PROCEEDINGS OF THE THIRTEENTH ANNUAL SYMPOSIUM ON

# COMPUTATIONAL GEOMETRY

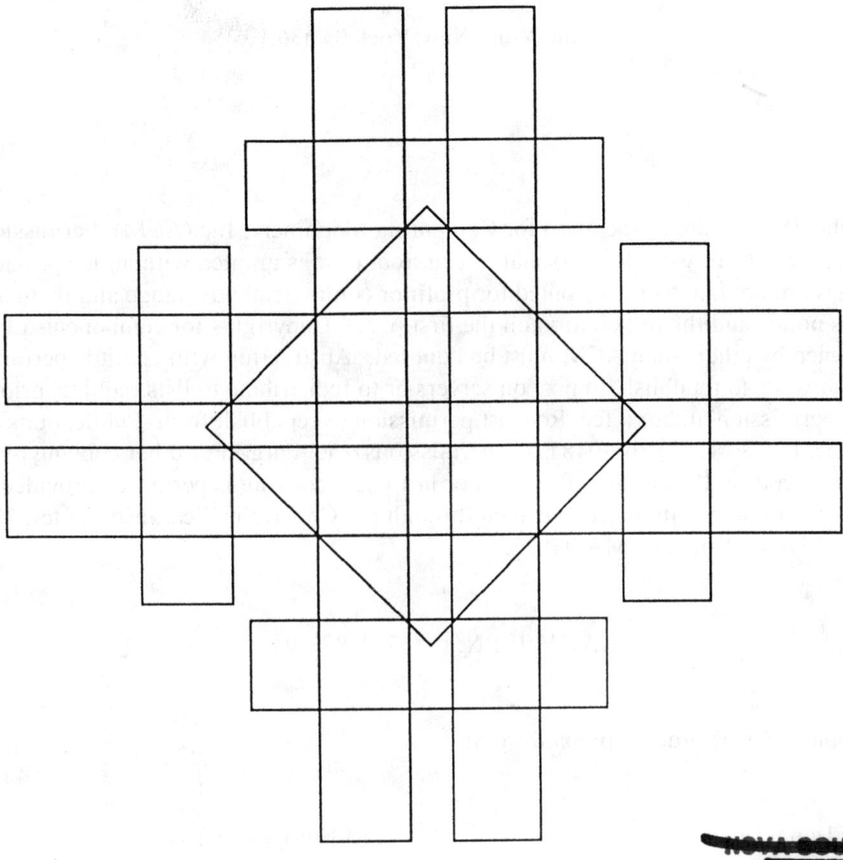



June 4-6, 1997
Centre Universitaire Méditerranéen
Nice, France

Sponsored by the ACM Special Interest Groups

for Graphics and Algorithms and Computation Theory

acm PRESS

**The Association for Computing Machinery**
**1515 Broadway**
**New York New York 10036**



**ACM ISBN:** 0-89791-878-9

Additional copies may be ordered prepaid from:

**ACM Order Department**
PO Box 12114
Church Street Station
New York, NY 10257

Phone: 1-800-342-6626
(US and Canada)
+1-212-626-0500
(all other countries)
Fax: +1-212-944-1318
E-mail: acmpubs@acm.org

ACM European Service Center
108 Cowley Road
Oxford OX 4 1 JF
UK

Phone: +44-1-865-382338
Fax: +44-1-865-381338
E-Mail: acm_europe@acm.org
URL:http://www.acm.org

**ACM Order Number:** 429970

Printed in the USA

# Foreword

This volume contains works presented at the *Thirteenth Annual Symposium on Computational Geometry,* held June 4–6, 1997 in Nice, France. The symposium was sponsored by the Special Interest Groups for Graphics (SIGGRAPH) and for Algorithms and Computation Theory (SIGACT) of the Association for Computing Machinery and was partially funded with the generous support of the following companies: Matra Marconi Space, Matra Datavision, Noesis, and Thomson Training and Simulation.

The symposium differed in significant ways from its twelve predecessors. At the business meeting of the 1996 symposium it was decided that an attempt should be made to broaden the scope of the conference towards more application oriented work. This mandate was realized by the adoption of a two-track format. Two independent program committees were formed, one for an "applied track" and one for a "theoretical track." A joint call-for-papers was issued with the instruction that submissions be explicitly directed to one of the two tracks. The committees decided independently about their respective submissions.

The community response to the new format was overwhelming: Each track, by itself, received roughly the same number of submissions as the entire 1996 symposium. While the submissions doubled, the conference duration had not, nor had the number of presentations. To alleviate the resulting problem of an unusually competitive selection process, the two committees were forced to ask the authors of many deserving paper submissions to let their work appear in the symposium proceedings as a *communication* with a poster presentation at the meeting. Almost all authors agreed. The committees would like to express their gratefulness for the authors' cooperativeness.

Some details: The applied track committee "met" electronically and selected 19 papers and 17 communications from a total of 96 paper and 7 communication submissions. The theoretical track committee met physically February 8–9, 1997 at Schloß Dagstuhl, Germany, and selected 21 papers and 18 communications from a total of 92 paper and 4 communication submissions. The selections were based on the quality and originality of the results and their relevance to computational geometry.

The program committees wish to thank all the people who helped in refereeing the papers — see the list on the following page. The papers and communciations generally present preliminary reports of ongoing research. It is expected that most papers will eventually appear in more polished form in refereed journals. It is the understanding of the program committees that appearance of a *communication* in these proceedings does *not* exclude submission of a corresponding paper to other conferences.

Descriptions of the accepted videos of the 6th Video Review of Computational Geometry appear in these proceedings. For details see the introduction on page V0.

The program committees wish to thank all authors who submitted abstracts in response to the call-for-papers.

**Program Committees:**

**Applied track:**

Bernard Chazelle (Princeton Univ.)
Jung-Hong Chuang (Natl. Chiao Tung U., Taiwan)
David Dobkin (Princeton Univ.)
Gerald Farin (Arizona State Univ.)
Christoph Hoffmann (**chair**) (Purdue Univ.)
Ming Lin (ARO)
Joe Mitchell (SUNY Stony Brook)
Nick Patrikalakis (MIT)
Franco Preparata (Brown Univ.)
Jarek Rossignac (Georgia Tech.)
Hans Peter Seidel (Erlangen, Germany)
Vadim Shapiro (Univ. of Wisconsin)
Tamas Varady (Natl Academy, Hungary)

**Theoretical track:**

Nina Amenta (Xerox PARC)
Tetsuo Asano (Osaka Electrocommunication Univ.)
Ken Clarkson (Bell Labs)
David Kirkpatrick (Univ. of British Columbia)
David Mount (Univ. of Maryland)
Stefan Näher (Univ. Halle)
Richard Pollack (New York University)
Claude Puech (Univ. Joseph Fourier, Grenoble)
Günter Rote (Technische Univ. Graz)
Otfried Schwarzkopf (Pohang U. of Science & Tech.)
Raimund Seidel (**chair**)(Univ. des Saarlandes)

**Conference Chair:** Jean-Daniel Boissonnat (INRIA Sophia-Antipolis)

**Symposium Sponsored by**

ACM Special Interest Groups for
Graphics (SIGGRAPH) and for
Algorithms & Computation Theory (SIGACT)

**With Support from**

Matra Marconi Space
Matra Datavision
Noesis
Thomson Training and Simulation

## Subreferees

P. Agarwal
O. Aichholzer
N. Amenta
S. Arya
F. Aurenhammer
S. Basu
P. Belleville
M. Bern
B. Bhattacharia
Th. Biedl
J.-D. Boissonnat
J. Bose
H. Bronniman
C. Burnikel
P. Cahoon
S. Campagna
F. Cazals
S.-W. Cheng
Y.-J. Chiang
J. Cohen
G. Csizmadia
O. Devillers
T. Dey
M. Dickerson
T. Dudás
F. Durand
H. Edelsbrunner
D. Eppstein
S. Felsner
S. Fortune
S. Frisch
M. Golin
P. Grabner
G. Greiner
S. Hahmann
M. Held
P. Hell
T. Hermann
J. Hershberger
J. Hobby
H. Hoppe
C.-Y. Hu
T.R. Jackson
P. Kaklis
T. Kameda
N. Katoh
M. Keil
C. Kenyon
S. Khuller
V. Klee
R. Klein
L. Kobbelt
G. Lukacs
T. Maekawa
D. Manocha
J. Matoušek
M. McAllister
K. Mehlhorn
L. Melissaratos
J. Mitchell
A. Mukhopadhey
N. Netanyahu
F. Nielsen
J. Pach
N. Pippenger
R. Pito
M. Pocchiola
F.-T. Pu
G. Renner
S. Rivière
K. Romanik
T. Sakkalis
N. Sapidis
S. Sarma
R. Sarraga
M. Sharir
E.C. Sherbrooke
C. Silva
Ph. Slusallek
M. Smid
J. Snoeyink
M. Stamminger
S. Suri
M. Taillaud
H. Tamaki
T. Tokuyama
S.T. Tuohy
A. Varshney
G. Vegter
D. Wagner
E. Welzl
G. Wilfong
S. Wismath
G. Wöginger
M. Yamashita
X. Ye
G. Yu
M. Yvinec
G. Ziegler
M. de Berg
M. van Kreveld

# Thirteenth Annual ACM Symposium on Computational Geometry

June 4–6, 1997
Nice, France

## Table of Contents

SESSION 10: **Theoretical Track: Computational Models**
CHAIRED BY DAVID KIRKPATRICK, University of British Columbia



SESSION 11: **Applied Track: Geometric Matching and Packing**
CHAIRED BY JOSEPH MITCHELL, SUNY Stony Brook



SESSION 12: **Theoretical Track: Proximity Problems**
CHAIRED BY DAVID MOUNT, University of Maryland

# Communications

## Applied Track

**Theoretical Track**

## Video Review

# Accelerated Occlusion Culling using Shadow Frusta *

T. Hudson D. Manocha J. Cohen M. Lin† K. Hoff H. Zhang
Department of Computer Science
University of North Carolina
Chapel Hill, NC 27599-3175
{hudson,manocha,cohenj,lin,hoff,zhangh}@cs.unc.edu

**Abstract:**
Many applications in computer graphics and virtual environments need to render datasets with large numbers of primitives and high depth complexity at interactive rates. However, standard techniques like view frustum culling and a hardware z-buffer are unable to display datasets composed of hundred of thousands of polygons at interactive frame rates on current high-end graphics systems. We add a "conservative" visibility culling stage to the rendering pipeline, attempting to identify and avoid processing of occluded polygons. Given a moving viewpoint, the algorithm dynamically chooses a set of *occluders*. Each occluder is used to compute a *shadow frustum*, and all primitives contained within this frustum are culled. The algorithm hierarchically traverses the model, culling out parts not visible from the current viewpoint using efficient, robust, and in some cases specialized interference detection algorithms. The algorithm's performance varies with the location of the viewpoint and the depth complexity of the model. In the worst case it is *linear* in the input size with a small constant. In this paper, we demonstrate its performance on a city model composed of 500,000 polygons and possessing varying depth complexity. We are able to cull an average of 55% of the polygons that would not be culled by view-frustum culling and obtain a commensurate improvement in frame rate. The overall approach is *effective* and *scalable*, is applicable to all polygonal models, and can be easily implemented on top of view-frustum culling.

## 1 Introduction

Interactive display of extremely large and complex geometric data sets has long been an important problem in computer graphics. Although throughput of graphics systems has improved considerably over the years, the size and complexity of models has grown even faster. In many walkthrough and virtual environment applications, models commonly consist of millions of primitives. Rendering such models at interactive rates is a major challenge. There is a great body of literature in computer graphics and computational geometry using techniques based on visibility culling and model simplification to render such large models. In this paper, we address visibility culling.

Given a large model and a viewpoint, the goal of visibility culling and hidden surface removal algorithms is to determine the set of primitives visible from that viewpoint. No general purpose, interactive algorithms are known for *exact* visibility determination on large models composed of hundred of thousands of polygons. Current graphics hardware provides support for visibility computations on a per-pixel basis with a z-buffer. However, high-end systems are only able to render a few tens of thousands of polygons at reasonable frame rates, due to bottlenecks in their hardware before the per-pixel analyses can be carried out. As a result, many applications use the following high-level software techniques to cull away a subset of the polygons which are not visible from the current viewpoint before they are needlessly sent to the rendering hardware:

- **View-Frustum Culling:** View-frustum culling uses a traversal of spatial data structures to cull out portions of the model not lying in the current *view frustum*. The viewing volume is represented as a frustum of six planes (near plane, far plane, and four sides). At runtime the display algorithm checks whether each node of the spatial structure overlaps this frustum; only nodes partially or completely contained within the frustum are rendered.
- **Occlusion Culling:** Hidden-surface removal algorithms and occlusion culling techniques are commonly used for models with high depth complexity to further reject the portion of geometry obscured by other objects in the scene. Some of the existing techniques are based on backface culling, binary space partition trees, or partitioning the models into cells and portals.

**Desiderata:** In designing algorithms for occlusion culling, we recognize the importance of *generality* and *robustness*. We make no assumptions about the structure of our input models – they may be arbitrary sets of polygons with no other topological information, also known as "polygon soup". The problem of 3D occlusion culling involves the computation of some geometric relationship between two or more objects. In our case, we reduce the problem to performing overlap tests between convex objects in 2D or 3D. Based on our experience in developing two interference detection systems, I-COLLIDE [CLMP95] and RAPID [GLM96], as well as that of other authors in implementing algorithms for interference detection [HKM95, BCG+96] and intersection computation for solid modeling [For96, HHK89], we have realized robustness is an important issue in the design and implementation of interference detection algorithms. Our goal is to develop algorithms which are relatively simple, efficient, and not prone to geometric degeneracies.

In this paper, we present object-space techniques for *occlusion culling*. This involves computation of a set of polygons that are within the view frustum but are not visible from the current viewpoint. We add this conservative visibility culling stage to the

*Supported in part by a Sloan fellowship, ARO Contract P-34982-MA, NSF grant CCR-9319957, NSF grant CCR-9625217, ONR Young Investigator Award, DARPA contract DABT63-93-C-0048, NSF/ARPA Science and Technology Center for Computer Graphics & Scientific Visualization NSF Prime contract No. 8920219.

†Also with U.S. Army Research Office

rendering pipeline in order to reduce the number of polygons sent to the graphics hardware.

As the viewpoint changes, the algorithm dynamically chooses a set of occluders. Each occluder is a convex polytope or a union of convex polytopes. For each occluder the algorithm computes a *shadow frustum* and uses fast interference detection and a hierarchical representation to find those portions of the model within the shadow frustum. The idea of shadow volumes was first introduced by Crow [Cro77] to generate shadows by creating for each object a shadow volume that the object blocks from the light source.

**Main contribution:** We present geometric algorithms for

1. Occluder selection using off-line and on-line techniques
2. Robust and efficient occlusion culling based on specialized interference detection algorithms, given the occluders and a hierarchical decomposition of scene geometry into bounding volumes.

The resulting algorithms have been implemented and we report their performance on a large model.

**Organization:** The rest of the paper is organized in the following manner. We survey related work in Section 2 and give an overview of the algorithm in Section 3. The occluder selection algorithm is presented in Section 4 and visibility culling based on occluders is described in Section 5. We present implementation details and the performance in Section 6 and analyse its complexity in Section 7.

## 2 Related Work

There has been significant amount of research on visibility and hidden surface removal in the fields of computer graphics and computational geometry. Many asymptotically efficient algorithms have been proposed for exact visibility and hidden surface removal [SSS74, Mul89, BDEG94, McK87]. [Dor94] provides a recent survey of object-space hidden surface removal algorithms. McKenna and Seidel [MS85] have presented an algorithm for computing optimal shadows of a convex polytope. For static models, it is possible to precompute the visibility from all the viewpoints in space based on aspect graphs [GCS91]. In the worst case, for input models composed of $n$ polygons, this algorithm can decompose the space into $O(n^9)$ regions, making it impractical for large models. The utility of all these algorithms for complex models is currently unclear.

In the last few years a number of techniques have been proposed for efficiently computing *conservative* visibility. These can be classified into two categories, object-space and image-space algorithms. More details on this classification are presented in [SSS74]. In object space, [Cla76] proposed view-frustum culling of a hierarchy of bounding volumes. Garlick et al. [GBW90] proposed using octree-based spatial subdivision to render polygons contained in the viewing frustum.

Several recent algorithms structure the database into *cells* or regions, and use a combination of off-line and on-line algorithms for cell-to-cell visibility and the conservative computation of the potentially visible set (PVS) of polygons [ARB90, TS91, LG95]. Such approaches have been successfully used in architectural walkthrough systems, where the division of a building into discrete rooms lends itself to a natural division of the database into cells. It is not apparent that cell-based approaches can be generalized to an arbitrary model, which may come with no structure information. Decomposing an arbitrary polygonal model into appropriate cells is rather difficult. Other algorithms for densely-occluded but somewhat less-structured models have been proposed by Yagel and Ray [YR96]. They use regular spatial subdivision to partition the model into cells and describe a 2D implementation. Some algorithms are based on binary space partition trees [FKN80, Nay92].

The hierarchical Z-buffer algorithm operates in both object-space and image-space [GKM93]. It combines spatial and temporal coherence with hierarchical structures. The algorithm exploits coherence by performing visibility queries on the Z-buffer. Currently, most graphics systems do not support this capability in hardware, and simulating the hierarchical Z-buffer in software is relatively expensive.

The work most directly related to our approach is that of Coorg and Teller [CT96, CT97]. Given two convex objects (an occluder and occludee), their early work required the construction and maintenance of a linearized portion of an aspect graph. They use this structure to track the viewpoint and determine whether one convex polytope occludes the other from a given viewpoint. This involves enumeration of all visual events and data structures for dynamic plane maintenance. In the worst case, the number of planes used to form a cell of the arrangement can be $O(m^2)$, where $m$ is the number of vertices of the convex polytopes, though dynamic and hierarchical data structures are used in [CT96] to speed-up the computation of relevant planes. Each arrangement cell classifies all polytopes as completely, partially, or un- occluded. This approach is similar to earlier shadow computation algorithms which, given a light source and an occluder, decompose space into penumbra and umbra volumes. In their newer work they reduce the amount of coherence used and simplify the structure of the arrangement. This yields a considerable speedup, eliminating the overhead cost of maintaining complex data structures.

## 3 Algorithm Overview

Occlusion culling can be divided into two subproblems. First, for a given viewpoint we must select a small set of good occluders to use. Second, given good occluders, we must use them to cull away occluded portions of the model. These two problems are partially independent, and so we treat them separately in this paper.

In general, any primitive in the model to be rendered or any combinations of such primitives can be used as an occluder. In this paper, we restrict occluders to be either convex objects or those which can be expressed as union of two convex objects.

### 3.1 Data Structures

To perform occlusion culling, we require that the model be represented both in a spatial partition and in a spatial hierarchy. The spatial partitioning structure is constructed as part of preprocessing and used for occluder selection, while the spatial hierarchy is used for fast culling of occluded geometry. These may be separate data structures, or may be united into a single structure, based on the implementation.

The spatial partition is a discrete structure that spans space, mapping every point in the space of the model onto one of a finite number of partition *regions*. If the space of the model has volume $v$, and no region is a volume larger than $v_r$, we have at least $\frac{v}{v_r}$ regions in the partition.

The hierarchy is not a data structure concerned with the space of the model, but rather with the geometry of the model. Every piece of geometry is mapped to exactly one leaf *volume* of the hierarchy, and every volume of the hierarchy encloses all the geometry mapped to it or its children. Let the entire model consist of $n$ pieces of geometry, and let each leaf volume contain no more than $n_l$ geometry elements. Then, if we want a binary hierarchy (each non-leaf volume having two children), we need at least $2\frac{n}{n_l} - 1$ volumes in the hierarchy.

### 3.2 Occluder Selection

Finding good occluders in real time for an arbitrary model is a difficult task. Among all of those objects in the database which we might use, we must select the few which will occlude the most geometry from the current viewpoint with the least computational overhead. We make this task feasible by using a preprocess to discard all occluders which are *not* likely to be good when viewed from a given set of viewpoints. Then, at runtime, the results of this preprocess give us a list of occluders that are good over the local region of space, and we further reduce this to select those occluders that are best from the current viewpoint.

### 3.3 Visibility Culling

Once we have selected a few good occluders, they can be used by our culling algorithm. For each occluder we construct a *shadow frustum*: a frustum with its apex at the viewpoint, near plane determined by the occluder, and sides determined by the occluder's silhouette (as shown in Fig. 1). The space contained within this frustum is that which is not visible from the viewpoint due to the occluder. The pixels rendered for objects in this space would be discarded during depth comparison by the hardware z-buffer. Thus, model geometry which is completely contained within any of these shadow frusta (as is object A) need not be rendered.

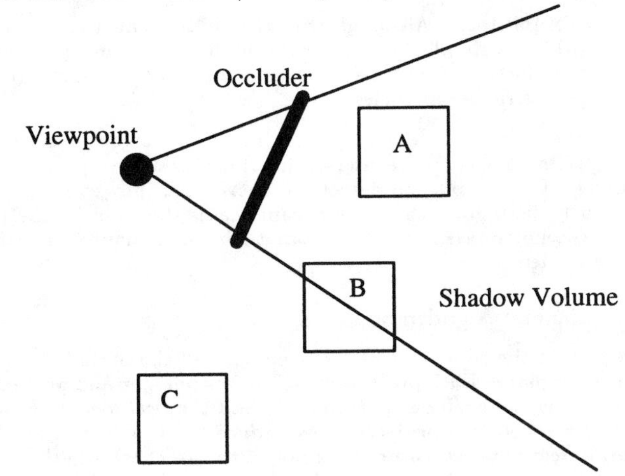


Figure 1: Relationship of bounding volumes to frusta.

## 4 Occluder Selection

Finding good occluders from a given viewpoint in a general unstructured model is a hard problem. By definition, a good occluder should occlude a large fraction of the geometry in the view frustum. For many viewpoints in the scene, no such good occluders may exist. In the worst case, computation of good occluders may correspond to computing the visible surface from that viewpoint using hidden surface removal algorithms. Some theoretically efficient algorithms of $O(n \log n + q)$ complexity, where $q$ is the number of edges in the visibility map, have been proposed by [Mul89]. However, not much is known about their practical performance.

The other alternative is to preprocess the entire model to determine a set of useful occluders at every viewpoint. However, the computation of such global visibility information is more difficult than hidden surface removal from a particular viewpoint. The fastest known algorithms known for computing the effects on global visibility due to a single polyhedron with $m$ vertices requires $O(m^6 \log m)$ time [GCS91].

Given the overall complexity of finding good occluders, we propose an approximation algorithm to find good occluders using a combination of online and offline techniques. Our algorithms work reasonably well in practice, but are not guaranteed to find good occluders all the time.

We pose occluder selection as an an optimization problem. Our goal is to use as few occluders as possible to keep CPU overhead down, but to occlude as many polygons as possible to keep the graphics pipeline lightly loaded. We take advantage of the empirical observation that for the data sets we are interested in, a few occluders cause most of the occlusion from most viewpoints, and using other occluders contributes little additional benefit[1].

[1] We performed experiments to confirm this. A typical result is shown in Figure 4. When we graph the fraction of polygons culled as a function of the number of occluders used, the optimum number of occluders is near the knee of the graph; in this case, approximately 8.

We use the following guiding principles in defining the optimization function to select good occluders:

1. **Solid Angle:** The viewed solid angle of a convex object is easily computable and measures the fraction of the visual field that it occupies. If we assume that the geometry of the scene is uniformly distributed about the viewpoint in all directions, the viewed solid angle is also directly proportional to the amount of geometry occluded.
2. **Depth Complexity:** We want to use occluders which occlude the maximum amount of geometry. In addition to using solid angle at runtime, we estimate the actual value of any occluder in the preprocess by random sampling. The algorithm selects some random viewpoints each partition region, constructs a shadow frustum from that viewpoint through each potentially good occluder, and determines the number of objects contained in the frustum. The average of several samples is a direct estimate of the value of the occluder.
3. **Coherence:** An occluder that does well or poorly at occluding for some frame will likely perform similarly for the next few frames. Similarly, objects that lie near a good occluder in screen space are themselves likely to be good occluders. The algorithm keeps track of the geometry culled by each occluder at each frame.

These criteria are integrated into our algorithm for both the online and offline processing of the geometry database. Rather than consider every polygon in the model as a potential occluder from every viewpoint, we use auxiliary data structures to quickly reduce the set of potential occluders to a manageable quantity. We can then further refine this set of potentially good occluders based on the exact viewing parameters of a particular frame.

### 4.1 Preprocess

The goal of the preprocess is to associate a set of potentially good occluders with every viewpoint in the model. We begin by constructing a spatial partition which divides the model into *regions*. Each region will store a list of potentially good occluders. At runtime, we determine which region contains the current viewpoint and use this region's associated list as a set of potentially good occluders. Thus, the region sizes control the granularity with which we sample the model's geometry for occlusion properties. Using small regions in our partition may shorten the list of potentially good occluders at each region, but this increases the number of regions in the data structure, thereby increasing its total memory usage.

Having constructed our spatial partition, we consider every potential occluder in the model[2]. For each potential occluder, we compute the set of viewpoints from which its viewed solid angle exceeds the threshold $\theta$ (a user-defined value). We use $\theta$ as the cutoff for consideration as an occluder. Objects whose viewed solid angles are less than the threshold cover only a small amount of screen space and, under the assumption that geometry is evenly distributed, are unlikely to occlude much geometry. For a single-sided ellipse, this set of viewpoints forms an ellipsoid and for a sphere, a concentric sphere. More complex occluders will have significantly more complex sets, which we approximate with the above. Each region of the partition which intersects this set of viewpoints adds the potential occluder to its list of potentially good occluders.

This solid-angle approximation is the first criterion of the runtime analysis we detail later. It is an *indirect* measure of the effectiveness of the occluder. We also pre-sample the second criterion, the actual amount of geometry behind the occluder, during the preprocess and store it in the region with the reference to the potentially good occluder. To do this, we choose a small number of viewpoints in the region and calculate the shadow frustum

[2] A separate preprocess may filter out small objects or simplify highly detailed objects to equivalently-occluding versions. Thus, the set of potential occluders need not be the same as the set of objects to be rendered.

cast by the occluder from each viewpoint. If we determine what fraction of the model geometry is actually occluded from each of these viewpoints and average our results together over all the viewpoints sampled, we have a *direct* estimate of the occluder's efficacy. Our selection of sample viewpoints be either random or guided by results on choosing optimal sampling patterns for antialiasing.

### 4.2 Runtime Computation

At every frame, we find the region of the spatial partition which contains the viewpoint. That region has an associated list of potentially good occluders obtained from the preprocess. The list is first narrowed by performing preliminary view-frustum culling to determine which potential occluders lie within the field of view. These potential occluders are then sorted based on our optimization function. The $n_{occ}$ occluders which produce the highest value of the function are used as that particular frame's occluders.

Throughout the process we exploit coherence. The algorithm exploits temporal coherence of occlusion. When the viewpoint only moves a little between frames, the value of a given occluder only changes a little. Using coherence, we improve our expected sorting performance to linear time. Similarly, due to spatial coherence and the fact we use convex occluders, the algorithm easily tracks the silhouette of each occluder during the overlap tests.

## 5 Visibility Culling Using Occluders

In order to use occluders, we must have constructed a hierarchy of *bounding volumes* that contains the entire model. Having selected $n_{occ}$ good occluders, we now proceed to perform occlusion culling with them. The critical requirement is to do so efficiently.

We begin by constructing a shadow frustum for each occluder. Taking the viewpoint as the apex of a frustum, we define the near plane of the frustum as a plane passing through the farthest point of the occluder's silhouette and whose normal points in the direction from that point towards the viewpoint. Each side of the frustum is a plane containing the viewpoint and an edge (two adjacent vertices) of the object's silhouette. This frustum contains the space which the occluder occludes from the viewpoint. Any geometry completely contained in the frustum is occluded and need not be rendered.

Using these shadow frusta requires an inorder traversal of the hierarchy. In fact, we incorporate view-frustum culling, shadow-frustum culling, and rendering into a single traversal of the hierarchy. We begin by marking all shadow frusta as active. As we encounter each volume during our traversal of the hierarchy, we first test for interference with the view frustum. If the volume is outside the view frustum, we are done with that sub-tree. Otherwise, we test for interference with all of the active shadow frusta. If the volume is entirely within any active frustum, the algorithm stops the current branch of the traversal without rendering any of the geometry in the volume. If the volume does not intersect any active frustum, we stop the current branch of the traversal and render all the geometry contained in the volume. Only if the volume partially overlaps some of the active frusta need we continue the traversal with that volume's children.

### 5.1 Overlap Tests

The algorithm determines the portions of the model occluded by a shadow frustum using interference tests with the bounding volume hierarchy. As the algorithm traverses the hierarchy, the intersection test between each bounding volume and the shadow frustum becomes a time-critical operation which directly influences the overall performance of occlusion culling. As shown in Figure 1, we need to quickly decide if a bounding volume is: (A) completely inside the shadow frustum, (B) partially overlapping the shadow frustum, or (C) completely outside the shadow frustum.

Consider the case when the occluder is a convex polytope. The shadow frustum must therefore be a convex volume. A number of efficient algorithms have been proposed in geometry and robotics literature for collision detection between convex polytopes. These are based on Minkowski sums [GJK88], closest features computation based on external Voronoi regions [LC91] and linear programming [Sei90]. All of them have been implemented and work reasonably well in practice. However, these algorithms are not directly applicable to our requirements for two reasons:

- **Limitation:** All these algorithms only check whether or not two objects are overlapping. They can not differentiate between partial overlap of two objects and complete containment of one object inside another, which is required for our algorithm.
- **Robustness:** Although the basic idea in all these algorithms is simple, their practical implementation are prone to robustness problems due to floating point arithmetic and geometric degeneracies.

In the geometry literature, other linear time algorithms have been proposed to compute intersections between convex polytopes [Cha89]. Unlike collision detection algorithms, they can unambiguously distinguish the three distinct cases shown in Figure 1. However, not much is known about their performance on real-world models.

### 5.2 General Algorithm

We project the silhouette of the occluder and the occludee onto the image plane. Each projection is a convex polygon and we refer to them as the occluder polygon ($A$) and the occludee polygon ($B$). Based on this projection, we reduce the problem to a 2D overlap test between two convex polygons. To check whether $B$ is totally or partially contained inside $A$, the algorithm initially checks whether they are overlapping or not. Our algorithm uses a modified Cyrus-Beck clipping algorithm [FDHF90], which quickly determines if two polygons intersect and robustly computes an edge and a common point (call it $O$) contained in the intersection of two polygons if they are overlapping. There are also other robust implementations available for computing the intersection of two planar polygons.

To check whether $B$ is totally contained inside $A$ or not, we need to check whether any of their edges intersect. Given $O$, we use a sweep-line approach to check whether or not $B$ is totally contained inside $A$. Therefore, the number of edge pairs we need to test for overlap is linear in the number of edges of the two polygons. In terms of robustness, this algorithm only requires a robust edge-edge overlap test.

### 5.3 Specialized Overlap Tests

Although the hierarchy of volumes may consist of arbitrarily-shaped convex volumes, this is rarely the case. Our work is a companion algorithm to view frustum culling, which normally uses hierarchies based on octrees, k-d trees, sphere-trees, OBB-Trees, R-trees etc. In all of these each node of the hierarchy is a sphere or a rectangular box. The rectangular box may be axis-aligned (an axis-aligned bounding box, or AABB) or be arbitrarily oriented (OBB).

In this section, we present robust and specialized overlap tests between a shadow frustum and an AABB or a OBB. Their overall running time is linear in the number of faces of the shadow frustum. However, we have also optimized the constant terms and overall operation count. The performance of the occlusion culling algorithm is dominated by these overlap tests.

The naive approach to exact interference detection requires an enormous amount of computation in the worst case: 54 "inside-outside" half-plane tests and 108 edge-face intersection tests (see Table 1). In the best case, it trivially rejects a box in 8 half-plane tests (dot products). We would like to maintain an exact test while drastically reducing the worst case cost and improving the best case cost. Our approach combines improved box-plane overlap tests and a fast edge-box intersection test using a parallel slabs representation for the bounding volume[Gre94, KK86].

Our box-plane overlap test uses the box polygon's normal to quickly find the box vertices which are closest to and farthest

| Interference Test | Dot Products | Scalar-Vector Mult/Divide | Vector Add/Sub | Cross Products |
|---|---|---|---|---|
| Naive with far plane | 270 (8) | 216 (0) | 1080 (0) | 432 (0) |
| Naive without far plane | 216 (8) | 168 (0) | 600 (0) | 336 (0) |
| Specialized with far plane | 294 (4) | 144 (0) | 144 (0) | 0 (0) |
| Specialized without far plane | 240 (4) | 144 (0) | 132 (0) | 0 (0) |

Table 1: Worst (and Best) case computational cost for the overlap tests

from the frustum's plane. Using these extremal points we need to check each plane against no more than two vertices, rather than eight. Oriented bounding boxes impose an overhead of only three additional dot products per plane tested at this stage.

We improve the worst case dramatically by speeding up expensive edge intersection tests. We can easily speed up the twelve tests between the frustum's edges and a box by representing our bounding volumes as three sets of slabs. This lets us avoid the edge-face test against each face individually by performing only a single edge-"solid" intersection test. Previously we required six expensive edge-face tests to test an edge against an entire box, but now we require only three edge-slab tests that completely eliminate the more expensive cross-product calculations.

Improving the twelve box-edge/frustum intersection tests proceeds similarly, using techniques from the ray-tracing literature. We represent the frustum as a set of planes forming a convex polyhedron and perform the edge test against the entire frustum at once. This also removes the expensive cross-product calculations entirely since we only have to perform six edge-plane intersection tests.

We further improve the performance by removing the far plane; this avoids having to perform the far-plane edge/box intersection tests and the view-frustum containment tests entirely (an infinite frustum never be contained in the bounding box). The improvements resulting from the new overlap test are shown in Table 1. Far planes are useful for view frusta, but are undesirable for shadow frusta.

## 6 Implementation and Performance

### 6.1 Data Structures

We have implemented the algorithms described in this paper. They scale well (efficiently and robustly) to large models. Our implementation has been interfaced within an unoptimized graphics viewer written using OpenGL. The viewer runs on top of UNC's view-frustum culling library and uses either exact or conservative versions of our specialized interference detection algorithm. We assume that *any* large model viewer will use view-frustum culling; results reported ignoring it are meaningless.

In our implementations, we also used the area-angle approximation presented by Coorg and Teller[CT96]. This is

$$-\frac{a\vec{\mathbf{N}}\cdot\vec{\mathbf{V}}}{|\vec{\mathbf{V}}|^2}$$

where $a$ is the area of the polygon (in object space), $\vec{\mathbf{N}}$ the polygon's normal vector, and $\vec{\mathbf{V}}$ the vector from the viewpoint to the center of the polygon. This gives a good approximation of the subtended solid angle of the polygon.

The results reported in this paper were obtained using area-angle as a goodness criterion; we have seen better results using the full three-part goodness criterion (solid angle, preprocess estimation, and coherence), but have not yet tested it on the large model.

We use a uniform division of the bounding box of the entire database into rectangular voxels as our spatial partition for occluder selection, and a tree of axis-aligned or oriented bounding boxes as our hierarchy for view-frustum culling and occlusion culling. We could alternatively use an octree, BSP tree, or other hierarchical spatial partition to contain the polygons of the model, thus using a single data structure to serve both purposes. However, in our experience that approach has two drawbacks:

1. Most spatial partition algorithms subdivide the polygons (or other primitives), such that each polygon lies in exactly one region. Polygon count is typically increased by a factor of two by this type of subdivision. Any increase in polygon count can be expected to slow rendering, which is the process we are fundamentally trying to speed up.
2. The two data structures capture two different properties of the model which may not be spatially similar. The spatial partition *should be* dense, i.e. have the smallest volumes, in regions of the model where the best set of occluders to change frequently. The bounding volume hierarchy needs to capture the geometry of the model as closely as possible for best results in culling. Occlusion culling as a field would benefit from further investigation of the tradeoffs in the number of spatial data structures used by algorithms.

### 6.2 Comparison with Other Algorithms

Our specialized shadow frustum/bounding box interference test has been compared against efficient algorithms and implementations for collision detection between convex polytopes [CLMP95, GJK88]. Our overlap test is at least two times faster and more robust than efficient implementations of these algorithms. Notice that some of the earlier algorithms for collision detection between convex polytopes utilize temporal and spatial coherence. Since the shadow frusta may change between frames, our algorithm uses no coherence in performing the overlap tests.

### 6.3 Performance on a Large Model

To test our algorithm in environments of varying high depth complexity, we built a composite model of more than 500,000 polygons by fusing together a model of central London, a model of the Aztec ceremonial center at Tenochtitlan, and a partially-developed region of hills in England. See plates I - IV at the end of the paper. The input model came to us as a collection of polygons with no adjacency or any other structure. We used uniform grids for occluder computation and a bounding volume hierarchy based on axis-aligned bounding boxes for view frusta and occlusion culling.

On our SGI Onyx RealityEngine 2, along a 2748-frame path through the central regions of this model we saw up to 80% of the model that passed view-frustum culling occluded (with roughly 40% of total polygons occluded on average). Our measured average speedup in rendering was 55%. (The theoretical maximum is 65%, with the discrepancy due to the overhead of managing the occluders and their shadow frusta.)

Occlusion culling is fundamentally a viewpoint-dependent optimization; Figures 2 and 3 show the frame time and rendered polygon count for each frame in our long path through the city model. On some frames occlusion culling gives extremely good results, while on others it does not obtain significant culling, depending on the depth complexity.

Occlusion culling yields good results with surprisingly few occluders. We rapidly see decreasing returns beyond five or six occluders active. Figure 4 shows the results of one experiment

| Method | Polygons per Frame | Time per Frame (us) | Frames per Second |
|---|---|---|---|
| View-Frustum Culling | 49196 | 276 | 3.6 |
| View-Frustum plus Occlusion Culling | 29749 | 178 | 5.6 |

Table 2: Comparison of average statistics over 2748-frame path down streets of the 500,000-polygon city model. Our implementation improves average frame rate by 55%.

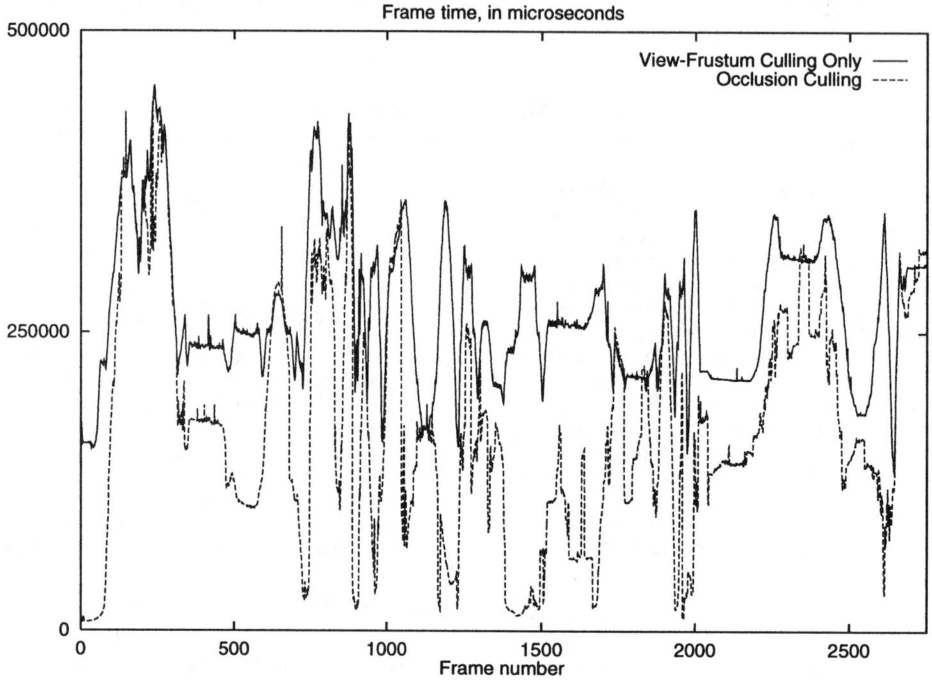


Figure 2: Time per frame, in microseconds, for each frame in the 2748-frame path through the large city model. Some viewpoints are good for occlusion culling, and at many of these points our implementation is seen to yield speedup.

with the city model, and we have seen similar results for other paths and other models.

### 6.4 Generality and Robustness

Generality and robustness have been main concerns in the design and implementation of our algorithm. The algorithm and implementation are applicable to all unstructured polygonal models. The algorithm requires no adjacency or connectivity information between the polygons. Only the silhouette tracking algorithm used for shadow frusta computation requires connectivity information. On unstructured models, we treat each convex polygon as a separate occluder.

Our specialized algorithms for overlap tests between the shadow frusta and bounding boxes are robust and not prone to degeneracies. They do not need to check for non-generic conditions such as parallel faces or edges. These are not special cases for the test and do not need to be handled separately. As a series of comparisons between linear combinations, the test is numerically stable. Our current implementation uses IEEE 64-bit floating point arithmetic instead of exact arithmetic for these overlap tests. Speed is an important concern in our system. We realize that floating point arithmetic can be a potential source of problem (though we have not experienced any so far), especially when one face of the box is just touching the shadow frusta. Many recent papers by Fortune, van Wijk, Clarkson, Boissonnat et al, have demonstrated that clever use of floating point arithmetic can yield very effective and yet correct implementations of geometric primitives. Many libraries like LEDA [MN95] have implementations of such algorithms for line segment intersections. As a result, it should be possible to have fast correct implementations of our specialized overlap test. For the general algorithm, we reduce the problem to 2D overlap tests between two convex polygons and use robust public domain implementations (available in Graphics Gems).

## 7 Algorithm Analysis

The overall asymptotic running time of the algorithm is at most linear in $n$, where $n$ is the total number of polygon in the model. The worst-case running times for the various phases of the algorithm are:

**Occluder Selection:** We select the $n_{occ}$ active occluders by sorting a list of $n_{pot}$ potentially good occluders. In practice, $n_{occ}$ is typically a small constant (between five and ten), and $n_{pot}$ is a small fraction of the input primitives. The total complexity of this phase is $O(n_{occ}n_{pot})$ or $O(n_{pot}log(n_{pot}))$.

**Shadow Frusta Computation:** Constructing shadow frusta based on silhouette tracking for these occluders requires time linear in the number of edges per occluder, which is a small constant $k$, for a phase complexity of $O(kn_{occ})$.

**Visibility Culling:** We use the $n_{occ}$ shadow frusta in the same manner as view frusta; each view-frustum cull is worst-case linear in the size of the model. The complexity of this phase is $O(n_{occ}n)$.

Since $n_{occ}$ is a small constant, and $n_{pot} << n$, our algorithm is in the worst case linear: $O(n)$. The spatial hierarchy constructed for occluder use keeps the constant in this term very small. We also use spatial and temporal coherence in resorting the list of potentially good occluders by using insertion sort. Since the ordering of this list only changes slightly between frames, that phase has an expected time linear in the number of potentially good occluders $n_{pot}$.

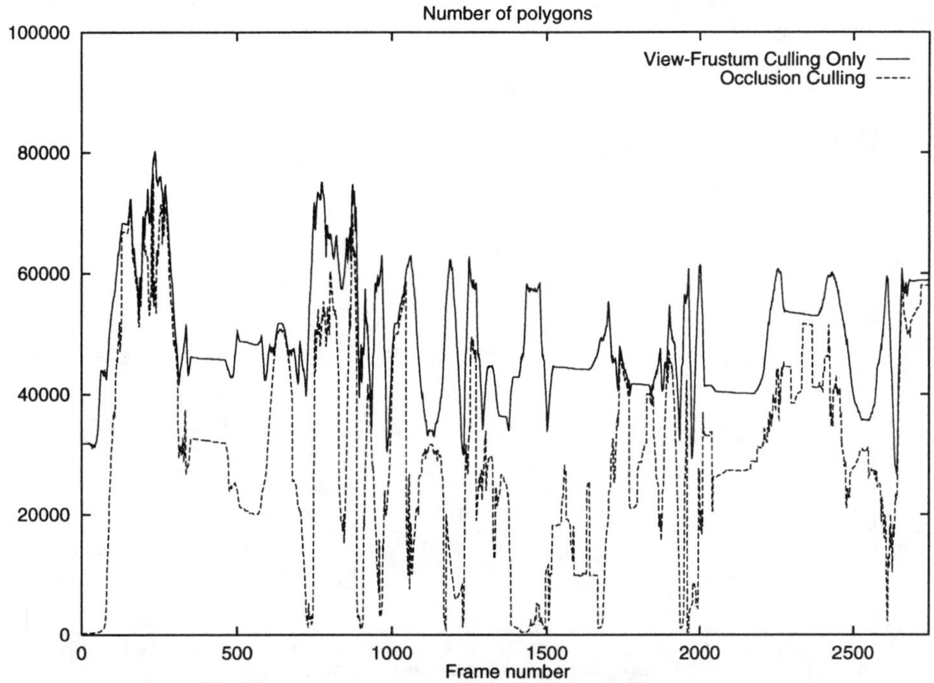


Figure 3: Polygons rendered per frame along a 2748-frame path through the large city model. Some viewpoints are good for occlusion culling, and at many of these points our implementation is seen to reduce the number of polygons rendered, which translates into less bus bandwidth consumed.

### 7.1 Main Features

Our main goals in developing these algorithms and systems have been robustness, applicability to large unstructured models, and scalability. None of the earlier conservative visibility algorithms has been applied to large and unstructured models composed of hundred of thousands of polygons. As a result, it is somewhat difficult to make exact comparisons in terms of robustness and efficiency.

Some of the main features of our approach are:

- **Robustness:** Since we use specialized interference detection algorithms for determining occluder-occludee relationships, our algorithm is not prone to robustness problems and geometric degeneracies.
- **Efficiency:** The additional CPU overhead due to occlusion culling varies with the viewpoint and is a linear function of the number of occluders chosen at that particular frame. In most cases, we use five to ten occluders. We use efficient and specialized algorithms for overlap tests between a shadow frustum and a hierarchy of bounding volumes. As a result, our algorithm performs interactively on large models composed of hundreds of thousands of polygons.
- **Generality:** The algorithm is applicable to all polygonal and unstructured models. In many CAD applications, the input models come as "polygon soup," with no structure or adjacency information available for constructing winged-edge or similar data structures. The presence of structure increases the effectiveness of our approach, but the results reported herein were obtained on unstructured data sets.
- **Ease of Implementation:** The implementation requires only simple data structures plus a general view frustum culling library. Since the occlusion culling algorithm sees the model as composed of simple bounding volumes (like spheres, rectangular boxes etc.), we use simple and easily-specialized overlap tests.

### 7.2 Ongoing Work

Often geometry is not occluded by any single convex occluder, but is occluded by the combination of several occluders. This is made more difficult to take advantage of in an object-space solution by the fact that the union of several convex occluders is not necessarily convex. We have developed and are implementing extensions to our interference detection algorithms to combine occluders in object space.

In [ZMHH97], Zhang et al. have successfully used the graphics pipeline to merge these occluders in the image space and form occlusion maps. Based on that they have proposed a two-pass algorithm. In the first pass, the algorithm constructs an occlusion map hierarchy and in the second pass used to cull away portions of the model not visible from the current viewpoint. This algorithm is able to cull away a higher percentage of the model than our object space approaches, but requires a high performance graphics pipeline. On the other hand, the algorithm presented in this paper does not use the graphics pipeline for visibility culling and should thus be useful for low-end hardware or personal computers.

## 8 Acknowledgements

Some of the models used in this work were originally constructed by:

- Dr. Vasilis Bourdakos of the Centre for Advanced Studies in Architecture, Bath University.
- Bob Galbraith Computer Graphics, East Longmeadow, MA.

## References

[ARB90] J. Airey, J. Rohlf, and F. Brooks. Towards image realism with interactive update rates in complex virtual building environments. In *Symposium on Interactive 3D Graphics*, pages 41–50, 1990.

[BCG+96] G. Barequet, B. Chazelle, L. Guibas, J. Mitchell, and A. Tal. Boxtree: A hierarchical representation of surfaces in 3d. In *Proc. of Eurographics'96*, 1996.

[BDEG94] M. Bern, D. Dobkin, D. Eppstein, and R. Grossman. Visibility with a moving point of view. *Algorithmica*, 11:360–78, 1994.

[Cha89] B. Chazelle. An optimal algorithm for intersecting three-dimensional convex polyhedra. *Proc. 30th Annu. IEEE Sympos. Found. Comput. Sci.*, pages 586–591, 1989.

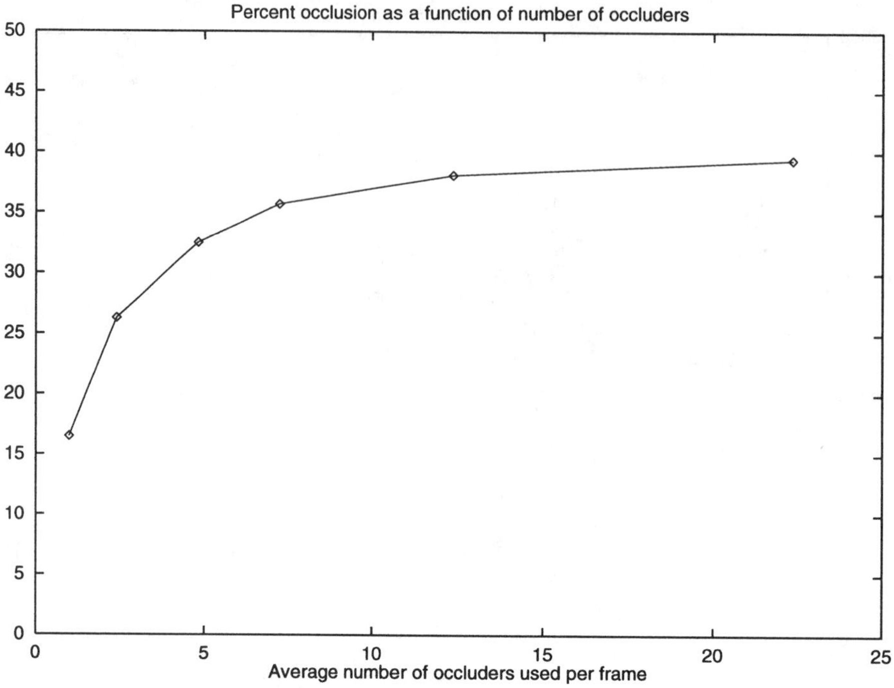


Figure 4: Average percentage of polygons within view frustum culled by occlusion culling as a function of average number of occluders active, along a 2748-frame path through the large city model. This supports the observation that for any given viewpoint a small number of occluders provide most of the occlusion.


[Cla76] J.H. Clark. Hierarchical geometric models for visible surface algorithms. *Communications of the ACM*, 19(10):547–554, 1976.

[Cla88] K. L. Clarkson. Applications of random sampling in computational geometry, II. In *Proc. 4th Annu. ACM Sympos. Comput. Geom.*, pages 1–11, 1988.

[CLMP95] J. Cohen, M. Lin, D. Manocha, and M. Ponamgi. I-collide: An interactive and exact collision detection system for large-scale environments. In *Proc. of ACM Interactive 3D Graphics Conference*, pages 189–196, 1995.

[Cro77] F. C. Crow. Shadow algorithms for computer graphics. *ACM Computer Graphics*, 11(3):242–248, 1977.

[CT96] S. Coorg and S. Teller. Temporally coherent conservative visibility. In *Proc. of 12th ACM Symposium on Computational Geometry*, 1996.

[CT97] S. Coorg and S. Teller. Real-time occlusion culling for models with large occluders. In *Proc. of ACM Symposium on Interactive 3D Graphics*, 1997.

[Dor94] S. E. Dorward. A survey of object-space hidden surface removal. *Internat. J. Comput. Geom. Appl.*, 4:325–362, 1994.

[FDHF90] J. Foley, A. Van Dam, J. Hughes, and S. Feiner. *Computer Graphics: Principles and Practice.* Addison Wesley, Reading, Mass., 1990.

[FKN80] H. Fuchs, Z. Kedem, and B. Naylor. On visible surface generation by a priori tree structures. *Proc. of ACM Siggraph*, 14(3):124–133, 1980.

[For96] S. Fortune. Robustness issues in geometric algorithms. In M.C. Lin and D. Manocha, editors, *Applied Computational Geometry*, pages 9–14. Springer-Verlag, 1996.

[GBW90] B. Garlick, D. Baum, and J. Winget. Interactive viewing of large geometric databases using multiprocessor graphics workstations. *Siggraph'90 course notes: Parallel Algorithms and Architectures for 3D Image Generation*, 1990.

[GCS91] Z. Gigus, J. Canny, and R. Seidel. Efficiently computing and representing aspect graphs of polyhedral objects. *IEEE Transactions on Pattern Analysis and Machine Intelligence*, 13(6):542–551, 1991.

[GJK88] E. G. Gilbert, D. W. Johnson, and S. S. Keerthi. A fast procedure for computing the distance between objects in three-dimensional space. *IEEE J. Robotics and Automation*, vol RA-4:193–203, 1988.

[GKM93] N. Greene, M. Kass, and G. Miller. Hierarchical z-buffer visibility. In *Proc. of ACM Siggraph*, pages 231–238, 1993.

[GLM96] S. Gottschalk, M. Lin, and D. Manocha. Obb-tree: A hierarchical structure for rapid interference detection. In *Proc. of ACM Siggraph'96*, pages 171–180, 1996.

[Gre94] N. Greene. Detecting intersection of a rectangular solid and a convex polyhedron. In *Graphics Gems IV*, pages 74–82. Academic Press, 1994.

[HHK89] C. Hoffmann, J. Hopcroft, and M. Karasick. Robust set operations on polyhedral solids. *IEEE Computer Graphics and Applications*, 9(6):50–59, 1989.

[HKM95] M. Held, J.T. Klosowski, and J.S.B. Mitchell. Evaluation of collision detection methods for virtual reality fly-throughs. In *Canadian Conference on Computational Geometry*, 1995.

[KK86] T. Kat and J. Kajiya. Ray tracing complex scenes. *Computer Graphics*, pages 269–278, 1986.

[LC91] M.C. Lin and John F. Canny. Efficient algorithms for incremental distance computation. In *IEEE Conference on Robotics and Automation*, pages 1008–1014, 1991.

[LG95] D. Luebke and C. Georges. Portals and mirrors: Simple, fast evaluation of potentially visible sets. In *ACM Interactive 3D Graphics Conference*, Monterey, CA, 1995.

[McK87] M. McKenna. Worst-case optimal hidden-surface removal. *ACM Trans. Graph.*, 6:19–28, 1987.

[MN95] K. Mehlhorn and S. Näher. LEDA: a platform for combinatorial and geometric computing. *Commun. ACM*, 38:96–102, 1995.

[MS85] M. McKenna and R. Seidel. Finding the optimal shadows of a convex polytope. In *Proc. 1st Annu. ACM Sympos. Comput. Geom.*, pages 24–28, 1985.

[Mul89] K. Mulmuley. An efficient algorithm for hidden surface removal. *Computer Graphics*, 23(3):379–388, 1989.

[Nay92] B. Naylor. Interactive solid geometry via partitioning trees. In *Proc. of Graphics Interface*, pages 11–18, 1992.

[Sei90] R. Seidel. Linear programming and convex hulls made easy. In *Proc. 6th Ann. ACM Conf. on Computational Geometry*, pages 211–215, Berkeley, California, 1990.

[SSS74] I. Sutherland, R. Sproull, and R. Schumaker. A characterization of ten hidden-surface algorithms. *Computing Surveys*, 6(1):1–55, 1974.

[TS91] S. Teller and C.H. Sequin. Visibility preprocessing for interactive walkthroughs. In *Proc. of ACM Siggraph*, pages 61–69, 1991.

[YR96] R. Yagel and W. Ray. Visibility computations for efficient walkthrough of complex environments. *Presence*, 5(1):1–16, 1996.

[ZMHH97] H. Zhang, D. Manocha, T. Hudson, and K. Hoff. Visibility culling using hierarchical occlusion maps. Technical Report TR97-004, Department of Computer Science, University of North Carolina, 1997. To Appear in Proc. of ACM Siggraph'97.

PLATE I

A view from within the streets of the London model. Note how all the outlying portions of the model are occluded by buildings on streets near the viewpoint.

PLATE II

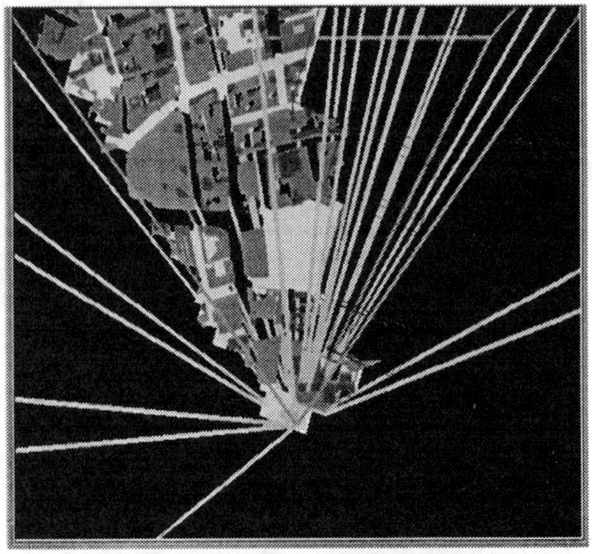

An overhead view of the same area, showing the view frustum, the shadow volumes of the occluders chosen by our algorithm, and the regions culled away as a result.

PLATE III

A close-up of the Aztec portion of the composite model, showing the great range of scales - from tiny detail polygons to huge blocks.

PLATE IV

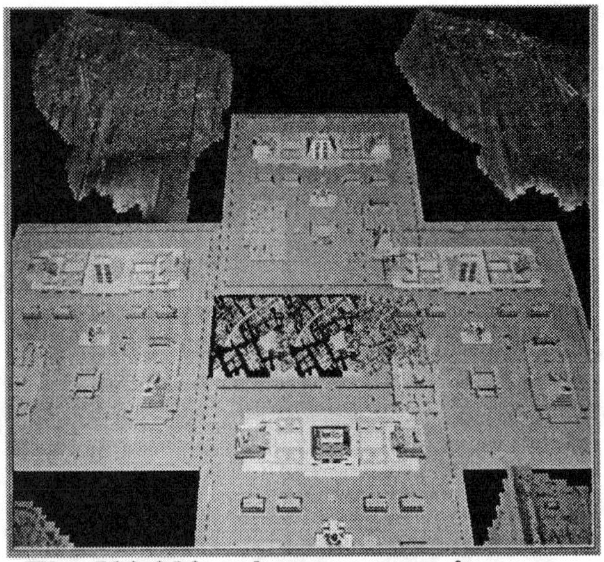

The 500,000-polygon composite model used in some of our tests, built out of London, Tenochtitlan, and central England.

# Bucket-like Space Partitioning Data Structures with Applications to Ray-Tracing

F. Cazals and C. Puech

iMAGIS/GRAVIR-IMAG
BP 53 - 38041 Grenoble cedex 09 - FRANCE
iMAGIS is a joint project of CNRS, INRIA, INPG and UJF
*email:* Frederic.Cazals@imag.fr

## Abstract

Data structures based on uniform subdivisions of the space —also known as bucketing— have the nice properties that they can be walked through very easily and can provide neighborhood relations at low cost. For data sets which are uniformly scattered in $2D$ or $3D$ space, this makes the implementation of algorithms such as ray tracing, nearest neighbors computation or Delaunay triangulation almost trivial. But should the processed data set admit dense clusters, the spatial partitioning does not result in data partitioning so that the performances are collapsing.

Although it has been known for a long time in dimension one that recursive bucket-sort admits a linear complexity for a wide range of probability densities, recursive bucket-like data structures have not received any attention in the computational geometry community. It has been observed in computer graphics that these were the fastest to ray-trace, but the question of understanding why they are not just another space partitioning data structure but rather the only data structure that succeeds in capturing the probabilistic properties of data distribution remains open.

This paper is a first step in this direction and investigates hierarchical recursive and non recursive data structures for ray-tracing. First, we show that precisely analyzing an optimized ray-tracer is a difficult task due to the context sensitivity of the calls costs of the functions called most often. Second, we exhibit statistics showing that if uniform grids are definitely not the right data structure to use for non-uniform distributions, recursive grids are very good at handling such distributions. Third, we present several improvements of the Hierarchy of Uniform Grids data structure, which result for the best cases in running times improved by up to a factor three with reference to the previously best known solution.

## 1 Introduction

### 1.1 Computational model, data structures and complexity

Although relatively new areas of computer science such as computer graphics or computational geometry focused primarily on the design and analysis of new algorithms — rendering, pixel based methods, animation, it is fair to say that the algorithms themselves received more attention that the underlying data structures. More precisely, after the pioneering work by Don Knuth in the early seventies, most of the research was carried out in the so called algebraic decision trees model defining a theoretical computer providing only algebraic operations (such as comparison, polynomial expressions evaluation) on the objects manipulated.

In dimension one, prototypes of such algorithms or data structures are quick-sort or balanced binary search trees. And in dimension two or three these are binary space partitioning trees or $k-d$-trees (see [dB$^+$96]). Their common feature is to recursively split the input data set around one of its elements, so that choosing this element as the median results in a balanced data structure whose height is logarithmic in the input size. When the splitting element is chosen at random, the worst case performance relies on the order of the items, and the analysis is generally done considering the $n!$ permutations equally likely.

Almost by then, another family of algorithms and data structures got some attention. Since the previous model is intrinsically binary, that is at any time a decision is made between a constant number of items (two for quick-sort, four for quad-trees, etc), it was noticed that using non constant *branching factors* could result in data structures of much reduced height. Splitting the data set was replaced by splitting the space containing it, which we illustrate with the example of bucket-sort. To sort the keys $\{x_1, \dots, x_n\}$, the algorithm (1)subdivides the interval $[inf\ x_i, sup\ x_i]$ into sub-intervals called *buckets* of equal length $\delta = (sup\ x_i - inf\ x_i)/n$ (2)assigns each $x_i$ to its sub-interval by computing $\lfloor (x_i - inf\ x_i)/\delta \rfloor$ (3)recurses in any bucket containing more than a constant $b$ (e.g. 5) number of points (4)sorts by insertion the buckets containing less than $b$ points. It is easy to see that if the probability density the $x_i$ are drawn from is uniform, each *bucket* contains $O(1)$ points after step (2) so that the sorting time is linear.

Two major differences with the algorithms/data structures of the former type are that the floor function is needed and that the running time of the algorithm does not depend on the order of the input data set but on the underlying probability density. If the first concern is not restrictive, being dependent on some possibly highly non uniform density is scary. Indeed, if at each step of recursion in bucket-sort all the points but one fall in the same bucket, the complexity becomes quadratic. Fortunately, such data sets must be built as follows $\{x_1 = 1!, x_2 = 2!, \dots, x_i = i!, \dots, x_n = n!\}$ and it is easily checked that the domain spanned by such


*Computational Geometry 97* Nice France

inputs gets huge even for very small samples: for $n = 70$ we already get $10^{100}$. (Remember that the diameter of the earth is just 6400 km so that a tiny object of say 1mm with reference to the whole earth would just result in a $10^{10}$ change of scale!) And more generally, it has been shown ([Dev86]) that forcing a recursive bucket-sort to recurse more than two levels requires fancy densities. The intuitive interpretation is that clusters of points that look very dense at some level of recursion appear to be almost uniform a few levels farther. Apart from bucket-sort, examples of such algorithms and data structures are interpolation-search in dimension one, grid-files, uniform and recursive grids in higher dimension [NH84, Dev86, CDP95]. Very interesting pieces of work are [SD95b, A+94] where space partitioning based algorithms are tested against divide-and-conquer and sweep-line algorithms for computing the Delaunay triangulation and reporting red/blue line segments intersections. In the two cases the former scheme appears to be the best for evenly distributed data sets and performs poorly on 'bumped' inputs. But as we shall see in this paper, recursive grids and hierarchies of uniform grids (HUG in the sequel) encompass uniform grids and provide a far better solution for such inputs.

More precisely, this paper addresses the question of building and evaluating (in the scope of a ray-tracer) data structures which are able to capture the objects repartition properties of complex real-world scenes. An example of such scenes is the kitchen from figure 5(a): whereas many tiny polygons are found in the neighborhood of the bowl teapot or stack of plates, the middle of the room is just empty! Before reviewing briefly the previous work and giving the paper overview, we want to emphasize the fact that we do not view bucket-like data structures as opposed to the classic ones. Indeed, since we are aiming at reproducing the efficiency of uniform grids for non uniform distributions, it seems natural to use algorithms whose complexity does not depend on the objects' spatial repartition. The construction of a recursive grid does not require such a step, but the one of a hierarchy of uniform grids does so for the clustering step. For applications such as the ones from [SD95b, A+94] where the data structure is built and used once, the cost of this pre-processing might be worth of consideration. But for ray tracing where the same data structure can be used to compute different views of say a walk-through within a complex environment, it does not matter. This is even more the case since as observed in [CDP95] the recursive grid and the HUG construction times are negligible with reference to the rendering one.

### 1.2 About ray tracing

Ray tracing primarily consists in finding for a given ray — that is a point and a direction in $3D$, the first object hit. From its early days it became clear that for complex scenes all the objects could not be tested for intersections which led to the development of spatial subdivision data structures. Still, if one can say that from a practical point of view a consensus considers the recursive grid as the best data structure to ray trace ([JdL92, CDP95, Hai96]), very little is known on the reasons why it is so and in particular the question of performing some average case analysis of ray tracing is open. This is a difficult concern especially since some notion of random scene is needed.

An attempt to find spatial partitioning taking into account the non uniformity of objects distribution was made in [Dev88] using integral geometry tools. Unfortunately, it is not clear how the kinematic density based formulas used to estimate the grids' optimal size are compatible with the non uniform distribution of objects. Another interesting piece of work is [MMS94] and focuses on characterizing the complexity of a particular directed segment shoot —which is like ray shooting excepted that the ray is replaced by a segment. More precisely, define a *strongly simple cover* of a segment $S$ as a set of balls (1)which individually do not intersect more than a constant number of obstacles edges (2 on figure 11(a)) even when grown of a factor $\varepsilon > 0$ (2)whose union covers $S$. Also, let the *simple cover complexity* of $S$ denoted $scc(S)$ be the cardinality of the smallest strongly simple cover of $S$. Then it is shown that it is possible to build a spatial subdivision of size $O(n)$ in $O(n \log n)$ such that segment shooting queries can be answered in $O(\log n + scc(s))$. Unfortunately, no experimental fact is reported about this data structure even if one can expect a quad-tree like behavior given the way it is built.

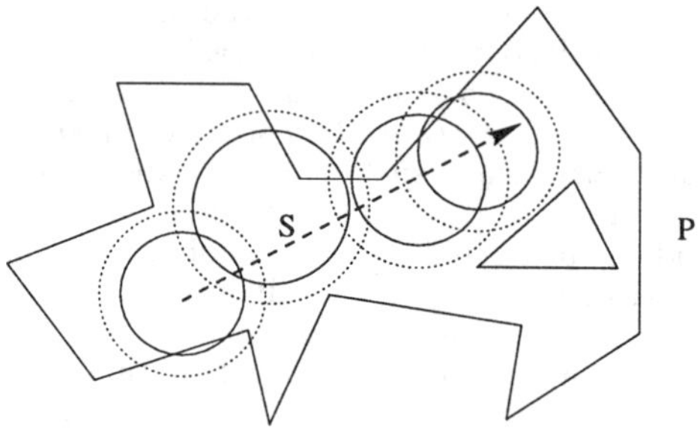


Figure 1: Strongly simple cover of a directed line segment $S$

### 1.3 Uniform grid, recursive grid and hierarchy of uniform grids

We briefly review the notions of uniform grid, recursive grid and hierarchy of uniform grids depicted on figures 2 and 3. More details on bucket-like space partitioning data structures can be found in [CDP95, A+94, AEII85, IBWY80, NH84]. We suppose we are given a set $\mathcal{O}$ of $n$ objects which can be polygons, implicit/algebraic patches, whatever. We just require each object to be bounded and by its length $d_i$ we refer to its axis aligned bounding box diameter length. The reason why we need so few hypothesis on the data manipulated is that we are interest in statistical properties under a huge numbers of objects and not in local properties attached to a particular object.

A *uniform grid* over the set $\mathcal{O}$ is a partition of its bounding box into $(n_x, n_y, n_z)$ subdivisions of equal length along the $x, y$ and $z$ axis. To get a memory requirement linear in the number of objects, the $n_i$ are usually taken so that $n_i = \alpha_i \sqrt[3]{n}$ which we call the $\sqrt[3]{n}$ criterion in the sequel. The $\alpha_i$ are positive constants and the simplest choice is $\alpha_x = \alpha_y = \alpha_z = 1$. Heterogeneous values based on the ratios of the dimensions of the box are also presented in [KS97].

Once a uniform grid is built of $\mathcal{O}$, one may find that some voxels are too populated. If $MAXP$ — standing for MAXimum number of Polygons — is some positive integer (50 e.g.) a *recursive grid* on $\mathcal{O}$ is a hierarchical data structure based on uniform grids such that whenever a voxel contains $N > MAXP$ items (the initial voxel being the scene bounding box) it is recursively split into a uniform grid of $N$ voxels.

Instead of letting the recursive partitioning figuring out which are the most populated areas of the scene, a *hier-*

archy of uniform grids* tries to do it more cleverly first by splitting $\mathcal{O}$ into subsets of objects of the 'same' size — the size $d_i$ of the object $o_i$ being for instance the length of its axis aligned bounding box diagonal, and second for each of these by looking for sub-subsets of neighbors. These two steps, respectively called the filtering and clustering steps, are defined as follows:

**Definition 1** ***Filter of lengths*** *Given a set $\mathcal{O}$ of objects, we call filter $\mathcal{F}$ a strictly increasing sequence of positive real numbers $\{f_1, ..., f_m\}$ such that $d_{inf} = inf_i d_i \in [f_1, f_2)$ and $d_{sup} = \sup_i d_i \in [f_{m-1}, f_m)$. A level of the filter $\mathcal{F}$ is an interval $[f_i, f_{i+1})$. Define also (i) $\bar{f}_k = (f_k + f_{k+1})/2$ the average length of level k (ii) $L_k$ the set of all objects from $\mathcal{O}$ the lengths of which $\in [f_k, f_{k+1})$.*

**Definition 2** ***$\delta$-connectivity - $\delta$-cSet*** *The distance between two objects is defined as the minimum distance between their bounding boxes. Two objects $o_i$ and $o_j$ are called $\delta$-connected where $\delta \in \mathbb{R}^{*+}$ if their distance $d(o_i, o_j) < \delta$. If they belong to the same level k of the filter $\mathcal{F}$, let $\delta_k$ be the $\delta$ associated to level k. We call* connectivity coefficient *$\alpha$ a strictly positive real number, and we set $\delta_k = \alpha\ \bar{f}_k$. A set $\mathcal{O}$ is said to be $\delta$-connected or a $\delta$-cSet if $\forall\ i \in$ 1..n there exists a $j \neq i$ such that $d(o_i, o_j) < \delta$.*

**Definition 3** ***Property $\pi_{xyz}$, potential cluster and cluster*** *A set $\mathcal{O}$ is said to be $\pi_x$ or to have property $\pi_x$ if the objects' projections along the x axis form a $\delta$-cSet. The opposite will be noted $\pi_{\bar{x}}$. Thus a set $\pi_{x\bar{y}z}$ will have the property $\pi_x$, $\pi_{\bar{y}}$ and $\pi_z$. $\mathcal{O}$ is said to be a potential cluster if it has property $\pi$ along one or two axis, and a cluster if it is $\pi_{xyz}$.*

The whole point is therefore to group objects of the same size and neighbors into clusters. The HUG construction algorithm does not assume anything about the clusters excepted that the highest filter level consists of a single cluster referred to as the *world* cluster in the sequel. Once a cluster has been isolated, its $N$ objects are stored into a uniform grid containing $\alpha^3 N$ voxels obtained by subdividing the cluster bounding box into $\alpha \sqrt[3]{N}$ subdivisions along each axis — as for the recursive grid, this is called the $\sqrt[3]{N}$ criterion in the sequel. The constant $\alpha$ is taken equal to one for every voxel but the world one. Indeed, as observed in [CDP95] taking a small value such as $\sqrt{2}$ for that particular cluster results in a substantial improvement of the rendering time with no significant memory requirement overhead. This observation which we call the $\sqrt{2}$ *criterion* is explained in section 6. The last step of the construction consists in building a hierarchy of clusters in a top-down fashion with respect to the filter levels: for a given grid, all the voxels intersected by a grid of lower level receive a pointer to that grid. The experiments of [CDP95] show that the value of the connectivity coefficient is not very important and that the practical running time of the hug construction is $O(n \log n)$ using $O(n)$ space.

Since most of this paper deals with experimental statistics measured on various test scenes under several data structures, we adopted the acronym

$$dataSetName.dataStructureName$$

to refer to a particular configuration. For example, *kit.r*50 refers to the kitchen model stored in a recursive grid with termination condition $MAXP$ = 50; *kit.g*30 refers to the same model stored in a uniform grid with $30^3$ voxels; and $kit.u.f3\sqrt{2}$ refers to a three levels HUG with the $\sqrt{2}$ criterion for the world cluster. Also we may use DS and bbox as a short cut for data structure and axis aligned bounding box.

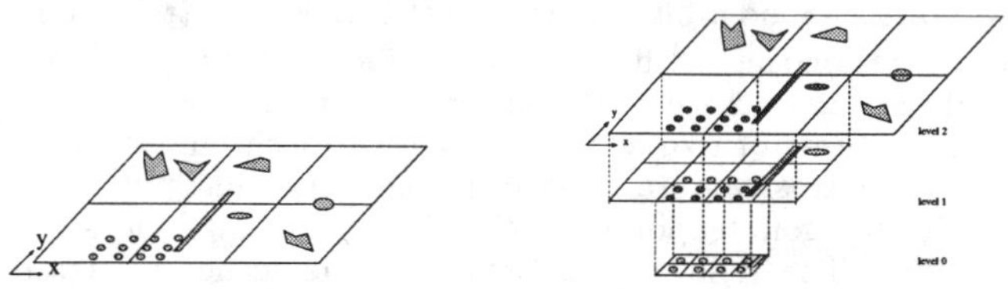

Figure 2: (a) Uniform grid (b) Recursive grid

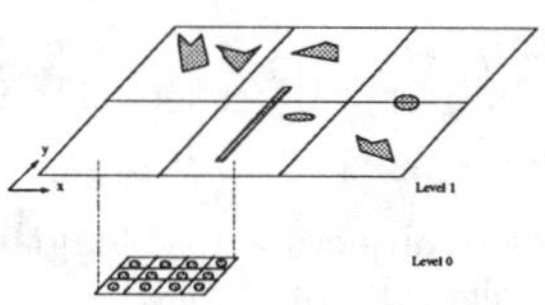

Figure 3: $\mathcal{H}$ierarchy of $\mathcal{U}$niform $\mathcal{G}$rids

### 1.4 Paper overview

In section 2 we present the test scenes; in section 3 we discuss some statistics of interest to characterize space partitioning data structures and the performances of a ray tracer; in section 4 we raise an important point about the context sensitivity of some functions calls costs; in section 5 we compare the ability of the uniform grid, recursive grid and HUG to partition non-uniform data sets; in section 6 we analyze the rendering statistics of section 3; in section 7 we present several improvements of the HUG data structure and in section 8 we conclude and list unsolved problems.

## 2 Scenes presentation

We performed tests on four real-world scenes: a kitchen model provided by Don Greenberg at the Cornell University program for Computer Graphics (USA); the Berkeley computer science building model provided by Seth Teller from MIT (USA); random cities generated by Phil Dench's computer program developed at the University of Curtin (Australia); and plants models generated by the AMAP computer program jointly developed by the CIRAD and the University of Montpellier (France). Pictures of these scenes are displayed on figures 5, 6, 7, 8 and 9.

The characteristics of these scenes are the following. The kitchen and the Cal CS building provide accurate models of data sets usually handled in architecture. The random cities are interesting because they are more $2,5D$ models. As for the plants models arranged to form a garden, they provide a more uniform distribution in $3D$ space since most of the polygons are leaves scattered all over the place.

The complexity of these scenes in terms of number of polygons can be made arbitrarily high. For the kitchen scene, just add dishes, forks, etc, and discretize them up to an arbitrarily precision. The full model of Berkeley CS building is already over 1,000,000 polygons ! The random cities can be parameterized to be as large as desired. As for the plants models, it is easy to generate plants of hundreds of thousands of polygons.

For each of the four scenes, we picked a configuration which we believe is representative of this kind of scene and does not require prohibitive computer resources to be interactively displayed and ray-traced. Indeed, all the test were performed on an Indigo2 extreme SGI workstation with 256

Mega-bytes of RAM and a MIPS R4400 processor at 200 MHZ. For example we did not work with the full Berkeley CS building model but either with the walls only (that is no furniture) or the $3^{rd}$ floor. Considering two or more floors would not have changed the spatial distribution of the polygons in terms of size as well as relative position. We also avoided excessive data clusters in the sense that putting a 20,000 polygon teapot in a 20,000 polygon kitchen is artificial ! The following statistics for the scenes used are displayed on figures 4, 5, 6, 7, 8 and 9:

- the number of polygons
- the ratios *min/max* and *min/diag*, with min (max) the smallest (largest) object length and diag the scene bounding box diagonal length
- the percentage of objects whose length is less than 20% of the largest size, denoted %.2
- the histogram of the objects lengths and the associated cumulated distribution function. These two functions are depicted on the same chart and are distinguished as follows: the leftmost scale corresponds to the histogram percentages while the rightmost one gives the values of the cumulated distribution function. The $x$ axis scale stops at .2 because most of the objects' sizes are smaller than the size of the largest objects over 5.

Of course these statistics do not provide any information on the relative position of the objects and this issue is discussed in a companion paper [CS97].

| | #polys | min/max | min/diag | %.2 |
|---|---|---|---|---|
| Kitchen | 20947 | 9,0e-4 | 6,5e-4 | 99,29 |
| CSB walls | 32811 | 5,7e-4 | 3,6e-4 | 99,49 |
| CSB $3^{rd}$ floor | 117605 | 1,0e-5 | 8,1e-6 | 99,99 |
| City | 52136 | 2,9e-3 | 1,0e-3 | 99,85 |
| Garden | 45293 | 4,3e-3 | 1,1e-3 | 99,72 |

Figure 4: Test scenes statistics

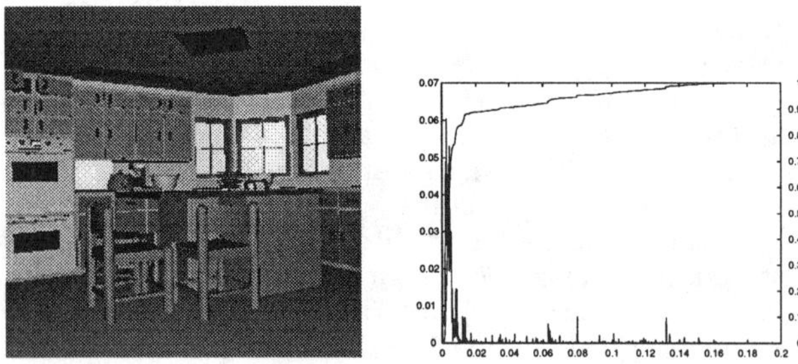

Figure 5: Kitchen scene

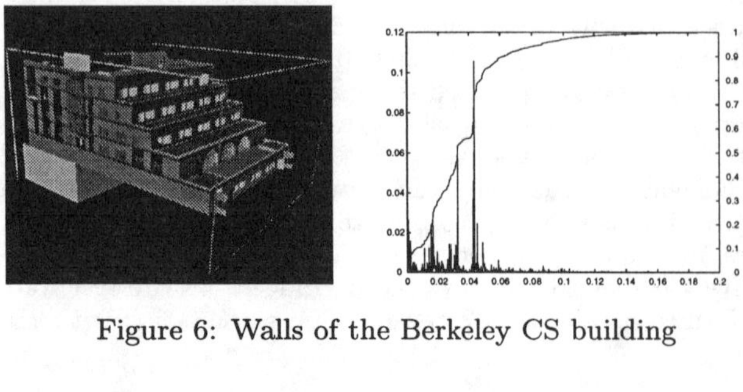

Figure 6: Walls of the Berkeley CS building

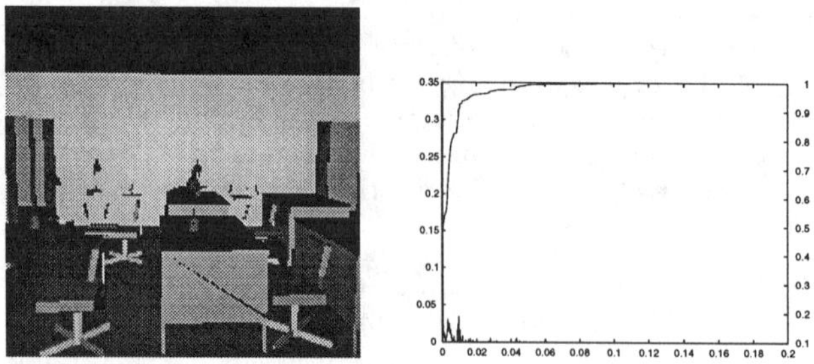

Figure 7: An office in the Berkeley CS building $3^{rd}$ floor

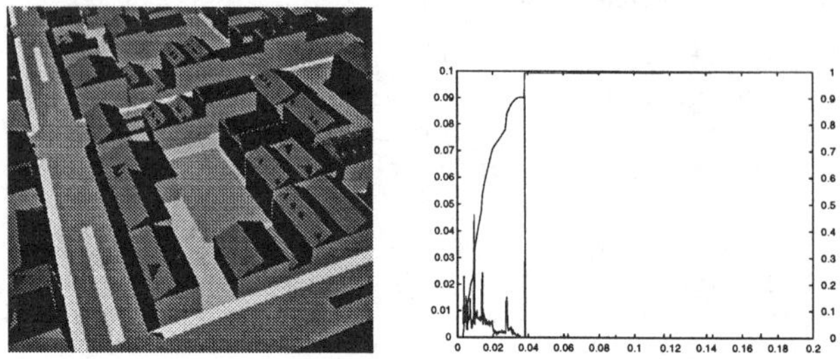

Figure 8: City model

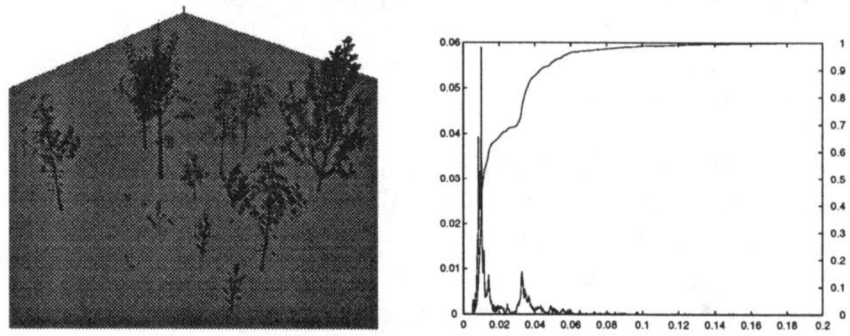

Figure 9: Garden scene

## 3 Ray tracing: statistics of interest

Pretty much all the information describing the ability of a data structure to ray-trace is captured by the number of ray-polygon and ray-bounding box intersection tests, the number of voxels traversed, the memory requirements expressed in mega-bytes, and the rendering time in seconds, which we respectively denote by #pit, #bbi, #vt, #mbs, and $t$. But to these high-level statistics correspond more precise metrics, some of which characterize the way the DS stores the scene primitives and are thus viewpoint independent, and some of which depend on the particular image rendered. More precisely, for each hierarchy level of our data structures we define the following statistics:

•**resources repartition statistics:** By %obj, %grids, %voxels, and %ptrs we respectively refer to the percentages of polygons stored at that level, grids voxels and pointers (to objects and sub-grids) allocated to that level with reference to the total number of such entities in the DS. By *%vol* we denote the sum of the volumes of the grids of that level with reference to the volume of the biggest grid. Note that as soon as there is more than one non empty level, $\sum_{all\ levels} \%vol > 100$. As we shall see, this is a good criterion to evaluate a HUG.

**•space occupancy statistics:** By $\mu o_g$ we refer to the average number of objects by grid. By $\%o$ we refer to the percentage of non empty voxels —that is containing objects. By $\mu o_v$, $\sigma o_v$, $\mu o'_v$, $\sigma o'_v$ we refer to the average and standard deviation of the number of objects by voxel, and to the average and standard deviation of the number of objects by *non-empty* voxel. The quantities $\mu g_v$, $\sigma g_v$, $\mu g'_v$, $\sigma g'_v$, have the same meaning but are related to the number of grids referenced by each voxel

**•rendering statistics:** By $\%pit$, $\%pi$, $\%bbi$, $\%vt$ we refer to the percentages of ray-polygon intersection tests, polygons intersected, bounding boxes intersection tests and voxels traversed with reference to the total number of such entities for the whole data structure. The ratio $\%pi/\%pit$ is equal to the number of polygons intersected over the number of tests. These viewpoint dependent statistics were computed while rendering 250x250 pixels images and averaged over a sequence of ten runs.

## 4 Heavy duties and functions calls costs

The rendering statistics depend on which functions are most often called and on how much each of these calls costs. An important point we want to raise here is that the running time of a function with more than one exit point can be context-sensitive. To see why, consider the following ray-polygon intersection algorithm:

**Intersect(ray, polygon)**
- [$test_1$] if the ray and the polygon are coplanar, return
- compute the intersection point $p$ between the ray and the polygon plane
- [$test_2$] if $p$ does not belong to the polygon $2D$ bounding box, return
- [$test_3$] compute the ray-crossing or winding number

Writing this routine this way is important since the two first tests are constant time whereas the third one depends linearly on the polygon number of vertices. But if one can expect $test_1$ to be passed most of the time, successing to $test_2$ obviously depends on the context from which the routine is called. Indeed, the ray-polygon intersection test is called whenever a ray is crossing a voxel referencing polygons. But the probability for a ray to intersect a polygon *given that* it intersects the voxel is proportional to the surfaces areas ratio ([S76, CS97]). Which shows that the test will be more expensive for polygons whose size matches the referencing voxel surface area.

The best one can look for is therefore average running times. Candidate tools to get these are the profiling programs such as `prof` and `pixie` on SGI. Unfortunately, the number of processor cycles spent within each function is computed by sampling the program counter every milli-second which is too sparse for functions whose running time order of magnitude is the micro-second. The only valid information one can get from a profile is therefore a rough idea of the functions called most often. In our case:

**ray polygon intersection:** see above

**ray bounding box intersection and grid walk through configuration:** intersection between the ray and a $3D$ bounding box (either the world bbox, or a cluster bbox for the HUG, or a sub-grid bbox for the recursive grid), and in the case of intersection, configuration of the parameters of the walk through across that grid

**next voxel computation:** given a location in the grid, computation of the next voxel traversed by the ray

To compute the costs $t_{pi}$, $t_{bi} + t_{rg}$ and $t_{nc}$ of these functions, we therefore timed each of them with two calls to the system `times()` function when stepping in and out. The average times are reported on the top part of figure 10 and confirm the context sensitivity of the calls costs. For example the average cost of the ray-polygon test varies from $1.19e-6$ seconds for the close view on the third floor of the Berkeley CSB with a $r50$ down-to $2.97e-7$ seconds for a recursive grid with recursion termination condition $MAXP = 500$. This drop corresponds to a lower percentage of tests reaching the winding number computation as shown by the $\%s_1$ and $\%s_2$ values expressing the percentages of successes to $test_1$ and $test_2$.

This variability makes the prediction of the expected performances difficult since as shown on the bottom part of figure 10 the numbers of calls of the ray-polygon, ray-bounding box and voxel crossing algorithms respectively expressed in million, thousand and million varies a lot from one data structure to another. In other words reducing the number of calls by the appropriate spatial subdivision results in more expensive calls and it is not clear where the optimum is.

| DS | $t_{pi}$ | $t_{rg} + t_{bi}$ | $t_{nc}$ | $\%s_1$ | $\%s_2$ |
|---|---|---|---|---|---|
| $csbc.u.f2.\sqrt{2}$ | 1.6e-6 | 3.2e-5 | 8.4e-6 | 72.3 | 9.35 |
| $csbc.u.f3.1$ | 8.6e-7 | 2.9e-5 | 9.3e-6 | 70.3 | 4.91 |
| $csbc.r50$ | 1.1e-6 | 4.6e-5 | 6.1e-6 | 84.4 | 6.63 |
| $csbc.r500$ | 2.9e-7 | 5.2e-5 | 7.1e-6 | 87.7 | 1.69 |
| $csbc.g80$ | 1.0e-6 | 5.0e-5 | 4.1e-6 | 90.6 | 6.29 |

| DS | $\#pit$ | $\#bbi$ | $\#vt$ | $\#mbs$ | $t$ |
|---|---|---|---|---|---|
| kit.u.f3.$\sqrt{2}$ | 5,3 | 509 | 2 | 4 | 57 |
| kit.u.f2.$\sqrt{2}$ | 5,3 | 367 | 2,3 | 3.63 | 47 |
| kit.r50 | 3,9 | 231 | 2,9 | 7.87 | 45 |
| kit.r500 | 18 | 138 | 2,4 | 4.38 | 86 |

Figure 10: Call costs, number of calls and overall performances

## 5 Voxels occupancy and spatial partitioning

To figure out how good at handling non uniform distributions the uniform grid, recursive grid and HUG are, we computed the statistics described in section 3 for the kitchen scene and the three configurations $g28$, $u.f3.\sqrt{2}$ and $r50$. In addition, we plotted for each hierarchy level of each data structure the histogram and the cumulated distribution function of the number of objects per voxel. The results are displayed below.

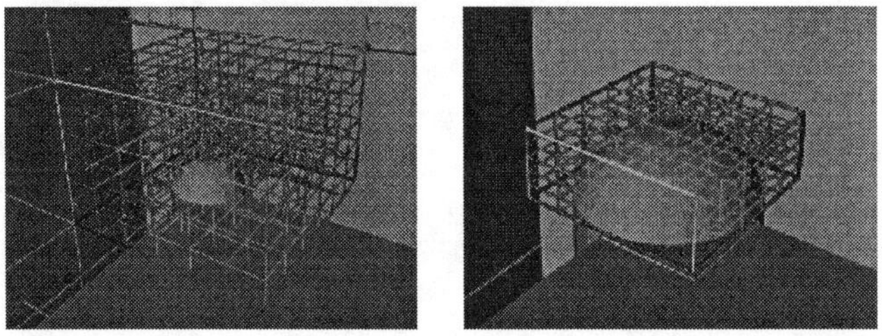

Figure 11: (a,b)Densely populated areas: chair foot top stored by a recursive grid and a HUG

| level | %obj | %grids | %voxels | %ptrs | %vol | #mbs |
|---|---|---|---|---|---|---|
| 0 | 100 | 100 | 100 | 100 | 100 | |
| % | 100 | 100 | 100 | 100 | 100 | |
| Σ | 20891 | 1 | 21952 | 58387 | | 2.11 |

| level | $\mu o_g$ | %o | $\mu o_v$ | $\sigma o_v$ | $\mu o_v'$ | $\sigma o_v'$ |
|---|---|---|---|---|---|---|
| 0 | 1 | 28.4 | 0.952 | 23 | 3.35 | 42.8 |

| level | %g | $\mu g_v$ | $\sigma g_v$ | $\mu g_v'$ | $\sigma g_v'$ |
|---|---|---|---|---|---|
| 0 | 0 | 0 | 0 | 0 | 0 |

Figure 12: Voxels occupancy for *kit.g28*

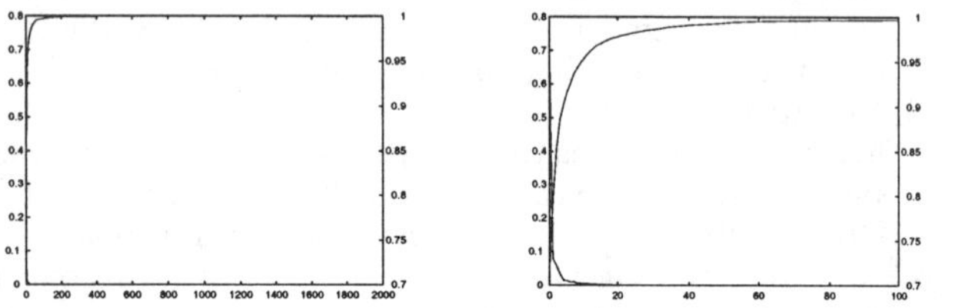

Figure 13: Distribution of the number of objects by voxel for *kit.g28*: (a)Full histogram (b)Zoom on the left part

| level | %obj | %grids | %voxels | %ptrs | %vol | #mbs |
|---|---|---|---|---|---|---|
| 0 | 82.6 | 70 | 65.3 | 66.1 | 0.152 | |
| 1 | 5.84 | 29.7 | 5.22 | 10.9 | 0.0218 | |
| 2 | 11.5 | 0.286 | 29.5 | 23 | 100 | |
| % | 100 | 100 | 100 | 100 | 100 | |
| Σ | 20891 | 350 | 31443 | 103150 | | 4.00 |

| level | $\mu o_g$ | %o | $\mu o_v$ | $\sigma o_v$ | $\mu o_v'$ | $\sigma o_v'$ |
|---|---|---|---|---|---|---|
| 0 | 1.18 | 43.6 | 3.32 | 6.87 | 7.62 | 8.7 |
| 1 | 0.197 | 81.4 | 6.01 | 6.1 | 7.38 | 5.96 |
| 2 | 40.4 | 34.2 | 2.49 | 7.08 | 7.28 | 10.6 |

| level | %g | $\mu g_v$ | $\sigma g_v$ | $\mu g_v'$ | $\sigma g_v'$ |
|---|---|---|---|---|---|
| 0 | 0 | 0 | 0 | 0 | 0 |
| 1 | 84.6 | 0.87 | 0.403 | 1.03 | 0.17 |
| 2 | 3.37 | 0.0669 | 0.449 | 1.99 | 1.48 |

Figure 14: Resources repartition and Voxels occupancy for *kit.u.f3.sqrt2*

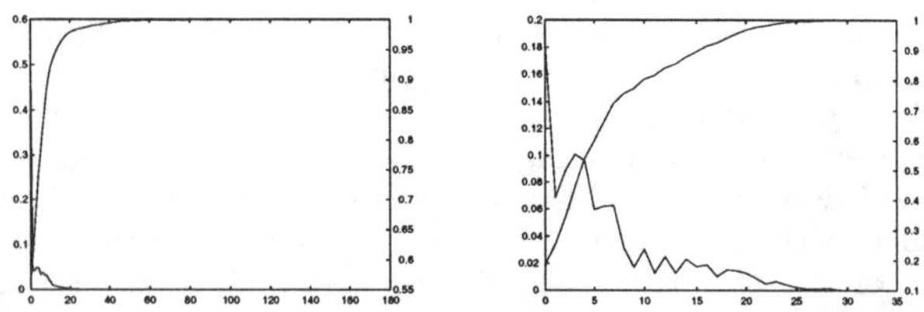

Figure 15: Distribution of the number of objects by voxel for *kit.u.f3.sqrt2*. levels 0 and 1

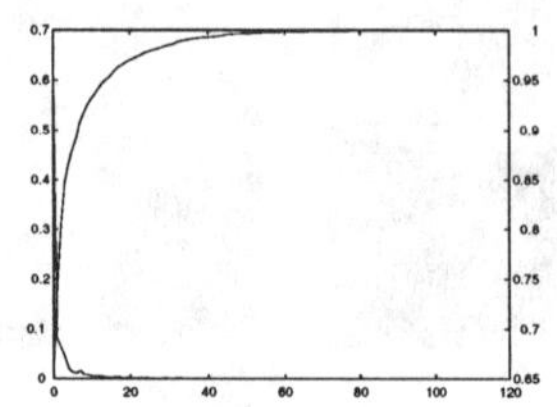

Figure 16: Distribution of the number of objects by voxel for *kit.u.f3.sqrt2*. level 2

| level | %obj | %grids | %voxels | %ptrs | %vol | #mbs |
|---|---|---|---|---|---|---|
| 0 | 0.11 | 0.813 | 0.21 | 0.713 | 3.37 | |
| 1 | 11.5 | 39 | 13.8 | 22 | 0.00144 | |
| 2 | 70.4 | 59.8 | 50.1 | 59.8 | 0.67 | |
| 3 | 18 | 0.407 | 36 | 17.4 | 100 | |
| % | 100 | 100 | 100 | 100 | 104 | |
| Σ | 20891 | 246 | 61030 | 196232 | | 7.87 |

| level | $\mu o_g$ | %o | $\mu o_v$ | $\sigma o_v$ | $\mu o_v'$ | $\sigma o_v'$ |
|---|---|---|---|---|---|---|
| 0 | 0.135 | 89.1 | 10.9 | 9.22 | 12.3 | 8.88 |
| 1 | 0.296 | 58.9 | 5.15 | 6.49 | 8.75 | 6.33 |
| 2 | 1.18 | 48.4 | 3.84 | 6.81 | 7.93 | 7.95 |
| 3 | 44.2 | 27.7 | 1.55 | 4.75 | 5.59 | 7.67 |

| level | %g | $\mu g_v$ | $\sigma g_v$ | $\mu g_v'$ | $\sigma g_v'$ |
|---|---|---|---|---|---|
| 0 | 0 | 0 | 0 | 0 | 0 |
| 1 | 0.0238 | 0.000238 | 0.0154 | 1 | 0 |
| 2 | 0.314 | 0.00314 | 0.056 | 1 | 0 |
| 3 | 0.67 | 0.0067 | 0.0816 | 1 | 0 |

Figure 17: Resources repartition and voxels occupancy for *kit.r50*

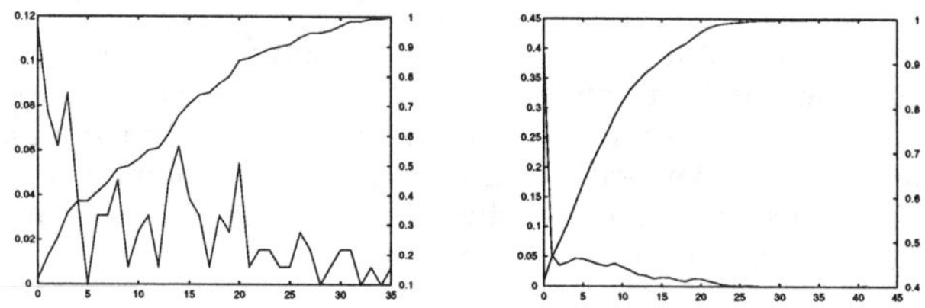

Figure 18: Distribution of the number of objects by voxel for *kit.r50*. levels 0 and 1

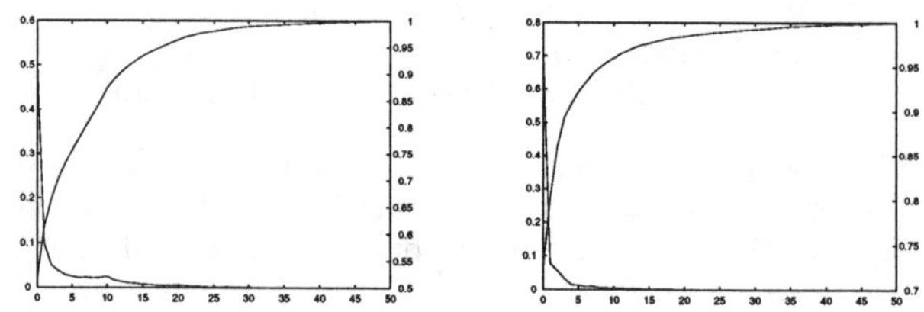

Figure 19: Distribution of the number of objects by voxel for *kit.r50*. levels 2 and 3

For the uniform grid the hierarchy has a single level but the way the polygons are referenced is disastrous: the average number of polygons per voxel is less than 1 and the standard deviation is 23. The cumulated distribution function actually shows that about 98% of the voxels contain a reasonable (less than 20) number of polygons. But the remaining ones admit a huge variability and are causing overall bad performances.

For the HUG, we first observe that the most resources consuming level is the level 0, which corresponds to the fact that most polygons of the scene are small polygons. This observation shows that the $\sqrt{2}$ criterion mentioned in section 1.3 is not too memory consuming since the big objects represent 11.5% of the scene. The voxels occupancy statistics are far better than for the uniform grid since $\mu o_g'$ and $\sigma o_g'$ are within a factor two or so. Also, the distribution of clusters is good since $\mu g_v'$ is bounded by 1 with a small standard deviation.

As for the recursive grid, we first note that two levels of recursivity are sufficient to capture about 88% of polygons within voxels containing less than $MAXP$ references. Also, the voxels occupancy statistics for these two levels are good, especially for the third. Indeed, in spite of a relatively low percentage of non empty voxels (27.7%), $\mu o_g$, $\sigma o_g$ $\mu o_g'$, and $\sigma o_g'$ are slightly better than for the HUG.

The HUG therefore performs pretty well at referencing the polygons. But the important fact is that the recursive grid which is built in a totally blind manner performs even better if one forgets the memory requirements. The point is that when a given volume containing $n$ objects is subdivided into $n$ voxels, this results in relatively small voxels so that most of the big polygons are referenced from scratch in voxels containing less than $MAXP$ polygons, which is also the case for the majority of the small polygons. And one level of recursion later, this is even more true for most of the small polygons, that is most of the polygons. Of course, one drawback is a relatively high percentage of empty voxels which affects the memory requirements. Another interesting fact is that the deeper in the hierarchy, the worse the statistics. This is a consequence of the above remark: at level 3, all the polygons take advantage of the thin subdivision induced by the small polygons scattered all over the place. But this property disappears when the grid recurses to handle dense spots. For example on figure 11, the 96 polygons modeling the top of the chair foot induce the subdivision of a voxel to eventually occupy just a corner. The HUG on the opposite figured out this dense spot through the filtering and clustering steps.

## 6 Rendering statistics: a detailed analysis

We now consider the viewpoint dependent statistics described in section 3 and averaged over a sequence of ten runs for 250x250 pixels pictures of the kitchen. We skip the uniform grid case whose running time is about 5 times slower than the two other data structures.

We start with the analysis of the performances for the viewpoint of figure 5 depicted on figure 20. First and for the two data structures, the percentage of intersected polygons with reference to the total number on intersection tests is pretty low (below 5%) which is not very satisfactory for the whole space partitioning paradigm ! Second, the majority of the ray-polygon intersection tests comes from the highest level of the hierarchy: 85.6% and 79.5% respectively for the recursive grid and the HUG. That's the reason of the efficiency of the $\sqrt{2}$ criterion mentioned in section 1.3, and that's also the reason why the recursive grid performs better than the HUG (see the discussion at the end of the previous section). As for the repartition of the number of bounding boxes intersected and the number of voxels traversed, the difference lies in the fact that the HUG inherently has more grids than the recursive grid (350 vs. 246, see section 5), and that these resources are allocated differently as already observed.

The discussion for the viewpoint of figure 21 is about the same. We observe that the rendering times are increased in a factor 1.37 and 1.29 for the recursive grid and HUG respectively, and that the number of ray-polygon tests is higher for the HUG while the opposite holds for the number of voxels traversed. For the HUG, the most consuming level in term of ray-polygon is now level 0, which is fine too since the voxels occupancy statistics are as good as the one of level 2 — fig. 14. And for the recursive grid, the most consuming level is now the second which is not very good since as shown on fig. 17, the deeper in the hierarchy, the higher the expected number of polygons per voxel.

| level | %pit | %pi | %(pi / pit) | %bbi | %vt | $t$ |
|---|---|---|---|---|---|---|
| 0 | 10.9 | 7.61 | 2.16 | 53.6 | 49.7 | |
| 1 | 3.51 | 3.74 | 3.29 | 21.8 | 22.2 | |
| 2 | 85.6 | 88.7 | 3.2 | 24.5 | 28.1 | |
| % | 100 | 100 | | 100 | 100 | |
| Σ | 5.3e6 | 1.6e5 | | 5.0e5 | 2.1e6 | 57 |

| level | %pit | %pi | %(pi / pit) | %bbi | %vt | $t$ |
|---|---|---|---|---|---|---|
| 0 | 6e-3 | 1e-3 | 0.735 | 9e-3 | 9e-3 | |
| 1 | 0.367 | 0.507 | 5.05 | 1.03 | 0.864 | |
| 2 | 20.2 | 15.9 | 2.87 | 44.9 | 42.2 | |
| 3 | 79.5 | 83.6 | 3.84 | 54.1 | 56.9 | |
| % | 100 | 100 | | 100 | 100 | |
| Σ | 3.9e6 | 1.4e5 | | 2.3e5 | 2.9e6 | 45 |

Figure 20: Rendering statistics for $kit.u.f3.sqrt2$ and $kit.r50$

Figure 21: Zoom on tiny polygons

| level | %pit | %pi | %(pi / pit) | %bbi | %vt | $t$ |
|---|---|---|---|---|---|---|
| 0 | 57.9 | 30.9 | 1.94 | 35.6 | 63.5 | |
| 1 | 1.15 | 2.03 | 6.37 | 46.6 | 31.4 | |
| 2 | 41 | 67 | 5.93 | 17.8 | 5.11 | |
| % | 100 | 100 | | 100 | 100 | |
| Σ | 10.1e6 | 3.6e5 | | 7e5 | 2.4e6 | 74.27 |

Figure 22: Rendering statistics for $kitchenBowlView.u.f3.sqrt2$

| level | %pit | %pi | %(pi / pit) | %bbi | %vt | $t$ |
|---|---|---|---|---|---|---|
| 0 | 0 | 0 | ∅ | 0 | 0 | |
| 1 | 3.86 | 3.79 | 3.43 | 7.45 | 14.4 | |
| 2 | 79.4 | 59.3 | 2.61 | 67.1 | 81.8 | |
| 3 | 16.7 | 36.9 | 7.69 | 25.5 | 3.77 | |
| % | 100 | 100 | | 100 | 100 | |
| Σ | 5.9e6 | 2e5 | | 4.9e5 | 3.3e6 | 62.73 |

Figure 23: Rendering statistics for $kitchenBowlView.r50$

## 7 Improvements of the HUG data structure

### 7.1 Some observations on filtering and clustering

Since the clustering step is intended to find closely fitting bounding boxes over sets of objects, an attractive idea is to use non axis aligned bounding boxes once the clusters have been isolated. This is even more so that, as already observed in section 4, the probability for a ray to cross a convex volume depends on the surface area delimiting this volume. So that a natural way to reduce the number of grids traversed per ray could consist in packing these objects into bounding boxes of minimal surface area. It should be emphasized that in doing so one can only expect to reduce the number of ray-box intersection tests, but not the number of ray-polygon tests. Indeed it is not clear that the orientation that minimizes the box surface area is also the one that makes the distribution of the number of polygons per voxel the best. This latter issue is actually related to the way a given box is partitioned into voxels and these two issues are discussed in section 7.2 and 7.3.

Another important question is the goodness of the original clustering algorithm. Applied to the kitchen scene, this algorithm does not result in surprising clusters: we indeed get separately the doors knobs, the chairs feet tops, etc. But some clusters as the one of figure 27(a) where the bowl, the knife handle and a part of the breadboard have been put together are more confusing since the configuration of figure 27(b) would be expected. In the same vein, this clustering algorithm on the smallest level of a three levels filter for the garden scene results in huge clusters containing leaves sparsely scattered in the cluster volume as depicted on figure 28, while a greater number of smaller clusters might appear as more natural as shown on figure 29. The problem here is that specifying a cluster as a set of objects whose projection on the three axis is (almost) connected is not sufficient. A symptom of this kind of problem is the $\sum_{all\ levels} \%vol$ statistic described in section 3. Getting a too big value for it —say over 120%, means that the number of voxels traversed during the rendering process is significantly increased and so is the rendering time. Section 7.4 presents a new clustering algorithm getting around this problem.

At last, a key issue is the synergy between the filtering and the clustering steps. Due to the intuition provided by the bucket-sort behavior, the original work of [CDP95] focussed on three levels hierarchies. But as seen in section 6, most of the intersection tests are performed at the highest level so that having a thinner grid for that one is better —remember the $\sqrt{2}$ criterion and the comment about the recursive grid. One way to achieve this goal is to merge the two first levels of the filter which corresponds to a binary partition of the objects' lengths histogram. But as mentioned above, a two levels filter should try to keep apart the big objects that would bridge the gaps from the small ones thus contributing to clusters of volume of the same order of magnitude as the whole scene. Since the filtering and clustering are performed in sequence, finding the threshold value inducing too high values the $\sum_{all\ levels} \%vol$ statistic is difficult. And that's even more the case that scenes as the city are not really prone to clustering — see fig. 8. In practice and due to the observation made in section 2, a two levels filter is just computed by putting together all the objects whose size is below a small percentage of the biggest object size — say 5%. As to decide whether or not a scene is amenable to clustering, some integral geometry based tools are proposed in [CS97].

### 7.2 Clusters orientation

For a given cluster whose polygons are defined by $n$ vertices $\{P_i\}_{i=1..n}$, the problem reduces to the computation of a box of minimal surface area containing the points $P_i$. Although this question has not received a specific attention in the computational geometry community, the related topic of computing bounding boxes of minimal volume has been shown to require sophisticated data structures such as the Voronoi diagram to be performed in optimal $O(n \log n)$ time. This led computer graphics programmers to look for approximate and more tractable solutions, and it was pointed out that the so-called Principal Components Analysis (PCA) used by statisticians could be useful ([Wu92, GLM96]). The goal of this method is to find a new frame such that the inertia of the points projection on the new axis is maximum. The calculation, which runs in linear time, is the following.

Let $X_{n,3}$ be the matrix containing the $3D$ coordinates of the $n$ points $\{P_i\}_{i=1..n}$, the coordinates of $P_i$ being noted $(v_{i,1}, v_{i,2}, v_{i,3})$, and let $C_{n,3}$ be the covariance matrix defined as follows:

$$C_{n,3} = \begin{bmatrix} C_{1,1} & C_{1,2} & C_{1,3} \\ C_{2,1} & C_{2,2} & C_{2,3} \\ C_{3,1} & C_{3,2} & C_{3,3} \end{bmatrix}$$

with

$$C_{i,j} = \frac{\sum_{k=1}^{n} v_{k,i} v_{k,j}}{n} - \mu_i \mu_j \text{ and } \mu_i = \frac{\sum_{k=1}^{n} v_{k,i}}{n}$$

The theory shows that the axis sought are the eigenvectors $(u, v, w)$ of $C_{n,3}$, which are orthogonal to one another since the matrix is symmetric. Once these new vectors have been computed, the coordinates of the projected points on say the axis defined by the vector $u$ are given by $X_{n,3} \cdot u$. Or equivalently, if we note $T$ the matrix $T = [u\ v\ w]$, the coordinates of the points in the new coordinate system are given by $T^{-1} X^t_{n,3}$, with $X^t_{n,3}$ the transposed matrix of $X_{n,3}$.

We implemented this algorithm and were expecting results as the one of figure 24(a), where the red box is obviously tighter than the yellow one. Unfortunately, configurations as the one of figure 24(b) are pretty frequent. And plotting the ratios of the volumes and surface areas for PCA computed bounding boxes over axis aligned ones for the bounding boxes of the clusters of the kitchen gives the graphics of figure 25 — as usual, the left scale refers to the histogram classes percentages, and the right one to the cumulated percentages.

Even if these diagrams are not presented in [GLM96], the reason why the PCA method does not work that well is mentioned: points in the interior of a cluster which ought not influence the selection of a bounding box, can have an arbitrary impact on the eigenvectors. The authors therefore propose a refined method based on the convex hull points. This works fine for their collision detection purposes, but the situation here is more involved as observed at the beginning of this section.

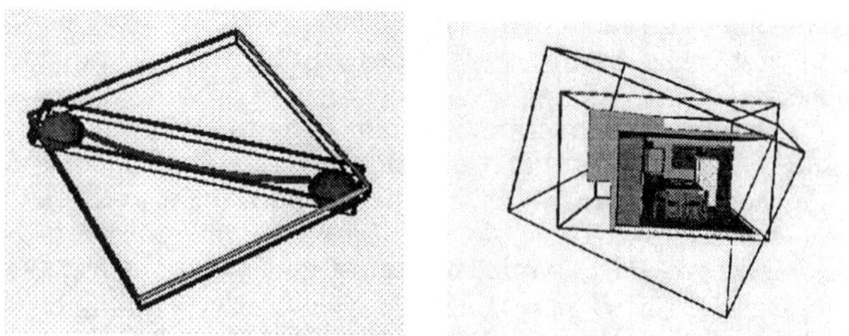

Figure 24: (a,b)Principal Components Analysis and bounding boxes: favorable and less favorable cases

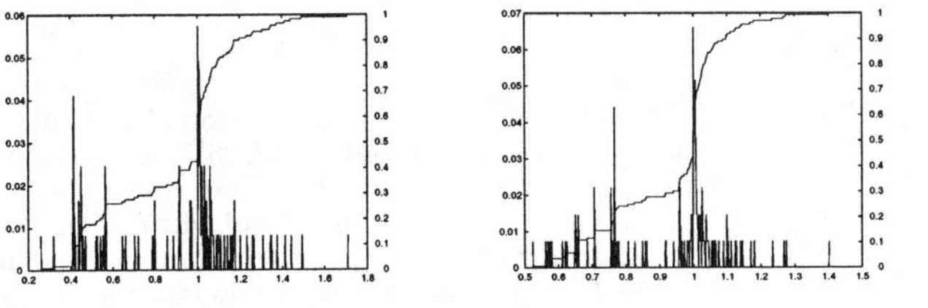

Figure 25: Volume and surface area ratios for the kitchen scene

### 7.3 Heterogeneous gridings

The second "natural" method to improve the way a cluster is stored is to make the distribution of the number of polygons referenced by the voxels as even as possible. And of course, choosing the same value for $n_x, n_y, n_z$ — see section 1.3 — is very unlikely to meet this need. In [KS97], an heterogeneous griding scheme based on the box dimensions is proposed, and the improvements reported are pretty good. We tried a different method based on the notion of discriminant axis. A discriminant axis is an axis which succeeds in separating the objects when subdivided into slabs of the same width. For example on fig. 26(a), the $y$ axis is more discriminant than $x$ axis since along that latter direction the line-segments projections overlap one-another all the way along. A measure of this overlapping introduced in [Caz97] is the following:

**Definition 4** *Let a one-dimensional arrangement of n line segments be a set of n line segments with distinct extremities on some line L. We define the* overlap number *of this arrangement as the sum over all the endpoints, of the number of line segments interiors these endpoints are contained in*

An example of arrangement of 3 line segments

$$\{(a_1, a_2), (b_1, b_2), (c_1, c_2)\}$$

is depicted on fig. 26(b), and the corresponding overlap number is $0+1+2+2+1+0=6$. Call $o_x, o_y$ and $o_z$ the overlap numbers of the three axis. The higher the overlap number $o_i$, the smaller the corresponding number of subdivisions $n_i$, with $i = x, y$ or $z$. A way to proceed is to set $n_i = n^{\beta_i}$ with $\beta_i = g(o_i)$, $g$ a decreasing function from $I\!N \rightarrow [0,1]$, and $\beta_x + \beta_y + \beta_z = 1$. This ensures exactly $n$ voxels for a cluster of $n$ objects and takes into account the ability of the $i$ axis to separate the objects. But so far, we did not find any expression of $\beta_i$ as a function of the $o_i$ leading to satisfactory improvements of the rendering time.

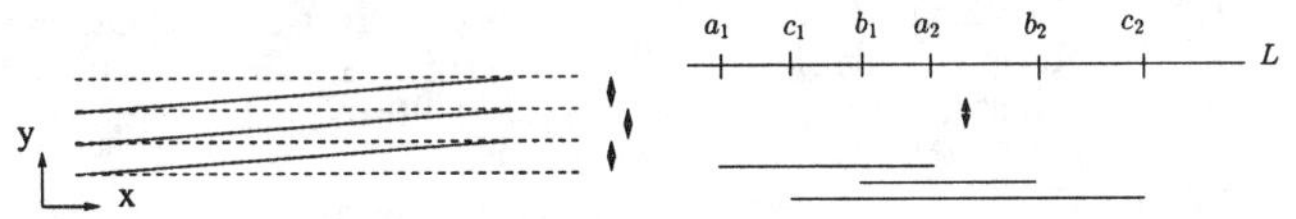


Figure 26: (a)Discriminant axis (b)Arrangement of line segments

### 7.4 Clustering by $2d$ sweeping

To remedy the problem mentioned in 7.1, we restrict the definition of a cluster as follows:

**Definition 5** ***Cluster*** *Let $\{b_1, \ldots, b_n\}$ be the axis aligned bounding boxes of the objects of the scene processed. We define the bounding boxes intersection graph $IG$ as the non directed graph whose nodes are the $b_i$ and two nodes $b_i$ and $b_j$ are linked by an arc if the two bounding boxes intersect one another. A cluster is then defined as a connected component of this graph.*

The use of this definition is to provide a straightforward two step algorithm to compute the clusters by first looking for all the pairs of intersecting bounding boxes and second running a depth first search on the resulting $IG$ graph. As far as we know, no equivalent data structure to the the interval tree (see [PS85, dB+96]) is known in dimension two, so that the first step can be done in sub-optimal time $O(n \log n + k')$ with $k'$ the number of intersecting bounding boxes projections in the $xy$ plane. Practically, it actually turned out that 3/4 of these intersections in the plane were denied along the $z$ axis. The second step just requires $O(n+k)$ with $k \leq k'$ the number of intersecting bounding boxes in $3D$. As shown on figure 27(b) and 29, the resulting clusters are more tight.

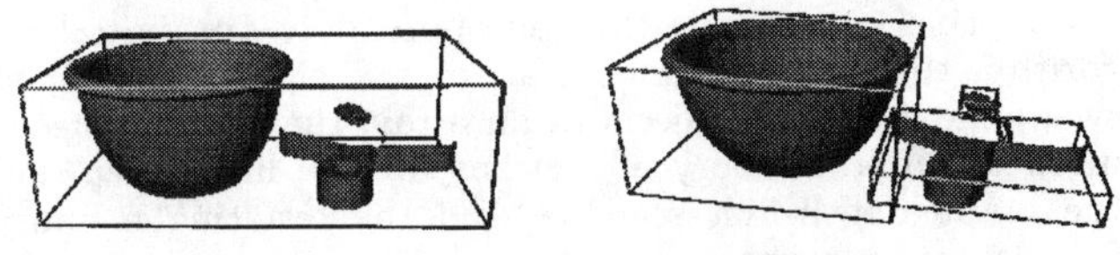

Figure 27: (a)(b)Kitchen stuff

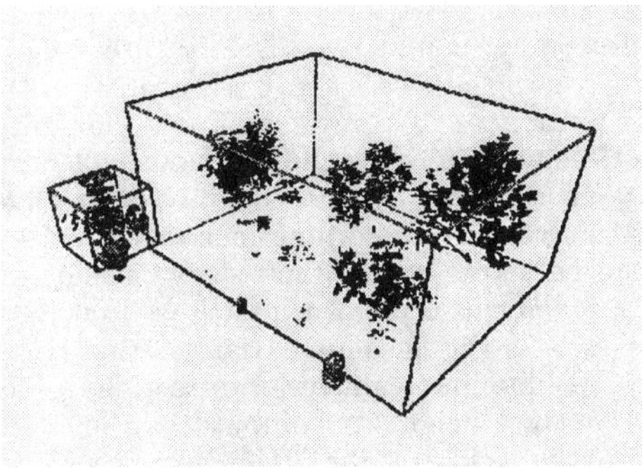

Figure 28: Huge tree leaves clusters

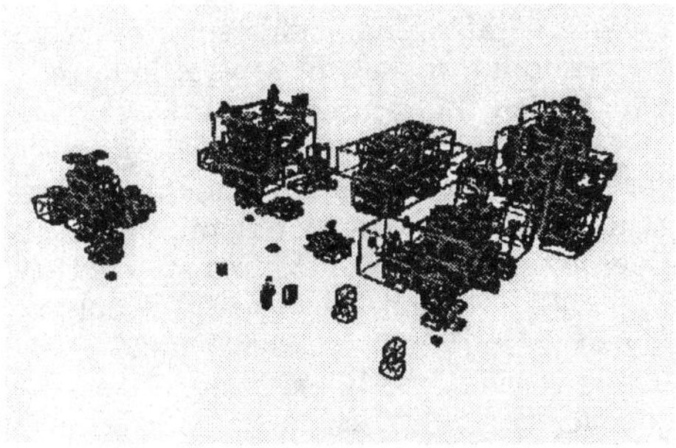

Figure 29: Tighter tree leaves clusters

### 7.5 Selected experimental results

The first improvement we made over the version of [CDP95] was to tune the subdivision parameter of the world cluster for a two-levels filter. For example for the kitchen scene it is easy to get a rendering time as good at the recursive grid that is around 45 seconds while using 3.99 mega-bytes instead of 7.87. On the opposite, restricting this resource to 4 mega-bytes for that data structure results in a performance of 103 seconds which is about 2.3 times slower than the HUG. The conclusions are the same for the Berkeley CS building, that is with the same memory requirement the HUG is between two and three times faster than the recursive grid. It should also be pointed out that for these scenes which are prone to clustering, using either clustering algorithm does not really matter.

On the opposite, for the garden and the city scenes where it was noticed that the original clustering algorithm was not performing that well, the situation is less favorable. The new clustering algorithm isolates houses in the city, sets of leaves in the trees, etc. But is a cluster of say 5 polygons still a cluster? The answer is probably not and due to the observation raised at the end of section 7.1 one could be tempted to pop up these polygons into the World grid ... which ends up being a uniform grid if there is no cluster left in the filter lower levels! This issue deserves more investigation.

Another problem is that as opposed to the initial algorithm that was performing an *in situ* clustering by just swapping the scene objects in an array, the new one needs more dynamic memory allocations for the line sweep and the graph depth-first search. And the resulting memory fragmentation turns out to require as much memory as the recursive grid.

## 8 Conclusion

The results achieved in this paper are threefold. First, we show that coming up with a precise average case analysis of the performances of a ray-tracer is a difficult question since the time consumed by the functions called most often is context-sensitive and depends within a factor five or more on the data structure these functions are called from. Second, we present experimental facts showing that the uniform grid which is known to perform pretty well for a wide set of problems such as the Delaunay triangulation computation, reporting red-blue line segments intersections as well as ray tracing or nearest neighbors computation, is outperformed by the recursive grid and the HUG in terms of data partitioning. Applied to ray-tracing this results in performances one order of magnitude better. Third, we exhibit an improved version of the hierarchy of uniform grids data structure which performs three times faster than the recursive grid for architectural scenes. This solution is particularly suitable for rendering models of several hundreds of megabytes and for which the recursive grid is too much memory consuming.

There are still important issues remaining. First, it would be important to have tools characterizing the ability of a particular model to be clustered and a first step in that direction is made in [CS97]. Second, the benefit of non axis-aligned bounding boxes as in [Wu92, GLM96] should be investigated further. Third it would be interesting to develop and analyze heterogeneous griding schemes. Fourth, our clustering algorithms may find interesting applications in the scope of hierarchical radiosity (see [SD95a]) or Geographic Information Systems.

**Acknowledgment:** The authors wish to thank the models providers as well as Mathieu Desbrun, Eric Haines and Stéphane Rivière for rereading this paper.

## References

[A+94] D.S. Andrews et al. Further comparison algorithms for geometric intersection problems. In $6^{th}$ *International Symposium on Spatial Data Handling*, 1994.

[AEII85] T. Asano, M. Edahiro, H. Imai, and M. Iri. Practical use of bucketing techniques in computational geometry. In G.T. Toussaint, editor, *Computational Geometry*. Elsevier, 1985.

[Caz97] F. Cazals. Combinatorial properties of one-dimensional arrangements. *Experimental Mathematics*, 6(1), 1997.

[CDP95] F. Cazals, G. Drettakis, and C. Puech. Filtering, clustering and hierarchy construction: a new solution for ray-tracing complex scenes. In *EG'95*, Maastricht, 1995.

[CS97] F. Cazals and M. Sbert. Some integral geometry tools to estimate the complexity of 3d scenes. *In preparation*, 1997.

[dB+96] M. de Berg et al. *Computational geometry by example*. Stanford University Press, draft version, 1996.

[Dev86] L. Devroye. *Lectures notes on bucket algorithms*. Birkhauser, 1986.

[Dev88] O. Devillers. *Méthodes d'optimisation du tracé de rayon*. PhD Thesis - Université Paris Sud, 1988.

[GLM96] S. Gottschalk, M.C. Lin, and D. Manocha. Obbtree: A hierarchical structure for rapid interference detection. In *ACM Siggraph 96*, New Orleans, 1996.

[Hai96] E. Haines. Personal communication. 1996.

[JdL92] Erik Jansen and Win de Leeuw. Recursive ray traversal. *Ray Tracing News (available at nic.funet.fi:/pub/graphics/misc/RTNews)*, 5(1), 1992.

[KS97] K. Klimaszewski and T. Sederberg. Faster ray-tracing using adaptative grids. *IEEE Computer Graphics and Applications*, Jan-Feb:42–51, 1997.

[lBWY80] J l. Bentley, B. Weide, and A. Yao. Optimal expected-time algorithms for closest point problems. *ACM Transactions on Mathematical Software*, 6, 1980.

[MMS94] J. Mitchell, D. Mount, and S. Suri. Query-sensitive ray shooting. In $10^{th}$ *ACM CG*, pages 359–368, Stony Brook, 1994.

[NH84] J. Nievergelt and H. Hinterberger. The grid file: an adaptable, symmetric multikey file structure. *ACM Transactions on Database Systems*, 9(1):38–71, 1984.

[O'R94] J. O'Rourke. *Computational Geometry in C*. Cambridge University Press, 1994.

[PS85] F. P. Preparata and M. I. Shamos. *Computational Geometry: An Introduction*. Springer-Verlag, New York, NY, 1985.

[S\'76] L. Sántalo. *Integral geometry and geometric probability*. Addison-Wesley, 1976.

[Sbe93] M. Sbert. An integral geometry based method for fast form-factor computation. *Computer Graphics Forum - Eurographics'93*, 12(3), 1993.

[SD95a] François X. Sillion and George Drettakis. Feature-based control of visibility error: A multi-resolution clustering algorithm for global illumination. In *SIGGRAPH 95*, Los Angeles, 1995.

[SD95b] P. Su and S. Drysdale. A comparison of sequential delaunay triangulation algorithms. In *Proc. of the* $11^{th}$ *ACM Symp. on Computational Geometry*, pages 61–70, 1995.

[Wu92] X. Wu. A linear-time simple bounding volume algorithm. *Graphics Gem*, III, 1992.

# On Fence Design and the Complexity of Push Plans for Orienting Parts *

*(extended abstract)*

Robert-Paul Berretty †‡ Ken Goldberg § Mark H. Overmars † A. Frank van der Stappen †

**Abstract**

A common task in automated manufacturing processes is to orient parts prior to assembly. We consider sensorless orientation of a polygonal part by a sequence of push actions. We show that any polygonal part can be oriented by a sequence of fences placed along a conveyor belt, thereby settling a conjecture by Wiegley *et al.* [25], and present the first polynomial-time algorithm to compute the shortest such sequence. The algorithm is simple and runs in time $O(n^3 \log n)$, where $n$ is the number of vertices of the part. Even though pathological parts can be constructed that require $\Omega(n)$ push actions, it turns out that almost all parts can be oriented using a small number of pushes. We deduce a new bound on the length of the shortest push plan that depends on the thinness – or *eccentricity* – of the part. The bound shows that only $O(1)$ pushes are required for the large class of parts with non-square minimum-width bounding boxes.

## 1 Introduction

Many automated manufacturing processes require parts to be oriented prior to assembly. A part feeder takes in a stream of identical parts in arbitrary orientations and outputs them in a uniform orientation. Part feeders often use data obtained from some kind of sensing to accomplish their task. We consider the problem of *sensorless orientation* of parts, in which the initial pose of the part is assumed to be unknown. In sensorless manipulation, parts are positioned and/or oriented using passive mechanical compliance. The input is a description of the part shape and the output is a sequence of open-loop actions that moves a part from an unknown initial pose into a unique final pose. Among the sensorless part feeders considered in literature are the parallel-jaw gripper [8, 12], a single pushing jaw [3, 14, 15, 19], a conveyor belt with a sequence of (stationary) fences placed along its sides [7, 20, 25], a conveyor belt with a single rotational fence (1JOC) [2], a tilting tray [11, 17], and vibratory plates and programmable vector fields [5, 6]. We focus on sensorless part feeders that rely on push actions to orient parts.

The pushing jaw [3, 14, 15, 19] orients a part by an alternating sequence of pushes and jaw reorientations. The problem of sensorless orientation by a pushing jaw is to find a sequence of push directions that will move the part from an arbitrary initial orientation into a single known final orientation. Such a sequence is referred to as a *push plan.* Akella *et al.* [2] have shown that finding a manipulation plan for the 1JOC part feeder is equivalent to finding a sequence of push actions, so any push plan immediately translates into a solution to the problem of sensorless orientation by the 1JOC part feeder. Goldberg [12] showed that any polygonal part can be oriented by a sequence of pushes. Chen and Ierardi [8] proved that any polygonal part with $n$ vertices can be oriented by $O(n)$ pushes. They showed that this bound is tight by constructing (pathological) $n$-gons that require $\Omega(n)$ pushes to be oriented. Goldberg gave an algorithm for computing the shortest push plan for a polygon. His algorithm runs in $O(n^2)$ time.

The problem of *fence design* is to determine a sequence of fence orientations (see Figure 1) such that the fences with these orientations align the part as it moves down a conveyor belt and slides along these fences [7, 20, 25]. The motion of the belt effectively turns each slide into a push action by the fence in the direction normal to the fence. The fact that the direction of the push, i.e., the normal at the fence, must have a non-zero component in the direction opposite to the motion of the belt imposes a restriction on successive push directions. Fence design can be regarded as finding a constrained sequence of push directions (see Subsection 2.2 for the actual constraints). The additional constraints make fence design considerably more difficult than sensorless orientation by a pushing jaw. Wiegley *et al.* [25] conjectured that a fence design exists for any polygonal part. They gave an exponential algorithm for computing the shortest sequence of fences for a given part.

In this paper, we prove the conjecture that a fence design exists for any polygonal part. In addition, we give a simple $O(n^3 \log n)$ algorithm for computing a fence design

*Research is supported by NATO Collaborative Research Grant CRG 951224,

†Dept. of Computer Science, Utrecht University, P.O. Box 80.089, 3508 TB Utrecht, The Netherlands.

‡Berretty's research is supported by the Dutch Organization for Scientific Research (N.W.O.)

§Industrial Engineering and Operations Research, University of California at Berkeley, Berkeley, CA 94720, USA. Goldberg's research is supported in part by the National Science Foundation under IRI-9612491 and by Presidential Faculty Fellow Award IRI-9553197.

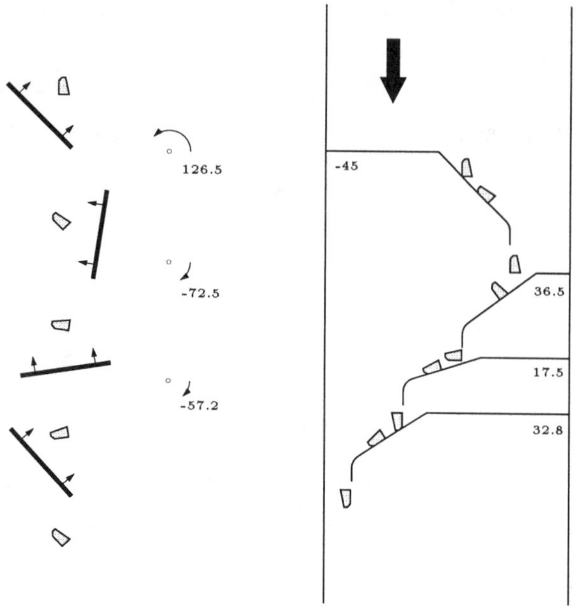


Figure 1: The left picture shows a plan for a pushing jaw. The jaw reorientations are (from top to bottom) 126.4536, −72.4762, and −57,1812. The conveyor belt on the right orients the same part. The fence angles, which are measured relative to the upward vector opposing the direction of belt motion, are −45, 36.4536, 17, 5238, and 32.8188.

of minimal length (in terms of the number of fences used). We show that fence designs of length $O(n)$ exist for a large class of parts. We have implemented the algorithm and the resulting program returns fence designs for input parts instantaneously. The program can be tuned to take into account certain quality requirements on the fence design, like minimum and maximum (successive) fence angles to prevent impractical steep and shallow fences and a long series of fences on a single side of the belt, which would require an impractically wide conveyor belt. The cost of the incorporation of quality measures is an increase of the algorithm's running time to $O(n^4)$.

Even though pathological polygons exist that give rise to push plans and fence designs of length $\Omega(n)$, experimental results indicate that in general only a few pushes or fences suffice to orient a part. In particular, we observe that long and thin parts are easier to orient than compact and fat ones. To formalize this intuition, we define the *geometric eccentricity* of a part based on the aspect ratio of its minimum-width bounding box. We deduce a bound on the length of a push plan that depends on the part's eccentricity rather than on its combinatorial complexity. The bound implies that a constant number of pushes is sufficient to orient a part with a non-square minimum-width bounding box. Our analysis also applies to curved parts, providing the first bound on the length of plans for orienting non-polygonal parts [21]. The results can be generalized to push-squeeze grasps by a parallel-jaw gripper, and apply to a large extent to fence designs. These results for part orienting are in contrast to computational geometric results in robot motion planning, hidden surface removal, depth orders, motion planning, point location, and range searching, where fatness often reduces combinatorial [4, 16, 23] and algorithmic [1, 10, 13, 18, 24] complexities. For part orienting, we show that thinness – or eccentricity – reduces complexity.

Throughout the paper, we assume zero friction between the part and the pushing device. Since any push action acts on the convex hull of the part rather than on the part itself, we assume without loss of generality that the part under consideration is convex.

This paper is organized as follows. In Section 2, we first review the key notion of a push plan, and identify the constraints on the relative push angles for fence design. In Section 3, we focus on a large class of asymmetric polygonal parts and show that these parts can be oriented by push plans and fence designs of linear length (in the number of vertices. We also show that, under the eccentricity assumption, the length of these plans and designs is actually constant. Section 4 discusses an $O(n^3 \log n)$ algorithm for computing the shortest fence design for a part. Section 5 generalizes the results of Section 3 to parts with some kind of symmetry.

## 2 Push plans and fence designs

### 2.1 The push function

In this section we focus on the push function of a part. Let $P$ a convex polygonal part with $n$ vertices and center-of-mass $c$. We assume that a fixed coordinate frame is attached to $P$. Directions are expressed relative to this frame. The contact direction of a supporting line $l$ of $P$ is uniquely defined as the direction of the normal of $l$ pointing into $P$. The radius function $r : [0, 2\pi) \to \mathbb{R}^+$ maps a direction $\phi$ onto the distance from $c$ to the supporting line of $P$ with contact direction $\phi$. For a polygonal part, the radius function is a continuous piecewise sinusoidal function, and can be computed in $O(n)$ time by checking each vertex [15]. The final orientation of a part that is being pushed can be determined from its radius function.

In most cases, parts will start to rotate when pushed. If pushing in a certain direction does *not* cause the part to rotate, then the contact normal at one of its points of contact with the jaw passes through the center-of-mass [15]. We refer to the corresponding direction of the contact normal as an *equilibrium* push direction or orientation. If pushing does change the orientation, then this rotation changes the orientation of the pushing device relative to the part. We assume that pushing continues until the part stops rotating and settles in a stable equilibrium pose, which is an equilibrium with a preimage of non-zero length.

The *push function* $p : [0, 2\pi) \to [0, 2\pi)$ links every orientation $\phi$ to the orientation $p(\phi)$ in which the part $P$ settles after being pushed by a jaw with initial contact direction $\phi$ (relative to the frame attached to $P$). The rotation of the part due to pushing causes the contact direction of the jaw to change. The final orientation $p(\phi)$ of the part is the contact direction of the jaw after the part has settled. The equilibrium push directions are the fixed points of $p$.

The push function $p$ of a polygonal part consists of steps, which are intervals $I \subset [0, 2\pi)$ for which $p(\phi) = C$ for all $\phi \in I$ and some constant $C \in I$. The steps of the push function are easily constructed from the radius function $r$. If the part is pushed in a direction corresponding to a point of non-horizontal tangency of the radius function then the part will rotate in the direction in which the radius decreases. The part finally settles in an orientation corresponding to a local minimum of the radius function. As a result, all points in the open interval $I$ bounded two consecutive local maxima

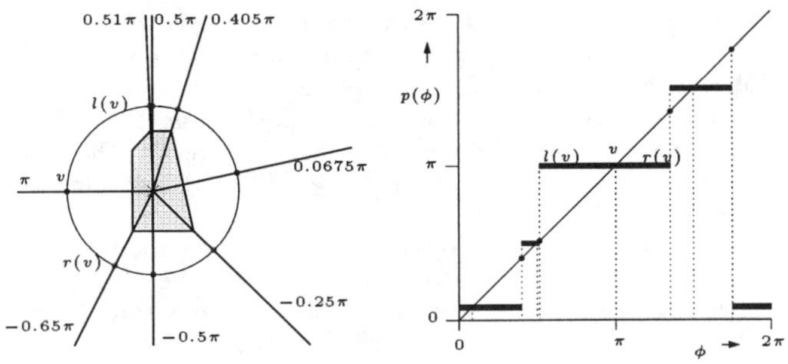


Figure 2: The push function for a polygonal part. The minima correspond to normals of edges that intersect the center-of-mass. The maxima correspond to a subset of the vertices of the polygon. The horizontal steps of the push function are angular intervals between two successive maxima.

of the radius function $r$ map onto the orientation $\phi \in I$ corresponding to the unique local minimum of $r$ on $I$. (Note that $\phi$ itself maps onto $\phi$ because it is a point of horizontal tangency.) This results in the steps of the push function. In preparation for the next section, we define two open intervals $l(v) = \{\phi < v | p(\phi) = v\}$ and $r(v) = \{\phi > v | p(\phi) = v\}$ for each fixed point – or equilibrium orientation – $v$ of the push function. We refer to these intervals as $v$'s left and right environment respectively. The interval $l(v)$ corresponds to the half-step left of $v = f(v)$ and $r(v)$ corresponds to the half-step right of $v = f(v)$ (see Figure 2). Note that an equilibrium $v$ is stable if $l(v) \cup r(v) \neq \emptyset$. The intervals $l(v)$ and $r(v)$ are the (maximal) intervals without equilibrium orientation.

Besides the steps, there are isolated points satisfying $p(\phi) = \phi$ in the push function, corresponding to local maxima of the radius function. Figure 2 shows a polygonal part and its push function. In the full version we show that any step function having only non-zero-length half-steps represents a polygonal part, and thus is a push function. A push function $p$ is said to have period $d$ if $p(\phi) = p((\phi + d) \bmod 2\pi)$ for all $\phi \in [0, 2\pi)$. Any part can only be oriented up to symmetry in its push function.

We use the abbreviation $p_\alpha$ to denote the (shifted) push function defined by

$$p_\alpha(\phi) = p((\phi + \alpha) \bmod 2\pi),$$

for all $\phi \in [0, 2\pi)$. Note that $p_\alpha(\phi)$ is the final orientation of a part in initial orientation $\phi$ after a reorientation by $\alpha$ followed by a push. We can now define a *push plan.*

**Definition 2.1** *A* push plan *is a sequence* $\alpha_1, \ldots, \alpha_m$ *such that* $p_{\alpha_m} \circ \ldots \circ p_{\alpha_1}(\phi) = \Phi$ *for all* $\phi \in [0, 2\pi)$ *and some fixed* $\Phi \in [0, 2\pi)$.

An important property of the push function is monotonicity, or the order preserving property [17]. A sequence $s_1, \ldots, s_n$ of elements of a set $S$ is *ordered* if $s_1, \ldots, s_n$ are encountered in order when the generating cycle of $S$ is traced *once*, starting from $s_1$. A function $p : S \to S$ is *monotonic* if for any ordered sequence $s_1, \ldots, s_n$ the sequence $p(s_1), \ldots, p(s_n)$ is also ordered.

### 2.2 Fence design

In this section we address the problem of designing a series of fences $f_1, \ldots, f_m$ that will orient $P$ when it moves down a

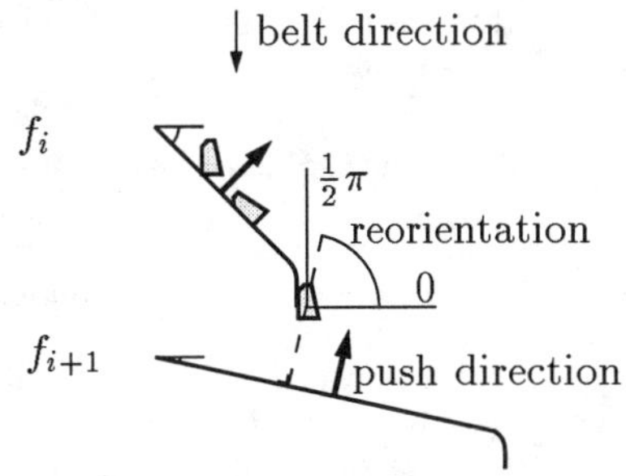


Figure 3: For two successive left fences, the reorientation of the push lies in the range $(0, \pi/2)$.

conveyor belt and slides along these fences $f_1, \ldots, f_m$. Let us assume that the conveyor belt moves in vertical downward direction as in Figure 1. We distinguish between left fences, which are placed along the left belt side, and right fences, which are placed along the right side. The fence angle $\beta_i$ of a fence $f_i$ denotes the angle between the vertical upward pointing vector opposing the motion of the belt and the normal to the fence with a positive component in upward direction. The motion of the belt turns the part's slide along a fence into a push by the fence. The direction of the push is – by the zero friction assumption – orthogonal to the fence with a positive component in the (upward) direction opposing the motion of the belt. Thus, the motion of the belt causes any push direction to have a positive component in the direction opposing the belt motion. We now translate this constraint on the push direction relative to the belt into a constraint on successive push directions relative to the part.

The slide along a fence $f_i$ causes one of $P$'s edges $e$ to align with the fence. The curved tip of the fence [7] guarantees that $e$ is aligned a with the belt sides as $P$ slides off the fence. If $f_i$ is a left fence then $e$ faces the left belt side (see Figure 3). If $f_i$ is a right fence, it faces the right side. Assume $f_i$ is a left fence. At the moment of leaving $f_i$, hence, after the push, the contact direction of $f_i$ is perpendicular to the belt and towards the right belt side. So, the reorientation of the push is expressed relative to this direction. Figure 3 shows that the reorientation $a_{i+1}$ is in the range $(0, \pi/2)$ if we choose $f_{i+1}$ to be a left fence. If we take a right fence $f_{i+1}$ then the reorientation is in the range $(\pi/2, \pi)$. A similar analysis can be done when $P$ slides off a right fence and the edge $e$ faces the left belt side. The result is given in the table.

| $f_i$ | $\alpha_{i+1}$ | $f_{i+1}$ |
|---|---|---|
| left | $(0, \pi/2)$ | left |
| left | $(\pi/2, \pi)$ | right |
| right | $(-\pi, -\pi/2)$ | left |
| right | $(-\pi/2, 0)$ | right |

The table shows that the type of fence $f_i$ imposes a bound on the reorientation $\alpha_{i+1}$. Application of the same analysis to fences $f_{i-1}$ and $f_i$ and reorientation $\alpha_i$ leads to the following definition of a valid fence design [25].

**Definition 2.2** *A* fence design *is a push plan* $\alpha_1, \ldots, \alpha_m$ *satisfying for all* $1 \leq i < m$*:*

- $\alpha_i \in (0, \frac{\pi}{2}) \cup (-\pi, -\frac{\pi}{2}) \Rightarrow \alpha_{i+1} \in (0, \frac{\pi}{2}) \cup (\frac{\pi}{2}, \pi)$
- $\alpha_i \in (-\frac{\pi}{2}, 0) \cup (\frac{\pi}{2}, \pi) \Rightarrow \alpha_{i+1} \in (-\frac{\pi}{2}, 0) \cup (-\pi, -\frac{\pi}{2})$.

The push plan on the left in Figure 1 satisfies the constraints of Definition 2.2, and is therefore also a fence design.

## 3 The complexity of push plans and fence designs

In this section we concentrate on parts with aperiodic push functions and a unique longest left environment $l(v)$ and a unique longest right environment $r(v')$. We prove that these *asymmetric* parts can be oriented by a push plan and a fence design of length $O(n)$. After that, we show that almost all parts can actually be oriented by a constant number of pushes or fences.

### 3.1 A linear bound

Chen and Ierardi [8] use a sequence of equivalent basic actions to orient a polygonal part $P$ with a (unique) longest right environment $r(a)$ of length $\alpha$ and a longest left environment $l(b)$ of length $\beta$. Each basic action consists of a reorientation of the jaw by an angle of $\alpha - \mu$, with $\mu > 0$ such that $\alpha - \mu > |r(a')|$ for any equilibrium orientation $a' \neq a$, and a subsequent push. (In a similar manner, we could use reorientations by $-(\beta - \mu)$). Note that a reorientation of the jaw by $\alpha - \mu$ corresponds to a change of the orientation of the part by $\alpha - \mu$. Every basic action puts the part into an equilibrium orientation. After each basic action, the part is therefore in one of a finite number of equilibrium orientations. Let us label the $m$ equilibrium orientations $a_0, \ldots, a_{m-1}$ in order of *decreasing* angle starting from $a_0 = a$. After the first push, the part $P$ can be in any of the equilibrium orientations $a_0, \ldots, a_{m-1}$. Chen and Ierardi show that every next basic action eliminates the last orientation in the sequence as possible orientation of the part. Assume that $P$ is in one of the orientations $a_0, \ldots, a_k$, for some $k \geq 1$. The key idea behind the proof is that a next basic action will cause $P$, when in $a_0$, to stay in $a_0$ because $\alpha - \mu < |r(a_0)| = |r(a)|$, and, when in $a_i$ for some $1 \leq i \leq k$, to move into some orientation $a_j$ with $0 \leq j \leq i-1$ because $\alpha - \mu > |r(a_i)|$. Upon completion of the basic action, the part will therefore be in one of the orientations $a_0, \ldots, a_{k-1}$. As a consequence, a total of $m+1$ basic actions suffices to put $P$ into orientation $a_0 = a$. In other words, the sequence $(\alpha - \mu)^{m+1}$ is a valid push plan for $P$. (If we use the longest left environment, we obtain the sequence $(-\beta + \mu)^{m+1}$.) The observation that $m = O(n)$ leads to the result of Theorem 3.1.

**Theorem 3.1** *[8] A polygonal part $P$ with $n$ vertices can be oriented by $N = O(n)$ pushes.*

The result of Chen and Ierardi can be further refined by monitoring the *length of the interval* containing the possible orientations of $P$ rather than the *number* of possible orientations while performing the basic actions described above. It can be shown that two subsequent basic actions shrink the interval of possible orientations from $[b, a]$ to $[\min\{b + (\alpha - \mu), a\}, a]$ (see [22] for a proof). Hence, the length of the interval decreases by $\alpha - \mu$ unless $b$ is close to $a$. This leads to an upper bound of $2\lceil 2\pi/\alpha \rceil + 1$ on the number of pushes.

**Lemma 3.2** *For part $P$, let $\alpha$ be the length of largest angular interval without equilibrium orientations. The part $P$ will be oriented after $N \leq 2\lceil 2\pi/\alpha \rceil + 1$ pushes.*

Lemma 3.2 serves as a tool in Subsection 3.2.

In order for the push plan $(\alpha - \mu)^{m+1}$ or $(-\beta + \mu)^{m+1}$ to be a valid fence design, we have to show that it satisfies the constraints formulated in Definition 2.2. We observe first of all that there can be no more than three environments of length at least $\pi/2$, because the longest two left environments have different lengths and the longest two right environments have different lengths. As a result, there is at most one left environment of size $\geq \pi/2$ or at most one right environment of size $\geq \pi/2$. Assume that there is at most one right environment of size $\geq \pi/2$. Although the length $\alpha$ of the longest right environment $r(a)$ can be at least $\pi/2$, the length $\alpha'$ of the second largest right enviroment $r(a')$ must be smaller than $\pi/2$. If we now choose $\mu$ such that $\alpha' < \alpha - \mu < \min\{\alpha, \pi/2\}$, then we get that $\alpha - \mu > |r(v)|$ for all equilibrium orientations $v \neq a$. In addition, we clearly have that $\alpha - \mu < \pi/2$, which makes it easy to verify that $(\alpha - \mu)^{m+1}$ is a fence design.

**Theorem 3.3** *An asymmetric polygonal part $P$ with $n$ vertices can be oriented by a fence design of length $N = O(n)$.*

### 3.2 Eccentric parts

The aim of this subsection is to show that, in practice, the length of the shortest push plan and fence design remains far below the the linear upper bound of the previous subsection, and is in fact constant for a large class of parts. To this end, we derive a bound that depends on the part's thinness – or eccentricity – rather than its combinatorial complexity. We use the aspect ratio of the minimum width bounding box as a measure for the eccentricity of the part. This bounding box is obtained by taking the two closest parallel supporting lines of the part and the two parallel supporting lines perpendicular to those. (The distance between its two closest parallel supporting lines is also referred to as the width of the part.) By definition, there exists no narrower bounding box of the part. The length-to-width ratio, or aspect ratio, of the minimum width bounding box can be expressed by a scalar of the form $1 + \epsilon$, with $\epsilon \geq 0$. Note that the minimum-width bounding box is a square if $\epsilon = 0$.

**Definition 3.4** *Let $P \subset \mathbb{R}^2$ be a part. Let $a \geq 1$ be the aspect ratio of the minimum width bounding box of $P$, which is defined by the two closest parallel supporting lines and the two perpendicular supporting lines of $P$.* The eccentricity $\epsilon$ *of $P$ is defined by $\epsilon = a - 1$.*

Figure 4 shows a part with eccentricity $\epsilon = (10.0/3.3) - 1 \approx 2.0$.

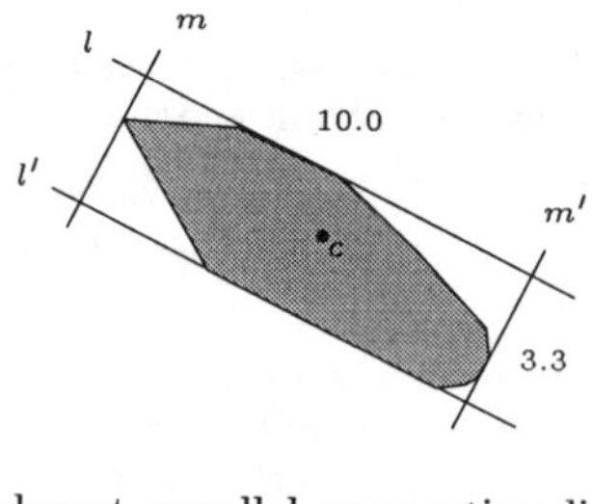


Figure 4: The closest parallel supporting lines $l$ and $l'$ and the orthogonal supporting lines $m$ and $m'$ define $P$'s minimum width bounding box.

As the number of push operations is (Lemma 3.2) inversely proportional to the length $\alpha$ of the longest interval without equilibrium push direction, we would like to have a lower bound on $\alpha$ given the eccentricity $\epsilon$ of the part. We derive the equivalent: an upper bound on $\epsilon$ given the length $\alpha$ of the longest interval without equilibrium push directions.

Consider a part $P$ with a largest angular interval without equilibrium push direction of length at most some $\alpha < \pi/4$. Now, there must exist an equilibrium push direction of $P$ in every interval of length $\alpha$. We show that achieving an equilibrium push direction for $P$ in every such interval requires an upper bound on its eccentricity $\epsilon$: the part cannot be arbitrarily thin.

In the considerations below, we use a property of the point of contact of the part $P$ with a jaw (or fence) pushing $P$ in an equilibrium push direction. The jaw touches the part $P$ either at a vertex or along an edge. We observe that $P$ will not rotate if the contact normal at the or one of the points of contact with the jaw passes through its center-of-mass $c$ [15, 22]. In other words, the normal at some point of contact of $P$ and a supporting line $l$ with contact direction equal to some equilibrium push direction of $P$ passes through $c$.

Let us consider the part $P$ and its two closest parallel supporting lines. Without loss of generality we assume that the distance between the closest parallel supporting lines $l$ and $l'$ equals 1. Let $l$ be supporting line that is at most a distance 0.5 away from $c$. We find it convenient to express vector and push directions relative to the contact direction of $l$. The direction of any other vector $v$ equals the clockwise angle between $v$ and the positive $y$-axis. Our objective is to show that there is a bound on the extension of the part in the direction parallel to $l$ and $l'$.

We extend a collection of rays from $c$. The first ray $\rho_0$ emanates from $c$ towards and perpendicular to $l$. Additional rays $\rho_i$ and $\rho_{-i}$ ($0 < i \leq \lfloor \pi/\alpha \rfloor$) are extended from $c$ in such a way that the clockwise angle from $\rho_i$ to $\rho_0$ and the counterclockwise angle from $\rho_{-i}$ to $\rho_0$ both equal $i\alpha$. A last ray referred to as both $\rho_\infty$ and $\rho_{-\infty}$ extends from $c$ towards and perpendicular to $l'$. Figure 5 shows the resulting collection of rays. Each pair of consecutive rays $\rho_i$ and $\rho_{i+1}$ (for $-\lfloor \pi/\alpha \rfloor \leq i < \lfloor \pi/\alpha \rfloor$) bounds a sector of angle $\alpha$ from the center-of-mass $c$. The boundary of the convex part $P$ intersects each such sector and each ray $\rho_i$.

The length $\alpha$ of the longest interval without equilibrium orientation implies that there must be an equilibrium push direction in every angular interval $[i\alpha, (i+1)\alpha]$. This in turn means that there must be a line with contact direction in $[i\alpha, (i+1)\alpha]$ such that the normal at its point of contact with $P$ passes through the center-of-mass $c \in P$. A simple observation is that in order for a point to have a vector with a direction in $[i\alpha, (i+1)\alpha]$ originating from it pass through $c$ that point must lie in the wedge bounded by the supporting lines of the vectors with directions $i\alpha$ and $(i+1)\alpha$ emanating from $c$. Using the fact that the vector originates at a point on the boundary of the convex part $P$, we can constrain the position of this point to lie in the half-wedge or sector bounded by the region bounded by the rays $\rho_i$ and $\rho_{i+1}$. Our assumption on the size of the longest interval without equilibrium orientation implies that *every* sector bounded by two consecutive rays $\rho_i$ and $\rho_{i+1}$ with $-\lfloor \pi/\alpha \rfloor \leq i < \lfloor \pi/\alpha \rfloor$ contains a point of $\partial P$ in which a supporting line of $P$ through that point has a contact direction in the range $[i\alpha, (i+1)\alpha]$. This implied fact is powerful enough to bound the extension of the part in each of the sectors by two polygonal chains.

We define a point $p_i$ on every ray $\rho_i$ and a point $p_{-i}$ on every ray $\rho_{-i}$ for $i \geq 1$. The construction of the points is incremental in the sense that the position of $p_{i+1}$ is determined from the position of $p_i$, and similarly for $p_{-(i+1)}$ from $p_{-i}$. Let $p_{-1}$, $p_0$, and $p_1$ be the points of intersection of $l$ and $\rho_{-1}$, $\rho_0$, and $\rho_1$ respectively. To find $p_{i+1}$ for all $1 < i < \lfloor \pi/\alpha \rfloor$, we take the line through $p_i$ at a counterclockwise angle of $(i-1)\alpha$ from $l$. The intersection of this line with $\rho_{i+1}$ defines $p_{i+1}$ The last point $p_\infty$ on $\rho_\infty$ lies at the intersection of $\rho_\infty$ and the line through $p_{\lfloor \pi/\alpha \rfloor}$ at a counterclockwise angle of $(\lfloor \pi/\alpha \rfloor - 1)\alpha$ from $l$. The resulting chain $p_0 p_1 \ldots p_\infty$ takes a left or counterclockwise turn of $\alpha$ at every intersection with a ray $\rho_i$, for $i > 1$. A second sequence of points $p_{-(i+1)}$ is constructed by a similar incremental scheme, using lines at a clockwise angle of $(\lfloor \pi/\alpha \rfloor - 1)\alpha$ from $l$. Figure 5 shows parts of the chains $p_0 p_1 \ldots p_\infty$ and $p_0 p_{-1} \ldots p_{-\infty}$. It can be

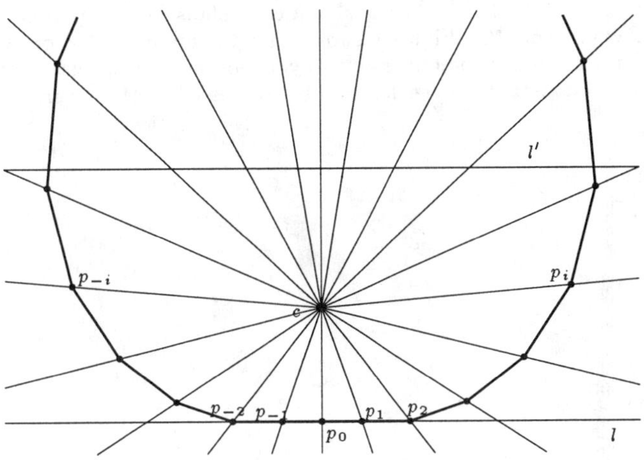


Figure 5: The chains $p_0 p_1 \ldots p_\infty$ and $p_0 p_{-1} \ldots p_{-\infty}$ which bound the extension of the part $P$.

shown that the part $P$ lies entirely inside the region bounded by the chains $p_0 p_1 \ldots p_\infty$ and $p_0 p_{-1} \ldots p_{-\infty}$. The proof is based on induction: assuming that the part $P$'s intersection with the sector bounded by $\rho_{i-1}$ and $\rho_i$ is restricted to the triangle $cp_{i-1}p_i$, we use the supporting line with (equilibrium) contact direction in $[(i-1)\alpha, i\alpha]$ tangent to $P$ inside $cp_{i-1}p_i$ to restrict $P$'s intersection with the sector bounded by $\rho_i$ and $\rho_{i+1}$ to the triangle $cp_ip_{i+1}$. The extension of the chains in the direction parallel to $l$ and $l'$ bounds the length of the minimum width bounding box [22]. Theorem 3.5 gives the resulting bound on $\epsilon$.

**Theorem 3.5** *Let $P$ be a part and let $\alpha$ be the length of the longest angular interval without equilibrium push directions. Then $P$'s eccentricity $\epsilon$ is bounded by*

$$\epsilon \leq \frac{\cos^{k-1}\alpha \cdot \sin(k+1)\alpha}{\cos^k 2\alpha} - 1,$$

*where $k = \lceil \pi/(2\alpha) \rceil$.*

In order to combine Lemma 3.2 and Theorem 3.5 into an upper bound on the number of pushes $N$ as a function of the eccentricity $\epsilon$, we would have to invert the upper bound on the eccentricity given $\alpha$. Given its complexity and its discontinuity, inverting seems impossible. Therefore, we settle for the less elegant formulation of the bound on the length of the push plan as a function of the part eccentricity given by Theorem 3.6.

**Theorem 3.6** *Let $P$ be a part with eccentricity*

$$\epsilon > \frac{\cos^{k-1}\alpha \cdot \sin(k+1)\alpha}{\cos^k 2\alpha} - 1$$

$(k = \lceil \pi/(2\alpha) \rceil)$, *for some* $\alpha \in (0, \pi/4)$. *Then,* $P$ *can be oriented by a push plan of length* $N$ *with*

$$N \leq 2\lceil \frac{2\pi}{\alpha} \rceil + 1.$$

Theorem 3.6 shows that the number of push actions needed to orient a part is a function of its eccentricity. As the entire analysis of this subsection applies to arbitrary parts, it provides the first bound on the length of a push plan for non-polygonal parts, and improves the bound of $2n - 1$ by Chen and Ierardi for a polygonal part with $n$ vertices to the minimum of $2n - 1$ and the bound based on the part's eccentricity provided by Theorem 3.6. The result in Theorem 3.6 can be extended [22] to push-squeeze plans for the parallel-jaw gripper [12]. Figure 6 shows that the upper bound on $N$ for a given eccentricity $\epsilon$ given by Theorem 3.6 is a relatively small constant – even for smaller values of $\epsilon$. Moreover, we

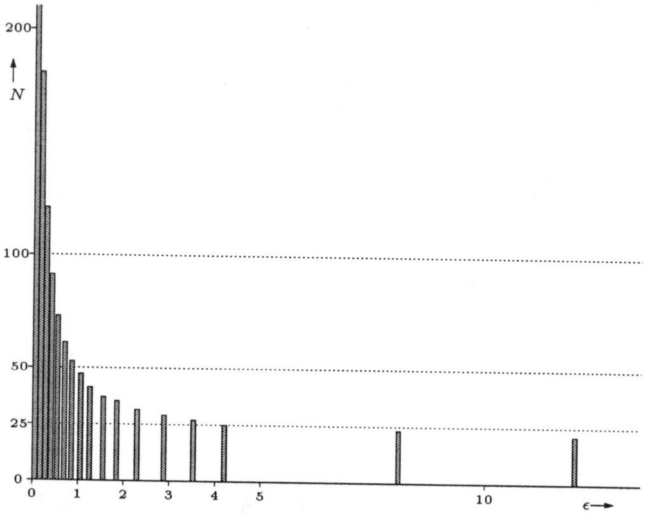


Figure 6: Upper bound on the length $N$ of the shortest push plan as a function of the eccentricity $\epsilon$.

expect the length of the shortest push plan to remain far below the bound of Theorem 3.6 – which actually bounds the length of some rigorous (non-optimal) push plan.

Let us now consider the consequences for fence design. We must use reorientations by $\alpha - \mu$ if we want to use no more than $2\lfloor 2\pi/\alpha \rfloor + 1$ fences. These reorientations can be either positive or negative, depending on whether the environment of length $\alpha$ is a right or left environment. Assume there exists a right environment $r(a)$ of length $\alpha$. If $\alpha \leq \pi/2$ or $r(a)$ is the only environment longer than $\pi/2$, then the considerations of the last paragraph of the previous subsection still apply and we end up with a fence design of length at most $2\lfloor 2\pi/\alpha \rfloor + 1$, which is a constant for parts with non-zero eccentricity. For parts with a second environment of length at least $\pi/2$, we do not yet have a fence design of constant length.

## 4 An algorithm for fence design

As every fence puts the part in a *stable* equilibrium orientation, the part is in one of these $m_s = O(n)$ orientations as it travels from one fence to another. Let us label these stable equilibria $a_0, \ldots, a_{m_s-1}$. After a first fence, the part can be in any of the stable orientations $a_0, \ldots, a_{m_s-1}$. The problem is – like in Subsection 3.1 – to reduce the set of possible orientations of $P$ to one stable orientation $a_i$ by a sequence of fences. We build a directed graph on all possible *states* of the part as it travels from one fence to a next fence. A state consists of a set of possible orientations of the part plus the type (left or right) of the last fence, as the latter imposes a restriction on the reorientation of the push direction. Although there are $O(2^{m_s})$ subsets of $\{a_0, \ldots, a_{m_s-1}\}$, we shall see that we can restrict ourselves to subsets consisting of sequences of adjacent stable equilibria. Any such sequence can be represented by the closed interval $s$ defined by the first and last stable orientation $a_i$ and $a_j$ of the sequence. The resulting graph has $O(m_s^2)$ nodes.

Consider two graph nodes $(s, t)$ and $(s', t')$, where $s = [a_i, a_j]$ and $s'$ are intervals of stable equilibria and $t$ and $t'$ are fence types. Let $I_{t,t'}$ be the open interval (of length $\pi/2$) of reorientations admitted by the successive fences of types $t$ and $t'$ according to the table in Subsection 2.2. There is a directed edge from $(s, t)$ to $(s', t')$ if there is an angle $\alpha \in I_{t,t'}$ such that a reorientation of the push direction by $\alpha$ followed by a push moves any stable orientation in $s$ into a stable orientation in $s'$. To check this condition, we determine the preimage $(u, w) \supseteq s'$ of $s'$ under the (monotonic) push function. Observe that if $|s| = a_j - a_i < w - u$, any reorientation in the open interval $r = (u - a_i, w - a_j)$ followed by a push will map $s$ into $s'$. We add an edge from $(s, t)$ to $(s', t')$ if the intersection of $r$ and the interval $I_{t,t'}$ of admitted reorientations is non-empty, and label this edge with $r \cap I_{t,t'}$. For convenience, we add a source and a sink to the graph. We connect the source to every node $(s, t)$ in the graph for which $s$ contains all $m_s$ stable equilibria, and every node $(s, t)$ with $s = [a_i, a_i]$ to the sink. Every path from the source to the sink now represents a fence design; a fence design of minimum length is found by a breadth first search from the source.

**Theorem 4.1** *The search for a path in the graph gives a fence design, if a fence design exists.*

**Proof:** We prove that all possible fence designs are represented by paths in the graph.
($\Rightarrow$) If there is a path, we prove there is a fence design. Since there is an edge from $(s, t)$ to $(s', t')$ if and only if the successive fences $t$ and $t'$ allow for a reorientation that maps $s$ into $s'$, this follows immediately from the construction of the graph.
($\Leftarrow$) If there is a fence design, we prove there is a path in the graph that represents this fence design. Let $f_1, \ldots, f_N$ be a fence design. We track the possible orientations of the fence design, and prove by induction that for every set of possible orientations, there is a node in the graph, and furthermore, there is a path from the source to such a node. Let $A_i$ denote the set of all possible orientations of $P$ leaving $f_i$. It is easy to see that for each $A_i$ there are multiple nodes that include the set of possible orientations.

After the first fence $f_1$, all stable orientations are possible. Since we added edges from the source to all nodes containing all stable orientations, all these nodes are reachable.

We now assume that for fence $f_i$ having type $t$ in our fence design the nodes $(s, t)$ with $s \supseteq A_i$ are reachable from the source. Let $t'$ be the fence type of $f_{i+1}$. Let $(s', t')$ be a node such that $s' \supseteq A_{i+1}$. Let $(u, w)$ denote the preimage of $s'$. Since the push function is monotonic and by existence of the fence design which maps $A_i$ onto $A_j$, there is an admitted reorientation $\alpha_{i+1}$ by $f_{i+1}$ such that $(u - \alpha_{i+1}, w - \alpha_{i+1}) \supset A_i$. Therefore, let $(s,t)$ be a node such that $(u - \alpha_{i+1}, w - \alpha_{i+1}) \supset s \supseteq A_i$. There is an edge from $(s,t)$ to $(s',t')$, and there is a path from the source to $(s,t)$. Since $s' \supseteq A_{i+1}$ is arbitrary, all $(s',t')$ with $s' \supseteq A_{i+1}$ are reachable from the source. □

An important observation is that some graph edges are redundant if we are just interested in a fence design of minimum length. Consider a node $(s,t)$ and all its outgoing edges to nodes $(s' = [a_i, a_j], t')$ for a fixed left endpoint $a_i$ and a fixed fence type $t'$. We show that only one of these outgoing edges is sufficient.

**Lemma 4.2** *Let $(s,t)$, $(s',t')$, and $(s'',t')$ be nodes with $s' = [a_i, a_j]$ and $s'' = [a_i, a_k]$ and let $s' \supseteq s''$. If there are edges from $(s,t)$ to both $(s',t')$ and $(s'',t')$, then the edge from $(s,t)$ to $(s',t')$ can be deleted without affecting the length of the shortest path in the graph.*

**Proof:** Assume that $\tau$ is a path from source to sink containing the edge $((s,t),(s',t'))$ and assume $((s',t'),(s''',t'''))$ succeeds this edge in $\tau$. Because $s' \supseteq s''$, there must also be an edge $((s'',t'),(s''',t'''))$ in the graph. Hence, we can replace the edges $((s,t),(s',t'))$ and $((s',t'),(s''',t'''))$ in $\tau$ by $((s,t),(s'',t'))$ and $((s'',t'),(s''',t'''))$ without affecting the length of $\tau$. □

Lemma 4.2 implies that every node in the graph has only one outgoing edge per set of nodes with intervals with a common left endpoint and a common fence type. The single edge that remains after application of Lemma 4.2 is the one to the node corresponding to the shortest interval. Since there are $O(m_s) = O(n)$ possible left endpoints and just two fence-types, the number of edges of one node can be reduced to $O(n)$. The total number of edges of the graph therefore becomes $O(n^3)$. A way to find the appropriate outgoing edge is to perform a binary search in each such set of nodes. Since a check for the existence of an edge between a pair of nodes takes $O(1)$ time, we can build the reduced graph in $O(n^3 \log n)$ time.

**Theorem 4.3** *The reduced graph contains $O(n^2)$ nodes and $O(n^3)$ edges and can be computed in $O(n^3 \log n)$ time.*

A breadth first search of the graph takes $O(n^3)$ time (see e.g. [9]). This yields the following theorem.

**Theorem 4.4** *The computation of the optimal fence design for a polygonal part takes $O(n^3 \log n)$.*

Let $\tau$ be a path in the graph from the source to the sink. Every edge of $\tau$ corresponds to a non-empty angular interval of possible reorientations of the push direction. We simply pick the midpoint of every such interval as the reorientation, and get a push plan which is a fence design. We can easily compute the fence angles from the reorientation angles on the path, using the properties of the fence framework [25].

Our implementation offers the user the additional possibility of adding costs to graph edges. By doing so, the user can prevent the algorithm from outputting certain types of fence designs. Assigning high costs to edges between any pair of nodes of the same fence type $t$, for example, will cause the algorithm to output a sequence of alternating (left and right) fences if such a sequence exists. Alternating sequences are often preferred over sequences containing cascades of left (or right fences), as they generally allow for narrower conveyor belts (see Figure 7). Different cost assignments can be found to prevent e.g. unwanted steep and shallow fences.

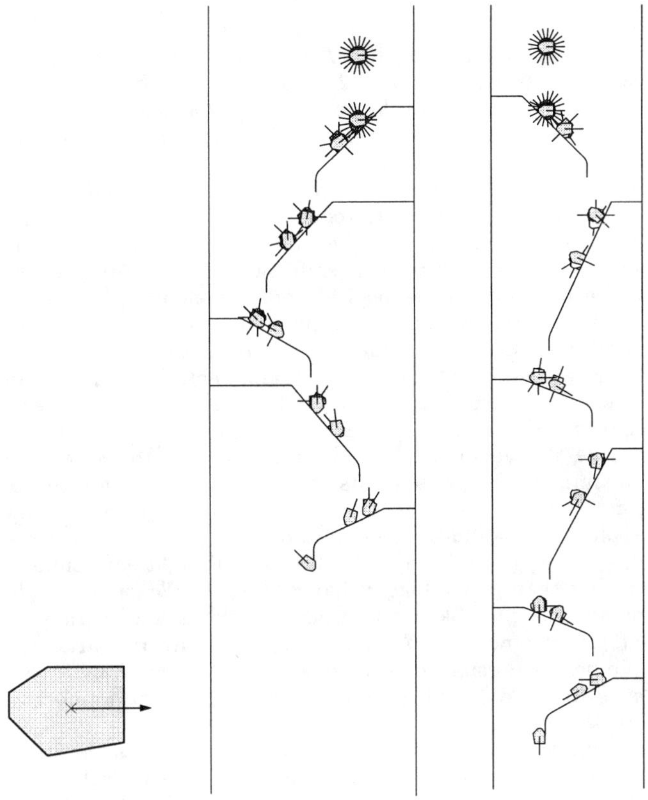

Figure 7: An optimal design of five fences, and a design of six alternating fences allowing for a narrower belt. Every line segment emanating from the part represents a possible orientation of the part.

The costs make it impossible to apply Lemma 4.2 to reduce the graph size, as this may cause the removal of equally long or longer paths with lower cost from the graph. The size of the resulting graph is therefore $O(n^4)$. Dijkstra's algorithm (see e.g. [9]) has been used to find the minimal cost path through the graph in time $O(n^4)$.

## 5 Arbitrary parts

In Section 3.1, we have shown that we can orient any part by a fence design if there is a unique largest left or right environment for which the second largest interval has size smaller than $\pi/2$. For asymmetric parts, there always exists such a large environment. If we deal with arbitrary parts, there can be several environments with the same size $\alpha$, even with size greater than $\pi/2$. In this section we show that for every polygonal part there is a fence design that orients the part up to symmetry in its push function. Due to lack of space, we do not give the complete proof. For more details, we refer to the full version. Let $P$ an arbitrary part. Let $m_s$ denote the number of stable orientations of $P$. We first show that we can orient $P$ if the period of the push function is $2\pi$. The plans can actually be used to orient any part up to the period in its push function as we show in the last part of this section. The method we use is similar to the method Chen and Ierardi introduced to generate push plans [8]. As mentioned, a fence design is a push plan satisfying certain constraints on the reorientations of the jaw. We try to produce push plans to that only use reorientations in either $(0, \pi/2)$ or $(-\pi/2, 0)$, as such plans clearly satisfy Definition 2.2. There are two problems with the implementation of the push plans of Chen and Ierardi as a fence design. If there are several right environments with size greater than $\pi/2$, we cannot orient the part. If there is no unique largest right environment, Chen and Ierardi apply their so called 'stretching lemma' which in general uses any reorientation of the pushing jaw. The stretching lemma shows that we can shift two possible orientations which both have a maximal left or right environment closer to each other with one single push; in other words: we can break the symmetry if there are multiple orientations with the same maximal left or right environment. For a push plan Theorem 3.1 and 3.2 remain valid [8, 22]. For fences over a conveyor this is not necessarily the case. We can, however, reduce the possible orientations of the part to those with maximal right environments or with right environments of length greater than $\pi/2$ by an alternating sequence of jaw applications and jaw reorientations by an angle $\min\{\alpha - \mu, \pi/2 - \mu\}$ ($\alpha$ and $\mu$ as in Section 3.1; no right environment smaller than $\pi/2$ is bigger than $\pi/2 - \mu$). (We can similarly reduce the possible orientations to orientations with such left environments.) As observed, for arbitrary parts, the number of possible orientations can be larger than 1. We have to use different ideas to further reduce the number of possible orientations.

The key to a reduction in the number of possible orientations is the cycle of the sequences of left and right environments. A sequence $\mathcal{S}$ has cycle $d$ if and only if for all $i \in 1 \ldots |\mathcal{S}|$, $\mathcal{S}_i = \mathcal{S}_{i+d \pmod{|\mathcal{S}|}}$. The sets of left and right environments are cyclically ordered sequences if we apply the order (by increasing or decreasing angle) of the fixed points to the environments. We classify the push functions, based on the cycle of their left and right environments, and the sizes of the environments.

1. Either the left or right environments have cycle $m_s$, and there is no environment of length greater than $\pi/2$ in that cycle.

2. Both left and right environments have a cycle smaller than $m_s$.

3. The left or right environments have cycle $m_s$ and there is more than one environment of length greater than $\pi/2$.

Let us first concentrate on push functions in the first and second class. The set of possible orientations is a cyclic sequence. We intuitively cut the cyclic sequence of orientations, which results in a sequence of orientations with a head and a tail. The head and tail are chosen such that the sequence of orientations is ordered from head to tail with decreasing angle. We move the possible orientations in such a way that the tail is placed on a orientation with maximal right environment and the head is not. During the move, we make sure that the number of stable orientations with maximal right environments between the orientation of the tail and the stable orientation before the head does not increase. If we have moved we reorient the jaw with angle $\alpha + \mu'$ ($\mu' > 0$, for all $v : l(v) < \mu'$ and $\alpha + \mu' < \pi/2$), push and again perform (at most $m_s$) reorientations by $\alpha - \mu$ and pushes. We now have reduced the number of orientations between the orientation of tail and the orientation before the head that have a maximal right environment. At this stage all possible orientations have maximal right environments.

For the first class of push functions, the move is based on the difference of sequences $\mathcal{B} = b_1 \ldots b_k$, such that $b_p b_1 \ldots b_k b_s$ is a subsequence of the right environments, $b_p, b_s$ maximal, and the elements of $\mathcal{B}$ are no maximal right environments. The sequence $\mathcal{B}$ is so to say an ordered set of right-intervals *between* two right-intervals of maximal size. Since the cycle of the right environments is $m_s$, the symmetry is broken in these $\mathcal{B}$'s. In the full version we describe the exact schemes, which only use reorientations in the angular interval $(0, \pi/2)$ (or $(-\pi/2, 0)$ if we consider the left environments). Thus, for the first class of push functions, we can generate push plans that are fence designs.

The second class of push functions is the class where both left and right environments have a cycle less than $m_s$. It is easy to see that the least common multiple of the two cycles has to be $n$, otherwise the period of the push function is smaller than $2\pi$. Since the cycle of the right environments is less than $m_s$, we can have to make use of unique combinations of left and right environments that break the symmetry. The reductio is very similar to the reduction for the first classes of push functions. Due to lack of space, the reduction is not included here. The schemes use only either reorientations in the angular interval $(0, \pi/2)$ or in $(-\pi/2, 0)$. Thus, for the second class of push functions, we can generate push plans that are fence designs.

The third class of push functions does not fit in the move and reduce framework. There are, however, at most three environments with size greater than $\pi/2$, which makes it possible to treat the different cases in the last class of push functions one by one, and provide dedicated push plans which use both left and right fences. Thus, for the third class of push functions, we can generate push plans that are fence designs.

If the period of the push function is less than $2\pi$, then the maximal size of the left and right environments only decreases. We can therefore use the fence designs of this section to orient the part, but only up to rotational symmetries. Theorem 5.1 summarizes the result.

**Theorem 5.1** *Any polygonal part can be oriented up to rotational symmetry by a fence design.*

As a result of Theorem 5.1 the algorithm we presented in Section 4 to compute fence designs always gives an answer. Since the graph used in the algorithm has $O(n^2)$ nodes, the length of the shortest path is at most $O(n^2)$.

**Theorem 5.2** *Any polygonal part with $n$ vertices can be oriented up to rotational symmetry by a fence design of length $O(n^2)$. The optimal fence design can be computed in $O(n^3 \log n)$ time.*

## 6 Conclusion

In this paper we have investigated the problem of sensorless part orientation by sequences of pushes. We have shown that any polygonal part can be oriented by a sequence of fences placed along a conveyor belt. We have presented the first polynomial-time algorithm for computing the shortest fence design for any given polygonal part. The algorithm is simple and runs in time $O(n^3 \log n)$. The structure of the algorithm yields an $O(n^2)$ bound on the length of the shortest fence design. It remains an open problem to extend the $O(n)$ bound on the length of push plans to fence designs.

Although pathological polygons can be constructed that lead to push plans and fence designs of length $\Omega(n)$, it turns out that the length of most plans almost always remains far below the worst-case length. We have shown that only $O(1)$ actions are required for parts with non-zero eccentricity, i.e., with non-square minimum-width bounding box. The analysis also applies to curved parts, providing the first complexity bound for non-polygonal parts. The results generalize to push-squeeze plans and to a large extent to fence designs. Similar results may be possible for other sensorless part feeders.

**Acknowledgements**

We thank Jeff Wiegley for useful feedback.

**References**

[1] P.K. Agarwal, M.J. Katz, and M. Sharir. Computing depth orders for fat objects and related problems. *Computational Geometry: Theory and Applications*, 5:187–206, 1995.

[2] S. Akella, W. Huang, K.M. Lynch, and M.T. Mason. Sensorless parts feeding with a one joint robot. *Algorithms for Robotic Motion and Manipulation* (J.-P. Laumond and M. Overmars (Eds.)), A.K. Peters, pages 229–238, 1996.

[3] S. Akella and M.T. Mason. Posing polygonal objects in the plane by pushing. In *IEEE International Conference on Robotics and Automation, Nice*, pages 2255–2268, 1992.

[4] H. Alt, R. Fleischer, M. Kaufmann, K. Mehlhorn, S. Näher, S. Schirra, and C. Uhrig. Approximate motion planning and the complexity of the boundary of the union of simple geometric figures. *Algorithmica*, 8:391–406, 1992.

[5] K.-F. Böhringer, V. Bhatt, and K.Y. Goldberg. Sensorless manipulation using transverse vibrations of a plate. In *Proc. IEEE Int. Conf. on Robotics and Automation, Nagoya, Japan*, pages 1989–1996, 1995.

[6] K.-F. Böhringer, B.R. Donald, and N.C. MacDonald. Upper and lower bounds for programmable vector fields with applications to mems and vibratory plate part feeders. *Algorithms for Robotic Motion and Manipulation,* J.-P. Laumond and M. Overmars (Eds.), A.K. Peters, pages 255–276, 1996.

[7] M. Brokowski, M.A. Peskin, and K. Goldberg. Curved fences for part alignment. In *IEEE International Conference on Robotics and Automation, Atlanta, GA*, volume 3, pages 467–473, 1993.

[8] Y.-B. Chen and D.J. Ierardi. The complexity of oblivious plans for orienting and distinguishing polygonal parts. *Algoritmica*, 14:367–397, 1995.

[9] T.H. Cormen, C.E. Leiserson, and R.L. Rivest. *Introduction to Algorithms.* The MIT Press, Cambridge, Massachusetts London, England, 1990.

[10] M. de Berg. Linear size binary space partitions for fat objects. In *Proc. 3rd Annual European Symp. on Algorithms (ESA'95), Lecture Notes in Computer Science*, volume 979, pages 252–263, 1995.

[11] M.A. Erdmann and M.T. Mason. An exploration of sensorless manipulation. *IEEE Journal of Robotics and Automation*, 4:367–379, 1988.

[12] K. Goldberg. Orienting polygonal parts without sensors. *Algoritmica*, 10(2):201–225, 1993.

[13] M.J. Katz, M.H. Overmars, and M. Sharir. Efficient hidden surface removal for objects with small union size. *Computational Geometry: Theory and Applications*, 2:223–234, 1992.

[14] K. M. Lynch and M.T. Mason. Stable pushing: Mechanics, controllability, and planning. *International Journal of Robotics Research*, 15:533–556, 1996.

[15] M. Mason. *Manipulator grasping and pushing operations.* PhD thesis, MIT, 1982. AI-TR-690.

[16] J. Matoušek, J. Pach, M. Sharir, S. Sifrony, and E. Welzl. Fat triangles determine linearly many holes. *SIAM Journal on Computing*, 23:154–169, 1994.

[17] B.K. Natarajan. An algorithmic approach to the automated design of parts orienters. In *IEEE Annual Symposium on Fundamentals of Computer Science*, pages 132–142, 1986.

[18] M.H. Overmars and A.F. van der Stappen. Range searching and point location among fat objects. *Journal of Algorithms*, 21:629–656, 1996.

[19] M.A. Peshkin and A.C. Sanderson. The motion of a pushed sliding workpiece. *IEEE Journal of Robotics and Automation*, 4(6):569–598, 1988.

[20] M.A. Peshkin and A.C. Sanderson. Planning robotic manipulation strategies for workpieces that slide. *IEEE Journal of Robotics and Automation*, 4(5):524–531, 1988.

[21] A. Rao and K. Goldberg. Manipulating algebraic parts in the plane. *IEEE Transactions on Robotics and Automation*, 11, 1995.

[22] A.F. van der Stappen, K. Goldberg, and M. Overmars. Geometric eccentricity and the complexity of manipulation plans. Technical report, UU-CS-1996-49, Dept. of Computer Science, Utrecht University, 1996.

[23] A.F. van der Stappen, D Halperin, and M. Overmars. The complexity of the free space for a robot moving amidst fat obstacles. *Computational Geometry: Theory and Applications*, 3:353–373, 1993.

[24] A.F. van der Stappen and M.H. Overmars. Motion planning amidst fat obstacles. In *Proc. 10th Ann. ACM Symp. on Computational Geometry*, pages 31–40, 1994.

[25] J. Wiegley, K. Goldberg, M. Peshkin, and M. Brokowski. A complete algorithm for designing passive fences to orient parts. In *IEEE International Conference on Robotics and Automation, Minneapolis, MN*, pages 1133–1139, 1996.

# On Levels in Arrangements of Lines, Segments, Planes, and Triangles*

Pankaj K. Agarwal† Boris Aronov‡ Micha Sharir§

**Abstract**

We consider the problem of bounding the complexity of the $k$-th level in an arrangement of $n$ curves or surfaces, a problem dual to, and extending, the well-known *$k$-set problem.* (a) We review and simplify some old proofs in new disguise and give new proofs of the bound $O(n\sqrt{k+1})$ for the complexity of the $k$-th level in an arrangement of $n$ lines. (b) We derive an improved version of Lovász Lemma in any dimension, and use it to prove a new bound, $O(n^2k^{2/3})$, on the complexity of the $k$-th level in an arrangement of $n$ planes in $\mathbb{R}^3$, or on the number of $k$-sets in a set of $n$ points in three dimensions. (c) We show that the complexity of any single level in an arrangement of $n$ line segments in the plane is $O(n^{3/2})$, and that the complexity of any single level in an arrangement of $n$ triangles in 3-space is $O(n^{17/6})$.

## 1 Introduction

**Background.** The *$k$-set problem* is one of the most challenging open problems in combinatorial geometry. The simplest variant of the problem is: Given a set $S$ of $n$ points in the plane in general position, and a parameter $0 \le k \le n-2$, what is the maximum possible number of lines that pass through a pair of points of $S$ and have exactly $k$ points of $S$ in one of the open halfplanes that they define? In a dual setting, we are given a set $\mathcal{L}$ of $n$ lines in the plane in general position, and want to bound the maximum possible number of vertices $v$ of the arrangement $\mathcal{A}(\mathcal{L})$, such that exactly $k$ lines pass below $v$. We denote this set of vertices by $V_k$. (Strictly speaking, $|V_k|$ is a slightly different quantity than the one defined above, as it corresponds to the number of lines passing through two of the given points and having exactly $k$ of the remaining points *below* them.) A slightly different variant of the dual problem is to define the *$k$-th level* of the arrangement, as the closure of the set of points that lie on the lines and have exactly $k$ lines below them, and seek a bound on the number of vertices of that level. (Each vertex of this closure may have either $k$ or $k-1$ lines below it.) See Figure 1 for an illustration.

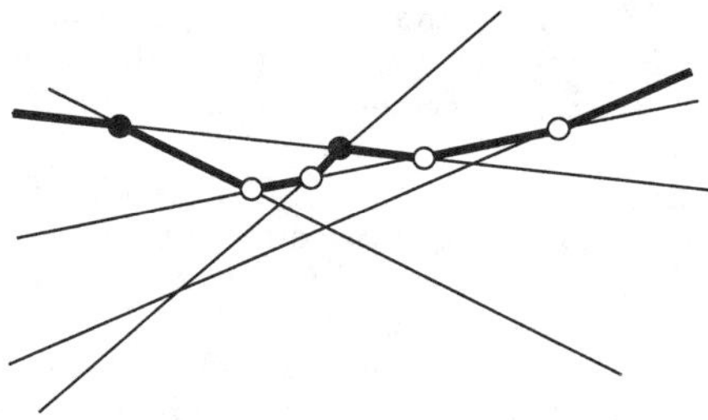

Figure 1: The third level in an arrangement of lines. The vertices of $V_2$ are indicated by empty circles and the vertices of $V_3$ are marked by black circles.

The $k$-set problem was first studied about 1970 by Erdős et al. and by Lovász [11, 19]. These papers have established an upper bound $O(n\sqrt{k+1})$ and a lower bound $\Omega(n\log(k+1))$ on the desired quantity, leaving a fairly big gap that is still mostly open. The only further progress on this problem is due to Pach et al. [21], where the upper bound is slightly improved to $O(n\sqrt{k+1}/\log^*(k+1))$; see [4, 10] for related results.

In the dual setting, the problem can be generalized in an obvious manner: In the plane, we are given a collection $\Sigma$ of $n$ $x$-monotone curves, each being the graph of a continuous totally or partially defined function, and a parameter $0 \le k < n$, and wish to bound the *complexity* (i.e., the number of vertices) of the $k$-th level in the arrangement $\mathcal{A}(\Sigma)$, defined exactly as in the case of lines. In this more general

*All three authors have been supported by a grant from the U.S.-Israeli Binational Science Foundation. Pankaj Agarwal has also been supported by a National Science Foundation Grant CCR-93-01259, by an Army Research Office MURI grant DAAH04-96-1-0013, by a Sloan fellowship, and by an NYI award and matching funds from Xerox Corporation. Boris Aronov has also been supported by NSF Grant CCR-92-11541 and a Sloan Research Fellowship. Micha Sharir has also been supported by NSF Grants CCR-94-24398 and CCR-93-11127, and by a grant from the G.I.F., the German-Israeli Foundation for Scientific Research and Development.

†Department of Computer Science, Box 90129, Duke University, Durham, NC 27708-0129, USA

‡Department of Computer and Information Science, Polytechnic University, Brooklyn, NY 11201-3840, USA

§School of Mathematical Sciences, Tel Aviv University, Tel Aviv 69978, Israel and Courant Institute of Mathematical Sciences, New York University, New York, NY 10012, USA



*Computational Geometry 97* Nice France

setting only two results are known: A recent seemingly weak, but elegant analysis by Tamaki and Tokuyama [23] yields the bound $O(n^{23/12})$ on the complexity of a level in an arrangement of $n$ *pseudo-parabolas*, which are graphs of total functions, each pair of which intersect at most twice. We also mention the case of *pseudo-lines*, which are $x$-monotone connected curves, unbounded at either end, each pair of which intersect exactly once, where a slightly larger lower bound of $\Omega(n \cdot 2^{c\sqrt{\log n}})$ for the complexity of the median level is established in [18]. Our proof techniques and upper bounds apply equally well to the case of pseudo-lines.

Similar extensions apply in higher dimensions. In the primal setting, we are given a set $S$ of $n$ points in $\mathbb{R}^d$ in general position, and wish to bound the number of hyperplanes passing through $d$ of the points such that one of the halfspaces that they bound contains exactly $k$ points of $S$. For $d = 3$, the best known upper and lower bounds are, respectively, $O(n^{8/3})$ and $\Omega(n^2 \log n)$ [3, 8]. For $d > 3$, the best known upper bound is $O(n^{d-c_d})$, for some exponentially small but positive constant $c_d$ [26]. Note that, in contrast to the planar case, these bounds depend only on $n$ and not on $k$.

We can formulate the problem in general dimension, in a dual setting: We consider an arrangement of hyperplanes, or, more generally, of surfaces that are graphs of continuous total or partial functions, and define the $k$-th level of the arrangement exactly as in the planar case. We now seek bounds on the maximum possible number of vertices (or of faces of all dimensions) of the level. Except for the case of hyperplanes, which is equivalent to the $k$-set problem mentioned in the preceding paragraph, no nontrivial bounds for the entire range of values of $k$ are known.

In spite of the sorry state of the problem, one can obtain nontrivial bounds when $k$ is small. The probabilistic analysis of Clarkson and Shor [7] (see also [22]) yields fairly sharp bounds on the combined complexity of the first $k$ levels in arrangements. For the case of hyperplanes, for example, the bound is $\Theta(n^{\lfloor d/2 \rfloor} k^{\lceil d/2 \rceil})$. For sufficiently small $k$, this gives a better upper bound on the complexity of a single level than the general bound stated above. The analysis of Clarkson and Shor [7] also implies, under fairly general assumptions that, for a constant $k > 0$, the worst-case number of vertices of the $k$-th level is asymptotically proportional to the maximum possible number of vertices on the lower envelope (i.e., the 0-th level) of the surfaces.

**New Results.** In this paper we make several contributions to these problems:

We first review some old proofs in new disguise, and present new proofs of the upper bound $O(n\sqrt{k+1})$ for the original planar $k$-set problem (or, dually, for the case of the $k$-th level in an arrangement of $n$ straight lines in the plane). We review the proof technique of Gusfield [13], which, as we perceive, is not well known within the combinatorial and computational geometry communities, and show its relationship to other proofs. We also give a simple proof of the dual version of what we call "Lovász Lemma" (see Lemma 2.3 below) that is used to prove the bound. As is well known, these techniques apply equally well to arrangements of pseudolines; see, for example, [12].

We adapt two of our proof techniques to yield the bound $O(n^{3/2})$ on the complexity of a single level in an arrangement of $n$ line segments (or "pseudo-segments," to be defined below). As far as we know, this bound is new.

We then proceed to study the problem in higher dimensions. First we obtain an improved version of Lovász Lemma that has a rather simple proof. Specifically, we show that no line can intersect more than $O(k^{d-1})$ $k$-set simplices, where a *$k$-set simplex* is a $(d-1)$-dimensional simplex, spanned by $d$ points of $S$, such that the hyperplane containing the simplex has exactly $k$ points of $S$ in one of its open halfspaces. The previous bound was $O(n^{d-1})$ (see [3, 19]), so this is a significant improvement when $k \ll n$. Plugging the new bound into the analysis technique of [8], we show that the complexity of the $k$-th level in an arrangement of $n$ planes in 3-space, and thus, the number of $k$-sets in a set of $n$ points in $\mathbb{R}^3$, is $O(n^2 k^{2/3})$. This is the first general bound for arrangements of planes that depends on $k$, besides the aforementioned $O(nk^2)$ bound on the overall complexity of the first $k$ levels. The new bound is an improvement when $k = \Omega(n^{3/4})$. A similar improved bound, of the form $O(n^{d-c_d-\varepsilon_d} k^{\varepsilon_d})$, can be obtained in any dimension $d > 3$, for appropriate constants $\varepsilon_d$, $c_d$, depending only on $d$, by combining the strengthened Lovász Lemma with the analysis of Živaljević and Vrećica [26]; see also [3].

Finally, we consider the case of triangles in 3-space, and show that the complexity of a single level in an arrangement of $n$ such triangles is $O(n^{17/6})$.

## 2 Arrangements of Lines

Let $\mathcal{L}$ be a collection of $n$ lines in the plane in general position. Let $V_k$, for $k = 0, \ldots, n-2$, denote, as in the introduction, the set of vertices of the arrangement $\mathcal{A}(\mathcal{L})$ that have exactly $k$ lines below them. Then the set of vertices of the $k$-th level is $V_k \cup V_{k-1}$ (or just $V_k$, for $k = 0$). When the level passes through a vertex of $V_{k-1}$ (resp. of $V_k$), it bends to the left (resp. to the right) as we traverse it in the positive $x$-direction. See Figure 1.

In this section we give four proofs of the following well-known result (which, as already pointed out, provides an estimate which is slightly larger than the best currently known upper bound of Pach *et al.* [21]):

**Theorem 2.1** *The complexity of the $k$-th level of $\mathcal{A}(\mathcal{L})$ is $O(n\sqrt{k+1})$.*

**Remark:** In most of the following proofs we will actually argue that $|V_{k-1}| = O(n\sqrt{k})$. The claimed bound on the number of vertices of the $k$-th level follows by repeating the argument for $|V_k|$ and combining the two estimates.

**First Proof (Potential Function):** This proof is not new, and is an adaptation of the analysis technique of Gusfield [13, 14]. We give it for the sake of completeness, and because we will shortly apply a variant of it to the case of segments. We note that the way it is presented below is somewhat different than Gusfield's own analysis; we will further comment on Gusfield's analysis later on.

Let the lines in $\mathcal{L}$ be $\ell_1, \ell_2, \ldots, \ell_n$, sorted in the order of decreasing slope, and let $k$ denote the given level. For any $\alpha \in \mathbb{R}$, we say that the level of a line $\ell \in \mathcal{L}$ is $j$ at $\alpha$ if exactly $j$ lines of $\mathcal{L}$ intersect the vertical line $x = \alpha$ below $\ell$. For each $x \in \mathbb{R}$, define the *potential function*

$$\Phi(x) = \sum \{ j \mid \text{ the level of } \ell_j \text{ at } x \text{ is } < k \}.$$

We clearly have $\Phi(-\infty)$, $\Phi(+\infty) = O(nk)$ (in fact, $\Phi(x) = O(nk)$ for each $x$). As we sweep $\mathcal{A}(\mathcal{L})$ with a vertical line from left to right, the value of $\Phi(x)$ can change only when $x$ equals the abscissa $v_x$ of a vertex $v \in V_{k-1}$ (refer to Figure 1; note that $\Phi(x)$ does not change at vertices of $V_k$.) Suppose that $v \in V_{k-1}$ is the intersection of lines $\ell_i$ and $\ell_j$, with $j > i$. Then, as easily checked, the change $\Delta\Phi(v_x) = \Phi(v_x + \varepsilon) - \Phi(v_x - \varepsilon)$, for a sufficiently small $\varepsilon > 0$, is $j - i > 0$. In other words

$$\Phi(+\infty) = \Phi(-\infty) + \sum_{v \in V_{k-1}} \Delta\Phi(v_x) = O(nk),$$

with each of these changes being a positive integer.

The number of vertices $v$ at which $\Delta\Phi(v_x) > \sqrt{k}$ is no more than $O(n\sqrt{k})$, as the sum of $\Delta\Phi(v_x)$ at these vertices is $O(nk)$, and each term in the sum is larger than $\sqrt{k}$. Concerning $v$ at which the change is at most $\sqrt{k}$, there are at most $n-1$ vertices with corresponding pairs of indices $(i, i+1)$, $n-2$ vertices with pairs $(i, i+2)$, etc., for a total of

$$(n-1) + (n-2) + \cdots + (n - \sqrt{k} + 1) < n\sqrt{k}$$

vertices. Combining the two estimates, we conclude that $|V_{k-1}| = O(n\sqrt{k})$. In fact, a more careful counting gives the bound $2n\sqrt{k}$. □

**Second Proof (Concave Chains):** Let $V_{k-1}$ denote, as above, the set of all vertices of the $k$-th level of $\mathcal{A}(\mathcal{L})$ at which the level makes a left turn, passing from a line with smaller slope to a line with larger slope. We associate with the $k$-th level a collection of $k$ *concave chains*, where each such chain is an unbounded $x$-monotone concave polygonal curve contained in the union of the lines of $\mathcal{L}$. (As will be seen below, the chains in a certain sense "cover" the portion of $\mathcal{A}(\mathcal{L})$ below the level.) This is done as follows. The desired chains, denoted $c_1, \ldots, c_k$, start at $x = -\infty$ along the $k$ lowest lines of the arrangement (these are the lines with the $k$ largest slopes). Whenever some chain $c_i$ reaches the $k$-th level, we are at a vertex $v \in V_{k-1}$, as is easily checked. We then continue $c_i$ to the right along the other line incident to $v$. The chains bend only at vertices of $V_{k-1}$; otherwise each chain follows the line it is on. See Figure 2 for an illustration.

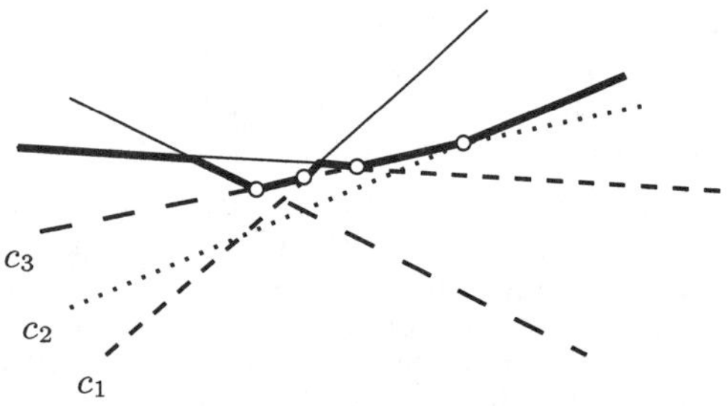


Figure 2: The concave chains associated with the third level; the level itself is drawn in bold, and the dashed paths denote the concave chains $c_1, c_2, c_3$.

It is easily seen that the resulting chains satisfy the following properties:

(i) The union of the chains is the closure of the portion of the union of the lines that lies below the $k$-th level. Except for the vertices of $V_{k-1}$, the union of the chains lies strictly below the $k$-th level.

(ii) The chains are vertex-disjoint and have non-overlapping edges, but they generally cross each other.

(iii) All the vertices of the chains lie on the upper envelope of the chains. Indeed, each chain, except for its vertices, lies fully below the $k$-th level, so any vertex of any chain lies above all the chains that are not incident to it.

Gusfield's analysis, with minor modifications, essentially establishes the following more general bound. The same result has been obtained independently by Halperin and Sharir [15], who were not aware of Gusfield's earlier work:

**Theorem 2.2 ([13, 14, 15])** *The overall number of vertices of k concave chains, which are vertex-disjoint and have non-overlapping edges, in an arrangement of n lines in the plane, is* $O(n\sqrt{k})$.

This result clearly yields another proof Theorem 2.1. We leave it to the reader to verify that the potential-function proof applies, almost verbatim, to the general case of concave chains. □

Note that the above bound does not count crossings between chains.

**Remarks:** (1) If the concave chains are not allowed to cross each other, then their overall complexity is only $O(k^{2/3}n^{2/3} + n)$, as shown in [15, 16], but the analysis in these papers crucially relies on the fact that the chains cannot cross.

(2) The chains associated with the $k$-th level have the additional property (iii) that all their vertices appear on their upper envelope. Can a sharper upper bound be proved for the complexity of $k$ concave chains with this extra property? Note, however, that property (iii) is not strong enough to improve the bound beyond $O(k^{2/3}n^{2/3} + n)$, as it is possible to produce a collection of $k$ chains in an arrangement of $n$ lines with a total of $\Omega(k^{2/3}n^{2/3} + n)$ vertices, so that all these vertices occur on the combined upper envelope of the chains [17].

**Third Proof (Concave Chains and Lovász Lemma):** One of the standard (and among the first) ways to prove the theorem is via the following statement, which we will refer to as "Lovász Lemma" [19]. It is usually stated in the primal plane, for a collection of $n$ points, but we will state a dual version of the lemma, for arrangements of lines, and give a simple proof that uses the concave chain structure, with the aim of extending it to other types of arrangements.

Let $\mathcal{L}$ be a collection of $n$ lines in the plane in general position, and let $1 \le k \le n-1$. For each vertex $v \in V_{k-1}$, let $W_v$ denote the double wedge formed by the two lines that meet at $v$ ($W_v$ is the region between the upper and lower envelopes of these two lines). See Figure 3.

**Lemma 2.3 (Dual Lovász Lemma in 2D)** *For any point z in the plane not lying on any line of* $\mathcal{L}$, *the number of double wedges* $W_v$, *for* $v \in V_{k-1}$, *that contain z is at most* $2\min\{k, j\} \le 2k$, *where j is the number of lines of* $\mathcal{L}$ *that pass below z. Actually, there are at most* $\min\{k, j\}$ *left wedges and at most* $\min\{k, j\}$ *right wedges that contain z.*

**Proof:** Let $c$ be one of the concave chains obtained in the previous proof, and consider the system of double wedges $W_v$, over all vertices $v$ of $c$. The concavity of $c$ is easily seen to imply that $z$ can lie in at most two of these double wedges (in at most one right wedge and in at most one left wedge). Since we have $k$ such chains, $z$ can lie in at most $2k$ double wedges $W_v$, for $v \in V_{k-1}$. Moreover, if $z$ lies above exactly

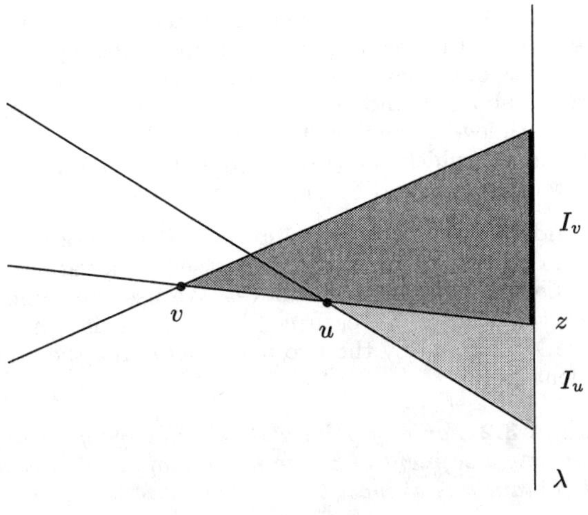


Figure 3: The setup for Lovász Lemma for the case of lines. The right wedges of $W_v, W_u$ are shown shaded.

$j < k$ lines of $\mathcal{L}$, then it lies only above $j$ concave chains, and can therefore only belong to double wedges corresponding to vertices of these chains. This easily implies the lemma. □

We can now complete the third proof of the theorem, using an analysis dual to that in the original proof in [19]. That is, fix a vertical line $\lambda$, and intersect each $W_v$, for $v \in V_{k-1}$, with $\lambda$, to obtain a system of $|V_{k-1}|$ intervals on $\lambda$, having a total of $n$ endpoints (which are the intersections of the lines in $\mathcal{L}$ with $\lambda$). It follows by a simple (and standard) counting argument (such as in [5, 19]) that $\lambda$ must contain a point that lies in at least $|V_{k-1}|^2/4n^2$ intervals. Since this number cannot be more than $2k$, we obtain $|V_{k-1}| \leq 2\sqrt{2}n\sqrt{k}$. With some care, this can be improved to $2n\sqrt{k}$.□

**Fourth Proof (Concave Chains and Cauchy-Schwarz Inequality):** Let $\mathcal{L} = \{\ell_1, \ldots, \ell_n\}$ be a set of $n$ lines, $k$ be an integer between 0 and $n-2$, and $c_1, \ldots, c_k$ be the concave chains associated with the $k$-th level of $\mathcal{A}(\mathcal{L})$, as defined above.

Let $w_{ij}$ denote the number of lines common to chains $c_i$ and $c_j$, for $1 \leq i < j \leq k$. For each $q = 1, \ldots, n$, let $M_q$ denote the number of chains that have an edge contained in $\ell_q$. Note that $|V_{k-1}| = \sum_{q=1}^{n} M_q - k$. On the other hand, we have

$$\sum_{i,j} w_{ij} = \sum_{q=1}^{n} \binom{M_q}{2}.$$

Hence,

$$\begin{aligned}
|V_{k-1}| &= \sum_{q=1}^{n} M_q - k = \sum_{q=1}^{n} (M_q - 1) + n - k \\
&\leq (2n)^{1/2} \left[ \sum_{q=1}^{n} \binom{M_q}{2} \right]^{1/2} + n - k \\
&= (2n)^{1/2} \left( \sum_{i,j} w_{ij} \right)^{1/2} + n - k.
\end{aligned}$$

Fix a pair of chains $c_i$, $c_j$. The concavity of the chains is easily seen to imply that both $c_i$ and $c_j$ must lie on or below each of the $w_{ij}$ common lines. That is, both $c_i$ and $c_j$ lie below (or on) the lower envelope $E$ of these lines, so each chain touches each of these lines only at its unique segment that appears on the lower envelope. It follows that $c_i$ and $c_j$ must intersect each other at least once below each of the $w_{ij}$ segments of $E$, so $t_{ij}$, the number of crossings between $c_i$ and $c_j$, must be at least $w_{ij}$. See Figure 4. We thus have $\sum_{i,j} w_{ij} \leq \sum_{i,j} t_{ij} \leq nk$, where the latter inequality follows from the fact that the number of chain-crossings is equal to the number of vertices of $\mathcal{A}(\mathcal{L})$ at level $< k$, and this number is known to be at most $nk$ [1]. This implies that

$$\begin{aligned}
|V_{k-1}| &\leq (2n)^{1/2} \left( \sum_{i,j} w_{ij} \right)^{1/2} + n - k \\
&\leq (2n)^{1/2} (nk)^{1/2} + n - k \\
&= 2^{1/2} n\sqrt{k} + n - k.
\end{aligned}$$

Note that the constant of proportionality is better than those yielded by the earlier proofs. □

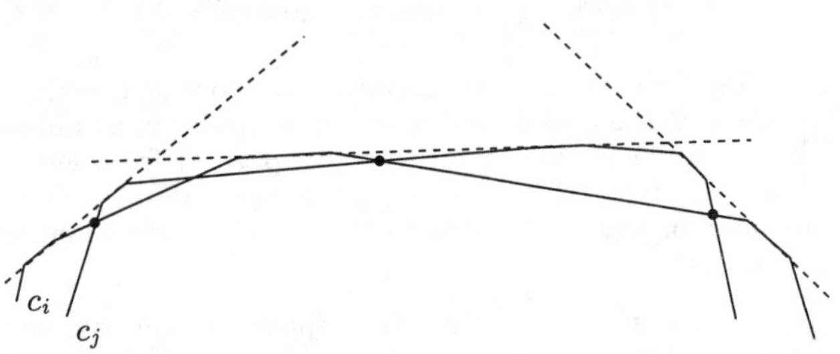


Figure 4: The chains $c_i$ and $c_j$ must cross at least $w_{ij}$ times.

## 3 Arrangements of Segments

In this section we extend some of the proofs given above to the case of line segments. Let $S$ be a collection of $n$ segments in the plane in general position. For $k = 0, \ldots, n-1$, the $k$-th level in the arrangement $\mathcal{A}(S)$ of $S$ is defined, as in the case of lines, to be the closure of the set of all points $w$ that lie on segments of $S$ and are such that the open downward-directed vertical ray emanating from $w$ intersects exactly $k$ segments of $S$ (that is, there are $k$ segments of $S$ below $w$). The *complexity* of a level is the number of vertices of $\mathcal{A}(S)$ that lie on the level plus the number of discontinuities of the level. (Unlike the case of lines, a level of $\mathcal{A}(S)$ is not necessarily connected, and it may involve vertical jumps from a segment to the segment lying directly above or below it, when a new segment starts or ends at a point below the level. Clearly, the number of such discontinuities is at most $2n$.) As in the case of lines, we define $V_k$, for $k = 0, \ldots, n-2$, to be the set of vertices of $\mathcal{A}(S)$ (excluding segment endpoints) that have exactly $k$ segments passing below them. The set of vertices of the $k$-th level, excluding segment endpoints and jump discontinuities, is $V_{k-1} \cup V_k$. The level bends to the left at vertices of $V_{k-1}$ and to the right at vertices of $V_k$. See Figure 5 for an illustration.

**Theorem 3.1** *The complexity of any single level in an arrangement of $n$ line segments in the plane in general position is $O(n^{3/2})$.*

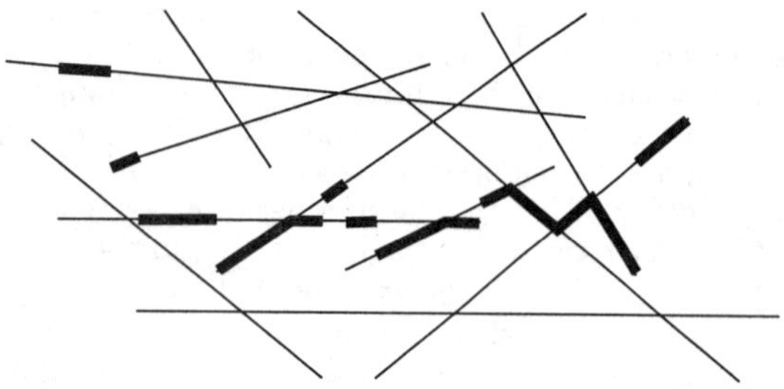

Figure 5: The second level in an arrangement of segments; here $|V_1| = 1$ and $|V_2| = 4$.

**First Proof (Potential Function):** This proof is an adaptation of the potential function proof for the case of lines, as given above. Let the segments be $s_1, s_2, \ldots, s_n$, sorted in the order of decreasing slope of their containing lines, and let $k$ denote the given level. For each $x$, define the potential function

$$\Phi(x) = \sum \{j \mid s_j \text{ lies at } x \text{ at level} < k\}.$$

Assuming that all the given segments are bounded, we have $\Phi(-\infty) = \Phi(+\infty) = 0$, and in any case we have, as above, $\Phi(x) = O(nk)$ for any $x$. As we sweep $\mathcal{A}(S)$ from left to right, we are interested in the signed changes $\Delta\Phi(x)$ in $\Phi(x)$. The value of $\Phi(x)$ can change only in one of the following three cases:

(i) $x$ is the abscissa of the left endpoint of some segment $s_i$, and this endpoint lies below the current $k$-th level: In this case we have $\Delta\Phi(x) = i - j$, where $s_j$ is the segment that currently lies directly below the level. (Here we have a discontinuity, where the level jumps down one segment.) Note that we may have $i = j$.

(ii) $x$ is the abscissa of the right endpoint of some segment $s_i$, and this endpoint lies on or below the current $k$-th level: In this case we have $\Delta\Phi(x) = j - i$, where $s_j$ is the segment that currently lies on the level. (Here we have a discontinuity, where the level jumps up one segment.) Again, it is possible that $i = j$.

(iii) $x$ is the abscissa of a vertex $v \in V_{k-1}$ (as in the case of lines, $\Phi(x)$ does not change at vertices of $V_k$): Suppose that $v$ is the intersection of segments $s_i$ and $s_j$, with $j > i$. Then $\Delta\Phi(x) = j - i > 0$.

The number of events of types (i) and (ii) is at most $2n$, and the change in the potential at each of these events has absolute value $O(n)$, for a total change of absolute value $O(n^2)$. We thus have

$$\sum_x \Delta\Phi(x) = O(n^2),$$

where the summation is taken over all $x$ that are the abscissae of a vertex of $V_{k-1}$, and each of these changes is a positive integer.

The proof now proceeds exactly as in the first proof of Theorem 2.1, and we leave it to the reader to fill in the straightforward details. The difference between the cases of segments and of lines is that, in the case of segments, we can only bound the total change in potential by $O(n^2)$, rather than by $O(nk)$. In fact, in the case of segments, the bound $O(n\sqrt{k+1})$ is too small for small values of $k$. For example, for $k = 0$ (that is, for the lower envelope of the segments) the complexity of the level can be $\Omega(n\alpha(n))$ [25], which is larger than the above bound. On the other hand, the complexity of the $k$-th level is smaller than the overall complexity of the first $k$ levels, which is $O(n(k+1)\alpha(n/(k+1)))$ [22]. This is a better upper bound for small values of $k$. □

**Second Proof (Concave Chains and Lovász Lemma):** We next present a second proof, based on a variant of the dual Lovász Lemma given above. We use the same setup as above. That is, for each $v \in V_{k-1}$, we define the double wedge $W_v$ formed by the two lines containing the segments incident to $v$.

**Lemma 3.2** *For any point $z \in \mathbb{R}^2$, not lying on any line containing a segment of $S$, the number of double wedges $W_v$ that contain $z$ is at most $4n$.*

**Proof:** Let us first extend the notion of concave chains to the case of segments. The chains are constructed as follows. We start a new chain at (i) the left endpoint of any segment, if that endpoint lies below the $k$-th level, and (ii) at any point of discontinuity of the level, when the level jumps *up* from a segment $s_i$ to a segment $s_j$ (the chain is started along the lower segment $s_i$). As $x$ increases, each chain $c$ follows the segment that it lies on, except when of the following situations occurs:

(i) $c$ reaches the right endpoint of that segment, and then $c$ terminates there;

(ii) $c$ follows a segment $s_i$ and reaches a discontinuity of the $k$-th level, where the level jumps *down* to $s_i$, in which case $c$ is terminated at that point; or

(iii) $c$ reaches a vertex $v \in V_{k-1}$, in which case $c$ bends to the right, and continues along the other segment incident to $v$.

We thus get a collection of at most $2n$ concave chains. It is easy to verify that these chains also satisfy (appropriate variants of) properties (i)–(iii) in the second proof of Theorem 2.1. Here the chains are graphs of partially-defined functions. Note that the domain of definition of some chains may also include intervals over which the $k$-th level is not defined (because there are fewer than $k + 1$ segments over such an interval).

The proof can now be completed as in the case of lines, because, for each of the at most $2n$ chains $c$, a point $z$ can belong to at most two double wedges $W_v$, for $v \in V_{k-1} \cap c$. □

The proof of Theorem 3.1 now proceeds along the same lines as the third proof of Theorem 2.1.

**Remarks:** (1) Both proofs presented in this section also apply to the cases of *pseudo-lines* and *pseudo-segments.* We have already defined the notion of a family of pseudo-lines. A collection $S$ of $n$ $x$-monotone connected arcs is *a family of pseudo-segments* if each of them can be extended to an $x$-monotone connected unbounded curve, so that this family of curves is a collection of pseudo-lines. (This is a much stronger definition than just requiring each pair of pseudo-segments to intersect at most once; see Figure 6.) We leave it to the reader to verify that both proofs go through in the case of pseudo-segments, with straightforward modifications.

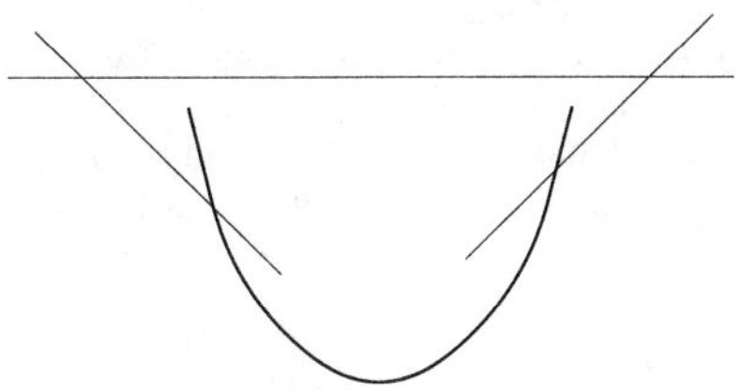

Figure 6: These five arcs do not form an arrangement of pseudo-segments.

(2) The immediate challenge is to improve Theorem 3.1, and obtain a better upper bound that also depends on $k$. As noted above, such a bound cannot be $O(n\sqrt{k+1})$, at least for small values of $k$.

(3) Another interesting open problem is to obtain an improved bound for the complexity of a single level in the arrangement of $q$ piecewise-linear functions, whose graphs consist of a total of $n$ segments. Of course, the bound $O(n^{3/2})$ applies here, but perhaps one can obtain an improved bound that is also a function of $q$ and is smaller when $q \ll n$.

## 4 Improved Lovász Lemma and Arrangements of Planes

Let $P = \{\pi_1, \ldots, \pi_n\}$ be a collection of $n$ planes in 3-space in general position, and let $\mathcal{A}(P)$ denote the arrangement of $P$. The $k$-th level of $\mathcal{A}(P)$ is defined as the closure of the set of all points that lie in the union of the planes and have exactly $k$ planes lying below them. The complexity of the level, regarded as a polyhedral surface, is the number of its vertices, edges and faces. This is clearly proportional to only the number of vertices, and we will focus on bounding this latter quantity.

**Theorem 4.1** *The number of vertices of the $k$-th level of $\mathcal{A}(P)$ is $O(n^2k^{2/3})$.*

As already noted, this improves the bound $O(n^{8/3})$ that was established in [8], when $k \ll n$, and is the first general bound for the case of planes that depends also on $k$, except for the $O(nk^2)$ bound on the combined complexity of the first $k$ levels [7]. Note that our new bound is smaller than this latter bound when $k = \Omega(n^{3/4})$.

The proof of Theorem 4.1 follows the previous proofs in [2, 8]. That is, it exploits a generalization of Lovász Lemma to three dimensions. We present here an improved version of this lemma, in arbitrary dimension, which leads to the improved bound of the theorem.

Let $H$ be a collection of $n$ hyperplanes in $\mathbb{R}^d$ in general position, and let $0 \le k \le n-d$. Let $V_k$ denote the set of those vertices $v$ of $\mathcal{A}(H)$ for which exactly $k$ hyperplanes of $H$ pass below $v$. For each $v \in V_k$, we denote by $H_v$ the set of the $d$ hyperplanes incident to $v$, and let $R_v$ denote the closed region ('corridor') lying between the upper and lower envelopes of the hyperplanes of $H_v$.

**Lemma 4.2 (Dual Lovász Lemma in $\mathbb{R}^d$)** *For any $(d-2)$-flat $f$ in $\mathbb{R}^d$, we have*

$$\left|\{v \in V_k \mid f \subset R_v\}\right| = O(k^{d-1}).$$

It will be more convenient to state and prove the primal version of this lemma. Fix a set $S$ of $n$ points in $\mathbb{R}^d$, in general position. A *$k$-set simplex* is a $(d-1)$-dimensional simplex spanned by $d$ points of $S$ with the property that its affine hull has precisely $k$ points of $S$ on one side of it.

**Lemma 4.3 (Primal Lovász Lemma in $\mathbb{R}^d$)** *Let $S$ be a finite point set in $\mathbb{R}^d$. Then, for any line $\ell$, the number of $k$-set simplices meeting $\ell$ is $O(k^{d-1})$.*

**Proof:** Note that this formulation of the lemma is independent of the choice of the coordinate system. Construct a coordinate system in which $\ell$ coincides with the $x_d$-axis. Dualize $S$ to a system $S^*$ of $n$ hyperplanes, using the standard duality that maps a point $(a_1, \ldots, a_d)$ to the hyperplane $x_d = -a_1x_1 - a_2x_2 - \cdots - a_{d-1}x_{d-1} + a_d$, and a hyperplane $x_d = b_1x_1 + b_2x_2 + \cdots + b_{d-1}x_{d-1} + b_d$ to the point $(b_1, \ldots, b_d)$ (see, e.g., [9]); this duality preserves incidences and above-below relationships between points and hyperplanes (that is, a point $p$ lies below, on, or above a hyperplane $h$ if and only if the dual hyperplane $p^*$ of $p$ lies below, on, or above the point $h^*$ dual to $h$). An application of such a duality also shows that this lemma and the preceding one are indeed dual versions of each other. It suffices to count the number of $k$-set simplices whose affine hulls have $k$ points of $S$ strictly below them. The remaining class of $k$-set simplices is handled by a symmetric argument.

The properties of the duality imply that the affine hull of a $k$-set simplex $\Delta$ as above is mapped into a vertex $\Delta^*$ of the arrangement of $S^*$ which has precisely $k$ hyperplanes below it (and $d$ hyperplanes passing through it). Hence $\Delta^*$ is a vertex of the $k$-th level of $\mathcal{A}(S^*)$. Moreover, $\Delta$ meets the $x_d$-axis $\ell$ if and only if the horizontal hyperplane through $\Delta^*$ is contained in $R_{\Delta^*}$, i.e., $\Delta^*$ is a local maximum of the $k$-th level of $\mathcal{A}(S^*)$.[1] Indeed, $\Delta$ meets $\ell$ if and only if every hyperplane that contains $\ell$ does not have all vertices of $\Delta$ on one side. The set of these hyperplanes is mapped by our duality to the set of all the points at infinity in horizontal directions. Hence $\Delta$ meets $\ell$ if and only if every point at infinity in a horizontal direction lies in $R_{\Delta^*}$, which is equivalent to the condition that the horizontal hyperplane through $\Delta^*$ is contained in $R_{\Delta^*}$, as asserted. As shown by Clarkson [6], the number of local extrema of the $k$-th level in an arrangement of hyperplanes in $d$-space is $O(k^{d-1})$, and this completes the proof of the lemma. □

**Proof of Theorem 4.1:** Dey and Edelsbrunner [8] have shown that if $T$ is a collection of $t$ triangles in 3-space, spanned by $n$ points in general position, then there exists a line $\ell$ that crosses $\Omega(t^3/n^6)$ triangles of $T$. Specifically, consider the collection of the $k$-set triangles of an $n$-point set $S$. Lemma 4.3 implies that no line can cross more than $O(k^2)$ of these triangles. Combining this with the result of [8], we have $|V_k|^3/n^6 = O(k^2)$, and the bound follows. □

## 5 Arrangements of Triangles

Let $\mathcal{T} = \{\Delta_1, \ldots, \Delta_n\}$ be a collection of $n$ triangles in 3-space in general position, and let $\mathcal{A}(\mathcal{T})$ denote the arrangement of $\mathcal{T}$. The $k$-th level of $\mathcal{A}(\mathcal{T})$ is defined, again, as the closure of the set of all points that lie in the union of the triangles and have exactly $k$ triangles lying below them (that is, the relatively open vertical downward-directed ray

[1] The connection between local extrema of $k$-levels and Lovász Lemma was first observed by Clarkson, as briefly remarked in the introduction of [6].

emerging from such a point intersects exactly $k$ triangles). As in the case of segments, the $k$-th level is not necessarily connected, and may have jump discontinuities at points that lie vertically above or on some triangle edge. The complexity of the level, regarded as a polyhedral surface, is the number of its vertices, edges and faces. Assuming general position, this is clearly proportional to the number of vertices only, and we will focus on bounding the number of *inner* vertices, which are incident to three distinct triangles. Any other, 'outer' vertex of the level lies in the vertical plane $H_e$ spanned by some triangle edge $e$. Moreover, if we intersect all the triangles with $H_e$, we get a collection of at most $n$ segments, and the vertices of the $k$-th level of $\mathcal{A}(\mathcal{T})$ that lie in $H_e$ are vertices of the $k$-th level of the 2-dimensional arrangement of these segments within $H_e$, where $e$ itself is also included. By Theorem 3.1, the number of such vertices is $O(n^{3/2})$. Repeating this analysis for each triangle edge $e$, we conclude that the number of outer vertices of the level is $O(n^{5/2})$.

We bound the number of inner vertices using a variant of the dual version of Lovász Lemma in 3-space. The bound that we obtain is considerably weaker than the one given in Lemma 4.2, but is still nontrivial. The proof of this version of the lemma is also different and somewhat more involved.

Let $v$ be an inner vertex of the $k$-th level, incident to three triangles $\Delta_1$, $\Delta_2$, $\Delta_3$; $v$ can be classified into three categories, depending on whether the $k$-th level in the neighborhood of $v$ coincides with

(a) the lower envelope of $\Delta_1$, $\Delta_2$, $\Delta_3$,

(b) the first level of the arrangement $\mathcal{A}(\{\Delta_1, \Delta_2, \Delta_3\})$, or

(c) the upper envelope of $\Delta_1$, $\Delta_2$, $\Delta_3$.

Note that vertices of type (b) have the property that all six edges of $\mathcal{A}(\mathcal{T})$ incident to the vertex lie on the $k$-th level, whereas for vertices of type (a) or (c), only three of these edges lie on the level, one edge on each segment of intersection of two of the triangles $\Delta_1$, $\Delta_2$, $\Delta_3$.

For each inner vertex $v$ of the $k$-th level of type (a) or (c), let $R_v$ be the closed region enclosed between the upper envelope and the lower envelope of the three planes containing the three triangles incident to $v$; see Figure 7 for a cross-section of such an $R_v$. We have the following weaker version of Lovász lemma:

**Lemma 5.1** *Any line in* $\mathbb{R}^3$ *is fully contained in at most* $O(n^{5/2})$ *regions* $R_v$ *of vertices of type (a) and (c).*

**Proof:** Let $\ell_1$ be a line in $\mathbb{R}^3$, and let $H$ be the vertical plane containing $\ell_1$. For a triangle $\Delta \in \mathcal{T}$, let $\pi_\Delta$ be the plane containing $\Delta$ and $\sigma_\Delta = \pi_\Delta \cap H$. Let $\mathcal{A}_H$ be the arrangement in $H$ of the lines $\{\sigma_\Delta \mid \Delta \in \mathcal{T}\}$. Let $\ell_0$ be a line contained in $H$, parallel to $\ell_1$, and lying below all vertices of $\mathcal{A}_H$. It is easily checked that no region $R_v$ contains $\ell_0$. We will move a line $\ell$ within $H$ upwards, parallel to itself, from the position when it coincides with $\ell_0$ until it coincides with $\ell_1$. We estimate the change in the number of regions $R_v$ that contain $\ell$ as it moves. Summing these changes yields the bound on the desired quantity for $\ell_1$.

The set of regions $R_v$ that fully contain $\ell$ can change only when $\ell$ passes through a vertex of $\mathcal{A}_H$. Clearly, the vertex $\chi = \sigma_{\Delta_1} \cap \sigma_{\Delta_2}$ has to be such that there is an inner type-(a) or type-(c) vertex $v$ in $\mathcal{A}$ incident to $\Delta_1$ and $\Delta_2$. Under these assumptions, for $\ell$ to become newly contained in a region $R_v$, or to stop being contained in $R_v$, as it sweeps past such a vertex $\chi$, it is necessary and sufficient that the slope of $\ell$ lie between the slopes of $\sigma_{\Delta_1}$ and $\sigma_{\Delta_2}$; see Figure 7(a). Let $\chi$ be such a vertex (where this latter condition also holds). Put $\lambda = \pi_{\Delta_1} \cap \pi_{\Delta_2}$. Let $s \subset \lambda$ be the segment $\Delta_1 \cap \Delta_2$. For all regions $R_v$ that either start or stop containing $\ell$ as it sweeps over $\chi$, $v$ is contained in $s$, so it suffices to concentrate only on such regions $R_v$.

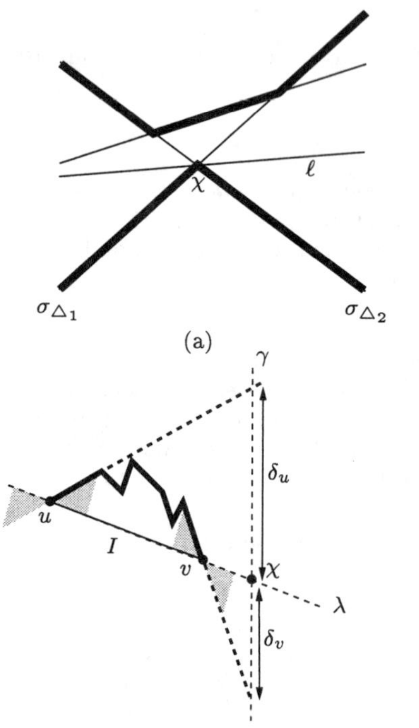


Figure 7: (a) Cross section of a region $R_v$ in $H$; the line $\ell$ just becomes contained in $R_v$; (b) cross section of $\mathcal{A}(\mathcal{T})$ by $\pi$; the intersections of $R_v, R_u$ with $\pi$ are shaded near the respective vertices.

We mark on $s$ all the inner vertices of the $k$-th level of $\mathcal{A}(\mathcal{T})$ of types (a) and (c), and consider the set of maximal subintervals of $s$ *not* contained in the $k$-th level. Each such subinterval $I$ is delimited by two points $u$, $v$, each of which is an inner vertex of the level of type (a) or (c) (it cannot be a vertex of type (b) because all six edges incident to a type-(b) vertex lie on the $k$-th level), a point of jump discontinuity of the level, or an endpoint of $s$. Let $q_s$ be the number of jump discontinuities of the $k$-th level along $s$. Note that each such discontinuity is an outer vertex of the $k$-th level. If an inner vertex $v$ is an endpoint of an interval along $s$ whose other endpoint $v'$ is either a jump discontinuity or an endpoint of $s$, we charge $v$ to $v'$. The number of such inner vertices (and therefore the change in the number of regions that contain $\ell$, corresponding to such vertices) is at most $q_s + 2$.

Next, consider an interval $I$, both of whose endpoints are inner vertices, say $u$ and $v$. Consider the vertical plane $\pi$ containing $\lambda$, and the cross-section of $\mathcal{A}(\mathcal{T})$ within $\pi$ (refer to Figure 7(b)). Clearly, the $k$-th level of this cross-section is contained in the $k$-th level of $\mathcal{A}(\mathcal{T})$, so it either lies fully above $I$ or fully below $I$. In the former case both $u$ and $v$ are of type (c), and in the latter case they are both of type (a). Let $\gamma$ be the vertical line $H \cap \pi$, and let $\delta_u = R_u \cap \gamma$ and $\delta_v = R_v \cap \gamma$. If $\chi \notin I$, then it is easily checked that

$\delta_u$ and $\delta_v$ lie on opposite sides of $\chi$ along $\gamma$ and thus are disjoint except at their common endpoint $\chi$. This fact, and our assumptions that the slope of $\ell$ is between the slopes of $\sigma_u = \Delta_u \cap H$ and $\sigma_v = \Delta_v \cap H$, imply that one of $R_u$, $R_v$ must be added, and the other one removed, from the set of regions containing $f$, as $\ell$ sweeps over $\chi$. Hence, as $\ell$ sweeps over $\chi$, $R_u$ and $R_v$ "cancel" out each other, in terms of containment of $\ell$.

To summarize, we have shown that as $\ell$ passes through $\chi$, the change in the number of regions $R_v$ containing $\ell$ is at most $4+q_s$. This implies that the number of regions $R_v$ that contain $\ell$ in its final position $\ell_1$ is at most $\sum_s(4+q_s)$, where the sum is over all $O(n^2)$ intersection segments between pairs of triangles in $\mathcal{T}$. Since the number of outer vertices on the $k$-th level is $O(n^{5/2})$, as argued above, and each is counted at most three times, $\sum_s q_s = O(n^{5/2})$. The number of regions containing $\ell$ is thus $O(n^2)+O(n^{5/2}) = O(n^{5/2})$, as asserted.

What if $\ell_1$ actually passes through a vertex $\chi = \sigma_{\Delta_i} \cap \sigma_{\Delta_j}$ of $\mathcal{A}_H$? Then the cancellation does not occur, which adds fewer than $n$ regions $R_v$ that can contain $\ell$—each such region corresponds to some vertex of $\mathcal{A}(\mathcal{T})$ on the segment $\Delta_i \cap \Delta_j$. □

**Theorem 5.2** *The complexity of any single level in an arrangement of n triangles in 3-space is* $O(n^{17/6})$.

**Proof:** Lemma 5.1 implies that no line $\ell$ is contained in more than $O(n^{5/2})$ regions $R_v$. Passing to the dual space, we obtain the following equivalent formulation, similar to the case of planes: The planes containing the triangles in $\mathcal{T}$ are mapped to a set of $n$ points. Each inner vertex $v$ of the $k$-th level is mapped to a triangle spanned by the three points dual to the planes containing the triangles incident to $v$. The line $\ell$ is mapped to another line $\ell^*$, and $\ell$ is contained in $R_v$ if and only if $\ell^*$ crosses the triangle dual to $v$. We now have a system of $X$ triangles in 3-space, spanned by a total of $n$ points, where $X$ is the number of inner vertices of the $k$-th level of types (a) and (c). As in the proof of Theorem 4.1, there exists a line that crosses at least $\Omega(X^3/n^6)$ such triangles [8]. On the other hand, by Lemma 5.1, this number is at most $O(n^{5/2})$. Combining these two inequalities yields $X = O(n^{17/6})$. We still need to bound the number of vertices of type (b). However, these vertices are vertices of type (a) of the $(k-1)$-st level, so, repeating the above analysis for this level, we obtain the bound asserted in the theorem. □

**Remark:** An open problem is to extend Lemmas 4.2 and 5.1 to the respective cases of pseudo-hyperplanes and pseudo-triangles, under appropriate definitions of these objects, and then to extend the proof of Theorem 5.2 to these cases. Note that there are two different problems to address: One is to extend Lovász Lemma, and the other calls for a dual and more general version of the analysis technique of [8] (that yields a line that stabs many triangles).

**Acknowledgments.** We wish to thank János Pach and Emo Welzl for useful discussions concerning these problems, and Ken Clarkson for pointing out a much simpler reduction from the number of local extrema of $k$-levels to Lovász Lemma, as presented in this paper.

## References

[1] N. Alon and E. Győri, The number of small semispaces of a finite set of points in the plane, *J. Combin. Theory Ser. A* 41 (1986), 154–157.

[2] B. Aronov, B. Chazelle, H. Edelsbrunner, L. Guibas, M. Sharir, and R. Wenger, Points and triangles in the plane and halving planes in space, *Discrete Comput. Geom.* 6 (1991), 435–442.

[3] I. Bárány, Z. Füredi, and L. Lovász, On the number of halving planes, *Combinatorica* 10 (1990), 175–183.

[4] I. Bárány and W. Steiger, On the expected number of $k$-sets, *Discrete Comput. Geom.* 11 (1994), 243–263.

[5] B. Chazelle, H. Edelsbrunner, L. Guibas, J. Hershberger, R. Seidel and M. Sharir, Selecting heavily covered points, *SIAM J. Comput.* 23 (1994), 1138–1151.

[6] K. Clarkson, A bound on local minima of arrangements that implies the Upper Bound Theorem, *Discrete Comput. Geom.* 10 (1993), 427–433.

[7] K. Clarkson and P. Shor, Applications of random sampling in computational geometry II, *Discrete Comput. Geom.* 4 (1989), 387–421.

[8] T. Dey and H. Edelsbrunner, Counting triangle crossings and halving planes, *Discrete Comput. Geom.* 12 (1994), 281–289.

[9] H. Edelsbrunner, *Algorithms in Combinatorial Geometry*, Springer-Verlag, Heidelberg 1987.

[10] H. Edelsbrunner, P. Valtr, and E. Welzl, Cutting dense point sets in half, *Proc. 10th Annu. ACM Symp. Comput. Geom.*, 1994, 364–476.

[11] P. Erdős, L. Lovász, A. Simmons, and E. G. Straus, Dissection graphs of planar point sets, In J. N. Srivastava *et al.*, editors, *A Survey of Combinatorial Theory*, North-Holland, Amsterdam, 1973, pp. 139–149.

[12] J. E. Goodman and R. Pollack, Allowable Sequences and Order Types in Discrete and Computational Geometry, in *New Trends in Discrete and Computational Geometry*, Algorithms and Combinatorics (J. Pach, Ed.), vol. 10, Springer-Verlag, 1993, 103–134.

[13] D. Gusfield, *Sensitivity Analysis for Combinatorial Optimization*, Ph.D. Thesis, University of California at Berkeley, 1980.

[14] D. Gusfield, Bounds for the parametric minimum spanning tree problem, *Proc. Humboldt Conference on Graph Theory, Combinatorics and Computing*, 1979, Utilitas Mathematica, pp. 173–183.

[15] D. Halperin and M. Sharir, On disjoint concave chains in arrangements of (pseudo)lines, *Inform. Process. Lett.* 40 (1991), 189–192.

[16] D. Halperin and M. Sharir, Corrigendum: On disjoint concave chains in arrangements of (pseudo) lines, *Inform. Process. Lett.* 51 (1994), 53–56.

[17] J. Hershberger and J. Snoeyink, Convex polygons made from few lines and convex decompositions of polyhedra, *Proc. 3rd Scand. Workshop on Algorithm Theory*, LNCS 621, Springer-Verlag, 1992, pp. 376–387.

[18] M. Klawe, M. Paterson, and N. Pippenger, unpublished manuscript.

[19] L. Lovász, On the number of halving lines, *Ann. Univ. Sci. Budapest, Eötvös, Sec. Math.* 14 (1971), 107–108.

[20] J. Pach and P.K. Agarwal, *Combinatorial Geometry*, Wiley Interscience, New York, 1995.

[21] J. Pach, W. Steiger, and M. Szemerédi, An upper bound on the number of planar $k$-sets, *Discrete Comput. Geom.* 7 (1992), 109–123.

[22] M. Sharir, On $k$–sets in arrangements of curves and surfaces, *Discrete Comput. Geom.* 6 (1991), 593–613.

[23] H. Tamaki and T. Tokuyama, How to cut pseudo parabolas into segments, *Prod. 11th ACM Symp. on Comput. Geom.*, 1995, 230–237.

[24] E. Welzl, More on $k$-sets of finite sets in the plane, *Discrete Comput. Geom.* 1 (1986), 95–100.

[25] A. Wiernik and M. Sharir, Planar realization of nonlinear Davenport–Schinzel sequences by segments, *Discrete Comput. Geom.* 3 (1988), 15–47.

[26] R. Živaljević and S. Vrećica, The colored Tverberg's problem and complexes of injective functions, *J. Combin. Theory Ser. A* 61 (1992), 309–318.

# Cylindrical Static and Kinetic Binary Space Partitions

Pankaj K. Agarwal* Leonidas J. Guibas† T. M. Murali‡ Jeffrey Scott Vitter§

**Abstract**

We describe the first known algorithm for efficiently maintaining a Binary Space Partition (BSP) for $n$ continuously moving segments in the plane. Under reasonable assumptions on the motion, we show that the total number of times the BSP changes is $O(n^2)$, and that we can update the BSP in $O(\log n)$ expected time per change. We also consider the problem of constructing a BSP for $n$ triangles in $\mathbb{R}^3$. We present a randomized algorithm that constructs a BSP of expected size $O(n^2)$ in $O(n^2 \log^2 n)$ expected time. We also describe a deterministic algorithm that constructs a BSP of size $O((n+k)\log n)$ and height $O(\log n)$ in $O((n+k)\log^2 n)$ time, where $k$ is the number of intersection points between the edges of the projections of the triangles onto the $xy$-plane.

## 1 Introduction

The Binary Space Partition (BSP, also known as BSP tree), originally proposed by Schumacker et al. [26] and further refined by Fuchs et al. [16], is a hierarchical partitioning of space widely used in several areas, including computer graphics (global illumination [7], shadow generation [10, 11], visibility determination [4, 28], and ray tracing [21]), solid modeling [22, 20, 29], geometric data repair [19], robotics [5], network design [18], and surface simplification [3]. Key to the BSP's success is that it serves both as a model for an object (or a set of objects) and as a data structure for querying the object.

Informally, a BSP $\mathcal{B}$ for a set of objects is a binary tree, where each node $v$ is associated with a convex region $\Delta_v$. The regions associated with the children of $v$ are obtained by splitting $\Delta_v$ with a hyperplane. If $v$ is a leaf of $\mathcal{B}$, then the interior of $\Delta_v$ does not intersect any object. The regions associated with the leaves of the tree form a convex decomposition of space. The faces of the decomposition induced by the leaves intersect the objects and divide them into fragments; these fragments are stored at appropriate nodes of the BSP. The efficiency of BSP-based algorithms depends on the number of nodes in the tree and on the height of the tree. As a result, several algorithms for constructing BSPs of small size and/or small height have been proposed; see [4, 8, 16, 24, 25, 28, 29].

In this paper, we study *cylindrical* BSPs in which all the cuts that do not contain any input objects are made by hyperplanes parallel to the same fixed direction. We address two problems. The first problem can be formulated as follows: Let $S$ be a set of $n$ interior-disjoint segments in the plane, each moving along a continuous path. We want to maintain the BSP for $S$ as the segments in $S$ move. We assume that the segments move in such a way that they never intersect, except possibly at their endpoints. Most of the work to date deals with constructing a BSP for a set of "static" segments, which do not move. Paterson and Yao propose a randomized algorithm that constructs a BSP of expected $O(n \log n)$ size for a set of $n$ segments in the plane [24]. They also propose a deterministic algorithm, based on a divide-and-conquer approach, that constructs a BSP of size $O(n \log n)$ in $O(n \log n)$ time [24]. Both of these algorithms are not "robust," in the sense that a small motion of one of the segments may cause many changes in the tree, or may cause non-local changes. Therefore, they are ill-suited for maintaining a BSP for a set of moving segments.

There have been a few attempts to update BSPs when the objects defining them move. Naylor describes a method to implement dynamic changes in a BSP, where the static objects are represented by a balanced BSP (computed in a preprocessing stage), and then the moving objects are inserted at each time step into the static tree [23]. Using the same assumption that moving objects are known *a priori*, Torres proposes the augmentation of BSPs with additional separating planes, which may localize the updates needed for deletion and re-insertion of moving objects in a BSP [30]. This approach does try to exploit (by introducing additional planes) the spatial coherence of the dynamic changes in the tree. Chrysanthou suggests a more general approach, which does not make any distinction between static and moving objects [12]. By keeping additional information about topological adjacencies in the tree, the algorithm performs insertions and deletions of a node in a more localized way. But all these prior efforts boil down to deleting moving objects from their earlier positions and re-inserting them in their current positions after some time interval has elapsed. Such approaches suffer from the fundamental problem that it is

---

*Support was provided by National Science Foundation research grant CCR-93-01259, by Army Research Office MURI grant DAAH04-96-1-0013, by a Sloan fellowship, by a National Science Foundation NYI award and matching funds from Xerox Corp, and by a grant from the U.S.-Israeli Binational Science Foundation. Address: Center for Geometric Computing, Department of Computer Science, Duke University, Box 90129, Durham, NC 27708-0129. Email: pankaj@cs.duke.edu

†Support was provided in part by National Science Foundation grant CCR-9623851 and by US Army MURI grant 5-23542-A. Address: Graphics Laboratory, Computer Science Department, Gates Building 3B, Room 374, Stanford University, Stanford, CA 94305. Email: guibas@cs.stanford.edu

‡This author is affiliated with Brown University. Support was provided in part by National Science Foundation research grant CCR-9522047 and by Army Research Office MURI grant DAAH04-96-1-0013. Address: Center for Geometric Computing, Department of Computer Science, Duke University, Box 90129, Durham, NC 27708-0129. Email: tmax@cs.duke.edu

§Support was provided in part by National Science Foundation research grant CCR-9522047, by Army Research Office grant DAAH04-93-G-0076, and by Army Research Office MURI grant DAAH04-96-1-0013. Address: Center for Geometric Computing, Department of Computer Science, Duke University, Box 90129, Durham, NC 27708-0129. Email: jsv@cs.duke.edu


*Computational Geometry 97* Nice France

very difficult to know how to choose the correct time interval size: if the interval is too small, then the BSP does not in fact change combinatorially, and the deletion/re-insertion is just wasted computation; if it is too big, then important intermediate events can be missed, which affect applications that use the tree.

Our algorithm, instead, treats the BSP as a *kinetic data structure*, as defined by Basch et al. [6]. We view the equations of the cuts made at the nodes of the BSP and the edges and faces of the subdivision induced by the BSP as functions of time. The cuts and the edges and faces of the subdivision change continuously with time. However, "combinatorial" changes in the BSP and in the subdivision (we precisely define this notion later) occur only at certain times. We explicitly take advantage of the continuity of the motion of the objects involved so as to generate updates to the BSP only when actual events cause the BSP to change combinatorially.

In Section 3, we describe a randomized kinetic algorithm for maintaining a BSP for moving segments in the plane. We assume that the segment motions are *oblivious* to the random bits used by the algorithm. Following Basch et al. [6], we assume that each moving segment has a posted flight plan that gives full or partial information about its current motion. Whenever a flight plan changes (possibly due to an external agent), our algorithm is notified and it updates a global event queue to reflect the change. We first derive a randomized algorithm for computing a BSP for a set of static segments, which combines ideas from Paterson and Yao's randomized and deterministic algorithms, but is also robust, in the sense described earlier. The "combinatorial structure" of the BSP constructed by this algorithm changes only when the $x$-coordinates of a pair of segment endpoints, among a certain subset of $O(n)$ pairs, become equal. We show that under the above assumption on the segment motions, the BSP can be updated in $O(\log n)$ expected time at each such event. We also show that if $k$ of the segments of $S$ move along "pseudo-algebraic" paths, and the remaining segments of $S$ are stationary, then the expected number of changes in the BSP is $O(kn \log n)$. As far as we know, this is the first nontrivial algorithm for maintaining a BSP for moving segments in the plane.

Next, we study the problem of computing a BSP for a set $S$ of $n$ interior-disjoint triangles in $\mathbb{R}^3$. Paterson and Yao [24] describe a randomized incremental algorithm that constructs a BSP of expected size $O(n^2)$ in time $O(n^3)$. They also show that their algorithm can be made deterministic without affecting its asymptotic running time. It has been an open problem whether a BSP for $n$ triangles in $\mathbb{R}^3$ can be constructed in near-quadratic time. Sub-quadratic bounds are known for special cases [2, 14, 25]. However, none of these approaches lead to a near-quadratic algorithm for triangles in $\mathbb{R}^3$. We present a randomized algorithm (in Section 4) that constructs a BSP for $S$ of expected size $O(n^2)$ in $O(n^2 \log^2 n)$ time.

The bottleneck in analyzing the expected running time of the Paterson-Yao algorithm is that no nontrivial bound is known on the number of vertices in the convex subdivision of $\mathbb{R}^3$ induced by the BSP constructed by the algorithm. Known techniques for analyzing randomized algorithms, such as the Clarkson-Shor framework [13] or backwards analysis [27], cannot be used to obtain a near-quadratic bound on the size of the convex subdivision corresponding to the BSP constructed by the Paterson-Yao algorithm, since the BSP constructed by the algorithm depends on the order in which triangles are added.

Our algorithm is a variant of the Paterson-Yao algorithm. We construct the BSP for $S$ in such a way that there is a close relationship between the BSP and the planar arrangement of lines supporting the edges of the $xy$-projections of the triangles in $S$. We use results from $\varepsilon$-net theory [17] and on arrangements of lines [15] to bound the expected number of vertices in the convex subdivision of $\mathbb{R}^3$ induced by the BSP and the expected running time of the algorithm.

Finally, we present a deterministic algorithm (Section 5) for constructing a BSP for a set $S$ of $n$ triangles in $\mathbb{R}^3$. If $k$ is the number of intersection points of the $xy$-projections of the edges of triangles in $S$, then the algorithm constructs a BSP of size $O((n + k) \log n)$ in time $O((n + k) \log^2 n)$; if $k \ll n^2$, the deterministic algorithm constructs a much smaller BSP than do Paterson and Yao's and our randomized algorithm. Another nice property of our deterministic algorithm is that the height of the BSP it constructs is $O(\log n)$, which is useful for ray-shooting queries, for example. It was an open problem whether BSPs of near-quadratic size and $O(\log n)$ height could be constructed for $n$ triangles in $\mathbb{R}^3$. The height of the BSP constructed by the randomized algorithms (both ours and the one by Paterson and Yao) can be $\Omega(n)$, e.g., when $S$ is the set of faces of a convex polytope. Due to lack of space, we omit many proofs and details from this abstract.

## 2 Definitions

A *binary space partition* $\mathcal{B}$ for a set $S$ of convex $(d-1)$-polytopes in $\mathbb{R}^d$ with pairwise-disjoint interiors is a tree defined as follows: Each node $v$ in $\mathcal{B}$ is associated with a convex $d$-polytope $\Delta_v$ and a set of $(d-1)$-polytopes $S_v = \{s \cap \Delta_v \mid s \in S\}$. The polytope associated with the root is $\mathbb{R}^d$ itself. If $S_v$ is empty, then node $v$ is a leaf of $\mathcal{B}$. Otherwise, we partition $\Delta_v$ into two convex polytopes by a *cutting hyperplane* $H_v$. At $v$, we store the equation of $H_v$ and the set $\{s \mid s \subseteq H_v, s \in S_v\}$ of polytopes in $S_v$ that lie in $H_v$. If we let $H_v^+$ be the positive halfspace and $H_v^-$ be the negative halfspace bounded by $H_v$, the polytopes associated with the left and right children of $v$ are $\Delta_v \cap H_v^-$ and $\Delta_v \cap H_v^+$, respectively. The left subtree of $v$ is a BSP for $S_v^- = \{s \cap H_v^- \mid s \in S_v\}$ and the right subtree is for $S_v^+ = \{s \cap H_v^+ \mid s \in S_v\}$. The size of $\mathcal{B}$ is the sum of the number of nodes in $\mathcal{B}$ and the total number of polytopes stored at all the nodes in $\mathcal{B}$.

At a node $v$ of $\mathcal{B}$, the cutting hyperplane $H_v$ may support a polytope $s \in S$ such that $H_v \cap \Delta_v \subseteq s$, i.e., the portion of $H_v$ that lies in the interior of $\Delta_v$ is contained in $s$. Such a cutting hyperplane will be referred to as a *free* cut. Free cuts will be critical in keeping the size of $\mathcal{B}$ small by preventing excessive fragmentation of the polytopes in $S$.

For our purposes, $S$ is either a set of $n$ segments in the plane or a set of $n$ triangles in $\mathbb{R}^3$. A unifying feature of all the BSPs constructed by our algorithms is that the region $\Delta_v$ associated with each node $v$ is a *cylindrical* cell in the sense that $\Delta_v$ may contain top and bottom faces that are contained in objects belonging to $S$, but all other faces are vertical. In the plane, $\Delta_v$ is a trapezoid; in $\mathbb{R}^3$, $\Delta_v$ may have large complexity, as it may contain many vertical faces.

## 3 Kinetic Algorithm for Segments

Let $S$ be a set of $n$ non-intersecting segments in the plane. We first describe a randomized algorithm for computing a BSP $\mathcal{B}$ for $S$ when the segments in $S$ are stationary, and then explain how to maintain $\mathcal{B}$ as each segment in $S$ moves along a continuous path.

Our algorithm makes two types of cuts: a *vertical* cut through an endpoint of a segment and an *edge* cut along a segment. Edge cuts are always contained totally within input segments; therefore, they are free cuts. Each face in the planar subdivision induced by $\mathcal{B}$ is a trapezoid; the left and right boundaries of a trapezoid are bounded by vertical cuts, and the top and bottom edges are bounded by edge cuts. At an interior node $v$ of $\mathcal{B}$, if $\Delta_v$ is split by a vertical cut through an endpoint $p$, we store $p$ at $v$; the left (resp., right) subtree of $v$ corresponds to the BSP for the trapezoid lying to the left (resp., to the right) of the cut; if $\Delta_v$ is split by an edge cut along a segment $s$, we store $s$ at $v$, and the left (resp., right) subtree of $v$ corresponds to the BSP for the trapezoid lying below (resp., above) of the cut.

For a node $v$ in $\mathcal{B}$, the *combinatorial structure* of the trapezoid $\Delta_v$ is the 4-tuple $(\lambda_v, \rho_v, \tau_v, \beta_v)$, where $\lambda_v$ (resp., $\rho_v$) is the endpoint of a segment in $S$ so that the vertical line passing it contains the left (resp., right) edge of $\Delta_v$, and $\tau_v$ (resp., $\beta_v$) is the segment in $S$ containing the top (resp., bottom) edge of $\Delta_v$. The *combinatorial structure* of $\mathcal{B}$ is a binary tree, each of whose nodes $v$ is associated with the set of segments $S_v$. We will use the combinatorial structure of the BSP crucially in our kinetic algorithm.

### 3.1 The static algorithm

We now describe our static algorithm. We choose a random permutation $\langle s_1, s_2, \ldots, s_n\rangle$ of $S$. We say that $s_i$ has a *higher priority* than $s_j$ if $i < j$. We add the segments in decreasing order of priority and maintain a BSP for the segments added so far. Let $S^{(i)} = \{s_1, s_2, \ldots, s_i\}$ be the set of the first $i$ segments in the permutation. Our algorithm works in stages. At the beginning of the $i$th stage, where $i > 0$, we have a BSP $\mathcal{B}^{(i-1)}$ for $S^{(i-1)}$; $\mathcal{B}^{(0)}$ consists of a single node $v$, where $\Delta_v$ is the entire plane. In the $i$th stage, we add $s_i$ and compute a BSP $\mathcal{B}^{(i)}$ for $S^{(i)}$ as follows: Suppose $p_{2i-1}$ and $p_{2i}$ are the left and right endpoints of $s_i$, respectively. Let $\Delta_v$ be the trapezoid containing $p_{2i-1}$ in the planar subdivision induced by $\mathcal{B}^{(i-1)}$. We store $p_{2i-1}$ at $v$ and create two children $w$ and $z$ of $v$. We partition $\Delta_v$ into two trapezoids by drawing a vertical segment through $p_{2i-1}$; $\Delta_w$ and $\Delta_z$ are the trapezoids lying to the left and right of the vertical line, respectively. The combinatorial structures of $w$ and $z$ are $(\lambda_v, p_{2i-1}, \tau_v, \beta_v)$ and $(p_{2i-1}, \rho_v, \tau_v, \beta_v)$, respectively. We then perform a similar step for $p_{2i}$. Finally, for each trapezoid $\Delta_x$ that intersects $s_i$, we split $\Delta_x$ into two trapezoids $\Delta_{x_1}$ and $\Delta_{x_2}$ by making an edge cut along $s_i$. We store $s_i$ at $x$, create two children $x_1$ and $x_2$ of $x$, and associate $\Delta_{x_1}$ (resp., $\Delta_{x_2}$) with $x_1$ (resp., $x_2$). If $\Delta_{x_1}$ is above $s_i$ and $\Delta_{x_2}$ is below $s_i$, then the combinatorial structure of $x_1$ and $x_2$ are $(\lambda_x, \rho_x, \tau_x, s_i)$ and $(\lambda_x, \rho_x, s_i, \beta_x)$, respectively. The resulting tree is the BSP $\mathcal{B}^{(i)}$ for $S^{(i)}$. See Figure 1 for an example of constructing $\mathcal{B}^{(i)}$ from $\mathcal{B}^{(i-1)}$.

The vertical segment drawn upwards (resp., downwards) from an endpoint $p_i$ will be referred to as the *upper* (resp., *lower*) *thread* of $p_i$. The segment containing the other endpoint of a thread is called the *stopper* of that thread. Note that the priority of the stopper of a thread of $p_i$ is higher than that of the segment containing $p_i$.

This completes the description of our algorithm. Note that once we fix the permutation, the algorithm is deterministic and constructs a unique BSP. Using an analysis similar to Paterson and Yao's [24], we can prove the following lemmas:

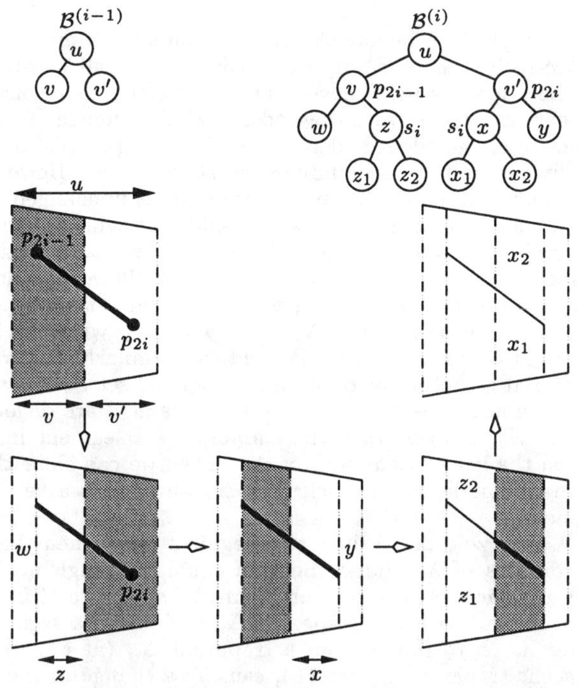

Figure 1: The BSP $\mathcal{B}^{(i-1)}$, the sequence of cuts made in the $i$th stage, and BSP $\mathcal{B}^{(i)}$. At each step, the shaded trapezoids are split. Portions of $s_i$ that lie in the interior of a trapezoid are drawn using thick lines.

**Lemma 3.1** *Let $p$ be an endpoint of a segment in $S$. The expected number of segments crossed by the threads of $p$ is $O(\log n)$.*

**Theorem 3.2** *The expected size of the BSP constructed by the above algorithm is $O(n \log n)$, and the expected height of the BSP is $O(\log n)$.*

### 3.2 The kinetic algorithm

We now describe how to maintain the static BSP as the segments in $S$ move continuously. The position of a segment $s_i$ with endpoints $p_{2i-1}$ and $p_{2i}$ can be specified by a point $(x, y, \tan(\theta/2)) \in \mathbb{R}^3$, where $(x, y)$ denotes the position of $p_{2i-1}$ in the plane and $\theta$ denotes the angle that the ray along the direction $\overrightarrow{p_{2i-1}p_{2i}}$ makes with the $(+x)$-axis, in the counterclockwise direction. Let $s_i(t)$ denote the segment $s_i$ at time $t$, and let $S(t)$ denote the set $S$ at time $t$. We assume that we choose a random permutation $\pi$ of $S$ once in the very beginning (at $t = 0$), and that $\pi$ does not change with time. Let $\mathcal{B}(t)$ denote the BSP of $S(t)$ constructed by the static algorithm, using $\pi$ as the permutation to decide the priority of the segments. We describe an algorithm that updates the BSP under the following assumption:

($\star$) *There is no correlation between the motion of the segments in $S$ and their priorities. Therefore, the chosen permutation $\pi$ always behaves like a random permutation, and Lemma 3.1 and Theorem 3.2 hold at all times.*

We parameterize the motion of the segments by time and use $t$ to denote time. For a given time instant $t$, we will use $t^-$ and $t^+$ to denote the time instants $t - \varepsilon$ and $t + \varepsilon$, respectively, where $\varepsilon > 0$ is an arbitrarily small constant. As the segments in $S$ move continuously, the equations of the cuts associated with the nodes of $\mathcal{B}$ also change. At the same time, the edges and vertices of the trapezoids in the subdivision of the plane induced by $\mathcal{B}$ also move. However, the combinatorial structure of a trapezoid is unchanged until two edges of the subdivision collide, at which point a trapezoid $\Delta_v$, where $v$ is a node in $\mathcal{B}$, shrinks to a vertical segment. Since the segments of $S$ are not allowed to intersect, the top and bottom edges of $\Delta_v$ cannot meet, so the combinatorial structure of $\Delta_v$ can change only when the left and right edges of $\Delta_v$ become identical. Similarly, the combinatorial structure of $\mathcal{B}$ changes when the set $S_v$ changes for some node $v \in \mathcal{B}$. Since the segments in $S$ are disjoint, the set $S_v$ changes when the endpoint of a segment in $S_v$ lies on the left or right edge of $\Delta_v$. Then we can show that there is a node $w \in \mathcal{B}$ such that $\Delta_w$ shrinks to a vertical segment.

A trapezoid $\Delta_v$ shrinks to a segment only when the $x$-coordinates of $\lambda_v$ and $\rho_v$ become equal, although not all such instances change the combinatorial structure of $\mathcal{B}$. At each instant $t$ when a trapezoid $\Delta_v$ shrinks to a segment, either $\lambda_v$ or $\rho_v$ moves from a trapezoid $\Delta_w$ (at $t^-$) to an adjacent trapezoid $\Delta_x$ (at $t^+$), causing a change in the set of segments $S_w$ or $S_x$. Hence, the combinatorial structure of $\mathcal{B}(t)$ also changes at $t$. Let $\mathcal{B}(t^-)$ and $\mathcal{B}(t^+)$ denote the trees at $t^-$ and $t^+$, respectively. We would like to ensure that the edit distance between $\mathcal{B}(t^-)$ and $\mathcal{B}(t^+)$ is small,[1] and that $\mathcal{B}(t^+)$ can be obtained from $\mathcal{B}(t^-)$ in time proportional to the edit distance between them.

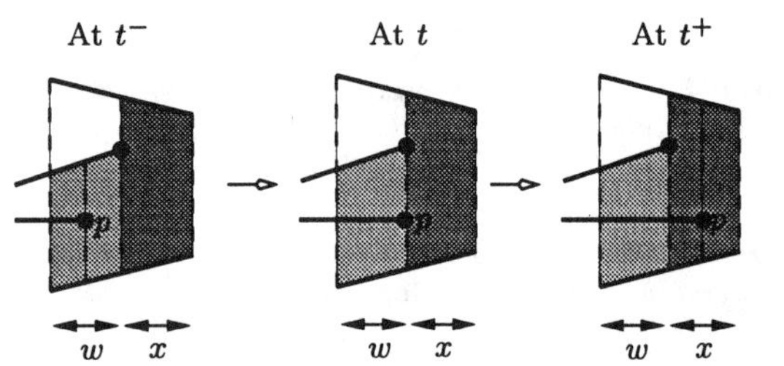


Figure 2: Endpoint $p$ moves from $\Delta_w$ at $t^-$ to $\Delta_x$ at $t^+$. The set $S_x$ changes at time instant $t$.

A node $v$ of $\mathcal{B}(t)$ is called *transient* if $\Delta_v$ does not contain any endpoint in its interior and a vertical cut was made at the parent $p(v)$ of $v$; $\Delta_v$ is called a transient trapezoid. See Figure 3. Using the above assumption and Lemma 3.1, we can prove that transient nodes have the following properties:

1. No proper ancestor or proper descendant of a transient node is a transient node.
2. The number of transient nodes in $\mathcal{B}(t)$ is $O(n)$.
3. Only edge cuts are made at the descendants of a transient node $v$ (including $v$ itself). The left and right edges of the trapezoids associated with all the descendants of $v$ are portions of the left and right edges of $\Delta_v$.
4. The expected number of descendants of a transient node is $O(\log n)$.

[1]The *edit distance* between $\mathcal{B}(t_1)$ and $\mathcal{B}(t_2)$ is the minimum number of insertions of nodes, deletions of nodes, and pointer changes required to obtain $\mathcal{B}(t_2)$ from $\mathcal{B}(t_1)$.

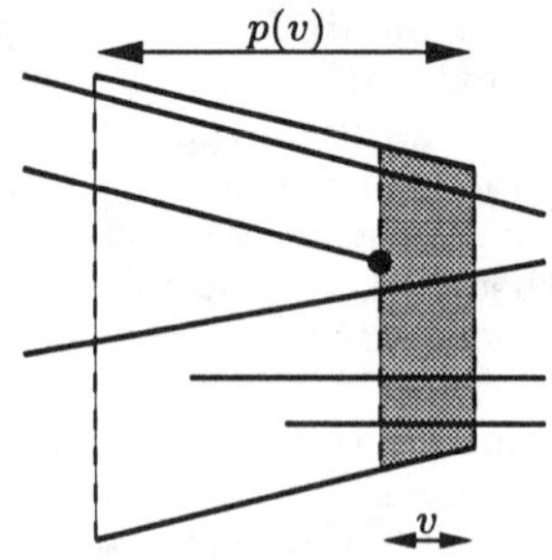


Figure 3: The shaded trapezoid $\Delta_v$ is transient.

Recall that for a node $v$ in $\mathcal{B}(t)$, $\lambda_v$ (resp., $\rho_v$) denotes the endpoint whose thread contains the left (resp., right) boundary of $\Delta_v$. The following lemma characterizes the time instances $t$ at which the combinatorial structure of $\mathcal{B}$ changes. Let $\Gamma(t) = \{(\lambda_v, \rho_v) \mid v \text{ is a transient node at time } t\}$; the properties of transient nodes imply that $|\Gamma(t)| = O(n)$.

**Lemma 3.3** *For any time instant $t$, $\mathcal{B}(t^-)$ and $\mathcal{B}(t^+)$ have different combinatorial structures if and only if there exists a transient node $v$ in $\mathcal{B}(t^-)$ such that the $x$-coordinates of the endpoints $\lambda_v$ and $\rho_v$ become equal at time $t$.*

Intuitively, transient nodes are the highest nodes in $\mathcal{B}(t)$ whose combinatorial structure can change next. If a trapezoid contains an endpoint in its interior, it cannot shrink to a segment; and if an edge cut is made at the parent $p(v)$ of a node $v$, then $\Delta_{p(v)}$ also shrinks to a segment whenever $\Delta_v$ shrinks to a segment and $\Delta_{p(v)}$ does not contain an endpoint. Hence, it suffices to keep track of transient nodes to determine all the combinatorial changes in $\mathcal{B}(t)$. To this end, we maintain the set $\Gamma(t)$. For each pair $(\lambda_v, \rho_v)$ in $\Gamma(t)$, we compute the time at which the $x$-coordinates of $\lambda_v$ and $\rho_v$ coincide, and store these time values in a global priority queue. We refer to these values of time as *critical events*.

We will prove that if the combinatorial structure of $\mathcal{B}$ changes at time $t$, then we can obtain $\mathcal{B}(t^+)$ from $\mathcal{B}(t^-)$ in $O(\log n)$ expected time. We will also show that at each event point, the expected number of changes in the global event queue is $O(\log n)$.

In order to expedite the updating of $\mathcal{B}$, we store some additional information with the nodes in $\mathcal{B}$: At each node $v$ of $\mathcal{B}$, we store the number $c_v$ of endpoints lying in the interior of $\Delta_v$ ($c_v$ helps determine the new transient trapezoids at $t^+$); and for each endpoint $p_j$, we maintain the list $T_j$ (resp., $B_j$) of segments that its upper (resp., lower) thread crosses; $T_j$ (resp., $B_j$) is sorted in the $(+y)$-direction (resp., $(-y)$-direction). As the endpoints move, these lists will be used to compute new stoppers of threads. Although we do not really need these lists, they simplify the description of the update procedure without affecting its expected running time.

We now describe the procedure for updating the tree at each critical event. Recall that at each such instant $t$, an endpoint $p$ of a segment in $S$ moves from a trapezoid $\Delta_w$ (at $t^-$) to an adjacent trapezoid $\Delta_x$ (at $t^+$). See Figure 2. At time $t^-$, the vertical cut made through $p$ divides $\Delta_w$ into two trapezoids. At time $t$, one of these two trapezoids shrinks to a segment, and at $t^+$, a new trapezoid appears inside $\Delta_x$. These changes inside $\Delta_w$ and $\Delta_x$ cause combinatorial changes in other trapezoids. We exemplify these ideas below. Let $\mathcal{B}^- = \mathcal{B}(t^-)$ and $\mathcal{B}^+ = \mathcal{B}(t^+)$. For a node $z \in \mathcal{B}^-$, let $\mathcal{B}_z^-$ denote the subtree of $\mathcal{B}^-$ rooted at $z$; define $\mathcal{B}_z^+$ similarly.

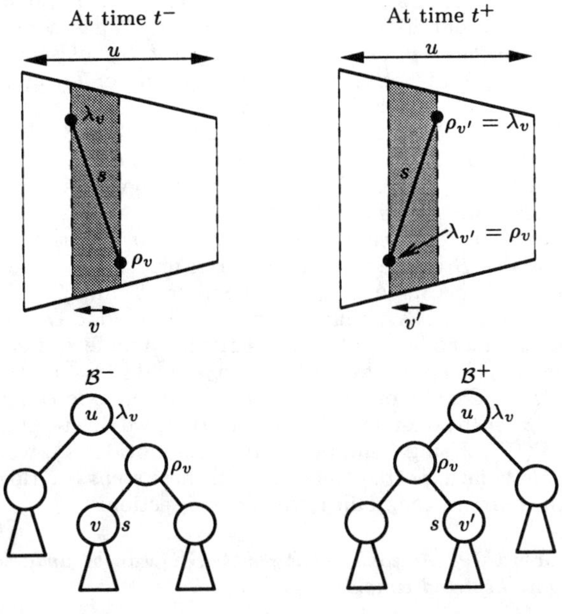


Figure 4: The case when $\lambda_v$ and $\rho_v$ belong to the same segment.

Suppose $v$ is a transient node in at time $\mathcal{B}^-$. Hence, the $x$-coordinates of the endpoints $\lambda_v$ and $\rho_v$ become equal at time $t$. Suppose $\lambda_v$ and $\rho_v$ are endpoints of the same segment $s \in S$; $s$ is vertical at time $t$. See Figure 4. Let $u$ be $v$'s grandparent in $\mathcal{B}^-$; the trapezoid $\Delta_u$ contains $s$. At time $t^+$, a new trapezoid $v'$ with combinatorial structure $(\rho_v, \lambda_v, \tau_v, \beta_v)$ appears inside $u$. We obtain $\mathcal{B}^+$ by simply swapping the left and right subtrees of $u$ in $\mathcal{B}^-$ and replacing $v$ with $v'$.

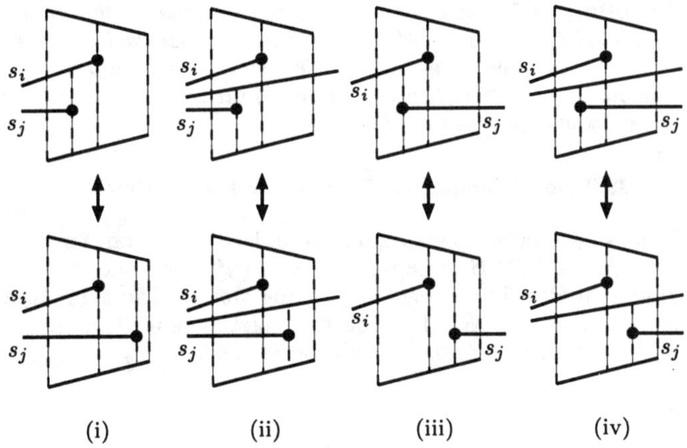


Figure 5: Some possible changes in the combinatorial structure of the trapezoids in the subdivision induced by the BSP.

Now suppose $\lambda_v$ and $\rho_v$ are endpoints of different segments. Figure 5 shows some of the ways in which the combinatorial structure of trapezoids can change at a critical event; the other cases can be reduced to these cases by taking a reflection with respect to one of the two axes, by going backward in time, or by doing both. Assume that $\rho_v = p_{2i}$ and $\lambda_v = p_{2j}$ are the right endpoints of the segments $s_i$ and $s_j$, respectively, that $s_i$ lies above $s_j$, that the priority of $s_i$ is higher than that of $s_j$, and that the $x$-coordinate of $p_{2j}$ is less than the $x$-coordinate of $p_{2i}$ at $t^-$; see Figures 5(i) and 5(ii). We now describe how we update $\mathcal{B}(t)$ for this case; we omit the other cases from this abstract, since they can be handled in a similar manner. Let $u$ and $w$ be the nodes in $\mathcal{B}^-$ at which the vertical cuts through $p_{2i}$ and $p_{2j}$, respectively, were made. Then, by our assumptions, $v$ is the right child of $w$, and $w$ lies in the left subtree of $u$. Let $u_L$ be the left child of $u$, and let $w_L$ be the left child of $w$. The segments $\tau_v$ and $\beta_v$ supporting the top and bottom edges of $\Delta_v$, respectively, are the stoppers of the top and bottom threads of $p_{2j}$; obviously, $\tau_v$ does not occur earlier than $s_i$ in the random permutation $\pi$ used to construct $\mathcal{B}$, and $\tau_v$ does not lie above $s_i$. There are two cases to consider depending on whether $\tau_v = s_i$.

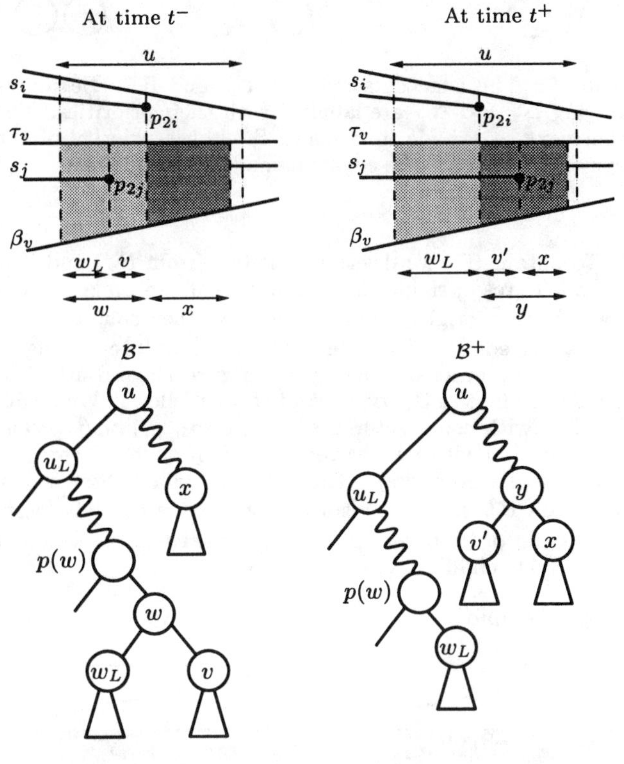


Figure 6: The case $\tau_v \neq s_i$: arrows on the bottom mark the horizontal extents of the shaded trapezoids.

*Case (i):* $\tau_v \neq s_i$. See Figure 6. Let $x$ be the highest node in the right subtree of $u$ (in $\mathcal{B}^-$) with the combinatorial structure $(p_{2i}, p_r, \tau_v, \beta_v)$, for some endpoint $p_r$. Then $\Delta_w$ and $\Delta_x$ share a common edge in $\mathcal{B}(t^-)$. At time $t^+$, as $p_{2j}$ leaves the trapezoid $\Delta_w$ and enters $\Delta_x$, $\Delta_{w_L}$ expands to $\Delta_w$, $\Delta_v$ disappears, and a new trapezoid $\Delta_{v'} = (p_{2i}, p_{2j}, \tau_v, \beta_v)$ appears (i..e, $\Delta_x$ is split at $t^+$ into two trapezoids: $\Delta_{v'}$ and the portion of $\Delta_x$ lying to the right of the cut through $p_{2j}$.) Moreover, $\Delta_{v'}$ (at time $t^+$) is intersected by $s_j$ and by the set of segments intersecting $\Delta_v$ (at time $t^-$). At time $t^-$, $\Delta_w$ is split by a vertical cut through $p_{2j}$ and $\Delta_{w_L}$ is split by an edge cut along $s_j$, while at time $t^+$, $\Delta_w$ is split by an edge cut along $s_j$. Therefore $\mathcal{B}_w^+$ is the same as $\mathcal{B}_{w_L}^-$.

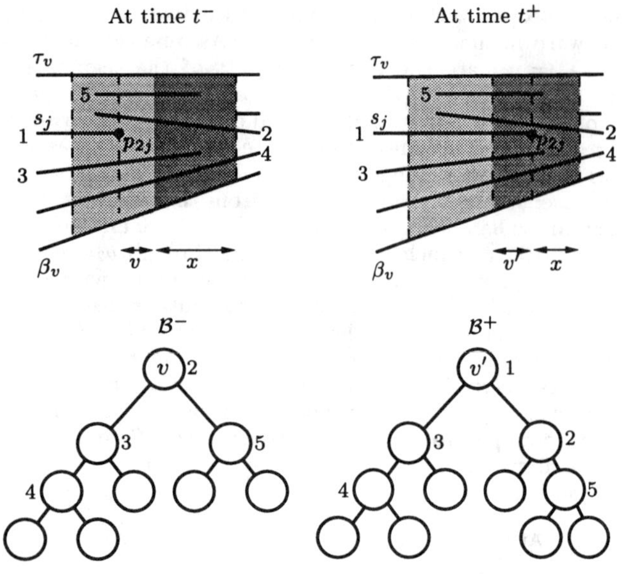


Figure 7: The edge cuts made in $\mathcal{B}_v^-$ and $\mathcal{B}_{v'}^+$. Segments crossing $\Delta_v$ and $\Delta_{v'}$ are labelled with their priorities. The label next to a node in $\mathcal{B}^-$ and $\mathcal{B}^+$ is the priority of the segment containing the edge cut made at that node.

To obtain $\mathcal{B}^+$, we delete the node $w$ from $\mathcal{B}^-$, and if $w$ was a left (resp., right) child of its parent, we make $w_L$ the new left (resp., right) child of $p(w)$. We then construct $\mathcal{B}_{v'}^+$ by making edge cuts through the segments intersecting $v'$ in decreasing order of priority (see Figure 7). and attach it to a descendant of the right child of $u$ as follows: We create a node $y$ with combinatorial structure $(p_{2i}, p_r, \tau_v, \beta_v)$. The cut associated with $y$ is the vertical cut through $p_{2j}$ (with $\tau_v$ and $\beta_v$ as its stoppers). The left and right subtrees of $y$ in $\mathcal{B}^+$ are $\mathcal{B}_{v'}^+$ and $\mathcal{B}_x^-$, respectively. If $x$ is the left (resp. right) child of its parent $p(x)$ in $\mathcal{B}^-$, then $y$ is the new left (resp. right) child of $p(x)$.

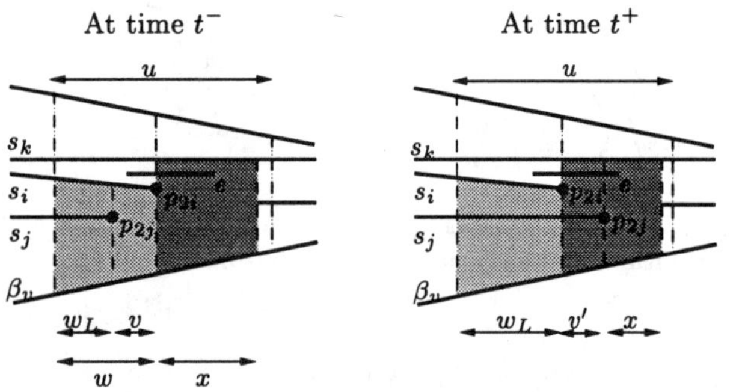


Figure 8: The case $\tau_v = s_i$: arrows on the bottom mark the horizontal extents of the shaded trapezoids. The segment $e$ is intersected by $T_{2j}$ at $t^+$ but not at $t^-$.

*Case (ii):* $\tau_v = s_i$. See Figure 8. In this case, at time $t$, the stopper of the upper thread of $p_{2j}$ switches from $s_i$ to a segment lying above $s_i$. The new stopper is the first segment $s_k$ in $T_{2i}$, the set of segments intersected by the top thread of $p_{2i}$, whose priority is higher than that of $s_j$. All segments that appear before $s_k$ in $T_{2i}$ are crossed by the top thread of $p_{2j}$ at $t^+$. The node $x$ now corresponds to the highest node in the right subtree of $u$ (in $\mathcal{B}^-$) whose combinatorial structure is $(p_{2i}, p_r, s_k, \beta_v)$, for some endpoint $p_r$. The new trapezoid $\Delta_{v'}$ that appears at $t^+$ has combinatorial structure $(p_{2i}, p_{2j}, s_k, \beta_v)$. The set $S_{v'}$ of segments crossing $\Delta_{v'}$ at $t^+$ is $s_j$, the set of segments in $T_{2j}$ and the segments in $T_{2i}$ stored before $s_k$ at $t^-$.

We first find $s_k$ by traversing the list $T_{2i}$. $\mathcal{B}^+$ is constructed in the same way as in the previous case. We again construct the subtree $\mathcal{B}_{v'}^+$ by adding the segments in $S_{v'}$ in decreasing order of their priority.

Finally, in both cases, we insert $s_j$ into $B_{2i}$, the list of segments in $S$ crossed by the lower thread of $p_{2i}$, and update $T_{2j}$. For a node $z$, $c_z$, the number of endpoints lying in the interior of $\Delta_z$, changes only if $z$ lies along the paths from $u$ to the nodes $p(w)$ and $y$. For such a node $z$, if $c_z = 0$ and if $\Delta_{p(z)}$ is split by a vertical cut, we add $(\lambda_z, \rho_z)$ to the list $\Gamma(t^+)$. On the other hand, if $c_z \neq 0$ but $z$ is transient at $t^-$ ($z$ must be an ancestor of $x$ in $\mathcal{B}^-$), we delete $(\lambda_v, \rho_z)$ from $\Gamma(t^+)$. Using Lemma 3.1 and assumption $(\star)$, we can show that the expected time spent in these steps is $O(\log n)$. We thus obtain the main result of this section:

**Theorem 3.4** *At each event point, $\mathcal{B}(t)$ can be updated in $O(\log n)$ expected time.*

Note that this theorem makes our BSP a kinetic data structure that is responsive, efficient, local, and compact, in the sense defined by Basch et al. [6].

We say that the trajectories followed by a set of segments are *pseudo-algebraic* if the segments move so that each pair of endpoints exchanges $y$-order only $O(1)$ times. A special case of pseudo-algebraic trajectories is when all the trajectories of the endpoints are constant-degree polynomials. If the trajectories of $k$ of the segments in $S$ are pseudo-algebraic and the remaining segments are stationary, then the total number of event points is $O(kn)$. We spend $O(\log n)$ expected time to maintain $\mathcal{B}(t)$ at each event point. Hence, we obtain the following corollary to Theorem 3.4:

**Corollary 3.5** *Let $S$ be a set of $n$ segments in the plane, and let $G \subseteq S$ be a set of $k$ segments. Suppose each segment of $G$ moves along a pseudo-algebraic trajectory and the remaining segments of $S$ are stationary, the total expected time spent in maintaining $\mathcal{B}$ is $O(kn \log n)$.*

## 4 BSPs for Triangles: A Randomized Algorithm

In this section we describe a randomized algorithm for constructing a BSP $\mathcal{B}$ of expected size $O(n^2)$ for a set of $n$ triangles in $\mathbb{R}^3$. The expected running time of the algorithm is $O(n^2 \log^2 n)$. We describe the algorithm in Section 4.1 and analyze its performance in Section 4.2.

### 4.1 Algorithm

For an object $s$ in $\mathbb{R}^3$, let $s^*$ denote the $xy$-projection of $s$. Let $E$ be the set of edges of the triangles in $S$, and let $E^*$ denote the set $\{e^* \mid e \in E\}$. Let $L$ be the set of lines in the $xy$-plane supporting the edges in $E^*$. We choose a random permutation $\langle \ell_1, \ell_2, \ldots, \ell_{3n} \rangle$ of $L$, and add the lines one-by-one in this order to compute $\mathcal{B}$. Let $L^{(i)} = \{\ell_1, \ell_2, \ldots \ell_i\}$. The algorithm works in $3n$ stages. The $i$th stage adds $\ell_i$ and constructs a top subtree $\mathcal{B}^{(i)}$ of $\mathcal{B}$ by refining the leaves of $\mathcal{B}^{(i-1)}$; $\mathcal{B}^{(0)}$ consists of one node (corresponding to $\mathbb{R}^3$) and $\mathcal{B}^{(3n)}$ is $\mathcal{B}$. As usual, we have a convex polytope $\Delta_v$

associated with each node $v$ of $\mathcal{B}^{(i)}$; $\Delta_v$ is a cylindrical cell, bounded by a set of vertical faces (i.e., faces parallel to the $z$-axis) and, possibly, top and bottom faces. If the top or bottom face of $\Delta_v$ exists, it is contained in a triangle of $S$. If $v$ is a leaf of $\mathcal{B}^{(i)}$ and no triangle in $S$ intersects the interior of $\Delta_v$, i.e., $S_v = \emptyset$, then $v$ is a leaf of $\mathcal{B}$ and we do not refine it further. Otherwise, we partition $\Delta_v$ into two cylindrical cells; these two cells are leaves of $\mathcal{B}^{(i+1)}$.

Before describing the $i$th stage of the algorithm in detail, we explain the structure of $\mathcal{B}^{(i)}$. We need a few definitions first. We say that a leaf $v$ of $\mathcal{B}^{(i)}$ (or the cell $\Delta_v$) is *active* if a triangle in $S$ intersects the interior of $\Delta_v$ (i.e. $S_v \neq \emptyset$); similarly, we say that a face $f$ in the line arrangement $\mathcal{A}(L^{(i)})$ is *active* if the $xy$-projection of some edge of a triangle in $S$ intersects the interior of $f$. For each active leaf $v$ in $\mathcal{B}^{(i)}$, $\Delta_v$ satisfies the following properties:

(P1) If a triangle $s \in S$ intersects the interior of $\Delta_v$, then the boundary of $s$ also intersects the interior of $\Delta_v$.

(P2) The cell $\Delta_v$ is a vertical section of the cylinder $\{(p,z) \mid p \in f, z \in \mathbb{R}\}$, for exactly one active face $f$ of $\mathcal{A}(L^{(i)})$; the vertical section may be truncated by triangles of $S$ at the top and bottom. See Figure 9.

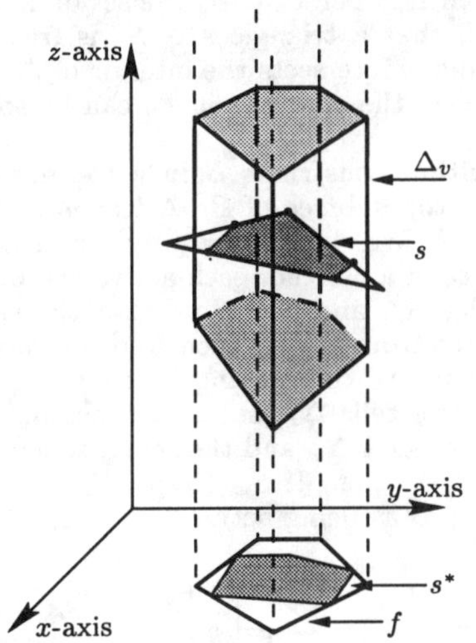


Figure 9: An active face $f$ and an active cell $\Delta_v \in \mathbf{\Delta}(f)$ that is a vertical section of the cylinder erected on $f$. The boundary of triangle $s$ intersects $\Delta_v$ and the boundary of $s^*$ intersects $f$.

In order to execute each stage efficiently, we maintain the following additional information:

(i) For each active cell $\Delta$ in $\mathcal{B}^{(i)}$, we store the set $S_\Delta \subseteq S$ of triangles that intersect the interior of $\Delta$.

(ii) We maintain the arrangement $\mathcal{A}(L^{(i)})$ as a planar graph; see [15]. For each active face $f$ in $\mathcal{A}(L^{(i)})$, we maintain the list $\mathbf{\Delta}(f)$ of those active cells $\Delta$ in $\mathcal{B}^{(i)}$ that lie inside the cylinder $\{(p,z) \mid p \in f, z \in \mathbb{R}^3\}$. Note that by Properties P(1) and P(2), a face $f \in \mathcal{A}(L^{(i)})$ is active if and only if $\mathbf{\Delta}(f) \neq \emptyset$.

In the $i$th stage, we make a cut along the vertical plane supporting $\ell_i$, and then make cuts contained in triangles of $S$ as follows: Let $h_i$ be the vertical plane supporting $\ell_i$ and let $h_i^+$ (resp., $h_i^-$) be the positive (resp., negative) halfspace supported by $h_i$.

1. We trace $\ell_i$ through the faces of $\mathcal{A}(L^{(i-1)})$. For each face $f \in \mathcal{A}(L^{(i-1)})$ intersected by $\ell_i$, we use $\ell_i$ to split $f$ into two faces $f^+$ and $f^-$. Let $\Delta$ be an active cell in $\mathbf{\Delta}(f)$. We partition $\Delta$ into two cells $\Delta^+ = \Delta \cap h_i^+$ and $\Delta^- = \Delta \cap h_i^-$.

2. We then compute the set $S_{\Delta^+} \subseteq S_\Delta$ of triangles that intersect the interior of $\Delta^+$. We also compute the set $F_{\Delta^+} \subseteq S_{\Delta^+}$ of triangles whose boundaries do not cross $\Delta^+$. Similarly, we compute the sets $S_{\Delta^-}$ and $F_{\Delta^-}$ for $\Delta^-$.

3. We split $\Delta^+$ into a set $\Psi$ of $|F_{\Delta^+}| + 1$ cells by making free cuts along each of the triangles in $F_{\Delta^+}$. For each triangle $s \in S_{\Delta^+} \setminus F_{\Delta^+}$, we perform a binary search to determine the unique cell in $\Delta' \in \Psi$ that intersects $s$ and add $s$ to $S_{\Delta'}$. ($\Delta'$ is unique because the triangles in $S$ are pairwise disjoint.) Finally, for each cell $\Delta' \in \Psi$, we add $\Delta'$ to the set $\mathbf{\Delta}(f^+)$ if $S_{\Delta'} \neq \emptyset$. Next, we perform a similar procedure for $\Delta^-$.

The resulting tree is $\mathcal{B}^{(i)}$. It is easily seen that $\mathcal{B}^{(i)}$ satisfies properties (P1) and (P2). Note that after the three lines supporting the $xy$-projections of the edges of a triangles $s \in S$ have been processed, $s$ does not intersect the interior of any cell.

**Remark:** The free cuts made in Step 3 are crucial in keeping the size of the BSP quadratic. Instead, if we simply erect vertical planes as we do in the algorithm, and make cuts along a triangle $s \in S$ only when all three lines supporting the $xy$-projections of $s$'s edges have been added, then there are instances of input triangles for which our algorithm will construct a BSP of $\Omega(n^3)$ size regardless of the initial random permutation.

## 4.2 Analysis of the algorithm

We first bound the expected size of $\mathcal{B}$. The cuts made in the algorithm partition each triangle in $S$ into a number of pieces. We can show that the size of $\mathcal{B}$ is boundedby the total number of pieces into which the triangles in $S$ are split by the cuts made in the algorithm. To bound the expected total number of pieces, we count the number of new pieces created in the $i$th stage, and sum the result over all stages. In the $i$th stage, we count the number of new pieces into which a triangle $s$ in $S$ is partitioned by the cuts made in the $i$th stage, and sum the resulting bound over all triangles in $S$.

Let $\nu_s$ be the total number of new pieces into which a triangle $s \in S$ is partitioned by the cuts made in the $i$th stage. Note that these cuts are contained in the vertical plane containing $\ell_i$. For $1 \leq k \leq i$, let $\lambda_k$ be the intersection of the vertical plane through $\ell_k$ with the triangle $s$, and let $\Lambda = \{\lambda_k, 1 \leq k \leq i\}$. To calculate $\nu_s$, consider the line arrangement $\mathcal{A}(\Lambda)$ on $s$. We call a face of $\mathcal{A}(\Lambda)$ a *boundary* face if it is adjacent to an edge of $s$; otherwise, it is an *interior* face. Recall that we partition a cell $\Delta_v$, for a leaf $v \in \mathcal{B}^{(i)}$, only if $\Delta_v$ is active. Property (P1) implies that the cuts made in the $i$th stage do not intersect the interior of any interior face of $\mathcal{A}(\Lambda)$, since such a face cannot intersect the interior of any active cell $\Delta_v$. Hence, $\nu_s$ is the number of boundary faces of $\mathcal{A}(\Lambda)$ that are intersected by $\lambda_i$.

For $1 \le k \le i$, let $\mu(\Lambda, k)$ denote the number of boundary faces in the arrangement $\mathcal{A}(\Lambda \setminus \{\lambda_k\})$ that are intersected by $\lambda_k$. Observe that the sum $\sum_{1\le k\le i} \mu(\Lambda, k)$ equals the total number of edges bounding the boundary faces of $\mathcal{A}(\Lambda)$. Each such edge lies in the zone (in $\mathcal{A}(\Lambda)$) of one of the edges of $s$. Hence, by the Zone Theorem [9, 15],

$$\sum_{1\le k\le i} \mu(\mathcal{A}(\Lambda), k) = O(i).$$

Since $\ell_i$ is chosen randomly from the set $L^{(i)}$, $\lambda_i$ can be any of the lines $\lambda_1, \lambda_2, \ldots, \lambda_i$ with equal probability. Therefore, the expected value $E[\nu_s]$ of $\nu_s$ is

$$E[\nu_s] = \frac{1}{i} \sum_{1\le k\le i} \mu(\mathcal{A}(\Lambda), k) = O(1).$$

Hence, the total number of pieces created in the $i$th stage is $O(n)$. Summing over $i$, we find that the total number of pieces into which the triangles in $S$ are partitioned into over the entire algorithm is $O(n^2)$. The following lemma is immediate:

**Lemma 4.1** *The expected number of nodes in the BSP constructed by the above algorithm is $O(n^2)$.*

For an active face $f$ in $\mathcal{A}(L^{(i-1)})$, let $k_f$ be the number of projected edges in $E^*$ that intersect the interior of $f$. By Property 1, if a triangle $s \in S$ intersects the interior of a cell $\Delta \in \mathbf{\Delta}(f)$, then the boundary of $s$ also intersects the interior of $\Delta$. Therefore, an edge of $s^*$ intersects the interior of $f$. Further, $s$ cannot intersect the interior of two different cells in $\mathbf{\Delta}(f)$; otherwise, $s$ intersects a non-vertical face $g$ of one of those cells, which is impossible, since each such face $g$ is contained in a triangle in $S$ and the triangles in $S$ are pairwise disjoint. As a result, we have

$$\sum_{\Delta\in\mathbf{\Delta}(f)} |S_\Delta| \le k_f. \tag{4.1}$$

We now analyze the expected running time of the algorithm. We first count the time spent during the $i$th stage in processing the line $\ell_i$. In Step 1, we spend $O(i)$ time to trace $\ell_i$ through $\mathcal{A}(L^{(i-1)})$. When processing an active face $f$ of $\mathcal{A}(L^{(i-1)})$ that is intersected by $\ell_i$, for each cell $\Delta \in \mathbf{\Delta}(f)$, we spend $|S_\Delta|$ time in Step 2. In Step 3, we can find the cell in the set $\Psi$ that intersects a triangle $s \in S_{\Delta^+} \setminus F_{\Delta^+}$ in $O(\log |F_{\Delta^+}|)$ time by performing a binary search. Hence, the total time spent in Step 3 for the face $f$ is $O(|S_\Delta| \log |S_\Delta|)$. Thus, (4.1) implies that the total time spent in processing $f$ is

$$\sum_{\Delta\in\mathbf{\Delta}(f)} O(|S_\Delta| \log |S_\Delta|) = O(k_f \log k_f).$$

Let $Z$ be the set of all active faces of $\mathcal{A}(L^{(i-1)})$ that are intersected by $\ell_i$. The total time spent in processing $\ell_i$ is $\sum_{f\in Z} O(k_f \log k_f)$. Using the Zone Theorem [9, 15] and the theory of random-sampling [13, 17], we can show that the expected value of $\sum_{f\in Z} k_f$ is $O(n \log n)$, which implies the following theorem:

**Theorem 4.2** *Let $S$ be a set of $n$ non-intersecting triangles in $\mathbb{R}^3$. We can compute a BSP for $S$ of expected size $O(n^2)$ in expected time $O(n^2 \log^2 n)$ time.*

**Remark:** Using a similar argument, we can also prove that the expected value of the total number of vertices of the nodes of $\mathcal{B}$ is $O(n^2)$. The height of $\mathcal{B}$ can be $\Omega(n)$, e.g., if the triangles in $S$ form a convex polytope.

## 5 BSPs for Triangles: A Deterministic Algorithm

In this section we describe a deterministic algorithm for computing a BSP for a set $S$ of $n$ triangles in $\mathbb{R}^3$. As in the previous section, let $E$ denote the set of edges of triangles in $S$, and let $E^* = \{e^* \mid e \in E\}$ be the set of $xy$-projections of the edges in $E$. Let $k$ be the number of intersection points in $E^*$. Our algorithm constructs in $O((n+k)\log^2 n)$ time a BSP $\mathcal{B}$ of size $O((n+k)\log n)$. As in the previous section, each node $v$ of $\mathcal{B}$ is associated with a cylindrical cell $\Delta_v$, but the top and bottom faces of $\Delta_v$ are now trapezoids. Let $\Delta_v^*$ denote the $xy$-projection of the top (or bottom) face of $\Delta_v$; two of the edges of $\Delta_v^*$ will be parallel to the $y$-axis. Let $E_v^*$ be the set of segments in $E^*$ that intersect the interior of $\Delta_v^*$ and are clipped within $\Delta_v^*$. A segment $\gamma \in E_v^*$ is called *anchored* if its endpoints lie on the two parallel edges of $\Delta_v^*$ and if it does not intersect any other edge of $E_v^*$. The anchored edges in $E_v^*$ can be linearly ordered from top to bottom (since they are disjoint). Let $A_v$ be the set of anchored edges in $E_v^*$. Let $F_v \subseteq S_v$ the set of all free triangles in $S_v$. Recall that a triangle $s \in S_v$ is free with respect to $\Delta_v$ if no edge $s$ intersects the interior of $\Delta_v$. Since $\Delta_v$ is a cylindrical cell, the triangles in $F_v$ can be sorted by their heights.

The algorithm constructs $\mathcal{B}$ in a top-down fashion by maintaining a top subtree of $\mathcal{B}$. A leaf $v$ of the subtree is *active* if $S_v \ne \emptyset$. We store the set of all active leaves of the current subtree in a list. For each active leaf $v$, we maintain three sets: $F_v$, $A_v$, and $S_v$. Note that the set $E_v^*$ can be easily computed from $S_v$. At each step, we choose an active leaf $v$ and compute the cutting plane $h_v$ that is used to split $\Delta_v$ into two cells $\Delta_z$ and $\Delta_w$. Once $h_v$ is chosen, we can compute $\Delta_z$ and $\Delta_w$ and the sets associated with them in $O(|S_v| + |F_v|)$ time. If $S_w$ (resp., $S_z$) is nonempty, we mark $w$ (resp., $z$) as being active.

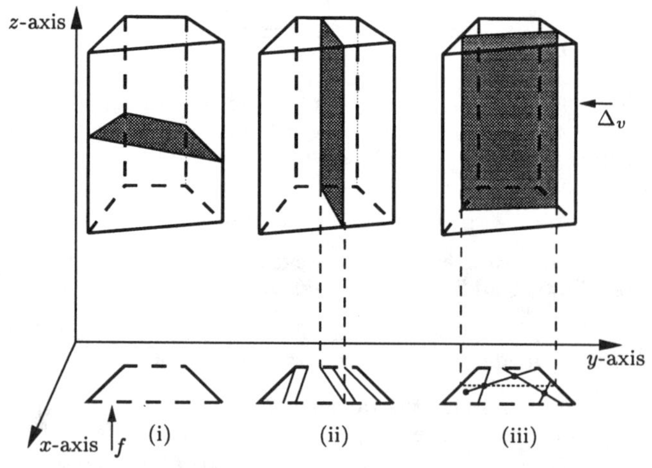


Figure 10: Cuts made in the deterministic algorithm: (i) Free cut, (ii) Cut parallel to the $z$-axis through an anchored edge, and (iii) Cut parallel to the $yz$-plane through a vertex of $\mathcal{A}(E_v^*)$

We choose the cutting plane $h_v$ as follows:

1. If $F_v \neq \emptyset$, we choose the median triangle $s$ of $F_v$ and set $h_v$ to be the plane supporting $s$. See Figure 10(i). Recall that the triangles in $F_v$ are ordered by height. Since $s$ does not intersect any triangle of $S_v$, each triangle of $S_v \setminus \{s\}$ belongs to either $S_w$ or $S_z$. Moreover, $F_w, F_z \subseteq F_v$ and $A_w, A_z \subseteq A_v$.

2. If $F_v = \emptyset$, i.e., there is no free triangle in $S_v$, we use $A_v$ to choose the cutting plane, as follows: If $A_v$ contains at least one anchored segment, let $\gamma$ be the middle anchored segment. Since the anchored segments can be totally ordered, the middle anchored segment is well defined. We choose the cutting plane $h_v$ to be the vertical plane containing $\gamma$. See Figure 10(ii). Since $h_v$ may intersect some triangles of $S_v$, a triangle of $S_v$ may belong to both $S_w$ and $S_z$. If a triangle $s \in S_v$ intersects $h_v$ and if none of its edges intersect the interior of $\Delta_w$ (resp., $\Delta_z$), then $s \in F_w$ (resp., $s \in F_z$).

3. Finally, if $A_v$ is empty, let $\sigma$ be a vertex in the arrangement $\mathcal{A}(E_v^*)$ with the median $y$-coordinate; $\sigma$ is either an endpoint of a segment of $E_v^*$ or an intersection point of two segments of $E_v^*$. We choose $h_v$ to be the plane $x = x(\sigma)$, i.e., $h_v$ is the plane parallel to the $yz$-plane and passing through $\sigma$. See Figure 10(iii). In this case too, $h_v$ can intersect the triangles of $S_v$. If $k_v$ is the number of intersection points of $E_v^*$, $\sigma$ can be computed in $O(|S_v| + k_v)$ time, after some initial preprocessing. We can again compute the required sets for $w$ and $z$ in $O(|S_v|)$ time.

This completes the description of the algorithm. We now give the analysis. At each node $v$, since we choose the middle free triangle, middle anchored edge, or a middle vertex of $\mathcal{A}(E_v^*)$, we can show that the height of $\mathcal{B}$ is $O(\log^2 n)$. However, if we assign appropriate weights to each free triangle in $F_v$ and to each anchored edge and choose a weighted median of the free triangles or the anchored edges, we can improve the height of $\mathcal{B}$ to $O(\log n)$; for example, see [24]. Following an analysis similar to the one given in [24], we can show that the number of nodes in $\mathcal{B}$ is $O((n+k)\log n)$, and that the running time of the algorithm is $O((n+k)\log^2 n)$. Hence, we obtain the following result:

**Theorem 5.1** *Let $S$ be a set of $n$ triangles in the plane, and let $k$ be the number of intersection points of the $xy$-projections of the edges of $S$. Then we can compute in $O((n+k)\log^2 n)$ time a BSP of size $O((n+k)\log n)$ for $S$.*

## 6 Conclusions

We have presented an efficient algorithm to maintain the BSP of moving segments in the plane. Currently, we do not know any non-trivial lower bounds for this problem. Recently, Agarwal et al. [1] have extended our result and developed an algorithm to maintain BSPs for moving triangles in $\mathbb{R}^3$.

We have also presented algorithms to construct BSPs for triangles in $\mathbb{R}^3$. The algorithms are (near-)optimal in the worst-case. No efficient algorithm is known for constructing a BSP of optimal or near-optimal size for triangles in $\mathbb{R}^3$. On the other hand, it is not known whether the problem is *NP*-hard.

## References


[1] P. K. Agarwal, J. Erickson, and L. J. Guibas, Kinetic binary space partitions in $\mathbb{R}^3$, In preparation, 1997.

[2] P. K. Agarwal, E. F. Grove, T. M. Murali, and J. S. Vitter, Binary space partitions for fat rectangles, *Proceedings of the 37th IEEE Annual Symposium on foundations of Computer Science (FOCS '96)*, October 1996.

[3] P. K. Agarwal and S. Suri, Surface approximation and geometric partitions, *Proc. 5th ACM-SIAM Sympos. Discrete Algorithms*, 1994, pp. 24–33.

[4] J. M. Airey, *Increasing Update Rates in the Building Walkthrough System with Automatic Model-space Subdivision and Potentially Visible Set Calculations*, Ph.D. Thesis, Dept. of Computer Science, University of North Carolina, Chapel Hill, 1990.

[5] C. Ballieux, Motion planning using binary space partitions, Tech. Rep. inf/src/93-25, Utrecht University, 1993.

[6] J. Basch, L. Guibas, and J. Hershberger, Data structures for mobile data, *Proc. 7th SIAM Symp. on Discr. Algorithms*, 1997, pp. 747–756.

[7] A. T. Campbell, *Modeling Global Diffuse Illumination for Image Synthesis*, Ph.D. Thesis, Dept. of Computer Sciences, University of Texas, Austin, 1991.

[8] T. Cassen, K. R. Subramanian, and Z. Michalewicz, Near-optimal construction of partitioning trees by evolutionary techniques, *Proc. of Graphics Interface '95*, 1995, pp. 263–271.

[9] B. Chazelle, L. J. Guibas, and D. T. Lee, The power of geometric duality, *BIT*, 25 (1985), 76–90.

[10] N. Chin and S. Feiner, Near real-time shadow generation using BSP trees, *Proc. SIGGRAPH '89, Comput. Graph.*, Vol. 23, ACM SIGGRAPH, 1989, pp. 99–106.

[11] N. Chin and S. Feiner, Fast object-precision shadow generation for areal light sources using BSP trees, *Comput. Graph. (Proc. 1992 Symp. on Interactive 3D Graphics)*, Vol. 25, 1992, pp. 21–30.

[12] Y. Chrysanthou, *Shadow Computation for 3D Interaction and Animation*, Ph.D. Thesis, Queen Mary and Westfield College, University of London, 1996.

[13] K. L. Clarkson and P. W. Shor, Applications of random sampling in computational geometry, II, *Discrete Comput. Geom.*, 4 (1989), 387–421.

[14] M. de Berg, Linear size binary space partitions for fat objects, *Proc. Third Europ. Symp. on Algorithms*, LNCS 979, Springer-Verlag, September 1995, pp. 252–263.

[15] H. Edelsbrunner, *Algorithms in Combinatorial Geometry*, Springer-Verlag, Heidelberg, West Germany, 1987.

[16] H. Fuchs, Z. M. Kedem, and B. Naylor, On visible surface generation by a priori tree structures, *Proc. SIGGRAPH '80, Comput. Graph.*, Vol. 14, ACM SIGGRAPH, 1980, pp. 124–133.

[17] D. Haussler and E. Welzl, Epsilon-nets and simplex range queries, *Discrete Comput. Geom.*, 2 (1987), 127–151.

[18] C. Mata and J. S. Mitchell, Approximation algorithms for geometric tour and network design problems, *Proc. 11th Annu. ACM Sympos. Comput. Geom.*, 1995, pp. 360–369.

[19] T. M. Murali and T. A. Funkhouser, Consistent solid and boundary representations from arbitrary polygonal data, *To appear in Proc. 1997 Symp. on Interactive 3D Graphics*, 1997.

[20] B. Naylor, J. A. Amanatides, and W. Thibault, Merging BSP trees yields polyhedral set operations, *Comput. Graph.*, 24 (1990), 115–124. Proc. SIGGRAPH '90.

[21] B. Naylor and W. Thibault, Application of BSP trees to ray-tracing and CSG evaluation, Technical Report GIT-ICS 86/03, Georgia Institute of Tech., School of Information and Computer Science, 1986.

[22] B. F. Naylor, SCULPT: an interactive solid modeling tool, *Proc. Graphics Interface '90*, 1990, pp. 138–148.

[23] B. F. Naylor, Interactive solid geometry via partitioning trees, *Proc. Graphics Interface '92*, 1992, pp. 11–18.

[24] M. S. Paterson and F. F. Yao, Efficient binary space partitions for hidden-surface removal and solid modeling, *Discrete Comput. Geom.*, 5 (1990), 485–503.

[25] M. S. Paterson and F. F. Yao, Optimal binary space partitions for orthogonal objects, *J. Algorithms*, 13 (1992), 99–113.

[26] R. A. Schumacker, R. Brand, M. Gilliland, and W. Sharp, Study for applying computer-generated images to visual simulation, Tech. Rep. AFHRL–TR–69–14, U.S. Air Force Human Resources Laboratory, 1969.

[27] R. Seidel, Backwards analysis of randomized geometric algorithms, in: *New Trends in Discrete and Computational Geometry* (J. Pach, ed.), *Algorithms and Combinatorics*, Vol. 10, Springer-Verlag, 1993, pp. 37–68.

[28] S. J. Teller, *Visibility Computations in Densely Occluded Polyhedral Environments*, Ph.D. Thesis, Dept. of Computer Science, University of California, Berkeley, 1992.

[29] W. C. Thibault and B. F. Naylor, Set operations on polyhedra using binary space partitioning trees, *Comput. Graph.*, 21 (1987), 153–162. Proc. SIGGRAPH '87.

[30] E. Torres, Optimization of the binary space partition algorithm (BSP) for the visualization of dynamic scenes, *Eurographics '90*, North-Holland, 1990, pp. 507–518.

# Primal–Dual Methods for Vertex and Facet Enumeration

PRELIMINARY VERSION

David Bremner*
School of Computer Science, McGill University, Montréal, Canada
bremner@cs.mcgill.ca

Komei Fukuda
Department of Mathematics, Swiss Federal Institute of Technology, Lausanne and
Institute for Operations Research, Swiss Federal Institute of Technology, Zurich, Switzerland
fukuda@ifor.math.ethz.ch

Ambros Marzetta
Institute for Theoretical Computer Science, Swiss Federal Institute of Technology, Zurich, Switzerland
marzetta@inf.ethz.ch

## Abstract

Every convex polytope can be represented as the intersection of a finite set of halfspaces and as the convex hull of its vertices. Transforming from the halfspace (respectively vertex) to the vertex (respectively halfspace) representation is called *vertex enumeration* (respectively *facet enumeration*). An open question is whether there is an algorithm for these two problems (equivalent by geometric duality) that is polynomial in the input size and the output size. In this paper, we extend the known polynomially solvable classes of polytopes by looking at the dual problems. The *dual* problem of a vertex (facet, respectively) enumeration problem is the facet (vertex) enumeration problem for the same polytope where the input and output are simply interchanged. For a particular class of polytopes and a fixed algorithm, one transformation may be much easier than its dual. In this paper, we propose a new class of algorithms that take advantage of this phenomenon. Loosely speaking, *primal–dual* algorithms use a solution to the easy direction as an oracle to help solve the seemingly hard direction.

## 1 Introduction

A *polytope* is the bounded intersection of a finite set of halfspaces in $\mathbb{R}^d$. The *vertices* of a polytope are those feasible points that do not lie in the interior of a line segment between two other feasible points. Every polytope $P$ can be represented as the intersection of a non-redundant set of halfspaces $\mathcal{H}(P)$ and as the convex hull of its vertices $\mathcal{V}(P)$. For the sake of a simpler duality, we assume that a polytope is full dimensional and contains the origin in its interior.

* Supported by NSERC Canada, FCAR Québec, and the J.W. McConnell Foundation


*Computational Geometry 97* Nice France


The problem of transforming from $\mathcal{H}(P)$ to $\mathcal{V}(P)$ is called *vertex enumeration*; transforming from $\mathcal{V}(P)$ to $\mathcal{H}(P)$ is called *facet enumeration* or *convex hull*.

An algorithm is said to be *polynomial* if the time to solve any instance is bounded above by a polynomial in the size of input and output. We consider the input (respectively, output) size to be the number of real (or rational) numbers needed to represent the input (respectively output); in particular we do not consider the dimension to be a constant. We assume each single arithmetic operation takes a constant amount of time. A *successively polynomial algorithm* is one whose $k$th output is generated in time polynomial in $k$ and the input size $s$, for each $k$ less than or equal to the cardinality of output. Clearly every successively polynomial algorithm is a polynomial algorithm.

It is easily seen from the duality of convex polytopes that the vertex and face enumeration problems are polynomially equivalent, that is, the existence of a polynomial algorithm for one problem implies the same for the other problem. Several polynomial algorithms (see e.g. [1, 4, 5, 7, 14, 15]) are known under strong assumptions of nondegeneracy, which restrict input polytopes to be simple in the case of vertex enumeration and simplicial in the case of facet enumeration. However, it appears to be extremely hard to determine whether there exists a polynomial algorithm in general.

In this paper, we extend the known polynomially solvable classes by looking at the dual problems. The *dual* problem of a vertex (facet, respectively) enumeration problem is the facet (vertex) enumeration problem for the same polytope where the input and output are simply interchanged. For a particular class of polytopes and a fixed algorithm, one transformation may be much easier than its dual. One might be tempted to explain this possible asymmetry by observing that the standard nondegeneracy assumption is not self-dual. Are the dual problems of nondegenerate vertex (face) enumeration problems harder? More generally, are the complexities of the primal and the dual problem distinct?

Here we will show in a certain sense that the primal and dual problems are of the same complexity. More precisely, we will show the following theorem: if there is a successively polynomial algorithm for the vertex (facet, respectively) enumeration problem for a hereditary class of problems, then there is a successively polynomial algorithm for the facet (vertex) enumeration problem for the same class, where a hereditary class contains all subproblems of any instance in the class. We propose a new class of algorithms that take advantage of this phenomenon. Loosely speaking, *primal–dual* algorithms use a solution to the easy direction as an oracle to help solve the seemingly hard direction.

From this general result relating the complexity of the primal and dual problems, and known polynomial algorithms for the primal nondegenerate case, we arrive at a polynomial algorithm for vertex enumeration for simplicial polytopes and facet enumeration for simple polytopes. We then show how to refine this algorithm to yield an algorithm with time complexity competitive with the algorithms known for the primal-nondegenerate case.

The only published investigation of the dual-nondegenerate case the authors are aware of is a paper by Peter Gritzmann and Victor Klee [10]. Their approach, most easily understood in terms of vertex enumeration, consists of intersecting the constraints with each defining hyperplane and, after removing the redundant constraints, finding the vertices lying on that facet by some brute force method. David Avis (private communication) has independently observed that this method can be extended to any polytope whose facets are simple (or nearly simple) polytopes. The method of Gritzmann and Klee requires solving $O(m^2)$ linear programs (where $m$ is the number of input halfspaces) to remove redundant constraints. Our approach does not rely on the polynomial solvability of linear programming.

### Notation

We start by defining some notation. Recall that $\mathcal{H}(P)$ (respectively $\mathcal{V}(P)$) is the non-redundant halfspace (respectively vertex) description of $P$. We use $m$ for $|\mathcal{H}(P)|$, $n$ for $|\mathcal{V}(P)|$, and $d$ for the dimension $\dim P$. The facets of $P$ are the intersection of the bounding hyperplanes of $\mathcal{H}(P)$ with $P$. We use $\mathbb{0}$ and $\mathbb{1}$ to denote the vector of all 0s and all 1s respectively. We treat sets of points and matrices interchangeably where convenient; the rows of a matrix are the elements of the corresponding set. Since we assume the origin is in the interior of each polytope, each facet defining inequality can be written as $ax \leq 1$ for some $a$. We use $\mathcal{P}(H)$ to denote the polyhedron $\{\, x \mid Hx \leq \mathbb{1} \,\}$. Similarly we use $\mathcal{H}(P)$ to mean the matrix $A$ where $P = \{\, x \mid Ax \leq \mathbb{1} \,\}$. For a set of points $V$ we use $\mathcal{H}(V)$ to mean $\mathcal{H}(\operatorname{conv} V)$; similarly for a set of halfspaces $H$, we use $\mathcal{V}(H)$ to mean $\mathcal{V}(\mathcal{P}(H))$. For a hyperplane $h = \{\, x \mid ax = 1 \,\}$, we use $h^+$ to mean the halfspace $\{\, x \mid ax \leq 1 \,\}$. We say that $h^+$ is a *valid inequality* for a set of points $X$ if $X \subseteq h^+$. We make extensive use of duality of convex polytopes in what follows. The *proper faces* of a convex polytope are the intersection of some set of facets. By adding the two *improper faces*, the polytope itself and the empty set, the faces form a lattice ordered by inclusion. Two polytopes are said to be *combinatorially equivalent* if their face lattices are isomorphic and *dual* if their face lattices are anti-isomorphic (i.e. isomorphic with the direction of inclusion reversed). The following is well known (see e.g. [2]).

**Lemma 1** *If $P = \mathcal{P}(H)$ is a polytope such that $\mathbb{0} \in \operatorname{int} P$, then $Q = \operatorname{conv} H$ is a polytope dual to $P$ such that $\mathbb{0} \in \operatorname{int} Q$.*

## 2 Primal–Dual Algorithms

In this section we first consider the relationship between the complexity of the primal problem and the complexity of the dual problem for vertex/facet enumeration. In Section 2.1 we show how to apply these results to design efficient algorithms for the dual-nondegenerate case. In Section 2.2 we modify the algorithms to work in the case where $P$ is moderately dual-degenerate. We will fix the primal problem as facet enumeration in this section, but the results can also be interpreted in terms of vertex enumeration. We call a family $\mathcal{F}$ of polytopes *facet-hereditary* if for any $P \in \mathcal{F}$, for any $H' \subset \mathcal{H}(P)$, if $\bigcap H'$ is bounded then $\bigcap H'$ is also in $\mathcal{F}$. The main idea of this paper is summarized by the following theorem.

**Theorem 1** *If there is a successively polynomial vertex enumeration algorithm for a facet-hereditary family of polytopes then there is a successively polynomial facet enumeration algorithm for the same family.*

Simple polytopes are not themselves facet-hereditary, but each simple polytope can be perturbed symbolically or lexicographically onto a combinatorially equivalent polytope whose defining halfspaces are in "general position", i.e. if we consider the natural arrangement of hyperplanes induced by the polytope, no $d+1$ meet in a point.

**Corollary 1** *There is a successively polynomial algorithm for facet enumeration of simple polytopes and for vertex enumeration of simplicial polytopes.*

Proof of Theorem 1 is constructive, via the following simple algorithm. The following algorithm takes as input a set $V$ of points in $\mathbb{R}^d$, and a subset $\mathcal{H}_0 \subset \mathcal{H}(P)$ such that $\bigcap \mathcal{H}_0$ is bounded. We show below how to compute such a set of halfspaces.

**Algorithm 1** *PrimalDualFacets*$(V, H_0 \subset \mathcal{H}(V))$

```
    Hc ← H0
    While ∃w ∈ V(Hc) \ V do                    FindWitness
        Find h+ ∈ H(V) s.t. w ∉ h+             DeleteVertex
        Hc ← Hc ∪ {h+}
    end while
    return Hc.
```

**end**

At every step of the algorithm we maintain the invariant that $\operatorname{conv} V \subseteq \mathcal{P}(H_c)$. When the algorithm terminates, we know that $\mathcal{V}(H_c) \subseteq V$. It follows that $\mathcal{P}(H_c) \subseteq \operatorname{conv} V$. There are two main steps in this algorithm that we have labeled FindWitness and DeleteVertex. The vertex $w \in \mathcal{V}(H_c) \setminus V$ is a *witness* in the sense that for any such vertex, there must be a facet of $\mathcal{H}(V)$ not yet discovered whose defining halfspace cuts off $w$. From the precondition of the theorem there exists a successively polynomial algorithm to enumerate the vertices of $H_c$. It follows that in time polynomial in $|V|$ we can find $|V|+1$ vertices of $\mathcal{P}(H_c)$, or discover $\mathcal{V}(H_c) = V$. If we discover $|V|+1$ vertices, one of these vertices must be a witness. In order to find the facet corresponding to a witness (the "DeleteVertex" step), we need to solve a separating hyperplane problem for a point and convex set. The separating hyperplane problem can be solved via the following linear program: maximize $yw$ subject to $Vy \leq \mathbb{1}$. If $y^*$ is the optimal value of the linear program then $y^* x \leq 1$ is the desired separating halfspace. While there are linear programming algorithms polynomial in the bit size of the input, there are not yet any known that are polynomial in $n = |V|$ and $d$, which is what we need for our theorem. It turns out that because we have a halfspace description of the union of our two sets, we can solve the separating hyperplane problem via a much simpler purely combinatorial algorithm. It is worth noting that there is no restriction of the input points to be in "convex position". Although the algorithm will be more efficient if there are no redundant interior points as part of the input, it will still function correctly if there are such points.

Our main tool here will be the pivot operation of the simplex method of linear programming. If we have a point in the interior of a polytope and a halfspace description, in at most $d$ pivot operations we can find a vertex of the polytope. If we ensure that each pivot does not decrease a given objective function, then we have the following.

**Lemma 2 (Raindrop Algorithm)** *Given $H \in \mathbb{R}^{m \times d}$, $c \in \mathbb{R}^d$ and $x_0 \in \mathcal{P}(H)$, in time $O(md^2)$ we can find $v \in \mathcal{V}(H)$ such that $cv \geq cz$.*

**Proof:** We start by translating our system by $-x_0$ so that our initial point is the origin. We represent the inequality system $Hx \leq \mathbb{1}$ in the standard "dictionary" form (see e.g. [5]) as follows. We transform each inequality into an equality by adding a slack variable, to arrive at the following system of linear equations or *dictionary*: $s = \mathbb{1} - Hx$. A solution to the dictionary is feasible for the system of inequalities iff $s \geq \mathbb{O}$. As a final row to our dictionary we add the the equation $z = cx$ (the *objective row*). The variables are naturally partitioned into two sets. Those on the left hand side (initially just the slacks) are called *basic*; the others are called *nonbasic*. Note that by construction, $x = \mathbb{O}$ is a feasible solution. We maintain the invariant that the nonbasic variables are set to 0. We start a pivot operation by choosing some nonbasic variable $x_j$ to become basic. Depending on the sign of the coefficient of $x_j$ in the objective row, we can always with increase or decrease $x_j$ without decreasing the value of $z$. As we change the value of $x_j$, some of the basic slack variables will decrease as we get closer to the corresponding hyperplane. By considering ratios of coefficients, we can find the first hyperplane reached. By moving that slack variable to the right hand side (making it nonbasic), and moving $x_j$ to the left hand side, we obtain a new dictionary in $O(md)$ time (see e.g. [5] for details of the simplex method). Note that since the old value of $x$ did not lie on the new hyperplane, the dimension of the intersection of all of the hyperplanes must decrease with each pivot, so that after at most $d$ pivots we have reached a vertex. ∎

By duality of convex polytopes we have the following.

**Lemma 3 (Dual Raindrop Algorithm)** *Given $V \in \mathbb{R}^{n \times d}$, $c \in \mathbb{R}^d$ and some $\alpha$ such that $V\alpha \leq \mathbb{1}$, in $O(nd^2)$ time we can find $a \in \mathcal{H}(V)$ such that $ac \geq \alpha c$.*

Essentially this is the same as the initialisation step of a gift-wrapping algorithm (see e.g. [4, 15]), except that we are careful that $c$ is on the same side of our final hyperplane as the one we started with.

We can now show how to implement the DeleteVertex step of Algorithm 1 without linear programming. A *basis* $B$ for a vertex $v \in \mathcal{V}(A)$ is a set of $d$ rows of $A$ such that $Bv = \mathbb{1}$ and $\mathsf{rank} B = d$. We can obviously find a basis in polynomial time; in the pivoting based algorithms in the next section we will always be given a basis for $v$.

**Figure 1** Illustrating the proof of Lemma 4.

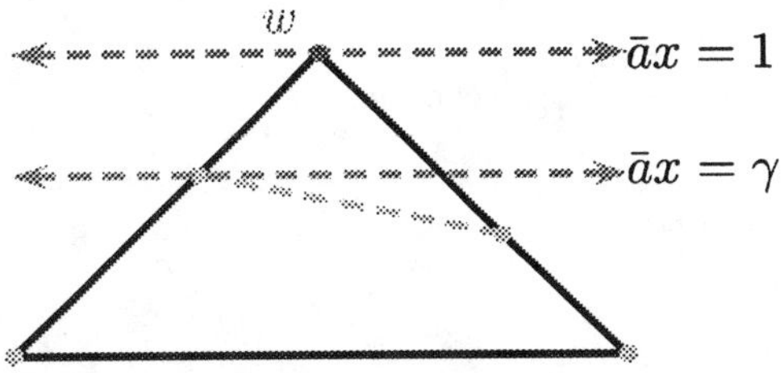


**Lemma 4 (DeleteVertex)** *Given $V \in \mathbb{R}^{n \times d}$, $H_0 \subset \mathcal{H}(V)$, $w \in \mathcal{V}(H_0) \setminus V$, and a basis $B$ for $w$, we can find $h^+ \in \mathcal{H}(V)$ such that $w \notin h^+$ in time $O(nd^2)$.*

**Proof:** Let

$$\bar{a} = \frac{1}{d} \sum_{b \in B} b\,.$$

The inequality $\bar{a}x \leq 1$ is satisfied with equality by $w$ and with strict inequality by every $v \in V$ (see Figure 1). Let

$$\gamma = \max_{v \in V} \bar{a}v\,.$$

Let $\alpha = \bar{a}/\gamma$. The constraint $\alpha x \leq 1$ is valid for $\operatorname{conv} V$, but $\alpha w > 1$. The lemma then follows from Lemma 3. ∎

It will prove useful below to be able to find a facet of $\operatorname{conv} V$ that cuts off a particular extreme ray or direction of unboundedness for our current intermediate polytope.

**Lemma 5 (DeleteRay)** *Given $V \in \mathbb{R}^{n \times d}$ and $r \in \mathbb{R}^d \setminus \{\mathbb{O}\}$, in $O(nd^2)$ time we can find $a \in \mathcal{H}(V)$ such that $ar > 0$.*

**Proof:** The proof is similar to that of Lemma 4. Let

$$\gamma = \max_{v \in V} rv\,.$$

Since $\mathbb{O} \in \operatorname{int} P$, $\gamma > 0$. Let $\alpha = r/\gamma$. The constraint $\alpha x \leq 1$ is valid for $\operatorname{conv} V$, but $\alpha r = (r \cdot r)/\gamma > 0$. By Lemma 3, in $O(nd^2)$ time we can compute $a \in \mathcal{H}(V)$ such that $ar \geq \alpha r > 0$. ∎

In order to initialize Algorithm 1, we need to find some subset $H_0 \subset \mathcal{H}(V)$ whose intersection is bounded. We start by showing how to find a subset whose intersection is pointed, i.e. has at least one vertex.

**Algorithm 2** *FindPointedCone*

$H \leftarrow \emptyset$. $\vec{r} \leftarrow \mathbb{1}$.
while $|H| < d$ do
    $H \leftarrow H \cup \text{DeleteRay}(\vec{r}, V)$
    $\mathcal{A} \leftarrow \bigcap_{h^+ \in H} h$
    Let $a$ and $b$ distinct points in $\mathcal{A}$.
    $\vec{r} \leftarrow a - b$.
end while
return $H$

**end**

**Lemma 6** *Given $V \in \mathbb{R}^{n \times d}$, in $O(nd^3)$ time, Algorithm 2 computes subset $H \subset \mathcal{H}(V)$ such that $\bigcap H$ defines a vertex.*

**Proof:** We can compute a parametric representation of the affine subspace $\mathcal{A}$ defined by the intersection of all hyperplanes found so far in $O(d^3)$ time by Gaussian Elimination. With each DeleteRay call in Algorithm 2, we find a hyperplane that cuts off some ray in the previous affine subspace. It follows that the dimension of $\mathcal{A}$ decreases with every iteration. ∎

**Figure 2** Using the Edmond's Oracle to find a set of points whose convex hull contains $z$.

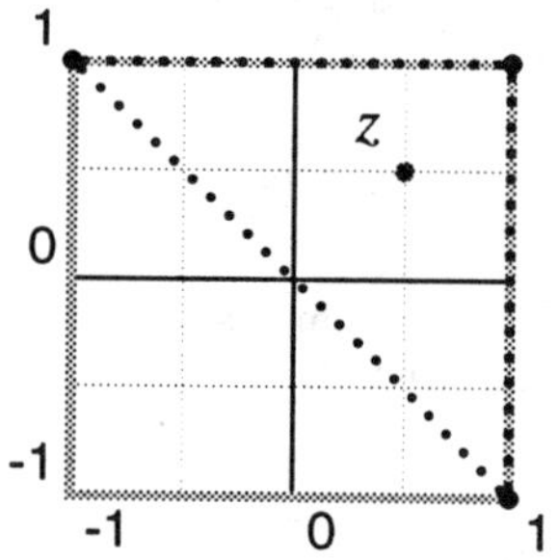


We now show how to augment the set of halfspaces computed by Algorithm 2 so that the intersection of our new set is bounded. To do so, we use an idea due to Jack Edmonds [8].

**Lemma 7 (Edmonds' Oracle)** *Given $H \in \mathbb{R}^{m \times d}$ and $z \in \mathcal{P}(H)$, in time $O(md^3)$ we can find $X \subset \mathcal{V}(H)$ such that $z \in \operatorname{conv} X$ and $|X| \leq d + 1$.*

**Proof:** (sketch) The idea is as follows. Let $P = \mathcal{P}(H)$. Apply Lemma 2 to find $v \in \mathcal{V}(H)$. Find the point $z'$ at which the ray $\overrightarrow{vz}$ exits $P$. If $z'$ is a vertex, we are done, otherwise intersect all constraints with the facet(s) containing $z'$ and recurse with $z'$ as the interior point. The recursively computed set, along with $v$, will contain $z$ in its convex hull. ∎

By duality of convex polytopes, we have the following:

**Lemma 8 (Dual Edmonds' Oracle)** *Given a d-polytope $P = \operatorname{conv} V$ and a valid inequality $ax \leq 1$ for $P$, we can find in time $O(|V|d^3)$, some $A \subset \mathcal{H}(V)$ such that $a \in \operatorname{conv} A$.*

**Figure 3** The dual interpretation of Figure 2 as the problem of finding a set of facets that imply a given valid constraint $zx \leq 1$.

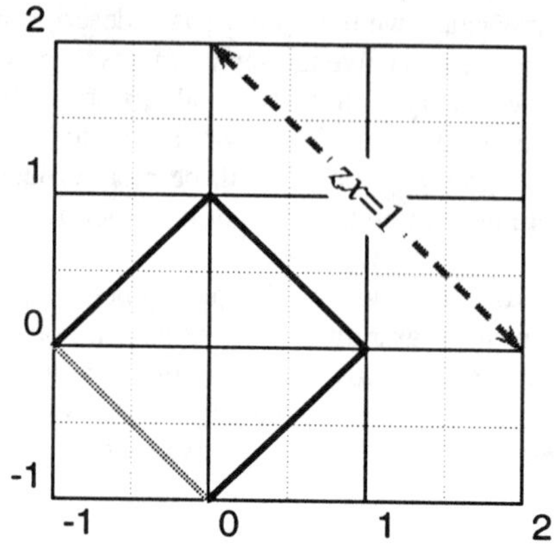


Figure 2 illustrates the application of Edmonds' oracle to find a subset of vertices of a cube containing an interior point $z$ in their convex hull. In Figure 3 the equivalent dual interpretation of finding a set of facets that imply a valid inequality is shown. In order to understand the application of Lemma 8, we note the following:

**Lemma 9** *Let $P = \{ x \mid Ax \leq 1 \}$ and $Q = \{ x \mid A'x \leq 1 \}$ be polyhedra such that each row $a'$ of $A'$ is a convex combination of rows of $A$. $P \subseteq Q$.*

Using Lemma 6 and Lemma 8, we can now find a subset of $\mathcal{H}(V)$ whose intersection is bounded.

**Lemma 10** *Given $V \in \mathbb{R}^{n \times d}$, in time $O(nd^3)$ we can compute a subset $H \subseteq \mathcal{H}(V)$ such that $\bigcap H$ is bounded.*

**Proof:** We start by computing set $B$ of $d$ facet defining halfspaces whose intersection defines a vertex, using Algorithm 2. The proof is then similar to that of of Lemma 4 and Lemma 5. Compute the mean vector $\bar{a}$ of the normal vectors in $B$. Let

$$\gamma = \max_{v \in V} -\bar{a}v\,.$$

Let $\alpha = -\bar{a}/\gamma$. Note that $h^+ \equiv \{ x \mid \alpha x \leq 1 \}$ is valid for $V$, but any ray feasible for $\bigcap B$ will be cut off by this constraint; hence $\mathcal{P}(B) \cap h^+$ is bounded. Now by applying Lemma 8 we can find a set of halfspaces $H \subset \mathcal{H}(V)$ such that $\alpha \in \operatorname{conv} H$. By Lemma 9 $\bigcap(B \cup H)$ is bounded. ∎

## 2.1 The dual-nondegenerate case

In this section we describe how the results of the previous section lead to a polynomial algorithm for facet enumeration of simple polytopes. We then give two refinements of this algorithm that will lead to an algorithm whose time complexity is competitive with the known algorithms for the primal nondegenerate case.

From the discussion above, we know that to achieve a polynomial algorithm for facet enumeration on a particular family of polytopes we need only have a polynomial algorithm for vertex enumeration for each subset of facet defining halfspaces of a polytope in the family. Dual-nondegeneracy (i.e. simplicity) is not quite enough in itself to guarantee this, but it is not difficult to see that the halfspaces defining any simple polytope can be perturbed so that they are in general position without affecting the combinatorial structure of the polytope. In this case each dual sub-problem is solvable by any number of pivoting methods (see e.g. [1, 5, 7]). Equivalently (and more cleanly) we can use lexicographic ratio testing (see e.g. [11]) in the pivoting method. A basis is a subset of $\mathcal{H}(P)$ whose bounding hyperplanes define a vertex of $P$. Although a pivoting algorithm may visit many bases (or perturbed vertices) equivalent to the same vertex, notice that any vertex of the input is simple hence will have exactly one basis. It follows that we can again guarantee to find a witness or find all vertices of $\mathcal{P}(H_c)$ in at most $n+1$ bases (where $n = |V|$, as before) output by the pivoting algorithm. In the case where each vertex is not too degenerate, say at most $d + \delta$ facets meet at every vertex for some small constant $\delta$, we may have to wait for as many as $n \cdot \binom{d+\delta}{\delta} + 1$ bases. Of course this grows rather quickly as a function of $\delta$, but is polynomial for $\delta$ constant. In the rest of this section we assume for ease of exposition that the polytope under consideration is simple.

It is not completely satisfactory to perform a complete vertex enumeration for each verification (FindWitness) step since each succeeding input to the vertex enumeration algorithm consists of adding exactly one halfspace to the previous input. Consider the following standard pivoting algorithm for vertex enumeration using depth first search. In what follows, we assume the halfspaces defining the current intermediate polytope are stored in some global dictionary (c.f. the proof of Lemma 2); we sometimes denote this set of halfspaces as $H_c$. We are given some subset $H_0 \subset \mathcal{H}(V)$ such that $\bigcap H_0$ is bounded and a starting vertex $v \in \mathcal{V}(H_0)$ (we can use the raindrop algorithm to find a starting vertex in $O(|H_0|d^2)$ time).

**Algorithm 3** *DFS(v)*

```
For j ∈ 1...d do
    v' ← ComputeNeighbour(v, j, H_c)
    if new(v') then
        DFS(v')
    end if
end for
```

**end**

The procedure ComputeNeighbour$(v, j, H_c)$ finds the $j$-th neighbour of $v$ in the the current intermediate polytope. This requires $O(md)$ time to accomplish using a standard simplex pivot. To check if a vertex is new (i.e. previously undiscovered by the depth first search) we can simply store the discovered vertices in some standard data structure such as a balanced tree, and query this structure in $O(d \log n)$ time.

We can modify Algorithm 3 so that it finds new facets as a side effect. We are given a subset $H_0 \subset \mathcal{H}(V)$ as before and a starting vertex $v \in \mathcal{V}(H_0)$ with the additional restriction that $v$ is a vertex of the input. In order to find a vertex of $\bigcap H_0$ that is also a vertex of the input, we find an arbitrary vertex of $\bigcap H_0$ by using Lemma 2. If this vertex is not a vertex of the input then we apply DeleteVertex to find a new halfspace which cuts it off, and repeat. We then modify Algorithm 3 by replacing the call to ComputeNeighbour with a call to the procedure ComputeNeighbour2, defined below. Call this modified version Algorithm 3a. The main modification is that in addition to the neighbouring vertex $v'$, the ComputeNeighbour2 routine computes any halfspaces defining $v'$ that are not already known. Suppose we have found $v$ (i.e. $v$ is a vertex of the current intermediate polytope). Since $P$ is simple we must have also found all of the halfspaces defining $v$. It follows that we have a halfspace description of each edge leaving $v$. Since we have a halfspace description of the edges, we can pivot from $v$. If we find a vertex $v' \in V$ by pivoting then we know it must be adjacent to $v$ in conv $V$; if $v' \notin V$ then we can cut it off using our DeleteVertex routine. If $P$ is simple, then no perturbation is necessary, since we will cut off degenerate vertices rather than trying to pivot away from them. Thus ComputeNeighbour2 can be implemented as follows:

**Algorithm 4** ComputeNeighbour2$(v, j, H_c)$

```
repeat
    w ← ComputeNeighbour(v, j)
    If w ∉ V then
        h⁺ ← DeleteVertex(w, H_c, V)
        AddToDictionary(h⁺, H_c)
    end if
until w ∈ V
return w
```

**end**

**Lemma 11** *With $O(mnd)$ preprocessing,* ComputeNeighbour2 *takes time $O(md + k(md + nd^2))$, where $k$ is the number of new halfspaces discovered.*

**Proof:** As mentioned above, ComputeNeighbour takes $O(md)$ time. The procedure AddToDictionary merges the newly discovered halfspace into the global dictionary. Since $P$ is simple, we know the new halfspace will be strictly satisfied by the current vertex $v$; it follows that we can merge it into the dictionary by making the slack variable basic. This amounts to a basis transformation of the bounding hyperplane, which can be done in $O(d^2)$ time.

Since the search problem is completely static (i.e. there are no insertions or deletions), it is relatively easy to achieve a query time of $O(d + \log n)$, with a preprocessing cost of $O(n(d + \log n))$ using e.g. *kd*-trees [13]. Suppose $\log n > md$. It follows that

$$\begin{aligned} n &> 2^{md} \\ &= (m^{d/2})^{md/(2\log m)}. \end{aligned}$$

But from the upper bound theorem for convex polytopes [12], we know $n \in O(m^{\lfloor d/2 \rfloor})$. It follows that $d + \log n \in O(md)$.

Since each pivot in ComputeNeighbour2 that does not discover a vertex of $V$ discovers a facet of conv $V$, we can charge the time for those pivots to the facets discovered. ■

Note that we no longer need an additional query step in Algorithm 3a to determine if the neighbour of the current vertex $v$ is new. We simply mark each vertex as discovered when we search in ComputeNeighbour2. Furthermore, for $P$ simple, $m \leq n$. Thus we have the following:

**Theorem 2** *Given $V \in \mathbb{R}^{n \times d}$, if* conv $V$ *is simple, we can compute $H = \mathcal{H}(V)$ in time $O(n|H|d^2)$.*

## 2.2 The dual-degenerate case

We would like an algorithm that is useful for moderately dual-degenerate polytopes. In this case each degenerate vertex will correspond to many vertices of the (halfspace) perturbed polytope. In the standard depth first search algorithm, previously discovered bases must be stored. Since the number of bases is not necessarily polynomially bounded in the dual-degenerate case[1] we turn to *reverse search* [1].

**Figure 4** The "simplex tree" induced by the objective $-\mathbb{1}$.

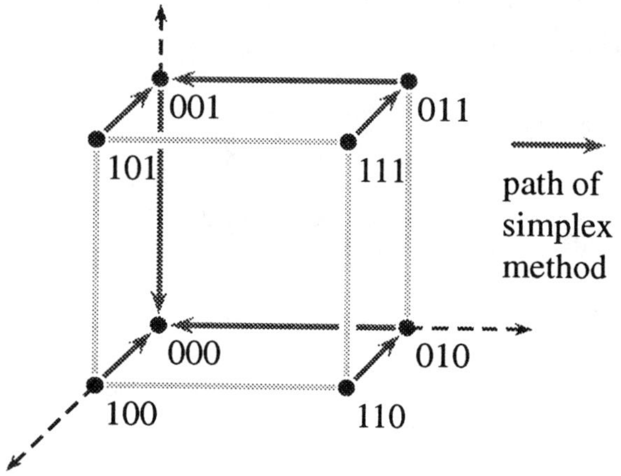


**Figure 5** The reverse search tree corresponding to Figure 4.

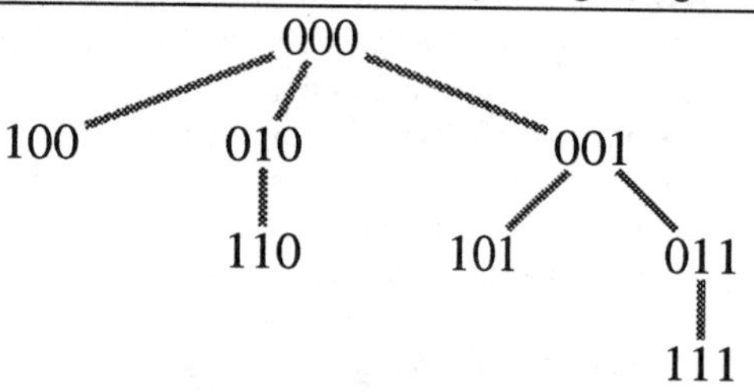


The essential idea of reverse search is quite simple, and requires only a basic knowledge of the simplex method of linear programming (see e.g. [5]). Choose an objective function (direction of optimization) so that there is a unique optimum vertex. Fix some arbitrary pivot rule. From any vertex in the polytope there is a unique sequence of pivots taken by the simplex method to that vertex (see Figure 4). If we take the union of these paths to the optimum vertex, it forms a tree, directed towards the root. It is easy to see algebraicly that the simplex pivot is completely reversible; in fact one just exchanges the roles of the leaving and entering variable. Thus we can perform depth-first search on the "simplex tree" by reversing the pivots from the root (see Figure 5). No storage is needed to backtrack, since we merely pivot towards the optimum vertex. In schematic form, the reverse search algorithm can be written as follows:

[1]Even if the number of bases is bounded by a small polynomial in the input size, any superlinear space usage may be impractical for large problems.

**Algorithm 5** *ReverseSearch*

```
v ← opt, j ← 1, AddToDictionary(H_0)
repeat
    while j ≤ d
        v' ← ComputeNeighbour2(v, j, H_c)
        if IsPivot(v', v) then
            v ← v', j ← 1                         down edge
        else
            j ← j + 1                             next sibling
        end if
    end while
    (v, j) ← PivotToOpt(v)                        up edge
    j ← j + 1.
until j > d and v = opt
```

**end**

The key to the correctness of Algorithm 5 is that halfspaces added later should not affect the earlier decisions of the pivoting rule, since reverse search relies on the pivoting rule making the same choices during backtracking as were made on the way down the tree. In the absence of degeneracy, there is no ambiguity since for each specified entering variable (edge leaving the vertex), there is only one possible leaving variable (new hyperplane). We use a small modification of the standard lexicographic pivoting technique (see e.g. [5], pp. 34–38) that we call "chronological lex pivoting". For each facet defining halfspace $a_i x \le 1$ discovered by Algorithm 5, we consider the "perturbed" halfspace $a_i x \le 1 + \varepsilon_i$. Each basis of the perturbed system can be viewed as a perturbed vertex $\tilde{v} = v + \Gamma\varepsilon$ where $v$ is a vertex of the original system and $\Gamma \in \mathbb{R}^{d \times m}$. Intuitively, we perturb the constant column of the inequality system by some vector $\varepsilon$ satisfying

$$0 < \varepsilon_1 \ll \varepsilon_2 \ll \ldots \varepsilon_m \ll 1 .$$

In fact, we treat expressions involving $\varepsilon$ purely symbolically, using lexicographic comparisons. By convention, set $\varepsilon_{m+1} = 1$. We say $a\varepsilon$ is lexicographically greater than $b\varepsilon$ if the nonzero entry in $a - b$ with largest index is positive. It is important to note that Algorithm 5 will never pivot down from a basis unless it is a basis of some input vertex. It follows that the reverse search tree may be viewed as consisting entirely of input vertices.

**Lemma 12** *Let $\tilde{v} = v + \Gamma\varepsilon$ be a perturbed vertex for some $v \in V$. Let $h^+$ be a perturbed constraint discovered after $\tilde{v}$. $\tilde{v}$ is strictly lexicographically feasible for $h^+$.*

**Proof:** Let $h^+ = \{\, x \mid ax \le 1 + \varepsilon_j \,\}$. Since $v$ is an input vertex, $av \le 1$. Furthermore, since $\tilde{v}$ is defined by halfspaces discovered earlier, each row $\gamma_i$ of $\Gamma\varepsilon$ is lexicographically less than $\varepsilon_j$. ∎

The chronological lex rule can implemented without any special symbolic computation techniques within the same $O(md)$ asymptotic time bounds for a pivot; see e.g. [11] for details.

From Lemma 12, we know we can add the new (perturbed) halfspaces in the same way as before, by making the slack variable basic. Equally importantly we know there is no ambiguity in the choices made by the pivot rule.

**Theorem 3** *Given* $V \in \mathbb{R}^{n \times d}$,

(a) *we can compute* $\mathcal{H}(V)$ *in time* $O(\beta m d^2)$ *and space* $O((m+n)d)$, *where* $m = |\mathcal{H}(V)|$ *and* $\beta$ *is the number of bases of* $\mathcal{H}(V)$.

(b) *we can decide if* conv $V$ *is simple in time* $O(n^2 d^2)$.

**Proof**

(a) Total cost for finding an initial set of halfspaces is $O(nkd^2)$, where $k$ is the size of the initial set. Since every DeleteVertex call finds a new halfspace, the total cost for these calls is $O(nmd^2)$. In every call to ComputeNeighbour2, each pivot except the last one discovers a new halfspace. Those which discover new halfspaces have total cost $O(m^2 d)$ which is $O(\beta m d)$ because the perturbed polytope is simple and $m \leq \beta$; the other pivots cost $O(\beta m d^2)$ as there are $\beta d$ calls to ComputeNeighbour2. The $\beta$ forward pivots (PivotToOpt) cost $O(\beta m d)$.

(b) At any step of the reverse search, we can read the number of halfspaces satisfied with equality by the current vertex off the dictionary in $O(m)$ time. If we reach a degenerate vertex, or discover more than $n$ facets, we stop. If the reverse search terminates, then in $O(nmd)$ time we can compute the number of facets meeting at each vertex. ■

Part b of Theorem 3 is of independent interest since the problem of given $H$, deciding whether $\mathcal{P}(H)$ is simple is known to be NP-Complete in the strong sense [9].

# 3 Experimental Results

We have implemented Algorithm 5 using rational arithmetic in C. For facet enumeration of hypercubes with dimension at least 5, it is faster than the reverse search implementation described in [3] (Table 1). We hope to include more extensive experimental results in the full version of the paper.

The memory requirements of our implementation are twice the input size plus twice the output size, as the program stores four dictionaries: a constant vertex dictionary $V$ and a growing halfspace dictionary $H_c$ in unpivoted form, and a working copy of both. The program uses an earlier version of the preprocessing step, with an upper bound of $O(nd^4)$, compared to the current bound of $O(nd^3)$. The source code is available at ⟨`http://wwwjn.inf.ethz.ch/ambros/pd.html`⟩.

**Table 1** Execution times (in seconds on a DEC 3000/700 AXP) of the primal–dual algorithm (`pd`) and reverse search algorithm (`prs`) for facet enumeration of hypercubes.

| Dimension | pd | prs |
|---|---|---|
| 4 | 0.06 | 0.04 |
| 5 | 0.27 | 0.40 |
| 6 | 0.87 | 6.07 |
| 7 | 2.62 | 91.57 |

# 4 Conclusions

An alternative approach to achieving an algorithm polynomial for the dual-nondegenerate case is to modify the method of Gritzmann and Klee [10]. An idea due to Clarkson [6] can be used to reduce the row size of each of these linear programs to $O(m')$ where $m'$ is the maximum number of facets meeting at a vertex. If we assume that $m' \leq d + \delta$ for some constant $\delta$ then we can solve each linear program by brute force in time polynomial in $d$. It seems that such an approach will be inherently quadratic in the input size since a quadratic number of linear programs may need to be solved.

It would be interesting to remove the requirement in Theorem 1 that the family be facet-hereditary, but it seems difficult to prove things in general about the polytopes formed by subsets of the halfspace description of a known polytope.

# References


[1] D. Avis and K. Fukuda. A pivoting algorithm for convex hulls and vertex enumeration of arrangements and polyhedra. *Disc. Comput. Geom.*, 8:295–313, 1992.

[2] A. Brøndsted. *Introduction to Convex Polytopes*. Springer-Verlag, 1981.

[3] A. Brüngger, A. Marzetta, K. Fukuda, and J. Nievergelt. The parallel search bench ZRAM and its applications. *Annals of Operations Research*, to appear.

[4] D. Chand and S. Kapur. An algorithm for convex polytopes. *J. ACM*, 17:78–86, 1970.

[5] V. Chvátal. *Linear Programming*. W. H. Freeman, New York, NY, 1983.

[6] K. L. Clarkson. More output-sensitive geometric algorithms. In *Proc. 35th Annu. IEEE Sympos. Found. Comput. Sci.*, pages 695–702, 1994.

[7] M. Dyer. The complexity of vertex enumeration methods. *Math. Oper. Res.*, 8(3):381–402, 1983.

[8] J. Edmonds. Decomposition using Minkowski. Abstracts of the 14th International Sysmposium on Mathematical Programming, Amsterdam, 1991.

[9] K. Fukuda, T. M. Liebling, and F. Margot. Analysis of backtrack algorithms for listing all vertices and all faces of a convex polyhedron. *Comput. Geom.: Theory and Appl.*, To appear.

[10] P. Gritzmann and V. Klee. On the complexity of some basic problems in computational convexity: II. volume and mixed volumes. In T. Bistztriczky, P. McMullen, and R. I. Weis, editors, *Polytopes: abstract, convex and computational (Scarborough, ON, 1993)*, NATO Adv. Sci. Inst. Ser. C Math. Phys. Sci., 440, pages 373–466. Kluwer Acad. Publ., Dordrecht, 1994.

[11] J. P. Ignizio and T. M. Cavalier. *Linear Programming*, pages 118–122. Prentice Hall International Series in Industrial and Systems Engineering. Prentice Hall, 1994.

[12] P. McMullen. The maximal number of faces of a convex polytope. *Mathematika*, 17:179–184, 1970.

[13] K. Mehlhorn. *Data Structures and Algorithms 3: Multi-dimensional Searching and Computational Geometry*, volume 3 of *EATCS Monographs on Theoretical Computer Science*. Springer-Verlag, Heidelberg, West Germany, 1984.

[14] R. Seidel. *Output-size sensitive algorithms for constructive problems in computational geometry*. Ph.D. thesis, Dept. Comput. Sci., Cornell Univ., Ithaca, NY, 1986. Technical Report TR 86-784.

[15] G. Swart. Finding the convex hull facet by facet. *J. Algorithms*, pages 17–48, 1985.

# Construction of 1-d lower envelopes and applications

Edgar A. Ramos*
Max-Planck-Institut für Informatik
Im Stadtwald, 66123 Saarbrücken, Germany
ramos@mpi-sb.mpg.de

## Abstract

We consider the problem of computing the lower envelope (the minimum) of $n$ constant degree algebraic functions of one variable. The lower envelope has size $O(n\beta(n))$ where $\beta(n)$ is a nearly constant function, and it can easily be computed in time $O(n\beta(n)\log n)$ by a simple deterministic divide-and-conquer algorithm [45]. We give an alternative simple (module a derandomization black box) approach using divide-and-conquer based on cuttings that results in a deterministic sequential algorithm that runs in the same time bound. This algorithm uses derandomization tools by now standard. This approach however allows us to obtain the following results:

• A deterministic sequential algorithm that is output sensitive and runs in time $O(n\log f)$ if $f \leq n^{\epsilon}$, or $O(n\beta(f)\log f) = O(n\beta(n)\log n)$ otherwise, where $f$ is the size of the output;
• a randomized parallel EREW algorithm that runs in time $O(\log n)$ and uses nearly optimal work $O(n\beta^2(n)\log n)$ with $n$-polynomial probability* .
• output sensitive deterministic parallel algorithms: in EREW, it runs in time $O(\log n(\log f + \log\log n))$ using work $O(n\log f)$ for $r \leq n^{\epsilon}$ or $O(n\beta(f)\log f)$ otherwise, in CRCW it runs in time $O(\log\log n(\log f + \log\log n))$ using work $O(n\log f)$ for $r \leq n^{\epsilon}$ or $O(n\beta^2(f)\log f)$ otherwise.

We use this algorithm as a component in the algorithm in [5] for a restricted class of planar algebraic Voronoi diagrams to obtain:

• A output sensitive deterministic sequential algorithm that runs in time $O(n\log f)$ where $f$ is the size of the output,
• an optimal work randomized parallel CREW algorithm that runs in time $O(\log n)$ with $n$-polynomial probability;
• deterministic parallel output sensitive algorithms using work $(n\log f)$ that run in time $O(\log n(\log f + \log^2\log n))$ for EREW and in time $O(\log\log n(\log f + \log^2\log n))$ for CRCW (in particular time $O(\log n\log\log n)$ in the worst case.

We consider a restricted subset of the type of optimization problems usually solved using parametric search, in which the algorithm that is run generically follows a divide-and-conquer approach based on cuttings, and in which the predicates involved (an object intersects a cell) are *linearizable*. We have the following results:

• an algorithm that solves the optimization problem for 1-d lower envelopes using $O(\log n)$ oracle queries,
• using a variation of the algorithm for planar Voronoi diagrams mentioned above, together with the optimization approach, we give a deterministic algorithm for computing the diameter of a set of $n$ points (furthest distance pair) that runs in time $O(n\log^2 n)$.

*A randomized algorithm on input of size $n$ is said to use time $t(n)$ and work $w(n)$ with $n$-polynomial probability if it fails to satisfy those bounds with probability at most $n^{-c}$ for some constant $c > 1$. Similarly, the bounds hold with $n$-exponential probability if it fails to satisfy them with probability at most $2^{-n^{\epsilon}}$ for some constant $\epsilon > 0$.



## 1 Introduction

Voronoi diagrams are a basic geometric structure and consequently their computation have received a great deal of attention [41, 18, 32, 39]. Specifically, in [5] a class of planar Voronoi diagrams was identified whose construction can be performed by a very general algorithm deterministically in worst case optimal work[1] $O(n\log n)$ both sequentially and in parallel. That was basically a generalization and derandomization of previous work in [14, 44, 3]. Such family of Voronoi diagrams includes, for example, Voronoi diagrams of line segments and the boundary of the intersection of unit balls in 3-d space. The latter was used in [3] to obtain a deterministic algorithm for the diameter problem in 3-d space running in time $O(n\log^3 n)$, still not matching the randomized performance $O(n\log n)$ [14].

Several questions remained: are output sensitive algorithms possible ? are faster parallel algorithms possible (both randomized and deterministic) ? can the running time

*This research was performed in part while at DIMACS/Rutgers University as a Postdoctoral Fellow. DIMACS is a cooperative project of Rutgers University, Princeton University, AT&T Research, Bell Labs and Bellcore. DIMACS is an NSF Science and Technology Center, funded under contract STC-91-19999; and also receives support from the New Jersey Commission on Science and Technology.

[1]We often refer to sequential time to treat uniformly both sequential and parallel algorithms.

of the diameter algorithm be improved ? In most cases, the bottleneck was the computation of the so called *contours*, the Voronoi diagram restricted to the boundary of a cell in a cutting of the diagram. This problem is basically the computation of the lower envelope (the minimum) of a set of functions of one variable. There is a very simple sequential and deterministic solution for this problem [45], and to a certain extent, it parallelizes [23]. However, its paralleliza-tion is not satisfactory, it cannot be made output sensitive, and it is a bottleneck in geometric optimization (diameter problem).

Thus, the objective of this work is to present an alternative algorithm for this problem and applications where it outperforms the other algorithm. The algorithm is a simple application of deterministic geometric sampling: a divide-and-conquer approach using *sensitive cuttings* [9, 40, 3] to control the total size of problems in all levels of the computation. The deterministic version of the algorithm however works only under certain restrictions for the functions which allow the efficient derandomization of the geometric sampling techniques. An important case for whch this is possible is when the functions are *piecewise algebraic*: the function is the solution to a polynomial of degree by a constant in each of a constant number of intervals covering $\mathbb{R}$. Although this seems a theoretical restriction, this is the case for all practical applications one encounters. The key implication of the algebraic restriction is the *linearizability* of the predicates that appear in the computations.

Using this algorithm, we can obtain sequential and parallel output sensitive algorithms for these algebraic functions. We extend and improve the results in [28] to a more general class of functions (there, only 2-d convex hulls are considered) and to deterministic algorithms. Using this new algorithm then we can obtain deterministic output sensitive construction for a restricted class (but again very general) of planar Voronoi diagrams (projections of lower envelopes of 2-d algebraic functions with a site having a simply connected face). These results also parallelize. The output sensitive results were not known previously even under randomization. An additional tool that we use is the fast construction of approximations and cuttings in parallel [24, 25, 27]. Even in the particular case of planar Euclidean Voronoi diagrams, we can improve previous work in [16] by obtaining a deterministic worst case optimal work CRCW algorithm that runs in time $O(\log n \log\log n)$ (this can also be obtained using the original divide-and-conquer 1-d lower envelope algorithm, but had not been noted before). The fast CRCW constructions use the recent work in [22].

Also using this 1-d lower envelope algorithm, and further insight into geometric optimization in an arrangement of surfaces, we can improve by a $\log n$ factor the running time of the diameter problem, thus obtaining an $O(n \log^2 n)$ time algorithm. The approach tries to transform the usual unintuitive parametric search to a more ordered search among critical points in an arrangement of surfaces.

The specific results are listed in the abstract and presented in the subsequent sections, but first we give a brief recount of the geometric sampling tools that are used.

## 2 Tools from geometric sampling

We present a minimum of definitions to facilitate the later description of our algorithms as approximations, cuttings, and linearization are essential for the algorithms. See [1, 34, 2, 36] for further information.

### 2.1 Approximations and cuttings

A *range space* is a pair $(X, \Gamma)$, where $X$ is a ground set whose elements are called *points* and $\Gamma$ is a set of subsets of $X$ called *ranges*. For $Y \subseteq X$, the *subspace of* $(X,\Gamma)$ *induced by* $Y$ is the range space $(Y, \Gamma_Y)$ where $\Gamma_Y = \{\gamma \cap Y : \gamma \in \Gamma\}$. $(X,\Gamma)$ has *bounded VC-exponent* if there is a constant $k$ such that for any $Y \subseteq X$, $|\Gamma_Y| = O(|Y|^k)$. All the range spaces we consider have bounded VC-exponent. For finite $X$, a $(1/r)$-*approximation* for $(X,\Gamma)$, is a set $A \subseteq X$ such that $||A \cap R|/|A| - |R|/|X|| \leq 1/r$ for every $R \in \Gamma$. A *range space with elementary cells* is a triple $S = (X, \Gamma, \mathcal{E})$ where $(X,\Gamma)$ is a range space and $\mathcal{E} = \mathcal{E}(\Gamma)$ is a set of subsets of $X$ called *elementary cells*. We assume there is a constant $d$ such that each $e \in \mathcal{E}$ is determined uniquely by a set of at most $d$ elements in $\Gamma'$, be the set of boundaries of the ranges $\Gamma$. Thus for finite $Q \subseteq \Gamma'$, the size of $\mathcal{E}(Q) = \{e \in \mathcal{E} : e$ is determined by elements in $Q\}$ has size $O(|Q|^d)$. We say $\gamma' \in \Gamma'$ *crosses* $e \in \mathcal{E}$ if $\gamma' \cap e \neq \emptyset$. For $Q \subseteq \Gamma'$ and $e \in \mathcal{E}$, $Q_{|e} = \{\gamma' \in \Gamma' : \gamma'$ crosses $e\}$ is called the *conflict list* of $e$. For the pair $(\Gamma', \mathcal{E})$, the *derived range space* is $(\Gamma', \Gamma'_{\mathcal{E}})$ where $\Gamma'_{\mathcal{E}} = \{\Gamma'_e : e \in \mathcal{E}\})$. A $(1/r)$-*cutting* for $(Y,Q)$ with cells from $\mathcal{E}$, where $Y \subseteq X$ and $Q \subseteq \Gamma'$ is finite, is a collection $\Xi \subseteq \mathcal{E}$ whose union contains $Y$ and with $|Q_{|e}| \leq |Q|/r$ for every $e \in \Xi$. $\Xi$ is an *elementary cell decomposition* of $(Y,Q)$ if the union of the cells in $\Xi$ is $Y$ and $Q_{|e} = \emptyset$ for every $e \in \Xi$.

### 2.2 Linearization

Two important families of range spaces are $(\mathbb{R}^d, \mathcal{H}_d)$ where $\mathcal{H}_d$ is the set of half-space in $\mathbb{R}^d$, and $(\mathbb{R}^d, \mathcal{L}_{d,s})$ where $\mathcal{L}_{d,s}$ is the set of *linear cells* of order $s$ in $\mathbb{R}^d$, that is, unions of at most $s$ cells that are intersections of at most $s$ halfspaces. For range spaces $(X,\Gamma)$ and $(Y,\Sigma)$, an injective function $\varphi : X \to Y$ is called an *embedding* if for each $\gamma \in \Gamma$, there is a $\sigma \in \Sigma$ such that $\gamma = \varphi^{-1}(\sigma)$. In our applications, we encounter pairs object–cell $(\Omega, \mathcal{E})$ such that $\omega \in \Omega$ and $e \in \mathcal{E}$ are described by $k + l$ real parameters $(x; a) \in \mathbb{R}^k \times \mathbb{R}^l$, and there is a first-order predicate $\Pi(x; a)$ in the theory or real closed fields (polynomial inequalities in the coordinates of $x$ and $a$ using Boolean connectives and quantifiers) expressing whether $\omega \cap e \neq \emptyset$. Using a quantifier elimination method, it follows that the derived range space is *linearizable*: There are integers $d$ and $s$, and mappings $\varphi : \mathbb{R}^k \to \mathbb{R}^d$ and $\psi : \mathbb{R}^l \to \mathcal{L}_{d,s}$ such that $\varphi(x) \in \psi(a)$ iff $\Pi(x; a)$ is true, and such that $\varphi$ is given by bounded degree polynomials in (the coordinates of $x$, and the functions describing the coefficients in the equations of the half-spaces defining $\psi(a)$ are given by bounded degree polynomial in (the coordinates of) $a$.

### 2.3 Deterministic constructions

**Approximations.** Let $(X,\Gamma)$ be a range space with constant VC-dimension. For $Y \subseteq X$ with $|Y| = n$, a $(1/r)$-approximation for $(Y, \Gamma_Y)$ of size $O(r^2 \log r)$ can be computed sequentially using work $O(nr^c)$ or $O(n \log r)$ if $r \leq n^\epsilon$ for certain constants $c > 0$ and $0 < \epsilon < 1$. In parallel the resulting size is $O(r^{2+\delta})$ and the construction can be performed using the same work and in times $O(\log n)$ and $O(\log n \log r)$ respectively for the EREW model, and in times $O(\log\log n)$ and $O(\log\log n \log r)$ respectively for the CRCW model.[2] See [33, 12, 24, 25].

**Cuttings.** Let $(X, \Gamma, \mathcal{E})$ be a range space with elementary cells and a linearizable derived range space $(\Gamma', \Gamma'_{\mathcal{E}})$. Let $Y(Q) \subseteq X$ for finite $Q \subseteq \Gamma'$ be a subset of $X$ determined by $Q$ and suppose there is a nondecreasing function $\xi$ so that $Y(Q) \subseteq X$ has elementary cell decomposition of size $\xi(|Q|)$. Then, there is a $(1/r)$-cutting for $(Y(Q), Q)$ of size $O(\xi(r))$. Let $n = |Q|$.

[2] These fast CRCW times are possible by allowing the output to be *padded* [37, 29], that is, in an array larger than the size of the output by a factor $(1+\Delta)$. In some of our CRCW algorithms these fast constructions are used only as intermediate steps, and the final output is not padded. In other cases, we can achieve a fast time by having the final output padded; these cases will be noted explicitly.

**Theorem 2.1:** *The above mentioned cutting can be computed deterministically in time $O(nr^c)$, or in time $O(n \log r)$ if $r \leq n^\epsilon$. In EREW PRAM the construction can be performed in times $O(\log n)$ and $O(\log n \log r)$ respectively; in CRCW PRAM the construction can be performed in times $O(\log \log n)$ and $O(\log \log n \log r)$ respectively (the output is padded, or add time $O(\log n)$ for not padded output).*

See [9, 33, 24, 4]. Only the CRCW construction has not been noted before, it uses the recent fast construction of approximation in [25, 27]. Briefly, for later reference, a cutting is computed by first (i) obtaining a $(1/2r)$-approximation $A$ for $(Q, Q_\mathcal{E})$, then (ii) computing a $(1/2r)$-cutting $\mathbf{T}$ for $(Y, A)$, and finally (iii) computing the conflict lists of $\mathbf{T}$ in $Q$. It is important to note that for the fast $O(n \log r)$ construction, for any constant $k > 0$, by choosing $\epsilon$ sufficiently small, we have that $|A|^k$ and $|T|^k$ are $O(|Q|)$; thus a polynomial amount of work in $|A|$ and $|T|$ is allowed. For our 1-d lower envelope algorithm, we need *sparse* cuttings, introduced in [9], and later used in [40, 3, 4]. In the next section, we state the specific constructions needed.

## 3 1-d lower envelopes

### 3.1 Statement of problem

Let $\mathcal{F}$ be the class of *1-d algebraic* functions, that is, functions $f : \mathbb{R} \to \mathbb{R}$ such that $f$ is defined in each of a constant number of intervals as an algebraic function of constant degree, and consider a set $F \subseteq \mathcal{F}$ of size $n$. The algebraic restriction implies that any two of these functions intersects at most a constant number $k$ of times and that certain operations on their intersections can be performed in constant time. It is also essential for our algorithms because the resulting predicates are linearizable. The *lower envelope* of $F$ is the function $L_F : \mathbb{R} \to \mathbb{R}$ defined by $L_F(x) = \min_{f \in F} f(x)$. This is determined by a sequence of points $p_0 = -\infty, p_1, \ldots, p_m, p_{m+1} = +\infty$ (the *vertices*) and indices $s_0, \ldots, s_m$ such that for each $i = 0, \ldots, m$, if $x \in [p_i, p_{i+1})$ then $f(x) = f_{s_i}(x)$ (the *edges*). Bounds on the *size* $m$ depending on $k$ have been extensively studied, and the interested reader is referred to the monograph [45]. For our purposes, we only use that $m$ is bounded in the worst case by $\lambda(n) = n\beta(n)$ where $\beta$ is a very slowly growing function ($\beta(n) = O(\log^* n)$, but better bounds are known). This problem is easily solved in time $O(n\beta(n) \log n)$ by a simple divide and conquer algorithm: compute recursively the lower envelopes for the collections $f_1, \ldots, f_{\lfloor n/2 \rfloor}$ and $f_{\lfloor n/2 \rfloor + 1}, \ldots, f_n$, merge the lists of their vertices, and compute the lower envelope in each resulting interval [6, 45]. In our applications (to parallelization and optimization), a divide-and-conquer algorithm based on cuttings turns out to be more appropriate.

### 3.2 Sparse cuttings

We identify a function with its graph. For a (partial) function $f$, let $f^- = \{(x, y) \in \mathbb{R}^2 : \exists y'(y \leq y', (x, y') \in f)\}$ (the closed region below the graph of $f$). $L_F^-$ has a natural decomposition into cells by extending its vertices downwards; thus, we let $\mathcal{E} = \mathcal{E}(\mathcal{F})$ consists of the cells of the form $e^-$ where $e$ is an edge in the lower envelope of any 3 functions in $\mathcal{F}$. We are interested in computing $L_F^- = \bigcap_{f \in F} f^-$ for a finite $F \subseteq \mathcal{F}$. An essential consequence of the algebraic assumption on $\mathcal{F}$ is that the range space derived from $(\mathcal{F}, \mathcal{E})$ is linearizable. The basic cutting result follows from 2.1:

**Theorem 3.1:** *Let $F \subseteq \mathcal{F}$ with $|F| = n$, and let $\sigma \in \mathcal{E}(\mathcal{F})$. A $(1/r)$-cutting for $(L_F^- \cap \sigma, F)$ of size $O(\lambda(r))$ can be computed deterministically with the same work and time bounds as theorem 2.1.*

Unfortunately, using recursively this cutting construction does not seem to lead to an efficient algorithm. A global mechanism to control the size of the subproblems is needed. Let $\mathrm{vert}(F)$ denote the set of vertices of the arrangement of $F$. Sparse cuttings have size that depend on the number of vertices of $F$ as in the following theorem (the proof uses standard techniques [3, 4] and will be given in the final version):

**Theorem 3.2:** *Let $F \subseteq \mathcal{F}$ with $|F| = n$, and let $\sigma \in \mathcal{E}(\mathcal{F})$. A $(1/r)$-cutting for $(L_F^- \cap \sigma, F)$ of size $O(1 + (r/n)^2 |\mathrm{vert}(F) \cap \sigma|)$ can be computed deterministically with the same work and time bounds as theorem 2.1.*

This follows because a random sample with probability $p = r/n$ (with appropriate *resampling*) is a $(1/r)$-cutting with good probability, and each vertex in $\sigma$ has probability $(r/n)^2$ of appearing in the arrangement of the sample. Unfortunately, for large $r$ this is a bad estimate. This needs to be corrected as large values of $r$ are necessary in some algorithms. Let $\mathrm{vert}(F, l)$ denote the set of vertices of the arrangement of $F$ of level at most $l$. The following result uses two stages: first a $(C/r)$-cutting, then a $(1/C)$-cutting, for an appropriate parameter $C$.

**Theorem 3.3:** *Let $F \subseteq \mathcal{F}$ with $|F| = n$, and let $\sigma \in \mathcal{E}(\mathcal{F})$. A $(1/r)$-cutting for $(L_F^- \cap \sigma, F)$ of size $O(\lambda(r/C) + (r/n)^2 |\mathrm{vert}(F, nC/r) \cap \sigma|)$ can be computed deterministically with the same work and time bounds as theorem 2.1.*

### 3.3 Basic algorithm

The algorithm is a straightforward construction of a cell decomposition of $L_F^-$ by recursively computing cuttings. However, the use of a sparse cutting is essential to control the total problem size over all levels of computation. Let $n_i$, $i = 0, \ldots, k$ be a sequence of integers such that $n_0 = n$ and $n_k = O(1)$. Let $n_i = n_{i-1}/r_i$. The algorithm works in $k$ stages. In the $i$th stage, $\mathbf{T}_i$ denotes the current cutting for $(L_F^-, F)$ which is an $(n_{i-1}/n)$-cutting. Let $C_i = |\mathbf{T}_i|$ and let $n_{|\sigma} = |F_{|\sigma}|$ (thus, $n_{|\sigma} \leq n_{i-1}$ for $\sigma \in \mathbf{T}_i$). The $i$-th stage is as follows:

For each $(\sigma, F_{|\sigma}) \in \mathbf{T}_i$ do
1. If $|F_{|\sigma}| \leq C$ then finish in time $O(1)$
2. Compute a $(1/r_\sigma)$-cutting $\mathbf{T}(\sigma)$ for $F_{|\sigma}$ with $1/r_\sigma = n_i/|F_{|\sigma}|$
3. Put each $(\tau, F_{|\tau}) \in \mathbf{T}(\sigma)$ in $\mathbf{T}_{i+1}$

Note that $1/r_\sigma = n_i/|F_{|\sigma}|$ implies that for $(\tau, F_{|\tau}) \in \mathbf{T}(\sigma)$, $|F_{|\tau}| \leq n_i$.

**Small size sampling.** Here $r_i = K$ where $K$ is an appropriate constant. Thus $k = O(\log n)$. In step 2, we use theorem 3.2. We have

$$\begin{aligned} C_i &\leq \sum_{\sigma \in \mathbf{T}_{i-1}} A(1 + (r_\sigma/|F_{|\sigma}|)^2 |\mathrm{vert}(F_{|\sigma}) \cap \sigma|) \\ &\leq AC_{i-1} + A(1/n_i)^2 |\mathrm{vert}(F, n_{i-1})| \\ &\leq AC_{i-1} + A(nr_i/n_i)\beta(n/n_{i-1}), \end{aligned}$$

where we have used the fact that the number of vertices of level at most $l$ is $O(l^2 \lambda(n/l))$ [14, 45]. Using induction, one can verify that $C_i \leq C'(nr_i/n_i)\beta(n/n_{i-1})$. The total conflict list size in the $i$th stage is $n_iC_i = O(nr_i\beta(n/n_{i-1})) = O(n\beta(n/n_{i-1}))$. Thus, for any $\sigma \in \mathbf{T}_i$, we can afford work $O(n_\sigma)$, and the total work is $O(n\beta(n)\log n)$. This is indeed possible, as the cutting and conflict lists can be computed in time $O(n_{|\sigma})$.

**Theorem 3.4:** *$L_F$ can be computed by a divide-and-conquer algorithm based on cuttings in time $O(n\beta(n)\log n)$.*

**Large size sampling.** Here $r_i = n_{i-1}^{\epsilon}$ for an appropriate constant $\epsilon > 0$, and so $n_i = n^{(1-\epsilon)^i}$ and $k = O(\log\log n)$. In step 2, use theorem 3.3 with $1/r = n_i/|F_{|\sigma}|$ and $C = K\beta(r_i)$ where $K$ is an appropriate constant. We have

$$\begin{aligned} C_i &\leq \sum_{\sigma\in\mathbf{T}_{i-1}} A(\lambda(r_i/C) + (1/n_i)^2|\mathrm{vert}(F_{|\sigma}, Cn_i)\cap\sigma|) \\ &\leq (A/C)r_i\beta(r_i/C)C_{i-1} + A(1/n_i)^2|\mathrm{vert}(F, n_{i-1}C/r_i)| \\ &\leq (A/K)r_iC_{i-1} + ABK\beta(r_i)(n/n_i)\beta(n/K\beta(r_i)n_i). \end{aligned}$$

Using induction, one verifies that $C_i \leq D(n/n_i)\beta(n/n_i)\beta(n)$ for an appropriate constant $D$. Thus, the total conflict list size in the $i$-th stage is $n_iC_i = O(n\beta(n/n_i)\beta(n))$. In this case in stage $i$, for any $\sigma \in \mathbf{T}_i$, we can afford work $O(n_{|\sigma}\log n_{|\sigma})$; thus the total work in the $i$-th stage is $O(n\beta^2(n)\log n_i)$ and the overall total work is $O(n\beta^2(n)\log n)$ since $\sum_i \log n_i = O(\log n)$.

**Remark.** Unfortunately, in the large size sampling case we cannot match the performance of other algorithms (note that $\beta(n)$ is very small). However, this approach leads to some interesting parallel algorithms, and in the case that $\beta(n) = O(1)$, the resulting algorithms are actually optimal (this is the case for our application to 2-d Voronoi diagrams).

### 3.4 Output sensitive algorithm

Using the basic algorithm above together with *pruning*, a very minor addition, we obtain an output sensitive algorithm with running time $O(n\beta(f)\log f)$. Adding a *filtering* stage improves the performance to $O(n\log f)$ for $f = O(n^\epsilon)$ for some $\epsilon > 0$ and provides fast parallel algorithms. Pruning and filtering have been used widely before to achieve output sensitive algorithms, beginning with the work of Clarkson and Shor [14]; for example [44, 11, 3, 28]. Also [31] is relevant.

#### 3.4.1 Pruning

The algorithm does not need to recurse on a cell $\sigma$ of the cutting $\mathbf{T}_i$ that does not contain a vertex of $L_F^-$. This can be detected by computing $L_F(a)$ and $L_F(b)$ where $a$ and $b$ are the $x$ coordinates of the left and right sides of $\sigma$. These are called the *contours*. Let $l_\sigma$ and $r_\sigma$ be the indices of the functions that determine $L_F(a)$ and $L_F(b)$. If $l_\sigma \neq r_\sigma$ or if some $f \in F_{|\sigma}$ intersects $f_I^-$ in $[a,b]$, where $I = l_\sigma = r_\sigma$, then $\sigma$ contains a vertex of $L_F^-$ and it must be retained, it is *nonredundant*. Otherwise, $L_F^-$ is completely determined by $I$ in $\sigma$, and need not be considered further. This can be determined in time $O(n_\sigma)$. Although this *pruning* by itself does not seem to control the total problem size, together with the use of sparse cuttings, it results in an output sensitive algorithm.

**Theorem 3.5:** *$L_F$ can be computed in time $O(n\beta(f)\log f)$ where $n = |F|$ and $f$ is the size of $L_F$.*

**Proof:** We use the small size sampling algorithm. The analysis above shows that the work at level is bounded by $O(n\beta(n/n_{i-1}))$. Furthermore, the size of each subproblem decreases at each level by at least a constant factor, and pruning enforces that the computation tree has at most $O(f)$ leaves. A lemma in [19, 8] implies that the total amount of work is $O(n\beta(f)\log f)$ (briefly, the work performed in the $O(\log f)$ stages until the subproblem size is $O(n/f)$ is $O(n\beta(f)\log f)$, and the work performed in the remaining stages is $O(n)$). ■

#### 3.4.2 Filtering

*Filtering* improves the previous result to $O(n\log f)$ for $f = O(n^\delta)$, some $\delta > 0$, and also makes possible faster parallel algorithms. Let $F_i$ be the filtered set at the beginning of the $i$-th iteration and let $n_{i-1} = |F_i|$ with $F_1 = F$ and $n_0 = n$. Let $r_i = 2^{2^i}$. In the $i$-th stage, the algorithm computes a $(1/r_i)$-cutting $\mathbf{T}_i$ for $F_i$ of size $O(r_i\beta(r_i))$ in time $O(n_{i-1}\log r_i)$ using theorem 3.1. Let $F_{i+1}$ be the union of the conflict lists of the nonredundant cells as determined by the contours as in the pruning algorithm. It is clear that $L_{F_i} = L_{F_{i+1}}$. Since there are at most $f$ nonredundant cells, then $n_i = |F_{i+1}|$ is at most $n_{i-1}f/r_i$. If $r_i > f$ then some size reduction is possible. The $i$-th stage is as follows:

1. Compute a $(1/r_i)$-cutting $\mathbf{T}_i$ for $F_i$
2. Determine nonredundant cells in $\mathbf{T}_i$
3. Let $F_{i+1}$ be the union of the conflict lists of nonredundant cells
4. If $n_i \leq n^{1-\delta}$ or $r_{i+1} \geq n^{4\delta}$ then end iterations

Here $\delta = \epsilon/4$ where $\epsilon$ is the constant in the cutting construction theorem. After $r_i > 4f^2$, $n_i$ decreases rapidly so the sum of terms $O(n_{i-1}\log r_i)$ is bounded by $O(n\log f)$ (since $n_i\log r_{i+1} \leq n_{i-1}\log r_i/2$). Similarly, before $r_i > f^2$ is achieved, at most work $O(n\log f)$ is performed (since $\log r_i$ in these stages adds up to $\log f$). After the last iteration, say the $I$-th one, use the ordinary algorithm performing work $O(n_I\beta(n_I)\log n_I)$. If $f \leq n^\delta$ then $n_I \leq n^{1-\delta}$, and this work is $O(n)$. Otherwise, if $f > n^\delta$, then the ordinary algorithm performs work $O(n\beta(n)\log n) = O(n\beta(f)\log f)$.

**Theorem 3.6:** *$L_F$ can be computed output sensitively using time $O(n\log f)$ if $f \leq n^\delta$, or time $O(n\beta(f)\log f)$ otherwise.*

**Remark.** It is not clear to us whether the techniques in [7] also produce such algorithm. If possible, however, it would result in a very sequential algorithm. Furthermore, the type of ray shooting that is needed seems much more difficult in the 2-d case. Our approach on the other hand extends to those cases as well.

### 3.5 Parallel algorithms

Our parallel algorithms use the fast constructions of approximations in [24, 25]. For the most part, the algorithms are direct parallelizations of the basic sequential algorithm.

#### Randomized algorithms

**Theorem 3.7:** *$L_F$ can be computed in EREW PRAM using time $O(\log n)$ and work $O(n\beta^2(n)\log n)$ with $n$-polynomial probability, and in CRCW using time $O(\log n\log^* n)$ and work $O(n\beta(n)\log n)$ with $n$-exponential probability.*

**Proof:** For the EREW algorithm we use large size sampling; the $i$-th stage takes time $O(\log n_i$. For the CRCW algorithm we use small size sampling; each stage takes time $O(\log^* n)$ using approximate counting and compaction techniques (but the final output is not padded). In order to achieve the $n$-exponential probability, we make use of the *failure sweeping* technique introduced in [21]. ■

The CRCW algorithm matches the time of an algorithm in [23], but it is considerably simpler. Although the EREW is not work optimal, we will use it to derive a work optimal algorithm for Voronoi diagrams as there $\beta(n) = O(1)$.

### Deterministic algorithms

**Theorem 3.8:** *$L_F$ can be computed deterministically in EREW PRAM using time $O(\log^2 n)$ and in CRCW using time $O(\log n \log\log n)$, both using work $O(n\beta(n)\log n)$.*

Here, in both cases we use small size sampling (in the CRCW algorithm padding is allowed in intermediate computations but not in the final result). These results are cetainly not new, they can easily be achieved by the basic divide-and-conquer algorithm.

### Output sensitive algorithms

Our approach is similar to that in [28], and to the sequential filtering algorithm. Here however, the filtering of $F_i$ using a $(1/r_i)$-cutting is repeated $k$ times and the result is $F_{i+1}$, where $k$ is an appropriate constant. As before $r_i = 2^{2^i}$. Let $p_i$ denote the number of processors available in the $i$-th iteration, with $p_0 = An$ for an appropriate constant $A$. We use two different algorithms for computing a $(1/r_i)$-cutting (stated in theorem 3.1: Algorithm **A** computes it using work $O(n_{i-1}\log r_i)$; algorithm **B** uses work $O(n_{i-1}r_i^c)$ for some constant $c > 0$. Algorithm **A** is slow, it takes time $O(\log n_{i-1}\log r_i)$ in EREW or $O(\log\log n_{i-1}\log r_i)$ in CRCW. Algorithm **B** is faster, it takes time $O(\log n_{i-1})$ in EREW or time $O(\log\log n_{i-1})$ in CRCW. The $i$-th iteration is as follows:

1. If $p_i < n_{i-1}r_i^c$
   then filter $F_i$ using algorithm **A** $k$ times
   $p_{i+1} = p_i$
   else filter $F_i$ using algorithm **B** $k$ times
   $p_{i+1} = p_i/2$
2. If $n_i \leq n^{1-\epsilon}$ or $r_{i+1} \geq n^\epsilon$ then end iterations

Suppose in the $(i-1)$-st stage $r_{i-1} \geq f^2$. Since there is a reduction by a factor at least $r_{i-1}/f$, after $k$ repetitions the reduction factor accumulates to at least $r_{i-1}^{k/2}$. So for $k$ a sufficiently large constant we have $p_i \geq n_{i-1}r_i^c$. Therefore, there are at most $O(\log\log f)$ iterations in which $p_i$ is not halved. In those iterations using algorithm **A** requires $O(n_{i-1}\log r_i)$ work, that is $O(p_i \log r_i)$, which in total is at most $O(n\log f)$ since $p_i = O(n)$ and $\log r_i$ grows geometrically with $r_i \leq f^2$. Because of the halving, the work performed in the remaining $O(\log\log n)$ stages is $O(n)$. Thus, the total work is $O(n\log f)$. If the filtering algorithm terminates because $n_i \leq n^{1-\epsilon}$, then $O(n)$ additional work computes $L_{F'}$ where $F'$ is the final $F_i$. If it terminates because $r_i \geq n^\epsilon$, then $O(n\beta(n)\log n) = O(n\beta(f)\log f)$ and the ordinary (non output sensitive) algorithm can finish the task. The running times are then as claimed in the following theorem, assuming a model in which we can allocate the number of processors in each iteration.

**Theorem 3.9:** *$L_F$ can be computed deterministically in the EREW model using time $O(\log n(\log f + \log\log n))$, and in the CRCW model (with padded output) using time $O(\log\log n(\log f + \log\log n))$, both using work $O(n\log f)$ for $r \leq n^\epsilon$ or ] $O(n\beta(f)\log f)$ otherwise.*

If the number of processors $p$ needs to be fixed, then by choosing its number appropriately, in the running time the additive term $\log\log n$ is changed into a multiplicative factor. Note however that we do not assume that the value of $f$ is known in advance.

**Remark.** For the EREW model, in the particular case of linear functions (a problem equivalent to convex hull computation), a faster time is possible [13]: $O(\log n(\log\log f + \log^* n))$ or, if the number of processors is fixed, $O(\log n \log f \log^* n)$ ($\log n \log f$ is trivially a lower bound in this case).

## 4 2-d lower envelopes and algebraic planar Voronoi diagrams

For $n$ algebraic functions of two variables, let $\psi(n)$ be an upper bound on the combinatorial complexity of their lower envelope. It is known that $\psi(n) = O(n^{2+\epsilon})$ [45]. The algorithm in the previous section can be extended to the 2-d case and it results in a deterministic algorithm that runs in time $O(\psi(n))$ as long as $\psi(n) = \Omega(n^{1+\epsilon})$ (the important point is that one can construct an elementary cell decomposition for the lower envelope whose size is also $O(\psi(n))$, this is not known for higher dimensions). So we get an optimal worst case algorithm under these conditions. In contrast, previous deterministic algorithms would result in $O(\psi(n)n^\epsilon)$ or $O(\psi(n)\log^c n)$ at best [45].

**Theorem 4.1:** *The lower envelope of $n$ 2-d algebraic functions can be computed deterministically in time $O(\psi(n))$ where $\psi(n)$ is an upper bound on the complexity of the lower envelope with $\psi(n) = \Omega(n^{1+\epsilon})$.*

Our main interest here is however a particular class of functions. It is well known that a planar Voronoi diagram is the projection of the lower envelope of a collection of surfaces (the distance functions from the sites) in 3-d. In [5], we gave algorithms for deterministic sequential construction and for parallel EREW construction of a particular but very general class of Voronoi diagrams, whose defining feature is that the bisectors are piecewise algebraic curves. That was a generalization and derandomization of the algorithm for halfspace intersections in [44]. A bottleneck in several applications of that algorithm was the computation of a 1-d lower envelope. Considering a more restricted class of Voronoi diagrams, but still including most important examples, the functions in those computations are algebraic and using the algorithm of the previous section we can obtain several interesting results: sequential and parallel output sensitive deterministic algorithms, fast CRCW and EREW algorithms, faster geometric optimization for 1-d lower envelopes, and faster computation of the diameter of a point set in $\mathbb{R}^3$.

### 4.1 Definition

We consider a set $S$ of $n$ *sites* in the plane. We identify a site $p \in S$ with its *distance function* $p : \mathbb{R}^2 \to \mathbb{R}$ (and also with its graph) which is assumed to be piecewise algebraic (the domain is divided into regions by $O(1)$ curves of bounded degree, and in each region the function is defined by a polynomial equation of bounded degree), Let $\mathcal{V}(S)$ denote the lower envelope of the functions in $S$, which we call its (lifted)

*Voronoi* diagram (the actual Voronoi diagram is its projection onto $\mathbb{R}^2$). $\mathcal{V}(S)$ consists of 0-, 1- and 2-dimensional *faces* called *vertices, edges and facets* respectively. We assume that each $p$ *supports* (coincides) with at most one facet in $\mathcal{V}(S)$, which is called the *Voronoi facet (or cell) of* $p$ and denoted by $V_p(S)$. This restriction is essential: First because it implies that the *size* of $\mathcal{V}(S)$ (the number of vertices, edges and facets) is $O(n)$; second because it allows to formulate a pruning rule for subproblems that result in an optimal work algorithm. In particular, given a cutting for $\mathcal{V}^-(S)$, if a site $p$ that does not contribute to $\mathcal{V}(S)$ on any of the boundaries of the cells of the cutting, then $p$ can contribute to $\mathcal{V}(S)$ in at most one cell $\tau$ of the cutting; we say then that $p$ is *interior* to $\tau$.[3] We assume that there are no degeneracies.[4] Also, we assume that there is a canonical way to decompose a Voronoi facet into elementary cells of constant size called *trapezoids* (one possibility is using *geodesics* from the vertices on its boundary) and so that the total number of resulting trapezoids is still $O(n)$. The collection of cells $e^-$, called *bricks*, where $e$ is a trapezoid on $\mathcal{V}(S)$ forms an elementary cell decomposition for $\mathcal{V}(S)^-$, denoted by $\mathcal{T}(S)$. Let $\mathcal{E}$ denote this class of elementary cells. Following our notation, for $\sigma \in \mathcal{T}(R)$, where $R \subseteq S$, $S_{|\sigma}$ is the conflict list for $\sigma$. In our algorithm, for a cell $\sigma$, we mantain a smaller conflict list $S_\sigma$ which is obtained from $S_{|\sigma}$ by *pruning*. For a brick $\tau = e^-$, the vertical portion of its boundary (that is, excluding $e$) is called its *contour* and denoted $C(\tau)$. We will need the details of the algorithm in [5] in the application to the 3-d diameter problem, so for reference we include it below with some minor revisions.

### 4.2 Basic algorithm

**Preliminaries.** Let $\tau \in \mathcal{E}$ be a brick in a cutting for $(\mathcal{V}^-(S), S)$. The restricted Voronoi diagram $\mathcal{V}_\tau(S)$ is $\mathcal{V}(S) \cap \tau$ and its faces are the connected components of all $f \cap \tau$ where $f$ is a face of $\mathcal{V}(S)$; similarly for $\mu = C(\tau)$. Note that a face in $\mathcal{V}(S)$ can originate more than one face in $\mathcal{V}_\tau(S)$. $\mathcal{V}_\mu(S)$ is called the *contour* of $\mathcal{V}(S)$ in $\tau$ (or of $\mathcal{V}_\tau(S)$) and denoted by $\mathcal{C}(S,\tau)$. An edge or a facet of $\mathcal{V}_\tau(S)$ which is not incident to vertices inside $\tau$ is said to be *green*, and the corresponding vertices and edges in $\mathcal{C}(S,\tau)$ are also said to be green. The green (or redundant) portion of $\tau$ is its intersection with the vertical extension of the green facets, and the non redundant portion, denoted $\tau_*$, is the closure of its complement. The non redundant portion of the contour $\mathcal{C}_*(S,\tau)$ is the restriction of $\mathcal{C}(S,\tau)$ to $\tau_*$. The (green) edges inside $\tau$ on the boundary of $\tau_*$ are called *attaching* edges and their projection onto $\mathbb{R}^2$ generate a subdivision $D(\tau)$ of the projection of $\tau$ into the redundant and non redundant portions. Suppose $\tau$ is a brick in the cutting of a brick $\sigma$. The portion $\tau_n = \tau \cap \sigma_*$ is called the *new* portion of $\tau$. When recursing on $\tau$, the algorithm is to construct $\mathcal{V}_{\tau_n}(S)$, as the remaining portion is already known. The corresponding new portion of the contour is denoted $\mathcal{C}_n(S,\tau)$. Point location in $D(\sigma)$ is used to identify vertices in $\mathcal{C}_n(S,\tau)$.

**General outline.** At the beginning of the $i$-th stage, $\mathbf{T}_i$ consists of subproblems $(\sigma, D(\sigma), S_\sigma)$ satisfying the invariant $\sum_{(\sigma,D(\sigma),S_\sigma)\in\mathbf{T}_i} |S_\sigma| = O(n)$. In the $i$-th stage, for $(\sigma, D(\sigma), S_\sigma) \in \mathbf{T}_i$, the algorithm constructs a $(1/r_\sigma)$-cutting $\mathbf{T}(\sigma)$ for $(\mathcal{V}^-(S_\sigma)\cap\sigma, S_\sigma)$ of size $O(r_\sigma)$, where $1/r_\sigma = n_i/|S_\sigma|$, using theorem 2.1. Then for each $(\tau, S_{\sigma|\tau}) \in \mathbf{T}(\sigma)$, it prunes $S_{\sigma|\tau}$ into $S_\tau$ so that the invariant holds for the next stage. Using large size sampling $r_i = n_{i-1}^\epsilon$, we get $n_i = n^{(1-\epsilon)^i}$ as an upper bound for $n_\sigma = |S_\sigma|$. Thus, if work $O(n_\sigma \log n_\sigma)$ is used for $\sigma$, the total work in the $i$-th iteration is $O(n \log n_i)$ and the total work of the algorithm is $O(n \log n)$. Let $n_{\sigma|\tau} = |S_{\sigma|\tau}|$. Since the cutting guarantees $\sum_{\tau\in\mathbf{T}(\sigma)} n_{\sigma|\tau} = O(n_\sigma)$ then an amount of work $O(n_{\sigma|\tau} \log n_\sigma)$ is allowed for each $\tau$. The steps in the $i$-th iteration are as follows (details are given below).

For each $(\sigma, D(\sigma), S_\sigma) \in \mathbf{T}_i$ do
1. If $|S_\sigma| \leq C$ then finish in time $O(1)$
2. Compute $(1/r_\sigma)$-cutting $\mathbf{T}(\sigma)$ where $1/r_\sigma = n_i/|S_\sigma|$
3. For each $(\tau, S_{\sigma|\tau}) \in \mathbf{T}(\sigma)$ do
4. Compute $\mathcal{C}_n(S,\tau)$ and determine $S_\tau^c$ and $S_\tau^{nc}$
5. Compute $\mathcal{V}(S_\tau^c)$ and its D-K hierarchy
6. Determine green faces and interior sites
7. Determine $\mathcal{C}_*(S,\tau)$, attaching edges and compute $D(\tau)$
8. Prune $S_{\sigma|\tau}$ into $S_\tau$ and put $(\tau, D(\tau), S_\tau)$ in $\mathbf{T}_{i+1}$

**Step 4.** To obtain $\mathcal{C}_n(S,\tau)$, the algorithm first computes $\mathcal{C}(S_{\sigma|\tau},\tau) = \mathcal{V}_\mu(S_{\sigma|\tau})$ (a corrupted version that includes also *spurious* edges and vertices) using our algorithm for 1-d lower envelopes in time $O(n_{\sigma|\tau} \log n_{\sigma|\tau})$ (note that $|\mathcal{V}_\mu(T))| = O(|T|)$, that is $\beta(\cdot) = O(1)$, because $|\mathcal{V}(T)| = O(|T|)$ and each edge of $\mathcal{V}(T)$ intersects $\mu$ at most a constant number of times; therefore, in the 1-d lower envelope we can use large size sampling and still obtain optimal algorithms). Then using a point location data structure for $D(\sigma)$ computed in the previous level, determine for each vertex in $\mathcal{C}(S_{\sigma|\tau},\tau)$ whether it is spurious (it is in the redundant portion). $D(\tau)$ supports queries in time $O(\log n_\sigma)$. Thus, $\mathcal{C}_n(S,\tau)$ is obtained in total time $O(\log n_\sigma)$ and work $O(n_{\sigma|\tau} \log n_\sigma)$. The contour sites $S_\tau^c \subseteq S_{\sigma|\tau}$ are those sites that touch $\mathcal{C}_n(S,\tau)$; the remaining $S_\tau^{nc} = S_{\sigma|\tau} - S_\tau^c$ are possible interior sites.

**Step 5.** Let $T = S_\tau^c$. To compute $\mathcal{V}(T)$, use this algorithm itself recursively, but notice there will be no interior sites: In the code above, step 5 and the determination of interior sites are omitted. A D-K hierarchy [17] for $\mathcal{V}(T)$ is a sequence $\mathcal{V}(T_i)$, $i = 0, \ldots, k$, where $T_0 = T$, $T_i \subseteq T_{i+1}$, $T_k = O(1)$ and $|T_{i+1}| \leq \alpha|T_i|$ for some constant $0 < \alpha < 1$ so that $k = O(\log|T|)$. $T_{i+1}$ is obtained from $T_i$ by removing sites whose cells are not adjacent and each with a number of adjacent cells $O(1)$ (in the dual graph, a large independent set of vertices of small degree). The D-K hierarchy can be computed sequentially in time $O(|T|)$, and in parallel (EREW) in time $O(\log|T|)$ using optimal work.

**Step 6.** A *quasi-edge* $e$ is an edge of $\mathcal{V}(S_\tau^c)$ in $\tau$ not incident to a vertex inside $\tau$. $e$ need not be an edge of $\mathcal{V}_\tau(S)$ as another site can intersect $e$ inside $\tau$. If $e$ is an edge of $\mathcal{V}_\tau(S)$, then it is green. Let $p \in S_\tau^{nc}$. The D-K hierarchy is used to detect whether the cell $V_p(S_\tau^c \cup \{p\})$ is nonempty by locating a vertex in $\mathcal{V}(S_\tau^c \cup \{p\})$ incident to $V_p(S_\tau^c \cup \{p\})$, if it exists. This can be done in time $O(\log|T|)$ (a standard search in a D-K hierarchy). If the vertex exists and is inside $\tau_n$, then $p$ is interior to $\tau$. A quasi-edge $e$ is an edge of $\mathcal{V}_\tau(S)$, and so a green edge, iff none of the vertices witnessing interior sites for $\tau$ lies on $e$.

**Step 7.** From the green edges, $\mathcal{C}_*(S,\tau)$ as well as the attaching edges can be determined. The structure of the subdivision $D(\tau)$ is particularly simple, since it has no vertices inside (the dual is a tree). A data structure for point location in $D(\tau)$ with the required performance can be constructed using standard techniques.

**Step 8.** Include in $S_\tau$ each site in $S_{\sigma|\tau}$ that is either interior to $\tau$ or incident to a non redundant edge in $\mathcal{C}_*(S,\tau)$. This enforces the global bound $O(n)$.

### 4.3 Output sensitive algorithm

We obtain an output sensitive algorithm for our class of Voronoi diagrams by using the previous basic algorithm, together with the output sensitive algorithm for 1-d lower envelopes and a filtering strategy similar to that for 1-d lower

---

[3] If we have an upper bound $\psi(n) = O(n)$, but a site can contribute more than one facet, our approach cannot be used to get an $O(n \log n)$ algorithm. We do not know of any deterministic approach that would achieve that running time (though it is possible with a randomized algorithm).

[4] They can certainly be handled through symbolic perturbation [20], but it would be interesting to handle them directly.

envelopes. Actually, we only need to modify the value of $r_i$. In the $i$-th iteration, contours are computed to determine redundant cells. Let $f_i'$ be the maximum size among the resulting contours. The value of $r_i$ is defined as the $\max(r_{i-1}^2, f_{i-1}'^2)$, following [28].

**Theorem 4.2:** *Voronoi diagrams in our restricted class can be computed output sensitively with a running time $O(n \log f)$ where $n$ is the number of sites and $f$ is the size of the output.*

This includes problems like Voronoi diagrams of line segments in the plane with aditive weights, and intersection of unit balls in the space, for which output sensitive algorithms were not previously known.

### 4.4 Parallel algorithms

Using the basic algorithm, together with our 1-d lower envelope algorithm, and appropriate tools for randomized construction of cuttings, we obtain the following theorems.

**Theorem 4.3:** *A restricted algebraic Voronoi diagram can be computed using randomization in the CREW model using time $O(\log n)$ and work $O(n \log n)$ with $n$-polynomial probability.*

The need for concurrent reads arises because of the simultaneous searches in the D-K hierarchy. This extends the work in [42] to a larger class of Voronoi diagrams (they only consider line segments in CRCW).[5] Using deterministic construction of cuttings, we obtain.

**Theorem 4.4:** *A algebraic Voronoi diagram can be computed deterministically using work $O(n \log f)$ in parallel. In the EREW model, using time $O(\log n(\log f + \log^2 \log n))$, and in the CRCW model using time $O(\log\log n(\log f + \log^2 \log n))$ (padded output).*

This theorem extends the results of [28] in several ways: to deterministic algorithms, EREW algorithms, and to a much larger class of Voronoi diagrams (they only consider convex hulls). For the case $f = n$, for example, we obtain an optimal work deterministic CRCW algorithm for computing our restricted Voronoi diagrams that runs in time $O(\log n \log\log n)$ (output not padded), improving over a very complicated non work optimal algorithm in [16] with the same running time, which applies only to Euclidean Voronoi diagrams.

Again, if the number of processors $p$ must be fixed, then by choosing $p$ appropriately, the additive term $\log^2 \log n$ becomes a multiplicative factor. For example in EREW, we obtain $O(\log n \log f \log^2 \log n)$ (a trivial lower bound is $O(\log n \log f)$).

## 5 Optimization

We consider optimization problems that can be expressed by a monotone predicate $P(r)$, where $r$ is a real parameter, and the value $r^*$ for which $P(r)$ changes from true to false is the optimal value. One is given an algorithm **Alg**$(r)$ that determines $P(r)$ and an oracle **Orc**$(r)$ that given $r$, replies whether $r < r^*$, $r = r^*$ or $r > r^*$. Parametric search is a powerful technique that can be used to obtain efficient algorithms for this type of problems [38]. To achieve efficiency, the technique calls for the use of parallel algorithms, and as a result often the algorithms seem somewhat unintuitive. A lot of effort has been made to remove the use of parametric search either by using randomization, or by some other deterministic approaches. In some cases, these efforts have led to a gain in the running time, usually just by a polylog factor. Our main aim here is the problem of computing the diameter of a 3-d point set which can be posed as an optimization problem of this type. We use an approach that although restricted, still it seems to apply in some interesting cases; it can be seen as an extension of the technique used in [10] to avoid parametric search in the solution to the slope selection problem.[6]

### 5.1 Approach

**Parametric search.** The parametric search technique makes use of **Alg** and**Orc**. It is required that the dependency of **Alg** on $r$ is through branchings such that for each there is a partition of $\mathbb{R}$ into a constant number of intervals, so that in each interval the action is the same, and the *critical points* defining the intervals can be determined in constant time. Usually, this holds because the action depends on the sign of a bounded degree polynomial in $r$ (when algebraic functions are involved). Thus, $r^*$ can be computed by running **Alg** "generically" on $r^*$, while using **Orc** to resolve its branchings (binary search on the set of critical points is used to reduce the number of oracle calls). Meggido [38] suggested that **Alg** be a parallel algorithm in order to reduce the number of oracles calls, and subsequently several authors (including [36]) have emphasized that what is needed is the presentation of oracle calls in parallel batches. Thus, a sequential algorithms that presents queries in parallel batches suffices: a batch of $m$ queries is resolved by $O(\log m)$ oracle calls by performing a binary search (with an additional cost $O(m)$ for computing medians).

**Parallelization via cuttings and linearization.** We consider an algorithm **Alg** that follows a divide-and-conquer approach based on cuttings. This directly provides the necessary parallelization of the oracle queries. Furthermore, each branching depending on $r$ in the algorithm is a test of the form $\omega \cap e \neq \emptyset$, for an object $\omega$ and cell $e$, that can be expressed as a first order predicate $\Pi(x; a; r)$ where $x$ and $a$ are the vectors of parameters of $\omega$ and $e$ respectively, other than $r$. In linearizing $\Pi$ (see section 2), there is the freedom of including $r$ in the $\varphi$ or in the $\psi$ functions: either (A) $\Pi(x; a, r)$ iff $\varphi(x; r) \in \psi(a)$ or (B) $\Pi(x; a; r)$ iff $\varphi(x) \in \psi(a; r)$ (of course for different functions). This freedom turns out to be very useful.

**Cole's trick.** Cole [15] pointed out that it is not necessary to wait to the resolution of all oracle queries in a batch to proceed with the algorithm: those parts of the algorithm that depend on resolved queries can proceed. To effectively reduce the number of oracle queries, a weigth $w(q)$ is assigned to each query $q$. The weights are assigned based on the directed graph $G$ of dependencies among the queries: $q$

[5] It is possible that their approach also extends to the more general case, but not clear whether to CREW.

[6] After an initial version of this manuscript, we found out that most of the elements in our approach that we regarded as new were already present in the work of Matoušek and Schwarzkopf [36]. We have now actually departed from our original presentation (more geometric and intuitive) and adopted theirs (more formal), which is more appropriate. Nevertheless, we have additional observations that lead to further improvement. Furthermore, by restating and emphasizing the approach, we expect to make it available for other possible applications.

points to the queries $q_1, \ldots, q_k$ that directly follow after $q$ (once $q$ is resolved). Under this condition, $w(q_i) = w(q)/2k$. Queries with no predecessor have weight 1. Given a batch of queries, the oracle resolves the weighted median, and as a result the *active* weight (the weight of the queries not yet resolved) is reduced by a factor at least 3/4 after each query is answered. So the number of oracle queries required is at most proportional to $\log m_0$, where $m_0$ is the number of queries with no predecessor, plus the maximum of $\sum_i \log(2d_i)$ over all sequences $d_1, d_2, d_3, \ldots$ of outdegrees along paths in $G$. A first useful example is the case of $n$ chains of length $l$ each: $O(\log n + l)$ queries are needed; another one is $n$ trees of constant degree and depth $l$, resulting in the same bound.

**Cutting computation.** We verify that under the assumption of linearization for queries $\gamma' \cap e \neq \emptyset$, the computation of an $\epsilon$-cutting for $(Y, Q)$, $|Q| = n$, with elementary cells from $\mathcal{E}(Q)$ (as reviwed after theorem 2.1, we follow the notation there) can be resolved using $O(\log n)$ oracle queries (and without affecting the amount of work). This is the same as in [36], but we take advantage of Cole's trick to reduce the number of oracle queries.
*Approximation.* Let $\gamma' \in Q$ and $e \in \mathcal{E}$. Option (B) results in $\gamma' \cap e \neq \emptyset$ iff $\varphi(x) \in \psi(a; r)$. Therefore, it suffices to compute an $(\epsilon/2)$-approximation $A$ for $\varphi(Q) = \{\varphi(x) : x$ is the parameter of some $q \in Q\}$ in $\mathbb{R}^d$ with respect to linear cells (or $(\epsilon/C)$-approximation with respect to simplices for an appropriate $C > 1$). This computation is independent of $r$.
*Cutting.* The set of critical points of $\gamma' \cap e \neq \emptyset$ for all $\gamma' \in A$ and $e \in \mathcal{E}(A)$ is of polynomial size in $|A|$ (and $O(n)$), and the corresponding queries can all be presented in a single batch. Resolving them requires $O(\log n)$ oracle queries. Once these queries are resolved, an $(\epsilon/2)$-cutting $\mathbf{T}$ for $A$ can be computed independently of $r$.
*Conflict lists.* Let $\gamma' \in Q$ and $e \in T$. Option (A) results in $\gamma' \cap e \neq \emptyset$ iff $\varphi(x; r) \in \psi(a)$. Thus, first construct a point location data structure $\mathcal{D}$ for the arrangement of all hyperplanes determining the linear cells $\psi(a)$ (this is independent of $r$) with search time $O(\log |T|)$ (preprocessing time is polynomial in $|T|$ which is $O(n)$). The conflict lists are determined by locating each $\varphi(x; r)$ in $\mathcal{D}$. Each test of $\varphi(x; r)$ against a hyperplane in $\mathcal{D}$ results in a constant number of critical points. If $|T|$ is $O(1)$, all the queries can be presented in a single batch. If $|T|$ is large (e.g. large size sampling is used) then we have $O(|Q|)$ parallel searches each of length $O(\log |T|)$, and then using Cole's trick, this is resolved using $O(\log |T| + \log |Q|) = O(\log n)$ queries to the oracle.

## 5.2 Optimization in 1-d lower envelopes

We consider an optimization problem in which **Alg** constructs the 1-d lower envelope of $n$ functions parametrized with $r$, and assume the availability of an oracle **Orc**. The framework explained above directly applies to our 1-d lower envelope algorithm, either with the slow or the large size sampling.The dependency of **Alg** on $r$ is only through the approximation and cutting computation. Let us assume small size sampling is used. Then, following the argument above, the critical values are presented in $O(\log n)$ batches, each of size $O(n)$. Using Cole's trick this results in $O(\log n)$ oracle queries while keeping the work $O(n\beta(n) \log n)$. The algorithm using large size sampling also leads to the same result.

**Theorem 5.1:** *The optimization problem for 1-d lower envelopes can be solved using $O(\log n)$ oracle queries.*

The best known previous result required $O(\log^2 n)$ oracle queries, using an algorithm for lower envelopes with parallel time $O(\log n)$ in a comparison model [23]; for that algorithm Cole's trick is not applicable. We apply this result for the diameter problem in the next section.

# 6 Diameter of a 3-d point set

Our algorithm is a further refinement in the sequence of results [10, 36, 3] that led to an algorithm with running time $O(n \log^3 n)$. Geometric sampling and parametric search were first used in [10]; then [36] introduced large size sampling and several tools to support it; finally [3] introduce pruning as a way to control the problem size in subproblems. They all used a relation between the diameter problem and the problem of intersecting unit balls in $\mathbb{R}^3$ [14]. The best known result not using sampling techniques is $O(n \log^5 n)$ [43] (but it uses parametric search). Our aim is to describe a deterministic algorithm that runs in time $O(n \log^2 n)$. The starting point is an application of the geometric optimization approach of the previous section to the ball intersection algorithm implicit in section 4. However, there are actually two levels in which cuttings are computed and that complicates the accounting of oracle queries. Fortunately, some observations lead to simplification in the algorithm.

## 6.1 Reduction to ball intersection

Let $P$ be a set of $n$ points in $\mathbb{R}^3$. The *diameter* $D(P)$ is the furthest (Euclidean) distance between any two points in $P$. For $p \in \mathbb{R}^3$, let $b(p; r)$ be the ball centered at $p$ and with radius $r$, and let $\mathcal{I}(P; r) = \bigcap_{p \in P} b(p; r)$. As observed in [14], an oracle oracle $\mathcal{O}(P)$ that for an $r > 0$ decides its relation to $D = D(P)$ can be implemented as follows: construct $\mathcal{I} = \mathcal{I}(P; r)$; if some $p \in P$ is outside $\mathcal{I}$ then $r < D$, if $P$ is contained in the interior to $\mathcal{I}$ then $r > D$, otherwise $r = D$ ($P$ is contained in $\mathcal{I}$ but not in its interior). Let $o$ be a point in the interior of $\mathcal{I}(P; D(P))$ (any point in the convex hull of $P$ will do, for example a point $p \in P$, but even the center of the minimum enclosing ball can be computed in time $O(n)$). Let $s(p; r)$ be the bounding sphere of $b(p; r)$, $S = S(P; r)$ be the set of bounding spheres, and $\mathcal{V}(P; r) = \mathcal{V}(S)$ be the boundary of $\mathcal{I}$. The requirements for the Voronoi diagram algorithm of section 4 are satisfied here: $\mathcal{V}$ is the lower envelope of the spheres in $S$ with respect to $o$ (hence the corresponding notation adopted); each $p \in P$ contributes at most one face in $\mathcal{V}$ [30]; a face of $\mathcal{V}(R; r)$ for $R \subseteq P$ is decomposed into *trapezoids* by drawing a *geodesic through the poles* for each vertex on the face, then $\mathcal{I}(R; r)$ is decomposed into *bricks* by joining these trapezoids to $o$ [36]. Since the transformation is clear, we will continue using here the notation of section 4, where appropriately $S$ corresponds to the set of spheres.

## 6.2 Oracle

The ball intersection algorithm almost for free implements the oracle $\mathcal{O}$. Let $Q$ be a copy of $P$. As $\mathcal{V}(S(P; r))$ is computed by successively refining a cutting, one can keep track of the position of each point $q \in Q$ in the cutting. At the end, a single test suffices to determine the facet of $\mathcal{I}$ hit by the ray $oq$, which we call its *antipodal facet* (and the point whose sphere supports that facet is its antipodal point), and their relative position. We need to verify that this can be done without affecting the time bound and within logarithmic number of oracle queries.

Let $Q_\sigma$ be the current subset of $Q$ inside $\sigma$. We follow the notation in the basic algorithm of subsection 4.2. Two point locations are needed to keep track of $q \in Q_\sigma$: (i) point location in the cutting $\mathbf{T}(\sigma)$ for $\sigma$, and this results in a $\tau \in \mathbf{T}(\sigma)$, and (ii) point location in the new redundant facets of $\tau$. Note that (ii) is necessary because those facets are removed from deeper subproblems as a result of pruning, so each point whose antipodal facet is among those must be identified at this level. Also note that as a result of this, $Q_\sigma$ is not necessarily equal to $Q_{|\sigma} = Q \cap \sigma$.

We deal with point location (i) similarly to the conflict list determination in the cutting construction: consider a linearization $q \in \tau$ iff $\varphi(p;r) \in \psi(a)$ where $a$ is the list of parameters of $\tau$ other than $r$. Then construct a point location data structure $\mathcal{D}$ for the hyperplanes bounding the linear cells $\psi(a)$, all $\tau \in \mathbf{T}(\sigma)$ (here we can afford polynomial size and processing time) and then search each point $\varphi(p;r)$ in $\mathcal{D}$. Again, using Cole's trick, this results in a logarithmic number of oracle queries.

Point location (ii) cannot be handled in this manner because its size $m$ could be large. Fortunately, it is a planar point location problem, and a data structure $D'(\tau)$ can be constructed using at most $O(m \log m)$ time and space. $D'(\tau)$ is actually an augmented version of $D(\tau)$ (it has further subdivisions), so we will use $D'(\tau)$ where we used $D(\tau)$ before. For $D'(\tau)$ we need to use a data structure that can be constructed using only $O(\log m)$ queries. This is possible following a divide-and-conquer approach based on cuttings much like the 1-d lower envelope and Voronoi diagram algorithms. First an appropriate decomposition into trapezoids is defined by extending *vertical* edges from vertices. Then the construction uses large size sampling together with a mechanism to enforce that the total size of subproblems at a level of the recursion be $O(m)$: say $\tau$ is a trapezoid in the cutting for $\sigma$; the edges crossing $\tau$ (incident to no vertex inside $\tau$) further decompose $\tau$ into a *slab* of smaller trapezoids; the construction recurses on these smaller trapezoids rather than on $\tau$ (point location in this slab is easy because there is a linear ordering among them). Using Cole's algorithm, one finds that $O(\log m)$ oracle queries are sufficient for the construction. Then the point location itself consists of some $m'$ independent searches of length $O(\log m)$, so it needs $O(\log m' + \log m)$ oracle queries to complete.

Finally, we extend this to $\mathbf{Orc}(P,Q)$ which implements the oracle for the furthest distance between points in $P$ and points in $Q$.

### 6.3 Diameter

The basic parametric search approach to determine $D(P)$ is to run $\mathbf{Orc}(P,P)$ generically while using $\mathbf{Orc}(P,P)$ with a fixed $r$ to resolve queries. There are complications in applying the basic approach and accounting the number of oracle calls. First, there are actually two levels: the main cutting $\mathbf{T}(\sigma)$, and the cuttings for computing contours for $\mu \in \mathbf{T}(\sigma)$. Second, there are other parts of the algorithm that need to resolve the position of $D$ among certain critical values (e.g. D-K hierarchy construction and search). By examining the algorithm we can make useful simplifications. We have the following observations:

- There is no need to maintain a single global oracle: Once $\sigma$ is split into a collection of subproblems $\mathbf{T}(\sigma)$, it suffices to determine the diameter in each subproblem, and the maximum among those will be the diameter of the original problem. (Actually, it is important that this is done *after* computing contours, because that guarantees the global $O(n)$ bound on conflict list sizes.)
- The determination of interior sites can be made with respect to a fixed radius (within the current invariant interval) once the contours have been computed. This eliminates of performing the parametric search on the construction of the D-K hierarchy (which would be a bottleneck) .
- The contour sites can be discarded from the subproblem, after $\mathcal{V}(S_\sigma^c)$ has been computed, because they have given already all their information; this results in maintaining a set of pairs $(P_\sigma, Q_\sigma)$ where each $p$ appears in at most one $P_\sigma$ and in at most one $Q_\sigma$.
- By reversing the roles of $P_\sigma$ and $Q_\sigma$ in each iteration, we can maintain pairs $(P_\sigma, Q_\sigma)$ in which both sets have a guaranteed size reduction.

The first and last modifications are not essential to achieve the $O(n \log^2 n)$ time, but we include as they may be useful for further refinement. These observations lead to the algorithm $\mathbf{diam}(P,Q)$ below which computes the furthest distance between point sets $P$ and $Q$. It uses $\mathbf{diam\text{-}bf}(P,Q)$ which also returns the furthest distance but using brute force (checking all pairs); this is acceptable if at least one of $P$ and $Q$ has constant size. Only steps 2,3 and 5 (marked with a *) need to query the oracle $\mathbf{Orc}(P,Q)$. Step 6 is performed for any fixed radius within the current invariant interval for $r$ without querying the oracle. In step 7, $D(\tau)$ is a trivial subdivision (no nonredundant regions). There is no need to eliminate redundant facets nor to determine points whose antipodal facets are redundant; **diam-bd** takes care of that. Note the roles of $P$ and $Q$ are reversed in the recursive call. We use notation so that $P_{|\tau}$ is such that $S(P)_{|\tau} = S(P_{|\tau})$, etc.

**diam**$(P,Q)$
1. If $|P| \leq C$ and $|Q| \leq C$ then return **diam-bf**$(P,Q)$
2.* Compute a $(1/r)$-cutting $\mathbf{T}$ for $S(P)$ where $r = |P|^\epsilon$
3.* Locate each $q \in Q$ in $\mathbf{T}$ resulting in $Q_{|\tau}$ for each $\tau \in \mathbf{T}$
4. For each $(\tau, P_{|\tau}) \in \mathbf{T}$ do
5.* Compute $\mathcal{C}(S(P_{|\tau}), \tau)$ and determine $S(P_\tau^c)$
6. Compute $\mathcal{V}(S(P_\tau^c))$, the D-K hierarchy and interior sites $P_\tau^i$
7. Return $\max_\tau$ max (**diam-bd**$(\tau, D(\tau), P_\tau^c, Q_{|\tau})$, **diam**$(Q_{|\tau}, P_\tau^i)$)

Algorithm **diam-bd** takes care of *boundary sites*. Here, the roles of $P$ and $Q$ cannot be reversed in the recursion. Again, the steps that use the oracle $\mathbf{Orc}(P_\sigma, Q_\sigma)$ are marked with a *. $d_\tau$ is the furthest distance for points whose antipodal facets are redundant.

**diam-bd**$(\sigma, D'(\sigma), P_\sigma, Q_\sigma)$
1. If $|P_\sigma| \leq C$ then return **diam-bf**$(P_\sigma, Q_\sigma)$
2.* Compute a $(1/r_\sigma)$-cutting $\mathbf{T}(\sigma)$ for $S(P_\sigma)$ where $r_\sigma = |P_\sigma|^\epsilon$
3.* Locate each $q \in Q_\sigma$ in $\mathbf{T}(\sigma)$ resulting in $Q_{\sigma|\tau}$ for each $\tau \in \mathbf{T}(\sigma)$
4. For each $(\tau, P_{\sigma|\tau}) \in \mathbf{T}(\sigma)$ do
5.* Compute $\mathcal{C}_n(S(P_{\sigma|\tau}), \tau)$ and determine $P_\tau^c$
6. Determine $\mathcal{C}_*(S(P_{\sigma|\tau}), \tau)$, attaching edges and $P_\tau$
7.* Compute $D'(\tau)$
8.* Locate each $q \in Q_{\sigma|\tau}$ in $D'(\tau)$ resulting in $Q_\tau$ and $d_\tau$
9. Return $\max_\tau$ max(**diam-bd**$(\tau, D'(\tau), P_\tau, Q_\tau)$, $d_\tau$)

The total oracle work at level $i$ is $O(n \log n_i^2)$, where $n_i = n^{(1-\epsilon)^i}$, and so the total work is $O(n \log^2 n)$.

**Theorem 6.1:** *There is a deterministic algorithm that solves the 3-d diameter problem in time $O(n \log^2 n)$.*

## References


[1] P. K. Agarwal. Geometric partitioning and its applications. In J. E. Goodman, R. Pollack, and W. Steiger, editors, *Computational Geometry: Papers from the DIMACS special year.* Amer. Math. Soc., 1991.

[2] P.K. Agarwal and J. Matoušek. On range searching with semialgebraic sets. *Discrete Comput. Geom.* **11** (1994), 393–418.

[3] N.M. Amato, M.T. Goodrich, and E.A. Ramos. Parallel algorithms for higher-dimensional convex hulls. In *Proc. 35th Annu. IEEE Sympos. Found. Comput. Sci. (FOCS 94)*, 683–694, 1994.

[4] N. M. Amato, M. T. Goodrich and E. A. Ramos. Computing faces in segment and simplex arrangements. In *Proc. 26th Annual ACM Sympos. Theory Comput.*, 672–682, 1995.

[5] N. M. Amato and E. A. Ramos. On computing diagrams by divide-prune-and-conquer. In *Proc. 12th Annual ACM Sympos. Comput. Geom.*, 1996. 672–682, 1995.

[6] M.J. Atallah. Some dynamic computational geometry problems. In *Comps. and Maths. with Appls.* **11** (1985) 1171–1181.

[7] T.M. Chan. Output-sensitive results on convex hulls, extreme points, and related problems In *Proc. 11th Annu. ACM Sympos. Comput. Geom.*, 10-19, 1995.

[8] T.M. Chan, J. Snoeyink, and C.-K. Yap. Output sensitive construction of polytopes in four dimensions and clipped Voronoi diagrams in three. In *Proc. 6th Annu. ACM-SIAM Sympos. Discrete Algorithms*, 282-291, 1995.

[9] B. Chazelle. Cutting hyperplanes for divide-and-conquer. *Discrete Comput. Geom.*, **9** (1993) 145–158.

[10] B. Chazelle, H. Edelsbrunner, L. Guibas and M. Sharir, Diameter, width, closest line pair, and parametric searching, *Discrete Comput. Geom.* **10** (1993), 183–196.

[11] B. Chazelle and J. Matoušek. Derandomizing an output sensitive convex hull algorithm in three dimensions. Technical Report, Dept. of Computer Science, Princeton University, 1992.

[12] B. Chazelle and J. Matoušek. On linear-time deterministic algorithms for optimization problems in fixed dimension. In *Proc. 4th ACM-SIAM Sympos. Discrete Algorithms*, pages 281–290, 1993.

[13] K.-W. Chong and E.A. Ramos. Manuscript in preparation. 1997

[14] K. L. Clarkson and P. W. Shor. Applications of random sampling in computational geometry, II. *Discrete Comput. Geom.*, **4** (1989) 387–421.

[15] R. Cole. Slowing down sorting networks to obtain faster sorting algorithms, *J. Assoc. Comput. Mach.* **34** (1987), 200–208.

[16] R. Cole, M.T. Goodrich and C.O'Dunlaing. Merging free trees in parallel for efficient Voronoi diagram construction. In *Proc. 17th Internat. Conf. on Automata, Languages and Programming*, 1990

[17] D. P. Dobkin and D. G. Kirkpatrick. Fast detection of polyhedral intersection. *Theoret. Comput. Sci.* **27** (1983) 241–253.

[18] H. Edelsbrunner. *Algorithms in Combinatorial Geometry*, volume 10 of *EATCS Monographs on Theoretical Computer Science.* Springer-Verlag, Heidelberg, West Germany, 1987.

[19] H. Edelsbrunner and W. Shi. An $O(n\log^2 h)$ time algorithm for the three-dimensional convex hull problem. *SIAM J. Comput.* **20** (1991) 259-277.

[20] H. Edelsbrunner and E. Mücke. Simulation of Simplicity: a technique to cope with degenerate cases in geometric algorithms. *ACM Trans. Graphics* **9** (1990) 66–104.

[21] M.R. Ghouse and M.T.Goodrich. Fast randomized parallel methods for planar convex hull construction. In *Proc. 3rd ACM Symposium on Parallel Algorithms and Architectures*, 1991, 192–203.

[22] T. Goldberg and U. Zwick. Optimal deterministic approximate parallel prefix sums and their applications. In *Proc. 4th IEEE Israel Symp. on Theory of Computing and Systems*, pages 220–228, 1995.

[23] M.T. Goodrich. Using approximations algorithms to design parallel algorithms that may ignore processor allocation. In *Proc. 32th Annu. IEEE Sympos. Found. Comput. Sci. (FOCS 91)*, 711–722, 1991.

[24] M.T. Goodrich. Geometric partitioning made easier, even in parallel. In *Proc. 9th Annu. ACM Sympos. Comput. Geom.*, 73–82, 1993.

[25] M.T. Goodrich. Fixed-dimensional parallel linear programming via relative $\epsilon$-approximations. In *Proc. 7th ACM-SIAM Symposium on Discrete Algorithms (SODA)*, 1996, 132–141.

[26] M.T. Goodrich, C. Ó'Dúnlaing and C.-K. Yap. Constructing the Voronoi diagram of a set of line segments in parallel. *Algorithmica* **9** (1993) 128–141.

[27] M.T. Goodrich and E.A. Ramos. Bounded independence derandomization of geometric partitioning with applications to parallel fixed-dimensional linear programming. To appear in *Discrete and Computational Geometry.*

[28] N. Gupta and S. Sen. Faster output-sensitive parallel convex-hulls for $d \leq 3$: optimal sublogarithmic algorithms for small outputs. In *Proc. 12th Annu. ACM Sympos. Comput. Geom.*, 176–185, 1996.

[29] T. Hagerup and R. Raman. Waste makes haste: tight bounds for loose parallel sorting. In *Proc. 33th Annu. IEEE Sympos. Found. Comput. Sci. (FOCS 92)*, 628–637, 1992.

[30] A. Heppes, Beweis einer Vermutung von A. Vázsonyi, *Acta Math. Acad. Sci. Hungar.* **7** (1956), 463–466.

[31] D. G. Kirkpatrick and R. Seidel. The ultimate planar convex hull algorithm? *SIAM J. Comput.* **15** (1986) 287–299.

[32] R. Klein. *Concrete and Abstract Voronoi diagrams.* LCNS 400, Springer-Verlag, 1988.

[33] J. Matoušek. Approximations and optimal geometric divide-and-conquer. In *Proc. 23rd Annu. ACM Sympos. Theory Comput.*, 505–511, 1991. Also in *J. Comput. Syst. Sci.* **50**, 203–208 (1995).

[34] J. Matoušek. Cutting hyperplane arrangements. *Discrete Comput. Geom.* **6** (1991) 385–406.

[35] J. Matoušek. Efficient partiton trees *Discrete Comput. Geom.* **8** (1992) 315–334.

[36] J. Matoušek and O. Schwarzkopf. A deterministic algorithm for the three-dimensional diameter problem, *Proc. 25th ACM Symposium on the Theory of Computing,* 1993, 478–484. Revised version in: *Computational Geometry: Theory and Applications.* **6** (1996) 253–262.

[37] P. MacKenzie and Q. Stout. Ultrafast expected time parallel algorithms. In *Proc. 2nd ACM-SIAM Symposium on Discrete Algorithms (SODA)*, 1991, 414–423.

[38] N. Meggido. Applying parallel computation algorithms in the design of serial algorithms, *J. Assoc. Comput. Mach.* **30** (1983), 852–865.

[39] K. Mulmuley. *Computational Geometry: An Introduction Through Randomized Algorithms.* Prentice Hall, Englewood Cliffs, NJ, 1993.

[40] M. Pellegrini. On point location and motion planning among simplices. In *Proc. ACM Sympos. Theory Comput.*, 1994, 95–104.

[41] F. P. Preparata and M. I. Shamos. *Computational Geometry: An Introduction.* Springer-Verlag, New York, NY, 1985.

[42] S. Rajasekaran and S. Ramaswami. Optimal parallel randomized algorithms for the Voronoi diagram of line segments in the plane and related problems. In *Proc. 10th Annu. ACM Sympos. Comput. Geom.*, 57–66, 1994.

[43] E.A. Ramos. An algorithm for intersecting equal radius balls in $\mathbb{R}^3$. Technical Report UIUCDCS-R-94-1851, Dept. of Computer Science. University of Illinois at Urbana-Champaign, 1994. To appear in *Computational Geometry: Theory and Applications.*

[44] J.H. Reif and S. Sen. Optimal parallel randomized algorithms for three-dimensional convex hulls and related problems. *SIAM J. Comput.* **21** (1992) 466–485.

[45] M. Sharir and P.K. Agarwal. *Davenport-Schinzel Sequences and Their Geometric Applications.* Cambridge University Press, 1995.

# Minkowski Operations For Satellite Antenna Layout*

Jean-Daniel Boissonnat[†], Eelco de Lange[†‡] and Monique Teillaud[†]

[†] INRIA

BP 93, 06902 Sophia Antipolis CEDEX, France.

[‡] Matra Marconi Space

Dept. AOII, 31 Avenue des Cosmonautes, Z.I. Du Palays, 31402 Toulouse CEDEX, France.

{Jean-Daniel.Boissonnat, Eelco.de_Lange, Monique.Teillaud}@sophia.inria.fr

**Abstract:** Satellite layout is a very hard task because available space for the equipments is small and the physical constraints on the layout are strong. In this article, we show how some physical layout constraints can be modeled geometrically and how Minkowski operations can be used to place equipments, and in particular antennas. Since antennas are supposed to be placed onto a satellite wall, search space is two-dimensional. We discuss an algorithm that efficiently calculates only the (planar) part we need of the admissible space and describe how we implemented this.

*Keywords: satellite layout, Minkowski operations, degeneracies, numerical stability*

## 1 Introduction

The layout of a satellite is a very hard problem in practice. Because a satellite is tried to be constructed as small as possible, the layout of the equipments is extremely constrained. One problem that is reputedly difficult to solve is the layout of the antennas and observation instruments on the outer walls of a satellite.

For instance, many feasibility studies are done to verify whether the specified set of antennas and instruments can be fitted on the outer walls. All along the development phase, specifications often change so one wants to know as soon as possible if a possibly existing layout is still realizable (or if not, how it should be modified) when the size or type of an equipment is changed or when an equipment is added. Typically, manual tools supplying placeability information are the most useful for these types of problems but automatic placement facilities are useful also.

An antenna, in general, consists of a parabolic reflector, a source that either emits or receives a signal, and their supporting structures. It also has a field of vision (from now on written as FOV), that can be shaped more or less as a cone. A FOV must not be obscured (not even partly) by an obstacle, which means that this cone is not intersected. An antenna may be mobile, which means that it can rotate along one or two predefined axes. There also exist deployable antennas which are folded in at the moment of launch and which are deployed once the satellite has entered into orbit.

We assume that we start the layout from scratch, adding the equipments one by one. Given an antenna, we want to place it *onto* a specified exterior wall of the satellite. Only when this isn't possible, we try to place it further from the wall but we try to stay as close as possible, to minimize the size and weight of the structure needed to support it. In practice, first the placements onto the wall are considered, then the placements at a certain distance of the wall, and so on. As well as the distance from the wall, we can also discretize the rotation of the antenna when we want to take it into account. So, basically, at every stage we only consider translations of the antenna in a plane, and the search space is contained in $\mathbb{R}^2$.

First, we will define the antenna layout problem and model it geometrically by defining an antenna's admissible space that is contained in a plane. We show how we can use planar sections of Minkowski differences between polyhedra to calculate this admissible space. Then, we present our algorithm that directly calculates this section and discuss the handling of degeneracies and numerical precision problems. Finally, we show some experimental results, and we compare them with the experimental results obtained by an algorithm that calculates a Minkowski difference in $\mathbb{R}^3$ and then finds the interesting planar section.

*This research was partially supported by the ESPRIT IV LTR Project No. 21957 (CGAL).



*Computational Geometry 97* Nice France

## 2 Definitions and Properties

The objects we handle are contained in $I\!R^3$. The parts of the equipments and the environment are modeled as convex polyhedra. Polyhedra are written as capitals, while planes, polygonal and polyhedral regions are written using calligraphic letters.

Finding an optimal solution for layout problems is known to be NP-hard: placing polygons in a rectangle of minimal length is an NP-hard problem [DM95]. In practice, fortunately, we are satisfied with a reasonable solution. By placing the objects one by one using a heuristic, one should be able to find good solutions in a reasonable time. However, while it is rather easy to find a good heuristic for placements of large numbers of relatively simple objects, the antenna layout problem consists of a small number of highly constrained objects and an efficient heuristic seems to be much harder to find.

We call $\mathcal{O}$ the set of *obstacles*, consisting of the satellite structure plus all equipments that have been placed on it at a certain moment. $\mathcal{P}$ is the equipment we are currently placing, and the set of all allowed translations that place $\mathcal{P}$ so that it does not collide with $\mathcal{O}$ (the *admissible space*) is written as $\mathcal{A}(\mathcal{O},\mathcal{P})$. The unions of all physical parts and of all FOVs of an equipment or obstacle $\mathcal{S}$ are denoted as $\mathcal{S}_P$ and $\mathcal{S}_F$ respectively.

$P^C = \{p \in I\!R^3 \,|\, p \notin P\}$ is the complement of $P$. The symmetric of $P$ is $\ominus P = \{-p \,|\, p \in P\}$.

The Minkowski sum between two polyhedra $A$ and $B$ is notated as $A \oplus B$. It is defined as $A \oplus B = \{a+b | a \in A, b \in B\}$. Note that the Minkowski sum is distributive and commutative over the union, so $(A_1 \cup A_2) \oplus B = B \oplus (A_1 \cup A_2) = (A_1 \oplus B) \cup (A_2 \oplus B)$.

## 3 Geometric Modelization

We shall describe how this problem can be modeled geometrically and we define the Gaussian diagram that allows us to calculate the admissible space efficiently.

### 3.1 Admissible Space

Given two convex polyhedra $O$ and $P$, translating $P$ by a vector $\vec{t}$ will make it intersect $O$ when there exist two points $o \in O$ and $p \in P$ such that $o = p+\vec{t}$, so, $\vec{t} = o-p$. Consequently, the set of all translations for $P$ such that it intersects $O$ is defined as $\{o-p | o \in O, p \in P\}$, which is exactly the Minkowski difference $O \ominus P$. Since $O$ and $P$ are convex, $O \ominus P$ also is convex. The boundary of $O \ominus P$ represents all translations that bring $P$ into contact with $O$ without intersecting its interior, while its complement represents all "free" translations for $P$.

Consequently, if we want to place $P$ without colliding with $O$, we must pick a translation $\vec{t} \in (O \ominus P)^C$. Since we want to allow contacts, we consider the boundary of the Minkowski difference to be part of the admissible space. If we want to place $\mathcal{P} = \{P \in \mathcal{P}\}$ among a set $\mathcal{O} = \{O \in \mathcal{O}\}$ of obstacles, we need to calculate $(\mathcal{O} \ominus \mathcal{P})^C$ where $\mathcal{O} \ominus \mathcal{P}$ is defined as $(\cup_{O \in \mathcal{O}} O) \ominus (\cup_{P \in \mathcal{P}} P)$ (see figure 1). Using the the distributivity of the Minkowski sum, this can be rewritten as $\cup_{O \in \mathcal{O}, P \in \mathcal{P}} O \ominus P$.

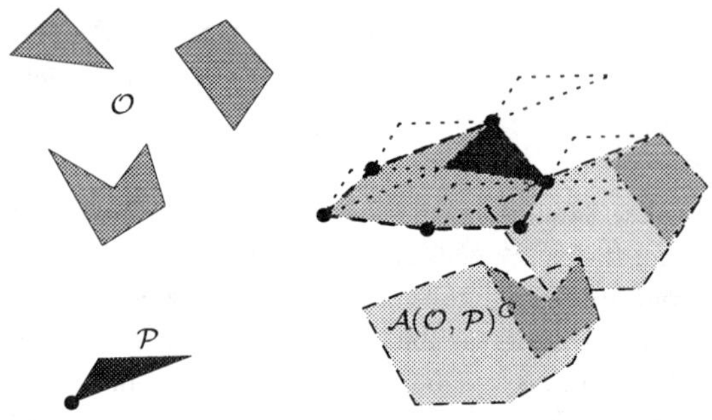


Figure 1: The admissible space for polygonal region $\mathcal{P} = \{P\}$ with respect to the polygonal region $\mathcal{O}$ is the complement of the polygonal region drawn in light gray.

This handles the physical parts of the equipment. FOVs can almost be treated as physical parts, because they may not intersect with "real" parts; the only difference is that two FOVs may freely intersect. Taking only in account parts that interfere, we can notate this as follows:

$$\mathcal{A}(\mathcal{O},\mathcal{P}) = (\ (\mathcal{O} \ominus \mathcal{P}_P) \cup (\mathcal{O}_P \ominus \mathcal{P}_F)\ )^C \tag{1}$$

We can rewrite this as the complementary of the union of $O \ominus P$ for all pairs $(O, P)$ such that $O$ is any element of $\mathcal{O}$ and $P \in \mathcal{P}_P$ or $O \in \mathcal{O}_P$ and $P \in \mathcal{P}_F$.

All the equipments and obstacles are decomposed as unions of convex polyhedra, so that in the sequel, $P$ and $O$ will always represent convex polyhedra.

We want to place the equipment onto a wall. Let us assume without loss of generality that this wall is horizontal and define a coordinate system $(O, X, Y, Z)$ where the $Z$-axis is perpendicular to the wall and the $X$- and $Y$- axes as well as $O$ are contained in it. Since we want to position the equipment $\mathcal{P}$ at a given height $z$, we are only interested in the translations $\vec{t} \in \mathcal{A}(\mathcal{O},\mathcal{P})$ that position $\mathcal{P}$ at this height. All these translations are contained in a horizontal plane $\mathcal{Z}$ defined by $Z = z$ in the admissible space so we are actually looking for the intersection of $\mathcal{A}(\mathcal{O},\mathcal{P})$ with $\mathcal{Z}$.

### 3.2 Gaussian diagrams of Convex Polyhedra

We shall define a mapping of a convex polyhedron $P$ onto the unit sphere of $I\!R^3$ denoted $S^2$ that is called the normal or Gaussian diagram (see [LP83, GS87, AB89, AB87]), and which can be used to calculate a Minkowski sum and bound its complexity.

For each facet $f \in P$, let $n_f$ denote its normal vector. The Gaussian diagram $P'$ on the unit sphere $S^2$ consists of the following elements (we intentionally use another terminology to avoid confusion between polyhedral elements and elements in the Gaussian diagram):

- Its *nodes* are the vectors $n_f$.
- Its *arcs* are great circular arcs, each being the locus of the normal vectors of all planes supporting $P$ at one of its edges.
- Its *cells* are regions, each being the locus of the normal vectors of all planes supporting $P$ at one of its vertices.

So, there is a one-to-one correspondence between all elements of $P$ and those of $P'$ (see figure 2) if $P$ has no coplanar facets.

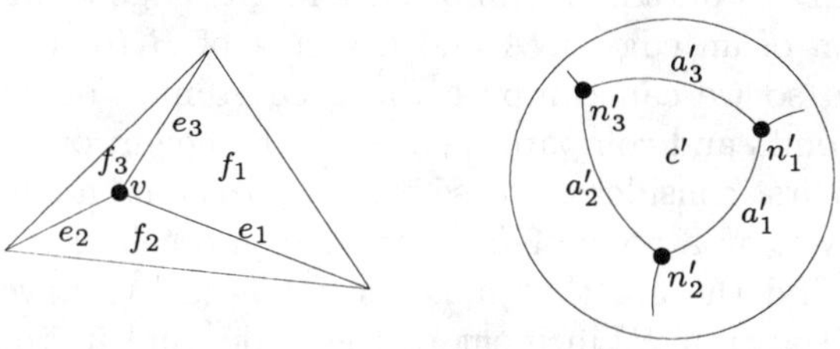


Figure 2: A convex polyhedron and its Gaussian diagram.

When a polyhedron has coplanar facets, the nodes associated with such coplanar adjacent facets coincide and the arcs and cells between such nodes reduce to points. Instead of trying to handle this type of degeneracy in the algorithm, we can work around it by eliminating coplanar adjacent facets in a preprocessing step. This can be achieved by merging possible coplanar facets in the input polyhedra.

So, from now on, we will assume that a convex polyhedron has no coplanar facets.

### 3.3 Gaussian Diagrams and Minkowski Sums

Let $A$ and $B$ be two convex polyhedra in general position and $M'$ be the superposition of their Gaussian diagrams. It is known that $M'$ is the Gaussian diagram of $M = A \oplus B$ [LP83, GS87]. Any facet of $M$ corresponds either to a node coming from $A'$ or $B'$ or to an intersection of two arcs of $A'$ and $B'$ (see figure 3).

If we assume that $A$ and $B$ are in general position, we distinguish three types of nodes of $M'$ associated with facets of $M$, all uniquely defined by an element of $A'$ and an element of $B'$:

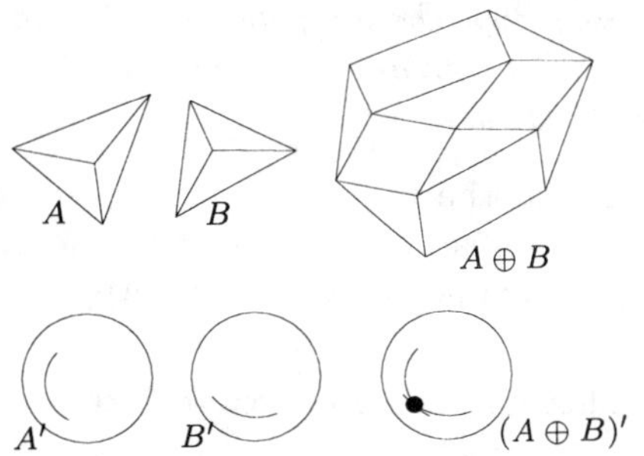


Figure 3: The intersection of two arcs of the Gaussian diagrams corresponds with an *aa*-facet of the Minkowski sum.

- Node $n_{A'} \in A'$ is contained in cell $c_{B'} \in B'$. We call this a *nc*-node.
- Symmetrically, a node $n_{B'} \in B'$ that is contained in a cell $c_{A'} \in A'$ is called a *cn*-node.
- An intersection of an arc $a_{A'} \in A'$ with an arc $a_{B'} \in B'$ is an *aa*-node.

When allowing degeneracies, we can encounter three other cases:

- Node $n_{A'} \in A'$ is contained in arc $a_{B'} \in B'$. We call this a *na*-node.
- Symmetrically, a node $n_{B'} \in B'$ that is contained in arc $a_{A'} \in A'$ defines a *an*-node.
- When two nodes $n_{A'} \in A'$ and $n_{B'} \in B'$ coincide, they define a *nn*-node.
- Arcs $a_{A'} \in A'$ and $a_{B'} \in B'$ overlap.

Knowing the elements of $A$ and $B$ that are associated with the two elements defining $n_{M'} \in M'$, we can construct the associated facet $f_M \in M$: it is the Minkowski sum of the two elements of $A$ and $B$.

### 3.4 Worst-case complexity

We will bound the worst-case size of the intersection of a Minkowski sum with a plane.

It is known that, given two convex polyhedra $A$ and $B$ of sizes $m$ and $n$ respectively, the size of their Minkowski sum $A \oplus B$ is $\Theta(mn)$ in the worst case. This result is easily verified by considering the superposition of the Gaussian diagrams of the two polyhedra: arcs of $A'$ and $B'$ can intersect $\Theta(mn)$ times, which means that as many *aa*-facets exist in the Minkowski sum. A new theorem, more relevant for our application, shows that, in the worst case, a section of $A \oplus B$ has the same worst-case size as $A \oplus B$ itself.

**Theorem 3.1** *Given two convex polyhedra $A$ and $B$, the intersection of $A \oplus B$ with any plane has a complexity of $\Theta(mn)$ in the worst case.*

**Proof** It is clear that the upper bound is quadratic, so we only have to show an example which is actually quadratic. Such an example is given in the Appendix.

## 4 A Direct Method to Compute a Minkowski Slice

While the boundary of the polyhedron $O \ominus P$ represents all translations that bring $P$ into contact with $O$ without intersecting it, the boundary of the polygon slice $(O \ominus P) \cap \mathcal{Z}$ is the subset of all such translations that position $P$ at a given distance from the wall. It corresponds with the displacement by translations that $P$ makes around $O$ when it keeps the same height and stays in contact with $O$.

The sequence of edges of $(O \ominus P) \cap \mathcal{Z}$ is given by the sequence of facets of $O \ominus P$ intersected by $\mathcal{Z}$. As $O \ominus P$ is convex ($P$ and $O$ are convex) if we project the normal vectors to this sequence of faces onto the horizontal plane, we notice that the sequence of polar angles of the projected vectors is monotonic. In fact, $(O \ominus P) \cap \mathcal{Z}$ is represented on $(O \ominus P)'$ as a path in the superposition of the Gaussian diagrams of $O$ and $\ominus P$. This path is monotonic in $\theta$, if $\theta$ denotes the polar angle of the projected points of the sphere onto the horizontal plane.

This expresses only the fact that, since $P$ and $O$ are convex, when $P$ is moving by translation at a fixed height while keeping a contact with an obstacle $O$, the contact point follows a path on $O$ describing a continuous sequence of adjacent facets, edges and vertices touched by $P$. If we project the normal vectors to this sequence of faces of $O$ onto the horizontal plane, the sequence of polar angles of the projected vectors is monotonic. The same holds on $P$.

### 4.1 Description of the Algorithm

We present here an algorithm that directly calculates the intersection of the Minkowski sum of two convex polyhedra $A$ and $B$ with a given horizontal plane $\mathcal{Z}$ by calculating the monotonic walk we just introduced, through the superposition of their Gaussian diagrams. We denote $M = A \oplus B$. It is assumed that $\mathcal{Z}$ intersects the interior of the Minkowski sum, because the output polygon is empty otherwise. We assume for the moment that $\mathcal{Z}$ does not contain any vertices of $M$ and that no degeneracies occur in the superposition of the Gaussian diagrams of $A$ and $B$; the degeneracies will be examined in section 4.2.

There exists a path through $M'$ (being the superposition of $A'$ and $B'$) that visits the arcs and nodes associated with all edges and facets of $M$ that intersect $\mathcal{Z}$ (and *only* those elements). We don't want to calculate the entire superposition and only construct the arcs and nodes of $M'$ we encounter along the way. We denote by $K_s$ the complexity of the output $(A \oplus B) \cap \mathcal{Z}$.

#### 4.1.1 How to get from a node to an arc

Assume that our walk has arrived at a certain node $n_{M'} \in M'$, associated with facet $f_M \in M$. Because $f_M$ cuts $\mathcal{Z}$ and it is convex, exactly two edges of $f_M$ cut $\mathcal{Z}$, one by whose associated arc we arrived at $n_{M'}$ and one by whose associated arc we will advance to the next node of the Gaussian diagram. We must determine which arc $a_{M'} \in M'$ incident to $n_{M'}$ to follow in order to get to the next node of $M'$ (it is either an arc from $A'$ or from $B'$) and we must also determine which cell it lies in: it is an arc of $A'$ that is contained in a cell of $B'$ or vice versa.

Note however that $M'$ does not contain the information needed to determine if an edge corresponding with an arc from $M'$ cuts $\mathcal{Z}$ or not, so we must look back at $A$ and $B$ to obtain this information: the edge of $M$ is the sum of an edge of $A$ and a vertex of $B$ (or the contrary), so we can compute the $z$ coordinate of its two endpoints and compare them with the height of $\mathcal{Z}$.

Let us first consider the case when $n_{M'}$ is a *cn*-node. $n_{M'}$ is given by a node of $B'$ contained in a cell of $A'$. Then to find the arc that must be followed, we have to test all arcs of $B'$ incident to the node, and it can take up to $O(n)$ time to identify the arc that must be walked. The case of a *nc*-node is symmetric and takes $O(m)$ time. We might visit all *cn*- and *nc*-nodes, but since arcs are incident to two nodes and the walk is monotonic, any arc from $A'$ or $B'$ is at most considered twice on the visited *cn*- and *nc*-nodes during the whole walk.

The case of *aa*-nodes is easier to analyze, since *aa*-nodes are incident to four arcs, and we visit at most $O(K_s)$ such nodes.

Consequently, $O(m + n + K_s)$ time is spent in total determining which arc to walk next.

#### 4.1.2 How to get from an arc to a node

Without loss of generality, suppose that we walk along arc $a_{M'}$ supported by $a_{A'} \in A'$, through cell $c_{B'} \in B'$. If $a_{A'}$ intersects an arc $a_{B'}$ incident to $c_{B'}$, the next node $n_{M'}$ is an *aa*-node, defined by $a_{A'}$ and $a_{B'}$. Otherwise, we encounter a *nc*-node $n_{M'}$, defined by $c_{B'}$ and the endpoint of $a_{A'}$ in the direction of the walk (see figure 4).

Notice that the boundary of $c_{B'}$ can be decomposed in two subsequences of arcs, both being $\theta$-monotonic as our path. Consequently, while following arc $a_{A'}$, we can follow at the same time the two sequences of arcs of the boundary of $c_{B'}$ that are respectively above and below

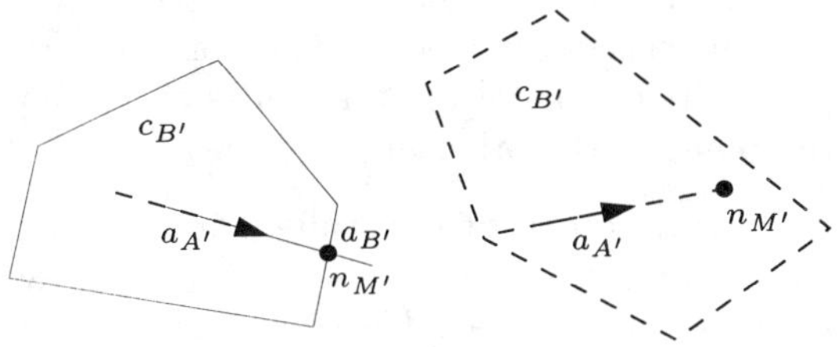


Figure 4: Identifying the next node $n_{M'} \in M'$. Arc $a_{A'}$ is walked through cell $c_{B'}$. (left) an $aa$-node. (right) a $nc$-node.

$a_{A'}$, until we reach either an intersection between $a_{A'}$ and an arc $a_{B'}$ or the endpoint of $a_{A'}$.

In the case we find an endpoint $n_{A'}$, we will follow a new arc of $A'$ and at the same time keep on traversing, starting from the current arcs, the ordered subsequences of arcs of $c_{B'}$. See figure 5(a).

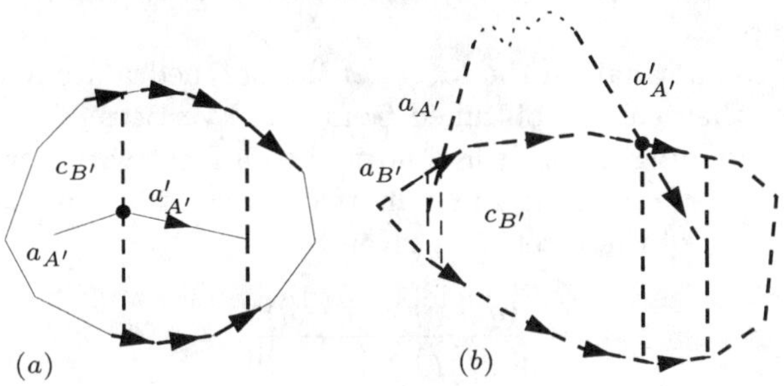


Figure 5: Following the boundary of $c_{B'}$ (a) after an endpoint of $a_{A'}$ (b) after an intersection between $a_{A'}$ and $a_{B'}$

In the case we reach an intersection between $a_{A'}$ and $a_{B'}$ we either now follow $a_{B'}$ through a cell $c'_{A'}$ and go on symmetrically. Or we keep on following $a_{A'}$ and we leave cell $c_{B'}$. Then we only have to remember the arcs of the two subsequences of the boundary of $c_{B'}$ at that point. Indeed, if the path re-enters the same cell $c_{B'}$ later with another arc $a'_{A'}$, since the path is monotonic, we will examine the two subsequences, starting from these remembered arcs, until we find the two arcs just above and below $a'_{A'}$, and then we go on as for $a_{A'}$. See figure 5(b).

$O(K_s)$ arcs of $M'$ are walked on the path, $O(K_s)$ nodes are computed. By following the boundaries of the cells of $A'$ and $B'$ crossed by the path, an arc of $A'$ or $B'$ can be examined once for each node of $M'$ that ends in one of the two cells incident to it, or intersects it, in which case the cost belongs to the cost of constructing node of $M'$; an arc of $A'$ or $B'$ can in addition be examined at most twice to locate the entry point of the path into each of its two incident cells. So in total, $O(m + n + K_s)$ time is spent.

#### 4.1.3 Finding an initial arc

The idea for starting the path is to find a first arc whose associated edge of $M$ is cut by $\mathcal{Z}$, belonging to a *shadow* of $M$. The shadow of $M$ is the boundary of the orthogonal projection of $M$ onto a plane. We will consider a vertical plane: this will ensure that the shadow of $M$ is cut by $\mathcal{Z}$. On $M'$, the chosen shadow of $M$ is associated with the sequence of arcs, cells and nodes of $M'$ cut by a given vertical plane containing the center of the sphere (the vertical direction on the sphere being given by the vector normal to the horizontal plane $\mathcal{Z}$).

Let us consider the great circle of the sphere contained in this vertical plane. Starting with the cell of $A'$ associated with the highest vertex of $A$, we can find the arcs of $A'$ intersected by the great circle by following the arcs incident to all the cells cut by this circle (see figure 6(a)). In this way, we directly obtain the ordered sequence of the arcs of $A'$ cut by the great circle. The same process is applied to $B$. Both arc-lists are associated to a sequence of connected edges from $A$ and $B$ respectively describing the shadow of $M$ from the direction normal to the great circle. The two lists are traversed at the same time, in such a way that the order in which the arcs cut the great circle is preserved, so we get an ordered list of arcs of $M'$ associated to the shadow of $M$ in $O(m + n)$ time.

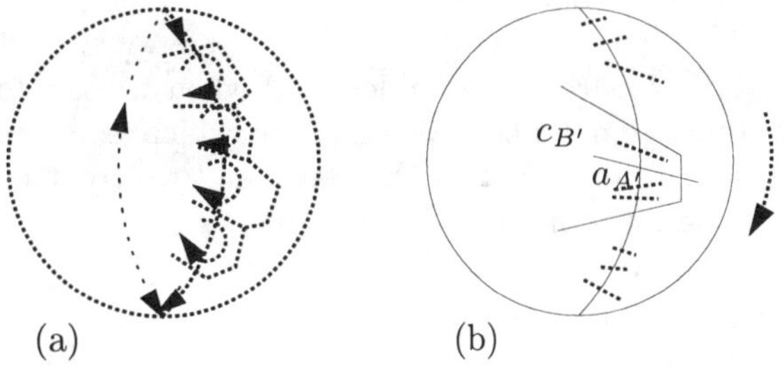


Figure 6: Finding a node $n_{M'}$ to start the walk with. (a) Constructing a sorted shadow-list in linear time. (b) Determining an arc whose associated edge cuts $\mathcal{Z}$.

In fact we can stop as soon as we encounter an arc corresponding in $M$ with an edge intersected by plane $\mathcal{Z}$: without loss of generality, assume that this arc is an arc $a_{A'}$ belonging to $A'$, associated with edge $[p, q]$ of $A$ (see figure 6(b)). It is contained in a cell of $B'$ associated with a vertex $v$. The edge $[p, q] \oplus v$ intersects $\mathcal{Z}$ if and only if its endpoints $p + v$ and $q + v$ lie on both sides of $\mathcal{Z}$.

#### 4.1.4 Global Complexity

Summing up, a total time of $O(m+n+K_s)$ is needed to run the algorithm, where $K_s$ is the size of the output polygon. This is better than the convolution algorithm because $K_s$ is at most equal to $K$.

### 4.2 Handling Degeneracies

We choose to properly handle the degeneracies. Remember that we decided to pre-process the input polyhedra to remove adjacent coplanar facets (Section 3.2).

As said in Section 3.3, the following degeneracies can be encountered while constructing $A \oplus B$:

1. a node of $A'$ might be contained in an arc or coincide with another node from $B'$ or inversely, so we have to be able to handle *na*-, *an*- and *nn*-nodes.

2. two arcs overlap: we have to make sure that if we walk one arc, we walk the other one at the same time.

The detection of the degeneracies occurs when getting from an arc to a node (Section 4.1.2).

In fact, the test whether two arcs intersect also checks if the intersection occurs at an endpoint of one (or both) arcs. When a *na*- or a *an*-node is detected in this way, we only have to add the arc in the ordered list of arcs incident to the node, and find the arc whose associated edge is cut by $\mathcal{Z}$, which will be followed next. When a *nn*-node is detected, we traverse the two lists of arcs incident to the nodes and walk the arc whose edge is cut by $\mathcal{Z}$.

Overlapping arcs are also detected when we test for their intersection. Then we compute which endpoint will be encountered first in the walk, and we treat the obtained *na*-, *an*- or *nn*-node as above.

### 4.3 Numerical Stability

Another problem often encountered when implementing geometric algorithms is the lack of numerical precision in geometric tests. This causes wrong decisions to be made leading to inconsistencies in the state data of the algorithm which causes it to malfunction.

Problems are encountered for example when testing for intersection between arcs when an endpoint of one of the arcs is nearly contained in the supporting plane of the other arc. So in fact, numerical problems occur in nearly degenerate cases.

The basic test for our algorithm is *which_side*, that tests on which side a node lies with respect to the supporting plane of an arc. All tests used in the direct algorithm can be written using one or more such tests:

- determining whether an arc contains a node: does the node lie on the plane that contains the arc, and does it lie between its two endpoints?

- testing for intersection between two (great circular) arcs: it must be verified whether for both arcs, the two endpoints lie at opposite sides of the supporting plane of the other arc.

- determining whether two (parallel) arcs overlap: do both endpoints of one arc lie on the plane defined by the other arc and is any endpoint contained in the other arc?

- determining the order between arcs from different Gaussian diagrams that are incident to a *nn*-node.

In fact, the *which_side* test can be written as the determinant of a matrix of size three. More precisely, let $n$ and $a$ be respectively the node and the arc to test. $n_1$ and $n_2$ will denote the endpoints of $a$.

$$\mathit{which\_side}(n, n_1, n_2) = sgn \begin{vmatrix} x & x_1 & x_2 \\ y & y_1 & y_2 \\ z & z_1 & z_2 \end{vmatrix}$$

where $(x, y, z)$ (resp $(x_i, y_i, z_i)$) are the coordinates of $n$ (resp $n_i$).

If the normals of the facets of the polyhedra are not given, they can be obtained from three vertices of the corresponding facet: if $n$ is normal to a facet with vertices $p, q, r$ appearing in counterclockwise order on the boundary of the facet, $n$ is given by

$$n = \frac{(q-p)\times(r-q)}{||(q-p)\times(r-q)||}.$$

However, by calculating this, we would introduce an error. So instead of calculating $n$, we represent it symbolically using $p, q, r$.

Note that, for the computation of the sign of the determinant, it is useless to normalize $n$. $n$ is proportional to

$$\left( \begin{vmatrix} y_q - y_p & z_q - z_p \\ y_r - y_q & z_r - z_q \end{vmatrix}, \begin{vmatrix} x_r - x_q & z_r - z_q \\ x_q - x_p & z_q - z_p \end{vmatrix}, \begin{vmatrix} x_q - x_p & y_q - y_p \\ x_r - x_q & y_r - y_q \end{vmatrix} \right)$$

So *which_side*$(n, n_1, n_2)$ is given by the sign of a determinant of a $3 \times 3$ matrix whose entries are $2 \times 2$ determinants.

*which_side*$(n, n_1, n_2)$ answers whether a node $n$ is lying on the supporting plane of an arc $n_1 n_2$. Then, to know whether $n$ is lying on the arc, the test reduces in fact to comparing the signs of the $3 \times 3$ determinants of the vectors $(n, n_1, q-p)$ and $(n, n_2, q-p)$, where $[p, q]$ is the edge common to the facets normal to $n_1$ and $n_2$. These determinants yield less restriction on the number of bits that can be allowed than the *which_side* test.

We want to make the direct method numerically stable, so, to get rid of the numerical problems we encountered, we decided to implement those tests using exact arithmetics. The determinant method described by Avnaim et al. in [ABD+97] can calculate the determinant sign of matrices of size three without error. If the coordinates of the vertices of the input polyhedra are represented by $b$ bit integers, the $2 \times 2$ determinants appearing in the *which_ side* test are represented by at most $2b+1$ bit integers. Then the exact determination of the sign of the $3 \times 3$ determinant will need $(2b+1)+1$ bits (in practice the implementation requires $(2b+1)+2$ bits). A standard IEEE double allows exact representation of 53-bit integers, so we can use 25-bit integers to represent our data, which is enough in practice.

## 5 Experimental Results

To compute the admissible space, we calculate all sections of Minkowski sums for pairs $(O, P)$ of polyhedra described in 3.1. Then we calculate the union of the resulting set of polygons, the complement of which is the wanted admissible space.

The implementation of our program allows the equipments to be placed one by one on a horizontal plane $\mathcal{Z}$. The user can request the equipment to be moved upwards or downwards, in this case $\mathcal{Z}$ is modified and the corresponding horizontal section of the three dimensional admissible space will be calculated again. Rotations can also be applied, but in this case the whole admissible space is modified and must be recomputed, not only its section by $\mathcal{Z}$.

We have tested the program on a realistic although fictional model of a satellite called PRISMESAT and present here the series of placements as well as the time it took to determine the Minkowski sums at each stage. The time it takes to calculate the union of the polygons is negligible and is not taken into account. The tests have been performed on a SUN SPARCstation20 running SunOS.

Figures 8, 9 give the sequence of placements performed. Table 1 gives the performances of the algorithm for all stages of the layout scenario. Time is measured in milliseconds CPU-time, and both total and average times are given, the average being calculated on all Minkowski sums computed.

We compare here the practical performances of our algorithm to compute the section of a Minkowski sum, with the performances of the algorithm proposed by Guibas and Seidel [GS87]. The last algorithm uses a topological sweep-line algorithm to determine all intersections of the superposed Gaussian diagrams of two convex polyhedra, constructing the Minkowski sum along the way. This is done in $O(m+n+K)$ time, where $m$ and $n$ are the sizes of the input polyhedra and $K$ the size of the resulting Minkowski sum. Then the time to slice the Minkowski sums by plane $\mathcal{Z}$ is added to the total time.

As can be seen, even though the direct method uses the exact evaluation of signs of determinants which is more costly than standard arithmetic, and is therefore much more robust than the convolution method, it is a lot faster.

In fact we also implemented the method that consists, for computing the Minkowski sum, in computing all vertices $v_A + v_B$ for all $mn$ pairs of vertices $v_A \in A, v_B \in B$ and then calculating their convex hull [AB89] with the Quick Hull algorithm [BDH93]. The performances were even poorer.

| Equipment | $N_p$ | $N_v$ | $N_m$ | $K$ | $N_s$ | $K_s$ |
|---|---|---|---|---|---|---|
| INIT | 54 | 837 | - | - | - | - |
| ESRI-1 | 3 | 32 | 162 | 30.8 | 44 | 11.0 |
| ESRI-2 | 3 | 32 | 171 | 30.3 | 46 | 10.8 |
| HORN-7-HST | 7 | 188 | 420 | 51.3 | 146 | 11.9 |
| SVS-ANTENNA | 10 | 727 | 670 | 144.5 | 257 | 22.3 |
| VOGO | 10 | 296 | 770 | 91.5 | 161 | 17.7 |
| TVGAVO-ENS | 4 | 70 | 348 | 62.3 | 107 | 19.3 |
| *Total* | *91* | *2082* | *2541* | *86.8* | *761* | *17.6* |

| | Convolution | | Direct Slice | |
|---|---|---|---|---|
| | *Average* | *Total* | *Average* | *Total* |
| WALL | - | - | - | - |
| ESRI-1 | 113.2 | 18339.1 | 26.6 | 4316.5 |
| ESRI-2 | 121.5 | 20772.3 | 25.7 | 4399.8 |
| HORN-7-HST | 161.3 | 67738.7 | 39.5 | 16599.3 |
| SVS-ANTENNA | 648.3 | 434371.9 | 85.8 | 57514.4 |
| VOGO | 387.5 | 298412.4 | 63.0 | 48481.4 |
| TVGAVO-ENS | 250.9 | 87318.2 | 54.8 | 19065.9 |
| *Total* | *364.8* | *926952.6* | *59.2* | *150377.3* |

Table 1: Performances of the used Minkowski algorithms in milliseconds CPU-time. $N_p$ is the number of polyhedra in the equipment and $N_v$ is the total number of vertices in the equipment. $N_m$ is the number of Minkowski sums calculated, $N_s$ the number of Minkowski sums that cut $\mathcal{Z}$, $K$ the average number of vertices of the Minkowski sums, and $K_s$ the average number of vertices of the (nonempty) sections with $\mathcal{Z}$.

## 6 Conclusions

We have proposed an optimal output-sensitive algorithm to compute a section of a three dimensional Minkowski sum by a plane without computing the whole sum. Degeneracies and numerical problems were solved.

The algorithm was used to compute the admissible space for the placement of antennas on a satellite. The implementation was compared to two methods that calculate the entire Minkowski sum on realistic satellite models and was shown to be very efficient in practice.

We are thinking about extending the functionalities of our program and will discuss here briefly some useful and interesting possibilities.

- Automatic layout. We were able to find a set of solutions using a basic branch-and-bound scheme for our example satellite. When a new equipment $\mathcal{P}$ is considered, the slices of the Minkowski sums of $\mathcal{P}$ with the previously considered equipments computed and stored. If the placement of $\mathcal{P}$ is changed the slices will only translate along with it, so they needn't be recalculated. Only the union must be computed. As a consequence, the admissible space can be constructed very efficiently for any configuration.

We plan to study heuristics to directly find solutions corresponding to the needs of the design office.

- A constraint that is more difficult to model is non-visibility. Instruments exist that serve to dissipate the heat generated by other equipments. When more than one such equipment is used then they may not "see" each other. The problem is that this constraint is not monotonic: placing an obstacle between two dissipators makes a previously inadmissible placement admissible again. This might make this constraint difficult to handle when we want to apply some automatic strategy. Another criterion to take into account could be the optimization of the inertia center of the satellite.

- One can also think about a post-processing step, that tries to compact the layout after all equipments have been placed, like is done for marker making in [LM93], for example.

## 7 Acknowledgments

The authors would like to thank Hervé Brönnimann for his example of a section of a Minkowski sum with quadratic complexity. Furthermore, thanks go to André Gasquet for his help on the specification of the antenna layout problem and on the implementation of the program. Thanks go also to our colleagues of the design office for answering many technical questions, supplying satellite models and testing our program.

## References

[AB87] F. Avnaim and J.-D. Boissonnat. Simultaneous containment of several polygons. In *Proc. 3rd Annu. ACM Sympos. Comput. Geom.*, pages 242–250, 1987.

[AB89] F. Avnaim and J.-D. Boissonnat. Polygon placement under translation and rotation. *RAIRO Inform. Theor.*, 23:5–28, 1989.

[ABD+97] F. Avnaim, J-D. Boissonnat, O. Devillers, F. Preparata, and M. Yvinec. Evaluating signs of determinants using single-precision arithmetic. *Algorithmica*, 17:111–132, 1997.

[BDH93] C. B. Barber, D. P. Dobkin, and H. Huhdanpaa. The Quickhull algorithm for convex hull. Technical Report GCG53, Geometry Center, Univ. of Minnesota, July 1993.

[DM95] K. Daniels and V. Milenkovic. Multiple translational containment: Approximate and exact algorithms. In *Proc. 6th ACM-SIAM Sympos. Discrete Algorithms*, pages 205–214, 1995.

[GS87] L. J. Guibas and R. Seidel. Computing convolutions by reciprocal search. *Discrete Comput. Geom.*, 2:175–193, 1987.

[LM93] Z. Li and V. Milenkovic. A compaction algorithm for non-convex polygons and its application. In *Proc. 9th Annu. ACM Sympos. Comput. Geom.*, pages 153–162, 1993.

[LP83] T. Lozano-Pérez. Spatial planning: A configuration space approach. *IEEE Trans. Comput.*, C-32:108–120, 1983.

## Appendix: a quadratic example

We construct a polyhedron $A$ by intersecting two open polyhedral vertical coaxial cones such that a pattern of $m$ zigzag edges is formed at the place where their boundaries intersect. $B$ is formed by taking the convex hull of a very small horizontal segment and of $n$ points on a small arc of a vertical circle centered at the midpoint of the segment. Then $A \oplus B$ is quadratic (see the Gaussian diagrams on figure 7(b)).

Let us denote $u$ (resp $v$) the upper (resp lower) vertex of $B$, and take the midpoint of the horizontal axis of $B$ as the origin on $B$. If any edge $[p,q]$ of the zigzag pattern on $A$ is sufficiently high with respect to $B$, then point $p+v$ of $A \oplus B$ will be above point $q+u$. So all the points of $[p,q] \oplus w$ defined as the sum $p+w$ of the upper endpoint $p$ of $[p,q]$ and a vertex $w$ of $B$ (except the two vertices of the axe of $B$) are above all the points defined as the sum $q+w$ of the lower endpoint $q$ of and a vertex $v$ of $B$. Which implies that an horizontal plane $\mathcal{Z}$ passing between $p+v$ and $q+u$ will intersect all the $O(mn)$ edges of the form $[p,q] \oplus w$ of $A \oplus B$.

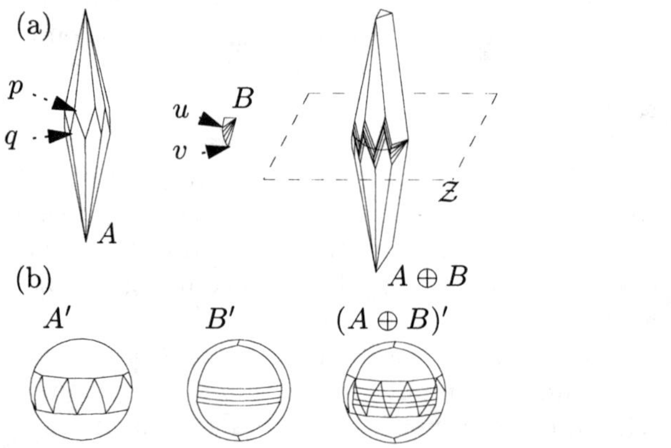


Figure 7: An example of a planar cross-section of a Minkowski sum with quadratic complexity.

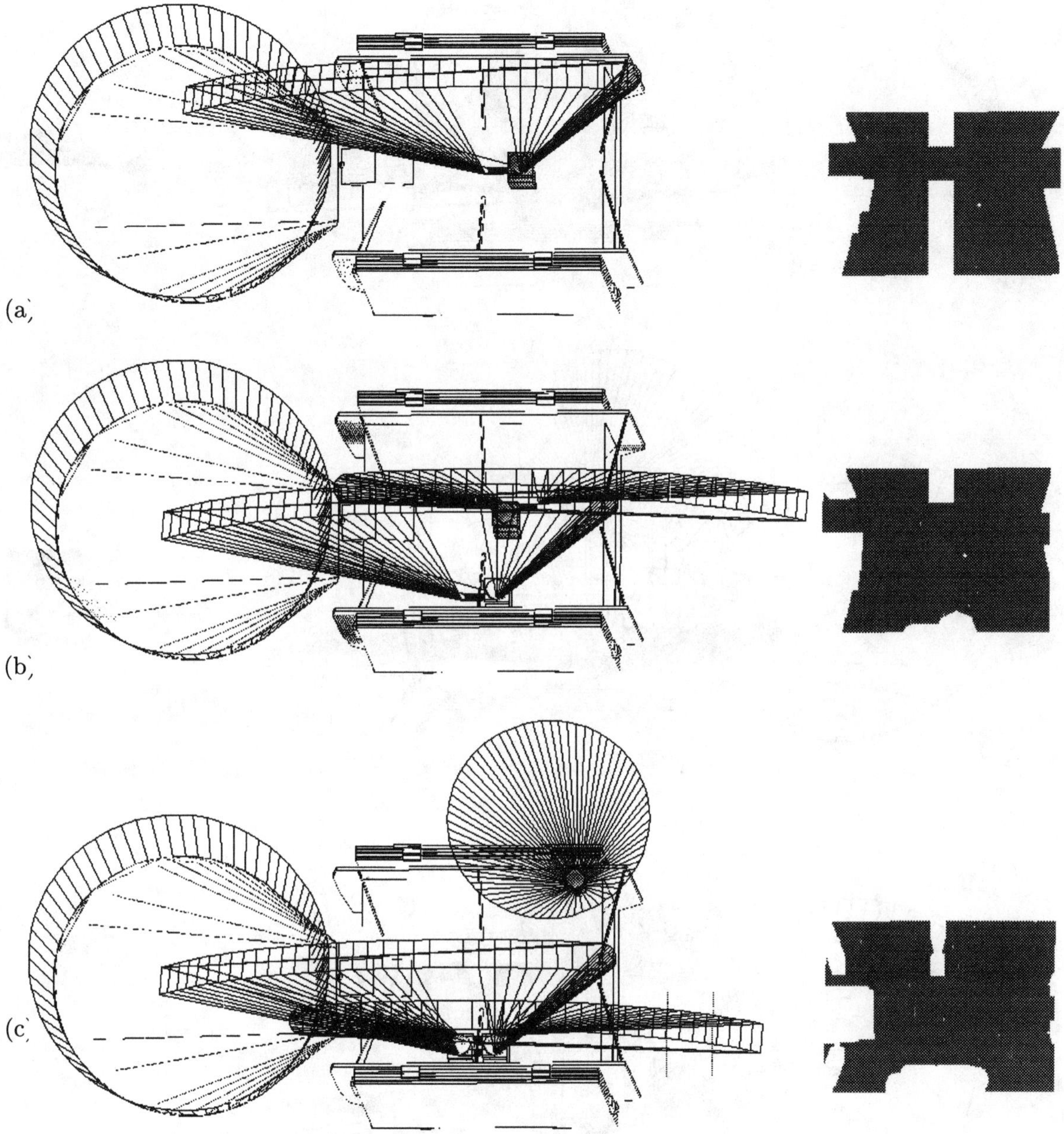


Figure 8: The sequence of placements performed for the imaginary satellite PRISMESAT: the equipment that is being placed is drawn in grey, the admissible space is drawn at the right side. (a) ESRI-1 (b) ESRI-2 (c) HORN-7-HST.

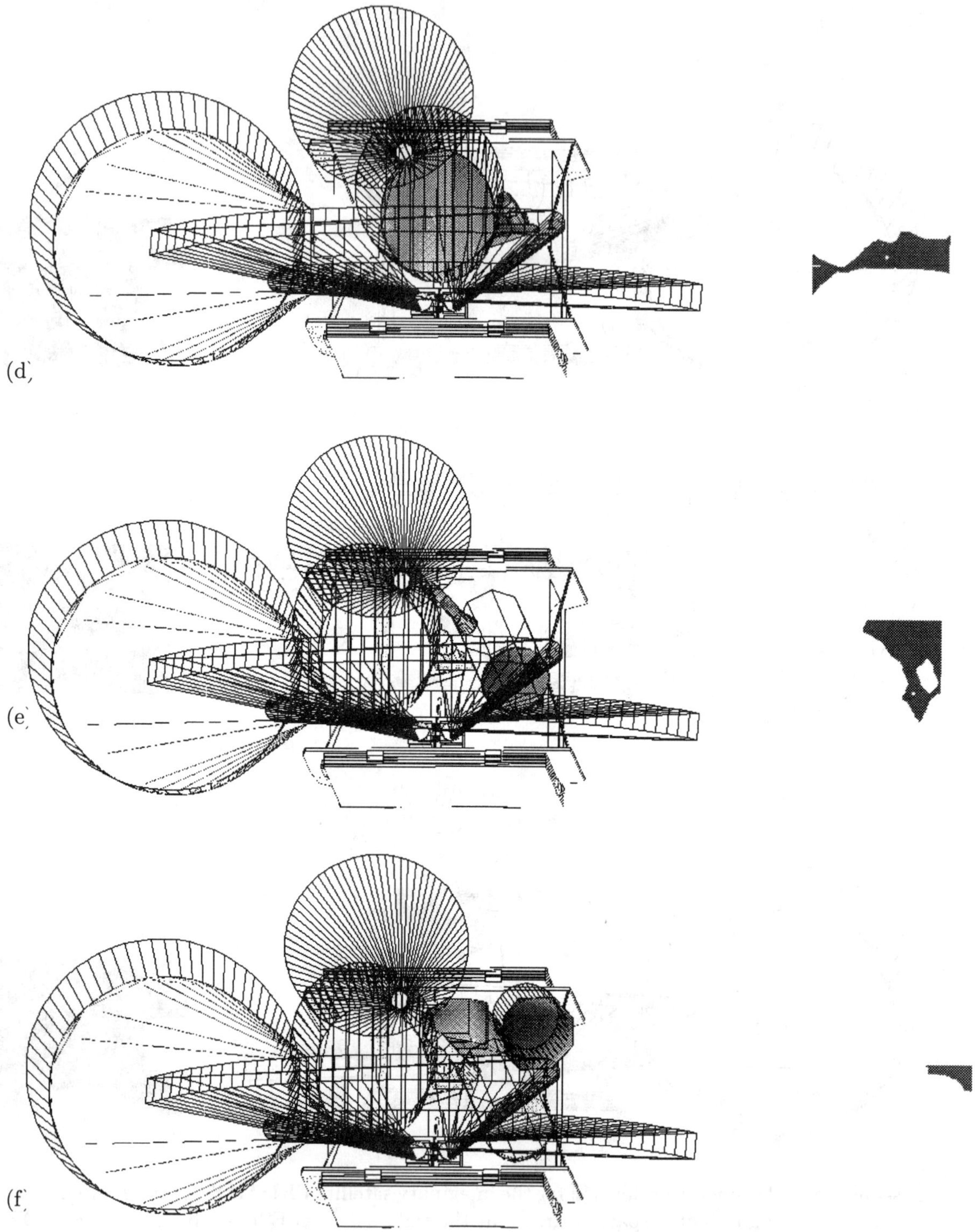


Figure 9: The sequence of placements performed for the imaginary satellite PRISMESAT (continued). (d) SVS-ANTENNA (e) VOGO (f) TVGAVO-ENS.

# Necessary Conditions for Boundary Representation Variance

Srinivas Raghothama Vadim Shapiro
Spatial Automation Laboratory*
University of Wisconsin-Madison, USA

**Abstract**

One of the key unsolved problems in parametric solid modeling is a robust update (regeneration) of the solid's boundary representation, given a specified change in solid's parameter values. The fundamental difficulty lies in determining the mapping between boundary representations for solids in the same parametric family, also known as "persistent naming." We formulate the notion of Boundary Representation (BR-)variance for solids in the same parametric family based on the assumption of continuity: small changes in solid's parameter values should result in small changes in solid's boundary representation, which may include local collapses of cells in boundary representations. We show that our formulation gives the necessary conditions that must be satisfied by any BR-variant mappings between boundary representations. These conditions are powerful enough to identify invalid updates in many practical situations and provide a formal criteria for the recently proposed heuristic approaches to the "persistent naming" problem.

## 1 Motivation

### 1.1 Ambiguity of parametric solid modeling

From its very inception solid modeling has been synonymous with unambiguous (informationally complete) representations of homogeneously $n$-dimensional subsets of Euclidean space [20, 9, 8]. On the other hand, the recent rise of solid modeling as a principal information medium first in engineering and now in consumer applications probably has to do more with development and successful marketing of new parametric ("feature-based" and "constraint-based") user interfaces than with the mathematical soundness of solid modeling systems. These parametric interfaces allow user to define and modify solid models in terms of high-level parametric definitions that are constructed to have intuitive and appealing meaning to the user and/or application (see [11] for examples and references). The success of the parametric solid modeling came at a high price: the new solid modeling systems no longer guarantee that the parametric models are valid or unambiguous, and the results of modeling operations are not always predictable.

Unpredictable behavior of the new systems is well documented in literature [10, 6, 24, 8], and we will consider some simple illustrative examples in the next section. The basic technical problem is that a parametric solid model corresponds to a *class* of solids, but there is no formal definition or standard for what this class is [24]. The lack of formal semantics manifests itself clearly through the so called "persistent naming" problem [10, 24]. In modern systems, every edit of a parametric definition is followed by a regeneration of the boundary representation (b-rep) for the resulting solid (boundary evaluation). Since parametric definitions and constraints explicitly refer to entities (faces, edges, vertices) in the boundary representation, every such regeneration must also establish the correspondence (a naming map $g$) between the pre-edit and post-edit boundary representations. The correspondence must persist over all valid edits, hence the term "persistent naming". But since the semantics of a parametric family is not well-defined, neither is the required correspondence.

Recently, several heuristic schemes for persistent naming have been proposed [13, 5, 22, 18] that appear to work well in a variety of situations. But the proposed techniques are not always consistent between themselves and none come with guarantees or with clearly stated limitations; they may be able to alleviate the observed problems in today's systems, but not solve them — be-

*Complete address: 1513 University Avenue, Madison, Wisconsin 53706, USA. Email: raghotha@cae.wisc.edu, vshapiro@engr.wisc.edu


*Computational Geometry 97* Nice France

cause the problems have not even been formulated.

### 1.2 Main contribution

A commonly cited obstacle to formalizing parametric editing stems from the belief that in the most general case parametric edits do not seem to preserve the basic structure of the boundary representation. Yet the naming mechanisms proposed in [13, 5] clearly rely on the generic structure of the boundary representation as a primary naming mechanism. Certainly, we would expect that the naming problem should have a unique solution when the parametric changes are small and the resulting boundary representation does not change very much. Accordingly, we propose that an intuitive notion of parametric class should be based on the following principle of *continuity:*

- *throughout a valid parameter range, small changes in a solid's parameter values result in small changes in the solid's representation.*

This requirement of continuity eliminates much of the unpredictable and erratic behavior of modern systems as described in [10, 24]. We use this principle to develop a formal definition for Boundary Representation (BR-)variance for solids in the same parametric family. Importantly, small changes may include limited local collapsing of entities in boundary representations. A special case of BR-variance is *BR-invariance* of a boundary representation, when it remains structurally unchanged.

Our definitions lead to the necessary conditions that must be satisfied by *any* naming map $g$ between boundary representations in the same parametric family. Loosely, the oriented boundary of every cell in the pre-edit b-rep must be mapped into the closure of the corresponding cell in the post-edit b-rep with the same orientation. This condition is powerful enough to identify invalid updates in many (but not all) practical situations, and the algorithms to check them are simple. In particular, the identified conditions give formal justification to the naming mechanisms proposed in [13, 5], but also identify techniques that are not acceptable, because they do not satisfy the assumption of continuity.

A notion of variational class is also prominent in mechanical tolerancing and robustness [21, 26]. Notably, Stewart proposed a definition of variational class in [26] which was later used in study of polyhedral perturbations that preserve topological form in [2]. The assumptions of continuity and continuous deformation are also mentioned informally in the naming techniques and discussion of [6]. Our definitions are consistent with the earlier definitions, but are stronger and stated in algebraic topological terms that are more appropriate for dealing with boundary representations and mappings between them. Additional discussion, details, and extensions to more general point sets and editing operations can be found in the full paper [25].

### 1.3 Examples

The following examples illustrate some of the erratic and inconsistent behavior during parametric updates of a solid model. These examples were created in several versions of two leading commercial solid modeling systems.[1]

A simple solid $S$ shown in Figure 1(a) is sufficient to illustrate many of the relevant issues. The shown boundary representation was produced from a parametric definition of $S$. The parametric definition uses a number of parameters, including parameters $t$ and $d$ that constrain the locations of features (slot and hole) with respect to the solid's edges as shown. The value of $t$ determines the location of edge $e_1$, which in turn determines the location of the hole constrained by parameter $d$. Consider what happens during a parametric edit of $t$. Every time $t$ is changed, the system attempts to regenerate the boundary representation of the solid and establish which of the edges should be called $e_1$. The result of this computation is easy to determine by simply observing what happens to the hole constrained by $d$, since it explicitly references edge $e_1$.

If the change in $t$ is small enough not to produce any changes in the combinatorial structure of the boundary representation, we would expect to see no abrupt changes resulting from the edit. In other words, the hole and constraint $d$ should move *continuously* with changes in the value of $t$. Instead, Figure 1(b) illustrates situations that have been observed in some earlier systems. In some cases the hole would "jump" on the face $f_2'$ and in other situations the hole would suddenly reappear on the other side of the edge $e_2'$ onto $f_1'$. In either case, the observed behavior would indicate that $e_1$ was mapped to $e_2'$; in the latter case the placement of the hole was also affected by the orientation of the mapped edge $e_2'$ (which is opposite to that of $e_1'$ with respect to face $f_3'$).

In both cases the assumption of continuity is violated, even though the correct mapping clearly exists. We will formally define such a mapping in section 4.1 as *BR-invariant*. Similar problems have been identified and illustrated with "jumping blends" by Hoffmann [5]. The techniques proposed in [13, 5, 6], as well as the latest versions of many commercial systems, are now able establish the proper mapping in this simple case, but not in many more general situations.

Figure 1(c) shows another parametric update of the same solid, where the value of $t$ is equal to the radius of the cylindrical slot, and the edge $e_1$ becomes coincident with $e_3$. Few (if any) of the latest commercial systems can handle such an update in a consistent and uniform fashion. A typical response in one system is shown in Figure 1(d) which indicates again that $e_1$ was mapped to $e_2'$; another system simply signaled an error indicating that it could not locate the proper edge

[1] We do not identify the specific systems, because similar technical problems are easily identified in *all* parametric modeling systems that rely on b-rep for internal representation of solids.

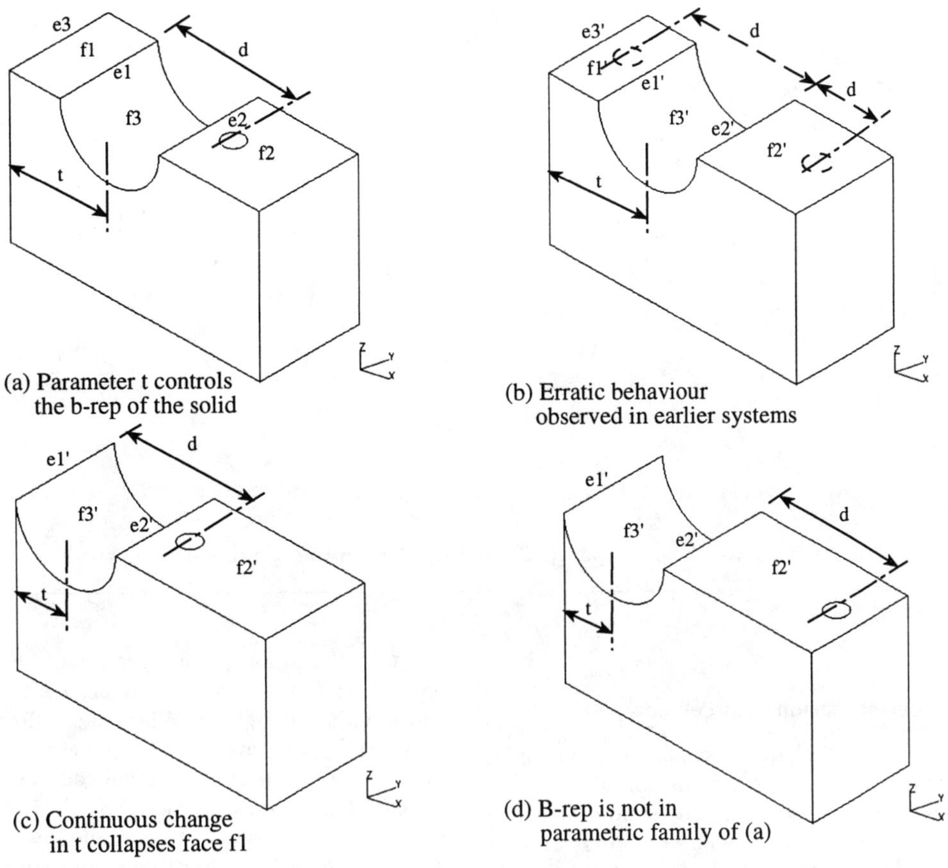


Figure 1: Parametric edits of a simple solid

(the naming map could not be found). When the slot was constructed as a "hole," the latter system gives no error and simply maps $e_1$ to $e'_2$, with the smaller hole reappearing on the other side of $e'_2$ (outside the solid). Yet another system might delete the constraint (and the hole) with an appropriate warning message. Again, intuitively it is clear that continuous change in $t$ should lead to the collapse of face $f_1$ and identification of edges $e_1$ and $e_3$ to $e'_1$. We will define such maps in section 4.1 as *BR-collapsing*.

This simple example clearly shows that the naming map is usually developed in *ad hoc* fashion, is very much dependent on the way the solid is created, and may be internally inconsistent between different features and schemes. The matching methods proposed in [13, 5] will not properly handle BR-collapsing maps, though it appears that the techniques in [5] could be further adapted.

One may also start with the solid shown in Figure 1(d) i.e., create the hole and constrain it with respect to $e'_2$, and then increase the value of parameter $t$. One would then expect the face $f'_3$ to gradually move to the right; we will call the corresponding mapping between pre-edit and post-edit boundary representations *BR-expanding*. Unfortunately, a more typical response from a system is the one shown in Figures 1(a), and the result will certainly differ from version to version and from system to system.

The final example is shown in Figure 2. Parameter $t$ controls the location of the cylindrical feature that is unioned to the cubical body, and one of the intersection edges is blended.[2] The change in value of $t$ moves the cylinder from one side of the block to the other. The system matches the blended edge $e_1$ with the edge $e'_1$ in the post-edit b-rep as shown. Is there a way to determine if this update is correct? Notice *there exists* a continuous map between pre-edit and post-edit boundary representations: it is a simple rigid body motion that flips $S$ upside-down. But this map is not consistent with the parametric change in $t$, which requires that face $f_1$ is mapped into face $f'_1$. While some of the proposed naming techniques will allow such edits, we will see in section 4.1 that no continuous change in $t$ could produce such a boundary representation.

[2] Blends provide a convenient way to track the names of the edges [5]. Throughout the paper we will ignore the geometric changes in b-rep resulting from blended edges.

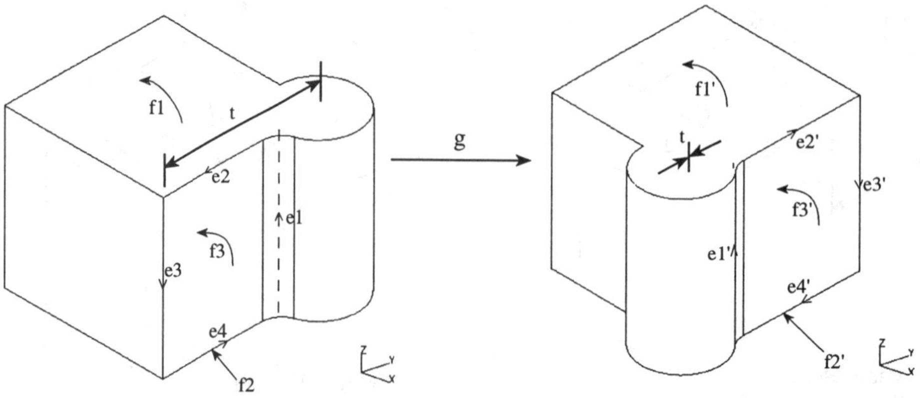


(a) Original solid's b-rep K

(b) Modified solid's b-rep L

Figure 2: Discontinuous update allowed by the parametric modeling system

## 2 Formulation

### 2.1 Boundary representation as a cell complex

Internally, every b-rep consists of an abstract (unembedded) cell complex that captures all combinatorial relationships between cells (vertices, edges, and faces), and geometric information that specifies the embedding for every cell in $E^3$. We do not deal with specific data structures in this paper, and therefore it is not necessary to distinguish between the abstract cell complex and its embedding. Thus, we will assume that boundary of a solid $S$ is represented by an embedded complex $K$ (b-rep of $S$).

Many different cell complexes have been used for boundary representations, including simplicial, polyhedral, and others with cells homeomorphic to disks. As example in Figure 1 illustrates, b-reps of many commercial systems may contain more general cells, such as faces with holes, circular edges, and so on; such cell complexes are also natural from the user's point of view. The corresponding formal definition, which we also adopt in this work, is given by Rossignac and O'Connor in [16]: a Selective Geometric Complex (SGC) is composed from cells that are relatively open submanifolds and are assembled together in a cell complex $K$. The solid's boundary $|K|$ is the *underlying space* of such a two-dimensional (boundary representation) complex.

### 2.2 Cell maps and cell homeomorphisms

The cellular structure of b-reps implies the proper mechanism for defining the notion of a variational class: we will consider the effect of a parametric change on the b-rep *cell by cell.* Informally, if $K$ is the pre-edit b-rep($t_0$) and $L$ is the post-edit b-rep($t_1$), then the naming map $g$ seeks to establish a correspondence between cells in $K$ and cells in $L$. So for example, when $t_1 = t_0$, we have $K = L$ and it should be clear that $g$ is a cell-by-cell identity mapping. When the difference $t_1 - t_0$ is very small, the correspondence between the cells does not change, even though some cells may continuously change their shapes. For example, in Figure 1(a), small changes in $t$ correspond to a small movement of the hole on face $f_2$ and therefore to a change in the shape of $f_2$ in the post-edit boundary representation. As the difference $t_1 - t_0$ gets larger, so does the difference between the shapes of the corresponding cells; and eventually some cells may actually *continuously* collapse into some other cells. This is exactly the case in Figure 1(c) where face $f_1$ and all of its bounding edges collapsed into the edge $e_3$ and its bounding vertices. This cell-by-cell continuous deformation is captured by the notion of a continuous *cell map* (modified from [12]).

**Definition 1 (Cell map)** *Let $K$ and $L$ be two b-reps, and let $g : |K| \to |L|$ be a map such that:*

1. *for each cell $\sigma \in K$, $g(\sigma)$ is a cell in $L$,*
2. *$dim(g(\sigma)) \leq dim(\sigma)$,*
3. *for every $\sigma \in K$, $g$ restricted to the closure of $\sigma$ is continuous.*

The first two conditions capture the notion of a "cell-by-cell" map. The third condition implies that $g$ can be extended to a continuous map $g : |K| \to |L|$ [15, 17, 1]. The cell map plays a dual role: on one hand, $g$ acts as a naming map and establishes the correspondence between the cells of two boundary representations; on the other hand $g$ guarantees the continuity of deformation of the underlying space (solid's boundary).

It may be instructive to check whether the naming maps in the examples of previous section are cell maps.

Consider the naming map $g$ between the two boundary representations of the solids in Figure 1(a) and (c):

$$g(f_1) = e_1{}', \quad g(e_3) = e_1{}', \quad g(e_1) = e_1{}', \quad g(e_2) = e_2{}',$$

and so on, i.e. the rest of the cells in Figure 1(a) map to their obvious images in Figure 1(c). It is easy to verify that $g$ is a cell map, because it satisfies all three conditions of the definition given above: every cell in (a) is continuously mapped to a cell in $L$ of same or lower dimension (note that face $f_1$ is mapped to edge $e_1{}'$) and the continuity condition is satisfied for the closure of every cell. In particular, note that the closure of the collapsed face $f_1$ is mapped into the closure of $e_1{}'$. Now consider the mapping of the same boundary representation in (a) to the one shown in Figure 1(d):

$$g(f_1) = e_1{}', \quad g(e_3) = e_1{}', \quad g(e_1) = e_2{}', \quad g(e_2) = e_2{}',$$

and the rest of the cells in (a) map to their appropriate images in (d). It is easy to check that $g$ is not a cell map; for example, the closure of face $f_1$ contains the edge $e_1$, while $e_2' = g(e_1)$ is not in the closure of $e_1' = g(f_1)$.

Referring to Figure 2, consider the mapping $g$ given by:

$$g(f_i) = f_i{}', \; i = 1, \ldots n; \qquad g(e_j) = e_j', \; j = 1, \ldots, m,$$

where $n$ is the number of faces and $m$ is the number of edges respectively in the boundary representation. This map $g$ trivially satisfies all three required conditions and is therefore a cell map. However, there is also another cell map $h$ between the two boundary representations:

$$h(f_1) = f_2{}', \quad h(f_2) = f_1{}', \quad h(f_3) = f_3{}', \quad h(e_4) = e_2{}',$$

$$h(e_1) = e_1{}', \quad h(e_3) = e_3{}', \quad h(e_2) = e_4{}'.$$

This cell map $h$ maps the loop of edges around face $f_1$ into the loop of edges around $f_2$ and vice versa, and corresponds to a rigid body rotation of 180 degrees about the shown $x$-axis. Thus given two boundary representations $K$ and $L$ there may be more than one valid cell map between them.

For sufficiently small parameter changes, we expect that there is also a one-to-one correspondence between cells in $K$ and $L$. In this case, we can strengthen the definition:

**Definition 2 (Cell homeomorphism)** *A bijective cell map $g : K \to L$ is called a cell homeomorphism between $|K|$ and $|L|$ [15].*

Notice that when $g$ is a cell homeomorphism, the second condition in the definition of the cell map can be also strengthened: the dimension of $g(\sigma)$ must be equal to that of $\sigma$ for every cell in $K$. For example, both cell maps $g$ and $h$ constructed above for the transformation in Figure 2 are cell homeomorphisms.

### 2.3 BR-variance and Invariance

For two boundary representations $K$ and $L$ to belong to the same parametric family there must exist a (not necessarily unique) cell map $g$ between $K$ and $L$. Unfortunately, the existence of such a map alone is not sufficient to assure the principle of continuity postulated in section 1.2. For example, even though the two boundary representations in Figure 2 are related by cell homeomorphism $g$, they cannot be deformed into another without leaving $E^3$. The reason that the deformation does not exist may not be obvious: the modified solid is the reflection of the original one. We will see in the next section such maps can be identified by considering the orientation of cells, which we have so far ignored. A more difficult situation is shown in Figure 3: the boundary representation of two linked tori is essentially the same as the boundary representation of the two disconnected tori. But despite of the apparent cell homeomorphism, it is not possible to deform one into another without either leaving $E^3$ or violating the continuity principle.

If $K$ is a b-rep($t_0$) and $L$ is a b-rep($t_1$), then the continuity principle requires that, as $t$ changes from $t_0$ to $t_1$, $K$ continuously deforms into $L$. These observations motivate the following definition.

**Definition 3 (BR-variance)** *Let $K$, $L$ be two boundary representations. A map $F : E^3 \times I \to E^3$ is called a* Boundary Representation Variance *or* BR-variance *of $K$ to $L$ if:*

1. *$F(|K|, 0) = |K|$,*
2. *$F(|K|, 1) = |L|$,*
3. *$F(|K|, t)$ is a cell map that is also continuous in $t \in [0, 1]$.*

In standard algebraic topological terms, BR-variance $F$ is a homotopy [1, 12] and can be also viewed as a continuous family of cell maps $F_t$, for every value $t \in [0, 1]$. A special case of BR-variance is a case when $F_t$ is a continuous family of cell homeomorphisms, in which case $F$ will be called *BR-invariance.* BR-variance captures the spirit of the postulated requirement of continuity and appears to be the weakest possible condition. But it may not be clear whether BR-variance exists for transformations shown in Figures 1, 2, and 3, and more generally how our definition can be used to distinguish b-reps which belong to the same parametric family and b-reps which do not.

**Definition 4 (BR-variant map)** *A map $g$ between two boundary representations $K$ and $L$ is called* BR-variant *if $F(|K|, 1) = g(|K|) = |L|$ for some BR-variance $F$.*

The last definition depends on the *existence* of some unknown BR-variance $F$, which may not be (and often is not) unique. In practice, only the mapping between two

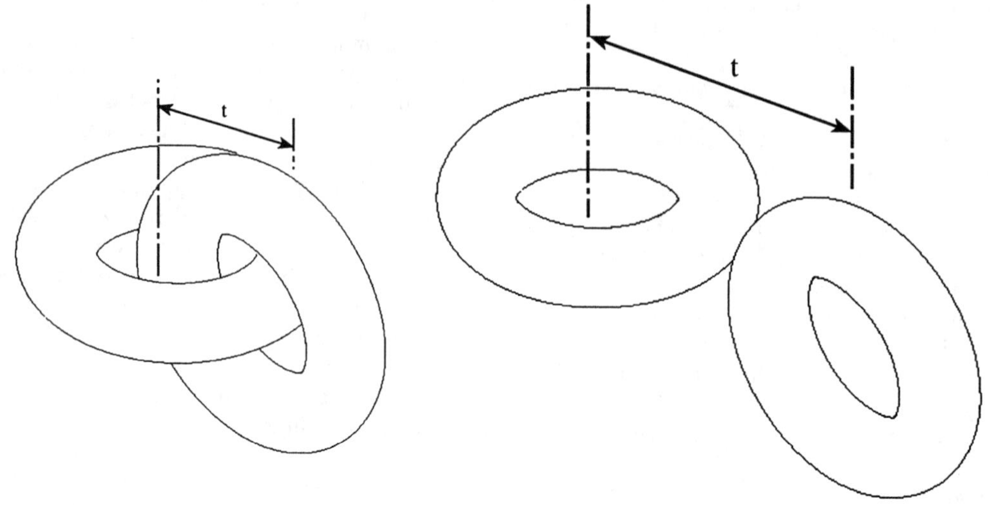


(a) Original cell complex K (b) Modified cell complex L

Figure 3: $K$ is homeomorphic to $L$, but $K$ cannot be deformed into $L$ without leaving $E^3$

boundary representations $g$ is known, and the problem is to determine whether an appropriate $F$ exists or not. We discuss the necessary conditions for existence of BR-variance in the next section and show through examples in section 4 how they can be used in practice.

## 3 Necessary Conditions for BR-Variance

### 3.1 Informal statement

In order to take advantage of the orientation information, we need to slightly modify the definition of the cell map by requiring that all cells in boundary representation are oriented, which is usually the case in most boundary representations. Orientation of a cell $\sigma \in K$ can be visualized as a sense of direction.

**Definition 5 (Oriented Cell Map)** *A map $g : K \to L$ is oriented cell map if it is a cell map and for every oriented cell $\sigma \in K$, $g(\sigma)$ is an oriented cell in $L$.*

The difficulty with the examples in Figures 2 is that the proposed cell maps between $K$ and $L$ do not preserve the relative orientation of certain cells, i.e. oriented cells of $K$ are mapped into oriented cells of $L$ with their relative orientation reversed. Since every boundary representation is orientable [19, 9], we will assume that 2-cells are always oriented counter-clockwise. We will also postulate that all 0-cells have positive orientation. The orientation of a 2-cell induces a orientation on its bounding 1-cells and an 1-cell orientation induces a orientation on its bounding vertices. Roughly, the naming map $g$ must preserve this relationship between oriented 0,1, and 2-cells. We did not explicitly require that BR-variance map is orientation preserving, because orientation is a homotopy invariant and is automatically preserved through continuity [3]. Thus, at the very least our definitions imply that any naming map $g$ must be an *orientation preserving cell map,* or as we explain below, the following equivalent conditions must be satisfied:

1. $K$ and $L$ are oriented consistently;
2. $g$ is an oriented cell map; and
3. if $dim(g(\sigma)) = dim(\sigma)$, then $g$ maps the *oriented boundary* of cell $\sigma \in K$ into the oriented boundary of the image cell $g(\sigma) \in L$.

### 3.2 Precise formulation

The first requirement is satisfied by requiring that all 2-cells are oriented counter-clockwise and all 0-cells have positive orientation. This assumption essentially takes care of the above requirement that two boundary representations $K$ and $L$ are oriented consistently. The second requirement that $g$ is a cell map was discussed in the last section; in the full paper [25] we also give equivalent combinatorial conditions that are easy to implement in any b-rep modeler. The third condition requires the notion of the *oriented* boundary operator on *p-chains* with coefficients $\{-1, 0, +1\}$ (sometimes called "elementary chains" [1, 7]). If $C_p(K)$ denotes the space of all $p$-chains on cell complex $K$, then boundary operator is a linear function $\partial : C_p(K) \to C_{p-1}(K)$. Any oriented cell map $g$ induces a family of chain maps $g_p$ in every dimension $p = 0, 1, 2$, defined as follows.

**Definition 6 (Chain map)** *Let $g : K \to L$ be an orientation preserving cell map. For each dimension $p$, define a chain map $g_p : C_p(K) \to C_p(L)$ as follows.*

1. *if $\sigma$ is a p-cell in $K$ and $g(\sigma)$ is a p-cell in $L$, $g_p(\sigma) = g(\sigma)$;*
2. *if $dim(g(\sigma)) < dim(\sigma)$, $g_p(\sigma) = 0$;*
3. *$g_p(a_1\sigma_1 + \ldots + a_n\sigma_n) = a_1 g_p(\sigma_1) + \ldots + a_n g_p(\sigma_n)$.*

Then it is known [1, 3, 12] that the family of such chain maps satisfy the following commutative diagram.

$$\begin{array}{ccc} C_p(K) & \xrightarrow{g_p} & C_p(L) \\ \downarrow \partial & & \downarrow \partial \\ C_{p-1}(K) & \xrightarrow{g_{p-1}} & C_{p-1}(L) \end{array}$$

And in particular, for every orientation preserving cell map the following simple condition is satisfied for each oriented $p$-cell $\sigma \in K$:

$$g_{p-1}(\partial(\sigma)) = \partial(g_p(\sigma)). \quad (1)$$

Together with the definition of an oriented cell map, (1) gives necessary and sufficient conditions for $g$ to be an orientation preserving cell map, which is in turn a necessary condition for $g$ to be BR-variant.

## 4 BR-variant naming

### 4.1 Examples revisited

Consider the naming map between the solids in Figure 1(a) and (c). The cells in the top view of the two solids are shown in Figure 4. The semantics of the update collapses the face $f_1$ and its boundary into the edge $e_1'$ and its bounding vertices. This BR-variance is reflected in the BR-variant naming map $g$ as follows:

$$g(f_1) = e_1', \quad g(e_3) = -e_1', \quad g(e_1) = e_1', \quad g(e_2) = e_2',$$

and the implied map for the respective vertices. Notice that $g$ is many-to-one. A many-to-one BR-variant map will be called **BR-collapsing**. Let's check the necessary conditions in terms of the induced chain maps $g_1$ and $g_2$:

$$g_2(f_1) = 0, \; g_2(f_2) = f_2', \; g_2(f_3) = f_3';$$
$$g_1(e_2) = e_2', \; g_1(e_6) = e_3', \; g_1(e_7) = e_4',$$
$$g_1(e_4) = 0, g_1(e_5) = 0, \; g_1(e_1) = e_1',$$
$$g_1(e_8) = e_5', \; g_1(e_3) = -e_1'.$$

A simple check shows that the necessary condition $\partial(g_2(f_1)) = g_1(\partial(f_1))$ holds, because

$$\partial(g_2(f_1)) = \partial(0) = 0, \text{and}$$

$$g_1(\partial(f_1)) = g_1(e_5 + e_1 + e_4 + e_3) = 0 + e_1' + 0 - e_1' = 0.$$

The rest of the cells remain essentially invariant under the naming map, and it is easy to check that the necessary condition holds for all other cells in the boundary representation.

Observe that the inverse map $g^{-1}$ from the solid in Figure 1(c) to (a) is *not* BR-variant; in fact it is not a valid cell map because it maps a one-dimensional edge $e_1'$ to a two-dimensional face $f_1$. Nevertheless, such maps appear useful and are allowed in many systems. Therefore we will say that a one-to-many relation $g$ is **BR-expanding** map if its inverse is BR-collapsing.

Now consider the apparent naming map between the solids in Figure 1(a) and (d):

$$g(f_1) = e_1', \quad g(f_3) = f_3', \quad g(e_3) = -e_1',$$
$$g(e_1) = e_2', \quad g(e_2) = e_2'.$$

We already know that this map $g$ is not BR-variant (because it is not a cell map), and it is easy to verify that the necessary condition does not hold either. The induced chain maps are:

$$g_2(f_1) = 0, \; g_2(f_2) = f_2', \; g_2(f_3) = f_3',$$
$$g(e_4) = 0, \; g_1(e_5) = 0, \; g_1(e_1) = e_2',$$
$$g_1(e_3) = -e_1', \; g_1(e_2) = e_2',$$

and the left and the right hand side of the equation (1) evaluate to different chains:

$$\partial(g_2(f_1)) = \partial(0) = 0,$$

$$g_1(\partial(f_1)) = g_1(e_5 + e_1 + e_4 + e_3) =$$
$$0 + e_2' + 0 - e_1' = e_2' - e_1'.$$

Clearly, $g$ is not BR-collapsing and the inverse map $g^{-1}$ cannot be BR-expanding.

Next consider the example in Figure 2. We described two distinct naming maps $g$ and $h$ in section 2.2, and argued that neither of them is appropriate — but for different reasons. We claimed that the naming map $g$ computed by a commercial system is not BR-variant because it is not orientation preserving. Indeed, $g_k(\sigma) = g(\sigma)$ for all cells in the boundary representation, because no cells have been collapsed, and in fact there is a one-to-one correspondence between the pre-edit and post-edit boundary representations. Yet, checking the necessary condition for face $f_3$ we see that

$$\partial(g_2(f_3)) = \partial(f_3') = -e_4' - e_3' - e_2' - e_1',$$

is not equal to

$$g_1(\partial(f_3)) = g_1(e_1 + e_2 + e_3 + e_4) = e_1' + e_2' + e_3' + e_4'.$$

Thus, $\partial(g_2(f_3)) = -g_1(\partial(f_3))$ and the similar reversal of orientation can be verified for all other faces in the boundary representation; this shows that $g$ is an orientation-reversing map and is not BR-variant.

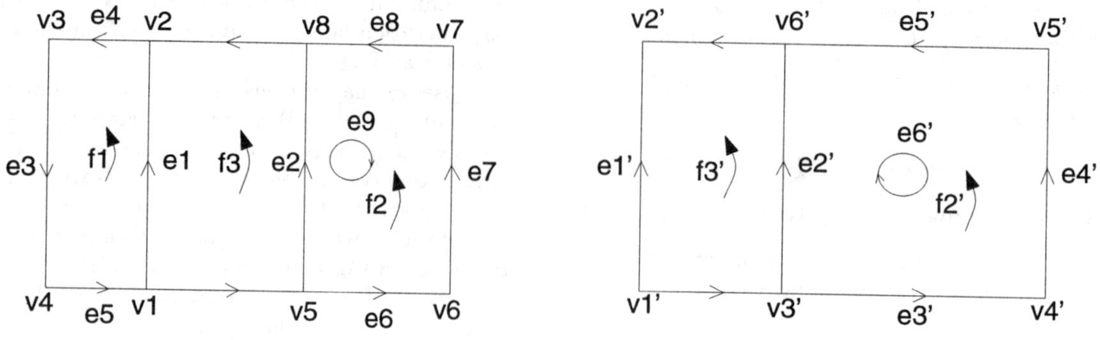


(a) Top view of K

(b) Top view of L

Figure 4: Top view of oriented boundary representation of the solid in Figure 1.

But now consider the necessary condition for the second naming map $h$ for the same solids:

$$h(f_1) = f_2{}', \quad h(f_2) = f_1{}', \quad h(f_3) = f_3{}',$$

$$h(e_4) = -e_2{}',\ h(e_1) = -e_1{}',\ h(e_3) = -e_3{}',\ h(e_2) = -e_4{}'.$$

and so on. Notice that $h$ reverses the orientation of all edges connecting $f_1$ and $f_2$. Checking the necessary condition on this oriented cell map:

$$\partial(h_2(f_3)) = \partial(f_3') = -e_4' - e_3' - e_2' - e_1' \quad =$$

$$h_1(\partial(f_3)) = h_1(e_4 + e_1 + e_2 + e_3) = -e_2' - e_1' - e_4' - e_3'.$$

Similarly, one can check that the map $h$ does satisfy the necessary condition for all other cells in the boundary representation and is a BR-variant map associated with the 180 degree rotation of the solid about X-axis. Furthermore, $h$ is **BR-invariant** map, because it is a cell homeomorphism. Unfortunately, this BR-invariance is not consistent with the intended semantics of the parametric edit. And in fact in this example, due to symmetry, there are many other BR-invariant maps (e.g. rotation about Z-axis by 90 degrees.) The guarantee of BR-variance (or invariance) should not be confused with the guarantee of the correct semantics of the parametric update.

### 4.2 Comparison with the proposed naming methods

In the typical scenario described in section 1, pre-edit b-rep $K$ and post-edit b-rep $L$ are given, and the system attempts to match the cells in $K$ with the cells in $L$, i.e. to construct the naming map $g : K \to L$. Several heuristic methods have been proposed for constructing the naming maps $g$ [13, 5, 6, 18, 22] in a variety of situations, including those where BR-variant maps may not exist. The naming techniques proposed by Kripac[13] and Hoffmann et al [6, 5] rely on several common methods for constructing the names for cells in boundary representations. In particular, both start with the names of the primitive surfaces appearing in the parametric definition and advocate constructing the names for individual cells using (1) list of adjacent cells in the boundary representation, and (2) relative orientation of the cell with respect to the adjacent cells. In general, these two techniques are not sufficient to uniquely match cells in two b-reps, and a number of additional techniques must be employed.

Focusing on the two common methods for naming cells, intuitively they correspond respectively to the two conditions that ensure an orientation preserving cell map: the adjacency information is a means for extending the continuity of the naming map to the whole boundary of the solid (as required by the third condition in the definition 1), and the matching of relative orientation of cells is directly related to the necessary condition (1) in section 3.

On closer examination, it appears that [13] may map a cell $\sigma \in K$ to $g(\sigma) \in L$ when all cells adjacent to $\sigma$ are mapped into cells adjacent to $g(\sigma)$, without considering the orientation information. But as clearly shown by example in Figure 2, such naming map $g$ does not have to preserve orientation and may not be BR-variant. By contrast, [6] matches $\sigma$ to $g(\sigma)$ only if the adjacent cells also preserve orientation. Both methods will find the BR-invariant map, if it exists and every cell has a unique name. It is not clear whether BR-invariance is always enforced when heuristic methods are used to resolve multiple matches between cells with the same names.

Neither of the above methods are able to handle BR-collapsing maps as defined in this paper, because both [13] and [6] require exact match of the adjacent cells. For example, edge $e_1$ in Figure 1(a) is adjacent to faces $f_1$ and $f_3$, while the image of face $f_1$ is not a face in the post-edit boundary representation of Figure 1(c). Thus, the proposed naming methods will never match $e_1$ to $e_1'$. More generally, all of the proposed methods preserve the dimension of the mapped cells, which is not the case for BR-collapsing and BR-expanding maps. Based on our results, it may make sense to redefine the adjacency conditions to be consistent with our definition 1 of cell map (additional details can be found in [25]).

Rossignac observed that when every cell in a b-rep can be represented by a Boolean (set-theoretic) expression on primitive surfaces and halfspaces, such expression can be used as a persistent name for the cells [22]. Boolean set-theoretic representations place no restrictions on topology of the resulting boundary representations, imply very different parametric family [24], and do not provide a mechanism for enforcing or even checking BR-variance.

A common method of matching cells in pre-edit and post-edit b-reps that can be observed in some commercial systems combines Boolean naming with the sequential ordering of the entities with respect to some external direction. For example, a system may attempt to distinguish between edges $e_1$ and $e_2$ in Figure 1(a) by indexing them along the length of the face ($x$-axis in the figure). This method is most likely responsible for producing the solid in Figure 1(d) instead of (c): the identification of $e_1$ with $e_3$ changes the adjacency information so that $e_2'$ appears to be the "first" edge to fit the old name of $e_1$.

## 5 Conclusion

### 5.1 Sufficient conditions for BR-variance

We have argued that the proposed principle of continuity and the implied notion of BR-variance should be accepted as the basis for formally defining the semantics of parametric family. In particular, a number of problems can be now formulated precisely, including:

1. Given a naming map $g$ between two b-reps $K$ and $L$, check if $g$ is BR-variant;
2. Determine if $K$ and $L$ belong to the same parametric family, or, equivalently, determine if there exists a BR-variant map between $K$ and $L$ corresponding to the particular parametric edit.
3. For $K$ and $L$ in the same parametric family, construct the naming BR-variant map $g$ that captures the semantics of the parametric edit.

Our formulation shows that it is easy to check whether a given map $g$ is an orientation preserving cell map, but this is not sufficient for solving the above problems. The main difficulty lies in the need to know what happens to the boundary representation not only at times $t_0$ and $t_1$ but also for all values of $t \in [t_0, t_1]$. To guarantee that $g$ is BR-variant, we must show that it can be extended to a continuous map on the whole of $E^3$. To be more precise, $g$ will satisfy all the requirements of BR-variant map if there exists a continuous orientation preserving map $h : E^3 \to E^3$ such that $h$ restricted to $K$ is a cell map $g$ from $K$ to $L$. [1]

This sufficient condition for BR-variance cannot be enforced by considering only the structure of the two-dimensional boundary representations. Other possible approaches to constructing BR-variant maps include constructing the corresponding map $h$ as a cell map on the three-dimensional decomposition of $E^3$ as suggested in [23, 24], or assuring the properties of $h$ by further restricting how $g$ modifies individual cells [26, 2].

### 5.2 Significance and limitations

The assumption of continuity makes sense from an engineering point of view and can serve as a starting point for developing universal standards in parametric modeling. The proposed definition of BR-variance and the implied necessary conditions can be employed in numerous other applications, including shape optimization, tolerancing, and constraint solving. For example, some heuristic methods proposed in [4] for identifying the correct solutions in constraint solving resemble the necessary conditions implied by our definitions, and the use of homotopy was proposed in [14].

It may appear that accepting the principle of continuity and BR-variance as a basis for parametric modeling is too limiting because a number of practically important parametric edits cannot be described by BR-variance. These include merging and splitting entities in boundary representations, (dis)appearance of holes, and other common edits. Recall that the main purpose of orientation preserving cell map was to enforce the continuity of the map $g : |K| \to |L|$ in a cell-by-cell fashion; a natural extension of this principle would allow structural changes in the cells of $K$ and $L$. For example, subdividing or merging cells in a solid's boundary representation, e.g. as proposed in [16], has no effect on the boundary set itself (the underlying space). In [25] we explain how more general edits can be defined by a sequence of splitting, merging, BR-collapsing, and BR-expanding mappings.


### Acknowledgements

This research is supported in part by the National Science Foundation CAREER award DMI-9502728 and grant DMI-9522806.


### References


[1] Max K. Agoston. *Algebraic Topology.* Marcel Dekker Inc, New York, 1976.

[2] L. E. Andersson, S. M. Dorney, T. J. Peters, and N. F. Stewart. Polyhedral perturbations that preserve topological form. *Computer-Aided Geometric Design*, 12:785–799, 1995.

[3] M. A. Armstrong. *Basic Topology.* Springer-Verlag, New York, 1983.

[4] W. Bouma, I. Fudos, C. Hoffmann, J. Cai, and R. Paige. A geometric constraint solver. Technical report, CSD-TR-93-054, Purdue University, Department of Computer Science, August 1993.

[5] V. Capoyleas, X. Chen, and C. M. Hoffmann. Generic naming in generative, constraint-based design. *Computer-Aided Design*, 28(1):17–26, 1996.

[6] X. Chen and C. M. Hoffmann. On editability of feature-based design. *Computer-Aided Design*, 27(12):905–914, 1996.

[7] J. G. Hocking and G. S. Young. *Topology*. Dover Publications, New York, 1961.

[8] C. H. Hoffmann and J. R. Rossignac. A road map to solid modeling. *IEEE Transactons on Visualization and Computer Graphics*, 2(1):3–10, March 1996.

[9] C. M. Hoffmann. *Geometric and Solid modeling*. Morgan Kaufman, USA, 1989.

[10] C. M. Hoffmann. On semantics of generative geometry representations. In *Proc. 19th ASME Design Automation Conference*, pages 411–420, 1993.

[11] J. Hoschek and W. Dankwort. *Parametric and Variational Design*. B. G. Teubner, 1994.

[12] Christine Kinsey. *Topology of Surfaces*. Springer-Verlag, New York, 1993.

[13] J. Kripac. A mechanism for persistently naming topological entities in history-based parametric solid models. In *3rd ACM Symposium on Solid Modeling and Applications, Salt Lake City, Utah*, May 1995.

[14] H. Lamure and D. Michelucci. Solving geometric constraints by homotopy. In *3rd ACM Symposium on Solid Modeling and Applications , Salt Lake City, Utah, May*, 1995.

[15] James R. Munkres. *Elements of Algebraic Topology*. Addison-Wesley, Reading, Massachusetts, 1984.

[16] M. A. O'Connor and J. R. Rossignac. SGC: A dimension independent model for pointsets with internal structures and incomplete boundaries. Technical report, IBM Research Report, RC14340, 1989.

[17] M. M Postnikov. *Lectures in Algebraic Topology*. Nauka, Moscow, 1983.

[18] A. Rappoport. Displayable-selectable models and boundary representations in geometric operation graphs. In *Theory and Practice of Geometric Modeling (Blaubeuren II)*. Tubingen, Germany, October 1996.

[19] A. A. G. Requicha. Mathematical models of rigid solid objects. Technical report, TM-28, PAP, University of Rochester, Rochester, NY, November 1977.

[20] A. A. G. Requicha. Representations for rigid solids: Theory, methods and systems. *ACM Computing Surveys*, 12:437–464, December 1980.

[21] A. A. G. Requicha. Representation of tolerances in solid modeling: Issues and alternative approaches. In M.S Pickett and Boyse J.W., editors, *Solid Modeling by Computers*. Plenum Publishing Corp, USA, 1984.

[22] J. R. Rossignac. CSG formulations for identifying and for trimming faces of CSG models. In *CSG'96, Winchester, UK*. Information Geometers Ltd, April 1996.

[23] V. Shapiro. Maintenance of geometric representations through space decompositions. *International Journal of Computational Geometry and Applications*, to appear.

[24] V. Shapiro and D. L. Vossler. What is a parametric family of solids? In *3rd ACM Symposium on Solid Modeling and Applications , Salt Lake City, Utah, May*, 1995.

[25] S.Raghothama and V.Shapiro. Boundary representation variance in parametric solid modeling. Technical report, SAL 1997-1, University of Wisconsin, Madison, WI, January 1997.

[26] N. F. Stewart. Sufficient condition for correct topological form in tolerance specification. *Computer-Aided Design*, 12:39–48, 1992.

# Choosing Corners of Rectangles for Mapped Meshing

Scott A. Mitchell[1]

## Abstract

Consider mapping a regular i x j quadrilateral mesh of a rectangle onto a surface. The quality of the mapped mesh of the surface depends heavily on which vertices of the surface correspond to corners of the rectangle. Our problem is, given an n-sided surface, chose as corners four vertices such that the surface resembles a rectangle with corners at those vertices. Note that n could be quite large, and the length and width of the rectangle, i and j, are not prespecified. In general, there is either a goal number or a prescribed number of mesh edges for each bounding curve of the surface. The goals affect the quality of the mesh, and the prescribed edges may make finding a feasible set of corners difficult.

The algorithm need only work for surfaces that are roughly rectangular, particularly those without large reflex angles, as otherwise an unstructured meshing algorithm is used instead. We report on the theory and implementation of algorithms for this problem.

We also give an overview of a solution to a related problem called interval assignment: Given a complex of surfaces sharing curves, globally assign the number of mesh edges or intervals for each curve such that it is possible to mesh each surface according to its prescribed quadrilateral meshing algorithm, and assigned and user-prescribed boundary mesh edges and corners. We also note a practical, constructive technique that relies on interval assignment that can generate a quadrilateral mesh of a complex of surfaces such that a compatible hexahedral mesh of the enclosed volume exists.

## 1 Introduction

Currently, practical mesh generation can be a very time consuming process. An experienced mesh generation software user may spend six months or more meshing a large industrial model. The user must supply goals (*soft-sets*) and constraints (*hard-sets*) on the sizes of mesh elements throughout the model, either by specifying a sizing function or actual values on various vertices, curves, surfaces, and volumes. The best software packages provide a toolbox of algorithms. These algorithms have complex trade-offs in terms of their automation, general applicability to complex geometries, quality of the produced mesh, and speed and memory requirements. Often the user must decide which algorithm to use for many subsets of the model, the order in which subsets are meshed, and additional parameters for each subset. Intelligent automation of these steps greatly speeds the overall mesh generation process. We report on an improved capability that automatically removes some order dependence and automatically specifies some parameters. (The automation of algorithm selection is also an important problem and is being researched.)

In particular, in the CUBIT[1] quadrilateral and hexahedral mesh generation toolkit, a common paradigm is an advancing front. This requires us to first mesh the curves, then the surfaces, and finally the interior of a volume. The user either specifies a *soft-set* goal or a fixed *hard-set* number of mesh edges (called *intervals*) for each curve. However, different quadrilateral surface meshing algorithms require a certain relationship between the number of intervals on each bounding curve. For meshing a surface with a *mapping algorithm*, these constraints depend on some additional parameters, namely which vertices are chosen as the *corners* of the mapping primitive. Typically a curve is contained in two surfaces, and its intervals are constrained in some way by both of them. If surfaces are meshed one by one, then often we paint ourselves into a corner and are left with a collection of surfaces that are unmeshable because their bounding curves are incompatibly meshed. Instead, we automatically and locally chose the corner parameters, then globally find the number of mesh edges (intervals) for each curve such that all surfaces can be meshed with their chosen algorithm and corners. These problems are called *corner picking* and *interval assignment*, respectively. Note that for a given choice of algorithms, corners, and hard-set intervals, interval assignment may be impossible (e.g. see Figure 1 and Figure 2). Simultaneously choosing corners and intervals to allow surface meshing is an NP-hard problem[2]. A structured or semi-structured volume meshing scheme may impose additional constraints.

We also note that if we cut surfaces into topological disks, and enforce that every curve has an even number of intervals, then any quadrilateral mesh of the surfaces that is compatible with the meshes of the curves admits a compatible hexahedral mesh of the enclosed volume. Actually finding a hexahedral mesh is another matter, particularly one with a quality guarantee.

Corner picking is a highly geometric problem, but is non-Euclidean in that the length of a curve matters less than its user-specified number of intervals, and whether these intervals are soft- or hard-set. The interval assignment problem comes from geometric data, but is essentially a combinatorial problem, usually solved by linear programming[3] or more recently by network flow algorithms[2].

## 2 Corner Picking

"Map meshing a surface" produces some (hopefully isomeric) map from an abstract structured quadrilateral mesh of an n-gon *primitive* to a surface. Typical n-gons include triangles, rectangles and pentagons. Mapping is valued as the fastest and least memory

---

1Parallel Computing Sciences Dept., MS 0441, Sandia National Laboratories, Albuquerque, NM 87185. E-mail *samitch@sandia.gov.* URL http://sass577.endo.sandia.gov:80/ 9225/Personnel/samitch/

Scott Mitchell was supported by the Mathematical, Information and Computational Sciences Division of the U.S. DOE, Office of Energy Research, and works at Sandia National Laboratories, operated for the U.S. DOE under contract No. DE-AC04-94AL85000. Sandia is a multiprogram laboratory operated by Sandia Corporation, a Lockheed Martin Company, for the U.S. DOE.

*Computational Geometry 97* Nice France
0-89791-878-9/97/06

intensive surface meshing algorithm, but it is only applicable to special geometries: The quality of the surface's mesh depends on how close the map is to a constant function. The quality of the map depends heavily on the interior angles of the vertices of the surface which are identified with corners of the abstract n-gon, the relative number of intervals on sides, and whether any interior-angles are reflex. Here for brevity we consider only the rectangle-primitive, which is a regular i x j mesh (see Figure 11 left for a 5x5 example). Our technique has been applied to triangle primitives and could be applied to other n-gons as well.

Often the surface is composed of four curves, and the corners of the rectangle primitive are uniquely determined by the geometry. However, we can rectangle-primitive map a surface that resembles a rectangle in a fuzzy sense, and is bounded by more or less than four curves. Some of the rectangle sides may be composed of a composite of several curves.

The interval assignment constraints are that opposite sides have exactly equal intervals. The intervals on hard-set curves are fixed, but the intervals on soft-set curves can be adjusted up or down to satisfy these constraints. In the case of some curves having hard-set intervals, the corner choice may also determine the feasibility of mapping the surface, as in Figure 1. There is even a global problem that is NP-complete[2] that we don't address here: Some corner choices may lead to the global interval assignment problem being infeasible, as in Figure 2.

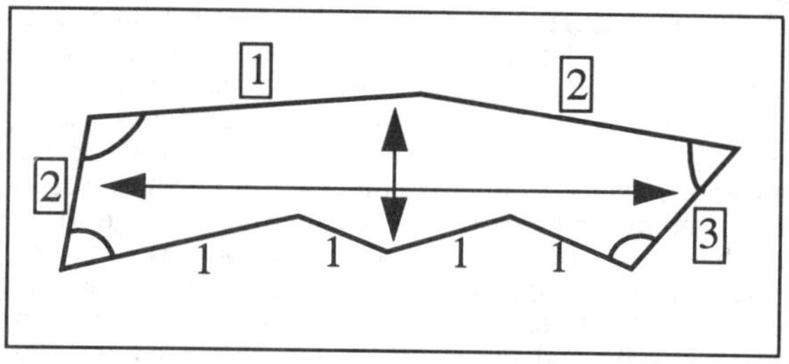


Figure 1. The hard-set (boxed) intervals and choice of corners leads to an infeasible interval assignment problem. First, the opposite right and left sides are hard-set with unequal values. More subtly, the top and bottom sides are also infeasible: Since each curve must have at least one interval the bottom side must have at least four intervals but the top side is hard-set to three.

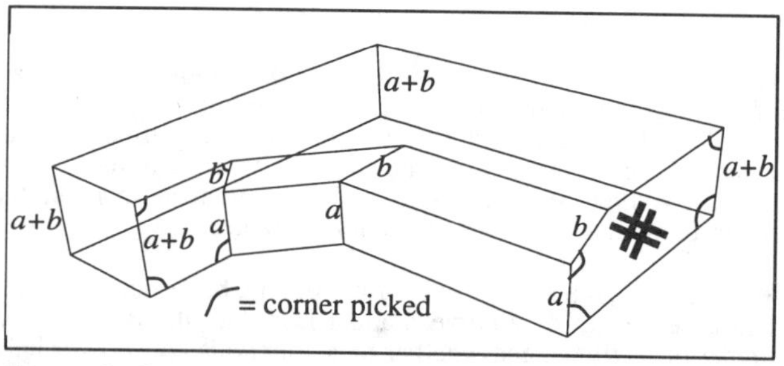


Figure 2. Local corner picking makes global interval assignment impossible on this real-world geometry. We get a system of equations that reduces to $a+b = a$, whose only solution has $b=0$, but a curve must have at least one interval!

A previous method used in CUBIT is to pick as corners the vertices with smallest interior angle. Obviously this can lead to a poor mesh. Arkin et. al[4] presents a polygon-matching algorithm that defines a metric for the shape-distance between two polygons in terms of an integral involving the two polygons' *turning-functions*. We can't immediately use this algorithm because one of our polygons is a rectangle of indeterminate height and width, and because interval assignment imposes different constraints and goals on which curves are on sides opposite each other.

However, our method of choice is a heuristic loosely based on the turning-function of Arkin et. al[4]. The heuristic has a provable running time of $O(n^2)$, where $n$ is the number of curves. It usually gives a good solution, but has trouble finding any solution in the case that many curves are hard-set and there are few feasible choices of corners. If the heuristic fails, we fall back on a set of algorithms that provably find a feasible set of corners, if one exists, in time $O(n^2 \log n)$. A post-processing heuristic shifts corners incrementally, singly and in pairs, keeping feasible at all times, searching for a local optima.

## 2.1 Heuristic Corner Picking

The algorithm makes two $O(n^2)$ passes. The first pass determines opposite corners 1 and 3, the second pass determines corners 2 and 4. In the first pass, all pairs of vertices $(a,b)$ are considered, and the pair with the best (smallest) *quality function* value is chosen. The quality function is a weighted sum of the corner_angle function, the turn_angle function, and the interval_ratio function. The corner_angle function depends on the interior angles at the corners (ideally $\pi/2$), the turn_angle function on the amount the vector tangent to the bounding curves turns strictly between successive corners (ideally $\pi/2$, to leave room for another corner), and the interval_ratio function depends on the ratio of intervals between sides (ideally 1). The parameters for corners $(a,b+1)$ can be determined in $O(1)$ time given their values for corners $(a,b)$. These functions are super-linear, the goal being to not vary too greatly from a rectangle in any one sense.

See Figure 4 for the effect of the turn_angle function. For certain pathological cases a possible improvement is to consider the maximum deviation of the tangent vector, rather than just comparing vectors adjacent to the corners. In addition to the interval ratio, a pair of vertices is excluded if the hard-set intervals preclude any possible mapping with those vertices as opposite corners. Since corner 2 must lie between 1 and 3, and 4 between 3 and 1, the choice of corners 1 and 3 restricts the choice of corners 2 and 4. Hence the quality function also takes into account the best corner-angles among the possible candidates for corners 2 and 4. Corner with interior angles greater than a user-specified threshold (usually $\pi$) can be excluded as well.

The second pass chooses corner 2 between 1 and 3, and corner 4 between 3 and 1. It is based on a similar quality function that considers the interior angles at corners 2 and 4 (ideally $\pi/2$), the ratio of intervals between opposite sides (ideally 1, for both pairs of opposite sides), and the amount of turning between successive corners (ideally 0).

If there are very few curves, say less than nine, then we forego the above heuristic and just exhaustively try all (< 210) combinations of four corners. We use similar functions to those described above, picking the combination with the best quality function. This quality function takes into account the fact that if a side must increase by some number of intervals to match the opposite side and many curves are hard-set, then the change must be distributed among the remaining soft-set curves. For illustration, the examples in section 2.3 and most other examples don't use the exhaustive search.

After intervals are assigned and curves are meshed, a similar heuristic is used to decide where to place the corners for the actual surface meshing. There is additional freedom, as any mesh vertex can be a corner, not just a curve vertex. In some volume meshing schemes, this freedom is not available. Usually the chosen corners are the same, but not always. For example, the circle in Figure 3 has only one bounding curve and no corners to chose before curve meshing, and the union of circles in Figure 3 has only bad reflex corners before curve meshing. Picking corners for surfaces with meshed curves takes time $O(m)$, where $m$ is the number of mesh edges, since in each pass only the $m/2$ pairs of mesh vertices that exactly divide the surface need be considered. (One could try all possible quadruples of corners in time $O(m^2)$.)

The following pseudocode describes the basics of the three main functions and their scaling with respect to each other.

```
turn_angle_function( turn_angle,
                     target_turn_angle )
    return
     | turn_angle - target_turn_angle |^1.7 / 2

corner_angle_function( corner_angle,
                       target_corner_angle)
    ratio =
        (corner_angle-target_corner_angle ) /
        target_corner_angle
    smoothly blow up ratio as corner_angle
    approaches a threshold (usually π).
    return 2 * ratio^1.7

interval_ratio_function( side_intervals,
                         opposite_side_intervals )
    ratio =
     side_intervals / opposite_side_intervals
    if (ratio < 1) then ratio = 1/ratio
    return ratio - 1
```

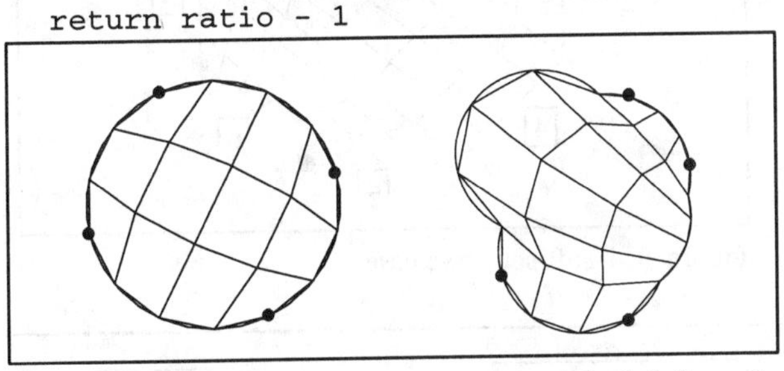

Figure 3. The left circle has no corners, and the left figure has only bad corners, until after intervals are assigned and the curves are meshed.

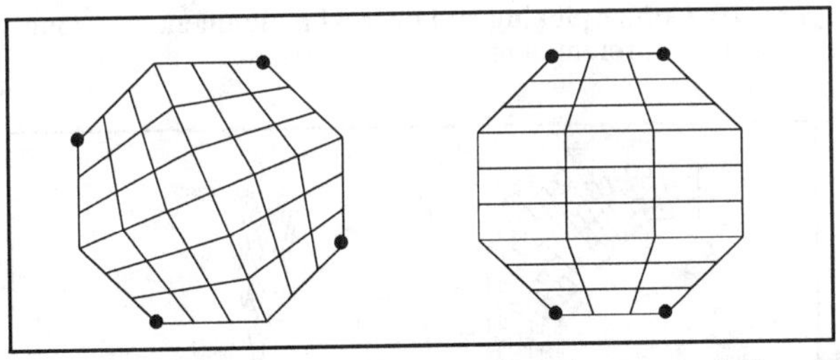

Figure 4. Without the turning function, the corner choices for both octagons have the same quality function value, as opposite-side intervals are equal and corner angles are all $3\pi/4$. But with the turning function the more regular mesh on the left is preferred.

## 2.2 Provable Corner Picking

Occasionally when there are many hard-set curves, the above heuristic fails to find a set of corners that admits a feasible solution to the interval assignment problem. In that case we can provably find a feasible a feasible set of corners, if any exist, in time $O(n^2 \log n)$, where $n$ is the number of curves bounding the surface. However we can't prove anything about the quality of the resultant mesh. For example the angles at the corners may be large. There are four cases, depending on the number of soft-set curves. We then apply an incremental shifting heuristic to get a good quality set of corners. Figures 6 to 9 illustrate these cases on an example.

**Case three or more soft-set curves.** We can find a feasible solution (one always exists) in $O(n)$ time. We put a soft-set curve by itself on a side, say opposite a side that's all hard-set, and the remaining two or more soft-set curves on sides opposite each other.

**Case two soft-set curves.** If there is a solution with both soft-set curves on the same side or on opposite sides, then there would also be a solution if one of the soft-set curves were hard-set to one interval. So we consider this a degenerate form of the one soft-set curve case below, and here only consider solutions with soft-set curves on adjacent sides.

Finding an adjacent-side solution if one exists takes time $O(n^2)$. If there is a solution with soft-set curves $S0$ and $S1$ on adjacent sides (sharing corner 2) then sliding corners 1 and 3 towards $S0$ and $S1$ is also a solution. Thus we reduce to the following form: $S0$ is immediately after corner 1, and $S1$ is immediately before corner 3, and the sum of the hard-set intervals from corner 3 to 1 is greater than the hard-set sum from 1 to 3. See Figure 5 left.

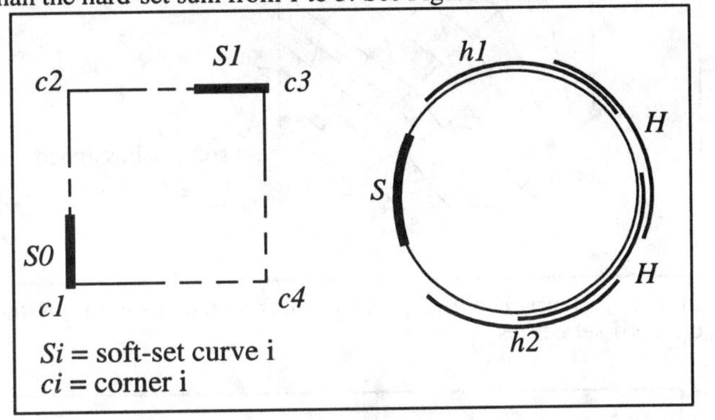


Figure 5. Left, the canonical form of a feasible solution with two soft-set curves $S0$ and $S1$ on adjacent sides. Right, the canonical form of a feasible solution with one soft set curve $S$. The $H$s are subsequences of hard-set edges, with interval count $A$. For a given $A$, if any pair of $H$ form a feasible choice of corners, then the pair $h1$ and $h2$ closest to $S$ on either side does.

Finding corners 1 and 3 satisfying the hard-set sum constraint takes linear time, and checking all possible pairs of corners $c2$ and $c4$ for feasibility takes only $O(n^2)$ time if we incrementally update parameters.

**Case one soft-set curve.** In this case any feasible corner solution has two opposite hard-set sides with exactly the same interval sum $A$. The remaining side $s1$ containing the soft-set curve $S$ has a smaller hard-set interval sum than its opposite side $s2$.

Finding a solution in time $O(n^2 \log n)$ time is somewhat tricky. Among all candidate sides $H$ with interval sum $A$, there is a side $h1$ closest to $S$ in the clockwise direction, and a side $h2$ closest in the counter-clockwise direction. See Figure 5 right. If there is any pair of candidate sides that do not overlap, then $h1$ and $h2$ do not overlap. In addition, among all candidate sides $H$, choosing $h1$ and $h2$ as opposite sides gives $s1$ with the smallest hard-set interval sum and $s2$ with the largest hard-set interval sum. Hence, for a given $A$, finding $h1$ and $h2$ is sufficient to determine if there is a solution. In practice we find $h1$ for various $A$ and as we encounter other $H$ with the same $A$, we stop when we have a feasible solution.

In particular, our algorithm is as follows: We consider all pairs of corners not straddling $S$ as candidate sides $H$. We first consider $H$ with smaller first corner. We keep a list of these $H$ sorted by hard-set interval sum $A$. If a second $H$ is found with a given $A$, then using stored data we can in constant time check if the two candidates form a feasible solution. If they do, we're done. If they don't, then we retain the first candidate (its $h1$ for $A$) and continue.

**Case no soft-set curves.** There is a $O(n \log n)$ algorithm in this case. Any two pairs of corners that exactly divide the interval sum in half are a feasible solution. All such pairs can be found by advancing opposite corners in lock-step manner. However, we have not implemented this since the heuristic (section 2.1) would have found a feasible solution if one existed, albeit in $O(n^2)$ time.

## 2.3 Corner Picking Examples

The first set of examples, Figures 6 to 9, illustrates heuristic corner picking, and each of the cases of the provable corner picking.

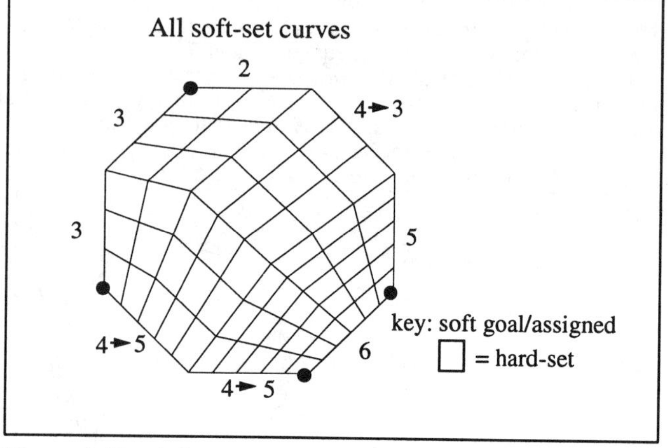


Figure 6. Heuristic corner picking works well on examples without hard-set curves.

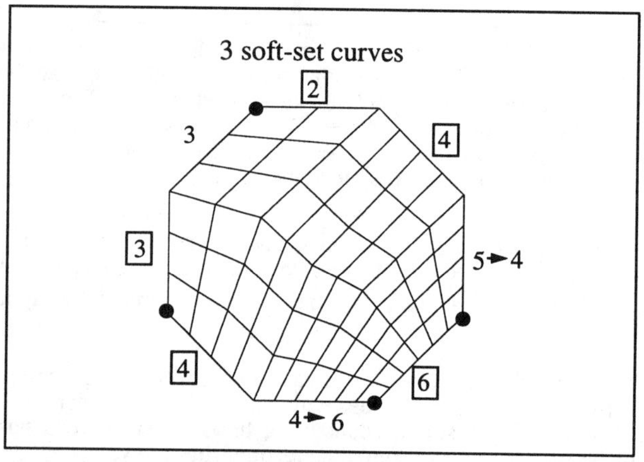


Figure 7. The heuristic and the exhaustive-search were turned off to demonstrate the provable algorithm for the 3+ soft-set curves case plus the incremental shifting heuristic.

The second set of examples, Figures 10 to 18, shows the results of corner picking and interval assignment on a sampling of surfaces.

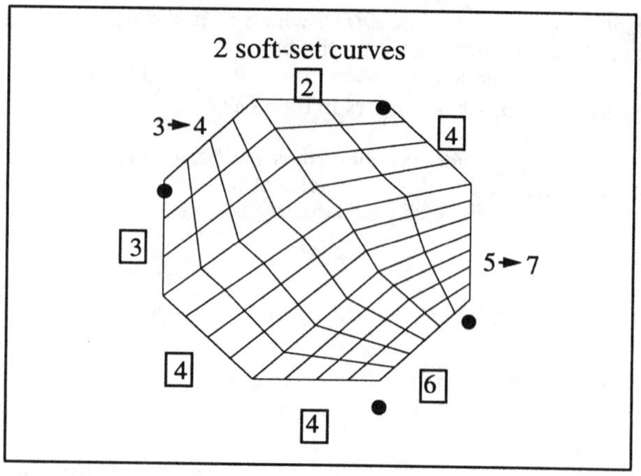


Figure 8. Provable corner picking plus heuristic improvement, 2-soft-set curves case.

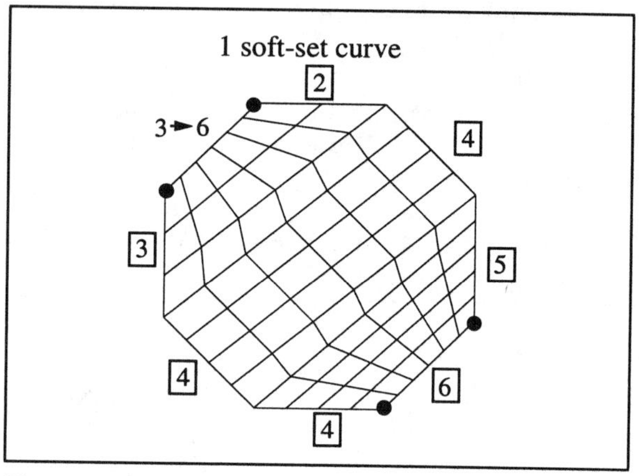


Figure 9. 1-soft-set curve case.

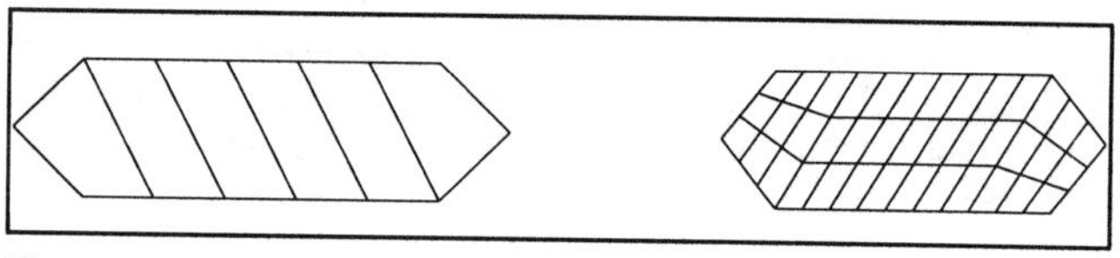

Figure 10. Corner picking and interval assignment on a long surfaces, at two different densities.

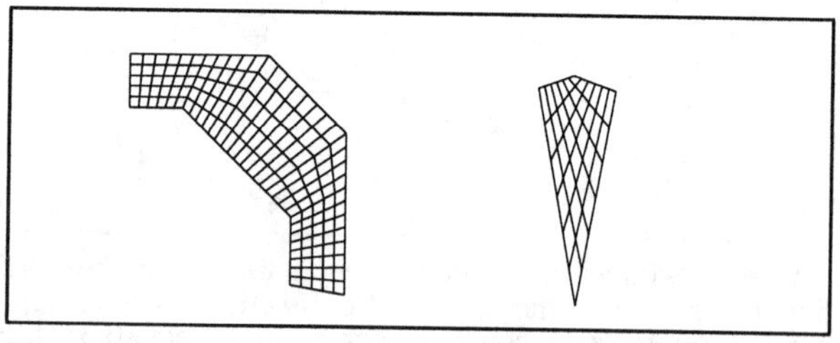

Figure 11. Some more corner picking and interval assignment examples.

# 3 Interval Assignment

We briefly give an overview of the interval assignment algorithm for a general advancing-front hexahedral meshing algorithm. First corners are picked for the primitive and semi-structured surface meshing algorithms. Then the interval constraints for all surfaces are assembled into a mixed-integer linear program (MILP). The various surface meshing algorithms create two types of constraints. First, more structured algorithms create constraints of the "intervals on opposite sides are equal", or "sum of intervals on two sides are greater than the third" variety. Second, a quadrilateral mesh of a surface must have an even number of mesh edges on its boundary, so the unstructured algorithms create "sum of intervals on all sides are even" constraints. The sum-even constraints force the problem

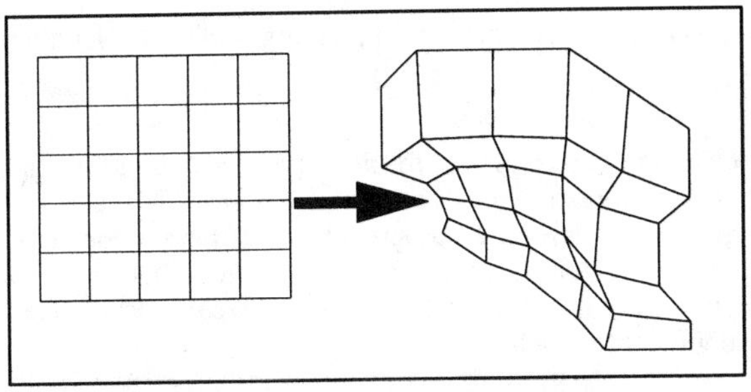

Figure 12. Abstract regular grid and mapping to planar surface geometry.

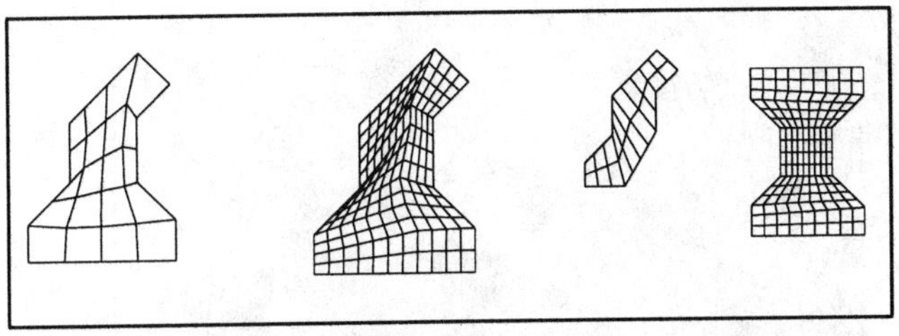

Figure 13. Corner picking and interval assignment examples.

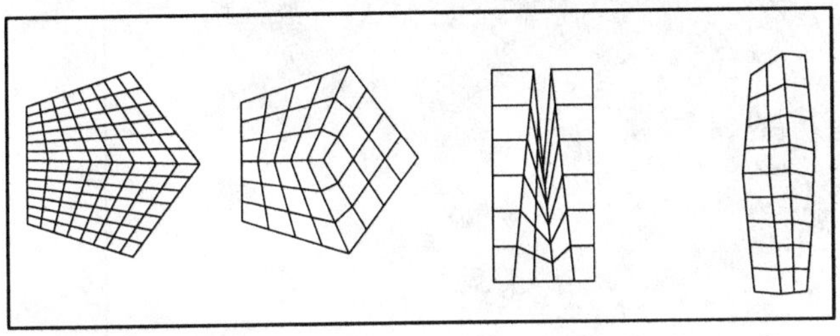

Figure 14. Corner picking and interval assignment examples. The center pentagon is triangle-primitive meshed.

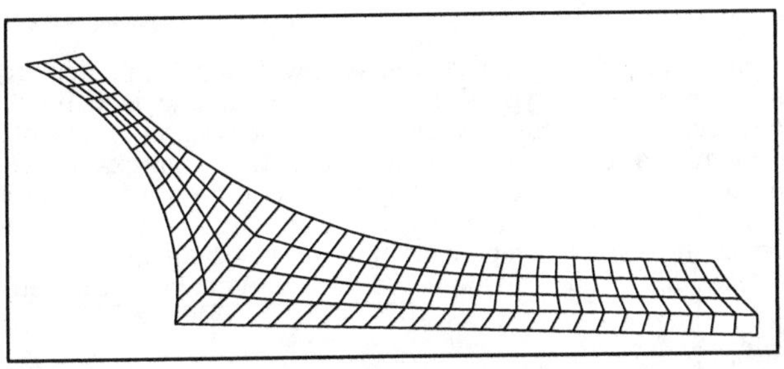

Figure 15. A real-world example where picking the four vertices with the smallest angles would have lead to a poorer mesh.

to be mixed-integer. The intervals themselves must also be integer, but this can often be obtained "for free" by careful MILP formulation.

The problem set-up is as follows: The soft-set curve intervals are the variables of the mixed-integer linear program (MILP). We add two extra *delta* variables for each curve, that compute the positive and negative difference of the assigned-interval to the goal-interval. We weight the deltas inversely proportional to the interval goal (to compute relative change). We add another variable $M$ that computes the maximum of the weighted deltas.

The solution process has two steps. In the first step we relax the integer variables: We solve the LP with objective minimizing $M$. We find a curve that forced $M$ to be as large as it is, and fix that curve's intervals to be the nearest integer value and remove it from

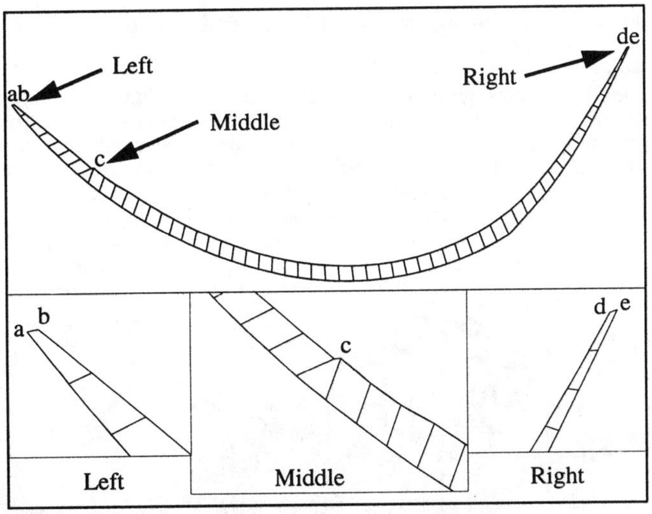


Figure 16. In this tire-section model, choosing the four vertices with smallest angles as corners would have lead to corners of (abce) and an unacceptable mesh, whereas our heuristic chose corners (abde).

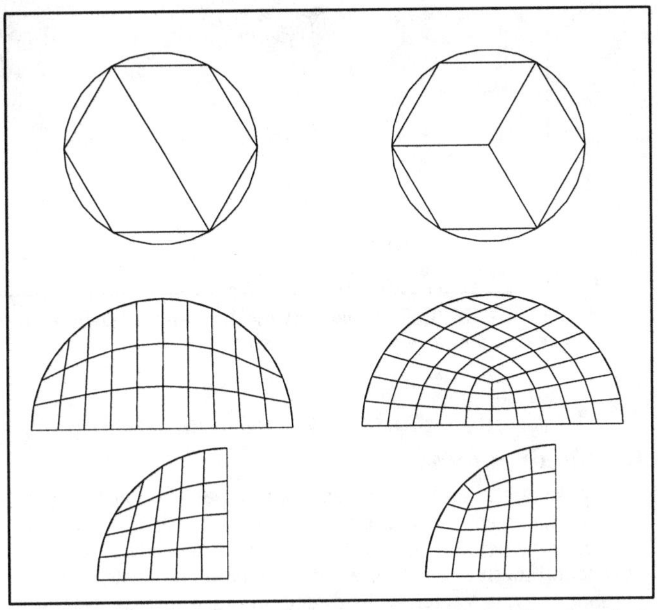

Figure 17. Left, rectangle-primitive meshes. Right, triangle primitive meshes on the same geometry with the same interval goals.

Figure 18. Another comparison of rectangle and triangle meshing.

the LP. We repeat until all curves are fixed or $M$ is zero. At the end of the first step we have an integer solution that satisfies all of the constraints except the sum-even constraints.

In the second step we set tight bounds on the sum-even "$k$" and curve variables, then solve the MILP minimizing the weighted sum of intervals (for speed). The Branch & Bound procedure may take too long (for a given set of bounds it may take exponential time). If we don't find a solution within an allotted amount of time, we try less tight bounds. We have four sets of bounds. There is guaranteed to be a solution for the least restrictive bounds, but since there is no guarantee on running time, we might not find a solution within the allotted time.

Our techniques give interval assignments that have very high fidelity to the user-desired goals: Most previous work relies on minimizing the sum of differences between assigned and goal intervals, coupled with local adjustments as a pre-process, whereas we essentially minimize the lexicographic vector of differences. We have noted cases where we succeed but some commercial packages fail to assign intervals compatible with meshing the surfaces. However, our techniques are slower because we iteratively solve the relaxed LP, taking time $O(n^3)$ rather than $O(n^2)$.

Our techniques are practical for models with up to a few thousand curves (or more if we're lucky and certain goals and meshing algorithms are chosen). To progress beyond that we conjecture that the problem should be divided up into subproblems, as is typically done with large LPs. Figure 19 shows some real-world models with about 500 curves each.

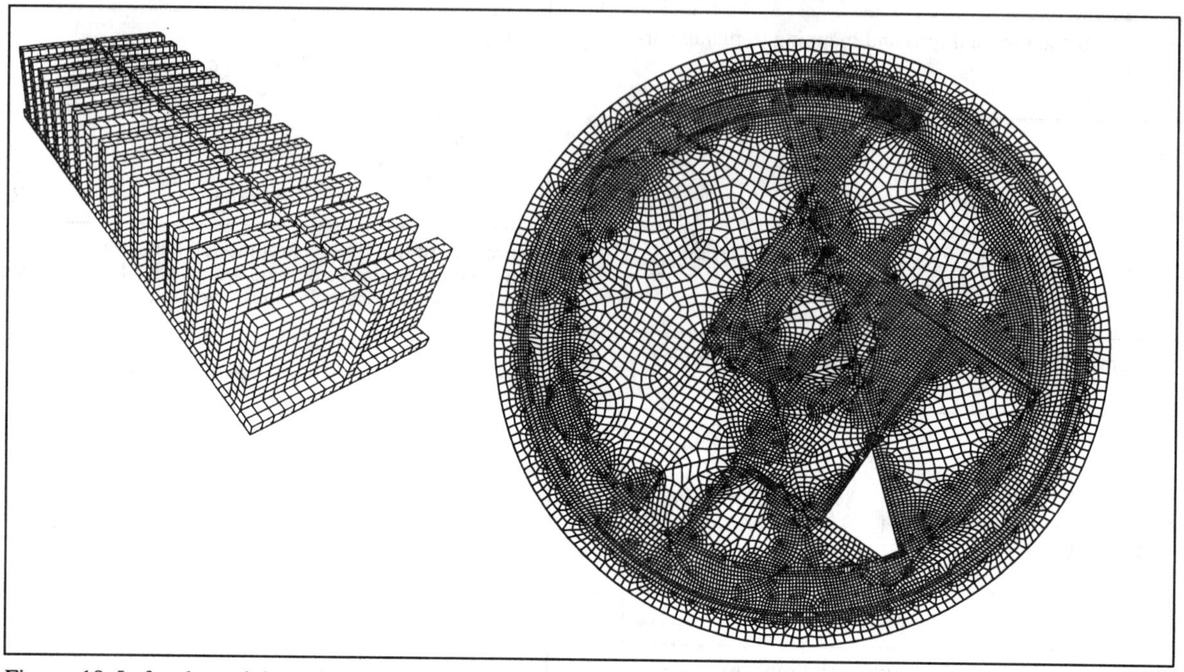

Figure 19. Left, a heat-sink meshed with a generalization of mapping. Right, a mostly unstructured mesh of a complex of 2d surfaces, courtesy of David Hensinger.

## 3.1 Ensuring the existence of a compatible hexahedral mesh

We consider the problem of generating a hexahedral mesh of a volume that exactly matches a quadrilateral mesh of the surfaces bounding the volume. Mitchell[5] notes that for such a compatible hexahedral mesh to exist for a handlebody, there must be an even number of quadrilaterals, and certain edge cycles (the ones contractible to a point in the volume) must be even. Being able to insert cutting disks along representative cycles is sufficient to show the existence of a hex mesh. See also Thurston[6]. Eppstein[7] gives a more constructive proof of the existence of a hexahedral mesh, in the case that the quadrilateral mesh is even and two-colorable. (Eppstein's sufficient condition is stronger in that two-colorable implies that **every** edge-cycle is even, yet weaker in that it is not restricted to handlebodies.)

Here we give a practical way to construct a quadrilateral mesh satisfying Eppstein's sufficient conditions, ensuring that a compatible hexahedral mesh exists. Thurston has also noted that a mesh with every quadrilateral divided into four admits a compatible hexahedral mesh. In essence, we reduce the existence problem to a surface cutting problem and a trivial interval assignment problem.

First, each surface bounding the volume must be "cut" by inserting two-sided curves, to reduce its topology to that of a disk; see Figure 20. This is somewhat analogous to inserting the cutting disks of Mitchell[5], but in a lower dimension and it does not restrict us to handlebodies. Also, finding these cutting curves is a much easier problem, perhaps solved by sweeping, particularly if we restrict to the typical case that the volume is defined by constructive solid geometry, and each surface is a subset of a plane, a sphere, or a cone. In the general case, finding cutting curves could be accomplished by forming a quadrilateral mesh of the surface, then retaining as cutting curves some chains of mesh edges connecting disjoint components of the surface boundary (one-sided surfaces don't bound a volume, so they need not be considered).

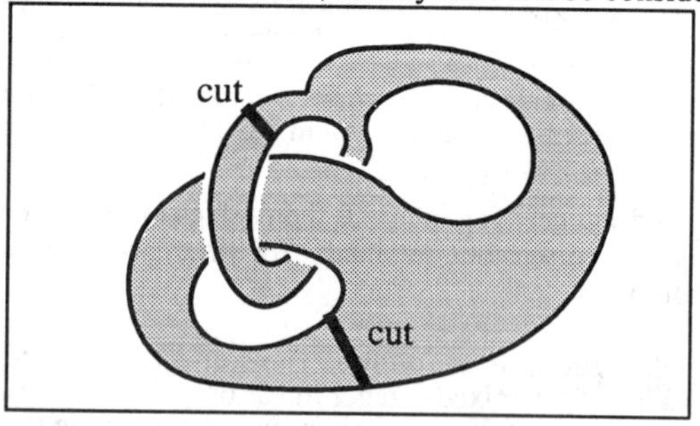


Figure 20. Cutting a surface into a topological disk by introducing two-sided cutting curves.

We can ensure that every curve (including the cutting curves) has an even number of intervals or mesh edges, and simultaneously satisfy the meshing algorithm requirements, by adding constraints to the interval assignment MILP (see section 3). We mesh the curves, then the surfaces with a quadrilateral meshing algorithm. (Every surface admits a quad mesh, since the number of bounding

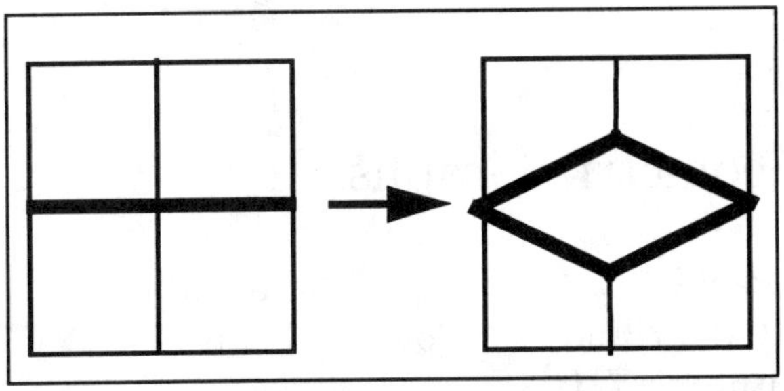

Figure 21. Adding a quad to a surface mesh.

mesh edges is even.) If there is an odd number of quadrilaterals, then we add one e.g. by opening a pair of mesh edges sharing a mesh node into a quadrilateral; see Figure 21. We claim the surface mesh is two-colorable (see Theorem 1), so by Eppstein[7] a hex mesh exists.

**Theorem 1** *If all bounding curves have an even number of intervals, and every bounding surface is topologically a disk, then any even quadrilateral mesh of the bounding surfaces admits a compatible hexahedral mesh.*

**Proof.** We claim the mesh of each surface is two-colorable red and green, with vertices of the surface red, hence the surfaces taken together are two-colorable. First, the mesh of the curves is two-colorable with all vertices red, since any path between two vertices is even (since curves only meet at vertices and curves have an even number of mesh edges). For each surface, there is a two-coloring that respects this coloring of the curves: First, each surface has the topology of a disk, and hence any quadrilateral mesh of each surface has a two-coloring[7]. Second, vertices are separated by an even length chain of edges and hence must all be the same color, so chose red.

# 4 Conclusions

First, we have demonstrated a practical and robust way of choosing which four vertices of a surface correspond to corners of a rectangle for the purposes of mapped meshing. This process takes into account the user-required and user-desired number of intervals on each bounding curve of the surface. This method has been extended to triangle-primitive mapped meshing, and could be extended to other n-gon primitive mapped meshing as well. Second, we have outlined an implementation of a practical and robust way to solve the problem of globally assigning intervals to a complex of surfaces, so that each surface may be meshed according to pre-set meshing schemes and corners if possible. Third, we have outlined a practical scheme for generating a surface mesh suitable as a starting point for an advancing-front hexahedral-meshing algorithm.

A number of open problems remain. Among them are: Is there a practical way to pick corners such that the interval assignment problem is feasible? Möhring[2] answers this for one set of meshing algorithms and notes that it is NP-complete. Also, corner picking and interval assignment may be infeasible due to hard-set intervals: automation would be increased if we could determine a small set of hard-set intervals to relax to make the problem feasible. Also, extending the interval assignment solution to take into account the constraints imposed by semi-structured hexahedral meshing algorithms such as sweeping is important. Another practical problem is determining and implementing a way to "cut" surfaces into topological disks, such that the cuts do not create small features or otherwise complicate the meshing problem.

# Acknowledgements

I wish to thank Sandia's Discrete Algorithms Group (DAG) and especially Jonathan E. Atkins for contributions to section 2.2.

# References

[1] T. D. Blacker, W. J. Bohnhoff, T. L. Edwards, J. R. Hipp, R. R. Lober, S. A. Mitchell, G. D. Sjaardema, T. J. Tautges, T. J. Wilson, W. R. Oakes, S. Benzley, J. C. Clements, L. Lopez-Buriek, S. Parker, M. Whitely, D. White, and E. Trimble. CUBIT mesh generation environment volume 1: users manual. SAND94-1100, Sandia National Laboratories, Albuquerque, New Mexico, May 1994.

[2] Rolf H. Möhring, Matthias Müller-Hannemann, and Karsten Weihe. Mesh refinement via bidirected flows: Modeling, complexity, and computational results. Technische Universität Berlin, Department of Mathematics, Report No. 520 / 1996. URL: http://www.math.TU-Berlin.DE/~mhannema/projecteng.html

[3] T. K. H. Tam and C. G. Armstrong. Finite element mesh control by integer programming, International Journal for Numerical Methods in Engineering, vol 36, 2581-2605, 1993.

[4] Esther M. Arkin, L. Paul Chew, Daniel P. Huttenlocher, Klara Kedem, and Joseph S. B. Mitchell. An efficiently computable metric for comparing polygonal shapes, Cornell University, Department of Computer Science, TR 89-1007, May 1989.

[5] Scott A. Mitchell. A characterization of the quadrilateral meshes of a surface which admit a compatible hexahedral mesh of the enclosed volume, Proc. 13th Annual Symposium on Theoretical Aspects of Computer Science (STACS '96), Lecture Notes in Computer Science 1046, pages 465-476, Springer, 1996. Also http://sass577.endo.sandia.gov/9225/Personnel/samitch/exist-abstract.html

[6] William Thurston. Hexahedral decomposition of polyhedra. Posting to sci.math, 25 Oct 1993. Available online at http://www.ics.uci.edu/~eppstein/gina/Thurston-hexahedra.html.

[7] David Eppstein. Linear complexity hexahedral mesh generation. 12th ACM Symp. Comp. Geom., Philadelphia (1996) 58-67. Also http://www.ics.uci.edu/~eppstein/

# Ramsey–Type Results for Geometric Graphs. II

*Gyula Károlyi*[*]
Eötvös University, Budapest
and ETH Zürich

*János Pach*[†]
City College, CUNY and
Courant Institute, NYU

*Géza Tóth*[‡]
Courant Institute, NYU

*Pavel Valtr*[§]
Charles University, Prague and
DIMACS Center, Rutgers University

## Abstract

We show that for any 2–coloring of the $\binom{n}{2}$ segments determined by $n$ points in the plane, one of the color classes contains non-crossing cycles of lengths $3, 4, \ldots, \lfloor\sqrt{n/2}\rfloor$. This result is tight up to a multiplicative constant. Under the same assumptions, we also prove that there is a non-crossing path of length $\Omega(n^{2/3})$, all of whose edges are of the same color. In the special case when the $n$ points are in convex position, we find longer monochromatic non-crossing paths, of length $\lceil\frac{n+1}{2}\rceil$. This bound cannot be improved. All of these cycles and paths can be found by $O(n^2)$ time algorithms. We also discuss some related problems and generalizations. In particular, we give sharp estimates for the largest number of disjoint monochromatic triangles that can always be selected from our segments.

[*]Supported by NSF grant DMS-9304580, the Alfred Sloan Foundation, and DIMACS under NSF grant STC-91-19999.

[†]Supported by NSF grant CCR-94-24398, PSC-CUNY Research Award 663472, and OTKA-T-020914.

[‡]Supported by OTKA-14220, OTKA-T-020914, and the Alfred Sloan Foundation.

[§]Supported by the Czech Republic grant GAČR 0194, Charles University grants GAUK 193/1996 and 194/1996, and by DIMACS Center.



*Computational Geometry 97* Nice France



## 1 Introduction

A *geometric graph* is a graph drawn in the plane so that every vertex corresponds to a point, and every edge is a closed straight-line segment connecting two vertices but not passing through a third. The $\binom{n}{2}$ segments determined by $n$ points in the plane, no three of which are collinear, form a *complete* geometric graph with $n$ vertices (see [PA95]). In classical Ramsey theory, we want to find large monochromatic subgraphs in a complete graph whose edges are colored with several colors [B74], [GRS90]. Most questions of this type can be formulated for complete geometric graphs, where the monochromatic subgraphs are required to satisfy certain geometric conditions. The investigation of these problems was initiated in [KPT96].

A subgraph of a geometric graph is said to be *non-crossing*, if no two of its edges have an interior point in common. In the present paper, we show how to find large non-crossing monochromatic paths and cycles in a geometric graph whose edges are colored with two colors.

**Theorem 1.1** *For any 2–coloring of the edges of a complete geometric graph with n vertices, there exist non-crossing cycles of length* $3, 4, \ldots, \lfloor\sqrt{n/2}\rfloor$, *all of whose edges are of the same color. The order of magnitude of this bound cannot be improved.*

**Theorem 1.2** *For any 2–coloring of the edges of a complete geometric graph with n vertices, there exists a non-crossing path of length* $\Omega(n^{2/3})$, *all of whose edges are of the same color.*

The last result improves the bound $\Omega(n^{1/2})$ established in [KPT96]. It is very likely that the $\Omega(n^{2/3})$

bound in Theorem 1.2 can be further improved to (nearly) linear. We can verify this conjecture for *convex geometric graphs*, i.e., for geometric graphs whose $n$ points form the vertex set of a convex polygon.

**Theorem 1.3** *For any 2-coloring of the edges of a complete convex geometric graph with $n \geq 3$ vertices, there exists a non-crossing monochromatic path of length $\lfloor \frac{n+1}{2} \rfloor$. This bound cannot be improved.*

The proof of Theorem 1.3 uses the following unpublished result of Micha Perles, whose proof is included.

**Theorem 1.4 (Perles)** *If a convex geometric graph of $n \geq k+1$ vertices has more than $\lfloor (k-1)n/2 \rfloor$ edges than it contains a non-crossing path of length $k$. This bound cannot be improved.*

A graph is called a *caterpillar*, if it is a tree containing no three edge disjoint paths of length two starting at the same vertex. In fact, Perles proved that under the conditions of Theorem 1.4, one can always find a non-crossing subgraph isomorphic to any given caterpillar of $k+1$ vertices.

The proofs of Theorems 1.1-1.4 can be turned into $O(n^2)$ time algorithms to find monochromatic cycles resp. paths with the required properties.

The above results can be rephrased using the following notation. Let $\mathcal{G}$ be a class of (so-called *forbidden*) geometric subgraphs. We want to determine the smallest number $R = R(\mathcal{G})$ with the property that every complete geometric graph with $R$ vertices, whose edges are colored with 2 colors, contains a monochromatic subgraph belonging to $\mathcal{G}$. If we restrict our attention to *convex* geometric graphs, then the corresponding function is denoted by $R_c(\mathcal{G})$. Clearly, $R_c(\mathcal{G}) \leq R(\mathcal{G})$.

For any positive integer $k$, let $k\mathcal{G}$ denote the class of all geometric graphs that can be obtained by taking the union of $k$ pairwise disjoint members of $\mathcal{G}$, any two of which can be separated by a straight line.

**Theorem 1.5** *Let $\mathcal{G}$ be any class of geometric graphs, each of which has at least two vertices.*

**(i)** *If $k$ is a power of 2 then*

$$R(k\mathcal{G}) \leq (R(\mathcal{G})+1)k - 1.$$

**(ii)** *For any $k > 0$*

$$R(k\mathcal{G}) \leq \left\lceil \frac{3(R(\mathcal{G})+1)}{2} \right\rceil k - \left\lceil \frac{R(\mathcal{G})+1}{2} \right\rceil .$$

**(iii)** *For any $k > 0$*

$$R_c(k\mathcal{G}) \leq (R_c(\mathcal{G})+1)k - 1.$$

In particular, if $\mathcal{G} = \mathcal{T}$ is the class of triangles, we have $R(\mathcal{T}) = 6$. Thus, by Theorem 1.5(i), we obtain

$$R(k\mathcal{T}) \leq 7k - 1,$$

provided that $k$ is a power of 2. The following theorem shows that this result cannot be improved.

**Theorem 1.6** *Let $\mathcal{T}$ denote the class of triangles and let $k$ be a positive integer. Then*

$$R_{\Delta}(k) := R(k\mathcal{T}) \geq (R(\mathcal{T})+1)k - 1 = 7k - 1 .$$

Using the approach of [KPTT97], one can design an $O(n^{\log\log n + 2})$-time algorithm to find $k$ pairwise disjoint triangles of the same color in any complete geometric graph of $n = 7k - 1$ vertices, whose edges are colored with two colors, provided that $k$ is a power of 2. (Note that in an abstract graph, one can always find $\lfloor n/5 \rfloor$ pairwise *vertex-disjoint* triangles of the same color [BES75].)

The paper is organized as follows. Sections 2 and 3 contain the proofs of Theorems 1.1 and 1.2, respectively. Theorems 1.3 and 1.4 are proved in Section 4, while Sections 5 and 6 contain the proofs of Theorems 1.5 and 1.6, respectively. At the end of each section, we give an efficient algorithm to find a subgraph with the properties guaranteed by the corresponding theorem(s).

## 2 Cycles – Proof of Theorem 1.1

Theorem 1.1 is an immediate consequence of the following result.

**Theorem 2.1** *Let $\mathcal{C}_k$ be the class of all non-crossing cycles of length $k$, and let let $\mathcal{D}_k$ be the class of all non-crossing cycles of length $k$, triangulated from a vertex. Then*

$$\Omega(k^2) = R_c(\mathcal{C}_k) \leq R(\mathcal{C}_k) \leq R(\mathcal{D}_k) = O(k^2).$$

**Proof.** First we prove that $R_c(\mathcal{C}_k) > (k-1)^2$.

Take $(k-1)^2$ points on a circle and partition them into $k-1$ groups, each containing $k-1$ consecutive points. Color with red all edges between points in different groups, and color with blue all edges between points belonging to the same group.

Any red non-crossing cycle contains at most one point from each group, hence it can not have more than $k-1$ points. On the other hand, all vertices of a blue cycle are from the same group, so there is no blue cycle with more than $k-1$ points.

Next we show that $R(\mathcal{D}_k) \leq 2(k-1)(k-2)+2$.

Let $P$ denote the vertex set of a complete geometric graph $G$ of $2(k-1)(k-2)+2$ vertices, whose edges are colored with red and blue. Let $p$ be a vertex of the convex hull of $P$. Since there are $2(k-1)(k-2)+1$ edges incident to $p$, at least $(k-1)(k-2)+1$ of them are of the same color, say, red. Let $p_1, p_2, \ldots, p_{(k-1)(k-2)+1}$ be vertices of $G$, listed in clockwise order of visibility from $p$, such that each edge $pp_i$ is red. A path $p_{i_1}p_{i_2}\cdots p_{i_j}$ is said to be *monotone* if $i_1 < i_2 < \ldots < i_j$.

Define a partial ordering of the vertices $p_1, p_2, \ldots, p_{(k-1)(k-2)+1}$, as follows. Let $p_i < p_j$ if $i < j$ and there is a monotone red path connecting $p_i$ to $p_j$. By Dilworth's theorem [D50], there are either $k-1$ elements that form a totally ordered subset, or $k$ elements that are pairwise incomparable. In the first case, there is a monotone red path $q_1, q_2, \ldots, q_{k-1}$ and we can complete it to a non-crossing red cycle $p, q_1, q_2, \ldots, q_{k-1}$ of size $k$, together with the corresponding diagonals from $p$. In the second case, there is a complete blue subgraph of $k$ vertices $q_1, q_2, \ldots, q_k$ because any two incomparable elements are connected by a blue edge. By a result of Gritzmann et al. [GMPP91, PA95], this contains a non-crossing cycle of length $k$, triangulated from a vertex. □

**Algorithm.** Based on this proof, in any 2-colored complete geometric graph of $n = 2(k-1)(k-2)+2$ vertices, one can find a monochromatic non-crossing cycle of length $k$, triangulated from a vertex, in $O(k^4) = O(n^2)$ time. We sketch the algorithm, using the notation in the above proof.

1. Find the vertices $p, p_1, p_2, \ldots, p_{(k-1)(k-2)+1}$. [This requires $O(k^2 \log k)$ steps.]

2. Build a data-structure to store for each vertex $p_i$ the longest clockwise monotone red path on $\{p_1, p_2, \ldots, p_i\}$ ending at $p_i$, together with its length $r_i$ ($i = 1, 2, \ldots, (k-1)(k-2)+1$). [$O(k^4)$ steps.]

3. Check if there is any $r_i \geq k-2$. If yes, then return the corresponding non-crossing red cycle of length $k$, triangulated from $p$. [$O(k^2)$ steps.]

4. If all $r_i < k-2$, then find $k$ vertices $p_{i_1}, p_{i_2}, \ldots, p_{i_k}$ such that $r_{i_1} = r_{i_2} = \ldots = r_{i_k}$. Return a non-crossing blue cycle of length $k$, triangulated from a vertex, induced by $\{p_{i_1}, p_{i_2}, \ldots, p_{i_k}\}$. [$O(k^2 \log k)$ steps.]

## 3 Paths – Proof of Theorem 1.2

Let $G$ be a complete geometric graph whose edges are colored red and blue. Suppose that $G$ contains no monochromatic non-crossing path of length $m$. To establish Theorem 1.2, it is enough to prove that $G$ has at most $O(m^{3/2})$ vertices.

Suppose without loss of generality that no two vertices lie on a vertical line. For each vertex $v$ of $G$, let $b(v)$ (resp. $r(v)$) denote the length of the longest $x$-monotone blue path (resp. red) in $G$ whose rightmost vertex is $v$.

Since every monotone path is non-crossing, $b(v)$ and $r(v)$ are integers between 0 and $m-1$. Observe that $(b(v), r(v)) \neq (b(v'), r(v'))$ for $v \neq v'$. Indeed, if $v \neq v'$ are two vertices of $G$ such that $v'$ lies to the right of $v$, then $b(v') > b(v)$ (if $vv'$ is colored blue) or $r(v') > r(v)$ (if $vv'$ is colored red). As a consequence, $G$ has at most $m^2$ vertices. The following more careful analysis gives a better bound.

Define $\mathcal{U}$ as the set of pairs with $(b, r) = (b(u), r(u))$ for some vertex $u$ of $G$. For $0 \leq b_1 \leq b_2 \leq m$ and $0 \leq r_1 \leq r_2 \leq m$, define a subset $\mathrm{Rect}(b_1, b_2, r_1, r_2)$ of $\mathcal{U}$ by

$$\mathrm{Rect}(b_1, b_2, r_1, r_2)$$
$$= \{(b, r) \in \mathcal{U} \mid b_1 \leq b < b_2, r_1 \leq r < r_2\}.$$

Every pair $(b, r)$, $0 \leq b \leq m$, $0 \leq r \leq m$, defines a partition of $\mathcal{U}$ into the following four subsets:

$$\begin{aligned} Q^{++}(b,r) &= \mathrm{Rect}(b, m, r, m), \\ Q^{+-}(b,r) &= \mathrm{Rect}(b, m, 0, r), \\ Q^{-+}(b,r) &= \mathrm{Rect}(0, b, r, m), \\ Q^{--}(b,r) &= \mathrm{Rect}(0, b, 0, r). \end{aligned}$$

The key observation is the following.

**Lemma 3.1** *For any $t \in \{1, 2, \ldots, 2m-1\}$, there are two integers $b(t), r(t) \in \{0, 1, \ldots, m\}$ such that $b(t) + r(t) = t$ and*

$$|Q^{+-}(b(t), r(t)) \cup Q^{-+}(b(t), r(t))| \leq 3m.$$

Before giving the proof of Lemma 3.1, we finish the proof of Theorem 1.2. Obviously, it suffices to show that $|\mathcal{U}| = O(m^{3/2})$.

Suppose for simplicity that $s = \sqrt{2m}$ is an integer. Somewhat inaccurately, we shorten the notation $Q^{++}(b(t), r(t))$ to $Q^{++}(t)$. We also shorten $Q^{+-}(b(t), r(t))$, $Q^{-+}(b(t), r(t))$, and $Q^{--}(b(t), r(t))$

analogously. We have

$$\mathcal{U} = \bigcup_{t=s,2s,\ldots,s^2-s} (Q^{+-}(t) \cup Q^{-+}(t))$$
$$\cup \bigcup_{t=0,s,2s,\ldots,s^2-s} (Q^{++}(t) \cap Q^{--}(t+s)),$$

because each pair $(b,r) \in \mathcal{U}$ with $ks \le b+r < (k+1)s$ lies in the first union (in a term with $t = ks$ or with $t = (k+1)s$) or in the term $Q^{++}(ks) \cap Q^{--}(ks+s)$ of the second union. Consequently,

$$|\mathcal{U}| \le \sum_{t=s,2s,\ldots,s^2-s} |Q^{+-}(t) \cup Q^{-+}(t)|$$
$$+ \sum_{t=0,s,2s,\ldots,s^2-s} |Q^{++}(t) \cap Q^{--}(t+s)|.$$

By Lemma 3.1, each term in the first sum is at most $3m$, and it is easy to check that each term in the second sum is at most $(s/2)^2 = m/2$. It follows that

$$|\mathcal{U}| \le (s-1)\cdot 3m + s\frac{m}{2} = O(m^{3/2}). \quad \square$$

It remains to prove Lemma 3.1.

**Proof of Lemma 3.1.** Suppose without loss of generality that $t \le m$. For each $b = 0, 1, \ldots, t$, denote

$$\Delta(b) = |Q^{-+}(b, t-b)| - |Q^{+-}(b, t-b)|.$$

It follows from

$$\Delta(0) \le 0 \le \Delta(m)$$

and from

$$|\Delta(b+1) - \Delta(b)|$$
$$= |\mathrm{Rect}(b, b+1, 0, m)| + |\mathrm{Rect}(0, m, t-b-1, t-b)|$$
$$\le m + m = 2m$$

that there is a $b_0 \in \{0, \ldots, m\}$ with

$$|\Delta(b_0)| \le m.$$

We now show that the lemma holds with $b(t) = b_0$ and $r(t) = t - b_0$. Suppose this is not true, i.e.,

$$|Q^{+-}(t)| + |Q^{-+}(t)| \ge 3m+1.$$

Then $|\Delta(b(t))| \le m$ gives

$$|Q^{+-}(t)| \ge m+1, \qquad |Q^{-+}(t)| \ge m+1.$$

Consequently, there is a vertical line $\ell$ such that a set $V_1$ of at least $\frac{m+1}{2}$ vertices to the left of $\ell$ corresponds to a subset of $Q^{+-}(t)$ or $Q^{-+}(t)$, and a set $V_2$ of at least $\frac{m+1}{2}$ vertices to the right of $\ell$ corresponds to a subset of $Q^{-+}(t)$ or $Q^{+-}(t)$, respectively. All edges of the complete bipartite geometric graph $(V_1 \cup V_2, V_1 \times V_2)$ are colored by the same color: red, if $V_1$ corresponds to elements of $Q^{+-}(t)$, and blue, otherwise. It is well-known (e.g., see [HS92]) that if $V_1$, $|V_1| \ge \frac{m+1}{2}$, is separated by a line from $V_2$, $|V_2| \ge \frac{m+1}{2}$, then the graph $(V_1 \cup V_2, V_1 \times V_2)$ contains a non-crossing path of length $m$, – a contradiction. $\square$

**Algorithm.** Based on the ideas of the proof of Theorem 1.2, we can design an algorithm to find a monochromatic path of length $m = \lfloor n^{2/3}/3 \rfloor$ in a complete 2-colored geometric graph on $n$ vertices in time $O(n^2)$, as follows.

1. Consider the vertices of $G$ from left to right, and for each of them find the number $b(v)$ (defined in the proof of Theorem 1.2), using the relation $b(v) = 1 + \max b(v')$, where the maximum is taken over all vertices $v'$ to the left of $v$ with the edge $vv'$ colored blue. Similarly, find the numbers $r(v)$. If any $b(v)$ or $r(v)$ is at least $m$, then return the corresponding path. [This takes $O(n^2)$ time.]

2. Find all entries of the matrix $M = (m_{b,r} : b, r \in \{0, 1, \ldots, m-1\})$ defined by

$$m_{b,r} = \begin{cases} 0, & \text{if } (b,r) \notin \mathcal{U}, \\ 1, & \text{if } (b,r) \in \mathcal{U}. \end{cases}$$

   Using this matrix, find the numbers $Q^{+-}(b,r), b, r \in \{0, 1, \ldots, m-1\}$, (defined in the proof of Theorem 1.2) as follows. For $b = m$ and for any $r$, set $Q^{+-}(m,r) = 0$. Similarly, for $r = 0$ and for any $b$, set $Q^{+-}(b,0) = 0$. For other entries $Q^{+-}(b,r)$, proceed by induction on $r - b$. Once we have found $Q^{+-}(b+1, r-1)$, $Q^{+-}(b+1, r)$, and $Q^{+-}(b, r-1)$, we can set

$$Q^{+-}(b,r) := Q^{+-}(b+1,r) + Q^{+-}(b,r-1) - Q^{+-}(b+1,r-1) + m_{b,r}.$$

   Analogously find the numbers $Q^{-+}(b,r)$. [$O(n^2)$ time.]

3. Find $1 \le b, r \le m-1$ such that

$$|Q^{+-}(b,r)| \ge m+1, \qquad |Q^{-+}(b,r)| \ge m+1.$$

   Following the proof of Theorem 1.2, find two $\lceil (m+1)/2 \rceil$-element sets separated by a vertical line, such that all edges between them have the same color. In this situation, it is known [HS92] how to find a monochromatic non-crossing path passing through all points of these subsets, in time $O(m \log m) = O(n \log n)$.

In Theorem 1.2, the edges of a complete geometric graph $G$ are colored by two colors. If they are colored by $r \geq 2$ colors, we obtain the following.

**Theorem 3.2** *Every complete geometric graph of $n$ vertices whose edges are colored by $r \geq 2$ colors, contains a non-crossing path of length $\Omega(n^{2/(2r-1)})$, all of whose edges are of the same color.*

**Proof.** Let the edges of $G$ be colored by $1, 2, \ldots, r$. Suppose that $G$ contains no non-crossing path of length $m$. For each color $i$, define a function $f_i$ on the vertex set of $G$ such that $f_i(v)$ is the length of the longest $x$-monotone monochromatic path of color $i$, whose rightmost point is $v$. Thus, for any vertex $v$, we have an $r$-tuple $f(v) = (f_1(v), \ldots, f_r(v))$ of integers between 0 and $m-1$. Of course, $f(v) \neq f(v')$ for $v \neq v'$, since $f(v)$ and $f(v')$ differ in the $i$-th coordinate, where $i$ is the color of the edge $vv'$. We now show by induction on $r$ that the number of vertices of $G$ is at most $O(m^{r-1/2})$. For $r = 2$, this was shown in the proof of Theorem 1.2. Suppose now that $r > 2$ and that the statement holds for $r-1$. Define $\mathcal{U}$ as the set of all $r$-tuples $f(v)$, where $v$ is a vertex of $G$. The set $\mathcal{U}$ is a disjoint union of $m$ sets $\mathcal{U}_j, j = 0, 1, \ldots, m-1$, where $\mathcal{U}_j$ consists of the $r$-tuples of $\mathcal{U}$ whose last coordinate is $j$. For each $j$, all edges between the vertices corresponding to elements of $\mathcal{U}_j$ are colored by colors $1, 2, \ldots, r-1$. By the inductive assumption, the size of $\mathcal{U}_j$ is at most $O(m^{r-3/2})$. Consequently, the size of $\mathcal{U}$ is at most $O(m^{r-1/2})$, and the result follows. □

# 4 Paths in convex geometric graphs – Proof of Theorems 1.3 and 1.4

The proof of Theorem 1.3 is based on the result of Perles (Theorem 1.4), whose proof is as follows.

**Proof of Theorem 1.4.** Let $G$ be a convex geometric graph with vertices $u_1, u_2, \ldots, u_n$ in clockwise order and with $|E(G)| = m > \lfloor (k-1)n/2 \rfloor$. For convenience, let $u_0 = u_n$ and $u_{n+i} = u_i$.

Fix $E_0 = \emptyset$, and define $E_1, E_2, \ldots, E_m$ inductively, as follows. There exists an edge $e_i \in \overline{E}_{i-1} = E(G) \setminus E_{i-1}$ such that one of the closed halfplanes determined by $e_i$ does not fully contain any element of $E(G) \setminus (E_{i-1} \cup \{e_i\})$. Such an edge will be called an *extreme* edge with respect to $\overline{E}_{i-1}$. Let $E_i = E_{i-1} \cup \{e_i\}$.

We also define some nonnegative integers $c_i(u_j)$ and $d_i(u_j)$. For any vertex $u_j$, let $I = I(u_j)$ be the smallest integer such that $E_I$ contains all edges of $G$ incident to $u_j$. Suppose first that $i < I$. Then, starting at $u_j$ and visiting the vertices of the $G$ in clockwise order, let $u_p$ and $u_q$ be the first and the last vertex such that $u_ju_p, u_ju_q \in \overline{E}_i$. Let $c_i(u_j)$ (resp. $d_i(u_j)$) denote the length of the longest non-crossing path in $E_i$ starting at $u_j$, all of whose vertices are in $\{u_j, u_{j+1}, \ldots, u_{p-1}\}$ (resp. $\{u_{q+1}, u_{q+2}, \ldots, u_j\}$). Next, if $i = I$, then $e_i = u_ju_p$ for some vertex $u_p$. Since $e_i$ is an extremal edge with respect to $\overline{E}_{i-1}$, we may suppose, by symmetry, that $\overline{E}_i$ has no element on the vertex set $\{u_j, u_{j+1}, \ldots, u_{p-1}, u_p\}$. If this is the case, then let $c_i(u_j)$ (resp. $d_i(u_j)$) denote the length of the longest non-crossing path in $E_i$ starting at $u_j$, all of whose vertices lie in $\{u_j, u_{j+1}, \ldots, u_p\}$ (resp. $\{u_{p+1}, u_{p+2}, \ldots, u_j\}$). (Otherwise we change the orientation.) Finally, if $i > I$ then put $c_i(u_j) = c_I(u_j)$ and $d_i(u_j) = d_I(u_j)$.

**Claim 4.1** *For every $0 \leq i \leq m$,*

$$\sum_{j=1}^{n} c_i(u_j) + \sum_{j=1}^{n} d_i(u_j) \geq 2|E_i|.$$

The claim is clearly true for $i = 0$. Let $0 < i \leq m$, and assume that the assertion has already been proved for $i-1$. Suppose that the endpoints of $e_i$ are $u_j$ and $u_l$. Since $c_i(u) \geq c_{i-1}(u)$ and $d_i(u) \geq d_{i-1}(u)$ for every fixed vertex $u$, it is enough to prove that

$$c_i(u_j) + d_i(u_j) + c_i(u_l) + d_i(u_l)$$
$$\geq c_{i-1}(u_j) + d_{i-1}(u_j) + c_{i-1}(u_l) + d_{i-1}(u_l) + 2.$$

However, we have either $c_i(u_j) > d_{i-1}(u_l)$ and $d_i(u_l) > c_{i-1}(u_j)$ or $c_i(u_l) > d_{i-1}(u_j)$ and $d_i(u_j) > c_{i-1}(u_l)$, depending on which side of $e_i = u_ju_l$ contains no edge of $\overline{E}_i$, and the statement follows.

Now it is easy to complete the proof of Theorem 1.4. Since $|E_m| > \lfloor (k-1)n/2 \rfloor$, there is $1 \leq j \leq n$ for with $c_m(u_j) + d_m(u_j) \geq k$. By the definition of $c_m$ and $d_m$, this means that there is a non-crossing path of length $k$, passing through $u_j$. The following construction shows that the bound in Theorem 1.4 is tight.

**Construction.** Let $0, 1, \ldots, n-1 \pmod{n}$ denote the vertices of a regular $n$-gon.

(i) If $n \equiv k \pmod{2}$, then connect each vertex to the $k-1$ vertices furthest from it.

(ii) If $n \not\equiv k \pmod{2}$, then connect each vertex $i$ to the $k-2$ vertices furthest from $i$, and for $0 \leq i \leq \lfloor \frac{n}{2} \rfloor - 1$, also to $i + \frac{n-k-1}{2}$.

In case (i), every edge has at least $\frac{n-k}{2}$ vertices on both sides. In particular, this holds for the first and last edges of any non-crossing path. This implies that any such path misses at least $n-k$ vertices, so it is of length at most $k-1$.

In case (ii), suppose that $k+3 \le n$. Consider any non-crossing path $P$. Call an edge of $P$ extremal if the whole path $P$ lies in one of the closed halfspaces determined by $P$. If $P$ has at least three extremal edges, then it has at most $n-3(\frac{n-k-1}{2}) \le k$ vertices, showing that its length is at most $k-1$. However, if $P$ has only two extremal edges, then $P$ is a "zigzag", in particular, for each $j$, $\lfloor\frac{n}{2}\rfloor \ge j \ge \frac{n-k+1}{2}$ it has at most two edges of the form $(i, i+j)$. If at most one edge of $P$ is of the form $(i, i+\frac{n-k+1}{2})$, then $P$ contains at most $k-1$ edges. Otherwise, the first and the last edges of $P$ are of the form $(i, i+\frac{n-k+1}{2})$ and $(j, j+\frac{n-k+1}{2})$ with $0 \le i < i+\frac{n-k+1}{2} < j \le \lfloor\frac{n}{2}\rfloor - 1$. In this case, the length of $P$ is at most

$$2\Big(j-\Big(i+\frac{n-k+1}{2}\Big)+1\Big) \le 2\Big(\Big\lfloor\frac{n}{2}\Big\rfloor - \frac{n-k+1}{2}\Big) \le k-1,$$

as required.

The case $n = k+1$ can be treated similarly. □

We prove the following equivalent form of Theorem 1.3.

**Theorem 4.2** *Let $\mathcal{P}_k$ be the class of all non-crossing paths of length $k$. Then $R_c(\mathcal{P}_k) = 2k-1$ unless $k = 1$.*

**Proof.** First, we show that $R_c(\mathcal{P}_k) > 2k-2$. Let $G$ be a convex geometric graph on $2k-2$ vertices, and let $p, v_1, v_2, \ldots, v_{k-2}, q, u_1, u_2, \ldots, u_{k-2}$ be its vertices listed in clockwise order. For all $i, j$, color all edges $(v_i, v_j)$, $(p, v_i)$, and $(q, v_i)$ blue; $(u_i, u_j)$, $(p, u_i)$, and $(q, u_i)$ red; $(v_i, u_j)$ red if $i+j$ is odd and blue if $i+j$ is even. The edge $(p, q)$ can have any color. It is not difficult to check that this graph contains no non-crossing monochromatic path of length $k$.

For the upper bound, first let $k$ be even and take a convex complete geometric graph on $2k-1$ vertices, all of whose edges are colored red or blue. This graph has $(2k-1)(k-1)$ edges, so one of the color classes (say, red) contains at least $\lceil(2k-1)(k-1)/2\rceil > \lfloor(2k-1)(k-1)/2\rfloor$ edges. Therefore, by Theorem 1.4, there is a non-crossing red path of length $k$.

If $k$ is odd, then we have to be more careful. Suppose there is a 2-coloring of the edges of a complete geometric graph $G$ on $2k-1$ vertices without a non-crossing path of length $k$. It follows from Theorem 1.4 that $(2k-1)(k-1)/2$ edges are red and the same number of edges are blue. Let $G_b$ denote the subgraph of $G$ obtained from $G$ by removing all red edges. Analyzing the proof of Theorem 1.4 for $G_b$, we obtain that (in the notation of the proof of Theorem 1.4)

$$\sum_{j=1}^{n} c_i(u_j) + \sum_{j=1}^{n} d_i(u_j) = 2|E_i| \qquad i = 1, 2, \ldots, m.$$

Consequently,

$$c_i(u_j) + d_i(u_j) + c_i(u_l) + d_i(u_l)$$

$$= c_{i-1}(u_j) + d_{i-1}(u_j) + c_{i-1}(u_l) + d_{i-1}(u_l) + 2.$$

(Otherwise, we would have $c_m(u_j) + d_m(u_j) \ge k$ for some $j$, and we could find a non-crossing path of length $k$.) This means, for example, that if $e_i = u_j u_l$ and $\overline{E}_i$ has no element on the vertex set $\{u_j, u_{j+1}, \ldots, u_{l-1}, u_l\}$, then $c_i(u_j) = d_{i-1}(u_l) + 1$, $d_i(u_l) = c_{i-1}(u_j) + 1$.

Looking at all possible ways how we may arrange the edges of $G_b$ in the sequence $e_1, e_2, \ldots, e_m$, we obtain the following proposition.

**Proposition 4.3** *Let $v, w \in \{1, 2, \ldots 2k-1\}$.*

(a) *If $u_v u_w$ is blue, then at least one of the longest blue non-crossing paths starting at $u_v$, all of whose vertices lie in $\{u_v, u_{v+1}, \ldots, u_w\}$, contains the edge $u_v u_w$.*

(b) *If $u_v u_w$ is blue, then at least one of the longest blue non-crossing paths starting at $u_w$, all of whose vertices lie in $\{u_v, u_{v+1}, \ldots, u_w\}$, contains the edge $u_v u_w$.*

(c) *If $u_v u_w$ is red, then at least one of the longest blue non-crossing paths starting at $u_v$, all of whose vertices lie in $\{u_v, u_{v+1}, \ldots, u_w\}$, does not contain $u_w$.*

(d) *If $u_v u_w$ is red, then at least one of the longest blue non-crossing paths starting at $u_w$, all of whose vertices lie in $\{u_v, u_{v+1}, \ldots, u_w\}$, does not contain $u_v$.*

Exchanging the roles of the colors, one can formulate a similar statement about red paths in the graph $G_r$ obtained from $G$ by deleting all blue edges.

Consider the $2k-1$ edges $u_i u_{i+1}$. At least $k$ of them are colored by the same color (blue, say). Let $u_p u_{p+1} \ldots u_q$ be the longest non-crossing blue path using some of the edges $u_i u_{i+1}$ (certainly, its length is at least 2). First, observe that Proposition 4.3 (c) yields that all edges within $\{u_p, u_{p+1}, \ldots, u_q\}$ are blue. Indeed, if $u_s u_t, p \le s < t \le q$, was a red edge such that all other edges $u_{s'}u_{t'}, s \le s' < t' \le t$ were blue, then Proposition 4.3 (c) would be false with $u = s$ and $w = t$. This implies that the blue path $u_p u_{p+1} \dots u_q$ cannot visit all or all but one of the vertices of $G$. By the maximality of this path, the edges $u_{p-1}u_p$ and $u_q u_{q+1}$ are red. It follows from the the dual of Proposition 4.3 (a) and (c) with $v = p-1$ and $w = q+1$ that $u_p u_{q+1}$ is blue. Analogously, $u_{p-1}u_q$ is also blue.

Suppose now that some of the edges $u_{p-1}u_r, r = p+1, p+2, \dots, q-1$, are red. Take one of them for which $r$ is maximum. Then Proposition 4.3 (b) would be violated for $v = p-1$ and $w = r+1$. Thus, all edges $u_{p-1}u_r, r = p+1, p+2, \dots, q$, are blue. Analogously, all edges $u_{q+1}u_r, r = p, p+1, \dots, q-1$, are also blue.

The edge $u_{p-2}u_{p-1}$ cannot be red. Otherwise, starting with the longest non-crossing red path $u_p u_{p-1} \dots$ and using a similar argument as above, we would conclude that $u_{p-1}u_{p+1}$ must be red, but we already know that it is blue. Thus, $u_{p-2}u_{p-1}$ is blue, and we get a contradiction with Proposition 4.3 (a) or (c) for $v = p-2$ and $w = p+1$. □

**Algorithm.** Suppose $G$ is a convex geometric graph with $n \ge k+1$ vertices $u_1, u_2, \dots, u_n$, $n \ge k+1$, and $G$ has more than $\lfloor (k-1)n/2 \rfloor$ edges. By Theorem 1.4, $G$ contains a non-crossing path of length $k$. It can be found in $O(n^2)$ time, or, if $G$ has $O(nk)$ edges, then also in $O(nk \log n)$ time, as follows. Let the *length* of an edge $u_i u_j$ be defined as $\min(|i-j|, n-|i-j|)$ for $1 \le i, j \le n$. So every edge has length at most $n/2$. We use the notation of the proof.

1. List the edges of $G$ in increasing order of length, $e_1, e_2, \dots, e_m$. (The order of the edges of the same length is arbitrary.) This can be done in two different ways:

   (i) For the *complete* graph on the same vertex set, such a list is instantly available. Go over this list and delete all edges missing from $G$. [This can be done in $O(n^2)$ time.]

   (ii) Order the edges of $G$ according to their lengths. [This can be done in $O(nk \log n)$ time, provided that $G$ has $O(nk)$ edges.]

2. At start, $E_0 = \emptyset$, $c_0(u_j) = d_0(u_j) = 0$, $1 \le j \le n$. For $1 \le i \le m$, at the $i$-th step we have $E_{i-1} = \{e_1, e_2, \dots, e_{i-1}\}$, $c_{i-1}(u_j)$, and $d_{i-1}(u_j)$, and we also have the corresponding paths of length $c_{i-1}(u_j)$ and $d_{i-1}(u_j)$, respectively, ending at $u_j$, for each $1 \le j \le n$. Observe that $e_i$ is an extreme edge with respect to $E_{i-1}$. So let $E_i = E_{i-1} \cup e_i$. Update $c(u), d(u)$, and the list of corresponding paths at the endpoints of $e_i$. Do the $(i+1)$-st step. [Steps 1 through $m$ can be done in a total of $O(nk)$ time.]

3. It follows from Theorem 1.4 that after the $m$-th step there will be a vertex $u$ with $c(u) + d(u) \ge k$. Return the union of the corresponding paths of lengths $c(u)$ and $d(u)$. [$O(n)$ time.]

To find a monochromatic non-crossing path of length $\lfloor n/2 \rfloor$ (ensured by Theorem 1.3) in a complete 2-colored graph, we first apply the above algorithm on the graphs $G_b$ and $G_r$ containing all blue and all red edges of $G$, respectively. If we do not find a monochromatic non-crossing path of length $\lfloor n/2 \rfloor$, then we use the proof of Theorem 4.2 to improve the performance of the algorithm (to increase at least one of the values $c(v)$, $d(v)$) so that it will find the desired path in time $O(n^2)$.

# 5 General estimates – Proof of Theorem 1.5

For any set of $n$ points $P$ in the plane, an $i$-element subset of $P$ is called an *i-set* if it can be obtained by intersecting $P$ with an open half-plane. It is easy to see that all $i$-sets can be generated by the following procedure [ELSS73]: Take an oriented line $\ell$ passing through precisely one point $p \in P$ and having $i$ elements of $P$ on its left side. Rotate $\ell$ around $p$ in the clockwise direction until it hits another point $q \in P$, and then continue the rotation around $q$, etc. Whenever $\ell$ passes through only one element of $P$, the points lying on its left side form an $i$-set.

Any geometric subgraph $K_{R(\mathcal{G})}$ induced by an $R(\mathcal{G})$-element subset of $P$ contains a subgraph of the class $\mathcal{G}$, all of whose edges are of the same color. If these edges are red (blue), we say that the *type* of the corresponding subset is *red (blue)*. Note that a set may have both types.

**Lemma 5.1** *Let $\mathcal{G}$ be any class of geometric graphs. Then $R(2\mathcal{G}) \le 2R(\mathcal{G}) + 1$.*

Let $N = 2R(\mathcal{G}) + 1$ and let $P$ be the vertex set of a complete geometric graph $K_N$ whose edges are colored by red and blue.

If all $R(\mathcal{G})$-sets of $P$ have the same type, then take two disjoint $R(\mathcal{G})$-sets. Both contain a monochromatic geometric subgraph belonging to $\mathcal{G}$, so their union is a monochromatic member of $2\mathcal{G}$.

On the other hand, if there are two $R(\mathcal{G})$-sets of different types, then we can also find two $R(\mathcal{G})$-sets, $A$ and $B$, of different types such that $|A \cup B| = R(\mathcal{G})+1$ and that $P \setminus (A \cup B)$ is also an $R(\mathcal{G})$-set. The type of $P \setminus (A \cup B)$ will coincide with the type of $A$ or $B$. □

**Proof of Theorem 1.5.** Part (i) is an immediate corollary of Lemma 5.1.

Let $\mathcal{G}$ be a (fixed) class of geometric graphs. Let $r = r(k,l)$ be the the smallest number with the property that every complete geometric graph with $r$ vertices, whose edges are colored by red and blue, contains either a subgraph belonging to $k\mathcal{G}$, all of whose edges are red, or a subgraph belonging to $l\mathcal{G}$, all of whose edges are blue. So, $R = R(\mathcal{G}) = r(1,1)$, $R(k\mathcal{G}) = r(k,k)$.

**Lemma 5.2** *For* $k \geq l \geq 1$

$$r(k,l) = r(l,k) \leq (R+1)k + \left\lceil \frac{R+1}{2} \right\rceil l - \left\lceil \frac{R+1}{2} \right\rceil.$$

**Proof.** By simmetry, clearly $r(k,l) = r(l,k)$. We proceed by induction on $k$ and $l$. First we show that $r(k,1) \leq kR$. Let $G$ be a complete geometric graph with $kR$ vertices, whose edges are colored by red and blue. Suppose that no two vertices determine a vertical line. Then we can choose $k-1$ vertical lines that divide the vertex set of $G$ into $k$ equal parts. Since each part contains $R$ of the vertices, the subgraph spanned by every part contains a monochromatic copy of a geometric graph from the class $\mathcal{G}$. Either all of them are red or one of them is blue. Therefore, the lemma holds for any pair $(k,l)$, where either $k=1$ or $l=1$.

Let $k \geq l$ be fixed, and suppose that we already know that the Lemma is true for every pair $(k',l') \neq (k,l)$, where $k' \geq k, l' \geq l$.

Consider a complete geometric graph $G$ with vertex set $V$, whose edges are colored by red and blue and

$$|V| = (R+1)k + \left\lceil \frac{R+1}{2} \right\rceil l - \left\lceil \frac{R+1}{2} \right\rceil.$$

By Theorem 1.5 (i), for any $n \geq 1$, if $f(n) = 2^n(R+1)-1 \leq |V|$, then any geometric subgraph induced by an $f(n)$-set contains a monochromatic subgraph from the class $2^n\mathcal{G}$. Its color will be the *type* of the corresponding $f(n)$-set.

If there are two $f(n)$-sets of different types, we can cut off an $(f(n)+1)$-set which contains an $f(n)$-set of both types. Since

$$(R+1)k + \left\lceil \frac{R+1}{2} \right\rceil l - 2^n(R+1)$$
$$\geq (R+1)(k-2^n) + \left\lceil \frac{R+1}{2} \right\rceil (l-2^n),$$

we can apply the induction hypothesis for the subgraph induced by the rest of the vertices, with $k' = k-2^n, l' = l-2^n$.

So we can assume that for all $n$, all $f(n)$-sets are of the same type, provided $f(n) = 2^n(R+1)-1 \leq |V|$. Suppose that for $n_1 < n_2$, the $f(n_1)$ and $f(n_2)$-sets are of different types. Then, there exists an $n_1 \leq n < n_2$ such that the $f(n)$ and $f(n+1)$-sets are also of different types. Cut off an $f(n+1)$-set. Since it contains an $f(n)$-set, the subgraph induced by the $f(n+1)$-set contains a blue (resp. red) subgraph from the class $2^{n+1}\mathcal{G}$ and a red (resp. blue) subgraph from the class $2^n\mathcal{G}$.

If the $f(n+1)$-sets are of type blue, then either $l \leq 2^{n+1}$ and we are done, or we can apply the induction hypothesis for the graph induced by the remaining vertices with $k' = k-2^n$ and $l' = l-2^{n+1}$ since it is easy to check that

$$(R+1)k + \left\lceil \frac{R+1}{2} \right\rceil l - 2^{n+1}(R+1)$$
$$\geq (R+1)(k-2^n) + \left\lceil \frac{R+1}{2} \right\rceil (l-2^{n+1}).$$

On the other hand, if the $f(n+1)$-sets are of type red then we can assume that $k > 2^{n+1}$ and $l > 2^n$ and we can apply the induction hypothesis with $k' = k-2^{n+1}$ and $l' = l-2^n$ since

$$(R+1)k + \left\lceil \frac{R+1}{2} \right\rceil l - 2^{n+1}(R+1)$$
$$\geq (R+1)(k-2^{n+1}) + \left\lceil \frac{R+1}{2} \right\rceil (l-2^n)$$

and

$$(R+1)k + \left\lceil \frac{R+1}{2} \right\rceil l - 2^{n+1}(R+1)$$
$$\geq (R+1)(l-2^n) + \left\lceil \frac{R+1}{2} \right\rceil (k-2^{n+1}).$$

Therefore, in the sequel we assume that for *all* $n$, all $f(n)$-sets are of the same type. For simplicity we assume that they are of type *red*. The case where these sets are of type *blue* can be settled very similarly. Write $k$ in base 2, that is, $k = 2^{n_1}+2^{n_2}+\ldots+2^{n_a}$, $n_1 < n_2 < \ldots < n_a$. Let $V_0 = V$. By the assumption, every $f(n_1)$- and $f(n_2)$-set of $V_0$ are of type red. Let $S_1$ be an $f(n_1)$-set and let $V_1 = V_0 \setminus S_1$. If every $f(n_2)$- and $f(n_3)$-set of $V_1$ are of type red, then let $S_2$ be such an $f(n_2)$-set and let $V_2 = V_1 \setminus S_2$. In general, if $V_i$ has already been defined by this procedure and every $f(n_{i+1})$- and $f(n_{i+2})$-set of $V_i$ are of type red, then let $S_{i+1}$ be any $f(n_{i+1})$-set and let $V_{i+1} = V_i \setminus S_{i+1}$. Otherwise, if not all $f(n_{i+1})$-sets or not all $f(n_{i+2})$-sets of $V_i$ are of type red, or $i = a-1$, then stop and let $j = i$.

If $j = a-1$, then $V_{a-1}$ still has an $f(n_a)$-set, $S_a$ of type red, and $S_1 \cup S_2 \cup \ldots \cup S_a$ contains a red subgraph from the class $k\mathcal{G}$ and we are done. Suppose $j < a-1$. We distinguish two cases.

**a.** *There is an $f(n_{j+1})$-set of $V_j$ of type blue.* Since all $f(n_{j+1})$-sets of $V_{j-1}$ were of type red, at least one of the $f(n_{j+1})$-sets of $V_j$ is of type red. Therefore, we can find an $(f(n_{j+1})+1)$-set, $S_{j+1}$, which contains an $f(n_{j+1})$-set of both types. $|S_1 \cup S_2 \cup \ldots \cup S_{j+1}| \leq (2^{n_1} + 2^{n_2} + \ldots + 2^{n_{j+1}})(R+1)$ and $S_1 \cup S_2 \cup \ldots \cup S_{j+1}$ contains a red subgraph from the class $(2^{n_1} + 2^{n_2} + \ldots + 2^{n_{j+1}})\mathcal{G}$ and a blue subgraph from the class $2^{n_{j+1}}\mathcal{G}$. If either $k \leq 2^{n_1}+2^{n_2}+\ldots+2^{n_{j+1}}$ or $l \leq 2^{n_{j+1}}$ then we are done. Otherwise, easy calculation shows that we can apply the induction hypothesis for for the graph induced by the remaining vertices with $k' = k-(2^{n_1}+2^{n_2}+\ldots+2^{n_{j+1}})$ and $l' = l - 2^{n_{j+1}}$.

**b.** *Every $f(n_{j+1})$-set of $V_i$ are of type red, but there is an $f(n_{j+2})$-set of $V_j$ of type blue.* In this case, there exists an $n_{j+1} \leq n < n_{j+2}$ such that every $f(n)$-set of $V_i$ are of type red, but there is an $f(n+1)$-set of $V_j$ of type blue. Then let $S_{j+1}$ be an $f(n+1)$-set of $V_j$ of type blue. Clearly, it contains an $f(n)$-set of $V_j$, which is of type red. $|S_1 \cup S_2 \cup \ldots \cup S_{j+1}| \leq (2^{n_1}+2^{n_2}+\ldots+2^{n_j}+2^{n+1})(R+1)$ and $S_1 \cup S_2 \cup \ldots \cup S_{j+1}$ contains a red subgraph from the class $(2^{n_1}+2^{n_2}+\ldots+2^{n_j}+2^n)\mathcal{G}$ and a blue subgraph from the class $2^{n+1}\mathcal{G}$. Again, easy calculation shows that we can proceed as in case a. □

Return to the proof of Theorem 1.5. Part (ii) is an immediate corollary of Lemma 5.2. In part (iii), if all $R_c(\mathcal{G})$-sets have the same type then, using the convexity of the geometric graph, we can take $k$ disjoint $R_c(\mathcal{G})$-sets and we get a monochromatic geometric graph from $k\mathcal{G}$.

If there are two $R(\mathcal{G})$-sets of different types, we can proceed by induction as in part (ii). □

**Algorithms.** Following the approach of [KPTT97], where $G$ was just an edge, one can obtain an $O(n^{\log\log n+2})$ time algorithm to find a monochromatic subgraph belonging to $k\mathcal{G}$ in a complete 2-colored geometric graph of $n = (R(\mathcal{G})+1)k-1$ vertices, provided that $k$ is a power of 2. The same approach yields an $O(n^{\log\log n+2})$ time algorithm to find a monochromatic subgraph belonging to $k\mathcal{G}$ in a complete 2-colored geometric graph of $n = \lceil 3(R(\mathcal{G})+1)/2 \rceil k - \lceil (R(\mathcal{G})+1)/2 \rceil$ vertices (see Theorem 1.5 (i) and (ii)).

However, it is much easier to find a monochromatic subgraph belonging to $k\mathcal{G}$ in a complete 2-colored *convex* geometric graph of $n = (R_c(\mathcal{G})+1)k-1$ vertices (see Theorem1.5 (iii)). This can be done in $O(n^2)$ time as follows.

Let $r = R_c(\mathcal{G})$ for simplicity. Let $v_1, v_2, \ldots, v_r$ be the vertices of a 2-colored complete convex geometric graph, listed in clockwise order. The algorithm will be described recursively. Its running time is denoted by $f(n)$.

1. Find a monochromatic copy of $G$ in the subgraphs induced by $\{v_{ir+1}, v_{ir+2}, \ldots, v_{(i+1)r}\}$, for $i = 0, 1, \ldots, n-1$. [This can be done in $O(n)$ time.]
2. If all of these monochromatic copies have the same color, then return their union. If not, then find two consecutive ones, say, in $\{v_1, v_2, \ldots, v_r\}$ and $\{v_{r+1}, v_{r+2}, \ldots, v_{2r}\}$, that have different colors. [$O(n)$ time.]
3. For $j = 0, 1, \ldots, r$, the subgraph spanned by $\{v_{j+1}, v_{j+2}, \ldots, v_{j+r}\}$ contains a monochromatic copy of $G$. Find two consecutive ones, say in $\{v_1, v_2, \ldots, v_r\}$ and $\{v_2, v_3, \ldots, v_{r+1}\}$, which have different different colors. [$O(1)$ time.]
4. Call the algorithm recursively for the subgraph induced by $\{v_{r+2}, v_{r+3}, \ldots, v_n\}$, to find a monochromatic subgraph which belongs to $(k-1)\mathcal{G}$. Depending on its color, augment it with the monochromatic copy of $G$ contained in $\{v_1, v_2, \ldots, v_r\}$ or $\{v_{r+1}, v_{r+2}, \ldots, v_{2r}\}$.
[$f(n-r)$ time.]

Thus, the running time $f(n)$ of this algorithm satisfies the inequality $f(n) \leq f(n-r)+O(n)$, which implies that $f(n) = O(n^2)$.

## 6 Triangles – Proof of Theorem 1.6

The proof is quite technical, therefore we sketch here only the main ideas.

It is sufficient to show that $R_\Delta(k) > 7k-7$ for every $k$. Indeed, if $R_\Delta(k) \leq 7k-2$ for some positive integer $k$, an application of Theorem 1.5 with $k = 8$ and $\mathcal{G} = k\mathcal{T}$ would yield $R_\Delta(8k) \leq 7(8k)-9$, a contradiction.

**Construction.** Let $k$ be a positive integer, $n = 7k-7$. Let $0, 1, \ldots, 7k-8$ denote the vertices of a regular $n$-gon, in clockwise order. If two vertices are separated by $m-1$ others, then we say that the *length* of the edge connecting them is $m$ ($m \le \lfloor n/2 \rfloor$). A vertex is colored red if $i \equiv 0, 1, 3$, or $5 \pmod 7$, and blue otherwise.

We color the edges $ij$ ($0 \le i \ne j \le n-1$) according to the following rules. Let $\lambda$ denote the length of $ij$.

(a) If $\lambda \equiv 1, 4$, or $6 \pmod 7$, let $ij$ be blue.

(b) If $\lambda \equiv 2$ or $3 \pmod 7$, let $ij$ be red.

(c) If $\lambda \equiv 5 \pmod 7$, let $ij$ be colored with the color of its starting point in the clockwise direction.

(d) If $\lambda \equiv 0 \pmod 7$, let $ij$ be colored with the color other than the color of $i+1 \pmod 7$.

To see that this graph does not contain $n$ pairwise non-crossing blue triangles, we need the following lemma that can be proved by induction. By an *arc* we mean a set of consecutive vertices.

**Lemma 6.1** *Suppose that an arc $I$ of $m \le n/2$ vertices contains the vertices of $h$ pairwise non-crossing blue triangles. Then $m \ge 7h-1$. Moreover, if $m = 7h-1$, then the starting point of $I$ is blue.*

Consider a maximal set of pairwise non-crossing blue triangles. Suppose first that the points can be partitioned into three arcs, $I_1, I_2, I_3$, with $m_i = |I_i| \le n/2$ such that the vertices of each of these triangles belong to the same arc. Let $k_i$ denote the number of triangles whose vertices belong to $I_i$. By Lemma 6.1, we have

$$\begin{aligned} 7n-7 &= m_1+m_2+m_3 \\ &\ge (7k_1-1)+(7k_2-1)+(7k_3-1) \\ &> 7(k_1+k_2+k_3)-7, \end{aligned}$$

and we can conclude that $k_1+k_2+k_3 < n$.

Otherwise, there is a blue triangle so that the convex hull of its vertex set contains the center of our regular $n$-gon. Its vertices, $p, q, r$, partition the remaining set of points into three arcs of lengths $m_1, m_2, m_3$ containing $k_1, k_2, k_3$ triangles, respectively. ¿From Lemma 6.1 we obtain

$$\begin{aligned} 7n-7 &= m_1+m_2+m_3+3 \\ &\ge (7k_1-1)+(7k_2-1)+(7k_3-1)+3 \\ &= 7(k_1+k_2+k_3), \end{aligned}$$

implying $n \ge k = k_1+k_2+k_3+1$. Suppose that $n = k$. Then we have equality in the previous inequalities, therefore, the length of each side of triangle $pqr$ is divisible by 7. On the other hand, by Lemma 6.1, all of the points $p+1, q+1, r+1$ are blue, so $pqr$ is a monochromatic red triangle, a contradiction.

The existence of $n$ non-crossing red triangles can be excluded by a similar argument.

# References


[B74] S.A. Burr, Generalized Ramsey theory for graphs – a survey, in: *Graphs and Combinatorics (R. Bari and F. Harary, eds.), Lecture Notes in Mathematics* **406**, Springer–Verlag, Berlin, 1974, 52–75.

[BES75] S.A. Burr, P. Erdős, and J.H. Spencer, Ramsey theorems for multiple copies of graphs, *Transactions of the American Mathematical Society* **209** (1975), 87–99.

[D50] R.P. Dilworth, A decomposition theorem for partially ordered sets, *Annals of Mathematics* **51** (1950), 161–166.

[ELSS73] P. Erdős, L. Lovász, A. Simmons, and E.G. Straus, Dissection graphs of planar point sets, in: *A Survey of Combinatorial Theory (G. Srivastava, ed.)*, North–Holland, Amsterdam, 1973, 139–149.

[GMPP91] P. Gritzmann, B. Mohar, J. Pach, and R. Pollack, Embedding a planar triangulation with vertices at specified points (solution to problem E3341), *American Mathematical Monthly* **98** (1991), 165-166.

[GRS90] R.L. Graham, B.L. Rothschild, and J.H. Spencer, *Ramsey Theory, 2nd ed.*, John Wiley, New York, 1990.

[HS92] J. Hershberger and S. Suri, Applications of a semidynamic convex hull algorithm, *BIT* **32** (1992), 249-267.

[KPT96] Gy. Károlyi, J. Pach, and G. Tóth, Ramsey–type results for geometric graphs. I, in: *Proceedings of the 12th Annual ACM Symposium on Computational Geometry* 1996, 359-365. Also in: *Discrete and Computational Geometry,* to appear.

[KPTT97] Gy. Károlyi, J. Pach, G. Tardos, and G. Tóth, An algorithm for finding many disjoint monochromatic edges in a complete 2-colored geometric graph, in: *Intuitive Geometry, (I. Bárány, K. Böröczky, eds.), Bolyai Soc. Math. Studies* **6**, to appear.

[PA95] J. Pach and P.K. Agarwal, *Combinatorial Geometry,* John Wiley, New York, 1995.

# On the Complexity of the Union of Fat Objects in the Plane*

Alon Efrat[†] Micha Sharir[‡]

**Abstract**

We prove a near-linear bound on the combinatorial complexity of the union of $n$ fat convex objects in the plane, each pair of whose boundaries cross at most a constant number of times.

## 1 Introduction

Let $C$ be a collection of $n$ compact convex sets in the plane, satisfying the following properties:

(i) The objects in $C$ are $\alpha$-*fat*, for some fixed $\alpha > 1$; that is, for each $c \in C$ there exist two concentric disks $D \subseteq c \subseteq D'$ such that the ratio between the radii of $D'$ and $D$ is at most $\alpha$.

(ii) For any pair of distinct objects $c, c' \in C$, their boundaries intersect in at most $s$ points, for some fixed constant $s$.

See [13] for more details concerning fat objects in the plane.

Our goal is to derive a near-linear upper bound on the *combinatorial complexity* of the union $U = \bigcup C$, where we measure the complexity by the number of intersection points between the boundaries of the sets of $C$ that lie on $\partial U$.

There are not too many results of this kind. If $C$ is a collection of $\alpha$-fat triangles,[1] then the complexity of $U$ is $O(n \log\log n)$ (with the constant of proportionality depending on $\alpha$) [9], and this bound improves to $O(n)$ if the triangles are nearly of the same size [1]. See also [14] for additional results concerning fat polygons. If $C$ is a collection of $n$ *pseudo-disks* (arbitrary simply-connected regions bounded by closed Jordan curves, each pair of whose boundaries intersect at most twice), then the complexity of $U$ is $O(n)$ [7]. Of course, without any additional conditions, the complexity of $U$ can be $\Omega(n^2)$, even for the case of (non-fat) triangles. Even for fat convex objects, something like condition (ii) must be assumed, or else the complexity of the union might be arbitrarily large, an easy observation that has been noted in [9].

The main result of this paper is

**Theorem 1.1** *The combinatorial complexity of the union of a collection $C$ that satisfies conditions (i)–(ii) is $O(n^{1+\varepsilon})$, for any $\varepsilon > 0$, where the constant of proportionality depends on $\varepsilon$, $\alpha$ and $s$.*

Theorem 1.1 constitutes a significant progress in the study of the union of planar objects, an area that has many algorithmic applications, such as finding the maximal depth in an arrangement of fat objects (see [3]), hidden surface removal in a collection of fat objects in 3-space [6], and point-enclosure queries in a collection of fat objects in the plane [5].

**Remark:** In an earlier version of this paper, the authors have proved the slightly better bound $O(\lambda_s(n))$ on the complexity of the union of $C$, under the additional assumptions that all the regions in $C$ are roughly of the same size and have bounded curva-

*Work by Micha Sharir has also been supported by a grant from the U.S.–Israeli Binational Science Foundation, by National Science Foundation Grants CCR-94-24398 and CCR-93-11127, and by a grant from the G.I.F. — the German Israeli Foundation for Scientific Research and Development.

†School of Mathematical Sciences, Tel Aviv University. *alone@math.tau.ac.il*

‡School of Mathematical Sciences, Tel Aviv University, and Courant Institute of Mathematical Sciences, New York University. *sharir@math.tau.ac.il*



[1] For triangles, there is an equivalent definition of fatness that requires all angles to be at least $\alpha_0$; in [9], this is called $\alpha_0$-fatness.

ture. The proof for this special case is considerably simpler than the one given here.

## 2 Analysis

### 2.1 Touching and shattering vertices

Let $\mathcal{C}$ be a collection of $n$ compact simply-connected sets in the plane, each bounded by a closed Jordan curve (we refer to the sets in $\mathcal{C}$ as *Jordan regions*), and let $U$ denote their union. We assume that these regions are in general position, so that each pair of boundaries intersect in a finite number of points and properly cross at each point of intersection, and no three boundaries have a common point. (In this subsection we make no other assumption on $\mathcal{C}$.) As already mentioned, we measure the combinatorial complexity of $U$ by the number of vertices of the *arrangement* $\mathcal{A}(\mathcal{C})$ of $\mathcal{C}$ (i.e., points of intersection between pairs of boundaries of regions in $\mathcal{C}$) that lie on its boundary. We classify the arrangement vertices into two categories:

*touching vertices:* these are intersections between pairs of boundaries that intersect at only two points.

*shattering vertices:* these are all the other boundary intersection points.

The *level* of a vertex of $\mathcal{A}(\mathcal{C})$ is the number of regions that contain it in their interior. Thus the vertices of $U$ are exactly the vertices at level 0.

Let $T(\mathcal{C})$ (resp. $S(\mathcal{C})$) denote the number of touching (resp. shattering) vertices of $U$. We use the following result of Pach and Sharir [10]:

**Theorem 2.1**

$$T(\mathcal{C}) \leq 2S(\mathcal{C}) + 6n - 12.$$

### 2.2 Caps, inscribed fat polygons, and their properties

We now return to the case where $\mathcal{C}$ satisfies the conditions (i) and (ii) in the introduction. Let $c \in \mathcal{C}$. We inscribe in $c$ a convex polygon $P_c$ defined as follows. We choose some constant integer parameter $t > 12$, which also satisfies

$$\alpha \sin \frac{2\pi}{t} \cdot \left(1 - \frac{\pi\alpha}{t} \tan \frac{\pi}{t}\right) < 1.$$

and define $\theta_j = 2\pi j/t$, for $j = 0, 1, \ldots, t-1$. For each $j$, let $w_j = w_j(c)$ denote the (unique) point on $\partial c$ that has a tangent (that is, a supporting line) at orientation $\theta_j$ (tangents are assumed to be oriented so that $c$ lies to their left). We also define $w'_j$, for $j = 1, \ldots, t-1$, to be the point on $\partial c$ such that the length of the portion of $\partial c$ extending counterclockwise from $w_0$ to $w'_j$ is $j/t$ times the perimeter of $c$.

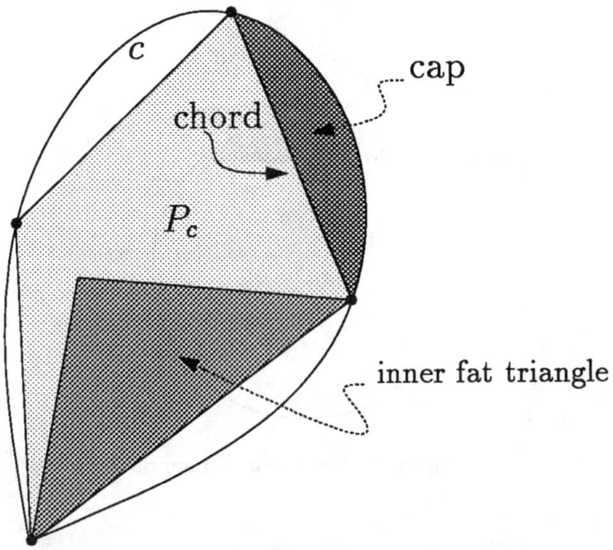


Figure 1: The inscribed polygon $P_c$ and the corresponding caps; one inner fat triangle is also illustrated.

$P_c$ is defined to be the convex polygon whose vertices are $w_0, \ldots, w_{t-1}, w'_1, \ldots, w'_{t-1}$. (Note that $P_c$ may have fewer than $2t-1$ vertices if $\partial c$ contains nonsmooth points whose tangent orientations span a sufficiently large interval.) The difference $c \setminus P_c$ is the union of at most $2t-1$ *caps* of $c$, where a cap is an intersection of $c$ with a halfplane. The *chord* of a cap is the intersection of $c$ with the line bounding the corresponding halfplane. An illustration of such an inscribed polygon and of the corresponding caps is shown in Figure 1.

**Lemma 2.2** *The polygons $P_c$ are $\alpha'$-fat, for*

$$\alpha' = \alpha\left(1 - \frac{\pi\alpha}{t} \tan \frac{\pi}{t}\right).$$

**Proof:** Since $c$ is $\alpha$-fat, there exist two concentric disks $D_1 \subseteq c \subseteq D_2$, with respective radii $r_1, r_2$, such that $r_2 \leq \alpha r_1$. Clearly, $P_c \subseteq D_2$. Let $K$ be one of the caps that constitute $c \setminus P_c$, and assume that $D_1$ intersects the chord $pq$ of $K$. It must do so at two points, or else its interior would have contained $p$ or $q$, contradicting the assumption that $D_1 \subseteq c$. By definition, there exist two tangents to $c$, $\tau_p$ at $p$ and $\tau_q$ at $q$, whose orientations differ by at most $2\pi/t$, and the distance $pq$ is smallet than $\sigma/t$, where $\sigma$ is the perimeter of $c$. Let $d$ denote the distance from the center $O$ of $D_1$ to $pq$. It is easy to see that $r_1 - d$ is at most the height to $pq$ in the triangle bounded by $pq$, $\tau_p$ and $\tau_q$ (see Figure 2), and a simple exercise shows that this height is at most $\frac{pq}{2} \tan \frac{\pi}{t}$. Hence

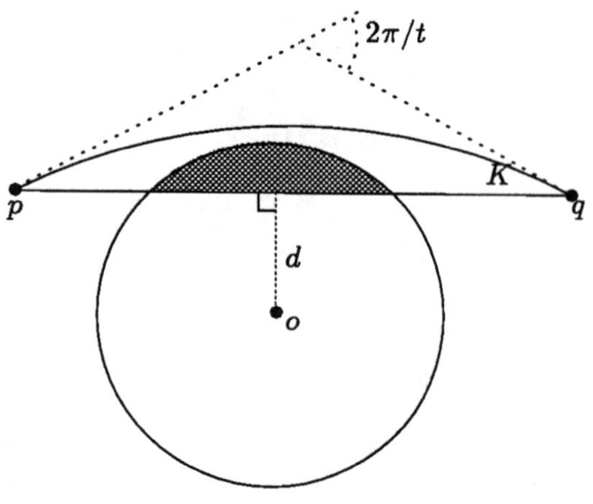


Figure 2: The proof of Lemma 2.2

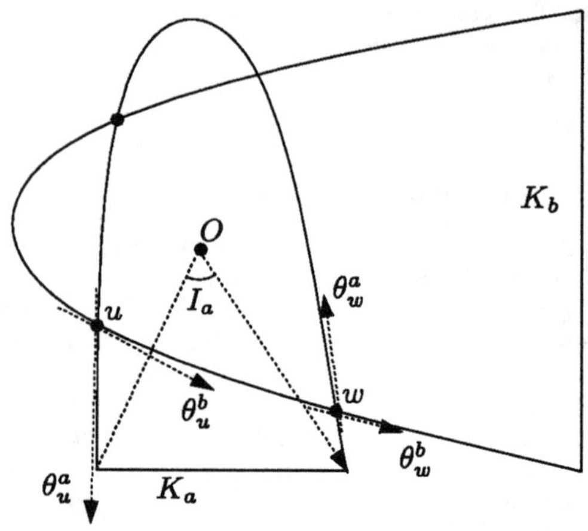


Figure 3: Two intersecting caps without a chordal intersection

$$
\begin{aligned}
r_1 - d &\le \frac{pq}{2}\tan\frac{\pi}{t} < \frac{\sigma}{2t}\tan\frac{\pi}{t} \\
&\le \frac{2\pi\alpha r_1}{2t}\tan\frac{\pi}{t} = \frac{\pi\alpha r_1}{t}\tan\frac{\pi}{t},
\end{aligned}
$$

where the last inequality follows from the fact that $c \subseteq D_2$. Hence

$$d \ge r_1\left(1 - \frac{\pi\alpha}{t}\tan\frac{\pi}{t}\right).$$

This implies that the disk concentric with $D_1$ and having radius $r_1\left(1 - \frac{\pi\alpha}{t}\tan\frac{\pi}{t}\right)$ is contained in $P_c$, and this completes the proof of the lemma. □

Let $c \in \mathcal{C}$, and let $O$ denote the common center of two disks $D_1 \subseteq P_c \subseteq D_2$, such that their respective radii $r_1$, $r_2$ satisfy $r_2 \le \alpha' r_1$. Let $pq$ be an edge of $P_c$. The convexity of $P_c$ and the fact that $D_1 \subseteq P_c$ are easily seen to imply that the angle $Opq$ must be at least the angle $\beta$ between $Op$ and the tangent to $D_1$ from $p$, which satisfies $\sin\beta = r_1/|Op| \ge r_1/r_2 \ge 1/\alpha'$. Similarly, the angle $Oqp$ must also be at least $\beta$. It follows that we can find a point $v$ inside $Opq$, such that all the angles of the triangle $vpq$ are at least

$$\beta_0 = \min\{\arcsin(1/\alpha'), \pi/3\}.$$

We repeat this analysis to each edge of each polygon, and replace the polygons $P_c$ by the collection of resulting triangles $vpq$. We refer to these triangles as ***inner fat triangles***. Let $\mathcal{T} = \mathcal{T}(\mathcal{C})$ denote the collection of inner fat triangles. Clearly, $|\mathcal{T}| \le (2t-1)n$. As an immediate consequence of [9], we have:

**Lemma 2.3** *The union $U_{\mathcal{T}}$ of the triangles in $\mathcal{T}$ has $O(n \log\log n)$ vertices.*

Let $v$ be a shattering vertex of $\partial U$, incident to two sets $a, b \in \mathcal{C}$. Let $K_a$, $K_b$ be the respective caps of $a$, $b$ that contain $v$, and let $p_a q_a$, $p_b q_b$ denote their respective chords. Consider the convex set $R = K_a \cap K_b$.

**Lemma 2.4** *At least one of the chords $p_a q_a$, $p_b q_b$ meets $\partial R$.*

**Proof:** Indeed, suppose to the contrary that both chords are disjoint from $R$. It follows that $R = a \cap b$, and that $\partial R$ contains at least four points of intersection between $\partial a$ and $\partial b$. Moreover, let $O$ be an interior point of $R$, and consider $\partial K_a$ and $\partial K_b$ as graphs of two respective functions $r = K_a(\theta)$, $r = K_b(\theta)$, in polar coordinates about $O$. Note that $\partial R$ is the graph of the pointwise minimum of $K_a$ and $K_b$. There is an angular interval $I_a$ over which $K_a(\theta)$ is attained at the chord of $K_a$, and a similar interval $I_b$ for the chord of $K_b$. These intervals must be disjoint, or else $\partial R$ would overlap one of these chords, contrary to assumption. See Figure 3.

Let $u$ (resp. $w$) denote the first vertex of $\partial R$ that we encounter as we rotate about $O$ clockwise (resp. counterclockwise) from $I_a$ (clearly, no vertex of $\partial R$ has an orientation in $I_a$). In the angular interval that runs counterclockwise from $u$ to $w$, the boundary of $R$ is attained by $\partial b$. Moreover, as we traverse, in counterclockwise direction, the portion of $\partial b$ that lies on $\partial K_b$, we first encounter $u$ and then $w$, and the reverse order is obtained along $\partial a$. See Figure 3.

Let $\theta_u^a$, $\theta_w^a$ denote the orientations of the tangents to $a$ at $u$ and $w$, respectively, and let $\theta_u^b$, $\theta_w^b$ denote the corresponding tangent orientations for $b$. (If any of these tangents is not unique, we fix an arbitrary tangent among those that are available.) The circular counterclockwise order of these four orientations is $(\theta_u^a, \theta_u^b, \theta_w^b, \theta_w^a)$, and they partition the circular range of orientations into four angular intervals that we denote by $(\theta_u^a, \theta_u^b)$, $(\theta_u^b, \theta_w^b)$,

$(\theta_w^b, \theta_w^a)$, and $(\theta_w^a, \theta_u^a)$. Each of the second and fourth intervals has length at most $2\pi/t$ (since the endpoints of any of these intervals are two tangent orientations within a single cap), and each of the first and third intervals has length at most $\pi$ (the total amount by which the tangent to a convex set can turn at a fixed point of its boundary is at most $\pi$). It follows that each of the lengths of the first and third intervals is at least $\pi - 4\pi/t > 2\pi/3$.

We now repeat the whole analysis in the last two paragraphs by interchanging $a$ and $b$. This yields two vertices $u'$, $w'$ of $\partial R$, such that the turning angle of the tangents to $R$ at each of these vertices is also greater than $2\pi/3$. It is easily verified that among the vertices $u, w, u', w'$ there exist at least three distinct vertices, or else $\partial a$ and $\partial b$ would have intersected at only two points, contrary to assumption. We have thus obtained at least three vertices of $R$ such that the turning angle of the tangents at each of them is greater than $2\pi/3$, which is impossible, because the overall turning angle for a convex set is $2\pi$. This contradiction completes the proof of the lemma. □

**Lemma 2.5** *Let $K_a$ be a cap of some set $a \in \mathcal{C}$, with chord $e_a$, and let $\Delta_b$ be an inner fat triangle in $\mathcal{T}$, obtained from the polygon $P_b$, for some $b \in \mathcal{C}$, such that the chord $e_b$ of $\Delta_b$ crosses $\partial K_a$. Then one of the following cases must occur:*

*(i) $e_a$ crosses $\partial\Delta_b$ (as in Figure 4(i)).*

*(ii) $K_a$ contains a vertex of $\Delta_b$ that is an endpoint of $e_b$ (as in Figure 4(ii)).*

*(iii) $\Delta_b$ contains a vertex of $K_a$ (as in Figure 4(iii)).*

*(iv) $\partial K_a$ and $\partial\Delta_b$ cross exactly twice, at two points that lie on $\partial a$ and on $e_b$, and $e_a$ is disjoint from $K_a \cap \Delta_b$. Furthermore, let $K_b$ denote the cap of $b$ that shares the same chord $e_b$ with $\Delta_b$. Then either $K_b$ contains an endpoint of $e_a$ (as in Figure 4(iv.a)), or $\partial a$ and $\partial b$ intersect only twice (as in Figure 4(iv.b)).*

**Proof:** Suppose that cases (i) and (ii) do not occur. That is, $e_a$ does not cross $\partial\Delta_b$ and no vertex of $\Delta_b$ lies in $K_a$. Then $e_b$ must intersect $\partial K_a$ at two points, $u$, $v$, both lying on $\partial a$. Therefore $e_b$ splits $K_a$ into two subregions, the region $K_a'$ that contains $e_a$, and the complementary region $K_a''$. Denote the range of the orientations of the tangents to $a$ at the points of $K_a$ by $(\theta_0, \theta_0')$, where $\theta_0 < \theta_0' \le \theta_0 + 2\pi/t$. Clearly, the orientations of $e_a$ and of $e_b$ also lie in this range. Two cases can arise:

(1) $\Delta_b$ overlaps $K_a'$ and is disjoint from $K_a''$ (see Figure 5(i)): If $K_a'$ is fully contained in $\Delta_b$ then $u$ and $v$ are the only two points of intersections between $\partial K_a$ and $\partial\Delta_b$, and, moreover, $\Delta_b$ contains both vertices of $K_a$, so we are in case (iii). Otherwise, since, by assumption, $\Delta_b$ does not intersect $e_a$ and does not have a vertex inside $K_a'$, one of its other

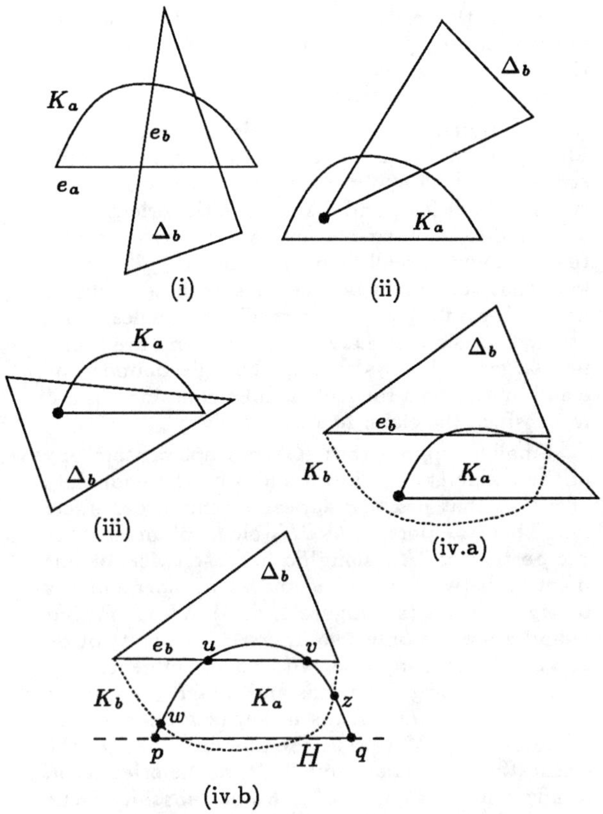


Figure 4: Illustrating the various cases in Lemma 2.5

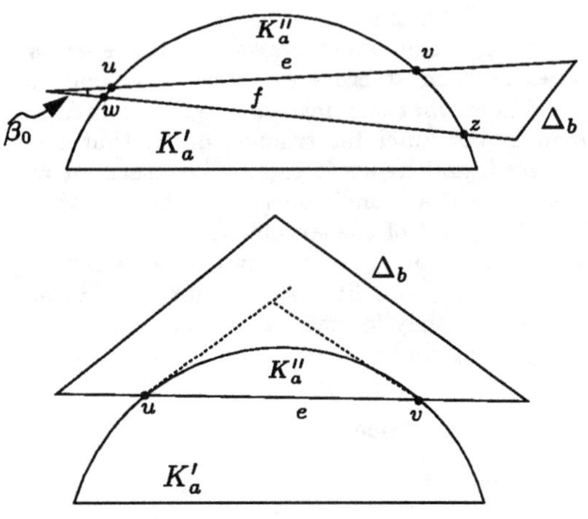


Figure 5: Two patterns of intersection of a cap $K_a$ and an inner fat triangle $\Delta_b$

edges, $f$, must also cross $\partial K_a$ twice, at two points $w, z$, lying on $\partial a$, so that the four points $w, u, v, z$ appear in this order along $\partial K_a$. In this case the orientation of $f$ also lies in the range $(\theta_0, \theta_0')$, and thus the angle between $e$ and $f$, which is $\geq \beta_0$, is at most $2\pi/t$, a contradiction.

(2) $\Delta_b$ overlaps $K_a''$ and is disjoint from $K_a'$ (see Figure 5(ii)): We claim that in this case $\Delta_b$ fully contains $K_a''$, so $u$ and $v$ are the only two intersection points of $\partial K_a$ and $\partial \Delta_b$. Since the orientations of $e_b$ and of the tangents (or, rather, any tangents) to $a$ at $u$ and at $v$ all lie in the range $(\theta_0, \theta_0')$, it follows that the angles between $e$ and these tangents are both at most $2\pi/t$. However, the angles of $\Delta_b$ at the endpoints of $e$ are both $\geq \beta_0$, and are therefore larger. It follows that the triangle bounded by $e$ and by two such tangents is fully contained in $\Delta_b$, from which the claim follows readily.

Finally, suppose that $K_b$ does not contain any of the endpoints $e_a$. Let $p$ and $q$ be the endpoints of $e_a$, so that $p, u, v, q$ appear in this order along $\partial a$. Then the portion of $\partial K_b$ along $\partial b$ must cross the portion of $\partial K_a$ along $\partial a$ at least twice, at one point $w$ between $p$ and $u$ and at another point $z$ between $v$ and $q$ (see Figure 4(iv.b)). We claim that $w$ and $z$ are the only two intersection points of $\partial a$ and $\partial b$. Indeed, suppose, with no loss of generality, that $e_a$ lies along the $x$-axis and that $K_a$ lies above it. Then $\gamma_a \equiv \partial a \cap K_a$ is a downward-concave $x$-monotone arc. Moreover, the absolute value of the orientation of $e_b$ is at most $2\pi/t$, so the orientation of any tangent to $\gamma_b \equiv \partial b \cap K_b$ has absolute value $\leq 4\pi/t$, which is easily seen to imply that $\gamma_b$ is also $x$-monotone and downward-convex. It follows that $\gamma_a$ and $\gamma_b$ cross each other exactly twice (at $w$ and $z$). We claim that there can be no other point of intersection between $\partial a$ and $\partial b$. Indeed, any such point must lie either in the halfplane below $e_a$ or in the halfplane above $e_b$. Consider the halfplane $H$ lying below $e_a$ (the second case is treated in a fully symmetric manner). It is easy to see that any such intersection must lie on $\gamma_b$. However, if $\gamma_b$ reaches $H$ it must cross $e_a$ twice. Arguing as above, it follows that the portion of $\gamma_b$ in $H$ is fully contained in the inner fat triangle of $P_a$ that has $e_a$ as a chord, and hence it cannot intersect $\partial a$ at all. This shows that condition (iv) holds, and thus completes the proof of the lemma. (Note that these arguments also imply that, in any configuration of case (iv), $\partial a \cap K_a$ and $\partial b \cap K_b$ can intersect in at most two points; they intersect in one or zero points if and only if $K_b$ contains an endpoint of $e_a$.) □

### 2.3 The proof of Theorem 1.1

The proof follows the technique used in the analysis of the complexity of lower envelopes of surfaces in higher dimensions and of related structures, as given in [4, 12].

Let $\mathcal{K} = \mathcal{K}(\mathcal{C})$ denote the collection of all caps of sets in $\mathcal{C}$, as defined above; recall that $|\mathcal{K}| \leq (2t-1)n$. Let $U_\mathcal{K}$ denote the union of these caps. The vertices of $U$ are also vertices of $U_\mathcal{K}$.

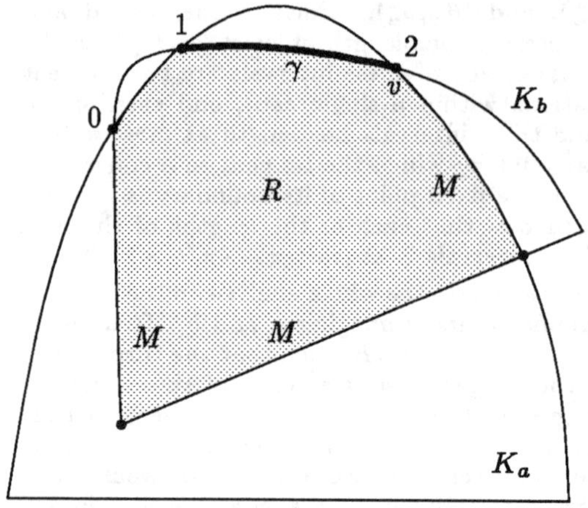


Figure 6: The region $R = K_a \cap K_b$; the marked arcs are labeled by $M$; $\partial R$ has three shattering vertices, with indices 0,1,2, as shown

Let $K_a \subseteq a$ and $K_b \subseteq b$ be two caps of two (distinct) regions $a, b \in \mathcal{C}$, such that $\partial K_a$ and $\partial K_b$ intersect in at least one vertex that is shattering in $\mathcal{A}(\mathcal{C})$ (in other words, this is a vertex incident to $\partial a$ and $\partial b$, and these boundaries cross at least four times; note also that a vertex can be shattering in $\mathcal{A}(\mathcal{C})$ and not in $\mathcal{A}(\mathcal{K})$ or vice versa). Put $R = K_a \cap K_b$. We call an arc of $\partial R$ *marked* if it contains a vertex of $\mathcal{A}(\mathcal{K})$ that lies on the chord of some cap, such that condition (iv) of Lemma 2.5 does not hold for that vertex (or, rather, for the cap and the triangle on whose boundaries the vertex lies), and *unmarked* otherwise; see Figure 6. We will refer to vertices of $\mathcal{A}(\mathcal{K})$ that lie on some chord as *chordal* vertices. Chordal vertices that satisfy condition (iv) of Lemma 2.5 will be called *special chordal vertices*, and all the other chordal vertices will be called *standard chordal vertices*.

Lemma 2.4 and Lemma 2.5 imply that at least one arc of $\partial R$ is marked. Actually, they imply that one of the vertices of $R$ is a standard chordal vertex. Indeed, Lemma 2.4 implies that $R$ has at least one chordal vertex. If all the chordal vertices of $R$ are special then each of the chords of $K_a$, $K_b$ contains zero or two such vertices, and, arguing as in the proof of Lemma 2.5, one can show that $\partial a$ and $\partial b$ intersect in exactly two vertices, contradicting the assumption that $\partial K_a$ and $\partial K_b$ intersect in at least one shattering vertex in $\mathcal{A}(\mathcal{C})$. Note also that if $\partial R$ has any special chordal vertex, then it must also have an endpoint of a chord as a vertex (see Lemma 2.5).

We define the *index* of a vertex $w$ of $a \cap b$ that lies on $\partial R$ to be the number of unmarked arcs of

$\partial R$ that we encounter before hitting a marked arc, as we traverse this boundary from $w$ in counterclockwise direction. However, if $\partial R$ has a special chordal vertex, it can have at most one vertex of $\partial a \cap \partial b$, as follows easily from the analysis of the proof of Lemma 2.5 (we must be at the situation shown in Figure 4(iv.a)), and we define the index of that vertex to be 0. The index of a vertex is an integer between 0 and $s-1$.

We also define the *level* of a vertex $v$ of the arrangement $\mathcal{A}(\mathcal{K})$ to be the number of caps of $\mathcal{K}$ containing $v$ in their interior. Clearly, vertices at level 0 are exactly the vertices of $\partial U_{\mathcal{K}}$.

We define the following quantities:

- $T(\mathcal{C})$ is the number of touching vertices in $\mathcal{A}(\mathcal{C})$ that lie on $\partial U$.
- $S(\mathcal{C})$ is the number of shattering vertices in $\mathcal{A}(\mathcal{C})$ that lie on $\partial U$.
- $S^{(j)}(\mathcal{C})$, for $j = 0, \ldots, s-1$, is the number of these shattering vertices whose index is at most $j$ (so $S(\mathcal{C}) = S^{(s-1)}(\mathcal{C})$).
- $Q^*(\mathcal{C})$ is the number of special chordal vertices that lie at level 0 in the arrangement $\mathcal{A}(\mathcal{K} \cup \mathcal{T})$ of the caps and inner fat triangles of the regions in $\mathcal{C}$.[2]
- $Q_0(\mathcal{C})$ is the number of standard chordal vertices that lie at level 0 in $\mathcal{A}(\mathcal{K} \cup \mathcal{T})$.
- $T(n)$ is the maximum of $T(\mathcal{C})$, over all collections $\mathcal{C}$ of $n$ sets satisfying (i) and (ii) (with fixed $\alpha$ and $s$).
- $S(n)$ is the maximum of $S(\mathcal{C})$, over all collections $\mathcal{C}$ as above.
- $S^{(j)}(n)$ is the maximum of $S^{(j)}(\mathcal{C})$, for $j = 0, \ldots, s-1$, over all collections $\mathcal{C}$ as above.
- $Q^*(n)$ is the maximum of $Q^*(\mathcal{C})$, over all collections $\mathcal{C}$ as above.
- $Q_0(n)$ is the maximum of $Q_0(\mathcal{C})$, over all collections $\mathcal{C}$ as above.

We will derive a (somewhat complex) system of recurrence relationships for the above quantities. Each of these recurrences involves a 'threshold parameter' $k < n/2$, which is arbitrary, and we will choose a different value of $k$ for each recurrence, in a manner detailed below.

First, using Theorem 2.1, we have:

$$T(n) \le c\,(n + S(n))\,, \qquad (1)$$

for some constant $c$ (for simplicity, we will use the same constant in all the recurrences).

We next estimate $S^{(j)}(n)$, for $j = 0, \ldots, s-1$. Let $v$ be a shattering vertex of $\mathcal{A}(\mathcal{C})$ that lies on $\partial U$, incident to two boundaries $\partial a$, $\partial b$, and contained in two respective caps $K_a \subseteq a$ and $K_b \subseteq b$, whose index is at most $j$. Let $R = K_a \cap K_b$, and let $\gamma$ denote the arc of $\partial R$ incident to $v$ and lying counterclockwise to it; see Figure 6. We fix a threshold parameter $k$, trace $\gamma$ from $v$, and examine the sequence $\sigma$ of vertices of $\mathcal{A}(\mathcal{K})$ that we encounter. Several cases can arise:

**(a)** $\sigma$ contains at least $k$ vertices, and none of the first $k$ vertices of $\sigma$ is a standard chordal vertex. (Note that $\sigma$ may contain a standard chordal vertex only if the index of $v$ is zero.) In this case we charge $v$ to the block of the first $k$ vertices of $\sigma$. Several important observations need to be made:

(i) Each of these charged vertices lies at level $\le k$ (in $\mathcal{A}(\mathcal{K})$, or in $\mathcal{A}(\mathcal{C})$ if it is a vertex of this latter arrangement).

(ii) Each charged vertex is charged at most twice, once along each of the boundaries containing it.

(iii) The portion of $\gamma$ that contains $\sigma$ may lie inside some inner fat triangles, but it does not cross any non-chordal edge of any such triangle (this follows from condition (iv) of Lemma 2.5).

(iv) Each of the charged special chordal vertices lies at level $\le k$ in $\mathcal{A}(\mathcal{K} \cup \mathcal{T})$ (this is a consequence of observation (iii)).

It follows that the overall number of shattering vertices of $\mathcal{A}(\mathcal{C})$ that lie on $\partial U$ and fall into this category is at most $2/k$ times the number of touching and shattering vertices of $\mathcal{A}(\mathcal{C})$ that lie at level at most $k$ (in either arrangement) plus $2/k$ times the number of special chordal vertices that lie at level at most $k$ in $\mathcal{A}(\mathcal{K} \cup \mathcal{T})$. Using the probabilistic analysis technique of Clarkson and Shor [2], this upper bound is at most $(2/k) \cdot O(k^2)$ times the expected number of touching and shattering vertices of the union of a random sample of $n/k$ regions of $\mathcal{C}$, plus $(2/k) \cdot O(k^2)$ times the expected number of special chordal vertices at level 0 in the arrangement of the caps and inner fat triangles of a random sample of $n/k$ regions of $\mathcal{C}$. In other words, the number of vertices $v$ of the present type is $O(kT(n/k) + kS(n/k) + kQ^*(n/k))$. Applying (1), this becomes $O(n + kS(n/k) + kQ^*(n/k))$.

**(b)** The first $k$ (or all, if $\sigma$ is shorter) vertices in $\sigma$ include at least one standard chordal vertex $w$ (in this case the index of $v$ must be zero). We charge $v$ to the first such $w$. It is easily verified that $w$ can be charged in this manner only once. Since $w$ is at level at most $k$ in $\mathcal{A}(\mathcal{K} \cup \mathcal{T})$ (see observation (iv)),

[2] One might be tempted to think that these vertices lie on the boundary of the union of these caps and triangles, but this is not the case, since each chordal vertex $z$ lies in the interior of the union of the cap and triangle that share the chord that contains $z$. This is why we use the more careful notion of level 0.

another application of the Clarkson-Shor technique implies that the number of vertices $v$ of this kind is at most $O(k^2 Q_0(n/k))$.

**(c)** $\sigma$ contains fewer than $k$ vertices of $\mathcal{A}(\mathcal{K})$, none of which is standard chordal. In this case we consider the other endpoint $w$ of $\gamma$. If $w$ is also a shattering vertex in $\mathcal{A}(\mathcal{C})$, then its index is at most $j-1$, and it lies at level at most $k$ in $\mathcal{A}(\mathcal{C})$. In this case we charge $v$ to $w$. Otherwise, $w$ is a special chordal vertex, in which case, as observed above, $\partial R$ must contain an endpoint $z$ of the chord of $K_a$ or of $K_b$. Moreover, suppose, without loss of generality, that $\gamma \subseteq \partial a$. Then the proof of Lemma 2.5 is easily seen to imply that $z$ is an endpoint of the chord of $K_a$, and that the entire counterclockwise portion of $\partial a$ from $v$ to $z$ is lies in the interior of $U_\mathcal{K}$ (see Figure 4(iv.a)). In this case we charge $v$ to $z$. The argument just given implies that $z$ can be charged at most once in this manner, and the number of such points $z$ is at most $(2t-1)n = O(n)$. Applying the Clarkson-Shor technique again to the former type of charging, we conclude that the number of vertices $v$ of this kind is $O(n + k^2 S^{(j-1)}(n/k))$.

Thus, summing up these bounds, we obtain the following system of recurrences:

$$\begin{aligned} S^{(j)}(n) \;&\le\; c\,(n + kS(n/k) + kQ^*(n/k) \\ &+\; k^2 S^{(j-1)}(n/k) + k^2 Q_0(n/k))\ , \end{aligned} \quad (2)$$

for $j = 0, \ldots, s-1$, and for some constant $c$ that depends on $\alpha$ (for $j = 0$, we put $S^{(-1)} = 0$ in the right-hand side).

Next we estimate $Q_0(\mathcal{C})$, using an analysis similar to the one just presented. Recall that we are counting standard chordal vertices at level 0 in $\mathcal{A}(\mathcal{K} \cup \mathcal{T})$. (Formally, we can also take each pair of a cap and a triangle with a common chord, and shift them slightly away from each other, so that each of the standard chordal vertices that lie on this chord and are counted in $Q_0(\mathcal{C})$ is split into two vertices, both of which lie on the boundary of the union of $\mathcal{K} \cup \mathcal{T}$.)

We analyze the number of these chordal vertices by considering them as vertices in $\mathcal{A}(\mathcal{K} \cup \mathcal{T})$, each lying on the boundary of the intersection of an inner fat triangle and a cap. So let $v$ be a standard chordal vertex, lying on the boundary of a cap $K_a \subseteq a$ and on the chord of an inner fat triangle $\Delta_b \subseteq b$, for two (distinct) sets $a, b \in \mathcal{C}$. By definition, $K_a$ and $\Delta_b$ satisfy one of the conditions (i)–(iii) of Lemma 2.5, which means that the boundary of the intersection $R = K_a \cap \Delta_b$ has at least one vertex that is a vertex of $\mathcal{A}(\mathcal{T})$, and none of its vertices is a special chordal vertex (see Figure 7). Observe that $\partial R$ consists of at most six arcs, since each edge of $\Delta_b$ can intersect $\partial K_a$ at most twice. We call an arc of $\partial R$ *marked* if it contains a vertex of $\mathcal{A}(\mathcal{T})$, and *unmarked* otherwise. We will refer to these vertices as *sharp* vertices. As just argued, at least one arc of $\partial R$ is marked. We define the index of

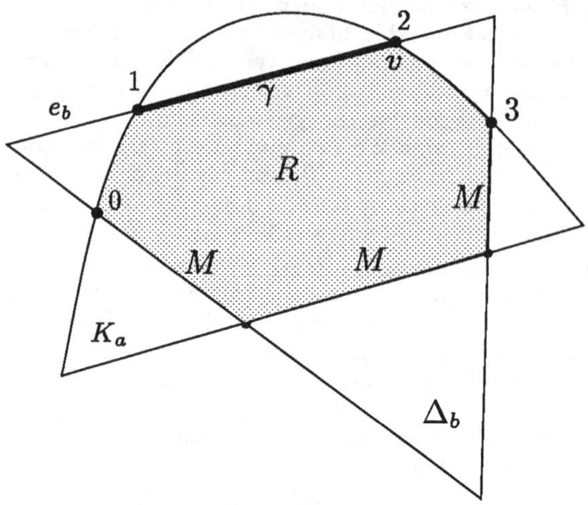


Figure 7: The region $R = K_a \cap \Delta_b$; the marked arcs are labeled by $M$; $\partial R$ has four semi-sharp vertices, with indices 0,1,2,3, as shown

any non-sharp vertex $w$ of $R$ (including non-chordal vertices as well, i.e., edges lying on the other edges of $\Delta_b$) to be the number of unmarked arcs of $\partial R$ that we encounter before hitting a marked arc, as we traverse this boundary from $w$ in counterclockwise direction. The index of a vertex is an integer between 0 and 5. As just mentioned, not all the vertices that we encounter during this traversal need be chordal; each non-chordal and non-sharp vertex lies on $\partial a$ and on some edge (other than the chord) of $\Delta_b$. We refer to all such vertices as *semi-sharp*; clearly, chordal vertices are also semi-sharp.

We introduce more quantities that we want to bound:

- $Q(\mathcal{C})$ is the number of semi-sharp vertices on $\partial U_{\mathcal{K}\cup\mathcal{T}}$ (perturbed as above), excluding special chordal vertices. Clearly, $Q_0(\mathcal{C}) \le Q(\mathcal{C})$.
- $Q^{(j)}(\mathcal{C})$, for $j = 0, \ldots, 5$, is the number of semi-sharp vertices on $\partial U_{\mathcal{K}\cup\mathcal{T}}$ (excluding special chordal vertices), whose index is at most $j$ (so $Q(\mathcal{C}) = Q^{(5)}(\mathcal{C})$).
- $Q(n)$ the maximum of $Q(\mathcal{C})$, over all collections $\mathcal{C}$ of $n$ regions satisfying (i) and (ii) (with fixed $\alpha$ and $s$).
- $Q^{(j)}(n)$ is the maximum of $Q^{(j)}(\mathcal{C})$, over all such collections $\mathcal{C}$, for $j = 0, \ldots, 5$.

Let $v$ be a semi-sharp vertex of $\partial U_{\mathcal{K}\cup\mathcal{T}}$, incident to the boundary of a cap $K_a \subseteq a$ (and to $\partial a$ itself) and to the boundary of an inner fat triangle $\Delta_b \subseteq b$, for two distinct sets $a, b \in \mathcal{C}$, whose index is at most $j$. Let $R = K_a \cap \Delta_b$, and let $\gamma$ denote the arc of $\partial R$ incident to $v$ and lying counterclockwise to it. We fix some threshold parameter $k$, trace $\gamma$ from $v$, and examine the sequence $\sigma$ of vertices of $\mathcal{A}(\mathcal{K} \cup \mathcal{T})$ that we encounter. Several cases can arise: **(a)** $\sigma$ contains at least $k$ vertices, and none of the first $k$ vertices of $\sigma$ is sharp. (Note that $\sigma$ may contain a sharp vertex only if the index of $v$ is zero.) In this case we charge $v$ to the block of the first $k$ vertices of $\sigma$. As above, we have the important observations that (i) each of these vertices lies at level $\leq k$ in $\mathcal{A}(\mathcal{K} \cup \mathcal{T})$, and (ii) each such vertex is charged at most twice, once along each of the boundaries containing it. It follows that the overall number of semi-sharp vertices of $\partial U_{\mathcal{K} \cup \mathcal{T}}$ that fall into this category is at most $(2/k)$ times the number of touching and shattering vertices of $\mathcal{A}(C)$ at level at most $k$, and of semi-sharp and special chordal vertices of $\mathcal{A}(\mathcal{K} \cup \mathcal{T})$ at level at most $k$. As above, the probabilistic analysis technique of Clarkson and Shor implies that this upper bound is at most $(2/k) \cdot O(k^2)$ times the expected number of touching and shattering vertices of the union of a random sample of $n/k$ regions of $\mathcal{C}$, plus $(2/k) \cdot O(k^2)$ times the expected number of semi-sharp and special chordal vertices of the union of the caps and triangles of a random sample of $n/k$ regions of $\mathcal{C}$. Hence the number of semi-sharp vertices of $\partial U_{\mathcal{K} \cup \mathcal{T}}$ of this kind is $O(kT(n/k)+kS(n/k)+kQ(n/k)+kQ^*(n/k))$. Applying (1), as above, this becomes $O(n+kS(n/k)+kQ(n/k)+kQ^*(n/k))$.

**(b)** The first $k$ (or all, if $\sigma$ is shorter) vertices in $\sigma$ include at least one sharp vertex $w$ (in this case the index of $v$ must be zero). We charge $v$ to the first such $w$. As above, it is easily seen that $w$ can be charged in this manner at most twice, and that it lies at level at most $k$ in $\mathcal{A}(\mathcal{K} \cup \mathcal{T})$. Hence, using Lemma 2.3 and the Clarkson-Shor analysis technique, the number of vertices $v$ of this kind is easily seen to be $O(k^2 \cdot \frac{n}{k} \log\log \frac{n}{k}) = O(nk \log\log n)$.

**(c)** $\sigma$ contains fewer than $k$ vertices of $\mathcal{A}(\mathcal{K} \cup \mathcal{T})$, none of which is sharp. In this case we charge $v$ to the other endpoint $w$ of $\gamma$. Clearly, $w$ is also a semi-sharp vertex (as noted above, no vertex of $R$ is special chordal), whose index is at most $j-1$, and it lies at level at most $k$ in $\mathcal{A}(\mathcal{K} \cup \mathcal{T})$. Applying the Clarkson-Shor technique again, we conclude that the number of vertices $v$ of this kind is $O(k^2 Q^{(j-1)}(n/k))$.

Thus, summing up these bounds, we obtain:

$$\begin{aligned} Q^{(j)}(n) \;&\leq\; c\,(nk \log\log n + kS(n/k) + kQ(n/k) \\ &+\; kQ^*(n/k) + k^2 Q^{(j-1)}(n/k))\,, \qquad (3) \end{aligned}$$

for $j = 0, \ldots, s$, and for some constant $c > 0$ that depends on $\alpha$ (for $j = 0$, we put $Q^{(-1)} = 0$ in the right-hand side).

Finally, we estimate $Q^*(\mathcal{C})$, that is, the number of special chordal vertices on $\partial U_{\mathcal{K} \cup \mathcal{T}}$ (perturbed as above). Note that the special chordal vertices are touching vertices in $\mathcal{A}(\mathcal{K} \cup \mathcal{T})$, but not all such touching vertices are necessarily special chordal. Nevertheless, Theorem 2.1, applied to $U_{\mathcal{K} \cup \mathcal{T}}$, implies that $Q^*(\mathcal{C})$ is at most proportional to $n$ plus the number of all other vertices of $\partial U_{\mathcal{K} \cup \mathcal{T}}$. That is, we have:

$$Q^*(n) = O\left(n + T(n) + S(n) + Q(n)\right),$$

or, using (1),

$$Q^*(n) \leq c\left(n + S(n) + Q(n)\right), \qquad (4)$$

for some constant $c$, as above. Substituting this into (2) and (3), we thus obtain

$$\begin{aligned} S^{(j)}(n) \;&\leq\; c\,(n + kS(n/k) \\ &+\; k^2 S^{(j-1)}(n/k) + k^2 Q(n/k))\,, \qquad (5) \end{aligned}$$

for $j = 0, \ldots, s-1$, and

$$\begin{aligned} Q^{(j)}(n) \;&\leq\; c\,(nk \log\log n + kS(n/k) \\ &+\; kQ(n/k) + k^2 Q^{(j-1)}(n/k))\,, \qquad (6) \end{aligned}$$

for $j = 0, \ldots, s$.

Following the analysis in [4, 12], the solution of the combined recurrences (5) and (6), with appropriate choice of the threshold parameters $k$, can be shown to imply that

$$T(n),\ S(n),\ Q(n),\ Q^*(n) = O(n^{1+\varepsilon}),$$

for any $\varepsilon > 0$, where the constants of proportionality depend on $\varepsilon$, $\alpha$, and $s$. Since these recurrences are somewhat more involved than those in [4, 12], we include, in the full version of the paper, an Appendix with a proof of these bounds, for the sake of completeness. This concludes the proof of Theorem 1.1. □

## Acknowledgment

We wish to thank János Pach for helpful discussions concerning this problem. His comments were especially helpful for the analysis given in Subsection 2.1.

## References

[1] P. Agarwal, M. Katz and M. Sharir, Computing depth order and related problems, *Comput. Geom. Theory Appls.* 5 (1995), 187–206.

[2] K. Clarkson and P. Shor, Applications of random sampling in computational geometry, II, *Discrete Comput. Geom.* 4 (1989), 387–421.

[3] A. Efrat, M. Sharir and A. Ziv, Computing the smallest $k$-enclosing circle and related problems, *Comput. Geom. Theory Appls.* 4 (1994), 119–136.

[4] D. Halperin and M. Sharir, Almost tight upper bounds for the single cell and zone problems in three dimensions, *Discrete Comput. Geom.* 14 (1995), 385–410.

[5] M. Katz, 3-D vertical ray shooting and 2-D point enclosure, range searching, and arc shooting amidst convex fat objects, to appear in *Comput. Geom. Theory Appls.* Also in a Tech. Report 2583, INRIA, Sophia-Antipolis, France, 1995.

[6] M. Katz, M. Overmars and M. Sharir, Efficient output sensitive hidden surface removal for objects with small union size, *Comput. Geom. Theory Appls.* 2 (1992), 223–234.

[7] K. Kedem, R. Livne, J. Pach and M. Sharir, On the union of Jordan regions and collision-free translational motion amidst polygonal obstacles, *Discrete Comput. Geom.* 1 (1986), 59–71.

[8] L. Lovász, J. Pach and M. Szegedy, On Conway's thrackle conjecture, *Proc. 11th ACM Symp. on Computational Geometry*, 1995, 147–151.

[9] J. Matoušek, J. Pach, M. Sharir, S. Sifrony and E. Welzl, Fat triangles determine linearly many holes, *SIAM J. Comput.* 23 (1994), 154–169.

[10] J. Pach and M. Sharir, On the boundary of the union of planar convex sets, manuscript, 1997.

[11] M. Sharir, On $k$-sets in arrangements of curves and surfaces, *Discrete Comput. Geom.* 6 (1991), 593–613.

[12] M. Sharir, Almost tight upper bounds for lower envelopes in higher dimensions, *Discrete Comput. Geom.* 12 (1994), 327–345.

[13] F. van der Stappen, *Motion Planning amidst Fat Obstacles*, Ph.D. Dissertation, Utrecht University, 1994.

[14] M. van Kreveld, On fat partitioning, fat covering, and the union size of polygons, *Proc. 3rd Workshop Algorithms Data Struct.*, Lecture Notes in Computer Science, Vol. 709, Springer-Verlag, 1993, 452–463.

# Some PL functions on surfaces are not height functions

Sergey P. Tarasov *
Computing center of RAS
Vavilova 40
117967 Moscow GSP-1, Russia.
e-mail: sergey@ccas.ru

Michael N. Vyalyi †
Computing center of RAS
Vavilova 40
117967 Moscow GSP-1, Russia.
e-mail: vyalyi@ium.ips.ras.ru

## Abstract

We find some obstructions for piecewise-linear (PL) functions defined on two-dimensional closed compact manifolds (surfaces) to be *height* functions in $\mathbf{R}^3$ and prove existence of some non-height PL functions. On the other hand we prove that all PL functions on a sphere are height functions.

Keywords: height function, index of singularity, Morse theory, upward drawing of an order.

## Introduction

It is well known that one can define on any compact closed two-dimensional manifold a smooth function $f(\cdot)$ having three critical points only: minimum, maximum and a degenerate saddlepoint (see, e.g. [8] and also [3]). Assume from now on that $M^2$ is oriented. An exercise attributed to H.Hopf (see [9], p.92) states that if $M^2$ is not a sphere then $f(\cdot)$ is not a height function, i.e. that it is not possible to represent $f(\cdot)$ in a form of a composition $zg(\cdot)$, where $g(\cdot) : M^2 \to \mathbf{R}^3$ is a smooth embedding of $M^2$ and $z(\cdot)$ is a linear functional on $\mathbf{R}^3$. (Piecewise-linear (PL) case is treated in Definition 2 below.)

On the other hand if we assume $g(\cdot)$ to be polyhedral embedding then $f(\cdot)$ can be a height function (see [3]). In this case $f(\cdot)$ is a piecewise-linear (PL) function, i.e. $f(\cdot)$ is defined on the manifold $M^2$ by specifying its values in the vertices of some triangulation of $M^2$ and continued onto $M^2$ by linearity. Let assume throughout the paper that values of $f$ in vertices are different. (If this assumption is dropped then it is easy to construct an example of a PL function on a sphere that is not a height function. Take for this the edge graph of a bipiramide with a triangle as a base. Let $f(\cdot)$ be 0 at both apexes and 1 at the vertices of the triangle-base. It is easy to check that $f(\cdot)$ is not a PL height function.)[1]

Then motivated by the facts stated above we get the following results.

We find necessary and sufficient condition for a PL function $h$ defined on $M^2$ to be a height function on a small enough neighborhood of its singular level line (Theorem 1).

* Supported in part by RFFI grant 96-01-00662 and Volkswagen Stiftung
† Supported in part by RFFI grant 96-01-00662
[1] We are grateful to Guenter Rote who attracted our attention to this case.



**Theorem 2.** *There exist PL functions that are not height functions.*

In some cases related e.g. to the problem of the visualization of the ordered structures or to *the upward drawing of orders on surfaces* (see, e.g. [1]) we are interested in sufficient conditions for a PL function $f(\cdot)$ to be a height function. Here is our result in this direction.

We find sufficient condition for a realization of a particular PL function on $M^2$ as height function (Theorem 3).

The paper is organized as follows. In Section 1 we briefly remind the notion of *singularity* of PL mapping. Next we introduce the notion of the *graph-type of a singular point* and prove that any graph-type can be realized by some PL-function. On the other hand to show that some PL-function $f(\cdot)$ defined on a manifold $M^2$ can be realized as a height function it is necessary to realize neighborhoods of all its singular level lines locally. Any local realization of a particular neighborhood of a singular level line can be formally described as a PL *admissible deformation* of a disk $B^2$ in a pure combinatorial form and --- what is essential--- all these admissible deformations correspond to the same graph-type of the singularity involved. Now to prove Theorem 1 it is enough to prove that the number of graph-types of singularities exceeds the number of admissible deformations when singularity is degenerate enough. This result is contained in Section 2. In Section 3 we propose some sufficient conditions for realization of PL functions as a height functions and even state them in a form of a *realization algorithm.* As a by product we prove that any PL function can be realized as a height function if we consider embeddings of the manifold $M^2$ into $\mathbf{R}^4$. In Section 4 we check conditions for the applicability of the realization algorithm for the spherical (genus 0) case (Theorem 4).

## 1 Graph-types of singularities of PL functions

It will be convenient for us to represent two-dimensional manifolds by standard polygons with identified edges and consider only straight line embeddings of graphs (that is always possible due to well-known result for *planar* graphs [6]).

Let's recall the definition of the *index of singularity* of the PL function $f(\cdot)$. Note that the edges of the triangulation incident to any vertex $u$ are cyclically oriented and form a kind of a circle on a manifold. Mark the vertex $v$ on this circle by "+" if $f(v) > f(u)$ or by "−" if $f(u) > f(v)$. Now separate the vertices on the circle into connected components of consecutive +'s and −'s and denote the number of positive (negative) components by $n_+$ ($n_-$), respectively. Call a vertex $u$ *nonsingular* if $n_+ + n_- = 2$ and set *index* $(u) = 0$, otherwise call it *singular* and set *index* $(u) = 1$ if either $n_+ = 0$ ($u$ is a *local maximum*) or $n_- = 0$ ($u$ is a *local minimum*) or finally set *index* $(u) = 1 - n_-$ if $n_+ + n_- > 2$ and this is the case of a *saddlepoint.* Unlike smooth case it is

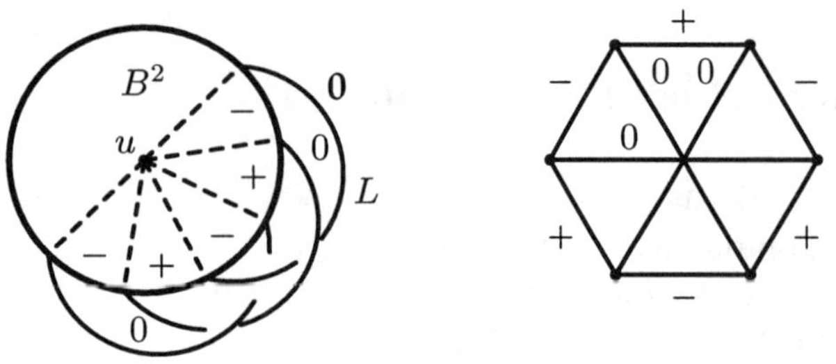


Figure 1: Singular point of *index* $(u) = -2$.

not possible to eliminate "degenerate" singularities of the large negative index by "small" perturbations.

The set $f^{-1}(u)$, where $u$ is a singular vertex, is called the *singular level line.*

**Definition 1.** Graph-type *of a singular point* $u \in M^2$ *of negative index of a function* $f(\cdot)$ *is a graph* $G$ *on* $2(1 - index(u))$ *vertices obtained by the following procedure. Choose any small disk* $B^2 \subset M^2$ *containing* $u$ *and mark points of intersection of its boundary and the singular level line* $L = f^{-1}(u)$. *These points will be vertices of* $G$. *They partition* $L$ *and the boundary circle of* $B^2$ *into disjoint components. Except the component containing* $u$ *all components are arcs. They are edges of* $G$ *that are colored into three colors* $\{0, +, -\}$. *0-edges correspond to the arcs of* $L$ *outside* $B^2$. $\pm$*-edges are arcs of the boundary of* $B^2$ *colored by* $+$ *or* $-$ *according to the sign of* $f(\cdot)$ *on the arc involved (see Fig. 1).*

By construction $G$ depends only on cyclic orientation of edges incident to $u$ and does not depend on the choice of a circle $B^2$ small enough.

It follows from orientability of $M^2$ [3] that 0-edges connect vertices of different parity (in the cyclic numbering of the vertices on the boundary of $B^2$). Lemma 1 below shows that this is the only nontrivial condition for PL realization of a particular graph-type of singularity

**Lemma 1.** *Let* $G$ *be a three-colored graph on* $2n$ *vertices so that edges of each color group form a perfect matching. Let two of these matchings form a hamiltonian cycle* $H \subset G$. *And let the edges of the third matching connect vertices of* $G$ *of different parity in the numbering induced by the hamiltonian cycle.*

*Then there exist a two-dimensional oriented closed compact manifold* $M$ *and a PL function* $f(\cdot)$ *on* $M$ *having singular point of graph-type* $G$.

**Proof.** Any embedding of the graph into the surface can be described by specifying cyclic orientations of the edges associated with each vertex (see, e.g. [4]).

Specify such a cyclic orientation in vertices that the hamiltonian cycle $H$ define a face $HF$ (i.e. $H$ is a rotational cycle in this orientation). Such an orientation always exists and is unique up to reflection. Denote the underlying manifold by $M$. Now consider an embedding of the *dual* graph $G^*$ into $M$. We may choose such an embedding that only dual pairs of edges from $G$ and $G^*$ have nonempty one point intersection. Denote the set of these points of intersection by $P$. Form a new embedded graph $\tilde{G}$ out of $G$ and $G^*$ adding the set $P$ to the vertices and updating resulting edges (formally we should add a new vertex for any pair of intersecting dual edges, then delete this pair and add four new edges instead).

Now we define a PL function $f(\cdot)$ on $M$ with a singularity of a prescribed graph-type $G$ in a vertex $u^* \in G^*$ corresponding to the hamiltonian face $HF$. It is enough to specify $f(\cdot)$ in the vertices of $\tilde{G}$ and continue it by linearity onto triangulation.

Let additionally partition the set $P$ of intersection points into two disjoint subsets $P = A \cup B$, where the vertices of $A$ result from intersection of edges of $G$ corresponding to hamiltonian cycle $H$. Set

$$f(x) = \begin{cases} 0, \; x \in u^* \cup V(G) \cup B; \\ 1, \; x \in A \text{ and the color of the} \\ \quad \text{underlying edge of } H \text{ is } +; \\ -1, \; x \in A \text{ and the color of the} \\ \quad \text{underlying edge of } H \text{ is } -; \\ 2, \; x \in V(G^*) \setminus u^*, \text{ and } f(\cdot) > 0 \\ \quad \text{in some neighbors of } x \text{ in } \tilde{G}; \\ -2, \; x \in V(G^*) \setminus u^*, \text{ and } f(\cdot) < 0 \\ \quad \text{in some neighbors of } x \text{ in } \tilde{G}. \end{cases}$$

By construction of $\tilde{G}$ consistency of this assignment is equivalent to proving that any face of $G$ except hamiltonian face $HF$ can contain only edges of the hamiltonian cycle of the same color. And this in turn is equivalent to the parity conditions for the edges of the third matching.

Also by construction $f(\cdot)$ is zero only on the edges of 0-colored matching and on the edges connecting $u^*$ and the vertices of $G$. In all other points $f(\cdot) \neq 0$ as any triangle of $\tilde{G}$ can contain at most one edge on which $f(\cdot)$ is identically zero (and thus there are no zero-triangles in the triangulation) and any triangle can't contain vertices of opposite signs.

To finish the proof we should eliminate edges on which $f(\cdot)$ is identically zero. This can be done by simple technical argument.

By construction graph-type of the singular point $u^*$ is $G$. The proof is completed. □

## 2 Local obstructions to embedding at a singular point

In this section we approach the problem locally, i.e. for the triangulated manifold $M^2$ and PL function $f(\cdot)$ on it we consider embeddings of some small neighborhood of a connected component of a singular level line containing singular point.

**Definition 2.** *PL function* $h(\cdot)$ *defined on a two-dimensional manifold is called (or equivalently can be realized as) a* PL height function *if it can be represented as a composition* $zg$, *where* $g(\cdot) : M \to \mathbf{R}^3$ *is a PL embedding of* $M$ *and* $z(\cdot)$ *is a linear functional on* $\mathbf{R}^3$.

**Definition 3.** Plane chord diagram *is such a perfect matching of* $2n$ *located on the circle vertices that chords corresponding to the edges of matching do not cross.*

Note that modulo cyclic order the precise positions of vertices on the circle are irrelevant for the intersection of chords. Thus two chords do not cross iff their ends are placed in the cyclic order alternatively with each other. To this extent plane chord diagram can be regarded as "pure" combinatorial object.

**Definition 4.** *A triple* $(D_-, D_0, D_+)$ *of plane chord diagrams on* $2n$ *cyclically ordered points is called* an admissible deformation *of the index* $n$. The graph *of the admissible deformation is a graph on* $2n$ *vertices with edges* $D_- \cup D_0 \cup D_+$ *colored in corresponding colors.*

It will be convenient to have a mental picture of an admissible deformation in which 0-labeled edges lie outside the underlying circle and the $\pm$-labeled edges lie inside the circle (see Fig 2).

Theorem 1 explains the definitions above.

**Theorem 1.** *Let* $h(\cdot)$ *be PL function on the manifold* $M^2$ *and* $u$ *is its singular point. Let* $M$ *be small enough neighborhood of the connected component of the singular level line of* $h(\cdot)$ *containing* $u$. $h(\cdot)$ *is PL height function on* $M$ *iff there exists*

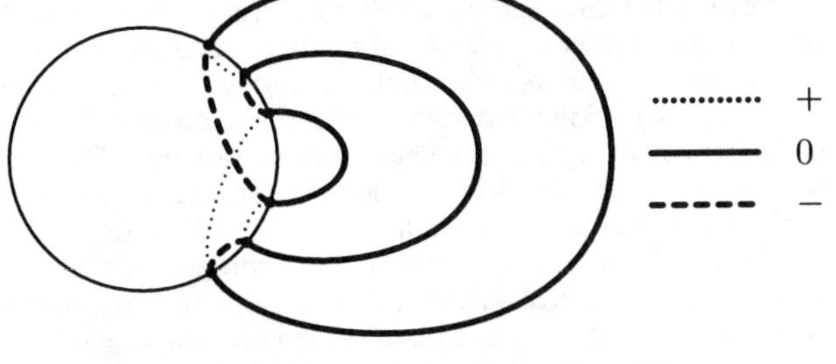


Figure 2: Picture of an admissible deformation.

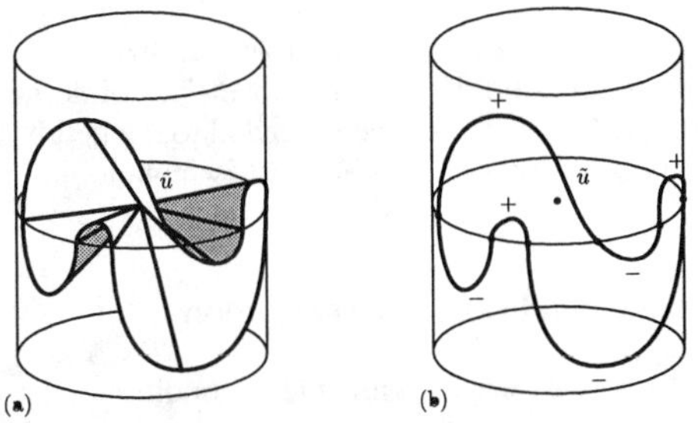


Figure 3: (a) Neighborhood of a singular point $u$ under embedding $g$ and (b) matchings $D_+$ and $D_-$.

*an admissible deformation $(D_-, D_0, D_+)$ whose graph coincides with graph-type of $u$.*

**Proof.** *Necessity.* Let $h(\cdot)$ be PL height function on $M$. Then by definition $h$ can be represented as a composition $h = zg$. We may assume that linear functional $z$ is simply standard $Z$-coordinate function and $h(u) = 0$. Let $\tilde{M} = g(M)$ and $g(u) = 0$.

The connected component $L \subset h^{-1}(u) \subset M^2$ of the level line $h^{-1}(u)$ containing singular point $u$ is a union of $(1 - index(u))$ circles with a unique point of intersection $u$. Small neighborhood of $L$ in $M^2$ (see [3]) is a disk $B^2$ centered at $u$ and $(1 - index(u))$ bands whose ends are attached to the boundary of $B^2$. In any embedding the disk is deformed and the bands become twisted (but the number of twists of any band should be even as manifold is oriented and does not contain Mobius band). If $h$ is a height function then bands in the embedding $g(M)$ can't twist at all.

We may assume (choosing smaller neighborhood of $M$ and slightly changing $g$) that all edges outgoing from the origin (the image of $u$ under $g$) have nonzero projections both on $0Z$-axis and on $0XY$ coordinate plane. Let $St(u)$ be the *star* of the singular vertex $u$, i.e. the union of all triangles incident to $u$ in the triangulation. We can take such a narrow vertical cylinder $C = \{(x, y, z) \in \mathbf{R}^3 | x^2 + y^2 \leq \delta\}$ that boundary of the image of $St(u)$ under $g$ lies outside $C$. We can also assume that the image of the connected component of the singular level line $L$ containing $u$ intersects the boundary $\partial C$ of the cylinder at exactly $2(1 - index(u))$ points. Let denote this set by $P$. Note that the curve $l = g(St(u)) \cap \partial C$ is a simple connected closed and not self-intersecting curve that intersects zero-level circle $S = \{Z = 0\} \cap \partial C$ exactly at points of $P$ (see Fig. 3). It is evident now how to construct admissible deformation. Its vertices correspond to $2n$ points of the set $P$. Edges of the perfect matching $D_0$ correspond to the arcs of the singular level line $L$ connecting points of $P$ and passing in plane $\{Z = 0\}$ *outside* circle $S$. $D_0$ is certainly plane chord diagram as the underlying arcs don't cross outside $S$. Edges of the matching $D_+ (D_-)$ correspond to the arcs of the curve $l$ connecting consecutive points of $P$ (in the cyclic ordering induced by $l$) which have positive (negative) $Z$-coordinates. Let prove that each of the perfect matchings $D_+$ and $D_-$ also forms a plane chord diagram in the cyclic order induced by $S$. Assume on the contrary that some chords of $D_+$ cross ($D_-$ is treated analogously). But then the corresponding arcs of $l$ should cross on $\partial C$, that is impossible. Thus we obtain an admissible deformation. By construction its graph coincides with the graph-type of the singularity involved.

*Sufficiency.* Let $(D_-, D_0, D_+)$ be the admissible deformation whose graph coincides with graph-type of the singular point $u$. We must construct an embedding of a small neighborhood of a singular level line into $\mathbf{R}^3$ whose height function is $h$. Take any planar embedding of the union of circles induced by $D_0$ into $0XY$-plane. Denote this set by $V$. Now let construct a cylinder $\tilde{C} = V \times [-\varepsilon, +\varepsilon]$ ($\varepsilon$ is small enough). Let cut from $\tilde{C}$ a small cylindrical neighborhood of the interval $\tilde{u} \times [-\varepsilon, +\varepsilon]$ ($\tilde{u}$ is a image of the singularity under embedding). Let denote by $v_i$, $(i = 1, \ldots 2n)$ the point of $V \cap \partial\tilde{C}$ corresponding to the vertex $i$ of the admissible deformation. Inside this neighborhood attach to the rest of $\tilde{C}$ the set of triangles $conv(\tilde{u} \times 0, v_i \times (-\varepsilon), v_i \times \varepsilon)$, $conv(\tilde{u} \times 0, v_i \times (-\varepsilon), v_j \times (-\varepsilon))$, $conv(\tilde{u} \times 0, v_k \times \varepsilon, v_l \times \varepsilon)$, where pairs $(i, j)$ pass through set of edges $D_-$ and pairs $(k, l)$ pass through set of edges $D_+$. This set of triangles corresponding to the admissible deformation on Fig. 2 is shown on Fig. 4.

As $+$ and $-$ edges in the graph-type of singularity form cycle we conclude that triangles involved have common center at point $\tilde{u} \times 0$ and form a disk and thus we get the required embedding. □

**Corollary 1.** *There exist forbidden graph-types of singularities and thus local obstructions for realization of PL functions on oriented two-dimensional manifolds as PL height functions.*

**Proof.** Any admissible deformation is uniquely up to rotation and reflection related to some graph-type of singularity. But there are not more then $2^{O(n)}$ (cube of the Catalan number) admissible deformations. And the number of different graph-types (the number of different graphs from Lemma 1) is at least $n!/4n = 2^{\Omega(n \log n)}$. Thus some graph-types are forbidden: no admissible deformations are related to them. □

**Theorem 2.** *There exist PL functions that are not height functions.*

**Proof.** To prove the theorem it's enough to take any singularity of a forbidden graph-type and construct function (and the underlying manifold) as in the proof of Lemma 1. □

We note in passing that computer experiment shows that there are no obstructions up to $n = 18$.

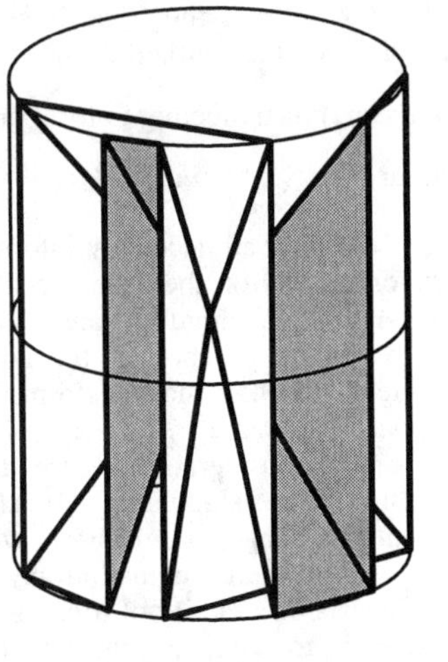

Figure 4: Embedding of the admissible deformation.

**Remark 1.** Theorem 2 can be slightly generalized:
There exists a function defined in the vertices of some embedded into $M^2$ graph $G$ that can not be extended to the height function of any PL embedding of $M^2$ into $\mathbf{R}^3$ that is monotonic on the edges of $G$.

## 3 Sufficient conditions for realization

In this section we propose sufficient conditions for realization of PL functions on manifolds as height functions using standard considerations of the elementary Morse theory.

At first we briefly discuss the main idea informally appealing to the elementary Morse theory for the two-dimensional manifolds.

Recall that the main objective of Morse theory (see, e.g.,[10]) is to deduce topological structure of the manifolds from studying the singular points and levels of smooth real-valued functions defined on them. Let $h(\cdot)$ be any smooth function defined on the manifold $M$. A point $x \in M$ is called *singular* for $h$ if the gradient of $h$ vanishes at it. The corresponding value $h(x)$ is called *singular value.* Moreover a singular point is called *nondegenerate* if the *Hessian* of $h$ (the determinant of the matrix of the second derivatives) is nonzero at $x$. The dimension of the maximal subspace on which Hessian is negatively definite is called the *Morse index* of the singular point (evidently this number is intrinsic to the point and does not depend on local coordinate system involved). Thus for the two-dimensional case Morse index can be 0, 1 or 2. Call $h(\cdot)$ *Morse function* if it has nondegenerate singular points only. We can assume that there is only one singular value on any singular level of Morse function. Evidently a nonsingular level set consists of a family of closed topological contours $\{C_i\}$.

Morse Theory states that topological structure of the level sets $h^{-1}(a) = \{x \in M \mid h(x) = a\}$ changes only while passing through singular values and there are four typical changes for the case of oriented closed compact two-dimensional manifolds.

If there are no singular values of $h$ on the interval $[a, b]$ then submanifold $h^{-1}([a, b])$ is homeomorphic and even diffeomorphic in the smooth case to the set $C_i \times [a, b]$.

After passing through singular value this family transforms depending on the index value *index* $(u)$ of the corresponding singular point.

1. If *index* $(u) = 0$, then one new contour is formed and this corresponds to passing a local minimum.
2. If *index* $(u) = 2$ then some contour disappears and this case corresponds to passing a local maximum.

   Finally, if *index* $(u) = 1$ then this situation corresponds to passing a saddlepoint and either
3. two contours are glued together to form one contour or
4. some contour breaks into two.

Note that we could infer all necessary information about these topological transformations of the level sets just tracing their evolution in the vicinity of singular value. Therefore we can distinguish the reasons by which e.g. the number of contours involved has changed, whether the transformation resulted from passing a saddlepoint or extrema.

We could use similar consideration in case of PL functions. We only need to use PL index of singularity instead of a smooth one. Thus for our case of the triangulated manifold with a function $f(\cdot)$ defined in the vertices of the triangulation any nonsingular level set $f^{-1}(a)$ also consists of a family of topological contours. It is evident that changes in topological structure of the level set can take place only after passing through a vertex (recall that only one vertex lies in any level set).

Level sets follow the same pattern of transformations as in the smooth case. Either we pass local minimum and one new contour appears or we pass local maximum and one contour vanishes or we pass a saddlepoint and some contours are dissolved into many and some contours are attached together to form one. The resulting transform can be completely deduced from the dynamics of the evolution of the level contours in the vicinity of the singular value.

Informally, *between singular values* manifold is homeomorphic to a system of vertical cylindrical tubes. *Passing singular value* corresponds to either emerging of some new tube or cutting some tube or pasting some tubes together to form a new tube or, finally, splitting some tube to form several new ones.

Moreover similarly to the smooth case stated above these changes essentially depend on the number $1 - index(u)$ of "negative directions" outgoing from that singular vertex $u$. But it was already pointed above that PL singularities of an arbitrary nonzero index are nondegenerate in the sense that it is not possible to eliminate them by small PL perturbation of the function $f$. And thus contrary to the nondegenerate smooth case considered above, in general, while passing through singular value more than one topological contour could be formed or destroyed and one contour could be simultaneously several times broken or attached to itself or to the other contours. And as Theorem 2 states under the assumption that level sets lie in the corresponding horizontal planes it is not always possible to realize all these changes as embedding. The straightforward way to partially avoid this difficulty is to consider more restrictive admissible deformations. For this purpose let recall the definition of an isotopic deformation.

**Definition 4.** *A family of homeomorphic transformations* $k^t$ $0 \leq t \leq 1$ *of the topological space* $X$ *is called an* isotopic deformation *of* $X$ *if* $k^0$ *is an identity map and the map* $K(t, \cdot) \stackrel{def}{=} k^t(\cdot)$ *is continuous in both variables. Two subspaces* $A^1, A^2 \subseteq X$ *are called* isotopic *if there is an isotopic deformation* $k^t$ *of* $X$ *such that* $k^1 A^1 = A^2$.

Note that sets $(D_-, D_0)$ and $(D_0, D_+)$ form two systems of contours. Loosely speaking, passing through singularity takes $(D_-, D_0)$ to $(D_0, D_+)$. To describe isotopic deformations of these systems of contours we need to define an *inclusion relation* on a pair $(C_1, C_2)$ of Jordan contours.

We say that $C_1$ and $C_2$ are *not included* into each other iff bounded components of their complements are disjoint (both these components are well-defined due to Jordan theorem).

Additionally let call a system of pairwise not included into each other disjoint Jordan contours *isotopically trivial.*

**Definition 5.** *An admissible deformation* $(D_-, D_0, D_+)$ *is called* trivial *if there exists an embedding of its graph into plane without inclusion of contours in* $(D_-, D_0)$ *and without inclusion of contours in* $(D_0, D_+)$. *(In other words, each of the two systems of cycles involved is isotopically trivial in this embedding.)*

**Lemma 2 [5, 11].** [2] *Any two closed Jordan contours* $C_1, C_2$ *in* $\mathbf{R}^2$ *are isotopic. In particular, if both contours are simple polygons then there exist PL isotopy.*

**Proof.** We prove the last assertion of Lemma 2. We can use induction on the number of edges of the first polygon $C_1$. Triangulate $C_1$. Then at least one triangle has two consecutive boundary edges of $C_1$ as sides thus producing an "ear" of $C_1$ that can be isotopically cut reducing the number of edges of the polygon. □.

**Remark 2.** There is a straightforward representation of an isotopy map $K(\cdot, \cdot)$ between two Jordan contours $C_1$ and $C_2$. A *tube* is a space homeomorphic to $S^1 \times [0, 1]$. Denote by $H^t = ((x, y, z) \in \mathbf{R}^3 | z = t)$ a family of parallel hyperplanes. Embed contours $C_1, C_2$

[2]The authors are grateful to M.A. Štan'ko and R.Seidel for pointing related references.

into two different hyperplanes of this family, say, $C_1 \subset H^0$ and $C_2 \subset H^1$, respectively. Then isotopy map $K(z, \cdot)$ can be naturally identified with a nonsingular w.r.t. height function $z$ embedding of a tube bounding contours $C_1, C_2$ into $\mathbf{R}^3$, i.e. $K(0, C_1) = C_1$, $K(1, C_1) = C_2$ and the set $K(t, C_1), 0 \le t \le 1$ is a tube.

**Lemma 3.** *Two isotopically trivial systems of Jordan contours $A_1, \ldots, A_n$ and $B_1, \ldots, B_n$ are isotopic.*

**Proof.** The proof immediately follows from Lemma 2 and the representation of an isotopy map above. □.

**Remark 3.** Arguing as above one can obtain a complete classification of systems of disjoint Jordan contours up to isotopy. Namely, two systems of disjoint Jordan contours $A_1, \ldots, A_n$ and $B_1, \ldots, B_n$ are isotopic iff their (oriented) inclusion graphs are isomorphic.

**Theorem 3.** *Let $h$ be a PL function on $M^2$. If graph-types of its singular points of the negative index admit trivial deformations then $h$ is a height function.*

**Proof.** The idea of the proof of the Theorem 3 is straightforward: we should trace evolution of the level sets of the function $h(\cdot)$ using trivial deformations in the small neighborhoods of the singular values unfolding possible intersections of tubes to form a homeomorphic embedding of the manifold $M^2$ into $\mathbf{R}^3$.

Let $h(\cdot)$ --- be PL function on the manifold $M^2$. Let $u_1, \ldots, u_k$ be its singular points. Choose such an $\varepsilon > 0$ that preimages of the $\varepsilon$-neighborhoods of the singular values ($t_1 = h(u_1) < \ldots < t_k = h(u_k)$) are disjoint.

One of the connected components $U_i$ of the set $h^{-1}([t_i - \varepsilon, t_i + \varepsilon])$, $i = 1 \ldots k$, contains the singularity $u_i$ and the rest are PL tubes. If $u_i$ is a local extremum then embedding of $U_i$ is a PL cup or cap (depending on the type of the extremum) with an apex in some point of plane $H^{t_i}$ and base in plane $H^{t_i \pm \varepsilon}$. For the singularities of negative index embedding of the corresponding component $U_i$ is constructed using trivial admissible deformation (that exists by assumption). The boundary of $U_i$ consists of two systems of disjoint topological contours $C_i(t_i + \varepsilon)$, $C_i(t_i - \varepsilon)$ The embedding maps them into isotopically trivial systems of simple polygons $\mathcal{B}_i(+)$ and $\mathcal{B}_i(-)$ in planes $H^{t_i + \varepsilon}$ and $H^{t_i - \varepsilon}$, respectively. Enlarge these systems (respecting isotopical triviality) by pairs of congruent triangles located in the same planes $H^{t_i + \varepsilon}$ and $H^{t_i - \varepsilon}$, respectively, that have equal projection on $0XY$-plane. The number of the triangles pairs involved should be equal to the number of the remaining connected components of $h^{-1}([t_i - \varepsilon, t_i + \varepsilon])$ and both systems of polygons in planes $H^{t_i + \varepsilon}$ and $H^{t_i - \varepsilon}$ are by construction isotopically trivial. Let define an embedding of these remaining components (different from $U_i$) as prisms with bases in corresponding pairs of triangles.

Thus we constructed PL embeddings of the sets $h^{-1}([t_i - \varepsilon, t_i + \varepsilon])$. It remains to continue them onto the embedding of the whole manifold. Note that the set $h^{-1}([t_i + \varepsilon, t_{i+1} - \varepsilon])$ is a union of PL tubes. By Lemma 3 there exists an isotopy between isotopically trivial and equal in size systems of polygons located in planes $H^{t_i + \varepsilon}$ and $H^{t_{i+1} - \varepsilon}$, respectively. In view of the Remark 2 this isotopy gives required embedding of the set $h^{-1}([t_i + \varepsilon, t_{i+1} - \varepsilon])$. □

**Remark 4.** All is greatly simplified if we consider realization of the height function in $\mathbf{R}^4$. Any admissible deformation is now related to an embedded into $\mathbf{R}^3$ graph $G$. Vertices of $G$ belong to some sphere. 0-colored ($\pm$-colored) edges are related to arcs connecting corresponding vertices outside (inside) the sphere. And the arcs of the same color do not cross. But this condition is fulfilled trivially as any graph can be embedded into $\mathbf{R}^3$ without edge crossings. Therefore there are no local obstructions to realization of height function in $\mathbf{R}^4$.

The existence of the trivial admissible deformation under embeddings into $\mathbf{R}^4$ is thus replaced by the condition that contours of the corresponding systems are not linked. Therefore any PL function on a two-dimensional manifold can be realized as a height function in $\mathbf{R}^4$. The algorithm of Theorem 3 is effective in this case.

## 4 Case of genus 0

If all admissible deformations were trivial then Theorems 1 and 3 would imply necessary and sufficient conditions on a PL function on a manifold to be a height function. Incidentally this is actually the case for the spheres (although we suspect that the answer to the general question is negative).

**Theorem 4.** *Any singular point of a PL function $f(\cdot)$ defined on a sphere has trivial admissible deformation.*

**Proof.** Let use induction to prove a stronger statement: there exists such a trivial admissible deformation in which ends of each 0-labeled edge are neighbors.

Under this condition the systems $(D_-, D_0)$ and $(D_0, D_+)$ are isotopically trivial (in each of these systems contours are disjoint and not included into each other).

The base for the induction ($n = 1$, i.e. there are 2 vertices) is evident.

Let prove the induction hypothesis.

First note that in the proof of Lemma 1 we construct an embedding of the graph-type of the singularity into some manifold $M'$. Let prove that the genus of this manifold is not greater than the genus of $M^2$. $M'$ can be obtained from $M^2$ by consecutive cutting along non-essential contours of the singular level line (not containing the singular point) and subsequent attaching of disks to the resulting boundary circles. After completing these cut-and-paste operations we will get union of some closed manifolds (including $M'$). Checking variation of the Euler characteristic we conclude that any indicated operation does not increase the genus.

Therefore the graph-type $G$ of the singular point is plane and the hamiltonian contour defines the face of the embedding. Therefore 0-labeled edges of $G$ form a trivial plane chord diagram when they are put simply along the hamiltonian contour.

Denote by $G_-, G_+$, respectively, graphs obtained from $G$ by deleting $e$ and the incident vertices $a$ and $b$ (pairs of other edges of graph incident to these vertices are glued together to form two connected components). By induction hypothesis our condition holds for $G_-, G_+$. This means that there exist trivial admissible deformations of graphs $G_-, G_+$ in which ends of 0-labeled edges are neighbors. Denote by $a_-, a_+$ and $b_-, b_+$ the neighbors of the vertices $a$ and $b$ in the graph $G$, respectively. (We assume that the edges $(a, a_-)$ and $(b, b_-)$ are $-$-labeled in $G$ and the edges $(a, a_+)$ are $(b, b_+)$ are $+$-labeled.)

Denote by $\overrightarrow{(x, y)}$ possibly empty sequence (*an interval*) of vertices between the vertices $x$ and $y$ in the cyclic order induced by trivial admissible deformation. Let

$a_-, \overrightarrow{(a_-, b_-)}, b_-, \overrightarrow{(b_-, a_-)}$ and $a_+, \overrightarrow{(a_+, b_+)}, b_+, \overrightarrow{(b_+, a_+)}$ be corresponding cyclic orders on the vertices of $G_-$ and $G_+$.

W.l.o.g. we may assume that the sequences $\overrightarrow{(a_-, b_-)}$ and $\overrightarrow{(a_+, b_+)}$ are nonempty. We need this assumption to guarantee that (possible) 0-edges $(a_- b_-)$ and/or $(a_+ b_+)$ would be placed properly in the resulting graph.

Now set the following cyclic order on the vertices of $G$:

$$a, b, b_-, \overrightarrow{(b_-, a_-)}, a_-, \overrightarrow{(a_-, b_-)}, b_+, \overrightarrow{(b_+, a_+)}, a_+, \overrightarrow{(a_+, b_+)}$$

(see Fig. 5).

In other words, the ends of the edge $e = (a, b)$ are neighbors and stand between the vertices in $G_-$ and $G_+$ that follow the cyclic order of their trivial admissible deformations. Hence, we

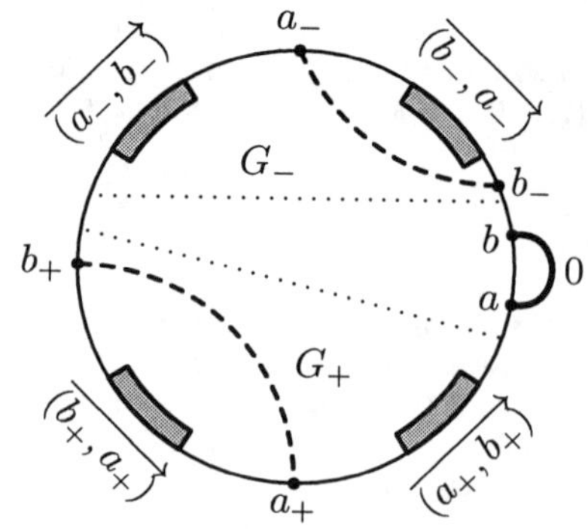


Figure 5: Construction of Theorem 4.

find the required admissible deformation. Indeed, the ends of the 0-labeled edges are neighbors. The edges of the same color inside $G_-$ and $G_+$ (not incident to $a$ and $b$) do not cross by induction hypothesis. Moreover, they do not cross (''virtual'', i.e. induced by the construction of graphs $G_-, G_+$) chords $(a_-, b_-)$ and $(a_+, b_+)$ because these chords are edges in graphs $G_-, G_+$. So, the edges of the same color incident to $a$ and $b$ do not cross the edges of this color inside $G_-$ and $G_+$. The edges of the same color incident to $a$ and $b$ do not cross by construction. □

Now an immediate consequence of Theorems 3 and 4.

**Corollary 2.** *Any PL function $f(\cdot)$ on a sphere is a height function.*

**Remark 5.**[3] Corollary 2 can be easily proved along the lines of ''lifting'' procedure suggested in [1]. Indeed, the underlying triangulation is planar.

Let $u$ be the unique minimum of $f(\cdot)$. Add two vertices $u_1$, $u_2$ on an arbitrary pair of neighboring edges incident to $u$ and the edge joining them. Then refine the triangulation. We may assume that these vertices are close enough to $u$ so that the values of $f(\cdot)$ in all remaining vertices are strictly greater. Now consider the embedding of the refined triangulation with outer triangle $(u, u_1 u_2)$. The graph of $f(\cdot)$ together with attached triangle $\Big((u, f(u)), (u_1, f(u_1)) (u_2, f(u_2))\Big)$ gives the desired embedding.

## 5 Concluding remarks

Our paper leaves at least three open issues.

What is the maximal genus of the manifold where all singular points of any PL function have trivial admissible deformations? (It seems that the answer is 1.)

What is the complexity status of finding an admissible deformation?

And last but not the least how to find explicit construction of nonrealizable singularity?

We have made limited computer experiment and found that all singularities up to the index $-8$ admit admissible deformation. This partly explains the fact that in all considered applications of the realization algorithm (e.g. for upward drawings of particular orders of small size) we succeeded to perform the algorithm by hand and have not encountered any difficulties.

## Acknowledgments

The first author is grateful to I.Rival who attracted his attention to the problem of upward drawing of orders on surfaces. We thank A. Yu. Kitaev for helpful discussions. We are grateful to the unknown referees for their valuable remarks.

And especially we are grateful to Guenter Rote for numerous insightful comments.

[3] We are grateful to Guenter Rote who called our attention to the existence of simple proof of the statement of Corollary 2.

## References

[1] Ewacha K., Li W., and Rival I. Order, genus and diagram invariance. Order, **8**, 1991, 107--113.

[2] Musin O.R. On some problems of computational geometry and topology. Lect. Notes Math., **1520**, 1992, 57--80.

[3] Banchoff T.F. and Takens F. Height functions with three critical points. Illinois J. Mathematics, **76**, 1975, 326--335.

[4] Ringel G. Map-color theorem. Springer, N.Y., 1974.

[5] Antoine L. Sur l'homeomorphisme des deux figures et de leurs voisinages. J. Math., Ser. 8. **4**, 1921, 221--235.

[6] Wagner K. Bemerkungen zum Vierfarbenproblem, *Jber. Deutsch. Math. Verein.* **46**, 1936, 26--32.

[7] Banchoff T.F. Critical points and curvature for embedded polyhedra, *J. Diff. Geom.* **1**, 1967, 245--256.

[8] Dubrovin B., Novikov S., Fomenko A. Modern Geometry. Moscow, 1979.

[9] H.Seifert and W.Threlfall. Variationsrechnung im Grossen. Teubner, Berlin, 1938.

[10] J.Milnor. Morse Theory. Princeton, NJ, Princeton University Press, 1963.

[11] B. Aronov and R. Seidel and D. Souvaine. On compatible triangulations of simple polygons. *Comput. Geom. Theory Appl.* **3**, No 1, 1993, 27--35.

# Vertex-Edge Pseudo-Visibility Graphs: Characterization and Recognition

Joseph O'Rourke
Ileana Streinu*

**Abstract**

We extend the notion of polygon visibility graphs to pseudo-polygons defined on *generalized configurations of points*. We consider both vertex-to-vertex, as well as vertex-to-edge visibility in pseudo-polygons.

We study the characterization and recognition problems for vertex-edge pseudo-visibility graphs. Given a bipartite graph $G$ satisfying three simple properties, which can all be checked in polynomial time, we show that we can define a generalized configuration of points and a pseudo-polygon on it, so that its vertex-edge pseudo-visibility graph is $G$. This provides a full characterization of vertex-edge pseudo-visibility graphs and a polynomial-time algorithm for the decision problem. It also implies that the decision problem for vertex visibility graphs of pseudo-polygons is in NP (as opposed to the same problem with straight-edge visibility, which is only known to be in PSPACE).

## 1 Introduction

Characterizing visibility graphs has remained an elusive problem [O'R93]. Ghosh [Gho88, Gho97] proposed a set of necessary conditions as a starting point. Everett [Eve90] proved their insufficiency and proposed new conditions. She also placed the recognition problem in PSPACE by reducing it to the existential theory of the reals. Abello and Kumar [AK95] expanded the set of conditions and first related the problem with oriented matroid theory. Their conditions, plus realizability (stretchability) of a certain oriented matroid associated with the graph, would allow a characterization of visibility graphs. But realizability of oriented matroids is a very strong condition. It has been shown by Mnëv [Mnë91] (see also Shor [Sho91]) that it is as hard as the existential theory of the reals, for which so far only exponential algorithms are known. In this context the problem seems too hard to attack.

In this paper we deal with the characterization and recognition problems for a class of polygon visibility graphs, the ve-graphs introduced in [OS97]. Elsewhere we deal with the reconstruction (drawing) problem [Str96a].

Our approach introduces two innovations. First, instead of the usual vertex-vertex visibility graph (v-graph), we study the ***vertex-edge visibility graph*** (ve-graph) $G_{VE}$ of a polygon. We introduced this concept in a previous paper [OS97]. There we showed that this combinatorial structure contains more geometric information than the vertex visibility graph for straightline polygons. Second, we generalize the notion of straightline visibility to visibility along pseudolines. Here we reinterpret and clarify some of the results of Abello and Kumar [AK95], as well as taking them a step farther. We mix our two ideas by starting with the ve-graph of a pseudo-polygon, and showing that such graphs can be characterized by three simple properties recognizable in polynomial time. It follows that recognition of vertex visibility graphs for pseudo-polygons is in NP. In a companion paper [Str96b], it is shown that the class of straightline ve-graphs is properly contained in the class of pseudoline ve-graphs, and similarly for v-graphs. In particular it follows that Abello and Kumar's oriented matroids may not be stretchable, and thus their characterization fails to completely capture straightline v-graphs. Similarly, our characterization of pseudo ve and v-graphs would be incomplete if taken to their straightline counterparts.

By stating the problem in the pseudo-visibility context, we isolate the combinatorial (oriented matroid) structure of the problem from the stretchability issue, in a manner similar to [AK95]. Focus on ve-graphs rather than v-graphs, and explicit focus on pseudo-

*Department of Computer Science, Smith College, Northampton, MA 01063, USA.
{orourke,streinu}@cs.smith.edu. Supported by NSF grant CCR-9421670.


*Computational Geometry 97* Nice France

visibility, result in a simpler set of conditions than have been obtained previously. The relative simplicity of these conditions has led, as just mentioned, to the first examples of non-stretchable visibility graphs [Str96b].

The paper has two main sections. In Section 2, we start with a pseudo-polygon and its ve-graph, and derive its properties. These properties are all relatively unsurprising, direct generalizations of those established for straight-line ve-graphs in [OS97]. The definition of vertex-edge visibility in the pseudoline context is, however, not completely straightforward. In order to concentrate on the more novel characterization, we do not include proofs in Section 2; see [OS96, OS97]. In Section 3, we start from abstract properties of the graph and construct a pseudo-polygon that realizes it. This involves the construction of an acyclic uniform rank-3 oriented matroid. We establish that the matriod has the claimed properties using Knuth CC-system axioms [Knu92]. Other systems of axioms (such as co-circuits) have been considered in previous versions of this paper, but they led to a much more involved case analysis.

## 2 Pseudo-Visibility in Pseudo-Polygons

### Generalized Configurations of Points

Our generalization of straightline visibility to pseudo-visibility depends the the notion of a "generalized configurations of points" introduced by Goodman and Pollack [GP84].[1] Recall that an *arrangement of pseudolines* $\mathcal{L}$ is a collection of simple curves, each of which separates the plane, such that each pair of lines of $\mathcal{L}$ meet in exactly one point, where they cross.

**Definition 2.1** *Let* $V = \{v_0, v_1, \ldots, v_{n-1}\}$ *be a set of points in the Euclidean plane* $\mathbb{R}^2$*, and let* $\mathcal{L}$ *be an arrangement of* $\binom{n}{2}$ *pseudolines such that every pair of points* $v_i$ *and* $v_j$ *lie on exactly one pseudoline* $l_{ij} \in \mathcal{L}$*, and each pseudoline in* $\mathcal{L}$ *contains exactly two points of* $V$*. Then the pair* $(V, \mathcal{L})$ *is a* generalized configuation of points *in general position.*

The phrase "in general position" indicates that no three points of $V$ lie on one line of $\mathcal{L}$.

### Pseudo-Polygon

Two points $a$ and $b$ on a pseudoline $l \in \mathcal{L}$ determine a unique (closed) *segment ab* consisting of those points on $l$ that lie between the two points. For $0 \le i \le n-1$, let $e_i = \overline{v_i v_{i+1}}$ be the segment determined by $v_i$ and $v_{i+1}$ on $l_{i,i+1}$.[2]

**Definition 2.2** *The segments* $e_i = v_i v_{i+1}$ *form a pseudo-polygon iff:*

1. *The intersection of each pair of segments adjacent in the cyclic ordering is the single point shared between them:* $e_i \cap e_{i+1} = v_{i+1}$*, for all* $i = 0, 1, \ldots, n-1$.
2. *Nonadjacent segments do not intersect:* $e_i \cap e_j = \emptyset$*, for all* $j \ne i+1$.

See Fig. 1a for an example. Throughout we let $P$ denote a pseudo-polygon, with $V$ its vertices labeled $V = (v_0, v_1, \ldots, v_{n-1})$ in counterclockwise (ccw) order, and $E$ its set of edges similarly labeled. Note edges are closed segments. We use the term *exterior* to designate points of the plane strictly exterior to $P$ (and so not on its boundary).

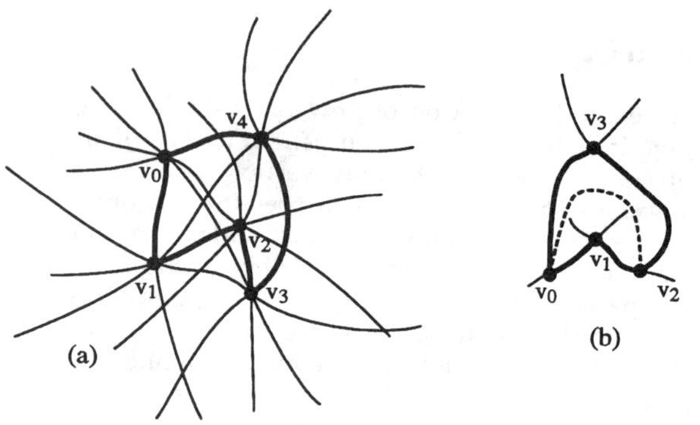


Figure 1: (a) A pseudo-polygon (not every intersection between pseudolines is shown); $|V| = 5$, $|\mathcal{L}| = 10$. (b) $v_0$ cannot see $v_2$.

### Vertex-vertex Pseudo-Visibility

Pseudo-visibility is determined by the underlying arrangement $\mathcal{L}$: lines-of-sight are along pseudolines in $\mathcal{L}$.

**Definition 2.3** *Vertex* $v_i$ *sees vertex* $v_j$ $(v_i \leftrightarrow v_j)$ *iff either* $v_i = v_j$*, or they lie on a line* $l_{ij} \in \mathcal{L}$ *and the segment* $v_i v_j$ *is nowhere exterior to* $P$.

Note that our definition of pseudo-visibility is dependent upon $\mathcal{L}$: it does not make sense to ask if two points of $V$ see one another without providing the underlying arrangement $\mathcal{L}$. Dependence upon $\mathcal{L}$ means there is not complete freedom to assign which vertex sees which. For example, in Fig. 1b, $v_0$ could not be arranged to see $v_2$, because the pseudoline $l_{02}$ would have to intersect $l_{01}$ (and $l_{12}$) twice, violating the definition of a pseudoline arrangement.

**Definition 2.4** *The* vertex pseudo-visibility graph $G_V(P)$ *of a polygon is a labeled graph with node set* $V$*, and an arc between two vertices iff they can see one another (according to Def. 2.3).*

[1] Their definition is for the projective plane, and includes a special line $l_\infty$. We use the Euclidian plane.

[2] All index arithmetic is mod $n$ throughout the paper.

We will often abbreviate $G_V(P)$ to $G_V$. Note that $G_V$ is Hamiltonian: the arcs corresponding to the polygon boundary form a Hamiltonian circuit $(v_0, \ldots, v_{n-1})$. And also note that since $G_V$ is labeled by $V$, which we assumed was labeled in a ccw boundary traversal order, the Hamiltonian circuit is provided by the labeling of the graph.

### Vertex-edge Pseudo-Visibility

We need to define when a vertex sees an edge. In [OS97], we defined $v$ to see $e$ in a straightline context if $v$ sees an open interval of $e$. Here we extend this notion to the pseudo-visibility context without adding any new points and pseudolines to the generalized configuration, to keep the definitions purely combinatorial. We start with the notion of a "witness." Let $r_j^i \subset l_{ij}$ be the ray along $l_{ij}$ starting at and including $v_j$, directed away from (and therefore excluding) $v_j$.

**Definition 2.5** *Vertex $v_j$ is a* witness *for the vertex-edge pair $(v_i, e)$ iff either*

1. *$v_i$ is an endpoint of $e$, and $v_j$ is also (here we permit $v_j = v_i$); or*
2. *$v_i$ is not an endpoint of $e$, and*
   (a) *$v_i$ sees $v_j$; and*
   (b) *the ray $r_j^i$ intersects $e$ at a point $p$,*
   (c) *either $v_j = p$, or the segment $v_jp$ is nowhere exterior.*

We will refer to the line $l_{ij}$ in the above definition as the *witness line*.

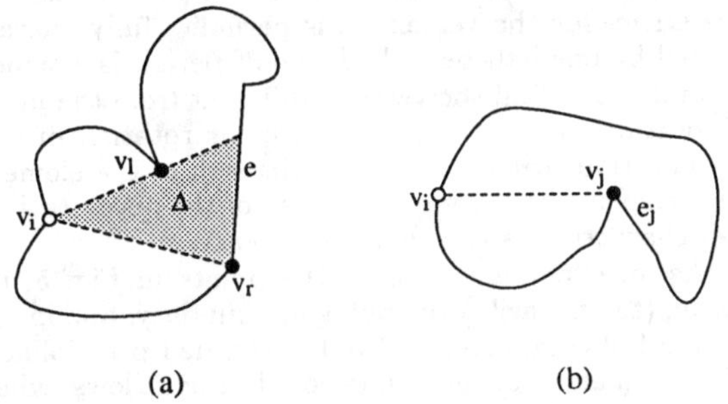


Figure 2: (a) $v_i$ sees $e$; (b) $v_i$ does not see $e_j$ although $v_j$ is a witness.

**Definition 2.6** *Vertex $v$* sees *edge $e$ ($v \to e$) iff there are at least two witnesses $v_r$ and $v_l$ for $(v, e)$.*

We call $v_r$ the *right witness* and $v_l$ the *left witness* if their ccw ordering is $(v_r, e, v_l)$. Thus $v_r$ is to the right from the viewpoint of $v$, and $v_l$ to the left. Note that an endpoint of an edge sees that edge, because then both endpoints are witnesses.

The intent of this definition is illustrated in Fig. 2a: $v_i$ sees an open interval of $e$. The reason we demand two witnesses is that one witness does not suffice, as is clear from Fig. 2b.

**Lemma 2.7** *Under the general position assumption, if $v_i$ sees $e$, there are* exactly *two witnesses for $(v_i, e)$.*

**Definition 2.8** *The* vertex-edge pseudo-visibility graph *(ve-graph) $G_{VE}$ of a polygon is a labeled bipartite graph with node node set $V \cup E$, and an arc between $v \in V$ and $e \in E$ iff $v$ can see $e$ (according to Def. 2.5).*

### Ve-graph Properties

Our aim now is to obtain characterizing properties of $G_{VE}$. The key property concerns how a gap in $v_k$'s view of the polygon's boundary can occur: it can only occur in one of the two ways illustrated in Fig. 3. We first state this condition using both v-v and v-e visibility, and later (Theorem 2.13) remove the v-v information.

We use the following notation to specify parts of the polygon boundary: $P[i,j]$ is the closed subset of the polygon boundary ccw from $v_i$ to $v_j$. $P(i,j]$ excludes $v_i$; $P[i,j)$ excludes $v_j$; and $P(i,j)$ excludes both.

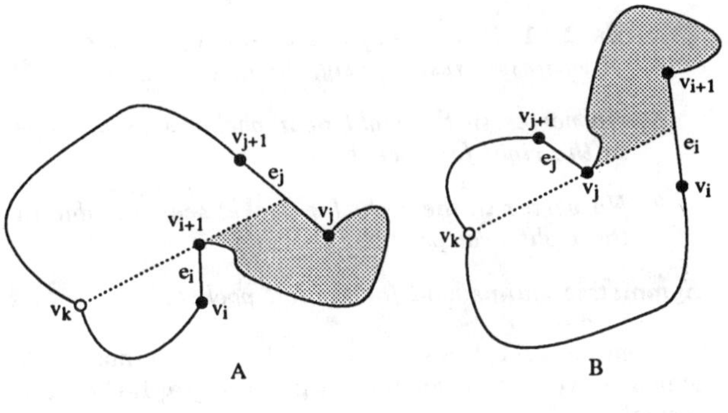


Figure 3: Two cases for $v_k$ seeing $e_i$ followed by $e_j$.

**Lemma 2.9** *If $v_k$ sees non-adjacent edges $e_i$ and $e_j$ and no edge between, $v_k \in P[j+1, i]$, then exactly one of Case A or B holds (see Fig. 3):*

**Case A**

1. *$v_k$ sees $v_{i+1}$ but not $v_j$; and*
2. *$v_{i+1}$ is the right-witness for $(v_k, e_j)$; and*
3. *$v_{i+1}$ sees $e_j$ but $v_j$ does not see $e_i$.*

**Case B**

1. *$v_k$ sees $v_j$ but not $v_{i+1}$; and*
2. *$v_j$ is the left-witness for $(v_k, e_i)$; and*
3. *$v_j$ sees $e_i$ but $v_{i+1}$ does not see $e_j$*

The structure established by the preceding lemma is best captured by the notion of "pockets":

**Definition 2.10** *If $v_i$ sees $e_j$ and $v_r$ and $v_l$ are the right and left witnesses respectively (cf. Fig. 4), then $P[i,r)$ and $P(l,i]$ are the right and left* near pockets *of $(v_i, e_j)$, and $P(r,j]$ and $P[j+1,l)$ are the right and left* far pockets *of $(v_i, e_j)$, respectively.*

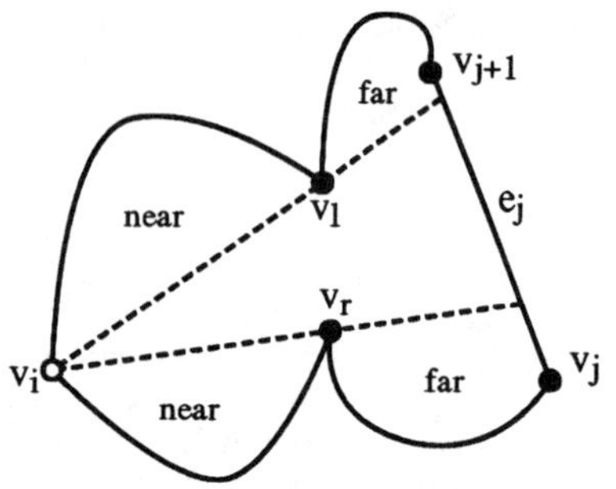


Figure 4: Definition of near and far pockets.

Note that: $v_i$ is in both near pockets; $e_j$ is not part of either far pocket, and the witnesses are not in any of the four pockets. If either witness is an endpoint of $e_j$, then the corresponding pocket is empty.

**Lemma 2.11** *If $v_i$ sees $e_j$ and $v_r$ and $v_l$ are the right and left witnesses respectively, then*

1. *No vertex in the right near pocket sees an edge in the right far pocket.*
2. *No vertex in the right far pocket sees an edge in the right near pocket.*

*Symmetric claims hold for the left pockets.*

Lemma 2.11 leads immediately to Lemma 2.12, which conveys the same import in more graph-theoretic terms.

**Lemma 2.12** *If $v_i$ sees $e_j$ and $v_r$ and $v_l$ are the right and left witnesses respectively, then $v_r$ is an articulation point of the subgraph of $G_{VE}$ induced by $P[i,j]$, and symmetrically, $v_l$ is an articulation point of the subgraph induced by $P[j+1,i]$.*

We may now state our characterization of ve-graphs by discarding from some of the previous lemmas all but vertex-edge visibility information.

**Theorem 2.13** *If $G_{VE}$ is the vertex-edge visibility graph of a pseudo-polygon $P$, then it satisfies these two properties:*

1. *If $v_k$ sees non-adjacent edges $e_i$ and $e_j$ and no edge between, $v_k \in P[j+1,i]$, then exactly one of these holds:*
   - **A.** *$(v_{i+1}, e_j) \in G_{VE}$, or*
   - **B.** *$(v_j, e_i) \in G_{VE}$.*
2. *In the two cases above, additionally:*
   - **A.** *$v_{i+1}$ is an articulation point of the subgraph of $G_{VE}$ induced by $P[k,j]$.*
   - **B.** *$v_j$ is an articulation point of the subgraph of $G_{VE}$ induced by $P[j+1,k]$.*

We will prove in Section 3 that these properties basically provide a complete characterization of vertex-edge visibility graphs.

## Information in the Vertex-Edge Visiblity Graph

We have established the key low-level properties of $G_{VE}$ in Theorem 2.13, but they give little insight into higher-level properties of the graph. To carry out the proof of Theorem 3.2 in the next section, we will need to derive from the lower-level concepts a number of additional combinatorial concepts, analogs of the geometric notions of convexity of polygon vertices, partial local sequences, and shortest paths trees. These have been shown to be derivable from $G_{VE}$ in the straight-line case in [OS97], and the generalization to pseudo-visibility is along similar lines. We will omit most of the details here and give just the definitions that are needed to understand the abstract counterparts introduced later in the proof of the main theorem. See [OS96] for details.

Given a set of points in the plane, rotate a directed line around each point and record the ordered list of the other points as they are encountered by the rotating line. In addition, assign a sign to each point: positive if it is encountered by the forward ray from the center of rotation, negative if by the backward ray. The infinite sequence thus obtained is called the *i-sequence* for the vertex. It is periodic, fully characterized by one half-period. The half-period is a signed permutation of all the vertices different from the point of rotation, $\alpha_1\alpha_2\cdots\alpha_{n-1}$. A circular rotation of this permutation, with a change of the sign of the element sent from the beginning to the end of the permutation, also characterizes the same $i$-sequence.

As an example, consider the points in Fig. 5, ignoring (temporarily) the polygon boundary, and imagining a full complement of fully-extended pseudolines. The $i$-sequences of the points could be as follows, where negative points are indicated with a bar:

$v_0$ : $2143\bar{6}5$
$v_1$ : $2\overline{506}34$
$v_2$ : $341506$
$v_3$ : $506412$
$v_4$ : $50\bar{3}612$
$v_5$ : $602143$
$v_6$ : $214305$

The collection of $i$-sequences for all the points in the set is a version of what Goodman and Pollack [GP84] called a *cluster of stars* (see also [Str96a]).

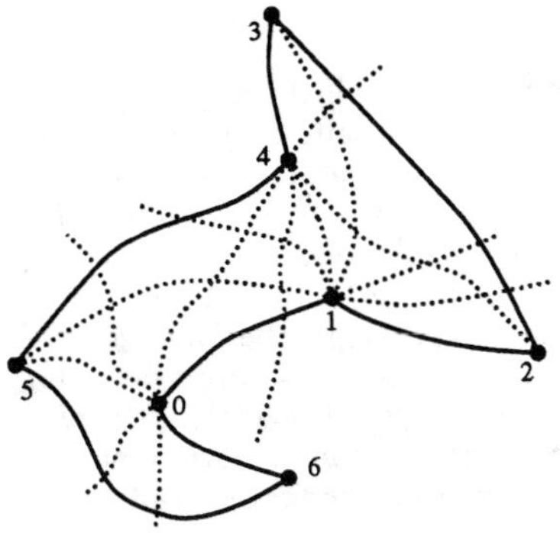


Figure 5: A pseudo-polygon used to illustrate $i$-sequences.

From an $i$-sequence, one can read off *chirotope* information, i.e., whether a triple $i, j, k$ makes a right or left turn at $j$. Using a different terminology (vertex-free tournaments, pre-CC and CC-systems), Knuth [Knu92] has given a system of axioms equivalent to uniform acyclic oriented matroids of rank 3. Knuth's CC-systems can be interpreted as characterizing the $i$-sequences of generalized configurations of points. We will make use of his results in the proof of our main theorem.

Rotating a directed pseudoline around a vertex in a polygon and recording in a circular (signed) sequence only the visible vertices induces a *partial i-sequence.* The partial $i$-sequences of a polygon are uniquely determined by the ve-graph, and can be easily computed from it as soon as the convexity properties of the vertices of the polygon are determined. For example, for the vertices of Fig. 5, the partial $i$-sequences for each vertex are as follows:

$v_0$ : $14\overline{6}5$
$v_1$ : $2\overline{50}34$
$v_2$ : 341
$v_3$ : 412
$v_4$ : $50\overline{3}12$
$v_5$ : 6014
$v_6$ : 05

The edges of the v-graph meeting at a common vertex may form convex or reflex angles. This information is uniquely determined from the ve-graph via computing an extended relation of visibility between vertices of the polygon and edges of the v-graph.

For any two distinct vertices of a simple polygon, there exists a unique shortest path between them. For a fixed source vertex $v_i$, the set of shortest paths from $v_i$ to all other vertices induces a *shortest-path tree* (sp-tree) rooted at $v_i$. The tree is ordered: there is a natural ordering between the subtrees rooted at each vertex, as given by the ccw traversal of the boundary of the polygon. Each internal node of a sp-tree is a reflex vertex and a *turn* is associated with it: left or right, according to how the shortest paths from the root through that vertex turn there.

Next we define two canonical circular orderings of vertices around each vertex of a polygon. These orderings will be used in the proof of the Theorem 3.2. The intuition behind them is that for every pseudo-polygon there exists another pseudo-polygon in a *normal form* having the same ve-graph. We will introduce a concept of $i$-sequences for a pseudo-polygon: the normal form has the property that the polygon $i$-sequences are identical to the $i$-sequences of the underlying generalized configuration of points. The existence of the normal form will be a consequence of the Theorem 3.2.

We define a total ordering on all the internal vertices of the sp-tree rooted at $v_i$ (and hence on all the shortest paths from the root to those vertices). If $v_j$ and $v_k$ are two children of the same node, with $v_i, v_j, v_k$ occuring in this order in a ccw traversal of the boundary of the polygon, then all the vertices in the tree rooted at $v_j$ are listed *before* all the vertices in the tree rooted at $v_k$. If $v_j$ is a right (left) turn, then vertex $v_j$ itself is listed *after* (*before*) all the children in the subtree rooted at $v_j$. For example, in Fig. 5, the sp-tree rooted at 1 makes a left turn at vertex 0. The tree rooted at 6 makes a right turn at 0, a left turn at 4, and a right turn at 1. The total order of the vertices of the polygon in the figure derived this way is as follows:

$v_0$ : 214356
$v_1$ : 234506
$v_2$ : 341506
$v_3$ : 506412
$v_4$ : 506123
$v_5$ : 621043
$v_6$ : 214305

Lastly, combining the total order of the shortest paths (vertices) around each vertex with the partial $i$-sequence information, one can define a canonical circular order of *signed* vertices around each vertex $v_i$ of the polygon, as follows. If the vertex $v_i$ is convex, the ordering is the same as the one induced by the sp-tree. If it is reflex, take the signed permutation representing its partial $i$-sequence. Each vertex in this permutation is a child of the root ($v_i$) in the sp-tree for the current vertex. Create a new signed permutation from the partial $i$-sequence by replacing each vertex with the ordered list of the vertices in the subtree rooted at $v_j$, with the same sign as the root $v_j$. For example, consider $v_i = v_4$ in Fig. 5, a reflex vertex whose partial $i$-sequence is $v_4$ : $50\overline{3}12$. We replace $v_j = v_0$ (0 in the sequence) with 06, resulting in $v_4$ : $506\overline{3}12$.

This canonical circular ordering will be called the *polygon i-sequence* for vertex $v_i$ in $P$; it is in general different from the $i$-sequence for vertex $v_i$ in the context of the point configuration of the vertices of the polygon. Note, for example, that the polygon $i$-sequence for $v_4$ just obtained is different from the $i$-sequence of $v_4$ in the point configuration ($50\overline{3}612$), because 6 was encountered before $\overline{3}$ when spinning about $v_4$ in the configuration, but all of 0's children are sorted along with 0 without regard to where blocked lines of visibility (e.g., $l_{43}$) might encounter them. The normal form mentioned previously forces, in this case, $v_6$ to lie left rather than right of $l_{43}$.

## 3 Abstract ve-graphs

We will work with bipartite graphs $G_{VE}$ defined on two circularly ordered lists $V = (v_0, \ldots, v_{n-1})$ of *vertices* and $E = (e_0, \ldots, e_{n-1})$ of *edges*. Most of our terminology and notation carries over from the geometric setting, but we repeat here to emphasize that in the abstract setting, all definitions must be combinatorial. Two edges $e_i$ and $e_j$ are *adjacent* if $j = i+1$ or $i = j+1$; vertex $v_i$ is *adjacent* to the edges $e_{i-1}$ and $e_i$; and edge $e_i$ is adjacent to its "endpoints" $v_i$ and $v_{i-1}$. The polygon boundary between vertices $v_i$ and $v_j$ is defined as a list $P[i,j] = (v_i, e_i, v_{i+1}, \ldots, e_{j-1}, v_j)$. As before, $P[i,j)$ and $P(i,j]$ exclude "endpoint" vertices. We use $v_i \to e_j$ as alternate notation for $(v_i, e_j) \in G_{VE}$: $v_i$ *sees* $e_j$. For a vertex $v_k \in P[j+1, i]$, two edges $e_i$ and $e_j$ are called *consecutive from* $v_k$ if $v_k$ sees $e_i$ and $e_j$ but does not see any $e_l \in P[i,j]$.

**Definition 3.1** *An* abstract ve-graph *is a bipartite graph $G_{VE}$ on sets of vertices $V$ and edges $E$, both circularly labeled as above, with $|V| = |E| \geq 3$, satisfying the two properties of Theorem 2.13, plus the additional condition that each vertex sees its adjacent edges: $\forall i$, $v_i \to e_i$ and $v_i \to e_{i-1}$.*

Our main goal is to prove that this definition captures ve-graphs of pseudo-polygons:

**Theorem 3.2 (Main Theorem)** *If $G_{VE}$ is an abstract ve-graph, then there exists a generalized configuration of points $\{p_0, \cdots, p_{n-1}\}$ and a pseudo-polygon specified by this ordering of the vertices, whose ve-graph is the same as $G_{VE}$.*

The plan of the proof is as follows. Our ultimate goal is to prove the configuration of points exists by constructing a corresponding uniform rank-3 acyclic oriented matroid. This requires specifying the orientation of all triples of points. We use shortest paths to define these signed triples. With these in hand, we will need to prove the collection of triples do constitute the appropriate type of matroid. We use Knuth's "counterclockwise" CC-axioms for this purpose. Several of his axioms can be established from the properties of $i$-sequences; the remaining we prove directly.

Note that we start only with a definition of v-e visibility. From here we will define a notion of v-v visibility, as well as abstract counterparts of convex and reflex vertices, pockets, shortest paths, shortest-path trees, and $i$-sequences. At this point we will be ready to define the triples and construct the matroid. Most of the definitions through shortest paths mimic in the abstract setting those in [OS97] for straight-line polygons. The additional factor here is that we have to prove that the definitions are consistent and lead to objects satisfying the properties one would expect from them.

We define v-v visibility according to which of Case A and B holds in the defining properties from Theorem 2.13: in Case A, we define $v_k \to v_{i+1}$, and in case B, we define $v_k \to v_j$, just as in Lemma 2.9. If $j = i+1$, we define $v_k \to v_{i+1}$. Moreover, if case $A$ holds, we say that $v_{i+1}$ is a *right* articulation point for $P[k, i+1)$ (the *near*) and $P(i+1, j]$ (the *far*) *right pockets*. If case $B$ holds, $v_j$ is the left articulation point for $P(j,k]$ (the near) and $P[i+1, j)$ (the far) *left pockets*. In either case, the pockets and articulation points are relative to visibility from $v_k$ (or to the triple $v_k, e_i, e_j$).

**Lemma 3.3** *$\forall i$, $v_i$ sees at least one other edge $e_j$ different from $e_{i-1}$ and $e_i$.*

**Lemma 3.4** *$\forall i$, exactly one of the following two cases holds:*

**A.** $v_{i-1} \to e_i$ *and* $v_{i+1} \to e_{i-1}$, *or*

**B.** $v_{i-1} \not\to e_i$ *and* $v_{i+1} \not\to e_{i-1}$.

In the first case we say $v_i$ is *convex*, and in the second *reflex*.[3]

**Proof:** Assume this is not true. Then either $v_{i-1} \to e_i$ but $v_{i+1} \not\to e_{i-1}$ or $v_{i-1} \not\to e_i$ but $v_{i+1} \to e_{i-1}$. Assume the first case holds (the other can be treated similarly). Then, taking $v_{i+1}$ to play the role of $v_k$ in property 1 from Theorem 2.13, we know that $v_{i+1} \to e_i$ by Def. 3.1. Because $v_{i+1} \not\to e_{i-1}$, there must be an edge $e_j$, $j \in P[i+2, i-1)$, such that $v_{i+1}$ sees $e_j$ and $e_i$ consecutively (such an edge must exist, by Lemma 3.3). See Fig. 6. Moreover, Case **B** of Theorem 2.13 must hold. Then $e_{i-1}$ lies in the far pocket for the triplet $v_{i+1}, e_j, e_i$. But then, by property 2, it cannot be the case that $v_{i-1}$, which is in the far pocket, sees $e_i$, which is in the near pocket. This contraction establishes the lemma. □

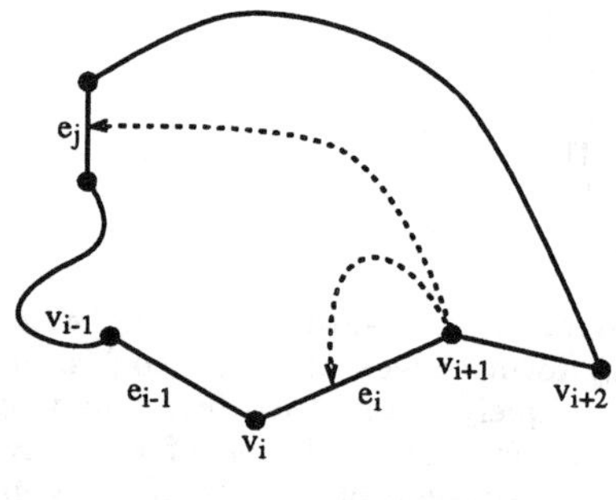


Figure 6: Lemma 3.4: $v_{i+1} \to e_i$ and $v_{i+1} \to e_j$ and $v_{i+1}$ sees no edge between.

The next lemma shows that the articulation point properties of the ve-graph carry through to v-v visibility.

[3] Recall our general position assumption removes worry about the intermediate case.

**Lemma 3.5** *No vertex in a near pocket can see any vertex in the corresponding far pocket and vice-versa.*

The next step is to show the v-v visibility is symmetric:

**Lemma 3.6** *If $v_i \to v_j$ then $v_j \to v_i$.*

This lemma guarantees that the v-graph associated to an abstract ve-graph is indeed an undirected graph, and is the base case for the inductive proof of the more general property of symmetry of paths between pairs of vertices $v_i$ and $v_j$ (Lemma 3.15).

As an aside, note that some of the properties that we prove can be found as axioms in [AK95] (e.g., path symmetry is their "Necessary Condition 2"): it is indeed surprising that our simple conditions are enough to imply these high level properties. The property in the following lemma is one of their defining axioms.

**Lemma 3.7** *If $v_k \to v_i$ and $v_k \to v_j$ consecutively, then $v_i \to v_j$.*

From the ve-graph we can also define an extended relation of visibility between vertices and v-graph edges, as well as an abstract angle (convex or reflex) between two v-edges with a common endpoint. We say that vertex $v_k$ sees visibility edge $v_iv_j$ if $v_k \to e_l$ and $i \in P(k,l]$, $v_j \in P[l+1,k)$. We define the abstract *angle* $\angle v_iv_kv_j$ between two adjacent v-edges $v_kv_i$ and $v_kv_j$ as being *convex* if $v_i \to v_kv_j$ and $v_j \to v_kv_i$, reflex if none of these two conditions holds. Then we can construct the abstract partial $i$-sequence for a vertex $v_k$ using the following algorithm.

**Algorithm 1** (Construction of the partial $i$-sequence for vertex $v_k$ from convex/reflex angle information)

*Start with vertex $v_{k+1}$ in a list.*

*Record in the list the consecutive vertices $v_i$ for which $\angle v_{k+1}v_kv_i$ is convex. Let $v_m$ be the last such vertex. If we have not yet reached $v_{k-1}$ (i.e., $m \neq k-1$), continue with the rest of the vertices $v_j$ for which the angle $\angle v_{k+1}v_kv_j$) is reflex. For each such vertex $v_j$, determine a pair $(v_{i_1}, v_{i_2})$ of previously listed consecutive vertices, $v_{i_1} \in P[k+1,m)$, $v_{i_2} \in P(i_1,m]$, so that $v_j$ does not see $v_{i_1}$ and sees $v_{i_2}$ consecutive to $v_k$. Then, if two consecutive vertices $v_{j_1}$ and $v_{j_2} \in P(m,k-1]$ lie between the same pair of consecutive vertices $v_{i_1}, v_{i_2}$, list them in the same order in the partial i-sequence, between $v_{i_1}$ and $v_{i_2}$. All the vertices $v_i$ with $v_{k+1}v_kv_i$ convex are listed as positive, the others as negative, in the signed partial i-sequence.*

Some simple lemmas guarantee that the above definition is consistent and the algorithm is correct.

The next result may seem rather trivial, although the proof is not.

**Lemma 3.8** *There exists at least one convex vertex.*

We have defined pockets and articulation points earlier. The following lemma shows that pockets embed nicely: this is the main tool used to define abstract shortest paths and sp-trees.

**Lemma 3.9** (Subpocket embedding)
*Let $v_i \to v_k$ with $v_k$ a right articulation point for visibility from $v_i$. Let $e_m$ be the next edge visible from $v_i$ after $v_k$ and let $v_j$ be in the far pocket $P(k,m]$. Assume that $v_k \not\to v_j$ and let $v_j$ be in a far pocket $P'$ from $v_k$. Then $P' \subset P(k,m]$. A similar embedding property holds when $v_k$ is a left articulation point.*

For example, in Fig. 5, consider $i=6$, $k=0$, $m=4$, and $j=3$: $v_i = v_6 \to v_k = v_0$, $v_k = v_0 \not\to v_j = v_3$, and $v_j = v_3 \in P(0,4]$. Then $P' = P[3,4) \subset P(0,4]$.

We need here some terminology which will help in formulating and proving symmetry properties for pockets and shortest paths. If $v_j$ is in the right far pocket $P(k,m]$ from $v_i$ and in the right far pocket $P(p,s]$ from $v_k$ we will say that the embedding pattern of the far pocket containing $v_j$ invisible from $v_i$ is of type $RR$, with first articulation point $v_k$ of type $R$ and second articulation point $v_p$ of type $R$. Similarly we define type $RL$, $LR$, $LL$ embedding patterns and articulation points, according to whether the first pocket is right and the second left, etc. The following lemma shows that there is a symmetry in the embedding patterns seen from the two endpoints. In what follows we will also use the notation $\overline{R} = L$ and $\overline{L} = R$. The following corrollary is a direct consequence of the proof of Lemma 3.9.

**Corollary 3.10** (Pocket symmetry) *If $v_i \not\to v_j$ and the embedding pattern of the far pockets from $v_i$ to $v_j$ is $AB$ ($A, B \in \{L,R\}$) with articulation points $v_k$ of type $A$ and $v_p$ of type $B$ in this order, then the embedding pattern of the pockets from $v_j$ to $v_i$ is reversed, $\overline{B}\overline{A}$, with articulation points $v_p$ of type $\overline{B}$ and $v_k$ of type $\overline{A}$ in this order.*

The previous lemmas allow us to correctly define the *pocket embedding tree* rooted at a vertex $v_i$. This is a tree with a fixed ordering on the children of each internal node and with a sign (corresponding to a right/left turn) associated to each node other than the root. The tree has exactly $n$ nodes. Each node is labelled with a signed vertex of $V$, and the set of vertices in a subtree correspond to the set of vertices in a subpocket with articulation point given by the vertex labelling the root of the subtree. Moreover, the sign of the root of the subtree is right/left, corresponding to whether it is a left or right articulation point for visibility from its parent.

The pocket embedding tree is defined recursively as follows. Its root is the vertex $v_i$ and its children correspond to the far pockets from $v_i$ in the order induced by the circular order of $V$. Each node is labelled by the corresponding articulation point and signed by its right/left type. Lemma 3.9 ensures that once we get into a pocket we can continue subdividing the pockets by looking at the far subpockets of invisibility from the root of the subtree, restricted only to the vertices in the current pocket (subtree). This in turn guarantees that this process generates a tree.

The labels of the paths from the root $v_i$ to the other nodes of this tree correspond intuitively (i.e., in the case of a polygon) to the shortest paths from $v_i$ to all the other vertices of the polygon. This motivates our using the terminology *shortest paths* for them in the combinatorial setting of abstract ve-graphs. Let $v_i$ and $v_j$ be two vertices of $G_{VE}$. We will define the *abstract shortest path* from $v_i$ to $v_j$, $sp(v_i, v_j)$ as an ordered list of vertices starting with $v_i$ and ending with $v_j$. In what follows, *concat* denotes the function that concatenates two lists and *reverse* the function that reverses an ordered list.

The following definition is just a rigorous formalization of this concept.

$sp(v_i, v_j)$ is constructed recursively as follows.

**if** $v_i \rightarrow v_j$ **then** $sp(v_i, v_j) = (v_i, v_j)$

**else** $sp(v_i, v_j) = (v_i)$ *concat* $sp(v_k, v_j)$, where $v_k$ is the articulation point of the far pocket from $v_i$ containing $v_j$.

The shortest paths contain more information that just the simple ordered list of vertices from a source to a destination. They also capture the *left* or *right* turns on the path, at articulation points along the way. First let us fix some notation for this. If $sp(v_i, v_j) = (v_i, v_k, \cdots, v_j)$ and $v_k$ is a right (left) articulation point for $v_i$, then we say that the shortest path $sp(v_i, v_j)$ *makes a right (left) turn* at $v_k$. From $v_k$ on, the left/right turns are defined on the subsequent subpaths towards $v_j$. We will denote by $ssp(v_i, v_j)$, the *signed shortest path* from $v_i$ to $v_j$, to be a signed list obtained from $sp(v_i, v_j)$ and attaching the appropriate signs to its vertices (+ for right turn, − for left turn). The following lemma shows that the signs are already determined by the shortest path.

**Lemma 3.11** *The right/left turns on a shortest path are determined by the indices of the articulation points. More precisely, if $v_k$ is a right articulation point from $v_i$ to the far pocket containing $v_j$, then all the vertices $v_{j'}$ in the far pocket have indices $j' > k$ (circularly, i.e. $v'_j \in P(k, i)$). If it is a left articulation point, all the vertices $v_{j'}$ in the far pocket have indices $j' < k$ (circularly, i.e. $v_{j'} \in P(i, k)$).*

**Lemma 3.12** *The signed list of vertices on a path from the root $v_i$ in a pocket tree rooted at $v_i$ to an internal node $v_j$ is equal to the $ssp(v_i, v_j)$.*

The following lemmas show that the shortest paths glue together nicely. We note that these properties were among the axioms in [AK95].

**Lemma 3.13** If $v_k \in sp(v_i, v_j)$ then $ssp(v_i, v_k)$ *is the signed sublist of* $ssp(v_i, v_j)$ *starting at* $v_i$ *and ending at* $v_k$.

**Lemma 3.14** If $v_k \in sp(v_i, v_j)$ then $sp(v_i, v_k)$ concat $sp(v_k, v_j) = sp(v_i, v_j)$.

**Lemma 3.15** $sp(v_i, v_j) =$ *reverse* $sp(v_j, v_i)$.

To summarize, up to this point we have defined for every vertex $v_i$ of $V$ a tree rooted at $v_i$ and labelled with all the vertices of $V$. There is an ordering of the children of all internal nodes and a sign (*turn*) associated with each internal node. The paths from the root to any vertex define *abstract shortest paths* from the root to that vertex. Two shortest paths trees are compatible in the sense of Lemma 3.13. We will now define a circular ordering of the shortest paths $sp(v_i, v_j)$ around a vertex $v_i$ by combining the partial $i$-sequences for $v_i$ with the shortest path tree rooted at $v_i$, exactly as described at the end of Section 2.

The last step is to define a predicate on any triple of vertices of $V$ and show that it satisfies Knuth's CC-system axioms. It has been shown by Knuth that CC-systems are equivalent to uniform rank-3 acyclic oriented matroids [Knu92, p. 40], and it is well known (see e.g., [BLW+93]) that these in turn are equivalent with Goodman and Pollack's generalized configurations of points (in general position).[4]

We will write $i < j < k$ iff the indices $i, j$ and $k$ occur in this order in $V$ (as usual, indices are taken mod $n$).

The predicate will be denoted as $ijk$. It is first defined for indices $i < j < k$ then extended as usual for other permutations of three points: odd number of inversions change the sign, even number keep the same sign (see Knuth's axioms on next page). We say that $ijk$ holds iff in the $i$-sequence for vertex $v_i$, the vertices $v_j$ and $v_k$ appear positively in this order in a half-period. For example, the polygon $i$-sequence for $v_1$ in Fig. 5 is $v_1$ : 234506. Thus $5 < 6$ occur in this order in the sequence, and so 156 holds.

The following lemma matches the chirotope definition in [AK95].

**Lemma 3.16** *If $i < j < k$ then $ijk$ holds iff $i, j, k$ do not occur on a common abstract shortest path. Equivalently, $\neg ijk$ iff $i, j, k$ belong to the same shortest path.* Continuing with the same example, 156 holds because $1 < 5 < 6$ and they do not occur on a shortest path; and $\neg 160$ because $1 < 6 < 0$ and they occur on the shortest path $(v_1, v_0, v_6)$.

**Proof:** We first prove sufficiency. Let $i < j < k$ and $ijk$, and assume for contradiction that $i, j, k$ are on the same shortest path. Then either $j \in sp(i, k)$, in which case $j$ has to be a right turn and contradicts the definition of the $i$-sequence for vertex $i$ ($k$ would occur before $j$ and not vice-versa); or $i \in sp(k, j)$ which again implies that in the $i$-sequence for $i$, $j$ is followed by $\bar{k}$, contradicting the hypothesis; or $k \in sp(j, i)$, which again implies that $k$ occurs after $j$ in the $i$-sequence for $i$, contradiction.

To prove the necessity, assume $i, j, k$ are not on the same shortest path. Then in the sp-tree rooted at $i$, $j$ and $k$ are on different branches, and if those branches originate at $i$, they do not form a reflex angle. The definition of the $i$-sequences then guarantees that a vertex of smaller index ($j$) is encountered in the $i$-sequence before a vertex of higher index ($k$). □

Note that the conditions given by this lemma constitute the definition in [AK95] for their oriented matroid defined via chirotope axioms. We have found that verifying these or any other equivalent oriented

[4] The modifier "uniform" indicates general position.

matroid axioms involves a tedious case analysis, which we circumvent via $i$-sequences and an appropriate interpretation of Knuth's CC-axioms [Knu92, p. 4]:

**Axiom 1.** (Cyclic symmetry): $ijk \Rightarrow jki$.

**Axiom 2.** (Antisymmetry): $ijk \Rightarrow \neg ikj$.

**Axiom 3.** (Nondegeneracy): $ijk \vee ikj$.

**Axiom 4.** (Interiority): $ijk \wedge ikl \wedge ilj \Rightarrow jkl$.

**Axiom 5.** (Transitivity): $ijk \wedge ijl \wedge ijm \wedge ikl \wedge ilm \Rightarrow ikm$.

To finish the proof we need to verify that Knuth's axioms hold, that we can define a pseudo-polygon on the corresponding generalized configuration of points (i.e., edges do not cross), and then show that its ve-graph is exactly the graph that we started with. First, we notice that the five axioms of Knuth can be grouped into three categories. Axioms 2, 3, and 5 are equivalent to the existence of circular $i$-sequences for each point of the system (this is a consequence the corrolary on [Knu92, p. 12]), which we have already established. Axiom 1 requires that the signs of triples do not change under circular permutations of the points, which is a direct consequence of Lemma 3.16. This leaves Axiom 4, the acyclicity condition, which ensures that the points can be realized as an affine generalized configuration of points.

**Lemma 3.17** *Axiom 4 holds:* $ijk \wedge ikl \wedge ilj \Rightarrow jkl$.
**Proof:** Let $a, b, c$ and $d$ be four arbitrary points occuring in this order in $V$: $a < b < c < d$. Without loss of generality we can assume that $i = a$. The symmetry in Axiom 4 tells that we have only two cases to verify: (1) $j = b, k = c, l = d$, or (2) $j = b, l = c, k = d$.

**Case 1.** $i < j < k < l$. The premiss, interpreted through Lemma 3.16, says that: $i, j, k$ and $i, k, l$ are not on the same shortest path, and, using Axiom 2 ($ilj \Rightarrow \neg ijl$), that $i, j, l$ are on the same shortest path. The consequent says that $j, k, l$ are not on the same path. Assume for the sake of a contradiction that $j, k, l$ are on the same path. Three cases are possible: $j$ on $sp(l, k)$ and is a right turn; $k$ on $sp(j, l)$ and is a right turn; or $l$ on $sp(k, j)$ and is a right turn. In the first case (Fig. 7a), $i$ is in the near right pocket from $l$, and $k$ in the far right pocket from $l$, contradicting the first premiss $ijk$. In the second case, we have to use the premiss $\neg ijl$ to get a contradiction. Either $j$ is on $sp(i, l)$ (Fig. 7b), in which case it follows that $k$ is on $sp(i, l)$, contradicting $ikl$; or $i$ on $sp(l, j)$, in which case either $ijk$ or $ikl$ have to be on the same sp, contradicting the first premisses; or, finally, $l$ on $sp((j, i)$. In this case we obtain a contradiction with $ikl$. The third case may be treated similarly.

**Case 2.** $i < j < l < k$ is treated similarly. □

This concludes the proof that the $i$-sequences defined from the ve-graph satisfy the CC-system axioms of Knuth, and hence they form an affine generalized configuration of points in general position. We now have to verify that the order of $V$ induces a pseudo-polygon.

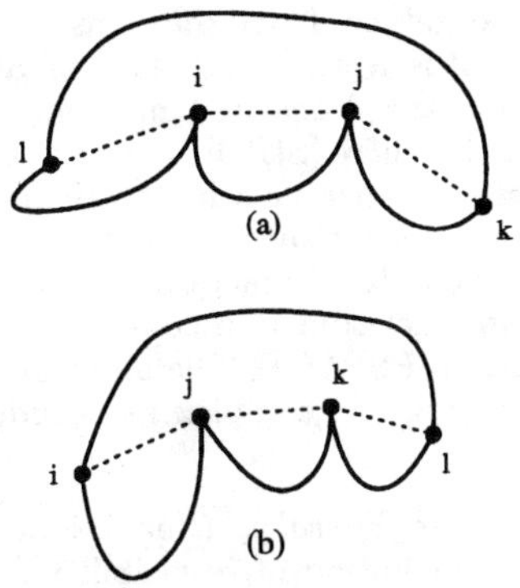


Figure 7: Lemma 3.17, $i < j < j < l$, $i, j, l$ on a shortest path: (a) $j \in sp(l, k)$; (b) $k \in sp(j, l)$ and $j \in sp(i.l)$.

**Lemma 3.18** *No two edges $v_i v_{i+1}$ and $v_j v_{j+1}$ of $\overline{P}$ cross.*
**Proof:** Assume the contrary. Then (1) $v_j$ and $v_{j+1}$ are on separate sides of $v_i v_{i+1}$, and (2) $v_i$ and $v_{i+1}$ are on separate sides of $v_j v_{j+1}$. There are two ways (1) can happen: either $v_i$ sees $e_j$ after $e_i$, in which case $v_{i+1} \rightarrow e_j$, so $v_{i+1} v_j v_{j+1}$ is a left turn, contradicting (2); or $v_{i+1}$ sees $e_j$, in which case a contradiction is similarly derived. □

We will denote by $\overline{P}$ the pseudo-polygon obtained this way from the generalized configuration of points associated with $G_{VE}$.

**Lemma 3.19** *The ve-graph of $\overline{P}$ coincides with $G_{VE}$.*

This is true since the construction of the generalized configuration of points preserved all the vertex-to-edge visibilities and invisibilities.

To conclude, let us notice that the conditions characterizing abstract ve-graphs can be easily verified in polynomial time. This proves the main result that the recognition problem for ve-graphs is in $P$. Also, as a consequence of the relationship between ve-graphs and v-graphs, if one is given a graph one can easily "guess" a ve-graph and verify that it matches the given one and satisfies the ve-graph properties. This places the pseudo v-graph recognition problem in $NP$.

Finally we mention some open problems raised by our work:

- *Given a v-graph $G_V$, in polynomial time find a ve-graph $G_{VE}$ compatible with it (in the sense that there is a pseudo-polygon $P$ such that $G_V$ and $G_{VE}$ are its pseudo-visbility v- and ve-graphs respectively), or show none exists.* Such a polynomial-time algorithm would place pseudo v-graph recognition problem in $P$.
- *Characterize the realizable rank-3 acyclic uniform oriented matroids produced by the construc-*

*tion in Theorem 3.2.* The class of oriented matroids obtained from the construction in Theorem 3.2 is a strict subclass of all the acyclic uniform rank-3 oriented matroids. It has been shown elsewhere [Str96b] that not all are stretchable (i.e., realizable with straight lines). However, this beeing such a restricted class, it might be possible to characterize or recognize them with an algorithm of a complexity smaller than the known PSPACE. The alternative is to show that they are as complex as pseudoline stretchablity.

- *Remove the general position assumption.* A more detailed definition of ve-visibility is necessary to deal with degeneracies. See [OS97] for a hint of the complications.


## Acknowledgements

We are grateful to James Abello for explaining his work to us.


## References


[AK95] J. Abello and K. Kumar. Visibility graphs and oriented matroids. In R. Tamassia and I. G. Tollis, editors, *Graph Drawing (Proc. GD '94)*, volume 894 of *Lecture Notes in Computer Science*, pages 147–158. Springer-Verlag, 1995.

[BLW+93] A. Björner, M. Las Vergnas, N. White, B. Sturmfe ls, and G. Ziegler. *Oriented Matroids.* Cambridge University Press, Cambridge, 1993.

[Eve90] H. Everett. *Visibility Graph Recognition.* PhD thesis, Dept. Comput. Sci., Univ. Toronto, Toronto, ON, 1990. Tech. Rep. 231/90.

[Gho88] S. K. Ghosh. On recognizing and characterizing visibility graphs of simple polygons. In *Proc. 1st Scand. Workshop Algorithm Theory*, volume 318 of *Lecture Notes in Computer Science*, pages 96–104. Springer-Verlag, 1988.

[Gho97] S. K. Ghosh. On recognizing and characterizing visibility graphs of simple polygons. *Discrete Comput. Geom.*, 17(2):143–162, 1997. Revision of [Gho88].

[GP84] J. E. Goodman and R. Pollack. Semispaces of configurations, cell complexes of arrangements. *J. Combin. Theory Ser. A*, 37:257–293, 1984.

[Knu92] D. E. Knuth. *Axioms and Hulls*, volume 606 of *Lecture Notes in Computer Science.* Springer-Verlag, Heidelberg, Germany, 1992.

[Mnëv91] N. E. Mnëv. The universality theorem on the oriented matroid stratification of the space of real matrices. In J. E. Goodman, R. Pollack, and W. Steiger, editors, *Discrete and Computational Geometry: Papers from the DIMACS Special Year*, volume 6 of *DIMACS Series in Discrete Mathematics and Theoretical Computer Science*, pages 237–243. AMS/ACM, 1991.

[O'R93] J. O'Rourke. Computational geometry column 18. *Internat. J. Comput. Geom. Appl.*, 3:107–113, 1993. Also in *SIGACT News* 24:1 (1993), 20–25.

[OS96] J. O'Rourke and I. Streinu. Pseudo-visibility graphs in pseudo-polygons: Part I. Technical Report 041, Dept. Comput. Sci., Smith College, Northampton, MA, January 1996. Revised Apr. 1996.

[OS97] J. O'Rourke and I. Streinu. The vertex-edge visibility graph of a polygon. *Comput. Geom. Theory Appl.*, 1997. To appear.

[Sho91] P. W. Shor. Stretchability of pseudolines is NP-hard. In P. Gritzmann and B. Sturmfels, editors, *Applied Geometry and Discrete Mathematics: The Victor Klee Festschrift*, volume 4 of *DIMACS Series in Discrete Mathematics and Theoretical Computer Science*, pages 531–554. AMS Press, 1991.

[Str96a] I. Streinu. Clusters of stars. Technical Report, Dept. Comput. Sci., Smith College, Northampton, MA, 1996. See also these proceedings.

[Str96b] I. Streinu. Non-stretchable pseudo-visibility graphs. Technical Report, Dept. Comput. Sci., Smith College, Northampton, MA, 1996.

# A Complete Roundness Classification Procedure

Kurt Mehlhorn[1] Thomas C. Shermer[2] Chee K. Yap[3]

**Abstract**

We describe a *roundness classification procedure*, that is, a procedure to determine if the roundness of a planar object $I$ is within some $\epsilon_0$ from an ideal circle. The procedure consists of a probing strategy and an evaluation algorithm working in a feedback loop. This approach of combining probing with evaluation is new in computational metrology. For several definitions of roundness, our procedure uses $O(1/qual(I))$ probes and runs in time $O(1/qual(I)^2)$. Here, the quality $qual(I)$ of $I$ measures how far the roundness of $I$ is from the accept-reject criterion. Hence our algorithms are "quality sensitive".

## 1 Introduction

A basic task of metrology is to decide whether a given planar object is "round" within some specified bound. We call this the *roundness classification problem.* The literature on roundness classification is fairly large; recent algorithmic papers include [10, 7, 3, 13, 11, 2, 6]. The area of computational metrology addresses this and similar problems [5, 4, 12, 9, 14].

[1]Max-Planck-Institute für Informatik, Im Stadtwald, 66123 Saarbrücken, Germany, mehlhorn@mpi-sb.mpg.de

[2]School of Computing Science, Simon Fraser University, Burnaby, B.C. V5A 1S6, Canada, shermer@cs.sfu.ca, supported by the Natural Sciences and Engineering Research Council of Canada.

[3]Courant Institute of Mathematical Sciences, New York University, New York, NY 10012, U.S.A., yap@cs.nyu.edu, supported by an National Science Foundation Grant #CCR-9402464.



*Computational Geometry* 97 Nice France



In this paper, we not only raise the issue of how to define roundness but also give one of the first "complete" roundness classification procedures. It consists of a *probing strategy* that plans and makes the measurements, and a *computation and decision strategy* which uses the measurements to ultimately classify the object as good or bad. Traditionally, these two strategies are decoupled. In our case, the two strategies are coupled in a feedback loop, giving a holistic approach to the classification problem. This approach can be applied to other problems in computational metrology as well.

Let us briefly illustrate the "standard approach": suppose the roundness of $I$ is defined to be the minimum width $w(I)$ of an annulus that contains the boundary of $I$. Assume that our classification problem is to accept $I$ iff $w(I) \leq \epsilon_0$. There are 3 steps:

**(i)** Some probing strategy takes a predetermined number $n$ of sample points on the boundary of $I$.

**(ii)** The set $S$ of sample points is then submitted to an algorithm to compute the minimum width $w(S)$ of an annulus that contains $S$.

**(iii)** Finally, a decision policy decides to accept or reject $I$ based on the computed value $w(S)$. Typically, we accept iff $w(S) \leq \epsilon_0$.

It is hardly clear that the policy (iii) gives a correct decision since we have substituted "$w(I)$" with "$w(S)$". Computational geometers have traditionally focused on step (ii) only. One of the few papers that discuss the decision policies in step (iii) is found in [15]. We say that the algorithm in this paper is "complete" in the sense that we integrate all three steps, and we prove that our decision policy is correct with respect to the original object $I$, not with respect to the sampled set $S$.

Let us fix some notations for the paper. An *object* $I$ is defined to be any compact simply-connected subset of the plane, with boundary denoted by bd $I$.

For a point $p$, we use $R(p, I)$ and $r(p, I)$ to denote the maximal and minimal distance from $p$ to a point in bd $I$ respectively[1], i.e.,

$$R(p, I) = \max\{dist(p, q); q \in \text{bd } I\}$$

$$r(p, I) = \min\{dist(p, q); q \in \text{bd } I\}.$$

Our distance measure is the Euclidean distance. When $I$ is understood, we write $R(p)$ instead of $R(p, I)$ and $r(p)$ instead of $r(p, I)$. Let $r_0 > 0$ and $\epsilon_0 > 0$ be parameters, which we consider fixed throughout the paper. An object $I$ is called $(r_0, \epsilon_0)$-*round* or *good*[2] if there is a point $p$ such that $r_0(1 - \epsilon_0) \leq r(p) \leq R(p) \leq r_0(1 + \epsilon_0)$. An object is called *bad* if it is not good.

In Figure 1, the shaded objects $C_1$ and $C_2$ are bad while $C_3$ and $C_4$ are good.

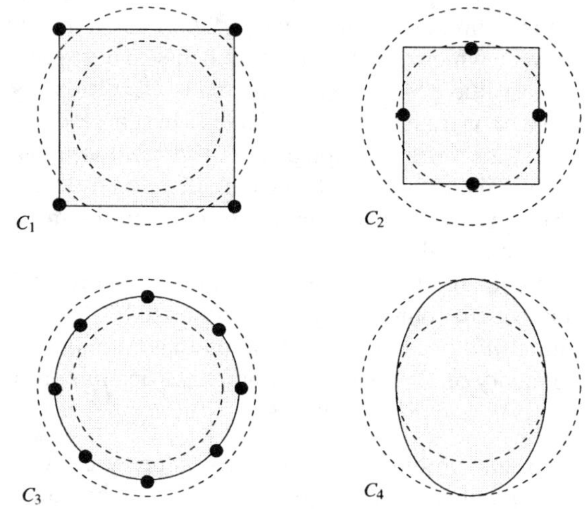


Figure 1: Objects $C_1$ and $C_2$ are "clearly" bad; $C_3$ is "clearly" good but the classification of $C_4$ as good is less evident. The four identical annuli indicated by dashed circles have radii $r_0(1 - \epsilon_0)$ and $r_0(1 + \epsilon_0)$.

We use the *finger probing* model of [1]. It postulates that the measurement device can identify a point in the interior of $I$ and can probe along any ray, i.e., determine the first point on the ray in bd $I$. Note that this is a simple mathematical model for Coordinate Measurement Machines (CMM's). CMM's are considered the state of the art in this area [14]. In this probing model, $n$ probes yield a set $S_n$ of at most $n$ points in bd $I$. A necessary condition for the classification procedure to be correct is that the set of all objects $I'$ with $S_n \subseteq \text{bd } I'$ is either collectively good or bad. Without additional assumption on $I$ this will not be the case for any finite $n$. In this paper, we use the following approximate roundness assumptions.

> **Minimum Quality Assumption (MQA and CMQA)**: $I$ is a body and there is a point $p$ and $\delta > 0$ such that $R(p) \leq r_0(1 + \delta)$ and $r(p) \geq r_0(1 - \delta)$, where $\delta = 1/20$ and $\epsilon_0 \leq \delta/2$. If we further assume that $I$ is convex, then we denote the assumption by CMQA.

The constant $1/20$ in MQA is arbitrary and is easily replaced by larger values. But we feel that the MQA assumption is not critical for practical purposes: modern processes for manufacturing round bodies can easily satisfy our requirement. But the convexity assumption in CMQA is limiting. However, we believe that our techniques can be extended to replace convexity by a bound on the maximum possible negative curvature

## 2 Main Results

We explain the intuitive idea of our main results, using the objects $C_1, \ldots, C_4$ in Figure 1. We shall assume that $C_i$'s are known a-priori to be convex. We say that $C_1$ is "clearly" bad because its violation of the outer circle of radius $r_0(1 + \epsilon_0)$ is easily witnessed by the four corners of $C_1$. (This depends on the convexity assumption.) So if our probe strategy yields a set of points that contains some approximation to each of these four corners, we can confidently reject $C_1$. Likewise, $C_2$ is bad as witnessed by the 4 indicated points on its boundary, and $C_3$ is witnessed to be good by the 8 indicated points on its boundary. Although $C_4$ is good, it cannot be witnessed by any finite set of points. Informally, the reason is that the roundness of $C_4$ is arbitrarily close to the boundary between "good" and "bad". We measure this closeness to the boundary by the "quality measure" (the quality of $C_4$ turns out to be 0).

For a point $p$, let

$$qual(p, I) = \min\{(r(p) - (1 - \epsilon_0)r_0, (1 + \epsilon_0)r_0 - R(p)\}/r_0,$$

and let

$$qual(I) = \max_p qual(p, I).$$

We call a point $p$ with $qual(p, I) = qual(I)$ a *center* of $I$ and use $c_I$ to denote a center of $I$. Let $A$ be the annulus with inner radius $(1 - \epsilon_0)r_0$ and outer radius $(1 + \epsilon_0)r_0$. For a good object, $c_I$ defines a

[1] Since bd $I$ is a closed bounded set the minimum and maximum exist.

[2] this definition of roundness is by far not the only conceivable one. It is called referenced rounding in [2]. Our results hold for several other definitions of roundness as we will show in section 5.

placement of $I$ such that bd $I \subseteq A$ and the distance between bd $I$ and bd $A$ is maximized, and for a bad object, $c_I$ defines a placement of $I$ that minimizes the distance between bd $I \cap$ cpl $A$ and bd $A$. Here, cpl $A$ denotes the complement of $A$. The minimum quality assumption implies that $qual(I) \geq -(\delta - \epsilon_0)$ and that $(1-\delta)r_0 \leq r(c_I) \leq R(c_I) \leq (1+\delta)r_0$.

We now state our main results. Basically, we show classification procedures whose complexity is *quality sensitive*:

**Theorem 1** *Under assumption CMQA there is a classification algorithm that classifies any body $I$ in time $O(|1/qual(I)| \log |1/qual(I)|)$ with $O(|1/qual(I)|)$ probes.*

Theorem 1 is based on the following two theorems. The following theorem has independent interest:

**Theorem 2** *Under assumption MQA, six probes and constant running time suffice to determine a point $c_0$ with $dist(c_0, c_I) \leq 2\delta r_0$.*

Let $c_0$ be as in the theorem above and let $n$ be any positive integer. For any integer $i$, $0 \leq i < n$, send a ray along the direction $2\pi i/n$, and let $S_n$ be the set of contact points. The quality of $S_n$ is a good approximation of the quality of $I$.

**Theorem 3** *Under assumption CMQA, $|qual(S_n) - qual(I)| = O(1/n)$ and $qual(S_n)$ can be computed in time $O(n \log n)$.*

The complete classification algorithm is given in Figure 2. By Theorem 3 there is a constant $c$ such that $|qual(S_n) - qual(I)| \leq c/n$. Therefore, $n$ cannot exceed $c \cdot qual(I)$ and Theorem 1 follows.

Note that this procedure runs forever if the quality is 0 (as when $I = C_4$ in Figure 1). In practice, we should stop the algorithm when the quality becomes sufficiently close to 0.

The structure of the paper is as follows. In section 3 we prove Theorem 2 and in section 4 we prove Theorem 3. In section 5 we extend our results to a larger class of quality measures and in section 6 we list open problems.

## 3 Initial Placement

We provide a simple strategy for finding a point $c_0$ close to the center $c_I$ of $I$. This result has independent and practical interest. For instance, there are highly specialized roundness-measuring machines. Traditionally, the placement of an object in such machines are carried out manually. Our result means that one can easily automate this process with just 6 initial probes.

The strategy is based on two simple subroutines: For any point $p$, the pair of horizontal probes directed towards $p$, one from the left and one from the right, is called the *horizontal pair* for $p$. If $p_\ell$ and $p_r$ are the two contact points returned by the horizontal pair for $p$, define $H(p) = H_I(p)$ to be $(p_\ell + p_r)/2$. Note that $y(p) = y(H(p))$. Also, $p_\ell = \infty$ iff $p_r = \infty$; in this case, $H(p)$ is defined to be $\infty$. Similarly, the two vertical probes directed towards $p$, one from above and one from below, is called the *vertical pair* for $p$. If the corresponding two contact points are $p_a$ and $p_b$, we define $V(p) = V_I(p)$ to be $(p_a + p_b)/2$. For completeness, define $H(\infty) = V(\infty) = \infty$.

**Theorem 4** *Let $c_I$ be a center of $I$ and let $\rho = R(c_I)/r(c_I) - 1$. Let $p_0$ be any point in the interior of $I$ and let $c_0 = H(V(H(p_0)))$. If $\rho \leq 0.1$ then $dist(c_0, c_I) \leq \rho \cdot r(c_I)$.*

**Proof:** In the following analysis, assume without loss of generality that the center $c_I$ is at the origin, and $r(c_I)$ is normalized to 1. Let

$$p_1 := H(p_0), \quad p_2 := V(p_1), \quad p_3 := H(p_2).$$

So $c_0 = p_3$. Without loss of generality, assume $x(p_1) \geq 0$ and $y(p_2) \geq 0$. First we show an upper bound on $x(p_1) = |x(p_1) - x(c_0)|$:

$$x(p_1) \leq \sqrt{2\rho + \rho^2}. \tag{1}$$

There are two cases to consider.

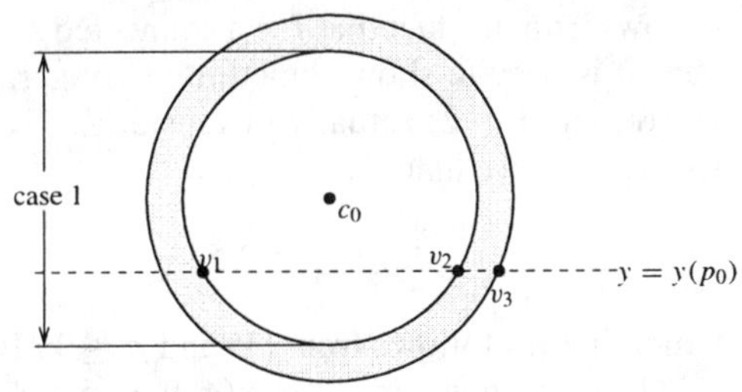


Figure 3: $-1 \leq y(p_1) \leq 1$

**Case 1:** $-1 \leq y(p_1) \leq 1$. Refer to Figure 3.

Let $v_1, v_2$ be the points where the horizontal line $y = y(p_0)$ intersects the unit circle centered at $c_0$. Let $x_i = x(v_i)$ $(i = 1, 2)$ and $x_1 < x_2$. Since the unit circle is contained in $I$, the left probe contacts $I$ at the $x$-coordinate $x_\ell$ where $x_\ell \leq x_1$. Let $x_3$ be the positive $x$-coordinate where this line intersects the circle of radius $1 + \rho$ centered at

```
Compute a point c0 as in Theorem 2.
n = 2;
Repeat
    n = 2 * n;
    Determine S_n and compute qual(S_n) and an interval [l, u] containing qual(I).
until 0 ∉ [l, u]
If (l > 0) declare I good;
If (u < 0) declare I bad;
```

Figure 2: The complete classification algorithm.

$c_0$. Since $I$ is contained in the circle of radius $1+\rho$, the right probe contacts $I$ at the $x$-coordinate $x_r$ where $x_r \le x_3$. So $x(p_1) = (x_\ell + x_r)/2 \le (x_1+x_3)/2$. Furthermore, since $x_1 = -x_2$, $x(p_1) \le (x_3 - x_2)/2$. The quantity $x_3 - x_2$ is largest when $y(p_0) = 1$ or $-1$; in this case, the Pythagorean theorem shows $x_3 - x_2 = \sqrt{(1+\rho)^2 - 1^2} = \sqrt{2\rho+\rho^2}$. Thus $x(p_1) \le \sqrt{2\rho+\rho^2}/2$.

**Case 2:** $y(p_1) < -1$ or $y(p_1) > 1$.

We may define $x_3, x_\ell$ and $x_r$ as in case 1. Clearly both $x_\ell$ and $x_r$ are no greater than $x_3$. Thus, $x(p_1) = (x_\ell + x_r)/2 \le x_3$. The coordinate $x_3$ is largest when $y$ equals 1 or $-1$, in which case $x_3 = \sqrt{2\rho+\rho^2}$. Thus $x(p_1) \le \sqrt{2\rho+\rho^2}$.

In either case, we obtain the desired bound (1). Notice that $p_1$ need not lie inside $I$. Hence it is important to argue that $p_2 = V(p_1)$ is finite. But this follows from the fact that $I$ is a connected set. But in fact, it is easy to show something stronger, namely $p_i$ (for any $i \ge 2$) actually lies inside $I$. To see this for $p_2$, observe that

$$x(p_1) < 0.5, \tag{2}$$

which in turn follows from (1) and $\rho \le 1/10$.

Next we show $|y(p_2) - y(c_0)| \le 5\rho/8$, which, under our assumptions, amounts to:

$$y(p_2) \le 5\rho/8. \tag{3}$$

Referring to Figure 4(i), let the line $x = x(p_1)$ intersect the unit circle centered at $c_0$ at the points $v_1, v_2$. Let $y_i := y(v_i)$ $(i = 1, 2)$ and $y_1 < y_2$. The line also intersects the circle of radius $1+\rho$ at the positive $y$-coordinate $y_3$. As in the proof of (1), Case 1, we have $y(p_2) \le (y_3 - y_2)/2$. Now the maximum value $y_3 - y_2$ obtains when $x(p_1)$ is maximized. By (1), $x(p_1) \le \sqrt{2\rho+\rho^2}$. Let $g := \sqrt{2\rho+\rho^2}$ and $e$ represent $y_3 - y_2$ when $x(p_1) = g$. So $y(p_2) \le e/2$ and its suffices to show $e \le 5\rho/4$. Define $h$ as in Figure 4(ii). A simple calculation shows that

$$e + h = 1, \qquad h = \sqrt{1-g^2}$$

and hence

$$e = 1 - \sqrt{1-g^2}.$$

Using the Taylor expansion

$$\begin{aligned}(1-x)^{1/2} &= 1 - \left(\frac{x}{2} + \frac{x^2}{8} + \frac{x^3}{16} + \cdots\right) \\ &> 1 - \left(\frac{x}{2} + \frac{x^2}{8} + \frac{x^3}{16(1-x)}\right),\end{aligned}$$

and on substituting $x = g^2$ in the expression for $e$,

$$e < \frac{g^2}{2} + \frac{g^4}{8} + \frac{g^6}{16(1-g^2)}.$$

Using $g^2 < 0.25$,

$$e < g^2\left(\frac{1}{2} + \frac{1}{32} + \frac{1}{128}\right) < \frac{9g^2}{16}.$$

From $g^2 = 2\rho + \rho^2 < 2.1\rho$, we conclude that $e < 5\rho/4$, which proves (3).

We next show $|x(p_3) - x(c_I)| \le 13\rho/25$. Assuming without loss of generality that $x(p_3) \ge 0$, this amounts to

$$x(p_3) \le 13\rho/25. \tag{4}$$

With reference to Figure 5(i), let the line $y = y(p_2)$ intersect the circles centered at $c_0$ with radii 1 and $1+\rho$ at the positive $x$-coordinates $x_2'$ and $x_3'$, respectively. The above analysis shows $x(p_3) \le (x_3' - x_2')/2$. Assuming without loss of generality that $y(p_3) \ge 0$, the value of $x_3' - x_2'$ is increasing with $y(p_2)$. Let $e'$ be the value of $x_3' - x_2'$ when $y(p_2) = 5\rho/8$, as in Figure 5(ii). Since

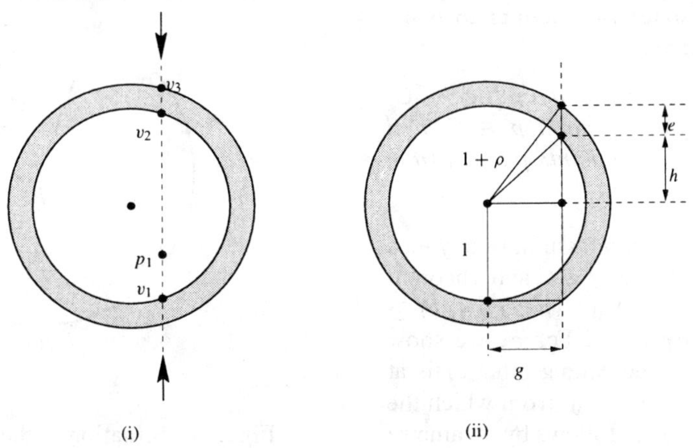

Figure 4: Second set of probes: bounding $y(p_2)$.

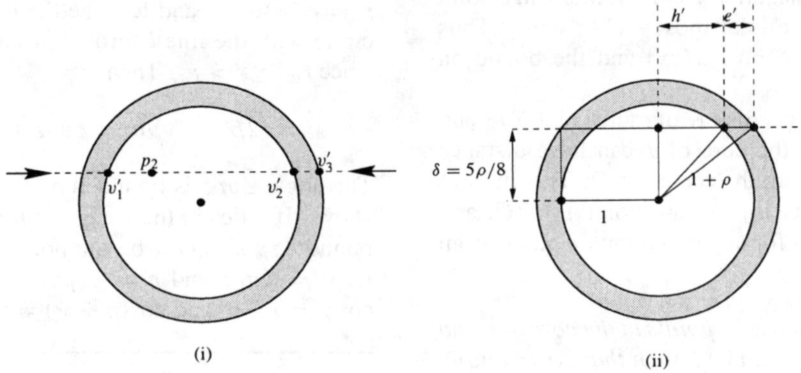

Figure 5: Third set of probes: bounding $x(p_3)$.

$y(p_2) \leq 5\rho/8$, this means $x(p_3) \leq e'/2$. It remains to bound $e'$. Putting $\delta = 5\rho/8$,

$$\begin{aligned} e' &= \sqrt{(1+\rho)^2 - \delta^2} - \sqrt{1-\delta^2} \\ &< (1+\rho) - 1 + \delta^2 \left( \frac{1}{2} + \frac{\delta^2}{8} \frac{1}{1-\delta^2} \right), \end{aligned}$$

using an abbreviated version of the previous Taylor expansion. Since $\delta^2 = 25\rho^2/64 < \rho/25$, we get $e' < 26\rho/25$, verifying (4).

From (3) and (4) we conclude that that $\|p_3 - c_I\| \leq \rho\sqrt{(5/8)^2 + (13/25)^2} < \rho$, and the theorem is proved. ∎

Theorem 2 is an easy consequence of Theorem 4 and MQA. MQA implies implies $\rho = R(c_I)/r(c_I) - 1 \leq (1+\delta)/(1-\delta) - 1 \leq 0.1$ and hence $dist(c_0, c_I) \leq \rho \cdot r(c_I) = R(c_I) - r(c_I) \leq 2\delta r_0$.

## 4 Uniform and Near-Uniform Probing

In this section we prove Theorem 3. Recall that we determine a set $S_n = \{v_0, \ldots, v_{n-1}\}$ of $n$ points in bd $I$ arranged in clockwise order by uniform probing about $c_0$, i.e., by probing along the directions $2\pi i/n$ for all $i$, $0 \leq i < n$. Let $c_S$ be the center of $S_n$ and define the core $C$ of $I$ by

$$C = \{p \in I;\ r(p, I) \geq (1 - 4\delta)r_0\}.$$

We show that $c_I$, $c_0$, and $c_S$ lie in the core and that if $p$ and $q$ are points in the core and $v$ and $w$ are points in bd $I$ then $\angle vpw \leq 6\angle vqw$. This implies that $\angle v_i c_S v_{i+1}$ is at most $12\pi i/n$ for all $i$, $0 \leq i < n$, i.e., the uniform probing about $c_0$ amounts to near-uniform probing about $c_S$ with about the same sampling density. Finally, we show that the near-uniform probing about $c_S$ implies that $|qual(I) - qual(S)| = O(1/n)$.

### 4.1 Uniform probing about $c_0$ amounts to near-uniform probing about $c_S$.

**Lemma 5** *The points $c_I$, $c_0$, and $c_S$ lie in $C$, $R(p, I) \leq (1 + 6\delta)r_0$ for any point $p \in C$, and $dist(p, q) \leq 10\delta r_0$ for any two points $p$ and $q$ from the core.*

**Proof:** The minimum quality assumption yields $r(c_I, I) \geq (1-\delta)r_0$ and hence $c_I \in C$ and Theorem 2 states that $dist(c_I, c_0) \leq 2\delta r_0$. Thus $r(c_0, I) \geq (1 - 3\delta)r_0$ and hence $c_0 \in I$. For $c_S$ we show $dist(c_S, c_I) \leq 3\delta r_0$. Assume w.l.o.g. that $c_I$ is at the origin. We show $x(c_S) \leq 2\delta r_0$, from which the claimed bound on $dist(c_I, c_S)$ follows by symmetry and elementary calculation.

First note that $qual(I) \leq qualS = qual(c_S, S)$ implies $R(c_S, S) \leq (1 + \delta)r_0$. Next note that $n$ is always a multiple of 4 and hence one of the probes is made along the negative $x$-axis. It hits $I$ in a point whose $x$-coordinate is at most $-(1 - \delta)r_0$. Thus $R(c_S, S) \geq (1 - \delta)r_0 + x(c_S)$ and the bound on $x(c_S)$ follows.

$I$ is contained in a disk $D$ of radius $(1+\delta)r_0$ and hence no point in the core of $I$ can have distance larger than $5\delta r_0$ from the center of $D$. This implies $R(p, I) \leq (1 + 6\delta)r_0$ for any point $p \in C$, and $dist(p, q) \leq 10\delta r_0$ for any two points $p$ and $q$ from the core. ∎

**Lemma 6** *Let $p$ and $q$ be points in the core of $I$ and let $v$ and $w$ be points in* bd $I$ *such that $2\alpha = \angle vqw$. Then $2\beta = \angle vpw \leq 12\alpha$.*

**Proof:** If $\alpha \geq \pi/12$ there is nothing to show. Assume, otherwise. Then $\arcsin 2\tan\alpha \leq 3\alpha$. For a point $x$ we use $r_x$ to denote $dist(p, x)$. We proceed in two steps.

First assume $r = r_v = r_w$. We show $\beta \leq 3\alpha$. Let $L = 2\beta r/(2\pi)$ be the length of the circular arc from $v$ to $w$. We bound $L$ from above. We may assume w.l.o.g. that $p$ is at the origin and that $q$ lies on the positive $x$-axis, cf. Figure 6.

We claim that $L$ is maximal if the axis of the wedge $W = \angle vqw$ aligns with the negative $x$-axis. Indeed consider any other orientation of $W$'s axis and consider turning $W$'s axis towards the negative $x$-axis. Then $L$ will grow as long as the "forward leg" of $W$ is longer than the "rear leg" of $W$. The worst case situation is therefore if $W$'s axis aligns with the negative $x$-axis, cf. Figure 7. We have $r \geq (1 - 4\delta)r_0$, $z \leq 10\delta r_0$, $\tan\alpha = y/(x + z)$, $\sin\beta = y/r$, and $x \leq r$. Thus,

$$\sin\beta = y/r = \frac{(x + z) \cdot \tan\alpha}{r} \leq 2\tan\alpha$$

and hence $\beta \leq 3\alpha$.

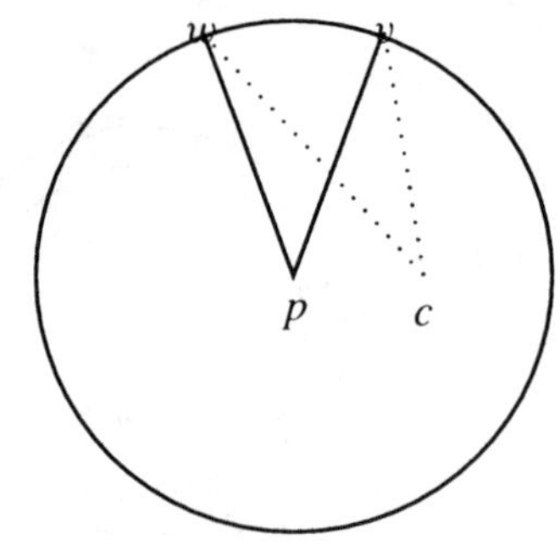


Figure 6: Situation in the proof of lemma 6

Now we come to the general case. Assume w.l.o.g, $r_v \leq r_w$. Consider the circle with radius $r = r_v$ about $v$ and let $z$ be the intersection of this circle with the line $l$ through $q$ and $w$; $z$ is unique since $r_w \geq r > r_q$. Then

$$2\beta = \angle vpw = \angle vpz + \angle zpw.$$

The angle $\angle vpz$ is bounded by $6\alpha$ by the argument above. If $z$ lies in the wedge $\angle vpw$ we also need to bound $\angle zpw$. Let $a$ be the point on $l$ closest to $p$, let $\gamma = \angle apz$ and $\eta = \angle zpw$, cf. Figure 8. Then $\cos\gamma = r_a/r_v$ and $\cos(\eta + \gamma) = r_a/r_w$.

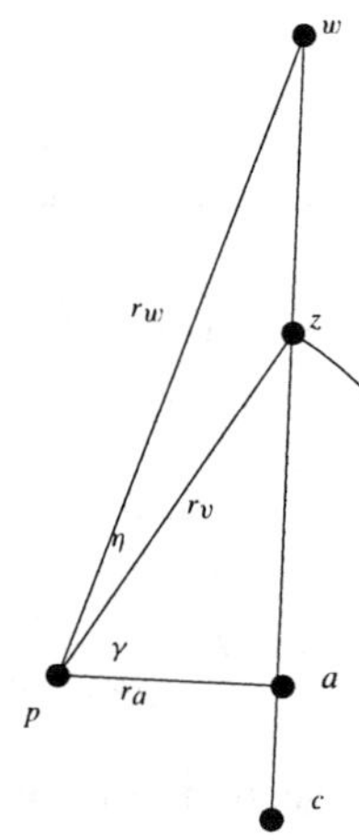


Figure 8: Bounding $\eta$

Since $\cos(\gamma + \eta) = \cos\gamma\cos\eta - \sin\gamma\sin\eta$ this implies

$$\begin{aligned} \sin\eta &= (\cos\gamma\cos\eta - \cos(\gamma + \eta))/\sin\gamma \\ &\leq (\cos\gamma - \cos(\gamma + \eta))/\sin\gamma \\ &= ((r_w - r_v)r_a)/(r_v \cdot r_w) \cdot (1/\sin\gamma). \end{aligned}$$

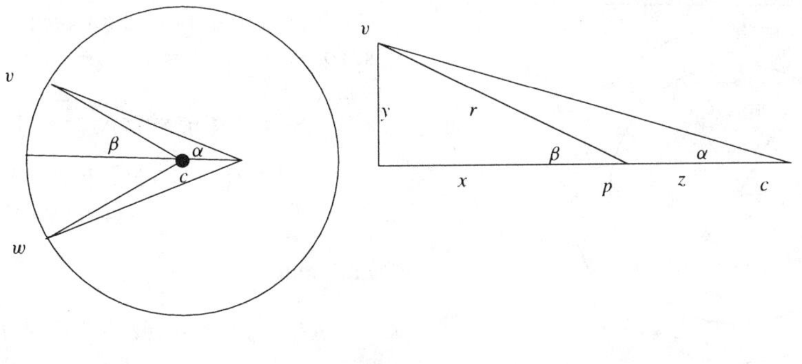


Figure 7: Worst Case

We have $\cos\gamma = r_a/r_v \leq 6\delta/(1-4\delta) \leq 1/10$ and hence $\sin\gamma = \sqrt{1-\cos^2\gamma} \geq \sqrt{99/100} \geq 9/10$. It remains to bound $(r_w - r_v)/r_w$. An obvious bound is $10\delta/(1-4\delta)$, which is, however, much too weak for our purposes.

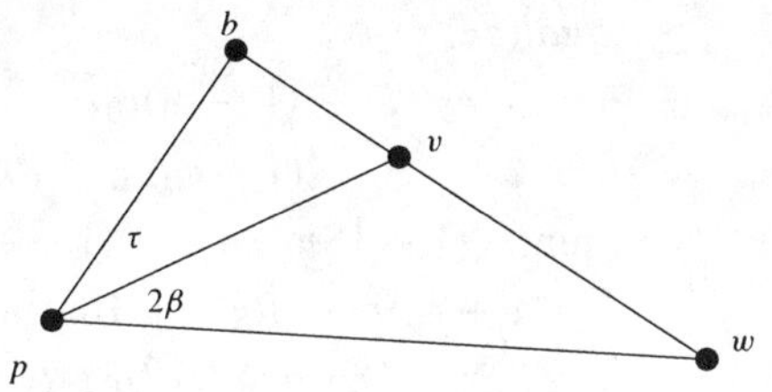


Figure 9: Bounding $rw$ minus $rv$

Consider the line $L$ through $v$ and $w$ and let $b$ be the point closest to $p$ on $L$; $b$ may or may not lie between $v$ and $w$. If $b$ lies between $v$ and $w$ then

$$\begin{aligned}(r_w - r_v)/r_w &\leq (r_w - r_b)/r_w \\ &\leq 1 - \cos 2\beta \\ &\leq 2\sin\beta^2 \\ &\leq 2\beta.\end{aligned}$$

If $b$ does not lie between $v$ and $w$ we have $r_b \geq (1-4\delta)r_0$ since otherwise a neighborhood of $b$ is contained in $I$ and hence $v \notin \mathrm{bd}\, I$ by the convexity of $I$. Let $\tau = \angle bpv$. We have $\cos\tau = r_b/r_v$ and $\cos(2\beta+\tau) = r_b/r_w$, cf. Figure 9, and hence

$$\begin{aligned}(r_w - r_v)/r_w &= 1 - \cos(2\beta+\tau)/\cos\tau \\ &= 1 - \cos 2\beta + (\sin\tau/\cos\tau)\sin 2\beta \\ &\leq 2\sin^2\beta + (1/\cos\tau)\sin 2\beta \\ &\leq 2\beta(1 + 1/\cos\tau).\end{aligned}$$

Since $\cos\tau = r_b/r_v \geq (1-4\delta)/(1+6\delta) \geq 1/2$ we conclude $(r_w - r_v)/r_w \leq 6\beta$ and hence $\sin\eta \leq 6\beta/9$. Substituting into $2\beta - 6\alpha \leq \eta$ yields $\sin(2\beta - 6\alpha) \leq 2\beta/3$. This implies $2\beta \leq 12\alpha$. Note that the equation $\sin(2\beta - 6\alpha) = 2\beta/3$ has a single solution with $2\beta \leq \pi/4$ and that the iteration $2\beta_0 = 6\alpha$, $2\beta_{i+1} = 6\alpha + \arcsin 2\beta_i/3$ converges monotonically towards this solution. A simple induction shows that all iterates stay bounded by $12\alpha$. Indeed, $2\beta_0 \leq 12\alpha$ and if $2\beta_i \leq 12\alpha$ and hence $\arcsin 2\beta_i/3 \leq \arcsin 4\alpha \leq 6\alpha$ then $2\beta_{i+1} \leq 12\alpha$. ∎

### 4.2 *qual*($S_n$) approximates *qual*($I$)

We now conclude our proof of Theorem 3. Let $\theta = \pi/n$ and let $S = \{v_1 \ldots v_n\} \subseteq \mathrm{bd}\, I$ be the points determined by probing along directions $2\theta i$, $0 \leq i < n$, from $c_0$. Then $\angle v_i c_0 v_{i+1} = 2\theta$ and $\angle v_i c_S v_{i+1} \leq 12\theta$, where $c_S$ is the center of $S$ by Lemma 6. Let $r_S = r(c_S, S)$ and $R_S = R(c_S, S)$.

**Lemma 7**

$$r_S(1 - 18\pi/n^2) \leq r_S \cos 6\theta \leq r(c_S, I) \leq r_S. \quad (5)$$

**Proof:** The upper bound on $r(c_S, I)$ is immediate. For the lower bound, choose $q \in \mathrm{bd}\, I$ at distance $r(c_S, I)$ from $c_S$. The point $q$ lies angularly between some two samples $v_i$ and $v_{i+1}$ around $c_S$. Let $2\sigma = \angle v_i c_S v_{i+1} \leq 12\theta$. Since $I$ is convex, the line segment $[v_i, v_{i+1}]$ must be contained in $I$. Convexity furthermore implies that the point $q$ does not lie on the same side of this segment as $c_S$ does. Thus, the distance $r(c_S, I)$ from $c_S$ to $q$ is at least the distance $D$ from $c_S$ to the segment. This distance $D$ is minimized when the distances from $c_S$ to $v_i$ and from $c_S$ to $v_{i+1}$ are both $r_S$; in this case $D = r_S \cos\sigma \geq \cos 6\theta$, as desired. See Figure 10. The lower bound in terms of $n$ follows from $\theta = \pi/n$ and $\cos\theta \geq 1 - \theta^2/2$. ∎

**Lemma 8** *Let* $z = R(c_S, I)/r(c_S, I)$. *Then*

$$\begin{aligned}R_S \leq R(c_S, I) &\leq \frac{R_S}{\cos\theta - \sin\theta\sqrt{z^2-1}} \\ &\leq R_S(1 + 18\pi/n).\end{aligned}$$

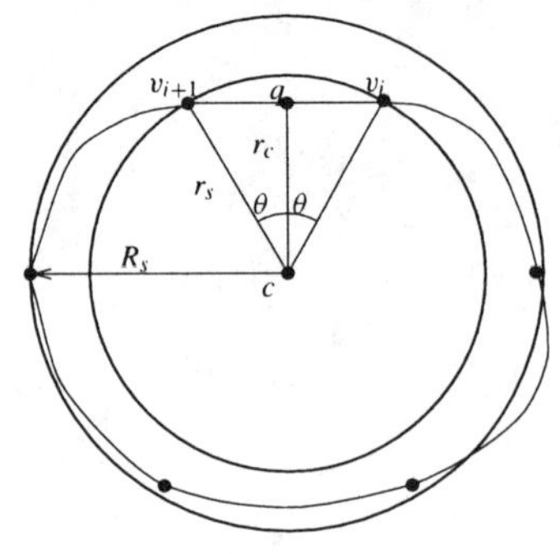


Figure 10: A lower bound on $r(c_S, I)$

**Proof:** The lower bound on $R(c_S, I)$ is immediate. For the upper bound, choose $q \in \text{bd}\, I$ at distance $R(c_S, I)$ from $c_S$, lying angularly between two samples $v_i$ and $v_{i+1}$ when seen from $c_S$. Let $2\sigma = \angle v_i c_S v_{i+1} \leq 12\theta$. The maximum value of $R(c_S, I)$ is obtained when $v_i$ and $v_{i+1}$ are both at distance $R_S$ from $c_S$, $q$ is on the bisector of $\angle v_i c_S v_{i+1}$, and the lines $qv_i$ and $qv_{i+1}$ are both tangent to the circle of radius $r(c_S, I)$ centered at $c_S$. This situation is illustrated in Figure 11.

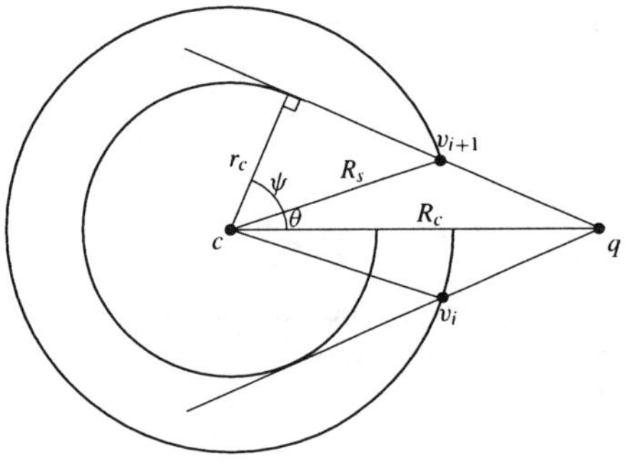


Figure 11: An upper bound on $R(c_S, I)$

Let $\psi$ be the angle about $c_S$ formed by $v_{i+1}$ and the point where $qv_{i+1}$ is tangent to the circle of radius $r(c_S, I)$. Then

$$\cos\psi = \frac{r(c_S, I)}{R_S}, \quad \cos(\psi + \sigma) = \frac{r(c_S, I)}{R(c_S, I)}.$$

Rearranging the latter equation,

$$\begin{aligned} R(c_S, I) &= \frac{r(c_S, I)}{\cos(\psi + \sigma)} \\ &= \frac{r(c_S, I)}{\cos\psi\cos\sigma - \sin\psi\sin\sigma} \\ &= \frac{R_S}{\cos\sigma - \sqrt{(R_S/r(c_S, I))^2 - 1}\,\sin\sigma}. \end{aligned}$$

For the upper bound in terms of $n$, observe that $\sin\sigma \leq \sigma$ and $z =\leq (1 + \delta)/(1 - \delta)$ implies

$$\begin{aligned} &\cos\sigma - \sin\sigma\sqrt{z^2 - 1} \\ &= 1 - \sin^2\sigma - \sin\sigma\sqrt{z^2 - 1} \\ &\geq 1 - \sigma(1 + \sqrt{2}) \end{aligned}$$

and hence

$$R(c_S, I) \leq R_S(1 + 3\sigma) \leq R_S(1 + 18\pi/n).$$

■

**Lemma 9** *$qual(S_n) - 20\pi/n \leq qual(I) \leq qual(S)$ and $qual(S_n)$ can be computed in time $O(n \log n)$.*

**Proof:** $qual((I) \leq qual(S_n)$ follows directly from $S \subseteq \text{bd}\, I$. For the lower bound, let $r_S = r(c_S, S)$ and $R_S = R(c_S, S)$ and observe

$$\begin{aligned} qual(I) &= \max_p qual(p, I) \\ &\geq qual(c_S, I) \\ &= \min(r(c_S, I) - (1 - \epsilon_0)r_0, \\ &\qquad (1 + \epsilon_0)r_0 - R_S)/r_0 \\ &\geq \min(r_S(1 - 18\pi^2/n^2) - (1 - \epsilon_0)r_0, \\ &\qquad (1 + \epsilon_0)r_0 - R_S(1 + 18\pi/n))/r_0 \\ &\geq qual(S) - 18\pi\max(r_S/r_0, R_S/r_0)/n \\ &= qual(S) - 20\pi/n. \end{aligned}$$

The bound on the running time follows from [2]. ■

## 5 Other Quality Measures

There are many ways to quantify the roundness of an object besides the one used above. For example, we may require

- that the inner radius $r(p)$ lies in some interval $[r_0 - \epsilon_1, r_0 + \epsilon_2]$ and the outer radius $R(p)$ in some interval $[r_0 - \epsilon_3, r_0 + \epsilon_4]$, or
- that the width $R(p) - r(p)$ lies in some interval $[0, \epsilon_5]$ or
- that the relative width $(R(p) - r(p))/r(p)$ lies in some interval $[0, \epsilon_6]$ or
- that the average distance of a point in $\text{bd}\, I$ from the center $c$ lies in some interval $[r_0 - \epsilon_7, r_0 + \epsilon_7]$.

The approach described in this paper is readily extended to the first three constraints above and in fact to a large class of constraints that can be formulated in terms of an annulus containing the boundary of the body to be classified. Let $\mathcal{F} = \{f_1, \ldots, f_k\}$ be a family of functions of two real arguments. We call $f_i$ the $i$-th *measure of quality*. For a point $p$ let

$$qual(p, I) = \min_{1 \le i \le k} f_i(r(p, I), R(p, I))$$

be the quality of the body $I$ with respect to $p$ and let

$$qual(I) = \max_p qual(p, I)$$

be the quality of $I$. We call a point $p$ with $qual(I) = qual(p, I)$ an *$\mathcal{F}$-center* of $I$ and use $c_{\mathcal{F},I}$ to denote an $\mathcal{F}$-center of $I$. The examples above lead to the quality functions $f_1(r, R) = r - (r_0 - \epsilon_1)$, $f_2(r, R) = r_0 + \epsilon_2 - r$, $f_3(r, R) = R - (r_0 - \epsilon_3)$, $f_4(r, R) = r_0 + \epsilon_4 - R$, $f_5(r, R) = \epsilon_5 - (R - r)$, and $f_6(r, R) = \epsilon_6 - (R - r)/r$. The specialization $\mathcal{F} = \{f_1, f_2, f_3, f_4\}$ and $\epsilon_1 = \epsilon_2 = \epsilon_3 = \epsilon_4 = \epsilon_0 r_0$ was considered in the preceding section.

Our approach hinges on the following two properties

- the quality of the sample $S = S_n$ can be computed efficiently and
- the quality of $S_n$ is a good estimate for the quality of $I$.

With respect to the first property we observe that the overlay of the furthest and nearest site Voronoi diagram partitions the plane into cells (= regions, edges, and vertices) with the property that for all $p$ in a fixed cell the radii $r(p, S)$ and $R(p, S)$ are determined by a small number of points in the sample $S$, namely two points for regions, three points for edges, and four points for vertices. For a cell $f$ let $S_f$ be the points in $S$ that determine $r(p, S)$ and $R(p, S)$ for all $p$ in $f$.
**Assumption BM**: Let $f$ be a cell and let $F$ be the affine hull of $f$; $F$ is either the plane, or a line, or a vertex. We assume that the function $p \longrightarrow qual(p, S_f)$ for $p \in F$ has only a bounded number of local minima and that these minima are computable in constant time.

Assumption BM is certainly satisfied for any quality measure based on the functions $f_1, \ldots, f_6$ above. We proceed under assumption BM. For each $f$, the equations $f_i = f_j$, $1 \le i < j < k$, partition $F$ into cells where in each cell we have a fixed ordering of the values of $f_1, \ldots, f_k$. Thus, determining the local minima within a cell amounts to a bounded size (since $|S_f| \le 4$) optimization problem.

We determine the local minima of $qual(p, S_f)$ in $F$ and then test these minima for containment in $f$. The "surviving" minima are candidates for the $\mathcal{F}$-center of $S$. The total number of candidates is $O(n^2)$. We conclude:

**Lemma 10** *Under assumption BM the quality of a sample $S$ of size $n$ can be computed in time $O(n^2)$.*

We next turn to the second property that $qual(S_n)$ is a good estimate of $qual(I)$.
**Assumption CD**: For all sufficiently large $n$, all $\mathcal{F}$-centers of $I$ and all $\mathcal{F}$-centers of $S_n$ lies in the core of $I$. The quality functions $f_1, \ldots, f_k$ are differentiable and have bounded derivative.

**Lemma 11** *Under assumptions CMQA, BM, and CD we have $|qual(I) - qual(S_n)| = O(1/n)$.*

**Proof:** By the first part of assumption CD any $\mathcal{F}$-center $c_S = c_{\mathcal{F},S}$ of $S$ lies in the core of $I$. Also $c_0$ lies in the core and hence $r(c_S, I) - O(1/n) \le r(c_S, I) \le r(c_S, S)$ and $R(c_S, S) \le R(c_S, I) \le R(c_S, S) + O(1/n)$ by Lemmas 7 and 8. Thus, $(|f_i(r(c_S, I), R(c_S, I)) - f_i((r(c_S, S), R(c_S, S))| = O(1/n)$ and hence

$$\begin{aligned} qual(I) &\ge qual(c_S, I) \\ &\ge qual(c_S, S) - O(1/n). \end{aligned}$$

The same argument with $c_{\mathcal{F},I}$ instead of $c_S$ shows

$$\begin{aligned} qual(I) &= qual(c_I, I) \\ &\le qual(c_I, S) + O(1/n) \\ &\le qual(S) + O(1/n). \end{aligned}$$

■

Assumption CD is satisfied for all quality measures based on functions $f_1, \ldots, f_5$. It is not satisfied for the measure $qual(p, I) = f_6(r(p, I), R(p, I))$. Under this measure the center may lie at infinity and not within $I$.

**Theorem 12** *Under assumptions CMQA, BM, and CD, there is a classification algorithm working in time $O(|1/qual(I)|^2)$ time with $O(|1/qual(I)|)$ probes.*

**Proof:** The algorithm was already given in section 1. Its correctness and performance follows immediately from Lemmas 10 and 11. ■

## 6 Discussion and Open Problems

Our work rises many open problems:

We have introduced several notions of roundness, and given a general procedure for their associated classification problems. It seems to us that the different notions of roundness are useful for different applications. It is unclear why the width measure $w(I)$ is dominant in practice.

What is the influence of the probing model? For example, what happens for a "line prober" that is able to determine the tangent to the object perpendicular to a specified direction. We conjecture that the error of approximation reduces from $O(1/n)$ to $O(1/n^2)$.

We claim only a running time of $O(n^2)$ for determining $qual(S_n)$. For two quality measures a time bound of $O(n \log n)$ can be obtained: For referenced roundness this is shown in [2] and for width this is shown in [6]. Can the time bound be improved for other quality measures.

Does the approach extend to more complicated classification tasks, e.g., determining roundness of a three-dimensional object?

In our approach we keep the probing center $c_0$ fixed although $c_{S_n}$ is known to be a much better approximation of $c_I$ as $n$ gets large. Can this be exploited?

Is an approach that classifies "clearly good" and "clearly bad" objects quickly and takes longer on "borderline objects" of interest to actual metrology? Of course, in practice, one should modify our decision procedure to simply accept or reject once we have determined that the quality is smaller than some constant. This is the decision policy issue studied in [15].

Measurement devices make errors [8]. How can this be taken into account?

## 7 Bibliography

### References


[1] Richard Cole and Chee K. Yap. Shape from probing. *Journal of Algorithms*, 8:19–38, 1987.

[2] Ch. Duncan, M.T. Goodrich, and E.A.Ramos. Efficient approximation algorithms and optimization algorithms for computational metrology. In *Proc. SODA 97*, pages 121–130, 1997.

[3] H. Ebara, N. Fukuyama, H. Nakano, and Y. Nakanishi. Roundness algorithms using the Voronoi diagrams. In *Abstracts 1st Canad. Conf. Comput. Geom.*, page 41, 1989.

[4] Shaw C. Feng and Theodore H. Hopp. A review of current geometric tolerancing theories and inspection data analysis algorithms. Technical Report NISTIR-4509, National Institute of Standards and Technology, U.S. Department of Commerce. Factory Automation Systems Division, Gaithersburg, MD 20899, February 1991.

[5] A. B. Forbes. Geometric tolerance assessment. NPL Technical Report DITC 210/92, National Physical Laboratory, Division of Information Technology and Computing, NPL, Teddington, Middlesex, U.K. TW11 0LW, October, 1992.

[6] J. Garcia and P.A.Ramos. Fitting a set of points by a circle. In *Proc. ACM Conference on Computational Geometry*, 1997.

[7] V. B. Le and D. T. Lee. Out-of-roundness problem revisited. *IEEE Trans. Pattern Anal. Mach. Intell.*, PAMI-13(3):217–223, 1991.

[8] S. D. Phillips, B. Borchardt, and G. Caskey. Measurement uncertainty considerations for coordinate measuring machines. NIST Technical Report NISTIR 5170, National Institute of Standards and Technology, NIST, Precision Engineering Division, Building 220, Room B113, Gaithersburg, MD 20899, April, 1993.

[9] U. Roy, C.R. Liu, and T.C. Woo. Review of dimensioning and tolerancing: representation and processing. *Computer-aided Design*, 23(7):466–483, 1991.

[10] U. Roy and X. Zhang. Establishment of a pair of concentric circles with the minimum radial separation for assessing rounding error. *Computer Aided Design*, 24(3):161–168, 1992.

[11] Michiel Smid and Ravi Janardan. On the width and roundness of a set of points in the plane. Department of Computer Science Report TR 94-62, University of Minnesota, 1994.

[12] Vijay Srinivasan and Herbert B. Voelcker, editors. *Dimensional Tolerancing and Metrology*, 345 East 47th Street, New York, NY 10017, 1993. The American Society of Mechanical Engineers. CRTD-Vol. 27.

[13] Kurt Swanson. An optimal algorithm for roundness determination on convex polygons. In *Proc. 3rd Workshop Algorithms Data Struct.*, volume 709 of *Lecture Notes in Computer Science*, pages 601–609, 1993.

[14] Chee K. Yap. Exact computational geometry and tolerancing metrology. In David Avis and Jit Bose, editors, *Snapshots of Computational and Discrete Geometry, Vol.3*. McGill School of Comp.Sci, Tech.Rep. No.SOCS-94.50, 1994. A Volume Dedicated to Godfried Toussaint.

[15] Chee K. Yap and Ee-Chien Chang. Issues in the metrology of geometric tolerancing. In M. Overmars, editor, *Proc. Workshop on Algorithmic Robotics*. Springer Verlag, 1996. (to appear) Lecture Notes in Computer Science.

# Fitting a set of points by a circle

Jesús García-López[†] Pedro A. Ramos[‡]

## Abstract

Given a set of points $S = \{p_1, \dots, p_n\}$ in the euclidean $d$-dimensional space, we address the problem of computing the $d$-dimensional annulus of smallest width containing the set. We give a complete characterization of the centers of annuli which are *locally minimal* in arbitrary dimension and we show that, for $d = 2$, a locally minimal annulus has two points on the inner circle and two points on the outer circle that interlace angle-wise as seen from the center of the annulus. Using this characterization, we show that, given a circular order of the points, there is at most *one* locally minimal annulus consistent with that order and it can be computed in time $O(n \log n)$ using a *simple* algorithm: it involves the computation of furthest and closest point Voronoi diagrams and a prune-and-search in the former one.

[†]**Dpto. Matemática Aplicada, E.U. Informática, U. Politécnica de Madrid, (jglopez@eui.upm.es).**

[‡]**Dpto. Matemática Aplicada, E.U.I.T. Aeronaútica, U. Politécnica de Madrid, (pramos@fi.upm.es).**



## 1 Introduction

Let $S = \{p_1, \dots, p_n\}$ be a set of $n$ points in $\mathbb{E}^d$ and let $d(x, y)$ denote the euclidean distance between points $x$ and $y$. If we denote by $\mathcal{F}$ the family of hyper-spheres in $\mathbb{E}^d$, the problem of finding the optimal Tchebyshev fit of $S$ with a hyper-sphere can be formulated as a min-max problem, that is, finding

$$\min_{\gamma \in \mathcal{F}} \max_{i=1,\dots,n} d(p_i, \gamma).$$

The problem can be formulated in a more geometric setting in the following way: we shall call $d$-annulus with *center* at $c$, *inner radius* $r$ and *outer radius* $R$, denoted $\mathcal{A}(c, r, R)$, to the locus of points between two concentric spheres of radius $r$ and $R$, that is,

$$\mathcal{A}(c, r, R) = \{x \in \mathbb{E}^d \mid r \leq d(x, c) \leq R\}.$$

The *width* of the annulus is $w = R - r$ and the problem that we address here is to compute the annulus of smallest width containing the set $S$. If we define the function $\mathcal{R}_S(x)$ as the difference of distances between the furthest and the closest neighbor of $x$ in $S$, we say that an annulus is locally minimal if its center is a local minimum of $\mathcal{R}_S(x)$ and the problem of finding the annulus of smallest width containing the set of points is equivalent to finding the global minimum of $\mathcal{R}_S(x)$.

The main motivation for this problem comes from Computational Metrology. Given an object that has to be tested for circularity ($d = 2$), take a sample of points from the object and measure the circularity of this sample set; then accept the object if the circularity is good enough and reject otherwise. The measure for circularity recommended by international standards is the width of the annulus of smallest width containing $S$ (see [5, pp. 40–42]). Further information about the relation between Computational Geometry and Metrology can be found in [13].

A related concept is the *width* of a set, defined as the width of the thinnest slab containing the set. We shall see that in some situations (that will be characterized) there is no annulus of smallest width because there is a *slab* (region between two parallel lines) of smaller width than any annulus containing $S$ and there is an infinite number of annulus which width converges to the width of the slab. We shall include this case into the general one by considering a slab as an annulus with center at $\infty$.

The problem of computing the annulus of smallest width containing a set of points has been extensively studied in the Computational Geometry literature. Up to our knowledge, the first non-trivial algorithm is due to Ebara *et al.* [3], where it is claimed that the center of the annulus is either an intersection of an edge of the closest point Voronoi diagram with an edge of the furthest point Voronoi diagram, either a vertex of one of these Voronoi diagrams, leading to an algorithm which complexity is $O(n^2)$. This characterization of the center of the annulus is used by Agarwal *et al.* [1] in order to give an $O(n^{8/5+\varepsilon})$ algorithm which is the unique $o(n^2)$ algorithm for solving this problem. This algorithm uses parametric search and other sophisticated programming techniques. Smid and Janardan [11] give a proof of that condition for the center of the annulus and point out that the optimal annulus can be a slab. The equivalent problem for simple polygons has been studied by Le and Lee [6] and has been solved in optimal $O(n)$ time for the special case of convex polygons by Swanson *et al.* [12].

We give necessary and sufficient conditions for a point to be the center of a locally minimal annulus in $\mathbb{E}^d$ which, when restricted to the planar case, show that there are always two points on the outer circle and two points on the inner circle of the annulus (thus the center is always an intersection of an edge of the closest point Voronoi diagram with an edge of the furthest point Voronoi diagram) and the points on the boundary interlace angle-wise as seen from the center of the annulus. Moreover, in Section 4 we will see that this characterization has important implications on the distribution of local minima for configurations of points arising in practice (namely, when the angular order of the points around the center of the annulus is known). We shall see that in the region where the circular order of $S$ does not change, there is at most *one* local minimum, which can be computed, using a simple algorithm, in $O(n \log n)$ time. Before going on, we state the formal definitions and introduce some notation in the next section.

## 2 Preliminaries

Let $S = \{p_1, \ldots, p_n\}$ be a set of points in $\mathbb{E}^d$. The convex hull of $S$ will be denoted *conv* $S$ and $\mathcal{V}_c(S)$ and $\mathcal{V}_f(S)$ will denote the closest and furthest point Voronoi diagram of $S$, respectively.

Given a point $x \in \mathbb{E}^d$, $CN(x)$ and $FN(x)$ will denote, respectively, the set of closest and furthest neighbors of $x$ in $S$, that is,

$$\begin{aligned} CN(x) &= \{p_j \in S \mid d(x,p_j) = \min_{i=1,\ldots,n} d(x,p_i)\} \\ FN(x) &= \{p_j \in S \mid d(x,p_j) = \max_{i=1,\ldots,n} d(x,p_i)\}. \end{aligned}$$

Then, the *roundness function* can be defined in the following way:

$$\mathcal{R}_S(x) = d(x, FN(x)) - d(x, CN(x)).$$

The problem of computing the annulus of smallest width containing $S$ is equivalent to the problem of finding the global minimum of $\mathcal{R}_S(x)$. The special case where the annulus of minimum width is a slab could be included into the general one by extending the definition of $\mathcal{R}_S(x)$ to the real projective $d$-space, $\mathrm{P}^d\mathbb{R}$. We recall that $\mathrm{P}^d\mathbb{R}$ can be obtained by adding to $\mathbb{R}^d$ a point at $\infty$ for each family of parallel lines.

Given a point $x \in \mathrm{P}^d\mathbb{R} \setminus \mathbb{R}^d$, we define $\mathcal{R}_S(x)$ as the limit, when $y$ goes to $\infty$ along a line of the corresponding family, of $\mathcal{R}_S(y)$. Clearly, this limit does not depend on the line and the value is the distance between the supporting hyperplanes of $S$ orthogonal to the family of lines.

If $p, q \in \mathbb{E}^d$, $\vec{pq}$ will denote the vector from $p$ to $q$ and, as usual, we shall identify point $p$ and vector $\vec{op}$ (for a fixed point $o$). The euclidean scalar product is denoted $\vec{u} \cdot \vec{v}$ and $\|\vec{u}\| = (\vec{u} \cdot \vec{u})^{1/2}$ is the length of the vector.

We shall call *k-plane* and *k-sphere* to a plane and a sphere of dimension $k$ and the unit $k$-sphere will be denoted $\mathcal{S}^k$.

If $S$ is included in a $(d-1)$-sphere the global minimun of $\mathcal{R}_S(x)$ is zero. This property can be checked in linear time so, from now on, we are going to exclude this case from our study.

## 3 Characterizing local minima of $\mathcal{R}_S(x)$

The following lemma, which proof follows easily from Taylor expansion, will be useful later:

**Lemma 1** *Let $p \in \mathbb{R}^d$, $\vec{v} \in \mathcal{S}^{d-1}$ and let $\alpha_p$ the angle determined by $\vec{op}$ and $\vec{v}$. Then*

$$d(\lambda\vec{v}, p) = d(o,p) - \lambda\cos\alpha_p + \frac{\lambda^2}{2}\frac{\sin^2\alpha_p}{d(o,p)} + o(\lambda^2).$$

Next result gives a complete characterization of the local minima of $\mathcal{R}_S(x)$ in arbitrary dimension. In what follows, we assume that $x_0$ is the origin of the coordinate system, and vector $\vec{x_0p}$ will be denoted $\vec{p}$.

**Theorem 1** *$x_0$ is a local minimum of $\mathcal{R}_S(x)$ if, and only if, $\forall\, \vec{v} \in \mathcal{S}^{d-1}$*

$$\min_{p_i \in FN(x_0)} \frac{\vec{v}\cdot\vec{p_i}}{\|\vec{p_i}\|} < \max_{p_i \in CN(x_0)} \frac{\vec{v}\cdot\vec{p_i}}{\|\vec{p_i}\|}. \tag{1}$$

**Proof:** For a given $\vec{v} \in \mathcal{S}^{d-1}$, let $p_{c(\vec{v})}$ denote the point of $CN(x_0)$ that minimizes the angle between $\vec{v}$ and $\vec{p_i}$ for $p_i \in CN(x_0)$ and let $\alpha_{c(\vec{v})}$ be that angle, that is,

$$\cos\alpha_{c(\vec{v})} = \frac{\vec{v}\cdot\vec{p}_{c(\vec{v})}}{\|\vec{p}_{c(\vec{v})}\|} = \max_{p_i \in CN(x_0)} \frac{\vec{v}\cdot\vec{p_i}}{\|\vec{p_i}\|}.$$

In a similar way, $p_{f(\vec{v})}$ is the point of $FN(x_0)$ that maximizes the angle between $\vec{v}$ and $\vec{p_i}$ for $p_i \in FN(x_0)$ and let $\alpha_{f(\vec{v})}$ be that angle, that is,

$$\cos\alpha_{f(\vec{v})} = \frac{\vec{v}\cdot\vec{p}_{f(\vec{v})}}{\|\vec{p}_{f(\vec{v})}\|} = \min_{p_i \in FN(x_0)} \frac{\vec{v}\cdot\vec{p_i}}{\|\vec{p_i}\|}.$$

Then, condition (1) says that

$$\forall\, \vec{v} \in \mathcal{S}^{d-1}, \quad \cos\alpha_{f(\vec{v})} < \cos\alpha_{c(\vec{v})}. \tag{2}$$

Using Lemma 1, it can be easily shown that, for $\lambda > 0$ small enough, $p_{c(\vec{v})} \in CN(x_0 + \lambda\vec{v})$ and $p_{f(\vec{v})} \in FN(x_0 + \lambda\vec{v})$ and, from Taylor expansion, it follows that

$$\mathcal{R}_S(x_0 + \lambda\vec{v}) = \mathcal{R}_S(x_0) + \lambda\left(\cos\alpha_{c(\vec{v})} - \cos\alpha_{f(\vec{v})}\right) + \frac{\lambda^2}{2}\left(\frac{\sin^2\alpha_{f(\vec{v})}}{\|\vec{p}_{f(\vec{v})}\|} - \frac{\sin^2\alpha_{c(\vec{v})}}{\|\vec{p}_{c(\vec{v})}\|}\right) + o(\lambda^2).$$

Therefore, if condition (1) holds, we have $\cos\alpha_{c(\vec{v})} - \cos\alpha_{f(\vec{v})} > 0$ for all $\vec{v} \in \mathcal{S}^{d-1}$ and $x_0$ is a local minimum. Conversely, if there exists $\vec{v} \in \mathcal{S}^{d-1}$ such that $\cos\alpha_{f(\vec{v})} \geq \cos\alpha_{c(\vec{v})}$, $\mathcal{R}_S(x)$ decreases in that direction and $x_0$ is not a local minimum. □

It is worth noticing that condition (1) can be stated in a more geometric setting: it is equivalent to say that there does not exist a circular cone $\mathcal{C}$ with apex at $x_0$ separating points of $CN(x_0)$ and $FN(x_0)$. We say that a cone $\mathcal{C}$ separates two sets of points if the interior and exterior of $\mathcal{C}$ include only points of one of the sets. Observing that the intersection of such a cone with a hypersphere is a $(d-2)$-sphere, condition (1) can also be stated saying that, if we project points of $CN(x_0)$ and $FN(x_0)$ onto a common sphere centered at $x_0$, then projections of points of $CN(x_0)$ cannot be separated from projections of points of $FN(x_0)$ by a hyperplane. Again, we say that a hyperplane separates two sets of points if the two open half-spaces determinated by the hyperplane include only points of one of the sets.

We also observe that condition (1) implies that $|CN(x_0)| + |FN(x_0)| \geq d+2$. To see this, we observe that, given two sets $R$ and $B$ in $\mathbb{E}^d$ with $|R| + |B| \leq d+1$, they can be separated by an

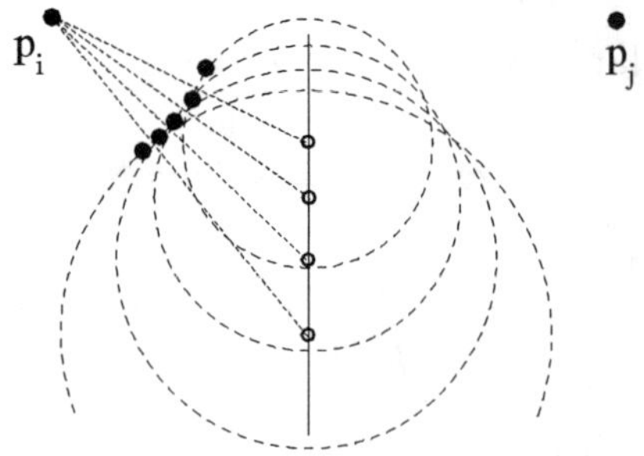


Figure 1: $\mathcal{R}_S(x)$ can have $\Omega(n)$ local minima.

hyperplane (this is easy to prove, using induction on $d$).

Finally, we point out that condition (1) implies $|CN(x_0)| > 1$ and $|FN(x_0)| > 1$.

For $d = 2$, this is also equivalent to say that there are two points on the outer circle and two points on the inner circle of the annulus that interlace angle-wise as seen from the center of the annulus.

This characterization of local minima of $\mathcal{R}_S(x)$ raises the open problem of determining the number of local minima that $\mathcal{R}_S(x)$ can have in the worst case. If we restrict ourselves to the two dimensional case, it was conjectured in previous works [1, 11] that $\mathcal{R}_S(x)$ can have as much as $\Omega(n^2)$ local minima because $\mathcal{V}_c(S)$ and $\mathcal{V}_f(S)$ can have a quadratic number of intersections (even for sets of points in convex position). Up to our knowledge, if we take into account the angular condition, there are no examples in the literature giving a lower bound for the number of local minima of $\mathcal{R}_S(x)$. Figure 1 gives an example of a configuration of points with $\Omega(n)$ local minima. Actually, all this minima are on the same edge of the furthest point Voronoi diagram, and all of them can be global minima of the function. However, it is not clear at all how this construction can be iterated and provide an example with more than a linear number of local minima. Therefore, the number of local minima that $\mathcal{R}_S(x)$ can have in the worst case is $\Omega(n)$ and $O(n^2)$ and there is a wide gap for future work.

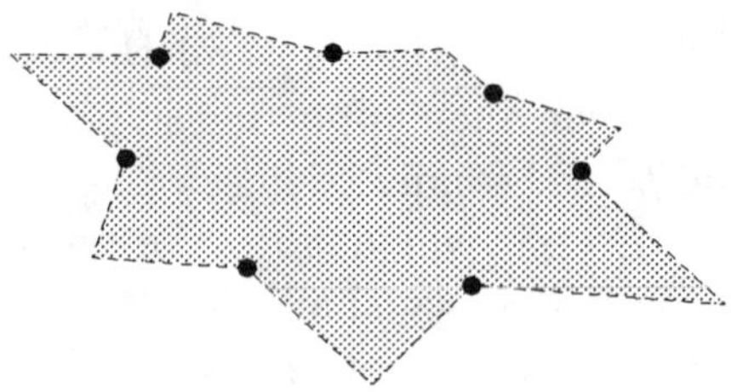

Figure 2: For a set of points in convex position, *conv* $S \subsetneq$ *ker* $S$.

# 4 Roundness in the plane

Let $S = \{p_1, \dots, p_n\}$ an *ordered* set of points in the plane. We define the *kernel* of $S$, denoted *ker* $S$, as the locus of points where points of $S$ are seen in the order $p_1, \dots, p_n$. Hereafter, we assume that points are ordered in positive direction (counterclockwise) and that indices are modulus $n$. We observe that, if points are in convex position, *conv* $S \subsetneq$ *ker* $S$ as shown in Figure 2 where *ker* $S$ is shaded.

Suppose that *ker* $S \neq \emptyset$ and consider the polygon $P$ defined by the sequence of vertices $p_1, p_2 \dots, p_n, p_1$. If $P$ is a simple polygon then *ker* $S$ is a star-shaped polygon and can be obtained as the intersection of the kernel of vertices of *conv* $S$ with the half-planes defined by two consecutive points of $S$ such that at least one of the points is not a vertex of *conv* $S$. If $P$ is not a simple polygon there is an edge $e$ of $P$ such that $P \setminus e$ is a simple chain and *ker* $S$ is a convex polygon. The intersection of *ker* $S$ with an arbitrary line has at most two connected components and, in this case, the line does not intersect *conv* $S$ (thus, it intersects two "ears" of *ker* $S$).

**Theorem 2** *Inside ker $S$ there is at most one local minimum of $\mathcal{R}_S(x)$.*

**Proof:** Assume that $x_0 \in$ *ker* $S$ is a local minimum of $\mathcal{R}_S(x)$. We are going to see that for an arbitrary $\vec{v} \in \mathcal{S}^1$, there is no other local minimum on the half-line $x_0 + \lambda\vec{v}$, $\lambda > 0$. Without loss of generality, we can assume that $x_0$ is the origin of the coordinate system and $\vec{v}$ points toward the positive $OX$ axis.

In what follows, we assume that $CN(x_0) = \{p_i, p_j\}$ and $FN(x_0) = \{p_k, p_l\}$. If this is not the case, the same arguments can be made choosing two points on each subset that satisfy the angular condition. Assume also that

$$\vec{v}\cdot\vec{p}_i \geq \vec{v}\cdot\vec{p}_j \quad (3)$$
$$\vec{v}\cdot\vec{p}_k \leq \vec{v}\cdot\vec{p}_l. \quad (4)$$

From condition (1), it follows that

$$\frac{\vec{v}\cdot\vec{p}_k}{\|\vec{p}_k\|} < \frac{\vec{v}\cdot\vec{p}_i}{\|\vec{p}_i\|}. \quad (5)$$

For a point $q \in \mathbb{E}^2$, let us define

$$\mathcal{C}(q) = \{\, x \in \mathbb{E}^2 \mid \frac{\vec{v}\cdot\vec{x}}{\|\vec{x}\|} \geq \frac{\vec{v}\cdot\vec{q}}{\|\vec{q}\|} \}.$$

We claim that, if $\vec{v}\cdot\vec{p}_k \leq 0$ then for every $\lambda > 0$, $FN(\lambda\vec{v}) \cap \mathcal{C}(p_k) = \emptyset$ and $CN(\lambda\vec{v}) \subset \mathcal{C}(p_k)$ and, therefore, there cannot be any other local minimum as long as the angular order is the same as in $x_0 = o$. To prove the claim, observe that, if $q \in \mathcal{C}(p_k) \cap S$ then $\|\vec{q}\| \leq \|\vec{p}_k\|$ and, because $\vec{v}\cdot\vec{p}_k \leq 0$, it holds

$$\|\vec{q}\|\Big(\|\vec{q}\| - 2\lambda\frac{\vec{v}\cdot\vec{q}}{\|\vec{q}\|}\Big) < \|\vec{p}_k\|\Big(\|\vec{p}_k\| - 2\lambda\frac{\vec{v}\cdot\vec{p}_k}{\|\vec{p}_k\|}\Big)$$

and therefore

$$\|\lambda\vec{v} - \vec{q}\|^2 < \|\lambda\vec{v} - \vec{p}_k\|^2,$$

and $q \notin FN(\lambda\vec{v})$.

Using a similar argument, it can be shown that $CN(\lambda\vec{v}) \subset \mathcal{C}(p_k)$ for every $\lambda > 0$ because, in this case, if $q \in S$ and $\vec{q} \notin \mathcal{C}(p_k)$, we have that

$$\frac{\vec{v}\cdot\vec{q}}{\|\vec{q}\|} < \frac{\vec{v}\cdot\vec{p}_k}{\|\vec{p}_k\|} < \frac{\vec{v}\cdot\vec{p}_i}{\|\vec{p}_i\|}$$

and therefore

$$\|\lambda\vec{v} - \vec{p}_i\|^2 < \|\lambda\vec{v} - \vec{q}\|^2.$$

Suppose now that $\vec{v}\cdot\vec{p}_k > 0$. Let $(x_i, y_i)$ the coordinates of point $p_i$. We consider separately the cases when $x_i \geq x_k$ and $x_i < x_k$. If $x_i \geq x_k$ it can be shown that we have, as before, that $CN(\lambda\vec{v}) \subset \mathcal{C}(p_k)$ and $FN(\lambda\vec{v}) \cap \mathcal{C}(p_k) = \emptyset$ for $\lambda > 0$ and thus there cannot be any other local minimum for any $\lambda > 0$ inside $ker\ S$.

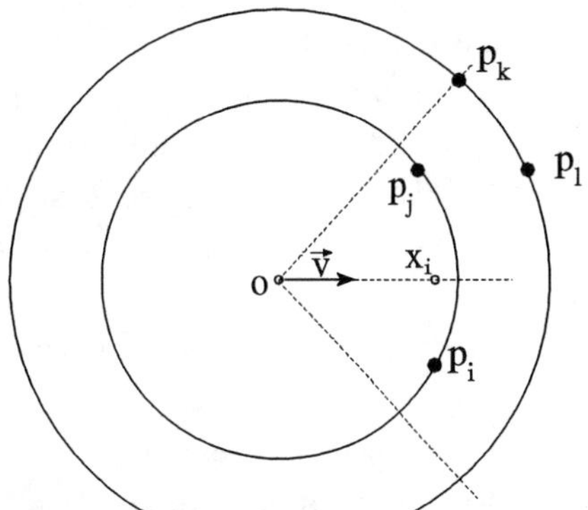


Figure 3: Illustrations of the proof of Theorem 2.

Finally, for the case when $x_i < x_k$, we claim that for $0 < \lambda < x_i$, $CN(\lambda\vec{v}) \subset \mathcal{C}(p_i)$ and $FN(\lambda\vec{v}) \cap \mathcal{C}(p_k) = \emptyset$ and, therefore, there is no local minimum in this segment. To prove that $CN(\lambda\vec{v}) \subset \mathcal{C}(p_i)$ we observe that if $q \in S$ and $q \notin \mathcal{C}(p_i)$ then $\|\vec{p}_i\| \leq \|\vec{q}\|$ and for $0 < \lambda < x_i$ it holds

$$\|\lambda\vec{v} - \vec{p}_i\| < \|\lambda\vec{v} - \vec{q}\|.$$

To see this, observe that this relation is equivalent to

$$2\lambda(\vec{v}\cdot\vec{q} - \vec{v}\cdot\vec{p}_i) < \|\vec{q}\|^2 - \|\vec{p}_i\|^2$$

and

$$2\lambda(\vec{v}\cdot\vec{q} - \vec{v}\cdot\vec{p}_i) < 2\lambda\frac{\vec{v}\cdot\vec{p}_i}{\|\vec{p}_i\|}(\|\vec{q}\| - \|\vec{p}_i\|)$$
$$< 2\|\vec{p}_i\|(\|\vec{q}\| - \|\vec{p}_i\|) < \|\vec{q}\|^2 - \|\vec{p}_i\|^2.$$

To see that $FN(\lambda\vec{v}) \cap \mathcal{C}(p_k) = \emptyset$, we show that if $q \in \mathcal{C}(p_k) \cap S$ then $\|\vec{p}_i\| \leq \|\vec{q}\| \leq \|\vec{p}_k\|$ and it holds

$$\|\lambda\vec{v} - \vec{q}\| < \|\lambda\vec{v} - \vec{p}_k\|.$$

Again, this expression is equivalent to show that

$$2\lambda(\vec{v}\cdot\vec{p}_k - \vec{v}\cdot\vec{q}) < \|\vec{p}_k\|^2 - \|\vec{q}\|^2$$

which is true because

$$2\lambda(\vec{v}\cdot\vec{p}_k - \vec{v}\cdot\vec{q}) < 2\,x_i\frac{\vec{v}\cdot\vec{q}}{\|\vec{q}\|}(\|\vec{p}_k\| - \|\vec{q}\|)$$
$$< \|\vec{p}_k\|^2 - \|\vec{q}\|^2.$$

In order to finish the proof, we only have to show that, for $\lambda > x_i$ points $\lambda\vec{v}$ are not in $ker\ S$, and this follows easily from Equation (5) and the alternating condition of the points of $CN(x_0)$ and $FN(x_0)$ around $x_0$: if $y_i$ and $y_k$ have the same sign, we have done and if they have different signs (as in Figure 3), we have that $p_l \in \mathcal{C}(p_k)$ and either $p_i$ or $p_j$ is between $p_k$ and $p_l$ and the same argument can be made with $p_k$ and the point between $p_k$ and $p_l$. □

If $S$ is a set of points in convex position (i.e. they are the vertices of a convex polygon), because $conv\, S \subset ker\, S$, we have:

**Corollary 1** *If $S$ is in convex position, $\mathcal{R}_S(x)$ has at most one local minimum inside $conv\, S$.*

It is worth noticing that Theorem 2 easily gives useful properties about distribution of local minima in practical examples. For instance, if we assume that we have taken a sample of $n$ points which are contained in an annulus of width $\varepsilon$ centered at the origin (we normalize the problem in such a way that the inner radius is one) and denote by $\theta$ the smallest angle between two consecutive points, a straightforward computation shows that the disk centered at the origin with radius

$$\delta = \frac{(1+\varepsilon)\sin\theta}{\sqrt{1+(1+\varepsilon)^2-2(1+\varepsilon)\cos\theta}}$$

is contained in ker $S$ and thus has at most one local minimum. If points are equally distributed and we assume a (huge) 5% error both in the angle and the width, for $n = 360$ we have $\delta \approx 0.33$. Then, this result can be used to explain why standard optimization methods perform well "in practice" to approximate the solution.

In the next section, we are going to see that additional structure of the problem will allow us to design a simple and efficient *discrete* algorithm which can find the *exact* solution to the problem.

# 5 The algorithm

Before stating the key result, we need some additional notation and definitions. Given two points $p_i$ and $p_j$, we denote by $S_{ij}$ the chain between $p_i$ and $p_j$ ordered in positive (counterclockwise) direction as seen from $ker\, S$. We denote by $e_{ij}$ the edge of $\mathcal{V}_f(S)$ contained in the bisector of segment $p_ip_j$. The end-points of $e_{ij}$ will be denoted by $c_{ij}$ (the closest one to $p_i$ and $p_j$) and by $f_{ij}$ (the furthest one). This is well defined except for, at most, one edge of $\mathcal{V}_f(S)$, which can be treated separately.

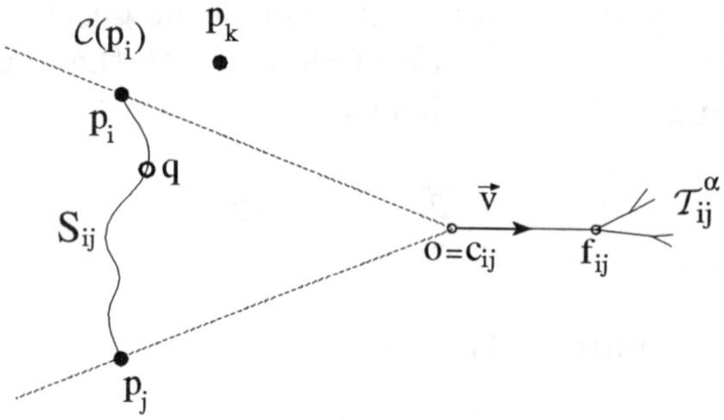


Figure 4: Illustration of Lemma 2.

We observe that $\mathcal{V}_f(S) \cap ker\, S$ is a forest and denote by $\mathcal{T}^1, \ldots, \mathcal{T}^m$ the trees of this forest.

We label the points in such a way that triangle $c_{ij}p_ip_j$ has positive orientation and denote by $\mathcal{T}^\alpha_{ij}$ the subtree of $\mathcal{T}^\alpha$ rooted at $c_{ij}$ containing the edge $e_{ij}$ (see Figure 4). We observe that this subtree is made up of segments of bisectors of points of $S_{ij}$.

**Lemma 2** *If $CN(c_{ij}) \cap S_{ij} = \emptyset$ there is no local minimum of $\mathcal{R}_S(x)$ in $\mathcal{T}^\alpha_{ij}$.*

**Proof:** Let $\vec{v}$ be the unit vector

$$\vec{v} = \frac{\overrightarrow{c_{ij}f_{ij}}}{||\overrightarrow{c_{ij}f_{ij}}||}.$$

We claim that, if $CN(c_{ij}) \cap S_{ij} = \emptyset$ then $CN(c_{ij} + \lambda\vec{v}) \cap S_{ij} = \emptyset$ for any $\lambda > 0$ and, therefore, there is no local minimum on $c_{ij} + \lambda\vec{v} \cap ker\, S$ because furthest and closest neighbors do not alternate angularly. Moreover, the lemma follows from the claim by iterating the argument for the set of edges of $\mathcal{T}^\alpha_{ij}$.

To prove the claim, assume that $c_{ij} = o$ and let $p_k = CN(c_{ij})$. If $q \in S$ and $q \notin \mathcal{C}(p_i)$ then it holds $||\vec{p}_k|| \leq ||\vec{q}||$ and $\vec{v} \cdot \vec{q} < 0$ and thus

$$||\vec{p}_k||\Big(||\vec{p}_k|| - 2\lambda\frac{\vec{v}\cdot\vec{p}_k}{||\vec{p}_k||}\Big) < ||\vec{q}||\Big(||\vec{q}|| - 2\lambda\frac{\vec{v}\cdot\vec{q}}{||\vec{q}||}\Big)$$

which is equivalent to say that

$$\|\lambda\vec{v} - \vec{p}_k\| < \|\lambda\vec{v} - \vec{q}\|$$

probing that $q$ is not the closest neighbor of $\lambda\vec{v}$.

□

Using the previous lemma the algorithm is straightforward: as a first step, we compute $\mathcal{V}_f(S)$, $\mathcal{V}_c(S)$ and preprocess $\mathcal{V}_c(S)$ for point location. This can be done in $O(n \log n)$ time. Because the number of intersections between $\mathcal{V}_f(S)$ and *ker S* is linear, the forest $\mathcal{T}^1, \ldots, \mathcal{T}^m$ can be computed in time $O(n \log n)$ with and standard sweepline algorithm.

Let us denote by $n_\alpha$ the size of $\mathcal{T}^\alpha$. We perform a depth first search in $\mathcal{T}^\alpha$ and locate a vertex with the property that the subtrees that it connects have size smaller than $cn_\alpha$ for $c < 1/2$ a positive constant and locate the closest neighbor of the vertex in $O(\log n)$ time. We apply the previous lemma and discard all the subtrees except the one corresponding to the chain containing the closest neighbor and discard a linear fraction of edges of $\mathcal{T}^\alpha$. Therefore, $\mathcal{T}^\alpha$ is procesed in $O(n_\alpha + \log n \log n_\alpha)$ time obtaining the edge of $\mathcal{T}^\alpha$ containing the local minimum (if such edge exists).

Because the total size of $\mathcal{T}^1 \cup \ldots \cup \mathcal{T}^m$ is $O(n)$, the algorithm runs in $O(n \log n)$ time. Then, we have shown:

**Theorem 3** *If there is a local minimum of $\mathcal{R}_S(x)$ inside ker $S$, it can be computed in $O(n \log n)$ time.*

## 6 Concluding remarks

In this paper we have dealt with the problem of computing the annulus of minimum width containing a set of points. We have characterized the set of local minima of $\mathcal{R}_S(x)$ (feasible centers for the annulus) in the $d$-dimensional case and we have shown that, inside the *kernel* of the set, there is at most one local minimum which can be computed in $O(n \log n)$. We remark that the algorithm is more efficient than previous algorithms solving the general problem and that it is useful to solve the problem in practice, because in the metrology application the sample points are given in angular order. We also remark that the algorithm is simple and, thus, easy to implement. The number of local minima in the whole plane remains as an open problem. It is also an interesting open problem to see whether or not a set of points in convex position in the space has only one local minimum inside its convex hull.

## 7 Acknowledgments

The authors thank Prof. G. Toussaint and C. K. Yap for bringing the problem to our attention during the Bellairs Workshop on Computational Geometry '95, colleagues M. Abellanas, G. Blanco, G. Hernández and F. Gómez for helpful discussions and P. Bose for let us know about reference [8].

## References

[1] P. K. Agarwal, M. Sharir, S. Toledo: Applications of parametric searching in geometric optimization, *Journal of Algorithms*, Vol. 17, pp. 292-318, 1994.

[2] A. Aggarwal, L. J. Guibas, J. Saxe, P. W. Shor: A linear time algorithm for computing the Voronoi diagram of a convex polygon, *Discrete and Computational Geometry*, Vol. 4, pp. 591-604, 1989.

[3] H. Ebara, N. Fukuyama, H. Nakano, Y. Nakanishi: Roundness algorithms using Voronoi diagrams, *Abstracts 1st Canadian Conference on Computational Geometry*, 1989.

[4] H. Edelsbrunner, L. J. Guibas, J. Stolfi: Optimal point location in a monotone subdivision, *SIAM Journal on Computing*, Vol. 15, pp. 317-340, 1986.

[5] L. W. Foster: *GEO-METRICS II: The application of geometric tolerancing techniques.* Addison-Wesley Publishing Co., 1982.

[6] V. B. Lee, D. T. Lee: Out-of-roundness problem revisited. *IEEE Transactions on Pattern Analysis and Machinery Intelligence*, Vol. 13, pp. 217-223, 1991.

[7] F. P. Preparata, M. I. Shamos, *Computational Geometry*, Springer Verlag, 1985.

[8] T. J. Rivlin: Approximation by circles, *Computing* Vol. 21, pp. 93-104, 1979.

[9] U. Roy, X. Zhang: Establishment of a pair of concentric circles with the minimum radial separation for assessing rounding error, *Computer Aided Design*, Vol. 24(3), pp. 161-168, 1992.

[10] T. Shermer, C. K. Yap: Probing for near-centers and approximate relative roundness, June 21-23, 1995. (Invited talk) ASME Workshop on Tolerancing & Metrology, University of North Carolina at Charlotte.

[11] M. Smid, R. Janardan: On the width and roundness of a set of points in the plane, *Proc. 7th. Canadian Conference on Computational Geometry*, pp. 193-198, 1995.

[12] K. Swanson, D. T. Lee, V. L. Wu: An optimal algorithm for roundness determination of convex polygons. *Computational Geometry: Theory and Applications* Vol. 5, pp. 225-235, 1995.

[13] C. K. Yap: Exact Computational Geometry and Tolerancing Metrology, *Snapshots on Computational Geometry*, Vol. 3, D. Avis and J. Bose (Eds.), McGill School of Computer Science, Tech. Rep. No. SOCS-94.50, 1994.

# The Discrete 2-Center Problem*

Pankaj K. Agarwal† Micha Sharir‡ Emo Welzl§

## Abstract

We present an algorithm for computing the *discrete 2-center* of a set $P$ of $n$ points in the plane; that is, computing two congruent disks of smallest possible radius, centered at two points of $P$, whose union covers $P$. Our algorithm runs in time $O(n^{4/3} \log^5 n)$.

## 1 Introduction

**Problem statement and previous results.** Let $P$ be a set of $n$ points in the plane. The *discrete 2-center* problem for $P$ is to cover $P$ by (the union of) two congruent closed disks whose radius is as small as possible, and whose centers are two points of $P$. This is a restricted version of the standard 2-center problem, where the centers of the two covering disks can be any pair of points in the plane. This latter problem has been studied extensively, where the best algorithm, due to Sharir [14] and slightly improved by Eppstein [6], runs in randomized expected $O(n \log^2 n)$ time.

The discrete 2-center problem has been studied in [7], where a near-quadratic algorithm is proposed (such an algorithm is briefly described later in this introduction). Before discussing it further, we note that the discrete 1-center problem, seeking the smallest disk centered at a point of $P$ and containing $P$, is much easier to solve, in time $O(n \log n)$, using the furthest-neighbor Voronoi diagram of $P$. That is, the diagram allows us to find, in $O(n \log n)$ time, the furthest neighbor $f(p)$ of each point $p \in P$. The point $p$ that minimizes the distance between $p$ and $f(p)$ is the center of the desired smallest enclosing disk.

The discrete 2-center appears to be more difficult than the standard 2-center problem. Both problems involve a 'decision procedure' that, given a fixed radius $r$, aims to determine whether $P$ can be covered by two disks of radius $r$. As an informal explanation of the additional difficulty of the discrete 2-center problem, suppose that we have already guessed one center $p$. The second center must then lie in ('pierce') each of the disks of radius $r$ centered at the points of $P$ and not containing $p$. In the standard 2-center problem we simply need to determine whether the intersection of all these disks is nonempty, whereas in the discrete 2-center problem we need to determine whether this intersection contains a point of $P$, which is a harder task.

**Main results and the overall approach.** In this paper we obtain an efficient solution to the discrete 2-center problem that runs in time $O(n^{4/3} \log^5 n)$. This is the first subquadratic algorithm for solving the problem. We note that a near-quadratic solution is rather easy: It suffices to show a near-quadratic solution to the fixed-size problem, and then follow (a simpler version of) the binary-search technique (on the radius $r$) that is described below in Section 5, to find the optimal solution. The fixed-size problem, for a given radius $r$, determines whether there exist $p, q \in P$ so that $P \subset D(p,r) \cup D(q,r)$, where $D(x,r)$ denotes the closed disk of radius $r$ centered at $x$. We try each point $p \in P$ as the first center, obtain the set $N_p \subset P$ of points not contained in $D(p,r)$, and solve the discrete 1-center problem on $N_p$ in $O(n \log n)$ time. The running time of the fixed-size procedure is therefore $O(n^2 \log n)$.

In order to improve the running time of the fixed-size problem, we proceed as follows: For each $p \in P$, let $K_p$ be the intersection of all the disks $D(q,r)$ centered at the points of $P$ and not containing $p$. If any set $K_p$ contains a point $q$ of $P$, then we are done: $p$ and $q$ are centers of two disks of radius $r$ whose union covers $P$. Conversely, if $p, q \in P$ are centers of two such disks, then $p \in K_q$ and $q \in K_p$. In other words, we need to compute the union $U$ of all the $K_p$'s, and determine whether $U \cap P \neq \emptyset$. The difficult step is to compute $U$ in time close to $n^{4/3}$.

---

*Pankaj Agarwal and Micha Sharir have been supported by a joint grant from the U.S.–Israeli Binational Science Foundation. Work by Pankaj Agarwal has also been supported by a National Science Foundation Grant CCR-93-01259, by an Army Research Office MURI grant DAAH04-96-1-0013, by a Sloan fellowship, and by an NYI award and matching funds from Xerox Corporation. Work by Micha Sharir has also been supported by National Science Foundation Grants CCR-94-24398 and CCR-93-11127, and by a grant from the G.I.F. — the German Israeli Foundation for Scientific Research and Development. Work by Micha Sharir and Emo Welzl has been supported by a Max-Planck Research Award.

†Center for Geometric Computing, Department of Computer Science, Box 90129, Duke University, Durham, NC 27708-0129, USA.

‡School of Mathematical Sciences, Tel Aviv University, Tel Aviv 69978, Israel, and Courant Institute of Mathematical Sciences, New York University, New York, NY 10012, USA.

§Institut für Theoretische Informatik, ETH Zürich, CH-8092 Zürich, Switzerland.



*Computational Geometry 97* Nice France

We consider a more general problem: Let $P$ be a set of $m$ points and $\mathcal{D}$ a set of $n$ congruent disks. For each $p \in P$, define $\mathcal{D}_p = \{D \in \mathcal{D} \mid p \notin D\}$, $K_p = \bigcap_{D \in \mathcal{D}_p} D$, $\mathcal{K} = \{K_p \mid p \in P\}$, and $U = \bigcup_{p \in P} K_p$. In Section 2 we present some important properties of $\mathcal{K}$, which we believe to be of independent interest. The main property is that $\mathcal{K}$ is a collection of *convex pseudo-disks*; i.e., these sets are compact and convex, and for any pair $K_p$, $K_q$ of such sets, both $K_p \setminus K_q$ and $K_q \setminus K_p$ are connected.

In Section 3 we show that the combinatorial complexity of $U$ is $O(m^{2/3}n^{2/3}\log^{1/3} n + n \log n)$. While this bound is non-trivial, and 'consistent' with the running time we are aiming at, we have so far been unable to exploit this bound to obtain an alternative simpler solution, of comparable complexity, of the discrete 2-center problem. The reasons for this are technical, and will be noted below.

In Section 4 we present an $O(n^{4/3}\log^4 n)$-time algorithm for computing $U$ (and for testing whether $U \cap P \neq \emptyset$) for the case in which $\mathcal{D} = \{D(p,r) \mid p \in P\}$. The algorithm constructs and searches in $U$ in a semi-implicit manner, using appropriate range-searching data structures and techniques similar to those used in parametric searching, for performing various primitive operations on the semi-implicit representation of $U$. Finally, we describe the overall algorithm for the discrete 2-center in Section 5. As mentioned, its running time is $O(n^{4/3}\log^5 n)$.

## 2 Structure of $\mathcal{K}$

In this section we prove some interesting properties of $\mathcal{K}$. These properties, besides being of independent interest, are crucial for making our algorithm efficient.

**Theorem 2.1** *Let $\mathcal{D}$ be a finite set of congruent disks in the plane, and let $P$ be a finite set of points. Let $\mathcal{K}$ be the same as defined in the introduction. Then $\mathcal{K}$ is a family of convex pseudo-disks; that is, each $K_p$ is a compact convex set, and for each pair of distinct sets $K_p$, $K_q$, both sets $K_p \setminus K_q$ and $K_q \setminus K_p$ are connected.*

We prove the theorem by a sequence of lemmas.

**Lemma 2.2** *For a point $p$, two distinct disks $D_1, D_2 \in \mathcal{D}$ that do not contain $p$, and another disk $D \in \mathcal{D}$ that contains $p$, the set $D \setminus (D_1 \cap D_2)$ is connected.*

**Proof:** Suppose to the contrary that $D \setminus (D_1 \cap D_2)$ is disconnected. Since $p \in D$ and $p \notin D_1, D_2$, all three disks $D, D_1, D_2$ are distinct. Since $D \setminus (D_1 \cap D_2)$ is disconnected, $\partial D$ and $\partial(D_1 \cap D_2)$ must cross at exactly 4 points, all lying on the boundary of $E = D \cap D_1 \cap D_2$. This however is impossible, since the intersection of three congruent disks can have at most three such intersections on its boundary. □

**Corollary 2.3** *For a point $p \in P$ and a disk $D \in \mathcal{D}$ that contains $p$, the set $K_p \setminus D$ is connected.*

**Proof:** Suppose to the contrary that $K_p \setminus D$ is disconnected. Since $D \notin \mathcal{D}_p$, it is distinct from any of the disks that form $K_p$, so any intersection of $\partial D$ with $\partial K_p$ must be a proper crossing. Moreover, since $K_p \setminus D$ is disconnected, the boundaries of $D$ and of $K_p$ must cross at least four times. This, however, implies that $D \setminus K_p$ is also disconnected (this follows from the convexity of $K_p$). But

$$D \setminus K_p = D \setminus \bigcap_{D' \in \mathcal{D}_p} D' = D \cap \bigcup_{D' \in \mathcal{D}_p} (D')^c = \bigcup_{D' \in \mathcal{D}_p} (D \setminus D').$$

If a union of a collection of sets is disconnected, then either one of the sets is disconnected, or there exist two disjoint sets in the collection. In our setting, $D \setminus D'$ is always connected, and the second case contradicts Lemma 2.2 (because for $D', D'' \in \mathcal{D}_p$, $(D \setminus D') \cup (D \setminus D'') = D \setminus (D' \cap D'')$). Hence, $K_p \setminus D$ is connected. □

**Lemma 2.4** *For a point $p$, two distinct disks $D_1, D_2 \in \mathcal{D}$ that do not contain $p$ and two other distinct disks $D_3, D_4 \in \mathcal{D}$ that contain $p$, the set $(D_1 \cap D_2) \setminus (D_3 \cap D_4)$ is connected.*

**Proof:** Suppose again to the contrary that $(D_1 \cap D_2) \setminus (D_3 \cap D_4)$ is disconnected. Lemma 2.2 implies (using the argument in the proof of Corollary 2.3) that $(D_1 \cap D_2) \setminus D_3$ is connected, and so is $(D_1 \cap D_2) \setminus D_4$. It follows that $\partial(D_1 \cap D_2)$ and $\partial D_3$ intersect at most twice, and the same holds for $\partial(D_1 \cap D_2)$ and $\partial D_4$. These conditions, along with our assumption, imply that $\partial(D_1 \cap D_2)$ and $\partial(D_3 \cap D_4)$ intersect exactly 4 times. Moreover, put $E = D_1 \cap D_2 \cap D_3 \cap D_4$, fix a point $o$ in the interior of $E$, and consider the boundaries of $D_1, \ldots, D_4$ and $E$ as graphs of functions $r = D_1(\theta), \ldots, r = E(\theta)$ in polar coordinates about $o$. Let $u, v, w, z$ be the four points of intersection between $\partial(D_1 \cap D_2)$ and $\partial(D_3 \cap D_4)$, appearing in this circular counterclockwise order along $\partial E$. Let $\theta_u < \theta_v < \theta_w < \theta_z$ be the polar orientations of $u, v, w, z$, respectively. Since $D_1, \ldots, D_4$ are congruent disks, each $\partial D_i$ appears along $\partial E$ in a single connected arc. Hence, with no loss of generality, we may assume that $\partial E$ is attained by $\partial D_1$ over $[\theta_u, \theta_v]$, by $\partial D_3$ over $[\theta_v, \theta_w]$, by $\partial D_2$ over $[\theta_w, \theta_z]$, and by $\partial D_4$ over $[\theta_z, \theta_u]$. See Figure 1. Let $\theta_p$ be the polar

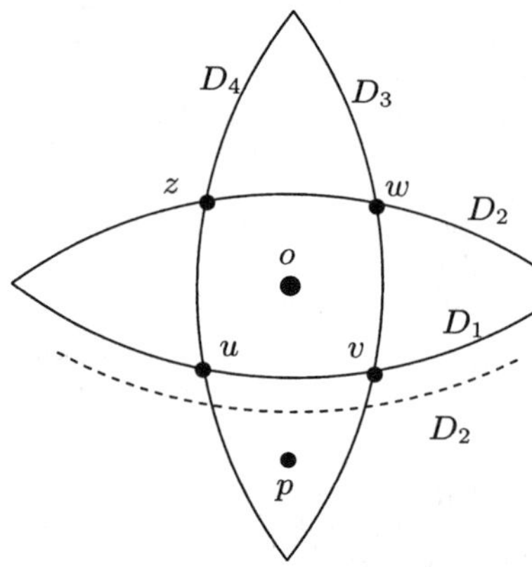


Figure 1: The proof of Lemma 2.4

orientation of $\vec{op}$. It is impossible that $\theta_p$ lies in $[\theta_v, \theta_w]$, for otherwise, since $p \in D_3$, we have $|op| \leq D_3(\theta_p) \leq D_1(\theta_p)$, implying that $p \in D_1$, contrary to assumption. Similarly, $\theta_p$ cannot lie in $[\theta_z, \theta_u]$. (We use the notation $[\theta, \theta']$ to denote the angular interval extending counterclockwise from $\theta$ to $\theta'$). Suppose then that $\theta_p \in [\theta_u, \theta_v]$. Let $F = D_3 \cap D_4$. Since $p \in F$ but $p \notin D_2$, we have $D_2(\theta_p) < F(\theta_p)$, and this inequality is reversed over the intervals $[\theta_v, \theta_w]$ and $[\theta_z, \theta_u]$. It follows that $\partial F$ and $\partial D_2$ intersect at least twice over the interval $[\theta_u, \theta_v]$, which, together with $w$ and $z$, yields 4 points of intersection between these boundaries, all lying along $\partial(D_2 \cap D_3 \cap D_4)$, which is impossible for congruent disks (cf. the proof of Lemma 2.2). A similar contradiction occurs when $\theta_p \in [\theta_w, \theta_z]$. All these contradictions establish the lemma. □

Following the same argument as in the proof of Corollary 2.3, we obtain:

**Corollary 2.5** *For a point $p \in P$ and two disks $D, D' \in \mathcal{D}$ that contain $p$, the set $K_p \setminus (D \cap D')$ is connected.*

We are now ready to prove Theorem 2.1.

**Proof of Theorem 2.1:** Suppose to the contrary that there exist $p, q \in P$ such that $K_p \setminus K_q$ is disconnected. As in the proof of Corollary 2.3, we can express $K_p \setminus K_q$ as

$$\begin{aligned} K_p \setminus K_q &= K_p \setminus \bigcap_{D \in \mathcal{D}_q} D = K_p \cap \bigcup_{D \in \mathcal{D}_q} D^c \\ &= \bigcup_{D \in \mathcal{D}_q} (K_p \setminus D) = \bigcup_{D \in \mathcal{D}_q \setminus \mathcal{D}_p} (K_p \setminus D). \end{aligned}$$

(The last equality follows from the fact that the disks in $\mathcal{D}_q \cap \mathcal{D}_p$ contribute empty sets to this union.) Again, as in the proof of Corollary 2.3, if $K_p \setminus K_q$ is disconnected, then either there exists a disk $D \in \mathcal{D}_q \setminus \mathcal{D}_p$ so that $K_p \setminus D$ is disconnected, or there exists two disks $D_1, D_2 \in \mathcal{D}_q \setminus \mathcal{D}_p$ so that $K_p \setminus (D_1 \cap D_2)$ is disconnected. The first condition contradicts Corollary 2.3 and the second contradicts Corollary 2.5. Hence, $K_p \setminus K_q$ is connected (and so is $K_q \setminus K_p$). □

Theorem 2.1 fails for non-congruent disks, as is illustrated in Figure 2. Nevertheless, the following variant of the theorem holds in even more generality:

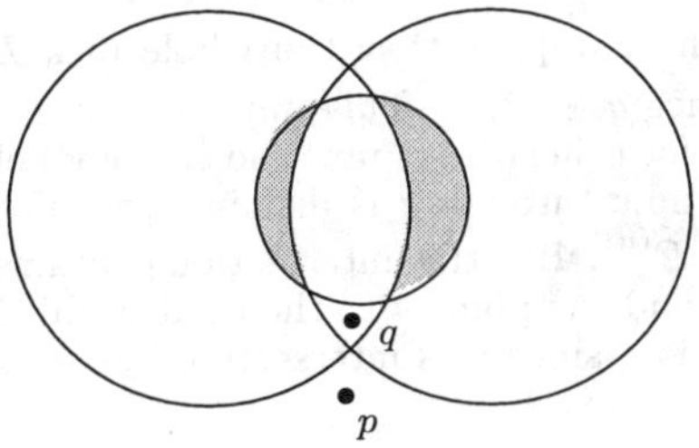


Figure 2: $K_q \setminus K_p$ (the shaded region) consists of two connected components.

**Theorem 2.6** *Let $\mathcal{D}$ be a finite set of convex pseudo-disks in the plane; that is, each $D \in \mathcal{D}$ is a compact convex set, and for each pair of distinct sets $D, D' \in \mathcal{D}$, both sets $D \setminus D'$ and $D' \setminus D$ are connected; we also assume that $\partial D$ and $\partial D'$ cross each other transversally at any point of intersection. Let $P$ be a finite set of points. For each $p \in P$, let $\mathcal{D}_p$ denote the set of pseudo-disks in $\mathcal{D}$ that do not contain $p$, and let $K_p$ denote their intersection. Then, for any $p, q \in P$, $\partial K_p$ and $\partial K_q$ can cross each other at most twice.*

(Note that for the sets $K_p$ and $K_q$ in Figure 2, their boundaries do not cross at all.)

**Proof:** We partition $\mathcal{D}$ into four subsets: the subset $\mathcal{D}_p \cap \mathcal{D}_q$ of pseudo-disks that contain neither $p$ nor $q$, the subset $\mathcal{D}_p \setminus \mathcal{D}_q$ of pseudo-disks that contain $q$ but not $p$, the subset $\mathcal{D}_q \setminus \mathcal{D}_p$ of pseudo-disks that contain $p$ but not $q$, and the subset $\mathcal{D} \setminus \mathcal{D}_p \setminus \mathcal{D}_q$ of pseudo-disks that contain both $p$ and $q$. We can ignore the last subset since the pseudo-disks in this set have no effect on $K_p$ or $K_q$. Let $\mathcal{I} = \bigcap(\mathcal{D}_p \cap \mathcal{D}_q)$. Clearly, both $K_p$ and $K_q$ are contained in $\mathcal{I}$, so any crossing between their boundaries must be interior to $\mathcal{I}$. In particular, if such a crossing occurs between a pseudo-disk $D \in \mathcal{D}_p$ and a pseudo-disk $D' \in \mathcal{D}_q$, then we must have $D \in \mathcal{D}_p \setminus \mathcal{D}_q$ and $D' \in \mathcal{D}_q \setminus \mathcal{D}_p$ (that is, $q \in D$ and $p \in D'$).

Now suppose that $\partial K_p$ and $\partial K_q$ cross each other three times, at points $u$, $v$, and $w$. By the above argument, there exist six (not necessarily all distinct) pseudo-disks, $D_u^{(p)}$, $D_u^{(q)}$, $D_v^{(p)}$, $D_v^{(q)}$, $D_w^{(p)}$, $D_w^{(q)}$, such that $D_u^{(p)}$, $D_v^{(p)}$, and $D_w^{(p)}$ are in $\mathcal{D}_p \setminus \mathcal{D}_q$, $D_u^{(q)}$, $D_v^{(q)}$, and $D_w^{(q)}$ are in $\mathcal{D}_q \setminus \mathcal{D}_p$, $D_u^{(p)}$ and $D_u^{(q)}$ cross at $u$, $D_v^{(p)}$ and $D_v^{(q)}$ cross at $v$, and $D_w^{(p)}$ and $D_w^{(q)}$ cross at $w$.

Let $o$ be a point in the interior of $K_p \cap K_q$. There must exist two of the crossing points, say $u$ and $v$, such that $p$ and $q$ appear between $u$ and $v$ in counterclockwise angular order about $o$. Without loss of generality, assume that $u$, $p$, $q$, and $v$ appear in this counterclockwise order about $o$, and let $\theta_u < \theta_p < \theta_q < \theta_v$ be the orientations of the vectors $\vec{ou}$, $\vec{op}$, $\vec{oq}$, and $\vec{ov}$, respectively.

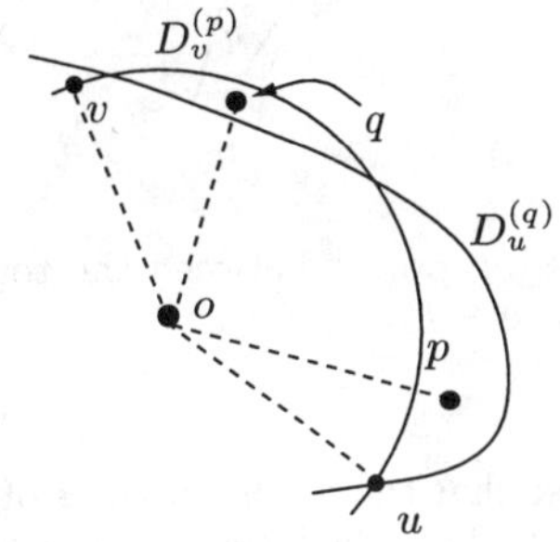


Figure 3: Proof of Theorem 2.6.

Now consider the two pseudo-disks $D_u^{(q)}$ and $D_v^{(p)}$, and regard their boundaries as functions $r = D_u^{(q)}(\theta)$ and $r = D_v^{(p)}(\theta)$ in polar coordinates about $o$. Then we have (see Figure 3):

$$\begin{aligned} D_u^{(q)}(\theta_u) &\le D_v^{(p)}(\theta_u) \\ D_u^{(q)}(\theta_p) &> D_v^{(p)}(\theta_p) \\ D_u^{(q)}(\theta_q) &< D_v^{(p)}(\theta_q) \\ D_u^{(q)}(\theta_v) &\ge D_v^{(p)}(\theta_v). \end{aligned}$$

These inequalities follow from the convexity of $K_p$ and $K_q$, from the fact that $u$ and $v$ lie on their boundaries, and from the fact that $D_v^{(p)} \in \mathcal{D}_p \setminus \mathcal{D}_q$ and $D_u^{(q)} \in \mathcal{D}_q \setminus \mathcal{D}_p$. However, this implies that $D_u^{(q)}$ and $D_v^{(p)}$ intersect at least three times, contradicting the assumption that $\mathcal{D}$ is a set of pseudo-disks. This completes the proof. □

The following corollary is an immediate consequence of the results of [10]:

**Corollary 2.7** *If $P$ has $m$ points, then the boundary of $\bigcup_{p \in P} K_p$ consists of $O(m)$ connected portions of the boundaries of the individual $K_p$'s.*

We now return to the assumption that $\mathcal{D}$ is a set of congruent disks. For a point $p \in P$, we say that $p$ lies *above* (resp. *below*) $K_p$ if the downward-directed (resp. upward-directed) vertical ray from $p$ intersects $K_p$, and $p$ lies to the *left* (resp. *right*) of $K_p$ if $p$ and $K_p$ are separated by a vertical line, so that $p$ lies to the left (resp. right) of the line. To facilitate our solution to the fixed-size decision problem, presented in Section 4, we need the following stronger property of the $K_p$'s: Define

$$\begin{aligned} P_T &= \{p \in P \mid p \text{ lies above } K_p\}, \\ P_B &= \{p \in P \mid p \text{ lies below } K_p\}, \\ P_L &= \{p \in P \mid p \text{ lies to the left of } K_p\}, \\ P_R &= \{p \in P \mid p \text{ lies to the right of } K_p\}. \end{aligned}$$

**Theorem 2.8** *Let $p, q$ be two distinct points in $P_T$. Then the top boundaries of $K_p$ and $K_q$ cross at most once, and the same holds for the bottom boundaries. The same properties hold for each of the other three sets $P_B$, $P_L$, and $P_R$.*

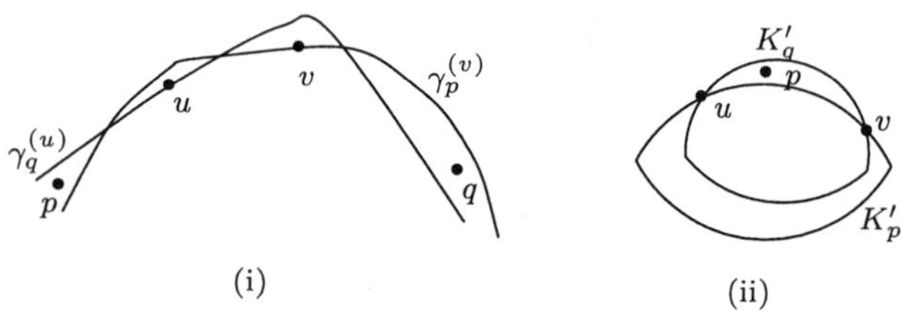


Figure 4: Crossing points between the top boundaries: (i) $p, q \in P_T$; (ii) $p, q \in P_B$.

**Proof:** Suppose that the top boundaries of $K_p$ and $K_q$ cross at two points $u$ and $v$. The intersection $u$ must be witnessed by two disks $D_p^{(u)}$ and $D_q^{(u)}$ with $u$ on the top boundaries of these disks, $D_p^{(u)} \cap \{p, q\} = \{q\}$ and $D_q^{(u)} \cap \{p, q\} = \{p\}$. Similarly, there exist witness disks $D_p^{(v)}$ and $D_q^{(v)}$, with similar properties, for the intersection $v$.

Let us first prove that two such intersections are not possible if $p, q \in P_T$.

We call the top boundary of a disk $D$ extended by vertical rays downwards at its endpoints the *top curve of $D$*. Since we are dealing with disks of equal radius, the top curves of $D_p^{(u)}$ and $D_q^{(u)}$ intersect in exactly one point (they have to intersect, since otherwise $K_p$ and $K_q$ are disjoint). Since $p$ is above $D_p^{(u)}$, and not above $D_q^{(u)}$, and vice versa for $q$, the $x$-coordinate of this unique intersection has to lie between the $x$-coordinates of $p$ and $q$. So we have shown that the $x$-coordinate of $u$ has to lie between $p$ and $q$, and the same is true for $v$. We may assume that the $x$-coordinates of $p$, $u$, $v$, and $q$ appear in this increasing order.

Now consider the top curves $\gamma_p^{(v)}$ and $\gamma_q^{(u)}$ of disks $D_p^{(v)}$ and $D_q^{(u)}$, respectively, and refer to Figure 4 (i). The curve $\gamma_p^{(v)}$ lies below $p$ (since $p \notin D_p^{(v)}$ and $p$ lies above $K_p$), lies above $u$ or passes through $u$ (otherwise $u$ cannot lie on the boundary of $K_p$), passes through $v$, and lies above $q$ (since $q \in D_p^{(v)}$). The curve $\gamma_q^{(u)}$ lies above $p$, passes through $u$, continues above $v$ and lies below $q$. It follows that the two curves switch sides three times: between $p$ and $u$, between $u$ and $v$, and between $v$ and $q$. (This also covers the case where, say, $\gamma_p^{(v)}$ passes through $u$, because the curves $\gamma_p^{(v)}$ and $\gamma_q^{(u)}$ must *cross* at this point, as is easily verified.) This gives three intersections of these curves, a contradiction, which concludes the argument for the case in which $p, q \in P_T$.

Suppose next that $p, q \in P_B$. Let $D_p^{(u)}$, $D_q^{(u)}$, $D_p^{(v)}$, and $D_q^{(v)}$ be four respective witness disks, defined as above. We exploit now the previously proved fact that $K_p' := D_p^{(v)} \cap D_p^{(u)}$ and $K_q' := D_q^{(v)} \cap D_q^{(u)}$ behave like pseudo-disks and thus their boundaries do not cross at any point other than $u$ and $v$. Let us assume that the top boundary of $K_q'$ lies above the top boundary of $K_p'$ in the range between the $x$-coordinates of $u$ and $v$. Now recall that $p$ must lie in $K_q' \setminus K_p'$, and thus it lies below the top boundary of $K_q'$ and *above* the top boundary of $K_p'$, in contradiction to the fact that $p$ lies above $K_p$ which is contained in $K_p'$; see Figure 4 (ii).

We now switch to the case of $P_L$, where $p$ and $q$ lie to the left of their regions $K_p$ and $K_q$. Without loss of generality, suppose that $p$ lies to the left of $q$. Since $K_q$ is to the right of $q$, any intersection of the boundaries of $K_p$ and $K_q$ must lie to the right of $q$. So let us assume that two such intersections $u$ and $v$ exist, both between the top boundaries of $K_p$ and $K_q$, and that the $x$-coordinates of $p$, $q$, $u$, and $v$ appear in this increasing order. Let $D_p^{(u)}$, $D_q^{(u)}$, $D_p^{(v)}$, and $D_q^{(v)}$ be four respective witness disks, defined as above.

First consider the top boundary of the disk $D_q^{(v)}$. It must lie above $p$ and $u$ and pass through $v$. Let us assume that it lies below $q$ (which we will lead to a contradiction). The top boundary of $D_p^{(u)}$ lies above $q$, goes through $u$ and lies above (or passes through) $v$, and so it must intersect the top boundary of $D_q^{(v)}$ twice, once between $q$ and $u$, and once between $u$ and $v$ (Figure 5 (i)); a contradiction (as in a preceding argument, this also covers the case where $\partial D_q^{(v)}$ passes through $u$). So the top boundary of $D_q^{(v)}$ must lie above $q$, which implies that the whole disk $D_q^{(v)}$ must lie above $q$, since $q \notin D_q^{(v)}$. This implies that $D_q^{(u)}$ must also lie above $q$, for otherwise it must lie entirely below $q$, and so the vertical line through $q$ is disjoint from the intersection of $D_q^{(v)}$ and $D_q^{(u)}$. But this intersection contains $p$ to the left of this line, and the point $u$ to the right of this line, which is a contradiction, since this intersection has to be connected.

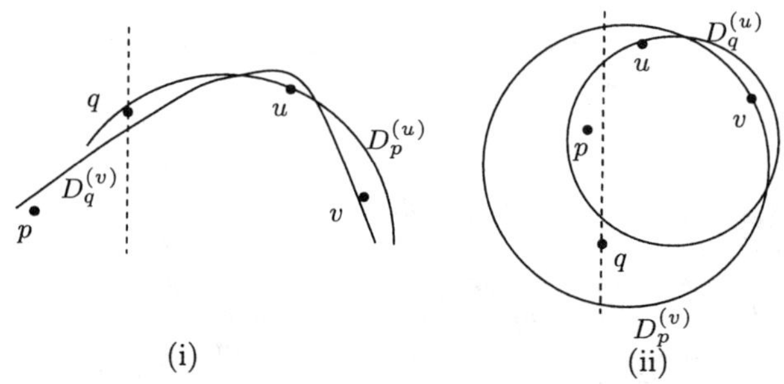


Figure 5: The proof of $p, q \in P_L$: (i) $D_q^{(v)}$ lies below $q$; (ii) the other situation.

Now we investigate the interplay between $D_q^{(u)}$ and $D_p^{(v)}$. Their top boundaries intersect between $u$ and $v$. Since the top boundaries of two congruent disks intersect at most once, the top boundary of $D_p^{(v)}$ lies above the top boundary of $D_q^{(u)}$ at the $x$-coordinate of $q$, which in turn lies above $q$, as just proved. Now the bottom boundary of $D_q^{(u)}$ must lie above $q$, while the bottom boundary of $D_p^{(v)}$ must lie below (one disk must not contain $q$, the other has to). So either the boundaries of these two disks intersect twice to the left of $q$ (which yields at least three intersections between these boundaries, including the one between $u$ and $v$), or they do not intersect there, which implies that $D_p^{(v)}$ contains $D_q^{(u)}$ to the left of the vertical line through $q$; see Figure 5 (ii). But $p \in D_q^{(u)}$, which implies $p \in D_p^{(v)}$; a contradiction.

The cases of bottom boundaries, and of $P_R$, are symmetric, which concludes the proof of the theorem. □

**Remark:** Two top boundaries (or two bottom boundaries) of two sets $K_p$, $K_q$, for two points $p, q$ in, say $P_T$, may also interact in somewhat more involved manners. First, we can have a *weak crossing* between two such top boundaries, in which the two boundaries have an overlapping portion, so that the top portion of $\partial K_p$ lies below the top portion of $\partial K_q$ to the left of the overlap, lies above it to the right of the overlap. See Figure 6(i) for an illustration. Another possibility is that these top boundaries meet twice, without crossing, and overlap between these two meeting points, as is illustrated in Figure 6(ii). Situations of the second type will not affect our algorithm, and we will have to exercise some care to accommodate situations of the first type in the algorithm; see Section 4 for details.

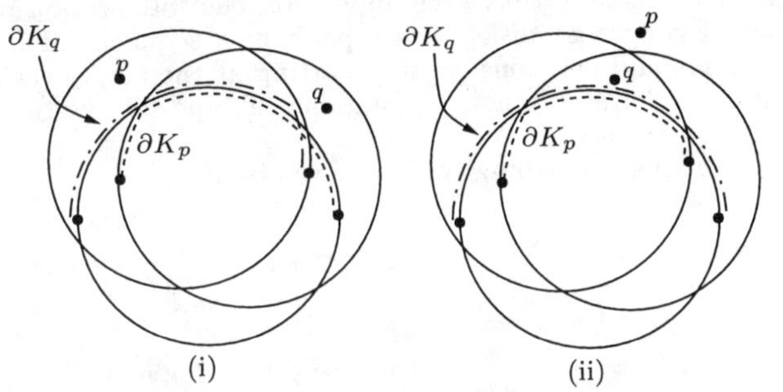


Figure 6: Counter-example for Theorem 2.8 if we just consider intersection points (including noncrossings).

## 3 Complexity of $\mathcal{K}$

Let $\mathcal{D}$ be a set of $n$ congruent disks and $P$ a set of $m$ points in the plane. Let $\mathcal{K}$ and $U$ be the same as defined in the introduction. In this section we obtain a bound on the combinatorial complexity of $\mathcal{K}$, which is defined as follows. Let $V(\mathcal{D}, P)$ be the set of intersection points of disks in $\mathcal{D}$ that lie on the boundary of some $K_p$. Set $\kappa(\mathcal{D}, P) = |V(\mathcal{D}, P)|$ and $\kappa(n, m) = \max \kappa(\mathcal{D}, P)$, where the maximum is taken over all sets of $n$ congruent disks and over all sets of $m$ points in the plane. Note that if a vertex appears on the boundaries of sets in $\mathcal{K}$, we count it only once. If we count the vertices with multiplicity, then $\kappa(n, m) \geq mn$ — take $n$ congruent disks all of whose boundaries appear on their common intersection and choose $m$ points in their common exterior. The main result of this section is the following theorem.

**Theorem 3.1** $\kappa(n, m) = O(m^{2/3} n^{2/3} \log^{1/3} n + n \log n)$.

The proof of the theorem is based on the random-sampling technique, and proceeds along the same lines as the proof by Clarkson et al. [4] for the bound on the complexity of many faces in an arrangement of lines in the plane. We first prove a technical lemma and a weaker bound on $\kappa(n, m)$, and then prove the theorem.

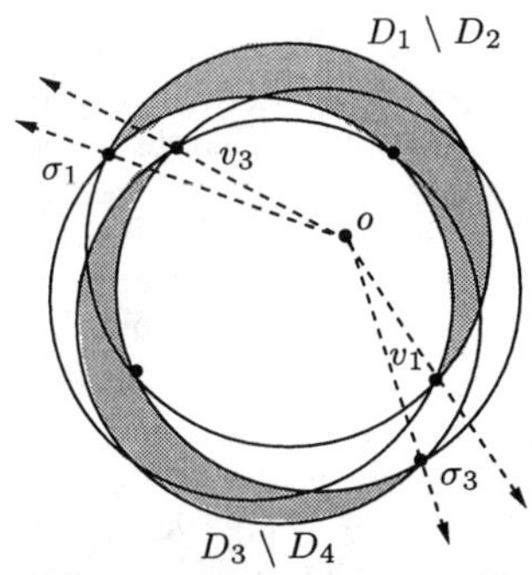


Figure 7: Proof of Lemma 3.2.

**Lemma 3.2** *Let $D_1, D_2, \ldots, D_k$ be a set of congruent disks, all of whose boundaries appear on their common intersection $\mathcal{I}$. Assume that $\partial D1, \partial D_2, \ldots, \partial D_k$ appear in this clockwise order along $\partial\mathcal{I}$. Then the sets $D_i \setminus D_{i+1}$, for $1 \leq i \leq k$ (where we put $D_{k+1} = D_1$), are disjoint, and the same holds for the sets $D_{i+1} \setminus D_i$.*

**Proof:** Suppose that there exist a pair of indices $1 \leq i < j \leq k$ so that $D_i \setminus D_{i+1}$ and $D_j \setminus D_{j+1}$ intersect. Note that $j$ must be at least $i+2$, and $i$ must be at least $j+2-k$; without loss of generality, we can assume that $i=1$, $j=3$, and $k \geq 4$. Consider the arrangement $\mathcal{A}(\{D_1, D_2, D_3, D_4\})$, and let $\mathcal{I}' = \bigcap_{i=1}^4 D_i$. Let us assume that the origin, $o$, lies in the interior of $\mathcal{I}'$. Let $v_i$ be the (unique) intersection point of $\partial D_i$ and $\partial D_{i+1}$ that appears on $\mathcal{I}'$, and let $\sigma_i$ be the other intersection point of these circles. Let $\theta_i$ (resp. $\alpha_i$) denote the orientation of $v_i$ (resp. $\sigma_i$). We will regard $\partial D_i$ as the graph of a univariate function $D_i(\theta)$ in polar coordinates. We will denote by $(\theta_1, \theta_2)$ the counterclockwise circular interval from $\theta_1$ to $\theta_2$.

By construction, $D_1(\theta) > D_2(\theta)$ for $\theta \in (\theta_1, \alpha_1)$, and therefore $D_1 \setminus D_2$ is nonempty only for $\theta \in (\theta_1, \alpha_1)$. Similarly, $D_3(\theta) > D_4(\theta)$ for $\theta \in (\theta_3, \alpha_3)$ and $D_3 \setminus D_4$ is nonempty only for $\theta \in (\theta_3, \alpha_3)$. Since $\partial D_1, \partial D_2, \partial D_3$, and $\partial D_4$ appear in this counterclockwise order along $\partial\mathcal{I}'$, it follows that $\theta_2 \in (\theta_1, \alpha_1)$ and $\theta_4 \in (\theta_3, \alpha_3)$, and that $\alpha_1 \in (\theta_2, \theta_4)$ and $\alpha_3 \in (\theta_4, \theta_2)$.

Let $\xi$ be a point in $(D_1 \setminus D_2) \cap (D_3 \setminus D_4)$, and let $\eta$ be the orientation of $\xi$. Then we have $\min\{D_3(\eta), D_1(\eta)\} > \max\{D_4(\eta), D_2(\eta)\}$. Moreover, $\eta \in (\theta_1, \alpha_1) \cap (\theta_3, \alpha_3)$. The order relationships noted at the preceding paragraph are easily seen to imply that only the following two cases can arise:

*Case (i)* $\eta \in (\theta_3, \alpha_1) \subseteq (\theta_3, \theta_4)$:

*Case (ii)* $\eta \in (\theta_1, \alpha_3) \subseteq (\theta_1, \theta_2)$:

In case (i), $\partial D_2$ appears along $\partial(D_1 \cap D_2 \cap D_3)$ in at least two disjoint arcs—the arc with angular range $(\theta_1, \theta_2)$ and another arc containing a point at orientation $\eta$ (recall that $\partial D_2$ does not appear on the boundary of the intersection in the angular range $(\theta_2, \theta_3)$). This however is impossible for congruent disks. Symmetrically, in case (ii), $\partial D_4$ appears along $\partial(D_1 \cap D_3 \cap D_4)$ in at least two disjoint arcs—the arc with angular range $(\theta_3, \theta_4)$ and another arc containing a point at orientation $\eta$. These two contradictions complete the proof that the regions $D_i \setminus D_{i+1}$, for $1 \leq i \leq k$, are disjoint. A symmetric argument shows that the regions $D_i \setminus D_{i-1}$ are also disjoint. This completes the proof of the lemma. □

**Lemma 3.3** *For $m, n \geq 1$, $\kappa(n,m) = O(m\sqrt{n} + n)$.*

**Proof:** Let $\mathcal{D}$ be a set of $n$ congruent disks and $P$ a set of $m$ points in the plane. It suffices to prove that $\kappa(n,m) = O(m^2 + n)$. Using a standard batching technique, the bound can be rewritten as $O(m\sqrt{n} + n)$; see e.g. [4].

We partition $\mathcal{D}$ into maximal subsets $\mathcal{D}_1, \ldots, \mathcal{D}_k$, so that all disks within each $\mathcal{D}_i$ contain the same subset of $P$. Let $V_i$ be the set of vertices on the boundary of $\bigcap \mathcal{D}_i$. Obviously $\sum_{i=1}^{k} |V_i| \leq \sum_i |\mathcal{D}_i| = n$. We partition the vertices in $V(\mathcal{D}, P)$ into two subsets $A$ and $B$, where $A$ consists of those vertices $v$ for which the two disks on whose boundaries $v$ lies belong to the same $D_i$, and $B$ consists of those vertices whose two associated disks belong to different $D_i$'s. Obviously $|A| \leq \sum_{i=1}^{k} |V_i| \leq n$. We next bound $|B|$.

For each point $p \in P$, let $B_p \subseteq B$ be the set of vertices in $B$ that appear on the boundary of $K_p$. Suppose $v \in B_p$ is an intersection point of the circles bounding two disks $D_1$ and $D_2$. Since $D_1$ and $D_2$ belong to two different subsets, the symmetric difference $D_1 \oplus D_2$ contains at least one point $q$ of $P$. We charge $v$ to $q$. If $v'$ is another vertex of $B_p$, which is an intersection point of the circles bounding two disks $D_3$ and $D_4$, then, by Lemma 3.2, $D_1 \setminus D_2$ and $D_3 \setminus D_4$ are disjoint, and the same holds for $D_2 \setminus D_1$ and $D_4 \setminus D_3$. Hence, each point $q \in P$ can be charged at most twice (once for lying in some $D_i \setminus D_{i+1}$ and once for lying in some $D_j \setminus D_{j-1}$, where $(D_1, D_2, \ldots)$ is the circular sequence of disks whose boundaries appear in this counterclockwise order along $\partial K_p$), thereby implying that $|B_p| \leq 2m$. Summing over all points $p \in P$, we obtain that $|B| \leq 2m^2$, and therefore

$$\kappa(\mathcal{D}, P) \leq 2m^2 + n,$$

as asserted. □

**Proof of Theorem 3.1.** Let $r \geq 1$ be a fixed parameter, to be specified later. We choose a random subset $R \subseteq \mathcal{D}$ of size $r$, where each subset of size $r$ is chosen with equal probability, and compute the *vertical decomposition* $\mathcal{A}^*(R)$ of the arrangement $\mathcal{A}(R)$ [2, 4]. For each cell $\Delta \in \mathcal{A}^*(R)$, let $\mathcal{D}_\Delta \subseteq \mathcal{D}$ be the set of disks whose boundaries intersect $\Delta$ (including the boundary of $\Delta$), let $E_\Delta \subseteq \mathcal{D}$ be the set of disks that are disjoint from $\Delta$, and let $P_\Delta \subseteq P$ be the set of points that lie in $\Delta$. Put $m_\Delta = |P_\Delta|$ and $n_\Delta = |\mathcal{D}_\Delta|$. We denote by $\mathcal{I}_\Delta$ the common intersection of the disks in $E_\Delta$.

Let $v$ be a vertex of $K_p$, for some $p \in P_\Delta$. Suppose that $v$ is an intersection point of the boundaries of two disks $D$ and $D'$. Since, by definition, none of these disks can fully contain $\Delta$, we can classify $v$ into three categories:

(i) Both $D$ and $D'$ belong to $\mathcal{D}_\Delta$,

(ii) $D \in \mathcal{D}_\Delta$ and $D' \in E_\Delta$ (or vice versa), or

(iii) both $D, D' \in E_\Delta$.

A vertex of type (i) is also a vertex of $V(\mathcal{D}_\Delta, P_\Delta)$, so the number of such vertices is at most $\kappa(\mathcal{D}_\Delta, P_\Delta) \leq \kappa(n_\Delta, m_\Delta)$. Since $E_\Delta \subseteq \mathcal{D}_p$ for every $p \in P_\Delta$, every vertex of type (ii) lies on the boundary of $\mathcal{I}_\Delta$. The boundary of each disk in $\mathcal{D}_\Delta$ intersects $\mathcal{I}_\Delta$ in at most two points, so the number of type (ii) vertices is at most $2n_\Delta$. Summing over all cells, the number of type (i) and type (ii) vertices is $\sum_{\Delta \in \mathcal{A}^*(R)} O(n_\Delta + \kappa(n_\Delta, m_\Delta))$.

Finally, each vertex of type (iii) is a vertex of $\mathcal{I}_\Delta$. Hence, in order to bound the number of (distinct) vertices of type (iii), we need an upper bound on the total number of distinct vertices of all the $\mathcal{I}_\Delta$'s, over all cells $\Delta \in \mathcal{A}^*(R)$. Let $G$ be the graph dual to $\mathcal{A}^*(R)$, that is, each node of $G$ corresponds to a cell of $\mathcal{A}^*(R)$, and two nodes corresponding to cells $\Delta, \Delta'$ are connected by an edge if the boundaries of $\Delta$ and $\Delta'$ overlap along (a portion of) an edge. We compute a path $\Pi$ in $G$ that visits each node of $G$ at least once and at most 4 times. The existence of such a path was proved in [2]. We traverse $\Pi$, and at each node corresponding to a cell $\Delta$, we maintain $\mathcal{I}_\Delta$, as follows. When we move from a node corresponding to $\Delta$ to the next node in $\Pi$, corresponding to a cell $\Delta' \in \mathcal{A}^*(R)$, we delete all the disks of $\mathcal{D}_\Delta \setminus \mathcal{D}_{\Delta'}$ from the intersection, and insert the disks of $\mathcal{D}_{\Delta'} \setminus \mathcal{D}_\Delta$ into the intersection. Since $E_\Delta \cup E_{\Delta'} \subseteq \mathcal{D}_\Delta \cup \mathcal{D}_{\Delta'}$, we now have the set $E'_{\Delta'}$. We thus perform at most $n_\Delta + n_{\Delta'}$ insertions and deletions as we move from one node of $\Pi$ to another. Summing over all nodes of $\Pi$, we perform $O(\sum_\Delta n_\Delta)$ insertions and deletions. We wish to bound the number of distinct vertices that ever appear on the intersection, we we traverse $\Pi$. Tamir [15] (see also Agarwal [1]) has shown that the number of distinct vertices that ever appear on the intersection of halfplanes, as we perform a mixed sequence of $k$ insertions and deletions (starting at the empty set), is $O(k \log k)$. Using the same argument, we can show that the total number distinct vertices that ever appear on the intersection of a set of congruent disks, as we perform a sequence of $k$ insertions and deletions (again, starting at the empty set), is also $O(k \log k)$. Hence, the number of type (iii) vertices is $O(\sum_\Delta n_\Delta \log n)$.

Putting everything together, we obtain

$$\begin{aligned}\kappa(\mathcal{D}, P) &\leq \sum_{\Delta \in \mathcal{A}^*(R)} \kappa(\mathcal{D}_\Delta, P_\Delta) + O\Big(\sum_{\Delta \in \mathcal{A}^*(R)} n_\Delta \log n\Big) \\ &= O\Big(\sum_{\Delta \in \mathcal{A}^*(R)} (m_\Delta \sqrt{n_\Delta} + n_\Delta \log n)\Big).\end{aligned}$$

Since $R$ is a random subset of $\mathcal{D}$, the random-sampling argument of Clarkson and Shor [5] implies that there exists $R$ for which

$$\begin{aligned}\sum_{\Delta \in \mathcal{A}^*(R)} n_\Delta &= O(nr), \\ \sum_{\Delta \in \mathcal{A}^*(R)} m_\Delta \sqrt{n_\Delta} &= O\left(m\sqrt{\frac{n}{r}}\right).\end{aligned}$$

Substituting these values and choosing $r = \left\lceil \frac{m^{2/3}}{n^{1/3} \log^{2/3} n} \right\rceil$, we obtain

$$K(n,m) = O(m^{2/3} n^{2/3} \log^{1/3} n + n \log n).$$

This completes the proof of the theorem. □

An immediate consequence of Theorem 3.1 and Corollary 2.7 is the following.

**Corollary 3.4** *The complexity of $U$ is $O(n^{4/3} \log^{1/3} n)$.*

Unfortunately, we have not been able to exploit this bound to obtain an efficient algorithm, of comparable complexity, that computes $U$ explicitly. The results of this section, although of interest in their own right, are not needed for the analysis of the algorithm that we prsent in the next two sections.

## 4 The Decision Algorithm

Let $P$ be a set of $n$ points in the plane, and let $\mathcal{D} = \{D(p,r) \mid p \in P\}$. Let $\mathcal{K}$ and $U$ be the same as defined in the introduction. We describe an $O(n^{4/3}\log^4 n)$-time algorithm to determine whether $U \cap P \neq \emptyset$. Our strategy is to construct separately each of the four sub-unions $U_T = \bigcup_{p\in P_T} K_p$, $U_B = \bigcup_{p\in P_B} K_p$, $U_L = \bigcup_{p\in P_L} K_p$, and $U_R = \bigcup_{p\in P_R} K_p$, and to test whether any of them contains any point of $P$. We next describe in detail the construction of one such sub-union, say $U_T$. As already mentioned, we do not know how to compute these unions efficiently in an explicit manner (for instance, it may be too expensive to construct each $K_p$ explicitly, for all $p \in P$). We will therefore represent the $K_p$'s and their unions implicitly; this implicit representation will be sufficient to determine whether $U_T \cap P \neq \emptyset$.

### 4.1 Representation of $\mathcal{K}$ and of its union

For a subset $A \subseteq P$, let $U^A$ denote the union $\bigcup_{p\in A} K_p$. For each connected component of $\partial U^A$, we store its concave vertices (points of crossing of the boundaries of two distinct $K_p$'s) and the points that are locally $x$-extremal along $\partial U^A$. If two top (or two bottom) boundaries, say of $K_p$ and of $K_q$, have a weak crossing along $\partial U^A$, as in the Remark following Theorem 2.8, we store the left endpoint of the common overlap between these boundaries, and think of it as a 'weakly-concave' vertex of $\partial U^A$. A maximal portion $\gamma$ of $\partial U^A$ that does not contain any of these points is $x$-monotone and lies on the boundary of a single $K_p$ (portions of $\gamma$ may overlap the boundaries of many $K_p$'s, but there is (at least) one point $p$ such that $\gamma$ is contained in $\partial K_p$). We refer to $\gamma$ as a *boundary arc* of $U^A$. We maintain $\gamma$ implicitly, by recording the point $p$ for which $\gamma \subset \partial K_p$ and a bit that indicates whether $\gamma$ is a portion of the top or bottom part of $\partial K_p$.

Next, to represent each $K_p$ implicitly, we compute a family $\{\mathcal{D}^{(1)}, \ldots, \mathcal{D}^{(s)}\}$ of "canonical" subsets of $\mathcal{D}$ such that $\sum_{i=1}^{s} |\mathcal{D}^{(i)}| = O(n^{4/3}\log n)$, and such that for any $p \in P$, $\mathcal{D}_p$ can be represented as the union of $O(n^{1/3}\log n)$ canonical subsets. Let $J_p$ be the set of indices of these canonical subsets (i.e., $\mathcal{D}_p = \bigcup_{i\in J_p} \mathcal{D}^{(i)}$). Katz and Sharir [9] have shown that the construction of such a family of canonical sets, and of the corresponding sets of indices $\{J_p\}_{p\in P}$, can be accomplished in time $O(n^{4/3}\log n)$. For each canonical subset $\mathcal{D}^{(j)}$, we compute the intersection $\mathcal{I}^{(j)} = \bigcap \mathcal{D}^{(j)}$ in $O(|\mathcal{D}^{(j)}|\log|\mathcal{D}^{(j)}|)$ time. We store the vertices of the top and bottom parts of $\mathcal{I}^{(j)}$ in separate lists, each sorted in increasing order of their $x$-coordinates. For each vertex $v \in \mathcal{I}^{(j)}$, we also store the disk whose boundary appears on $\partial\mathcal{I}^{(j)}$ immediately to its right. Finally, we store the vertices of all the $\mathcal{I}^{(j)}$'s in a single master list $\Lambda$, sorted in increasing order of their $x$-coordinates. The total time spent in computing this implicit representation of the $K_p$'s is $\sum_{i=1}^{s} O(|\mathcal{D}^{(j)}|\log n) = O(n^{4/3}\log^2 n)$.

### 4.2 Basic operations on $\mathcal{K}$

In order to compute the implicit representation of $U_T$, we will need subroutines for the following basic operations on the boundaries of the sets in $\mathcal{K}$.

**(S1) Leftmost and rightmost points.** *Given a point $p$, compute the leftmost and the rightmost points of $K_p$.*

This requires computing the leftmost and the rightmost points of $\bigcap_{j\in J_p} \mathcal{I}^{(j)}$. Reichling [13] has shown that the leftmost (or rightmost) point of the intersection of $k$ convex polygons with a total of $n$ vertices can be computed in time $O(k\log^2 n)$. In fact, his algorithm can also be applied to a family of intersections of congruent disks. Applying Reichling's algorithm to the set $\{\mathcal{I}^{(j)} \mid j \in J_p\}$, we can compute the leftmost (or rightmost) point of $K_p$ in time $O(|J_p|\log^2 n) = O(n^{1/3}\log^3 n)$.

**(S2) Intersection points with a vertical line.** *Given a vertical line $\ell$ and a point $p \in P$, determine the intersection points of $\ell$ with $\partial K_p$.*

For each $j \in J_p$, we can compute $\ell \cap \mathcal{I}^{(j)}$ in $O(\log n)$ time. Repeating this step for all $j \in J_p$, we obtain a collection of $O(n^{1/3}\log n)$ intervals along $\ell$. We can compute the endpoints of the intersection of these intervals (or detect that the intersection is empty) in an additional $O(n^{1/3}\log n)$ time. Hence, the total running time of this procedure is $O(n^{1/3}\log^2 n)$. This procedure can also be used to determine whether a query point in the plane lies above, below, or on a boundary arc $\gamma$.

**(S3) Crossing points of two top (or two bottom) boundary arcs.** *Given two points $p, q \in P_T$ and an $x$-interval $[a,b]$ contained in the $x$-span of both $K_p$ and $K_q$, determine whether the top boundaries of $K_p$ and $K_q$ cross in $[a,b]$. If so, return their crossing point. If they weakly cross in $[a,b]$, then return the leftmost endpoint of their common overlap in $[a,b]$. A similar operation is prescribed for the bottom boundaries of $K_p$ and $K_q$.*

Let $\gamma_p$ (resp. $\gamma_q$) be the portion of the top boundary of $K_p$ (resp. $K_q$) in the $x$-interval $[a,b]$, and let $\ell_a : x = a$ and $\ell_b : x = b$. By computing the intersection points of $\partial K_p$ and $\partial K_q$ with $\ell_a$, we can determine, in $O(n^{1/3}\log n)$ time, whether $\gamma_p$ lies above or below $\gamma_q$ at $\ell_a$. Suppose $\gamma_p$ lies above $\gamma_q$ at $\ell_a$. We repeat the same procedure at $\ell_b$. Note that, by Theorem 2.8, $\gamma_p$ lies below $\gamma_q$ at $\ell_b$ too if and only if $\gamma_p$ and $\gamma_q$ cross. If they do cross, then, by performing a binary search over the points stored in the master list $\Lambda$, we obtain two consecutive vertices $\alpha, \beta \in \Lambda$ so that the crossing point (or, in case of weak crossing, the leftmost point of the common overlap) of $\gamma_p$ and $\gamma_q$ lies in the $x$-interval $I$ between $\alpha$ and $\beta$. Each step of the binary search involves determining whether $\gamma_p$ lies above $\gamma_q$ at a vertical line $\ell : x = x_0$, for some $x_0 \in \Lambda$, and is performed using subroutine (S2). Hence the total cost of the binary search is $O(n^{1/3}\log^3 n)$. The top boundary of each $\mathcal{I}^{(j)}$, for $j \in J_p$, is composed of a single circular arc in the $x$-interval $I$. We therefore collect the $O(n^{1/3}\log n)$ corresponding disks, and compute, in $O(n^{1/3}\log^2 n)$ time, the top boundary $\hat{\gamma}_p$ of their intersection within $I$. Similarly, we compute $\hat{\gamma}_q$, the top boundary of $K_q$ over $I$. We can now compute the crossing point (or the leftmost point of the common overlap) of $\hat{\gamma}_p$ and $\hat{\gamma}_q$ in an additional $O(n^{1/3}\log n)$ time, by merging the lists of vertices of $\hat{\gamma}_p$ and $\hat{\gamma}_q$, and by inspecting each 'atomic' interval formed by this merge. The total time spent by this procedure is thus $O(n^{1/3}\log^3 n)$. A symmetric procedure can compute the unique crossing point (or the leftmost point of the common overlap of a weak crossing of these boundaries) of the bottom boundaries of $K_p$ and $K_q$, within the same time bound.

**(S4) Crossing points of a top boundary arc and a bottom boundary arc.** *Given two points $p, q \in P_T$ and an $x$-interval $[a, b]$ contained in the $x$-span of both $K_p$ and $K_q$, determine whether the top boundary of $K_p$ crosses the bottom boundary of $K_q$ in the interval $[a, b]$. If so, return their crossing point(s).*

Let $\gamma_p$ be the portion of the top boundary of $K_p$ lying in the interval $[a, b]$, and let $\gamma_q$ be the portion of the bottom boundary of $K_q$ lying in the interval $[a, b]$. Note that, by convexity, $\gamma_p$ and $\gamma_q$ can cross in at most two points. By comparing the $y$-coordinates of the endpoints of $\gamma_p, \gamma_q$, using subroutine (S2), we can determine whether they cross exactly once. In this case, we can determine their unique crossing point (or the leftmost point of overlap of a weak crossing) using an appropriate variant of subroutine (S3). Suppose that we determine that $\gamma_p$ and $\gamma_q$ cross at zero or two points, and that $\gamma_q$ lies above $\gamma_p$ at the vertical line $x = a$ (and also at $x = b$); if $\gamma_q$ lies below $\gamma_p$ at these points, the arcs do not intersect. If we regard $\gamma_p$ and $\gamma_q$ as graphs of univariate, partially defined functions $\gamma_p(x), \gamma_q(x)$, respectively, then $\Delta\gamma(x) = \gamma_q(x) - \gamma_p(x)$ is a convex function. Therefore, by a binary search through $\Lambda$, each step of which requires determining the intersection points of a vertical line with $\gamma_p$ and $\gamma_q$, we can determine, in overall $O(n^{1/3} \log^3 n)$ time, the unique $x$-value $x_0$ at which $\Delta\gamma$ attains its minimum. If $\Delta\gamma(x_0) > 0$, then $\gamma_p$ and $\gamma_q$ do not intersect. If $\Delta\gamma(x_0) = 0$, then the minimum of $\Delta\gamma$ is the unique point of intersection (actually, of tangency) of $\gamma_p$ and $\gamma_q$. If $\Delta\gamma(x_0) < 0$, then $\gamma_p$ and $\gamma_q$ have two crossings, one of which lies in the interval $[a, x_0]$ and the other lies in the interval $[x_0, b]$. Now we can compute both crossing points in $O(n^{1/3} \log^3 n)$ time, using subroutine (S3).

### 4.3 Computing $U_T$

We now describe an algorithm for computing the implicit representation of $U_T$ described above, and for determining whether $U_T \cap P \neq \emptyset$. We first compute, using subroutine (S1), the leftmost and rightmost points, $l_p, r_p$, of each $K_p$, for $p \in P$. This, combined with calls to subroutine (S2), allows us to compute the sets $P_T$, $P_B$, $P_L$, and $P_R$, in overall $O(n^{4/3} \log^3 n)$ time. Next, we compute $U_T$, using a divide-and-conquer algorithm. If $|P_T| = 1$, then $U_T = K_p$, where $p$ is the only point in $P_T$. In this case, we output $\partial U_T$ as consisting of two boundary arcs, both connecting $l_p$ and $r_p$, where the top (resp. bottom) arc is the top (resp. bottom) boundary of $K_p$. If $|P_T| > 1$, we partition $P_T$ into two subsets $P_T^1$ and $P_T^2$, each of size at most $\lceil |P_T|/2 \rceil$. We recursively compute $U_T^1 = \bigcup_{p \in P_T^1} K_p$ and $U_T^2 = \bigcup_{p \in P_T^2} K_p$, and then compute $U_T = U_T^1 \cup U_T^2$, using a sweep-line algorithm. This 'merge' step computes the implicit representation of $U_T$ from those of $U_T^1, U_T^2$, which are output by the respective recursive calls.

The sweep line scans the plane from left to right, stopping at the concave vertices and the locally $x$-extremal points of $U_T^1, U_T^2$, and $U_T$. By Corollary 2.7, the number of such "event points" is only $O(n)$. The algorithm maintains those arcs of $U_T^1, U_T^2$ that currently intersect the sweep line in a height-balanced tree $T$, sorted in the increasing order of the $y$-coordinates of their intersection points with the line. At each event point, the algorithm inserts a new arc, deletes an arc, or swaps two adjacent arcs in the tree $T$. In order to insert a new arc into $T$, the algorithm has to perform $O(\log n)$ comparisons of the following form: given a point $q$ and a boundary arc $\gamma$, determine whether $q$ lies above, below, or on $\gamma$. Using subroutine (S2), such a comparison can be performed in $O(n^{1/3} \log^2 n)$ time. The time spent in inserting an arc is thus $O(n^{1/3} \log^3 n)$. The deletion of an arc follows a standard deletion procedure of a height-balanced tree. After having inserted or deleted an arc, we obtain the new $O(1)$ adjacent pairs of arcs in $T$, compute their (leftmost) intersection points to the right of the current sweep line, and insert them into the event queue. We thus need to perform $O(1)$ calls to the subroutines (S3) and (S4), each of which takes $O(n^{1/3} \log^3 n)$ time. Omitting all the other straightforward and standard details of the sweep-line algorithm, we conclude that the algorithm spends $O(n^{1/3} \log^3 n)$ time at each event point, therefore the total time spent by the sweep-line algorithm is $O(n^{4/3} \log^3 n)$. The overall time spent in computing the implicit representation of $U_T$ is thus $O(n^{4/3} \log^4 n)$. (Note that the computation of the sets $\mathcal{D}^{(j)}$, $\mathcal{I}^{(j)}$, and $J_p$ is performed only once, before starting the recursive construction of $U_T$.)

We next have to determine whether $U_T \cap P \neq \emptyset$. This can easily be done, at no increase in the asymptotic running time, during the topmost sweep of the recursion, in which the entire $U_T$ is constructed. We include the points of $P$ as additional event points of the line sweep. Whenever we encounter a point $p \in P$, we find the arc $\gamma$ of $U_T$ lying immediately above $p$. If $\gamma$ is a portion of the top boundary of some $K_q$, then $p \in U$. Moreover, $D(p, r) \cup D(q, r)$ covers all points of $P$, so we can return $p, q$ as the solution to the fixed-size problem; if $\gamma$ is a portion of the bottom boundary of some $K_q$, then $p \notin U$. The arc $\gamma$ can be determined by searching the tree with $p$, where each step of the search determines whether $p$ lies above, below, or on an arc $\gamma'$. Since each such step can be performed in $O(n^{1/3} \log^2 n)$ time, using subroutine (S2), we can determine in $O(n^{1/3} \log^3 n)$ time whether $p \in U_T$. Summing this cost over all points $p \in P$, the total time spent by this stage is $O(n^{4/3} \log^3 n)$.

We now construct and search in $U_B$, $U_L$ and $U_R$, using the algorithm just described. If at least one of these unions contains a point of $P$ then we have found, and can output, two points $p, q \in P$ such that $D(p, r) \cup D(q, r)$ covers $P$; otherwise, no two such points exist. The overall running time is $O(n^{4/3} \log^4 n)$. Hence, we obtain the following result.

**Theorem 4.1** *Given a set $P$ of $n$ points in the plane and a real value $r > 0$, we can determine, in $O(n^{4/3} \log^4 n)$ time, whether there are two points $p, q \in P$ so that $P \subset D(p, r) \cup D(q, r)$. If so, we can also find such a pair within the same time bound.*

## 5 The Overall Algorithm

The overall algorithm for the discrete 2-center problem proceeds as follows. We note that the optimum radius $r^*$ is a distance between two points of $P$, so we will run a binary search over these $\binom{n}{2}$ distances, using the fixed-size decision procedure given in the preceding section to determine whether the optimum $r^*$ is larger than, smaller than, or equal to a distance $r$. (Testing for an equality is somewhat trickier; we give more details in the full version of the paper.) Of course, running this binary search explicitly will require quadratic time, so we use instead the distance-selection algorithm of [9] (see also [2]), which computes the $k$-th smallest distance in a set of $n$ points in the plane in time $O(n^{4/3}\log^2 n)$. Since we need to invoke this procedure, and also the fixed-size decision procedure, only $O(\log n)$ times, the overall running time of the algorithm is $O(n^{4/3}\log^5 n)$.

**Theorem 5.1** *The discrete 2-center problem for a set of $n$ points in the plane can be solved in time $O(n^{4/3}\log^5 n)$.*

## 6 Conclusion

We have presented an $O(n^{4/3}\log^5 n)$-time algorithm for the planar discrete 2-center problem. Our decision algorithm relies heavily on the properties of the combinatorial structure of $\mathcal{K}$ that we have proved in Section 2. Although we have shown in Section 3 that the complexity of $U$ is roughly $n^{4/3}$, we do not have an algorithm with comparable running time that computes $U$ explicitly. There are, in fact, a number of substructures in an arrangement of a set of congruent disks, whose worst-case complexity has the same asymptotic upper bound as that of the corresponding structure in an arrangement of lines. For example, the number of incidences between points and congruent disks, the complexity of many faces, and the complexity of $U$ (one can define a structure analogous to $U$ for a set of halfplanes in the plane). However, unlike the case of lines, no efficient algorithm is known for computing most of these substructures. It is an open problem whether $m$ distinct faces in an arrangement of $n$ congruent disks can be computed in time close to $m^{2/3}n^{2/3}+n$. A solution to any of these problems will most likely offer insights for developing a simpler algorithm (still with running time close to $n^{4/3}$) for the discrete 2-center problem. A more challenging open problem is whether a near-linear-time algorithm can be developed for the discrete 2-center problem.


## Acknowledgments

The authors wish to thank Arie Tamir for helpful discussions concerning this problem.


## References


[1] P.K. Agarwal, On stabbing lines for convex polyhedra in 3D, *Comput. Geom: Theory Appls.* 4 (1994), 177–189.

[2] P.K. Agarwal, B. Aronov, M. Sharir and S. Suri, Selecting distances in the plane, *Algorithmica* 9 (1993), 495–514.

[3] P.K. Agarwal and M. Sharir, Planar geometric location problems, *Algorithmica* 11 (1994), 185–195.

[4] K. Clarkson, H. Edelsbrunner, L. Guibas, M. Sharir, and E. Welzl, Combinatorial complexity bounds for arrangements of curves and spheres, *Discrete Comput. Geom.* 5 (1990), 99–160.

[5] K. Clarkson and P. W. Shor, Applications of random sampling in computational geometry, II, *Discrete Comput. Geom.* 4 (1989), 387–421.

[6] D. Eppstein, Faster construction of planar 2-centers, *Proc. 8th ACM-SIAM Symp. on Discrete Algorithms*, 1997, 131–138.

[7] J. Hershberger and S. Suri, Finding tailored partitions, *J. Algorithms* 12 (1991), 431–463.

[8] J. Jaromczyk and M. Kowaluk, An efficient algorithm for the Euclidean two-center problem, *Proc. 10th ACM Symp. on Computational Geometry*, 1994, 303–311.

[9] M. Katz and M. Sharir, An expander-based approach to geometric optimization, *Proc. 9th ACM Symp. on Computational Geometry*, 1993, 198–207.

[10] K. Kedem, R. Livne, J. Pach and M. Sharir, On the union of Jordan regions and collision-free translational motion amidst polygonal obstacles, *Discrete Comput. Geom.* 1 (1986), 59–71.

[11] N. Megiddo, Applying parallel computation algorithms in the design of serial algorithms, *J. ACM* 30 (1983), 852–865.

[12] N. Megiddo and K. Supowit, On the complexity of some common geometric location problems, *SIAM J. Comput.* 13 (1984), 1182–1196.

[13] M. Reichling, On the detection of a common intersection of $k$ convex objects in the plane, *Inform. Process. Lett.* 29 (1988), 25–29.

[14] M. Sharir, A near-linear algorithm for the planar two-center problem, *Discrete Comput. Geom.* (in press).

[15] A. Tamir, Improved complexity bounds for center location problem on networks by using dynamic data structures, *SIAM J. Discrete Math.* 1 (1988), 377-396.

# Robust Proximity Queries: an Illustration of Degree-driven Algorithm Design*

*Giuseppe Liotta*[†] *Franco P. Preparata*[‡] *Roberto Tamassia*[§]

**Abstract** In the context of methodologies intended to confer robustness to geometric algorithms, we elaborate on the exact computation paradigm and formalize the notion of degree of a geometric algorithm, as a worst-case quantification of the precision (number of bits) to which arithmetic calculation have to be executed in order to guarantee topological correctness. We also propose a formalism for the expeditious evaluation of algorithmic degree. As an application of this paradigm and an illustration of our general approach, we consider the important classical problem of proximity queries in 2 and 3 dimensions, and develop a new technique for the efficient and robust execution of such queries based on an implicit representation of Voronoi diagrams. Our new technique gives both low degree and fast query time, and for 2D queries is optimal with respect to both cost measures of the paradigm, asymptotic number of operations and arithmetic degree.

## 1 Introduction

The increasing demand for efficient and reliable geometric software libraries in key applications such as computer graphics, geographic information systems, and computer-aided manufacturing is stimulating a major renovation in the field of computational geometry. The inadequacy of the traditional simplified framework has become apparent, and it is being realized that, in order to achieve an effective technology transfer, new frameworks and models are needed to design and analyze geometric algorithms that are efficient in a practical realm.

The real-RAM model with its implicit infinite-precision requirement, has proved unrealistic and needs to be replaced with a realistic finite-precision model where geometric computations can be carried out either exactly or with a guaranteed error bound. This has motivated a great deal of research on the subject of robust computational geometry (see, e.g., [6, 7, 12, 13, 25, 27]). For an early survey of the different approaches to robust computational geometry the reader is referred to [16].

To a first, rough, approximation, robustness approaches are of two main types: perturbing and nonperturbing. Perturbing approaches transform the given problem into one that is intended not to suffer from well-identified shortcomings; nonperturbing approaches are based on the notion of "exact" (rather than "approximate") computations, with the assumption that (bounded-length) input data are error free. In this category falls the exact geometric computation paradigm independently advocated by Yap [28] and by the Saarbrücken school [3, 4], and so does our approach. Within this paradigm, we introduce the concept of *degree* of a geometric algorithm, which characterizes, up to a small additive constant, the arithmetic precision (i.e., number of bits) needed by a large class of geometric algorithms. Namely, if the coordinates of the input points of a degree-$d$ algorithm within this class are $b$-bit integers, then the algorithm may be required on some instances to perform arithmetic computations with bit precision $d(b + O(1))$.

Theoretical analysis and experimental results show that multiprecision numerical computations take up most of the CPU time of exact geometric algorithms (see, e.g., [17]). Thus, we believe that, in defining the efficiency of a geometric algorithm, the degree should be considered as important as the asymptotic time complexity and should correspondingly play a major role in the design stage. In fact, the principal thrust of this paper is to present algorithm degree as a major design criterion for geometric computation. Research along these lines involves re-examining the entire rich body of computational geometry as we know it today.

---

*Research supported in part by the National Science Foundation under grant CCR–9423847, by the U.S. Army Research Office under grant DAAH04–96–1–0013, and by the EC ESPRIT Long Term Research Project ALCOM-IT under contract 20244.

[†]Dipartimento di Informatica e Sistemistica, Universitá di Roma "La Sapienza", Via Salaria 113, Roma 00198, Italy. Work by this author performed when he was with the Center for Geometric Computing at Brown University. liotta@dis.uniroma1.it

[‡]Center for Geometric Computing, Department of Computer Science, Brown University, 115 Waterman Street, Providence, RI 02912–1910, USA. franco@cs.brown.edu

[§]Center for Geometric Computing, Department of Computer Science, Brown University, 115 Waterman Street, Providence, RI 02912–1910, USA. rt@cs.brown.edu



*Computational Geometry 97* Nice France

In this paper, we consider as a test case a problem area, geometric proximity, which plays a major role in applications and has recently attracted considerable attention because, due to its demands of high precision for exact computation, it is particularly appropriate in assessing effectiveness of robust approaches (see, e.g., [13]). In particular, we shall illustrate the role played by the degree criterion if one wishes to comply with the exact-computation paradigm.

To illustrate the approach, we recall that Voronoi diagrams of 2D point sites are the search structures which permit answering a proximity query without evaluating all query-point/site distances. Therefore, given the set of sites, their Voronoi diagram is computed and supplied as a planar subdivision to a point-location procedure. Assuming that the coordinates of all input data (also called *primitive* points) are $b$-bit integers, the coordinates of the points computed by the algorithm (referred to here as *derived* points, e.g., the vertices of a Voronoi diagram of points and segments) must be stored with a representation scheme that supports rational or algebraic numbers as data types (through multiprecision integers). Specifically, the coordinates $(x, y)$ of a Voronoi vertex are rational numbers given by the ratio of two determinants (of respective orders 3 and 2) whose entries are integers of well-defined maximum modulus. The fundamental operation used by any point location algorithm is the point-line discrimination, which consists of determining whether the query point $q$ is to the left or to the right of an edge between vertices $v_1$ and $v_2$. For the case of the Voronoi diagram $V(S)$, this is equivalent to evaluating the sign of a $3 \times 3$ determinant whose rows are the homogeneous coordinates of $q$, $v_1$, and $v_2$, a computation that needs about $6b$ bits of precision. This should be compared with the $O(n)$-time brute-force method that computes the (squares of the) distances from $q$ to all the sites of $S$, and executes arithmetic computations with only $2b$ bits of precision (which is optimal).

Guided by the low-degree criterion, in this paper we present a technique — complying with the exact-computation paradigm — which uses a new point-location data structure for Voronoi diagrams, such that the test operations executed in the point location procedure are distance comparisons, and are therefore executed with optimal $2b$ bits of precision. Hence, our approach reconciles efficiency with robustness, and advocates an object-oriented programming style where access to the geometry of Voronoi diagrams in point-location queries is encapsulated in a small set of geometric test primitives. It must be pointed out that distance comparisons have already been used nontrivially for proximity search (extremal-search method [11]). However, we shall show that the latter method fails to achieve optimal degree because the search is based on predicates requiring $4b$ bits of precision; moreover, the high overhead of the search technique (which uses the hierarchical polytope representation [8]) casts some doubts on the practicality of the method.

The main results of this work are summarized in Table 1. We show that previous methods for proximity queries exhibit a sharp tradeoff between degree and query time. Namely, low degree is achieved by the slow brute-force search method, while fast algorithms based on point-location in a preprocessed Voronoi diagram or on the extremal-search method have high degree. Our new technique gives instead both low degree and fast query time, and is optimal with respect to both cost measures for queries in sets of 2D point sites.

For proofs and details omitted in this extended abstract, the reader is referred to the full paper [20].

# 2 Degree of Geometric Problems

The numerical computations of a geometric algorithm are basically of two types: tests (predicates) and constructions. The two types of computations have clearly distinct roles. Tests are associated with branching decisions in the algorithm that determine the flow of control, whereas constructions are needed to produce the output data of the algorithm. Approximations in the execution of constructions are acceptable, since approximate results are perfectly tolerable, provided that the error magnitude does not exceed the resolution required by the application. On the other hand, approximations in the execution of tests may produce an incorrect branching of the algorithm. Such event may have catastrophic consequences, giving rise to *structurally* incorrect results. Te exact-computation paradigm therefore requires that tests be executed with total accuracy. We shall therefore characterize geometric algorithms on the basis of the complexity of their test computations. Any such computation consists of evaluating the sign of an *algebraic expression* over the input variables, constructed using an adequate set of operators, such has $\{+, -, \times, \div, \sqrt[k]{\ }, \ldots\}$. As we shall show below, the expressions under considerations are equivalent to multivariate polynomials.

We make here the reasonable assumption that input data be *dimensionally consistent*, i.e., that if an entity with the physical dimension of a length is represented with $b$ bits, then one with the dimension of an area be represented with $2b$ bits, and so on. Under the hypothesis of dimensional consistency ( where point coordinates are $b$-bit entries), a polynomial expressing a test is homogeneous because all of its monomials must have the same physical dimension. A *primitive variable* is an input variable of the algorithm and has conventional arithmetic degree 1. The arithmetic degree of

| Query | Method | Degree | Time |
|---|---|---|---|
| nearest neighbor | brute-force distance comparison | 2 * | $O(n)$ |
| | point location in explicit Voronoi diagram | 6 | $O(\log n)$ * |
| | extremal-search method | 4 | $O(\log n)$ * |
| | *point location in implicit Voronoi diagram* | 2 * | $O(\log n)$ * |
| $k$-nearest neighbors and circular range search | brute-force distance comparison | 2 * | $O(n)$ |
| | point location in explicit order-$k$ Voronoi diagram | 6 | $O(\log n + k)$ * |
| | *point location in implicit order-$k$ Voronoi diagram* | 2 * | $O(\log n + k)$ * |
| nearest neighbor among points and segments | brute-force distance comparison | 6 | $O(n)$ |
| | point location in explicit Voronoi diagram | 64 | $O(\log n)$ * |
| | *point location in implicit Voronoi diagram* | 6 | $O(\log n)$ * |
| 3D nearest neighbor | brute-force distance comparison | 2 * | $O(n)$ |
| | point location in explicit 3D Voronoi diagram | 8 | $O(\log^2 n)$ |
| | *point location in implicit 3D Voronoi diagram* | 3 | $O(\log^2 n)$ |

Table 1: Comparison of the degree and running time of algorithms for some fundamental proximity query problems. A * denotes optimality. The technique introduced in this paper (*point location in an implicit Voronoi diagram*) outperforms previous methods and is optimal for 2D queries.

a polynomial expression $E$ is the common arithmetic degree of its monomials. The arithmetic degree of a monomial is the sum of the arithmetic degrees of its variables.

**Definition 1** *An algorithm has* degree $d$ *if its test computations involve the evaluation of multivariate polynomials of arithmetic degree at most $d$. A problem $\Pi$ has* degree $d$ *if any algorithm that solves $\Pi$ has degree at least $d$.*

Recently, Burnikel [2] has independently defined the notion of *precision of an algorithm*, which is equivalent to our notion of degree of an algorithm. Also, our definition of degree is related to that of *depth of derivation* proposed by Yap [27, 28].

Motivated by a standard feature of geometric algorithms, we also make the assumption that every multivariate polynomial of degree $d$ used in tests depends upon a set of size $s$ (a small constant) of primitive variables. Therefore a multivariate polynomial has $O(s^d)$ distinct monomials with bounded integer coefficients, so that the maximum value of the multivariate polynomial is expressible with at most $db + d \log s$ bits. A consequence of Definition 1 and of the above assumption is the following fact, which justifies our use of the degree of an algorithm to characterize the precision required in test computations.

**Lemma 1** *If an algorithm has degree $d$ and its input variables are $b$-bit integers, then all the test computations can be carried out with $d(b + O(1))$ bits.*

Typically the support of a geometric test is not naturally expressed by a multivariate polynomial, but, rather, by a pair $(E_1, E_2)$ of expressions involving the four arithmetic operations, powering, and the extraction of square roots, and the test consists of comparing the magnitudes of $E_1$ and $E_2$. Such expressions always have a physical dimension (a length, an area, a volume, etc.), so that if they have the form of ratios, the degree of the numerator exceeds that of the denominator.

Expressions such as $E_1$ and $E_2$ can be viewed as a binary tree, whose leaves represent input variables and whose internal nodes are of two types: binary nodes, which are labeled with an operation from the set $\{+, -, \times, \div\}$, and unary nodes, which are labeled either with a power or with a square root extraction (notice that we restrict ourselves to this type of radicals). If no radical appears in the trees of $E_1$ and $E_2$, then the test is trivially equivalent to the evaluation of the sign of a polynomial, since $E_i$ is a rational function of the form $\frac{N_i}{D_i}$ ($i = 1, 2$, $N_i, D_i$ are not both trivial polynomials and $D_i \neq 0$) and $E_1 \geq E_2 \iff (-1)^{\sigma(D_1)+\sigma(D_2)}(N_1 D_2 - N_2 D_1) \geq 0$, where $\sigma(E) = 1$ if $E < 0$ and $\sigma(E) = 0$ if $E \geq 0$. (Note that the above predicate implies the inductive assumption that the signs of lower degree expressions $N_1$,$N_2$, $D_1$, and $D_2$ are known.) Suppose now that at least one of the trees of $E_1$ and $E_2$ contains radicals. We prune the tree so that the pruned tree contains no radicals except at its leaves (notice that pruned subtrees may themselves contain radicals). Then $N_i$ and $D_i$ ($i = 1, 2$) can be viewed as polynomials whose variables are the radicals and whose coefficients are (polynomial) functions of non-radicals. Given a polynomial $E$ in a set of radicals, for any radical $R$ in this set, we can express $E$ as $E = E''R + E'$ where neither $E''$ nor $E'$ contains $R$. Then $E \geq 0 \iff E''R \geq -E'$. The resulting expression $(E''^2 R^2 - E'^2)$ does not contain $R$. Therefore by this device we can eliminate one radical. This shows that by the four rational operations we can reduce the sign test to the computation of the signs of a collection of multivariate polynomials.

We now present a very simple, but useful, formalism that enables us to rapidly evaluate the degree of the multivariate polynomial which uniquely determines the sign of the original algebraic expression.

The terms involved in the formal manipulations are of two types, generic and specific. *Generic* terms have the form $\alpha^s$ (for a formal variable $\alpha$ and an integer $s$), representing an unspecified multivariate polynomial of degree $s$ over primitive variables. *Specific* terms have the form $\rho_j$, for some integer index $j$, representing a specified expression involving the operators $\{+, -, \times, \div, \sqrt{}\}$. The terms are members of a free commutative semiring, i.e., addition and multiplication are associative and commutative, addition distributes over multiplication, and specific terms can be factored out. Beside these conventional algebraic rules, we have a set of *rewriting rules* of the form $A \rightarrow B$, meaning that the sign of $A$ is *uniquely determined* by the sign of $B$. Note, the two signs not always coincide: indeed the correspondence between the signs of the sides depends upon the evaluated signs of other expressions of lower degree.

We have seven rules, whose correctness can be proved with elementary algebra. The rules are:

$$
\begin{array}{lccc}
(1) & \rho_j & \longrightarrow & \alpha^s \\
(2) & \alpha^s\alpha^r & \longrightarrow & \alpha^{s+r} \\
(3) & \alpha^s + \alpha^s & \longrightarrow & \alpha^s \\
(4) & -\alpha^s & \longrightarrow & \alpha^s \\
(5) & \frac{\rho_j}{\rho_i} \pm \frac{\rho_h}{\rho_i} & \longrightarrow & \rho_j \pm \rho_h \\
(6) & \frac{\rho_j}{\rho_i} \pm \frac{\rho_h}{\rho_k} & \longrightarrow & \rho_j\rho_k \pm \rho_i\rho_h \\
(7) & \rho_i \pm \rho_j & \longrightarrow & \rho_i^2 - \rho_j^2.
\end{array}
$$

The preceding discussion establishes the following:

**Theorem 1** *Rules* (1) – (7) *are adequate to evaluate the degree of a multivariate polynomials which uniquely determine the sign of an arbitrary algebraic expression involving square roots.*

While the above rules represent an adequate formalism for obtaining an upper bound to the degree of an algorithm, more subtle is the corresponding lower-bound question. In other words, given a predicate $\mathcal{P}$ that is essential to the solution of a given problem, what is the inherent degree of $\mathcal{P}$? Suppose that predicate $\mathcal{P}$ is expressed by a polynomial $P$ of degree $d$, and we must decide whether the value of $P$ is positive, negative, or zero. Can we answer this question by computing a discrete (ternary) function $f$ of analogous evaluations of irreducible polynomials $P_1, \ldots, P_k$ of maximum degree smaller than $d$? Clearly, $f$ changes value only when some $P_j$ changes sign (exactly, when the value of $P_j$ passes by 0). Thus, a 0 of $P$ corresponds to a 0 of some $P_j$. Moreover, as the arguments of $P_j$ vary while $P_j$ remains 0, so does $f$ and hence $P$. Therefore, $P$ vanishes at all points for which $P_j$ vanishes and, for a well-known theorem of polynomial algebra (see, e.g., [1] pp. 212–216), we conclude that $P_j$ is a factor of $P$. This is summarized as follows:

**Theorem 2** *The degree of the problem of evaluating a predicate expressed by the sign of a polynomial $P$ is the maximum arithmetic degree of the factors of $P$ that are not always positive or always negative.*

As an application of Theorem 2, we can prove a lower bound on the degree of the nearest neighbor search problem.

**Theorem 3** *The nearest neighbor search problem for a point set has degree 2 in any fixed dimension $d \geq 2$.*

Observe that an optimal-degree algorithm for the nearest neighbor search problem in a planar point set can be easily obtained with the brute force approach, where one computes all the distances between the query point and all other points and reports the point at minimum distance. However, such algorithm is both computationally inefficient (it requires quadratic time) and does not support repetitive-mode queries. In Section 3, we present an optimal degree algorithm whose query time and space are optimal.

Let $q$ be a query point, let $p_1$ and $p_2$ be two points, and let $r_1$ and $r_2$ be two lines on the plane. The point-to-points distance test determines whether $q$ is closer to $p_1$ or to $p_2$. The point-to-point-line distance test determines whether $q$ is closer to $p_1$ or to $r_1$. The point-to-lines distance test determines whether $q$ is closer to $r_1$ or to $r_2$.

**Lemma 2** *The* point-to-points distance test, point-to-point-line distance test, point-to-lines distance test *can be solved with degree 2, 4, and 6, respectively.*

Another fundamental proximity primitive is the incircle test, that is testing whether the circle determined by three distinct sites (points and segments) of the plane contains a given query site. An incircle test is expressed as a quadruple $(a_1, a_2, a_3; a_4)$, where each $a_i \in \{p, l\}$ $(i = 1, \ldots, 4)$ is either a point or a line on the plane and we test whether $a_4$ intersects the circle determined by $a_1$,$a_2$, and $a_3$. The incircle test can be generalized to the insphere test in any dimension $d \geq 2$. The insphere test $(p_1, \ldots, p_{d+1}; p_{d+2})$, where points $p_1, \ldots, p_{d+1}$ determine a $d$-dimensional sphere and $p_{d+2}$ is the query point.

**Lemma 3** [2] *The* incircle test $(l_1, l_2, l_3; l_4)$ *can be solved with degree* 40. *The* insphere test $(p_1, \ldots, p_{d+1}; p_{d+2})$ *in any fixed dimension $d \geq 2$ can be solved with degree $d + 2$.*

## 3 Point Sites in the Plane

In this section, under our standard assumption that all input parameters — such as coordinates of sites and query points — are represented by $b$-bit integers, we consider the following proximity queries on a set $S$ of point sites in the plane:

*nearest neighbor search:* given query point $q$, find a site of $S$ whose Euclidean distance from $q$ is less than or equal to that of any other site;

*k-nearest neighbors search:* given query point $q$, find $k$ sites of $S$ whose Euclidean distances from $q$ are less than or equal to that of any other site;

*circular range search:* given query points $q$ and $r$, find the sites of $S$ that are inside the circle with center $q$ passing through $r$.

It is well known that such queries are efficiently solved by performing point location in the Voronoi diagram (possibly of higher order) $V(S)$ of the sites [22].

The *chain method* [19], the *bridged chain method* [10], the *trapezoid method* [21], the *subdivision refinement method* [18], and the *persistent search tree method* [24] are popular deterministic techniques for point location in a planar map that combine theoretical efficiency with good performance in practice (see, e.g., [22]). The *randomized-incremental method* [14] also exists, that is specialized for point location in Voronoi diagrams.

By a careful examination of the query algorithms used in the point location methods presented in the literature, it is possible to clearly separate the primitive operations that access the geometry of the map from those that access only the topology. We say that a point location method is *native* for a class of maps if it performs point locations queries in a map $M$ of the class by accessing the geometry of $M$ exclusively through the following three geometric test primitives that discriminate the query point with respect to the vertices and edges of $M$:

above-below$(q, v)$ determine whether query point $q$ is vertically above or below vertex $v$.

left-right$(q, v)$ determine whether query point $q$ is horizontally to the left or to the right of vertex $v$.

left-right$(q, e)$ determine whether query point $q$ is to the left or to the right of edge $e$; this operation assumes that edge $e$ is not horizontal and its vertical span includes $q$.

Test primitive left-right$(q, v)$ is typically used only in degenerate cases (e.g., in the presence of horizontal edges).

Some point location methods work on modified versions of the original subdivision by means of auxiliary geometric objects introduced in the preprocessing (e.g., triangulation or regularization edges). We say that a point location method is *ordinary* for a class of maps if it performs point locations queries in a map $M$ of the class by accessing the geometry of $M$ through the three geometric test primitives described above for the native methods and through left-right$(q, e)$ tests such that $e$ is a fictitious edge connecting two vertices of $M$.

More specifically, we have:

**Lemma 4** *The trapezoid method and the persistent search tree method are native for general maps. The chain method and the bridged chain method are ordinary for general maps and native for monotone maps. The subdivision refinement method is ordinary for general maps. The randomized-incremental method is ordinary for Voronoi diagrams.*

Hence, all the known planar point location methods described in the literature are ordinary for Voronoi diagrams.

Let $S$ be a set of $n$ point sites in the plane, where each site is a primitive point with $b$-bit integer coordinates. The Voronoi diagram $V(S)$ of $S$ is said to be *explicit* if the coordinates of the vertices of $V(S)$ are computed and stored with exact arithmetic, i.e., as rational numbers (pairs of integers).

**Lemma 5** *The* left-right$(q, e)$ *test primitive in an explicit Voronoi diagram of point sites in the plane has degree* 6.

An algorithm for proximity queries on a set $S$ of point sites in the plane is said to be *conventional* if it computes the explicit Voronoi diagram $V(S)$ of $S$ and then performs point location queries on $V(S)$ with an ordinary method. Note that the class of conventional proximity query algorithms includes all the efficient algorithms presented in the literature. A conventional proximity query algorithm needs to perform test primitive left-right$(q, e)$. Thus, by Lemma 5 we have:

**Theorem 4** *Conventional algorithms for the following proximity query problems on a set of point sites in the plane have degree at least* 6*: (i) nearest neighbor query; (ii) k-nearest neighbor query; and (iii) circular range query.*

We observe that a degree-6 algorithm implies that a $k$-bit arithmetic unit can handle with native precision queries for points in a grid of size at most $2^{k/6} \times 2^{k/6}$. For example, if $k = 32$, the points that can be treated with single-precision arithmetic belong to a grid of size at most $64 \times 64$.

The extremal-search method [11], also designed for proximity queries, reduces the nearest neighbor search problem for a set $S$ of 2D point sites to an extremal search problem.

**Theorem 5** *The extremal-search method for the nearest neighbor query problem on a set of point sites in the plane has degree at least* 4.

Let $S$ be a set of $n$ point sites in the plane, and recall our assumption that each site or query point is a primitive point with $b$-bit integer coordinates. We say that a number $s$ is a *semi-integer* if it is a rational number of the type $s = m/2$ for some integer $m$. The *implicit Voronoi diagram* $V^*(S)$ of $S$ is a representation of the Voronoi diagram $V(S)$ of $S$ that consists of a topological component and of a geometric component. The topological component of $V^*(S)$ is the planar embedding of $V(S)$, represented by a suitable data structure (e.g., doubly-connected edge lists [22] or the data structure of [15]). The geometric component of $V^*(S)$ stores the following geometric information for each vertex and edge of the embedding:

- For each vertex $v$ of $V(S)$, $V^*(S)$ stores semi-integers $x^*(v)$ and $y^*(v)$ that approximate the $x$- and $y$-coordinates $y(v)$ of $v$, We provide the definition of $y^*(v)$ below. The definition of $x^*(v)$ is analogous.

$$y^*(v) = \begin{cases} y(v) & 0 \le y(v) \le 2^b - 1,\ y(v) \text{ integer} \\ \lfloor y(v) \rfloor + \frac{1}{2} & 0 \le y(v) \le 2^b - 1,\ y(v) \text{ not integer} \\ 2^b - \frac{1}{2} & y(v) > 2^b - 1 \\ 0 & y(v) < 0 \end{cases}$$

  Note that semi-integers $x^*(v)$ and $y^*(v)$ can be stored with $(b+1)$-bits.

- For each non-horizontal edge $e$ of $V(S)$, $V^*(S)$ stores the pair of sites $\ell(e)$ and $r(e)$ such that $e$ is a portion of the perpendicular bisector of $\ell(e)$ and $r(e)$, and $\ell(e)$ is to the left of $r(e)$.

Equipped with the above information, the three test primitives for point location can be performed in the implicit Voronoi diagram $V^*(S)$ as follows:

above-below$(q, v)$ compare the $y$-coordinate of $q$ with $y^*(v)$;

left-right$(q, v)$ compare the $x$-coordinate of $q$ with $x^*(v)$;

left-right$(q, e)$ compare the Euclidean distances of point $q$ from sites $\ell(e)$ and $r(e)$.

Since the query point $q$ is by assumption a primitive point whose coordinates are $b$-bit integers, we have that $y(q) \le y(v)$ if and only if $y(q) \le y^*(v)$, where testing the latter inequality has degree 1. Similar considerations apply to testing $x(q) \le x(v)$. This proves the correctness of our implementation of above-below$(q, v)$ and left-right$(q, v)$.

The correctness of the above implementation of test left-right$(q, e)$ follows directly from the definition of Voronoi edges. Thus, in an implicit Voronoi diagram, test left-right$(q, e)$ can be implemented with a point-to-points distance test that has degree 2 (Lemma 2).

Hence, we obtain the following lemma:

**Lemma 6** *In an implicit Voronoi diagram of point sites in the plane test primitives* above-below$(q, v)$ *and* left-right$(q, v)$ *can be performed in $O(1)$ time with degree 1, and test primitive* left-right$(q, e)$ *in $O(1)$ time with degree 2.*

In order to execute a native point location algorithm in an implicit Voronoi diagram, we only need to redefine the implementation of the three test primitives. By having encapsulated the geometry in the test primitives, no further modifications are needed. Hence, by Lemma 6 and Theorem 3 we obtain:

**Lemma 7** *For any native method on a class of maps that includes Voronoi diagrams, a point location query in an implicit Voronoi diagram has optimal degree 2 and has the same asymptotic time complexity as a point location query in the corresponding explicit Voronoi diagram.*

In order to compute the implicit Voronoi diagram $V^*(S)$, we begin by constructing the Delaunay triangulation of $S$, denoted $DT(S)$, by means of the $O(n \log n)$-time algorithm of [15], which has degree 4 because its most expensive operation in terms of the degree is the incircle test (see Lemma 3). The topological structure of $V(S)$ and the sites $\ell(e)$ and $r(e)$ for each edge $e$ of $V(S)$ are immediately derived from $DT(S)$ by duality. Next, we compute the approximations $x^*(v)$ and $y^*(v)$ for each vertex $v$ of $V(S)$ by means of integer division. Let $a$, $b$, and $c$ be the three sites of $S$ that define vertex $v$. Adopting the same notation as in the proof of Lemma 5, the $y$-coordinate $y(v)$ of $v$ is given by the ratio $y(v) = \frac{Y_1}{2W_1}$, where $Y_1$ is a variable of arithmetic degree 3 and $W_1$ is a variable of arithmetic degree 2, and similarly for $x(v)$. Hence, the computation of $x^*(v)$ and $y^*(v)$ involves integer represented by at most $3(b+O(1))$ bits. We summarize the above analysis as follows.

**Lemma 8** *The implicit Voronoi diagram of $n$ point sites in the plane can be computed in $O(n \log n)$ time, $O(n)$ space, and with degree 4.*

**Theorem 6** *Let $S$ be a set of $n$ point sites in the plane. There exists an $O(n)$-space data structure for $S$, based on the implicit Voronoi diagram $V^*(S)$, that can be computed in $O(n \log n)$ time with degree 5, and supports nearest neighbor queries in $O(\log n)$ time with optimal degree 2.*

We can define implicit higher order Voronoi diagrams for point sites in the plane, and extend the above results to $k$-nearest neighbors and circular range search queries. We have

**Theorem 7** *Let $S$ be a set of $n$ point sites in the plane and $k$ an integer with $1 \le k \le n-1$. There exists an $O(k(n-k))$-space data structure for $S$, based on the implicit order-$k$ Voronoi diagram $V_k^*(S)$, that can be computed in $O(k(n-k)\sqrt{n} \log n)$ time with degree 5 and supports $k$-nearest neighbors queries in $O(\log n + k)$ time with optimal degree 2.*

**Theorem 8** *Let $S$ be a set of $n$ point sites in the plane. There exists an $O(n^3)$-space data structure for $S$, based on implicit order-$k$ Voronoi diagrams, that can be computed in $O(n^3)$ time with degree 5 and supports circular range search queries in $O(\log n \log\log n + k)$ time with optimal degree 2.*

# 4 Point Sites in 3D Space

In this section, we consider the following proximity query on a set $S$ of point sites in three-dimensional (3D) space:

*nearest neighbor search:* given query point $q$, find a site of $S$ whose Euclidean distance from $q$ is less than or equal to that of any other site;

We recall our assumption that the sites and query points are primitive points represented by $b$-bit integers.

As for the two-dimensional case, such query is efficiently answered by performing point location in the 3D Voronoi diagram of $S$.

There are only two known efficient spatial point location methods for cell-complexes that are applicable to 3D Voronoi diagrams: the *separating surfaces method* [5, 26], which extends the chain-method [19], and the *persistent planar location method* [23], which extends the persistent search tree method [24]. Let $N$ be the number of facets of a cell-complex $C$. The query time is $O(\log^2 N)$ for both methods. The space used is $O(N)$ for the separating surfaces method and $O(N \log^2 N)$ for the persistent planar location method. Both methods are restricted to convex cell-complexes. The separating surfaces method is further restricted to acyclic convex cell-complexes, where the dominance relation among cells in the $z$-direction is acyclic.

We can separate the primitive operations that access the geometry of the cell-complex from those that access only the topology. We say that a point location method is *native* for a class of 3D cell-complexes if it performs point locations queries in a cell complex $C$ of the class by accessing the geometry of $C$ exclusively through the following three geometric test primitives that discriminate the query point with respect to the vertices and edges of $C$:

above-below$(q, v)$ compare the $z$-coordinate of the query point $q$ with the $z$-coordinate of vertex $v$.

left-right$(q, v)$ compare the $x$-coordinate of the query point $q$ with the $x$-coordinate of vertex $v$.

front-rear$(q, v)$ compare the $y$-coordinate of the query point $q$ with the $y$-coordinate of vertex $v$.

left-right$(q_{xy}, e_{xy})$ compare the $xy$-projection $q_{xy}$ of the query point $q$ with the $xy$-projection of edge $e_{xy}$. This operation assumes that $e_{xy}$ is not parallel to the $x$-axis and its $y$-span includes $q_{xy}$.

above-below$(q, f)$ determine whether query point $q$ is above or below a facet $f$; this operation assumes that facet $f$ is not parallel to the $z$-axis and that the $xy$-projection of $f$ contains the $xy$-projection of $q$.

Test primitives above-below$(q, v)$ and left-right$(q, v)$ are used only in degenerate cases (e.g. in the presence of facets parallel to the $z$-axis and in cases where $e_{xy}$ is horizontal).

**Lemma 9** *The separating surfaces method is native for acyclic convex cell-complexes. The persistent planar location method is native for convex cell-complexes.*

Hence, all the known spatial point location methods described in the literature are native for 3D Voronoi diagrams.

Let $S$ be a set of $n$ point sites in 3D, where each site is a primitive point with $b$-bit integer coordinates. The 3D Voronoi diagram $V(S)$ of $S$ is said to be *explicit* if the coordinates of the vertices of $V(S)$ are computed and stored with exact arithmetic, i.e., as rational numbers (pairs of integers).

**Lemma 10** *The* left-right$(q_{xy}, e_{xy})$ *test primitive in an explicit Voronoi diagram of point sites in 3D space has degree 8.*

An algorithm for nearest neighbor queries on a set $S$ of point sites in 3D space is said to be *conventional* if it computes the explicit 3D Voronoi diagram $V(S)$ of $S$ and then performs point location queries on $V(S)$ with a native method. Recall that the class of conventional nearest neighbor query algorithms includes the two efficient algorithms presented in the literature. A conventional proximity query algorithm needs to perform test primitive left-right$(q_{xy}, e_{xy})$. Thus, by Lemma 10, we have:

**Theorem 9** *Conventional algorithms for the nearest neighbor query problem on a set of point sites in 3D space have degree at least 8.*

The definition of the implicit 3D Voronoi diagram $V^*(S)$ of a set of $S$ of point sites in 3D space is a straightforward extension of the definition for two-dimensional Voronoi diagrams given in Section 3. Namely $V^*(S)$ stores the topological structure of the 3D Voronoi diagram $V(S)$ of $S$ (e.g., the data structure of [9]) and the following geometric information for each vertex and facet:

- For each vertex $v$ of $V(S)$, $V^*(S)$ stores the semi-integer $(b+1)$-bit approximations $x^*(v)$, $y^*(v)$ and $z^*(v)$ of the $x$-, $y$-, and $z$-coordinates of $v$.
- For each facet $f$ of $V(S)$ that is not parallel to any of three Cartesian planes, $V^*(S)$ stores the pair of sites $\ell(f)$ and $r(f)$ such that $f$ is a portion of the

perpendicular bisector of $\ell(f)$ and $r(f)$, and $\ell(f)$ is below $r(f)$.

The tests $\mathsf{above\text{-}below}(q, v)$, $\mathsf{left\text{-}right}(q, v)$, $\mathsf{front\text{-}rear}(q, v)$ can be implemented comparing the $x$-, $y$- and $z$-coordinate of query point $q$ with $x(v)^*$, $y(v)^*$, and $z(v)^*$ respectively. With the same reasoning as for the two-dimensional case (see Section 3), it is easy to see that such implementations are correct. Test primitive $\mathsf{above\text{-}below}(q, f)$ is implemented by comparing the Euclidean distances of point $q$ from the two sites $\ell(e)$ and $r(e)$ of which $f$ is the perpendicular bisector, with a point-to-points distance test. The implementation is correct by the definition of Voronoi facet. Thus, by Lemma 2, we have.

**Lemma 11** *In an implicit Voronoi diagram of 3D point sites test primitives* $\mathsf{above\text{-}below}(q, v)$, $\mathsf{left\text{-}right}(q, v)$, $\mathsf{front\text{-}rear}(q, v)$ *can be performed in* $O(1)$ *time with degree* 1, *and test primitive* $\mathsf{above\text{-}below}(q, f)$ *can be performed in* $O(1)$ *time and with degree* 2.

Finally, test $\mathsf{left\text{-}right}(q_{xy}, e_{xy})$ is implemented by determining the sign of the equation of the line that contains edge $e_{xy}$ when computed at point $q_{xy}$.

**Lemma 12** *Test primitive* $\mathsf{left\text{-}right}(q_{xy}, e_{xy})$ *in an implicit Voronoi diagram of 3D point sites can be performed in* $O(1)$ *time and with degree* 3.

In order to execute a native point location algorithm in an implicit 3D Voronoi diagram, we only need to redefine the implementation of the five test primitives. By having encapsulated the geometry in the test primitives, no further modifications are needed. Hence, by Lemmas 11–12 we obtain:

**Lemma 13** *For any native method on a class of cell-complexes that includes 3D Voronoi diagrams, a point location query in an implicit 3D Voronoi diagram has degree 3 and has the same asymptotic time complexity as a point location query in an explicit 3D Voronoi diagram.*

The Voronoi diagram of $n$ point sites in 3D space is an acyclic convex cell complex with $N = O(n^2)$ facets. Hence, using the separating surfaces method on the implicit 3D Voronoi diagram yields the following result:

**Theorem 10** *Let* $S$ *be a set of* $n$ *point sites in 3D space. There exists an* $O(n^2)$*-space data structure for* $S$ *that can be computed in* $O(n^2)$ *time with degree* 7 *and supports nearest neighbor queries in* $O(\log^2 n)$ *time with degree* 3.

Although the algorithm for nearest neighbor queries proposed in this section has nonoptimal degree 3, it is a practical approach for the important application scenario where the primitive points are pixels on a computer screen. On a typical screen with about $2^{10} \times 2^{10}$ pixels, our nearest neighbor query can be executed with the standard integer arithmetic of a 32-bit processor.

# 5 Point and Segment Sites in the Plane

In this section, we consider the following proximity query on a set $S$ of point and segment sites in the plane:

*nearest neighbor search:* given query point $q$, find a site of $S$ whose Euclidean distance from $q$ is less than or equal to that of any other site.

As for the other queries studied in the previous sections, such query is efficiently solved by performing point location in the Voronoi diagram of the set of point and segment sites [22]. The Voronoi diagram $V(S)$ of a set $S$ of point and segment sites is a map whose edges are either straight-line segments or arcs of parabolas. Hence, in general $V(S)$ is neither convex nor monotone. In order to perform point location in $V(S)$, we refine $V(S)$ into a map with monotone edges as follows. If edge $e$ of $V(S)$ is an arc of parabola whose point $p$ of maximum (or minimum) $y$-coordinate is not a vertex, we split $e$ into two edges by inserting a fictitious vertex at point $p$. We call the resulting map the *extended Voronoi diagram* $V'(S)$ of $S$. The persistent search tree method and the trapezoid method can be used as native methods on the extended Voronoi diagram, where the test primitives are the same as those defined in Section 3 for point sites. If we want to use the chain method or the bridged chain method, we need to do a further refinement that transforms the map into a monotone map by adding vertical fictitious edges emanating from the fictitious vertices previously inserted along the parabolic edges.

**Lemma 14** *The trapezoid method and the persistent search tree method are native, and the chain method and the bridged chain method are ordinary for extended Voronoi diagrams of point and segment sites.*

Let $S$ be a set of $n$ points and segment sites. The extended Voronoi diagram $V'(S)$ of $S$ is said to be *explicit* if the coordinates of the vertices of $V'(S)$ are computed and stored with exact arithmetic, i.e., as algebraic numbers [3, 28]. In the following lemma, we analyze the degree of test primitive $\mathsf{left\text{-}right}(q, e)$ for a straight-line edge $e$ of an explicit extended Voronoi diagram.

**Lemma 15** *The* $\mathsf{left\text{-}right}(q, e)$ *test primitive for a straight-line edge* $e$ *in an explicit extended Voronoi diagram of point and segment sites in the plane has degree* 64.

An algorithm for proximity queries on a set $S$ of point and segment sites in the plane is said to be *conventional* if it computes the explicit extended Voronoi diagram $V'(S)$ of $S$ and then performs point location queries on $V'(S)$ with a native method. Note that the class of conventional proximity query algorithms includes all the efficient algorithms presented in the literature. A conventional proximity query algorithm needs to perform test primitive left-right$(q, e)$. Thus, by Lemma 15 we conclude:

**Theorem 11** *Conventional algorithms for the nearest neighbor query problem on a set of point and segment sites in the plane have degree at least* 64.

Our analysis shows that performing point location in an explicit Voronoi diagram of points and segments is not practically feasible due to the high degree.

The definition of the implicit Voronoi diagram $V^*(S)$ of a set of $S$ of point and segment sites is a straightforward extension of the definition for Voronoi diagrams of point sites given in Section 3. Namely $V^*(S)$ stores the topological structure of the extended Voronoi diagram $V'(S)$ of $S$ (e.g., the data structure of [9]) and the following geometric information for each vertex and edge:

- For each vertex $v$ of $V'(S)$, $V^*(S)$ stores the semi-integer $(b+1)$-bit approximations $x^*(v)$ and $y^*(v)$ of the $x$- and $y$-coordinates of $v$.
- For each non-horizontal edge $e$ of $V'(S)$, $V^*(S)$ stores the pair of sites $\ell(e)$ and $r(e)$ such that $e$ is a portion of the bisector of $\ell(e)$ and $r(e)$, and $\ell(e)$ is to the left of $r(e)$.

In the implicit Voronoi diagram $V^*(S)$ of $S$, test left-right$(q, e)$ is implemented by comparing the distances of query point $q$ from sites $\ell(e)$ and $r(e)$ with one of the following tests, depending on the type (point or line) of sites $\ell(e)$ and $r(e)$: point-to-lines distance test, point-to-point-line distance test, or point-to-points distance test. Thus, by Lemma 2, we have.

**Lemma 16** *For any native method on a class of maps that includes extended Voronoi diagrams of point and segment sites in the plane, a point location query in an implicit Voronoi diagram has degree 6 and has the same asymptotic time complexity as a point location query in an explicit Voronoi diagram.*

**Theorem 12** *Let $S$ be a set of $n$ point and segment sites in the plane. There exists an $O(n)$-space data structure for $S$ that can be computed in $O(n \log n)$ expected time with degree* 40 *and supports nearest neighbor queries in $O(\log n)$ time with degree* 6.

# 6 Simplified Implicit Voronoi Diagrams

In this section, we describe a modification of implicit Voronoi diagrams of point sites that allows us to reduce the degree of the preprocessing task from 5 to 4 when the sites are in the plane (see Theorems 6—8), and from 7 to 5 when the sites are in three-dimensional space (see Theorem 10). This modification also has a positive impact on the space requirement of the data structure and on the running time of point location queries. Let $V(S)$ be the Voronoi diagram of a set $S$ of point sites in the plane. We recall our standard assumption that all input parameters — such as coordinates of sites and query points — are represented as $b$-bit integers. An *island* of $V(S)$ is a connected component of the map obtained from $V(S)$ by removing all the vertices with integer $y$-coordinate and all the edges containing a point with integer $y$-coordinate. Note that for any two vertices $v_1$ and $v_2$ of an island, $y^*(v_1) = y^*(v_2) = m + \frac{1}{2}$ for some integer $m$, where $y^*(v)$ is the semi-integer approximation defined in Section 3. The *simplified implicit Voronoi diagram* $V^\circ(S)$ of $S$ is a representation of the Voronoi diagram $V(S)$ of $S$ that consists of a topological component and of a geometric component. The topological component of $V^\circ(S)$ is the planar embedding obtained from $V(S)$ by contracting each island of $V(S)$ into an *alias vertex*. The geometric component of $V^\circ(S)$ stores the following geometric information for each vertex and edge of the embedding:

- For each vertex $v$ that is also a vertex of $V(S)$, $V^\circ(S)$ stores the $(b+1)$-bit semi-integers approximations $x^*(v)$ and $y^*(v)$.
- For each alias vertex $a$, which is associated with an island of $V(S)$, $V^\circ(S)$ stores semi-integer $y^*(a)$ such that $y^*(a) = y^*(v)$ for each vertex $v$ of the island.
- For each non-horizontal edge $e$ that is also an edge of $V(S)$, $V^\circ(S)$ stores the pair of sites $\ell(e)$ and $r(e)$ such that $e$ is a portion of the perpendicular bisector of $\ell(e)$ and $r(e)$, and $\ell(e)$ is to the left of $r(e)$.

The space requirement of the simplified implicit Voronoi diagram is less than or equal to that of the implicit Voronoi diagram, since each island is represented by a single alias vertex storing only its semi-integer $y$-approximation. We can show examples where the simplified implicit Voronoi diagram of $n$ point sites has $O(n)$ fewer vertices and edges than the corresponding implicit Voronoi diagram.

**Lemma 17** *For any native method on a class of maps that includes monotone maps, a point location query in a simplified implicit Voronoi diagram has optimal degree 2 and executes a number of operations less than or equal to a point location query in the corresponding explicit Voronoi diagram.*

**Lemma 18** *The simplified implicit Voronoi diagram of $n$ point sites in the plane can be computed in $O(n \log n)$ time, $O(n)$ space, and with degree 4.*

The main advantage of the simplified implicit Voronoi diagram with respect to the degree cost measure is that the additional test primitive needed in the preprocessing that consists of comparing the $y$-coordinates of two Voronoi vertices (see the proof of Theorem 6) is now reduced to the comparison of two $(b+1)$-bit semi-integers, and thus has degree 1. Hence, the preprocessing for point location using a native method for monotone maps has degree 1.

**Theorem 13** *Let $S$ be a set of $n$ point sites in the plane. There exists an $O(n)$-space data structure for $S$, based on the simplified implicit Voronoi diagram $V^\circ(S)$, that can be computed in $O(n \log n)$ time with degree 4 and supports nearest neighbor queries in $O(\log n)$ time with optimal degree 2.*

Using a similar approach, we can define simplified implicit order-$k$ Voronoi diagrams for point sites in the plane and simplified implicit Voronoi diagrams for point sites in three-dimensional space. This reduces the degree of the preprocessing from 5 to 4 in Theorems 7 and 8, and from 7 to 5 in Theorem 10.

# References


[1] M. Bocher. *Introduction to higher algebra.* Macmillan, 1907.

[2] C. Burnikel. *Exact Computation of Voronoi Diagrams and Line Segment Intersections.* Ph.D thesis, Universität des Saarlandes, Mar. 1996.

[3] C. Burnikel, J. Könnemann, K. Mehlhorn, S. Näher, S. Schirra, and C. Uhrig. Exact geometric computation in LEDA. In *Proc. 11th Annu. ACM Sympos. Comput. Geom.*, pages C18–C19, 1995.

[4] C. Burnikel, K. Mehlhorn, and S. Schirra. On degeneracy in geometric computations. In *Proc. 5th ACM-SIAM Sympos. Discrete Algorithms*, pages 16–23, 1994.

[5] B. Chazelle. How to search in history. *Inform. Control*, 64:77–99, 1985.

[6] K. L. Clarkson. Safe and effective determinant evaluation. In *Proc. 33rd Annu. IEEE Sympos. Found. Comput. Sci.*, pages 387–395, 1992.

[7] D. P. Dobkin. Computational geometry and computer graphics. *Proc. IEEE*, 80(9):1400–1411, Sept. 1992.

[8] D. P. Dobkin and D. G. Kirkpatrick. Fast detection of polyhedral intersection. *Theoret. Comput. Sci.*, 27:241–253, 1983.

[9] D. P. Dobkin and M. J. Laszlo. Primitives for the manipulation of three-dimensional subdivisions. *Algorithmica*, 4:3–32, 1989.

[10] H. Edelsbrunner, L. J. Guibas, and J. Stolfi. Optimal point location in a monotone subdivision. *SIAM J. Comput.*, 15:317–340, 1986.

[11] H. Edelsbrunner and H. A. Maurer. Finding extreme points in three dimensions and solving the post-office problem in the plane. *Inform. Process. Lett.*, 21:39–47, 1985.

[12] S. Fortune. Stable maintenance of point set triangulations in two dimensions. In *Proc. 30th Annu. IEEE Sympos. Found. Comput. Sci.*, pages 494–505, 1989.

[13] S. Fortune and C. J. Van Wyk. Efficient exact arithmetic for computational geometry. In *Proc. 9th Annu. ACM Sympos. Comput. Geom.*, pages 163–172, 1993.

[14] L. J. Guibas, D. E. Knuth, and M. Sharir. Randomized incremental construction of Delaunay and Voronoi diagrams. *Algorithmica*, 7:381–413, 1992.

[15] L. J. Guibas and J. Stolfi. Primitives for the manipulation of general subdivisions and the computation of Voronoi diagrams. *ACM Trans. Graph.*, 4:74–123, 1985.

[16] C. M. Hoffmann. The problems of accuracy and robustness in geometric computation. *IEEE Computer*, 22(3):31–41, Mar. 1989.

[17] M. Karasick, D. Lieber, and L. R. Nackman. Efficient Delaunay triangulations using rational arithmetic. *ACM Trans. Graph.*, 10:71–91, 1991.

[18] D. G. Kirkpatrick. Optimal search in planar subdivisions. *SIAM J. Comput.*, 12:28–35, 1983.

[19] D. T. Lee and F. P. Preparata. Location of a point in a planar subdivision and its applications. *SIAM J. Comput.*, 6:594–606, 1977.

[20] G. Liotta, F. P. Preparata, and R. Tamassia. Robust proximity queries: an illustration of degree-driven algorithm design. *SIAM J. Comput.* (to appear). Also available as Technical Report CS-96-16, Center for Geometric Computing, Comput. Sci. Dept., Brown Univ., Providence, RI, 1996. URL: http://www.cs.brown.edu/cgc/cgc-papers/cgc-papers.html.

[21] F. P. Preparata. A new approach to planar point location. *SIAM J. Comput.*, 10:473–482, 1981.

[22] F. P. Preparata and M. I. Shamos. *Computational Geometry: An Introduction.* Springer-Verlag, New York, NY, 1985.

[23] F. P. Preparata and R. Tamassia. Efficient point location in a convex spatial cell-complex. *SIAM J. Comput.*, 21:267–280, 1992.

[24] N. Sarnak and R. E. Tarjan. Planar point location using persistent search trees. *Commun. ACM*, 29:669–679, 1986.

[25] K. Sugihara and M. Iri. Construction of the Voronoi diagram for 'one million' generators in single-precision arithmetic. *Proc. IEEE*, 80(9):1471–1484, Sept. 1992.

[26] R. Tamassia and J. S. Vitter. Optimal cooperative search in fractional cascaded data structures. *Algorithmica*, 15(2), 1996.

[27] C. K. Yap. Symbolic treatment of geometric degeneracies. *J. Symbolic Comput.*, 10:349–370, 1990.

[28] C. K. Yap. Toward exact geometric computation. *Computational Geometry: Theory and Applications* (to appear).

# Efficient Exact Evaluation of Signs of Determinants *

Hervé Brönnimann[†] Mariette Yvinec[‡]

**Abstract**

This paper presents a theoretical and experimental study on two different methods to evaluate the sign of a determinant with integer entries. The first one is a method based on the Gram-Schmidt orthogonalisation process which has been proposed by Clarkson [Cla92]. We review the analysis of Clarkson and propose a variant of his method. The second method is an extension to $n \times n$ determinants of the ABDPY method [ABD$^+$94] which works only for $2 \times 2$ and $3 \times 3$ determinants. Both methods compute the signs of a $n \times n$ determinants whose entries are integers on $b$ bits, by using an exact arithmetic on only $b + O(n)$ bits. Furthermore, both methods are adaptive, dealing quickly with easy cases and resorting to the full-length computation only for null determinants.

## 1 Introduction

Geometric algorithms are known to be highly sensitive to numerical inaccuracy. Those algorithms generally rely on building discrete combinatorial structures whose actual state depend on the outcomes of some numerical tests. In this context, roundoff errors leads quickly to fatal inconsistencies and failure of the programs. Robustness has now become one of the major issue in the field of computational geometry [C$^+$96, chap. 10].

Some attempts have been made to design geometric algorithms such that robust implementations can be obtained using only the inaccurate but fast arithmetic provided by floating point processors (see for examples [SI89, SI94, Mil89, LM90, Hof89, HHK88, For92]). Such solutions, although very useful in some domains like solid modeling and CSG applications, are still painful to design and known only for a few geometric problems. Another approach is to turn to exact arithmetic which makes robustness a non issue. The use of exact arithmetic has been recently advocated by Fortune and Van Wyk [FV93], Yap [Yap93, YD95] , Burnikel and coll. [BKM$^+$95] and many others. However, as reported for example by Karasick and coll. [KLN91], naive implementation of exact arithmetic can be quite slow and many works are now devoted to speed up the paradigm of exact geometric computing. Fortunately exact geometric computing does not imply computing everything exactly. Most of the numerical tests arising in geometric algorithm amount to determine only the sign of a determinant or of a polynomial expression. Thus arithmetic filters based on a fast floating point evaluation of the expression and of a bound on the error allow very often to make safe decision. This approach is used for example in the LN package by Fortune and Van Wyk [FV93] and has been shown experimentally to provide a substantial speedup. Devillers and Preparata investigate the theoretical behavior of some filters [DP96]. In degenerate or near degenerate cases, however, exact arithmetic has to be carried out in full. Burnikel and coll. [BKM$^+$95] and Yap [Yap93] provide powerful software to perform exact arithmetic on algebraic numbers. But numerical tests arising in geometric algorithms are not arbitrary. In fact, most geometric algorithms rely on a few number of geometric predicates such as which-side or orientation test, in-circle or in-sphere tests, all of which amount to compute the sign of a determinant. Thus designing a specialized implementation for evaluating exactly the sign of a determinant can in many cases avoid to pay the price of a general purpose multi-precision package. Clarkson [Cla92] propose an efficient method to compute the sign of a determinant whose entries are

*This research was partially supported by the ESPRIT IV LTR Project No. 21957 (CGAL).

[†]INRIA Sophia-Antipolis, BP 93, 06902 Sophia-Antipolis cedex (France).

[‡]INRIA Sophia-Antipolis and CNRS, Laboratoire I3S, BP 145, 06903 Sophia-Antipolis cedex (France)

integers. The so called ABDPY method, due to Avnaim and col. [ABD+94], is an alternative solution for $2 \times 2$ and $3 \times 3$ determinants. Using an uncommon multiple-term arithmetic, Schewchuck [She96] designs an adaptative implementation for low dimensional geometric predicates on floating point entries.

In this paper, we propose a theoretical and experimental study on the exact evaluation of the sign of a determinant with integer entries. Mainly, we have revisited the method of Clarkson and extended to higher dimensions the ABDPY method.

Clarkson's method is based on the Gram-Schmidt orthogonalization process and will be called hereafter the reorthogonalization method. We propose a variant of his algorithm which is somewhat simpler to analyze. This variant allows to compute the sign of an $n \times n$ determinant whose entries are $b$-bit integers, that is, integers whose absolute values are smaller than $2^b$, using an exact arithmetic on $b + \lceil 1.32(n-1) + 0.5 \log n \rceil$ bits, and with a worst case complexity of $O(bn^3 + n^3 \log n)$.

We call lattice method the extension of ABDPY method to higher dimensions. In fact the lattice method, borrows from both the ABDPY method and Clarkson's method. Like ABDPY, it mostly consists in locating one of the column vectors of the determinant with respect to a region approximating the hyperplane spanned by the others column vectors. But, when the vector is found to lie within the approximating region, the algorithm resorts to an iterative doubling technique which was suggested to us by the study of Clarkson's method. For a $n \times n$ determinant with $b$-bit integer entries the lattice method requires an exact arithmetic on $b + n - 2 + \lceil \log n \rceil$ bits and, though its worst case complexity is exponential, its behavior is $O(bn^3)$ in most cases.

The efficiency of both methods comes mostly from the small number of extra bits they require. Indeed, in most practical cases, the exact arithmetic required will stay within the 53 bits of precision available in the mantissa of standard IEEE doubles. Then all numerical computations entailed by the evaluation of the sign of a determinant can be performed exactly using the floating point processor of any computer conforming to the IEEE norm. Another factor of efficiency of those methods is their adaptivity. Both are iterative methods and the number of iterations performed depends on the actual value of the determinant. This number is quite small in easy cases where the determinant is far from zero.

The next two sections respectively present the reorthogonalization and the lattice method. Experimental evidence of their efficiency is given in the last section.

## 2 The reorthogonalization method

**Overview.** The goal is to compute $\det(\mathcal{A})$, where $\mathcal{A}$ is a matrix whose coefficients are $b$-bit integers. One may use Gram-Schmidt reductions to compute an orthogonal matrix $\mathcal{C}$ whose determinant is the same as that of $\mathcal{A}$, and such that each column $C_k$ of $\mathcal{C}$ is $A_k + \mathrm{LC}(A_1, \ldots, A_{k-1})$, where LC denotes any linear combination of its arguments. Computing the determinant of $\mathcal{C}$ can then be done by standard Gaussian elimination. In practice, roundoff errors result in a matrix $\mathcal{B}$ that approximates $\mathcal{C}$. The sign of $\det(\mathcal{B})$ is correct if $\mathcal{A}$ is well-conditioned. It could be wrong, however, if $\mathcal{A}$ is ill-conditioned. Clarkson's idea amounts to preconditioning the matrix $\mathcal{A}$ without introducing errors, before doing Gram-Schmidt reductions. The invariant enforced by preconditioning is that each column $B_k$ of $\mathcal{B}$ obtained by reducing the preconditioned $A_k$ verifies $2{B_k}^2 \geq {A_k}^2$.

Clarkson's preconditioning applies a variant of Gram-Schmidt reductions in which the vector subtracted from $A_k$ is a linear combination of $A_k^{(1)}, \ldots, A_{k-1}$ (rather than $C_k^{(1)}, \ldots, C_{k-1}$ for standard Gram-Schmidt reductions). Thus, whatever the coefficients in this combinations be, $\det(\mathcal{A})$ is not affected as long as the linear combination is computed exactly. Nevertheless, the norm of $A_k$ decreases in the process, so we may multiply it by a suitable factor $s \geq 2$ without overflowing. Iterating for a given $k$ until the invariant is satisfied leads to a transformed matrix $\mathcal{A}'$ whose determinant has been multiplied by a factor $\prod s > 0$. Therefore the sign of $\det(\mathcal{A}')$ is the same as that of $\det(\mathcal{A})$.

The iterative preconditioning process is depicted in figure 1, and the code for preconditioning and reducing is given below. It is the same as that of [Cla92]. Our variant is in the choice of $s$ that leads to a much simpler analysis.

**for** $k := 1$ **to** $n$
  **loop**
  $B_k^{(k)} := A_k$
  **for** $j := k-1$ **downto** 1
    $B_k^{(j)} := fl\left(B_k^{(j+1)} - fl\left(fl\left(\frac{A_k}{B_j}\right) B_j\right)\right)$
  **if** $fl(A_k \cdot A_k) \leq 2 fl\left(B_k^{(1)} \cdot B_k^{(1)}\right)$
    $B_k := B_k^{(1)}$, **exit loop**
  Compute some integer $s$
  $A_k^{(k)} := s \times A_k$
  **for** $j := k-1$ **downto** 1
    $A_k^{(j)} := A_k^{(j+1)} - \left\lceil fl\left(\frac{A_k^{(j+1)}}{B_j}\right)\right\rfloor A_j$
  $A_k := A_k^{(1)}$
  **end loop**

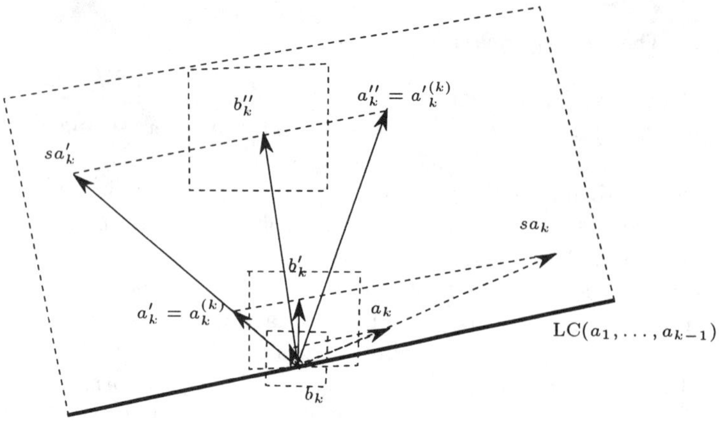


Figure 1: The process of reducing vector $A_k$ is shown. At any time, $A_k$ remains in the bounding box, but its component along the orthogonal of $(A_1, \ldots, A_{k-1})$ increases at each step. The computed vector $B_k$ lies in the square box. At the first and second steps, the box does not ensure that the sign of the determinant is known safely, but it becomes so after the third reduction of $A_k$.

Notation *fl* (expr) means to evaluate an expression within machine floating point precision. Lines 3–5 compute the Gram-Schmidt reductions, lines 6–7 check if the invariant is satisfied, and if not, lines 10–13 compute the modified Gram-Schmidt reductions and loop.

When the determinant is null, some vector $C_k$ is null and the invariant $2\|B_k\|^2 \geq \|A_k\|^2$ is never satisfied. In this case, the algorithms stops and concludes that the determinant is null after a certain number of iterations have been performed (this number is given in the complexity analysis below). Otherwise, let $\mathcal{B}'$ be the floating point approximation of $\mathcal{C}'$, obtained from $\mathcal{A}'$ by Gram-Schmidt reductions. Since the preconditioning invariant is satisfied, it is not hard to show that $\det(\mathcal{B}')$ can be evaluated with relative precision less than 1 (using standard Gaussian elimination [Cla92, FM67]). Hence the sign of $\det(\mathcal{A})$ is computed exactly, and a good approximation $(\prod s)^{-1} \det(\mathcal{B}')$ is available.

**Needed arithmetic.** To bound the number of bits needed by the algorithm, we first obtain a bound on the error $\mathcal{B} - \mathcal{C}$ and deduce a bound on the norms of the coefficients of $\mathcal{A}$ after several reduction steps.

To bound the error $\mathcal{B} - \mathcal{C}$, we assume that the algorithm has conditioned $A_1, \ldots, A_{k-1}$ (therefore the preconditioning invariant is valid for $j < k$), and has performed a number of preconditioning steps on $A_k$. When we exit the reduction of $A_k$, the invariant is satisfied for $j = k$. In [Cla92], it is shown that this implies that $\|B_j - C_j\| \leq \delta_j \|B_j\|$ for all $j \leq k$, where the $\delta_j$'s depend on $j$, $n$, and the precision $\mathbf{u}$ of the floating point computations. We can assume that $\delta_n \leq 0.01$, for all practical values of $n$ and $\mathbf{u}$, which implies loose limits on $n$ as a function of $\mathbf{u}$.

Having bounded the error $\mathcal{B} - \mathcal{C}$, we must now bound the coefficients of $\mathcal{A}$ after several reduction steps. Let $A'_k$ be the vector $A_k$ after several modified Gram-Schmidt reductions. As is done in [Cla92], we can bound the norms of all the vectors $A_k^{(j)}$ computed in line 11 by

$$\left\| A_k^{(j)} \right\|^2 \leq 0.55\, s^2 \|A'_k\|^2 + 0.51\, S_k^c\,, \tag{1}$$

where $S_k^c = \sum_{j=1}^{k-1} \|C_j\|^2$. It is now clear that, to control the growth of $A_k$ after several reductions, we must choose $s$ carefully in terms of $S_k^c$ and $\|A'_k\|$. The value $S_k^c$ is unknown to the algorithm, however, and only an approximation $S_k^b$ of $S_k^c$ is known. Assuming that $\delta_n \leq 0.01$, we can show (as in done in [Cla92]) that $\left|S_k^b - S_k^c\right| \leq 0.07 S_k^c$.

Clarkson's choice [Cla92] is given in terms of $S_k^b$ and $\|B_k\|$, and yields a complicated analysis. Instead, we pick (with a parameter $\lambda$ to be set later)

$$s = \left\lceil fl\left(\sqrt{1.3 + \frac{S_k^b}{\lambda \left\|A'_k\right\|^2}}\right)\right\rfloor.$$

If $s = 1$, we must ensure that $A'_k$ is indeed shrinking, otherwise the algorithm could loop infinitely. But if $s = 1$, the square root is (conservatively) smaller than 1.51, hence $S_k^b \geq \lambda \|A'_k\|^2$, and we know that $S_k^c \leq 1.07 S_k^b$. Plugging into (1) shows that $\left\|A_k^{(1)}\right\|^2 \leq 0.57(1+\lambda)\|A'_k\|^2$. We now set $\lambda = 0.42$ to obtain $\left\|A_k^{(1)}\right\| \leq 0.9\|A'_k\|$. Thus, if $s = 1$, $A'_k$ shrinks in norm by at least 10%.

Otherwise, put $s' = 1.3 + S_k^b/(\lambda\|A'_k\|^2)$ so that $s = \lceil fl\,(s')\rfloor$. Estimating rounding errors conservatively, we can safely say that $s'$ is less than $(s-0.51)^2$, and hence that

$$s^2\|A'_k\|^2 \leq \frac{s^2}{(s-0.51)^2 - 1.3} \frac{1.07}{\lambda} S_k^c \leq 11.05, S_k^c, \tag{2}$$

since $s$ is greater than 2 and $\lambda$ is set at 0.42. Plugging this bound into (1), we finally get that $\left\|A_k^{(1)}\right\|^2 \leq 6.59\, S_k^c$, which bounds the norm of the next vector $A'_k$.

From all this, we gather (conservatively) that

$$\|A'_k\|^2 \leq \max\left(\|A_k\|^2, 6.59 S_k^c\right)$$

where $A'_k$ is after any number of reductions of $A_k$. The key here is that $S_k^c$ depends only $C_1 \ldots, C_{k-1}$ and remains fixed during the $k$-th reduction loop (no matter how many times we enter the loop).

To obtain a bound on how many bits are needed by the algorithm is now routine. If $\mathcal{A}$ has $b$-bit integer coefficients, the maximum norm of a vector $A_j$ is $\sqrt{n}2^b$. Then $\left\|A_k^{(j)}\right\|^2 \leq \max\left(n2^{2b}, 6.59S_k^c\right)$ holds for all $k \geq n$, for all $j \leq k$, and after any number of reductions of $A_k$. By induction, we find that $S_k^c \leq n\,6.59^{k-2}2^{2b}$. Hence $b + \lceil 1.36(n-1) + 0.5\log n \rceil$ bits suffice to express all the vectors $A_k$ and $A_k^{(j)}$ occurring in the algorithm.

**Complexity.** It is also simple to bound the number of iterations. Indeed, each iteration in which $s = 1$ decreases the norm of $A'_k$ by 10% while keeping the determinant constant, and the others multiply the determinant by a value $s \geq 2$ while keeping the coefficients of $\mathcal{A}$ in the allowable range. When the determinant is not null, each iteration of the first kind decreases the orthogonality defect $\prod_{1 \leq j \leq k} \|A_j\|/\|C_j\|$ by 10%, and the defect is not increased by iterations of the second kind; since it is greater than 1 and smaller than $\det(\mathcal{A})$, there cannot be more than $t_1 = O(\log\det(\mathcal{A}))$ such iterations. After $t_2$ iterations of the second kind, we have $2^{t_2}|\det(\mathcal{A})| \leq |\det(\mathcal{A}')| \leq n^{n/2}2^{bn}$. Hence $t_1 + t_2 = O(bn + n\log n)$. When the determinant is null, the algorithms stops after $O(bn+n\log n)$ iterations and concludes that the determinant is null. Each iteration performs $O(n^2)$ operations, and the final Gaussian elimination performs $O(n^3)$ operations. Hence the complexity of the algorithm is $O(bn^3 + n^3\log n)$.

In fact, the algorithm does not usually perform so many iterations. When reducing $A_k$, the product $P_k^c = \prod_{1 \leq j \leq k} \|C_j\|$ has been multiplied by a quantity which we denote by $\prod s$. Since $P_k^c$ is either 0 or greater than 1 to start with, it must be zero or at least $\prod s$. The following upper bound on $P_k^c$ can be computed by the algorithm:

$$P_k^b = \prod_{j=1}^{k-1}(1+\delta_j^2)\|B_j\|^2 \times (\|B_k\|^2 + \delta_k^2\|A'_k\|^2).$$

As soon as $P_k^b \leq \prod s$, we can rest assured that the first $k$ columns are linearly dependent and that the determinant is 0. Since $\mathcal{B}'$ is close in practice to $\mathcal{C}'$, this is usually how the loop is exited when the determinant is null.

## 3 The lattice method

**Overview.** Here and in the following we assume that we want to evaluate the sign of the determinant $D = \det(U_1, U_2, \ldots, U_n)$ where $U_1, U_2, \ldots, U_n$ are $n$-dimensional column vectors whose coordinates are supposed to be $b$-bit integers.

The lattice method considers in a special way the last column and last row of determinant D. We note $z_1, z_2 \ldots z_n$ the last components of the input vectors and $u_1, u_2, \ldots, u_n$ their orthogonal projections on $\mathbb{R}^{n-1}$.

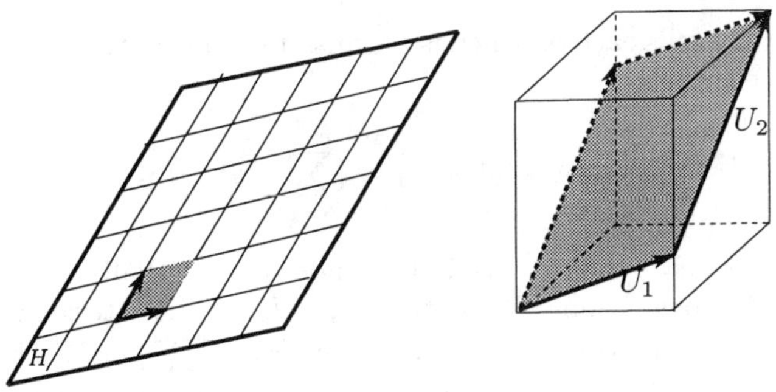


Figure 2: The cell $\mathcal{C}_H$ and the box $\mathcal{B}$

Without loss of generality, $z_1, z_2, \ldots, z_n$ are assumed to be non negative. We choose the same point $O$ as origin for $\mathbb{R}^n$ and $\mathbb{R}^{n-1}$, and we note with the same capital letter $U$ either a vector in $\mathbb{R}^n$ or the point $O + U$, and with the same small letter $u$ either a vector in $\mathbb{R}^{n-1}$ or the point $O + u$.

Let $H$ be the hyperplane through $\{O, U_1, U_2, \ldots, U_{n-1}\}$. The basic underlying idea is that computing the sign of the $n \times n$ determinant $D$ reduces to compute the sign of a $(n-1) \times (n-1)$ determinant if the position of the point $U_n$ with respect to the hyperplane $H$ is known. Indeed, assuming that vectors $U_1, U_2, \ldots, U_{n-1}$ together with $E_n$, the unit vector along the $n$th axis, span $\mathbb{R}^n$, we have $U_n = \mathrm{LC}(U_1, U_2, \ldots, U_{n-1}) + \alpha E_n$ and $D = \alpha\det(u_1, u_2, \ldots, u_{n-1})$ where the sign of $\alpha$ only depends of the position of $U_n$ with respect to the hyperplane $H$.

Locating $U_n$ with respect to $H$ amounts in turn to evaluate the sign of $D$, and this might look like going round in circle. In fact, in a first phase, point $U_n$ is located with respect to some region $\mathcal{H}$ of $\mathbb{R}^n$ which contains $H$ and can be considered as an approximation of $H$. If $U_n$ is found to be outside region $\mathcal{H}$, the position of $U_n$ with respect to $H$ is known and the problem is reduced by one dimension. Otherwise (and this is where the lattice method mostly departs from ABDPY), the algorithm enters a second phase in which the last column vector and therefore the numerical value of the determinant are iteratively doubled, which amounts to iteratively refine the approximation $\mathcal{H}$ of $H$.

Before being more precise on those two phases of the algorithm, let us describe region $\mathcal{H}$. We consider the lattice $\mathcal{L}_H$ formed by vectors in $H$ that are linear combination of $\{U_1, U_2, \ldots, U_{n-1}\}$ with integer coefficients, and the lattice $\mathcal{L}$ which is the projection of $\mathcal{L}_H$ in $\mathbb{R}^{n-1}$.

$$\mathcal{L}_H = \left\{\sum_{i=1}^{n-1} l_i U_i, l_i \in \mathbb{N}\right\}, \quad \mathcal{L} = \left\{\sum_{i=1}^{n-1} l_i u_i, l_i \in \mathbb{N}\right\}.$$

The lattice $\mathcal{L}$ induces a partition of $\mathbb{R}^{n-1}$ into elementary cells each of which is a translated copy of the origin cell :

$$\mathcal{C} = u_1 \oplus u_2 \oplus \ldots \oplus u_{n-1} = \{\sum_{i=1}^{n-1} \alpha_i u_i,\ \ 0 \leq \alpha_i < 1\}$$

, where the $\oplus$ symbol stands for a (half-closed) Minkowski sum. We note $\mathcal{C}(l_1, \ldots l_{d-1})$ the lattice cell that is the translated of $\mathcal{C}$ by the vector $\sum_{i=1}^{d-1} l_i u_i$. The point $c(l_1, \ldots l_{d-1}) = \sum_{i=1}^{d-1} l_i u_i$ is called the reference point of the cell $\mathcal{C}(l_1, \ldots l_{d-1})$. Analogous definitions and notation hold for the lattice $\mathcal{L}_H$ whose elementary cells partition $H$ : the cell $\mathcal{C}_H(l_1, \ldots l_{n-1})$ of $\mathcal{L}_H$ is the copy of the origin cell $\mathcal{C}_H = U_1 \oplus U_2 \oplus \ldots \oplus U_{n-1}$ translated by the vector $\sum_{i=1}^{n-1} l_i U_i$. The cell $\mathcal{C}(l_1, \ldots l_{d-1})$ of $\mathcal{L}$ is the orthogonal projection of the cell $\mathcal{C}_H(l_1, \ldots l_{d-1})$ of $\mathcal{L}_H$.

To describe the region $\mathcal{H}$, we consider the $n$-dimensional box (see figure 2)

$$\mathcal{B} = u_1 \oplus u_2 \oplus \ldots \oplus u_{n-1} \oplus \left(\sum_{i=1}^{n-1} z_i\right) E_n.$$

Box $\mathcal{B}$ has the origin cell $\mathcal{C}$ of lattice $\mathcal{L}$ as orthogonal projection and contains the origin cell $\mathcal{C}_H$ of $\mathcal{L}_H$. Then, region $\mathcal{H}$ is defined as the union of all the boxes $\mathcal{B}(l_1, \ldots l_{n-1})$ which are copies of $\mathcal{B}$ translated by the vectors of lattice $\mathcal{L}_H$. Since $\mathcal{B}$ contains the origin cell $\mathcal{C}_H$, $\mathcal{B}(l_1, \ldots l_{n-1})$ contains the cell $\mathcal{C}_H(l_1, \ldots l_{d-1})$ and $\mathcal{H}$ contains $H$.

Now, the overview of the algorithm is a follows.

- *First phase.* The algorithm determines the position of point $U_n$ with respect to $\mathcal{H}$.
  - If point $U_n$ is found to be above or below $\mathcal{H}$, the relative position of $U_n$ with respect to $H$ is known and the algorithm ends up with evaluating the sign of a $(n-1) \times (n-1)$ determinant.
  - Otherwise, $U_n$ lies within some box $\mathcal{B}(k_1, \ldots k_{n-1})$ of $\mathcal{H}$ and the algorithm computes the vector $R = U_n - \sum_{i=1}^{n-1} k_i U_i$ which satisfies both following conditions $D = \det(U_1, U_2, \ldots, R)$ and $R \in \mathcal{B}$. Then, the second phase is entered.
- *Second phase.* A variable number of the following iterations are performed. Iteration $t$ takes as input a vector $R_t$ of $\mathbb{R}^n$ such that

$$D_t = \det(U_1, U_2, \ldots, R_t) = 2^t D \text{ and } R_t \in \mathcal{B}. \quad (3)$$

(Initially, $D_0 = D$ and $R_0 = R$.) The algorithm sets $R' = 2R_t$ and considers the determinant $D_{t+1} = \det(U_1, U_2, \ldots, R') = 2D_t$. Point $R'$ is located with respect to the union of boxes $\mathcal{H}$.

  - If $R'$ is found outside $\mathcal{H}$ the relative position of $R'$ and $H$ is known and the algorithm ends up getting the sign of $D_{t+1}$ which is also the sign of $D$ from the sign of a $(n-1) \times (n-1)$ determinant.

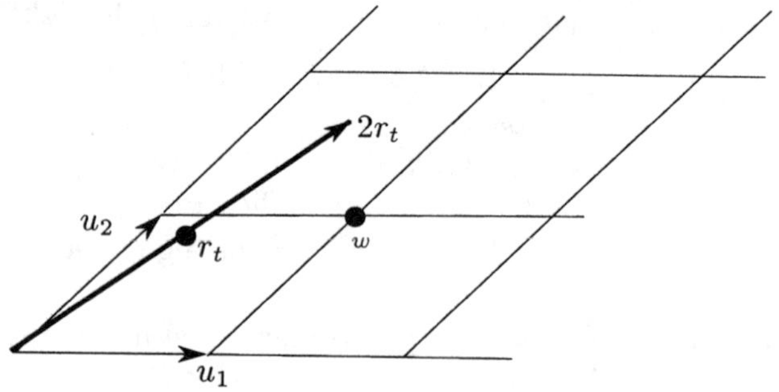


Figure 3: Locating $2r_t$ in the lattice $\mathcal{L}$

  - Otherwise $R'$ belongs to a box $\mathcal{B}(k_1, \ldots k_{n-1})$ of $\mathcal{H}$ and the algorithm computes the vector $R_{t+1} = R' - \sum_{i=1}^{n-1} k_i U_i$ which verifies both $D_{t+1} = \det(U_1, U_2, \ldots, R_{t+1}) = 2^{t+1} D$ and $R_{t+1} \in \mathcal{B}$. Then one proceeds to the next iteration with $R_{t+1}$

The algorithm ends up evaluating the sign of a $(n-1) \times (n-1)$ determinant as soon as one of the points $R' = 2R_t$ is located outside $\mathcal{H}$. If the determinant $D$ is null, this will never happen but the algorithm can stop and be sure than $D$ is zero after at most $bn + \lceil (\frac{n}{2} + 1) \rceil \log n$ iterations. Indeed, at each iteration the value of the determinant is multiplied by a factor 2. Thus, $D_t = 2^t D$ which is more than $2^t$ if $D$ is not zero. On the other end, each entry of determinant $D_t$ is at most $2^b$ except those of the last column (the components of $R_t$) which are at most $n2^b$ because $R_t$ belongs to box $\mathcal{B}$. Thus $D(t)$ is less than $n\sqrt{n}^n 2^{bn}$. Therefore, when $2^t > n\sqrt{n}^n 2^{bn}$, the algorithm can stop and conclude that $D$ is null.

**Further details.** Due to lack of space, we skip the detailed description of the first phase.

At each iteration of the second phase, the algorithm is given a point $R_t(r_t, z_t)$ in $\mathcal{B}$ and has to locate the endpoint of $R' = 2R_t$ with respect to $\mathcal{H}$. For this the algorithm determines the reference point $c_{t+1}$ of the cell of lattice $\mathcal{L}$ containing the endpoint of the projection $2r_t$ of $R'$. Owing to the fact that $r_t$ belongs to the origin cell $\mathcal{C}$, point $c_{t+1}$ is necessarily one of the $2^{d-1}$ vertices of $\mathcal{C}$ (see figure 3) and finding which one amounts to determine the sign of the following $(n-1) \times (n-1)$ determinants, for $i = 1, \ldots, n-1$:

$$a_i = [u_1, \ldots, u_{i-1}, (2r_t - w), u_{i+1}, \ldots, u_{n-1}],$$

where $w = \sum_{i=1}^{n-1} u_i$. Let $C_{t+1}$ be the lattice point of $H$ that projects on $c_{t+1}$. We compute $R_{t+1} = 2R_t - C_{t+1}$. If the last component $z_{t+1}$ of $R_{t+1}$ is negative or greater than $\sum_{i=1}^{n-1} z_i$, the position of $R_{t+1}$ with respect to $H$ is known. Else $D_{t+1} = \det(U_1, U_2, \ldots, R_{t+1}) = 2^{t+1} D$ and $R_{t+1} \in \mathcal{B}$ and the algorithm proceeds to iteration $t+1$.

| Test | Det3x3 | | | Det4x4 | | | Det5x5 | | | Det6x6 | | |
|---|---|---|---|---|---|---|---|---|---|---|---|---|
| | $b = 50$ | | | $b = 49$ | | | $b = 47$ | | | $b = 46$ | | |
| Determinant | R | Q | N | R | Q | N | R | Q | N | R | Q | N |
| Gauss | 9 | 9 | 9 | 13 | 13 | 12 | 23 | 21 | 22 | 36 | 35 | 34 |
| Leda | 339 | 334 | 337 | 1661 | 1648 | 1650 | 5503 | 5487 | 5250 | 9800 | 8600 | 8300 |
| Reorth. | 4 | 30 | 403 | 77 | 386 | 1100 | 149 | 604 | 2207 | 263 | 848 | 3354 |
| Lattice | 11 | 25 | 345 | 88 | 187 | 1232 | 352 | 635 | 2243 | 888 | 1491 | 4051 |
| Filt.+Lat. | 3 | 13 | 355 | 13 | 104 | 1171 | 13 | 288 | 2284 | 26 | 1060 | 4137 |
| LN | 3 | 53 | 53 | 4 | 295 | 296 | na | na | na | | | |

Figure 4: Timings in microseconds on a Sun Sparc5, 110MHz, running Solaris (na: not available). R, Q, N stand respectively for random, quasi-null, and null determinants

**Needed arithmetic.** To evaluate the sign of an $n \times n$-determinant, the above algorithm performs different operations among which are the evaluations of signs of $(n-1) \times (n-1)$-determinants. To evaluate the signs of those determinants, the algorithm calls the $(n-1)$-dimensional version of the same algorithm which in turn involves evaluating signs of $(n-2) \times (n-2)$ determinants and so on until dimension 2 is reached, where the original ABDPY method is used. We call $k$-th level of computation the set of numerical computations performed in calls to the algorithm evaluating the sign of $k \times k$-determinants except the computations involved in evaluating the sign of $(k-1) \times (k-1)$-determinants. In the following, we first focus on the computations performed at the $n$-th level and then consider the lower levels.

At the $n$-th level of computations, the input vectors $\{U_1, U_2, \dots U_n\}$ are all vectors with $b$-bit integer entries. However to be able to generalize our conclusions to lower levels, we shall here assume a weaker hypothesis. Namely, we consider the $L_\infty$ norm of the input vectors (the $L_\infty$ norm $\|U\|_\infty$ of a vector $U$ is the maximum absolute value of any components of $U$), and define the *norm* of determinant $\det(U_1, U_2, \dots U_n)$ as the sum $\sum_{i=1}^{d} \|U_i\|_\infty$ of the $L_\infty$ norm of its column vectors. The following fact is proved in the full length version of this paper.

**Fact 1** *If the norm of $D$ is less than a constant $S_n$, any of the $(n-1)\times(n-1)$ determinants considered by the algorithm has a norm less than $S_{n-1} = 2S_n$. Furthermore computations performed at the nth level do not require an exact integer arithmetic on more than $\lceil \log S_n \rceil + 1$ bits.*

An easy recurrence using the above fact shows that if the norm of the $n \times n$ determinant $D$ is less than $S_n$, the computations performed at level $k$ require an exact integer arithmetic on $\lceil \log S_k \rceil + 1$ bits with $S_k = 2^{n-k} S_n$. For a $n \times n$-determinant with $b$-bits integer entries, the norm hypothesis is satisfied with $S_n = n2^b$. Since the original ABDPY algorithm yields the sign of a $2 \times 2$ determinant with $b'$-bits integer entries using only a $b'$-bit arithmetic [ABD+95], the whole algorithm requires only an exact integer arithmetic on $n - 2 + \lceil \log n \rceil + b$ bits.

**Complexity.** During the first phase, the algorithm requires at most $O(bn)$ evaluations of signs of $(n-1)\times(n-1)$ determinants (this is clear from the full description of the first phase of the algorithm). During the second phase, the algorithm performs at most $O(bn + n \log n)$ iterations each of which involves $(n-1)$ evaluation of signs of $(n-1) \times (n-1)$ determinants. The other computations at the $n$-th level take time $O(n^2)$. Thus the complexity $t_n$ obeys a recurrence equation of the form

$$t_n = O(bn^2 + n^2 \log n) t_{n-1} + O(n^2),$$

which leads to an exponential complexity $t_n = O((b + \log n)^{n-1} (n!)^2)$. But this bound is very pessimistic: indeed, even if the determinant is null and requires a full-fledged first phase followed by $bn$ iterations in the second phase, the $(n-1) \times (n-1)$ encountered determinants have no reason to be close to zero. Therefore they are very likely to be caught by some floating point filter. As the different determinants arising in the loops of the first phases or during iterations of the second phase differ only by a single column, the floating point evaluations involved in filtering are reduced to a scalar product (using the minors relative to the changing column). Thus in practice, evaluating the sign of each encountered $(n-1) \times (n-1)$ determinant takes only $O(n)$ time and the time required by the whole algorithm should be close to $O(bn^3 + n^3 \log n)$ even for a null determinant. Of course, if the determinant is not even close to zero, few iterations are performed and the algorithm is very fast.

## 4 Experimental results

The reorthogonalization method and the lattice method have both been implemented in $C$. The lattice method is at present time available up to dimension five only but will be soon available in dimension 6 and higher. To be able to compare the efficiency of those methods with respect to others we have also implemented a floating point Gaussian elimination (which of course does not always yields the right sign) and an exact computation

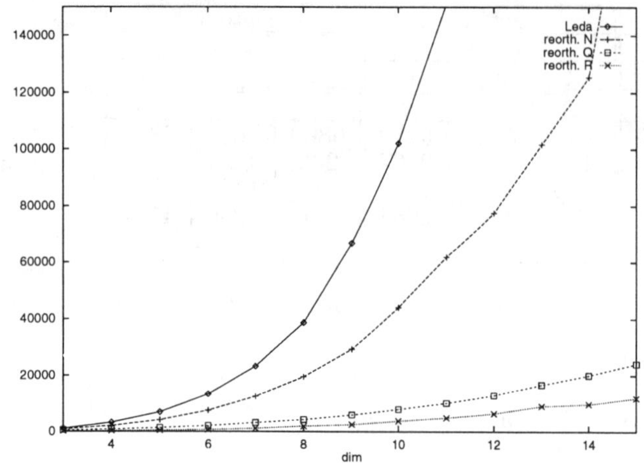


Figure 5: Timing as function of the dimension for the exact computation of signs of $n \times n$ determinants. The timing are in microsecond. The solid curve correspond respectively to an exact computation using Leda. The three dotted curves correspond to the reorthogonalization method on respectively null, quasi-null and random determinants. Using the lattice method instead would yield almost identical behavior.

of the determinant using the exact integer arithmetic provided by LEDA. In order to show the practical efficiency of our methods and to get a fair comparison with the code produced by LN [FV93] for sign of determinants computation, we have also implemented the lattice method combined with a floating point filter (the results would have been quite similar for the reorthogonalization method).

We have experimented on determinants of dimensions $n$ from 2 to 6 with $b$-bit integers entries, where $b = 53-(n-2+\lceil \log n \rceil)$. (This is the precision bound allowed for the lattice method and slightly over the bound allowed for reorthogonalization, however these bounds are pessimistic and, in practice the reorthogonalization method works and yields the correct sign up to this precision on the entries). In each dimension, three types of determinants respectively called *random*, *null* and *quasi-null* have been used. Random determinants have as entries random signed integers numbers on $b$ bits and their value is of the the order of $2^{bn}$. Null determinants are formed by $n-1$ column vectors of the form $k_i U_i$ and a last column of the form $\sum_{i=1}^{n-1} l_i U_i$ where the components of vectors $U_i$ are random signed integers numbers on $\lceil b/2 \rceil$ bits while the coefficients $k_i$ and $l_i$ are random signed integers numbers on $\lfloor b/2 \rfloor$ bits. Therefore null determinants have rank $n-1$. Quasi-null determinants are obtained by a small perturbation of null determinants adding to each entry a random sign integer on two bits. Timing results appear in figures 4, 4.

Lattice and reorthogonalization method show about the same performances at least up to dimension 6. Both appear to be highly adaptive method being very fast for random determinants (less than a factor 10 above the floating point calculation) and also fast enough for quasi-null determinants which are not caught by usual floating point filters. Both method are still faster than the exact computation for null determinants. Furthermore the speeding factor indeed increases as the dimension rise. This is confirmed by tests performed on the reorthogonalization method up to dimension 12 and whose results are gathered in the graph of figure 4. At least for dimension $n$ up to 6, the timing results for null determinants agree with the predicted $O(n^3)$ law.

In fact, the lattice method can be very slow, turning to the exponential behaviour expected on the theoretical point of view, in some special cases for instance when two column vectors of the input determinant are colinear. However a simple pretraitment allows to avoid these bad cases. This pretraitment consist in first making sure of the independance of the first two column vectors (computing the $2 \times 2$ minors of these two columns) and then multiplying the input matrix both on the left and on the right side by a $n \times n$ matrix $M$ whose entries are either 1 or $-1$ and whose determinant is not null. Such a pretraitement is very fast and only substract $\lceil 2 \log n \rceil$ from the allowed bit complexity of the entries. Another caveat to be mentioned : any of the above method except Leda will fail in dimension $n$ equal to 20 (possibly less) due to overflow of the range of exponents in IEEE double precision. At last, we can also mention that reorthogonalization in dimension 6 has been successfully used to detect the singular configurations of a parallel robot with six degrees of freedom an application where lots of null and quasi-null determinants have to be tested.

## 5 Conclusions

Both reorthogonalization and lattice method appear to be practically feasible methods to compute exactly signs of determinants with integer entries at least for dimension up to 15. This should provided an affordable way of performing exact tests in most geometric algorithms which is a decisive step toward robustness.

## References

[ABD+94] F. Avnaim, J.-D. Boissonnat, O. Devillers, F. Preparata, and M. Yvinec. Evaluating signs of determinants using single-precision arithmetic. Research Report 2306, INRIA, BP93, 06902 Sophia-Antipolis, France, 1994.

[ABD+95] F. Avnaim, J.-D. Boissonnat, O. Devillers, F. Preparata, and M. Yvinec. Evaluation of a new method to compute signs of determinants. In *Proc. 11th Annu. ACM Sympos. Comput. Geom.*, pages C16–C17, 1995.

[BKM+95] Christoph Burnikel, Jochen Könnemann, Kurt Mehlhorn, Stefan Näher, Stefan Schirra, and Christian Uhrig. Exact geometric computation in LEDA. In *Proc. 11th Annu. ACM Sympos. Comput. Geom.*, pages C18–C19, 1995.

[C+96] B. Chazelle et al. Application challenges to computational geometry: CG impact task force report. Technical Report TR-521-96, Princeton Univ., April 1996.

[Cla92] K. L. Clarkson. Safe and effective determinant evaluation. In *Proc. 33rd Annu. IEEE Sympos. Found. Comput. Sci.*, pages 387–395, 1992.

[DP96] O. Devillers and F. Preparata. A probabilistic analysis of the power of arithmetic filters. Rapport de recherche 2971, INRIA, 1996. also report CS96-27 Brown University.

[FM67] G. Forsythe and C. Moler. *Computer solutions of linear algebraic systems.* Prentice Hall, 1967.

[For92] S. Fortune. Numerical stability of algorithms for 2-d Delaunay triangulations and Voronoi diagrams. In *Proc. 8th Annu. ACM Sympos. Comput. Geom.*, pages 83–92, 1992.

[FV93] S. Fortune and C. J. Van Wyk. Efficient exact arithmetic for computational geometry. In *Proc. 9th Annu. ACM Sympos. Comput. Geom.*, pages 163–172, 1993.

[HHK88] C. M. Hoffmann, J. E. Hopcroft, and M. S. Karasick. Towards implementing robust geometric computations. In *Proc. 4th Annu. ACM Sympos. Comput. Geom.*, pages 106–117, 1988.

[Hof89] C. Hoffmann. *Geometric and Solid Modeling.* Morgan Kaufmann, San Mateo, California, 1989.

[KLN91] M. Karasick, D. Lieber, and L. R. Nackman. Efficient Delaunay triangulations using rational arithmetic. *ACM Trans. Graph.*, 10:71–91, 1991.

[LM90] Z. Li and V. Milenkovic. Constructing strongly convex hulls using exact or rounded arithmetic. In *Proc. 6th Annu. ACM Sympos. Comput. Geom.*, pages 235–243, 1990.

[Mil89] V. Milenkovic. Double precision geometry: a general technique for calculating line and segment intersections using rounded arithmetic. In *Proc. 30th Annu. IEEE Sympos. Found. Comput. Sci.*, pages 500–505, 1989.

[She96] Jonathan R. Shewchuk. Robust adaptive floating-point geometric predicates. In *Proc. 12th Annu. ACM Sympos. Comput. Geom.*, pages 141–150, 1996.

[SI89] K. Sugihara and M. Iri. A solid modelling system free from topological inconsistency. *J. Inform. Proc.*, 12(4):380–393, 1989.

[SI94] K. Sugihara and M. Iri. A robust topology-oriented incremental algorithm for Voronoi diagrams. *Internat. J. Comput. Geom. Appl.*, 4:179–228, 1994.

[Yap93] C. K. Yap. Towards exact geometric computation. In *Proc. 5th Canad. Conf. Comput. Geom.*, pages 405–419, 1993.

[YD95] C. K. Yap and T. Dubhe. The exact computation paradigm. In D. Du and F. Hwang, editors, *Computing in Euclidean Geometry.* World Scientific Press, 1995.

# Computing exact geometric predicates using modular arithmetic with single precision

Hervé Brönnimann* Ioannis Z. Emiris* Victor Y. Pan‡ Sylvain Pion*

**Abstract:** We propose an efficient method that determines the sign of a multivariate polynomial expression with integer coefficients. This is a central operation on which the robustness of many geometric algorithms depends. The method relies on modular computations, for which comparisons are usually thought to require multiprecision. Our novel technique of *recursive relaxation of the moduli* enables us to carry out sign determination and comparisons by using only floating point computations in single precision. The method is highly parallelizable and is the fastest of all known multiprecision methods from a complexity point of view. We show how to compute a few geometric predicates that reduce to matrix determinants. We discuss implementation efficiency, which can be enhanced by good arithmetic filters. We substantiate these claims by experimental results and comparisons to other existing approaches. This method can be used to generate robust and efficient implementations of geometric algorithms, including solid modeling, manufacturing and tolerancing, and numerical computer algebra (algebraic representation of curves and points, symbolic perturbation, Sturm sequences and multivariate resultants).

**Keywords:** computational geometry, exact arithmetic, robustness, modular computations, single precision, (RNS) Residue Number Systems

*INRIA Sophia-Antipolis, B.P. 93, 2004, Route des Lucioles, 06902 Sophia-Antipolis Cedex, FRANCE. This research was partially supported by the ESPRIT IV LTR Project No. 21957 (CGAL).

‡ Department of Mathematics and Computer Science, Lehman College, City University of New-York, Bronx NY 10468, USA. Work on this paper was performed while visiting the second author and was supported by NSF Grants CCR 9020690 and 9625344 and PS-CUNY Awards 666327 and 667340.



*Computational Geometry 97* Nice France



## 1 Introduction

Most of geometric predicates can be expressed as computing the sign of an algebraic expression. In principle, one may compute such expressions by using floating-point arithmetic with a fixed finite precision (f.p. arithmetic), but then the roundoff errors may easily lead to the wrong sign. This problem is often referred to as the *robustness* problem in computational geometry [17].

One solution to the robustness problem is to explicitly handle numerical inaccuracies, so as to design an algorithm that does not fail even if the numerical part of the computation is done approximately [19, 29], or to analyze the error due to the f.p. imprecision [12]. Such designs are extremely involved and have only been done for a few algorithms. The general solution, it has been widely argued, is to compute the predicates exactly [9, 6, 13, 33, 11]. This can be achieved in many ways: computing the algebraic expressions with infinite precision [31], with a finite but much higher precision that can be shown to suffice [14], or by using an algorithm that performs a specific test exactly. In the last category, much work has focused on computing the sign of the determinant of a matrix with integer entries [7, 2, 4], which applies to many geometric tests (such as orientation tests, in-circle tests, comparing segment intersections) as well as to algebraic primitives (such as resultants and algebraic representations of curves and surfaces). Recently, some techniques have been devised for handling arbitrary expressions and f.p. representation [28].

In our present paper, we propose a method that determines the sign of a multivariate polynomial expression with integer coefficients, using no operations other than modular arithmetic and f.p. computations with a fixed finite (single) precision. The latter operations can be performed very fast on usual computers. The Chinese remainder algorithms enable us to perform rational algebraic computations modulo several primes, that is, with a lower precision, and then to combine them together in order to recover the desired output value. The latter stage of combining the values modulo smaller primes, however, was always considered a bottleneck of this approach, because higher precision computations were required at this stage. Our paper proposes a new technique, which we call *recursive relaxation of the moduli* and which enables us to resolve the latter problem. Due to this technique, we correctly recover the sign of an integer from its value reduced modulo several smaller primes, and we only use some simple lower precision computations at the recovery stage. (This should make our algorithms of some independent interest also for the theory and practice of algebraic computing.) Our deterministic algorithms 1 and 2 of sections 3 and 4, respectively, specify our approach and our technique based on Lagrange's and Newton's interpolation formulae, respectively. Our algorithm 4 of section 5 gives a probabilistic simplification of algorithm 3. Preliminary experimental results and running times are discussed in section 7. In general, our methods are comparable in speed to other exact methods and even faster for particular inputs.

**Related work.** Performing exact arithmetic is usually expensive. Thus, it is customary to resort to arithmetic filters [14]: those filters safely evaluate a predicate in most cases, in order to avoid performing a more expensive exact implementation. The difficult cases arise when the expression whose sign we wish to compute is very small. For typical filters, the smaller this quantity, the slower the filter [7, 2, 28]: this is referred to as *adaptivity.* Modular arithmetic displays an opposite kind of adaptivity: with a smaller quantity, fewer moduli have to be computed, hence the test is faster. Typically, when filters fail, they also provide an upper bound on the absolute value of the expression whose sign we wish to compute (see many details and estimates in [25]). This bound can then be used to determine how many moduli should be taken. Modular arithmetic is therefore complementary to the filtering approach. We also observe this in section 7.

Residue Number Systems (RNS) express and manipulate numbers of arbitrary precision by their moduli with respect to a given set of numbers. They have been popular because they provide a cheap and highly parallelizable version of multiprecision arithmetic. It is impossible here to give a fair and full account on RNS, but Knuth [22] and Aho, Hopcroft, and Ullman [1] provide a good introduction to the topic. From a complexity point of view, RNS allows to add and multiply numbers in linear time. Its weak point is that sign computation and comparisons are not easily performed and seem to require full reconstruction in multiple precision, which defeats its purpose. This is precisely the issue that our paper handles.

The closest predecessors of our work are apparently [10] and [20]. The algorithm of Hung and Parhami [20] corresponds to single application of the second stage of our recursive relaxation of the moduli. Such a single application suffices in the context of the goal of [20], that is, application to divisions in RNS, but in terms of the sign determination of an integer, this only works for an absolutely larger input. The paper [10] gives probabilistic estimates for early termination of Newton's interpolation process, which we apply in our probabilistic analysis of our algorithm 4. Its main subject is an implementation of an algorithm computing multidimensional convex hulls. The paper [10] does not use our techniques of recursive relaxation of the moduli, and it does not contain the basic equations (1)–(3) of our section 3.

## 2 Exact sign computation using modular arithmetic

**Modular computations.** Our model of a computer is that of a f.p. processor that performs operations at unit cost by using $b$-bit precision (e.g., in the IEEE 754 double precision standard, we have $b = 53$). It is a realistic model as it covers the case of most workstations used in research and industry. We will use mainly one basic property of f.p. arithmetic on such a computer: for all four arithmetic operations (and for computing a square root too but we will not need it), the computed result is always the f.p. representation that best approximates the exact result [22, 28]. This means that the relative error incurred by an operation returning $x$ is at most $2^{-b}$, and that the absolute error is at most $2^{\lfloor \log |x| - b \rfloor}$. (All logarithms in this paper are base 2.) In particular, operations performed on pairs of integers smaller than $2^b$ are performed exactly as long as the result is also smaller than $2^b$.

Let $m_1, \ldots, m_k$ be $k$ pairwise relatively prime integers and let $m = \prod_i m_i$. For any number $x$ (not necessarily an integer), we let $x_i = x \bmod m_i$ be the only number in the range $\left[-\frac{m_i}{2}, \frac{m_i}{2}\right)$ such that $x_i - x$ is a multiple of $m_i$. (This operation is always among the standard operations because it is needed for reducing the arguments of periodic functions.)

To be able to perform arithmetic modulo $m_i$ on integers by using f.p. arithmetic with $b$-bit precision, we will assume that $m_i \leq 2^{b/2+1}$. Performing modular multiplications of two integers from the interval $\left[-\frac{m_i}{2}, \frac{m_i}{2}\right)$ can be done by multiplying these numbers and returning their product modulo $m_i$. (The product is smaller than $2^b$ in magnitude and hence is computed exactly.) Performing additions can be done very much in the same way, but since the result is in the range $\left[-\frac{m_i}{2}, \frac{m_i}{2}\right)$, taking the sum modulo $m_i$ is more easily achieved by adding or subtracting $m_i$ if necessary. Integral divisions modulo $m_i$ can be computed using Euclid's algorithm; we will need them in this paper only in section 6.

Therefore, arithmetic modulo $m_i$ can be performed using f.p. arithmetic with $b$-bit precision, provided that $m_i \le 2^{b/2+1}$.

**Exact sign computation.** In this paper, we consider the following computational problem.

**Problem 1** *Let $k$, $b$, $m_1, \ldots, m_k$ denote positive integers, $m_1, \ldots, m_k$ being pairwise relatively prime, such that $m_i \le 2^{b/2+1}$, and let $m = \prod_{i=1}^k m_i$. Let $x$ be an integer whose magnitude is smaller than $\lfloor (m/2)(1-k2^{-b}) \rfloor$. Given $x_i = x \bmod m_i$, compute the sign of $x$ by using only modular and floating-point arithmetic both performed with $b$-bit precision.*

We will solve this problem, even though $x$ can be huge and, therefore, not even representable by using $b$ bits.

## 3 Lagrange's method

According to the Chinese remainder theorem [8], $x$ is uniquely determined by its residues $x_i$, that is, Problem 1 is well defined and admits a unique solution. Moreover, this solution can be derived algorithmically from the following formula, due to Lagrange [22, 23]. If $x$ is an integer in the range $\left[-\frac{m}{2}, \frac{m}{2}\right)$, $x_i$ stands for the residue $x \bmod m_i$, $v_i = m/m_i = \prod_{j \ne i} m_j$, and $w_i = v_i^{-1} \bmod m_i$, then

$$x = \left( \sum_{i=1}^{k} ((x_i w_i) \bmod m_i)\, v_i \right) \bmod m.$$

Trying to determine the sign of such an integer, we computed the latter sum approximately in fixed $b$-bit precision. Computing a linear combination of large integers $v_i$ with its subsequent reduction modulo $m$ can be difficult, so we prefer to compute the number

$$S = \text{frac}\left( \sum_{i=1}^{k} \frac{(x_i w_i) \bmod m_i}{m_i} \right),$$

where frac$(z)$ is the fractional part of a number $x$ that belongs to $\left[-\frac{1}{2}, \frac{1}{2}\right)$.

If $S$ were computed exactly, then we would have $S = \text{frac}(x/m)$, due to Lagrange's interpolation formula. In fact, $S$ is computed with a fixed $b$-bit precision. Nevertheless, we prove in the full version that exact rounding and summing terms pairwise in a tree-like fashion introduces an absolute error $\varepsilon_k = k\,2^{-b-1}$ in the sum $S$. Therefore, if $S$ is greater than $\varepsilon_k$, the sign of $x$ is the same as the sign of $S$, and we are done. Otherwise, $|x| < \varepsilon_k m$. Since $m_k \le 2^{b/2+1}$, we can say conservatively that for all practical values of $k$ and $b$, this is smaller than $\frac{m}{2m_k}(1-\varepsilon_{k-1})$, and hence we may recover $x$ already from $x_i = x \bmod m_i$ for $i = 1, \ldots, k-1$, that is, it suffices to repeat the computation using only $k-1$, rather than $k$ moduli. Recursively, we will reduce the solution to the case of a single modulus $m_1$ where $x = x_1$. We will call this technique *recursive relaxation of the moduli*, and we will also apply it in section 4.

We will present our resulting algorithm by using additional notation:

$$\begin{aligned} m^{(j)} &= \prod_{1 \le i \le j} m_i, \\ v_i^{(j)} &= \prod_{\substack{1 \le i \le j \\ i \ne j}} m_i, \\ w_i^{(j)} &= \left(v_i^{(j)}\right)^{-1} \bmod m_i, \\ S^{(j)} &= \text{frac}\left( \sum_{i=1}^{j} \frac{x_i w_i^{(j)} \bmod m_i}{m_i} \right), \end{aligned}$$

so that $m = m^{(k)}$, $w_i = w_i^{(k)}$ and $S = S^{(k)}$. All the computations in this algorithm are performed by using f.p. arithmetic with $b$-bit precision.

**Algorithm 1** : Compute the sign of $x$ knowing $x_i = x \bmod m_i$

> **Precomputed data:** $m_j$, $w_i^{(j)}$, $\varepsilon_j$, *for all* $1 \le i \le j \le k$
> **Input:** *integers* $k$ *and* $x_i \in \left[-\frac{m_i}{2}, \frac{m_i}{2}\right)$, *for all* $1 \le i \le k$
> **Output:** *sign of* $x$, *the unique solution of* $x_i = x \bmod m_i$ *in* $\left[-\frac{m^{(k)}}{2}, \frac{m^{(k)}}{2}\right)$
> **Precondition:** $|x| \le \frac{m^{(k)}}{2}(1-\varepsilon_k)$
>
> *1. Let* $j \leftarrow k+1$
> *2. Repeat* $j \leftarrow j-1$,
> $S^{(j)} \leftarrow \textit{frac}\left( \sum_{i=1}^{j} \frac{x_i w_i^{(j)} \bmod m_i}{m_i} \right)$
> *until* $|S^{(j)}| > \varepsilon_j$ *or* $j = 0$
> *3. If* $j = 0$ *return* "$x = 0$"
> *4. If* $S^{(j)} > 0$ *return* "$x > 0$"
> *5. If* $S^{(j)} < 0$ *return* "$x < 0$"

**Lemma 3.1** *Algorithm 1 computes the sign of $x$ knowing its residues $x_i$ by using at most $\frac{k(k-1)}{2}$ f.p. modular multiplications, $\frac{k(k-1)}{2}$ f.p. divisions, $\frac{k(k-1)}{2}$ f.p. additions, and $k+2$ f.p. comparisons.*

**Proof.** The $m_i$'s and the $w_i^{(j)}$'s are computed once and for all and placed into a table, so they are assumed to be available to the algorithm at no cost. In step 2, a total of $j$ modular multiplications, $j$ f.p. divisions, and $j$ f.p. additions (including taking the fractional part) are performed. □

By using parallel implementation of the summation of $k$ numbers on $\lceil k/\log k \rceil$ arithmetic processors in

$2\lceil \log k \rceil$ time (cf. e.g. [3, ch.4]), we may perform algorithm 1 on $\lceil k/\log k \rceil$ arithmetic processors in $O(k \log k)$ time, assuming each $b$-bit f.p. operation takes constant time. Furthermore, if $\lceil k^2/\log k \rceil$ processors are available, we may compute all the $S^{(j)}$ and compare $|S^{(j)}|$ with $\varepsilon_j$, for all $j = 1, \ldots, k$ concurrently. This would require $O(\log k)$ time on $\lceil k^2/\log k \rceil$ processors. Finally, if $\lceil tk/\log k \rceil$ processors are available for some parameter $1 \le t \le k$, we may perform algorithm 1 in $O((k \log k)/t)$ time by batching $\lceil t \rceil$ consecutive values of $j$ in parallel. In practice, the algorithm needs to examine only a few values of $j$, so $O(\log k)$ time suffices even with $\lceil k/\log k \rceil$ processors.

**Remark 1.** If actually $x = 0$, the algorithm can be greatly sped up by testing if $x_j = 0$ in step 2, in which case we may directly pass to $j - 1$. Furthermore, stage 3 is not needed unless $x = x_j = 0$ for all $j$, which can be tested beforehand. Of course, if the only answer needed is "$x = 0$" or "$x \ne 0$", then it suffices to test if all the $x_i$'s are zero.

**Remark 2.** If $|x|$ is not too small compared to $m^{(k)}$, then only step $k$ is performed, involving only $k$ f.p. operations of each kind. Also, we note that the costly part of the computation is likely to be the determination of the $x_i$'s. For these reasons, we should try to minimize the number $k$ of moduli $m_i$ involved in the algorithm. This can be done by getting better upper estimates on the magnitude of the output or by using the probabilistic technique of section 5.

## 4 A generalization of Lagrange's method

We will show that Lagrange's method is in fact a particular case of the following method. Let

$$\Sigma^{(0)} = S^{(k)} = \mathrm{frac}\left( \sum_{i=1}^{k} \frac{(x_i w_i) \bmod m_i}{m_i} \right).$$

This quantity is computed in the first step of algorithm 1. If the computed value of $\Sigma^{(0)}$ is smaller than $\varepsilon_k$, it implies that $\Sigma^{(0)} < 2\varepsilon_k$. Thus, $|x|$ is smaller than $2m\varepsilon_k$. We can then multiply $x_i w_i$ by

$$\alpha_k = \lfloor \frac{\frac{1}{2}(1 - \varepsilon_k)}{2\varepsilon_k} \rfloor,$$

to obtain $(x_i w_i \alpha_k) \bmod m_i$ for all $i = 1, \ldots, k$. This can be easily done by precomputing $\alpha_k$ modulo each $m_i$. We then compute

$$\Sigma^{(1)} = \mathrm{frac}\left( \sum_{i=1}^{k} \frac{(x_i w_i \alpha_k) \bmod m_i}{m_i} \right),$$

and more generally,

$$\Sigma^{(j)} = \mathrm{frac}\left( \sum_{i=1}^{k} \frac{(x_i w_i \alpha_k^j) \bmod m_i}{m_i} \right),$$

where we assume $\alpha_k \bmod m_i$ precomputed for all $i = 1, \ldots, k$. This leads to the following algorithm:

**Algorithm 2** : Generalized Lagrange's method. Compute the sign of $x$ knowing $x_i = x \bmod m_i$.

> **Precomputed data:** $m_i$, $w_i$, $\varepsilon_k$, $\alpha_k \bmod m_i$, *for all* $i = 1, \ldots, k$
> **Input:** *integers* $k$ *and* $x_i \in \left[-\frac{m_i}{2}, \frac{m_i}{2}\right)$ *for all* $i = 1, \ldots, k$
> **Output:** *sign of* $x$*, the unique solution of* $x_i = x \bmod m_i$ *in* $\left[-\frac{m}{2}, \frac{m}{2}\right)$
> **Preconditions:** $|x| \le \frac{m}{2}(1 - \varepsilon_k)$ *and* $x \ne 0$
>
> 1. *Let* $j \leftarrow -1$
> 2. *Repeat* $j \leftarrow j + 1$,
>    $\Sigma^{(j)} \leftarrow \mathit{frac}\left( \sum_{i=1}^{k} \frac{x_i w_i \bmod m_i}{m_i} \right)$
>    $x_i \leftarrow x_i \alpha_k \bmod m_i$ *for all* $1 \le i \le k$,
>    *until* $|\Sigma^{(j)}| > \varepsilon_k$ *or* $j = k$
> 3. *If* $j = k$ *return* "$x = 0$"
> 4. *If* $\Sigma^{(j)} > 0$ *return* "$x > 0$"
> 5. *If* $\Sigma^{(j)} < 0$ *return* "$x < 0$"

It is easy to see that the number of iterations in step 2 is $O(\log m/\log \alpha_k) = O(k)$, because $|x|$ is no less than 1 and no more than $m^{(k)} \le 2^{k(b/2+1)}$, and is multiplied by $\alpha_k$ at each iteration.

**Remark 3.** Algorithm 1 corresponds to a choice of $\alpha_k = m_j$ in step $j$, this is why we call algorithm 2 a generalization. Applying Lagrange's method, we do not multiply by the maximum at each step, but by a smaller number chosen so as to simplify the computations.

**Remark 4.** To yield the parallel time bounds such as $O(\log k)$ using $\lceil k^2/\log k \rceil$ processors for algorithm 2, we need to precompute ${\alpha_k}^j$ for all $i, j = 1, \ldots, k$.

## 5 An incremental variant

A recursive incremental version of the Chinese remainder algorithm, named after Newton, is described in this section. Its main advantage is that it does not require an *a priori* bound on the magnitude of $x$.

Let $x^{(j)} = x \bmod m^{(j)}$, for $j = 1, \ldots, k$, so that $x^{(1)} = x_1$ and $x = x^{(k)}$. Let $y_1 = x_1$, and for all $j = 2, \ldots, k$,

$$\begin{aligned} t_j &= w_j^{(j)} = (m^{(j-1)})^{-1} \bmod m_j, \\ y_j &= \left(x_j - x^{(j-1)}\right) t_j \bmod m_j \in \left[-\frac{m_j}{2}, \frac{m_j}{2}\right). \end{aligned}$$

Then (see, e.g., [22, 23]), for all $j = 2, \ldots, k$,

$$x^{(j)} = x^{(j-1)} + y_j m^{(j-1)}.$$

Clearly, this leads to an incremental computation of the solution $x = x^{(k)}$ to problem 1; we see below how this can be exploited for an early termination of the interpolation. A further advantage is that all computation can be kept modulo $m_j$, and no floating-point computation is required, in contrast to sections 3 and 4 where $S^{(j)}$ or $\Sigma^{(j)}$ are computed. It is obvious, that when $y_j \neq 0$, then the sign of $x^{(j)}$ is the same as the sign of $y_j$ since $|x^{(j-1)}| \leq m^{(j-1)}/2$. If $y_j = 0$, the sign of $x^{(j)}$ is the same as that of $x^{(j-1)}$, for $j \geq 2$, whereas the sign of $x^{(1)} = x_1 = y_1$ is known. If $y_j = 0$ for all $j$, then this is precisely the case when $x = 0$.

For $1 \leq i < j \leq k$, we introduce integers

$$u_i^{(j-1)} = \left(m^{(i-1)} t_j\right) \bmod m_j = \left(\prod_{l=i}^{j-1} m_l\right)^{-1} \bmod m_j.$$

Then $t_j = u_1^{(j-1)}$. In the full version of the paper, we show that the quantities $y_j$ verify the following equality for all $j = 2, \ldots, k$,

$$y_j = \left((x_j - x_1)u_1^{(j-1)} - \sum_{i=2}^{j-1} y_i u_i^{(j-1)}\right) \bmod m_j.$$

Therefore, they can be computed by using modular arithmetic with bit-precision given by the maximum bit-size of the $m_j^2$. Here it suffices to assume that the absolute value of $x$ is bounded by $m^{(k)}/2$.

**Algorithm 3** : Compute the sign of $x$, knowing $x \bmod m_i$, by Newton's incremental method

> **Precomputed data:** $m_j$, $u_i^{(j-1)}$, *for all* $1 \leq i < j \leq k$
>
> **Input:** *integers* $k$ *and* $x_i \in \left[-\frac{m_i}{2}, \frac{m_i}{2}\right)$ *for all* $i = 1, \ldots, k$
>
> **Output:** *sign of* $x$*, where* $x$ *is the unique solution of* $x_i = x \bmod m_i$ *in* $\left[-\frac{m^{(k)}}{2}, \frac{m^{(k)}}{2}\right)$
>
> **Precondition:** *None.*

1. *Let* $y_1 \leftarrow x_1$, $j \leftarrow 1$. *Depending on whether* $y_1$ *is negative, zero or positive, set* $s$ *to* $-1, 0$ *or* $1$*, respectively.*
2. *Repeat* $j \leftarrow j + 1$,

$$y_j \leftarrow \left((x_j - x_1)u_1^{(j-1)} - \sum_{i=2}^{j-1} y_i u_i^{(j-1)}\right) \bmod m_j,$$

*until* $j = k$. *For every* $j$*, if* $y_j$ *is strictly negative or positive, then set* $s$ *to* $-1$ *or* $1$*, respectively.*

3. *Depending on whether* $s$ *is* $-1, 0,$ *or* $1$*, return "*$x < 0$*," "*$x = 0$*," or "*$x > 0$*", respectively.*

**Remark 5.** As in remark 1, we can test beforehand if all $x_i = 0$, which is precisely the case when $x = 0$.

**Lemma 5.1** *Algorithm 3 computes the sign of* $x$ *knowing its residues* $x_i$ *using at most* $\frac{k(k-1)}{2}$ *f.p. modular multiplications,* $\frac{k(k-1)}{2}$ *f.p. modular additions, and* $k$ *f.p. comparisons.*

**Proof.** For every $j = 2, \ldots, k$, there are $j - 1$ f.p. modular additions and multiplications. There is one comparison for each $j = 1, \ldots, k$. □

Algorithm 3 requires $k$ recursive steps in the worst case (though we expect to have it terminate earlier in practice), so its parallel time cannot be decreased below $\Omega(k \log k)$. Nevertheless the algorithm can be implemented in $O(k \log k)$ time on $\lceil k/\log k \rceil$ processors, assuming each $b$-bit f.p. operation takes constant time.

To compare with algorithm 1, realistically assume that a modular addition is equivalent to 3/2 f.p. additions and one comparison, on the average. Then, algorithm 1 requires $\frac{k(k-1)}{2}$ f.p. divisions (which are essentially multiplications with precomputed reciprocals) more than algorithm 3, whereas the latter requires $\frac{k(k-1)}{4}$ extra f.p. additions and $\frac{k(k-1)}{2}$ additional comparisons.

The principal feature of this approach, based on Newton's formula for recovering $x$, is its incremental nature. This may lead to faster termination, before examining all $k$ moduli. Informally, this should happen whenever the magnitude of $x$ is significantly smaller than $m^{(k)}/2$, in which case we would save the computation required to obtain $x_j$ for all larger $j$. This saves a significant amount of computation if termination occurs earlier than the static bound indicated by $k$. A quantification of this property in the case of convex hulls can be found in [10].

We propose below a probabilistic variant of algorithm 3 which, moreover, removes the need of an *a priori* knowledge of $k$. Step 2 is modified to include a test of $y_j$ against zero. Clearly, $y_j = 0$ precisely when $x^{(j)} = x^{(j-1)}$. Then we may deduce that $x^{(j)} = x^{(k)} = x$, with a very high probability, and terminate the iteration.

**Algorithm 4** : Yield earlier termination of algorithm 3 for absolutely smaller input. Algorithm 3 is modified exactly as shown.

> **Input:** *integers* $x_i \in \left[-\frac{m_i}{2}, \frac{m_i}{2}\right)$ *for* $i = 1, \ldots$ *as required in the course of the algorithm; no need for* $k$
>
> **Output:** *sign of* $x$ *with very high probability*

2. *Terminate the loop also if* $y_j = 0$

By lemma 3.1 of [10], this algorithm fails with probability bounded by $(k-2)/m_{\min}$, where

$$m_{\min} = \min\{m_1, m_2, \ldots, m_k\}.$$

For $k \leq 12$, $m_{\min} \geq 2^{25}$, the error probability is less than $10^{-6}$. A more careful analysis can reduce this probability by exploiting the correlation of failure at different stages. For experimental support of this claim, we refer to [10].

## 6 Applications

### 6.1 Exact geometric predicates

Exact geometric predicates is the most general way to provide robust implementations of geometric algorithms [9, 13, 33, 11]. In particular, orientation tests can be implemented by looking for the sign of a determinant. Modular arithmetic becomes increasingly interesting when the geometric tests (e.g. determinants) are of higher order and complexity. They are central in, notably:

- Convex hull computations: this is a fundamental problem of computational geometry and of optimization for larger dimensions. Among numerous practical applications, one may note collision detection in dynamic simulation and animation [24], prediction of poses for industrial parts on a conveyer belt, and computation of stable grasps by robots [30].
- Voronoi diagrams: for points, their computation reduces to convex hulls. The sweepline algorithm in 2D is relatively simple, but involves tests of degree 20 and modular arithmetic can be of substantial help in conjunction with arithmetic filters [14]. For segments, the tests become of even higher degree and complexity [6], and f.p. computation is likely to introduce errors, so exact arithmetic is often a must.
- Mixed subdivisions used in solving systems of nonlinear equations. Sparse elimination theory is a relatively new area of computational algebraic geometry, which exploits the geometric structure of polynomial systems in order to obtain tighter bounds and faster algorithms for their manipulation [15]. The algebraic questions are formulated in terms of *Newton polytopes* and their mixed volume, each polytope being the convex hull of the exponent vectors appearing in a polynomial.

Even for small dimensions, the nature of the data may force the f.p. computation to introduce inconsistencies, for instance, in:

- Planarity testing in geometric tolerancing [32]. Here, one must determine if a set of points sampling a plane surface can be enclosed in a slab whose width is part of the planarity requirements. The computation usually goes by computing the width of the convex hull, and the data is usually very flat, hence prone to numerical inaccuracies.

In geometric and solid modeling, traditional approaches have employed finite precision floating point arithmetic, based on bounds on the roundoff errors. Although certain basic questions in this domain are now considered closed, there remain some fundamental open problems, including boundary computation [18]. Tolerance techniques and symbolic reasoning have been used, but have been mostly restricted to polyhedral objects; their extension to curved or arbitrary degree sculptured solids would be complicated and expensive. More recently, exact arithmetic has been proposed as a valid alternative for generating boundary representations of sculptured solids, since it guarantees robustness and precision even for degenerate inputs at a reasonable or negligible performance penalty [21].

Furthermore, exact arithmetic allows the use of a variety of algebraic and symbolic methods, including algebraic representation of curves and points, symbolic perturbation, Sturm sequences and multivariate resultants; for an introduction to these methods, see [5]. The critical operation is deciding the sign of a multivariate polynomial expression with rational coefficients on a set of points. Recent data structures that exploit structure of algebraic objects, such as *straight-line programs*, also use precisely this kind of primitive operation [27].

### 6.2 Sign of the determinant of a matrix

As mentioned, computing the sign of a matrix determinant is a basic operation in computational geometry, applied to many geometric tests (such as orientation tests, in-circle tests, comparing segment intersections) [7, 2]. Sometimes, the entries to the determinant are themselves algebraic expressions. For instance, the in-circle test can be reduced to computing a $2 \times 2$ determinant, whose entries have degree 2 and thus require $2b + O(1)$-bit precision to be computed exactly [2]. Computing these entries by using modular arithmetic enables in-circle tests with $b$-bit precision while still computing exactly the sign of a $2 \times 2$ determinant.

To compute an $n \times n$ determinant modulo $m_k$, we may use Gaussian elimination with a single final division. At step $i < n$ of the algorithm, the matrix is

$$\begin{pmatrix} \vdots & \cdots & \cdots \\ 0 & \alpha_{i,i} & \cdots \\ \vdots & \vdots & \vdots \\ 0 & \alpha_{n,i} & \cdots \end{pmatrix}$$

and we assume that the pivot $\alpha_{i,i}$ is invertible modulo $m_k$. Then we change line $L_j$ to $\alpha_{i,i}L_j - \alpha_{j,i}L_i$ for all

$j = i+1, \ldots, n$. At step $n$ of the algorithm, we multiply the coefficient $\alpha_{n,n}$ by the modular inverse of the product $\prod_{i=1}^{n-1} \alpha_{i,i}^{n-i}$. This gives us the value of the determinant modulo $m_k$. Note that the same method but with non-modular integers and a final division would have involved exponentially large integers and several slow divisions at each step. Nevertheless, it is only the range of the final result that matters for modular computations. This shows a big advantage of modular arithmetic over other multiprecision approaches.

The pivots should be invertible modulo $m_k$. If $m_k$ is prime, the pivot simply has to be non-zero modulo $m_k$. The algorithm can be also easily implemented if $m_k$ is a power of a prime, or if $m_k$ is the product of two primes. This would be desirable mainly for taking $m_k = 2^{b_k}$ for which modular arithmetic is done naturally by integer processors, though here, special care must be taken about even output. Other choices of $m_k$ do not seem to bring any improvement.

With IEEE double precision ($b = 53$), we choose moduli smaller than $2^{27}$, so that $2(\frac{m_k}{2})^2 \leq 2^{53}$: Gaussian elimination intensively uses $(ad-bc)$-style operations; here we may apply one final modular reduction, instead of two for each product before subtracting.

This algorithm performs $O(n^3)$ operations for each modulus $m_i$. With Hadamard's determinant bound and $m_k$ greater than $2^{b/2}$, only $k = \lceil 2n \log n \rceil$ finite fields need to be considered. Hence the complexity of finding the sign of the determinant is $O(n^4 \log n)$ single precision operations. On a $O(n^3 \log n)$-processor machine, the time complexity drops to $O(n)$, if we use custormary parallelization of the Gaussian elimination routine for matrix triangulation (cf. [16]), which gives us the value of the determinant. (We apply this routine in modular arithmetic, with simplified pivoting, concurrently for all $m_i$'s.) Theoretically, substantial additional parallel acceleration can be achieved by using randomization [3, ch. 4], [26], yielding the time bound $O(\log^2 n)$ on $\lceil n^3 \log n \rceil$ arithmetic processors, and the processor bound can be decreased further to $O(n^{2.376})$, by applying asymptotically fast algorithms for matrix multiplication.

## 7 Experimental results

We present several benchmark results of our implementations of the described methods for computing a determinant in C, and compare them with different existing packages. Method FP is a straightforward f.p. implementation of Gaussian elimination. Method LEDA uses the routine `sign_of_determinant(integer_matrix)` of Leda [6]. Method CL has been implemented by us based on [7, 4]. As we compare with methods that handle arbitrary dimensions, we did not specialize the implementation for small dimensions as is done

| $n$ | FP | MOD | CL | GMP | LEDA |
|---|---|---|---|---|---|
| 2 | 3.5 | 28 | 47 | 12 | 243 |
| 3 | 9 | 71 | 131 | 61 | 1032 |
| 4 | 18 | 151 | 276 | 226 | 2847 |
| 5 | 32 | 525 | 503 | 660 | 6370 |
| 6 | 55 | 857 | 850 | 1820 | 12330 |
| 7 | 85 | 1410 | 1213 | 3810 | 23400 |
| 8 | 130 | 2150 | 1954 | 6700 | 38700 |
| 9 | 182 | 3370 | 2669 | 11080 | 60500 |
| 10 | 250 | 4810 | 3671 | 17100 | 89000 |
| 11 | 330 | 6340 | 5063 | 27200 | 135800 |
| 12 | 420 | 8410 | 6559 | 39100 | 191900 |
| 13 | 550 | 11600 | 9272 | 53000 | 210750 |
| 14 | 700 | 14930 | 11461 | 72400 | 255750 |

Table 1: Performance on random determinants.

| $n$ | FP | MOD | CL | GMP | LEDA |
|---|---|---|---|---|---|
| 2 | 3.5 | 30 | 344 | 12 | 240 |
| 3 | 9 | 75 | 832 | 60 | 1025 |
| 4 | 18 | 155 | 1472 | 220 | 2605 |
| 5 | 32 | 532 | 2598 | 655 | 5880 |
| 6 | 55 | 871 | 4419 | 1810 | 11370 |
| 7 | 85 | 1420 | 7040 | 3800 | 20700 |
| 8 | 130 | 2160 | 9590 | 6690 | 38300 |
| 9 | 182 | 3390 | 13610 | 11060 | 60300 |
| 10 | 250 | 4840 | 18590 | 17080 | 94000 |
| 11 | 330 | 6360 | 24850 | 27170 | 130000 |
| 12 | 420 | 8450 | 33800 | 39050 | 195000 |
| 13 | 550 | 11670 | 42325 | 52940 | 205700 |
| 14 | 700 | 15110 | 52800 | 72330 | 240500 |

Table 2: Performance on small determinants.

| $n$ | FP | MOD | CL | GMP | LEDA |
|---|---|---|---|---|---|
| 2 | 3.5 | 31 | 343 | 12 | 230 |
| 3 | 9 | 76 | 912 | 57 | 890 |
| 4 | 18 | 157 | 2195 | 220 | 2500 |
| 5 | 32 | 540 | 4647 | 650 | 5820 |
| 6 | 55 | 885 | 8270 | 1800 | 12290 |
| 7 | 85 | 1440 | 13790 | 3790 | 20780 |
| 8 | 130 | 2180 | 21110 | 6670 | 34420 |
| 9 | 182 | 3420 | 31610 | 11040 | 55090 |
| 10 | 250 | 4870 | 44800 | 17050 | 92640 |
| 11 | 330 | 6400 | 62920 | 27150 | 136520 |
| 12 | 420 | 8480 | 87300 | 39020 | 184000 |
| 13 | 550 | 11700 | 119800 | 52920 | 201500 |
| 14 | 700 | 15150 | 143650 | 72250 | 248000 |

Table 3: Performance on zero determinants.

in [4] (this would provide an additional speedup of approximately 3). Method GMP is an implementation of Gaussian elimination using the GNU Multiprecision Package, for dimension lower than 5, and an implementation of Bareiss' extension of Gaussian elimination, for higher dimensions. Method MOD is an implementation of modular Gaussian elimination as described in section 6 using our recursive relaxation of the moduli. Of the other methods available, the lattice method of [4] has not yet been implemented in dimensions higher than 5; LN [14] provides a very fast implementation in dimensions up to 5 but was not available to us in higher dimensions.

Among the methods that guarantee exact computation, our implementations are at least as efficient as the others, and for certain classes of input they outperform all available programs. Furthermore our approach applies to arbitrary dimensions, whereas methods that compute a f.p. approximation of the determinant value are doomed to fail in dimensions higher than 15 because of overflow in the f.p. exponent.

All tests were carried out on a 85MHz Sun Sparc 5 workstation, using the `clock()` function. Each program is compiled with the compiler that gives best results. Each entry in the following tables represents the average time of one run in microseconds, with a maximum deviation of about 10%. We concentrated on determinant sign evaluation and considered three classes of matrices: random matrices, whose determinant is typically away from zero, in table 1, almost-singular matrices with single-precision determinant in table 2, and lastly singular matrices with zero determinant in table 3. The coefficients are integers of bit-size $53 - n$ (due to restrictions of Clarkson's method).

Our results suggest that our approach is comparable, and for certain classes of input significantly faster than the examined alternatives that guarantee exact results. The running times are displayed in tables 1–3. (For small dimensions, specialized implementations can provide an additional speedup for all methods.) Our code is reasonably compact and easy to maintain. As an obvious improvement, with a reasonably accurate f.p. filter, the penalty of exact arithmetic can be paid only for small determinants (tables 2 and 3). Another improvement we plan on exploring is to use parallelization.

Some side effects may occur, due to the way we generate matrices. The code of the modular package is free, and anyone can benchmark it on the kind of matrices that he uses. It is available via the URL `http://www.inria.fr/prisme/personnel/pion/progs/modular.html`

## 8 Conclusion

RNS systems have been used in number systems because they provide a highly parallelizable technique for multiprecision. As parallel computers are becoming more available, RNS provide an increasingly desirable implementation of multiprecision. This comes in sharp contrast with other multiprecision methods that are not easily parallelizable. Perhaps the main problem with RNS is that comparisons and sign computations seem to require full reconstruction and, therefore, use standard multiprecision arithmetic. We show that one may in fact use only single precision and still perform these operations exactly and efficiently. In some applications, the number of moduli may be large. Our algorithms may be easily implemented in parallel with a speedup depending almost linearly on the number of processors.

As an application, we show how to compute the sign of a determinant. This problem has received considerable attention, yet the fastest techniques are usually iterative and do not parallelize easily. Moreover, they usually only handle single precision inputs. Our techniques are comparable in speed or even faster than other techniques (e.g. [2, 6, 7]), and can easily handle arbitrarily large inputs.

A central problem we plan to explore further is to design algorithms that compute upper bounds on the quantities involved to determine how many moduli should be taken. For determinants, the static bounds we use seem to suffice for applications in computational geometry [14]. They might be overly pessimistic in other areas (such as tolerancing or symbolic algebra) where the nature of the data or algebraic techniques might imply much better bounds. A valid approach we will further study and implement is Newton's incremental method of section 5, provided that we are willing to afford some small probability of error.

## References


[1] A. V. Aho and J. E. Hopcroft and J. D. Ullman. *The Design And Analysis Of Computer Algorithms.* Addison-Wesley, Reading, MA, 1974.

[2] F. Avnaim, J.-D. Boissonnat, O. Devillers, F. Preparata, and M. Yvinec. Evaluation of a new method to compute signs of determinants. In *Proc. 11th Annu. ACM Sympos. Comput. Geom.*, pages C16–C17, 1995.

[3] D. Bini, V. Y. Pan. *Polynomial and Matrix Computations. Vol. 1: Fundamental Algorithms.* Birkhäuser, Boston, 1994.

[4] H. Brönnimann, M. Yvinec. Efficient exact evaluation of signs of determinants. These proceedings.

[5] B. Buchberger, G.E. Collins, and R. Loos, editors.

*Computer Algebra: Symbolic and Algebraic Computation.* Springer, Wien, 2nd edition, 1982.

[6] C. Burnikel, J. Könnemann, K. Mehlhorn, S. Näher, S. Schirra, and C. Uhrig. Exact geometric computation in LEDA. In *Proc. 11th Annu. ACM Sympos. Comput. Geom.*, pages C18–C19, 1995. Package available at http://www.mpi-sb.mpg.de/LEDA/leda.html.

[7] K. L. Clarkson. Safe and effective determinant evaluation. In *Proc. 33rd Annu. IEEE Sympos. Found. Comput. Sci.*, pages 387–395, 1992.

[8] T. H. Cormen, C. E. Leiserson, and R. L. Rivest. *Introduction to Algorithms.* The MIT Press, Cambridge, Mass., 1990.

[9] H. Edelsbrunner and E.P. Mücke. Simulation of simplicity: A technique to cope with degenerate cases in geometric algorithms. *ACM Trans. Graphics*, 9(1):67–104, 1990.

[10] I.Z. Emiris. A complete implementation for computing general dimensional convex hulls. *Intern. J. Computational Geom. & Applications*, 1997. To appear. Preliminary version as Tech. Report 2551, INRIA Sophia-Antipolis, France, 1995.

[11] I.Z. Emiris and J.F. Canny. A general approach to removing degeneracies. *SIAM J. Computing*, 24(3):650–664, 1995.

[12] S. Fortune. Numerical stability of algorithms for 2-d Delaunay triangulations and Voronoi diagrams. In *Proc. 8th Annu. ACM Sympos. Comput. Geom.*, pages 83–92, 1992.

[13] S. Fortune and C. J. Van Wyk. Efficient exact arithmetic for computational geometry. In *Proc. 9th Annu. ACM Sympos. Comput. Geom.*, pages 163–172, 1993.

[14] S. Fortune, C.J. Van Wyk. Static analysis yields efficient exact integer arithmetic for computational geometry. *ACM Trans. on Graphics*, 1996.

[15] I.M. Gelfand, M.M. Kapranov, and A.V. Zelevinsky. *Discriminants and Resultants.* Birkhäuser, Boston, 1994.

[16] G. H. Golub and C. F. van Loan *Matrix computations.* Johns Hopkins University Press, Baltimore, Maryland, 1996.

[17] C. M. Hoffmann. The problem of accuracy and robustness in geometric computation. Report CSD-TR-771, Dept. Comput. Sci., Purdue Univ., West Lafayette, IN, 1988.

[18] C.M. Hoffmann. How solid is solid modeling? In *Applied Computational Geometry*, volume 1148 of *LNCS*, pages 1–8. Springer, 1996.

[19] C. M. Hoffmann, J. E. Hopcroft, and M. T. Karasick. Robust set operations on polyhedral solids. *IEEE Comput. Graph. Appl.*, 9(6):50–59, November 1989.

[20] C.Y. Hung, B. Parhami. An approximate sign detection method for residue numbers and its application to RNS division. *Computers Math. Applic.* 27(4):23–35, 1994.

[21] J. Keyser, S. Krishnan, and D. Manocha. Efficient B-rep generation of low degree sculptured solids using exact arithmetic. Technical Report 40, Dept. Computer Science, Univ. N. Carolina, Chapel Hill, 1996.

[22] D.E. Knuth. *The Art of Computer Programming: Seminumerical Algorithms*, volume 2. Addison-Wesley, Reading, Massachusetts, 1981 and 1997.

[23] M. Lauer. Computing by homomorphic images. In B. Buchberger, G.E. Collins, and R. Loos, editors, *Computer Algebra: Symbolic and Algebraic Computation*, pages 139–168. Springer, Wien, 2nd edition, 1982.

[24] M.C. Lin, D. Manocha, and J. Canny. Efficient contact determination for dynamic environments. In *Proc. IEEE Conf. Robotics and Automation*, pages 602–608, 1994.

[25] V. Y. Pan, Y. Yu, and C. Stewart. Algebraic and numerical techniques for the computation of matrix determinants. *Computers & Math. (with Applications)*, 1997, to appear.

[26] V. Y. Pan. Parallel computation of polynomial GCD and some related parallel computations over abstract fields. *Theoretical Computer Science*, 162:2, 173–223, 1996.

[27] A. Rege. A complete and practical algorithm for geometric theorem proving. In *Proc. ACM Symp. on Computational Geometry*, pages 277–286, Vancouver, June 1995.

[28] J.R. Shewchuk. Robust adaptive floating-point geometric predicates. In *Proc. 12th Annu. ACM Sympos. Comput. Geom.*, pages 141–150, 1996.

[29] K. Sugihara and M. Iri. A robust topology-oriented incremental algorithm for Voronoi diagrams. *Internat. J. Comput. Geom. Appl.*, 4:179–228, 1994.

[30] J. Wiegley, A. Rao, and K. Goldberg. Computing a statistical distribution of stable poses for a polyhedron. In *Proc. 30th Annual Allerton Conf. on Comm. Control and Computing*, Univ.Ill. Urbana-Champaign, 1992.

[31] C. Yap. Towards exact geometric computation. *Comput. Geom. Theory Appl.*, 7:3–23, 1997.

[32] C. K. Yap. Exact computational geometry and tolerancing metrology. In D. Avis and J. Bose, editors, *Snapshots of Computational and Discrete Geometry, Vol. 3, Tech. Rep. SOCS-94.50.* McGill School of Comp. Sci., 1995.

[33] C. K. Yap and T. Dubhe. The exact computation paradigm. In D. Du and F. Hwang, editors, *Computing in Euclidean Geometry.* World Scientific Press, 1995.

# A Perturbation Scheme for Spherical Arrangements with Application to Molecular Modeling*

DAN HALPERIN†
Tel Aviv University

CHRISTIAN R. SHELTON‡
Massachusetts Institute of Technology

**Abstract**

We describe a software package for computing and manipulating the subdivision of a sphere by a collection of (not necessarily great) circles and for computing the boundary surface of the union of spheres. We present problems that arise in the implementation of the software and the solutions that we have found for them. At the core of the paper is a novel perturbation scheme to overcome degeneracies and precision problems in computing spherical arrangements while using floating point arithmetic. The scheme is relatively simple, it balances between the efficiency of computation and the magnitude of the perturbation, and it performs well in practice. We report and discuss experimental results. Our package is a major component in a larger package aimed to support geometric queries on molecular models; it is currently employed by chemists working in 'rational drug design.' The spherical subdivisions are used to construct a geometric model of a molecule where each sphere represents an atom. We also give an overview of the molecular modeling package and detail additional features and implementation issues.

## 1 Introduction

Implementing geometric algorithms can be a difficult task. It has been found out and repeatedly rediscovered that there is a huge gap between geometric algorithms as they are described in most theoretical papers and their implementation in software. Two issues that are often ignored in the theoretical approach turn out to be critical in practice: Degeneracies and numerical precision. These issues are collectively referred to as "robustness". Surveys on the topic can be found in [8],[15],[17].

In theory degeneracies are often handled by assuming *general position*, namely assuming that degeneracies do not occur. The general position assumption had contributed significantly to the advancement of geometric algorithms by letting the researchers focus on the key (theoretical) problems while ignoring many technical issues. When implementing a geometric algorithm however, one must take degeneracies into consideration.

The numerical precision problem was solved in the theory of geometric algorithms by assuming infinite precision real arithmetic [18]. For certain algorithms and geometric objects this assumption is realizable in practice by using *exact arithmetic* [5],[9],[20],[24]. Computing with exact arithmetic is in general more costly than using floating point arithmetic, and in certain cases not realizable because of the geometric primitives that need to manipulated. Here again there is a gap between what could in theory be handled by exact arithmetic and what current technology offers [8].

The software package that we describe in this paper computes the boundary of a union of spheres, the surface area of the boundary, and the intersection pattern of any sphere with all the other spheres in a given set. We therefore have to compute the intersection of pairs and triples of spheres, various tangency points of great circles and *little circles*[1] on a sphere, and so on. Such computations would be difficult to carry out using exact arithmetic (all these operations require the solution of polynomial equations, so theoretically symbolic schemes could possibly be used here). Also floating point arithmetic has the obvious advantages of availability and efficiency. Our goal here is to devise robust algorithms that deal with intersecting spheres in $\mathbb{R}^3$ while using floating point arithmetic. Some examples of previous and related work on robust floating point geometric algorithms can be found in [11],[17],[21], and [22].

Our motivating application is geometric modeling of molecules. Our software package is part of a toolbox aimed to support the chemist in the drug design process

*This work has been supported in part by a grant from Pfizer Central Research. Dan Halperin has also been supported by an Alon Fellowship, and by ESPRIT IV LTR Project No. 21957 (CGAL).

†Department of Computer Science, Tel Aviv University, Tel Aviv 69978, Israel. E-mail: halperin@math.tau.ac.il.

‡Department of Computer Science, MIT, Cambridge, MA 02139. E-mail: cshelton@ai.mit.edu. Part of the work on this paper was carried out while Christian Shelton was at the Department of Computer Science, Stanford University

[1]Throughout the paper we use the term *little circle* to mean any circle on a sphere $S$ that is the intersection of $S$ with another sphere.



*Computational Geometry 97* Nice France

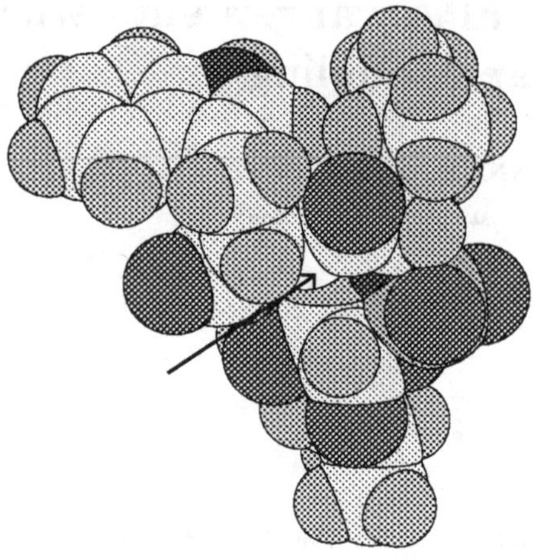

Figure 1: The sphere model of a molecule. The arrow points to the sphere for which the spherical arrangement is drawn in Figure 2(a).

[6], [7]. The basic geometric model of a molecule that we use is the so-called *hard sphere model* where every atom is represented by a sphere at some fixed position relative to the other atom spheres in the molecule. Since the hard sphere model is an approximate model to begin with, we have the freedom to perturb the spheres slightly without much effect on the relevance of the model.

When computing with floating point arithmetic we cannot precisely determine degeneracies; we can only know that we are in a potentially degenerate situation. For example, we may not be able to determine for sure that four spheres in our collection meet at a single point, but we could detect that the point of intersection of three out of the four spheres is less than some $\varepsilon > 0$ away from a point of intersection of a different triple of these spheres. For a small parameter $\varepsilon > 0$ which we call the *resolution parameter*, we regard such a pair of intersection points as a *degeneracy*. (A "potential degeneracy" may be a more appropriate term, but for brevity we will refer to such a situation as degeneracy.) Our new scheme guarantees that for a given parameter $\varepsilon > 0$, all the features of the spherical arrangement are at least $\varepsilon$ apart. The resolution parameter $\varepsilon$ depends on the floating point precision and on the type of operations (e.g., computing the intersection points of three spheres). We assume here that $\varepsilon$ is given. There are numerical analysis methods to compute useful bounds on $\varepsilon$ (see [14, Chapter 4] for examples concerning linear objects). We point out that in our algorithms the 'depth of operations' (namely how many times in a row the result of one operation is the operand in another operation) is bounded by a small constant. Therefore one can obtain a bound on the resolution parameter that does not depend on the input size $n$—the number of spheres, in our case.

We present an efficient perturbation scheme for a collection $M$ of $n$ spheres in $\mathbb{R}^3$ that makes our geometric algorithms robust. We call the decomposition of $\mathbb{R}^3$ induced by the spheres the *arrangement of the spheres*; the subdivision of each sphere by the circles of intersection with other spheres we call a *spherical arrangement*. We denote the maximum number of spheres in $M$ intersecting any single sphere in $M$ by $k$ (as was shown in [13], $k$ is a constant for the hard sphere model of molecules; see also the next section). For any given resolution value $\varepsilon > 0$, we determine a parameter $\delta$ that depends on $\varepsilon$, on $k$, and on the maximum radius $R$ of a sphere in $M$. We then present a scheme that perturbs each sphere by at most $\delta$, resolves all the degeneracies in the arrangement of the spheres, and runs in $O(n)$ time.

We also take care of degeneracies that result from a refinement of the spherical arrangements known as the *trapezoidal decomposition* (see Section 2.1). Since in the trapezoidal decomposition we are free to choose a direction for the 'poles' (two antipodal points on a sphere, such that all the added arcs in the refinement are portions of great circles through these poles), we choose the poles so that the angular separation of the added arcs will be above a certain threshold $\omega$. Here also, we choose $\omega$ such that determining the poles could be done fast.

We have implemented this perturbation scheme and we report experimental results below. For example, letting $\varepsilon = 10^{-11}$, $\delta = 10^{-9}$, and $\omega = \arccos(1 - 10^{-11})$, loading a molecule with 2034 atoms (namely computing the boundary surface and all the individual spherical arrangements) takes 146.8 seconds (including the perturbation) on an SGI Indigo 250Mhz RS4400. Our experimental results show that in most cases the time taken up by the perturbation scheme is negligible (Section 6).

Our software is part of a toolbox of data structures and algorithms aimed to support the chemist in the *rational drug design* process [1], and it is currently being used by chemists in a pharmaceutical company. Further details on the larger software (the toolbox) can be found in [6]. Ours is definitely not the first implementation of an algorithm for computing molecular surfaces (see, e.g., [2],[19],[23] and the recent survey by Connolly [3]). We believe that our approach stands out in its treatment of robustness issues and in the ease and flexibility of computing substructures of the collection of the spheres, such as the union of a subset of the spheres of one molecule consumed by another molecule.

The rest of the paper is organized as follows. In the next section we give more background details on spherical arrangements and on the hard sphere model of a molecule. In Section 3 we review our software package. The key ideas underlying our new perturbation scheme are presented in Section 4. Algorithmic details of the scheme and running time analysis are given in Section 5. In Section 6 we report experimental results. More practical issues concerning our implementation are discussed in Section 7. A brief summary and suggestions for future research on the topic are given in Section 8. In the Appendix we complete technical details concerning shapes and volumes that arise in the perturbation scheme.

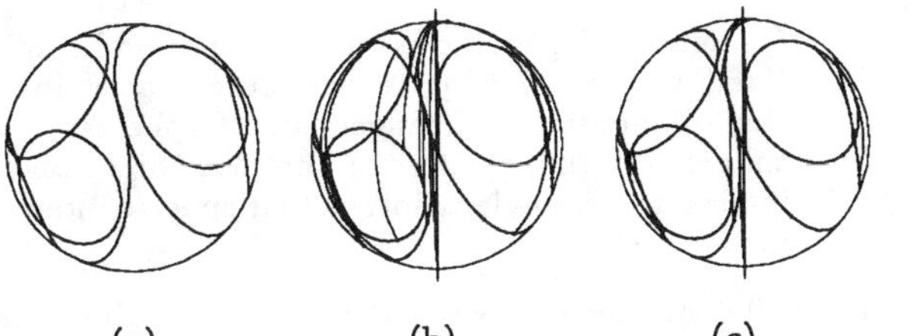

Figure 2: A spherical arrangement (a), its full trapezoidal decomposition (b), and partial trapezoidal decomposition (c).

## 2 Preliminaries: Spheres and Molecules

### 2.1 Little Circles on a Sphere

A collection of circles on a sphere $S$ induces a partitioning of the sphere into vertices, edges and faces. We call such a partitioning a ***spherical arrangement***; see Figure 2(a). If all the circles are great circles, then one can easily transform the spherical arrangement into a ***planar arrangement of lines*** [4], which is a simpler object to handle [10]. However, in our application the circles are not necessarily great circles. Each of the circles is the result of the intersection of $S$ with another sphere. We refer to these more general circles as ***little circles***.

The faces in the spherical arrangement need not be simply connected, and each face may have a large number of edges on its boundary. We can apply a standard refinement procedure, called a ***trapezoidal decomposition*** that will make each face homeomorphic to a disc and have at most four edges on its boundary. See Figure 2(b). We fix a pair of antipodal points as ***poles***. We call the great circles through the poles ***polar circles***, arcs of polar circles ***polar arcs***, and any point on a little circle that is tangent to a polar circle we call a ***polar tangency***. For every polar tangency of every circle in our collection, we extend a polar arc in either direction until it hits another little circle or reaches a pole. We do the same from every intersection point of a pair of little circles. We call this refinement the ***full trapezoidal decomposition***.

If we are only concerned with making each face simply connected and making the graph of all the edges of the arrangement connected then using a ***partial trapezoidal decomposition***, in which polar arcs are only extended from polar tangency points and not from intersections, suffices. See Figure 2(c).

### 2.2 The Hard Sphere Model

A common approach to representing the three-dimensional geometric structure of a molecule, is to represent each of its atoms by a "hard" sphere. In certain applications it is also assumed that the relative displacement of the spheres is fixed. There are recommended values for the radius of each atom sphere and for the distance between the centers of every pair of spheres. In this model, the spheres are allowed to interpenetrate one another, therefore it is sometimes referred to as the "fused spheres" model (see Figure 1). The envelope surface of the fused spheres may be regarded as a formal ***molecular surface***. It is evident that various properties of molecules are disregarded in this simple model. However, in spite of its approximate nature, it has proven useful in many practical applications. For more background material and references, see for example the survey paper by Mezey [16].

In [13] the hard sphere model is studied from a ***computational geometry*** point of view. Several observations are made in that paper showing that, because of certain special properties, the spheres in this model can be efficiently manipulated. We cite below the results that will be needed in later sections. Theorem 2.1, states the conditions that make the sphere model of a molecule favorable. Theorem 2.2 summarizes a hash-table based data structure that is constructed exploiting these conditions.

**Theorem 2.1** [13] *Let $M = \{B_1, \ldots, B_n\}$ be a collection of $n$ balls in 3-space with radii $r_1, \ldots, r_n$ and centers at $c_1, \ldots, c_n$. Let $r = \min_i r_i$ and let $R = \max_i r_i$. Also let $S = \{S_1, \ldots, S_n\}$ be the collection of spheres such that $S_i$ is the boundary surface of $B_i$. If there are positive constants $\rho_1, \rho_2$ such that $\frac{R}{r} < \rho_1$ and for each $B_i$ the ball with radius $\rho_2 \cdot r_i$ and concentric with $B_i$ does not contain the center of any other ball in $M$ (besides $c_i$), then:* **(i)** *for each $B_i \in M$, the maximum number of balls in $M$ that intersect it is bounded by a constant, and* **(ii)** *the maximum combinatorial complexity of the boundary of the union of the balls in $M$ is $O(n)$.*

**Theorem 2.2** [13] *Given a collection $M$ of $n$ balls as defined in Theorem 2.1, one can construct a data structure using $O(n)$ space, to answer intersection queries for balls whose radii are not greater than $r_{max}$, in $O(1)$ time. The expected preprocessing time of the structure is $O(n)$.*

## 3 Overview of the Software Package

Our software package is divided into two main parts: the core computation engine and the interface. The computation engine consists of the code necessary to compute the surface area of any arbitrary subset of the spheres, to answer intersection queries, and to create decompositions of the spheres. A number of different interfaces have been constructed. Two primary ones are currently being used. One is a batch processor which reads in a list of molecules (or a list of sets of molecules) to process and outputs a surface file for each molecule (or set) consisting of general statistics such as name and number of atoms, the surface area and void areas, and the surface area of one set of atoms in one molecule consumed by another set of atoms in another molecule (for multiple molecule sets).

The other interface is a graphical interface. It displays the general statistics of the molecule along with boundary surface and void area computations. In a separate window, it displays a rendering of a three-dimensional model of the spheres which can be rotated, translated, and scaled. This model can further be colored and shaded per atom by various calculations from the computational engine such as surface area, surface percentage, element type, molecule, and number of intersecting atoms. This interface also allows the user to select any atoms in the molecule. The selected atoms

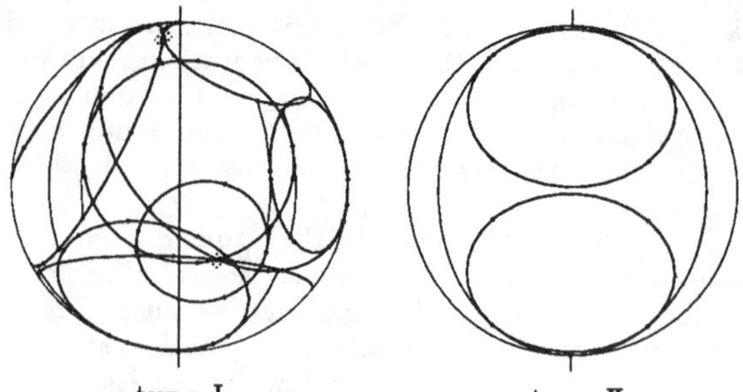


Figure 3: Examples of the two types of degeneracies. Type I degeneracies are marked by small shaded circles.

are displayed separately with a full view of the spherical arrangement on the atom (we also call it the *atom map*) and the exact statistics on the atom's contribution in terms of surface area to the molecule's outer boundary and inner voids.

This paper will detail the computational engine of the package. In particular, we will concentrate on robust computation of the spherical decomposition. Higher level data structures for keeping track of atoms within multiple molecules and lower level code for computing intersections and keeping track of the geometry of points and arc segments will be assumed for our discussion.

For representing the atom map, we chose a variant of the quad-edge data structure [12]. This structure makes updating the subdivision simple. Each arc is stored as a plane and two points (thus the arc is the intersection of the plane and the sphere between the two points). Four operations are needed: adding an arc, removing an arc, splitting an arc, and merging two connected arcs (namely two arcs that belong to and are adjacent along a single circle). The maintenance of this data structure is fairly standard and we omit the details here.

The data structure representing the spherical arrangement assumes that the graph of arcs is connected. This means that all of the regions must be simply connected. To ensure this, we break up the surface with additional arcs in order to create a trapezoidal decomposition, as described in Section 2.1.

We have presented so far the ingredients that are necessary to understanding the perturbation scheme. Further details on our software are given below in Section 7.

## 4 The Perturbation Scheme

We distinguish two types of degeneracies that arise in a collection of intersecting spheres and atom maps as we compute them. Recall that we are concerned with floating point arithmetic and that we define a degeneracy using a resolution parameter $\varepsilon > 0$. In this section we give a precise definition of the various degeneracies. We describe the two types as they occur on a single sphere. See Figure 3.

**Type I** A little circle is too small (see below for numerical values), two little circles are tangent (or almost tangent), or two intersection points each being the intersection of three spheres are close together. Had we been using exact arithmetic these (potential) degeneracies would correspond respectively to the following (exact) degeneracies in the three-dimensional arrangement of spheres: Two spheres are tangent to one another, three spheres intersect in a single point, or four spheres intersect in a point.

**Type II** The angle between the planes containing two distinct polar arcs is too small (it is below some threshold $\omega$).

Type I is inherent to the arrangement of spheres, whereas type II is an artifact of our decomposition method.

We wish to perturb the spheres slightly so that all the features of the arrangement will be at least $\varepsilon$ apart, for a given resolution parameter $\varepsilon$. We would like the perturbation procedure to be efficient and at the same time that the perturbation magnitude will be as small as possible. For a given $\varepsilon$ our procedure will determine a perturbation value $\delta$ that will guarantee that the procedure will take $O(n)$ time for $n$ spheres.

We use a two-step perturbation scheme. At the first step we remove type I degeneracies by an incremental procedure where we add the spheres one by one and if a degeneracy occurs we only perturb the last sphere that has been added. We then choose the orientation of the poles (namely the direction to which all the planes containing polar arcs will be parallel) that will eliminate degeneracies of type II.

### 4.1 Removing Type I Degeneracies

Let $S_1, S_2, \ldots, S_n$ be an ordering of the spheres in $M$. For a pair of spheres $S_i$ and $S_j$ let $\rho_{ij}$ denote the distance between their centers. Let $M_t$ denote the set $\cup_{i=1}^{t}\{S_i\}$. After the completion of stage $t$, the incremental procedure will maintain the following invariants:

$I_1$ The center $C_i$ of any sphere in $M$ has been moved by at most $\delta$ from its original placement ($\delta$ is a constant to be determined below).

$I_2$ For every pair of spheres $S_i$ and $S_j$, $i \neq j \leq t$, and $r_j \geq r_i$, we have $|\rho_{ij} - r_i - r_j| > \varepsilon$ and $|\rho_{ij} + r_i - r_j| > \varepsilon$.

$I_3$ For every triple of spheres $S_i, S_j$ and $S_k$, $i \neq j \neq k \leq t$, the circle $C_{ij} := S_i \cap S_j$ and the sphere $S_k$ are not tangent, and are at least $\varepsilon$ away from being tangent. (The formal definition of this invariant is given in the Appendix.)

$I_4$ Let $\tau_i$ and $\tau_j$ each be an intersection point of three spheres in $M_t$; then $d(\tau_i, \tau_j) > \varepsilon$.

We call a perturbation scheme that satisfies the above invariants at the end of each stage (and in particular after the $n$th stage) a *valid perturbation scheme*.

Suppose the procedure has been carried out successfully for the first $t$ stages. We next describe how we add the sphere $S_{t+1}$ so that $M_{t+1}$ will maintain the invariants above. We denote the center of sphere $S_i$ for $i \leq t$ after the completion of step $t$ by $C_i^t$. Let $B(C_{t+1}, \delta)$ be the ball of radius $\delta$ around the original placement of the center $C_{t+1}$ of $S_{t+1}$. We will place the center of $S_{t+1}$ inside $B(C_{t+1}, \delta)$ and this will guarantee the invariant $I_1$. The invariants $I_2, I_3$ and $I_4$ define forbidden loci

$F_2, F_3$ and $F_4$ respectively for the center of $S_{t+1}$. We will choose the new placement $C'_{t+1}$ of the center of $S_{t+1}$ to be in $G_{t+1} := B(C_{t+1}, \delta) \setminus (F_2 \cup F_3 \cup F_4)$. If such a placement exists, we call it a ***valid placement*** of the center of $S_{t+1}$. If the original placement of $S_{t+1}$ lies in $G_{t+1}$ then we do not move $S_{t+1}$.

The region $F_2$ is the union of ***spherical shells***[2] of two types. Each sphere $S_i \in M_t$ induces two spherical shells according to whether the tangency is external (namely, at the tangency point the interior of the spheres are disjoint) or internal. One shell has its center in $C'_i$ and its radii are $r_i + r_t - \varepsilon$ and $r_i + r_t + \varepsilon$. This spherical shell represents the loci of placement of the center of $S_t$ that result in the spheres $S_i$ and $S_t$ being tangent, or almost tangent. The other type of shells is defined similarly.

We postpone a detailed explanation of the shape of the objects contributing to $F_2, F_3$, and $F_4$ to the Appendix. We just mention that $F_4$ is also the union of spherical shells, whereas $F_3$ is the union of ***toric shells***. In the Appendix we derive upper bounds on the volume taken up by the regions $F_2, F_3$, and $F_4$. Let $VF :=$ Volume$(F_2 \cup F_3 \cup F_4)$ and let $VB :=$ Volume$(B(C_{t+1}, \delta))$.

Our goal is to choose $\delta$ such that $VB > 2 \cdot VF$. If we do that, we are guaranteed that a point chosen uniformly at random inside $B(C_{t+1}, \delta)$ has probability $> \frac{1}{2}$ to be a valid placement for the center of $S_{t+1}$. Our calculations show that we need to choose $\delta = f(k, \varepsilon, R) = 2k\varepsilon^{1/3}R^{2/3}$, where $k$ is the maximum number of spheres in $M$ intersecting any single sphere in $M$, $R$ is the maximum radius of a sphere in $M$, and assuming $k \geq 10$. Not surprisingly, our experimental results show that this is a very conservative bound. The theoretical bound is a crude worst-case bound, and even as such it shows that our approach does not conceal any very large constant.

### 4.2 Removing Type II Degeneracies

After we have removed all type I degeneracies, we wish to choose a direction for the poles such that no type II degeneracy arises, when using the **partial** vertical decomposition. (As we remark below the method can be extended to deal with full vertical decomposition.) We will handle each sphere separately, namely choose the direction for the poles independently for each sphere.

The situation here is significantly different from the situation with type I degeneracies: It is impossible to guarantee an $\varepsilon$-separation of the polar arcs. Polar arcs can get arbitrarily close to one another, and they can coincide in the poles. Therefore our goal is to obtain a good angular separation, which we call ***$\omega$-separation***, and is defined as follows: Once the direction of the poles has been determined, for any pair of distinct polar arcs the angle between the planes containing them is at least $\omega$. We wish to determine the biggest $\omega$ that will allow us to run Step 2 in $O(n)$ time. If $\omega$-separation has been obtained for an $\omega$ value which is greater than the floating point resolution, polar arcs can be safely and consistently distinguished during computation.

We next describe this procedure for a single sphere $S_t$, after Step 1 has been completed (and type I degeneracies have been removed). Fix two little circles $C_{it}$ and $C_{jt}$ on $S_t$ (each being the intersection of $S_t$ with another sphere in $M$), and let $\Phi$ be a great circle on $S_t$ that is tangent to $C_{it}$ and $C_{jt}$ (it is one of at most four great circles tangent to this two little circles). The circle $\Phi$ is the locus of pole locations that will cause the planes containing the polar arcs extending from the tangency points of $\Phi$ with $C_{it}$ and $C_{jt}$ to coincide (namely have zero angular separation).

[2] A spherical shell is the region enclosed between two concentric spheres. It is the three-dimensional analog of an annulus.

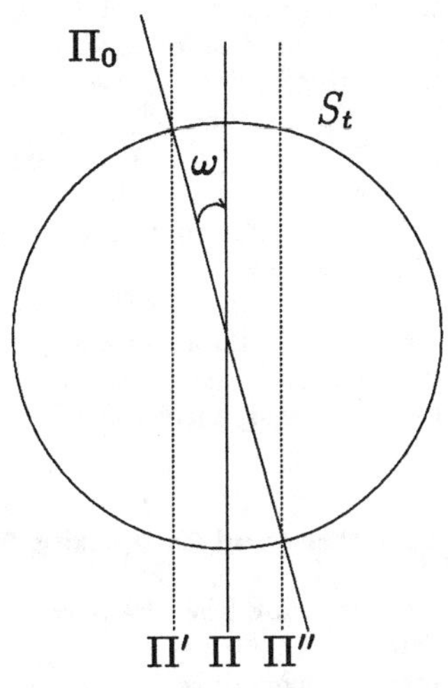


Figure 4: A cross-section of $S_t$.

Let $\Pi := \Pi(\Phi)$ denote the plane containing $\Phi$. Let $\Pi'$ and $\Pi''$ be two planes parallel to $\Pi$, each on a different side of $\Pi$, and such that any plane $\Pi_0$ passing through the center of $S_t$ and tangent to the circle $\Pi' \cap S_t$ (or the circle $\Pi'' \cap S_t$) makes an angle $\omega$ with $\Pi$. See Figure 4 for an illustration. We call the portion of $S_t$ between $\Pi' \cap S_t$ and $\Pi'' \cap S_t$ the ***$\omega$-strip of $\Phi$***, and denote it by $\sigma(\omega, \Phi)$. It is easily verified that if the poles are chosen outside $\sigma(\omega, \Phi)$ then the polar arcs that correspond to the tangencies of $\Phi$ with $C_{it}$ and $C_{jt}$ are at least $\omega$ apart.

Every tangent circle $\Phi$ of every pair of circles on $S_t$ defines an $\omega$-strip $\sigma(\omega, \Phi)$. Assume without loss of generality that the radius of $S_t$ is 1. The area of any such strip is $4\pi \sin \omega$. The maximum number of strips that need to be considered on $S_t$ is $4\binom{k}{2}$. For Step 2 to run in $O(n)$ time we wish that at least half of the surface area of $S_t$ will not be covered by $\omega$-strips (any constant fraction will do; the smaller the uncovered fraction the higher the constant factor in the expected running time bound). Therefore we choose $\omega$ such that $2 \cdot 4\binom{k}{2} 4\pi \sin \omega < 4\pi \implies \sin \omega < 1/(4k(k-1))$ .

**Remarks.** (1) The $\omega$-strip is only an inscribing region of the forbidden region corresponding to any $\Phi$. The actual forbidden region induced by any tangency is in fact smaller, and it has a more complicated boundary. (2) Extending the method to the full vertical decomposition requires handling several more cases involving up to four circles $C_{it}$ each. This extension is straightforward and we omit further details here. We have implemented this extension and we report experimental results for it below. (3) In practice, it is convenient to impose the same pole direction for all the spheres in the arrangement (for ease of coding, debugging and visualization). To this end we draw all the $\omega$-strips on the (unit) sphere of directions $S^2$, and the constraint on $\omega$ becomes $\sin\omega < 1/(4nk(k-1))$. The experimental results described below are for this choice of pole directions. (4) The construction of the forbidden regions on $S_t$ is diametrically symmetric, namely a point on $S_t$ is free (i.e., represents a pole that will induce an $\omega$-separation) if and only if its antipodal point is free. (5) In our implementation we also check to make sure that a polar arc does not pass within $\omega$ of any intersection point.

## 5 Algorithmic Details and Complexity Analysis

In this section we describe how the perturbation scheme is carried out efficiently.

The incremental procedure of Step 1 will move the center of each sphere by at most $\delta$ from its original placement. We expand each of the original spheres $S_t$ into a sphere $S_t''$ whose radius is $r_t + \delta$, and let $M''$ be the set of the expanded spheres $S_t''$. We construct a data structure as described in Theorem 2.2 to support range queries on the spheres in $M''$. This structure enables us to find in $O(1)$ time all the spheres in $M''$ that intersect a given query sphere $S_t''$.

When looking for a perturbation of the center of $S_t$ during Step 1 of the procedure we proceed as follows. Although constructing the subdivision of $B(C_t, \delta)$ into free and forbidden regions could be done in maximum constant time, it would be an extremely difficult task that may introduce new precision problems. Instead we do something much simpler. We choose a point $p$ uniformly at random in the ball $B(C_t, \delta)$. We next check whether $p$ is a valid perturbation by checking that it lies outside all the forbidden regions $F_2, F_3$ and $F_4$. For example, to check whether $p$ lies outside $F_2$ we check whether it lies outside each of the at most $2k$ spherical shells defining $F_2$. Each of these tests is simple and fast and their overall number is bounded by a constant. See below for experimental results. Recall that $\delta$ was chosen such that with probability $> \frac{1}{2}$, the point $p$ will represent a valid perturbation. Thus the expected number of trials before we find a valid $p$ is $\leq 2$, and the overall running time of Step 1 is $O(n)$.

Similar arguments apply to Step 2 procedure. For any given sphere $S_t$, we choose a point $p$ uniformly at random on the boundary of $S^2$. We then test all the $\omega$-strips relevant to $S_t$ to check whether $p$ lies outside all those strips. By the choice of $\omega$ the expected number of trials before we find a valid $p$ is $\leq 2$, and the overall running time of Step 2 is $O(n)$ as well.

We summarize the performance of the perturbation scheme in the following theorem.

**Theorem 5.1** *Given a collection $M$ of $n$ spheres as described in Theorem 2.1, and a resolution parameter $\varepsilon > 0$, a valid perturbation of the spheres in $M$ can be computed in $O(n)$ time, by moving each sphere by at most $\delta$ from its original placement, where $\delta$ is a parameter that depends on $\varepsilon$, on the maximum number of spheres in $M$ intersecting any single sphere in $M$, and on the maximum radius of a sphere in $M$, and such that all the degeneracies are resolved. In $O(n)$ time we can also find a direction for the poles on each sphere so that all the polar arcs in the trapezoidal decomposition of the spherical arrangements lie on planes the angle between any pair of which is at least $\omega$, for $\omega$ that depends on the maximum number of spheres in $M$ intersecting any single sphere in $M$.*

| number | file name | size | max $k$ | mean $k$ |
|---|---|---|---|---|
| 1 | estradiol.mol2 | 44 | 17 | 6.81 |
| 2 | clofazimine.mol2 | 55 | 13 | 6.67 |
| 3 | michellamine_b.mol2 | 104 | 13 | 6.81 |
| 4 | 288d.pdb | 120 | 9 | 6.25 |
| 5 | conocurvone.mol2 | 127 | 14 | 6.67 |
| 6 | 245d.pdb | 240 | 9 | 6.17 |
| 7 | 1ppt.pdb | 301 | 10 | 5.92 |
| 8 | 4pti.pdb | 454 | 10 | 5.79 |
| 9 | 1bzm.pdb | 2034 | 10 | 5.74 |
| 10 | 2pka.pdb | 3598 | 11 | 5.79 |
| 11 | 2ace.pdb | 4143 | 12 | 6.05 |
| 12 | 1sdk.pdb | 4384 | 12 | 6.00 |
| 13 | 1nok.pdb | 6759 | 11 | 5.73 |
| 14 | 7at1.pdb | 7106 | 12 | 5.70 |

Table 1: Listing of the molecules used for testing. Molecules with file names ending in pdb can be found at the Protein Databank at http://www.pdb.bnl.gov/. Molecules with file names ending in mol2 can be found at the Center for Molecular Modeling at http://cmm.info.nih.gov/modeling/. The size column gives the number of atoms in the molecule, and $k$ is the number of atom spheres intersecting any single atom sphere.

We emphasize again that the theoretical bounds that we obtain on $\delta$ and $\omega$ are crude and in practice (as shown below) these quantities are much smaller.

## 6 Experimental Results

For timing experiments, we chose fourteen molecules as listed in Table 1. The molecules from the Protein Databank did not specify the positions of hydrogen atoms whereas the others did. This accounts for most of the difference in $k$ values. All timings were done on an SGI Indigo with a 250Mhz MIPS RS4400 CPU.

For our timings, we ran the program on each of the molecules in our set for varying $\epsilon$ and $\delta$ ($\delta$ on $[10^{-6}, 10^{-10}]$ and $\epsilon$ on $[10^{-9}, 10^{-12}]$). Figure 5 shows the results of these timings for the portion of the code not involved with computing perturbations. Figure 6 details the time spent in the perturbation code.

It should be noted that almost all of the perturbation time $(90-98\%)$ is spent finding the pole direction. Table 2 shows the fraction of the $\delta$-spheres which are free and Table 3 shows the fraction of the pole directions that are free for a typical molecule. To obtain the former chart, we modified the program to attempt (and discard) 1000 perturbations each time the center of an atom was positioned. The values shown in the table are the average over all center placements of the fraction of these perturbations that were valid. The latter was obtained with a similar sampling method for pole directions.

The reason for the dramatic difference between the time spent on pole location and the time spent on cen-

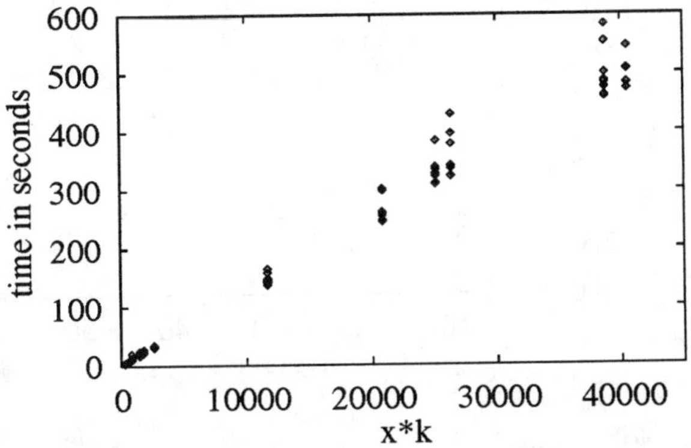


Figure 5: Time to build the molecules shown in Table 1 excluding perturbation time. The horizontal coordinate is size-of-molecule × average-$k$.

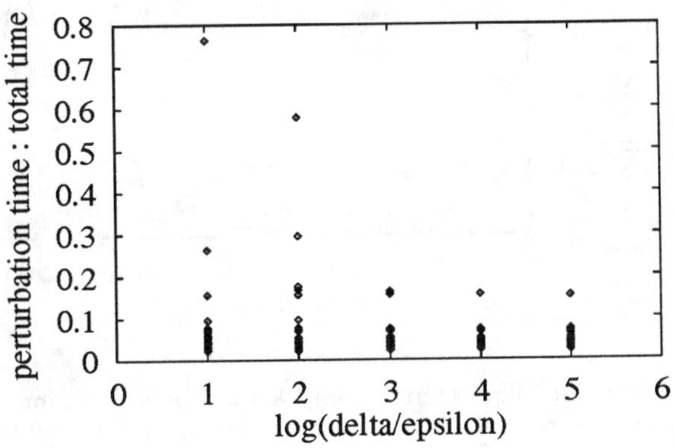


Figure 6: Ratio of perturbation time to total time versus ratio of $\delta$ to $\epsilon$.

| | 1e-3 | 1e-4 | 1e-5 | 1e-6 | 1e-7 | 1e-8 | 1e-9 |
|---|---|---|---|---|---|---|---|
| 1e-2 | 0.993 | 0.999 | 0.999 | 0.999 | 0.999 | 1.000 | 1.000 |
| 1e-3 | | 0.999 | 0.999 | 0.999 | 1.000 | 1.000 | 1.000 |
| 1e-4 | | | 0.999 | 0.999 | 0.999 | 0.999 | 1.000 |
| 1e-5 | | | | 1.000 | 1.000 | 1.000 | 1.000 |
| 1e-6 | | | | | 1.000 | 1.000 | 1.000 |
| 1e-7 | | | | | | 1.000 | 1.000 |
| 1e-8 | | | | | | | 1.000 |

Table 2: Valid fraction of the $\delta$-sphere for molecule 10 from Table 1. $\epsilon$ varies across the table and $\delta$ varies downwards.

| $1-\cos(\omega)$ | | | | |
|---|---|---|---|---|
| 1e-9 | 1e-10 | 1e-11 | 1e-12 | 1e-13 |
| 0.285 | 0.675 | 0.890 | 0.950 | 0.975 |

Table 3: Valid fraction of pole positions for molecule 8 from Table 1.

| | 1e-3 | 1e-4 | 1e-5 | 1e-6 | 1e-7 | 1e-8 | 1e-9 |
|---|---|---|---|---|---|---|---|
| 1e-2 | 0.625 | 0.937 | 0.993 | 0.999 | 0.999 | 1.000 | 1.000 |
| 1e-3 | | 0.626 | 0.937 | 0.993 | 0.999 | 0.999 | 1.000 |
| 1e-4 | | | 0.626 | 0.937 | 0.993 | 0.999 | 0.999 |
| 1e-5 | | | | 0.626 | 0.937 | 0.993 | 0.999 |
| 1e-6 | | | | | 0.626 | 0.937 | 0.993 |
| 1e-7 | | | | | | 0.626 | 0.937 |
| 1e-8 | | | | | | | 0.626 |

Table 4: Valid fraction of the $\delta$-sphere for the degenerate set of spheres described in Section 6. $\epsilon$ varies across the table and $\delta$ varies downwards.

ter location can be attributed to a few additional constraints we placed on the poles. In order to simplify certain coding and debugging procedures, we insisted that all spheres share the same pole direction. In addition, the poles were required to be chosen such that no intersection point of a polar arc and a little circle would be within $\epsilon$ of any other vertex of the arrangement, and that no polar arc be within $\omega$ of the special great circle that is added as part of our "simplified point location" procedure (which we describe in the next section).

We also reperformed the experiment of Table 2 for a purposefully degenerate collection of spheres (Table 4). For this set, we arranged 27 spheres in a cube on a unit regular grid. Each sphere had a radius of 1.05 thus producing 2 to 12 degeneracies where 3 little circles intersect at one point on each sphere. In this highly degenerate case, most of the $\delta$-sphere was free even when the ratio of $\delta$ to $\epsilon$ was 10.

## 7 Additional Issues in Computing Spherical Arrangements

### 7.1 Simplified Point Location

We construct the spherical arrangement on each sphere by sequentially adding one sphere at a time to the collection. Each sphere induces a number of little circles on both itself and other spheres. These are added in corresponding pairs. When a new circle is added to a sphere, we must locate in the previously constructed arrangement a point on the little circle. For circles which do not encompass a pole, we choose to locate the tangency points. Since polar arcs must be extended from each tangency to the arc segments "above" and "below" the tangency, finding these "upper" and "lower" bounds on the polar arc is sufficient for point location.

In order not to have to search through all of the arcs on a sphere to find the two arcs needed, we use a simple dividing scheme. Each sphere is divided into sections along the equator. All arcs on the sphere are projected onto the plane of the equator and then outwards to the

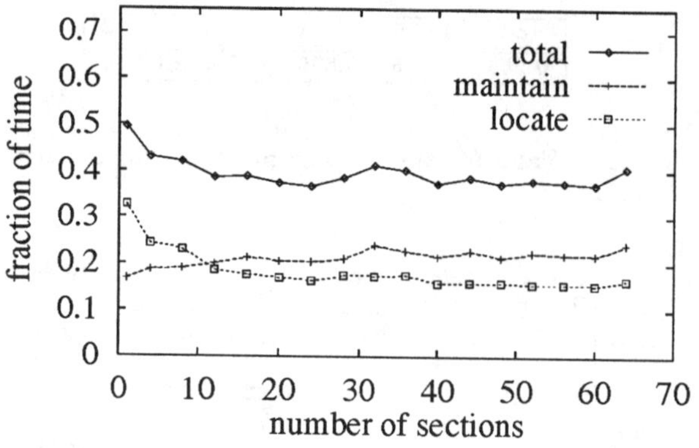


Figure 7: Fraction of the total time of the construction spent in locating points, divided into time spent maintaining the data structure and the time spent querying and locating points on the decomposition.

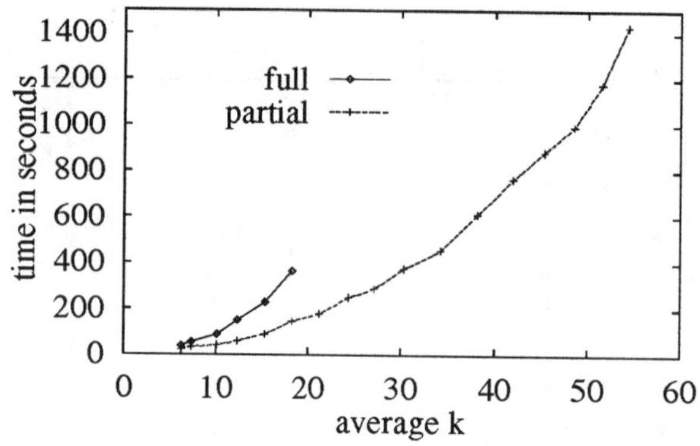


Figure 8: Time spent constructing the decompositions after the perturbation for full and partial decompositions against $k$ (the average number of little circles per sphere)

equator. Each section on the equator stores references to the arcs whose projections pass through it. Furthermore, since all little circles are broken at polar tangency points, the equatorial extent of any arc segment can be found by projecting only its endpoints.

Thus, for point location, the point is similarly projected onto the equator and then into a section. Half of a great circle passing from one pole to the other through the point is constructed and this polar arc is then intersected with the arcs listed in the found section to find the arc immediately "above" and "below" the point. For our implementation, we chose to divide the equator into a fixed number of equal-size sections.

For little circles which surround one of the poles, such a scheme will not work since there are no such tangency points. To ensure simple point location in this case, one great circle passing through both poles is maintained. Thus each little circle without tangencies will intersect this great circle twice. These two points are then used as starting locations for adding the rest of the little circle just as the tangency points would otherwise be used.

Figure 7 shows timings of the code involved in point location during a load of molecule 10 from table 1. The plot is not as conclusive as we might hope, but it certainly shows an initial negative slope for small numbers of sections and hints at a positive slope for larger values.

### 7.2 Full vs. Partial Decomposition

In order to evaluate the experimental advantage of either the partial or full decompositions, we took molecule 4 and timed the algorithm while successively adding to the radii table used to convert element types to radii. Thus the radii were expanded by a constant each time and the calculations redone. This provided a number of different molecules with different $k$ values, with a somewhat "natural" spread of the spheres in space.

Figures 8, 9, and 10 detail the amount of time spent in each of the phases (decomposition construction, pole perturbation, and center perturbation respectively). The last figure only has one graph since the algorithm for

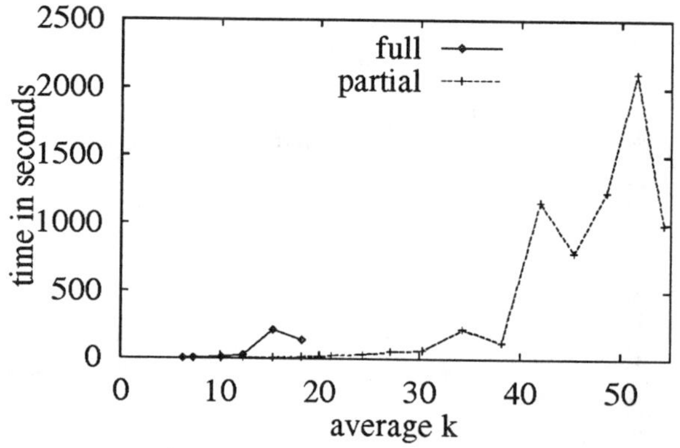


Figure 9: Time spent finding a valid pole direction for full and partial decompositions against $k$ (the average number of little circles per sphere)

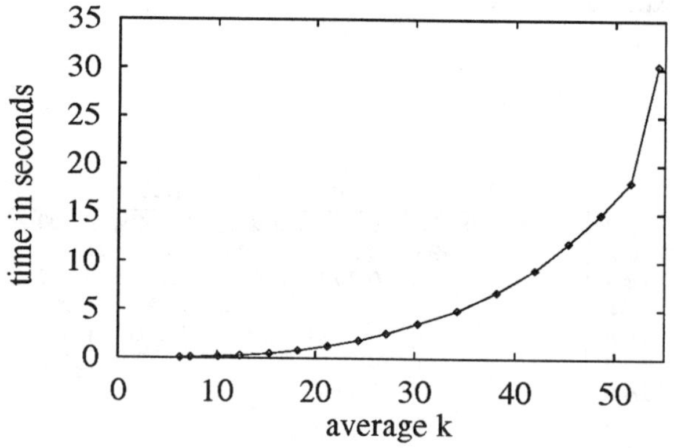


Figure 10: Time spent perturbing the centers of the spheres against $k$ (the average number of little circles per sphere)

finding centers does not change based on the decomposition method. The timings were stopped at different points ($k = 18$ for the full decomposition and $k = 55$ for the partial decomposition) since any further increase in the radii caused the pole perturbation to run for hours with no success.

The difference in the perturbation times can be explained in terms of the constraints put on the pole direction. In the full decomposition case, due to the added polar arcs, it is necessary to check additional plane conflicts, some of these involving five spheres. This adds significantly to the time necessary to find a pole direction (or subtracts significantly from the sphere of possible pole directions). As mentioned above our code also enforces that all spheres have the same pole direction. For the molecules normally considered by the program, this does not have a serious effect on running time. However this restriction does explain the difference between the perturbation time for centers and poles since the centers are chosen separately.

## 8 Conclusion

We have presented a perturbation scheme for a collection of spheres in three-dimensional space. Our scheme is suitable for computing with finite precision arithmetic and we have presented experimental results while using standard floating point arithmetic. For a given resolution parameter $\varepsilon$ we perturb the spheres such that features of the three dimensional arrangement of the spheres (and hence of the two-dimensional spherical arrangement on each sphere) are at least $\varepsilon$ apart. Our scheme balances between the size of the perturbation (which we aim to minimize) and the expected running time of the scheme (the smaller the magnitude of the perturbation the longer the expected time it may take to compute a valid perturbation).

Our motivation to develop this scheme is a software package that we have devised aimed to support geometric queries on molecular models. The new scheme has made our algorithms and data structures robust with hardly any effect on the running and reaction time of the system. We have also presented experimental results showing this small effect. Our software package is currently used by chemists working in rational drug design.

The main direction for further research that we propose is to extend the scheme to other types of arrangements of geometric objects. An obvious limitation of our approach, that may make it unsuitable for certain applications, is that we actually move the input geometric objects from their given placement. However, we believe that there are applications where a small bounded perturbation of the input objects is permissible, since often the precision of the input objects is limited to start with (due to measurement limitations, for example). Moreover, applications that use floating point arithmetic without further adaptations, already make approximate computations, and a small enough perturbation may not have any noticeable effect.

## Acknowledgment

The molecular modeling package described in the paper has been developed as part of a joint project between the Computer Science Robotics Laboratory at Stanford University and Pfizer Central Research in Sandwich, England. The authors acknowledge the assistance of Jean-Claude Latombe who initiated and supervised the project, Paul Finn of Pfizer Central Research who provided useful guidance, and the help and support of the other participants at Stanford: Lydia Kavraki (now at Rice University), Rajeev Motwani, and Suresh Venkatasubramanian.

## References

[1] L. Balbes, S. Mascarella, and D. Boyd. A perspective of modern methods for computer aided drug design. In *Reviews in Computational Chemistry*, volume 5, pages 265–294. VCH Publishers Inc., 1994.

[2] M. Connolly. Analytical molecular surface calculation. *Journal of Applied Crystallography*, 16:548–558, 1983.

[3] M. Connolly. Molecular surfaces: A review. *Network Science*, 1996. http://www.awod.com/netsci/Issues/Apr96/feature1.html.

[4] H. Edelsbrunner. *Algorithms in Combinatorial Geometry*, volume 10 of *EATCS Monographs on Theoretical Computer Science*. Springer-Verlag, Heidelberg, West Germany, 1987.

[5] H. Edelsbrunner and E. P. Mücke. Simulation of simplicity: a technique to cope with degenerate cases in geometric algorithms. *ACM Trans. Graph.*, 9:66–104, 1990.

[6] P. Finn, D. Halperin, L. Kavraki, J.-C. Latombe, R. Motwani, C. Shelton, and S. Venkatsubramanian. Geometric manipulation of flexible ligands. In M. C. Lin and D. Manocha, editors, *Applied Computational Geometry: Towards Geometric Engineering*, volume 1148 of *Lecture Notes in Computer Science*, pages 67–78. Springer, 1996.

[7] P. Finn, L. Kavraki, J.-C. Latombe, R. Motwani, C. Shelton, S. Venkatsubramanian, and F. Yao. Rapid: randomized pharmacophore identification for drug design. This proceedings.

[8] S. Fortune. Ronustness issues in geometric algorithms. In M. C. Lin and D. Manocha, editors, *Applied Computational Geometry: Towards Geometric Engineering*, volume 1148 of *Lecture Notes in Computer Science*, pages 9–14. Springer, 1996.

[9] S. Fortune and C. J. Van Wyk. Static analysis yields efficient exact arithmetic for computational geometry. *ACM Trans. on Graphics*, 25(3):223–248, 1996.

[10] M. Goldwasser. An implementation for maintaining arrangements of polygons. In *Proc. 11th Annu. ACM Sympos. Comput. Geom.*, pages C32–C33, 1995.

[11] D. H. Greene and F. F. Yao. Finite-resolution computational geometry. In *Proc. 27th Annu. IEEE Sympos. Found. Comput. Sci.*, pages 143–152, 1986.

[12] L. J. Guibas and J. Stolfi. Primitives for the manipulation of general subdivisions and the computation of Voronoi diagrams. *ACM Trans. Graph.*, 4:74–123, 1985.

[13] D. Halperin and M. H. Overmars. Spheres, molecules, and hidden surface removal. In *Proc. 10th Annu. ACM Sympos. Comput. Geom.*, pages 113–122, 1994.

[14] C. Hoffmann. *Geometric and Solid Modeling*. Morgan Kaufmann, San Mateo, California, 1989.

[15] C. Hoffmann. The problems of accuracy and robustness in geometric computation. *IEEE Computer*, 22(3):31–41, Mar. 1989.

[16] P. Mezey. Molecular surfaces. In K. Lipkowitz and D. Boyd, editors, *Reviews in Computational Chemistry*, volume 1, pages 265–294. VCH Publishers Inc., 1990.

[17] V. Milenkovic. Verifiable implementations of geometric algorithms using finite precision arithmetic. *Artif. Intell.*, 37:377–401, 1988.

[18] F. P. Preparata and M. I. Shamos. *Computational Geometry: An Introduction.* Springer-Verlag, New York, NY, 1985.

[19] M. F. Sanner, A. J. Olson, and J.-C. Spehner. Fast and robust computation of molecular surfaces. In *Proc. 11th Annu. ACM Sympos. Comput. Geom.*, pages C6–C7, 1995.

[20] J. R. Shewchuk. Robust adaptive floating-point geometric predicates. In *Proc. 12th Annu. ACM Sympos. Comput. Geom.*, pages 141–150, 1996.

[21] K. Sugihara. On finite-precision representations of geometric objects. *J. Comput. Syst. Sci.*, 39:236–247, 1989.

[22] K. Sugihara and M. Iri. Two design principles of geometric algorithms in finite-precision arithmetic. *Appl. Math. Lett.*, 2(2):203–206, 1989.

[23] A. Varshney, F. B. Jr., and W. Wright. Computing smooth molecular surfaces. *IEEE Computer Graphics and Applications*, 14:19–25, 1994.

[24] C. Yap and T. Dubé. The exact computation paradigm. In D. Du and F. Hwang, editors, *Computing in Euclidean Geometry*, pages 452–492. World Scientific, 1995.

## Appendix: Shells of Degenerate Placements

In this appendix we give more details on the shape and volume of the shells defining the regions $F_2$, $F_3$ and $F_4$ as mentioned in Section 4; we use the notation established there.

**The region $F_2$.** $F_2$ consists of placements of the center of $S_{t+1}$ that induce tangency or near-tangency of $S_{t+1}$ and another sphere. There are two types of tangencies: external and internal. We first describe the shell for external tangency. For a sphere $S_i$, an exact tangency is induced by placing the center $C_{t+1}$ of $S_{t+1}$ at distance exactly $r_i + r_{t+1}$ away from the center of $S_i$, namely, $\rho_{i(t+1)} - r_i - r_{t+1} = 0$. We define the potential degeneracy of this type when using floating point with resolution $\varepsilon$ as $-\varepsilon \leq \rho_{i(t+1)} - r_i - r_{t+1} \leq \varepsilon$, which is a spherical shell (i.e. the region sandwiched between two concentric spheres) with center at $C_i$ and radii $r_i + r_{t+1} - \varepsilon$ and $r_i + r_{t+1} + \varepsilon$. Its volume is $\frac{4}{3}\pi[(r_i + r_{t+1} + \varepsilon)^3 - (r_i + r_{t+1} - \varepsilon)^3]$. Similarly the volume of a shell corresponding to internal tangency is (assuming $r_{t+1} > r_i$) $\frac{4}{3}\pi[(r_{t+1} - r_i + \varepsilon)^3 - (r_{t+1} - r_i - \varepsilon)^3]$.

**The region $F_3$.** The region $F_3$ is the union of *toric shells* each defined for a pair of distinct sphere $S_i, S_j \in M_t$. Let $C_{ij}$ denote $S_i \cap S_j$ after both spheres have been possibly perturbed. The toric shell for the pair $S_i, S_j$ is the loci of placements of the center of $S_{t+1}$ that will cause $S_{t+1}$ to be tangent, or almost tangent to $C_{ij}$. Thus it represents the placements of $S_{t+1}$ that may cause two circles on a spherical arrangement to be tangent or almost tangent.

We denote the center of circle $C_{ij}$ by $c_{ij}$ and its radius by $r_{ij}$. At a point $p$ of tangency between $C_{ij}$ and $S_{t+1}$ there is a plane $\Pi(p)$ that is tangent to $S_{t+1}$, and such that its intersection with the plane containing $C_{ij}$ is a line $L(p)$ tangent to $C_{ij}$. If we fix the point $p$ on $C_{ij}$ and hence the line $L(p)$ there is a pencil of planes through $L(p)$ where each plane defines a possible osculation placement of the sphere and the circle at $p$. The induced loci of forbidden placements for the center of $S_{t+1}$ are a circle with center at $p$, lying on a plane

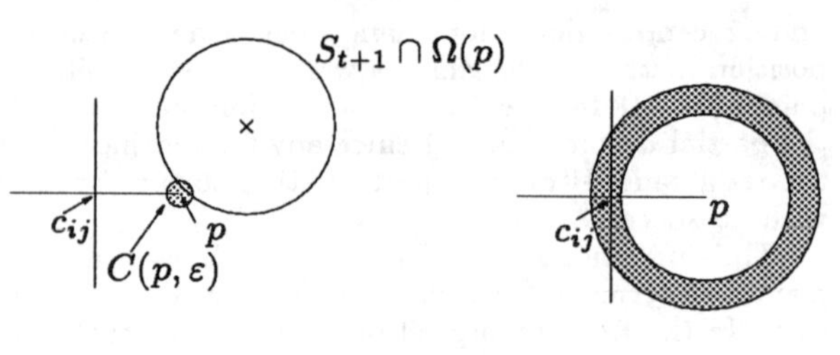


Figure 11: A cross-section of $C_{ij}$ and $S_{t+1}$ on the plane $\Omega(p)$.

$\Omega(p)$ orthogonal to $L(p)$ and having radius $r_{t+1}$. We now extend the definition to near-tangency at $p$ and restrict ourselves to the plane $\Omega(p)$.

We assume that $C_{t+1}$ lies on $\Omega(p)$ and hence $S_{t+1} \cap \Omega(p)$ is a great circle of $S_{t+1}$. We expand a circle $C(p, \varepsilon)$ of radius $\varepsilon$ around $p$. The forbidden placements for $C_{t+1}$ are now all the placement where $S_{t+1} \cap \Omega(p) \cap C(p, \varepsilon) \neq \emptyset$—the shaded annulus in Figure 11.

We repeat this for every point $p$ on $C_{ij}$ and obtain a toric shell. In fact it is only toric-like but for brevity we call it a toric shell. It would have been a real toric shell (the region sandwiched between two tori with the same center, axis of rotation, and major radius) were it not self intersecting as it may possibly be here. Assuming that both $r_{ij}$ and $r_{t+1}$ are bounded by $R$, we get that the volume of the toric shell is bounded by $4\pi(2\pi + 1)R^2\varepsilon + \frac{16}{3}\pi\varepsilon^3$. For lack of space, we omit the details of how this bound is obtained. The details can be found in the full version of the paper.

**The region $F_4$.** The region $F_4$ is the union of spherical shells each centered at a point of intersection $p$ of three distinct spheres in $M_t$ and having radii $r_{t+1} - \varepsilon$ and $r_{t+1} + \varepsilon$. The spherical shell here is the loci of placements of the center of $S_{t+1}$ that will cause $S_{t+1}$ to pass through an intersection point of three distinct spheres in $M_t$ or very close to this intersection point.

Let $p$ denote such an intersection point. The loci of placements of $C_{t+1}$ that will cause $S_{t+1}$ to go through $p$ constitute the sphere centered at $p$ with radius $r_{t+1}$. It is easily verified that if we wish that $p$ will be $\varepsilon$ away from $S_{t+1}$ then the loci of forbidden placements are the spherical shell centered at $p$ with radii $r_{t+1} - \varepsilon$ an $r_{t+1} + \varepsilon$. Its volume is $\frac{4}{3}\pi[(r_{t+1} + \varepsilon)^3 - (r_{t+1} - \varepsilon)^3]$.

**Bounding the volume of $F_2, F_3$, and $F_4$** Let $M$ be a collection of spheres as defined in Theorem 2.1. Also let $R := \max_{i=1}^{n} r_i$, and let $k$ denote the maximum number of spheres of $M$ intersecting a single sphere of $M$. In the full version of the paper we show that for a given parameter $\varepsilon > 0$, if we choose $\delta = f(k, \varepsilon, R) = 2k\varepsilon^{1/3}R^{2/3}$ then we get that the volume of the forbidden region inside $B(C_{t+1}, \delta)$ is less than half the volume of $B(C_{t+1}, \delta)$, as desired.

# Minimum Strictly Convex Quadrangulations of Convex Polygons

Matthias Müller–Hannemann[1] Karsten Weihe[2]

**Abstract**

We present a linear–time algorithm that decomposes a convex polygon conformally into a minimum number of *strictly* convex quadrilaterals. Moreover, we characterize the polygons that can be decomposed without additional vertices inside the polygon, and we present a linear–time algorithm for such decompositions, too. As an application, we consider the problem of constructing a minimum conformal refinement of a mesh in the three–dimensional space, which approximates the surface of a workpiece. It turns out that this problem is $\mathcal{NP}$–hard, and we present a linear-time algorithm with a constant approximation ratio of 4.

## 1 Introduction

In this paper, we introduce a new approach to the decomposition of convex polygons and, as an application, to the conformal refinement of finite element meshes into strictly convex quadrilaterals. The latter research has resulted from a cooperation with an engineering company that sells CAD packages.

Much work has been done on decompositions into *triangles*; see [Ho88] and [BE92] for surveys. However, there is much less work on quadrilaterals, although quadrilaterals are more appropriate in certain situations. Ref. [Joe95] states one possible reason for that: "good quadrilateral meshes are harder to generate than good triangular meshes."

**Conformal decompositions of polygons.** Conformal quadrangulations of polygons is a fundamental problem and has applications in finite element methods, watchguard problems, and scattered data interpolation. See [Tou95] for a survey of this topic and its applications.

[1] Technische Universität Berlin, Sekr. MA 6–1, Str.d. 17.Juni 136, 10623 Berlin, Germany, mhannema@math.tu–berlin.de, http://www.math.tu–berlin.de/~mhannema.

This author was supported by the special program "Efficient Algorithms for Discrete Problems and Their Applications" of the Deutsche Forschungsgemeinschaft under grant Mo 446/2-2.

[2] Universität Konstanz, Fakultät für Mathematik und Informatik, Postfach 5560/D188, 78434 Konstanz, Germany, karsten.weihe@uni–konstanz.de, http://www.informatik.uni–konstanz.de/~weihe.



*Computational Geometry 97* Nice France



A *polygon* is a closed and connected region in the plane or, more generally, of a smooth surface in the three–dimensional space, bounded by a finite, closed sequence of straight line segments (*edges*). A polygon is *convex* if the internal angle at each vertex is at most $\pi$. A vertex is a *corner* if its internal angle is strictly less than $\pi$. A polygon is *strictly convex* if all its vertices are corners.

Throughout this paper, a *conformal decomposition* of a polygon is a decomposition of its enclosed area into quadrilaterals such that the following holds (see Fig. 1):

1. The quadrilaterals are strictly convex.
2. The quadrilaterals are openly disjoint.
3. Each edge of the polygon is an edge of exactly one quadrilateral. In particular, additional vertices on the boundary of the polygon are not allowed (nonetheless in the interior).
4. If two quadrilaterals share more than one corner, they share exactly one edge as a whole.

There is a conformal decomposition for a polygon $P$ if and only if the number of vertices of $P$ is even (cf. Lemma 2.2). This fact will be proved only for convex polygons, because this is our focus in this work, but the proof can be easily extended to general polygons.

We call a conformal decomposition of a convex polygon $P$ *perfect* if it has no internal vertices. In other words, all internal edges are chords of $P$. If $P$ admits a perfect conformal decomposition, we call $P$ *perfect*, too. In Sect. 2, we will prove a useful characterization of perfect polygons (Lemmas 2.6 and 2.8). Based on this result, we will present a linear–time algorithm that tests whether a given polygon $P$ is perfect and, if so, constructs a perfect decomposition.

In Sect. 3, we will present a linear–time algorithm that constructs a conformal decomposition of a convex polygon with a minimum number of quadrilaterals. This problem is closely related to

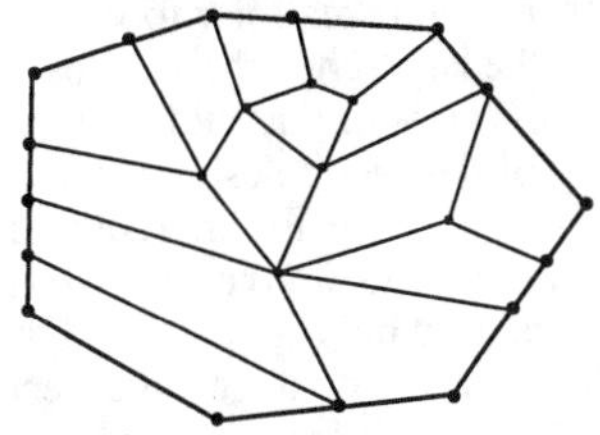

Figure 1: A convex polygon with 7 corners and 16 vertices and a conformal decomposition with 7 additional, internal vertices.

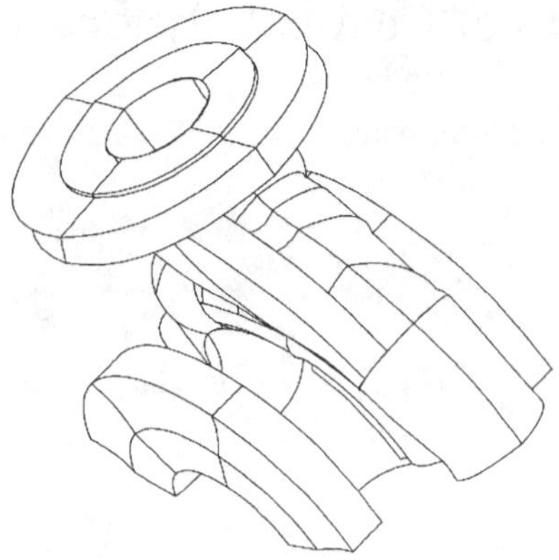

Figure 2: A mesh modeling a pump. This mesh has been constructed by a German car company.

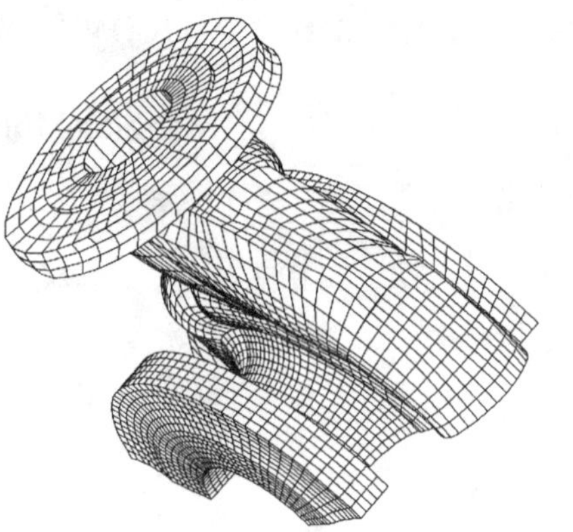

Figure 3: The conformal refinement produced by the algorithm in [MMW95].

perfect decompositions, since for perfect polygons, the perfect decompositions are exactly the minimum decompositions (cf. Corollary 3.3).

Neither problem has been dealt with so far. Nonetheless, a few related problems have found some attention. Ref. [ELOSU92] gives lower and upper bounds on the number of quadrilaterals in a conformal decomposition of non-convex polygons (even with holes), but in contrast they also allow additional vertices to be placed on the boundary. Refs. [Sac82, ST81] investigate perfect decompositions of (star-shaped) rectilinear polygons into *non–strictly* convex quadrilaterals, and [Lub85] considers perfect decompositions of non-convex polygons but even allows overlapping internal edges. See [Tou95] for a systematic survey.

**Conformal mesh refinement.** Section 4 is devoted to an application, which has motivated the research on decompositions of polygons: Conformal refinements of *finite–element meshes*, which approximate the surface of a workpiece, are important for the computer aided design of machines, vehicles, and many other kinds of technical devices; see Figs. 2 and 3.

A *mesh* is a complex of openly disjoint, convex polygons. In a *conformal refinement* of a mesh, each polygon is decomposed into strictly convex quadrilaterals, and if two quadrilaterals share more than a corner, they share exactly one edge as a whole. Note that the polygons of the input mesh need not be refined conformally in a conformal refinement of the mesh. In fact, the third condition for decompositions of polygons to be conformal must be fulfilled for the mesh as a whole, not for the individual original polygons; see Fig. 3.

Workpieces are modeled interactively as meshes. However, these meshes are usually very coarse and not conformal. To be suitable for the finite–element method, the mesh has to be refined into a conformal mesh in a preprocessing step.

Previous work puts emphasis on the shape of the quadrilaterals (angles should neither be too small nor too large; the *aspect ratio*, i.e. the ratio between the largest and the smallest side of a quadrilateral, should be small). This is important for the numerical accuracy in the later iterations of the cyclic design process, when the model has become mature and exact results are required for fine–tuning.

In this paper now, we focus on the early stages of this process, where the model is designed only roughly, and the numerical accuracy must only suffice to indicate the general tendency. Hence, the development time is crucial, which in turn is determined by the run time of the finite–element method. This raises the following problem: Given a mesh, find a conformal refinement with a *minimum* number of quadrilaterals.

Unfortunately, it is not easy to find conformal decompositions into a small number of quadrilaterals. Therefore, algorithms often refine workpieces into too many quadrilaterals, which makes the finite–element method very costly or even infeasible. In all previous work on this problem (see [MMW95] and [TA93]), the number of quadrilaterals is considered only heuristically or not at all. The reason is that all previous work is based on the *template model*. In that model, the possibilities of decomposing a single polygon are restricted to a few classes of *templates*, which are designed to achieve good angles and aspect ratios heuristically. For example, the most important template for quadrangular polygons is an $(m \times n)$–grid, where $m$ and $n$ are variable. However, in the template model it is $\mathcal{NP}$–hard to construct a conformal refinement even when all objectives are ignored. See [MMW95] for a proof in the case of standard templates; this proof can be easily extended to other variants. Hence, it is not surprising that, in practice, algorithms working on the template model often fail to refine all quadrilaterals conformally.

In Sect. 4 we introduce a linear–time algorithm that does not rely on the template model and is guaranteed to construct a feasible refinement for every input workpiece. Based on the algorithm presented in Sect. 3 for optimal refinements of convex polygons, our algorithm can simply be stated as follows: Subdivide each edge of the mesh once, and then refine each polygon separately.

It turns out that the problem is $\mathcal{NP}$–hard when (i) all template restrictions are dropped, but (ii) the number of quadrilaterals is to be minimized (cf. Theorem 4.1). However, we will prove that the number of quadrilaterals produced by our approximation algorithm is at most four times the minimum. This implies that the asymptotic complexity of the finite–element method is still best possible, and the constant of four is reasonably small.

Finally, we would like to note that our results also apply to the more general case where one looks for a minimal conformal refinement subject to *lower bounds* on the number of additional vertices to be placed on each edge of the given mesh.

## 2 Perfect Polygons

In this section, we sketch the proof of the following theorem.

**Theorem 2.1** *There is a linear–time algorithm which tests whether or not a polygon is perfect and, if so, constructs a perfect conformal decomposition.*

To prove Theorem 2.1, we will need some general notation. An *interval* on a polygon $P$ is a path on its boundary. An interval is *trivial* if it consists of exactly one edge of $P$. A *segment* $S$ is an interval between two successive corners of $P$. Let $e_1$ and $e_2$

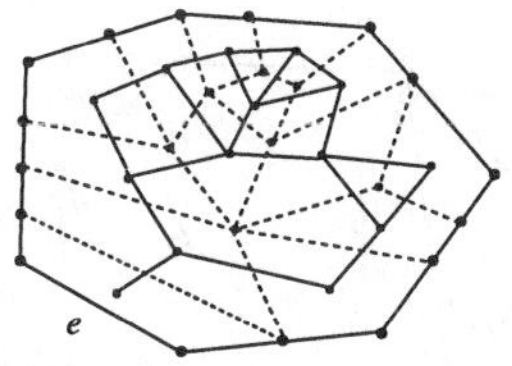


Figure 4: The polygon $P$ (solid) and the conformal decomposition $G$ (dashed) of Fig. 1, and the graph $G^*$ (solid). The unique degree–one vertex of $G^*$ points to the trivial segment $e$.

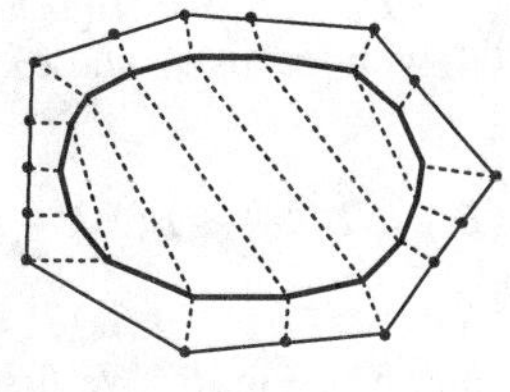

Figure 5: Proof of Lemma 2.2. First the convex polygon is reduced to a strictly convex core (thick solid line), then the core can be refined in a straightforward manner. Obviously, this procedure does not minimize the number of quadrilaterals in general.

be two different edges of $P$. Then $I[e_1, e_2]$ denotes the interval counterclockwise from $e_1$ to $e_2$, including neither $e_1$ nor $e_2$.

A conformal decomposition of $P$ is usually identified with the planar, embedded graph $G = (V, E)$ whose outer face is $P$ and whose internal faces are the quadrilaterals. For such a graph $G = (V, E)$, the *dual graph* $G^d = (V^d, E^d)$ is defined as usual [NC88]: $V^d$ is the set of all faces of $G$, and two vertices in $V^d$ are adjacent if and only if they are incident to a common edge of $G$. Let $G^* = (V^*, E^*)$ be the graph arising from $G^d$ by removing the vertex corresponding to the outer face of $G$. See Fig. 4. A planar graph whose vertices all lie on a common face is called *outerplanar*. Hence, a perfect decomposition corresponds to an outerplanar graph.

The following lemma is well-known (see for example [Joe95]).

**Lemma 2.2** *A polygon $P$ admits a conformal decomposition if and only if the number of vertices of $P$ is even.*

For completeness, we sketch an illustrative proof for the special case of convex polygons.

**Proof:** The *if* part follows from the construction in Fig. 5.

Fig. 4 demonstrates an obvious fact: that the degree of a face $v^* \in E^*$ has in $G^*$ the same parity as the number of edges of $P$ incident to $v^*$. Now the *only–if* part follows from the fact that the sum of all degrees in an undirected graph $G^*$ is always even. □

**Assumption 2.3** *Throughout Sects. 2 and 3, we assume that the number of vertices of $P$ is even.*

Note that $G^*$ is always connected. The following observation is well-known for outerplanar graphs and is also obvious from Fig. 4.

**Observation 2.4** *$G$ is a perfect decomposition if and only if $G^*$ is a tree.*

Each degree–one vertex $v^*$ of $G^*$ points to a trivial segment of $P$ (see Fig. 4). In the quadrilateral corresponding to $v^*$, this is the edge opposite to the edge crossed by the unique incident dual edge. We will sometimes identify such a vertex with this trivial segment. Since each tree has at least two leaves, we obtain:

**Corollary 2.5** *Every perfect polygon contains at least two trivial segments.*

The length $L(I)$ of an interval $I$ is the number of its edges. Moreover, $K(I)$ denotes the maximum size of a choice of strictly convex internal vertices of $I$ such that no two of them are neighbored on $P$. We often denote $(L - 2K)(I) := L(I) - 2 \cdot K(I)$. Note that $(L - 2K)(I)$ is always nonnegative.

Roughly speaking, Lemmas 2.6 and 2.8 show that, if a polygon is perfect, it even admits a perfect decomposition of a very restricted shape: The reduced dual graph $G^*$ is either a path, or it has a single branching vertex, which has degree three. In the former case, it is called a *path decomposition*, and in the latter case, a *$K_{1,3}$–decomposition*, because then $G^*$ is a subdivision of the complete bipartite graph $K_{1,3}$ on $1 + 3$ vertices.

**Lemma 2.6** *Let $P$ be a polygon with exactly two trivial segments $e_1$ and $e_2$, and let $I_1 := I[e_1, e_2]$ and $I_2 := I[e_2, e_1]$. Assume that we have $L(I_1) \geq L(I_2)$. Then $P$ is perfectly decomposable if and only if $(L - 2K)(I_1) \leq L(I_2)$. In this case, it admits a path decomposition.*

**Proof:** As each tree with two leaves is a path, any perfect decomposition is a path decomposition. When $G^*$ is a path, each vertex of $G^*$ corresponds to a quadrilateral that shares exactly two edges with $I_1 \cup I_2$. If these two edges are on the same interval $I_1$ or $I_2$, they are incident and enclose a corner of $P$. (Otherwise, they would enclose a third leaf of $G^*$.) Therefore, we have $L(I_1) - L(I_2) \leq 2 \cdot K(I_1)$ and $L(I_2) - L(I_1) \leq 2 \cdot K(I_2)$. Obviously, these two inequalities are also sufficient for perfectness. The latter inequality is always true because of $L(I_1) \geq L(I_2) \geq (L - 2K)(I_2)$. Hence, the former inequality determines perfect decomposability, which proves Lemma 2.6. □

In Lemma 2.8, we assume the following scenario.

**Scenario 2.7** *Let $P$ be a polygon with at least three trivial segments. Let $e_1$, $e_2$, and $e_3$ be three trivial segments such that the counterclockwise order around $P$ is $e_1 \prec e_2 \prec e_3 \prec e_1$. Let $I_1 := I[e_1, e_2]$, $I_2 := I[e_2, e_3]$, $I_3 := I[e_3, e_1]$, and w.l.o.g. $L(I_1) \geq L(I_2)$ and $L(I_1) \geq L(I_3)$. Assume that $L(I_1)$ is minimum subject to all these conditions.*

Fig. 6 shows an example for this scenario.

**Lemma 2.8** *In Scenario 2.7, $P$ is perfectly decomposable if and only if $(L - 2K)(I_1) \leq L(I_2) + L(I_3) + 1$. In this case, there is a perfect decomposition that is either a path decomposition with leaves $e_1$ and $e_2$, or it is a $K_{1,3}$–decomposition with leaves $e_1$, $e_2$, and $e_3$.*

**Proof:** We consider four different cases separately.

*Case I*: $(L - 2K)(I_1) = L(I_2) + L(I_3) + 1$.

Then $L(I_1) \geq L(I_2) + L(I_3) + 1 = L(I_2 \cup e_3 \cup I_3)$, and the existence of a path $G^*$ with leaves $e_1$ and $e_2$ follows analogously to Lemma 2.6.

*Case II*: $L(I_2) = 0$ or $L(I_3) = 0$.

Without loss of generality, let $L(I_2) = 0$. By Assumption 2.3, $L(I_1) + L(I_2) + L(I_3)$ is odd. Hence, $L(I_1)$ and $L(I_3)$ have different parities, and the overall assumption $L(I_1) \geq L(I_3)$ can be strengthened to $L(I_1) \geq L(I_3)+1 = L(I_2 \cup e_3 \cup I_3)$. Again the existence of a path $G^*$ with leaves $e_1$ and $e_2$ follows analogously to Lemma 2.6.

*Case III*: $(L-2K)(I_1) < L(I_2)+L(I_3)+1$ and $L(I_2) > 0$ and $L(I_3) > 0$.

This is the case shown in Fig. 6. We first define three values $b_1$, $b_2$, and $b_3$. Recall the overall assumption $L(I_1) \geq L(I_2)$ and $L(I_1) \geq L(I_3)$. In particular, this yields $L(I_1) \geq |L(I_2)-L(I_3)|$. Together with the condition defining Case III, we obtain that the interval $\mathcal{I} :=$

$$\Big[(L-2K)(I_1), L(I_1)\Big] \cap \Big[|L(I_2)-L(I_3)|, L(I_2)+L(I_3)\Big]$$

is not empty. If $|\mathcal{I}| = 1$, the unique element is either $(L-2K)(I_1)$ or $L(I_1)$, because $L(I_2)$, $L(I_3) > 0$ implies $|[|L(I_2)-L(I_3)|, L(I_2)+L(I_3)]| > 1$. In this case, we define $b_1$ as this unique element of $\mathcal{I}$, and hence $b_1 + L(I_2) + L(I_3)$ is odd. On the other hand, if $|\mathcal{I}| > 1$, we choose $b_1 \in \mathcal{I}$ so that $b_1 + L(I_2) + L(I_3)$ is odd again. Next define $b_2 := L(I_2) - 1$ and $b_3 := L(I_3)$ and consider the following linear equation system:

$$\left(\begin{array}{ccc|c} 0 & 1 & 1 & b_1 \\ 1 & 0 & 1 & b_2 \\ 1 & 1 & 0 & b_3 \end{array}\right)$$

The unique solution is

$$\begin{aligned} x_1 &= (b_2+b_3-b_1)/2, \\ x_2 &= (b_1+b_3-b_2)/2, \\ x_3 &= (b_1+b_2-b_3)/2. \end{aligned}$$

By definition of the $b_i$, $b_1+b_2+b_3$ is even. Hence, the $x_i$ are integral. Next we show that the $x_i$ are nonnegative. Because of $b_1 \in \mathcal{I}$, we have $b_1 \leq L(I_2)+L(I_3)$, and since $b_1$ and $L(I_2)+L(I_3)$ have different parities, we even have $b_1 \leq b_2+b_3$, which means $x_1 \geq 0$. Moreover, $b_1 \in \mathcal{I}$ immediately yields $b_1 \geq |L(I_2)-L(I_3)|$, and by the same parity argument, we even have $b_1 \geq |b_2 - b_3|$. This implies $x_2 \geq 0$ and $x_3 \geq 0$.

All these facts together suffice, because we can construct a subdivision of $K_{1,3}$ from the values $x_i$ (cf. Fig. 6): $x_1$ is the number of quadrilaterals sharing one edge with $I_2$ and one with $I_3$. Analogously, $x_2$ quadrilaterals share an edge with $I_1$ and $I_3$, and $x_3$ with $I_1$ and $I_2$.

There are as many as $\frac{1}{2}\Big(L(I_1)-(x_2+x_3)\Big)$ quadrilaterals that share two edges with $I_1$. Note that the number of quadrilaterals that share two edges with $I_1$ is geometrically feasible, because we obtain $\frac{1}{2}\Big(L(I_1)-(x_2+x_3)\Big) \leq K(I_1)$ from $(L-2K)(I_1) \leq b_1 = x_2 + x_3$.

As $b_2 = L(I_2) - 1$, there is one quadrilateral that shares one edge with $I_2$ and no edge with $P \setminus I_2$. This is the branching vertex of the subdivision of $K_{1,3}$ (no. 5 in Fig. 6).

*Case IV*: $(L-2K)(I_1) > L(I_2)+L(I_3)+1$.

First we show that $I_1$ contains no trivial segment. To see this, assume $I_1$ does contain a trivial segment $e_4$. Let $I_4 := I[e_1, e_4]$ and $I_5 := I[e_4, e_2]$. Clearly, we have $L(I_4) < L(I_1)$ and $L(I_5) < L(I_1)$. In Case IV, we also have $L(I_2 \cup e_3 \cup I_3) < L(I_1)$. However, this contradicts the specific choice of $\{e_1, e_2, e_3\}$ in the scenario of Lemma 2.8, because $\{e_1, e_4, e_2\}$ would be a strictly better choice.

Now we know that $I_1$ contains no trivial segment. Suppose that nonetheless $P$ admits a perfect decomposition $G = (V, E)$ of $P$.

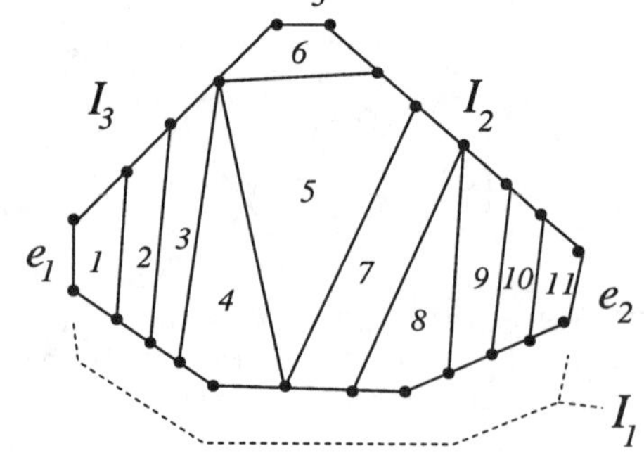


Figure 6: An example for Scenario 2.7. Here we have $L(I_1) = 11$, $(L-2K)(I_1) = 7$, $L(I_2) = 6$, $L(I_3) = 4$, and $\mathcal{I} = [7, 11] \cap [2, 10] = [7, 10]$. For this decomposition, we have chosen $b_1 = 7$, $b_2 = 5$, and $b_3 = 4$. This yields $x_1 = 1$ (inducing quadr. no. 6), $x_2 = 3$ (nos. 1,2,3), and $x_3 = 4$ (nos. 7,9,10,11).

Then all leaves of $G^*$ belong to $e_2 \cup I_2 \cup e_3 \cup I_3 \cup e_1$. In other words, the quadrilaterals incident to $I_1$ form one single path $p$ of $G^*$.

Let $Q_1$ denote the quadrilaterals sharing one edge with $I_1$ and one or more edges with $I_2 \cup e_3 \cup I_3$, let $Q_2$ be the quadrilaterals sharing exactly one edge with $I_1$ and none with $I_2 \cup e_3 \cup I_3$, and let $Q_3$ be the quadrilaterals sharing at least one edge with $I_2 \cup e_3 \cup I_3$ and none with $I_1$. Since $I_1$ contains no trivial segments, no quadrilateral shares more than two edges with $I_1$. Note that at most $K(I_1)$ quadrilaterals share exactly two edges with $I_1$. Therefore, we have $|Q_1| + |Q_2| \geq (L-2K)(I_1)$.

Next note that each quadrilateral in $Q_2$ is a branching vertex of the tree $G^*$. The branch that does not belong to $p$ contains at least one leaf, and this leaf belongs to $Q_3$. This implies $|Q_3| \geq |Q_2|$. Clearly, we have $|Q_1| + |Q_3| \leq L(I_2) + L(I_3) + 1$. Altogether, we obtain the contradiction $(L-2K)(I_1) \leq |Q_1| + |Q_2| \leq |Q_1| + |Q_3| \leq L(I_2) + L(I_3) + 1$. □

Now we are going to introduce the linear–time algorithm for perfect convex polygons. Recall from Corollary 2.5 that polygons with less than two trivial segments are not perfect. Lemma 2.6 immediately translates into a linear–time algorithm for polygons with exactly two trivial segments.

So consider the case that $P$ has more than two trivial segments. Clearly, the only problem is to find three trivial segments $e_1$, $e_2$ and $e_3$ as described in Scenario 2.7, because the proof of Lemma 2.8, Cases I–III, immediately yields a linear–time algorithm for the rest.

To find suitable trivial segments $e_1, e_2, e_3$, we apply the following strategy. Let $n$ denote the number of edges of $P$. We number all edges counterclockwise $0 \dots n-1$, starting anywhere. In the following, all additions are performed modulo $n$. In particular, a subtraction $x - y$ denotes the counterclockwise distance from $y$ to $x$ around $P$.

For a real number $x \in [0, n]$, $\ell(x) \in \{0, \dots, n-1\}$ denotes the first value $i$ in the sequence $(\lfloor x \rfloor, \lfloor x \rfloor - 1, \lfloor x \rfloor - 2, \dots)$ such that the edge no. $i$ is a trivial segment. Analogously, $r(x) \in \{0, \dots, n-1\}$ denotes the first value $i$ in the sequence $(\lceil x \rceil, \lceil x \rceil + 1, \lceil x \rceil + 2, \dots)$ such that the edge no. $i$ is a trivial segment. Clearly, all these values can be computed in linear time during one single pass around $P$.

For an $i \in \{0, \dots, n-1\}$ that is the number of a trivial segment, let $M(i)$ denote the minimum possible value of $\max\{i' - i, i'' - i', i - i''\}$ such that $i'$ and $i''$ are numbers of trivial segments and $i < i' < i'' < i$ is the cyclic order modulo $n$. It suffices to compute $M(i)$ for each $i$ and to take the overall minimum. For real numbers $x$, $y \in [0, n]$ let

$$A(x, y) := \begin{cases} \frac{x+y}{2}, & x \le y; \\ \frac{x+y}{2} + \frac{n}{2}, & x > y. \end{cases}$$

Intuitively, $A(x, y)$ is the middle of the interval of $P$ that goes counterclockwise from $x$ to $y$. The following lemma fills in the remaining gap.

**Lemma 2.9** *For $i \in \{0, \ldots, n-1\}$, $M(i)$ is attained by the following choices of $i'$ and $i''$.*

1. *If all trivial segments except $i$ belong to $[i+1, \lfloor i + \frac{1}{3}n \rfloor]$, we choose $i'' = \ell(i + \frac{1}{3}n)$, and $i'$ is an arbitrary trivial segment different from $i$ and $i''$.*
2. *Analogously, if all trivial segments except $i$ belong to $[\lceil i - \frac{1}{3}n \rceil, i-1]$, we choose $i' = r(i - \frac{1}{3}n)$ and $i''$ arbitrarily.*
3. *Otherwise, one of the following choices is appropriate:*

| $i'$ | $i''$ |
|---|---|
| $\ell(i + \frac{1}{3}n)$ | $\ell(A(i', i))$ or $r(A(i', i))$ |
| $\ell(A(i, i''))$ or $r(A(i, i''))$ | $r(i - \frac{1}{3}n)$ |
| $r(i + \frac{1}{3}n)$ | $\ell(i - \frac{1}{3}n)$ |

**Proof:** In the first case of Lemma 2.9, we have $i' - i \le n/3$, $i'' - i' \le n/3$, and $i - i'' \ge 2n/3$ for any choice of $i'$ and $i''$. Therefore, $i''$ is best chosen counterclockwise "farthest away" from $i$, which means $i'' = \ell(i + \frac{1}{3}n)$. Clearly, $i'$ may be chosen arbitrarily. The second case is mirror–symmetric.

In the third case, we apply a further case distinction. For $i' \in [i+1, \lfloor i + \frac{1}{3}n \rfloor]$, we have $i' - i \le \max\{i'' - i', i - i''\}$, because $i' - i \le \frac{1}{3}n$ and all three differences sum up to $n$. Hence, the best choice for $i'$ is $i' = \ell(i + \frac{1}{3}n)$. To minimize $\max\{i'' - i', i - i''\}$, we clearly choose $i''$ as close to $A(i', i)$ as possible, that is, $i'' = \ell(A(i', i))$ or $i'' = r(A(i', i))$. The case $i'' \in [\lceil i - \frac{1}{3}n \rceil, i-1]$ is mirror–symmetric.

It remains the case $i', i'' \in [i + \frac{1}{3}n, i - \frac{1}{3}n]$ (note that $i - \frac{1}{3}n = i + \frac{2}{3}n$). Then we have $i'' - i' \le \frac{1}{3}n$, $i' - i \ge \frac{1}{3}n$, and $i - i'' \ge \frac{1}{3}n$, which means that $i'$ and $i''$ should be chosen "farthest away" from each other, that is, $i' = r(i + \frac{1}{3}n)$ and $i'' = \ell(i - \frac{1}{3}n)$. □

## 3 Non–Perfect Polygons

Remember Assumption 2.3. For a conformal decomposition $G = (V, E)$ of polygon $P$ let $q(G)$ denote the number of internal, quadrangular faces. In this section, we will sketch the proof of the following result.

**Theorem 3.1** *There is a linear–time algorithm that constructs, for a convex polygon, a conformal decomposition $G$ that minimizes $q(G)$.*

Let $i(G)$ denote the number of internal vertices, that is, the members of $V$ that do not lie on $P$. In particular, a decomposition is perfect if and only if $i(G) = 0$. The following lemma relates $i(G)$ to $q(G)$.

**Lemma 3.2** *For a fixed polygon $P$, $q(G)$ grows strictly monotonously with $i(G)$. Therefore, to minimize $q(G)$ it suffices to minimize $i(G)$.*

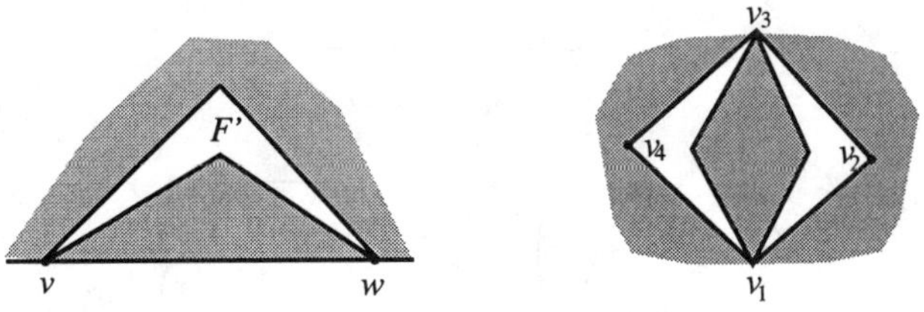


(a) (b)

Figure 7: The schematic situations prevented by the fourth and fifth condition in Lemma 3.4, respectively.

**Proof:** Euler's formula says $|V| + (q(G)+1) = |E| + 2$. Let denote $(V_1, E_1) = P$, and let $E_2 := E \setminus E_1$ denote the set of all internal edges of $G$. Then we have $|V_1| = |E_1|$, and since all internal faces of $G$ are quadrilaterals, we also have $4 \cdot q(G) = |E_1| + 2 \cdot |E_2|$. Together, this yields $q(G) = i(G) + |E_1|/2 - 1$. □

**Corollary 3.3** *If existing, a perfect decomposition $G$ of a polygon $P$ minimizes $q(G)$.*

Therefore, the algorithm from Sect. 2 solves the problem for perfect polygons. For non–perfect polygons, we abstract from all geometrical aspects and restrict our attention to purely structural aspects. More precisely, in Lemma 3.4 we first characterize the planar graphs induced by conformal decompositions.

**Lemma 3.4** *Let $P$ be a convex polygon and $G = (V, E)$ a biconnected planar graph together with a designated face $F$ such that*

1. *$F$ is isomorphic to $P$;*
2. *all other faces of $G$ are quadrilaterals;*
3. *the following vertices have degree at least 3: all internal vertices and all vertices on $F$ that do not correspond to corners of $P$;*
4. *if $F$ shares two vertices $v, w \in G$ with another face $F'$ such that $v$ and $w$ correspond to vertices on the same segment of $P$, then $v$ and $w$ are adjacent in $G$ (see Fig. 7(a) for the situations prevented by this condition);*
5. *when two non–adjacent vertices, $v_1$ and $v_3$, on a cycle $(v_1 - v_2 - v_3 - v_4 - v_1)$ of length 4 are removed, each connected component of $G \setminus \{v_1, v_3\}$ contains $v_2$ or $v_4$ (see Fig. 7(b) for the situations prevented by this condition).*

*Then (and only then) there is a planar embedding of $G$ such that the outer face is mapped onto $P$ and all internal faces are strictly convex.*

**Proof:** See Appendix A. □

It is easy to see that all these preconditions are necessary. Our proof of sufficiency is constructive and yields a linear–time algorithm for a geometrically feasible embedding of such a graph $G$ (all details omitted for brevity). Hence, from now on we may restrict our attention to the structural aspects of $G$. It is also easy to see from Fig. 7 that the last two conditions of Lemma 3.4 are fulfilled for each optimal decomposition. Thus, we may even restrict our attention to the first three conditions. These conditions will be trivially fulfilled by our construction.

Lemma 3.5 below is the basis of the linear–time algorithm for non–perfect polygons. To state this lemma, we need some more terminology.

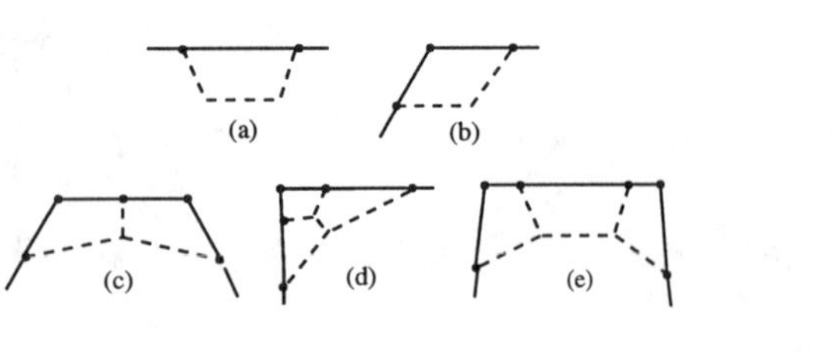


Figure 8: The first class of cut components in Lemma 3.5(2). The solid lines belong to $P$, and the dashed lines are internal edges. By Lemma 3.4, only the structure of the graph matters; the concrete lengths and angles are only exemplary.

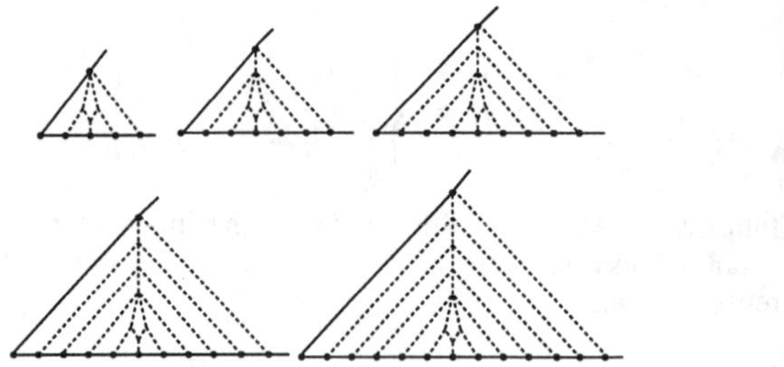

Figure 9: The five smallest cut components of the second class. The definition of the whole (infinite) class might be obvious.

We will denote a polygon by the counterclockwise sequence of the lengths of its segments. For example, $(1,1,1,1)$ denotes the strictly convex quadrilateral, $(1,1,2) = (1,2,1) = (2,1,1)$ the quadrilateral degenerated to a triangle, and $(4,1,2,3,2,2,2) = (1,2,3,2,2,2,4) = \cdots$ the polygon in Fig. 1. By Lemma 3.4, this notation encodes all information that we need for our purpose.

Recall the definition of $G^* = (V^*, E^*)$ from the beginning of Sect. 2. Let $G = (V, E)$ be an undirected planar, embedded graph. An *area component* of $G$ is a subgraph $G'$ induced by a connected component of $G^*$. More precisely, $G'$ consists of all vertices and edges incident to the polygons that correspond to this component of $G^*$. An *area decomposition* of $G$ is a collection of area components such that the inducing components of $G^*$ partition all vertices in $V^*$. Intuitively, this means that the internal faces of $G$ are partitioned and covered by closed, but openly disjoint, connected areas.

**Lemma 3.5** *For $P \notin \{(2,1,1), (4,2,2), (4,3,3), (3,3,3,3)\}$, there is a conformal decomposition $G$ with minimum $q(G)$ such that there is an area decomposition of $G$ with the following properties:*

1. *The area decomposition consists of at most four area components.*
2. *All area components except one are isomorphic to one of the components depicted in Figs. 8 and 9. These area components are henceforth called the* cut components.
3. *The remaining area component is outerplanar. This area component is henceforth called the* core component.
4. *No two cut components share an edge.*
5. *All cut components except at most one are of type (c), (d), or (e) in Fig. 8.*
6. *If a cut component of type (a) or (b) in Fig. 8 or a cut component in Fig. 9 occurs, the core component admits a path decomposition.*
7. *If a cut component of the type in Fig. 9 occurs, the area decomposition contains at most two cut components; if a cut, component of type (a) or (b) in Fig. 8 occurs, this is the only cut component.*

**Proof:** See Appendix B for a sketch of the proof. □

In the following, we will only sketch the linear–time algorithm that performs this reduction. See Figs. 10 and 11 for examples of optimal solutions constructed by the algorithm.

By Lemma 3.4, constructing an outerplanar decomposition of an area component amounts to applying the algorithm from Sect. 2. Hence, by Lemma 3.2, it suffices to find a collection of cut components so that the remainder is a suitable core component and the number of internal vertices introduced by these cut components is minimized.

We need some more terminology. Let $G'$ be a *partial conformal decomposition* of polygon $P$. This means that the outer face of $G'$ is isomorphic to $P$, and that all internal faces except one are quadrilaterals. (The exceptional face is yet to be refined.) Let $F'$ denote this face. We call a vertex of $F'$ *relatively convex* if it is either a corner of $P$, or it has degree at least three in $G'$. A *relative segment* of $F'$ is an interval of $F'$ between two successive relative corners. By Lemma 3.4, we can extend $G'$ to a full conformal decomposition of $P$ by decomposing $F'$ and thereby treating relatively convex vertices as corners and every other vertex of $F'$ as a vertex having internal angle $\pi$.

Note that all cut components except those in Fig. 8(a) and (b) introduce trivial relative segments for the core component. The idea is to construct a collection of cut components such that the induced trivial relative segments allow a path decomposition or a $K_{1,3}$–decomposition.

Consequently, the algorithm consists of two stages: The first stage tries to find such a collection so that the core component admits a $K_{1,3}$–decomposition; in the second stage, the core component shall instead be a path decomposition. If succeeding, each stage determines a solution with a minimum number of vertices. By Lemmas 2.6, 2.8, and 3.5(3), at least one stage will succeed, and if both stages succeed, the better result is the overall optimum.

Remember the definition of $r(\cdot)$ and $\ell(\cdot)$ for Lemma 2.9. For a real number $x \in [0, n]$, we next introduce the values $\ell_j(x)$ and $r_j(x)$ for $j = 0, 1, 2$. This is a generalization of the values $\ell(x)$ and $r(x)$ introduced in Sect. 2. Roughly speaking, $\ell_j(x)$ ($r_j(x)$, resp.) denotes the clockwise (counterclockwise) distance from $x$ to the next possibility to create a trivial relative segment spending $j$ additional internal vertices. In particular, we have $\ell_0(x) = \ell(x)$ and $r_0(x) = r(x)$.

More formally, $\ell_1(x)$ is the first value $i$ in the sequence $(\lfloor x \rfloor, \lfloor x \rfloor - 1, \lfloor x \rfloor - 2, \ldots)$ such that the edges no. $i, \ldots, i-3$ may belong to a cut component of type (c) in Fig. 8. In other words, the vertices between the edges $i$ and $i-1$ and between the edges $i-2$ and $i-3$ are corners of $P$. Analogously, $\ell_2(x)$ is the first value $i$ in this sequence such that the edges no. $i, \ldots, i-3$ may belong to a cut component of type (d), or edges no. $i, \ldots, i-4$ by a component of type (e). This means that either the vertex between the edges $i-1$ and $i-2$ is a corner or the vertices between the edges $i$ and $i-1$ and the edges $i-3$ and $i-4$ are corners. The definitions of $r_1(x)$ and $r_2(x)$ are mirror–symmetric. Clearly, we can compute all these values in linear time during one single pass around $P$.

Next we explain the first stage of the algorithm in detail, that is, finding a minimum conformal decomposition such that the core component admits a $K_{1,3}$–solution. See Fig. 10 for an example where this stage succeeds and yields the overall optimum. Like in

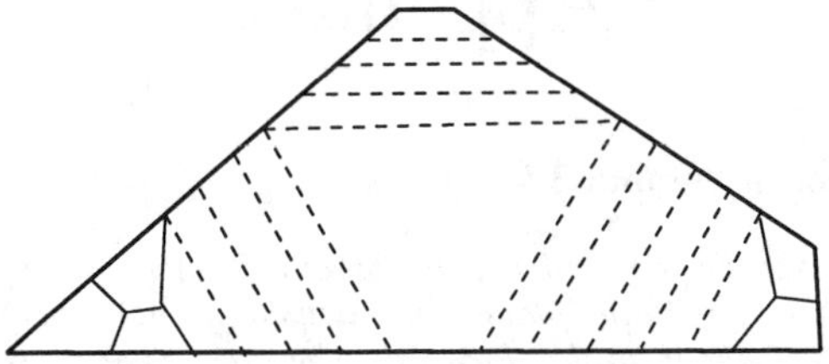

Figure 10: An optimal decomposition with one cut component of the type in Fig. 8(c) and one of the type in Fig. 8(d). The dashed lines are the chords of the (outerplanar) core component. In this example, the core component is decomposed according to a dual subdivision of $K_{1,3}$.

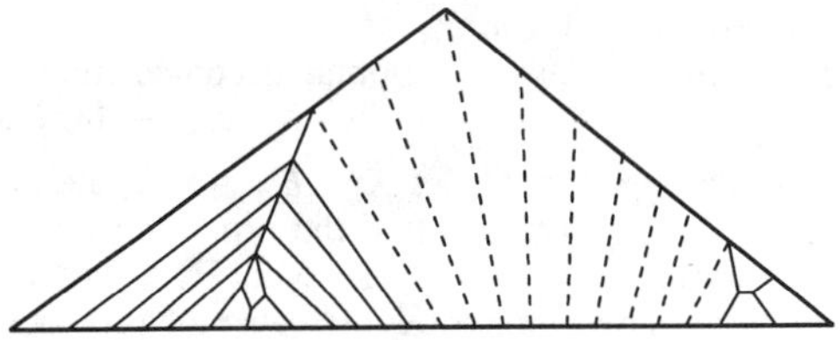

Figure 11: An optimal decomposition with one cut component of the type in Fig. 8(d) and one of the type in Fig. 9. By Lemma 3.5(6), the core component admits a path decomposition.

Sect. 2, we try to find optimal $i'$ and $i''$ for each trivial segment $i$. However, here we do the same also for each possible trivial relative segment that may result from cutting off an area component of type (c),(d) or (e) in Fig. 8. For each such trivial segment or trivial relative segment $e$, we apply 9 variants on the procedure in Lemma 2.9: one for each combination $j_1, j_2 \in \{0, 1, 2\}$. For $j_1 = j_2 = 0$, this is exactly the algorithm in Lemma 2.9. The other variants differ (up to minor details) from this one in that $\ell_0$ and $r_0$ are replaced by $\ell_{j_1}$ and $r_{j_1}$ for $i'$, and by $\ell_{j_2}$ and $r_{j_2}$ for $i''$. The best result of all these variants is the optimum for $e$, and the best result for all $e$ is the overall optimum.

It remains to explain the second stage of the algorithm in detail, that is, finding a minimum conformal decomposition such that the core component admits a path solution. See Fig. 11. Because of Lemma 3.5(7), all possibilities to apply type (a) or (b) in Fig. 8 may be tested separately. On the other hand, all possible reductions that need only components of types (c), (d) or (e) in Fig. 8 may be examined analogously to the first stage of the algorithm.

Lemma 3.5(5) and (7) say that we need at most one cut component of the type in Fig. 9, and that there is at most one further cut component, which is of type (c), (d), or (e) in Fig. 9. In other words, we may assume that one leaf of the decomposition according to Lemma 2.6 is either an original trivial segment or is created by type (c), (d) or (e) in Fig. 8. Let $i$ be the index of this (relatively) trivial segment. Then it is easy to see that, in an optimal solution, a component of Fig. 9 can only be applied to a corner incident to the segment to which the index $i + n/2$ belongs. Therefore, at most two possibilities of this type must be examined after fixing the first trivial segment.

This concludes the sketch of the algorithm.

## 4 Conformal Mesh Refinements

As an application for the decomposition of polygons, we consider in this section the problem of constructing a minimum conformal refinement of a mesh in the three–dimensional space. This problem has resulted from a cooperation with an engineering company that sells CAD packages.

Throughout this section, we identify a mesh with the undirected graph $G = (V, E)$ that is induced by all corners of these polygons. In other words, each polygon is a "hole" of $G$, and since a side of a polygon may consist of more than one edge of $G$ (e.g., in Figs. 2 and 12), these polygons of $G$ are not strictly convex, but still convex.

Note that there may be holes of $G$ that do not belong to the approximation of the surface, namely holes of the surface itself. (These holes may not be convex.) Note further that the graph $G$ of a mesh need not be planar; for example, a mesh approximating the surface of a torus has genus one. Even more, a mesh may contain *folding edges*, that is, edges incident to more than two polygons (Fig. 12). We call a mesh *homogeneous* if it does not contain folding edges.

In this paper now, we attack the minimum conformal mesh refinement problem from a theoretical point of view and introduce a new approach that does not rely on template restrictions.

Unfortunately, this problem is highly intractable:

**Theorem 4.1** *The minimum conformal mesh refinement problem is $\mathcal{NP}$–hard even for homogeneous meshes.*

**Proof:** by a reduction from *Exact Cover by 3–Sets* [GJ79]; see Appendix C for further details. □

However, for the asymptotic complexity of the numerical analysis it is not necessary to find the exact minimum of quadrilaterals, but to achieve a *constant* approximation ratio.

The following lemma shows that there is an algorithm that runs in linear time and achieves a small constant approximation ratio.

**Theorem 4.2** *There is a linear–time algorithm that constructs a conformal refinement of a mesh such that the number of quadrilaterals is at most four times the optimum.*

**Proof (sketch):** Let $G = (V, E)$ denote the mesh. We subdivide each edge in $E$ exactly once. Obviously, each polygon is even afterwards, and we may apply the algorithm from Sect. 3 separately to each polygon. This already completes the description of the algorithm.

To prove the approximation ratio, we need some further terminology. For a convex polygon $P$ with an even number of vertices, $\min(P)$ denotes the minimum number of quadrilaterals required by any conformal decomposition of $P$.

For an arbitrary convex polygon $P$ with edge set $E_P$, a mapping $X : E_P \to \mathbb{N}_0$ is called *feasible* if $\sum_{e \in E_P} X(e)$ has the same parity as $E_P$. In particular, if $|E_P|$ is even, $X \equiv 0$ is possible, too. The polygon $P_X$ is constructed from $P$ by subdividing each edge $e \in E_P$ exactly $X(e)$ times. Hence, $X$ feasible means that $P_X$ admits a conformal decomposition. Moreover, $\text{Min}(P)$

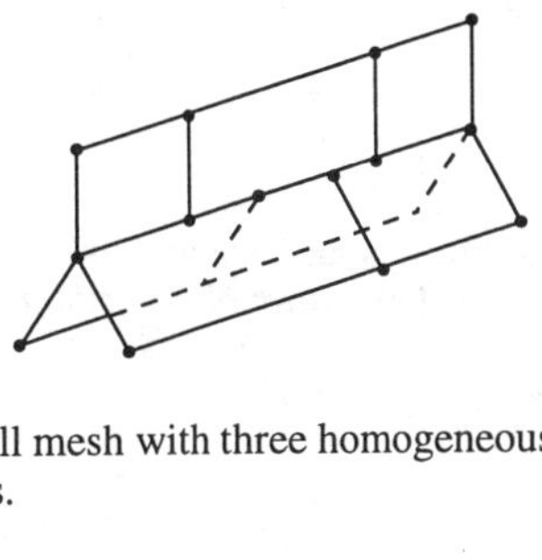

Figure 12: A small mesh with three homogeneous components and five folding edges.

denotes the minimum number of quadrilaterals in any conformal decomposition of any polygon $P_X$,

$$\mathrm{Min}(P) := \min\{q(G) \mid G \text{ conformal decomposition of } P_X,\ X : E_P \to \mathbb{N}_0 \text{ feasible}\}.$$

Let $P$ be a polygon of $G$, let $X : E_P \to \mathbb{N}_0$ be uniformly equal to 1, $X \equiv 1$. The heart of the proof (all details omitted) is to show that $\min(P_X) \leq 2 \cdot \mathrm{Min}(P) + |E_P| - 2$ for this special class of weightings $X$. Because of this fact, it suffices to prove $|E_P| \leq 2 \cdot \mathrm{Min}(P) + 2$. However, this follows from the proof of Lemma 3.2: Let $Y : E_P \to \mathbb{N}_0$ such that $\mathrm{Min}(P) = \min(P_Y)$, and let $i(Y)$ denote the number of internal vertices of an optimal conformal decomposition of $P_Y$. Then we have $\min(P_Y) = i(Y) + |E_P|/2 - 1$, and the claim follows. □

## References


[BE92] M. Bern and D. Eppstein. Mesh generation and optimal triangulation. In D.-Z. Du and F. Hwang, editors, *Computing in Euclidean Geometry*, pages 23–90. World Scientific, 1992.

[ELOSU92] H. Everett, W. Lenhart, M. Overmars, T. Shermer, and J. Urrutia. Strictly convex quadrilateralizations of polygons. In *Proceedings Fourth Canadian Conference on Computational Geometry*, pages 77–83, 1992.

[GJ79] M. R. Garey and D. S. Johnson. *Computers and Intractability: A Guide to the Theory of NP–Completeness.* W H Freeman & Co Ltd, 1979.

[Ho88] K. Ho-Le. Finite element mesh generation methods: a review and classification. *Computer–Aided Design*, 20:27–38, 1988.

[Joe95] B. Joe. Quadrilateral mesh generation in polygonal regions. *Computer–Aided Design*, 27:209–222, 1995.

[Lub85] A. Lubiw. Decomposing polygonal regions into convex quadrilaterals. In *Proceedings of the ACM Symposium on Computational Geometry*, pages 97–106, 1985.

[MMW95] R. H. Möhring, M. Müller-Hannemann, and K. Weihe. Using network flows for surface modelling. In *Proceedings of the 6th Annual ACM-SIAM Symposium on Discrete Algorithms, SODA'95*, pages 350–359, 1995.

[NC88] T. Nishizeki and N. Chiba. *Planar Graphs: Theory and Algorithms*, volume 32 of *Annals of Discrete Mathematics.* North–Holland, 1988.

[Sac82] J. R. Sack. An $\mathcal{O}(n \log n)$ algorithm for decomposing simple rectilinear polygons into convex quadrilaterals. In *Proceedings 20th Conference on Communications, Control, and Computing*, pages 64–74, 1982.

[ST81] J. R. Sack and G. Toussaint. A linear–time algorithm for decomposing rectilinear star–shaped polygons into convex quadrilaterals. In *Proceedings 19th Conference on Communications, Control, and Computing*, pages 21–30, 1981.

[TA93] T. Tam and C. Armstrong. Finite element mesh control by integer programming. *Int. J. Numer. Methods Eng.*, 36:2581-2605, 1993.

[Tou95] G. Toussaint. Quadrangulations of planar sets. In *Proceedings 4th International Workshop on Algorithms and Datastructures, WADS'95, Kingston, Canada, August 16-18, 1995*, pages 218–227, 1995.


# Appendix

## A Proof of Lemma 3.4

**Remark A.1** *For convenience, the proof is formulated only for the case of polygons in the plane; the extension to polygons in other smooth surfaces is straightforward. The proof disregards further interesting aspects such as internal angles of embedded quadrilaterals. However, the decompositions constructed in Sects. 2 and 3 are of a very restricted shape, and further criteria may be fulfilled to some extent, too.*

**Proof:** The idea of the proof is to use induction on the number of internal faces $|F|$ of $G$. For technical reasons, however, we use a slightly different induction hypothesis. Let us relax the degree constraints of Item 3 in the following way:

3' Every internal vertex of $G$ has degree 3.

4' If $F$ shares two or more vertices with another face $F'$, these vertices form a common interval of both $F$ and $F'$.

Hence we allow non-corners of $P$ to have degree 2.

**Claim A.2** *Let $P$ and $G$ and the Items 1,2, and 5 as in Lemma 3.4, but replace Item 3 by 3' and Item 4 by 4'.*

*Then there is a planar embedding of $G$ such that the outer face is embedded like $P$ and all internal faces are strictly convex except those which are incident to non-corners of degree 2. Nonetheless, the latter faces are still convex.*

Clearly, it suffices to prove this claim, as Item 3 implies that all faces of $G$ can be embedded strictly convex.

The case $|F| = 1$ is trivial. Hence, assume $|F| > 1$.

The conditions in Lemma 3.4 imply that there is a path $P$ which connects two vertices $v$ and $w$ on the boundary of $P$ such that all edges of $p$ are internal edges. In particular, we assume that the vertices $v, w$ and the path $p$ are chosen such that $p$ is shortest subject to this condition.

As $v$ and $w$ belong to different segments of $P$ and $P$ is convex, we can embed $p$ on a straight line between $v$ and $w$ (such that the vertices of $p$ are equally spaced, say). Regard $p$ as oriented from $v$ to $w$.

Now consider the polygons $P_\ell = p \cup I_P(w, v)$, $P_r = p \cup I_P(v, w)$ ($I_P(\cdot, \cdot)$ is the counterclockwise interval with respect to $P$), and the subgraphs $G_\ell$ and $G_r$ which are the restrictions of $P$ and $G$ to the left hand and right hand side of $p$, respectively.

We want to apply the induction hypothesis to the convex polygon $P_\ell$ and $G_\ell$, and to $P_r$ and $G_r$. However, to this end we have to modify $p$ to ensure all preconditions. Clearly, $G_\ell$ and $G_r$ fulfill Items 1 and 2 of the preconditions, and each has fewer internal faces than $G$. It is also easy to see that the property stated in Item 5 carries over to them.

At this point it becomes clear why we used the relaxed version of Item 3: We cannot be sure that each internal vertex of $p$ has degree 3 or more in $G_\ell$ and $G_r$.

Suppose Item 4' is not valid anymore for $P_\ell$ or $P_r$, say for $P_\ell$. Then there is a face $f$ with two non-adjacent vertices $v_1, v_2$ on $p$. Denote by $v_3, v_4$ the two other vertices of this face, and let $v_3$ be the one which is enclosed by the cycle formed by the edges $(v_4, v_1), (v_4, v_2)$ and the interval between $v_1$ and $v_2$ on $p$ in the embedding of $G$. The distance of $v_1$ and $v_2$ on $p$ must be equal to 2, as otherwise $p$ would not be a shortest path between $v$ and $w$. Let $v_5$ be the vertex between $v_1$ and $v_2$ on $p$. Then remove from $p$ the edges $(v_1, v_5), (v_5, v_2)$ and replace them by $(v_1, v_3), (v_3, v_2)$ to yield a new path $p'$ between $v$ and $w$. Assume that we have chosen

$p'$ instead of $p$ in the beginning, and iterate this modification step if necessary. Observe that Item 5 now guarantees that the change from $p$ to $p'$ cannot cause a (new) violation of Item 4' in the other subgraph. (In fact, this is the place where we need Item 5.)

Hence, we finally find $P_\ell$ and $P_r$ to which the induction hypothesis applies. Thus, both polygons admit the desired strictly convex embeddings with the exception of only those faces incident internal vertices of $p$ which have degree 2 in $G_\ell$ or $G_r$.

But now we use the fact that for any strictly convex polygon there is an $\epsilon > 0$ such that we can shift any of its vertices from its current position into an arbitrary direction within some disk of radius $\epsilon$, and the "shifted" polygon remains strictly convex. Hence, if necessary, we can readjust internal vertices of $p$ (and keep all other vertices stationary) such that all faces of $G$ are simultaneously strictly convex. □

## B Proof of Lemma 3.5 (Sketch)

In this section, we outline the proof of Lemma 3.5, by stating the crucial technical lemmas from which the proof is derived. We do not give all details, because the detailed proof is much too long and includes several extensive case distinctions.

Basically, the proof consists of three steps. In the first step we consider *internal quadrilaterals*, that is, quadrilaterals that share no edge with the boundary of the polygon (though possibly vertices). As a by–product (Lemma B.3), we characterize the unique optimal conformal decompositions for polygons of the type $(2k, 1, 1)$, $k \in \mathbb{N}$.

Then, in the second step, we show that we may also restrict our attention to optimal conformal decompositions where internal vertices appear only in very specific situations. Based on this characterization, we prove that there is an optimal conformal decomposition with an area decomposition such that all but at most one area components belong to one of the types depicted in Figs. 8 and 9, the cut components.

Finally, we prove that there is an optimal conformal decomposition such that an area decomposition contains only a few of these cut components.

Let us start with the preparation for the first step. The following two lemmas show that internal quadrilaterals appear only in very restricted situations.

**Lemma B.1** *No two internal quadrilaterals are adjacent in an optimal conformal decomposition if at least one of them is incident to an internal vertex.*

**Lemma B.2** *Let $P \neq (2, 1, 1)$ be a convex polygon, and let $G$ be an optimal conformal decomposition of $P$ with a minimum number of internal quadrilaterals. Assume that $G$ contains an internal quadrilateral $V$ that is incident to at least one internal vertex, and let $V$ be adjacent to the quadrilaterals $V_1, V_2, V_3, V_4$ in cyclic order. Then (for an appropriate choice of $V_1$) $V_i$ and $V_{i+1}$ are adjacent for $i = 1, 2, 3$, and all four quadrilaterals are incident to consecutive edges of the same segment of $P$.*

Fig. 13 illustrates the cases where internal quadrilaterals cannot be removed in optimal solutions. For "triangle–shaped" polygons of the form $(2k, 1, 1)$ we can prove even more:

**Lemma B.3** *The optimal conformal decomposition of the triangle-shaped polygon $(2k, 1, 1)$ is unique for all $k \geq 1$. The polygon $(2, 1, 1)$ needs 4 internal vertices, and each polygon $(2k, 1, 1)$ needs $k + 1$ internal vertices, for $k \geq 2$.*

**Proof:** by induction on $k$; details are left out. □

For a conformal decomposition $G$ of polygon $P$ let $G^i$ denote the subgraph induced by all internal vertices. We call an internal vertex *isolated* if it is not incident to another internal vertex. Analogously, an edge of $G^i$ is *isolated* if it is not incident to any other edge of $G^i$. Lemma B.4 characterizes the connected components of $G^i$ that are not isolated vertices or edges. Afterwards, this result is used to characterize the area components of $G$.

**Lemma B.4** *For any polygon $P \notin \{(3,3,3,3), (4,3,3), (4,2,2)\}$, there is an optimal conformal decomposition $G$ such that for any two incident edges $(v_1, v_2)$, $(v_2, v_3)$ in $G^i$ one of the following holds:*

- *Either at least one of $(v_1, v_2)$ and $(v_2, v_3)$ is incident to an internal quadrilateral, or*
- *all four incident quadrilaterals are pairwise different, and the four edges opposite to $(v_1, v_2)$ and $(v_2, v_3)$ belong to the same segment of $P$.*

**Corollary B.5** *There is a conformal decomposition $G$ of the polygon $P$ with minimum $q(G)$ such that a connected component of $G^i$ which contains more than two vertices induces a subgraph in $G$ which is isomorphic to the optimal decomposition of some $(2k, 1, 1)$ in G, and one of the trivial segments of the $(2k, 1, 1)$ is an edge of $P$.*

**Lemma B.6** *There is an optimal decomposition such that $degree(v) = 3$ for all isolated internal vertices, unless $P = (2, 2, \ldots, 2)$. This decomposition has also the properties of Corollary B.5.*

**Lemma B.7** *There is an optimal decomposition such that for each component of $G^i$ which is a single edge, say $e = (v_1, v_2)$, we have $degree(v_1) = degree(v_2) = 3$, unless $P = (3, 3, 2, 2, \ldots, 2)$. This decomposition has also the properties of Corollary B.5 and Lemma B.7.*

With the previous lemmas we have completely characterized which kind of internal components have to be considered in optimal conformal decompositions.

The final task will be to show that we may restrict our attention essentially to the area components described in Lemma 3.5.

**Lemma B.8** *For any polygon $P \notin \{(3,3,3,3), (2,1,1), (4,3,3), (4,2,2)\}$, there is a conformal decomposition $G$ with minimum $q(G)$ which has an area decomposition such that there is at most one outerplanar area component and all other area components are isomorphic to one of the cut components in Figs. 8 or 9. Furthermore, no two cut components share an edge in that area decomposition.*

With these lemmata at hand it is not too difficult to prove the remaining statements of Lemma 3.5. The proof technique is as follows: We suppose for contradiction that there is no optimal conformal decomposition such that $G$ has the form according to Lemma 3.5.

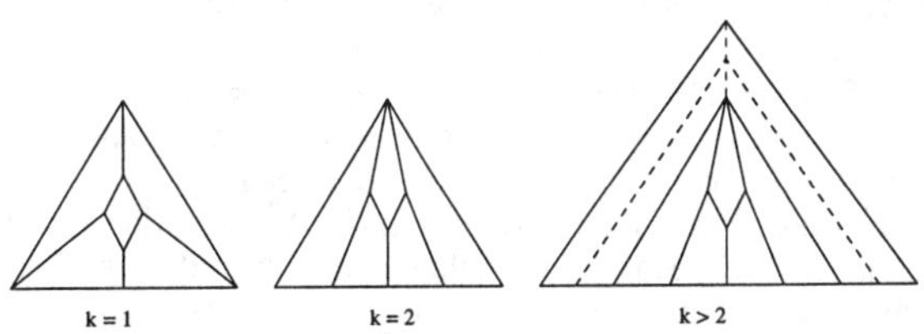


Figure 13: Illustration of optimal conformal decompositions of $(2k, 1, 1)$

Since we know so far that there is an optimal conformal decomposition with an area decomposition into cut components plus at most one single core component, this means that we have applied too many cut operations, or equivalently, $G$ has too many internal components. Hence, we have to show that there is another optimal conformal decomposition $G'$ which uses at least one internal component less, and then we are done by induction. The other possibility is to derive a contradiction to optimality. This is achieved in a case analysis for the different kinds of operations. Roughly speaking, the idea is to "unite" two internal components to a single larger one (with not more internal vertices in total) and to rebuild the conformal decomposition afterwards.

## C Proof of Theorem 4.1

**Proof:** *Exact Cover by 3-Sets* is the following problem. Given a ground set $U = \{u_1, ..., u_{3m}\}$ and a family $S_1, S_2, ..., S_n$ of subsets of $U$, each of cardinality three, is there a subfamily of $m$ subsets that covers $U$ ? This problem is well-known to be strongly $\mathcal{NP}$–complete [GJ79]. The reduction from *Exact Cover by 3-Sets* to the minimum conformal mesh refinement problem is as follows.

For each subset $S_i$ we create a polygon as in Fig. 14. For simplicity we will refer to it as polygon $S_i$. Edge $e_{lS_i}$ with $l \in \{1,2,3\}$ is to correspond with the $l$-th element of $S_i$. If an edge $e_{lS_i}$ is subdivided this will mean that subset $S_i$ covers the corresponding element.

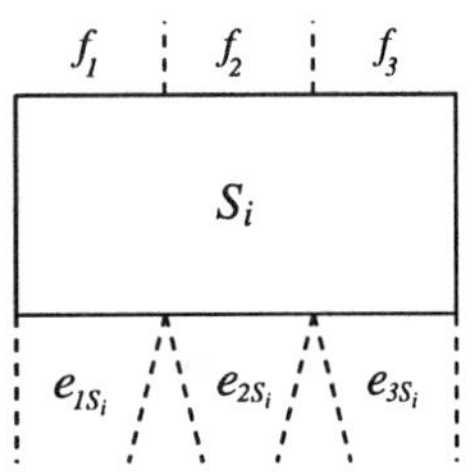


Figure 14: Polygon for a subset $S_i$.

Suppose that the edges $f_1, f_2, f_3$ cannot be subdivided in any optimal solution. This can be assumed by the following construction. Add a homogeneous bunch of conforming quadrilaterals to such an edge $e$ such that the shortest path from it to the mesh boundary has a length of at least $k$, where $k$ is some constant to be specified later. Each bunch needs at most $O(k^2)$ quadrilaterals. By parity, the subdivision of $f_i$ by one unit would imply at least $k$ additional quadrilaterals in any feasible solution. If no edge of $S_i$ is subdivided at all, then we clearly need exactly 3 quadrilaterals. If one subdivides each of the edges $e_{lS_i}$ and $x_{S_i}$ exactly once, then we need exactly 7 quadrilaterals. If only one edge $e_{lS_i}$ is subdivided once, then $x_{S_i}$ must be subdivided by parity reasons, and we need 5 quadrilaterals. If exactly two edges $e_{lS_i}$ are subdivided once, one needs exactly 6 quadrilaterals.

Hence, in an exact cover one needs exactly $7m + 3(n - m) = 3n+4m$ quadrilaterals for all subsets $S_i$ together. If an exact cover exists, it is not worth to cover three times only one element, as this needs 15 quadrilaterals compared to $7 + 2 \cdot 3 = 13$. Furthermore, it is more expensive to cover once only one element and once two elements (and once no element), as this needs $5 + 6 + 3 = 14$ quadrilaterals. Finally, it is more expensive to cover three times two elements, as this needs 18 quadrilaterals compared with 17 quadrilaterals in an exact cover. If no exact cover exists one needs more than $3n + 4m$ quadrilaterals for the refinement of all subsets $S_i$.

Let $k_i$ denote the total number of appearances of $u_i$ in the subsets $S_j$. Define for each element $u_i$ a polygon as in Fig. 15.

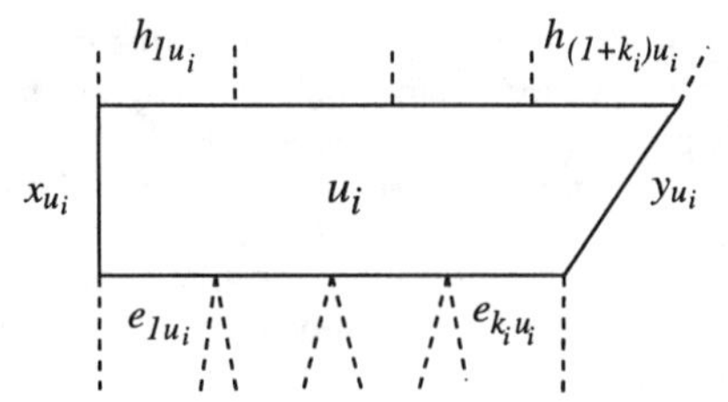


Figure 15: Polygon for an element $u_i$.

The parity of such an "element" polygon $u_i$ is odd. Suppose that none of its edges $h_{1u_i}, ..., h_{k_iu_i}, h_{(1+k_i)u_i}, x_{iu_i}, y_{iu_i}$ is subdivided. This can be achieved by the same trick with additional quadrilaterals as above. Then a conformal refinement of $u_i$ requires that an odd number of additional points is placed on the edges $e_{1u_i}, ..., e_{k_iu_i}$. In particular, the optimum refinement for $u_i$ is achieved if we add exactly one additional point on one of these edges. In this case, $u_i$ becomes perfect and needs exactly $1 + k_i$ quadrilaterals.

Now we establish the relationship between elements $u_i$ and subsets $S_j$. If $u_i$ is an element of $S_j$, we connect an edge $e_{lS_j}$ with a corresponding edge $e_{ru_i}$ by a chain of $q$ conforming quadrilaterals ($q$ is some small constant) as in Fig. 16 which we call channels.

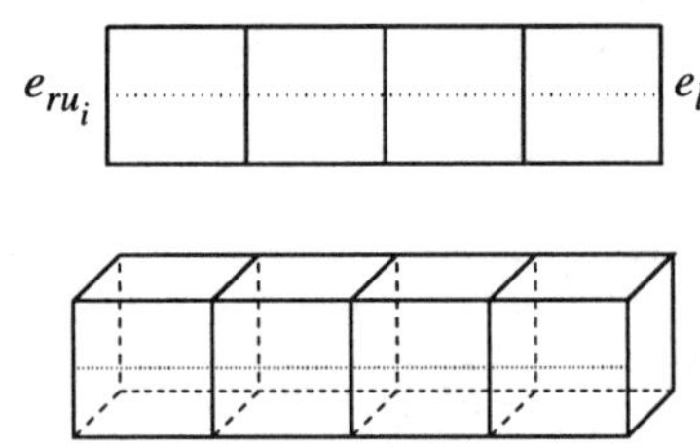


Figure 16: "Channel" between element $u_i$ and subset $S_i$ before and after blowing up. The pointed line denotes the split if this channel is used.

Clearly, all these channels can easily be embedded into the three-dimensional space such that they do not cross each other. The purpose of these connections is to make sure that each element $u_i$ may be covered by any subset $S_j$ it belongs to. All channels have the same length such that it makes no difference which one is chosen to cover an element.

We also want that if $e_{ru_i}$ is subdivided once then also $e_{lS_j}$ is subdivided once in any optimal solution. Hence, it must be cheaper to split each quadrilateral on the channel into two quadrilaterals than to use any other feasible refinement. To this end, we "blow up" each channel: we take four copies of each channel and glue them together as in Fig. 16. As all these quadrilaterals are conforming, a subdivision by one unit of the original edge $e_{ru_i}$ implies by parity arguments an odd subdivision of either the original edge $e_{lS_j}$ as intended or of one of the copies of $e_{ru_i}$ or $e_{lS_j}$. In the first case, the original path is split into two chains. To rule out the latter case, we have to make such a subdivision expensive. This can be done by the same trick with additional bunches of conforming quadrilaterals as above. If the constant $k$ is chosen larger than $q$, then it is easy to see that any optimal solution uses exactly one channel per element $u_i$. From the above discussion it is also clear that an exact cover is cheaper than any other cover of $U$.

Observe that no folding edge occurs in our construction. Hence, the theorem remains true in the special case of homogeneous meshes. □

# Average-case ray shooting and minimum weight triangulations

*Boris Aronov*[1]
Computer and Information Science
Polytechnic University
Brooklyn, NY 11201
aronov@ziggy.poly.edu

*Steven Fortune*
Bell Laboratories
Murray Hill, New Jersey 07974
sjf@research.bell-labs.com

**Abstract**

Consider an environment filled with polyhedral obstacles. The answer to a *ray-shooting query* is the first obstacle encountered by the ray. A simple way to answer ray-shooting queries is to triangulate space in a manner compatible with the obstacles, and then walk through the triangulation along the ray until the first obstacle is encountered. We show that the average walk length can be reduced by choosing a triangulation with weight (i.e., length in two dimensions or area in three dimensions) as small as possible. In two dimensions, we observe that the length of the minimum-length triangulation can be estimated by the total length of the obstacles plus the total length of a minimum spanning tree of the obstacles, up to a logarithmic factor; this gives an *a priori* estimate of the average-case walk length. Our main result is in three dimensions. We give a polynomial-time algorithm that computes a triangulation compatible with a set of polyhedral obstacles; the area of the triangulation is within a multiplicative constant of the smallest possible.

## 1 Introduction

The *ray-shooting* problem is to preprocess a set of polyhedral obstacles so that subsequent queries of the form "What obstacle is hit first by a query ray?" can be answered efficiently. The ray-shooting problem has been studied extensively in computational geometry (see [1] for a survey of theoretical results) and has many practical applications.

Ray-shooting queries can be answered in a particularly simple way given a triangulation of space compatible with the obstacles. Starting in the triangle containing the endpoint of the query ray, it suffices to walk through the triangulation along the ray until an obstacle is encountered. The walk takes constant time per visited triangle, so the total cost of the walk is proportional to the number of triangle faces crossed. In the worst case, the walk can be long.

[1]Work of the first author was performed while he was visiting Bell Laboratories, Murray Hill, NJ.


*Computational Geometry 97* Nice France
Copyright 1997 ACM 0-89791-878-9/97/06 ..$3.50

For example, Agarwal *et al* [2] describe a configuration of $n$ obstacle triangles in $\mathbb{R}^3$ so that there is a line missing all the obstacles yet hitting $\Omega(n)$ faces of any triangulation compatible with the obstacles.

We consider instead the average-case cost of ray-shooting. Let $S$ be a set of obstacles (line segments in two dimensions, triangles in three dimensions) containing all faces of its convex hull $H$. The distribution on rays can be described as follows: the ray direction is chosen uniformly at random from the space of all directions; the ray endpoint is chosen uniformly at random from the obstacle set $S$. To analyze the cost of ray-shooting with this distribution we first consider the average-case cost of line-stabbing queries.

A *line-stabbing query* reports all obstacle facets intersected by a query line in order. Let $\mathcal{T}$ be a triangulation compatible with the obstacles $S$. For line $\ell$, let $s(\ell)$ and $t(\ell)$ be the number of obstacle facets and triangulation facets intersected by $\ell$, respectively. Let $\mu$ be the standard measure on lines invariant under rigid motions [14]. Let $L(U)$ be the set of lines that intersect a set $U$. Then

$$\frac{\int t(\ell)\, d\mu}{\mu(L(H))}$$

is the average walk length to answer a line-stabbing query that meets the convex hull $H$. The ratio

$$c(\mathcal{T}) = \frac{\int t(l)\, d\mu}{\int s(l)\, d\mu}$$

is the average walk length per reported obstacle facet. A basic fact from integral geometry [14] is that for a facet $f$, $\mu(L(f)) = c_d|f|$, where $|f|$ is the *weight* of $f$ (length in two dimensions and area in three dimensions) and the constant $c_d$ depends upon dimension, with $c_2 = 2$ and $c_3 = \pi$. It follows easily that

$$c(\mathcal{T}) = \frac{\sum_{f \in \mathcal{T}} |f|}{\sum_{f \in S} |f|} = \frac{|\mathcal{T}|}{|S|}.$$

Now for a ray $r$, let $t(r)$ be the number of triangulation facets encountered in a walk along $r$ up to an obstacle. Formally, $t(r)$ should include the facet containing the ray endpoint and should count all facets up to but not including the first obstacle facet. (Thus if the endpoint of $r$ is on the boundary of the convex hull and $r$ does not enter the hull, $t(r)$ is 1.) It is easy to verify that $c(\mathcal{T})$ is also the average value of $t(r)$ for rays chosen from the distribution described above.

Let

$$c(S) = \inf_{\hat{\mathcal{T}}} c(\hat{\mathcal{T}}),$$

where $\hat{\mathcal{T}}$ ranges over all triangulations compatible with $S$. Clearly, to answer ray-shooting or line-stabbing queries it is desirable to choose a triangulation $\mathcal{T}$ with $c(\mathcal{T})$ as close as possible to $c(S)$. It is not obvious that the lower bound $c(S)$ can be attained [5]; for example, it is conceivable that it is always possible to decrease weight and hence $c(\mathcal{T})$ by adding Steiner points.

We consider triangulations whose weight approximates the minimum, in two and three dimensions. Section 2 briefly considers the two-dimensional case. Let $MST$ be the minimum Steiner spanning tree of the obstacles and $M$ be the greatest lower bound on triangulation length. We observe that $|S| + |MST| \approx M$; specifically, they differ by at most a logarithmic factor. If the $MST$ is very short, such as when the obstacle set is already connected, then the ratio $M/|S|$ is small, and we would expect that ray-shooting by walking through a triangulation would perform well. Such good behavior has been reported experimentally [9] (even without explicitly minimizing the weight of the triangulation).

The main result, in section 3, is a polynomial-time algorithm to compute a three-dimensional Steiner triangulation compatible with a set of polyhedral obstacles. The total area of the triangulation is within a constant factor of the smallest possible. The first step of the algorithm produces a depth-bounded octtree from the obstacles, where the rule is that an octtree cube is split if it meets an obstacle vertex or edge and is not too small. The second step of the algorithm triangulates each leaf cube in a manner compatible with the obstacles and neighboring leaf cubes. In the worst case, the algorithm runs in time $O(n^6)$ and produces as many tetrahedra; an improvement of an order of magnitude in both running time and number of tetrahedra is possible if the obstacle set is well-shaped (see section 3.3). A further improvement is possible if obstacles are just points (though this is not discussed in this paper). The algorithm is a generalization of Eppstein's algorithm [8] to compute a constant-factor approximation to the minimum-length Steiner triangulation of a set of points in two dimensions. As with Eppstein's analysis, the approximation ratio is large, though constant.

The proof of the approximation ratio has two parts. The first part is to show that the total surface area of the octtree is at most a constant factor times the area of an arbitrary triangulation $\hat{\mathcal{T}}$. To do this we charge the surface area of each leaf cube to some local feature of $\hat{\mathcal{T}}$. There are essentially two cases: if the central subcube of the leaf cube meets a vertex, edge, or face of $\hat{\mathcal{T}}$, then the area of $\hat{\mathcal{T}}$ within the leaf cube must be at least proportional to its surface area. Otherwise the central subcube must be contained in a tetrahedron of $\hat{\mathcal{T}}$, some face of which must have area at least proportional to the surface area of the leaf cube. (The actual argument is more complex, to guarantee that a single tetrahedral face is not charged by too many leaf cubes.) The second part of the approximation-ratio bound is to triangulate all leaf cubes with total area at most a constant times the tree surface area plus the obstacle area.

**Other work.** Assuming roughly linear data-structure storage, the best theoretical algorithms for ray-shooting have worst-case query time $O(\log n)$ for a simple polygon [10], approximately $O(\sqrt{n})$ for a set of planar line-segments [4], and approximately $O(n^{3/4})$ for a set of triangles in three dimensions [3]. Agarwal *et al* [2] consider the line-stabbing number of triangulations that are consistent with a set of obstacles. Mitchell *et al* [11] consider segment-shooting, a variant of ray-shooting. They show that the cost of a segment-shooting query in an octtree can be bounded up to a constant factor by the "cover complexity" of the segment.

## 2 The two-dimensional case

Throughout this section $S$ is a set of $n$ planar obstacle segments that meet only at endpoints; $S$ must include segments partitioning the boundary of its convex hull. A triangulation $\mathcal{T}$ is *compatible* with $S$ if any edge in $S$ is the union of closed edges of $\mathcal{T}$. Vertical bars $|\cdot|$ denote length, thus $|S|$ is the sum of the lengths of the segments in $S$. Let $M$ be $\inf_{\mathcal{T}} |\mathcal{T}|$, where $\mathcal{T}$ varies over triangulations compatible with $S$.

**Lemma 2.1** *In polynomial time it is possible to compute a polynomial-sized Steiner triangulation $\mathcal{T}$ compatible with $S$ so that $|\mathcal{T}| = O(M)$.*

We omit a detailed proof of lemma 2.1; it can be proven using techniques of Eppstein [8] or of the proof in section 3. The basic strategy is to build a depth-bounded balanced quadtree using only the vertices of $S$, and then triangulate each square in a fashion compatible with the edges of $S$ meeting the square. We remark that lemma 2.1 depends upon including the length of the obstacle set $S$ in the length of the triangulation; if the length is not included, a constant-factor polynomial-time approximation algorithm is not known [8].

Let $MST$ be a *minimum (Steiner) spanning tree* of $S$, i.e., a set of segments of minimum total length so that $S \cup MST$ is connected. An easy compactness argument shows that the minimum length can actually be attained.

**Lemma 2.2** $|S| + |MST| \leq M \leq O((|S| + |MST|) \log n)$.

*Proof:* The first inequality is obvious, since any triangulation compatible with $S$ must be connected. For the second, notice that $MST \cup S$ partitions the convex hull of $S$ into simple polygons (recall that $S$ includes convex hull edges). The total number of vertices is $O(n)$, since the $MST$ is a forest with no vertices of degree 2 and the number of leaves is the number of connected components of $S$. By a result of Clarkson [6], each simple polygon can be triangulated with weight proportional to $\log n$ times the perimeter of the polygon. □

**Corollary 2.3** $(|MST| + |S|)/|S| \leq c(S) \leq O(((|MST| + |S|)/|S|) \log n)$.

Let $D$ be the diameter of $S$, i.e., the largest distance between two points of $S$. The minimum spanning tree of the vertices of $S$ has length at most $O(D\sqrt{n})$, hence $|MST| \leq O(D\sqrt{n})$. The following lemma improves the worst-case bound that can be obtained from lemma 2.2 and this estimate by a factor of $\log n$. Again we omit a detailed proof of the lemma (see section 3.6 for a similar proof in three dimensions).

**Lemma 2.4** $M \le O(|S| + D\sqrt{n})$.

**Corollary 2.5** $c(S) \le O(1 + D\sqrt{n}/|S|))$.

## 3 The three-dimensional case

This section describes an algorithm that produces a Steiner triangulation compatible with a set of polyhedral obstacles in three dimensions. The triangulation has area within a constant factor of the smallest possible. Section 3.1 reviews some basic definitions. Section 3.2 states the main theorem in the case that the obstacles are "wide," a condition on the aspect ratio of their convex hull. (The full paper indicates how to remove the wideness assumption.) The main theorem is proved in sections 3.3 through 3.5. Section 3.3 gives an algorithm that constructs an octtree from a set of polyhedral obstacles. In section 3.4 we prove that the surface area of the octtree is at most a constant factor times the area of any Steiner triangulation compatible with the obstacles, while section 3.5 gives an algorithm to triangulate the octtree with total area proportional to the area of the octtree plus the area of the obstacles. Finally in section 3.6 we show some worst-case bounds on the ratio of triangulation area to obstacle area.

### 3.1 Definitions

We use terminology from the theory of convex polyhedra [7]. For example, a *polyhedral set* is any set obtained from open and closed halfspaces by a finite number of unions and intersections. A *face* of a convex polyhedron $C$ is the relative interior of the intersection of the closure of $C$ with a hyperplane supporting $C$; $\partial C$ is the relative boundary of $C$.

A *polyhedral subdivision* $\mathcal{Q}$ of a polyhedral set $Q$ is a finite partition of $Q$ into relatively open convex polyhedral *cells* so that every face of every cell is a union of cells in $\mathcal{Q}$. A subdivision is *proper* if every face of every cell is a cell in the subdivision; note that not all subdivisions are proper. If $f$ is a face of a cell $C$, we will occasionally refer to $f$ as a *polyhedral face* to distinguish it from a cell contained in $\partial C$ that is not a face of $C$. The $k$-skeleton $\mathcal{Q}^{(k)}$ of $\mathcal{Q}$ is the subdivision consisting of all cells of $\mathcal{Q}$ of dimension at most $k$. If $C$ is a convex polytope that is the union of cells of $\mathcal{Q}$, then bdry$(C, \mathcal{Q})$ is the subdivision consisting of the cells in $\mathcal{Q}$ whose union is $\partial C$. For convenience we define bdry$(\mathcal{Q})$ to be bdry$(Q, \mathcal{Q})$. The *area* of a subdivision $\mathcal{Q}$, area$(\mathcal{Q})$, is the sum of the areas of the 2-cells in $\mathcal{Q}$; similarly the *length* of $\mathcal{Q}$, length$(\mathcal{Q})$, is the sum of the lengths of the 1-cells in $\mathcal{Q}$. A *triangulation* of a polyhedral set is a proper polyhedral subdivision so that all cells are simplices. A triangulation $\mathcal{T}$ is *compatible* with a subdivision $\mathcal{S}$ if every cell in $\mathcal{S}$ is a union of cells in $\mathcal{T}$.

If $C$ is a square or cube and $k$ a positive real, then $kC$ is the square or cube with the same center and orientation as $C$ and side length $k$ times the side length of $C$. We write for example $C/2$ for $(1/2)C$.

### 3.2 The main theorem for wide obstacles

The *obstacle set* $S$ is a subdivision in $\mathbb{R}^3$ consisting only of 0-, 1-, and 2-simplices. For simplicity we assume that there are no isolated edges and vertices in $S$, i.e., every edge or vertex lies on the boundary of a 2-simplex; with minor modifications the algorithm can be extended to handle isolated edges and vertices. We let $H$ be the convex hull of $S$; we assume that $S$ contains a triangulation of $\partial H$. Throughout $n$ is the total number of cells in $S$.

Let $B$ be the smallest cube containing $S$, and let its side length be $b$. Obstacle set $S$ is *wide* if the area of $\partial H$ is $\Omega(b^2)$. Informally, $H$ can look like a ball or pancake, but not like a pencil. The following theorem holds for wide obstacles; it follows immediately from theorem 3.2 in section 3.4 and theorem 3.9 in section 3.5, using the fact that the area of $\partial H$ and hence $S$ is $\Omega(b^2)$. (The full paper shows how to triangulate obstacles that are not wide, at a cost of an additional factor of $n$ in time and number of tetrahedra.)

**Theorem 3.1** *Let $S$ be a wide set of obstacles with a total of $n$ simplices. In time $O(n^5)$ it is possible to compute a triangulation compatible with $S$ whose area is within a constant factor of the smallest possible. The triangulation has $O(n^5)$ tetrahedra and partitions the convex hull of $S$.*

### 3.3 The octtree algorithm

We now describe how to build an octtree $T$ from the obstacle set $S$. If necessary, perturb the minimal enclosing cube $B$ slightly so that no obstacle face is parallel to a face of $B$. Recall that $b$ is the side length of $B$. Choose $s_0$ so that $b/s_0$ is a power of 2 and $b/(2n) < s_0 \le b/n$.

We first build an octtree $T_0$. Every node of the tree is a cube. $B$ is the root node. A cube is *skewered* if it meets the 1-skeleton of $S$. A cube $C$ is subdivided (i.e., is not a leaf) if it is skewered and has side length greater than $s_0$; its children are the eight subcubes obtained by cutting it by the three planes through its center parallel to its facets. This process is repeated until no cube can be further subdivided. Clearly, each leaf cube in the octtree has side length $s_0 2^i$, for some integral $i \ge 0$, and each skewered cube is a leaf with side length $s_0$.

Octtree $T_0$ must then be made balanced, resulting in the octtree $T$. Two cubes are *adjacent* if they have overlapping 2-faces; an octtree is *balanced* if the side lengths of any two adjacent leaf cubes differ by at most two. If an octtree is unbalanced, it can be balanced by repeatedly choosing a pair of adjacent cubes that violate the balance property and subdividing the larger cube. Standard results [12] imply that balancing $T$ increases the number of cubes by at most a constant factor.

An obstacle triangle $\Delta$ *fully cuts* a cube $C$ if $\Delta$ intersects $C$ and $\partial\Delta$ avoids $C$. Notice that a cube may be fully cut by an obstacle triangle independently of whether or not it is skewered.

### 3.4 Area of the octtree

The *area* of octtree $T$, area$(T)$, is $\sum_C \text{area}(\partial C \cap H)$, where $C$ varies over the leaf cubes of $T$. Let $M = \inf_{\hat{\mathcal{T}}} \text{area}(\hat{\mathcal{T}})$, where $\hat{\mathcal{T}}$ ranges over triangulations compatible with $S$.

**Theorem 3.2** area$(T) \le c \cdot M$, *for some absolute constant* $c$.

*Proof:* Let $\hat{\mathcal{T}}$ be an arbitrary triangulation compatible with $S$ and let $C$ be a leaf cube of $T$. We can assume $C \cap H$ is

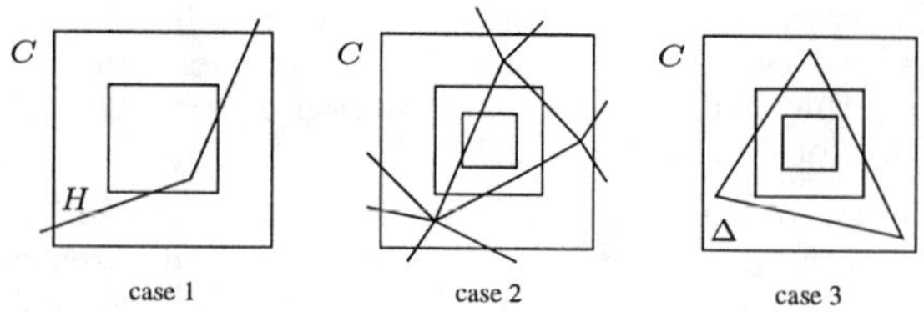


Figure 1: Cases of thereom 3.2. In case 1, $\partial H$ meets $C/2$; in case 2, the 2-skeleton of $\hat{\mathcal{T}}$ meets $C/4$; in case 3, the 2-skeleton of $\hat{\mathcal{T}}$ avoids $C/4$.

not empty. We charge area($\partial C \cap H$) to features of $\hat{\mathcal{T}}$, with cases as follows (see figure 1):

1. $C/2 \not\subseteq H$. Necessarily $\partial H$ meets $C$. By lemma 3.3 below, area($\partial C \cap H$) is $O$(area($\partial H \cap C$)). We charge area($\partial C \cap H$) to $\partial H \cap C$.

2. $C/2 \subseteq H$ and $C/4$ meets the 2-skeleton of $\hat{\mathcal{T}}$. By lemma 3.4 below, area($\partial C \cap H$) is $O$(area($\hat{\mathcal{T}} \cap C$)). We charge area($\partial C \cap H$) to $\hat{\mathcal{T}} \cap C$.

3. $C/2 \subseteq H$ and $C/4$ avoids the 2-skeleton of $\hat{\mathcal{T}}$. This is the most complex case. Let $\alpha$, $\beta$, and $\gamma$ be constants (chosen in the proof of Lemma 3.6 below, with $0 < \alpha, \beta < 1$ and $\gamma > 1$). Let the side length of $C$ be $s$. A *charging pair* $(\Delta, e)$ for $C$ is a triangle $\Delta$ of $\hat{\mathcal{T}}$ and an edge $e$ of $\Delta$ so that area($\Delta \cap \gamma C$) $\geq \alpha s^2$ and length($e \cap \gamma C$) $\geq \beta s$. We charge the area of $\partial C \cap H$ (which is $O(s^2)$) to the triangle of a charging pair, whose existence is guaranteed by Lemma 3.6 below. By lemma 3.8 the total charge to any triangle $\Delta$ of $\hat{\mathcal{T}}$, over all leaf cubes of $\mathcal{T}$, is at most $O$(area($\Delta$)).

In cases 1 and 2, the charges are to disjoint portions of $\partial H$ and $\hat{\mathcal{T}}$, respectively, so the total charge is $O$(area($\partial H$)+ area($\hat{\mathcal{T}}$)) which is $O$(area($\hat{\mathcal{T}}$)). In case 3, the total charge to any triangle $\Delta$ of $\hat{\mathcal{T}}$ is $O$(area($\Delta$)), so the total charge over all triangles is $O$(area($\hat{\mathcal{T}}$)). □

**Lemma 3.3** *Suppose $C$ is a leaf cube and $C/2 \not\subseteq H$. Then* area($\partial C \cap H$) *is* $O$(area($\partial H \cap C$)).

*Proof:* Let $s$ be the side length of $C$. Pair each corner of $C/2$ with the corresponding corner of $C$, and consider the 8 subcubes whose opposite corners are formed by the pairs. At least one such subcube $D$ must be entirely outside $H$, else $H$ would contain $C/2$. Let $d$ be the center of $D$. See figure 2.

Let $S_0$ and $S_1$ be the spheres of radius $s/8$ and $\sqrt{3}s$, respectively, with center $d$. $S_0$ is contained in $D$ and is outside $H$, and $S_1$ contains $C$. Let mappings $\pi_0$ and $\pi_1$ be central projections from $d$ into $S_0$ and $S_1$, respectively. Any ray from $d$ to a point of $\partial C \cap H$ must first hit a point of $\partial H \cap C$, since $d$ is outside $H$ and inside $C$. Hence $\pi_1(\partial C \cap H) \subseteq \pi_1(\partial H \cap C)$. For some constant $c$, we have

$$c \cdot \text{area}(\partial C \cap H) \leq \text{area}(\pi_1(\partial C \cap H))$$

since for any point $p \in \partial C \cap H$, the angle that a ray $\vec{dp}$ makes with $\partial C \cap H$ is bounded away from zero, as $d$ is bounded

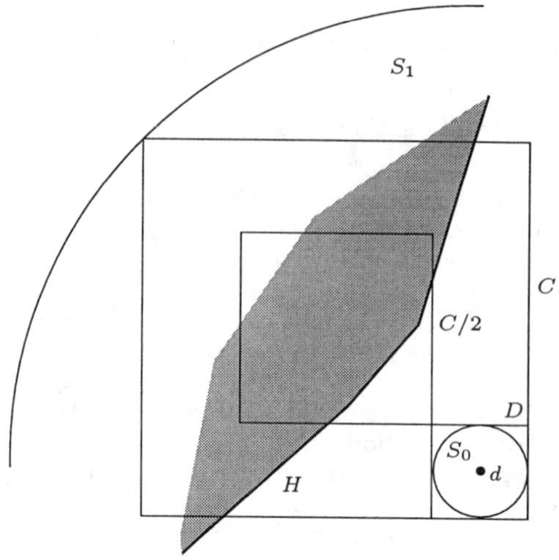


Figure 2: Proof of Lemma 3.3.

away from $\partial C$. We also have

$$\begin{aligned} \text{area}(\pi_1(\partial C \cap H)) &\leq \text{area}(\pi_1(\partial H \cap C)) \\ &= (8/\sqrt{3})^2 \cdot \text{area}(\pi_0(\partial H \cap C)) \\ &\leq 64/3 \cdot \text{area}(\partial H \cap C), \end{aligned}$$

proving the lemma. □

**Lemma 3.4** *If $C$ is a leaf cube and $B = C/2 \subseteq H$ and $B/2$ meets the 2-skeleton of $\hat{\mathcal{T}}$, then* area($\partial C \cap H$) *is* $O$(area($\hat{\mathcal{T}} \cap C$)).

*Proof:* Let $s$ be the edge length of $B$. Clearly area($\partial C \cap H$) is $O(s^2)$; we show area($\hat{\mathcal{T}} \cap B$) is $\Omega(s^2)$. Let $D$ be the open cube of side length $s/2$ centered at a point $x$ of the 2-skeleton of $\hat{\mathcal{T}}$ in $B/2$, with the same orientation as $B/2$. Choose three orthogonal edges $e_1$, $e_2$, $e_3$ of $D$, and for any $p \in D$ let $l_i = l_i(p)$, $i = 1, 2, 3$, be the segment through $p$ parallel to $e_i$ connecting opposite faces of $D$. At least one of $l_1$, $l_2$, $l_3$ must meet the 2-skeleton of $\hat{\mathcal{T}}$; for if not, they must all lie in the same tetrahedron $\Delta$ of $\hat{\mathcal{T}}$, since they share point $p$, and hence the convex hull of $l_1$, $l_2$, $l_3$ is contained in $\Delta$, a contradiction since the convex hull contains $x$, which is in $\hat{\mathcal{T}}^{(2)}$. Let $D_i$, $i = 1, 2, 3$, be the set of points $p \in D$ for which $l_i(p)$ meets the $\hat{\mathcal{T}}^{(2)}$. Since $D = D_1 \cup D_2 \cup D_3$, we can assume, say, that the volume of $D_1$ is at least $(s/2)^3/3 = s^3/24$. Now $D_1$ must be the Cartesian product of a segment of length $s/2$ parallel to $e_1$ with the projection of $\hat{\mathcal{T}}^{(2)} \cap D$ onto a facet of $D$ perpendicular to $e_1$. Hence the projection of $\hat{\mathcal{T}}^{(2)}$ has area at least $s^2/12$, and $\hat{\mathcal{T}}^{(2)}$ itself has at least the same area. □

**Lemma 3.5** *For any leaf cube $C$ of $T$, $7C$ meets the 1-skeleton of $S$.*

*Proof:* Let a *refinement step* during balancing be the replacement of a leaf node by an interior node with 8 children. We show that the lemma holds for $T_0$ and is maintained by every refinement step.

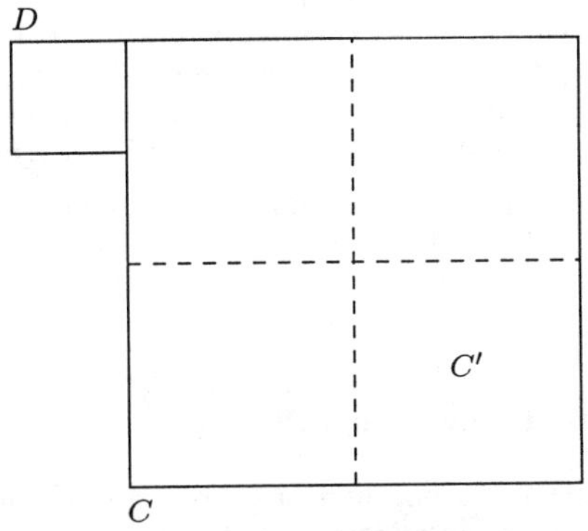


Figure 3: Proof of lemma 3.5.

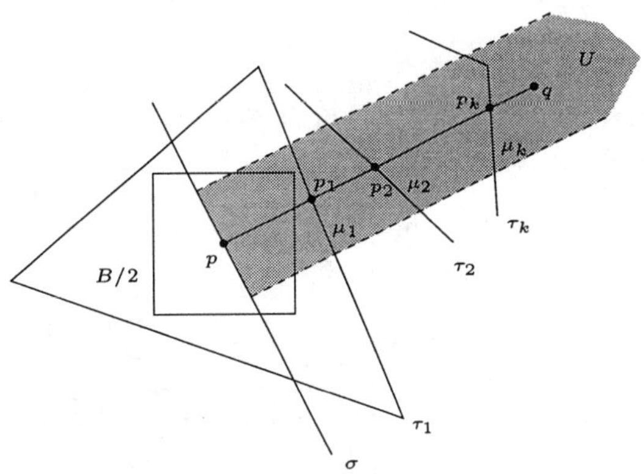


Figure 4: Proof of Lemma 3.6.

By construction, every skewered leaf in $T_0$ meets the 1-skeleton of $S$. Every unskewered leaf $C$ in $T_0$ has a skewered parent, so certainly $3C$ meets the 1-skeleton of $S$.

A refinement step maintains the condition of the lemma for every unchanged cube. So suppose $C$ is refined into 8 subcubes, with $s$ the original side length of $C$ and $C'$ an arbitrary child of $C$. See figure 3. $C$ must share a common 2-face with a cube $D$ whose side length is at most $s/4$. By inductive hypothesis, $7D$ meets the 1-skeleton of $S$. Easy calculations show that the $L_\infty$ distance between the centers of $C'$ and $D$ is at most $7s/8$ and that $7D \subset 7C'$. Hence $7C'$ meets the 1-skeleton of $S$. □

**Lemma 3.6** *If $C$ is a leaf cube, $B = C/2 \subseteq H$, and $B/2$ avoids the 2-skeleton of $\hat{T}$, then there is a charging pair for $C$.*

*Proof:* Let $s$ be the side length of $B$, so $B/2$ has side length $s/2$. By lemma 3.5 there is a point $q$ on the 1-skeleton of $\hat{T}$ within $7C = 14B$, hence distance $7\sqrt{3}s < 14s$ from the center $p$ of $B$. We assume that open segment $pq$ meets the 2-skeleton of $\hat{T}$ only at interior points of 2-cells; otherwise a slightly modified argument with a perturbed segment $p'q$ is necessary.

Let $\sigma$ be the plane orthogonal to $pq$ through $p$ and let $\pi_\sigma$ be orthogonal projection onto $\sigma$. Let $U$ be the infinite cylinder with axis $pq$ and radius $s/4$; then the disc $D = U \cap \sigma$ is contained in $B/2$. Let $p_1, \ldots, p_{l-1}$ be the intersections of 2-cells of $\hat{T}$ with open segment $pq$ in order from $p$ to $q$, and set $p_0 = p$ and $p_l = q$. For $i = 1, \ldots, l$, let $\tau_i$ be the 3-simplex of $\hat{T}$ containing the open segment $p_{i-1}p_i$. See figure 4. We have that $B/2 \subseteq \tau_1$, since $p_0 = p \in B/2$ and the 2-skeleton of $\hat{T}$ avoids $B/2$.

Let $\mu_i$ be the connected component of $\partial\tau_i \cap U$ containing $p_i$. Choose $k$ minimal so that $\mu_k$ contains a point of the 1-skeleton of $\hat{T}$; such $k$ must exist since $p_l = q$ is on the 1-skeleton of $\hat{T}$.

We claim that $\mu_i$ *fully cuts* $U$, i.e., $\pi_\sigma(\mu_i) = D$, for $i = 1, \ldots, k$. Now $\mu_1$ fully cuts $U$, since $D \subseteq B/2 \subseteq \tau_1$. For $1 < i \leq k$, $\mu_{i-1}$ must be a portion of a single 2-face of $\tau_i$, since it contains no point on the 1-skeleton of $\hat{T}$. Hence there must be two connected components to $U \cap \partial\tau_i$, specifically $\mu_{i-1}$ and $\mu_i$, and $\mu_i$ must fully cut $U$.

We now show that $\mu_k$ lies at distance at most $28s$ from $\sigma$. Let $f_{k-1}$ and $f_k$ be the faces of $\tau_k$ containing $p_{k-1}$ and $p_k$, and let $P_{k-1}$ and $P_k$ be the planes containing $f_{k-1}$ and $f_k$, respectively. Now $P_{k-1} \cap D = P_{k-1} \cap U \cap \sigma$ must be empty, since $P_{k-1} \cap U = \mu_{k-1}$ lies entirely in face $f_{k-1}$ but $D = U \cap \sigma$ lies entirely in the interior of simplex $\tau_1$. Also $P_k \cap D$ must be empty: if not, choose $u \in P_k \cap D$; then $p_k$ and $u$ lie on opposite sides of $P_{k-1}$ within $U$, so segment $up_k$ meets an interior point of face $f_{k-1}$ of $\tau_k$; however $up_k$ also lies entirely in the plane $P_k$ of face $f_k$ of $\tau_k$, a contradiction. The intersection of $P_k$ with $\partial U$ is an ellipse, whose center $p_k$ is at distance at most $14s$ from $\sigma$. The ellipse avoids $D$, since $P_k$ avoids $D$. Hence the furthest point of the ellipse is at most $28s$ from $\sigma$. Since the closed half-space of $P_k$ containing $D$ also contains $\tau_k$ and $\mu_k$, $\mu_k$ is at most distance $28s$ from $\sigma$.

We set the charging-pair constants $\alpha = \pi/48$, $\beta = 1/8$, and $\gamma = 43$. To finish the proof, notice that $\text{area}(\mu_k) \geq \text{area}(\pi_\sigma(\mu_k)) = \pi(s/4)^2$. Since $\mu_k$ consists of portions of at most three triangular faces of $\tau_k$, at least one such triangle $\Delta$ satisfies $\text{area}(\Delta \cap U) \geq \pi s^2/48 = \alpha s^2$. $\Delta$ is the triangle of $\hat{T}$ promised by the lemma statement. We have $\Delta \cap U$ within distance $28s$ from $\sigma$, in fact within distance $28s$ from $\sigma \cap U \subset B/2 \subset B$. Since the side length of $B$ is $s$, $\Delta \cap U \subset (2 \cdot 28 + 1)B \subset \gamma C$.

It remains to find a piece of an edge of $\Delta$ of length at least $\beta s$ within $\gamma C$. Let $U'$ be the cylinder coaxial with $U$ but with double the radius. We have $\Delta \cap U'$ within distance $42s$ from $\sigma \cap U' \subset B$, so $\Delta \cap U' \subset (2 \cdot 42 + 1)B = \gamma C$. If $\Delta$ is completely contained within $U'$, then by the isoperimetric inequality [14] the perimeter $L$ of $\Delta$ satisfies $L \geq \sqrt{4\pi \cdot \text{area}(\Delta)} \geq \pi s/2\sqrt{3}$, so one edge of $\Delta$ has length at least $\pi s/6\sqrt{3} \geq \beta s$. Otherwise $\partial\Delta$ intersects both $\partial U'$ and the interior of $U$. Since $\partial U$ and $\partial U'$ are separated by $s/4$, there is a portion of an edge of $\partial\Delta$ lying within $U$ of length at least $s/8 = \beta s$. □

The *width* of a compact planar set $S$, $\text{width}(S)$, is the smallest distance $w$ so that $S$ is contained in the closed region between two parallel lines $w$ apart. The *diameter* of $S$, $\text{diam}(S)$, is the largest distance between two points of $S$. It is easy to see that the area of $S$ is at most $\text{width}(S) \cdot \text{diam}(S)$.

**Lemma 3.7** *Let $\Delta$ be an obstacle triangle and $r$ the radius of its inscribed circle. If leaf cube $C$ with side length $s$ has* *charging pair* $(\Delta, e)$, *then* $s$ *is* $O(r)$.

*Proof:* The width of $\Delta$ is at most $3r$, by some elementary geometry. Since $(\Delta, e)$ is a charging pair for $C$, we have

$$\begin{aligned} \alpha s^2 &\leq \text{area}(\Delta \cap \gamma C) \\ &\leq \text{width}(\Delta \cap \gamma C) \cdot \text{diam}(\Delta \cap \gamma C) \\ &\leq 3r \cdot \sqrt{3}\gamma s. \end{aligned}$$

Hence $s \leq 3\sqrt{3}\gamma r/\alpha$, which is $O(r)$. □

**Lemma 3.8** *The total charge to triangle* $\Delta$ *of* $\hat{T}$ *from all leaf cubes in case 3 above is* $O(\text{area}(\Delta))$.

*Proof:* Let $p$ and $r$ be the perimeter and inscribed-circle radius of $\Delta$, respectively. Some elementary geometry gives $\text{area}(\Delta) = pr/2$.

Consider the set $C_s$ of all leaf cubes with a fixed side length $s$. Since no two cubes in $C_s$ overlap, no point in space is covered by more than $\gamma^3$ open supercubes $\gamma C$, for $C \in C_s$. Since the intersection of $\gamma C$ with $\partial\Delta$ has length at least $\beta s$, no more than $\gamma^3 p/(\beta s) = O(p/s)$ cubes in $C_s$ can be charged to $\Delta$. The charge for each cube is $O(s^2)$, so the total charge to $\Delta$ for cubes in $C_s$ is $O(ps)$.

Let $\hat{s}$ be the largest cube side length for which a cube is charged to $\Delta$; by lemma 3.7, $\hat{s}$ is $O(r)$. The total charge over all cube sizes is $O(p(s_0 + 2s_0 + \ldots + \hat{s})) = O(p\hat{s}) = O(pr)$, which is $O(\text{area}(\Delta))$. □

## 3.5 Triangulating the tree

**Theorem 3.9** *In time* $O(n^5)$ *it is possible to construct a triangulation* $\mathcal{T}$ *from the octtree* $T$ *so that* $\text{area}(\mathcal{T}) \leq c' \cdot (\text{area}(T) + \text{area}(S) + b^2)$, *where* $c'$ *is an absolute constant and* $b$ *is the side-length of the root cube of* $T$. $\mathcal{T}$ *has* $O(n^5)$ *tetrahedra and partitions the convex hull* $H$ *of* $S$.

For $C$ a leaf cube of $T$, the *clipped cube* $C_H$ is $C \cap H$. We choose below a subdivision $\mathcal{P}$ of $H$ whose 3-cells are clipped-cube interiors and whose 2-skeleton is a triangulation. In sections 3.5.2 and 3.5.3 we show how to triangulate a clipped cube so that its boundary triangulation matches $\mathcal{P}$. The desired triangulation $\mathcal{T}$ is then obtained simply by pasting the clipped-cube triangulations into $\mathcal{P}$.

In octtree $T$, two adjacent cubes may have overlapping but distinct $k$-faces, $k = 1, 2$, if the cubes are of different sizes. A clipped-cube $k$-face is *minimal* if no other clipped-cube face of the same dimension is properly contained within it; a minimal $k$-face is a *tree-partitioning* face if it is properly contained in another clipped-cube face (possibly of higher dimension). It is easy to see that any clipped-cube $k$-face is the union of minimal faces of dimension at most $k$. In fact, since $T$ is balanced, any 1-face is the union of at most four minimal 1-faces and three vertices, and any 2-face is the union of at most four minimal 2-faces, four minimal 1-faces, and a vertex (forming a '+'-shape in the middle of the 2-face).

Let $\mathcal{P}'$ consist of clipped-cube interiors plus all minimal 0-, 1-, and 2-faces. Subdivision $\mathcal{P}$ is obtained from $\mathcal{P}'$ by replacing 1- and 2-faces. A 1-face $e$ of $\mathcal{P}'$ is replaced with the chain of 0- and 1-cells formed by subdividing $e$ at each point of intersection with an obstacle triangle. Let $f$ be a 2-face of $\mathcal{P}'$. Form the set of line segments obtained by intersecting $f$ with all obstacle triangles, then triangulate this set together with the set of 0- and 1-cells partitioning $\partial f$. The resulting triangulation replaces $f$ in $\mathcal{P}$. It is easy to check that $\mathcal{P}$ is a subdivision of $H$ whose 2-skeleton has been triangulated.

The *combinatorial complexity* of leaf cube $C$, $n_C$, is number of edges of $C$ plus the number of obstacle triangles meeting $C$. It is easy to check that $\text{bdry}(C_H, \mathcal{P})$ has $O(n_C)$ edges and vertices.

### 3.5.1 Central triangulations

Let $Q$ be a 3-cell in a polyhedral subdivision $\mathcal{Q}$ and suppose $\text{bdry}(Q, \mathcal{Q})$ is a triangulation. Let $q$ be an interior point of $Q$. The *central triangulation* of $Q$ from $q$ consists of the tetrahedra that can be formed by $q$ and the triangles in $\text{bdry}(Q, \mathcal{Q})$, and all tetrahedral faces.

**Proposition 3.10** *If* $\text{bdry}(Q, \mathcal{Q})$ *has* $k$ *vertices, then the area of any central triangulation of* $Q$ *is at most* $3k/2$ *times the area of* $\partial Q$.

*Proof:* Let $q$ be the central triangulation vertex. There are at most $3k$ edges of $\text{bdry}(Q, \mathcal{Q})$. Each new tetrahedral 2-face is formed by $q$ and such an edge, and has area at most half the area of $\partial Q$. □

### 3.5.2 Triangulating an unskewered cube

For this section, $C$ is an unskewered leaf cube (i.e., $C$ avoids the 1-skeleton of $S$). A triangulation $\mathcal{T}_C$ of $C_H$ is obtained as follows. Start with the subdivision consisting of $C_H$ and $\text{bdry}(C_H, \mathcal{P})$. Subdivide $C_H$ by all obstacle triangles that meet it. Notice that all such obstacle triangles must cut $C$ fully and cannot meet within $C$; furthermore $\text{bdry}(C_H, \mathcal{P})$ already contains all edges of intersection between obstacle triangles and $\partial C_H$. Now triangulate any new 2-cells, and centrally triangulate each 3-cell.

**Lemma 3.11** $\mathcal{T}_C$ *has area* $O(\text{area}(S \cap C) + \text{area}(\partial C \cap H))$.

*Proof:* Let $\mathcal{T}_0$ be the subdivision before central triangulation, and let $C_0$ be a 3-cell of $\mathcal{T}_0$. We show that the number of vertices in $\text{bdry}(C_0, \mathcal{T}_0)$ is constant. The lemma then follows using Proposition 3.10 and the fact that $\text{area}(\mathcal{T}_0)$ is $\text{area}(S \cap C) + \text{area}(\partial C \cap H)$.

First notice that $C_0$ has at most 14 polyhedral 2-faces, 6 faces which are subsets of a 2-face of $C$ and at most 8 faces separating a vertex of $C$ from $C_0$. Hence $C_0$ has a constant number of polyhedral vertices. Any vertex of $\text{bdry}(C_0, \mathcal{T}_0)$ that is not a polyhedral vertex must be a cube partitioning vertex, of which there are at most a constant number. □

### 3.5.3 Triangulating a skewered cube

For this section, $C$ is a skewered leaf cube (i.e., $C$ meets the 1-skeleton of $S$). A triangulation $\mathcal{T}_C$ of $C_H$ is obtained in two steps. We first compute a triangulation $\mathcal{T}_C$ so that $\text{bdry}(\mathcal{T}_C) = \text{bdry}(C_H, \mathcal{T}_C)$ is a refinement of $\text{bdry}(C_H, \mathcal{P})$. Using lemma 3.15, we then update $\mathcal{T}_C$ to remove any vertex of $\text{bdry}(C_H, \mathcal{T}_C)$ not in $\text{bdry}(C_H, \mathcal{P})$.

The first step has four substeps. The "vertical" direction can be chosen to be any direction not parallel to a face of $C_H$ or a face of $S$ and so that no vertical line meets three obstacle edges that do not already meet at a common vertex.

1. Orthogonally project each edge of $\text{bdry}(C_H, \mathcal{P})$ onto a horizontal plane (i.e., a plane orthogonal to the vertical direction), and extend the projection to a line. Similarly project and extend each obstacle edge meeting $C$.
2. Let $\mathcal{T}_2$ be a (two-dimensional) triangulation of the resulting arrangement, truncated to the projection of $C_H$. Lift $\mathcal{T}_2$ to a set of infinite vertical triangular prisms.
3. Subdivide $C_H$ using both the vertical prisms and any obstacle triangles meeting $C$, forming a subdivision $\mathcal{T}_3$.
4. Extend the triangulation of $\text{bdry}(C_H, \mathcal{P})$ to form a triangulation of the 2-skeleton of $\mathcal{T}_3$. Then centrally triangulate each 3-cell, forming the triangulation $\mathcal{T}_C$.

**Lemma 3.12** *The total length of $\mathcal{T}_2$ is $O(n_C s_0)$.*

*Proof:* Let $\mu$ be the rigid-motion invariant measure on sets of lines in the plane [14] and $t(\ell)$ be the number of edges of $\mathcal{T}_2$ met by line $\ell$. Up to a constant multiple, the length of $\mathcal{T}_2$ is $\int t(\ell)\, d\mu$. Consider the arrangement before triangulation and truncation. By the zone theorem for lines [7], the total combinatorial complexity of the cells of the arrangement intersected by a line $\ell$ is $O(n_C)$, so $t(\ell)$ is $O(n_C)$. Hence $\int t(\ell)\, d\mu$ is bounded by $O(n_C)$ times the measure of the set of lines intersecting $\mathcal{T}_2$, which is $O(s_0)$ since the perimeter of $\mathcal{T}_2$ is $O(s_0)$ [14]. □

**Lemma 3.13** *Triangulation $\mathcal{T}_C$ has area $O(\text{area}(S \cap C) + \text{area}(\partial C \cap H) + n_C s_0^2)$.*

*Proof:* The total area of the vertical 2-cells in the subdivision $\mathcal{T}_3$ is $O(n_C s_0^2)$, since the total length of $\mathcal{T}_2$ is $O(n_C s_0)$ and the height of $C_H$ at most $O(s_0)$. Any nonvertical 2-cell is a portion of either $\partial C \cap H$ or $S \cap C$, hence the total area of all nonvertical 2-cells is $\text{area}(S \cap C) + \text{area}(\partial C \cap H)$. Let $P$ be a 3-cell in $\mathcal{T}_3$; we show that the number of vertices in $\text{bdry}(P, \mathcal{T}_3)$ is constant. The lemma then follows using proposition 3.10 and the bound on the sum of the areas of all 2-cells.

Cell $P$ is a section of a triangular prism and has at most 5 polyhedral faces, hence a constant number of polyhedral vertices. We claim $\text{bdry}(P, \mathcal{T}_3)$ has at most two additional vertices per vertical polyhedral edge of $P$. To see this, note that by construction of $\mathcal{T}_3$, any vertex of $\text{bdry}(P, \mathcal{T}_3)$ must either be a vertex of $\text{bdry}(C_H, \mathcal{T}_3)$, and hence a polyhedral vertex of $P$, or must lie in the interior of a collinear vertical chain of edges of $\mathcal{T}_3$. In the latter case, the vertex must either lie in the interior of an obstacle triangle, and hence must be a polyhedral vertex of $P$, or on the closure of an obstacle edge. However, there can be at most two such vertices of this last type per vertical chain, since the vertical direction was chosen so that no three obstacle edges lie on a common vertical line. □

It is easy to see that $\mathcal{T}_C$ has $O(n_C^3)$ tetrahedra, has $O(n_C^2)$ vertices on $\text{bdry}(\mathcal{T}_C)$, and can be computed in time $O(n_C^3)$.

We now update the triangulation to remove the vertices of $\text{bdry}(C_H, \mathcal{T}_C)$ not in $\text{bdry}(C_H, \mathcal{P})$. Let $G = (V, E)$ be the 1-skeleton of $\text{bdry}(C_H, \mathcal{T}_C)$ as a graph, and similarly let $G_P = (V_P, E_P)$ be the 1-skeleton of $\text{bdry}(C_H, \mathcal{P})$. A subset of $V$ is *independent* if no two vertices are connected by an edge of $E$.

**Lemma 3.14** *There is an independent subset of $V \setminus V_P$ whose size is a constant fraction of $V \setminus V_P$ so that each vertex has constant degree in $G$.*

*Proof:* The average degree of vertices in the subgraph of $G$ induced by $V \setminus V_P$ is at most 6, by planarity. Any vertex $v$ in $V \setminus V_P$ is incident to at most 4 vertices of $V_P$: either $v$ lies in the interior of a triangle of $\text{bdry}(C_H, \mathcal{P})$, in which case it can be incident to at most 3 vertices of $V_P$, or $v$ lies on an edge of $\text{bdry}(C_H, \mathcal{P})$, in which case it can be incident to at most 4 vertices of $V_P$. Hence the average degree of vertices $V \setminus V_P$ in $G$ is at most ten. By standard techniques [13], it follows that there is a bounded-degree independent set whose size is a constant fraction of $V \setminus V_P$. □

**Lemma 3.15** *$\mathcal{T}_C$ can be updated to remove any vertex in* $\text{bdry}(C_H, \mathcal{T}_C)$ *not in* $\text{bdry}(C_H, \mathcal{P})$. *The update takes time $O(n_C^2 \log n_C)$ and adds area $O(s_0^2 \log n_C + \text{area}(S \cap C))$ and $O(n_C^3)$ tetrahedra.*

*Proof:* $\mathcal{T}_C$ is updated in stages. Each stage removes the independent set of vertices guaranteed by lemma 3.14. Clearly, the number of stages is logarithmic in the number of vertices of $\mathcal{T}_C$, i.e., $O(\log n_C)$.

We assume that no vertex to be removed lies on an obstacle triangle. (The case that a vertex lies on an obstacle triangle is more complex and is described in the full paper.) Then a vertex $v$ in the independent set can be removed as follows. Suppose $v$ does not lie on an edge of $\text{bdry}(C_H, \mathcal{P})$. Let $S_v$ be the polygon formed by the edges opposite $v$ in triangles of $\text{bdry}(C_H, \mathcal{T}_C)$ incident to $v$. It is possible to push $v$ slightly inside $C_H$ maintaining the combinatorial structure of $\mathcal{T}_C$ (in particular, $v$ must not be pushed through the plane of the face opposite $v$ in any tetrahedron incident to $v$). Pushing $v$ leaves a dimple in the triangulation of $C_H$. The dimple can be filled by triangulating $S_v$ and then adding to $\mathcal{T}_C$ the tetrahedra formed by $v$ and the new triangles. If $v$ does lie on an edge $e$ of *bdry*$(C_H, \mathcal{P})$ the approach is similar, except that the triangulation of $S_v$ must use a piece of edge $e$.

The area added to $\mathcal{T}_C$ at each stage is at most $O(s_C^2)$. To see this, first note that when a vertex $v$ is pushed, the change in area of the tetrahedra incident to $v$ is negligible, since the perturbation of $v$ can be made arbitrarily small. The area of the new tetrahedral faces is at most a constant times the area of $S_v$, since $S_v$ has at most a constant number of edges. Since the vertices at each stage form an independent set, the polygons $\{S_v\}$ have disjoint interiors and their total area is at most the surface area of $C$, i.e., $O(s_C^2)$. Hence the total area added at all stages is $O(s_C^2 \log n_C)$.

The number of tetrahedra required to fill each dimple is constant, hence only $O(n_C^2)$ tetrahedra are added for all vertices. (In the full paper, we show that $O(n_c^3)$ tetrahedra have to be added to fill dimples that result from perturbing vertices lying on obstacle boundaries.) □

### 3.5.4 Accounting

**Lemma 3.16** *$\sum_C n_C s_0^2 = O(\text{area}(S) + b^2)$, where the sum runs over all skewered leaf cubes $C$ and $b$ is the side length of the root bounding cube.*

*Proof:* Let $m_C$ be the total number of obstacle triangles that meet cube $C$. Since every skewered cube meets some obstacle triangle, $n_C$ is $O(m_C)$. We can write $\sum_C m_C s_0^2$ as

$$\sum_{\Delta \in S} |S_\Delta| s_0^2$$

where $S_\Delta$ is the set of skewered leaf cubes intersected by obstacle triangle $\Delta$. We can write

$$|S_\Delta| = a_\Delta + m_\Delta$$

where $a_\Delta$ and $m_\Delta$ are respectively the number of cubes $C$ in $S_\Delta$ so that the doubled cube $2C$ avoids or meets the boundary of $\Delta$.

Consider first cubes $C$ in $S_\Delta$ so that $2C$ avoids the boundary of $\Delta$. Plainly the area of $2C \cap \Delta$ is $\Theta(s_0^2)$. Since no point in space is covered by more than 8 open doubled skewered cubes,

$$a_\Delta \leq 8 \cdot \text{area}(\Delta)/\Omega(s_0^2) = O(\text{area}(\Delta)/s_0^2).$$

It follows that $a_\Delta s_0^2$ is $O(\text{area}(\Delta))$, and hence $\sum_{\Delta \in S} a_\Delta s_0^2$ is $O(\text{area}(S))$.

Now consider cubes $C$ in $S_\Delta$ so that $2C$ meets the boundary of $\Delta$. The boundaries of the doubled cubes lie on the planes of a cubic grid of step size $s_0$. An obstacle triangle edge $e$ meets at most $3\lceil \text{length}(e)/s_0 \rceil$ planes, which is $O(b/s_0)$ (recall that $b$ is the side length of the box bounding the obstacle set $S$). Hence each edge meets $O(b/s_0)$ doubled cubes, so $m_\Delta$ is $O(b/s_0)$. We have

$$\sum_{\Delta \in S} m_\Delta s_0^2 = O(n \cdot b/s_0 \cdot s_0^2) = O(nbs_0) = O(b^2)$$

using $b/(2n) < s_0 \leq b/n$. □

*Proof of Theorem 3.9:* By lemmas 3.11, 3.13, and 3.15, the total area of $\mathcal{T}$ is

$$\sum_C \text{area}(\partial C \cap H) + \sum_C \text{area}(S \cap C) + \sum_C n_C s_0^2$$

where the first two summations run over all leaf cubes and the last over skewered leaf cubes. The first two summations add to $O(\text{area}(T) + \text{area}(S))$. Using lemma 3.16, the last summation is $O(\text{area}(S) + b^2)$.

Since the minimum cube size is about $b/n$, the total number of leaf cubes is $O(n^3)$. Both the running time to compute $\mathcal{T}$ and the number of tetrahedra are bounded by $O(\sum_C n_C^3) = O(\sum_C (m_C + 1)^3) = O(\sum_C m_C^3) + O(n^3)$, where $m_C$ is the number of triangles meeting cube $C$. All leaf cubes $C$ have faces lying on a cubic grid of planes with step size $s_0$; a triangle can hit at most $O(n^2)$ such grid cubes and thus at most $O(n^2)$ leaf cubes of $T$. Hence the number of incidences between leaf cubes and triangles, $\sum_C m_C$ is $O(n^3)$. The sum $\sum_C m_C^3$ is maximized if $m_C$ is as large as possible, i.e., $n$, for as many cubes $C$ as possible, i.e., $O(n^2)$, yielding an upper bound of $O(n^3 \cdot n^2) = O(n^5)$. □

### 3.6 Worst-case bounds

Recall that $c(S) = \inf_{\mathcal{T}} \text{area}(\mathcal{T})/\text{area}(S)$, where $\mathcal{T}$ varies over all triangulations compatible with $S$. The following lemma gives worst-case bounds on $c(S)$. As before, $S$ must contain faces partitioning the boundary of its convex hull.

**Lemma 3.17** *For any wide obstacle set $S$ in $\mathbb{R}^3$, $c(S) = O(\sqrt{n})$, where $n$ is the number of simplices in $S$. There is a wide set $\hat{S}$ of $O(n)$ obstacles with $c(\hat{S}) = \Omega(\sqrt{n})$.*

*Proof:* Let $B$ be the minimum-size bounding cube of $S$, perturbed slightly so that no obstacle face is parallel to a face of $B$. Let $B$ have side length $b$. Split $B$ into a grid of identical cubes, where each cube has side length $s \approx b/\sqrt{n}$, so there are about $n^{3/2}$ cubes altogether. Using the algorithms of section 3.5, triangulate each clipped cube and paste the triangulations together, yielding a triangulation $\mathcal{T}$ compatible with $S$. $\mathcal{T}$ has area $O(n^{3/2} \cdot s^2) = O(b^2\sqrt{n})$ (the surface area of the cubes) plus $O(\text{area}(S))$ plus $O(\sum_C n_C s^2)$, where $n_C$ is the combinatorial complexity of the cube. An analysis similar to lemma 3.16 shows that $\sum_C n_C s^2$ is $O(\text{area}(S) + n \cdot (b/s) \cdot s^2) = O(\text{area}(S) + b^2\sqrt{n})$. Hence the ratio $\text{area}(\mathcal{T})/\text{area}(S)$ is $O(\sqrt{n})$, as $\text{area}(S)$ is $\Omega(b^2)$.

For the second statement of the lemma, choose an axis-aligned unit cube $B$. Subdivide $B$ into a cubical grid of about $\sqrt{n} \times \sqrt{n} \times \sqrt{n}$ identical subcubes. For each one-dimensional row of subcubes parallel to the $x$-, $y$-, or $z$-axis, choose a very thin obstacle triangle that covers all the subcube centers in the row. Slightly perturb the resulting set of $\Theta(n)$ triangles so that no two intersect. Add to the obstacle triangles a triangulation of their convex hull, forming the obstacle set $\hat{S}$. Since the 1-skeleton of $\hat{S}$ passes very near the center of each subcube, Lemma 3.4 implies that any triangulation $\mathcal{T}$ compatible with $\hat{S}$ must have area $\Omega((1/\sqrt{n})^2) = \Omega(1/n)$ within each subcube. Since there are $n^{3/2}$ subcubes, the total area of $\mathcal{T}$ is $\Omega(\sqrt{n})$. Since the area of the convex hull is $O(1)$ and the area of the remaining obstacles can be made arbitrarily small, $c(\hat{S}) = \Omega(\sqrt{n})$. □

An argument similar to this proof shows that if $S$ contains $n$ points (and the faces of the convex hull) then $c(S) = O(n^{1/3})$. Furthermore there is a set $\hat{S}$ of $n$ points with $c(\hat{S}) = \Omega(n^{1/3})$. These results contrast with the results of Agarwal *et al* [2]. They show there is a set $S$ of $n$ points in $\mathbb{R}^3$ so that in any triangulation of $S$, some line meets $\sqrt{n}$ triangulation faces. Similarly there is a set $S$ of $n$ obstacle triangles in $\mathbb{R}^3$ so that in any triangulation compatible with $S'$, some line meets $O(n)$ triangulation triangles, even though it misses all obstacles.

## 4 Discussion

We have not tried to estimate the approximation ratio for the construction in section 3. Our algorithm is based on Eppstein's algorithm, which approximates the minimum-length Steiner triangulation of a set of points in two dimensions. Eppstein is able to prove an approximation ratio of 316 (though he suspects the true ratio is much smaller, perhaps around 20); our proof is much less careful about constants than his. A challenging open problem is to construct triangulations of approximately minimum weight in two or three dimensions with reasonable constants and with a reasonable number of vertices.

In two dimensions, the minimum spanning tree provides an intrinsic measure of the minimum weight triangulation, in the sense that their weights differ by at most a logarithmic factor (lemma 2.2). In three dimensions, the surface area of the octtree constructed in section 3 is an intrinsic measure of the minimum weight triangulation (theorem 3.2). It would be of interest to obtain a more natural intrinsic measure.

The analysis of ray-shooting-by-walking can be extended to other subdivisions besides triangulations. For example, consider the leaf cubes of the octtree constructed in section 3, with the modification that each unskewered cube is partitioned by all obstacles that cut it fully. Label each skewered leaf cube with the number of obstacle triangles that meet it, and label all the other 3-cells 1. Then it is possible to walk through the partitioned octtree along a line $\ell$ with total cost proportional to $w(\ell)$, where $w(\ell)$ is the sum of the labels of the 3-cells intersected by $\ell$. The analysis in section 3 shows that $\int w(\ell) d\mu$ is the area of the minimum weight triangulation, to within a constant factor. Two problems arise naturally when considering alternative subdivisions for ray-shooting queries: first, to determine if the area on the minimum weight triangulation is always a relevant bound; second, to provide an analytic comparison of the constants that arise from different subdivisions.

## References


[1] P. Agarwal, Range searching, *CRC Handbook of Discrete and Computational Geometry*, J. E. Goodman and J. O'Rourke, eds., CRC Press, Boca Raton, to appear, 1997.

[2] P. Agarwal, B. Aronov, and S. Suri, Stabbing triangulations by lines in 3D, *Proc. Eleventh Ann. Symp. Comp. Geometry*, 1995, pp. 267–276.

[3] P. Agarwal and J. Matoušek, On range searching with semialgebraic sets, *Discrete and Comput. Geom.* **11**:393–418, 1994.

[4] P. Agarwal and M. Sharir, Applications of a new partition scheme, *Discrete and Comput. Geom.* **9**:11–38, 1993.

[5] M. Bern and D. Eppstein, Approximation algorithms for geometric problems, in *Approximation Algorithms for NP-hard Problems*, D. Hochbaum, ed., PWS Publishing Co., 1997, pp. 296–345.

[6] K. Clarkson, Approximation algorithms for planar traveling salesman tours and minimum-length triangulations, *Proc. 2nd ACM-SIAM Symp. on Discrete Algorithms*, 1991, pp. 17–23.

[7] H. Edelsbrunner, *Algorithms in Combinatorial Geometry*, Springer-Verlag, Berlin, 1987.

[8] D. Eppstein, Approximating the minimum weight Steiner triangulation, *Discrete and Computational Geometry*, **11**:163–191, 1994.

[9] S. Fortune, A beam-tracing algorithm for prediction of indoor radio propagation, *Proc. First ACM Workshop on Applied Computational Geometry*, 1996, pp. 76–81.

[10] J. Hershberger and S. Suri, A pedestrian approach to ray shooting: shoot a ray, take a walk, *Proc. 4th ACM-SIAM Symp. on Discrete Algorithms*, 1993, pp. 54–63.

[11] J.S.B. Mitchell, D.M. Mount, and S. Suri, Query-sensitive ray shooting, *Proc. Tenth Ann. Symp. Comp. Geometry*, 1994, pp. 359–368.

[12] D.W. Moore, *Simplicial mesh generation with applications*. Ph.D. Thesis, Dept. of Computer Science, Cornell Univ., 1992. Report 92-1322.

[13] J. O'Rourke, *Computational Geometry in C*, Cambridge University Press, Cambridge, 1994.

[14] L. Santaló, *Integral Probability and Geometric Probability*, Encyclopedia of Mathematics and its Applications, Volume 1, Addison-Wesley, 1979.

# Contour Trees and Small Seed Sets for Isosurface Traversal*

Marc van Kreveld[†]
marc@cs.ruu.nl

René van Oostrum[†]
rene@cs.ruu.nl

Chandrajit Bajaj[‡]
bajaj@cs.purdue.edu

Valerio Pascucci[‡]
pascucci@cs.purdue.edu

Dan Schikore[‡]
drs@cs.purdue.edu

## Abstract

For 2D or 3D meshes that represent a continuous function to the reals, the contours—or isosurfaces—of a specified value are an important way to visualize it. To find such contours, a seed set can be used for the starting points from which the traversal of the contours can start. This paper gives the first methods to obtain seed sets that are provably small in size. They are based on a variant of the contour tree (or topographic change tree). We give a new, simple algorithm to compute such a tree in regular and irregular meshes that requires $O(n \log n)$ time in 2D for meshes with $n$ elements, and in $O(n^2)$ time in higher dimensions. The additional storage overhead is proportial to the maximum size of any contour (linear in the worst case, but typically less). Given the contour tree, a minimum size seed set can be computed in polynomial time and storage. Since in practice at most linear storage is allowed, we develop a simple approximation algorithm giving a seed set of size at most twice the size of the minimum. It requires $O(n \log^2 n)$ time in 2D and $O(n^2)$ time otherwise, and requires linear storage. We also give experimental results, showing the size of the seed sets and supporting the claim that sublinear storage is used.

## 1 Introduction

Scalar data defined over the plane or 3-space is quite common in fields like medical imaging, scientific visualization, and geographic information systems. Such data can be visualized after interpolation by showing one or more contours or isosurfaces: the sets of points having a specified scalar value. For example, scalar data over the plane are used to model elevation in the landscape, and a contour is just an isoline of elevation. In atmospheric pressure modelling, a contour is a surface in the atmosphere where the air pressure is constant, an isobar. In medical imaging, isosurfaces are used to show reconstructed scans of the brain or parts of the body. The scalar data can be seen as a sample of some real-valued function, which is called a terrain or elevation model in GIS, and a scalar field in imaging.

A real-valued function over 2D or 3D can be represented in a computer using a 2D or 3D mesh, which can be regular (all cells have the same size and shape) or irregular. A terrain (mountain landscape) in GIS is commonly represented by regular square grid or an irregular triangulation. The elements of the grid, or vertices of the triangulation, have a scalar function value associated to them. The function value of non-vertex points in the 2D mesh can be obtained by interpolation. An easy form of interpolation for irregular triangulations is linear interpolation over each triangle. The resulting model is known as the TIN model for terrains (Triangulated Irregular Network) in GIS. In computational geometry, it is known as a polyhedral terrain. In this paper we will consider the interpolation issues only as long as may they affect the isocontour computation. In particular in Section 2 we will state briefly the properties we assume for the interpolating function (satisfied by the most commonly used linear interpolation over simplicial complexes or multi-linear interpolation over regular grids). More on interpolation of spatial data and references to the literature can be found in the book by Watson [24].

One can expect that the complexity of the contours with a single function value in a mesh with $n$ elements is roughly proportional to $\sqrt{n}$ in the 2D case and to $n^{2/3}$ in the 3D case [15]. Therefore, it is worthwhile to have a search structure to find the mesh elements through which the contours pass. This will be more efficient than retrieving the contours of a single function value by inspecting all mesh elements.

There are basically two approaches to find the contours more efficiently. Firstly, one could store the 2D or 3D domain of the mesh in a hierarchical structure and associate the minimum and maximum occurring scalar values at the subdomains to prune the search. For example, octrees have been used this way for regular 3D meshes [25].

The second approach is to store the *scalar range*, also called *span*, of all the mesh elements in a search structure. Kd-trees [15], segment trees [3], and interval trees [5, 23] have been suggested as the search structure, leading to a contour retrieval time of $O(\sqrt{n} + k)$ or $O(\log n + k)$, where $n$ is the number of mesh elements and $k$ is the size of the output. A problem with this approach is that the search structure can be a serious storage overhead, even though an interval tree needs only linear storage. Still, one doesn't want to store a tree with a few hundred million intervals

*The research of the first and second authors was partially supported by the ESPRIT IV LTR Project No. 21957 (CGAL).The research of the third, fourth and fifth authors was partially supported by AFOSR grant F-49620-94-1-0080, NSF grant CCR 92-22467 and ONR grant N00014-94-1-0370



*Computational Geometry 97* Nice France

that would arise from regular 3D meshes. It is possible to reduce the storage requirements of the search structures by observing that a whole contour can be traced directly in the mesh if one mesh element through which the contour passes is known. Such a starting element of the mesh is also called a *seed.* Instead of storing the scalar range of all mesh elements, we need only store the scalar range of the seeds as intervals in the tree, and a pointer into the mesh. There are a few papers that take this approach [3, 13, 23]. The tracing algorithms to extract a contour have been developed before, and they require time linear in the size of the output [2, 12, 13].

The objective of this paper is to present new methods for seed set computation. Of a seed set, we require that any possible connected component of any contour in the mesh pass through at least one seed. Otherwise we could miss a (portion of a) contour. To construct such a small size seed set, we use a variation of the *contour tree*, a tree that captures the contour topology of the function represented by the mesh. It has been used before in image processing and GIS research [9, 10, 14, 21, 22]. Another name in use is the *topographic change tree*, and it is related to the *Reeb graph* used in Morse Theory [18, 19, 20, 22]. It can be computed in $O(n \log n)$ time for piecewise linear functions over 2D [6].

This paper includes the following results.

- We give a new, simple algorithm that constructs the contour tree. For 2D meshes with $n$ elements, it runs in $O(n \log n)$ time like a previous algorithm [6], but the new method is much simpler and needs less additional storage. For meshes with $n$ faces in $d$-space, it runs in $O(n^2)$ time. In typical cases, less than linear temporary storage is needed during the construction, which is important in practice. Also, the higher-dimensional algorithm requires subquadratic time in typical cases.
- We show that the contour tree is the appropriate structure to use when selecting seed sets. We give a polynomial time and storage algorithm for minimum size seed sets by using min-cost flow in a DAG [1].
- In practice one can use at most linear storage when computing seed sets. We give a simple approximation algorithm that requires $O(n \log^2 n)$ time and linear storage, and gives at most twice as many seeds as the minimum size seed set. In $d$-space, the algorithm takes $O(n^2)$ time.
- The approximation algorithm has been implementated, and we supply test results of various kind.

Previous methods to find small size seed sets didn't give any guarantee on their size [3, 13, 23].

## 2 Preliminaries on scalar functions and the contour tree

On a function $\mathcal{F}$ from $d$-space to the reals, the *criticalities* can be identified. These are the local maxima, the local minima, and the saddles (or passes). If we consider all contours of a specific function value, we have a collection of lower-dimensional regions in $d$-space (typically, $(d-1)$-dimensional surfaces of arbitrary topology). If we let the function value take on the values from $+\infty$ to $-\infty$, a number of things may happen to the contours. Contour shapes deform continuously, with changes in topology only when a criticality is met (i.e., its function value is passed). A new contour component starts to form whenever the function value is equivalent to a locally maximal value of $\mathcal{F}$. An existing contour component disappears whenever the function value is equivalent to a locally minimal value.

At saddle points, various different things can happen. It may be that two (or more) contour components adjoin, or one contour component splits into two (or more) components, or that a contour component gets a different topological structure (e.g., from a sphere to a torus in 3D). The changes that can occur have been documented well in texts on Morse theory or differential topology [11, 16]. They can be described by a structure called the contour tree, which we describe shortly.

For example, consider 2D triangular meshes with linear interpolation and note how the contour tree relates to such meshes. For simplicity, we assume that all vertices have a distinct function value. If we draw the contours of all critical vertices of the mesh, then we get a subdivision of the 2D domain into regions (see Figure 1). Since all saddle points must be vertices in our setting, one can show that every region between contours is bounded by exactly two contours. We let every contour in this subdivision correspond to a node in a graph, and two nodes are connected (from max to min) if there is a region bounded by their corresponding contours. This graph is a tree, which is easy to show [6, 23], and it is called the contour tree. All nodes in the tree have degree 1 (corresponding to local extrema), degree 2 (normal vertices), or at least 3 (saddles). In other words, every contour of a saddle vertex splits the domain into at least three regions. For each vertex in the triangulation, one can test locally whether it is a saddle. This is the case if and only if it has neighboring vertices around it that are higher, lower, higher, and lower, in cyclic order around it. If one would take the approach outlined above to construct the contour tree, $\Omega(n^2)$ time may be necessary in the worst case, because the total complexity of all contours through saddles may be quadratic [6]. An $O(n \log n)$ time divide-and-conquer algorithm exists, however [6].

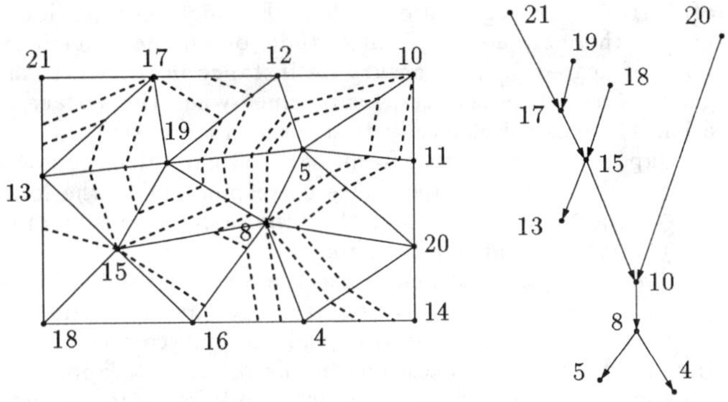


Figure 1: 2D triangular mesh with the contours of the saddles, and the contour tree.

In a general framework, we define the contour tree without assumptions on the type of mesh, interpolant, and dimension of the space over which function $\mathcal{F}$ is defined. The input data is assumed to be:

- a mesh $M$ of size $n$ embedded in $\mathbb{R}^d$;
- a continuous real-valued function $\mathcal{F}$ defined over all cells of $M$.

We define the contour tree $\mathcal{T}$ as follows.

- Take each maximal connected contour component which contains a criticality.
- These components correspond to the *supernodes* of $\mathcal{T}$ (the tree will be augmented later with additional nodes, hence we use the term supernodes here). Each supernode is labeled with the function value of its contour.
- For each region bounded by two contour components, we add a superarc between the corresponding supernodes in $\mathcal{T}$, oriented from the the higher to the lower function value.

The contour tree is well defined, because each region is bounded by two and only two contour components which correspond to supernodes. In fact, it is easy to see that the contour tree is a special case of the more general Reeb graph in the $(d+1)$-dimensional space obtained from the domain (the mesh) extended with the function image space [18, 19, 20, 22]. Furthermore, one can show that the contour tree is indeed a tree.

For 2D meshes, all criticalities correspond to supernodes of degree 1, or degree 3 or higher. For higher-dimensional meshes there are also criticalities that correspond to a supernode of degree 2. This occurs for instance in 3D when the genus of a surface changes, for instance when the surface of a ball changes topologically to a torus (Figure 2(b)).

Superarcs are directed from higher scalar values to lower scalar values. Thus, supernodes corresponding to the local maxima are the sources and the supernodes corresponding to the local minima are the sinks.

Since the type of the mesh and the function used to interpolate the associated discrete data may have some impact on the the seed cells selection and contour tree computation we define the weakest conditions required to apply the present approach. Note that such conditions are satisfied by the most common simplicial decompositions with linear interpolant and regular grids with multi-linear interpolant.

To be able to compute the contour tree, we make the following assumptions:

- Inside any face of any dimension of $M$, all criticalities and their function values can be determined.
- Inside any face of any dimension of $M$, the range $(min, max)$ of the function values taken inside the face can be determined.

We assume that in facets and edges of 2D meshes, the items above can be computed in $O(1)$ time. For vertices, we assume that the first item takes time linear in its degree. Similarly, in 3D meshes we assume that both items take $O(1)$ to compute in cells and on facets, and time linear in the degree on edges and at vertices.

In 3D, a saddle point $p$ is a point such that for any sufficiently small $\epsilon$-sphere around $p$, the contour of $p$'s value intersects the $\epsilon$-sphere in at least two components. Possible criticalities are shown in Figure 2. When sweeping the function value from $\infty$ to $-\infty$, they correspond to (a) two contours merging or splitting, but not containing the other, (b) an increment or decrement of the genus of one contour surface, and (c) two contours merging or splitting, and one containing the other. More cases can occur when a criticality causes several of these changes at once, or when the contour ends at the boundary of the mesh.

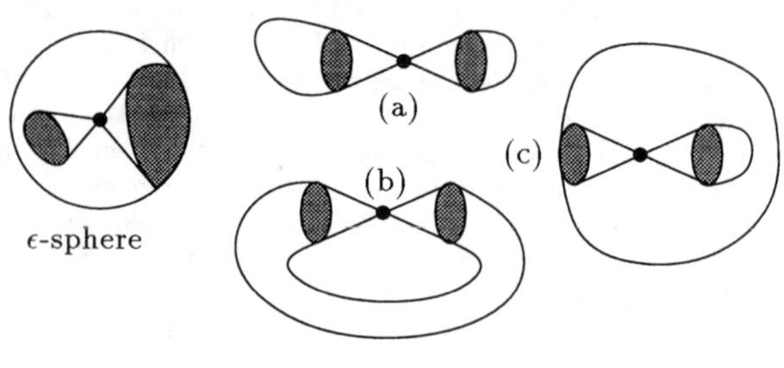


Figure 2: Citicalities in 3D.

## 3 Contour tree algorithms

In this section we assume for simplicity that the mesh $M$ is a simplicial decomposition with $n$ cells, and linear interpolation is used. As a consequence, all critical points are vertices of the mesh $M$. Instead of computing the contour tree as defined in the previous section, we compute a variation that includes nodes for all the vertices of $M$, also the non-critical ones. So supernodes correspond to critical vertices and normal nodes correspond to other vertices. Superarcs are now sequences of arcs, and connect two supernodes. It is easy to determine the contour tree with only the supernodes using the same technique. From now on, we call the contour tree with nodes for all vertices the contour tree $\mathcal{T}$. We'll need this augmented tree for seed selection in the next section.

The critical nodes of $\mathcal{T}$ that have in-degree 1 and out-degree greater than 1 are called *bifurcations*, and the nodes with in-degree greater than 1 and out-degree 1 are called *junctions*. We'll assume that all bifurcations and junctions have degree exactly 3, that is, out-degree 2 for bifurcations and in-degree 2 for junctions. This assumption can be removed, but it facilitates the following descriptions considerably. Basically, other critical nodes can be seen as clusters of critical nodes of degree 3. For example, a node with in-degree 2 and out-degree 2 can be treated as a junction and a bifurcation, with a directed arc from the junction to the bifurcation.

### 3.1 The general approach

To construct the contour tree $\mathcal{T}$ for a given mesh in $d$-space, we let the function value take on the values from $+\infty$ to $-\infty$ and we keep track of the contours for these values. In other words, we sweep the scalar value. For 2D meshes, one can image sweeping a polyhedral terrain embedded in 3D and moving down a horizontal plane. The sweep stops at certain event points: the vertices of the mesh. During the sweep, we keep track of the contour components in the mesh at the value of the sweep function, and the set of cells of the mesh that cross these components. The cells that contain a point with value equivalent to the present function value are called *active*. The tree $\mathcal{T}$ under construction during the sweep will be growing at the bottom at several places simultaneously. Each such part of $\mathcal{T}$ that is still growing corresponds to a unique contour component at the current sweep value. We group the cells into contour components by storing a pointer at each active cell in the mesh to the corresponding superarc in $\mathcal{T}$. The contours can only change structurally at the event points, and the possible changes are the following:

- At a local maximum of the mesh (more correctly: function), a new contour appears. This is reflected in $\mathcal{T}$ by creating a new supernode and a new arc incident to it. This arc is also the start of a new superarc, which will be represented. Each cell incident to the maximum becomes active, and we set their pointer to the new superarc of $\mathcal{T}$. At this stage of the algorithm, the new superarc has no lower node attached to it yet.
- At a local minimum of the mesh, a contour disappears; a new supernodenode of $\mathcal{T}$ is created, and the arc corresponding to the disappearing component at the current value of the sweep is attached to the new supernode. It is also the end of a superarc. The cells of the mesh incident to the local minimum are no longer active.
- At a non-critical vertex of the mesh, a new node of $\mathcal{T}$ is created, the arc corresponding to the contour containing the vertex is made incident to the node, and a new arc incident to the node is created (there is no new superarc). Some cells stop being active, while others incident to the vertex start being active. Their pointers are set to the current superarc of the contour. For the cells that remain active, nothing changes: their pointer keeps pointing to the same superarc.
- At a saddle of the mesh, there is some change in topology in the collection of contours. It may be that two or more contours merge into one, one contour splits into two or more, or one contour changes its topological structure. A combination of these is also possible in general. The first thing to do is to determine what type of saddle we are dealing with. This can be decided by traversing the whole contour on which the saddle lies.

  If two contours merge, a new supernode (junction) is created in $\mathcal{T}$ for the saddle, and the superarcs corresponding to the two merging contours are made incident to this supernode. Furthermore, a new arc and superarc are created for the contour that results from the merge. The new arc is attached to the new supernode. All cells that are active in the contour after the merge set their pointer to the new superarc in $\mathcal{T}$.

  If a contour splits, then similar actions are taken. If the saddle is because of a change in topology of one single contour, a new supernode is made for one existing superarc, and a new arc and superarc are created in $\mathcal{T}$. All active cells of the contour set their pointers to the new superarc.

For the sweep algorithm, we need an event queue and a status structure. The event queue can be implemented with a standard heap structure, such that insertions and extractions take logarithmic time per operation. The status structure is implicitly present in the mesh with the additional pointers from the cells to the superarcs in the contour tree.

**Theorem 1** *Let $M$ be a mesh in $d$-space with $n$ faces in total, representing a continuous, piecewise linear function over the mesh elements. The contour tree of $M$ can be constructed in $O(n^2)$ time and $O(n)$ storage.*

**Proof:** The algorithm clearly takes time $O(n \log n)$ for all heap operations. If the mesh is given in an adjacency structure, then the traversal of any contour takes time linear in the combinatorial complexity of the contour. Any saddle of the function is a vertex, and any contour can pass through any mesh cell only once. Therefore, the total time for traversal is $O(n^2)$ in the worst case, and the same amount of time is needed for setting the pointers of the active cells. ◇

The quadratic running time shown above is somewhat pessimistic, since it applies only when there are a linear number of saddles for which the contour through them has linear complexity. We can also state that the running time is $O(n \log n + \sum_{i=1}^{m} |C_i|)$, where the $m$ saddles lie on contours $C_1, \ldots, C_m$ with complexities $|C_1|, \ldots, |C_m|$. Also, besides the mesh (input) and the contour tree (output), the additional storage required can be made linear in the maximum number of active cells and the number of local maxima. So this is $O([\text{no. maxima}] + \max_{1 \leq i \leq m} |C_i|)$ additional storage. We can avoid the pointers from the mesh cells to the superarcs by copying the active part of the mesh into a separate, temporary structure. Then these pointers don't form a permanent storage overhead.

### 3.2 The two-dimensional case

In the 2D case, the time bound can be improved to $O(n \log n)$ time in the worst case by a few simple adaptations. First, a crucial observation: for 2D meshes representing continuous functions, all saddles correspond to nodes of degree at least 3 in $\mathcal{T}$. Hence, at any saddle two or more contours merge, or one contour splits into at least two contours, or both. The main idea is to implement a merge in time linear in the size of the *smaller* of the two contours, and similarly, to implement a split in time linear in the size of the *smaller resulting contour*. We also show how to maintain the pointers with the active cells efficiently.

In the structure, each active cell has a pointer to a *name* of a contour, and the name has a pointer to the corresponding superarc in $\mathcal{T}$. Maintaining pointers now comes down to changing names. We consider the active cells and names as a union-find like structure that allows the following operations:

- *Merge*: given two contours about to merge, combine them into a single one by renaming them to a single name,
- *Split*: given one contour about to split, split it into two separate contours by renaming one subset of the contour cells to a new name.
- *Find*: given one active cell, report the name of the contour it is in.

Like in the simplest union-find structure, a *Find* takes $O(1)$ time since we have a pointer to the name explicitly. A *Merge* is best implemented by changing the name of the cells in smaller contour to the name of the larger contour. Lets' say that contour $C_i$ and $C_j$ are about to merge. Determining which of them is the smallest takes $O(\min(|C_i|, |C_j|))$ time if we traverse both contours simultaneously. We alternatingly take one "step" in $C_i$ and one "step" in $C_j$. After a number of steps twice the size of the smaller contour, we have traversed the whole smaller contour. This technique is sometimes called a *tandem search*. To rename for a *Merge*, we traverse this smaller contour again and rename the cells in it, again taking $O(\min(|C_i|, |C_j|))$ time.

The *Split* operation is analogous: if a contour $C_k$ splits into $C_i$ and $C_j$, the name of $C_k$ is preserved for the largest of $C_i$ and $C_j$, and by tandem search starting at the saddle in two opposite directions we find out which of $C_i$ and $C_j$ will be the smaller one. This will take $O(\min(|C_i|, |C_j|))$ time. Note that we cannot keep track of the size in an integer for each contour instead of doing tandem search, because a *Split* cannot be supported efficiently.

**Theorem 2** *Let M be a mesh in 2D with n faces in total, representing a continuous, piecewise linear scalar function. The contour tree of this function can be computed in $O(n \log n)$ time and linear storage.*

Sketch of the proof for the claimed time bound:

- Determining for each vertex of what type it is (min, max, saddle, normal) takes $O(n)$ in total.
- The operations on the event queue take $O(n \log n)$ in total.
- Creating the nodes and arcs of $\mathcal{T}$, and setting the incidence relationships takes $O(n)$ time in total.
- When a cell becomes active for the first time, the name of the contour it belongs to is stored with it; this can be done on $O(1)$ time, and since there are $O(n)$ such events, it takes $O(n)$ time in total.
- At the saddles of the mesh, contours merge or split. Updating the names of the contours stored with the cells takes $O(\min(|C_i|, |C_j|))$, where $C_i$ and $C_j$ are the contours merging into one, or resulting from a split, respectively. It remains to show that summing these costs over all saddles yields a total of $O(n \log n)$ time.

We prove the bound on the summed cost for renaming by transforming $\mathcal{T}$ in two steps into another tree $\mathcal{T}'$ for which the construction is at least as time-expensive as for $\mathcal{T}$, and showing that the cost at the saddles in $\mathcal{T}'$ are $O(n \log n)$ in total.

Consider the cells to be additional *segments* in $\mathcal{T}$ as follows. Any cell becomes active at a vertex and stops being active at another vertex. These vertices are nodes in $\mathcal{T}$, and the cell is represented by a segment connecting these nodes. Note that any segment connects two nodes one of which is ancestor of the other. A segment can be seen as a shortcut of a directed path in $\mathcal{T}$, where it may pass over several nodes and supernodes.

The number of cells involved in a merge or split at a saddle is equivalent to the number segments that pass over the saddle node in $\mathcal{T}$; the size of the smallest set at this node determines the costs for processing the saddle (since we do tandem search).

The first transformation step is to *stretch* all segments; we simply assume that a segment starts at some source node that is an ancestor of the original start node, and ends at a sink that is a descendant of the original end node. It is easy to see that the number passing segments at all saddles cannot decrease by the stretch.

The second transformation step is to repeatedly *swap* superarcs, until no supernode arising from a split (bifurcation) is an ancestor of a supernode arising from a merge (junction). Swapping a superarc from a bifurcation to a junction is illustrated in Figure 4; the integers $a, b, c$ represent the number of segments passing over the superarcs shown. Before the swap, the time spent in the merge at $u$ and the split at $v$, is $O(\min(a,b) + \min(b,c))$ where $a, b, c$ denote the number of segments passing these superarcs. After the swap, this becomes $O(\min(a, b+c) + \min(a+b, c))$,

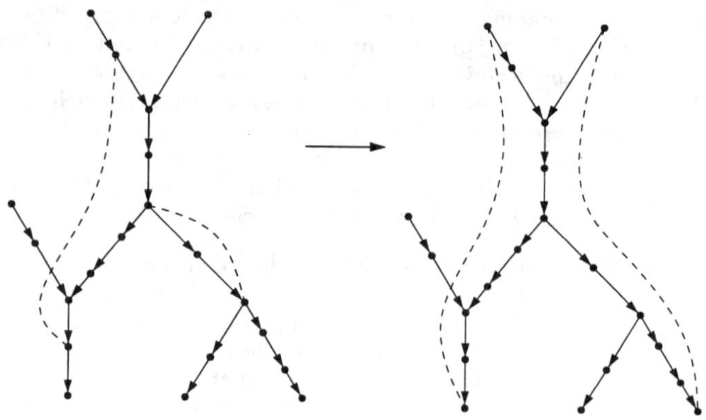

Figure 3: Stretching two segments (dashed) in $\mathcal{T}$.

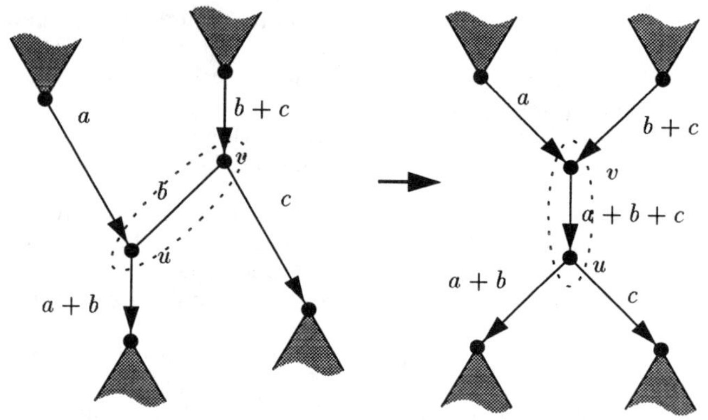


Figure 4: Swapping a superarc.

which is at least as much. No segment ends, because all of them were stretched.

It can easily be verified that by repeatedly swapping superarcs, $\mathcal{T}'$ can be derived such that no junction in $\mathcal{T}'$ has a bifurcation as an ancestor.

Now, every segment can pass $O(n)$ junctions and bifurcations, but no segment can be more than $O(\log n)$ times in the smallest set. Summing this over the $O(n)$ segments, this results in a total of $O(n \log n)$ time for all renaming of the cells.

## 4 Seed set selection

A seed set is a subset of the cells (triangles or tetrahedra) the mesh. A seed set is *complete* if every possible contour passes through at least one seed. Since we assume linear interpolation over the cells, the function values occurring in one cell is exactly the range between the lowest and the highest valued vertices. This range is simply a one-dimensional interval, and is also called the *span*. For each cell, its span can be located in the contour tree, where it is represented by two nodes. We add the span to the contour tree as a directed arc, which we call a *segment* to distinguish it from the arcs of the contour tree. Let $\mathcal{T}$ denote the contour tree, and let $\mathcal{G}$ denote the DAG that is the contour tree extended with the segments of all mesh elements. Observe that each segment is a shortcut in $\mathcal{T}$ for one or more arcs on a directed path. We say that the segment *passes*, or *covers*, these arcs of $\mathcal{T}$ (see Figure 4.1(a)). The small seed set problem now is the following graph problem: find a small subset of the segments such that each arc of $\mathcal{T}$ is passed by some segment of the subset.

In this section we give two methods to obtain complete seed sets. The first gives a seed set of minimum size, but it requires polynomial time and storage. The second method requires $O(n \log^2 n)$ time and linear storage in 2D, and gives a seed set at most twice the size of the minimum. In $d$-space, this approximation algorithm takes $O(n^2)$ time and linear storage.

### 4.1 Minimum seed sets in polynomial time and storage

We define a bipartite graph $\mathcal{B} = (U \cup V, A)$ as follows. The set $U$ of nodes corresponds to the set of segments of the mesh cells, and the set $V$ corresponds to the set of arcs of $\mathcal{T}$. An arc $(u, v) \in A$ if the segment corresponding to $u \in U$ passes the arc corresponding to $v \in V$ ((see Figure 4.1(b))). A complete seed set corresponds to a subset of $U$ that dominates all nodes in $V$, that is, each node in $V$ should have a neighbor in the chosen subset. The smallest cardinality subset $U' \subseteq U$ corresponds to a minimum seed set. The graph $\mathcal{B}$ can have a number of arcs quadratic in the size of $\mathcal{G}$.

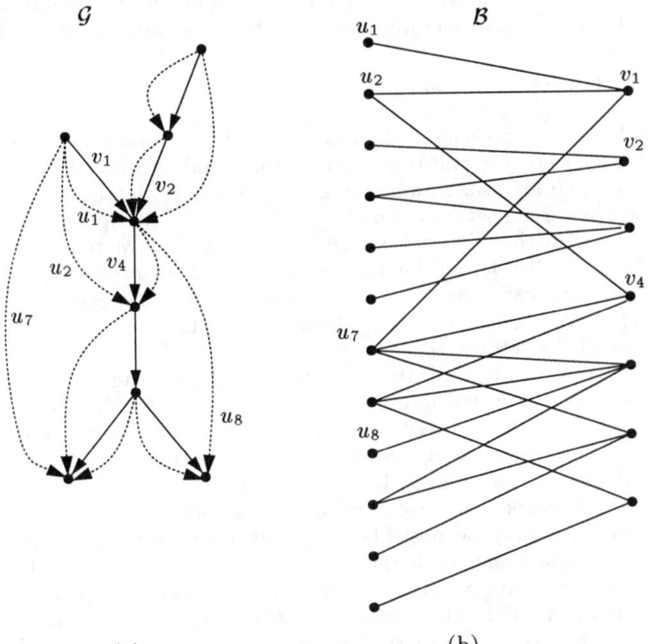


Figure 5: (a) A DAG $\mathcal{G}$ with the segments shown dashed. (b) The bipartite graph $\mathcal{B}$ assiciated with $\mathcal{G}$.

Observe that $\mathcal{B}$ is strongly chordal: every cycle of even length exceeding 5 has an odd chord. This is for the following reason. Since the nodes in $U$ represent segments in $\mathcal{G}$, and three segments in $\mathcal{G}$ cannot have pairwise overlap of an arc in $\mathcal{T}$ (or $\mathcal{G}$) without having a triple overlap as well, every cycle of even length exceeding 5 has a node in $U$ and in $V$ that are connected but aren't in the cycle (a chord).

We can augment $\mathcal{B}$ by turning $U$ into a clique; $\mathcal{B}$ remains strongly chordal. A minimum dominating clique on strongly chordal graphs can be computed in linear time, because the same result holds for the superclass of dually chordal graphs [4, 8]. This gives a minimum seed set.

**Theorem 3** *Given a contour tree $\mathcal{T}$ for a continuous function defined by a mesh with $n$ cells, an optimum size seed set can be computed in polynomial time and storage.*

### 4.2 Approximation of small seed sets in linear storage

The excessive time and storage requirements for optimal seed sets makes it nearly useless in practical applications. We therefore developed an approximation algorithm to compute a seed set using linear storage and $O(n \log^2 n)$ time in the 2D case. It yields a seed set of size no more than twice the size of the smallest seed set. In higher dimensions, the running time is $O(n^2)$.

Our approximation algorithm is a simple greedy method that operates quite similarly to the contour tree construction algorithm. We first construct the contour tree $\mathcal{T}$ as shown before. We add the segments to form the graph $\mathcal{G}$. Then we sweep again, now in the mesh and in the graph $\mathcal{G}$ simultaneously. During the sweep, greedy choices are made in the set of segments in $\mathcal{G}$; these segments correspond to cells that will be seeds. The greedily chosen segments are stored in a data structure $\mathcal{D}$ that allows for insertions of newly chosen segments and for queries. A query specifies a node $v$ of $\mathcal{T}$, and asks whether some chosen segment in $\mathcal{D}$ ends at $v$, and whether $v$ is passed by some chosen segment in $\mathcal{D}$. The segments are grouped by superarc they pass as usual, corresponding to cells intersected by the same contour. To make the greedy choice at any superarc, we store all active segments at a superarc sorted by function value of the lower node of $\mathcal{G}$ in a binary search tree. The following events can occur:

- Source: Choose the segment leaving it with the lowest value at the other end node, the greedy choice. Initialize a set of active segments for this superarc of $\mathcal{T}$.
- Normal node: Update the currently active segments.

  Query with the node in $\mathcal{D}$ to decide if any chosen segment passes it. If not, it must be the end node of a chosen segment. Choose a new segment greedily and add it to $\mathcal{D}$.
- Sink: Remove the group of active segments.
- Junction: Update the active segments, merge the two groups of active segments into one, and merge the corresponding two binary search trees.

  Query with the node as for a normal node.
- Bifurcation: Split the group of active segments into two, and update them.

  For each highest node $v$ below the bifurcation, test by querying in $\mathcal{D}$ if any chosen segment ends at it or passes it. If neither is true, choose a segment active at the bifurcation and into this superarc greedily. Add the chosen segments (zero or more) to $\mathcal{D}$.

**Lemma 1** *The greedy strategy has an approximation factor of 2.*

We give a sketch of the proof. Note that if $\mathcal{T}$ only contains junctions, then a greedy strategy is optimal. At a bifurcation, some segment leading into one superarc may be the chosen one, whereas the greedy choice for the other superarc would have been optimal. We can bound the number of segments chosen by the greedy strategy from above by assuming that for *both* superarcs the greedy choice was taken. Then we continue on both parts of the tree below these superarcs in the same way. We can bound the minimum number from below by assuming that only *one* segment was taken, and we continue the argument on the same two subtrees as in the greedy algorithm. So the subtrees are the same, and the greedy method chooses at most twice minimum.

We treat junctions and bifurcations efficiently as before, by tandem search in the contours on the mesh. We can traverse in time linear in the size of the smaller group to decide how to merge or split the segments efficiently. A merge or split of components $C_i$ and $C_j$ takes $O(\min(|C_i|, |C_j|) \log n)$ time, because it involves merging two binary search trees, or splitting one. So we need $O(n \log^2 n)$ time for manipulating the binary search trees.

To define the data structure $\mathcal{D}$, we first need a transformation of $\mathcal{T}$ and its segments. Give $\mathcal{T}$ some fixed, left-to-right order of the children and parents of each supernode. Then perform a left-to-right topological sort to number all nodes. Then perform a right-to-left topological sort to give each node a second number. The numbers are such that one node $u$ is an ancestor of another node $v$ if and only if the first number and the second number of $u$ is smaller than the corresponding numbers of $v$ (see Figure 6). These numbers can be seen as coordinates in the plane. Any segment from a start node $u$ to an end node $v$ in $\mathcal{T}$ transforms to a rectangle by using the numbers as coordinates. The lower left corner of the rectangle has the coordinates of $v$, and the upper right corner has the coordinates of $u$. The greedily chosen segments are stored as such rectangles in the data structure $\mathcal{D}$. A query with a node of $\mathcal{T}$ asks if the point is contained in some rectangle; this corresponds to some chosen segment passing the node. $\mathcal{D}$ is an interval tree with associated structures [17]. The main tree is defined on the intervals of the first coordinate. Each node stores a vertical split line and two segment trees, one for query points to the left of the split line and one for query points to the right. We store the rectangles that intersect the vertical split line in the left segment tree by storing the left envelope of the rectangles only (see Figure 6). This segment tree requires only linear storage, and queries and insertions take logarithmic time. For the whole interval tree with associated structures, the storage is still linear, the query time is $O(\log^2 n)$ and

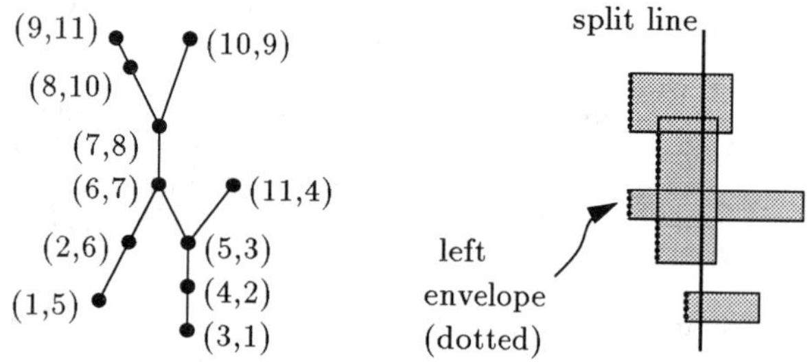


Figure 6: Left, the numbering of $\mathcal{T}$. Right, the left envelope of the rectangles intersecting a split line.

insertions take $O(\log n)$ time. The number of queries and insertions is linear in the number of nodes in the contour tree. Therefore, all operations on $\mathcal{D}$ also take $O(n \log^2 n)$ time together. More details are in the full paper.

**Theorem 4** *Let $M$ be a 2D mesh with $n$ cells representing a real function. A seed set of size at most twice the optimum can be determined in $O(n \log^2 n)$ time and linear storage. For a mesh in d-space, the running time is $O(n^2)$.*

## 5 Test results

In this section we present empirical results for generation of seed sets within bounds of optimality. Given in Table 1 are results collected from six datasets, both 2d and 3d. Presented are the total number of cells in the mesh, in addition to seed extraction statistics and comparisons to previously known efficient approximation methods. The methods presented here, shown to be within a factor of 2 of optimal, represent an improvement of 2 to 20 times over the method of [3], which had no claim on the seed set size. The presented storage statistics account only for the number of stored items, and not the size of each storage item (a constant). Note that the bounded seed set method presented here has, in general, greater storage demands, though storage remains sublinear in practice. Such tradeoffs are considered acceptable for the benefit of seed sets within guaranteed bounds of optimality. Sample images from the test function and LAMP datasets are given in the Appendix.

## 6 Further research

This paper presented the first methods to obtain seed sets for contour retrieval that are provably small in size. We gave a polynomial time and storage algorithm to determine the smallest seed set, and we also gave a factor two approximation algorithm that takes $O(n \log^2 n)$ time for functions over 2D and $O(n^2)$ time for functions over 3D. In typical cases, the worst case quadratic time bound seems too pessimistic. The algorithms make use of new methods to compute the so-called contour tree.

Test results indicate that seed sets resulting from the methods described improve on previous methods by greater than an order of magnitude in seed set size for some cases. Storage requirements in the seed set computation remain sublinear, as evidenced by the test results.

Our work can be extended in the following directions. Firstly, it may be possible to give worst case subquadratic time algorithms for higher-dimensional meshes. Secondly, it is important to study what properties an interpolation scheme on the mesh should have to allow for efficient contour tree construction and seed set selection. We are currently studying these extensions.

**Acknowledgements.** The authors thank Dieter Kratsch, Hans Bodlaender and Dirk Siersma for their helpful comments.

Sample images from the some of the test data is given in the Appendix. The heart data appears courtesy Tsuyoshi Yamamoto and Hiroyuki Fukuda of Hokkaido University. The LAMP data is courtesy the Space Science and Engineering Center at the University of Wisconsin. The bullet data is courtesy Lawrence Livermore National Laboratory. The terrain data is a triangulated irregular network from USGS DEM data.

| Data | total cells | #seeds | storage | time (s) | #seeds by method of [3] | storage req of [3] | time (s) |
|---|---|---|---|---|---|---|---|
| | | | | | | | |
| Heart | 65025 | 5631 | 30651 | 32.68 | 12214 | 255 | 0.87 |
| Function | 3969 | 80 | 664 | 1.23 | 230 | 63 | 0.15 |
| Bullet | 20000 | 8 | 964 | 2.74 | 47 | 1000 | 0.30 |
| LAMP 3d | 19040 | 172 | 9267 | 6.82 | 576 | 1360 | 0.33 |
| | | | | | | | |
| LAMP 2d | 2720 | 73 | 473 | 0.69 | n/a | n/a | n/a |
| Terrain | 95911 | 188 | 2078 | 13.67 | n/a | n/a | n/a |

Table 1: Seed cell statistics for regular (top) and irregular (bottom) data

## References


[1] R. K. Ahuja, T. L. Magnanti, and J. B. Orlin. *Network Flows: Theory, Algorithms, and Applications.* Prentice Hall, Englewood Cliffs, NJ, 1993.

[2] E. Artzy, G. Frieder, and G. T. Herman. The theory, design, implementation, and evaluation of 3-d surface detection algorithms. *Comput. Graph. Image Process.*, 15:1–24, 1981.

[3] C.L. Bajaj, V. Pascucci, and D.R. Schikore. Fast isocontouring for improved interactivity. In *Proc. 1996 IEEE Symposium on Volume Visualization*, pages 39-46, San Francisco, Oct 7-8, 1996.

[4] A. Brandstädt, V. D. Chepoi, and F. F. Dragan. The algorithmic use of hypertree structure and maximum neighbourhood orderings. In *20th International Workshop "Graph-Theoretic Concepts in Computer Science"(WG'94)*, volume 903 of *Lecture Notes in Computer Science*, pages 65–80. Springer-Verlag, Berlin, 1995.

[5] P. Cignoni, C. Montani, E. Puppo, and R. Scopigno. Optimal isosurface extraction from irregular volume data. In *Proc. IEEE Volume Visualization*, 1996.

[6] Mark de Berg and Marc van Kreveld. Trekking in the alps without freezing or getting tired. In *1st Annual European Symposium on Algorithms (ESA '93)*, volume 726 of *Lecture Notes in Computer Science*, pages 121–132. Springer-Verlag, 1993.

[7] F. F. Dragan and A. Brandstädt. Dominating cliques in graphs with hypertree structure. In *International Symposium on Theoretical Aspects of Computer Science (STACS'94)*, Lecture Notes in Computer Science number 775 , pages 735–746. Springer-Verlag, 1994.

[8] F. F. Dragan and A. Brandstädt. Dominating cliques in graphs with hypertree structure. In *International Symposium on Theoretical Aspects of Computer Science (STACS'94)*, volume 775 of *Lecture Notes in Computer Science*, pages 735–746, Berlin, 1994. Springer-Verlag.

[9] H. Freeman and S.P. Morse. On searching a contour map for a given terrain profile. *Journal of the Franklin Institute*, 248:1–25, 1967.

[10] C. Gold and S. Cormack. Spatially ordered networks and topographic reconstructions. In *Proc. 2nd Int. Sympos. Spatial Data Handling*, pages 74–85, 1986.

[11] M.W. Hirsch. *Differential Topology*, volume 33 of *Graduate Texts in Mathematics.* Springer-Verlag, Berlin, 1976.

[12] C.T. Howie and E.H. Blake. The mesh propagation algorithm for isosurface construction. *Computer Graphics Forum*, 13:65–74, 1994.

[13] T. Itoh and K. Koyamada. Automatic isosurface propagation using an extrema graph and sorted boundary cell lists. *IEEE Trans. on Visualization and Computer Graphics*, 1:319–327, 1995.

[14] I.S. Kweon and T. Kanade. Extracting topographic terrain features from elevation maps. *CVGIP: Image Understanding*, 59:171–182, 1994.

[15] Y. Livnat, H.-W. Shen, and C.R. Johnson. A near optimal isosurface extraction algorithm using the span space. *IEEE Transactions on Visualization and Computer Graphics*, 2:73–84, 1996.

[16] J. Milnor. *Morse Theory*, volume 51 of *Annals of Mathematics Studies.* Princeton University Press, 1963.

[17] F. P. Preparata and M. I. Shamos. *Computational Geometry: An Introduction.* Springer-Verlag, New York, NY, 1985.

[18] G. Reeb. Sur les points singuliers d'une forme de pfaff completement integrable ou d'une fonction numerique. *Comptes Rendus Acad. Sciences Paris*, 222:847–849, 1946.

[19] Y. Shinagawa and T.L. Kunii. Constructing a Reeb graph automatically from cross sections. *IEEE Computer Graphics and Applications*, 11:44–51, November 1991.

[20] Y. Shinagawa, T.L. Kunii, and Y.L. Kergosien. Surface coding based on morse theory. *IEEE Computer Graphics and Applications*, 11:66–78, September 1991.

[21] J.K. Sircar and J.A. Cerbrian. Application of image processing techniques to the automated labelling of raster digitized contours. In *Proc. 2nd Int. Symp. on Spatial Data Handling*, pages 171–184, 1986.

[22] S. Takahashi, T. Ikeda, Y. Shinagawa, T.L. Kunii, and M. Ueda. Algorithms for extracting correct critical points and constructing topological graphs from discrete geographical elevation data. *Eurographics '95*, 14:C–181–C–192, 1995.

[23] M. van Kreveld. Efficient methods for isoline extraction from a TIN. *Int. J. of GIS*, 10:523–540, 1996.

[24] David F. Watson. *Contouring: A Guide to the Analysis and Display of Spatial Data.* Pergamon, 1992.

[25] J. Wilhelms and A. Van Gelder. Octrees for faster isosurface generation. *ACM Transactions on Graphics*, 11:201–227, 1992.


## A Appendix

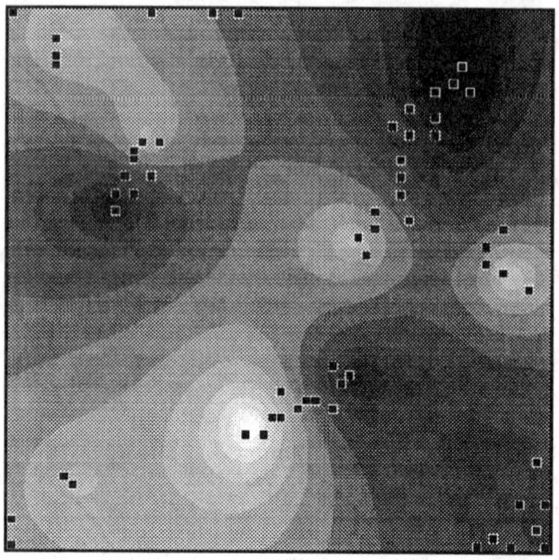

(a)

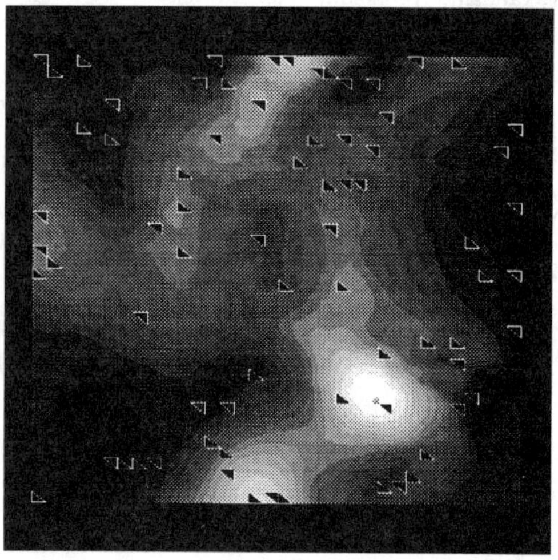

(b)

Figure 7: Seed sets from 2d scalar data
(a) seed set from a synthetic smooth function
(b) seed set for a slice of wind speed data (LAMP)

# Separating an Object from its Cast

Hee-Kap Ahn[1] Mark de Berg[2] Prosenjit Bose[3] Siu-Wing Cheng[4]
Dan Halperin[5] Jiří Matoušek[6] Otfried Schwarzkopf[1]

## Abstract

In casting, liquid is poured into a cast that has a cavity with the shape of the object to be manufactured. The liquid then hardens, after which the cast is removed. We consider the case where the cast consists of two parts and address the following problems: (1) Given a cast for an object and a direction $\vec{d}$, can the cast be partitioned into two parts such that the parts can be removed in directions $\vec{d}$ and $-\vec{d}$, respectively, without colliding with the object or the other cast part? (2) How to find a direction $\vec{d}$ such that the above cast partitioning can be done? We give necessary and sufficient conditions for both problems, as well as algorithms to decide them for polyhedral objects. We also present a surprising result on the case where the cast parts need not be removed in opposite directions.

[1]Department of Computer Science, Pohang University of Science and Technology, Hyoja-Dong, Pohang 790-784, South Korea. Supported by the non-directed research fund of the Korean Ministry of Education, and partially by Postech Grant 96F004.

[2]Department of Computer Science, Utrecht University, P.O.Box 80.089, 3508 TB Utrecht, the Netherlands. Supported by the Netherlands' Organization for Scientific Research (N.W.O.) and by ESPRIT Basic Research Action No. 7141 (project ALCOM II: *Algorithms and Complexity*).

[3]Département de Mathématiques et d'Informatique, Université du Québec à Trois-Rivières, C.P. 500, Trois-Rivières, Québec G9A 5H7, Canada. Research supported by NSERC Grant OGP0183877 and a FIR Grant.

[4]Department of Computer Science, HKUST, Clear Water Bay, Hong Kong. Supported partly by the Research Grant Council CERG HKUST 650/93E and HKUST 650/95E.

[5]Department of Computer Science, Tel Aviv University, Tel Aviv 69978, Israel. Research supported by an Alon Fellowship, and by ESPRIT IV LTR Project No. 21957 (CGAL).

[6]Department of Applied Mathematics, Charles University, Malostranské nám. 25, 118 00 Praha 1, Czech Republic. Supported by Czech Republic Grant GAČR 0194 and by Charles University Grants No. 193,194.


*Computational Geometry 97* Nice France


## 1 Introduction

**Background.** To facilitate manufacturing and assembly, it is desirable to design parts in such a way that manufacturing and assembly can be performed easily, and therefore cheaply (*design for assembly, design for manufacturing*). To make this feasible, computer-aided design systems have to be augmented with a component that verifies *on-line* that a part being designed can actually be manufactured using the intended techniques. Such algorithms are required to work purely on the basis of a CAD model of the part.

Algorithms have been proposed for a number of manufacturing processes, such as injection molding [4, 8, 21], NC-machining [11], and stereolithography [1]. For an overview of this application area, the reader is referred to the survey by Bose and Toussaint [2, 5].

In assembly, the emphasis is on planning tasks so as to put parts together to form the final product. Interesting geometric problems arise in almost every step of automatic assembly planning. Assembly sequencing [24], part orienting [9], fixturing [18], and welding [17], are just a few of many examples.

Although these manufacturing and assembly processes may be totally different from one another, some of them raise similar geometric problems that can be generally termed *separability* problems [22]. This is the case in the manufacturing process that we study in this paper: the process of casting [7, 23].

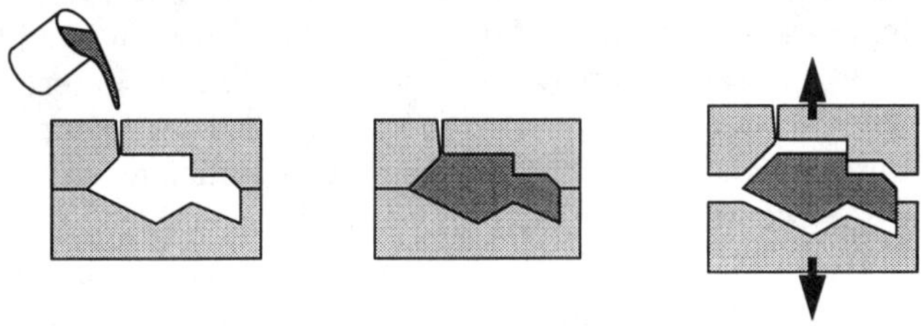

Figure 1. The casting process.

The casting process consists of filling the cavity between two cast parts with some molten liquid. After the liquid has solidified, the cast parts are removed and one is left with an object whose shape is that of the cavity. Figure 1 illustrates the process on a 2-dimensional example. The key property necessary for casting is that the cast parts can be removed from the object without destroying either cast parts or the object. This ensures that the given object can be mass produced by re-using the same cast parts. Observe that the cast parts in the example are removed in opposite directions. This need not always be possible; sometimes it may be necessary to remove them in non-opposite directions.

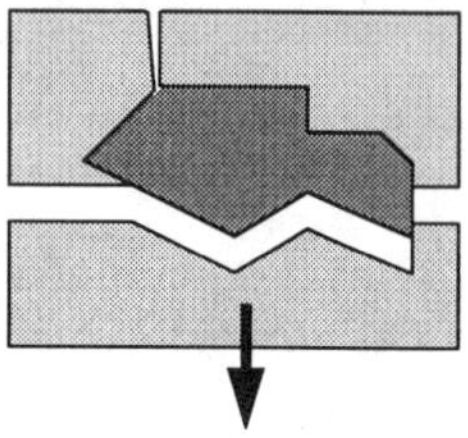

Figure 2. The top part of the cast is stuck.

The casting process may fail in the removal of the cast parts: if the cast is not designed properly, then one or more of the cast parts may be stuck during the removal phase, as in Figure 2. The problem we address here concerns this aspect: Given a 3-dimensional object, is there a cast for it whose parts can be removed after the liquid has solidified? An object for which this is the case is called *castable*.

We note that more complicated objects may be constructed through the use of cores and inserts [7, 20, 23]. Cores and inserts are appendages to the cast parts. These appendages are removed after the liquid has hardened and before the cast parts themselves are removed. They allow the possibility of building more complicated objects, in the sense that if the cores and inserts were not removed, then the cast could not be removed without breaking. However, their use slows down the manufacturing process and makes it more costly. We do not study the extra possibilities of cores and inserts.

**Related work.** The 2-dimensional version of our problem has been studied by Rappaport and Rosenbloom [21]. They presented an $O(n)$ time algorithm to determine whether a simple $n$-vertex polygon can be decomposed into two monotone chains, which is a sufficient and necessary condition for the polygon to be castable.

Hui and Tan [13] gave a heuristic approach to the 3-dimensional problem. Some candidate parting directions are heuristically chosen and ordered, and for each direction in order, a sampled set of boundary points are checked if they are obscured by other parts of the product, If not, that parting direction is assumed to be feasible.

Kwong [14] gave an algorithm to determine the feasibility of a given parting direction. He reduced the problem to the hidden surface removal problem in computer graphics by observing that if all the facets can be completely illuminated from the parting direction and its opposite, then the parting direction is feasible.

Chen et al. [6] first computed the convex hull of a polyhedral object, and then obtained the *pockets* of the object by subtracting it from its convex hull. They observed that if all pockets are completely visible in either the parting direction or its opposite, then the parting direction is feasible, and gave an algorithm that returns the parting direction that maximizes the number of completely visible pockets. However, as the converse of the above observation is not necessarily true, their algorithm is not complete and may not find a good parting direction even though it exists.

Hui [12] built on the previous paper [6] and obtained exponential time incomplete algorithms that also take into account cores and inserts.

Finally, Bose et al. [3] considered a special model of casting, the *sand casting model*, where two parts of the cast must be separated by a plane. Given a simple polyhedron with $n$ vertices, they presented an $O(n^{3/2+\varepsilon})$ time and space algorithm[1] (based on partition trees [16]), for any fixed $\varepsilon > 0$, as well as an $O(n^2)$ time and $O(n)$ space algorithm (based on linear programming) that decides whether there is a cast whose constituent parts can be removed in opposite directions. When arbitrary directions are allowed, their partition tree based algorithm still takes $O(n^{3/2+\varepsilon})$ time and space, however their linear programming based algorithm increases slightly to $O(n^2 \log n)$ time and $O(n)$ space. Note that even convex polyhedra are not castable in general if the two cast parts must be separated by a plane [3].

**Summary of results.** In this paper we give a complete characterization of castability for general objects, and obtain an algorithm to verify this condition for polyhedral objects. We also present an algorithm that reports all *combinatorially distinct directions* in which there is a good cast (we postpone a formal definition of "distinct directions" to Section 4). We do not assume any special separability of the two cast parts, and, generalizing previous work, allow objects of arbitrary genus. The running time of our algorithm for determining the legal parting directions is $O(n^4)$, where $n$ is the combinatorial complexity of the polyhedral model. We show that there exist polyhedra for which there are $\Omega(n^4)$ combinatorially distinct directions in which there is a good cast. This implies that our algorithm is optimal in the worst case if we want to report all such directions.

In sand casting, the problem of finding a good cast reduces to the problem of finding a plane subdividing the

[1] Bose et al. remarked that this can be improved to $O(n^{4/3+\varepsilon})$.

boundary of the part into two terrains. Extending the boundary subdivision into two cast parts is trivial: one simply splits the cast with this plane. In our more general setting, the object boundary must also be split into terrains, but extension of the boundary subdivision to a legal cast is no longer straightforward. In fact, we show that that this extension is not always possible for the case of arbitrary removal directions: there are polyhedra whose boundary can be decomposed into two terrains but that are not castable. This implies that Hui and Tan's [13] and Kwong's [14] approaches are not sufficient for finding a legal cast with arbitrary removal directions. (For opposite cast removal we prove that such an extension is always possible.) It has to be noted, though, that we are not aware of any industrial setting in which non-opposite directions are being used for cast removal, and we therefore do not study this case in more detail.

**Outline.** In Section 2, we introduces more terminology and certain assumptions on the objects. In Section 3, we present criteria for testing whether a given object admits opposite cast removal in a given direction $\vec{d}$, and give an algorithm to test this condition for polyhedral objects. In Section 4, we give an algorithm to compute all possible directions of opposite cast removal for a given polyhedron. We also construct an example showing that our algorithm is optimal in the worst case. Experimental results are presented and discussed in Section 5. The case of non-opposite cast removal is briefly addressed in section 6. We conclude in Section 7 with some open problems.

## 2 Preliminaries: Assumptions and Terminology

Throughout this paper, $\mathcal{P}$ denotes a polyhedron which is a body, not necessarily convex, bounded by a piecewise linear surface. The union of vertices, edges, and facets on this piecewise linear surface forms the boundary of $\mathcal{P}$ denoted by $\partial\mathcal{P}$. We assume that $\partial\mathcal{P}$ is a connected 2-manifold. Each facet of $\mathcal{P}$ is a connected planar polygon which is allowed to have polygonal holes. We assume that two adjacent facets cannot be coplanar, but allow coplanar non-adjacent facets. We also assume that $\mathcal{P}$ is *simple*, namely there is no pair of non-adjacent facets sharing a point. Our assumptions imply that $\mathcal{P}$ may contain tunnels, but no voids—a polyhedron with a void is not castable anyway. For a more thorough description of polyhedra and some of their properties, the reader is referred to the book by Preparata and Shamos [19].

For generality—and since we are aware that many industrial parts are not polyhedral—we characterize castability for more general objects. We assume that the bounding surface of an object $\mathcal{B}$ consists of a finite number of patches of algebraic surfaces of bounded degree, and the patches meet along algebraic curve segments of bounded degree. $\mathcal{B}$ has to be topologically equivalent to a polyhedron $\mathcal{P}$ as defined above. We call an object $\mathcal{B}$ *$\vec{d}$-monotone* if every line with direction $\vec{d}$ intersects the interior of $\mathcal{B}$ in at most one connected component.

We will assume that the outer shape of a cast equals the boundary of an axis-parallel box $B$. A cast for an object $\mathcal{B}$, denoted by $\mathcal{C}$, consists of a *red part* $\mathcal{C}_r$ and a *blue part* $\mathcal{C}_b$ such that each is a subset of $B$, the interior of $\mathcal{C}_b$ and $\mathcal{C}_r$ do not intersect, and $B \setminus (\mathcal{C}_r \cup \mathcal{C}_b)$ equals $\mathcal{B}$.

In our discussion we will refer to the subdivision of the unit sphere $\mathcal{S}^2$ by a collection $\mathcal{Q}$ of great circles and arcs of great circles. We call this subdivision the *arrangement* of $\mathcal{Q}$ on $\mathcal{S}^2$ and denote it by $\mathcal{A}(\mathcal{Q})$. This arrangement consists of *faces* of dimensions 0, 1 and 2 called vertices, edges and cells respectively. A vertex of $\mathcal{A}(\mathcal{Q})$ is either an intersection point of two curves in $\mathcal{Q}$ or an endpoint of an arc in $\mathcal{Q}$. An edge of $\mathcal{A}(\mathcal{Q})$ is a maximal connected component of a curve in $\mathcal{Q}$ not intersecting any other curve in $\mathcal{Q}$. A cell of $\mathcal{A}(\mathcal{Q})$ is a maximal connected region of $\mathcal{S}^2$ not intersecting any curve in $\mathcal{Q}$.

The *combinatorial complexity* of the arrangement $\mathcal{A}(\mathcal{Q})$ is the total number of faces (of all dimensions) in the arrangement. If $Q$ consists of $q$ curves, each being a great circle or an arc of a great circle, then the complexity of the arrangement is $O(q^2)$ and it can be $\Omega(q^2)$. We say that the curves in $\mathcal{Q}$ are in *general position* if no three curves in $\mathcal{Q}$ meet at a single point.

A *polyhedral terrain* is the graph of a (possibly partially defined) continuous piecewise linear function with domain $\mathbb{R}^2$ and range $\mathbb{R}$. A typical example is a polyhedral surface with the property that every vertical line intersects it in at most one point.

## 3 Opposite Cast Removal—Testing a Direction

In this section we present a criterion for testing whether a given object $\mathcal{B}$ admits opposite cast removal in a given direction $\vec{d}$. In other words, we give a way to determine whether a cast $\mathcal{C} = (\mathcal{C}_r, \mathcal{C}_b)$ can be built for $\mathcal{B}$ such that $\mathcal{C}_r$ and $\mathcal{C}_b$ can be translated to infinity in directions $\vec{d}$ and $-\vec{d}$ respectively so that the interior of $\mathcal{C}_r$, $\mathcal{C}_b$, and $\mathcal{B}$ do not intersect during the translation. If this is the case, we say that $\mathcal{C}_r$ and $\mathcal{C}_b$ can be removed *without collision* and $\mathcal{B}$ *can be cast* in direction $\vec{d}$. The order of removing the cast parts is irrelevant in this situation.

Throughout this section we assume that $\vec{d}$ is the vertical direction (the positive $z$ direction). We say that $\mathcal{B}$ is castable if it is castable in the vertical direction. The red cast part has to be translated upward, the blue cast part downward.

**Lemma 3.1** *An object $\mathcal{B}$ is castable if and only if it is vertically monotone.*

**Proof:** Assume that $\mathcal{B}$ is castable. Suppose that a vertical line intersects the interior of $\mathcal{B}$ in more than one connected component. Between these connected components, there must be a point $p$ that belongs to either $\mathcal{C}_r$ or $\mathcal{C}_b$. However, $p$ can be translated neither upward nor downward without intersecting the interior of $\mathcal{B}$, a contradiction.

Assume now that $\mathcal{B}$ is vertically monotone, and recall that the cast $\mathcal{C}$ is made from a rectangular axis-parallel box $B$. Let now $\mathcal{B}^*$ be the solid obtained by sweeping $\mathcal{B}$ upward to infinity. By the assumption, $\mathcal{B}^* \setminus \mathcal{B}$ can be translated upward to infinity without intersecting the interior of $\mathcal{B}$. We can therefore set $\mathcal{C}_r := B \cap (\mathcal{B}^* \setminus \mathcal{B})$ and $\mathcal{C}_b := B \setminus (\mathcal{B} \cup \mathcal{C}_r)$. Clearly, the red cast part $\mathcal{C}_r$ can be removed upward without collision. Let now $q$ be a point in $\mathcal{C}_b$, and consider the vertical ray emanating downward from $q$. This ray cannot contain a point of $\mathcal{C}_r$, and it cannot intersect the interior of $\mathcal{B}$. Therefore, $\mathcal{C}_b$ can be removed downward without collision. ◻

Let's turn our attention to the special case of a polyhedral object $\mathcal{P}$. The red and blue cast parts induce a partition of $\partial\mathcal{P}$ into a red part and a blue part. We call a facet of $\mathcal{P}$ an *up-facet* if its outward normal points upward, and a *down-facet* if its outward normal points downward. Vertical facets are neither up- nor down-facets. If $\mathcal{P}$ is castable, then clearly every up-facet must be completely red, while every down-facet must be blue. The object shown in Figures 3 and 4 illustrates that vertical facets sometimes need to be assigned both colors. The facet *abcd* is coplanar with the vertical facets incident to the edges $uv$ and $rs$. The construction in the proof clearly colors all up-facets red and all down-facets blue. The vertical facets are all colored blue as well, except for the facet *abcd* which is subdivided by *pq* with *abpq* colored red and *pqcd* colored blue. In fact, to obtain a cast for this object in the vertical direction, the facet *abcd* must be assigned two different colors.

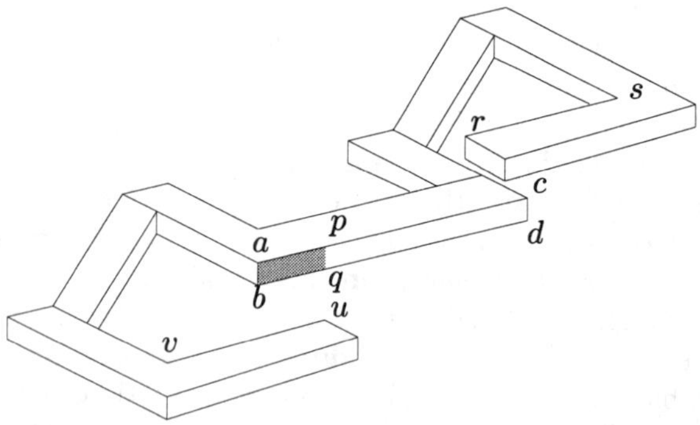


Figure 3. The facet *abcd* is coplanar with the vertical facets incident to $uv$ and $rs$. The line segment *pq* is the intersection of *abcd* with a ray shooting vertically upward from $u$.

The construction used in Lemma 3.1 does not result in practically useful casts, since it generates vertical walls between the red and blue cast parts. Sometimes these are unavoidable (as for the object in Figure 3). However, we next present an alternative construction that does not create any vertical walls if they are not necessary.

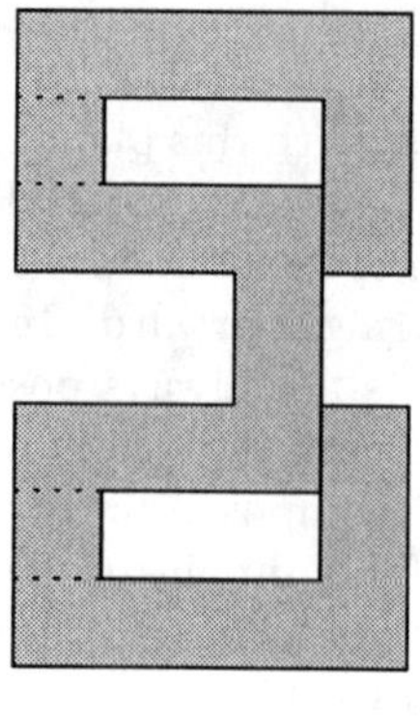

Figure 4. Projection onto the $xy$-plane of the object in Figure 3, with the interior shaded.

**Theorem 3.2** *We can construct a vertical cast for a vertically monotone polyhedron $\mathcal{P}$ with $n$ vertices in time $O(n \log n)$. The two cast parts do not meet along vertical facets, if no vertical line avoiding the interior of $\mathcal{P}$ touches two non-adjacent facets of $\mathcal{P}$.*

**Proof:** Let $h$ be an arbitrary plane that is parallel to the $xy$-plane and cuts the box $B$ into two halves, and let $R$ be the rectangle $R := h \cap B$. We project $\mathcal{P}$ onto $h$ and obtain a polygon $\overline{\mathcal{P}}$ with holes as in Figure 4. This can be done in time $O(n \log n)$ using a simple plane sweep algorithm.

Since $\mathcal{P}$ is vertically monotone, every point on $\partial\overline{\mathcal{P}}$ must be the projection of a point on $\partial\mathcal{P}$. Indeed, with each vertex $\overline{v}$ of $\overline{\mathcal{P}}$, we associate a vertex $v$ of $\mathcal{P}$ that projects onto $\overline{v}$. If there is more than one such vertex, we choose the topmost one.

It follows that the complexity of $\overline{\mathcal{P}}$ is $O(n)$ and hence $R \setminus \overline{\mathcal{P}}$ can be triangulated in $O(n)$ time, as shown in Figure 5. Every triangle is now "lifted" into three dimensional space by replacing every vertex $\overline{v}$ by its associated vertex $v$. We obtain a triangulated terrain $\sigma$ inside the box $B$.

Let now $\mathcal{U}$ be the union of all up-facets of $\mathcal{P}$ and $\sigma$, and let $\mathcal{U}^*$ be the solid obtained by sweeping $\mathcal{U}$ upward. We can now let $\mathcal{C}_r := B \cap \mathcal{U}^*$ be the red cast part. The blue cast part is defined as $\mathcal{C}_b := B \setminus (\mathcal{P} \cup \mathcal{C}_r)$.

By its definition, $\mathcal{C}_r$ can be removed upward without collision. Consider a point $q$ in $\mathcal{C}_b$. Clearly, a vertical downward ray emanating from $q$ cannot intersect $\mathcal{P}$. It also cannot intersect $\sigma$, and hence $\mathcal{C}_b$ can be removed downward without collision.

It remains to argue that when no vertical line avoiding the interior of $\mathcal{P}$ touches two non-adjacent facets of $\mathcal{P}$, then the separation of $\mathcal{C}_r$ and $\mathcal{C}_b$ has no vertical facets.

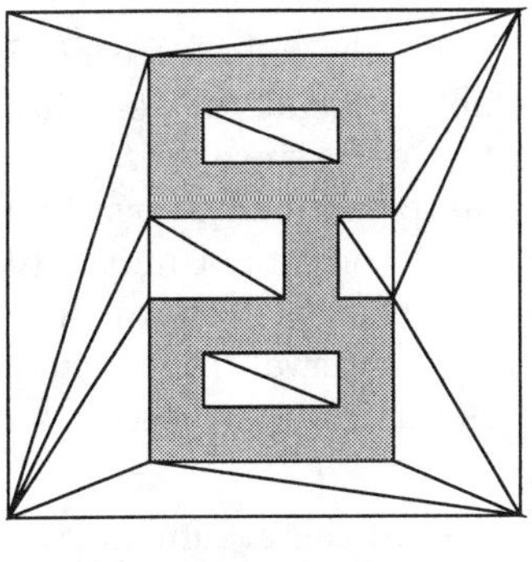

Figure 5.

This follows from the fact that in that case $\mathcal{U}$ is a terrain (without vertical facets) that splits $B$ into two parts. ◻

To check a polyhedron $\mathcal{P}$ for castability, we can use Lemma 3.1 and the following simple observation.

**Observation 3.3** *A polyhedron $\mathcal{P}$ is vertically monotone if and only if the union of open up-facets forms a terrain.*

This observation is essentially the basis for Kwong's algorithm [14], see the Introduction. We prefer a somewhat different approach, that decides monotonicity by looking at the silhouette edges of the polyhedron only. This is the key behind the efficient algorithm presented in the next section where we go over all possible directions and check in which of them the given polyhedron is castable.

We first give a precise definition of the silhouette of an object. This turns out to be somewhat tricky if the object can have vertical facets.

Let $\mathcal{B}$ be an object, and consider a vertical line $\ell$. $\ell$ intersects $\partial\mathcal{B}$ in a number of maximal closed intervals. These intervals separate open intervals lying either completely inside or outside $\mathcal{B}$. A boundary interval that is surrounded on both sides by intervals outside $\mathcal{B}$ is called a *convex silhouette interval.* A boundary interval that is surrounded on both sides by intervals in the interior of $\mathcal{B}$ is called a *reflex silhouette interval.* Over all vertical lines, the union of all convex silhouette intervals forms the *convex silhouette* and the union of all reflex silhouette intervals forms the *reflex silhouette.* The union of both is called the *silhouette* of $\mathcal{B}$.

We visualize these definitions for the case of a polyhedron $\mathcal{P}$. If $\mathcal{P}$ has no vertical facets, then the silhouette consists exactly of the *silhouette edges* of $\mathcal{P}$, namely the edges $e$ where the two facets incident to $e$ lie on one side of the unique vertical plane through $e$. A silhouette edge is convex if the dihedral angle between the two facets in the interior of $\mathcal{P}$ is smaller than $\pi$, otherwise it is reflex.

If $\mathcal{P}$ has vertical facets, then the silhouette is no longer 1-dimensional. For instance, the silhouette of the object in Figure 3 consists of *all* vertical facets of the polyhedron. Figure 6 shows another object with part of its quite complicated silhouette shown shaded. Note that the segments $a$ and $b$ are also part of the silhouette.

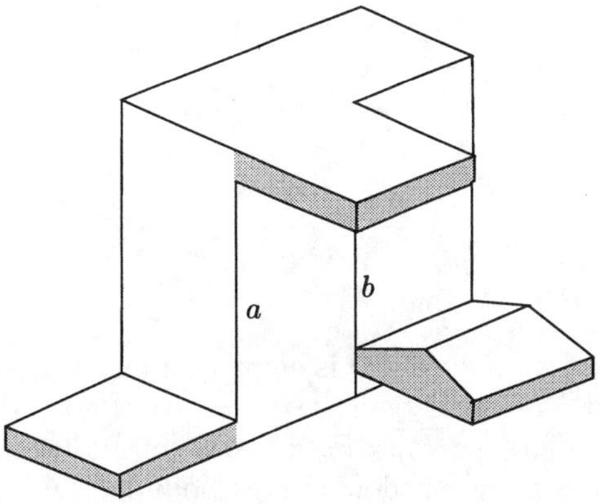


Figure 6.

The following lemma is straightforward:

**Lemma 3.4** *If the reflex silhouette of an object $\mathcal{B}$ is not empty, then $\mathcal{B}$ is not vertically monotone.*

From now on, we will therefore only consider objects whose reflex silhouette is empty. The convex silhouette (the silhouette, for short) of such an object consists of a finite number of disjoint curves or "bands" on $\partial\mathcal{B}$, the *silhouette curves.* Let $\pi(p)$ denote the vertical projection of $p$ on the $xy$-plane. The silhouette curves project on closed curves in the plane. We call these projected curves *shadow curves.* The key step in our argument is the following lemma.

**Lemma 3.5** *Let $\mathcal{B}$ be an object with empty reflex silhouette and such that no vertical line contains two silhouette intervals. Then $\mathcal{B}$ is vertically monotone.*

**Proof:** Let $S$ be the silhouette of $\mathcal{B}$. Since no vertical line contains two silhouette intervals, the shadow curves of $\mathcal{B}$ are mutually disjoint, simple, closed curves in the $xy$-plane. Hence we can express $\pi(S)$ as the disjoint union of topological circles $\gamma_1, \gamma_2, \ldots, \gamma_n$, which partition the $xy$-plane into open regions $R_1, R_2, \ldots, R_k$. Every region $R_i$ is topologically equivalent to a disc with a finite number of holes, as in Figure 7.

For a point $p$ in the $xy$-plane, $\pi^{-1}(p) \cap \mathcal{B}$ is the disjoint union of closed intervals. We number those intervals from bottom to top as $I_1(p)$, $I_2(p)$, $\ldots$, $I_m(p)$. Within each region $R_i$, the number $m$ here is a constant.

Let now

$$P_{ij} := \bigcup_{p \in R_i} I_j(p),$$

and let $\overline{P}_{ij}$ denote the topological closure of $P_{ij}$. We observe that every $\overline{P}_{ij}$ is a closed, arc-connected set. Since

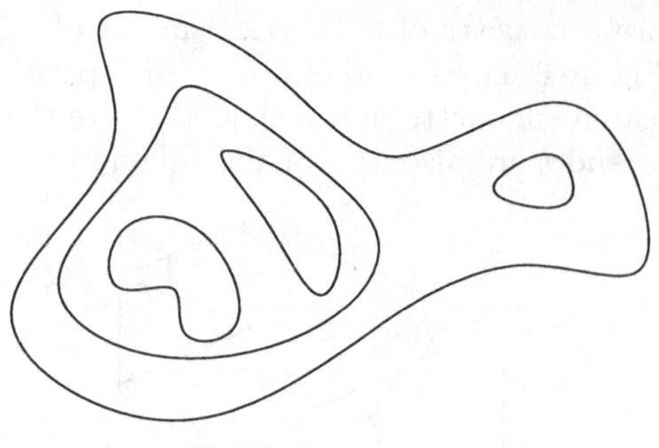

Figure 7.

the reflex silhouette of $\mathcal{B}$ is empty, for any $i$, $j$, and $j' \neq j$, $\overline{P}_{ij}$ and $\overline{P}_{ij'}$ are disjoint.

The difference $\overline{P_{ij}} \setminus P_{ij}$ consists of additional "cylinders" above the shadow curves bounding $R_i$. Consider now such a shadow curve $\gamma_t$ between $R_i$ and another region $R_{i'}$. Each piece of $\pi^{-1}(\gamma_t) \cap \mathcal{B}$ belongs to at most one $\overline{P}_{ij}$ and at most one $\overline{P}_{i'j'}$. It follows that every $\overline{P}_{ij}$ is adjacent to at most one $\overline{P}_{i'j'}$ in every $R_{i'}$ that is adjacent to $R_i$.

By the condition on $\pi(S)$, the adjacency relationship between the regions $R_i$ has tree structure. This implies that there cannot be a path connecting $\overline{P}_{ij}$ to $\overline{P}_{ij'}$ for $j \neq j'$. Since $\mathcal{B}$ is arc-connected, this is only possible if there is at most one such $P_{ij}$ above every region $R_i$, and hence $\mathcal{B}$ is vertically monotone. □

The lemma does not apply to an object such as in Figure 3, since its shadow curve is not simple. To extend the lemma to the more general setting, we have to define clearly what kind of non-simplicity we allow. We orient all silhouette curves such that the interior of $\mathcal{B}$ lies locally to the left of the curve. This induces an orientation in the shadow curves. We say that a set of shadow curves are *non-crossing* if a slight shrinking of every curve—obtained by moving every point slightly to the left—results in a set of mutually disjoint, simple curves.

We can now prove the generalization of the previous lemma.

**Lemma 3.6** *Let $\mathcal{B}$ be an object with empty reflex silhouette and non-crossing shadow curves. Then $\mathcal{B}$ is vertically monotone.*

**Proof:** Assume that $\mathcal{B}$ is not vertically monotone. Then there is a vertical line $\ell$ containing two points $p$ and $q$ in the interior of $\mathcal{B}$, and a point $r$ between $p$ and $q$ on the boundary or outside $\mathcal{B}$. Let $\varepsilon > 0$ be such that balls of radius $2\varepsilon$ centered at $p$ and $q$ are contained in $\mathcal{B}$. We now shrink $\mathcal{B}$ to obtain a new object $\mathcal{B}'$ by removing from $\mathcal{B}$ every point that has distance at most $\varepsilon$ to the complement of $\mathcal{B}$. If $\varepsilon$ is chosen small enough, this results in a legal object $\mathcal{B}'$. Since $p$ and $q$ lie in the interior of $\mathcal{B}'$ and $r$ lies between $p$ and $q$ outside $\mathcal{B}'$, $\mathcal{B}'$ is not vertically monotone.

Let $S$ and $S'$ be the silhouette of $\mathcal{B}$ and $\mathcal{B}'$ respectively. It is not difficult to see that $\mathcal{B}'$ has empty reflex silhouette, and since $\pi(S')$ is obtained from $\pi(S)$ precisely as prescribed above in the definition of non-crossing, $\pi(S')$ consists of mutually disjoint, simple curves. It follows that no vertical line intersects more than one silhouette interval in $\mathcal{B}'$. But then Lemma 3.5 implies that $\mathcal{B}'$ is vertically monotone, a contradiction. □

Finally, to wrap up the argument, we need the following lemma.

**Lemma 3.7** *Let $\mathcal{B}$ be an object with empty reflex silhouette whose shadow curves are crossing. Then $\mathcal{B}$ is not vertically monotone.*

**Proof:** Let $\overline{p}$ be a point in the $xy$-plane where a shadow curve crosses itself or another shadow curve. By definition, this means that the vertical line through $\overline{p}$ touches the boundary of $\mathcal{B}$ in two points $p$ and $p'$ lying in two different silhouette intervals. Therefore, there is a point $r$ between $p$ and $p'$ that lies outside $\mathcal{B}$.

By the definition of crossing, there must be a point $\overline{q}$ in the neighborhood of $\overline{p}$ that lies locally to the left of the two shadow curve segments. Indeed, otherwise a slight shrinking of the two shadow curves to the left would separate them.

By the construction, the vertical line $\ell$ through $\overline{q}$ passes through points $q$ and $q'$ close to $p$ and $p'$ in the interior of $\mathcal{B}$. If the distance between $\overline{p}$ and $\overline{q}$ is chosen small enough, then $\ell$ also intersects a neighborhood of $r$ lying completely outside $\mathcal{B}$. It follows that $\mathcal{B}$ is not vertically monotone. □

We summarize Lemmas 3.4, 3.6, and 3.7 in the following theorem.

**Theorem 3.8** *An object $\mathcal{B}$ is vertically monotone (and therefore castable) if and only if its reflex silhouette is empty and its shadow curves are non-crossing.*

Before we outline our algorithm for testing castability, we have to examine the silhouette of a polyhedron $\mathcal{P}$ in more detail. This silhouette consists of silhouette edges, vertical edges of the polyhedron, and parts of vertical facets. To find the silhouette on the vertical facets correctly, we use the vertical decomposition (or trapezoidal map) of these facets. Every trapezoid $\Delta$ of the decomposition is bounded from above and from below by (parts of) edges $e_1$ and $e_2$ of $\mathcal{P}$. If both $e_1$ and $e_2$ are convex edges, then $\Delta$ is part of the convex silhouette of $\mathcal{P}$. If both $e_1$ and $e_2$ are reflex edges, then $\Delta$ is part of the reflex silhouette, and $\mathcal{P}$ is not castable. If one edge is convex while the other is reflex, $\Delta$ does not belong to the silhouette. Note that certain vertical extensions produced by the vertical decomposition also belong to the silhouette. We can ignore them, however, as their projection coincides with the projection of the endpoints of the incident non-vertical edges of trapezoids. We call the projection of every silhouette edge and silhouette trapezoid a *shadow edge*. By Theorem 3.8, the polyhedron $\mathcal{P}$ is castable if and only if it has no reflex silhouette elements, and its shadow edges form a set of non-crossing curves.

If we examine the possible intersections of two shadow edges, we find that there are four cases that have to be treated as crossings, as shown in Figure 8. Two shadow

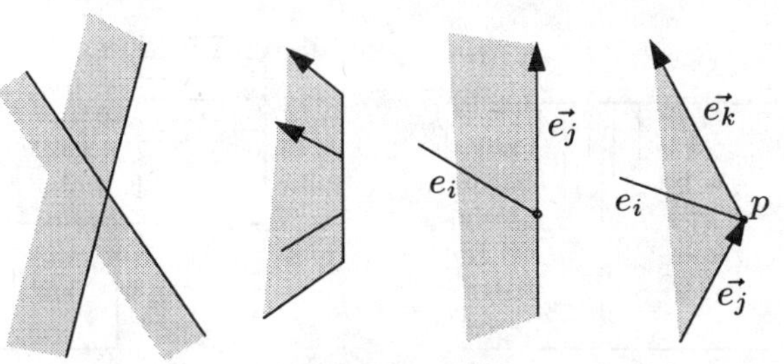


Figure 8.

edges cross if they intersect in their interior. They cross if they overlap with the same orientation (compare with Figure 4, where they overlap with opposite orientation—that is not a crossing). If an endpoint of edge $e_i$ lies on $e_j$, and $e_i$ lies to the left of $e_j$, this is a crossing. Finally, if an endpoint of $e_i$ coincides with the destination of $e_j$, we have to test whether $e_i$ is contained in the wedge formed by $e_j$ and the next shadow edge $e_k$ on the shadow curve of $e_j$.

**Theorem 3.9** *Given a polyhedron $\mathcal{P}$ with $n$ vertices, we can test in time $O(n \log n)$ whether $\mathcal{P}$ is vertically monotone (and therefore castable).*

**Proof:** We first form the vertical decomposition of all vertical facets in time $O(n \log n)$. Then, local analysis suffices to identify the silhouette curves. We project them into the plane to find the shadow curves. A plane sweep algorithm is used to compute the planar subdivision induced by the shadow curves. The algorithm finds all places where shadow edges meet. We test the edges for crossing. If no crossing exists, the subdivision has linear complexity and the algorithm terminates in time $O(n \log n)$. As soon as a crossing is found, we can also terminate, so the running time is $O(n \log n)$. □

## 4 Opposite Cast Removal—Finding a Direction

We have seen how to test whether a polyhedron $\mathcal{P}$ is castable in a given direction $\vec{d}$. In this section we describe an algorithm to solve the following problem: Given a polyhedron $\mathcal{P}$, decide whether it can be cast in some direction $\vec{d}$. In fact, we will solve the more general problem of finding all directions $\vec{d}$ for which $\mathcal{P}$ can be cast.

We represent every possible direction by a point on the unit sphere $\mathcal{S}^2$ (centered at the origin): a point $p$ on $\mathcal{S}^2$ represents the direction from the origin to $p$. Our goal is to identify the region of $\mathcal{S}^2$ corresponding to directions in which $\mathcal{P}$ is castable. By Lemma 3.1 and Observation 3.3 $\mathcal{P}$ is castable in direction $\vec{d}$ if and only if the union of open up-facets forms a terrain relative to $\vec{d}$.

If we imagine the direction $\vec{d}$ changing continuously, there are two events that may influence the terrain-property of the up-facets: First, an up-facet may become a down-facet, or vice versa—the set of these directions forms $O(n)$ great circles on $\mathcal{S}^2$. Second, the projection of a vertex $v$ may cross the projection of an edge $e$—the set of these directions can be described by $O(n^2)$ great circle arcs. Let $\mathcal{Q}$ denote the union of these curves which represent "critical events".

Consider the arrangement $\mathcal{A}(\mathcal{Q})$ of $O(n^2)$ great circles and great circle arcs on $\mathcal{S}^2$. It is easily verified that inside every face of the arrangement, $\mathcal{P}$ is either $\vec{d}$-monotone for every direction $\vec{d}$, or for none. We say that two directions are *combinatorially distinct* if they lie in two different faces of the arrangement. We aim to compute all combinatorially distinct directions in which $\mathcal{P}$ is castable.

By our definition of castability, if a cell or an edge of $\mathcal{A}(\mathcal{Q})$ represents directions in which $\mathcal{P}$ is castable, then any vertex on its boundary represents a direction in which $\mathcal{P}$ is castable. This suggests the following simple algorithm. We compute all the vertices of the arrangement $\mathcal{A}(\mathcal{Q})$ by simply computing the intersections of all pairs of curves in $\mathcal{Q}$. This takes $O(n^4)$ time. Then for each computed vertex we test, in $O(n \log n)$ time whether $\mathcal{P}$ is castable in the corresponding direction using Observation 3.3 and Theorem 3.9. The total running time of this algorithm is $O(n^5 \log n)$.

There are several ways in which this straightforward approach can be improved. Note that any vertex $v$ in $\mathcal{A}(\mathcal{Q})$ represents a degenerate casting direction $\vec{d_v}$: either one facet or more are parallel to $\vec{d_v}$, or a line parallel to $\vec{d_v}$ intersects $\mathcal{P}$ in more than one connected component. Therefore, we may be better off searching for directions in the interior of cells of $\mathcal{A}(\mathcal{Q})$. However, we need to consider all faces of $\mathcal{A}(\mathcal{Q})$ instead of just cells, because there may be directions of castability that appear only along edges (such as the situation depicted in Figure 3) or vertices of the arrangement.

Note also that the difference between two adjacent cells in $\mathcal{A}(\mathcal{Q})$ is quite small, provided that $\mathcal{A}(\mathcal{Q})$ is in general position. When going from one face of the arrangement to another, either one facet of $\mathcal{P}$ changes its status (among down-facet, up-facet or parallel relative to a given direction) or the projection of a vertex of $\mathcal{P}$ crosses into (or over) the projection of an edge of $\mathcal{P}$. We defer the details of handling degeneracy in $\mathcal{A}(\mathcal{Q})$ to the full paper.

To exploit the coherence between adjacent cells we proceed as follows. First we compute the arrangement $\mathcal{A}(\mathcal{Q})$. We do this with an output-sensitive algorithm. Let $m$ denote combinatorial complexity of $\mathcal{A}(\mathcal{Q})$. The algorithm is a straightforward adaptation of the plane-sweep paradigm to the sphere and its running time is $O(m \log m)$. Its output is a data structure that allows for a traversal of the arrangement cell by adjacent cell (we could use, say, the quad-edge data structure for the purpose [10]).

If we are only concerned with worst-case running time, and since the complexity of the arrangement $\mathcal{A}(\mathcal{Q})$ can be $\Theta(n^4)$ in the worst case (see below), we can compute the arrangement in $\Theta(n^4)$ time by substituting each arc in $\mathcal{Q}$ by the great circle containing it and computing the arrangement of the resulting collection of great circles. (The latter arrangement is a refinement of $\mathcal{A}(\mathcal{Q})$ from which we could easily obtain the required output.) Using central projection from the sphere onto two parallel planes tangent to $\mathcal{S}^2$ at two antipodal points we obtain two arrangements of straight lines. Such arrangements can be computed in $\Theta(n^4)$ time each. In fact computing the arrangement on one plane suffices since the two arrangements are symmetric, provided some caution is exercised in choosing the projection planes.

After $\mathcal{A}(\mathcal{Q})$ has been computed, we choose a point $p$ inside a face of the arrangement arbitrarily and compute the silhouette elements of $\mathcal{P}$ corresponding to the direction $\vec{d_p}$. For each silhouette element we check whether it is convex or reflex. We initialize two counters: how many silhouette elements are reflex and how many pairs of shadow edges cross one another. For the direction $\vec{d_p}$, this computation takes $O(n^2 \log n)$ time by a simple plane-sweep algorithm. By Theorem 3.8, the polyhedron is castable in direction $\vec{d_p}$ if both counters are zero. We move to an adjacent cell of the arrangement. By the edges of the arrangement that are involved in the move, we know how to update the counters at a constant time per edge involved. At the end of the move we check the counters and report castability if they are both zero. We conclude that all directions for which there is a good cast can be computed in $O(n^4)$ time, or in time $O(m \log m)$.

There exist polyhedra that have as many as $\Omega(n^4)$ distinct cast directions. Figure 9 shows a polyhedron with two horizontal "legs" and there is a row of small (resp. large) "teeth" positioned along the upper leg (resp. lower leg). The schematic diagram on the left in Figure 10 gives the top view showing the interaction of the large and small teeth. In the top view, if we move from left to right, one large tooth will appear in each of the gaps among the small teeth before an adjacent large tooth appears in any gap among the small teeth. If there are $n/4$ small teeth and $n/4$ large teeth, then there are $\Omega(n^2)$ distinct and good parting directions. The schematic diagram on the right in Figure 10 shows the top view of a composite object consisting of two copies of the polyhedron in Figure 9: the lower leg of the left copy and the upper leg of the right copy are at the same level, and the two shaded boxes represent the projection of the two rows of small teeth. The number of distinct and good parting directions for the composite object can be as large as $\Omega(n^4)$, which shows that our algorithm is optimal in the worst case.

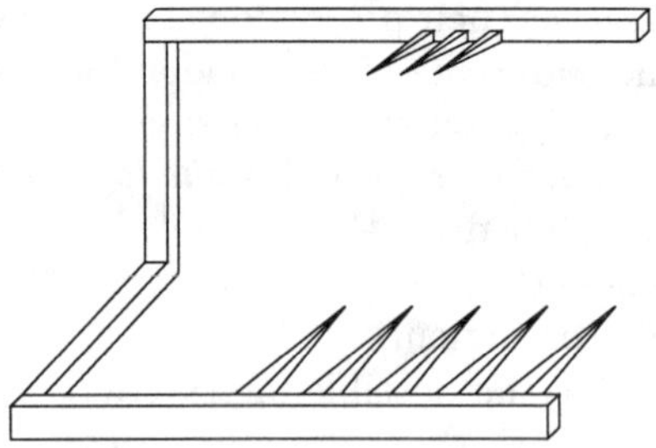

Figure 9.

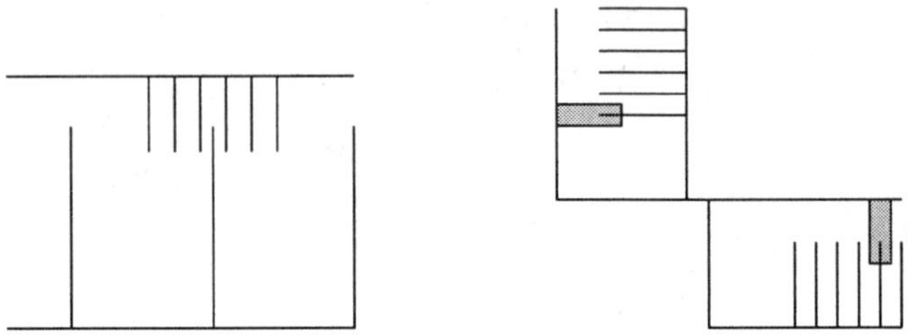

Figure 10.

**Theorem 4.1** *Let $\mathcal{P}$ be a simple polyhedron with $n$ vertices. All directions in which there is a good cast can be computed in $O(n^4)$ time. Moreover, there are polyhedra for which there are $\Omega(n^4)$ distinct directions in which there is a good cast. Alternatively, all the directions in which there is a good cast can be computed in $O(m \log m)$ time, where $m$ is the combinatorial complexity of the arrangement $\mathcal{A}(\mathcal{Q})$.*

## 5 Experimental Results

We have implemented a simplified version of the algorithm of Theorem 3.9 (instead of a plane sweep, we simply test all pairs of shadow edges for crossings). Given an object to be cast, we test a random set of directions, as well as heuristically chosen directions (currently the directions of all edges). This is simple to implement and seems to work fine in practice. Figure 11 shows a heart-shaped object. The sphere represents the sphere of directions. A black stipple is plotted on the sphere for every direction in which the object has been found castable. Two directions have been specially marked, one of these is an edge direction, the other one a randomly chosen direction. The heart has 75 edges and 7 edge directions were

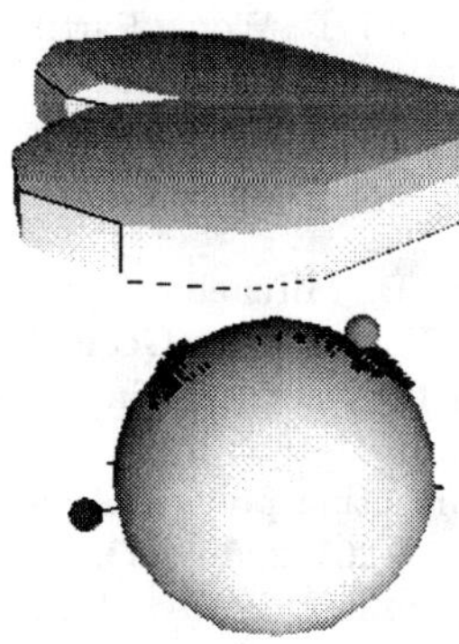

Figure 11.

found to be feasible. Of the 32,000 randomly chosen directions, 803 were found to be feasible. The black line on the heart shows the silhouette for one of the two marked directions (the randomly chosen one).

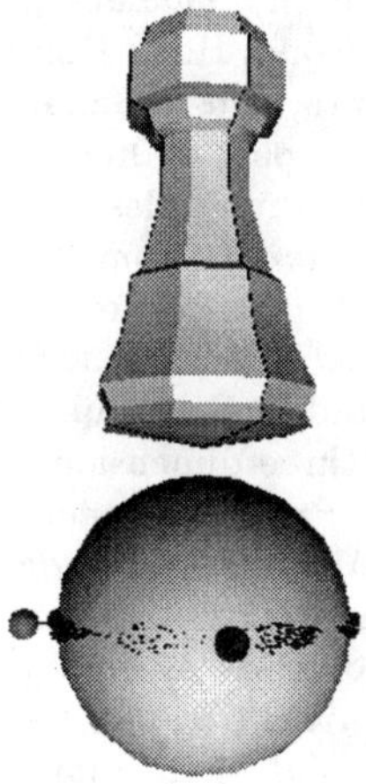

Figure 12.

The rook from a chess game in Figure 12 had 264 edges and the directions of 8 of them are feasible. Of the tested 32,000 random directions, 576 are feasible. Silhouettes for the two marked directions are shown on the object.

Clearly, further experimentation is necessary to improve the heuristics. Although the program runs only a few seconds on these parts, it would have to be faster to allow on-line warnings inside a CAD system (we imagine a system that automatically warns the designer as soon as the object becomes uncastable).

A natural extension would be to test directions that are parallel to a pair of facet planes. Obviously we could also test all $O(n^4)$ vertices of the arrangement, but so far it seems that that would make the program slower without helping much in practice.

## 6 Arbitrary Cast Removal

We now briefly discuss the case where the removal directions for the cast parts are not necessarily opposite. We call a polyhedron $\mathcal{P}$ castable with cast $\mathcal{C} = (\mathcal{C}_r, \mathcal{C}_b)$ and removal directions $(\vec{d_r}, \vec{d_b})$ if $\mathcal{C}_r$ and $\mathcal{C}_b$ can be translated to infinity in directions $\vec{d_r}$ and $\vec{d_b}$ respectively so that the interior of $\mathcal{C}_r$, $\mathcal{C}_b$, and $\mathcal{P}$ do not intersect during the translation. Contrary to the case of opposite removal directions, the order of removing the casts can be important. The cast induces partition of $\partial\mathcal{P}$ into two parts, a red part touching $\mathcal{C}_r$ and a blue part touching $\mathcal{C}_b$. When looking at $\mathcal{P}$ from infinity in direction $-\vec{d_r}$ every point on the red part must be visible, and similar for the blue part. The reverse of this statement, however, is not true, as the next theorem shows.

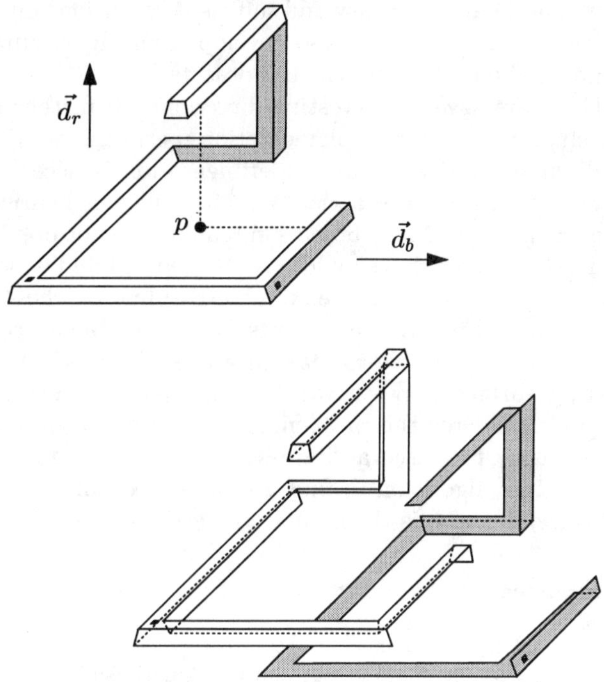


Figure 13. Illustration for the proof of Theorem 6.1

**Theorem 6.1** *In the case of arbitrary cast removal, there are polyhedra that are not castable even though their boundary can be split into two terrains with respect to the removal directions, and which are completely visible from those directions.*

**Proof:** Consider the polyhedron $\mathcal{P}$ and the removal directions depicted in Figure 13. The shaded facets of $\mathcal{P}$ together with the bottom facets form the blue terrain on the boundary; the remaining facets form the red terrain. Both the blue and the red terrain are actually terrains in their respective removal directions. There is no good cast, however, for these removal directions for $\mathcal{P}$: The point $p$ will intersect the interior of $\mathcal{P}$ when moved in direction $\vec{d_b}$ and when moved in $\vec{d_r}$, so it can neither be in the blue nor in the red cast part. This means that there is no good cast for the directions $(\vec{d_b}, \vec{d_r})$. By poking two thin holes into the object, as in Figure 13, we can ensure that there cannot be a good cast for any other pair of directions either. The hole in the blue terrain, for example, makes $\vec{d_b}$ the only direction for which this hole is part of a terrain. □

## 7 Concluding Remarks

We have studied the problem of determining whether there is a two-part cast for a given object such that the two cast parts can be removed without collision. We considered the case where the removal directions must be opposite, and gave necessary and sufficient conditions under which a cast exists. We also developed an algorithm to compute the cast for polyhedral objects.

There are several interesting directions for further research. While our implementation performs well on medium size models, more experimentation is necessary to develop a robust, practically useful, efficient heuristic implementation. Many objects in real life are not polyhedral, so the algorithm should be extended to handle more general object boundary (e.g. cubic B-spline patches) as well as to allow cores and inserts for uncastable objects. Finally, it is desirable to maximize the "flatness" of the parting surface between the two cast parts. Majhi et al. [15] considered this problem for convex polyhedral objects. They proposed a "flatness" measure and gave an $O(n^2)$ time algorithm to find a cast that optimizes this measure, where $n$ is the number of vertices.

## References


[1] B. Asberg, G. Blanco, P. Bose, J. Garcia, M. Overmars, G. Toussaint, G. Wilfong, and B. Zhu. Feasibility of design in stereolithography. In *Proc. 13th Symp. on FST & TCS. Lecture Notes in Computer Science*, vol. 761, pages 228–237. Springer-Verlag, 1993. To appear in Algorithmica.

[2] P. Bose. *Geometric and Computational Aspects of Manufacturing Processes.* PhD thesis, McGill University, 1994. Also available as UBC Tech Rep 95-02, Dept. of Comp. Sci, Univ. of British Columbia, 1995.

[3] P. Bose, D. Bremner, and M. van Kreveld. Determining the castability of simple polyhedra. In *Proc. 10th Annu. ACM Sympos. Comput. Geom.*, pages 123–131, 1994. To appear in Algorithmica.

[4] P. Bose, M. van Kreveld, and G. Toussaint. Filling polyhedral molds. In *Proc. 3rd Workshop Algorithms Data Struct. Lecture Notes in Computer Science*, vol. 709, pages 210–221. Springer-Verlag, 1993.

[5] P. Bose and G. Toussaint. Geometric and computational aspects of manufacturing processes. *Comput. & Graphics*, 18:487–497, 1994.

[6] L. Chen, S. Chou, and T. Woo. Parting directions for mould and die design. *Computer-Aided Design*, 25:762–768, 1993.

[7] R. Elliot. *Cast Iron Technology.* Butterworths, London, 1988.

[8] S. Fekete and J. S. B. Mitchell. Geometric aspects of injection molding. Workshop on Geometric and Computational Aspects of Injection Molding, Bellairs Research Institute, February 5–12, 1993.

[9] K. Y. Goldberg. Orienting polygonal parts without sensors. *Algorithmica*, 10:201–225, 1993.

[10] L. J. Guibas and J. Stolfi. Primitives for the manipulation of general subdivisions and the computation of Voronoi diagrams. *ACM Trans. Graph.*, 4:74–123, 1985.

[11] M. Held. *On the Computational Geometry of Pocket Machining. Lecture Notes in Computer Science*, vol. 500. Springer-Verlag, June 1991.

[12] K. Hui. Geometric aspects of mouldability of parts. *Computer Aided Design*, 29:197–208, 1997.

[13] K. Hui and S. Tan. Mould design with sweep operations—a heuristic search approach. *Computer-Aided Design*, 24:81–91, 1992.

[14] K. K. Kwong. *Computer-aided parting line and parting surface generation in mould design.* PhD thesis, The University of Hong Kong, Hong Kong, 1992.

[15] J. Majhi, P. Gupta, and R. Janardan. Computing a flattest, undercut-free parting line for a convex polyhedron, with application to mold design. In *ACM Workshop on Applied Computational Geometry*, pages 39–45, 1996.

[16] J. Matoušek. Efficient partiton trees. *Discrete & Computational Geometry*, 8:315–334, 1992.

[17] M. McAllister and J. Snoeyink. Two-dimensional computation of the three-dimensional reachable region for a welding head. In *Proc. 5th Canad. Conf. Comput. Geom.*, pages 437–442, Waterloo, Canada, 1993.

[18] M. Overmars, A. Rao, O. Schwarzkopf, and C. Wentink. Immobilizing polygons against a wall. In *Proc. 11th Annu. ACM Sympos. Comput. Geom.*, pages 29–38, 1995.

[19] F. P. Preparata and M. I. Shamos. *Computational Geometry: An Introduction.* Springer-Verlag, 1985.

[20] W. Pribble. Molds for reaction injection, structural foam and expandable styrene molding. In J. DuBois and W. Pribble, editors, *Plastics Mold Engineering Handbook.* Van Nostrand Reinhold Company, New York, 1987.

[21] A. Rosenbloom and D. Rappaport. Moldable and castable polygons. In *Proc. 4th Canad. Conf. Comput. Geom.*, pages 322–327, 1992.

[22] G. T. Toussaint. Movable separability of sets. In G. T. Toussaint, editor, *Computational Geometry*, pages 335–375. North-Holland, Amsterdam, Netherlands, 1985.

[23] C. Walton and T. Opar, editors. *Iron Castings Handbook.* Iron casting society, Inc., 1981.

[24] R. H. Wilson and J.-C. Latombe. Geometric reasoning about mechanical assembly. *Artificial Intelligence*, 71:371–396, 1994.

# A Data Front-End for Layered Manufacturing*

Gill Barequet†  Yuval Kaplan‡

## Abstract

In this paper we describe a geometric software package which serves as a data front-end for rapid prototyping systems. This package can be customized for every layered-manufacturing technique. We survey the full path of data in the system, starting as data files originated by external tools, in most cases CAD systems, going through some repairing operations, then editing a job composed of several objects, and finally processing the job for actual production.

**Keywords:** geometric software, rapid prototyping, layered manufacturing, polyhedra.

## 1 Introduction

Layered manufacturing has become during the last decade the major method for Rapid Prototyping (RP) of models. Physical models are needed in almost every industry for design verification, simulation, integration of complex systems, etc. RP has also applications in medicine (e.g., in surgery planning), in chemistry (e.g., for verifying docking sites of macromolecules), in architecture, civil engineering, and so on. RP has emerged as a supporting tool for shortening the time-to-market of a product. In most stages of the design a prototype of a model does not necessarily have to be as accurate as the final product or to have the same mechanical properties, color, functionality, etc. Rapid prototypes are expected above all to be produced *fast* and with low cost. While consuming relatively short time, layered manufacturing overcomes many limitations of traditional milling techniques, especially the limitation on the geometries of the produced models. RP is also lately used for final-product fabrication, e.g., in some casting processes.

RP techniques need to interface with the CAD world. The first player in the RP market, 3D Systems, Inc., set in 1988 the de-facto file-format standard in the RP industry [3DS]. This standard is a syntax for describing the boundary of a model in a *polyhedral* (in fact, triangular) format, that is, as a piecewise-linear surface which contains only triangular faces. Objects however are rarely designed as polyhedra but as smooth objects described by more involved mathematical entities. The polyhedral boundary representation is actually an *approximation* of the original object. Later file formats extended this standard by allowing non-triangular faces, possibly perforated, but the whole rapid prototyping industry is still based on polyhedral object descriptions, where every model is described by a collection of *facets* (planar polygonal faces) which approximates the object boundary surface.

The emerging RP industry created the need for supporting software whose main features are:

1. Computing polyhedral approximations of smooth CAD files.
2. Cleaning and repairing ill-defined CAD files (or rather, their approximations), whose defects may not disturb the graphics display or external applications (e.g., finite element analysis), but spoil the production of the models.
3. Designing a production job (a "batch" of several models to be produced simultaneously).
4. Simulating the production, estimating its cost and duration, and predicting errors in it.
5. Converting polyhedral formats to data recognized by an actual RP technique.

Here is a brief (and incomplete) review of the currently available RP technologies (a more detailed review is given in the full version of the paper). In 1988, 3D Systems, Inc.,

*Work of the first author on this paper has been supported in part by the Israeli Ministry of Science and the Arts under an Eshkol Grant 0562-2-95 and by the U.S. Army Research Office under Grant DAAH04-96-1-0013.

†Center for Geometric Computing, Dept. of Computer Science, Johns Hopkins University, Baltimore, MD 21218. E-mail: barequet@cs.jhu.edu

‡Responsible for software development, Cubital Ltd., 13 HaSadna St., Industrial Zone North, Ra'anana 43650, Israel. E-mail: cubital@netvision.net.il



*Computational Geometry 97* Nice France

introduced the stereolithography technology. Stereolithography builds parts from a 3D CAD file in "slices," layer-by-layer from the bottom up, by curing layers of a liquid photopolymer resin with a laser in a slow serial process, one dot at a time. Other stereolithography-like machines are produced by EOS GmbH, Germany, and by CMET, Inc., D-MEC, Ltd., and Denken Engineering, all from Japan. In 1991 Cubital Ltd. from Israel introduced its additive bi-material process, SGC (Solid Ground Curing), which uses a photo-curable acrylate or epoxy polymer resin and solid wax support material. Each layer image is first produced on a glass plate to form a photomask. An ultraviolet (UV) light, projected through the photomask of the cross-sectional layer onto the liquid ultraviolet-sensitive photo-polymer resin, solidifies the entire layer within seconds. Unsolidified excess resin is vacuumed away and molten liquid wax fills in all cavities, surrounding the cured resin. The wax is then cooled to become a solid support media. DTM Corporation's Selective Laser Sintering (SLS) process uses data from a CAD file to direct deposition of thin layers of heat-fusible powder in a pattern that is heated (sintered) with a $CO_2$ laser. RP manufacturer Helisys, Inc., offers Laminated Object Manufacturing (LOM), a process wherein a laser beam cuts through a series of sheets of heat-sensitive paper, layered upon one another to build a part with the resulting consistency similar to that of wood. A similar technology is available from Kira Corp., Japan. In Fused Deposition Modeling of Stratasys, CAD file data directs part-building in slices by extruding a thermoplastic material from a temperature-controlled head in layers that fuse to one another to build a solid part. Sanders Prototype, BPM Technology, Inc., and Soligen Corp. each use a form of ink-jet type spray to build parts. Sanders calls its process "3-D plotting," while BPM terms its methodology "Ballistic-particle manufacturing." Soligen's Direct-shell production casting (DSPC) builds a ceramic mold which can be used to create parts from molten metal. 3D Systems has also introduced its own ink-jet type modeling process called "Multi-jet modeling." Other processes deposit a stream of metallic powder into the path of a laser beam aimed at a target position on the built model. Such processes include CMB (Controlled Metal Build-up) of Fraunhofer IPT (Aachen, Germany) and LENS (Laser Engineered Net Shaping) currently under development at Sandia National Labs (Albuquerque, New Mexico).

Three important facts characterize the flow of data from CAD systems to RP technologies:

1. CAD systems transfer to RP polyhedral *approximations* of the original objects, which contain in many cases errors or ambiguities [BW, BS1].
2. CAD systems are able to output such files but are unable in most cases to read them back for examining modifications and corrections [Wo].
3. It is much more economical to produce several models during one "run" of an RP machine. For this purpose several objects are positioned in a "job" so as to use the full capacity of the machine.

These three facts, together with the growth of the popularity of RP (as a result of the increase in its availability and decrease in production time and cost), led to evolving implementation of software packages which allow the user to view RP files, edit them, *repair* them, customize them for layered manufacturing, edit production jobs, simulate the manufacturing (for estimating the production time, accuracy, etc.), and finally for actually feeding the RP machines with data.

Supporting software is supplied by most of the RP vendors listed above and also by independent firms. A recent review of Wohlers [Wo] lists most of these packages and their major features. In our opinion the most efficient packages (out of those listed in this reference) are Rapid Tools of DeskArtes Oy (mainly due to its ability to fix defects in files) and SolidView of Solid Concepts, Inc. (mainly because of its friendliness). The reader is referred to the reference cited above for details and commercial information. Recently, Morvan and Fadel [MF] and Krause, Stiel, and Luddermann [KSL] claimed systems for processing and repairing of CAD data.

The DFE (Data Front-End) system described in this paper is a geometric software package which supports all stages of preparing CAD data files for production. It is customizable for every layered-manufacturing technique. We present the various components of the system along with detailed descriptions of the implemented algorithms. The system currently runs on DEC stations (CPU 5000/200-260) and SGI Indy and Indigo2 (CPU R5000) under the Ultrix 4.3 and Irix 5.3 or 6.2 operating systems, respectively. The underlying graphics package and required card are Phigs 2.3a and a Z-buffer (on DEC), and OpenGL and 24-bit XL (on SGI).

The paper is organized as follows. In Section 2 we describe how data files are imported into the system. In Section 3 we review the main geometric tools of the Academy, the package for editing and repairing data files. In Section 4 we present the job editor, and in Section 5 we describe the production tool. We end in Section 6 with some concluding remarks.

## 2 Importing Data Files

The data front-end is capable of reading files that contain polyhedral descriptions in many commonly-used file formats. Most of the formats describe the *boundary representation* (B-Rep) of a polyhedron, which is typically an approximation of a smooth object designed by using a CAD system. These formats usually belong to one of the following two classes:

I. Only one list of polygons, each consists of a sequence of vertices, where each vertex is specified by its 3-dimensional coordinates.

II. Two lists: The first list consists of indexed points, each specified by its 3-dimensional coordinates, and the second list consists of polygons, each specified by a sequence of indices referring to the point list.

The advantages of the second class are obvious: vertex coordinates need not be repeated and adjacency between polygons (that share an edge of the polyhedron) is implicitly specified by the existence of two successive vertex indices $(i, j)$ in one polygon and a matching successive pair $(j, i)$ in the other polygon. Note that the neighborhood information is not provided explicitly in either class.

The following is a partial list of file formats recognized by the data front-end.

STL This is the de-facto standard of the rapid-prototyping industry. It is output by practically all CAD systems and is recognized by every rapid-prototyping system. This file format belongs to class I and allows triangles only. The file format contains for each triangle the unit vector normal to it. The full specifications of this format are given in [3DS].

BSTL This is the binary form of STL.

CFL This file format belongs to class II and is used solely by Cubital's DFE and the SGC rapid-prototyping technique. The file syntax allows additional pieces of information, e.g., details of the designer, geometric attributes, and manufacturing instructions.

SOLI This is the binary form of CFL.

RF This is another class II Ascii format used by the RobCad (Robotic CAD) system of Tecnomatics Ltd. [Te].

UNV This is yet another class II Ascii format used by the I-Deas CAD system of SDRC Ltd.

The front-end can also write data into CFL, STL, and BSTL files. When STL or BSTL files are output the boundary polygons need to be triangulated. Since the vertices of every polygon are embedded in a 3-dimensional space, they are not necessarily all coplanar. (In most practical cases the vertices of a facet are almost coplanar, but the algorithm does not make any use of this assumption.) The system triangulates the spatial polygons so as to minimize the total area of the triangles. For this purpose it uses a dynamic-programming technique which closely follows Klincsek's algorithm [Kl] for triangulating a 2D polygon. The running-time of this algorithm is cubic in the complexity of the input polygon. The full details of this algorithm are found in [BS1, BS2].

## 3 Repairing Data Files

After importing a data file, the DFE allows the user to analize it, to modify it, and to fix defects in it. This is done via a software package called the *Academy*.

The system automatically performs an analysis of the connected components of the object boundary (by performing a depth-first search on the dual graph of the boundary) and provides the user with detailed statistics. Valid objects may consist of more than one connected component, where each one is either a solid or an "anti-solid"—a void in a solid. Figure 1 shows an object with two solid components: a ball trapped in a "cage." The graphics display of the Academy shows each connected component in a different color. Some CAD systems compute inaccurately the polyhedral approximation of a smooth object as is explained later in this section. Thus the boundary of the polyhedron may contain flaws ("holes") or may even be broken into several connected components which may not necessarily enclose volumes. Connected components which are not true solids are bounded by one or more closed polygonal chains in space

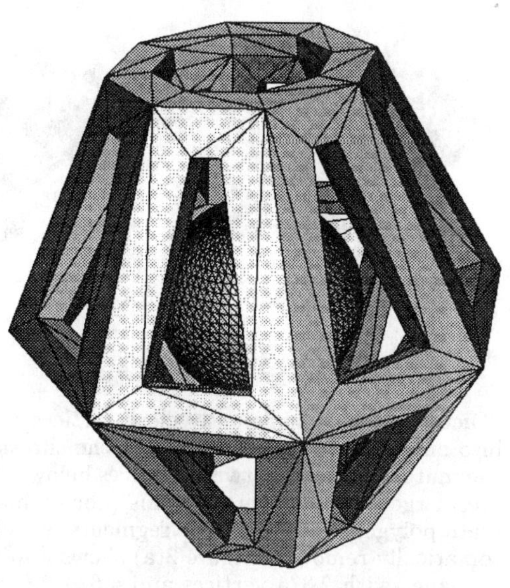

Figure 1: A ball in a cage

which the system denotes as *flaws*. The statistics which the Academy provides for each connected component includes the number of vertices and facets, number of flaws, an indication whether it encloses a volume or not (that is, whether it is closed or open), and an indication whether the component is *orientable* or not. (The facets of a non-orientable manifold cannot be oriented consistently, such that each edge appears exactly twice and in opposite directions in the two neighboring facets that share the edge.)

The system provides some tools for handling the connected components. The user may select some connected components of the object, then delete them, or extract them into a new file, or split the object into two files: one that contains the selected components and one that contains the remaining components. The user may also separate the object into multiple objects, each containing one connected component of the original object. In many cases the orientations of the polygons that describe the boundary of the object are unreliable. As a rule, each facet should be oriented counter-clockwise when viewed from outside the object. The user may reorient the facets of each connected component. When a component is orientable, the system indicates for each facet its correct orientation and whether it is currently oriented correctly or not. For each connected component the user may choose to invert all the polygons, to orient them correctly, or to orient them in their incorrect orientations. The last option is used for creating "anti-objects," voids in correctly-oriented surrounding objects.

In practice, both the resolution and the addressability of the manufacturing process are far coarser than the accuracy of the CAD system or that of the process that produced the polyhedral approximation of the original object. In many cases the object is an overdetailed approximation of the original object in which too small tolerance was allowed. The result is a tessellation of the model, which, although very similar to the original model, contains too many (close) vertices and very small facets. The Academy provides two major tools for reducing the complexity of the description of the object:

(a) Original object (b) $\varepsilon = 5$in (c) $\varepsilon = 10$in (d) $\varepsilon = 25$in

Figure 2: Unifying $\varepsilon$-close vertices

1. "Diet," in which sufficiently close vertices are unified into one vertex. The user supplies the threshold $\varepsilon$ for the mutual distance between vertices below which they are merged into one vertex. This process may degenerate polygons into points or segments, which are automatically removed. Figure 2(a) shows a model of an airplane (with 2,375 vertices and 4,629 facets) of size 580in × 520in × 220in. Figures 2(b,c,d) show its "dieted" versions by $\varepsilon = 5$in (with 2,025 vertices and 3,886 facets), by $\varepsilon = 10$in (with 1,584 vertices and 3,017 facets), and by $\varepsilon = 25$in (with 614 vertices and 1,243 facets), respectively. (Typically the user applies this operation with a value of $\varepsilon$ which is less than half of the minimum feature size of the layered-manufacturing technology.) When the user does not supply a threshold, the default value $\varepsilon = 0$ is taken. In this case the tool simply unifies vertices which appear more than once in the polyhedral description of the object. In fact, data files with syntax of class I (see Section 2) automatically undergo this operation when they are imported into the system.

2. "Trim," in which almost coplanar neighboring facets are united to form facets with larger complexities. The user supplies the threshold $\alpha$ for the deviation between the directions of the normal vectors to the facets, below which the facets are united. In other words, neighboring facets are united when the dihedral angle between them is less than $\alpha$ away from $\pi$ (exact coplanarity). Figure 3(a) shows a rotor of a real car engin, which consists of 2,636 facets (and of 1,312 vertices). Figure 3(b) shows its "trimmed" version (with $\alpha = 0.5^o$) which contains only 872 facets (the number of vertices is unchanged). Similarly, Figures 4(a,b) show a leg-bone implant (with 7,700 facets and 3,852 vertices) and its trimmed version by $\alpha = 2^o$ (with only 2,621 facets), respectively. When the user does not supply a threshold, the default value $\alpha = 0$ is taken. In this case the tool simply unites only coplanar neighboring facets. Figures 5(a,b) show a mechanical part (with 1,164 facets and 572 vertices) and its simplified version (with only 337 facets), respectively. When the system imports data files which allow triangles only (e.g., STL—see Section 2) the user is prompted whether he or she wishes this operation be performed as part of the conversion.

Yet the most useful ability of the Academy is to fix flaws in the boundary of an object. These flaws usually appear in polyhedral approximations of CAD objects whose boundaries are described by curved entities [BS1, BW, DM, MD,

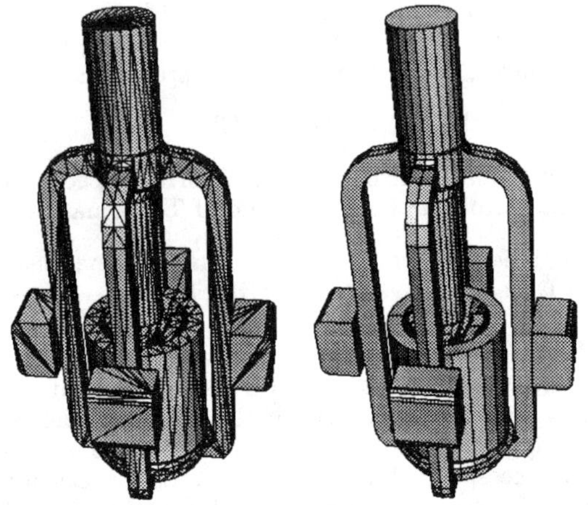

(a) Original object (b) Trimmed object

Figure 3: Trimming a car rotor with $\alpha = 0.5^o$

(a) Original object

(b) Trimmed object

Figure 4: Trimming a leg-bone implant with $\alpha = 2^o$

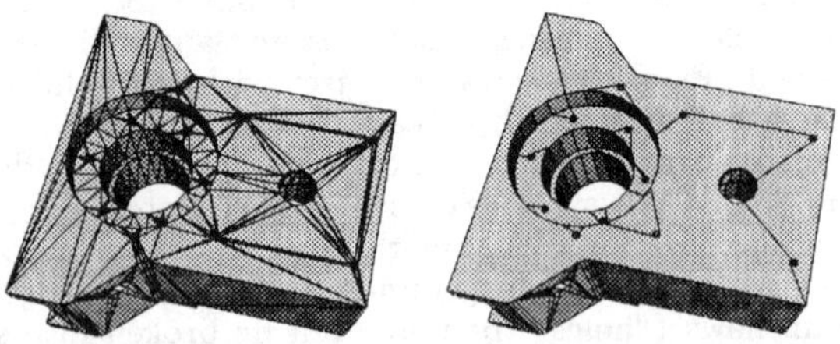

(a) Original object (b) Simplified object

Figure 5: Uniting coplanar neighboring facets of an object

ST]. In CSG (Constructive Solid Geometry) the object is described as the unions and intersections of cubes, spheres, cones, etc., whereas in using its B-Rep (Boundary Representation) it may be described by NURBs (Non-Uniform Rational Bisplines), Bézier surfaces, etc. Most flaws are caused by incorrect handling of trimming curves defined by intersections of surfaces. Other flaws are caused by missing surfaces or patches within a surface, or by incorrect handling of adjacencies between them. Flaws do not usually disturb graphics displays where the gaps between the surfaces are often too small to be seen or are handled straightforwardly [SW]. However, it causes severe problems in applications which rely on the continuity of the boundary, such as rasterization algorithms [FD] which play a crucial role in layered manufacturing (see Section 5).

Flaws seem to appear in most large CAD files, so their detection and elimination is indeed a rather important practical problem. Sheng and Tucholke [ST] refer to these errors as one of the most severe software problems in rapid prototyping. Dolenc and Mäkelä [DM] and Sheng and Hirsch [SH] try to avoid it already in the surface fitting triangulation process. Others attempt to solve this problem given only the polyhedral approximation of the object. Some works use only local information and do not check for any global consistency violations. Bøhn and Wozny [BW] treat local flaws by iteratively triangulating them: at each step they eliminate the vertex which spans the smallest angle with its two neighboring vertices. Mäkelä and Dolenc [MD] apply a minimum distance heuristic in order to locally fill flaws in the boundary of the object. Barequet and Sharir [BS1] attempt to solve this problem while considering the global consistency of the resulting polyhedron. Their algorithm uses a geometric hashing technique for matching portions of the flaws (which are then "stitched") and an optimal triangulation of 3D polygons for resolving the unmatched portions; it also ensures that the resulting polyhedron is *orientable*.

We now proceed to describe the Academy feature of detecting and fixing flaws. The Academy first displays the list of flaws of each connected component of the object boundary. Each flaw is classified according to whether it is a real *border* of a surface patch, a thin *crack* in it, a *missing triangle*, or a *degenerate gap* in the boundary. (A degenerate gap occurs when edges do not match properly in pairs but yet the boundary is continuous.) For each flaw the system provides its complexity (number of vertices), its 3-dimensional bounding-box, the total length of its edges, and in case of a thin crack, the two spatial angles spanned by the pairs of edges that form its two endpoints and the ratio between the lengths of the two polygonal chains between the two endpoints. Degenerate gaps may be ignored, or added (in the opposite orientation) to the model, so as to topologically fill in the gaps. Missing triangles are also simple to resolve by simply adding them (or rather, the inverted versions of the holes in the boundary) to the object. Cracks and borders may however be much more involved and user intervention is needed in most cases.

Cracks are usually fairly easy to fix: the two sides of a crack need to be "stitched" to each other by a sequence of triangles that follows the crack in space. This procedure resembles the "merging" of two sorted lists of numbers. The Academy provides such a patching tool which differs from that cited above only in the restriction that the two stitched polygonal chains must share their endpoints. The user may set the maximum allowed length of a chord (edge of a triangle) added in this process. Portions of the crack in which its "width" exceeds this value are not stitched and are left for further repairing operations.

It is also possible that portions of *different* borders should be stitched together. This is often the case when the boundary of an object is broken into pieces due to inconsistent approximations of the surfaces that describe the original object. Barequet and Sharir [BS1] discuss a heuristic method, based on a computer-vision technique, for identifying such matching portions of borders of connected components of the object. When automating this process one must take care not to create a non-orientable surface by inconsistently stitching between components of the object. They show that when matches are given scores (say, according to the lengths of the matches and their widths—the distances between the matching portions) the problem of selecting a subset of *consistent* matches (which does not result in a non-orientable manifold) with a maximum total score is $\mathcal{NP}$-Hard. The cited reference hence uses an approximation scheme which reduces the time complexity of choosing the optimal consistent subset of matches from practically exponential in $c$, the number of components, and in $m$, the number of matches between them, to only $O(m \log m + m\alpha(m, c))$.[1]

The Academy provides a "manual" version of the algorithm described in [BS1]. The user selects a subset of borders on which the binding (matching and stitching) operation is to be performed and a maximum allowed distance between matched portions of borders. Then the system iteratively attempts to match all pairs of selected borders and to stitch sufficiently close matching portions of them. Typically, the user invokes this tool several times with a slightly increasing sequence of limits on the inter-border distance. In difficult instances of the problem the user may select only two borders at a time. Identifying matching portions of borders is performed by searching for a sufficiently close pair of vertices (one on each border) and expanding the match along the borders. No global surface-orientability is checked. If we denote by $b$ the number of selected borders and by $\ell_i$ the complexity (number of vertices) of the $i$th border (for $1 \le i \le b$), the running time of the implemented version is $O(\Sigma_{1 \le i < j \le b} \ell_i \ell_j)$. This term is quadratic in the worst case in the complexity of the whole object, but in practice the user selects a small number of close borders in each iteration. Figures 6(a,b) show a CAD object and the borders of its connected components, respectively. Figures 6(c,d) show closeups of some flaws in the boundary of the object before and after fixing these flaws, respectively.

Two additional operations on objects are provided by the Academy:

1. Widening a surface into a solid object. Occasionally an object is designed as a zero-width open surface. The user specifies the desired thickness and the side of the original surface in which the offset surface is to be generated. Then the system generates an offset surface and the two surfaces are bound so as to create a solid. In case the offset size is comparable to that of

[1] $\alpha(m, c)$ is an extremely slowly growing functional inverse of Ackermann's function. For all practical values of $m$ and $c$, the function $\alpha(m, c)$ does not exceed a very small constant [Ta].

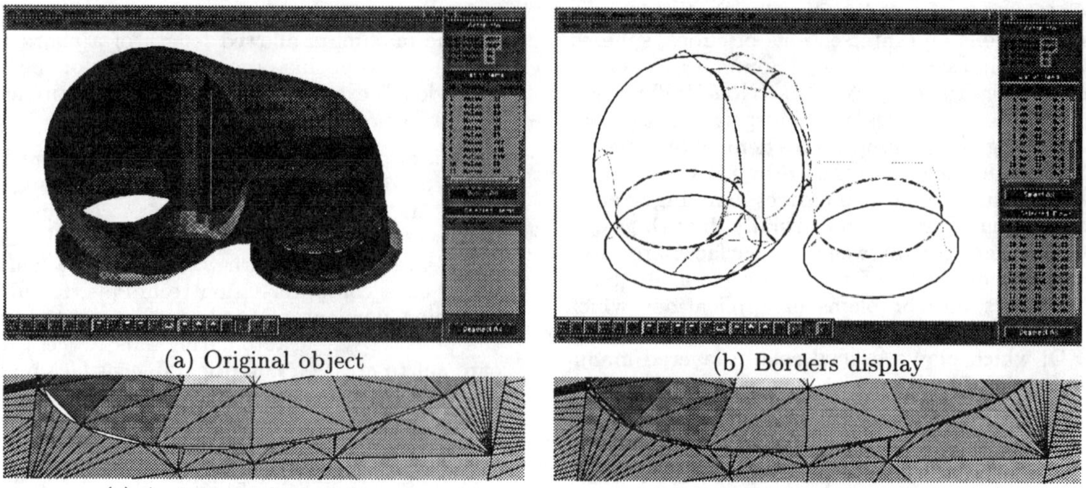

(a) Original object

(b) Borders display

(c) A closeup of some flaws in the object (before fixing)

(d) A closeup of some flaws in the object (after fixing)

Figure 6: Repairing flaws in the boundary of a CAD object

the features of the original surface (especially when it has sharp dihedral angles between neighboring facets) the offset surface may intersect itself or the original surface. A future release of the system will identify such self-intersections and will notify the user, which will then be able either to ignore the intersections or to perform local corrections of the object.

2. Cutting an object by a plane into two objects. This tool computes all the intersections between the plane and facets of the object: each hit facet is split into a number of pieces (this number is proportional to the number of edges of the facet intersected by the plane). Then the original object is split into two parts. Each facet not hit by the cutting plane is put in its respective part, while the pieces of the hit facets are located in the appropriate part according to the side of the cutting plane in which they lie. Finally, the two parts of the original object are closed by additional facets which lie within the cutting plane. The human interface of the system allows the user to specify a series of $k$ parallel planes used to cut the object into $k+1$ new objects.

## 4 Designing a Job

In most cases the capacity of the layered-manufacturing machine is by far larger than the size of any single built model. For example, the capacity of Cubital's Solider 5600 machine is 20in × 14in × 20in. Therefore, it is much more economical to produce a huge "job" which contains many small models than to build each model separately. The DFE system provides a package for editing such a production job whose main features are:

- Positioning objects at different locations. Each object may have multiple instances in a job.
- Rotating (by specifying Euler angles [Go, p. 608]—a triple of rotations around the $x$, $y$, and $z$ axes, or by some angle $\theta$ about some axis $\ell$) an object participating in a job, as well as scaling it (uniformly in all directions, or with different scaling factors in each direction), or mirroring it (reflecting it through some plane in 3-space). In fact, the user may apply any affine transformation to an object.
- Creating multiple copies of the same object, either along a line, or on a 2- or 3-dimensional grid, or along a 2-dimensional arc parallel to the $xy$-plane.

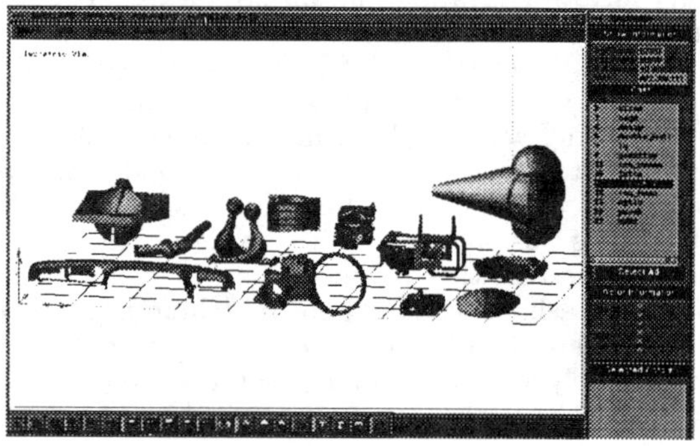

Figure 7: A production job

Figure 7 shows an example of a production job.

The entire job description is kept in a simple Ascii file. The sophisticate user may edit this file by using a text editor. Except that, of course, by doing that the user is unable to *visualize* the job. The purpose of visualizing a job is 3-fold:

1. Positioning models in the desired orientations. Most layered-manufacturing techniques perform differently along the different axes (especially the $z$ axis), e.g., with respect to the amount of model *shrinkage* during and after production.

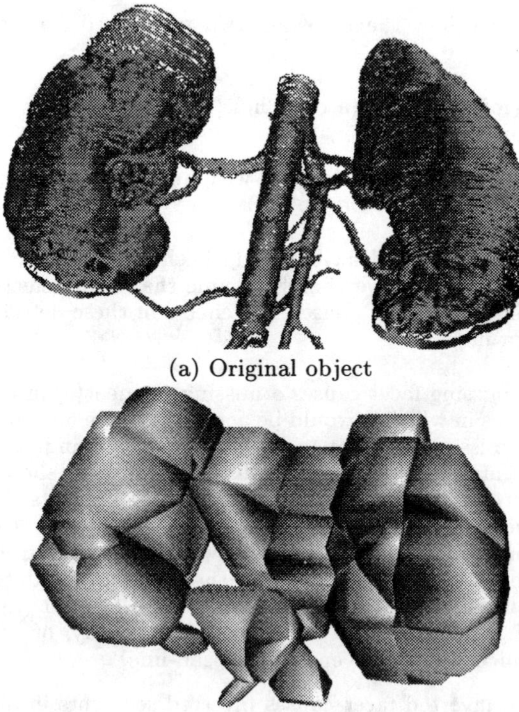

(a) Original object

(b) Displayed object

Figure 8: A "ghost" of an object

2. Positioning models economically. That is, avoiding as much as possible voids in the job volume not occupied by models. Usually the cost of the production of a job is proportional to the volume of its axis-aligned bounding-box.

3. Positioning models so that they do not intersect each other. Such intersections spoil in most cases only the involved models but are nevertheless undesirable.

The collection of objects that participate in a job may sum up to a huge amount of data which require a large amount of memory resources and may slow down the display substantially. The user is therefore advised not to display every single object in its full details but to display a "ghost," a rough approximation of the object which is yet descriptive enough for visualization and for designing the job. The system provides five approximation levels; in the higher levels the object is first subdivided into parts and then a ghost is computed for each part separately. Figure 8(a) shows a model of human kidneys which contains 192,062 vertices and 96,031 facets. Its ghost version, shown in Figure 8(b), contains only 964 vertices and 924 facets.

Finally, the ultimate goal of the job editor is to "consolidate" all the objects participating in the job into one object which will then be submitted for production. The consolidated job is saved in a file with the same structure as any other object. Thus, this operation may also be performed for saving a layout of multiple objects into one object to be further referred to as a whole for relocation, rotation, duplicating, etc.

# 5 Job Previewing and Manufacturing

The rear-end of the data front-end may be connected to a layered-manufacturing machine. The system allows the user to first preview the course of the manufacturing process without actually producing anything, and then, when the user is satisfied (and/or when the machine is ready), to perform the manufacturing job. Otherwise the user is able to return to the Academy (for repairing some object(s)) or to the job editor (for modifying the job), and then to invoke the production tool again.

As described above, all layered-manufacturing technologies produce the job layer by layer, and the different technologies vary in *how* they produce the layers. As explained earlier in this paper (see Section 4), the user designs a production job which contains multiple objects and then "consolidates" it into a single object—input to the production tool. This tool computes a series of intersections between $xy$-parallel planes (hereafter denoted *slices*), usually equally spaced along the $z$ axis, and the job object. Since the job is actually a polyhedron (even though it consists of many connected components), computing its intersection with a plane is straightforward: Each facet that intersects the plane contributes one or more directed segments, e.g., so that its right side is on the "material" side. The collection of all these segments at some $z$ (the *height* of the slice above the bottom of the job) forms a set of closed polygonal chains—simple polygons within the slice—which describe areas of "material" (inside some object) or non-material (outside any object). The computation of the series of such polygonal slices is performed efficiently: Since every facet is connected, it may intersect $xy$-parallel planes only within some range of heights ($z$ values) which may be precomputed only once by scanning all the vertices of the facet and by finding the minimum and maximum $z$ coordinates of the facet. Maintaining the active facets along the sequence of slices may be implemented, e.g., by two priority queues, whose keys are the minimum and maximum $z$ coordinates of the facets, respectively.

The process described above is common to all layered-manufacturing techniques. Now the system invokes the customizable part: converting the slice data into a format recognized by the actual production machine. Some machines behave like "pen-plotters": they need only the set of closed polygons per slice, so the production tool has to pack these data in the appropriate format and send them to the target machine. Other machines (e.g., the SLA series of 3D Systems) make some "hatching" process, where a laser beam first solidifies liquid along the polygons themselves, then makes a hatch at the interiors of these polygons.

The DFE customization described here is for the Solider 5600 MPM (Model-Production Machine) of Cubital Ltd., which requires each slice data to be given as a raster image (pixel bitmap). The production tool is customized to produce slices at some $z$ intervals, and then to *rasterize* each slice according to the required $xy$-resolution. Before sending the slices raster data to the machine, the user may wish to preview it in order to detect errors, as explained below. Figure 9 shows a typical slice of a job, in which the cross-sections of several objects are seen, all surrounded by machine-required protection aids. This specific machine requires the creation of a "fence" around the entire job, where

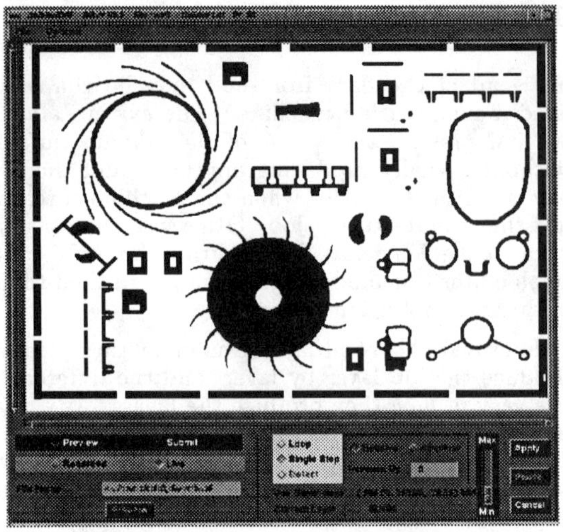

Figure 9: A raster image of a job slice

the walls of the fence are broken at predefined locations for releasing pressures that arise in the process of solidifying liquid.

The rasterization process is based on a well-known scan-line algorithm [FD]. The polygons are scanned along the $y$ axis (from top to bottom), and for each particular value of $y$, each active segment (a segment within the slice that intersects with the scan-line at that value of $y$) contributes a point where the scan-line either enters or exits the "material" area. The type of the point (*entry* or *exit*) is determined according to whether the scan-line crosses the edge from its "left" side to its "right" side (recall that the segments are directed) or vice versa. Note that for this purpose the segments within a slice need not be collected into closed polygons: they can appear in random order but still the algorithm performs correctly. After all the crossing points are computed for some scan-line, they are ordered according to their $x$ coordinates. An entry point means that the scan-line enters a material region, and conversely, an exit point means that the scan-line leaves the material region. When the original objects do not overlap (as in most cases) all the crossing points along a scan-line appear in alternate order, starting with an entry point and ending with an exit point. However the job editor allows objects to intersect. Overlapping objects are reflected by some permutation of the crossing points: the scan-line may enter two objects in a row (the second entry-point is the entry to the overlapping area) and then exit from both (the first exit-point occurs when the scan-line leaves the overlapping area). More generally, any valid parentheses-sequence of entry and exit points is legal; the permutation of the points reflects the overlaps between the various objects. The scan-line needs only to count the "level" of material it is in. Initially (at $-\infty$ along the $x$ axis) it starts with level 0 (outside any object), then every entry point increments the level by 1 and every exit point decrements the level by 1. Finally (at $+\infty$ along the $x$ axis) the scan-line should return to level 0. The simple observation is that the material lies wherever the level is greater than 0 (level $k$ means that $k$ objects overlap at this portion of the scan-line).

Now consider the following two possible defects of an object:

1. A missing facet (a real "hole" in the boundary of the object).
2. An inverted facet (a facet with an incorrect indication of the inside of the object).

These defects should have already been detected and fixed in the Academy. Nevertheless assume that such defects did reach the production phase. The effects of these defects are the following:

1. A missing facet causes a missing segment(s) in all the slices in which it would be active if it were not missing. A missing segment in a slice causes missing crossing points along all the scan-lines in which it should be active. Such a missing crossing point prevents the increase or decrease of the material level. Thus, a scan-line fails to "detect" an entry to or an exit from an object. After processing all the crossing points the final level is not 0, as expected, but $+1$ or $-1$ (in case of a single defect). Multiple defects may, of course, cancel out at the end of the scan-line.
2. An inverted facet causes inverted segments in all the slices in which it is active. An inverted segment in a slice causes crossing points of the wrong type (exit instead of entry and vice versa) in all the scan-lines in which it is active. Such an erroneous crossing point causes decreasing the material level instead of increasing it or vice versa. After processing all the crossing points the final level is not 0, as expected, but $+2$ or $-2$ (in case of a single defect). Multiple defects may also cancel out at the end of the scan-line.

These defects cause the so-called *stray-lines*: scan-lines which do not have, starting at some defected point, the correct level of material. From this point the result of the rasterization is incorrect. The rest of the scan-line erroneously becomes either "material" (a "black" hair that connects between seemingly separate objects) or non-material (a "white" hair that erases subsequent objects). The production previewing tool offers automatic detection of stray-lines which verifies that all the scan-lines terminate with level 0. It is obvious from the discussion above that this simple mechanism misses the rare (and unrealistic) situations where such defects cancel out in all the scan-lines within all the slices affected by the defects. This may happen, for example, when two facets, where one is an exact $x$-translation of the other, are inverted. In this pathological case only the spatial region bounded by the convex hull of these two facets will have a level either greater or less than by 2 from the correct level.

For completeness we mention that stray-lines may also be the result of (in)robustness issues. Round-off errors may cause the algorithm to miss segments (intersections of facets with the cutting planes) or crossing points. Furthermore, when a scan-line intersects with an original vertex of a polygon in the slice or when it contains a segment, wrong handling of these degeneracies may lead to an incorrect level update. Such events are inevitably present in our system too. Nevertheless most of these situations are avoided by perturbing the vertices such that no scan-line occurs at a vertex.

Upon detecting "hairs," the user may either return to the Academy in attempt to repair the defected object, or return to the job editor for repositioning the object (either in a position where it will not hurt other objects, or for rotating it a bit—such operation solves in most cases stray-lines generated by robustness problems!), or to ignore them. The user may choose to ignore stray-lines either because he or she is unable to fix the problem, or because the stray-lines are thin (i.e., there are only few successive scan-lines within one slice with the same problem), there are not too many of them, and they do not appear in subsequent slices.

The user may preview all the slices in a sequence, display every $k$th slice (for $k > 1$), or jump between them at random. Once the user is satisfied with the slices, he or she may submit the job to the production machine. The sliced data is then sent through communication channels by using the appropriate machine-dependent protocol.

## 6 Summary

In this paper we describe the DFE, a data front-end for layered manufacturing of 3-dimensional models out of computer data. The system accepts data files from a variety of sources, chief of which is CAD systems, analizes them, and allows the user to modify and repair them individually, and to design a production job for rapid prototyping. The DFE may be customized for any layered-manufacturing technique. The system is commercially available from Cubital Ltd.; details can be obtained from the second author. In our opinion, the next generation of the DFE and similar systems will handle not only polyhedra but general smooth objects which are typically designed by using CAD systems. The approximation of such smooth objects by polyhedra was imposed on this industry only because the first RP technologies were accompanied by software that insisted on inputing polyhedra. The de-facto standard of this industry (STL—see Section 2) is widely considered inadequate for accurate yet efficient production. However, direct slicing of smooth objects may be a much more computationally involved (and error-prune) task to perform.

## Acknowledgments

The geometric data structures of the system [Ba] were developed by the first author as a Ph.D. student at Tel-Aviv University under the supervision of Prof. M. Sharir. The authors wish to thank L. Werner from A-Z PR for helpful discussions on the RP industry.

## References

[3DS] 3D SYSTEMS, INC. (Valencia, CA), Stereolithography Interface Specification, p/n 50065-S01-00, 1989.

[Ba] G. BAREQUET, DCEL: A polyhedral database and programming environment, submitted for publication, 1996.

[BS1] G. BAREQUET AND M. SHARIR, Filling gaps in the boundary of a polyhedron, *Computer-Aided Geometric Design*, 12 (1995), 207–229.

[BS2] G. BAREQUET AND M. SHARIR, Piecewise-linear interpolation between polygonal slices, *Computer Vision and Image Understanding*, 63 (1996), 251–272.

[BW] J.H BØHN AND M.J. WOZNY, Automatic CAD-model repair: Shell-closure, *Proc. Symp. on Solid Freeform Fabrication*, Dept. of Mech. Eng., Univ. of Texas at Austin, 1992, 86–94.

[DM] A. DOLENC AND I. MÄKELÄ, Optimized triangulation of parametric surfaces, Technical Report TKO-B74, Helsinki University of Technology, 1991, *Mathematics of Surfaces IV*.

[FD] J.D. FOLEY AND A. VAN DAM, *Fundamentals of Interactive Computer Graphics*, Addison Wesley, 1984.

[Go] H. Goldstein, *Classical Mechanics*, Addison-Wesley, Reading, MA, 1980.

[Kl] G.T. KLINCSEK, Minimal triangulations of polygonal domains, *Annals of Discrete Mathematics*, 9 (1980), 121–123.

[KSL] F.-L. KRAUSE, C. STIEL, AND J. LUDDERMANN, Processing of CAD-data—conversion, verification and repair, *Proc. 4th ACM Symp. on Solid Modeling*, Atlanta, GA, 1997, to appear.

[MD] I. MÄKELÄ AND A. DOLENC, Some efficient procedures for correcting triangulated models, *Proc. Symp. on Solid Freeform Fabrication*, Dept. of Mech. Eng., Univ. of Texas at Austin, 1993.

[MF] S.M. MORVAN AND G.M. FADEL, IVECS, interactive correction of .STL files in a virtual environment, *Proc. Symp. on Solid Freeform Fabrication*, Dept. of Mech. Eng., Univ. of Texas at Austin, 1996, 491–498.

[SW] H. SAMET AND R.E. WEBBER, Hierarchical data structures and algorithms for computer graphics; Part II: Applications, *IEEE Computer Graphics & Applications*, 8 (1988), 59–75.

[SH] X. SHENG AND B.E. HIRSCH, Triangulation of trimmed surfaces in parametric space, *Computer-Aided Design*, 24 (1992), 437–444.

[ST] X. SHENG AND U. TUCHOLKE, On triangulating surface model for SLA, *Proc. 2nd Int. Conf. on Rapid Prototyping*, Dayton, OH, 1991, 236–239.

[Ta] R.E. TARJAN, *Data Structures and Network Algorithms*, SIAM, Philadelphia, 1983.

[Te] TECNOMATICS, N.V., Robface: The universal Rob-Cad programming interface, 1988.

[Wo] T.T. WOHLERS, STL viewers and editors, 1996 (available at `http://lamar.colostate.edu/~wohlers/stl_view.html`).

# Wall Thickness Control in Layered Manufacturing

Seth Allen
swallen@math.lsa.umich.edu

Debasish Dutta
dutta@engin.umich.edu

Department of Mechanical Engineering and Applied Mechanics
University of Michigan, Ann Arbor, MI 48109–2125

**Abstract**

The layered manufacture of sheet (surface) parts is considered for processes that use support structure. In particular, for a certain class of surfaces, a new method for construction that eliminates the need for "traditional" support structure, hence reducing wasted material and decreasing build time, is proposed. This method selectively thickens the surface's walls and is a natural compliment to adaptive slicing. Four related goals that a designer using these techniques might want to achieve are carefully analyzed. Instructive examples illustrate each goal. Finally, the application of these methods to general surfaces and to solids whose interiors are unimportant is considered.

## 1 Introduction

Layered manufacturing is a method of fabrication where objects are constructed layer by layer. This method offers a significant reduction in protptyping time compared to convention manufacturing techniques, and remains fast even when building geometrically complex objects. On the software side, many issues in layered manufacturing depend on computational geometry, e.g., build time, surface quality, nozzle path, etc.

Although many different processes are available for layered manufacturing, we are primarily concerned with those processes that might require the use of external support structure during the object's formation. Stereolithography as found in 3D Systems' machines and material deposition processes as found in machine made by Stratasys are two examples of processes that make use of support structure. For more on support structure, see [1].

Support structures and their effective use is an important problem in layered manufacturing. Since these have to be physically/manually removed from the finished part, time and material is wasted. Some (commercial) machines have employed novel techniques to facilitate support structure material identification and removal, e.g., using different materials–mechanical/thermal properties and color–than the build material. This strategy also requires hardware modifications such as dual heads to deposit the different materials.

The ability to build a hollow box/sphere, directly, has often been cited as a novel and distinguishing feature of layered manufacturing. However, the removal of trapped support material is the caveat. To the best of our knowledge, the necessity of support structures in layered manufactured parts has not been investigated, in the literature, thus far. Support structures can be in the inside (for hollow objects) or the outside (for hollow and solid objects). The latter has often been included as a criterion in the part orientation determination [1]. The focus of this paper is on how to avoid using support structures in the inside, thereby saving material and reducing build time. We propose achieving this by selectively thickening the surface's walls.

Even greater savings are possible when wall thickness control is used with adaptive slicing. Also, by employing the techniques presented here, a designer will be able to specify one measure of part strength.

For general surfaces, we apply our methods only to those regions that benefit from these techniques. Further, for solids whose exact interior specification is unimportant, the methods presented here are immediately applicable by thinking of the solid as a closed (hollow) surface. In these two cases, a hybrid approach that uses both traditional support structure and thickened walls is employed. Even here, a significant savings is still possible.

Although the methods developed here work for every surface type, we focus our attention in the examples to surfaces of revolution built in the direction of the axis of rotation. For these objects, the analysis is greatly simplified and so our methods are completely specified. Additional work remains to fully analyze the general, free-form surface cases.

### 1.1 Overview

§1 continues by defining the Wall Constructibility Property (WCP), and concludes with an example of a hemisphere built using "traditional" support structure.

In §2, we prove that any surface with WCP can be built without external support by thickening its walls. An algorithm for determining the required wall thickness is presented, and an example of a hemisphere build using the algorithm is considered.

We generalize the problem of constructing parts with WCP and no traditional support structure to include adaptively sliced parts and adaptively sliced parts with a minimum specified part strength in §3 and §4 respectively. Again algorithms and examples are given.

In §5, we consider building an object with WCP using thickened walls and a given specified maximum wall thickness. This is the first case where a hybrid approach that uses both traditional support structure and thickened walls may be required. An example illustrating the problem is presented.

§6 shows how the techniques developed in the preceding sections extend to objects that do not possess WCP. In these cases, the hybrid approach using thickened walls and tradition supports must


*Computational Geometry 97* Nice France

again be employed. An example demonstrates the methods used in this section.

§7 describes physical models of some of the previous examples that were built on a Stratasys 3D Modeler layered manufacturing machine.

Finally, §8 contains our conclusions.

### 1.2 Setup

Suppose a user would like to build a sheet object $S$ using a material deposition layered manufacturing process. For each point $p \in S$, let $N_S(p)$ be the outward pointing normal to $S$ at $p$. Let $\nu$ be the build direction. The surfaces that we consider first will have the Wall Constructibility Property (WCP) defined below.

**Definition 1.1 (Wall Constructibility Property).** Given a surface $S$ and build direction $\nu$, $S$ has the *Wall Constructibility Property* if $S$ can be and is oriented so that $\forall p \in S, N_S(p) \cdot \nu \geq 0$.

Note, every surface has an outer side that is to be built and an inner side which is, for us, unimportant. Definition 1.1 says that the surface's outer side can be safely thought of as the upper side. So for a surface $S$ with WCP, all the traditional support structure required to build $S$ lies below the surface. Examples of parts with WCP include molds and dies.

Given a surface $S$, we refer to the manufactured part as $LM(S)$ and the $i$-th layer of $LM(S)$ as $L_i$.

When discussing over-hang in the context of layered manufacturing, two meanings come to mind:

1. the over-hang between layers, and

2. the structural or global over-hang of the object as it is constructed.

In this paper, we are primarily concerned with the first meaning, and so, we define the over-hang $\triangle o_i$ between $L_i$ and $L_{i-1}$ is the maximum of the minimum distances from each point $p$ on the bottom of $L_i$ to a point on the top of $L_{i-1}$. Note, the bottom of $L_i$ and the top of $L_{i-1}$ lie in the same plane.

Further suppose the deposition process has the following characteristics. The maximum distance that one layer can over-hang the layer directly below without requiring support is $\triangle o_{\max}$. The thickness (height) of a layer is $\triangle l$ and the road width is $\triangle r$. Since $S$ is a surface, the nozzle path on each layer follows the intersection of $S$ with a plane at the height of the given layer and offsets of this intersection, i.e. the nozzle path uses the contour pattern. So the thickness of $LM(S)$'s walls (before they are thickened) at a given layer is determined by the number of contours used times $\triangle r$ and the thinnest possible wall is $\triangle r$.

We remark that support structure is required at points $p$ on $S$ where $|N_S(p) \cdot \nu| > 1 - \epsilon$ for some small $\epsilon$. Since these are the points where the normal to the surface lies along the build direction, at these points the over-hang between layers is the greatest and often gaps around these points will appear in the sliced object. More generally, given a point $p$ on $S$, the over-hang at $p$, $\triangle o_{(p)}$, can be estimated with the following formula,

$$\triangle o_{(p)} \approx \triangle l \cot \theta, \tag{1}$$

where $\theta = \cos^{-1}(N_S(p) \cdot \nu)$. See Figure 1. Note, as the slice thickness is decreased, over-hang amounts also decrease and the approximation in Equation (1) improves. Now define the minimum angle $\theta_{\min}$ between the surface normal and build direction at which no support structure is required. This angle is given by

$$\theta_{\min} = \tan^{-1}\left(\frac{\triangle l}{\triangle o_{\max}}\right). \tag{2}$$

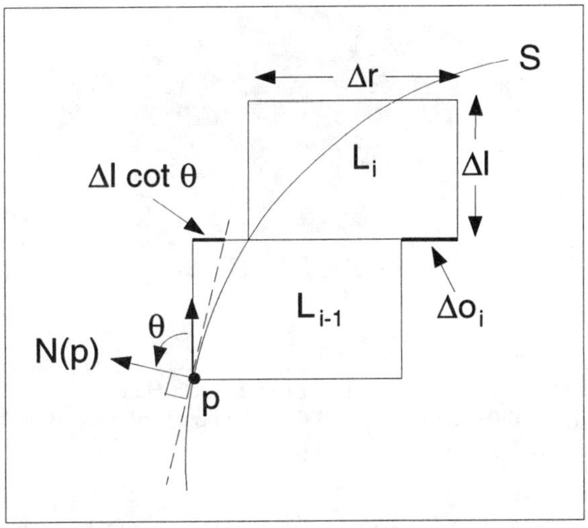


Figure 1: Demonstrates the approximation $\triangle o_{(p)} \approx \triangle l \cot \theta$.

So at a point $p \in S$, support is required if and only if $N_S(p) \cdot \nu \geq \cos(\theta_{\min})$.

By "traditional" support structure, we refer to the scaffolding built along with a part in a primarily vertical direction. This scaffolding is used both to support large over-hangs and to keep the object from toppling as it is built, and is removed (with some difficulty) and discarded at build completion.

### 1.3 Traditional Support Example

Consider the problem of building the hemisphere $S$ given by $x^2 + y^2 + z^2 = R^2, z \geq 0$ using a material deposition process and traditional support structure. Suppose that the build direction $\nu$ is $(0,0,1)$. Since the normal to $S$ at the point $p = (x,y,z)$ is $(x/R, y/R, z/R)$, we get $N_S(p) \cdot \nu = z/R$. So support is required at all layers whose height is greater than $z_{\min}$ given by:

$$z_{\min} = R\cos\left(\tan^{-1}\left(\frac{\triangle l}{\triangle o_{\max}}\right)\right). \tag{3}$$

Using the traditional method of support, we build a thin hemisphere with an internal (solid) cylinder of support with radius $\sqrt{R^2 - z_{\min}^2}$. See Figure 2. Notice the large relative size of the support structure compared to the desired surface. Constructing these supports wastes both material and time.

In §7, we discuss this example further and show pictures of the hemisphere built with traditional support structure on a Stratasys 3D Modeler using the fused deposition modeling process.

## 2 Problem I - Constant Layer Thickness

Given a surface $S$ with WCP, we want to determine the thickness of the $S$'s walls so that $S$ can be built without support structure. In other words, how do we thicken the walls of $S$ (as little as possible) so that $\forall i, \triangle o_i \leq \triangle o_{\max}$. By Theorem 2.1, we see that this goal is always possible.

Recall, support structure is required during layered manufacturing in the following two cases:

1. To support large over-hangs between layers. For solids, these over-hangs can only occur on faces with downward pointing normals, either because the angle between the surface normal

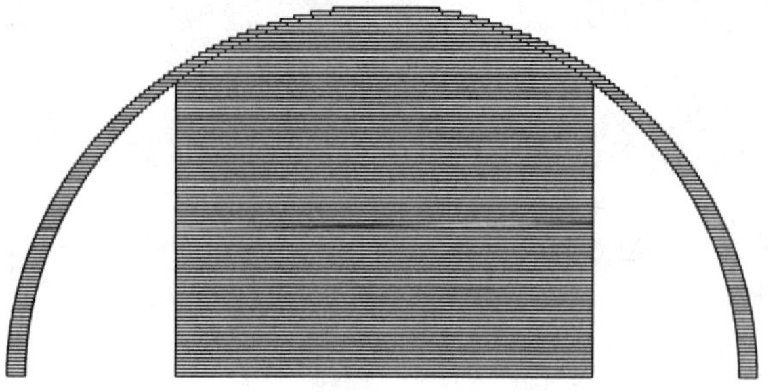

Figure 2: A cross-section of the hemisphere $LM(S)$ with radius 1 created with traditional support structure. Here, $\triangle l = 0.01$, $\triangle o_{\max} = 0.015$, and $\triangle r = 0.05$. The height of the lowest layer requiring support is $z_{\min} \approx 0.83$.

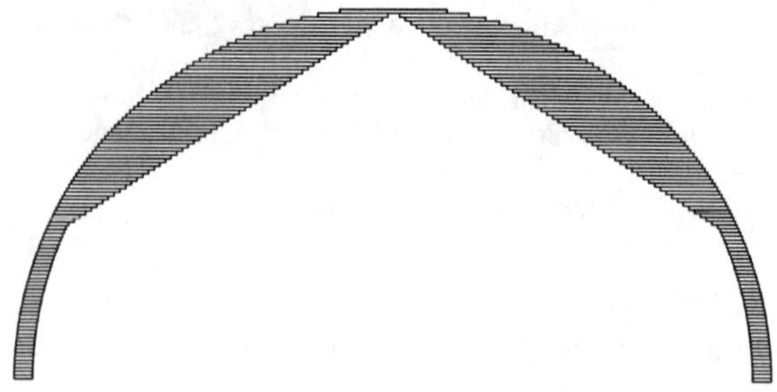

Figure 3: A cross-section of the hemisphere $LM(S)$ from Example I with radius 1 created with thickened walls for support. Here, $\triangle l = 0.01$, $\triangle o_{\max} = 0.015$, and $\triangle r = 0.05$.

and the build direction is small or because the face lies on a floating component.

2. To prevent the object from toppling as it is built.

For more on determining when support structure is necessary, we refer the interested reader to [1].

**Theorem 2.1.** *Let $S$ be any surface with WCP. Then $S$ is constructible* without any external support *by sufficiently thickening $S$'s walls.*

*Proof.* Define $B(S)$ be the projection of $S$ in direction $-\nu$ into the plane on which $S$ rests. So $B(S)$ can be thought of as a base for $S$. Then let $Cyl(S)$ be the set of all points lying between $S$ and $B(S)$. Since $S$ has WCP, $Cyl(S)$ is a solid whose boundary has top face $S$, bottom face $B(S)$, and, possibly, vertical side walls. Note, by thickening $S$'s walls enough, we are always able to get the solid $Cyl(S)$. We will consider building this solid.

Note, since $S$ has WCP, the only downward pointing normals on $Cyl(S)$ are on its base, and so no support structure is required support over-hangs between layers. Further, $Cyl(S)$ is stable at every stage of the construction, since the definition of $B(S)$ ensures that the center of mass of the "growing" part always lies above the convex hull of its base. So no support structure is required to keep $Cyl(S)$ from toppling. This completes the proof. □

Suppose $S$ has WCP. Let $\widetilde{S}$ be the union of $S$ and the vertical boundary walls of $Cyl(S)$. Since the center of mass of $\widetilde{S}$ as it is constructed always lies above the convex hull of $B(S)$, $\widetilde{S}$ is a stable surface that will not topple. So by using $\widetilde{S}$ instead of $S$, we are able to assume that every surface $S$ with WCP has the additional property that $S \cup B(S)$ encloses a solid. Therefore, none of our thickened wall algorithms for surfaces with WCP need to supply support to prevent the surfaces from toppling. However, some waste may result from the possibly unnecessary construction of $Cyl(S)$'s vertical boundary walls.

### 2.1 Algorithm for Thickened Walls Construction

Let $S$ be a surface with WCP. Suppose $LM(S)$ has $n$ layers where $L_n$ is the top layer. With this setup, the algorithm for constructing $S$ with thickened walls is as follows:
**Algorithm 1:**

1. Let $i = n$.
2. Measure the over-hang $\triangle o_i$ between layers $L_i$ and $L_{i-1}$. If $\triangle o_i > \triangle o_{\max}$, thicken $L_{i-1}$'s walls (as little as possible) so that this condition is false.
3. $i = i - 1$.
4. If $i = 1$, then stop. Otherwise, goto Step 2.

Note, the work in this algorithm is all in Step 2. It may be difficult both to measure the over-hang between layers, and to decide the best way to thicken the lower layer once the over-hang is determined. This computation is greatly simplified for a surface of revolution built in the direction of its axis $\nu$ of rotation, since each layer becomes a collection of concentric annuli. In this case, it is enough to calculate the over-hang at any point $p$ on one of the annuli by measuring the distance between layers in the plane determined by $\nu$ and $p$. Thickening the layer amounts to reducing the radius of the annulus' hole.

### 2.2 Example I

Consider again the problem of building the hemisphere $S$ given by $x^2 + y^2 + z^2 = R^2$, $z \geq 0$ using a material deposition process, but this time *without* external support structure. Notice that $S$ clearly has WCP. At the point $(0, 0, R)$, the surface normal and build direction coincide, therefore we concentrate on this point.

In the worst case, the final layer built $L_{\text{top}}$ in the hemisphere lies at a height $R - \triangle l$. By plugging this height into the equation of the hemisphere, we see that a disk of radius $\sqrt{2R\triangle l - (\triangle l)^2}$ and center $(0, 0, R - \triangle l)$ must be deposited at this layer. Therefore, the wall at $L_{\text{top}-1}$ must be thickened so that $L_{\text{top}-1}$ becomes an annulus with interior radius $\triangle o_{\max}$ centered at $(0, 0, R - 2\triangle l)$. Similarly, $L_{\text{top}-2}$ will be an annulus with interior radius $2\triangle o_{\max}$. Continuing in this manner, we see that $L_{\text{top}-n}$ is an annulus with interior radius $n\triangle o_{\max}$. This process stops when either the bottom of the hemisphere is reached, or the thickness of the wall is $\triangle r$. So $LM(S)$ is a solid hemisphere with a (stepped) cone removed from its interior. Figure 3 shows a cross-section of $LM(S)$ after thickening its walls using the above algorithm. See §7 for more on this example including a picture of the hemisphere built with thickened walls and no traditional supports.

## 3 Problem II - Variable Layer Thickness

Geometric inaccuracies due to the stair-case effect in layered manufacturing can be quantified by the cusp height between a layer and the boundary of the original surface. Since using thicker layers decreases the time for the build, and since the cusp height is a function

Figure 4: A cross-section of the hemisphere $LM(S)$ from Example II with radius 1 with variable slice thickness chosen so that the cusp height is always 0.006. The other parameters are the same as in Example I, $\triangle o_{\max} = 0.015$, and $\triangle r = 0.05$.

of both the surface shape and the slice thickness, using variable slice thicknesses allows a user to maximize build speed for a specified part accuracy. For more on this, see [2].

Now we solve Problem I, but with the layer thicknesses specified to lie within some range allowing $\triangle l$ to vary between $\triangle l_{\min}$ and $\triangle l_{\max}$, where the thickness of $L_i$ is $\triangle l_i$. The maximum over-hang constructible without support depends on the layer thickness (since the reduced weight of material deposited causes less droop, surface tension is better able to support itself, etc), on the overlap between layers, and on other factors. So, in place of $\triangle o_{\max}$ on $L_i$, we use $\triangle o_{i,\max}$ which varies with $\triangle l_i$. Further, since the amount of over-hang along $L_i$ depends on the thickness of $L_{i-1}$, we get

$$\theta_{i,\min} = \tan^{-1}\left(\frac{\triangle l_{i-1}}{\triangle o_{i,\max}}\right). \tag{4}$$

So at a point $p$ on the bottom of $L_i \cap S$, support is required if and only if $N_S(p) \cdot \nu \geq \cos(\theta_{i,\min})$.

Note, the algorithm to compute how surface's walls should be thickened in the variable layer thickness case, is exactly the same as Algorithm 1 presented in §2.1, except now in Step 2, we check that $\triangle o_i < \triangle o_{i,\max}$.

### 3.1 Example II

Consider again the hemisphere example as presented in §2.2, but this time allow layer thickness to vary. In particular, suppose the layer thicknesses are chosen to keep the cusp height constant and hence improve the quality of the constructed surface. Suppose the cusp height is the constant $c$, and the height of the base of layer $L_i$ is $z_i$. Then the layer thickness of $L_i$ is given by:

$$\triangle l_i = \sqrt{(R+c)^2 - (R^2 - z_i^2)} - z_i. \tag{5}$$

Note, $\triangle l_{i-1} \geq \triangle l_i$. To simplify the analysis, we assume $\triangle o_{i,\max}$ is constant for all $i$ (the material properties do not depend on the thickness of the material deposited). Now using the same reasoning as before, we see that $LM(S)$ is a solid hemisphere with a rounded cone removed. See Figure 4.

Note, since the cusp height and the over-hang at a layer are directly related, the thinner layers are also the layers at which the walls are most thickened. So less material (and time) is required for this build compared to that of Example I in §2.2.

For more on this example, including a picture of a physical model, see §7.

## 4 Problem III - Specified Part Strength

Now we solve Problem II, but further take into account a sense of $S$'s part strength as measured by the minimum depth of $S$'s walls at each point $p$ on $S$. The depth $D_S(p)$ of a point $p \in S$ is the distance measured along $-N_S(p)$ that a ray with end point $p$ travels through $LM(S)$ before first leaving the manufactured part. Note, to achieve a desired part strength, $S$'s walls may have to be thickened beyond the point where construction without traditional support structure first becomes possible.

### 4.1 Algorithm for a Specified Part Strength

Let $S$ be a surface with WCP. Suppose $LM(S)$ has $n$ layers where $L_n$ is the top layer and that the minimum depth allowed on $LM(S)$ is $d_{\min}$. With this setup, the algorithm is as follows:
**Algorithm 2:**

1. Let $i = n$.
2. Let $j = i$.
3. Let $d_i$ be the minimum of all depths $D_S(p)$ calculated at each point $p$ on the top of $L_i$. If $d_i \geq d_{\min}$, goto Step 7.
4. $j = j - 1$.
5. If $j = 0$, goto Step 7.
6. Thicken $L_j$'s walls (as little as possible) so that $d_i$ is increased. Do not thicken $L_j$'s walls past the point where $d_i = d_{\min}$. If it is not possible to increase $d_i$ by thickening $L_j$, stop since the desired part strength cannot be achieved. Otherwise, goto Step 3.
7. Measure the over-hang $\triangle o_i$ between layers $L_i$ and $L_{i-1}$. If $\triangle o_i > \triangle o_{\max}$, thicken $L_{i-1}$'s walls (as little as possible) so that this condition is false.
8. $i = i - 1$.
9. If $i = 1$, then stop. Otherwise, goto Step 2.

Note, in this algorithm, the base is thought of as a final infinitely thick layer that contributes to part strength. An alternative is to stop in Step 5 if $j = 0$, since the desired depth has not been achieved and no more layers remain to be thickened.

Again, the algorithm works by thickening $LM(S)$'s walls from the top down. Here however, more than one layer below the current layer may need be thickened so that the desired strength is achieved. After, the proper depth is assured, the layer below the current layer may need to be thickened further to support any large over-hangs. In the domain of surfaces of revolution, these computations are greatly simplified and may be performed in any plane containing the axis of revolution.

This algorithm may fail if the part depth is chosen to be too large. These are cases where even the solid $Cyl(S)$ is not strong enough. For example, consider making a tall, thin tube (cylinder) with radius $r$. The maximum depth at every point on the tube cannot be greater than $2r$. So for $r$ small, the tube is necessarily weak.

### 4.2 Example III

Consider the variably sliced hemisphere $S$ presented in §3.1, but now thicken $S$'s walls to ensure a minimum depth $d_{\min}$ on the surface. Output from the algorithm for this example is shown in Figure 5.

Notice that before ensuring a minimum depth, $LM(S)$ is very weak at the point $p = (0,0,1)$, since the depth at this point is only one (thin) layer thickness, see Figure 4. But not only are the upper

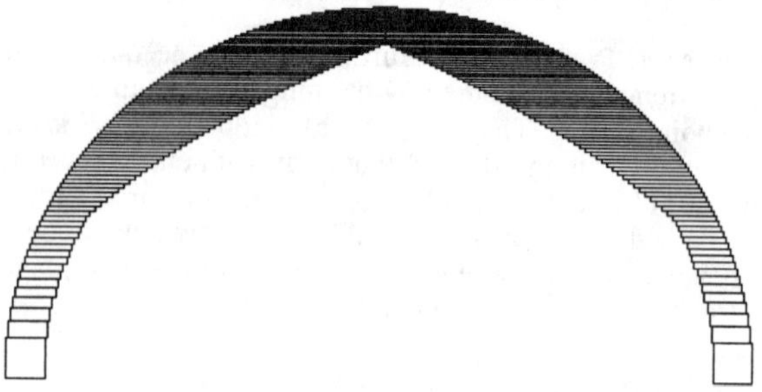

Figure 5: A cross-section of the variably sliced hemisphere $LM(S)$ from Example III with minimum depth $d_{\min} = 0.1$, radius 1, cusp height 0.006, $\triangle o_{\max} = 0.015$, and $\triangle r = 0.05$.

layers thickened to maintain strength near $p$, the lower layers are also thickened since $\triangle r < d_{\min}$.

Observe that the inner surface obtained in this example is not a simple offset of the inner surface from Example II, since the inner surface is offset more near $p$ than near points closer to the surface's base. However, an equivalent way to generate this example is to offset the outer surface from Example II by the specified depth forming a solid and then to thicken the solid's walls further to support any large over-hang.

## 5 Problem IV - Maximum Wall Thickness

Solve Problem III with a further restriction on the maximum (and possibly minimum) wall thickness. In this case, we may be forced to use a hybrid of traditional supports and thickened walls.

Note, increasing part strength often requires increasing wall thickness. For example, consider a cylinder with its axis in the build direction. Here, wall thickness and depth are equal. Since part strength and the maximum allowable wall thickness are contradictory specifications, no solution may be possible when both are specified. Note, this problem could be cast as an optimization problem that determines a maximum wall thickness while minimizing the volume of material used in the build. Our approach here is not one of optimization but of geometric analysis.

The algorithm to compute how a surface's walls should be thickened in this case, is exactly the same as Algorithm 2 presented in §4.1, except for the following two changes.

1. When Algorithm 2 calls for a wall thickness that is greater than the specified maximum, the thickness we choose is the smallest thickness less than the specified maximum so that the current layer supports as much of the upper layer as possible.

2. An additional step is added to the end of Algorithm 2. This step adds traditional support structure to support any remaining over-hangs that are too large for manufacture. In practice, this final step amounts to using commercial support software to generate the necessary remaining support structure. (In cases where the walls are adequately thickened, no additional support would be generated in this step.)

### 5.1 Example IV

To get a meaningful example, we must use a surface that is slightly more complicated than a hemisphere. We consider the surface of revolution $S$ that is generated by rotating the curve $z(x)$ for

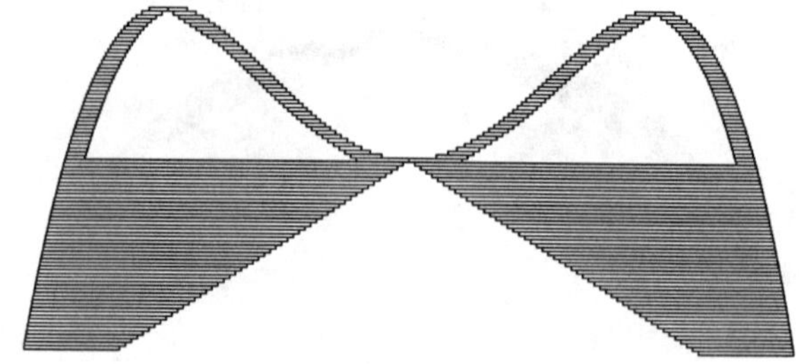

Figure 7: A cross section of $LM(S)$ from Example IV built using thickened walls with no restriction on maximum wall thickness. Here, $\triangle l = 0.01$, $\triangle o_{\max} = 0.015$, and $\triangle r = 0.05$.

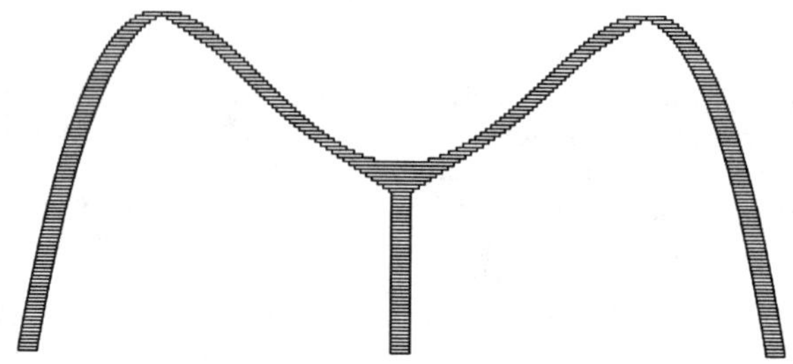

Figure 8: A cross section of $LM(S)$ from Example IV built using a hybrid of thickened walls and traditional support. Here the maximum wall thickness is 0.4. The other parameters are $\triangle l = 0.01$, $\triangle o_{\max} = 0.015$, and $\triangle r = 0.05$.

$0 \leq x \leq 1$ around the $z$-axis, where $z(x)$ is defined by:

$$z(x) \stackrel{\text{def}}{=} -\frac{1}{2}(x+1)(x-1)(1+5x^2). \qquad (6)$$

Figure 6 contains a picture of $S$ and of $z(x)$. For comparison, we consider constructing $S$ both without any restrictions on maximum wall thickness (using no traditional support) and with reasonable restrictions on the maximum wall thickness. In the latter case, a hybrid approach using both thickened walls and traditional support is employed.

As will be shown below, using variable slice thicknesses does not change the analysis in this case. Therefore, in the following examples, we assume constant slice thickness. Further, we assume that no part strength is specified.

First consider constructing $S$ without traditional support structure using uniform slice thickness and no restrictions on depth or wall thickness. Because $S$ has a local minimum at the point $p = (0, 0, 1/2)$, the section of the layer of $LM(S)$ containing $p$ lies on a floating component. Note, an equivalent way to think of this is that near $p$ there is an extremely large over-hang between layers. To support $p$, $LM(S)$'s walls with height less than $1/2$ must be greatly thickened. See Figure 7 for the picture of a cross section of $LM(S)$ in this case. Note, since we are only interested how $p$ is supported, the thickness of the layers does not matter.

Now suppose we wish to build the same surface $S$, except that we specify a maximum wall thickness that is less than 0.88. In this case to support $LM(S)$ near $p$, we must use traditional support structure beneath $p$. See Figure 8 for a picture of $LM(S)$'s cross section.

Note, if a minimum depth that is larger than one slice thickness

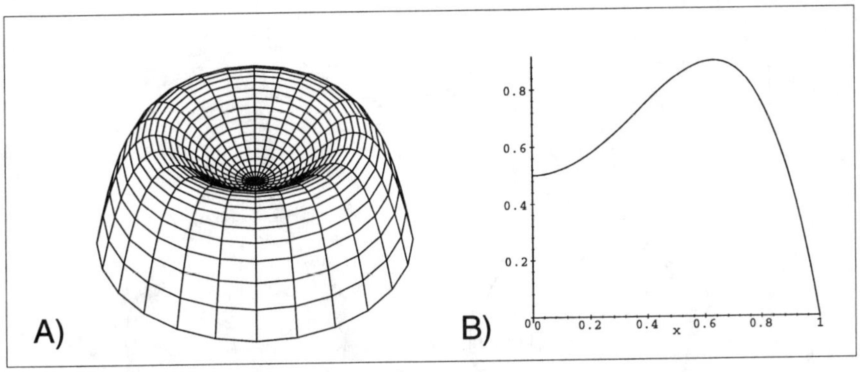


Figure 6: A) A surface of revolution $S$. B) The curve $z(x)$ that generates $S$ when rotated about the $z$-axis.

is specified, solid discs must be added to the slices directly below $p$. Traditional support structure would then be used to support the new material after the desired depth is achieved.

## 6 Extension to Non-WCP Surfaces and Solids

Now we consider the problem of building a surface $S$ that does not possess WCP but still has one "important" side and one side whose shape does not matter. We will assume that the important side is always the outer side of the surface. (For a surface with WCP, the important side is the upper side. This characterization can no longer be made.) Our goal will be to minimize the use of traditional support structure by selectively thickening $S$'s walls. This problem is solved by noticing that at points $p$ where $N_S(p) \cdot \nu < 0$, traditional support must be used to support large overhangs, and at $p$ where $N_S(p) \cdot \nu \geq 0$, the techniques of the preceding sections can be applied.

The method in these cases is to first use Algorithm 1 or 2 to support large over-hangs in those regions where the surface normal points up. Note, this first step may create new over-hangs in regions where the surface normal points down. These newly created over-hangs can be (and should be) supported by thickening $S$'s walls. So as before, we thicken $S$'s upper walls first and propagate the results downward, being careful only to thicken a wall in the direction away from $S$'s important side. The second step is to use traditional support structure to carry any remaining large over-hangs. The remaining over-hangs will include all large over-hangs in regions where surface normals points down and over-hangs that could not be supported due to restrictions on maximum wall thickness. This algorithm is summarized below:
**Algorithm 3:**

1. Apply Algorithm 1 or 2 to $S$ making sure to thicken $S$'s walls only in the direction away from $S$'s outer side.

2. Support any remaining large over-hangs using traditional support structure.

In the case of a surface $S$ of revolution built so that the axis of revolution is parallel to the build direction $\nu$, $S$ can be divided into disjoint regions $R_i$ separated by planes perpendicular to $\nu$, where $\forall p \in R_i$, $N_S(p)$ always points either up or down, i.e., into regions that either have or do not have WCP.

### 6.1 Example V

Now we consider building an artifact that resembles a light bulb. This object is formed from CSG primitives (two spheres, a cone, a cylinder, and a torus) using boolean operations. Figure 9 shows a cross section of the bulb in a plane containing its axis of rotation. This figure also indicates, with vertical lines, the regions where traditional support structure is required in order to build the bulb. Figures 9A and 9B show the bulb before and after its walls are thickened respectively. Notice:

1. the overall savings in material when the bulb is built using thickened walls, and

2. that in both cases, some traditional support structure lying outside the bulb is required.

Figure 9 suggests how a designer could use software that automatically thickens a surface's walls for support as a manufacturing aid. The material volume (and/or an estimation of build time) of both the object built using only traditional support structure and the object built using thickened walls for (some) internal support, could be performed. With regard to the results of these computations, the designer chooses the support generation technique.

## 7 Implementation and Builds

A suite of software tools to assist in layered manufacturing related tasks is currently under development in the University of Michigan CAD/CAM Laboratory. These tools include modules for variable slicing, nozzle path planning, part orientation, constructing CAD models from slice data, etc. The software that thickens a surface's walls for support will be a part of that suite.

Figure 10 contains pictures of hemispheres built using the Stratasys 3D Modeler layered manufacturing machine. The left hemisphere was built using traditional support structure with constant slice thickness (where the support structure has not been removed). The bottom and top hemispheres were built using thickened walls for support and constant slice thickness. The top hemisphere is turned to show its outer surface. The right hemisphere was adaptively sliced and built with thickened walls for support. Notice that the right hemisphere has a deeper hollow than the bottom hemisphere, and therefore required less material to build. All hemispheres use a road width of 0.02 inches, minimum wall thickness of 0.05 inches, and maximum over-hang of 0.01 inches. The constantly sliced hemispheres have a slice thickness of 0.01 inches and each took approximately three hours to build. The adaptively sliced hemisphere has a cusp height of at most 0.006 inches with layer thicknesses between 0.006 and 0.014 inches and a build time of about 2.5 hours.

Figure 11 contains two pictures of the bulb shaped object built with the Stratasys machine. This model was build in about 1.8 hours

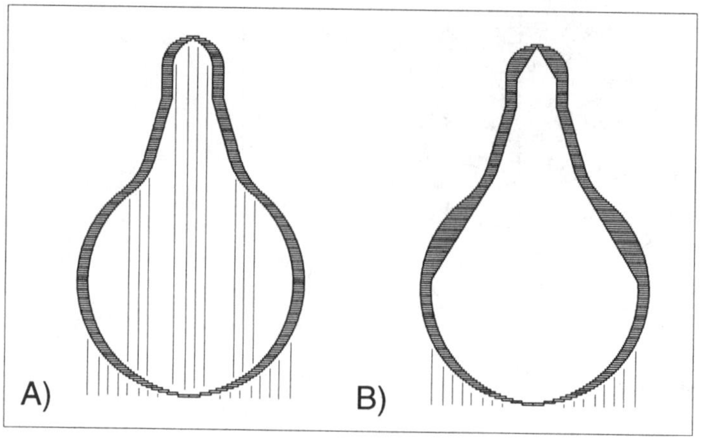


Figure 9: A) The cross section of $LM(S)$ from Example V without thickened walls. B) The same cross section with thickened walls. In both pictures, traditional support structure is required in the regions indicated by vertical lines, and the parameters used are $\triangle l = 0.01$, $\triangle o_{\max} = 0.006$, and $\triangle r = 0.05$.

Figure 10: Four hemispheres.

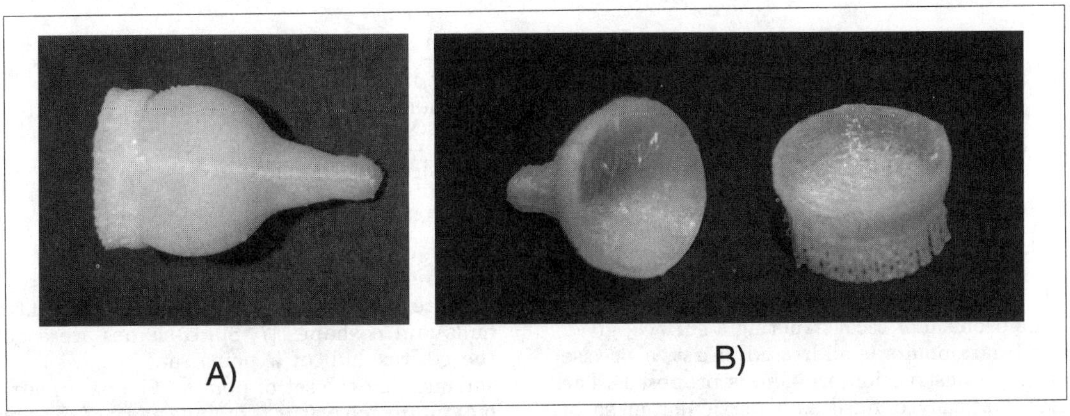


Figure 11: A) The bulb from Example V built using thickened walls for internal support. B) The same bulb after being cut to show that it is hollow. Notice the external traditional support structure remaining near the bulb's base.

using the following parameters: a road width of 0.02 inches, a constant layer thickness of 0.01 inches, a minimum wall thickness of 0.05 inches, and a maximum over-hang of 0.006 inches.

## 8 Conclusions

In this paper, we develop algorithms for wall thickness control in layered manufacturing. These algorithms provide a designer with a previously unconsidered alternative to the generation of traditional support structure in the layered manufacture of thin-walled parts. The algorithms presented in this paper enable a manufacturer to save both material and time when building an object compared to the methods commonly employed today. An additional benefit of these techniques is that the designer can specify a minimum overall strength for the constructed part. These results are achieved by selectively thickening the part's walls. Examples and actual physical prototypes demonstrating these techniques are given. Further, by thinking of solid objects as hollow, closed surfaces, these methods are immediately applicable to a very large class of objects. Although the algorithms developed here work for all surface types, in this paper, we focus on surfaces of revolution. Our ongoing work involves fully analyzing the general, free-form surface cases.

## 9 Acknowledgments

We gratefully acknowledge the financial support received from ONR grants N00014-94-1-0022 and N00014-95-1-0767.

## References

[1] Seth Allen and Deba Dutta, *Determination and evaluation of support structures in layered manufacturing*, Journal of Design and Manufacturing **5** (1995), 153–162.

[2] Prashant Kulkarni and Deba Dutta, *An accurate slicing procedure for layered manufacturing*, Computer Aided Design **28** (1996), no. 9, 683–697.

# $r$-Regular Shape Reconstruction from Unorganized Points

Dominique Attali

Communication Technology Laboratory, Swiss Federal Institute of Technology ETH,
Gloriastr. 35, CH-8092 Zurich, Switzerland
E-mail: Dominique.Attali@vision.ee.ethz.ch

## Abstract

In this paper, the problem of reconstructing a surface, given a set of scattered data points is addressed. First, a precise formulation of the reconstruction problem is proposed. The solution is mathematically defined as a particular mesh of the surface called the *normalized mesh*. This solution has the property to be included inside the Delaunay graph.

A criterion to select boundary faces inside the Delaunay graph is proposed. This criterion is proven to provide the exact solution in 2D for points sampling a $r$-regular shapes with a sampling path $\epsilon < 0.38r$. In $3D$, this results cannot be extended and the criterion cannot retrieve every faces. Some heuristics are then proposed in order to complete the surface.

## 1 Introduction

This paper addresses the problem of meshing a surface only known by an unorganized set of points. Such a problem may occur in many domains including pattern recognition, computer vision, and graphics. Meshing the boundary of an object is useful to study its geometry (volume, axes of inertia, area), to decide if a point is inside or outside the object, to provide a smooth interpolation of the boundary, or to display the object.

The initial data is a set of points with no structure, which sample a surface. Such points can be measured directly on the boundary of an object using 3D sensors or laser range scanning systems. They can also result from the segmentation of a 3D volumetric image. Starting from this mere set of points only known by their coordinates, we are interested in building a mesh induced by the proximity relationships of the points on the surface.

To solve this problem, it is common to compute first a neighbourhood graph of the data points and then to restrict the search for a mesh inside this neighbourhood graph. Many neighbourhood graphs for which efficient algorithms from computational geometry exist have been tried. In [1], the $k$-nearest neighbour graph is used to estimate a signed distance to the surface. After this step, a set of cubes crossing the surface is computed and a marching-cube algorithm provides the searched triangulation. Another appealing idea consists in starting from the convex hull of the points and to sculpture this convex hull until it passes through all the points [2]. The sculpture of the convex hull can be simplified by computing the Delaunay graph [3]. Thus, the sculpture consists in removing tetrahedra located on the current surface in such a way that at each step, the current surface remains a polyhedron. A drawback of such an approach is that no change of the topology is allowed and consequently, it is impossible to get a surface formed of several connected components or having holes. The Voronoi graph is sometimes used to get additional information on the skeleton of the object [4, 5, 6].

More complex graphs have also been introduced like $\alpha$-hulls and $\alpha$-shapes [7, 8]. $\alpha$-shapes are a generalization of the convex hull of a point set. An $\alpha$-shape is a polytope surrounding the set of points. The parameter $\alpha$ controls the maximum "curvature" of any cavity of the polytope. Several $\alpha$-shapes with different values of $\alpha$ are presented in figure 1. The choice of the parameter $\alpha$ might be tricky.

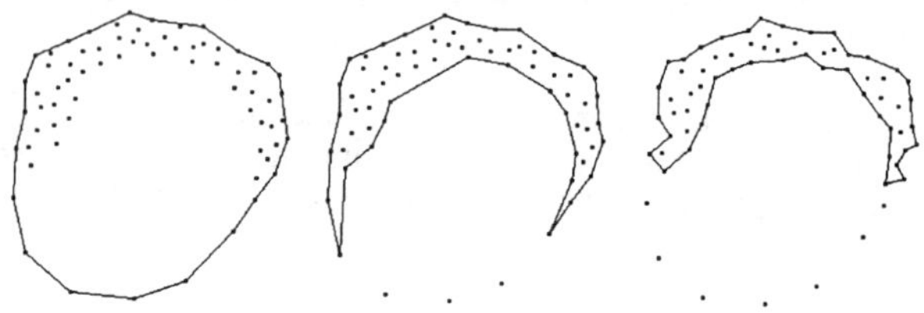

Figure 1: $\alpha$-shapes

Although a large amount of work has been done on this topic, few theoretical results [9] exist mainly due to the difficulty formalizing the problem. In most papers, no mathematical definition of the searched mesh is given. In section 2, we try to fill this deficiency by formulating the problem more precisely. In particular, the searched solution is defined as a particular mesh of the surface called the *normalized mesh*. In section 3, a method is proposed to build this solution. The convergence of our approach is proven in the two-dimensional case for *r-regular shapes*. Unfortunately, the demonstration cannot be extended to the three-dimensional case. It results that some faces of the surface cannot be retrieved by our algorithm. In section 4, some heuristics are proposed to complete the surface.

## 2 Problem statement

In this section, definitions are given in three-dimensional space but they can easily be extended in a space of any dimension. $d$ represents the Euclidean distance and $\overline{X}$ designates the closure of the set $X$.

### 2.1 General case

Let $S$ be the unknown surface to mesh and $E$ a finite set of points located on $S$. To measure the quality of the sample, we introduce the notion of sampling path:

**Definition 1 (Sampling path)** *$E \subset S$ is said to sample $S$ with the sampling path $\epsilon$ if any sphere with radius $\epsilon$ and center in $S$ contains at least one sample point in $E$ (Fig. 2).*

The end of this section is concerned with building up a *normalized mesh* from certain elementary spaces called *cells*.


*Computational Geometry 97* Nice France

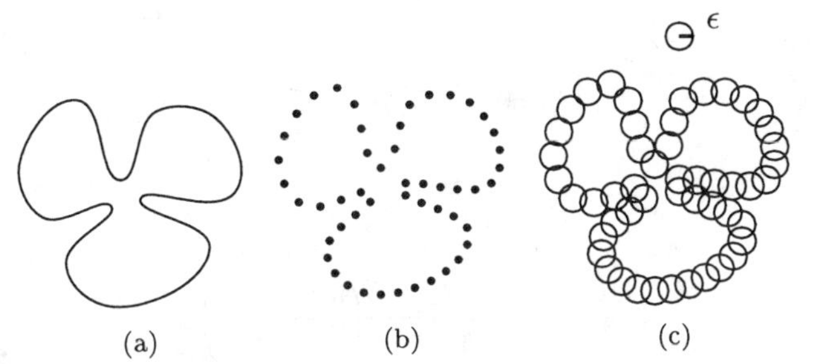


Figure 2: (a) Curve. (b) Sample of the curve with sampling path $\epsilon$. (c) Disks containing the curve.

A cell is just a generalization to an arbitrary number of vertices of a simplex.

**Definition 2 (cell)** *A $k$-cell $[a_1, \cdots, a_k]$ is the set of points $\sum_{i=1}^{k} \lambda_i a_i$, where $a_1, \ldots, a_k$ are points, and the $\lambda_i$ are real numbers such that $\lambda_i \geq 0$ for all $i$ and $\sum_{i=1}^{k} \lambda_i = 1$. The points $a_1, \ldots, a_n$ are called the vertices of the cell.*

The normalized mesh is a particular mesh of the surface $S$ defined as a union of cells with vertices at the data points $E$:

**Definition 3 (Normalized mesh)** *The normalized mesh of $S$ associated to $E$ is the set of cells $[p_1, \cdots, p_k]$ with $p_i \in E$ for which exists a point $m \in S$ such that $d(m, p_1) = \cdots = d(m, p_k) = d(m, E)$.*

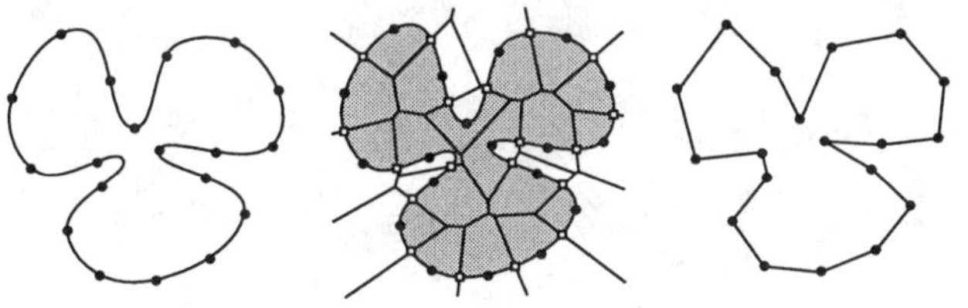

Figure 3: Construction of the normalized mesh from the Voronoi graph of the sample points.

By definition, the normalized mesh is included in the Delaunay diagram of $E$. It is formed of the Delaunay elements (edges, faces and simplices) whose dual intersects the surface (Fig. 3). Generally, the normalized mesh is formed of edges in $2D$ and triangular facets in $3D$ but it may also happen that the normalized mesh has a non-zero thickness if it includes a cell with not linearly independent vertices. One can remark that if $E$ samples $S$ with the sampling path $\epsilon$, then each face of the normalized mesh has diameter lower than $2\epsilon$.

Normalized mesh are particularly attractive as they provides a piecewise linear interpolant of the surface that converges to the surface when the sampling path tends to 0:

**Proposition 1 (Convergence)** *Let $S$ be a surface and $E_n$ be a sample of $S$ with sampling path $\frac{1}{n}$. Let $S_n$ be the normalized interpolant of $S$ associated to $E_n$. Then:*

$$\lim_{n \to \infty} S_n = \overline{S}$$

Using the previously defined notions, it is now possible to give a precise formulation of the initial reconstruction problem. Given a surface $S$ and a set of points $E$ located on $S$, the problem can be expressed as finding the normalized mesh of $S$ associated to $E$. Such a mesh provides a natural solution. It is a piecewise linear interpolant of the surface which tends to $S$ when the sampling path tends to 0.

### 2.2 $r$-regular shapes

In order to simplify the search for the normalized mesh, we will assume in the next section that $S$ is the boundary of an $r$-regular shape.

**Definition 4 ($r$-regular shape)** *Let $B_0$ be the unit ball. A shape $X$ is said to be $r$-regular if it is morphologically open and closed with respect to a disk of radius $r > 0$:*

$$X = (X \ominus rB_0) \oplus rB_0 = (X \oplus rB_0) \ominus rB_0$$

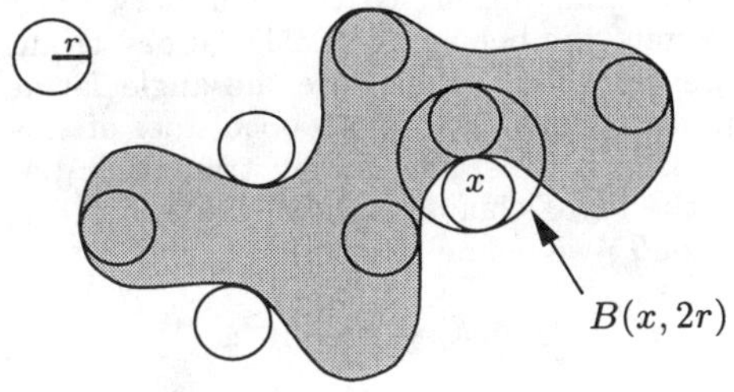


Figure 4: $r$-regular shape

The concept of $r$-regular shapes was first introduced in the field of mathematical morphology [10]. Since the parameter $r$ can be arbitrarily small, $r$-regular shapes offer a relevant model to represent objects arising in image analysis [11]. $r$-regular shapes have many nice geometric properties (Fig. 4):

1. The boundary of an $r$-regular shape has at each point a tangent and a radius of curvature greater or equal to $r$.

2. The boundary of an $r$-regular shape divides any ball with radius $2r$ and center on the boundary in exactly two connected componants.

3. From the previous property, it follows that if $\epsilon < 2r$, the normalized mesh provides a tiling of the surface. In other words, the normalized mesh retains all the topological properties of the surface.

4. In the two-dimensional space, any circle passing through three distinct boundary points has radius greater than $r$. This property will be crucial in the next section to validate our algorithm. Unfortunately, this property has no $3D$ equivalent.

## 3 Characterizing boundary faces

In this section, our goal is to characterize faces of the normalized mesh. From the previous section, we know that this search can be restricted inside the Delaunay diagram of the sample points. Therefore, the adressed problem is to determine which faces of the Delaunay diagram belong to the normalized mesh. Afterwards, a *Delaunay sphere* refers to a

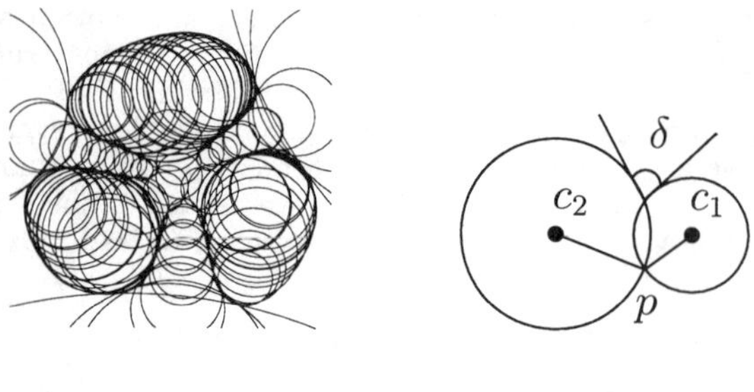


Figure 5: Angle formed by Delaunay disks

sphere circumscribed to a Delaunay simplex. The surface is assumed to be the boundary of an $r$-regular shape.

To detect boundary faces, one can take advantage of the following remark. In 2D, Delaunay discs tend to maximal discs of the object and of the complement of the object when the sampling density tends to 0 [12]. Consequently, Delaunay discs become tangent to the boundary (Fig. 5). In 3D, to evaluate the belonging of a Delaunay triangle to the surface, we are going to measure the angle formed by the two Delaunay spheres passing on both sides of this triangle. More precisely, if $T$ is a Delaunay triangle, $c_1$ and $c_2$ the centers of the two Delaunay spheres passing through $T$ and $p$ a vertex of $T$, we define (Fig. 5):

$$\delta(T) = \pi - \widehat{c_1 p c_2}$$

Broadly speaking, when the angle $\delta(T)$ is near 0, there is every chance that the triangle $T$ belongs to the boundary. On the contrary, if this angle is near $\pi$, the triangle has every chance to be inside or outside the object.

Let us remark that an analogous criterion was proposed in [3]. The distance from a face $T$ to the Delaunay sphere passing through $T$ was used in order to measure the belonging of this face to the surface. One advantage of our criterion is to be symmetric with respect to the boundary.

To reconstruct the surface, we are going to consider the set of triangles $T$ for which $\delta(T) \leq \delta_0$. This set is called afterwards $S_{\delta_0}$.

### 3.1 2D case

In $2D$, $S_{\delta_0}$ is a set of edges selected from the Delaunay diagram of $E$. The main result of this paper is that if $\epsilon < \sin(\frac{\pi}{8})r$, then the set of edges $S_{\frac{\pi}{2}}$ is the normalized mesh of $S$ associated to the points $E$. To demonstrate this result, one has first to establish the following proposition:

**Proposition 2** *Let $X$ be an $r$-regular shape and $E$ be a sample of $\partial X$ with sampling path $\epsilon$. For each edge $(pq)$ of the Delaunay diagram of $E$, we have :*

1. *if $(pq)$ belongs to the normalized mesh and $\epsilon < \frac{r}{2}$, then $\delta(pq) < \frac{\pi}{2}$*
2. *if $(pq)$ does not belong to the normalized mesh and $\epsilon < \sin(\frac{\pi}{8})r$, then $\delta(pq) > \frac{\pi}{2}$*

*Proof.* A complete proof may be found in [13]. In this paper, we only demonstrate the first point of the proposition. The second point can be established in a similar way.

Let $(pq)$ be an edge of the normalized mesh (Fig. 6). Let $B(x_1, r_1)$ and $B(x_2, r_2)$ be the two Delaunay disks intersecting at points $p$ and $q$. If $m$ designates the middle point of $[pq]$, we have :

$$\delta = \pi - \widehat{x_1 p m} - \widehat{m p x_2}$$

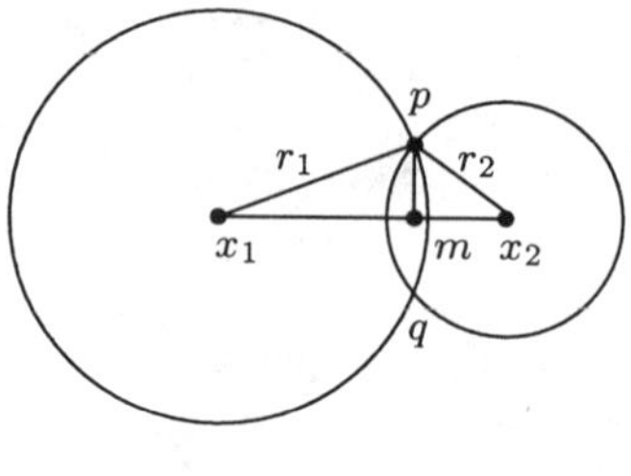


Figure 6: Illustration of the proof.

A disk intersecting the boundary of an $r$-regular shape in 3 distinct points has radius greater than $r$. Consequently, $r_1 \geq r$ and $r_2 \geq r$. Furthermore, edges of the normalized mesh have length smaller than $2\epsilon$. It follows that $d(p,m) < \epsilon$. Thus :

$$\cos(\widehat{x_1 p m}) = \frac{d(p,m)}{r_1} < \frac{\epsilon}{r} < \frac{1}{2}$$
$$\cos(\widehat{m p x_2}) = \frac{d(p,m)}{r_2} < \frac{\epsilon}{r} < \frac{1}{2}$$

Consequently, $\widehat{x_1 p m} > \frac{\pi}{3}$, $\widehat{m p x_2} > \frac{\pi}{3}$ and $\delta < \frac{\pi}{3} < \frac{\pi}{2}$.

**Proposition 3** *Let $X$ be an $r$-regular shape and $E$ be a sample of $\partial X$ with sampling path $\epsilon$. If $\epsilon < \sin(\frac{\pi}{8})r$, then $S_{\frac{\pi}{2}}$ is the normalized mesh*

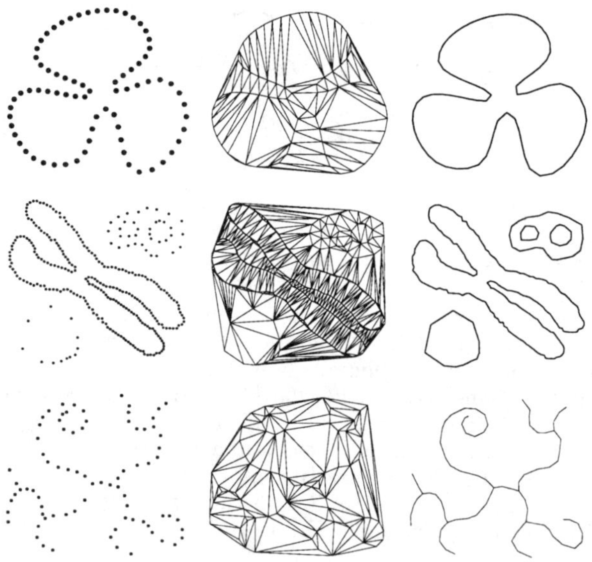

Figure 7: Sample points, Delaunay triangles and normalized mesh.

The beauty of this result lies in the fact that there is no need to know the two parameters $r$ and $\epsilon$ in order to compute the normalized mesh. Moreover, the sampling path has no need to be close to zero in order to find the correct result. It is enough that the sampling path $\epsilon$ and the constant $r$ characterizing the searched shape have approximately the same order. The computational time is also very attractive. The detection of boundary faces requires only the computation of a Voronoi graph. Given $n$ sample points in a $N$-dimensional space, the computational time is $O(n \log n + n^{\lceil N/2 \rceil})$. Figure 7 illustrates the robustness of the proposed method on several examples. The method was able to detect curves even if not closed and even when the points were not regularly spaced on the curves.

### 3.2 3D case

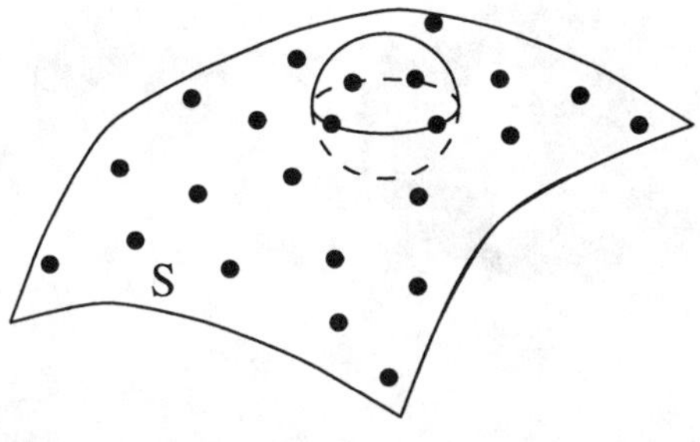


Figure 8: In 3D, a Delaunay sphere may intersect the surface

In $3D$, contrary to the 2D case, it may happen that some Delaunay spheres intersect the boundary without being tangent to it as illustrated in figure 8. This new phenomenon makes it impossible to minorate the radius of the Delaunay spheres. Consequently, proposition 2 cannot be generalized.

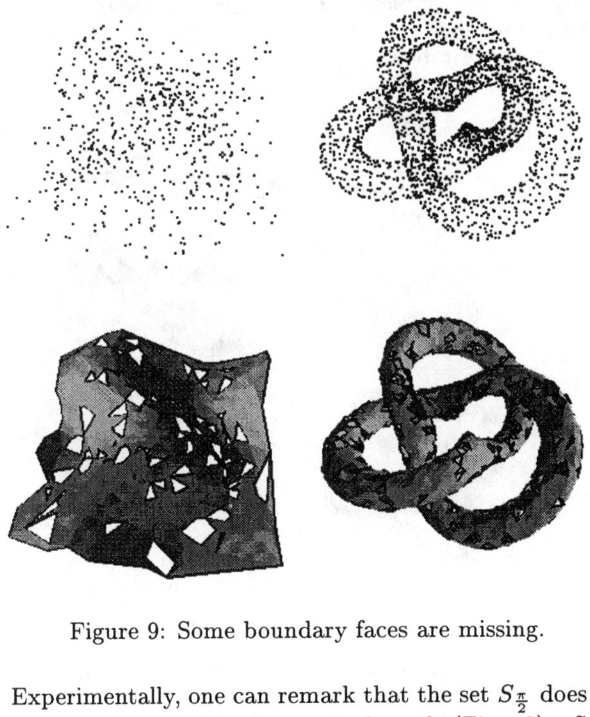

Figure 9: Some boundary faces are missing.

Experimentally, one can remark that the set $S_{\frac{\pi}{2}}$ does not capture all faces of the normalized mesh (Fig. 9). Some triangles are missing and the surface has more or less holes, depending on the distribution of sample points (Fig. 10). Nevertheless, as the proposed criterion does not produce erroneous triangles, one can start from this first approximation of the surface in order to find the solution.

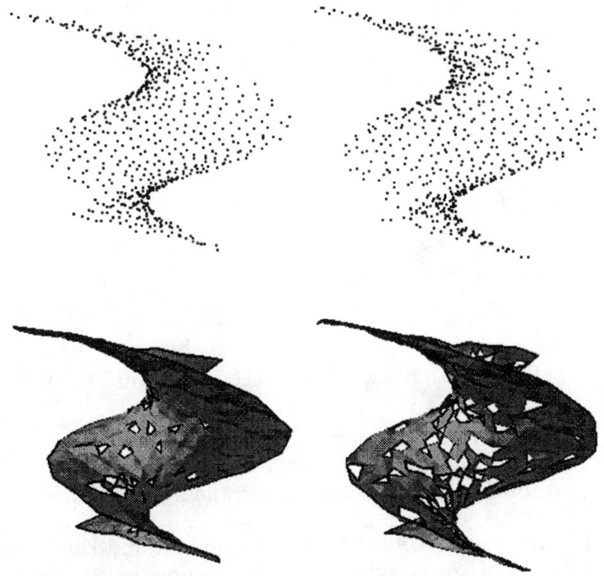

Figure 10: The number of missing triangles depends on the distribution of sample points.

## 4 Closing the surface

In this section, two different heuristics are proposed in order to complete surfaces from the previous section. Afterwards, the term "surface" refers to the set $S_{\frac{\pi}{2}}$.

### 4.1 Triangulation of the border

The border of the surface is formed of polygons. A first idea is to triangulate each polygon of the border independently from the others (Fig. 17). Given a simple polygon on the border of the surface, a triangulation of this polygon can be performed by adding one after the other Delaunay triangles sharing two or three edges with this polygon. If the triangulation fails, the surface cannot be closed by this method. Border polygons that are not simple must first be splitted up.

This approach has the advantage to make absolutely no assumption on the distribution of sample points. It deals with closed as well as open surfaces. Nevertheless, as illustrated on figure 11, it may sometimes not provide the expected solution. This approach has not yet been implemented.

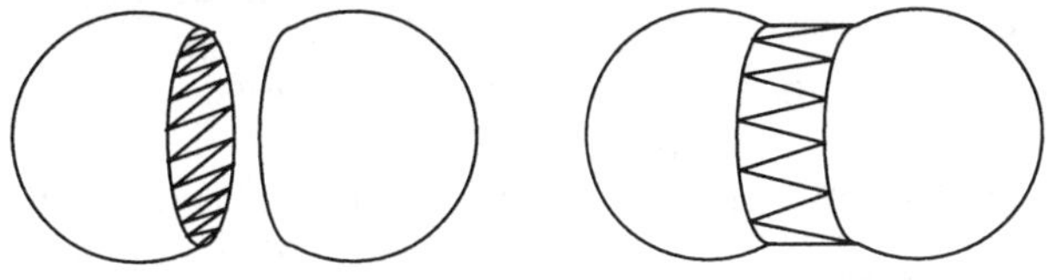

Figure 11: Two possible triangulations of the border.

### 4.2 Tessellation of the volume

In this section, a volume-based approach is proposed. Instead of computing directly the surface, we first try to find a collection of objects whose boundary provides the searched surface. This approach assumes surfaces to be closed, and consequently, is less general than the previous one.

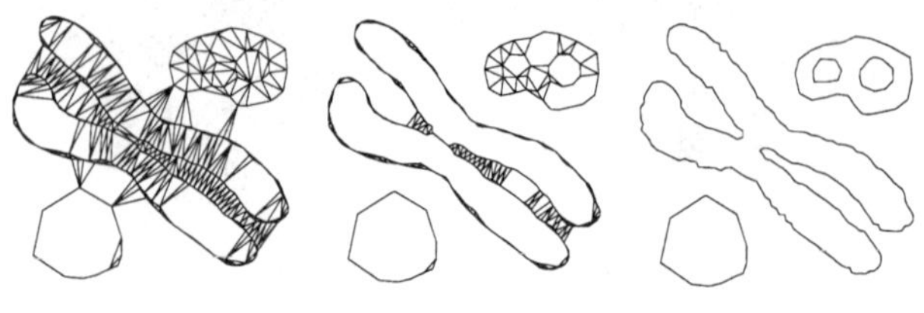

Figure 12: Illustration of the algorithm in 2D.

The main idea is to merge Delaunay tetrahedra until space is partitionned into a satisfactory number of objects. Our algorithm aggregates Delaunay tetrahedra as follows :

1. INITIALIZATION. Delaunay tetrahedra and the complement of the convex hull of the sample points form the initial set of objects. Delaunay triangles are put in a list $\mathcal{L}$ that is sorted according to the diameter of triangles.

2. MERGE. While $\mathcal{L}$ is not empty, the triangle $T$ having greatest diameter is taken out from $\mathcal{L}$. If $T$ separates two different objects $O_1$ and $O_2$, those two objects are merged providing that :

    - no triangle from $S_{\frac{\pi}{2}}$ disappears,
    - the merge does not isolate sample points inside $O_1 \cup O_2$. In other words, after the merge, every sample point must still belong to the boundary of at least one object (Fig. 13).

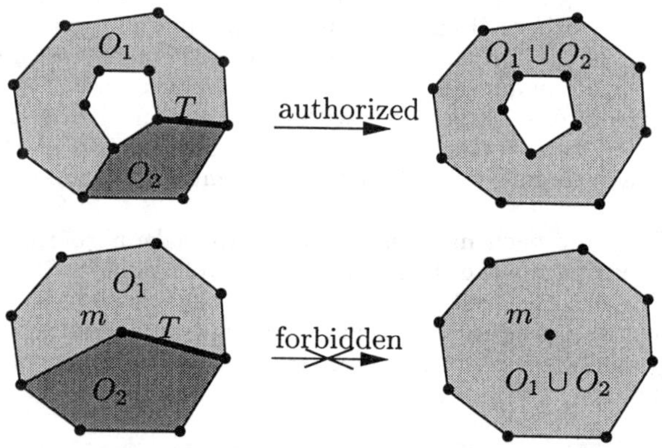


Figure 13: Rules to merge objects.

As soon as two objects share an important number of boundary triangles, they cannot merge anymore. Rules chosen to merge tetrahedra differ from rules used in [3] to sculpture the convex hull. In particular, more operations are permitted in order to allow topological change of the surface. The result is a set of closed surfaces of arbitrary topology.

If $n$ designates the number of sample points and $k$ the number of Delaunay faces in the list at the beginning of the algorithm, the complexity is $O(k \log k + kn^2)$. Indeed, the sort is in $O(k \log k)$. To detect a possible merge between two objects, it is necessary to consider the neighbours of each sample point, which can be $n$ in the worst case. Consequently, the complexity of the algorithm is $O(n^3)$ in 2D and $O(n^4)$ in 3D.

Figure 14 presents some results. The number of objects detected by our algorithm remains small. The percentage of tetrahedra not belonging to the main components is less than 0.5%.

Figure 14: Uncomplete surface from the previous section. First, second and other components resulting from the merge of Delaunay tetrahedra.

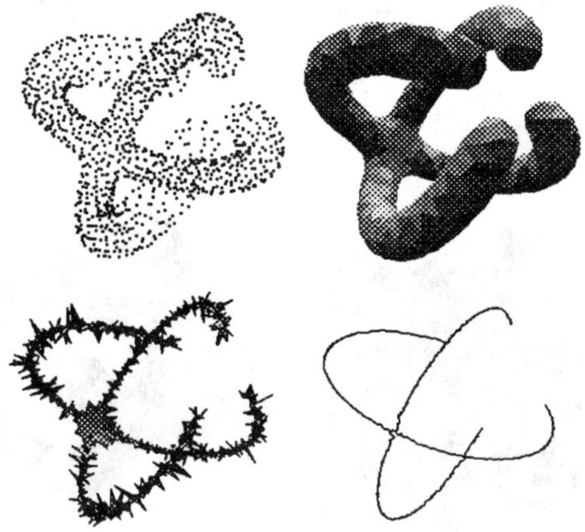

Figure 15: Sample points, reconstructed object, initial and simplified skeleton.

One advantage of this approach is that the skeleton of objects can easily be deduced (Fig. 15). The skeleton is a very famous representation of objects from image analysis and computer vision. It is a thin figure which retains all the topological properties of the object. It is useful to describe and analyse objects. In 3D, it is made up of surfaces and curves passing through the middle of the object. The skeleton can be computed by considering the dual of the Delaunay tetrahedra included in the object.

## 5 Conclusion

In this paper, a precise formulation of the reconstruction problem has been given. We defined the solution as a particular subset of the Delaunay diagram. A first method has been proposed to construct this solution. The convergence of this method has been established in 2D. For 3D objects, the method does not detect every boundary face. Two methods have been proposed for completing surfaces and detect missing faces.

A weak point of our study is that points must be located on or near the boundary of the objects. In the case of real data, this assumption cannot always be fulfilled. Sample points may form a dense cluster around the objects. A new formulation of the reconstruction problem must then be proposed. Current research is directed towards this direction. An idea would be to combine $\alpha$-shapes and the set of boundary faces detected with the criterion $\delta$ (Fig. 16). Indeed, the two approaches seem to complement one another. While $\alpha$-shapes can capture noisy set of points, our approach enables to connect points located on curves and surfaces.

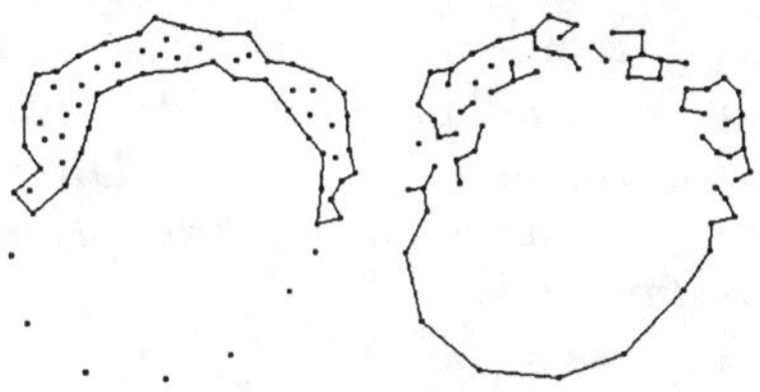

Figure 16: comparison of $\alpha$-shape and $S_{\frac{\pi}{2}}$.

## References


[1] H. Hoppe, T. DeRose, T. Duchamp, J. McDonnald, and W. Stuetzle. Surface reconstruction from unorganized points. *SIGGRAPH'92, Computer Graphics*, 26(2):71–77, 1992.

[2] J. O'Rourke. Polyhedra of minimal area as 3*d* objects models. In *Proc. of the International Joint Conference on Artificial Intelligence*, pages 664–666, 1981.

[3] J. D. Boissonnat. Geometric structures for three-dimensional shape representation. *ACM Transactions on Graphics*, 3(4):266–286, October 1984.

[4] J. Fairfield. Segmenting dot patterns by voronoi diagram concavity. *IEEE Trans. on PAMI*, 5(1):104–110, 1983.

[5] J. O'Rourke, H. Booth, and R. Washington. Connect-the-dots: A new heuristic. *Computer Vision, Graphics and Image Processing*, 39:258–266, 1987.

[6] A. C. F. Colchester, G. P. Robinson, and L. D. Griffin. A unified approach to the segmentation of grey-level and dot-pattern. In *Proc. of the 11th International Conference on Pattern Recognition*, pages 319–322, The Netherlands, 1992.

[7] H. Edelsbrunner, D. G. Kirkpatrick, and R. Seidel. On the shape of a set of points in the plane. *IEEE Transactions on Information Theory*, 29(4):551–559, 1983.

[8] H. Edelsbrunner and E. P. Mucke. Three-dimensional alpha shapes. *ACM Transactions on Graphics*, 13(1):43–72, 1994.

[9] S. Rippa. Minimal roughness property of the delaunay triangulation. *Computer Aided Geometric Design*, 7:489–497, 1990.

[10] J. Serra. *Image Analysis and Mathematical Morphology.* Academic Press, London, 1982.

[11] J. W. Brandt and V. R. Algazi. Continuous skeleton computation by Voronoi diagram. *CVGIP: Image Understanding*, 55(3):329–337, 1992.

[12] J. W. Brandt. Convergence and continuity criteria for discrete approximations of the continuous planar skeletons. *CVGIP : Image Understanding*, 59(1):116–124, 1994.

[13] D. Attali. *Squelettes et graphes de Vorono 2D et 3D.* PhD thesis, Universit Joseph Fourier- Grenoble I, October 1995.


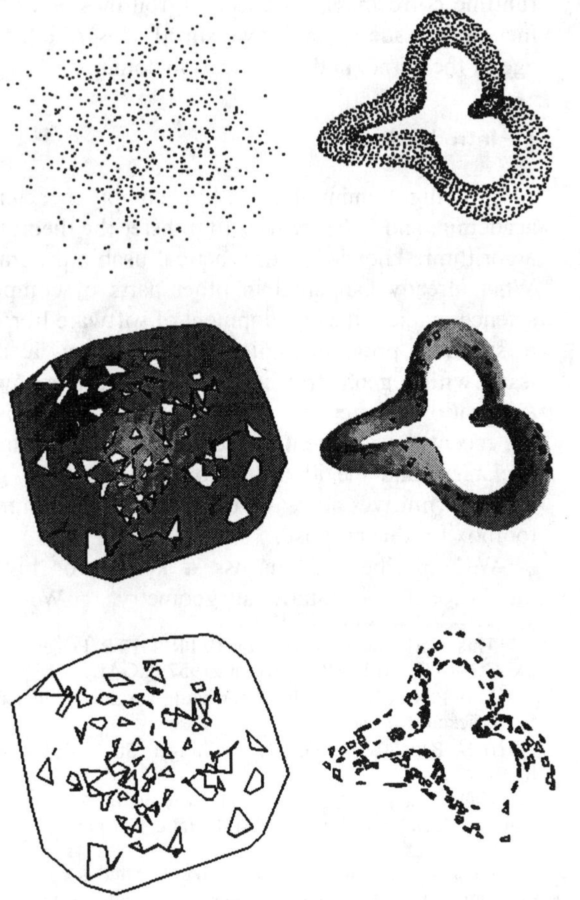

Figure 17: Sample points, reconstructed surface and border of the surface.

# A Computational Basis for Higher-Dimensional Computational Geometry and Applications[1]

Kurt Mehlhorn[2] Michael Müller[3] Stefan Näher[4] Stefan Schirra[2] Michael Seel[2] Christian Uhrig[2] Joachim Ziegler[2]

**Abstract**

In this paper we describe and discuss a kernel for higher-dimensional computational geometry and we present its application in the calculation of convex hulls and delaunay triangulations. We introduce the basic data types like points, vectors, directions, hyperplanes, segments, rays, lines, spheres, affine transformations, and operations connecting these types. The description consists of a motivation for the basic class layout as well as topics like layered software design, runtime correctness via checking routines and documentation issues. Finally we shortly describe the usage of the kernel in the application domain.

## 1 Introduction

A growing community within computer science, academia, and industry tries to transfer the theoretical algorithmic knowledge to practical usable programs. What already happened in other parts of computer science, namely the development of software libraries to speed up program implementation is now also an issue within geometric computing. We are now in a situation where the available computing power, the recent developments concerning exact arithmetic packages, and the identification of reasonable geometric primitives allow us to design a programming toolbox for this purpose.

We describe and discuss a kernel for higher-dimensional computational geometry. We have implemented all basic data types like points, vectors, directions, hyperplanes, segments, rays, lines, spheres, affine transformations, and operations connecting these types. The kernel is structured into three layers.

- arbitrary precision integer and rational arithmetic (classes *integer* and *rational*),
- exact linear algebra (classes *integer_vector* and *integer_matrix*), and
- basic geometric objects (classes *dd_rat_point*, *dd_rat_vector*, *dd_rat_direction*, *dd_rat_hyperplane*, *dd_rat_segment*, *dd_rat_ray*, *dd_rat_line*, *dd_rat_sphere*, and *aff_transformation*)

On top of the kernel we implemented some classical computation tasks in higher-dimensional CG like convex hulls and Delaunay triangulations. To make the kernel a toolbox for a wide user community and to give the whole project some paedagocical value we aimed for the following:

**Ease of use** — we aimed for a natural and intuitive interface as far as construction of objects, the conversion between objects, and the interaction of the classes and the operations are concerned. The naming scheme tries to achieve a compromise between mnemonics and word length. We followed a clean, complete and adaptable documentation scheme which provides all necessary information for the user of the kernel and at the same time is integrated into the implementation to enforce consistency between implementation and documentation. The information provided by this manual production toolset consists of prototype information, semantic preconditions, helpful implementation details and runtime information.

**Functionality** — we tried to provide a comprehensive functionality of the objects while avoiding interface bloating. The identification of the set

---

[1]This work was supported by ESPRIT LTR Project 20244 (ALCOM IT) and ESPRIT Project 21957 (CGAL)

[2]Max-Planck-Institute für Informatik, Im Stadtwald, 66123 Saarbrücken, Germany

[3]RIB Bausoftware GmbH, Vaihinger Str. 151, 70507 Stuttgart

[4]Fachbereich Mathematik und Informatik, Martin-Luther Universität Halle-Wittenberg, 06120 Halle, Germany



*Computational Geometry 97* Nice France
Copyright 1997 ACM 0-89791-878-9/97/06 ..$3.50

of primitives for higher level geometric applications was partly a dynamic process influenced by application design (see part 7).

**Layered Design** — we designed the kernel in a layered fashion for several reasons. First, the functionality of the lower levels is interesting in its own right. In particular, it can be used to realize additional geometric primitives. Second, the fact that the linear algebra layer provides extensive testing and checking routines considerably simplified the development of the geometry layer.

**Efficiency** — the code is designed to be as fast as possible respecting our primary goal: to develop modular, reusable and maintainable components which are not prone to arithmetic shortcomings like rounding errors. To optimize the runtime behaviour we use the LEDA memory management and a handle-rep scheme to improve memory consumption and to allow identity tests on objects.

The design of our kernel was mainly influenced by three sources: the experiences with the two-dimensional LEDA geometry kernel [MNU95], our experiences with an experimental higher-dimensional kernel [MZ94], and discussions with the group developing the CGAL-kernel [FGK+96].

In this short paper we can only give an overview of the kernel and the applications on top of it. We refer the reader to our website `http://www.mpi-sb.mpg/~seel` for the complete set of manual pages and for the complete documentation of the kernel and to `http://www.mpi-sb.mpg.de/~mehlhorn/Programs.html` for the application packet.

In the following parts we describe the three software layers of the kernel and the application layer, give implementation details, and report about experience with the two software packets.

## 2 Arithmetic

The bottom layer of our kernel is exact integral computing. We use the LEDA datatypes *integer* and *rational*. Any other bignum package providing the required functionality could be used instead.

The LEDA type *integer* realizes the mathematical type integer. The arithmetic operations $+$, $-$, $*$, $/$, $+=$, $-=$, $-$ (unary), $++$, $--$, the modulus operation ($\%$, $\% =$), bitwise AND ($\&$, $\&=$), bitwise OR ($|$,$|=$), the complement operator ($\sim$), the shift operators ($\ll$, $\gg$), the comparison operators $<$, $\leq$, $>$, $\geq$, $==$, $!=$, and the stream operators are all available. These operations never overflow and always yield the exact result. Of course, they may run out of memory.

Integers are essentially implemented by a vector of unsigned longs. The sign and the size are stored in extra variables. Some time critical functions are implemented in Sparc assembler code. The running time of addition is linear and the running time of multiplication is $O(L^{\log 3})$, where $L$ is the length of the operands. (Karazuba-Offman for multiplication)

A LEDA *rational* is essentially a pair of integers. The arithmetic operations $+$, $-$, $*$, $/$, $+=$, $-=$, $*=$, $/=$, $-$ (unary), $++$, $--$ are available on rationals. In addition, there are functions to extract the numerator and denominator, to cancel out the greatest common divisor of numerator and denominator, to compute squares and powers, to round rationals to integers, and many others. LEDA's rational numbers are not necessarily normalized, i.e., numerator and denominator of a rational number may have a common factor. A call *p.normalize*( ) normalizes $p$. This involves a gcd-computation to find the common factor $m$ in numerator and denominator and two divisions to remove the $m$. Since normalization is a fairly costly process we do not do it automatically.

We have run some runtime tests to compare the LEDA *integers* with the GNU *Integers* included in GNU's C++ library. The tests have been executed on an Ultrasparc 140. We tested the four basic operations $+$, $-$, $*$, $/$ iteratively on a table of $k$-bit random numbers.

| #Bits | #Ops | LEDA \| GNU in seconds | | | |
|---|---|---|---|---|---|
| | | + | | − | |
| 32 | $10^6$ | 1.1 | 5.1 | 1.3 | 5.5 |
| 64 | $10^6$ | 1.2 | 5.0 | 1.3 | 5.4 |
| 500 | $10^6$ | 1.8 | 7.3 | 1.86 | 7.8 |
| 1000 | $10^6$ | 2.7 | 10.4 | 2.8 | 13.4 |
| | | $*$ | | / | |
| 32 | $10^6$ | 1.6 | 7.5 | 4.7 | 26.6 |
| 64 | $10^6$ | 4.2 | 10.9 | 24.8 | 36.6 |
| 500 | $10^6$ | 158.8 | 272.7 | 225.7 | 375.3 |
| 1000 | $10^6$ | 591.6 | 1063.4 | 736.5 | 1281.3 |

## 3 Linear Algebra

The second layer of our kernel provides the standard operations of linear algebra encapsulated in the two classes *integer_vector* and *integer_matrix*. These are vectors and matrices with integer entries (= entries of type *integer*). The possible operations on matrices and vectors include

**construction mechanisms** — we can construct $m$-vectors, $m \times n$-matrices from a tuple of equidimensional vectors, identity matrices etc.

**data access operations** — we can access the integral and rational components, the dimensions of matrices and vectors, and the rows and columns of matrices.

**arithmetic operations** — we provide the inner product, scalar multiplication to vectors and matrices, and vector- and matrix-addition and -multiplication.

**Figure 1:** The Manual Page of class integer_matrix, an excerpt

**1. Definition**
An instance of data type *integer_matrix* is a matrix of integer variables. The types integer_matrix and integer_vector together realize many functions of basic linear algebra. All functions on integer matrices compute the exact result, i.e., there is no rounding error. Most functions of linear algebra are *checkable*, i.e., the programs can be asked for a proof that their output is correct.

**2. Creation**
*integer_matrix* $M(int\ n = 0,\ int\ m = 0)$;

creates an instance $M$ of type integer_matrix, $M$ is initialized to the $n \times m$ - zero matrix.

*integer_matrix* *M(array<integer_vector> A)*;

creates an instance $M$ of type integer_matrix. Let $A$ be an array of $m$ column-vectors of common dimension $n$. $M$ is initialized to an $n \times m$ matrix with the columns as specified by $A$.

**3. Operations**

| | | |
|---|---|---|
| *int* | *M*.dim1( ) | returns $n$, the number of rows of $M$. |
| *integer_vector&* | *M*.row(*int i*) | returns the $i$-th row of $M$ (an $m$-vector).<br>*Precondition*: $0 \leq i \leq n-1$. |
| *integer_matrix* | $M + M1$ | Addition of two matrices.<br>*Precondition*: $M$.dim1() = $M1$.dim1() and $M$.dim2() = $M1$.dim2(). |
| *bool* | linear_solver(*integer_matrix M, integer_vector b, integer_vector& x, integer& D, integer_matrix& spanning_vectors, integer_vector& c*) | determines the complete solution space of the linear system $M \cdot x = b$. If the system is unsolvable then $c^T \cdot M = 0$ and $c^T \cdot b \neq 0$. If the system is solvable then $(1/D)x$ is a solution, and the columns of *spanning_vectors* are a maximal set of linearly independent solutions to the corresponding homogeneous system.<br>*Precondition*: $M$.dim1() = $b$.dim(). |

**4. Implementation**
The datatype integer_matrix is implemented by two-dimensional arrays of integers. Operations *determinant*, *inverse*, *linear_solver*, and *rank* use $O(n^3)$ arithmetic operations, *col* takes time $O(n)$, *row*, *dim1*, *dim2*, take constant time, and all other operations take time $O(nm)$. The space requirement is $O(nm)$.

**operations based on solving a linear system** — we can determine the solution of a linear system $Ax = b$, calculate the determinant, rank, inverse, and independent columns of a matrix

The core operation of the last category is a Gaussian elimination scheme for a non-homogeneous linear system $Ax = b$ as described by J. Edmonds [Edm67]. For a recent reference see the books of A. Schrijver [Sch86, part I] or C. Yap [Yap93, lecture X]. Basically we transform the original matrix into a diagonalized matrix of integral entries which encodes the numerators of the solution vector and accumulate in parallel a common denominator of the numerator entries.

Figure 1 shows an excerpt of the manual page of class *integer_matrix*. The same style is used for all other types of the kernel. Each implemented class is documented by four major sections **Definition**, **Creation**, **Operations**, and **Implementation**. The first part gives an overview of the class specification like intended usage of the class and implementation features which influence this usage like exact arithmetic and proof features. The second and third part describe the user accessible operations of the class like constructors, member operations, and functions which work on the class. The description of each operation gives the semantics of it, describes the result transfer by return values and reference parameters, and states preconditions for the usage of the operation. The fourth part of our manual page completes the description by helpful implementation details and runtime and space bounds.

The cited description of the friend function *linear_solver* shows nicely the proof feature of our core operation. Either a solution $x$ is calculated which can be easily checked by substitution into the linear system $Mx = b$ or a vector $c$ is provided which proves the unsolvability of the system. Of course there's also a selftest incorporated in the code which can be switched on by a compilation flag and thus the testing can be done permanently.

In a second runtime test we compared the LEDA linear algebra module with commercial math packages like *Maple V* and *Mathematica 2.0*. We solved randomly generated non-homogeneous linear systems with *dim* rows and columns and 32 bit entries. The tests were executed on a Sun Sparc 4 with 40 MHz.

| *dim* | LEDA | Maple V R3 | Mathematica 2.0 |
|---|---|---|---|
| 20 | 1.3 s | 15.8 s | 31.6 s |
| 30 | 7.6 s | 92.4 s | 211.3 s |
| 40 | 28.4 s | 363.7 s | 840.3 s |
| 100 | 32 min | > 1 h | > 1 h |

## 4 Geometric Classes

As mentioned above we provide the geometric classes *dd_rat_point*, *dd_rat_vector*, *dd_rat_direction*, *dd_rat_hyperplane*, *dd_rat_segment*, *dd_rat_ray*, *dd_rat_line*, *dd_rat_sphere*, and *aff_transformation*.

We first give a motivation for our interface design and review briefly the basics of analytical geometry.

We use $d$ to denote the dimension of the ambient space and assume that our space is equipped with a standard Cartesian coordinate system. The basic object within this space is a point $p$, which we identify with its cartesian coordinate vector $p = (p_0, \ldots, p_{d-1})$, where the $p_i$, $0 \leq i < d$, are rational numbers. We store a *dd_rat_point* by homogeneous coordinates $(h_0, \ldots, h_d)$ where $p_i = h_i / h_d$ for all $i$, $0 \leq i < d$, and the $h_i$'s are integer (LEDA type *integer*). The homogenizing coordinate $h_d$ is always positive.

Points, vectors, and directions are closely related but nevertheless clearly distinct types. In order to work out the relationship, it is useful to identify a point with an arrow extending from the origin (= an arbitrary but fixed point) to the point. In this view a point is an arrow attached to the origin. A vector is an arrow that is allowed to float freely in space, more precisely, a vector is an equivalence class of arrows where two arrows are equivalent if one can be moved into the other by a translation of space. Points and vectors can be combined by some arithmetical operations. For two points $p$ and $q$ the difference $p - q$ is a vector (= the equivalence class of arrows containing the arrow extending from $q$ to $p$) and for a point $p$ and a vector $v$, $p + v$ is a point.

All operations of linear algebra apply to vectors, i.e., vectors can be stretched and shrunk (by multiplication with a scalar) and inner and cross product applies to them. On the other hand, geometric tests like collinearity or orientation only apply to points. Note that we distinguish the vector type *dd_rat_vector* in this scenario of geometric objects from the data type *integer_vector*, which we use to formulate calculations in our arithmetic linear algebra layer. We cannot identify both because their role and thereby their functionality within their respective code module is quite different and we don't want to have this mixed up.

A direction is also an equivalence class of arrows, where two arrows are equivalent if one can be moved into the other by a translation of space followed by stretching or shrinking. Alternatively, we may view a direction as a point on the unit sphere. In two-dimensional space directions correspond to angles. As in the case of *dd_rat_point* we store *dd_rat_vector*, and *dd_rat_direction*, respectively, as a homogeneous tuple of integers with positive homogenizing component.

The common one-dimensional straight-line objects in $d$-space like lines, rays and segments (which we allow to be trivial) are implemented in the classes *dd_rat_line*, *dd_rat_ray* and *dd_rat_segment* and determined by a pair of points.

With respect to the user interface we can group together points, vectors, directions, and on the other hand segments, rays, and lines. For the first group there are common operations to access Cartesian and homogeneous coordinates. Conversions within the first group can be made by explicit operations. For the second group there are similar operations to access the coordinates of the determining pair of points.

Oriented hyperplanes in the class *dd_rat_hyperplane* can be used to model halfspaces and affine hulls of $(d - 1)$-dimensional point sets. They are internally stored as a $(d + 1)$-tuple of integer coefficients. Finally oriented spheres of type *dd_rat_sphere* are helpfull in proximity calculations like Voronoi diagrams or Delaunay triangulations. They are stored as a tuple of $d + 1$ *dd_rat_points*.

For all of our basic geometric types we have affine transformations, which can be used by a call of a common member operation which gets an *aff_transformation*-instance as an argument and delivers a transformed object.

All object classes mentioned use a common handle-rep scheme which is already used in many modules of LEDA. We distinguish between a front-end object which is created by the constructor and a storage object of concrete geometric information which is referenced from the front-end object. The advantages of this scheme emerge in case of frequent copy construction and assignment where only references have to be redirected and no geometric information has to be copied. For large objects this lessens memory consumption and allows us to improve equality checks by testing the reference addresses before a comparison of geometric coordinate information. As in the case of our linear algebra module we use LEDA's improved memory management module which gives us a certain speed-up compared to the standard C++ allocation scheme.

We now take a closer look at two example manual pages, by which we elaborate further on the features of our data types. Figure 2 shows an excerpt from the point class manual page.

The default dimension of all objects is 2. This means that a call to the standard constructor *dd_rat_point*( ) delivers an instance of type *dd_rat_point* for planar geometry. A point in $d$-dimensional space is constructed by $p$(*integer_vector c, integer D*) or $p$(*integer_vector c*). In addition, there are standard initialization operations which allow comfortable creation of objects for the dimensions 2 and 3, which are not shown here. The data access operations allow access to the dimension and to Cartesian and homogeneous coordinates. Operator overloading allows the intuitive calculation of *dd_rat_point* difference, which results in a *dd_rat_vector* and the translation of *dd_rat_points* by adding a *dd_rat_vector*. The orientation predicate and other affine operations use LEDA *arrays* as a container type for a tuple of points.

To represent the second group of straight line objects we show a piece of the *dd_rat_line* documentation in figure 3. The basic operations provided on these objects are mainly position checks of points with respect to the objects, like *contains*( ) and intersection calculation between all higher dimensional objects like segments, rays, lines, and hyperplanes.

**Figure 2:** The Manual Page of class dd_rat_point, an excerpt

**1. Definition**
An instance of data type *dd_rat_point* is a point with rational coordinates in an arbitrary dimensional space ...

**2. Creation**

*dd_rat_point p(int d = 2);* introduces a variable $p$ of type *dd_rat_point* in d-dimensional space.

*dd_rat_point p(integer_vector c, integer D);*

introduces a variable $p$ of type *dd_rat_point* initialized to the point with homogeneous coordinates $(\pm c_0, \ldots, \pm c_{d-1}, \pm D)$, where $d$ is the dimension of $c$ and the sign chosen is the sign of $D$.
*Precondition*: $D$ is non-zero.

**3. Operations**

| | | |
|---|---|---|
| *int* | p.dim( ) | returns the dimension of $p$. |
| *rational* | p.coord(*int i*) | returns the $i$-th cartesian coordinate of $p$. |
| *integer* | p.hcoord(*int i*) | returns the $i$-th homogeneous coordinate of $p$. |
| *dd_rat_vector* | $p - q$ | returns $p - q$. *Precondition*: *p.dim( ) == q.dim( )*. |
| *dd_rat_point* | $p +$ *dd_rat_vector v* | returns $p + v$. *Precondition*: *p.dim( ) == v.dim( )*. |

*int* orientation(*array<dd_rat_point> A*)

determines the orientation of the points in $A$. This is the sign of the determinant

$$\begin{vmatrix} 1 & 1 & 1 & 1 \\ A[0] & A[1] & \ldots & A[d] \end{vmatrix}$$

where $A[i]$ denotes the cartesian coordinate vector of the $i$-th point in $A$.
*Precondition*: $A$ consists of $d + 1$ points of dimension $d$.

*bool* contained_in_affine_hull(*array<dd_rat_point> A, dd_rat_point x*)

determines whether $x$ is contained in the affine hull of the points in $A$.

**4. Implementation**
Points are implemented by arrays of integers as an item type. All operations like ...

**Figure 3:** The Manual Page of class dd_rat_line, an excerpt

**1. Definition**
An instance of data type *dd_rat_line* is an oriented line in $d$-dimensional Euclidian space ...

**2. Creation**

*dd_rat_line l(int d = 2);* introduces a variable $l$ of type *dd_rat_line* and initializes it to some line in $d$-dimensional space

*dd_rat_line l(dd_rat_point p, dd_rat_point q);*

introduces a line through $p$ and $q$ and oriented from $p$ to $q$.
*Precondition*: $p$ and $q$ are distinct and have the same dimension.

**3. Operations**

| | | |
|---|---|---|
| *dd_rat_point* | l.point1( ) | returns a point on $l$. |
| *dd_rat_direction* | l.direction( ) | returns the direction of $l$. |

*bool* l.contains(*dd_rat_point p*)

returns true if $p$ lies on $l$ and false otherwise.

*int* l.intersection(*dd_rat_segment s, dd_rat_point& i1, dd_rat_point& i2*)

returns the intersection set $l \cap s$ by the following means. The return value is one of the constants $\{$*IS_EMPTY, IS_POINT, IS_SEGMENT, IS_RAY, IS_LINE*$\}$. The corresponding set is determined by the two points *i1* and *i2*:

| return value | intersection set |
|---|---|
| *IS_EMPTY* | empty |
| *IS_POINT* | *dd_rat_point(i1)* |
| *IS_SEGMENT* | *dd_rat_segment(i1, i2)* |
| *IS_RAY* | *dd_rat_ray(i1, i2)* |
| *IS_LINE* | *dd_rat_line(i1, i2)* |

**4. Implementation**
Lines are implemented by a pair of points as an item type ...

## 5 Applications: Convex Hulls and Delaunay Triangulations

The convex hull and the Delaunay triangulation problem are traditionally specified as functions, i.e., given a set of points compute their convex hull or their Delaunay triangulation in some representation. We specify both problems as data types that support insertions and a large variety of query operations. In the case of convex hulls we support navigation through the interior and the boundary of the hull and we support membership and visibility queries. In the case of Delaunay triangulations we support navigation through the triangulation, we support locate[1] and nearest neighbor queries, and we support range queries with spheres and simplices. For two-dimensional convex hulls and Delaunay triangulations we also support an interface to the LEDA graph and window classes [MNU95, MN95]. In this way one can, for example, construct two-dimensional nearest and furthest site Voronoi diagrams and minimum spanning trees, display hulls and Delaunay triangulations.

The next two sections present parts of the specifications of convex hulls and Delaunay triangulations, respectively.

### Convex Hulls

An instance $C$ of type *chull* is the convex hull of a multi-set $S$ of points in $d$-dimensional space. We call $S$ the underlying point set and $d$ or *dim* the dimension of the underlying space. We use *dcur* or *dcurrent* to denote the affine dimension of $S$. The data type supports incremental construction of hulls.

The closure of the hull is maintained as a simplicial complex, i.e., as a collection of simplices the intersection of any two is a face of both[2]. In the sequel we reserve the word simplex for the simplices of dimension *dcur*. For each simplex there is an item of type *ch_simplex* and for each vertex there is an item of type *ch_vertex*. Each simplex has $1 + dcur$ vertices indexed from 0 to *dcur*; for a simplex $s$ and an index $i$, $C.vertex(s, i)$ returns the $i$-th vertex of $s$. For any simplex $s$ and any index $i$ of $s$ there may or may not be a simplex $t$ opposite to the $i$-th vertex of $s$. The function $C.opposite\_simplex(s, i)$ returns $t$ if it exists and returns *nil* otherwise. If $t$ exists then $s$ and $t$ share *dcur* vertices, namely all but the vertex with index $i$ of $s$ and the vertex with index $C.index\_of\_vertex\_in\_opposite\_simplex(s, i)$ of $t$. Assume that $t$ exists and let $j = C.index\_of\_vertex\_in\_opposite\_simplex(s, i)$. Then $s = C.opposite\_simplex(t, j)$ and $i = C.index\_of\_vertex\_in\_opposite\_simplex(t, j)$. Again, the specification and semantics of all operations connected to *chull* is documented in a manual page shown partly in figure 4.

[1] a locate query finds the simplex of the triangulation containing the query point

[2] The empty set is a facet of every simplex.

### Delaunay Triangulations

A instance *DT* of type *dd_delaunay* is the nearest and furthest site Delaunay triangulation of a set $S$ of points in some $d$-dimensional space. We call $S$ the underlying point set and $d$ or *dim* the dimension of the underlying space. We use *dcur* or *dcurrent* to denote the affine dimension of $S$. The data type supports incremental construction of Delaunay triangulations and various kinds of query operations (in particular, nearest and furthest neighbor queries and range queries with spheres and simplices).

A Delaunay triangulation is a simplicial complex. All simplices in the Delaunay triangulation have dimension *dcur*. In the nearest site Delaunay triangulation the circumsphere of any simplex in the triangulation contains no point of $S$ in its interior. In the furthest site Delaunay triangulation the circumsphere of any simplex contains no point of $S$ in its exterior. If the points in $S$ are co-circular then any triangulation of $S$ is a nearest as well as a furthest site Delaunay triangulation of $S$. If the points in $S$ are not co-circular then no simplex can be a simplex of both triangulations. Accordingly, we view *DT* as either one or two collection of simplices. If the points in $S$ are co-circular there is just one collection: the set of simplices of some triangulation. If the points in $S$ are not co-circular there are two collections. One collection consists of the simplices of a nearest site Delaunay triangulation and the other collection consists of the simplices of a furthest site Delaunay triangulation.

For each simplex of maximal dimension there is an item of type *dt_simplex* and for each vertex of the triangulation there is an item of type *dt_vertex*. Each simplex has $1 + dcur$ vertices indexed from 0 to *dcur*. For any simplex $s$ and any index $i$, $DT.vertex\_of(s, i)$ returns the $i$-th vertex of $s$. There may or may not be a simplex $t$ opposite to the vertex of $s$ with index $i$. The function $DT.opposite\_simplex(s, i)$ returns $t$ if it exists and returns *nil* otherwise. If $t$ exists then $s$ and $t$ share *dcur* vertices, namely all but the vertex with index $i$ of $s$ and the vertex with index $DT.index\_of\_vertex\_in\_opposite\_simplex(s, i)$ of $t$. Assume that $t = DT.opposite\_simplex(s, i)$ exists and let $j = DT.index\_of\_vertex\_in\_opposite\_simplex(s, i)$. Then $s = DT.opposite\_simplex(t, j)$ and $i = DT.index\_of\_vertex\_in\_opposite\_simplex(t, j)$. In general, a vertex belongs to many simplices.

Any simplex of *DT* belongs either to the nearest or to the furthest site Delaunay triangulation or both. The test $DT.simplex\_of\_nearest(dt\_simplex\ s)$ returns true if $s$ belongs to the nearest site triangulation and the test $DT.simplex\_of\_furthest(dt\_simplex\ s)$ returns true if $s$ belongs to the furthest site triangulation.

### Further Implementation Issues

The implementation of type *chull* follows [CMS93] and Delaunay triangulations are reduced to convex

**Figure 4:** The Manual Page of class chull, an excerpt

**1. Definition**
An instance $C$ of type *chull* is ...

**2. Creation**

| | |
|---|---|
| *chull* $C(int\ d\ =\ 2)$; | creates an instance $C$ of type *chull*. The dimension of the underlying space is $d$ and $S$ is initialized to the empty point set. |

The data type *chull* offers neither copy constructor nor assignment operator.

**3. Operations**

| | | |
|---|---|---|
| *int* | C.dim( ) | returns the dimension of ambient space |
| *dd_rat_point* | C.associated_point(*ch_vertex v*) | returns the point associated with vertex $v$. |
| *ch_vertex* | C.vertex_of_simplex(*ch_simplex s, int i*) | returns the vertex corresponding to the $i$-th vertex of $s$. *Precondition*: $0 \leq i \leq dcur$. |
| *dd_rat_hyperplane* | C.hyperplane_supporting(*ch_facet f*) | returns a hyperplane supporting facet $f$. The hyperplane is oriented such that the interior of $C$ is on the negative side of it. *Precondition*: $f$ is a facet of $C$ and $dcur > 1$. |
| *list<ch_facet>* | C.all_facets( ) | returns a list of all facets of $C$. |

⋮

**4. Implementation**
The time and space requirement is input dependent. Let $C_1, C_2, C_3, \ldots$ be the sequence of hulls constructed and for a point $x$ let $k_i$ be the number of facets of $C_i$ that are visible from $x$ and that are not already facets of $C_{i-1}$. Then the time for inserting $x$ is $O(dim \sum_i k_i)$ and the number of new simplices constructed during the insertion of $x$ is the number of facets of the hull which were not already facets of the hull before the insertion.
The data type *chull* is derived from *regl_complex*. The space requirement of regular complexes is essentially $12(dim + 2)$ Bytes times the number of simplices plus the space for the points. *chull* needs an additional $8 + (4 + x)dim$ Bytes per simplex where $x$ is the space requirement of the underlying number type and an additional 12 Bytes per point. The total is therefore $(16+x)dim + 32$ Bytes times the number of simplices plus $28 + x \cdot dim$ Bytes times the number of points.

hulls through the well-known lifting map, see for example [Ede87]. Based on our kernel a class *regl_complex* was implemented that can represent so-called regular simplicial complexes. A simplicial complex is called regular if all maximal simplices, i.e., simplices that are not a subsimplex of another simplex of the complex, have the same dimension. The class *regl_complex* provides operations for navigation through the complex and update operations. The class *chull* is derived from *regl_complex* and the class *dd_delaunay* is derived from *chull*.

The work horse for the query operations on convex hulls and Delaunay triangulations is a method

```
C.visibility_search(dd_rat_point x,
  list<ch_facet>& visible_facets,
  int& location, ch_facet& f);
```

that constructs the list of all $x$-visible hull facets in *visible_facets*, returns the position of $x$ with respect to the current hull in *location* ($-1$ for inside, 0 for on the the boundary, and $+1$ for outside) and, if $x$ is contained in the boundary of $C$, returns a facet incident to $x$ in $f$.

The membership query and the visible facets query for hulls are easily realized by this method and the nearest neighbor and the range query for Delaunay triangulations use it in an essential way. The nearest neighbor query for Delaunay triangulations lifts the query point (using the lifting map), then determines all visible facets of the hull, and then selects the best vertex by linear search through their vertices[3]. The range query with spheres lifts the sphere (using the lifting map) and then finds all vertices of the hull that lie below the resulting hyperplane[4].

We use program checking [BLR90, MNS+96] in our implementation. In particular,

- the class *regl_complex* provides a method *RC.check_topology*( ) that partially checks whether *RC* is an abstract simplicial complex[5], and a method *RC.is_Delaunay*(*kind*) that checks whether *RC* is a nearest (*kind* = *nearest*) or furthest (*kind* = *furthest*) site Delaunay triangulation of its vertex set.
- the class *chull* provides a method *C.check*( ) that verifies convex hulls as described in [MNS+96].
- we have algorithms that check whether a graph is a nearest or furthest site Delaunay diagram of its vertex set, whether a graph is a triangulation of its vertex set, and whether a graph is a nearest or furthest site Voronoi diagram.

The representation of convex hulls and Delaunay triangulations in data types *chull* and *dd_delaunay* is simplex-based, i.e., simplices are the main objects and

[3]This method is only efficient in low-dimensional space

[4]The lifing map turns a sphere into a hyperplane.

[5]The method checks whether the neighborhood relationship on simplices is symmetric, whether all vertices of a simplex are distinct, and whether two neighboring simplices share all but one of their vertices. It does not check whether simplices that share all but one of their vertices are actually neighbors in the complex.

lower dimensional faces are only implicitely represented. In two-dimensional space there is an alternative representation which makes the vertices and edges the primary objects and represents simplices (= triangles) implicitely as faces of a planar graph. This is the representation used for two-dimensional Delaunay triangulations (type *delaunay*) in LEDA. If *DT* has type *dd_delaunay*, *DTG* has type *GRAPH<POINT, int>*[6], and *kind* is one of *nearest* or *furthest*, then

```
DT.graphrep(DTG,kind);
```

constructs the graph representation of *DT* in *DTG*. All LEDA graph algorithms can now be applied to *DTG*. For example,

```
compute_voronoi(DTG,VD,kind);
```

will construct the graph representation of the Voronoi diagram in *VD*, and

```
edge_array<rational> dist(DTG);
forall_edges(e,DTG)
  dist[e] = DTG[source(e)].
            sqr_dist(DTG[target(e)]);
list<edge> L =
  MIN_SPANNING_TREE(DTG,dist);
```

will construct in *L* the set of edges comprising a minimum spannning tree of *DTG*.

## 6 Documentation

The full documentation of the kernel consists of about 240 pages[7] and the documentation of the application layer comprises about 100 pages[8].

Figure 5 shows the implementation of the *orientation*-predicate of the class *dd_rat_point* and one member operation of class *chull*. For each class the documentation and the implementation are collected in a noweb-file[9] and different tools are used to give different views of the noweb-file: the noweb tool **notangle** extracts the code, i.e., the view needed by the C++ compiler, and the LEDA tools **Lman** and **Ldoc** give the manual view and the documentation view, respectively[10].

Figure 5 also illustrates the vertical interaction between the software layers. In case of the *orientation*-predicate the determinant calculation is done by *determinant*( ) provided as a friend of *integer_matrix*. Most affine operations on point tuples map to the solution of a corresponding linear system calculated by our linear algebra layer. Notice that precondition checks are formulated as preprocessor macros which can be switched off by the flag `-DLEDA_CHECKING_OFF`.

## 7 Experimental Experiences

Both programs rely heavily on the fact that the kernel is exact. For example, the insertion routine for convex hulls distinguishes cases according to whether the newly inserted point lies in the affine hull of the points already present or not, and the checking programs would hardly make sense without exact primitives.

In the early stages of program development the checking feature of the kernel was particulary useful. For example, the convex hull program needs to compute the hyperplane defined by a set of points. This can be done by solving a linear system. In the first version of the program we set up the wrong linear system. It was very useful that the linear system solver gives a proof of unsolvability and does not just claim unsolvability. This located the error fairly quickly.

We have used classes *chull* and *dd_delaunay* on problems up to dimension 10. We have also compared it to the *qhull*-program of Barber, Dobkin, and Hudhanpaa[BDH96] and the *hull*-program of Clarkson. The first method computes approximate convex hulls and the latter method computes exact hulls but works only for a limited (albeit large) range of coordinate values. Both methods are significantly faster than ours. This is mostly due to their use of floating point arithmetic. Neither of the algorithms provides the rich functionality that we provide.

## 8 Availability and Extensions

The whole kernel has been used internally since June 96 and will become part of the next major LEDA release in form of a so called LEDA extension package. The complete specification of the listed data types and the code projects can be obtained via WWW from the LEDA home page `http://www.mpi-sb.mpg.de/LEDA`.

In cooperation with the CGAL-project the code base was extended with number type templatization. This allows runtime comparisons depending on the plugged in arithmetic components. We want to evaluate the use of modular integer data types within the linear algebra packet as well as in different geometric application scenarios. An adapted version of the templatized code will become part of the CGAL-kernel.

## 9 Conclusions

We described the layered design of a kernel for higher dimensional computational geometry and gave two ex-

---

[6] In LEDA, *GRAPH<POINT, int>* is the type of graphs where each node has an associated information of type *POINT* and each edge has an associated information of type *int*; the edge information is not used in this code.

[7] see `http://www.mpi-sb.mpg.de/~seel/geo.tar.gz`

[8] see `http://www.mpi-sb.mpg.de/~mehlhorn/Programs.html`

[9] see `http://www.cs.purdue.edu/homes/nr/noweb` for an introduction to noweb

[10] see `http://www.mpi-sb.mpg.de/~mehlhorn/LEDAbook.html` for an introduction to this tools

**Figure 5:** The implementation style

### Orientation

⟨*dd_rat_point.h*⟩≡

```
int orientation(const array<rat_point> & A);
```

**Semantics**: determines the orientation of the points in $A$, where $A$ consists of $d+1$ points in $d$-space. This is the sign of the determinant

$$\begin{vmatrix} 1 & 1 & 1 & 1 \\ A[0] & A[1] & \ldots & A[d] \end{vmatrix}$$

where $A[i]$ denotes the cartesian coordinate vector of the $i$-th point in $A$.

We are given an array $A$ of $d+1$ points in $d$-space and compute their orientation. Multiplying the $i$-th column of the above matrix by the homogenizing coordinate of $A[i]$ leaves the sign of the determinant unchanged. We set up this matrix and return its determinant. Actually, it is more convenient to transpose it and to make the first row the last. This changes the sign if the number of rows is even, i.e., if $d$ is odd.

⟨*implementing dd_rat_point*⟩+≡

```
int
orientation(const array< dd_rat_point > & A)
{
  TUPLE_DIM_CHECK(A,orientation)

  int al = A.low(); // the lower index start of |A|
  int d = A.high() - al; // A contains d + 1 points

  LEDA_OPT_PRECOND((A[al].dim() == d),"orientation: \
  needs A[].dim() + 1 many input points.")

  integer_matrix M(d+1); // quadratic

  for (int i = 0; i <= d; i++)
  {
    for (int j = 0; j <= d; j++)
      M(i,j) = A[al + i].hcoord(j);
  }

  int row_correction = ( (d % 2 == 0) ? +1 : -1 );
  // we invert the sign if the row number is even i.e. d is odd
  return row_correction * sign_of_determinant(M);
}
```

### Facets visible from a point

⟨*public members of class chull*⟩+≡

```
list<ch_facet> facets_visible_from(const dd_rat_point& x);
```

**Semantics**: returns the list of all facets that are visible from $x$.
*Precondition*: $x$ is contained in the affine hull of $S$.

⟨*Member Impls*⟩+≡

```
list<ch_facet> chull::facets_visible_from(const rat_point& x)
/* returns the list of all facets that are visible from |x|. */
{ list<ch_simplex> visible_simplices;
  int location = -1;                       // intialization is important
  int number_of_visited_simplices = 0;     // irrelevant
  ch_facet f;                              // irrelevant

  visibility_search(origin_simplex, x, visible_simplices,
                    number_of_visited_simplices, location, f);
  return visible_simplices;
}
```

amples for applications. The primary goal was to develop a useful base for geometric application design including checkable correctness, efficiency based on existing software library concepts and a clear and comprehensive documentation scheme.

#### References


[BDH96] C. Barber, D. Dobkin, and H. Hudhanpaa. The quickhull program for convex hulls. *ACM Transactions on Mathematical Software*, 22:469–483, 1996.

[BLR90] M. Blum, M. Luby, and R. Rubinfeld. Self-testing/correcting with applications to numerical problems. In *Proc. 22nd Annual ACM Symp. on Theory of Computing*, pages 73–83, 1990.

[CMS93] Kenneth L. Clarkson, Kurt Mehlhorn, and Raimund Seidel. Four results on randomized incremental constructions. In *Computational geometry : theory and applications*, volume 3, pages 185–212, Amsterdam, 1993. Elsevier.

[Ede87] H. Edelsbrunner. *Algorithms in Combinatorial Geometry*. Springer, 1987.

[Edm67] J. Edmonds. Systems of distinct representatives and linear algebra. *Journal of Research of the National Bureau of Standards*, 71(B):241–245, 1967.

[FGK+96] Andreas Fabri, Geert-Jan Giezeman, Lutz Kettner, Stefan Schirra, and Sven Schönherr. The CGAL kernel : a basis for geometric computation. In Ming C. Lin and Dinesh Manocha, editors, *Applied Computational Geometry : Towards Geometric Engineering : Workshop (FCRC-96; WACG-96), Philadelphia, PA, USA, May 27-28, 1996; selected paper*, volume LNCS 1148, pages 191–202S., Berlin, 1996. Springer.

[MN95] K. Mehlhorn and S. Näher. LEDA, a platform for combinatorial and geometric computing. *Communications of the ACM*, 38:96–102, 1995.

[MNS+96] K. Mehlhorn, S. Näher, T. Schilz, S. Schirra, M. Seel, R. Seidel, and Ch. Uhrig. Checking Geometric Programs or Verification of Geometric Structures. In *Proc. of the 12th Annual Symposium on Computational Geometry*, pages 159–165, 1996.

[MNU95] K. Mehlhorn, S. Näher, and C. Uhrig. *The LEDA User Manual*, 1995.

[MZ94] M. Müller and J. Ziegler. An implementation of a convex hull algorithm. Technical Report MPI-I-94-105, Max-Planck-Institut für Informatik, Saarbrücken, 1994.

[Sch86] A. Schrijver. *Theory of Linear and Integer Programming*. John Wiley and Sons, Chichester, New York, Brisbane, Toronto, Singapore, 1986.

[Yap93] C. K. Yap. *Fundamental Problems in Algorithmic Algebra*. Princeton University Press, 1993.

# A New Algorithm for Computing Shortest Paths in Weighted Planar Subdivisions

(Extended Abstract)

Cristian S. Mata *      Joseph S. B. Mitchell †

**Abstract**

We present a practical new algorithm for the problem of computing low-cost paths in a weighted planar subdivision or on a weighted polyhedral surface. The algorithm is based on constructing a relatively sparse graph, a "pathnet", that links selected pairs of subdivision vertices (and "critical points of entry") with locally optimal paths. The pathnet can be searched for paths that are provably close to optimal and approach optimal, as one varies the parameter that controls the sparsity of the pathnet.

We analyze our algorithm both analytically and experimentally. We report on the results of a set of experiments comparing the new algorithm with other standard methods.

## 1 Introduction

For a given weight function, $F : \Re^2 \rightarrow \Re$, the *weighted length* of an $s$-$t$ path $\pi$ in the plane is the path integral, $\int_\pi F(x,y)d\sigma$, of the weight function along the path $\pi$, linking the *start* $s$ to the *goal* $t$. The *weighted region metric* associated with $F$ defines the distance $d_F(s,t)$ to be the infimum over all $s$-$t$ paths $\pi$ of the weighted length of $\pi$. The *weighted region problem* (WRP) is to find an optimal $s$-$t$ path according to the weighted region metric, $d_F$, induced by a given piecewise-constant weight function, $F$. This problem is a natural generalization of the shortest-path problem in a polygonal domain: Consider a weight function that assigns weight 1 to $P$ and weight $\infty$ (or a sufficiently large constant) to the obstacles (the complement of $P$).

The weighted region problem models the minimum-time path problem for a point robot moving in a terrain of varied types (e.g., grassland, brushland, blacktop, bodies of water, etc), where each type of terrain has an assigned weight equal to the reciprocal of the maximum speed of traversal for the robot. It also arises in numerous other applications involving route planning in geographic data ([5, 15, 23]), military mission planning and decision support ([1, 10, 11, 13, 14, 15, 21, 22]), and fluid flow in injection molding [8].

We assume that $f$ is specified by a triangulation having $n$ vertices, with each face assigned an integer weight $\alpha \in \{0, 1, \ldots, W, +\infty\}$. We can allow edges of the triangulation to have a weight that is possibly distinct from that of the triangular facets on either side of it; in this way, "linear features" such as "roads" can be modeled.

This paper describes an algorithm to compute nearly optimal paths between two points in weighted polygonal subdivisions. We call the data structure built by the algorithm a *pathnet.* The pathnet can be used to answer path queries between pairs of points in the subdivision.

The pathnet algorithm is strongly motivated by results on approximating Euclidean shortest paths among obstacles (as in Clarkson [3] and Mitchell [16]), $t$-spanners [9], and approximation methods devised for minimum spanning trees [24]. In essence, the method is simply a generalization of the "fixed orientation" metric approximation to Euclidean distances, as employed in [3, 16], with appropriate modifications and complications that arise in the weighted region metric; e.g., a "straight" ray eminating from a point is now a locally optimal path, which refracts according to Snell's Law.

The new algorithm has been fully implemented (in C++) and tested against various other approaches on a large set of data. We report here on our experimental results.

### Summary

(1) We present a new algorithm for the weighted region problem, and analyze its effectiveness in approximating an optimal path. The algorithm is based on constructing a graph, a *pathnet,* that is guaranteed to contain an approximately optimal path between two query points. By varying the parameter, $k$, that controls the density of the graph, we are able to get arbitrarily close to optimal.

(2) We report on an implementation in C++ of our pathnet algorithm. The implementation is part of a more extensive system, called *WRP-Solve,* that has been built, with a convenient graphical user interface, to compare various approaches to route planning. In ad-

*Department of Computer Science, State University of New York, Stony Brook, NY 11794-4400, **email: cristian@cs.sunysb.edu**; Supported by NSF (CCR-9204585 and CCR-9504192) and by a grant from Hughes Aircraft.

†Department of Applied Mathematics and Statistics, State University of New York, Stony Brook, NY 11794-3600, **email: jsbm@ams.sunysb.edu**; Partially supported by NSF grants CCR-9204585 and CCR-9504192, and by grants from Boeing Computer Services and Hughes Aircraft.

dition to the pathnet algorithm, the system contains implementations of at least four other approaches to solving the WRP, and the user can freely select among them.

The software has been integrated within an extensive military simulation system, ModSAF, by our industrial collaborators at Hughes Aircraft. We expect to release the *WRP-Solve* code to the public domain.

(3) We provide experimental results comparing the pathnet approach to other approaches to the WRP. We present results obtained in both real and simulated datasets.

## Related Work

There are numerous papers on optimal route planning in weighted terrain maps. We refer the reader to the survey [15] for pointers to the literature.

The main theoretical results known for the weighted region problem are presented in Mitchell and Papadimitriou [17]. They give an algorithm, based on the "continuous Dijkstra method", to find a path whose weighted length is guaranteed to be within a factor $(1+\epsilon)$ of optimal, where $\epsilon > 0$ is any user-specified degree of precision. The time complexity of the algorithm is $O(E \cdot S)$, where $E$ is the number of "events" in the continuous Dijkstra algorithm, and $S$ is the complexity of performing a numerical search to solve the following subproblem: Find a $(1+\epsilon)$-shortest path from $s$ to $t$ that goes through a given sequence of $k$ edges of the triangulation. They show that $E = O(n^4)$ and that there are examples where $E$ can actually achieve this upper bound. The numerical search can be done using a form of binary search that exploits the local optimality condition: An optimal path bends according to "Snell's Law of Refraction" when crossing a region boundary. (The earliest reference we have found to the use of Snell's Law in optimal route planning applications is to the work of Warntz [23].) This leads to a bound of $S = O(k^2 \log(nNW/\epsilon))$ on the time needed to perform a search on a $k$-edge sequence, where $N$ is the largest integer coordinate of any vertex of the triangulation. Since one can show that $k = O(n^2)$, this yields an overall time bound of $O(n^8 L)$, where $L = \log(nNW/\epsilon)$ can be thought of as the bit complexity of the problem instance.

The algorithm of [17] assumes that the start (source) point is fixed, and computes a representation of optimal paths from the start to all other points (a form of "shortest path map"). If the start point is moved, the algorithm would need to be run all over again from the new start point. In contrast, our new algorithm computes a data structure that can be used for path queries between *pairs* of points. (The query time, though, is not logarithmic, as is the case if one builds a shortest path map and queries with a new goal point.) Further, the algorithm of [17] has worst-case running time that is *logarithmic* in $1/\epsilon$, while our new algorithm is linear in $1/\epsilon$; however, the worst-case dependence on $n$ is much better for the new algorithm: $O(n^3)$ versus $O(n^8)$. Finally, in contrast with the algorithm of [17], our new algorithm has been fully implemented and its practicality has been shown.

Various special cases of the weighted region problem admit faster and simpler algorithms. For example, if the weighted subdivision is rectilinear, and path length is measured according to weighted $L_1$ length, then efficient algorithms for single-source and two-point queries can be based upon searching a path-preserving graph [2]. Similarly, if the region weights are restricted to $\{0, 1, \infty\}$ (while edges may have arbitrary (nonnegative) weights), then an $O(n^2)$ algorithm can be based on constructing a path-preserving graph similar to a visibility graph [6]. This also leads to an efficient method for performing *lexicographic* optimization, in which one prioritizes various types of regions according to which is most important for path length minimization.

## 2 Preliminaries

We let $S$ denote a planar polygonal subdivision having convex faces. (In the event that the input subdivision has nonconvex faces, we first convexify (e.g., by triangulation).) We will assume that $S$ is given to us in a convenient data structure that allows the usual basic operations to be done in constant time; e.g., the quad-edge data structure [7] is one possibility (a variant of which is used by our code).

Let $n$ denote the total number of vertices of $S$. Then $S$ also has $O(n)$ faces and edges. Each face $f$ has an associated integer weight $\alpha_f \in \{0, 1, \ldots, W, +\infty\}$. The weight of an edge $e$ is also an integer $\alpha_e \in \{0, 1, \ldots, W, +\infty\}$ and is assumed to be at most $\min\{\alpha_f, \alpha_{f'}\}$, where $f$ and $f'$ are the faces incident on $e$. (In the event that the input does not specify a weight for $e$, we define $\alpha_e = \min\{\alpha_f, \alpha_{f'}\}$.)

A path is said to be *locally optimal* (an *LO-path*) if an infinitesimal perturbation of it does not result in its weighted length decreasing. The following properties of LO-paths in weighted regions are important to our algorithm; see [17] for proofs and further details.

(1) An LO-path is piecewise-linear, with bend points only at vertices or on edges of the subdivision.

(2) If an LO-path bends at a point on an edge, while crossing that edge, then it does so in a manner that obeys Snell's Law of Refraction: $\alpha_f \sin\theta = \alpha_{f'} \sin\theta'$ where $\theta$ and $\theta'$ are the angles of incidence and refraction respectively, and $\alpha_f$ and $\alpha_{f'}$ denote the corresponding face weights. (The angle of incidence (refraction) is defined to be the angle between the incoming (outgoing) segment and the normal to the edge.)

(3) If an LO-path bends at a point on an edge, while *entering* the edge at that point (turning to follow the edge for some distance), then it enters the edge at the critical angle of refraction, $\theta_c = \sin^{-1}(\alpha_e/\alpha_f)$.

(4) An LO-path cannot be incident on an edge at an angle greater than the critical angle, $\theta_c$.

## 3 The Pathnet Algorithm

The pathnet algorithm constructs a graph, a *pathnet* $\mathcal{G} = (V, E)$, on the vertices $V$ of the subdivision. The basic idea is simple: We discretize the continuum of orientations about each vertex $v \in V$, using $k$ evenly-spaced directions to approximate the full range $[0, 2\pi]$. Within each of the $k$ cones determined at $v$, we determine a possible edge (a *link*, which is a chain of segments, crossing edges of the subdivision) joining $v$ to another vertex, $u$, of $S$, or to a critical point of entry on some edge of $S$. It is possible that no link is created for a given cone; but there is *at most one* link per cone.

The (possible) link is determined by tracing out a locally-optimal path (*refraction ray*) from $v$ at each of the two orientations bounding the cone, advancing both paths (rays) in lock-step across faces of $S$, obeying Snell's Law of refraction with each crossing, until the two refraction rays first encounter a "topological fork" – the edge sequence of one ray first starts to differ from the edge sequence of the other ray. (Actually, for the sake of efficiency, we advance all $k$ rays out of $v$ in lock step, so that each ray is traced only once, not twice.) In other words, we determine the longest common edge sequence prefix of the refraction rays bounding each cone. This fork event must be of one of the following types:

(a) The two refraction rays are incident on distinct edges of $S$ – then there is at least one *splitting vertex*, $u$, on the face $f$ where they first fork, such that $u$ "splits" the two paths. We now create a link (an edge in $E$) between $v$ and (any) one such vertex $u$. Associated with this link is a pointer to the polygonal chain consisting of the line segments (along either of the two rays) that brought us to face $f$, and the segment joining the entry point onto $f$ with the vertex $u$. This chain represents an approximation to a refraction path from $v$ to $u$. We store its length as the weight of the edge $(v, u) \in E$. Refer to Figure 3.

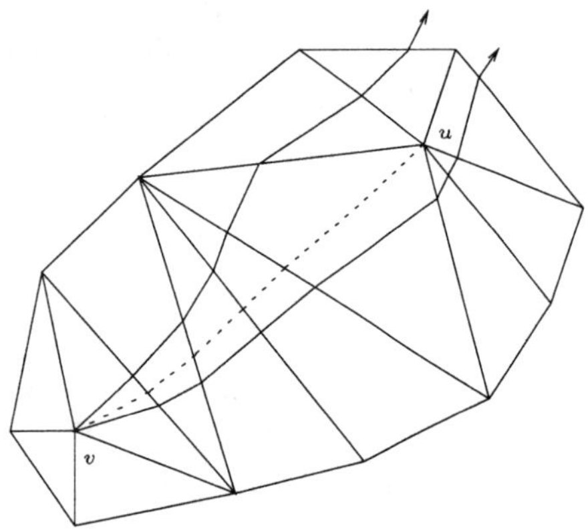


Figure 1: A cone determined by two refraction rays from vertex $v$ gets split by vertex $u$.

**Lemma 1** *There exists a unique locally-optimal path from $v$ to $u$ that stays within the refraction cone that is split by $u$.*

*Proof.* The proof follows from the monotonicity lemmas proven in [17]. □

*Remark.* Optionally, we can search for a chain linking $v$ and $u$ that is arbitrarily close to being a refraction (locally optimal) path; our code does this by a simple binary search, or a coordinate descent method. In practice, this may lead to somewhat shorter paths, and better "looking" paths in some cases. However, we see no way to make the worst-case error analysis take advantage of this extra optimization. We omitted it from the subset of experiments reported in this abstract.

(b) The two rays are incident on the same edge, $e$, but one or both rays are incident at an angle whose magnitude is greater than that of the critical angle at $e$.

**Lemma 2** *There exist at most two locally-optimal paths (corresponding to incidence angles $+\theta_c$ and $-\theta_c$) from $v$ to a critical entry point on edge $e$ that stays within the refraction cone.*

*Proof.* The proof follows from the monotonicity lemmas proven in [17]. □

We can trace a critical refraction path from a critical entry point $u \in e$ back to $v$, through the cone, since the edge sequence is known, and the critical angle at $e$ will therefore determine all of the angles of incidence along the edge sequence (even though the point $u$ is not discovered until we trace the path forward again). (Refer to Figure 3.) A pointer to this chain is stored with the link $(v, u) \in E$, and its length is recorded as the weight of the pathnet edge. Furthermore, the critical entry point $u$ is instantiated as a new vertex of the pathnet (it is added to $V$), and it is linked to neighboring such critical points along $e$ (or the endpoints of $e$). Note that, in total, no more than $O(kn)$ critical entry points will ever be added to the pathnet.

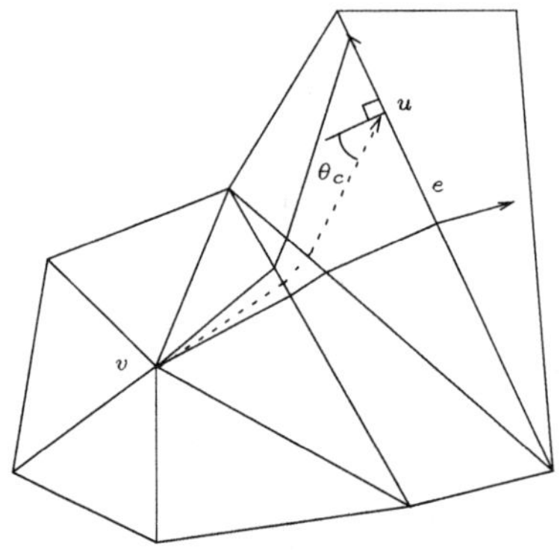


Figure 2: A cone determined by two refraction rays from vertex $v$ gets split at edge $e$, due to a critical entry point $u \in e$.

*Remark.* If one ray is incident on $e$ at an angle greater than $\theta_c$ and the other ray is incident at an angle less than $-\theta_c$, then it is possible that within the refraction cone there are *two* critical refraction paths incident on $e$. However, if both rays are incident at angles greater than $\theta_c$ (or both $< -\theta_c$), then there will be *no* critical refraction path within the cone.

(c) Both rays encounter the outer boundary of the subdivision. In this case, we simply cease the tracing of the cone, as no link of the pathnet will be made within it.

Once the pathnet is constructed, we search for a path between two vertices simply by running Dijkstra's shortest-path algorithm within the pathnet. (Alternatively, we can apply standard heuristic search variants of Dijkstra's shortest path algorithm, such as the $A^*$ algorithm [15]; however, our experiments reported here are all based on using Dijkstra's algorithm directly, not with the $A^*$ algorithm.) For arbitrary start/goal points, we first do point location queries to locate them in the subdivision, and then temporarily link them into the pathnet, by applying the same refraction ray tracing that was done from each vertex of $S$, in constructing $\mathcal{G}$. Then, a search of the augmented pathnet yields the desired path.

Alternatively, as we mention again later, our implementation also has an "on-demand" feature, which allows one to build the relevant portion of the pathnet "on the fly", during path search for a given query, without first constructing the full pathnet.

## 4 Analysis

The pathnet is a graph with $O(kn)$ nodes, since each of the $k$ cones at each of the $n$ vertices can produce at most one link. The propagation of the two rays defining a cone can be truncated after at most $O(n^2)$ steps (and typically it does not require more than about 5-10, in practice). This follows from Lemma 7.1 of [17], which proves that a locally optimal path that does not go through vertices or critical points can cross each edge at most $O(n)$ times, implying an edge sequence of at most $O(n^2)$ crossed edges. Thus, we get an overall worst-case time complexity of $O(kn^3)$, to build a data structure of size $O(kn)$.

It now remains to prove that a shortest path in the pathnet approximates the shortest weighted path between two points. The full details of the proof are deferred to the full paper, as they are somewhat tedious and are not central to the main ideas and applicability of the pathnet algorithm.

We begin by observing that if a large enough number $k$ of cones is used, then any vertex to vertex path (or vertex to critical entry point path) is contained in a single cone and the pathnet algorithm provides exact answers to shortest path queries.

Next, we observe that if we use $k$ cones per vertex, then the angle between two rays that define a cone begins at angle $2\pi/k$, and, after multiple refractions of the rays along a common edge sequence, never gets larger than $O(W/kw)$, where $W$ is the maximum finite weight and $w$ is the minimum nonzero weight. (This follows from the simple geometry of Snell's Law.)

Consider an optimal path, $\pi^*$, from a vertex $s$ to a vertex $t$. We will prove the existence of a path, $\pi$, that lies in the pathnet whose (weighted) length is at most $(1+\epsilon)$ times that of $\pi^*$, with an appropriate choice of $\epsilon$, which depends on $k$. Now, $\pi^*$ is partitioned into subpaths by the vertices and critical entry points along $\pi^*$. Consider one such subpath of $\pi^*$, call it $\pi^*_{u,v}$, joining vertex $u$ with vertex $v$. (The case of a subpath joining two critical entry points, or a vertex and a critical entry point, is handled similarly.) In the proposition below, we prove that there exists a path, $\pi_{u,v}$, from $u$ to $v$, within the pathnet, whose weighted length is at most $(1+\epsilon)$ times that of $\pi^*_{u,v}$, for an appropriate choice of $\epsilon$, as a function of $k$. It is then easy to conclude that any optimal path $\pi^*$ can be approximated by paths within the pathnet, within the same approximation factor, since each of the subpaths can be approximated.

**Proposition 1** *There exists a path, $\pi_{u,v}$, from $u$ to $v$, within the pathnet, whose weighted length is at most $(1+\epsilon)$ times that of $\pi^*_{u,v}$, where $\epsilon = O(\frac{W/w}{k\theta_{min}})$, $\theta_{min}$ is the minimum among the internal face angles of the subdivision, and $W/w$ is the ratio of the maximum non-infinite weight to the minimum non-zero weight.*

*Proof.* (Sketch) Consider the subpath $\pi^*_{u,v}$. Consider the refraction cone with apex $u$ that locally contains $\pi^*_{u,v}$. If this cone is split by vertex $v$, then we appeal to the claim below, and we are done. Otherwise, if the cone is split by some $r \neq v$, then we show that we can make a detour to $r$ (at a "small" cost), and then we can appeal to induction. (Specifically, we do induction on the index $i$ of the sorted lengths $\lambda_i$, which are the lengths of the links on the pathnet: $\lambda_1 \leq \lambda_2 \leq \cdots \leq \lambda_N$, where $N \leq \min\{nk, \binom{n}{2}\}$.)

Let $P$ be the polygon determined by the two refraction rays bounding the cone, stopping the rays at the points where they exit the face $f$ where the rays are first split, and closing the polygon $P$ using the boundary of $f$ (which contains the splitting vertex $r$). When the cone is split, the bounding refraction rays are at angle $O(W/kw)$. This fact, together with the lower bound on the face angles, allows one to show

*Claim:* The detour necessary to make $\pi^*_{u,v}$ go visit $r$ (by a detour from the point of exit from face $f$) is at most $O(\frac{W/w}{k\theta_{min}})$ times the length of $\pi^*_{u,v}$ within $P$. □

**Theorem 3** *In time $O(kn^3)$ a pathnet of size $O(kn)$ can be constructed such that, when searched, it yields paths that are within factor $(1+\epsilon)$ of optimal, where $\epsilon = O(\frac{W/w}{k\theta_{min}})$.*

## 5 Other Algorithms

In order to determine the practicality and relative effectiveness of our new algorithm, we implemented several other methods, for experimental comparison.

### 5.1 Grid-Based Algorithm

The most basic and most popular practical approach to solving path planning problems in varied terrain is to us a discretization into a uniform grid, and then to search the (weighted) *grid graph* for shortest paths. This method has many advantages over other more complex algorithms: (1) it is simple to implement; (2) it is particularly well suited for problems in which the input data structure comes as a grid; and (3) it is highly versatile, allowing for many other heuristic cost criteria (e.g., charging for turns) to be added quite easily.

We assume as input an array of sampled values of the underlying weight function $F$, giving values, $F_{i,j}$, at regular grid points indexed by $(i,j)$. Since our pathnet algorithm assumes as input a polygonal subdivision, not a weight array, our grid algorithm initializes a weight array by sampling the polygonal subdivision at grid points (whose spacing, $\gamma$, is a user-specified parameter).

The grid data points determine a *grid graph*, whose vertices are the grid points and whose edges link "nearby" grid points. Most grid-based algorithms assume a 4- or 8-connectivity of the grid, joining point $(i,j)$ to its four immediate neighbors $((i,j+1), (i,j-1), (i-1,j), (i+1,j))$, plus possibly the diagonals.

One of the main drawbacks to a grid graph solution is that, even for a trivial (constant) weight function $F$, there is an inherent error in distance estimate, since we measure grid distance instead of Euclidean distance. For 4-connectivity this "digitization bias" (aliasing) results in an error factor of $\sqrt{2}$; for 8-connectivity, it results in a factor of $(\sqrt{2}+1)/\sqrt{5} = 1.08$. See [18, 20] for discussions on the digitization bias problem, and on different distance transforms that can be used to address it.

One approach to reducing grid bias is to increase the connectivity of the grid. Thus, the grid-based algorithm that we implemented includes an enhancement to the usual methods, in that we permit higher degrees of connectivity: We connect $(i,j)$ to each of the grid points $(i+I, j+J)$, for $I, J \in \{-K, -K+1, \ldots, K-1, K\}$, except those that can be reached directly (without error) by two shorter edges (e.g., we do not connect $(i,j)$ to $(i+2, j+2)$, since this is effected by two diagonal connections already). Thus, for $K = 1,2,3,4$, we get connectivity of 8, 16, 32, 48, respectively. Allowing this $K$ as a parameter implies that, as $K$ goes to infinity, and the resolution $\gamma$ goes to zero, the grid-based solution will in fact go to the true optimum. (The dependence on $K$ is $O(1/K^2)$.) This will allow a fair comparison with our pathnet algorithm, for which we can also get arbitrarily close to optimum by increasing a parameter ($k$, the number of cones).

Since our underlying input function $F$ is given in polygonal subdivision form, we assign weights to the edges in the grid graph by computing the actual weighted cost of the direct (straight) line segment joining the endpoints. This is done by a simple ray tracing (walk) through $S$, integrating $F$ along the segment.

### 5.2 Edge Subdivision Algorithm

We also devised and implemented another natural method for searching a weighted subdivision for paths, based on discretizing the *edges* of the subdivision (rather than discreting the *faces*, as is the case with a grid-based method). This method has been implemented and used by several others, including Mitchell (who implemented it as part of a system built at Hughes, for the DARPA ALV project in 1985), Johansson [8] (who implemented it for use in fluid flow computations for injection molding), and Lanthier, Maheshwari, and Sack [12] (who, in independent work closely related work our own, have conducted experiments on this, and related, practical methods for the WRP).

The basic *edge subdivision algorithm* is as follows. For each edge $e$ of the subdivision, we place $\lfloor \alpha_e/\delta \rfloor$ new (Steiner) nodes, evenly spaced along $e$. Here, $\delta$ is a parameter of the algorithm, and $\alpha_e$ denotes the weight of edges $e$. We construct (weighted) edges joining each such Steiner node with all other nodes (both original vertices and Steiner nodes) that appear on the boundary of each of the two faces incident on $e$. In this way, each face of the subdivision corresponds to a complete graph on the set of vertices and Steiner nodes that appear on its boundary. Since faces are assumed to be convex, the line segment joining any pair, $u$ and $v$, of boundary points lies entirely within the face, so we assign the weight of $(u,v)$ to be $|uv| \cdot \alpha$, where $\alpha$ is the weight of the face. (Note that the shortest path joining $u$ and $v$ need not be the single edge, $(u,v)$, especially if the weight $\alpha$ of the face is very high.)

The worst-case error in the approximation is easy to estimate: For an optimal subpath that crosses $m$ edges, we can perturb it onto the graph induced by the subdivision nodes, adding weighted length $O(m\delta)$ to the path. This shows that the shortest path in the network constructed will be at most an additive term $O(m\delta)$ longer than optimal; as shown in [17], $m = O(n^2)$.

The edge subdivision algorithm has the following design choices: (1) the parameter $\delta$ (or one can specify a maximum number of Steiner points on any one edge); and (2) how to space the points on an edge, e.g., evenly (as we do), in geometric progression (as Papadimitriou [19] does for shortest paths in $\Re^3$), or in some other way. We have not yet seen how to use uneven spacing to achieve provably better approximation bounds; this may be a subject of future experimental analysis.

### 5.3 Other Approaches

Our *WRP-Solve* system also has a couple of other heuristic algorithms implemented:

(1) Search the (weighted) edge graph of $S$ for a shortest path between two vertices. (If the start/goal are not vertices of $S$, then we simply augment the graph by connecting them to the vertices of the convex containing cell.) This method has the advantage of being very simple and fast, but of course it can produce paths that are arbitrarily bad with respect to optimal.

(2) Search the dual graph of $S$, using edge weights that are the average of the two adjacent face weights, times the distance between the centroids of the faces. Again, this method is fast, but can yield arbitrarily bad paths with respect to optimum. (Note that if the faces are all unit squares – i.e., $S$ is a regular grid – then we are simply doing a 4-connectivity grid graph method, and the approximation error factor is then bounded by $\sqrt{2}$.)

Other approaches being currently designed for the *WRP-Solve* system include a simulated annealing approach (using, e.g., a starting seed path determined by either of the fast heuristics (1) or (2) above, together with a post-processing local optimality stage), and more sophisticated grid-based or quadtree-based algorithms. It will also be interesting to implement and test *hybrid* methods of computing optimal paths, using the best aspects of pathnets, grid graphs, edge subdivisions, etc.

## 6 Implementation and Experimental Setup

The *WRP-Solve* system is implemented in C++. Experiments were run on a set of SGI Indigos with R5000 processors at 150MHz running IRIX6.2. Internal memory size was 64MB for all, 512KB level 2 and 32KB separate data and instruction level 1 caches. Times quoted in the experiments were measured only on the SGI. For memory intensive experiments we used an Intel Pentium Pro computer at 180MHz with 128MB of memory.

Th system has several options for the user to set parameters, read/write data, time experiments, and select different algorithms. A screen shot showing one example is given in Figure 3.

### 6.1 Design choices

All of the algorithms model weighted subdivisions and grid sampled data with an undirected network where the nodes of the network correspond to points in the plane and the edge weights represent the weighted distance between nodes. The nodes are associated to a feature of the subdivision – vertex, edge, face – or to a Steiner point added by sampling an edge or a face of the subdivision.

Network representation is dependent on the search algorithm. For example, the edge subdivision algorithm does not require an explicit representation of the links between nodes on the same face. Determining the neighbors of a node during the search algorithm and the weighted length of the links can be done while the search algorithm is executed. This network representation would require no more storage than storage for the nodes but it would be inefficient in terms of execution time because dynamically determining the neighbors of a node has an impact on performance. On

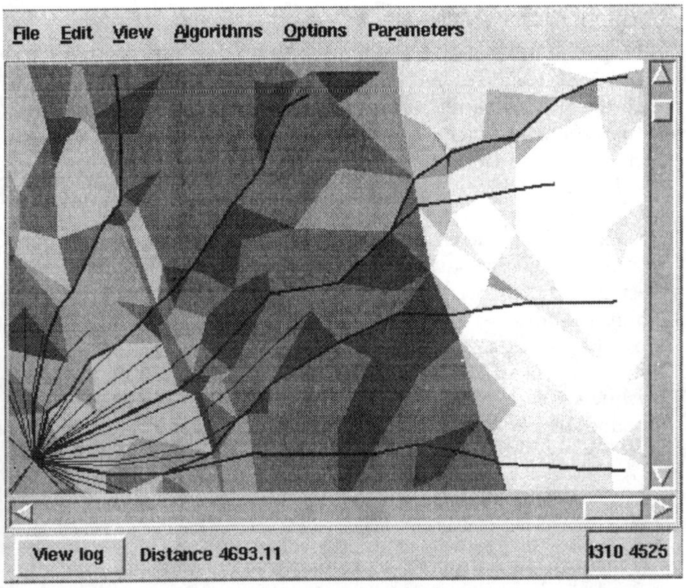


Figure 3: A screen shot from *WRP-Solve*. The starting point is in the lower left corner. Pathnet links to the start are shown with thin segments. Thicker segments (shown in "red" when viewed on a color workstation) are used to highlight the results of various shortest path queries to other points in the subdivision. Shading is used to indicate weight, with whiter regions having higher weights.

the other hand, building a complete representation – nodes and adjacency lists – for a large sampled grid is not practical given the large number of nodes and links.

The requirements that each algorithm imposes upon the representation of the search network had to be balanced with the fact that the experimental setup must compare the performance of algorithms. The choice was made to use the the same data structures for all algorithms.

The advantage of this solution is that the code used to implement the search algorithm for all the approximation algorithms is the same. This means that the number of elementary operations to compute a shortest path can be used to compare the performance of the algorithms. On the other hand, a specialized implementation for any one of the search methods would result in better performance – in terms of space and time used.

### 6.2 On-demand behavior

For all of the algorithms implemented, we have the option in *WRP-Solve* to do the preprocessing (e.g., pathnet construction) off-line or "on demand". In the "on-demand" mode, which is the default, the search network is constructed only as needed during the processing of a shortest path query. This implies that a search network may not contain all the links and nodes for the dataset. This on-demand feature is for some practical applications in which it is not necessary to preprocess the entire dataset. Once a portion of the search network is built, it is used by all later queries. This efffectively amortizes the cost of building the search network over multiple shortest path queries without requiring any preprocessing time.

For the experiments described in this paper, however, we chose to separate the preprocessing times from the query times. The search network is therefore created off-line – by selecting a start point and by building the shortest path tree for that point. This provides, on one hand, a complete data structure and, on the other hand, the results can be used to compare preprocessing time. (The same data is collected during preprocessing as for the shortest path queries.)

### 6.3 Datasets

The Weighted Regions Problem has practical applications in multiple domains, including decision support systems for the military [13, 14], manufacturing [8] and Geographic Information Systems. In practice, datasets – weighted subdivisions – have up to 10000 vertices. Most applications use smaller subdivisions – for example in the manufacturing application the size of the dataset is less than a 1000 vertices.

The experiments were run with a mix of real terrain and artificially generated datasets. The drawback of using real datasets is the presence of features that are degenerate or topologically inconsistent. In the implementation these problems are accounted for by letting the algorithms abandon computing a link – locally optimal path – if the underlying data structure is inconsistent.

**Real World Datasets.** The subdivisions used for the experiments are extracted from the "microterrain" elements in the Hunter-Ligett Compact Terrain Database (CTDB). The terrain database is used in the ModSAF[13, 14] training simulator for visibility and shortest path computation. The weighted subdivision corresponding to microterrain elements is obtained by assigning weights to the faces of the subdivision using a heuristic that assigns to each face integer weights between 3 and 12. The number of different weights used is higher than that which is typically used in practice by ModSAF; it was chosen to make shortest path planning even more challenging for our experiments.

The sizes of the weighted subdivisions vary from less than 100 vertices to around 1000 vertices. The faces of the subdivision are mostly triangles, although buildings and other man-made features have more vertices. The complicating problem with this data is the relatively large number of topologically incorrect features (e.g. overlaping faces and dangling edges).

**Simulated Datasets.** Artificial datasets are generated using two methods:

1. One method starts with a regular rectilinear grid and randomly perturbs the positions of the vertices, while constraining the perturbation so that all faces of the subdivision remain convex.
2. A second method uses Delaunay triangulations of point sets (based on the algorithm and code of Devillers [4]). In one set of datasets the vertices of the subdivision are generated uniformly at random within a square. Another set of triangulations is designed to yield nonuniform point distributions, by concentrating a majority of the points in a few small disks; the goal is to simulate the irregular distribution of points encountered in applications.

Each of the artificial datasets is generated with a variable number $n$ of vertices, where $n$ = 100, 200, 500, 1000, 2000, 5000, 10000 and 20000. Larger subdivisions are possible, but these datasets incur run-time penalties – for some of the algorithms we compare, these subdivisions lead to data structures that exceed the amount of internal memory available. It is an open question how the shortest path algorithms fare experimentally in external memory settings.

### 6.4 Generation of Start/Goal Pairs

The *WRP-Solve* system allows start and goal points to be chosen arbitrarily. A simple (worst-case linear-time) algorithm is used for point location, with special code to handle start/goal points that fall on edges/vertices of the subdivision.

Start/goal point pairs are picked at random with a minimum Euclidean distance between start and goal of 20% of the diameter of the dataset.

Start and goal points act as special vertices of the search network. These vertices are inserted when the search begins and deleted when the search ends. Time measurements include the time to insert the special vertices because this time (exclusive of point location) is partially dependent on the search algorithm. For example, inserting a vertex into a pathnet takes more time than the time to insert a vertex in a grid. In contrast, deleting vertices is a standard operation that involves the same operations for all algorithms implemented.

### 6.5 Data structures

The choice of representation for the subdivision and network data structures influences directly the performance of the algorithms. This effect is lessened by measuring the time performance for the each algorithms measured on the CPU clock and by counting the number of elementary operations performed by each algorithm during a shortest path search.

**Subdivision.** The natural choice for representing the input subdivisions is the *quad-edge* [7] data structure. Because of the limited operations needed, and the fact that the datasets are terrains, the implementation is simplified: *Rot* (edge-dual) and *Flip* (orientation) are not implemented.

Faces and vertices of the planar subdivision are objects with attributes. The vertices have $(x, y)$-coordinates and faces have weights. In all experiments the edges have the default weight which is the minimum weight of the incident faces.

It is possible to define obstacles in the subdivision when the weight of a face is over a user-defined threshold. In this case, no shortest path is allowed to cross an obstacle face. Free (no cost) regions are implemented by using low unit weight faces. Most applications use only a small number of weights. For example, in applications of *WRP-Solve* to air traffic control, Seagull Technologies, Inc. uses heuristics to assign five different weights to regions. The experiments use $\{1, \ldots, 15\}$ for weight values.

**Search Network.** The search network is a data structure designed to answer efficiently membership queries like "Is there a node at vertex $v$ of the planar subdivision?" Hash tables are used to store the nodes because they are efficient, simple and a robust implementation is available from the Tcl library. The average size of the buckets used in the hash tables was always less than two.

To speed things up separate hash tables are maintained for each of the objects making up a weighted subdivision: vertices, edges and faces.

### 6.6 Robustness issues

The pathnet algorithm assumes a topologically correct and consistent weighted subdivision. Subdivisions can be input interactively with click and drag. After an input operation, *WRP-solve* performs a topological check to verify consistency before a new edge is added to the data structure.

The procedure that solves Snell's Law of Refraction for each crossing is particularly sensitive to numerical errors. An exact solution requires solving a $4^{th}$ degree polynomial. If the refraction path is sufficiently close to a subdivision vertex and an incorrect decision is made – because of floating point rounding – the program will go into an infinite loop. To address this issue, each refraction case subproblem is solved with the following constraint: if the crossing point is within a small distance – a constant fraction of edge length, divided by the number of cones – of an endpoint, then the path "snaps" to the edge endpoint.

## 7 Results

**Measured parameters** We measured the following parameters for all shortest path queries:

- The number of operations performed during each shortest path search; here, "operations" counts the number of nodes visited by the search algorithm and the number of insertion and update operations to the heap.
- The time spent during each search. We measured two times: (1) the elapsed time which measures the interval of time elapsed between the start of the search and the end of the search - as returned by the function *gettimeofday* and (2) the user and system times as measured by *getrusage*. The reason was that the experiments were run on hardware in use by other users and the elapsed time might prove an incorrect measure.
- The length of the shortest path approximation and the actual Euclidean length between the start and goal parameters.
- The number of nodes in the network.

In addition to the above values, which were measured for all methods, we also measured for the pathnet algorithm some additional parameters, including the number of critical entry points, the number of *simple* and the number of *chain* links, the average degree of the nodes of the search network, and the average number of edge crossings in a chain.

**Average query time vs. approximation factor** The pathnet achieves consistently approximation factors of less than 5 percent, for all choices of parameters that we tested, and for all datasets. Figure 4 presents a plot of the average time required to process a query versus the approximation factor. The edge subdivision algorithm achieves the same approximation factors using more processing time. The dataset being used for the experiment is a Delaunay triangulation with 1000 vertices. The grid used in the experiment is 8-connected. Results show that the contenders for accurate results are the edge and the pathnet algorithms. The approximation factor obtained with the grid algorithm shows the digitization bias, although by using a higher degree of grid connectivity, results for the grid algorithm can be improved (see the full paper).

**Preprocessing time and memory use for pathnet algorithm** Preprocessing times required to build pathnets for different parameter values are measured. Results show that building the search network takes time proportional to the number of cones. Build times depend on the number of nodes in the search network, see Figure 6. Triangulations – terrain and Delaunay datasets – have larger size pathnets than grid subdivisions.

The pathnet algorithm achieves good approximation results for shortest paths even when the number of cones used is relatively small – less than 20. The data shows that the time to preprocess a subdivision with 1,000 nodes is less than 10 seconds. Note that in practical use the pathnet is built on demand and the preprocessing penalty is correspondingly lower. For the same subdivision the edge algorithm is on average 2.5 times slower for the same approximation factor. The grid – 8-connected – for a weaker approximation factor uses 10 times more.

The amount of memory used by the pathnet algorithm is proportional to the number of cones, see Figure 5. Artificial datasets have less variation than terrain data.

**Number of operations** The edge and pathnet algorithms perform fewer operations than the grid algorithm to achieve the same approximation factor. The pathnet has the advantage of links spanning across multiple faces. Figure 7 presents the average number of nodes visited per unit distance for different approximation factors. Scaling to unit distance allows an estimate of the time required to complete a shortest path query – a useful feature in an application where time resources are limited.

Complementary to the average number of operations is the average time required to plan a shortest path for a unit distance, see Figure 8.

**Approximation factors for pathnets** The experiment looks at the correlation between approximation factors and the number of cones used to build a pathnet. Figure 9 links the parameter – number of cones – necessary to achieve a given approximation factor, with respect to the optimal path.

The $x$-axis represents the number of vertices. For artificial datasets the number of cones remains relatively constant and decreases as the number of vertices increases.

For a given approximation factor the number of cones is the smallest number such that *all* shortest path queries tested are within the approximation factor. For example, a pathnet with 15 cones will in most cases achieve an approximation factor of 2 percent. A pathnet built with 20 cones achieves on average a 1 percent approximation factor.

The peaks in the plot indicate that some datasets require a large number of cones for a good approximation factor. A useful observation is that the "behavior" of the dataset remains the same over several values of the parameter. In this case, the approximation factor achieved with a small number of cones is a predictor on the approximation factor once the number of cones increases.

## 8 Conclusion

In this paper, we have taken a practical look at the problem of computing low-cost paths in weighted terrains. Our main contribution is a new method to compute paths in weighted terrains. The method has been fully implemented and tested in planar weighted regions, and comparisons have been made with several other simple and natural heuristic algorithms, such as regular grids and edge subdivision. Our experiments show that the pathnet method performs very well in practice and is highly competitive with other methods, generally yielding a shorter path in less query time. One other advantage of the pathnet algorithm is that its form of discretization is independent of the scale of the dataset. The grid and the edge subdivision algorithms depend in their choice of parameters on the (Euclidean) sizes of the features (edges and faces) in the map.

While the worst-case dependence on $\epsilon$ is worse than that of Mitchell and Papadimitriou [17], the worst-case dependence on $n$ is substantially lower, both in terms of time and space complexity: $O(n^3)$ versus $O(n^8)$ time and $O(n)$ versus $O(n^4)$ space.

Several extensions to our work are natural:

1. Our method applies directly to polyhedral surfaces with weighted facets. The only change to the implementation necessary is to modify the ray tracing procedure to compute the refraction path in the "unfolded" version of the local geometry. This code exists already, and will soon be fully tested.
2. The same method also applies to cases of weight functions that are not necessarily constant within faces of the subdivision (or facets of the polyhedron). For example, we could consider piecewise-linear weight functions, or other simple-to-describe weight functions for which we can compute the local optimality condition when a path crosses a region boundary. We also plan to consider weight functions that vary with time.
3. At the other extreme, we can apply our method to the unweighted (constant weight) case too, obtaining an approximation for shortest paths on polyhedral surfaces. For this problem, we know that a locally optimal path can cross only a linear number of edges (since each edge can be crossed at most once), so the complexity becomes $O(kn^2)$. Further, the actual performance in practice may be close to linear, since the average running time is $O(knm)$, where $m$ is the average number of edges crossed by a cone before its defining rays get split; in practice, we expect $m$ to be much less than $n$.

In conclusion, we should add that there are many other interesting experiments still to be conducted. In particular, we plan to conduct direct comparisons between the methods presented here and the methods recently developed in a parallel effort by Lanthier, Maheshwari, and Sack [12]. Further, it would be interesting to compare our methods' performance to the simulated annealing approaches implemented by Kindl, Shing, and Rowe [10, 11]. By combining different heuristic methods, we expect that the ultimate winner in our quest for better heuristics will be a "hybrid" method, which can take advantage of data given both in terms of regular grids and in terms of polygonal subdivisions.

## References


[1] R. Alexander and N. Rowe. Path planning by optimal-path-map construction for homogeneous-cost two-dimensional regions. In *Proc. IEEE Internat. Conf. Robot. Autom.*, 1990.

[2] D. Z. Chen, K. S. Klenk, and H-Y. T. Tu. Shortest path queries among weighted obstacles in the rectilinear plane. In *Proc. 11th Annu. ACM Sympos. Comput. Geom.*, pages 370–379, 1995.

[3] K. L. Clarkson. Approximation algorithms for shortest path motion planning. In *Proc. 19th Annu. ACM Sympos. Theory Comput.*, pages 56–65, 1987.

[4] O. Devillers. Robust and efficient implementation of the Delaunay tree. Report 1619, INRIA Sophia-Antipolis, Valbonne, France, 1992.

[5] D. H. Douglas. Least cost path in geographic information systems. Research note No. 61, Department of Geography, University of Ottawa, Ottawa, Ontario, August 1993.

[6] L. Gewali, A. Meng, J. S. B. Mitchell, and S. Ntafos. Path planning in $0/1/\infty$ weighted regions with applications. *ORSA J. Comput.*, 2(3):253–272, Summer 1990.

[7] L. J. Guibas and J. Stolfi. Primitives for the manipulation of general subdivisions and the computation of Voronoi diagrams. *ACM Trans. Graph.*, 4:74–123, 1985.

[8] P. Johansson. On a weighted distance model for injection moulding. Linköping Studies in Science and Technology, Thesis No. 604 LiU-TEK-LIC-1997:05, Division of Applied Mathematics, Linköping University, Linköping, Sweden, February 1997.

[9] J. M. Keil and C. A. Gutwin. Classes of graphs which approximate the complete Euclidean graph. *Discrete Comput. Geom.*, 7:13–28, 1992.

[10] M. Kindl, M. Shing, and N. Rowe. A stochastic approach to the weighted-region problem, I: The design of the path annealing algorithm. Technical report, Computer Science, U.S. Naval Postgraduate School, Monterey, CA, 1991.

[11] M. Kindl, M. Shing, and N. Rowe. A stochastic approach to the weighted-region problem, II: Performance enhancement techniques and experimental results. Technical report, Computer Science, U.S. Naval Postgraduate School, Monterey, CA, 1991.

[12] M. Lanthier, A. Maheshwari, and J. Sack. Approximating weighted shortest paths on polyhedral surfaces. In *Proc. 13th Annu. ACM Sympos. Comput. Geom.*, page To appear, 1997.

[13] M. J. Longtin. Cover and concealment in ModSAF. In *Proc. Fourth Conference on Computer Generated Forces and Behavioral Representation*, pages 239–247. STRICOM-DMSO, 1994.

[14] M. J. Longtin and D. Megherbi. Concealed routes in ModSAF. In *Proc. Fifth Conference on Computer Generated Forces and Behavioral Representation*, pages 305–313. STRICOM-DMSO, 1995.

[15] J. S. B. Mitchell. An algorithmic approach to some problems in terrain navigation. In S. Sitharama Iyengar and Alberto Elfes, editors, *Autonomous Mobile Robots: Perception, Mapping, and Navigation*, pages 408–427. IEEE Computer Society Press, Los Alamitos, CA, 1991.

[16] J. S. B. Mitchell. $L_1$ shortest paths among polygonal obstacles in the plane. *Algorithmica*, 8:55–88, 1992.

[17] J. S. B. Mitchell and C. H. Papadimitriou. The weighted region problem: Finding shortest paths through a weighted planar subdivision. *J. ACM*, 38:18–73, 1991.

[18] J. S. B. Mitchell, D. W. Payton, and D. M. Keirsey. Planning and reasoning for autonomous vehicle control. *Internat. J. Intell. Syst.*, II:129–198, 1987.

[19] C. H. Papadimitriou. An algorithm for shortest-path motion in three dimensions. *Inform. Process. Lett.*, 20:259–263, 1985.

[20] I. Ragnemalm. The Euclidean distance transform. Linköping Studies in Science and Technology, Ph.D. Dissertation 304, Department of Electrical Engineering, Linköping University, Sweden, 1993.

[21] R. F. Richbourg, N. C. Rowe, M. J. Zyda, and R. McGhee. Solving global two-dimensional routing problems using Snell's law. In *Proc. IEEE Internat. Conf. Robot. Autom.*, pages 1631–1636, Raleigh, NC, 1987.

[22] N. C. Rowe and R. F. Richbourg. An efficient Snell's law method for optimal-path planning across multiple two-dimensional, irregular, homogeneous-cost regions. *Internat. J. Robot. Res.*, 9:48–66, 1990.

[23] W. Warntz. Transportation, social physics, and the law of refraction. *The Professional Geographer*, 9(4):2–7, 1957.

[24] A. C. Yao. On constructing minimum spanning trees in $k$-dimensional spaces and related problems. *SIAM J. Comput.*, 11:721–736, 1982.

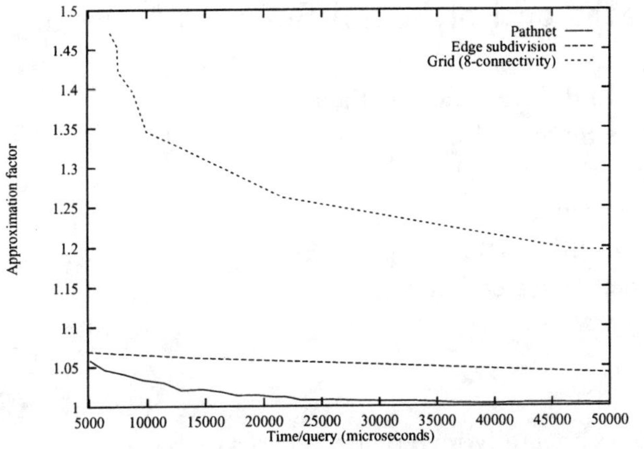


Figure 4: Approximation factor for the pathnet, grid and edge subdivision algorithms with respect to shortest path query time

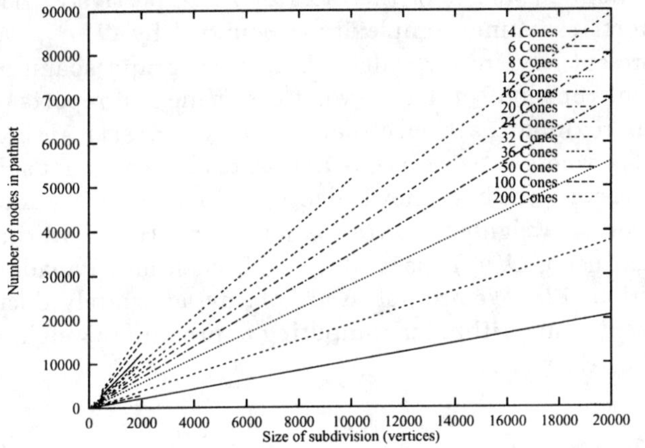


Figure 5: Number of nodes in pathnets with respect to the size of the weighted subdivision

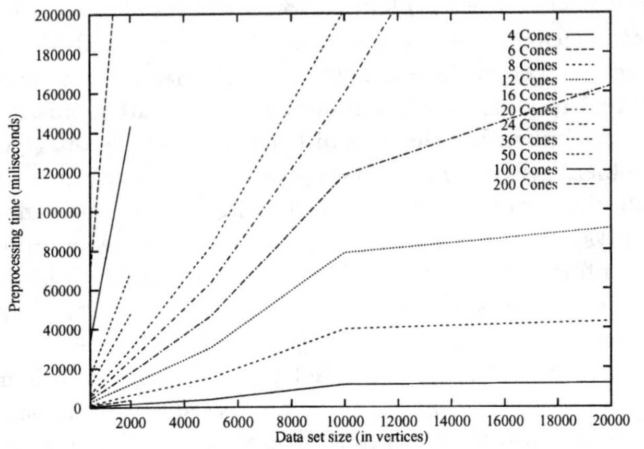


Figure 6: Time required to build a pathnet search graph versus number of cones for each vertex of the weighted subdivision

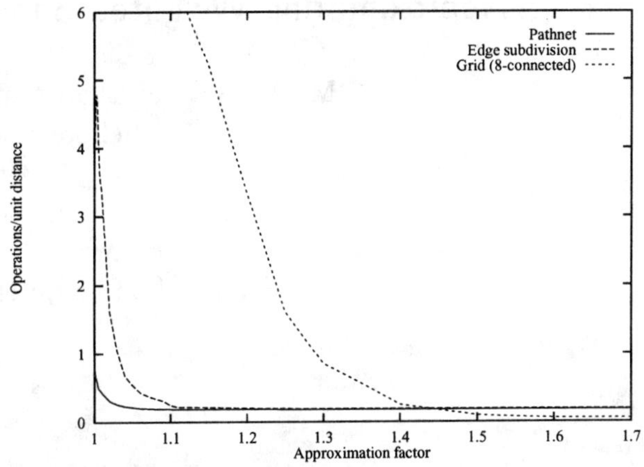


Figure 7: Number of nodes visited with respect to the approximation factor

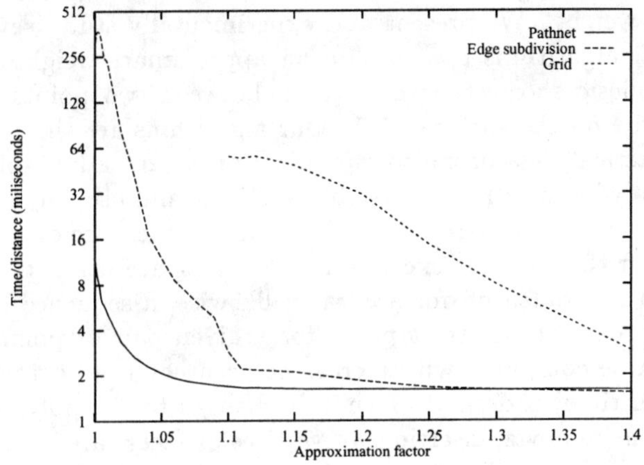


Figure 8: Average search time versus approximation factor, normalized to unit euclidean distance.

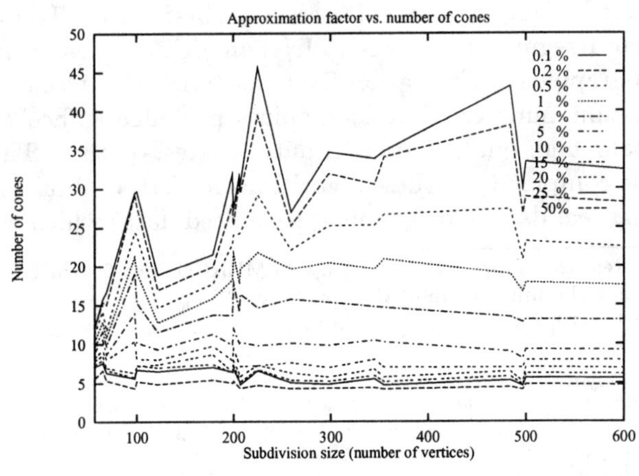


Figure 9: Maximum approximation factors for the pathnet algorithm versus number of cones.

# Approximating Weighted Shortest Paths on Polyhedral Surfaces *†‡

Mark Lanthier, Anil Maheshwari and Jörg-Rüdiger Sack
School of Computer Science
Carleton University
1125 Colonel By Drive
Ottawa, ON, CANADA K1S 5B6
E-mail: {lanthier,maheshwa,sack}@scs.carleton.ca

**Abstract**

Consider a simple polyhedron $\mathcal{P}$, possibly non-convex, composed of $n$ triangular regions (faces), in which each region has an associated positive weight. The cost of travel through each region is the distance traveled times its weight. We present and experimentally study several algorithms to compute an approximate weighted geodesic shortest path, $\pi'(s,t)$, between two points $s$ and $t$ on the surface of $\mathcal{P}$. Our algorithms are simple, practical, less prone to numerical problems, adaptable to a wide spectrum of weight functions, and use only elementary data structures. An additional feature of our algorithms is that execution time and space utilization can be traded off for accuracy; likewise, a sequence of approximate shortest paths for a given pair of points can be computed with increasing accuracy (and execution time) if desired. Dynamic changes to the polyhedron (removal, insertions of vertices or faces) are easily handled. The key step in these algorithms is the construction of a graph by introducing Steiner points on the edges of the given polyhedron and compute a shortest path in the resulting graph using Dijkstra's algorithm. Different strategies for Steiner point placement are examined. Our experimental results obtained on Triangular Irregular Networks (TINs) modeling terrains in Geographical Information Systems (GIS) show that a constant number of Steiner points per edge suffice to obtain high-quality approximate shortest paths. The time complexity of these algorithms for TINs (obtained using real data and randomly generated data) which we experimentally investigated is $\mathcal{O}(n \log n)$.

Our analysis bounds the approximate shortest path cost, $\pi'(s,t)$, by $\pi(s,t) + w_{max} \cdot |l_e|$, where $\pi(s,t)$ denotes the geodesic shortest path between $s$ and $t$ on the boundary of $\mathcal{P}$, $l_e$ is the longest edge and $w_{max}$ is the maximum weight of the faces of $\mathcal{P}$, respectively. The worst case time complexity is bounded by $\mathcal{O}(n^5)$. We present an alternate algorithm, using graph spanners, that runs in $\mathcal{O}(n^3 \log n)$ worst case time and reports an approximate path such that $\pi'(s,t) \leq \beta(\pi(s,t) + w_{max} \cdot |l_e|)$, where $\beta > 1$ is a constant. Already, for planar subdivisions, the best known algorithm for computing exact geodesic weighted shortest path runs in $\mathcal{O}(n^8 \log n)$ time and using $\mathcal{O}(n^4)$ space, due to Mitchell and Papadimitriou [12]. We are not aware of any adequately documented algorithm for computing approximate weighted shortest paths.

*Research supported in part by ALMERCO Inc. & NSERC.
†For the full version of this paper see [9].
‡See [10] for an accompanying video.


*Computational Geometry 97* Nice France


## 1 Introduction

Shortest path problems are among the fundamental problems studied in computational geometry. In this paper, we consider the problem of computing a shortest cost path between two points $s$ and $t$ on a polyhedral surface $\mathcal{P}$. The surface is composed of triangular regions (faces) in which each region has an associated positive weight indicating the cost of travel in that region. This problem arises in numerous applications in areas such as robotics, traffic control, search and rescue, water flow analysis, road design, navigation, routing, geographical information systems. Most of these applications demand simple and efficient algorithms to compute approximate shortest paths as opposed to a complex algorithm that computes an exact path. Polyhedra arising in these applications approximate real surfaces and thus an approximate path will typically suffice. Our interest is also motivated by our research and development on a parallel system for GIS and spatial modeling.

Research work on computing a shortest geodesic path on unweighted polyhedral surfaces began with the special case of convex polyhedra [14, 17]. Approximate computations of Euclidean paths on unweighted polyhedra have recently been reported by [1, 7]. The computation of Euclidean shortest paths on non-convex polyhedra has been investigated by [3, 11, 15, 19]; currently, the best known algorithm due to Chen and Han [3] runs in $\mathcal{O}(n^2)$ time. * The best known algorithm for the query problem, where both source and destination are unknown, requires $\mathcal{O}(n^6 m^{1+\delta})$, for $\delta > 0$ and $1 \le m \le n^2$, to answer queries in $\mathcal{O}((n/m^{1/4}) \log n)$ time [2].

Mitchell and Papadimitriou [12] introduced the "weighted region problem" and presented an algorithm that computes a shortest weighted cost path between two points in a planar subdivision; it requires $\mathcal{O}(n^8 \log n)$ time in the worst case. They state that their algorithm applies to non-convex polyhedral terrains with modifications. (See the survey paper by Mitchell and Suri [13] for several results on shortest path problems.)

Motivated by the practical importance of these problems and very high complexities for computing exact shortest paths, we investigate algorithms for computing approximated shortest paths. In this paper we propose several simple and practical algorithms (schemes) to compute an approximated weighted shortest path $\pi'(s,t)$ between two points $s$ and $t$ on the surface of a (possibly, non-convex) polyhedron $\mathcal{P}$. The accuracy of the approximation varies with the length of the longest edge $l_e$ of $\mathcal{P}$ and the maximum weight $w_{max}$ of the faces of $\mathcal{P}$. We present two variations on the algorithm. The first ensures that $|\pi'(s,t)| \le |\pi(s,t)| + w_{max} \cdot |l_e|$. The second, based on graph spanners, ensures that $|\pi'(s,t)| \le \beta(|\pi(s,t)| + w_{max} \cdot |l_e|)$ for some fixed constant $\beta > 1$. Our experimental results indicate that the schemes presented here perform very well in practice and guarantee excellent results for polyhedra in which $|\pi(s,t)| >> |l_e|$. The typical running time for our algorithms is $\mathcal{O}(n \log n)$ although the worst case running times for our algorithms are $\mathcal{O}(n^5)$ and $\mathcal{O}(n^3 \log n)$, respectively.

Our methodology for computing approximate shortest paths in unweighted TINs involves computing the sequence of faces through which the approximate path passes and then constructing the path by performing a "sleeve computation". In some cases the set of faces intersected by our approximate shortest path is identical to that of the actual shortest path, and hence the "sleeve computation" reports the actual shortest path.

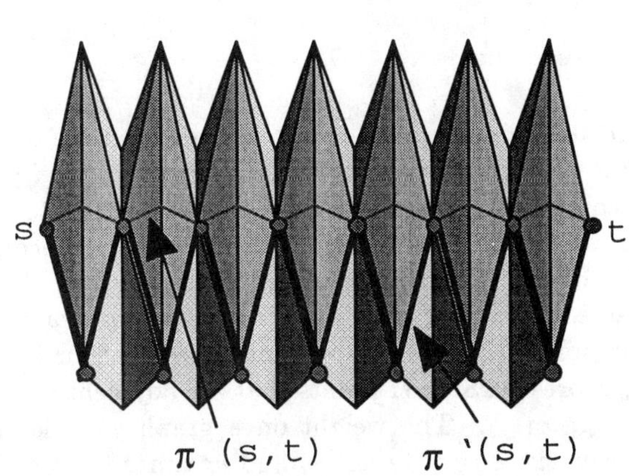

Figure 1: A worst case example

## 2 Shortest Path Approximations

### 2.1 A Simple Approach

Given a polyhedron $\mathcal{P}$, we could compute a graph $G$ (with edge weights) as follows. The vertices in $G$ correspond to the vertices of $\mathcal{P}$ and there is an edge between two vertices in $G$ if the corresponding vertices in $\mathcal{P}$ are connected by an edge.

For simplicity, assume that the source, $s$, and the target, $t$, are vertices of $\mathcal{P}$. Let $\pi(s,t)$ be composed of a sequence of $k$ adjacent straight line segments $s_1, s_2, ..., s_k$. The cost of a segment $s_j, 1 \le j \le k$ passing through face $f_i, 1 \le i \le n$ is assumed to be $w_{fi} \cdot |s_j|$. Similarly, an approximated path is denoted by $\pi'(s,t)$ with cost $|\pi'(s,t)|$ and segments $s'_1, s'_2, ..., s'_{k'}$.

$\pi'(s,t)$ can be computed using Dijkstra's algorithm [5]. This scheme confines the path to traveling on edges of $\mathcal{P}$, the quality of the approximation depends on the given triangulation, which could be bad in the worst case (see Figure 1). The approximated path $\pi'(s,t)$ with $k'$ links may have length $k' \cdot |l_e|$ and this could be much larger than $|\pi(s,t)|$.

### 2.2 Our Approximation Algorithms

We improve on the previous scheme by introducing Steiner points on the edges of the polyhedron. The set of vertices of $G$ will now consist of vertices and Steiner points of $\mathcal{P}$. We designed several strategies for placing vertices and edges in the graph $G$; this distinguishes the schemes examined here.

*Very recently, Varadarajan and Agarwal [19] propose an algorithm that computes an unweighted approximated path which is at most 13 times the actual path length and requires $\mathcal{O}(n^{\frac{5}{3}} \log^{\frac{5}{3}} n)$ time. They provide another algorithm that requires $\mathcal{O}(n^{\frac{8}{5}} \log^{\frac{8}{5}} n)$ time and produces a path that is at most 15 times the optimal.

### 2.2.1 Fixed Scheme

We add $m$ Steiner points evenly along each edge of $\mathcal{P}$, for some positive integer $m$. For each face $f_i, 1 \le i \le n$ of $\mathcal{P}$, compute a subgraph $G_i$ as follows. The Steiner points, along with the original vertices of $f_i$, become vertices of $G_i$. Connect a vertex pair $v_a, v_b$ of $G_i$ to form an edge of $G_i$ if and only if a) $v_a$ and $v_b$ represent Steiner points that lie on different edges of $f_i$ or b) $v_a$ and $v_b$ represent Steiner points that are adjacent on the same edge of $f_i$. The weight on a graph edge is the Euclidean distance between $v_a$ and $v_b$ times the weight of $f_i$. Denote $G_i$ as a complete graph although some edges are missing (i.e., those connecting vertices of non-adjacent Steiner points of the same polyhedral edge). A graph $G$ is then computed by forming the union of all $G_i, 1 \le i \le n$.

The approximated shortest path in $\mathcal{P}$ is computed by first determining a shortest path in $G$ using Dijkstra's algorithm and then transforming this path to a corresponding path in $\mathcal{P}$. It can easily be shown that all edges of $G$ lie on the surface of $\mathcal{P}$. Hence, any path in $G$ (i.e., our approximation) must also be a path on the surface of $\mathcal{P}$.

In order to analyze the cost of an approximated path, we first consider how each segment of a shortest path is approximated. If we compute a path $\pi'(s,t)$ in $G$ that passes through the same sequence of faces as $\pi(s,t)$, then the following claim is made:

**Claim 1** We can approximate a segment $s_j, 1 \le j \le k$ of $\pi(s,t)$ passing through face $f_i, 1 \le i \le n$ with an edge $s'_j$ of $G_i$ such that $w_{fi} \cdot |s'_j| \le w_{fi} \cdot |s_j| + w_{fi} \cdot \frac{|l_e|}{m+1}$, where $m$ is the number of Steiner points added to each edge of the face $f_i$.

**Proof:** Each edge is divided into $m+1$ intervals which have length at most $\frac{|l_e|}{m+1}$. From the properties of weighted paths as described by Mitchell and Papadimitriou [12], it follows that $s_j$ either crosses $f_i$ completely as shown in Figure 2 or it travels along an edge of $f_i$ (i.e., they show that the weighted shortest path obeys Snell's law of refraction and bends only at the edges of $\mathcal{P}$). Assume that $s_j$ begins at a point $a$ inside an interval of an edge of $f_i$ and ends at the point $b$ inside an interval of another edge (possibly the same one) of $f_i$. Without loss of generality assume that $s'_j$ begins at an interval endpoint (i.e., a Steiner point), say $c$, that is closest to $a$ and ends at an interval endpoint, say $d$, which is closest to $b$. Now if $c$ and $d$ are on different edges we know that there is a Steiner edge joining them. If they lie on the same edge, then there exists a series of adjacent collinear Steiner edges joining them, and we view these collinear edges as a single segment $s'_j$.

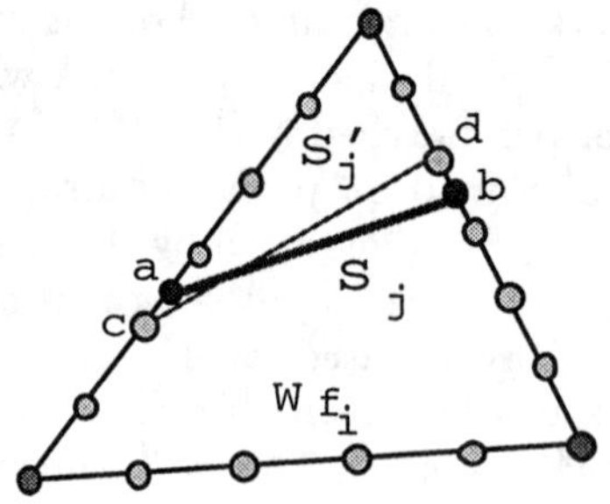


Figure 2: A weighted shortest path segment $s_j$ that crosses a face.

In Figure 2, the triangle inequality ensures that $|s'_j| \le |\overline{ca}| + |s_j| + |\overline{bd}|$. Since we chose the closer interval endpoints, then $|\overline{ca}| \le \frac{l_e}{2(m+1)}$ and $|\overline{bd}| \le \frac{l_e}{2(m+1)}$. Hence

$$|s'_j| \le |s_j| + \frac{l_e}{m+1}. \quad (1)$$

Now, multiplying by $w_{f_i}$ we have

$$w_{f_i} \cdot |s'_j| \le w_{f_i} \cdot |s_j| + w_{f_i} \cdot \frac{|l_e|}{m+1}. \quad (2)$$

In the case where a segment of the path travels along an edge of $\mathcal{P}$, we know that $|\overline{ca}|$ and $|\overline{bd}|$ have the same bounds as in 2. It is easily shown that equations 1 and 2 still hold. The proof also applies to the first and last segments of $\pi(s,t)$ in which $s$ and $t$ may be internal to faces.□

**Lemma 1** There exists an approximated path $\pi'(s,t)$ in $\mathcal{P}$ such that $|\pi'(s,t)| \le |\pi(s,t)| + \frac{|l_e|}{m+1} \cdot k \cdot w_{max}$.

**Proof:** Let $\pi'(s,t)$ be an approximated path with $k$ segments that passes through the same sequence of faces as $\pi(s,t)$. We write: $|\pi(s,t)| = \sum_{i=1}^{k}(w_{f_{s_i}} \cdot |s_i|)$ and $|\pi'(s,t)| = \sum_{i=1}^{k}(w_{f_{s_i}} \cdot |s'_i|)$ where $w_{f_{s_i}}$ is the weight of the face that $s_i$ passes through. By applying the results of Claim 1 to each segment of $\pi(s,t)$ we have: $\sum_{i=1}^{k}(w_{f_{s_i}} \cdot |s'_i|) \le \sum_{i=1}^{k}(w_{f_{s_i}} \cdot |s_i| + w_{f_{s_i}} \cdot \frac{|l_e|}{m+1})$.

We can rewrite this as

$$|\pi'(s,t)| \le |\pi(s,t)| + \tfrac{|l_e|}{m+1} \cdot \textstyle\sum_{i=1}^{k}(w_{f_{s_i}}).$$

Since $w_{f_{s_i}} \le w_{max}$ for all $f_{s_i}$ by definition, then

$$|\pi'(s,t)| \le |\pi(s,t)| + \frac{|l_e|}{m+1} \cdot k \cdot w_{max}. \quad (3)$$

□

Since a shortest weighted path on $\mathcal{P}$ may cross an edge $\mathcal{O}(n)$ times, we obtain the following theorem:

**Theorem 1** Using the fixed scheme, we can compute an approximation $\pi'(s,t)$ of the weighted shortest path $\pi(s,t)$ between two points $s$ and $t$ on a polyhedral surface $\mathcal{P}$ such that $|\pi'(s,t)| \leq |\pi(s,t)| + w_{max} \cdot |l_e|$, where $l_e$ is the longest edge of $\mathcal{P}$ and $w_{max}$ is the maximum weight of any face of $\mathcal{P}$. Moreover, we can compute this path in $\mathcal{O}(n^5)$ time.

**Proof:** We know (from Mitchell and Papadimitriou [12]) that a shortest weighted path on $\mathcal{P}$ may cross an edge $\mathcal{O}(n)$ times. Hence, a weighted shortest path may have $\mathcal{O}(n^2)$ segments. In Lemma 1, set $k = n^2$ and $m = n^2$, then we obtain:

$$|\pi'(s,t)| \leq |\pi(s,t)| + w_{max} \cdot |l_e|. \quad (4)$$

Using the fixed scheme, each edge contributes $\mathcal{O}(n^2)$ graph vertices and each face contributes $\mathcal{O}(n^4)$ graph edges. Thus, the algorithm requires $\mathcal{O}(n^3)$ space for graph vertices and $\mathcal{O}(n^5)$ space for graph edges. Dijkstra's shortest path algorithm takes $\mathcal{O}(n^5)$ time on this graph, using a fibonacci heap. □

### 2.2.2 Interval Scheme

In the fixed scheme, we made an assumption in our analysis that each edge crossed by the shortest path was of length $|l_e|$. In practice there are many edges of $\mathcal{P}$ with small length compared to $|l_e|$ (we have examined edge-length histograms for all of our data). This means that the addition of $m$ Steiner points to the smaller edges resulted in very small intervals with length much less than $\frac{|l_e|}{m+1}$.

We can improve the fixed scheme by forcing the intervals between adjacent Steiner points on an edge to be of length $\frac{|l_e|}{m+1}$. As a result, we can typically reduce the number of Steiner points added per edge considerably.

Since the maximum length of an interval is at most $\frac{|l_e|}{m+1}$, the proofs of Claim 1 and Lemma 1 still apply and hence Theorem 1 holds for the interval scheme. Although, the worst case analysis is the same for both schemes, the reduction of Steiner points has the advantage of reducing the number of graph vertices and edges that are created and processed by the graph shortest path algorithm.

### 2.2.3 Spanner Schemes

The time complexities of the previous two schemes can be further improved as described next; though the approximation achieved is not as good. Intuitively, we should be able to eliminate some edges joining Steiner points without a drastic reduction in our approximation factor. We eliminate these edges by using the notion of a *spanner*. Let $G_i, 1 \leq i \leq n$ be the complete graph formed by applying the edge decomposition scheme on a face $f_i$ of $\mathcal{P}$ with $m$ Steiner points per edge. $G_i$ has $3(m+1)$ vertices and $3(m+1)^2$ edges (including those along the face boundaries). We construct a *$\beta$-spanner* of $G_i$ and call it $G'_i$. The vertices of $G'_i$ are the vertices of $G_i$ which we sort in CCW order around any interior point of $f_i$ to form a sorted list $V_{G'_i}$.

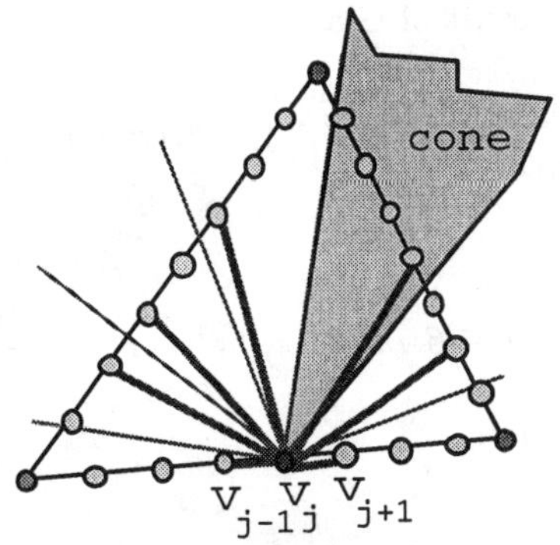


Figure 3: The spanner edges added from a vertex $v_j$ with $\theta = 30°$.

Let $\mathcal{C}$ be the set of 2-d cones with apex at $v_j, 1 \leq j \leq 3(m+1)$ and the conical angle $\theta = \frac{\pi}{\rho}$ for $\rho > 4$. For each vertex $v_j$ perform a radial sweep of the vertices $v_r, (j+1)\text{mod}(3(m+1)) \leq r \leq (j-1+3(m+1))\text{mod}(3(m+1))$. During this sweep compute the vertex $v_{min}$ that has minimal distance to $v_j$ in each of the $\rho$ cones and add an edge to $G'_i$ connecting $v_j$ and $v_{min}$ (see Figure 3).

Results of Clarkson [4], ensure that $G'_i$ is a $\beta$-spanner for $G_i$ where $\beta = \frac{1}{cos\theta - sin\theta}$. Since $G'_i$ is a $\beta$-spanner with $\mathcal{O}(m)$ vertices, it has $\mathcal{O}(m)$ edges. We create a similar spanner for each face of $\mathcal{P}$ individually and then merge each $G'_i, 1 \leq i \leq n$ to form the union $G'$. The approximate shortest path $\pi'(s,t)$ is computed by computing a shortest path in $G'$.

**Claim 2** A segment $s_j, 1 \leq j \leq k$ of $\pi(s,t)$ passing through $f_i$ can be approximated by a path $p_j$ in $G'_i$ such that $w_{f_i} \cdot |p_j| \leq \beta \cdot w_{f_i} \cdot |s_j| + \beta \cdot w_{f_i} \cdot \frac{|l_e|}{m+1}$, where $m$ is the number of Steiner points added per edge.

**Proof:** From Claim 1 it follows that $w_{f_i} \cdot |s'_j| \leq w_{f_i} \cdot |s_j| + w_{f_i} \cdot \frac{|l_e|}{m+1}$ for some segment $s'_j$ in $G_i$. From the definition of $G'_i$, it follows that $s'_j$ can be approximated by a path $p_j$ in $G'_i$ such that $|p_j| \leq \beta \cdot |s'_j|$. By substituting this into the result of Claim 1,

$$w_{f_i} \cdot \frac{|p_j|}{\beta} \leq w_{f_i} \cdot |s'_j| \leq w_{f_i} \cdot |s_j| + w_{f_i} \cdot \frac{|l_e|}{m+1} \quad (5)$$

Hence,

$$w_{f_i} \cdot |p_j| \leq \beta \cdot w_{f_i} \cdot |s_j| + \beta \cdot w_{f_i} \cdot \frac{|l_e|}{m+1}. \quad (6)$$

□

**Lemma 2** By applying the $\beta$-spanner scheme to $\mathcal{P}$ using $m$ Steiner points per edge, an approximate path $\pi'(s,t)$ is obtained for which $|\pi'(s,t)| \leq \beta(|\pi(s,t)| + \frac{|l_e|}{m+1} \cdot k \cdot w_{max})$.

**Proof:** The proof uses the results of Claim 2 and the techniques of Lemma 1. Let $\pi'(s,t)$ be the approximated path with $k$ subpaths $p_1, p_2, ..., p_k$ that passes through the same faces as $s_1, s_2, ..., s_k$ of $\pi(s,t)$ respectively. From the definition of the path cost it follows that

$|\pi(s,t)| = \sum_{i=1}^{k}(w_{f_{s_i}} \cdot |s_i|)$ and

$|\pi'(s,t)| = \sum_{i=1}^{k}(w_{f_{s_i}} \cdot |p_i|)$.

By applying the results of Claim 2 to each segment of $\pi(s,t)$ we have

$\sum_{i=1}^{k}(w_{f_{s_i}} \cdot |p_i|) \leq \sum_{i=1}^{k}(\beta \cdot w_{f_{s_i}} \cdot |s_i| + \beta \cdot w_{f_{s_i}} \cdot \frac{|l_e|}{m+1})$.

Rewrite this as

$|\pi'(s,t)| \leq \beta(|\pi(s,t)| + \frac{|l_e|}{m+1} \cdot \sum_{i=1}^{k}(w_{f_{s_i}}))$

Since $w_{f_{s_i}} \leq w_{max}$ for all $f_{s_i}$ by definition, then

$$|\pi'(s,t)| \leq \beta(|\pi(s,t)| + \frac{|l_e|}{m+1} \cdot k \cdot w_{max}). \quad (7)$$

□

**Theorem 2** An approximate weighted shortest path $\pi'(s,t)$ between two points $s$ and $t$ on a polyhedron of $n$ faces can be computed in $\mathcal{O}(n^3 \log n)$ time such that $|\pi'(s,t)| \leq \beta(|\pi(s,t)| + w_{max} \cdot |l_e|)$, $\beta > 1$, where $l_e$ is the longest edge of $\mathcal{P}$ and $w_{max}$ is the maximum weight of any face of $\mathcal{P}$.

**Proof:** The correctness follows from Lemma 2. We apply the $\beta$-spanner scheme to obtain $G'_i$ for each face $f_i$ of $\mathcal{P}$ where $1 \leq i \leq n$. Using $\mathcal{O}(n^2)$ Steiner points per edge, each subgraph $G'_i$ contains $\mathcal{O}(n^2)$ graph vertices and each vertex has a constant (depending upon $\beta$) degree since $G'_i$ is a spanner. Each $G'_i$ is computed by applying a radial sweep to each vertex in order to add its incident edges. Using a simple approach, this sweep could take $\mathcal{O}(n^2)$ time per vertex. However, since there are a constant number of cones ($\approx 2\rho$), we can divide $\mathcal{O}(n^2)$ vertices into an equal number of partitions. Since these vertices are sorted radially, we can apply a binary search for each of the cones to obtain the partitions in $\mathcal{O}(\log n)$ time. For each partition, we can again apply a binary search to obtain the closest vertex ($v_{min}$) in $\mathcal{O}(\log n)$ time as well. Since there are a constant number of partitions, each of the $\mathcal{O}(n^2)$ vertices can be processed in $\mathcal{O}(\log n)$ time and hence $G'_i$ can be computed in $\mathcal{O}(n^2 \log n)$ time. Since there are $n$ faces, $G'$ can be computed in $\mathcal{O}(n^3 \log n)$ time. A shortest path in $G'$ can be computed by using Dijkstra's algorithm (with a Fibonacci heap) and it runs in $\mathcal{O}(n^3 \log n)$ time. No further improvement on the graph construction is stated as the cost matches that of Dijkstra's shortest path algorithm. □

### 2.2.4 Sleeve Based Schemes

The scheme discussed next applies only to unweighted polyhedra. We first compute the approximate shortest path $\pi'(s,t)$ either using the fixed or the interval scheme. We then determine a sleeve by unfolding the faces along the edge sequence of $\pi'(s,t)$ and compute the shortest path that lies within this sleeve. We have implemented an algorithm similar to the algorithm of Guibas and Hershberger [6] and have applied this to our approximated path. (We omit the details of our sleeve computation due to space constraints from this version.) Section 3 shows that these approximated paths are much more accurate with this additional computation at a negligible increase in execution time. In most of our cases, the edge sequence of the approximated path is identical to that of an exact shortest path. The sleeve computation then produces the exact shortest path.

There is no efficient algorithm for computing shortest paths in weighted sleeves. For the weighted case however, we can perform a second approximation based on the outcome of the first. To do this, we select a *buffer* of faces from $\mathcal{P}$ to be used in the computation of the second approximation by selecting all faces that were "passed through" by the first approximated path. If this approximated path passed through a vertex, we add all incident faces of this vertex to the buffer. We then apply the approximation scheme on the buffer faces with an increased number of Steiner points per edge. As a result, we obtain a refined path. The refinement can be repeated as many times as desired.

### 2.3 Implementation Related Issues

**A Variation of Dijkstra's Algorithm:** In place of using Dijkstra's shortest path graph algorithm, we implemented a known variation of the algorithm. In this scheme we associate an additional weight to each vertex, namely its Euclidean distance to the target vertex. In this algorithm, at any iteration, the least cost vertex which we choose is the one which minimizes the sum of its cost from the source vertex plus the Euclidean distance to the target vertex, over all possible vertices.

**Space Complexity:** In applications, the space requirements for algorithms on TINs are often important. To address this, in any of our schemes, we can store the graph as it pertains to Steiner points implicitly. In an iteration of Dijkstra's algorithm, adjacency information can be computed on the fly with little penalty.

**Numerical Issues:** Suri [18] points out that the Chen and Han algorithm [3] based on unfolding is sensitive to numerical problems. This is due to the fact that 3D rotations are performed and errors accumulate along the paths and geometric structures computed. In our schemes the paths go through vertices or Steiner points (with the exception of the variation using as final step a sleeve computation); thus reducing the accumulation of numerical errors.

## 3 Experimental Results

A natural subclass of non-convex polyhedra is a *Triangular Irregular Network* or TIN which has triangular faces. A TIN is often constructed from a triangulated point set in the plane in which each point is assigned a height. In Geographic Information Systems, Cartography and related areas, shortest path problems arise frequently on terrains which are often modeled using TINs as shown in Figure 4. Due to their practical relevance, and in the context of our R&D [8], our experimental results are for TINs although the algorithms presented here also apply to any non-convex polyhedra.

One of the main difficulties in presenting experimental results is to choose *benchmark* TINs. It is conceivable that different TIN characteristics could affect the performance of an algorithm. We have attempted to accommodate different characteristics by performing our tests with TINs that have different sizes (i.e. number of faces), height characteristics (i.e. smooth or spiky (modeled by accentuating the heights)), and data sources (i.e. random or sampled from Digital Elevation Models (DEM)). Table 1 shows the attributes of the TINs that we've tested. TINs with stretched heights were created by multiplying their heights by five.

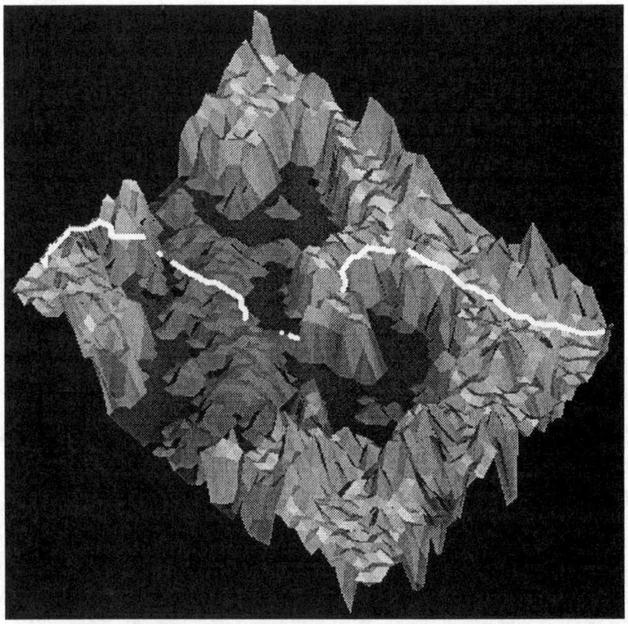

Figure 4: A weighted shortest path on a terrain in which traveling on water is expensive.

| FACES | STRETCHED HEIGHTS | DATA SOURCE |
|---|---|---|
| 1012 | NO | DEM |
| 1012 | YES | DEM |
| 5000 | NO | RANDOM |
| 5000 | YES | RANDOM |
| 10082 | NO | DEM |
| 10082 | YES | DEM |

Table 1: The various TINs and their attributes.

For each TIN, we computed a set of 100 random vertex pairs. We then tested each of the approximation schemes as shown in Table 2. We give the id of each scheme as they appear in the graphs. For each test, we computed the path cost between each of the 100 vertex pairs and then obtained an "average path cost" for these pairs. The tests were performed in iterations based on the number of Steiner points per edge. Each scheme was tested for both weighted and non-weighted scenarios (with the exception of the sleeve computations which were only computed in the unweighted case; a second approximation using a buffer was applied in the weighted case).

For the weighted domain, we use the same TINs and

| LEGEND ID | SCHEME | SLEEVE or BUFFER |
|---|---|---|
| INT | INTERVAL | NO |
| FIX | FIXED | NO |
| INTSLV | INTERVAL | YES |
| FIXSLV | FIXED | YES |

Table 2: The different approximation schemes.

we assign a weight to each face as the slope of the face. Thus, steeper faces have higher weight. Each edge of the TIN is given weight equal to the minimum of its adjacent faces. We determined experimentally that the spanner schemes provide slightly worse approximations and do not provide adequate improvement in running time; we omit the graphs here. Our analysis for the approximation, bounds the number of Steiner points by $\mathcal{O}(n^2)$ per edge in the worst case, this is far from the values required in our test suites using typical TINs.

### 3.1 Path Accuracy

We first ran tests on a terrain where the weight was homogeneous throughout all faces, i.e., we examined the Euclidean shortest path problem. The graphs in Figure 6 depict the results of these tests for variations of our approximation schemes. We can see that by using only a small number of Steiner points per edge, the approximated path length quickly converges to the actual path length (as computed using Chen and Han algorithm [3]). The graph shows that only a small number of Steiner points (i.e., 6) per edge suffice to obtain close-to-optimal approximations. Furthermore, the path accuracy observed is far better than the theoretical bound derived. The graphs also illustrate that the additional sleeve computation helps to obtain even more accurate approximations. Therefore, the best of our unweighted schemes is the interval scheme with the sleeve computation (IntSlv).

The graphs on the right of Figure 6 represent the approximations we obtained on the TINs with the heights of the vertices multiplied by a factor of 5. The graphs show that the results remain very good in that the approximate path length converges after only a few Steiner points per edge are added. We do see however, that the convergence is not as quick. This is mainly due to the fact that Steiner points are placed further apart along the now longer edges. Therefore it requires more Steiner points to reduce the interval size to that of the flatter TIN. The interval scheme performs better than the fixed scheme, since interval scheme favors placement of Steiner points on longer edges and longer edges are more likely to be crossed by the set of our random paths.

### 3.2 Computation Time

Since we are adding only a constant number of Steiner points on the average per edge, the running time of our algorithms is $\mathcal{O}(n \log n)$. In general, the algorithms' running times depend on the number of Steiner edges since we are evoking Dijkstra's graph shortest path algorithm.

The graphs of Figure 7 depict the running time for our 10,082 face TIN. We can see that our algorithms are substantially faster than Chen and Han[3]. The main reason is that our algorithms do not require any complex data structures nor do they perform expensive computations (i.e., 3D rotation and unfoldings). We precompute the graph $G$ and then perform a search for each query; a query is for a pair (source, destination), and we measure the time it takes to answer a query. From the graphs, we can see that the time required for the additional sleeve computation is negligible.

### 3.3 Weighted Paths

One problem in determining the accuracy of the algorithm in the weighted scenario is that we do not have an implementation of any algorithm that determines the actual shortest weighted path. The graphs of Figure 8 show the accuracy obtained through experimentation. As with the unweighted case, our path costs converge to some value after only a few Steiner points were added per edge. We therefore conjecture, that the cost of the paths converge similar to the unweighted scenario. From the graphs we can see that the second approximation based on the buffer technique provides a similar increase in accuracy as with the unweighted sleeve computations.

Since our algorithm is the same for unweighted and weighted scenarios, we obtained almost identical running time as shown in Figure 9. We can see however, that the second approximation resulted in a significant increase in computation time with respect to the increase shown for the unweighted sleeve computations. This increase is mainly due to the construction of a new refined graph which is necessary for each query.

## 4 Conclusion and Future Work

Shortest path problems belong to a class of geometric problems that are fundamental and of significant practical relevance. While realistic shortest path problems frequently arise in applications where the cost of travel is not uniform over the domain, the time, space and implementation complexities of existing algorithms even for the planar case are extremely large which motivates our study of approximation algorithms. Our experimental results show that high-quality approximations can be obtained with very good run-times. More precisely, we have provided empirical results showing that typical terrain data requires only a few (constant) Steiner points per edge. This reduces the running time to $\mathcal{O}(n \log n)$ in practice which is orders of magnitude smaller than the best known exact shortest path algorithm. The solutions are simple and of practical value.

We also theoretically establish bounds on the approximation quality and give worst-case bounds on the run-time of our algorithms.

For the unweighted scenario, we compared our accuracy to that of Chen and Han [3] and gave results indicating that our algorithm performs up to 50 times faster with a mimimum observed speedup of 14 times and produces nearly identical path results. We claim that our algorithm is efficient w.r.t. accuracy versus running time and is simple to implement. Our algorithm is of particular interest also for the case of queries with unknown source and destination.

Currently, we are investigating scenarios that involve other realistic weights taking into consideration physical properties of vehicles (see Figure 5 ). We are also working on other schemes and a parallel implementation of our algorithms.

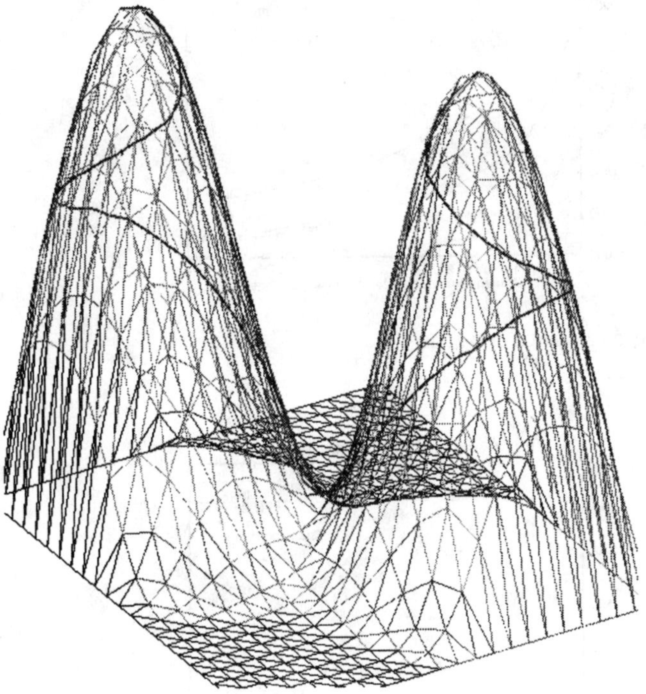

Figure 5: Shortest weighted path taking into consideration the maximum slope a vehicle can travel, as well as the turning angle.

**Acknowledgments:** We would like to thank Doron Nussbaum for useful discussions.

## References


[1] S. Har-Peled, M. Sharir, and K.R. Varadarajan, "Approximating Shortest Paths on a Convex Polytope in Three Dimensions", *Proc. 12th Annual Symp. on Computational Geometry*, Philadelphia, PA, 1996, pp. 329-338.

[2] P.K. Agarwal, B. Aronov, J. O'Rourke, and C. Schevon, "Star Unfolding of a Polytope with Applications", to appear in *SIAM Journal on Computing*, 1997.

[3] J. Chen and Y. Han, "Shortest Paths on a Polyhedron", *Proc. of the 6th ACM Symposium on Computational Geometry*, 1990, pp. 360-369.

[4] K.L. Clarkson, "Approximation algorithms for shortest path motion planning", *Proc. 19th Annual ACM Symp. Theory of Computing*, 1987, pp. 56-65.

[5] E.W. Dijkstra, "A Note on Two Problems in Connection with Graphs", *Numerical Mathematics 1*, 1959, pp.269-271.

[6] L. Guibas and J. Hershberger, "Optimal Shortest Path Queries in a Simple Polygons", *Proc. of the 3rd ACM Symposium on Computational Geometry*, 1987, pp. 50-63.

[7] J. Hershberger and S. Suri, "Practical Methods for Approximating Shortest Paths on a Convex Polytope in $\Re^3$", *Proc. of the 6th Annual ACM-SIAM Symposium on Discrete Algorithms*, 1995, pp. 447-456.

[8] D. Hutchinson, M. Lanthier, A. Maheshwari, D. Nussbaum, D. Roytenberg, J.-R. Sack, "Parallel Neighbourhood Modeling", *Proc. of the 4th ACM Workshop on Advances in Geographic Information Systems*,Minnesota, 1996, pp. 25-34.

[9] M. Lanthier, A. Maheshwari and J.-R. Sack, "Approximating Weighted Shortest Paths on Polyhedral Surfaces", Tech. Report TR-96-32, Carleton University, Dec. 96.

[10] M. Lanthier, A. Maheshwari and J.-R. Sack, "Approximating Weighted Shortest Paths on Polyhedral Surfaces", video in this proceedings, *13th Annual ACM Symposium on Computational Geometry*, June 1997.

[11] J.S.B. Mitchell, D.M. Mount and C.H. Papadimitriou, "The Discrete Geodesic Problem", *SIAM Journal of Computing*, **16**, August 1987, pp. 647-668.

[12] J.S.B. Mitchell and C.H. Papadimitriou, "The Weighted Region Problem: Finding Shortest Paths Through a Weighted Planar Subdivision", *Journal of the ACM*, **38**, January 1991, pp. 18-73.

[13] J.S.B. Mitchell and S. Suri, "A Survey of Computational Geometry", M.O. Ball et al. Eds., *Handbooks in Operations Research and Management Science*, Elsevier Science B.V., Vol. 7, Chapter 7, 1995, pp. 439-458.

[14] D.M. Mount, "On Finding Shortest Paths on Convex Polyhedra", Tech. Report 1495, Department of Computer Science, University of Maryland, Baltimore, MD, 1985.

[15] J. O'Rourke, S. Suri and H. Booth, "Shortest Paths on Polyhedral Surfaces", extended abstract, Dept. Electrical Engineering and Computer Science, Johns Hopkins University, Baltimore, Maryland, September 1984.

[16] N.C. Rowe, and R.S. Ross, "Optimal Grid-Free Path Planning Across Arbitrarily Contoured Terrain with Anisotropic Friction and Gravity Effects", *IEEE Transactions on Robotics and Automation*, Vol. 6, No. 5, October 1990, pp. 540-553.

[17] M. Sharir and A. Schorr, "On Shortest Paths in Polyhedral Spaces", *SIAM Journal of Computing*, **15**, 1986, pp. 193-215.

[18] S. Suri, personal communication, 1996.

[19] K.R. Varadarajan and P.K. Agarwal, "Approximating Shortest Paths on a Polyhedron", unpublished report, November 1996.

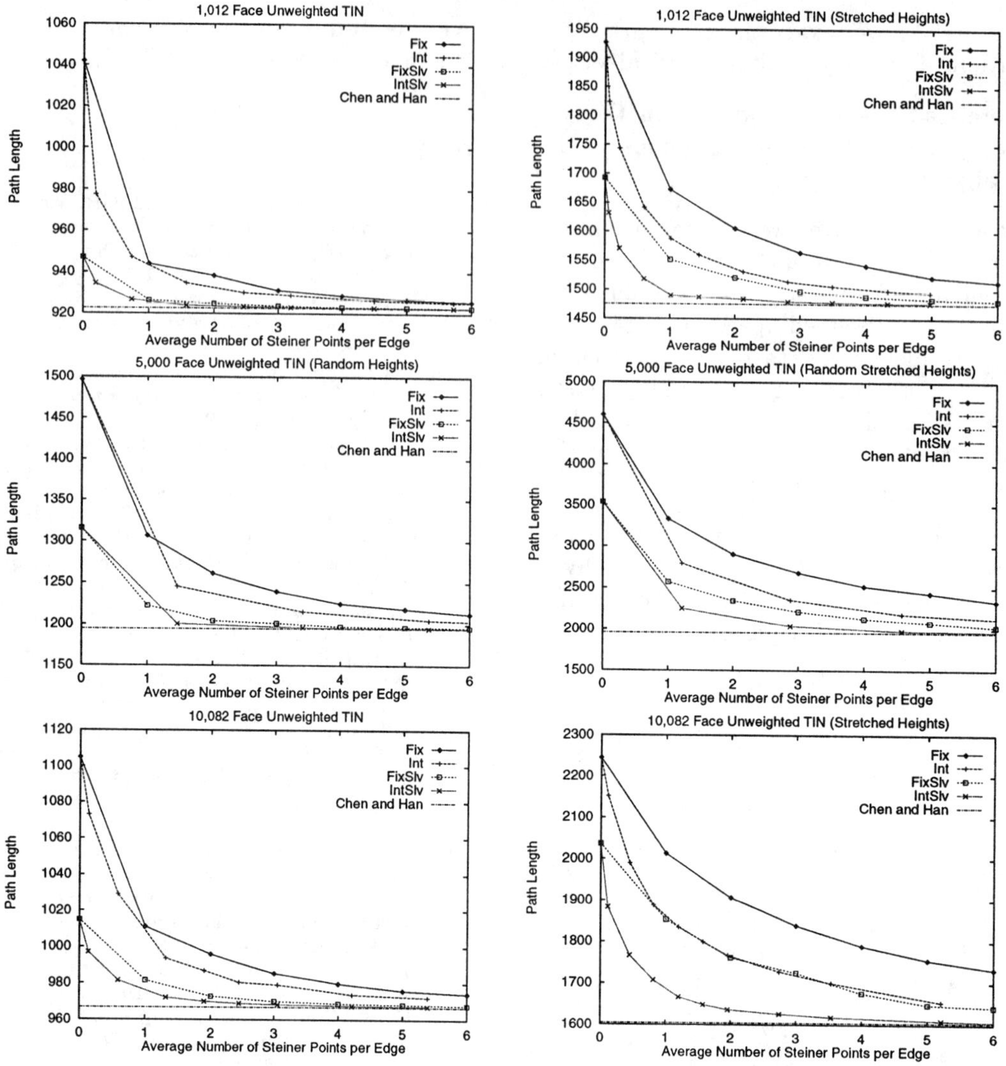


Figure 6: Graphs showing avg. path length.

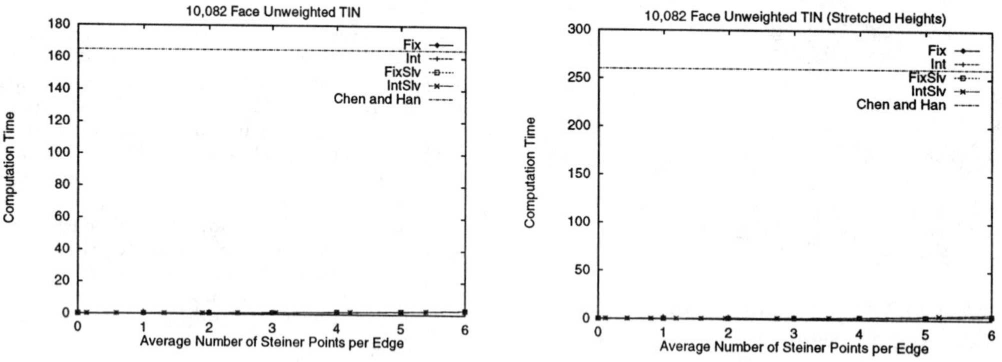


Figure 7: Graphs showing avg. computation time.

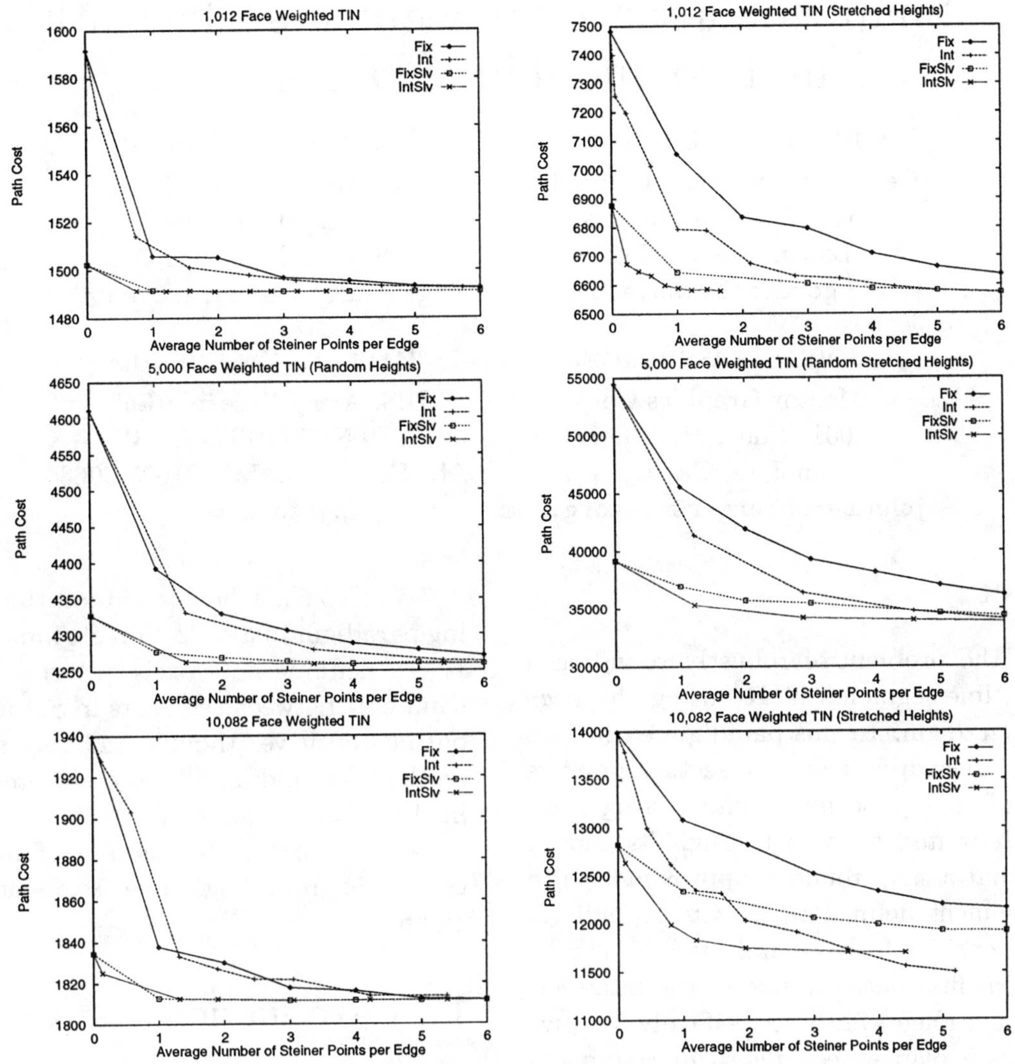


Figure 8: Graphs showing avg. path cost.

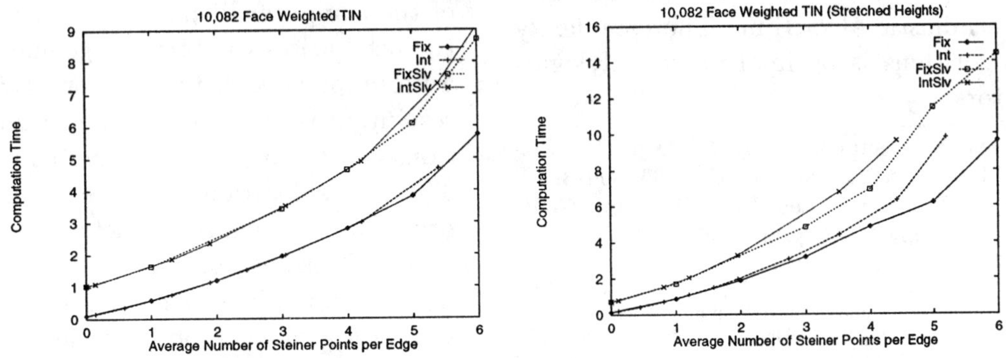


Figure 9: Graphs showing avg. computation time.

# Snap Rounding Line Segments Efficiently in Two and Three Dimensions

Michael T. Goodrich*
Cent. for Geometric Computing
Johns Hopkins University
Baltimore, MD 21218
goodrich@jhu.edu

Leonidas J. Guibas†
Dept. of Computer Science
Stanford University
Stanford, CA 94305
guibas@cs.stanford.edu

John Hershberger
Mentor Graphics Corp.
1001 Ridder Park Drive
San Jose, CA 95131
john_hershberger@mentorg.com

Paul J. Tanenbaum
U.S. Army Research Lab.
ATTN: AMSRL-SL-BV
Ab. Prv. Gnd., MD 21005-5068
pjt@arl.mil

## Abstract

We study the problem of robustly rounding a set $S$ of $n$ line segments in $\mathbf{R}^2$ using the *snap rounding* paradigm. In this paradigm each pixel containing an endpoint or intersection point is called "hot," and all segments intersecting a hot pixel are re-routed to pass through its center. We show that a snap-rounded approximation to the arrangement defined by $S$ can be built in an output-sensitive fashion, and that this can be done *without* first determining all the intersecting pairs of segments in $S$. Specifically, we give a deterministic plane-sweep algorithm running in time $O(n \log n + \sum_{h \in H} |h| \log n)$, where $H$ is the set of hot pixels and $|h|$ is the number of segments intersecting a hot pixel $h \in H$. We also give a simple randomized incremental construction whose expected running time matches that of our deterministic algorithm. The complexity of these algorithms is optimal up to polylogarithmic factors.

We also show how to extend the snap rounding paradigm to a collection $S$ of line segments in $\mathbf{R}^3$ by defining hot voxels in terms of "close encounters" between segments in $S$, and we give an output-sensitive (though probably sub-optimal) method for finding all close encounters determined by the segments in $S$.

**Key words.** Robustness, finite precision, geometric rounding, line segments, arrangements.

*This research is supported by NSF grant CCR-9625289 and by U.S. ARO grant DAAH04-96-1-0013.

†This research is supported by NSF grant CCR-9623851 and US Army MURI grant 5-23542-A.



## 1 Introduction

Geometric objects typically live in a continuous geometric space such as $\mathbf{R}^2$ or $\mathbf{R}^3$. Yet computer representations of such objects are necessarily discrete, both because of the digital nature of computers themselves, and of the raster nature of the computer displays currently in use. Thus the need arises to represent geometric objects in a finite, prespecified resolution. Even if variable resolution is allowed, repeated geometric operations can rapidly increase the required precision. These considerations give rise to the fundamental problem of *rounding* geometric objects to a prespecified resolution. Though individually rounding the numerical attributes of the representation of a geometric object is usually straightforward, such rounding may violate various structural properties which the geometric object is supposed to satisfy, such as convexity, simplicity, etc. In this paper we are interested in rounding collections of line segments so as to guarantee certain topological consistency properties between the ideal collection and its rounded counterpart. Failure to guarantee this consistency can lead to erroneous results and algorithmic failures in further processing. We are interested in roundings that perturb and fragment the original input as little as possible, and which can be computed efficiently.

## 1.1 Some Approaches to Rounding

The problem of dealing with finite precision and robustness in geometric algorithms is fundamental, and there has been considerable work done on developing good approaches to this problem (e.g., see [2, 8, 9, 10, 11, 13, 15, 16, 17, 19, 22, 23]). Of particular relevance to this paper is the previous work done on producing rounded versions of arrangements of line segments. At a high level, of course, the goal of such a method is to round the given set of line segments so that each rounded version of a segment is "close" to the original segment and the important topological properties of the original configuration are preserved as much as possible. It is of prime importance in performing such a computation that the rounding be done efficiently, both in terms of the combinatorial size of the representation and in terms of the running time of the algorithm that performs this rounding.

Greene and Yao [10] introduced the framework of rounding line segments to a pixel grid. They gave a method that preserves the topology of a segment arrangement sufficiently, but at the expense of converting each individual line segment into a polygonal chain containing many subsegments. In addition, the running time of their method depends on both the combinatorial complexity of the output (the number of subsegments needed to represent rounded segments) and also on the number of actual intersections among the original segments. Subsequent to this early work on segment rounding, there have been several papers that have examined the arithmetic complexity (in terms of bits of accuracy) needed to construct arrangements [9, 16, 19]. The general framework of these approaches still involves the computation of all segment intersections, although possibly at a reduced bit complexity than a naive method might use.

One approach to the segment rounding problem that has been shown to be very promising, from the standpoint of the combinatorial complexity of the rounded representation, is the *snap rounding* paradigm introduced by Greene and Hobby [15] and studied in more detail by Guibas and Marimont [11]. Given a set $S$ of $n$ line segments in the plane and a regular pixel grid $\mathcal{G}$, this approach involves defining pixels in $\mathcal{G}$ as being "hot" if they contain segment endpoints or segment intersection points (the point features of the arrangement). Guibas and Marimont show that if one "snap rounds" each segment passing through a hot pixel to the center of that pixel, then one is guaranteed not to introduce any new crossings (although one may, of course, introduce new incidences). (See Figure 1.) This has the benefit of being combinatorially efficient, generating a minimal fragmentation of the input segments consistent with topological consistency. In addition, the size of the rounded arrangement will be almost always smaller than that of the ideal arrangement (though counterexamples are possible), and much more so with coarse pixel sizes.

In addition, Guibas and Marimont give an algorithm for constructing such a snap-rounded arrangement of $S$, including its vertical decomposition (i.e., its trapezoidal decomposition), in expected time $O(n \log n + A + \sum_{h\in H} |h| \log |h| + \sum_{w\in W} |w|)$, where $A$ is the total number of pairs of intersecting segments in $S$, $H$ is the set of hot pixels, $|h|$ denotes the number segments intersecting a hot pixel $h \in H$, $W$ is the set of all pixels containing a vertical attachment (defined in the vertical decomposition), and $|w|$ is the number of segments intersecting a pixel $w \in W$. Their algorithm is dynamic, allowing also for efficient segment insertions and deletions. Alternatively, Hobby [15] describes a deterministic batch algorithm that first constructs the actual segment arrangement and then snap rounds it, resulting in an algorithm that runs in $O((n + A) \log n + \sum_{h\in H} |h|)$ time. Ideally,

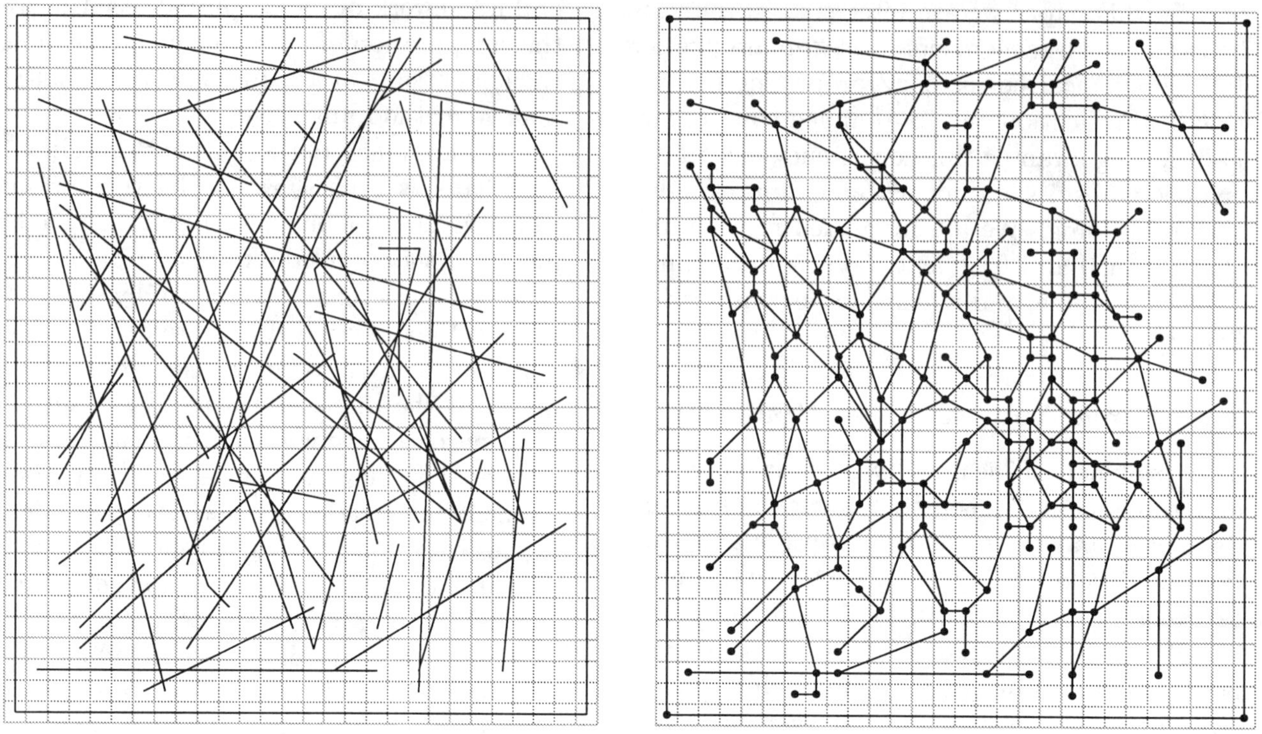

Figure 1: Left, an arrangement of 50 segments; right, the snap-rounded form of the arrangement.

however, one would desire an algorithm whose running time does not depend so heavily on $A$, reflecting the total number of intersecting pairs of segments in the ideal arrangement. In addition, one would like to eliminate any dependence upon other factors that do not contribute to the output size, such as the term $\sum_{w \in W} |w|$. A more desirable time bound would, for example, just depend upon $n$ and the complexities of the hot pixels in $H$. This would give a method for rounding line segments that is efficient in terms of both its combinatorial and computational complexity.

## 1.2 Our Results

In this paper we give two output-sensitive algorithms for efficiently performing snap rounding of line segments in the plane. The first method is deterministic and runs in time $O(n \log n + \sum_{h \in H} |h| \log n)$. It is based upon a plane-sweep strategy; it avoids computing the full arrangement of $S$ by erasing segments of $S$ inside hot pixels. We also give a simple randomized incremental method whose expected running time matches that of our deterministic method.

In addition to providing efficient methods for 2-dimensional snap rounding, we also develop a framework for 3-dimensional snap rounding. In this context the $n$ line segments in $S$ are embedded in $\mathbf{R}^3$ and we wish to round them to a regular 3-dimensional voxel grid. Unlike the 2-dimensional case, however, these segments will typically not intersect at all. Thus, we characterize of the combinatorial complexity of a 3-dimensional set of segments in terms of $n$ and the number of *close encounters* (or "near misses") between pairs of segments in $S$. We then propose a "hot" voxels definition in terms of those voxels containing segment endpoints and close encounters, and we propose a 3-dimensional rounded set of segments to be defined by taking all segments that intersect a hot voxel and snap-rounding them to the center of each such voxel. This retains the proximity of snap-rounded segments with original segments, but can be represented in the finite representation imposed by an integer-grid discretization of space. We also describe an output-sensitive method for finding all close encounters between the segments in $S$, which can then be used immediately to identify all the hot voxels determined by the segments in $S$.

We outline the main ideas behind our results in the sections that follow.

## 2 2-Dimensional Snap Rounding

Let $S$ be a collection of $n$ line segments in the plane, and let there be a regular grid $\mathcal{G}$ defining pixels in $\mathbf{R}^2$. In this section we describe two efficient methods for performing two-dimensional snap rounding of $S$, one deterministic and the other randomized. Both of these methods produce a vertical trapezoidal decomposition $\mathcal{S}$ of the snap-rounded arrangement of the segments in $S$, based upon the same strategy: first intersect segments with hot pixels and then collapse hot pixels to single points. This contrasts with the method of Guibas and Marimont [11], which is based upon a randomized incremental construction of $\mathcal{S}$ directly.

We say that a point in $\mathbf{R}^2$ is *critical* if it is the endpoint of a segment in $S$, or is the intersection point of two segments in $S$; we say that a pixel in $\mathcal{G}$ is *hot* if it contains a critical point. To *snap round* a segment $s$ in $S$, then, we convert $s$ into a polygonal chain (or *polyline*) that connects the centers of all the hot pixels that $s$ intersects (in order of their intersection along $s$). Let $H$ denote the set of all hot pixels, and let $\mathcal{H}$ denote the regions in $\mathbf{R}^2$ covered by hot pixels in $H$. In addition, let $\text{int}(\mathcal{H})$ denote the interiors of all the hot pixels in $H$ and let $\delta(\mathcal{H})$ denote the set of all line segments that bound hot pixels in $H$ (so that, when viewed as sets of points, $\mathcal{H} = \text{int}(\mathcal{H}) \cup \delta(\mathcal{H})$). For each segment $s$ in $S$, define the *external fragments* of $s$ to be the set of segments in $s \backslash \text{int}(\mathcal{H})$. That is, the fragments of $s$ are determined by "slicing" away all the portions of $s$ that intersect hot pixel interiors. If a segment $s$ in $S$ has a non-empty fragment, then we form the polyline for $s$ by moving the two endpoints of each fragment $f$ of $s$ to the centers of the two hot pixels that $f$ is incident upon (if a fragment consists of just a single point, then we expand this to the segment joining the centers of the two abutting hot pixels upon which this point is simultaneously incident).

Let $\bar{S}$ denote the set of external fragments for the segments in $S$. The method we use to implement the above strategy is to construct a representation of $\mathcal{S}'$, the vertical decomposition of $\bar{S} \cup \delta(\mathcal{H})$ (using, say, the quad-edge data structure of Guibas and Stolfi [14]). We call $\mathcal{S}'$ the *pixel-clipped arrangement* of the segments in $S$. Note that none of segments in $\bar{S} \cup \delta(\mathcal{H})$ cross (although there will be intersections defined by fragment endpoints and hot pixel boundaries). Moreover, the combinatorial complexity of $\mathcal{S}'$ is proportional to the snap-rounded representation $\mathcal{S}$ of the arrangement of the segments in $S$. Since converting $\mathcal{S}'$ to $\mathcal{S}$ is a straightforward linear-time computation, we concentrate on the problem of constructing the pixel-clipped arrangement $\mathcal{S}'$ from the set of segments $S$.

### 2.1 Deterministic Snap Rounding

We present a deterministic plane-sweep method for constructing $\mathcal{S}'$, as described above. In what follows we refer to the original, unrounded segments as *ursegments*, to distinguish them from other segments used by the algorithm.

Our approach is based on the plane sweeping algorithm of Bentley and Ottmann [3] for constructing (unrounded) segment arrangements. We sweep over the ursegments from left to right, processing events, some of which are critical points. However, we dynamically modify the set of segments so that the sweepline processes only the leftmost critical point in each hot pixel. The sweepline is also used to detect all intersections between ursegments and hot pixel boundaries. The key idea is the following: when we detect a critical point in a pixel, we erase all the ursegment portions that intersect the (now known to be hot) pixel's interior. Simultaneously, we introduce the pixel's top/bottom boundaries into the sweep structure as horizontal line segments. This eliminates all ursegment intersections inside the pixel except the leftmost, and adds some ursegment/pixel boundary intersection events.

Our algorithm uses three data structures, which we describe in a syntax based loosely on C. First is a Bentley-Ottmann sweepline, which consists of a current $x$ position `xpos`, a searchable list[1] storing the segments that intersect the vertical line $x$ = `xpos` in their $y$ order, and an $x$-ordered priority queue of segment insertions, deletions, and intersections to the right of `xpos`—intersection events in the priority queue involve only adjacent segments on the sweepline [3, 4]. We denote the sweepline by `SL`. Second, each hot pixel `pix` with a critical point left of `xpos` has two searchable $x$-ordered lists of the ursegments that intersect its top and bottom left of `xpos`; these lists are `pix.toplist` and `pix.botlist`. Third is a searchable $y$-ordered list of the hot pixels that intersect $x$ = `xpos` and have a critical point left of `xpos`—call this list `Hcur`.

The output of the algorithm is an $x$-ordered list of hot pixels `H`, and for each hot pixel `pix` a set of ursegments `pix.segs` that intersect `pix`. The pseudocode below uses two subroutines: `pixel(Point p)` rounds point `p` to its containing pixel, and `heat(Pixel pix)` makes `pix` hot. In particular, `heat(pix)` finds all the ursegments that intersect `pix` left of the current `xpos`, erases the portion of these segments inside `pix`, initializes `pix.toplist` and `pix.botlist`, and introduces horizontal segments into `SL` at the top and bottom boundaries of `pix`. Whenever `heat(pix)` is called, there are no critical points between `xpos` and the left side of `pix`.

The algorithm processes five different kinds of events:

1. Ursegment endpoints. N.B. Right endpoints sort before left endpoints at the same $x$.
2. Ursegment intersections.
3. Ursegment/pixel boundary intersections.
4. Ursegment re-insertions. (These occur when an ursegment exits a hot pixel.)
5. Right ends of hot pixel boundaries. N.B. Type (5) events sort before type (4) events at the same $x$.

[1] A searchable list is any data structure that stores an ordered list and supports logarithmic-time insertions, deletions, and searches, and constant (amortized) time successor/predecessor queries. Balanced binary trees and skip lists are standard examples.

The algorithm begins by initializing the sweepline `SL` with events of types (1) and (2). No events of types (3), (4), and (5) exist yet. We process events in left-to-right order, as follows:

```
1. Urseg_endpt(Point p, Urseg u)
   if (pixel(p) ∉ Hcur) heat(pixel(p)).
   if (p is the left end of u)
      Insert u into SL.
      Proceed as in case (3), beginning at (*).
   else
      Remove u from SL.
2. Urseg_intersection(Point p)
   heat(pixel(p)).
3. Urseg_pixbdy_intersection(Point p,
   Pixel pix, Urseg u)
   if (p is on the top of pix)
      append u to pix.toplist;
   else
      append u to pix.botlist.
(*) Add u to pix.segs, if not already there.
   Remove u from SL, as if u ended at p.
   if (u intersects the boundary of
      pix right of p at pp)
      Insert the event
      Urseg_reinsertion(pp, pix, u)
      into the priority queue.
4. Urseg_reinsertion(Point p, Pixel pix,
   Urseg u)
   Insert u into SL at p.
   if (p is on the top of pix)
      append u to pix.toplist;
   else if (p is on the bottom of pix)
      append u to pix.botlist.
   Let ppix be the pixel adjacent to pix with
   p on their common boundary.
   if (ppix ∈ Hcur)
      Process an event
      Urseg_pixbdy_intersection(p, ppix, u
      (A careful implementation—making pixel
      boundaries lie infinitesimally inside their
```

pixels—could avoid this test.)

5. `Pixbdy_end(Pixbdy bdy, Point p)`
   Remove `bdy` from `SL` at `p`.
   Append all pixels in `Hcur` to `H`, leaving `Hcur` empty.

The implementation of `pixel(Point p)` is trivial, so we focus on `heat(Pixel pix)`. We need the following lemma.

**Lemma 1:** *Given a pixel* `pix` $\notin$ `Hcur` *that intersects* $x$ = `xpos`, *let* `pixup` *and* `pixdown` *be the pixels in* `Hcur` *above and below it, if any. Let* `rect` *be the (possibly infinite) rectangle bounded by* `pixup`, `pixdown`, `xpos`, *and the* $x$ *coordinate of the left side of* `pix`. *All the ursegments that intersect* `rect` *belong to* `pixup.botlist`, `pixdown.toplist`, *or the* $y$*-ordered list of segments on the sweepline* `SL`. *In each of these lists, the segments that intersect* `pix` *form a contiguous subsequence, which can be found in* $O(\log n + k)$ *time, where* $k$ *is the length of the subsequence.*

**Proof:** By the definition of `Hcur`, no ursegment intersections or endpoints lie in `rect`. Any ursegment that intersects `pix` also intersects `rect`, and must intersect its top, bottom, or right side. Each of the three lists stores ursegments that intersect a particular line segment. If two ursegments $s$, $s'$ in the list for a segment $e$ intersect `pix`, then there is a line segment $e'$ inside `pix` joining them. The four segments $s$, $s'$, $e$, and $e'$ form a quadrilateral. Any ursegment between $s$ and $s'$ in the list intersects $e$ and enters the quadrilateral; it does not intersect $s$ or $s'$, so it must intersect $e'$, and hence `pix`. We can find the subsequence of each list that intersects `pix` by searching in the list: for each segment $s$ in the list that does not intersect `pix`, we can tell in constant time whether the portion of the list that intersects `pix` lies before or after $s$. □

We are now ready to present the implementation of `heat(pix)`.

```
heat(Pixel pix)
Insert pix into Hcur.
Find the ursegments that intersect
pix left of xpos, using Lemma 1.
For each ursegment u that intersects
pix left of xpos
  Add u to pix.segs.
  if (u intersects pix right of xpos)
    Remove u from SL.
    if (u intersects the boundary of pix
      right of xpos at a point p)
      Schedule an event
      Urseg_reinsertion(p, pix, u).
Build pix.toplist and pix.botlist
by sorting pix.segs.
Insert horizontal segments into SL at
the top and bottom boundaries of
pix, extending from xpos to the right
end of pix. For each pixel boundary
bdy inserted, with right endpoint p,
schedule an event Pixbdy_end(bdy,
p).
```

**Theorem 2:** *Given a set* $S$ *of* $n$ *line segments in the plane and a regular pixel grid* $G$, *one can snap-round the segments in* $S$ *to* $G$ *in time* $O(n \log n + \sum_{h \in H} |h| \log n)$, *where* $H$ *is the set of hot pixels and* $|h|$ *is the number of segments intersecting a hot pixel* $h \in H$.

**Proof:** A pixel is hot iff it contains an ursegment endpoint or intersection. The algorithm detects these events in cases (1) and (2). Ursegments are erased from `SL` only inside pixels already known to be hot, so all hot pixels are detected.

Ursegment/hot pixel incidences are found in case (3) and in `heat(pix)` (called from cases (1) and (2)). This finds all the incidences, because any ursegment that intersects a pixel `pix` without starting or ending there must intersect either the top of `pix`, the bottom of `pix`, or all vertical segments connecting top and bottom. The subroutine `heat(pix)` finds all incidences with the top and bottom of `pix` left of `xpos`, as well as all incidences with the vertical segment spanning `pix` at `xpos`. Case (3) finds all incidences with top and bottom to the right of the `xpos` where `heat(pix)` was called.

Let $m = \sum_{h \in H} |h|$. It is not hard to see that the running time of `heat(pix)` is $O(k \log n)$, where $k$ is the number of segments added to `pix.segs`. This sums to $O(m \log n)$ over all invocations of `heat()`. The rest of the algorithm requires only $O(\log n)$ time per event. The number of events of type (1) is $2n = O(m)$; the number of events of types (2) and (5) is $O(h)$, where $h$ is the number of hot pixels, which is $O(m)$; and the number of events of types (3) and (4) is $O(m)$.

Once all the ursegment/hot pixel incidences are known, the order in which the segments intersect the pixel boundary can be obtained in $O(m \log n)$ time by sorting. (The order is already known for the intersections on the top and bottom of each hot pixel.) Given this information, it is straightforward to compute the snap-rounded arrangement $\mathcal{S}$. □

In the next subsection we describe a simple randomized algorithm whose expected running time matches this bound.

## 2.2 Randomized Snap Rounding in $\mathbf{R}^2$

In this section we give a randomized incremental construction (RIC) for building the pixel-clipped arrangement $\mathcal{S}'$ of the segments in $S$. The basic approach is similar to that of previous RIC's for constructing segment arrangements (e.g., see [5, 6, 18, 20, 21]). We again maintain a trapezoidal decomposition of $\mathcal{S}'$, except that here we dynamically "clip" the current subdivision each time we discover a new hot pixel.

In the usual RIC of line segment arrangements, two different operations need to be addressed. One is the *point location* of (typically the left) endpoint of a new segment $s$ in the vertical trapezoidal decomposition of the arrangement of the segments inserted so far. The other is the *propagation* of $s$ through the trapezoidal decomposition, in order to discover the intersections of $s$ with existing segments, and to update the trapezoidal decomposition in the process. The point location step is normally handled by maintaining a conflict graph between (uninserted) segments and trapezoids [5], or by the "history-dag" technique of [12].

Like these methods, we also build the trapezoidal decomposition of the pixel-clipped arrangement $\mathcal{S}'$ in an incremental manner, by inserting all the segments in $S$ into this arrangement one after the other, in a random sequence. But we differ from these schemes for ideal line segments in several important ways. Fist of all, because we are dealing with a fixed pixel grid, we can completely finesse the point location issue. We simply initialize our vertical decomposition of the pixel-clipped arrangement to be the vertical decomposition of all the hot pixels containing segment endpoints. We can easily compute this decomposition by a line sweep in time $O(n \log n)$. We also initialize each of these hot pixels with a dynamic "by-pass" structure that allows us to trace other segments though them efficiently. Secondly, during the propagation stage in the insertion of a new segment $s$, we may discover new intersections between $s$ and other existing segments that lie outside of the currently known hot pixels. Whenever that happens, we need to create a new hot pixel $h$ corresponding to the new intersection, clip out from $\mathcal{S}'$ the portion corresponding to $h$, and initialize a new bypass structure for $h$ containing all the current segments intersecting $h$.

Let us now be more precise about these operations; call $s$ the next segment in $S$ to be added. We locate the (existing) hot pixel $h$ containing the left endpoint of $s$ through a simple indexing operation. This can be done in time $O(\log n)$ by keeping on the side a binary tree of all the pixel grid columns containing hot pixels (due to endpoints) in the initial $\mathcal{S}'$, and for each node of that tree another binary tree of all the hot pixels in that column. In practice, of course, we would simply index into the pixel array using the endpoint coordinates. Starting now from where $s$ exits the boundary of $h$, we trace $s$ into the pixel-clipped decomposition $\mathcal{S}'$. The segment $s$ may have to be propagated through a trapezoid of the decomposition or through an existing hot pixel, and in the process it may cross a vertical attachment, another segment, or a hot pixel boundary.

For every hot pixel $h$, we will maintain an ordered list of the intersections of the boundary of $h$ by the current segment as a dynamic binary search tree (such as a red-black tree). By traversing this tree we are able to propagate a new segment, such as $s$, through $h$ in $O(\log n)$ time; we can also update the representation to include

$s$ in $O(\log n)$ time. Thus the traversal by $s$ of a standard trapezoid in $\mathcal{S}'$ takes as usual constant time, while the traversal of a hot pixel uses the above by-pass structure. If $s$ exits a standard trapezoid by crossing a vertical attachment, then that attachment has to be shortened to end at $s$.

The most interesting situation arises when $s$ exits a standard trapezoid by crossing an existing segment $r$ at a point $P$. In this case we have discovered a new hot pixel $h$, and there is more work to do. In order to clip out from $\mathcal{S}'$ the part corresponding to $h$, we walk in $\mathcal{S}'$ from $P$ to the (say) north-east corner of $h$ (in a straight-line fashion), and then counterclockwise around the entire boundary of $h$. In this process we will discover all the current segments crossing $h$ (each at most two or three times)[2]. Notice that in this traversal of $\mathcal{S}'$ we will not encounter any other hot pixels (obviously), and that the number of vertical attachments we can meet is at most four. Thus the work involved is proportional to the number of segments currently intersecting $h$. We can now excise $h$ from $\mathcal{S}'$ and replace it by a by-pass structure reflecting the segment crossings with its boundary. We also add any necessary new vertical attachments emanating from the corners of $h$ in $O(1)$ time.

This finishes the description of our RIC algorithm for the pixel-clipped arrangement. The analysis is very straightforward. Clearly the point-location cost is $O(\log n)$ per segment. Also, the cost of detecting the new hot pixels, of forming the hot pixel by-pass structures, and of tracing segments through them and then adding these segment to the by-pass structure is all accounted for by the term $O(\sum_{h \in H} |h| \log n))$. Note also that these are both worst-case bounds. The remaining cost of our algorithm is that of propagating segments through trapezoids by crossing vertical attachments. Each such step "clips," or shortens, such an attachment. Because the segments are added in a random sequence, any specific vertical attachment will be clipped in this way an expected $O(\log n)$ times (this is just the left-to-right minima in a random permutation problem). Note that a vertical attachment can also be clipped by the creation of a new hot pixel, but we can charge this clipping to the hot pixel responsible for it. This additional clipping can only help the expected $O(\log n)$ bound above. Thus we have:

**Theorem 3:** *Given a set $S$ of $n$ line segments in the plane and a regular pixel grid $G$, one can snap-round the segments in $S$ to $G$ in expected time $O(n \log n + \sum_{h \in H} |h| \log n)$ by a randomized incremental construction.*

## 3 3-Dimensional Snap Rounding

Let us now consider the problem of snap rounding a set $S$ of $n$ line segments in $\mathbf{R}^3$. In this case we assume a bounding box in $\mathbf{R}^3$ containing $S$ has been partitioned into unit-cube voxels centered on the points with integer coordinates. We propse a notion of 3-dimensional snap rounding that is defined as follows. For distinct input segments $r$ and $s$, consider the distance between them and define their *connector* to be the segment $rs$ between their points of closest approach. We assume that no two input segments are parallel, so these points of closest approach are unique.

A connector will be called *short* if its length in the $L_\infty$ metric is 1 or less. In this case, there is said to exist a *close encounter* between the two input segments. We define a voxel to be *hot* if it contains any endpoints either of input segments or of short connectors.

The rounding process proceeds analogously to the two-dimensional case described above. Every input segment $s$ is transformed into a polyline $\sigma$ such that the endpoints of $\sigma$ are the centers of the hot voxels containing the corresponding endpoints of $s$ and the bends in $\sigma$, corresponding to transits of $s$ across hot voxels between its endpoints, are the centers of those hot voxels. Let us, then, address the computational issues involved with snap-rounding a set of $n$ line segments in $\mathbf{R}^3$ using this notion of rounding.

[2]Some tiny segments may be entirely inside $h$, but these contract to a point after snap-rounding.

### 3.1 Determining all close encounters

Suppose we are given a set $S$ of $n$ segments in $\mathbf{R}^3$. In this section we describe an efficient output-sensitive method for determining all pairs $(s,t)$ of segments in $S$ that are within an $L_\infty$ distance of 1 from each other (which are all segments that determine a close encounter). From each segment $s$ let us form the *tube* $\tau(s)$, which is the Minkowski sum of $s$ with an axis-oriented unit cube centered at the origin. The following facts are immediate:

1. each tube $\tau(s)$ is a zonotope; it has at most twelve facets which are parallelograms;

2. segments $s$ and $t$ are within an $L_\infty$ distance of 1 from each other if and only if their tubes $\tau(s)$ and $\tau(t)$ intersect.

We assume that no two segments are parallel; symbolic perturbation techniques can be used to guarantee that this is so [7]. It follows that two tubes $\tau(s)$ and $\tau(t)$ will intersect if and only if an edge of one of the tubes pierces (or touches) a face of the other. Therefore, let us form two sets from the collection of tubes for all the segments in $S$: the set $E$ of all edges of the tubes, and the set $F$ of all faces of the tubes. Each of these has clearly size $O(n)$. Note also that any intersection or contact between an edge $e$ in $E$ and a parallelogram $f$ in $F$ implies an intersection or contact between two tubes (except if $e$ and $f$ belong to the same tube), and any pairwise tube intersection will be captured this way.

We now use the range searching techniques for semi-algebraic varieties of Agarwal and Matoušek [1] to develop an efficient algorithm for reporting all the edge/face intersections. Consider a particular edge $e$ and face $f$; orient the four edges of $f$ consistently around $f$. Then the condition that $e$ intersects $f$ can be expressed by asserting that the line supporting $e$ has positive orientation with respect to the four lines supporting the edges of $f$, and that that the plane supporting $f$ separates the endpoints of $v$. By using the techniques of [1] we now preprocess all the faces in $F$ so that, given a query edge $e$, we can quickly report all the faces that $e$ intersects. This requires a six-level partition tree: four levels for the four sides of a face and two for the two endpoints of the query edge. Assuming that we want to use only linear space, the dominant query cost of this structure comes from the levels of the tree dealing with the line orientation conditions. By using Plücker coordinates, Agarwal and Matoušek [1] show how such a structure can be developed whose query time will be $O(n/s^{1/4}+k)$, where $k$ is the number of reported faces intersecting $e$, using space and preprocessing of $O(s^{1+\delta}$, for any $\delta > 0$. By querying with all the edges in $E$ and balancing the preprocessing and query costs, we can obtain all the edge/face intersections in total time $O(n^{8/5+\delta}+K)$ and space $O(n^{8/5+\delta})$, where here $K$ denotes the number of intersecting tubes. Further details will be given in the full paper.

## 4 Conclusion

We have given output-sensitive methods for two-dimensional segment snap-rounding, in both deterministic and randomized settings. In both cases our methods have a running time that is sensitive to both the number of input line segments and also to the number of segments in a snap-rounded representation. We have also given an extension of the snap-rounding notion to three-dimensional segments and we have given an output-sensitive method for snap-rounding segments in $\mathbf{R}^3$ as well. We feel that an interesting direction for future work is to explore the variety of topological properties that are preserved by this notion of three-dimensional snap rounding.

### Acknowledgement

We would like to thank Pankaj Agarwal for helpful discussions involving three-dimensional range searching.

## References

[1] P. K. Agarwal and J. Matoušek. On range searching with semialgebraic sets. *Discrete Comput. Geom.*, 11:393–418, 1994.

[2] F. Avnaim, J.-D. Boissonnat, O. Devillers, F. Preparata, and M. Yvinec. Evaluation of

a new method to compute signs of determinants. In *Proc. 11th Annu. ACM Sympos. Comput. Geom.*, pages C16–C17, 1995.

[3] J. L. Bentley and T. A. Ottmann. Algorithms for reporting and counting geometric intersections. *IEEE Trans. Comput.*, C-28:643–647, 1979.

[4] K. Q. Brown. Comments on "Algorithms for reporting and counting geometric intersections". *IEEE Trans. Comput.*, C-30:147–148, 1981.

[5] K. L. Clarkson. Randomized geometric algorithms. In D.-Z. Du and F. K. Hwang, editors, *Computing in Euclidean Geometry*, volume 1 of *Lecture Notes Series on Computing*, pages 117–162. World Scientific, Singapore, 1992.

[6] K. L. Clarkson, K. Mehlhorn, and R. Seidel. Four results on randomized incremental constructions. *Comput. Geom. Theory Appl.*, 3(4):185–212, 1993.

[7] H. Edelsbrunner and E. P. Mücke. Simulation of simplicity: a technique to cope with degenerate cases in geometric algorithms. *ACM Trans. Graph.*, 9:66–104, 1990.

[8] S. Fortune. Numerical stability of algorithms for 2-d Delaunay triangulations. *Internat. J. Comput. Geom. Appl.*, 5(1):193–213, 1995.

[9] S. Fortune and V. Milenkovic. Numerical stability of algorithms for line arrangements. In *Proc. 7th Annu. ACM Sympos. Comput. Geom.*, pages 334–341, 1991.

[10] D. H. Greene and F. F. Yao. Finite-resolution computational geometry. In *Proc. 27th Annu. IEEE Sympos. Found. Comput. Sci.*, pages 143–152, 1986.

[11] L. Guibas and D. Marimont. Rounding arrangements dynamically. In *Proc. 11th Annu. ACM Sympos. Comput. Geom.*, pages 190–199, 1995.

[12] L. J. Guibas, D. E. Knuth, and M. Sharir. Randomized incremental construction of Delaunay and Voronoi diagrams. *Algorithmica*, 7:381–413, 1992.

[13] L. J. Guibas, D. Salesin, and J. Stolfi. Epsilon geometry: building robust algorithms from imprecise computations. In *Proc. 5th Annu. ACM Sympos. Comput. Geom.*, pages 208–217, 1989.

[14] L. J. Guibas and J. Stolfi. Primitives for the manipulation of general subdivisions and the computation of Voronoi diagrams. *ACM Trans. Graph.*, 4:74–123, 1985.

[15] J. Hobby. Practical segment intersection with finite precision output. Technical Report 93/2-27, Bell Laboratories (Lucent Technologies), 1993.

[16] C. M. Hoffmann, J. E. Hopcroft, and M. S. Karasick. Towards implementing robust geometric computations. In *Proc. 4th Annu. ACM Sympos. Comput. Geom.*, pages 106–117, 1988.

[17] G. Liotta, F. P. Preparata, and R. Tamassia. Robust proximity queries in implicit Voronoi diagrams. Technical Report CS-96-16, Center for Geometric Computing, Comput. Sci. Dept., Brown Univ., Providence, RI, 1996.

[18] J. Matoušek and R. Seidel. A tail estimate for Mulmuley's segment intersection algorithm. In *19th International Colloquium on Automata, Languages, and Programming*, volume 623 of *Lecture Notes in Computer Science*, pages 427–438. Springer-Verlag, 1992.

[19] V. Milenkovic. Double precision geometry: a general technique for calculating line and segment intersections using rounded arithmetic. In *Proc. 30th Annu. IEEE Sympos. Found. Comput. Sci.*, pages 500–505, 1989.

[20] K. Mulmuley. A fast planar partition algorithm: part I. Technical Report 88-007, Dept. Comput. Sci., Univ. Chicago, Chicago, IL, 1988.

[21] R. Seidel. Backwards analysis of randomized geometric algorithms. In J. Pach, editor, *New Trends in Discrete and Computational Geometry*, volume 10 of *Algorithms and Combinatorics*, pages 37–68. Springer-Verlag, 1993.

[22] K. Sugihara and M. Iri. Geometric algorithms in finite-precision arithmetic. Technical Report 88-10, Math. Eng. and Physics Dept., U. of Tokyo, Japan, Sept. 1988.

[23] K. Sugihara and M. Iri. Two design principles of geometric algorithms in finite-precision arithmetic. *Appl. Math. Lett.*, 2(2):203–206, 1989.

# Realistic Input Models for Geometric Algorithms

Mark de Berg* Matthew Katz† A. Frank van der Stappen* Jules Vleugels*

## Abstract

Many algorithms developed in computational geometry are needlessly complicated and slow because they have to be prepared for very complicated, hypothetical inputs. To avoid this, realistic models are needed that describe the properties that realistic inputs have, so that algorithms can de designed that take advantage of these properties. This can lead to algorithms that are provably efficient in realistic situations.

We obtain some fundamental results in this research direction. In particular, we have the following results.

- We show the relations between various models that have been proposed in the literature.
- For several of these models, we give algorithms to compute the model parameter(s) for a given scene; these algorithms can be used to verify whether a model is appropriate for typical scenes in some application area.
- As a case study, we give some experimental results on the appropriateness of some of the models for one particular type of scenes often encountered in GIS, namely certain triangulated irregular networks.

## 1 Introduction

Many of the problems studied in computational geometry come from application areas such as computer graphics, robotics, and GIS. These problems are studied in an abstract setting. There are two dangers in the abstraction process. The first, well-known danger is that the abstract version is a too-much-simplified version of the real problem, so that the obtained solution is not applicable in practice. Most researchers are well aware of this danger, and they try to get rid of unrealistic assumptions as much as possible. The second danger is that the abstract version of the problem is *too* general, so that useful properties are lost. As a result, the algorithms have to be prepared for very complicated, hypothetical inputs. Actually, the situation is even worse: because one is usually interested in the *worst-case* performance, the algorithms are geared to these unrealistic inputs. Thus they become needlessly complicated and slow.

Consider as an example the construction of a binary space partition (BSP) tree for a set of disjoint triangles in 3-space. Paterson and Yao [12] showed that it is always possible to construct a BSP of size $O(n^2)$. Moreover, there are sets of triangles for which any BSP must have quadratic size, so the $O(n^2)$ upper bound is the best we can hope for *in the worst case*. The lower bound example, however, is very artificial: it consists of $n$ lines arranged in a very special way, namely in a grid-like fashion on a ruled surface. Hence, it does not say anything about what can be achieved in practice; perhaps realistic scenes have some special property that makes it possible to construct a BSP of linear size. Indeed, de Berg [6] showed that so-called *uncluttered scenes*—and it is claimed that in practical applications most scenes are uncluttered—admit a linear size BSP. (See Section 2 for a definition of "uncluttered".)

Another example comes from ray tracing. Agarwal and Matoušek [2] developed a data structure for 3-dimensional objects with roughly $O(n^{3/4})$ query time, which uses roughly linear storage. It is believed that this is theoretically close to optimal. But octrees, for instance, have been observed to behave well in many practical situations. Apparently, ray-shooting queries are not as difficult in practice as they are in theory. For this reason, Mitchell et al. [10] developed a data structure for 2-dimensional objects based on quad trees, and they analyzed its performance in terms of the so-called *simple-cover complexity*. They showed that the storage of the data structure is small if the simple-cover complexity of the scene is small, and that the query time is small if the simple-cover complexity of the query ray is small. (See Section 2 for a definition of "simple-cover complexity".) Again, it is expected that this will usually be the case in practice.

These two examples point in a promising research direction, where problems are studied under realistic assumptions, and algorithms analyzed using different, more practical complexity measures. This type of research can lead to algorithms and data structures that are *provably efficient* in *realistic situations*. Hence, the research is interesting from a theoretical as well as from a practical point of view. The crucial factor in this approach lies in the development of re-

---

*Department of Computer Science, Utrecht University, P.O.Box 80.089, 3508 TB Utrecht, the Netherlands. Supported by the Dutch Organisation for Scientific Research (N.W.O.) and by the ESPRIT IV LTR Project No. 21957 (CGAL).

†Departments of Industrial Engineering & Management and Mathematics & Computer Science, Ben-Gurion University of the Negev, Beer-Sheva 84105, Israel.



*Computational Geometry 97* Nice France

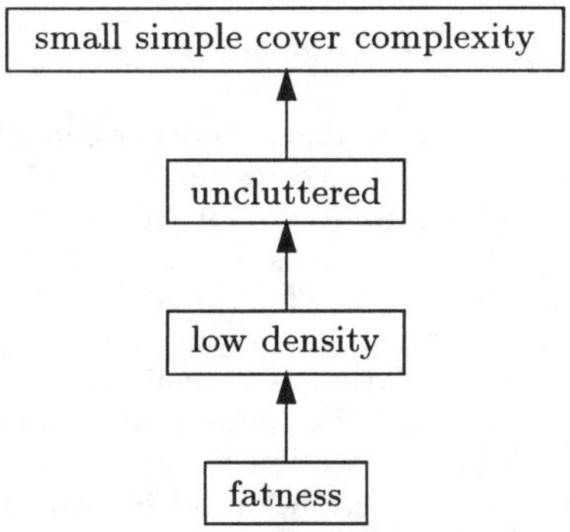


Figure 1: Relations between the models.

alistic models for the input scenes. The models should be general enough so that they are realistic, while still precluding complicated hypothetical inputs.

In this paper we take a first step in this direction, by studying the relations between various models that have been proposed in the literature for sets of non-intersecting objects. In particular, we study the following models: fatness, low density, clutteredness, and simple-cover complexity. (Definitions of the models are given in Section 2.) In Section 3 we show that these models are related as follows: fatness implies low density, which in turn implies unclutteredness, which implies small simple-cover complexity. Hence, the models form a hierarchy as depicted in Figure 1.

The relations between fatness and low density, and fatness and unclutteredness, were already known [6, 15]. One of the main contributions of our paper is that we prove that any low-density scene is uncluttered, and that any uncluttered scene—and hence any low-density scene, or any scene consisting of fat objects—has small simple-cover complexity.

Each of the models assigns one (or two) parameter(s) to a scene. If a model is realistic in a certain setting, then its parameter should have a favourable value—usually, a small, or for some models not-too-large, constant—for a typical scene in that setting. This has to be verified, of course. Therefore we develop, in Section 4, algorithms to compute the model parameters. This is also useful because some algorithms—for instance, the range searching data structure for fat objects developed by Overmars and van der Stappen [11]—may require the value of the model parameter as input in order to work correctly.

What constitutes a good model depends on the application domain: architectural models have different properties than molecular models, for instance. For each application domain, it should be verified experimentally which models are appropriate. As a case study, we report in Section 6 some initial experimental results indicating which models are realistic for certain triangulated irregular networks (TINs), which are encountered frequently in GIS. More precisely, we have taken DEM files giving the elevations of points in various areas in Canada and the US. The data points are arranged in a grid. We selected the most important points using the so-called VIP method [5], and computed the 2D Delaunay triangulation of the selected points, thus obtaining a TIN. We then computed the model parameters for the resulting set of 2D triangles, which indicate whether or not the models are realistic. Our experimental results are still preliminary. In the full version of the paper we hope to add some more results, also on architectural models.

The results stated above form the main contribution of our paper. The paper has one extra section, Section 5, in which we study the relation between the normal fatness and the so-called boundary fatness of an object. We prove that the boundary fatness is within a constant factor of the normal fatness; the constant factor we show is tight in the worst case for disconnected objects, and almost tight for connected ones. Determining this relation is not very useful from a practical point of view, but we found it a challenging theoretical problem.

## 2 The models

Below we define the models we study in this paper. Each model assigns one (or two) parameter(s) to a scene or to the objects in it. Algorithms can then be analyzed not only in terms of the number of objects in the scene, but also in terms of these model parameters. The hope is that, for realistic inputs, the model parameters have favorable values (not-too-large, or for some models: not-too-small, constants), so that the performance of the algorithm is good in practice.

Before we define the models we need to introduce some terminology and notation.

We say that two objects are *non-intersecting* if they have disjoint interiors. A *$d$-dimensional scene* is defined to be any collection of non-intersecting, constant-complexity objects in $\mathbb{E}^d$.

We denote the volume of an object $\mathcal{P}$ by $\mathrm{vol}(\mathcal{P})$, and the Euclidean distance between two points $p$ and $q$ by $\mathrm{dist}(p,q)$. For sets $A$ and $B$, we define $\mathrm{dist}(A,B) := \inf_{a\in A, b\in B} \mathrm{dist}(a,b)$. The minimal enclosing ball of an object $\mathcal{P}$ is denoted by $\mathrm{meb}(\mathcal{P})$, and the radius of the minimal enclosing ball is denoted $\rho_{\mathrm{meb}}(\mathcal{P})$. We define $\omega_d$ to be the volume of the $d$-dimensional unit ball. (For even dimensions, $\omega_d = \pi^{d/2}/(d/2)!$, and for odd dimensions, $\omega_d = 2(2\pi)^{\lfloor d/2\rfloor}/d!!$.) We use $B_{p,r}$ to denote the ball of radius $r$ centered at $p$.

Finally, we shall sometimes work with bounding boxes or bounding cubes of the objects. These are always axis-parallel. Thus the *bounding box* $\mathrm{bb}(\mathcal{P})$ of an object $\mathcal{P}$ is the smallest axis-parallel hyperrectangle that contains the object, and a minimal enclosing hypercube $\mathrm{mec}(\mathcal{P})$ of $\mathcal{P}$ is an axis-parallel hypercube of minimal size that contains the object. We use $\rho_{\mathrm{mec}}(\mathcal{P})$ to denote the radius of $\mathrm{mec}(\mathcal{P})$, which we define to be half its edge length.

**Fatness.** The first, and best known, model we consider is fatness. Intuitively, an object is called fat if it contains no long and skinny parts. There are many different definitions of fatness [1, 3, 9, 17], which are all more or less equivalent—at least for convex objects. We shall use a definition similar to the one used in van der Stappen's thesis [15], where the motion planning problem for scenes consisting of fat objects is studied.

**Definition 2.1** *Let $\mathcal{P} \subseteq \mathbb{E}^d$ be an object and let $\beta$ be a constant with $0 \leqslant \beta \leqslant 1$. Define $U(\mathcal{P})$ as the set of all balls centered inside $\mathcal{P}$ whose boundary intersects $\mathcal{P}$. We say that the object $\mathcal{P}$ is $\beta$-fat if for all balls $B \in U(\mathcal{P})$, $\mathrm{vol}(\mathcal{P} \cap B) \geqslant \beta \cdot \mathrm{vol}(B)$. The fatness of $\mathcal{P}$ is defined as the maximal $\beta$ for which $\mathcal{P}$ is $\beta$-fat.*

(Van der Stappen's original definition is slightly different: what he calls $k$-fat for some $k \geqslant 1$, is $(1/k)$-fat in our definition.) In the rest of the paper, we shall sometimes omit the fatness-parameter $\beta$, and call an object *fat* if it is $\beta$-fat for some not too small constant $\beta$. In addition, the fatness of a scene of objects is defined as the maximal $\beta$ for which every individual object is $\beta$-fat.

**Low density.** The model of low density was introduced by Overmars and van der Stappen [16] and refined by Schwarzkopf and Vleugels [14]. It forbids any ball $B$ to be intersected by many objects whose minimal-enclosing-ball radius is at least as large as the radius of $B$.

**Definition 2.2** *Let $\mathcal{S} := \{\mathcal{P}_1, \ldots, \mathcal{P}_n\}$ be a d-dimensional scene, and let $\lambda \geqslant 0$ be a parameter. We call $\mathcal{S}$ a $\lambda$-low-density scene if for any ball $B$, the number of objects $\mathcal{P}_i \in \mathcal{S}$ with $\rho_{\text{meb}}(\mathcal{P}_i) \geqslant \text{radius}(B)$ that intersect $B$ is at most $\lambda$. The density of $\mathcal{S}$ is defined to be the smallest $\lambda$ for which $\mathcal{S}$ is a $\lambda$-low-density scene.*

We say that a scene has *low density* if its density is a small constant.

**Clutteredness.** This model was introduced by de Berg [6] under the name bounding-box-fitness condition. It is defined as follows.

**Definition 2.3** *Let $\mathcal{S}$ be a d-dimensional scene. We call $\mathcal{S}$ a $\kappa$-cluttered scene if any hypercube whose interior does not contain a vertex of one of the bounding boxes of the objects in $\mathcal{S}$ is intersected by at most $\kappa$ objects in $\mathcal{S}$. The clutter factor of a scene is the smallest $\kappa$ for which it is $\kappa$-cluttered.*

We sometimes call a scene whose clutter factor is a small constant *uncluttered.* Of the models we consider, this is the only one not invariant under rotations, since it uses bounding boxes and axis-parallel hypercubes. Although invariance under rotations is a desirable property from a mathematical point of view, it is not appropriate in all applications; architectural models, for instance, are not invariant under rotations.

**Simple-cover complexity.** The last model we consider is a variant of the simple-cover complexity as defined by Mitchell et al. [10]. Given a scene $\mathcal{S}$, we call a ball *$\delta$-simple* if it intersects at most $\delta$ objects in $\mathcal{S}$.

**Definition 2.4** *Let $\mathcal{S}$ be a d-dimensional scene, and let $\delta > 0$ be a parameter. A $\delta$-simple cover for $\mathcal{S}$ is a collection of $\delta$-simple balls whose union covers $\text{bb}(\mathcal{S})$. We say that $\mathcal{S}$ has $(\sigma, \delta)$-simple-cover complexity if there is a $\delta$-simple cover for $\mathcal{S}$ of cardinality $\sigma n$. The $\delta$-simple-cover complexity of $\mathcal{S}$ is the smallest $\sigma$ for which $\mathcal{S}$ has $(\sigma, \delta)$-simple-cover complexity.*

(In the original definition, there was an additional, somewhat unnatural, parameter $\epsilon$. Furthermore, the balls only had to cover the complement of the objects within the bounding box, not the entire bounding box.) We will say that a scene has *small simple-cover complexity* if there are small constants $\sigma$ and $\delta$ such that it has $(\sigma, \delta)$-simple-cover complexity. One of the somewhat unpleasant properties of this model is that the simple-cover complexity depends heavily on the choice of the parameter $\delta$. In general, this parameter should be at least as large as the maximum number of objects touching in a single point, otherwise no $\delta$-simple cover exists.

## 3 Relations between the models

We say that one model *implies* another model if, for any given constant model parameter(s) for the first model, there exist constant model parameters for the second model that only depend on the constants for the first model, such that any input satisfying the first model for the given constants also satisfies the second model for the derived constants. Clearly, this relation is transitive: if model $A$ implies model $B$, and model $B$ implies model $C$, then model $A$ implies model $C$. In this section we prove that the models defined in the previous section form a hierarchy as depicted in Figure 1, and as made precise in the theorem below. The hierarchy is strict, that is, the relations stated in the theorem and the ones implied by transitivity are the only relations between the models. For instance, an uncluttered scene does not necessarily have low density. For lack of space, we omit the examples showing that the hierarchy is strict.

There are two obvious reasons why establising the relations between the models is useful. First, in this initial stage of the development of a model-based theory of geometric algorithms, it is important that we understand the properties and relations between the models. Second, if we know that model $A$ implies model $B$, and we have an efficient data structure or algorithm for model $B$, then we can safely use that structure or algorithm in application domains where model $A$ is appropriate. For example, if we have an application domain where the objects are fat, then algorithms based on the octree-like data structure of Mitchell et al. [10] will be efficient.

**Theorem 3.1** *The following relations exist between the models:*

*(i) Any d-dimensional scene consisting of $\beta$-fat objects has density at most $2^d/\beta$.*

*(ii) Any d-dimensional $\lambda$-low-density scene has clutter factor at most $2^d\lambda$.*

*(iii) Any $\kappa$-cluttered scene has $(\sigma, \delta)$-simple-cover complexity for $\sigma = O(d2^{5d+3})$ and $\delta = O(6^d\kappa)$.*

*Proof.* (i) This was proved by van der Stappen [15]; the global approach is as follows. For any region $R$, he defines a region $R'$ that results from growing $R$ by an amount that is proportional to the size of $R$. Each object that intersects $R$ and is larger than $R$ contains at least some constant fraction of the volume of $R'$. The combination of the volume of $R'$ and a lower bound on the volume of $E \cap R'$ results in the claimed bound on the number of objects that can intersect $R$.

(ii) Let $\mathcal{S}$ be a $d$-dimensional $\lambda$-low-density scene, and let $C$ be a hypercube. We will show that the number of objects $\mathcal{P} \in \mathcal{S}$ such that $\mathcal{P}$ intersects $C$, and none of the vertices of $\text{bb}(\mathcal{P})$ lie inside $C$, is bounded by $2^d\lambda$. If $\mathcal{P}$ intersects $C$ but the vertices of $\text{bb}(\mathcal{P})$ lie outside $C$, then $\text{bb}(\mathcal{P})$ must have a side length of at least $l$, where $l$ is the side length of $C$; therefore, the radius of the minimal enclosing ball $\text{meb}(\mathcal{P})$ of $\mathcal{P}$ is at least $l/2$. On the other hand, we can cover $C$ by $2^d$ balls of radius $l/2$. Since $\mathcal{P}$—and therefore $\text{meb}(\mathcal{P})$—intersects $C$, it must also intersect at least one of these balls; the number of such objects $\mathcal{P}$ is therefore bounded by $2^d\lambda$.

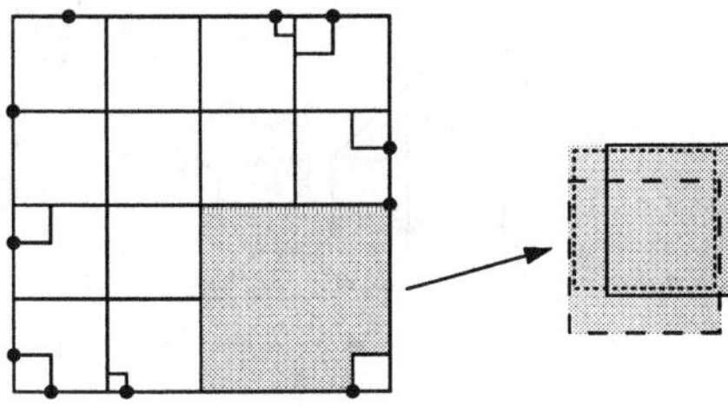

Figure 2: The refinement of a cube with markers on its boundary.

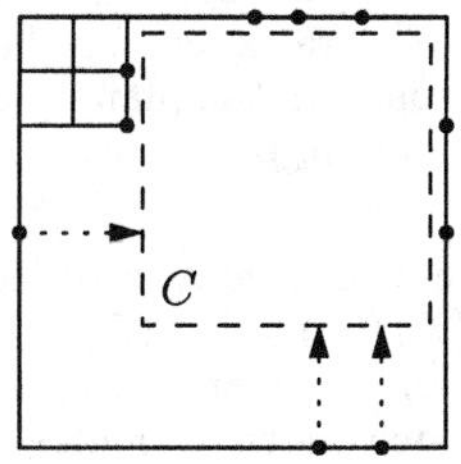


Figure 3: The cube $C$ has ten markers, three of which are obtained by projection.

(iii) Let $\mathcal{S}$ be a $d$-dimensional $\kappa$-cluttered scene consisting of $n$ objects. Our proof uses the fact that bb($\mathcal{S}$) can be partitioned into a collection $\mathcal{C}_1$ of $O(2^{d+2}n)$ cells with the following properties [6]:

- Every cell is either a cube, or the difference of a cube with another smaller cube that is contained in it and that shares a vertex with it. We call the latter type of cells L-shapes, because in the plane they have the shape of an L.
- Every cell can be covered by $O(2^d)$ cubes that have no bounding-box vertices in their interior. Since the scene is $\kappa$-cluttered this means that every cell is intersected by $O(2^d\kappa)$ objects.

We construct a simple cover from this partitioning as follows. Consider a cell $C$ in $\mathcal{C}_1$. Assume for the moment that $C$ is a cube. We would like to use meb($C$), the minimal enclosing ball of $C$, in the simple cover. However, if $C$ is adjacent to many smaller cells in $\mathcal{C}_1$ then meb($C$) may intersect too many objects. Hence, we refine $C$ as follows.

Let $V$ denote the set of vertices of the cells in $\mathcal{C}_1$. We call the vertices of $V$ that are located on the boundary of a cell but not vertices of the cell itself, the *markers* of that cell. We shall cover $C$ with a set $S(C)$ of subcubes without markers. The collection $S(C)$ is obtained recursively, as follows. Suppose that at some moment we have a subcube $C'$. If $C'$ does not have a marker then $C' \in S(C)$. Otherwise, consider the partitioning of $C'$ into $2^d$ 'octants'. If two or more of these octants contain markers then we replace $C'$ by the octants, with which we proceed recursively. If all markers are contained in one octant, however, we shrink that octant—meanwhile growing the other octants, which now start to overlap—until it hits a marker. We proceed recursively with the octants thus obtained. Figure 2 shows the result of this process in the planar case. For clarity we have not shown the cubes that are grown octants, because these overlap; we have only indicated them for the grey area. This recursive procedure results in a collection of subcubes that together cover $C$ and that have no markers. Moreover, the cardinality of $S(C)$ is $O(d2^{2d+1}m_C)$, where $m_C$ is the number of markers of $C$; the proof of this fact is omitted in this extended abstract. We claim that the minimum enclosing ball meb($C'$) of any cube $C' \in S(C)$ intersects $O(3^d)$ cells of $\mathcal{C}_1$, so that it intersects $O(6^d\kappa)$ objects. To prove this claim, we show in the full paper that the union of the cells of $\mathcal{C}_1$ whose closure intersects the closure of $C'$ contains meb($C'$), and that the number of such neighboring cells for $C'$ is $O(3^d)$.

The L-shapes are first covered by $O(2^d)$ cubes, which are then subjected to the recursive refinement procedure just described. There are some technical problems, however, which complicate things slightly. In particular, a cube $C$ that is used to cover an L-shape has to inherit the markers from the L-shape. For this we may have to project certain markers from the L-shape boundary onto the boundary of the cube—see Figure 3. The details of this are described in the full paper.

A vertex of any cell in $\mathcal{C}_1$ is a marker for $O(2^d)$ other cells in $\mathcal{C}_1$. Since the L-shapes in $\mathcal{C}_1$ are covered by $O(2^d)$ cubes before the partitioning into subcubes starts, and these cubes inherit the markers of their L-shape, this means that a vertex of a cell in $\mathcal{C}_1$ is a marker for $O(2^{2d})$ unrefined cubes. Hence, $\sum_C m_C = O(2^{2d}|\mathcal{C}_1|)$, where the sum is over all unrefined cubes.

Define $\mathcal{C}_2 := \bigcup_{C \in \mathcal{C}_1} S(C)$ to be the set of all subcubes resulting from the refinement process, and let $\mathcal{B} := \{\text{meb}(C) \mid C \in \mathcal{C}_2\}$ be the set of minimal enclosing balls of the subcubes. The total number of balls is

$$\begin{aligned} |\mathcal{B}| &= \sum_{C \in \mathcal{C}_1} O(d2^{2d+1}m_C) \\ &= O(d2^{4d+1}|\mathcal{C}_1|) \\ &= O(d2^{5d+3}n). \end{aligned}$$

As stated above, we can show that any ball in $\mathcal{B}$ intersects at most $O(3^d)$ cells from $\mathcal{C}_1$ and, hence, $O(6^d\kappa)$ objects. It follows that the $O(6^d\kappa)$-simple-cover complexity of the scene is $O(d2^{5d+3})$.

□

## 4 Computing the model parameters

To check whether a given model is realistic in a certain application, one needs to determine the model parameter for typical scenes in that application. Computing model parameters is also useful for other reasons. Some algorithms may need to know the value of the parameter in order to work correctly. (The range searching data structure of Overmars and van der Stappen [11] is an example.) In addition, the ability to compute the model parameters for a given scene can be used to establish which model best fits the scene, allowing us to choose "the right algorithm for the job". For example, if the scene consists of relatively fat objects, a simple algorithm that is very efficient for fat objects might be preferred to a more general algorithm for uncluttered scenes. (Such an algorithm-selection strategy is commonplace in commercial LP applications.) In this section we develop algorithms to compute the model parameters for the models defined in Section 2. We only consider the planar case, and we assume that the objects are simple polygons.

**Fatness.** The fatness of a convex object $\mathcal{P}$ can easily be computed: it is $\text{vol}(\mathcal{P})/(\omega_d \cdot \text{diam}(\mathcal{P})^2)$, where $\text{diam}(\mathcal{P})$ denotes the diameter of $\mathcal{P}$. Computing the fatness of non-convex objects is much more difficult—see Vleugels's thesis [18] for some results. Since in our setting the individual objects in a scene have constant complexity, we do not go into the computation of the fatness here.

**Low density.** To compute the density of a planar scene $\mathcal{S}$, we have to determine the smallest $\lambda$ such that any disc $D$ is intersected by at most $\lambda$ objects $\mathcal{P}_i \in \mathcal{S}$ with $\rho_i \geqslant \text{radius}(D)$, where $\rho_i := \rho_{\text{meb}}(\mathcal{P}_i)$. It is easy to see that we can restrict our attention to discs with radius $\rho_i$ for some $1 \leqslant i \leqslant n$.

To simplify the description, we assume from now on that all radii $\rho_i$ are distinct. We first order the objects according to the radius of their minimal enclosing disc, such that $\rho_1 > \ldots > \rho_n$. Define $\mathcal{S}_i := \{\mathcal{P}_1, \ldots, \mathcal{P}_i\}$. Let $\mathcal{M}_j(\rho)$ denote the Minkowski sum of $\mathcal{P}_j$ and disc of radius $\rho$ centered at the origin. Our algorithm is based on the following observation. A disc $D$ of radius $\rho_i$ intersects $\lambda$ objects $\mathcal{P}_j$ with $j \leqslant i$ if and only if the center of $D$ is covered by $\lambda$ of the Minkowski sums $\mathcal{M}_j(\rho_i)$ with $j \leqslant i$.

Our algorithm globally works as follows. We consider the radii $\rho_i$ in decreasing order. By the observation above, we can compute the maximum number of objects in $\mathcal{S}_i$ intersected by any disc of radius $\rho_i$ as follows: compute the Minkowski sums $\mathcal{M}_j(\rho_i)$ for $j \leqslant i$, and find the cell in the arrangement defined by the Minkowski sums that is contained in the maximum number of Minkowski sums. Implemented this straightforward manner, the algorithm is not very efficient. We can make it more efficient by using the fact that we treat the radii in order of decreasing value, as follows. Suppose that the maximum number of objects intersected by any disc of radius $\rho_j$ for any $j < i$ is $\lambda$. (Of course, we only consider the objects in $\mathcal{S}_j$ when we consider discs of radius $\rho_j$). Suppose furthermore that there is a disc of radius $\rho_i$ intersecting more than $\lambda$ objects from $\mathcal{S}_i$, so that we have to increase the current value of the density. Because $\rho_j > \rho_i$ for all $j < i$, this disc must intersect $\lambda + 1$ objects, one of which is the object $\mathcal{P}_i$ that now for the first time comes into consideration. Hence, we can restrict our attention to the part of the arrangement defined by the Minkowski sums $\mathcal{M}_j(\rho_i)$ lying inside $\mathcal{M}_i(\rho_i)$. To determine the objects whose Minkowski sums intersect $\mathcal{M}_i(\rho_i)$, we do a range query with $\mathcal{M}_i(2\rho_i)$ in $\mathcal{S}_{i-1}$. We thus obtain the following algorithm.

1. $\lambda_1 := 1$.
2. **for** $i := 2$ to $n$
3. **do** Determine the set $S$ of objects in $\mathcal{S}_{i-1}$ intersecting $\mathcal{M}_i(2\rho_i)$. Compute $\mathcal{M}_j(\rho_i)$ for each $\mathcal{P}_j \in S$, and compute the arrangement defined by these Minkowski sums inside $\mathcal{M}_i(\rho_i)$. Check whether there is a cell in this arrangement covered by $\lambda_{i-1}$ of these Minkowski sums. If so, set $\lambda_i := \lambda_{i-1}+1$, otherwise set $\lambda_i := \lambda_{i-1}$.
4. **return** $\lambda_n$

The range searching query we have to perform in Step 3 to find the objects in $\mathcal{S}_{i-1}$ intersecting $\mathcal{M}_i(2\rho_i)$ is done using a generalized version of the data structure of Schwarzkopf and Vleugels [14], as described by Vleugels [18]. The preprocessing time of the data structure for a $\lambda$-low-density scene is $O(n \log^3 n + \lambda n \log^2 n)$, and the time to perform a query is $O(\log^2 n + \lambda)$. This structure allows queries with all possible radii, so it has to be built only once at the start of the algorithm.

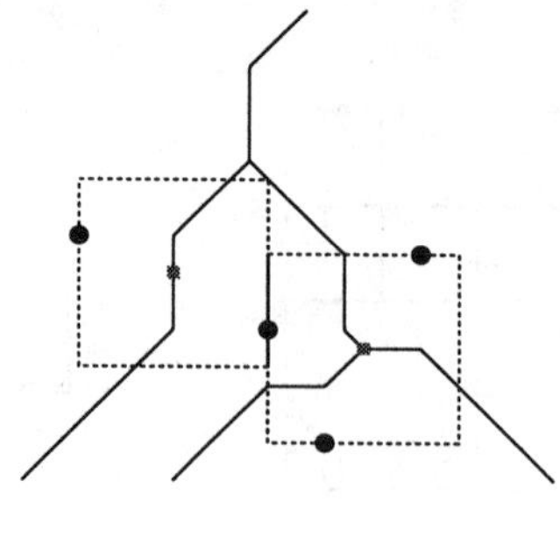

Figure 4: Maximal empty cubes are centered on the $L_\infty$-Voronoi diagram.

To compute the Minkowski sums and their arrangement, we spend quadratic time in the number of objects reported by the range query. Because the range with which we query has size $O(\rho_i)$, we know that the number of reported objects is $O(\lambda_{i-1})$. Hence, the time spent in stage $i$ is $O(\log^2 n + \lambda_i^2)$, leading to a total running time of

$$O(n \log^3 n + \lambda n \log^2 n) + \sum_{i=2}^{n} O(\log^2 n + \lambda_i^2),$$

which is $O(n \log^3 n + \lambda n \log^2 n + \lambda^2 n)$. Note that the running time is $O(n \log^3 n)$ if $\lambda$ is a constant.

**Theorem 4.1** *The density a planar scene $\mathcal{S}$ consisting of $n$ polygonal objects can be computed in $O(n \log^3 n + \lambda n \log^2 n + \lambda^2 n)$ time, where $\lambda$ is the density.*

**Clutteredness.** Let $\mathcal{S}$ be a planar scene consisting of $n$ objects, and let $V$ be the set of vertices of the bounding boxes of the objects in $\mathcal{S}$. We call a square *empty* if its interior does not contain any vertex in $V$. Recall that the clutter factor of $\mathcal{S}$ is the maximum number of objects in $\mathcal{S}$ intersected by any empty square. To compute the clutter factor we can restrict our attention to empty squares that are maximal in the sense that they are not contained inside some other empty square. Any maximal empty square has either three points from $V$ on its boundary, or it has two points on opposite sides. Hence, the center of such a square must either lie on a vertex of the $L_\infty$-Voronoi diagram of $V$, or it must lie on the vertical or horizontal part of a $L_\infty$-bisector that occurs in the Voronoi diagram—see Figure 4.

Our algorithm to compute the clutter factor is thus as follows.

1. Compute the $L_\infty$-Voronoi diagram of the set of bounding-box vertices of the objects.

2. For each vertex of the Voronoi diagram, compute the number of objects intersecting the maximal empty square centered at that vertex.

3. For each vertical or horizontal part of a $L_\infty$-bisector that occurs in the Voronoi diagram, sweep a maximal empty square along that part to compute the maximum

number of objects intersected by any maximal empty square centered on that part.

Step 1 can be done in $O(n \log n)$ time [8].

For Step 2 we store the vertices of the objects in a range tree, and the edges in a segment tree [13]. We apply fractional cascading [4] to these data structures, so that we can report the number of objects intersecting any square in $O(\log n + A)$ time, where $A$ is the number of answers.

To implement Step 3, we first determine the objects intersected by the sweep area of the square along the $L_\infty$-bisector part we are dealing with, using the range searching data structure. If there are $A$ such objects, this takes $O(\log n + A)$ time. Because the sweep we have to perform on the objects intersecting the sweep area is basically one-dimensional, it can be performed in $O(A \log A)$ time.

To analyze the final time complexity of the algorithm it remains to observe that for a $\kappa$-cluttered scene, the number of objects reported for each query in Step 2 of the algorithm is obviously bounded by $\kappa$. Moreover, the number of objects reported for each query in Step 3 is bounded by $2\kappa$, because each sweep area can be covered by two empty squares. This leads to the following result.

**Theorem 4.2** *The clutter factor of a planar scene $\mathcal{S}$ consisting of $n$ polygonal objects can be computed in $O(n \log n + n\kappa \log \kappa)$ time, where $\kappa$ is the clutter factor.*

Note that the running time is $O(n \log n)$ if the clutter factor is a constant.

**Simple-cover complexity.** For this model we would like to compute for a given parameter $\delta$ and a given scene $\mathcal{S}$, the $\delta$-simple-cover complexity of $\mathcal{S}$. Currently we have no efficient algorithm to compute for this; the best algorithm we have runs in exponential time. We also do not have a proof that computing the $\delta$-simple-cover complexity is NP-complete, although the seemingly-simpler DISC-PACK problem—which consists of covering a set of $n$ planar points by a given set of discs—is NP-hard [7], and a restricted version, in which the discs are identical, is NP-complete [7]. We leave the determination of the computational complexity as an open problem.

## 5 Intermezzo: boundary fatness

In the definition of fatness given in Section 2, we used the family $U(\mathcal{P})$ of all balls centered inside $\mathcal{P}$ and whose boundary intersects the boundary of $\mathcal{P}$; the fatness of $\mathcal{P}$ was then defined as the minimum over all such balls of the fraction of the ball covered by $\mathcal{P}$. At first sight it seems that the ball achieving this minimum, which we call the *emptiest ball*, would have its center on the boundary of $\mathcal{P}$. In other words, it seems that the fatness of $\mathcal{P}$ is equal to its *boundary fatness*, which is defined in the same way as the fatness, except that we now use the family $U(\partial\mathcal{P})$ consisting of all balls whose center lies on $\partial\mathcal{P}$ and whose boundary intersects the boundary of $\mathcal{P}$. However, this need not be the case: it can happen that the emptiest ball has its center in the interior of the object. Nevertheless, the fatness and the boundary fatness of an object are within a constant factor of each other, as the next theorem states.

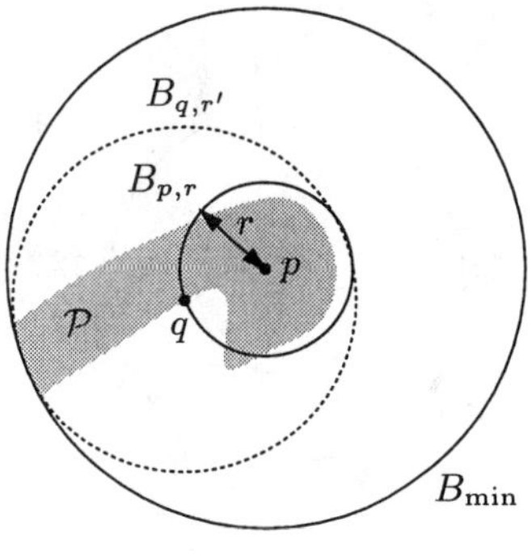


Figure 5: Illustration to the proof of Theorem 5.1.

**Theorem 5.1** *Let $\mathcal{P}$ be a (not necessarily connected) object in $\mathbb{E}^d$, then*

$$1 \leqslant \frac{\partial\text{-fatness}(\mathcal{P})}{\text{fatness}(\mathcal{P})} \leqslant \xi,$$

*where*

$$\xi = \min\left( (3/2)^d, \left(1 - \sqrt[d]{\text{fatness}(\mathcal{P})}\right)^{-d} \right).$$

*If $\mathcal{P}$ is disconnected, these bounds are tight. Moreover, in any dimension there exist connected objects $\mathcal{P}$ such that*

$$\frac{\partial\text{-fatness}(\mathcal{P})}{\text{fatness}(\mathcal{P})} \geqslant \left(\frac{2^d}{2^d+1}\right)\xi.$$

*Proof.* Let $\beta_\partial := \partial$-fatness$(\mathcal{P})$ and $\beta :=$ fatness$(\mathcal{P})$. The fact that $\beta_\partial \geqslant \beta$ follows immediately from the definitions. To prove the second inequality, we first prove that $\beta_\partial/\beta \leqslant (1-\sqrt[d]{\beta})^{-d}$.

Consider the emptiest ball $B_{\min} \in U(\mathcal{P})$. Let $p$ denote the center of $B_{\min}$—see Figure 5. We call the volume of the intersection of a ball with $\mathcal{P}$ the *weight* of the ball. We can assume without loss of generality that $B_{\min}$ has radius 1. Hence, the weight of $B_{\min}$ is $\omega_d\beta$. Now consider a second ball $B_{p,r}$ centered at $p$ with radius $r := \sqrt[d]{\beta}$. The volume of $B_{p,r}$ is exactly $\omega_d\beta$. We have $B_{p,r} \in U(\mathcal{P})$ because $p \in \mathcal{P}$ and $B_{p,r}$ cannot be completely inside $\mathcal{P}$. The latter is true because on the one hand $\mathcal{P}$ has a point outside $B_{p,r}$ (namely on $\partial B_{\min}$), and on the other hand $\text{vol}(B_{p,r}) = \text{vol}(\mathcal{P} \cap B_{\min})$. If $\mathcal{P}$ is connected, this implies that $\partial B_{p,r}$ must intersect $\partial\mathcal{P}$; if $\mathcal{P}$ is disconnected $B_{p,r}$ must still contain parts of $\partial\mathcal{P}$ in its interior and in its exterior, but it is no longer necessarily the case that $\partial B_{p,r}$ intersects $\partial\mathcal{P}$. In either case, there is a point $q$ in $\partial\mathcal{P} \cap B_{p,r}$.

Let $r'$ be the largest radius for which the ball $B_{q,r'}$ is contained in $B_{\min}$. Since $q \in \partial\mathcal{P}$ and $B_{q,r'}$ does not fully contain $\mathcal{P}$, we have $B_{q,r'} \in U(\partial\mathcal{P})$. Because $q$ lies inside $B_{p,r}$, we have $\text{dist}(q,p) \leqslant r$ and hence $r' \geqslant 1 - r = 1 - \sqrt[d]{\beta}$. The volume of $B_{q,r'}$ is at least $\omega_d(1-\sqrt[d]{\beta})^d$ and its weight is no more than the weight of $B_{\min}$ which is $\omega_d\beta$. Hence,

$$\begin{aligned} \beta_\partial/\beta &\leqslant \left(\frac{\omega_d\beta}{\omega_d(1-\sqrt[d]{\beta})^d}\right)/\beta \\ &= \frac{\beta}{(1-\sqrt[d]{\beta})^d}\cdot\frac{1}{\beta} \\ &= (1-\sqrt[d]{\beta})^{-d}. \end{aligned}$$

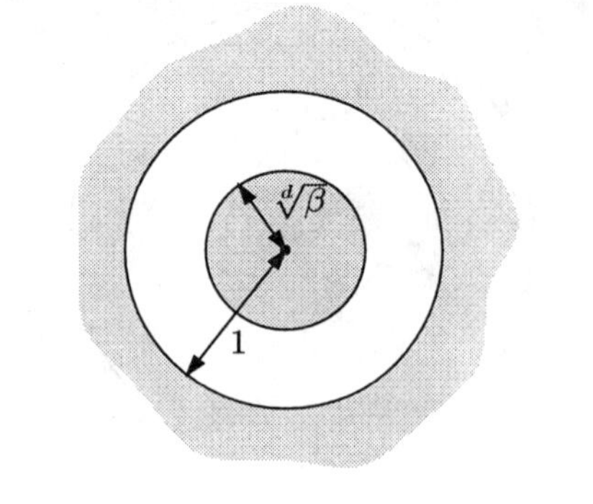


Figure 6: The lower bound example.

The lemma claims that the ratio between the boundary fatness of an object and its fatness is the minimum of $(3/2)^d$ and $(1-\sqrt[d]{\beta})^{-d}$. To prove this, we note that $(1-\sqrt[d]{\beta})^{-d} \geqslant (3/2)^d$ implies that $1-\sqrt[d]{\beta} \leqslant 2/3$ and hence that $\beta \geqslant (1/3)^d$. Since the boundary fatness of any object is at most $(1/2)^d$, this implies that $\beta_\partial/\beta$ is always less than or equal to $(1/2)^d/(1/3)^d = (3/2)^d$. This finishes the proof of the first part of the lemma.

The second part of the lemma states that the bound we just proved is tight if the object is not required to be connected. The example showing this is illustrated in Figure 6.

It consists of a spherical shell with a spherical hole of radius 1 at its center; the shell is sufficiently large in comparison to the hole so that any ball centered inside the shell achieves a fatness of close to $(1/2)^d$. Centered at the center of the hole is the second part of the object, which is a ball $B_1$ of radius $\sqrt[d]{\beta}$. Some easy calculations show that the ratio of the boundary fatness and the fatness for this object is $\xi = \left(1-\sqrt[d]{\text{fatness}(\mathcal{P})}\right)^{-d}$.

If we require the object to be connected, we can use almost the same construction, except that we now have to connect the ball $B_1$ to its surrounding shell. If this is done carefully (with a small but not too small crescent-shaped bridge) then the resulting object has a boundary-fatness-to-fatness ratio of $\left(\frac{2^d}{2^d+1}\right)\xi$. □

Filling in $d=2$, we get the following worst-case bounds for the planar case:

$$1 \leqslant \frac{\partial\text{-fatness}(\mathcal{P})}{\text{fatness}(\mathcal{P})} \leqslant 9/4.$$

Moreover, there are connected objects $\mathcal{P}$ for which

$$\frac{\partial\text{-fatness}(\mathcal{P})}{\text{fatness}(\mathcal{P})} \geqslant 9/5.$$

It is possible to improve these bounds. More precisely, we can prove that for any connected object $\mathcal{P}$ in the plane, the ratio between the boundary fatness of $\mathcal{P}$ and its fatness is less than 9/4. The best exact bound that is obtained from our construction (see below) has not been determined yet. We now describe the main idea behind this construction. As before, let $B_{\min} \in U(\mathcal{P})$ be the emptiest disc (we may assume that its radius is 1). Assume we can find $k$ discs $B_1,\dots,B_k$ in $U(\partial\mathcal{P})$ such that (i) they are contained in $B_{\min}$, and (ii) they are pairwise disjoint. Then $\beta_\partial \leq \min_{1\leq i\leq k}\beta_\partial^i$, where $\beta_\partial^i = \text{weight}(B_i)/\text{area}(B_i)$ is the emptiness of disc $B_i$. But,

$$\begin{aligned}\min_{1\leq i\leq k}\beta_\partial^i &= \min_{1\leq i\leq k}\frac{\text{weight}(B_i)}{\text{area}(B_i)}\\ &\leq \frac{\sum_{i=1}^k \text{weight}(B_i)}{\sum_{i=1}^k \text{area}(B_i)}\\ &\leq \frac{\text{weight}(B_{\min})}{\sum_{i=1}^k \text{area}(B_i)}\\ &= \frac{\omega_2\beta}{\sum_{i=1}^k \text{area}(B_i)},\end{aligned}$$

and therefore

$$\beta_\partial/\beta \leq \frac{\omega_2}{\sum_{i=1}^k \text{area}(B_i)}.$$

Thus, if we could find $k$ such discs such that their total area is a large enough fraction of the area of $B_{\min}$, then we would improve the upper bound of 9/4. Indeed, in our construction $k=2$, and we can find two disjoint discs $B_1, B_2 \in U(\partial\mathcal{P})$ that are contained in $B_{\min}$, such that $\text{area}(B_1)+\text{area}(B_2)$ is greater than $2.1\cdot\omega_2$, and therefore $\beta_\partial/\beta < 2.1$. The complete description of our construction appears in the full version of this paper.

## 6 Experimental verification of the models: a case study

Below we give some initial experimental results on the applicability of the models for one particular type of scenes: polyhedral terrains, as encountered frequently in GIS. In the full version of the paper we hope to report some more results, also on architectural models.

Because many algorithms working on terrains in fact work with its 2D projection, we shall do the same. Thus our test scenes are collections of triangles in the plane. The test scenes were generated as follows. We used so-called DEM files, describing certain areas in Canada and the US. Such a DEM file specifies the elevation of a set of sample points in the area at hand, where the sample points form a regular grid. Using the so-called VIP method [5] the $n$ most important points were extracted for the following values of $n$: 100, 200, 500, 1000, 2000, and 5000. The test scenes were then generated by computing the Delaunay triangulation of the extracted sample points. Below we describe the outcome of our tests on test scenes taken from Death Valley and from San Bernardino; the results we got for other areas were similar. Figure 7 lists the numerical data that we acquired from our experiments on these test scenes; Figure 8 shows the actual scenes we got for Death Valley and for San Bernardino, both with $n = 1000$.

**Fatness.** The fatness of a scene is determined by the least fat object in that scene. The presence of a single outlier in the scene—that is, a considerably less fat object—can have a dramatic impact on the performance of algorithms with a running time depending on the fatness of the scene. The results in the table—particularly those for the Death Valley scenes—confirm that this is indeed a potential danger: the fatness of the scene is sometimes a factor of 1,000 smaller than the average fatness of its triangles. Over 95 percent of the triangles in the Death Valley scenes, and over 99.5 percent of the triangles in the San Bernardino scenes, have a fatness that is within a factor ten of the average

| test scene | $n$ | fatness $\beta$ | avg. | density $\lambda$ | avg. | clutter factor $\kappa$ | avg. | simple-cover compl. $\sigma$ |
|---|---|---|---|---|---|---|---|---|
| Death Valley | 100 | $2.44 \cdot 10^{-4}$ | $3.63 \cdot 10^{-2}$ | 20 | 10.5 | 22 | 6.52 | 1.01 |
| | 200 | $2.20 \cdot 10^{-5}$ | $3.96 \cdot 10^{-2}$ | 16 | 9.87 | 22 | 5.63 | 0.622 |
| | 500 | $1.60 \cdot 10^{-4}$ | $4.45 \cdot 10^{-2}$ | 19 | 9.24 | 17 | 4.75 | 0.471 |
| | 1000 | $1.42 \cdot 10^{-5}$ | $4.96 \cdot 10^{-2}$ | 21 | 9.05 | 17 | 4.57 | 0.366 |
| | 2000 | $6.34 \cdot 10^{-6}$ | $5.37 \cdot 10^{-2}$ | 20 | 8.82 | 21 | 4.49 | 0.335 |
| | 5000 | $4.53 \cdot 10^{-6}$ | $5.73 \cdot 10^{-2}$ | 27 | 8.82 | 20 | 4.44 | 0.315 |
| San Bernardino | 100 | $2.65 \cdot 10^{-4}$ | $4.65 \cdot 10^{-2}$ | 13 | 8.32 | 11 | 4.32 | 0.332 |
| | 200 | $2.65 \cdot 10^{-4}$ | $4.66 \cdot 10^{-2}$ | 16 | 8.83 | 14 | 4.49 | 0.264 |
| | 500 | $2.91 \cdot 10^{-5}$ | $5.37 \cdot 10^{-2}$ | 15 | 8.75 | 15 | 4.26 | 0.270 |
| | 1000 | $1.81 \cdot 10^{-4}$ | $5.61 \cdot 10^{-2}$ | 17 | 8.61 | 17 | 4.26 | 0.252 |
| | 2000 | $1.67 \cdot 10^{-6}$ | $5.59 \cdot 10^{-2}$ | 23 | 8.80 | 21 | 4.33 | 0.265 |
| | 5000 | $2.03 \cdot 10^{-6}$ | $5.80 \cdot 10^{-2}$ | 25 | 8.76 | 19 | 4.30 | 0.275 |

Figure 7: Experimental results for two of the test scenes.

fatness. The planar range searching algorithm by Overmars and van der Stappen [11] has a quadratic dependence on the fatness. In the Death Valley scenes, exclusion of the least fat five percent of the triangles results in an overall fatness that is approximately one tenth of the average fatness of the complete scene; the fatness of the left-out five percent is a factor 100 smaller than this. Therefore, including these five percent least-fast triangles would increase the running time of the algorithm by a factor up to 10,000.

The experiments suggest that algorithms for this type of TINs, whose running time depends on the minimal fatness of the scene, will perform badly, whereas algorithms whose running time depends on the average fatness may perform well. Furthermore, it is interesting to see that the average fatness is fairly constant over different test scenes generated from the same area.

**Low density.** In the algorithm described in Section 4, we add the objects one by one, maintaining the density of the objects added so far. This is done by computing the arrangement of Minkowski sums of the objects. In our experiments, the candidate regions consists of discs that are centered at the vertices of this arrangement.

For our test scenes, the density typically varies between approximately fifteen and twenty-five. Unlike fatness, an algorithm that requires the input to have low density is not likely to have a running time whose dependence on the density is more than linear. (In most cases, the density will determine the number of objects for which the problem under consideration is 'easy enough' to solve it in constant time.) The average density does not differ much from this number. This suggests that, unlike with fatness, the density of a scene supplies a good estimation of the average number of objects intersecting a region.

The density of our test scenes appears to increase somewhat with a growing number of points. This is hardly surprising because the density of a scene is determined by the single worst candidate region. In contrast, the average density seems to either decrease with a growing number of points—as in the Death Valley scenes—or remain relatively constant, which is the case for San Bernardino. This suggests that regions with high density are introduced by the need to approximate the terrain with only few data points, especially if the terrain contains steep ridges. The approximation provided by the triangulation grows more accurate as the number of data points increases.

**Clutteredness.** The planar clutter factor of a scene is determined by the square without bounding-box vertices in its interior that intersects the largest number of objects. In the scenes we consider, the clutter factor typically lies between eleven and twenty-two; although the exact clutter factor varies somewhat with the number of points, it does not seem to increase significantly with a growing number of points.

As in the case of low density, the clutter factor will in most cases determine the number of objects for which the problem under consideration is sufficiently easy. Although the clutter factor lies in the same range as the density, the average clutter factor is in all cases lower than the average density. In our test scenes, the average clutter factor is typically less than five, and varies little for different scenes. This suggests that only few squares are relatively cluttered. Indeed, our experiments show that only approximately five percent of the squares intersect nine or more objects, while typically less than 0.5 percent intersect twelve or more objects. Like the density, the average clutter factor either decreases or remains relatively constant for larger scenes.

**Simple-cover complexity.** Currently we have no algorithm to compute the simple cover complexity of a scene. Instead we build smoothed quad trees for the scenes, the leaves of which are cells that intersect at most $t$ objects, where $t$ is the maximum number of objects touching in a single point. We believe the size of this tree to be closely related to the simple cover complexity, as we do not expect situations in which a small crowded corner of an otherwise empty cell would lead to many recursive splits of that cell without actually reducing the maximum number of objects in its subcells. The resulting squares are $\delta$-simple for a small constant $\delta$. The fact that the tree is smoothed makes it possible to cover the squares by a similar number of $\delta'$-simple discs, where $\delta'$ lies within a constant factor of $\delta$. Again, details can be obtained from the full version. Figure 7 shows that the ratio of the number of quad-tree leaves and the number of objects is indeed a small constant.

## 7 Conclusions

We have studied the relations between various scene models that have been proposed in the computational geometry literature, we have developed algorithms to compute the model parameters, and we have experimentally investigated

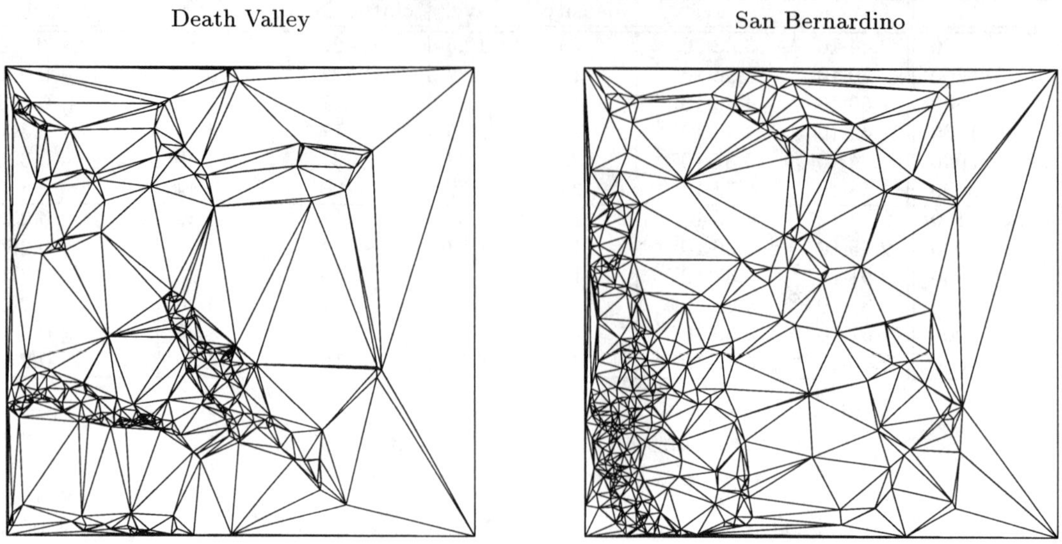


Figure 8: The Death Valley and San Bernardino test scenes for $n = 1000$.

the appropriateness of some of the models for certain types of TINs. We believe that our paper points into an important research direction, which can help to bridge the gap between theory (computational geometry) and practice (its application areas). Future work in this direction should include:

- For a number of important application domains, there should be models that are on the one hand general enough to include (almost all) scenes in the domain, and on the other hand restrictive enough so that complicated, hypothetical input scenes are precluded. Ideally, these models should be widely accepted as being realistic for their respective application areas. To achieve this, experimental research in the spirit of Section 6 is needed.

- New algorithms should be developed that solve a problem from a given application domain efficiently within a model appropriate for that domain. Here the focus should be on problems for which no efficient solution is available for general, unrestricted scenes.

We do not claim that the models we have studied in this paper are the best or most intuitive ones. Nevertheless, we believe that scenes in many areas will be uncluttered and have small simple-cover complexity. Solutions based on octree-like structures as described by de Berg [6] and Mitchell et al. [10] will perform well in that case. For many application domains, there may well be models that are more intuitive or more appropriate than these models. If this is the case, one would of course like to work with these models. We want to emphasize, however, that it is important to stick to a few, generally accepted models, in order to be able to develop a coherent model-based theory.

## References

[1] P. Agarwal, M. Katz, and M. Sharir. Computing depth orders for fat objects and related problems. *Comput. Geom. Theory Appl.*, 5:187–206, 1995.

[2] Pankaj K. Agarwal and J. Matoušek. On range searching with semialgebraic sets. In *Proc. 17th Internat. Sympos. Math. Found. Comput. Sci.*, volume 629 of *Lecture Notes in Computer Science*, pages 1–13. Springer-Verlag, 1992.

[3] H. Alt, R. Fleischer, M. Kaufmann, K. Mehlhorn, S. Näher, S. Schirra, and C. Uhrig. Approximate motion planning and the complexity of the boundary of the union of simple geometric figures. *Algorithmica*, 8:391–406, 1992.

[4] B. Chazelle and L. J. Guibas. Fractional cascading: I. A data structuring technique. *Algorithmica*, 1:133–162, 1986.

[5] Z. Chen and J. A. Guevara. System selection of very important points (VIP) from digital terrain models for constructing triangular irregular networks. In *Proc. 8th Internat. Sympos. Comput.-Assist. Cartog. (Auto-Carto)*, pages 50–56, 1988.

[6] Mark de Berg. Linear size binary space partitions for fat objects. In *Proc. 3rd Annu. European Sympos. Algorithms (ESA '95)*, volume 979 of *Lecture Notes in Computer Science*, pages 252–263, 1995.

[7] R. J. Fowler, M. S. Paterson, and S. L. Tanimoto. Optimal packing and covering in the plane are NP-complete. *Inform. Process. Lett.*, 12(3):133–137, 1981.

[8] D. T. Lee and C. K. Wong. Voronoi diagrams in $L_1$ ($L_\infty$) metrics with 2-dimensional storage applications. *SIAM J. Comput.*, 9:200–211, 1980.

[9] J. Matoušek, N. Miller, J. Pach, M. Sharir, S. Sifrony, and E. Welzl. Fat triangles determine linearly many holes. In *Proc. 32nd Annu. IEEE Sympos. Found. Comput. Sci.*, pages 49–58, 1991.

[10] J. S. B. Mitchell, D. M. Mount, and S. Suri. Query-sensitive ray shooting. In *Proc. 10th Annu. ACM Sympos. Comput. Geom.*, pages 359–368, 1994.

[11] Mark H. Overmars and A. Frank van der Stappen. Range searching and point location among fat objects. *J. Algorithms*, 21:629–656, 1996.

[12] M. S. Paterson and F. F. Yao. Efficient binary space partitions for hidden-surface removal and solid modeling. *Discrete Comput. Geom.*, 5:485–503, 1990.

[13] F. P. Preparata and M. I. Shamos. *Computational Geometry: An Introduction.* Springer-Verlag, New York, NY, 1985.

[14] Otfried Schwarzkopf and Jules Vleugels. Range searching in low-density environments. *Inform. Process. Lett.*, 60:121–127, 1996.

[15] A. F. van der Stappen. *Motion Planning amidst Fat Obstacles.* Ph.D. thesis, Dept. Comput. Sci., Utrecht Univ., Utrecht, the Netherlands, October 1994.

[16] A. F. van der Stappen and M. H. Overmars. Motion planning in environments with low obstacle density. Technical Report UU-CS-1995-33, Dept. Comput. Sci., Utrecht Univ., Utrecht, the Netherlands, October 1995.

[17] Marc van Kreveld. On fat partitioning, fat covering, and the union size of polygons. In *Proc. 3rd Workshop Algorithms Data Struct.*, volume 709 of *Lecture Notes in Computer Science*, pages 452–463. Springer-Verlag, 1993.

[18] Jules Vleugels. *On Fatness and Fitness—Realistic Input Models for Geometric Algorithms.* Ph.D. thesis, Dept. Comput. Sci., Utrecht Univ., Utrecht, the Netherlands, March 1997.

# Space-Time Tradeoffs for Emptiness Queries*

**(Extended abstract)**

Jeff Erickson

Center for Geometric Computing
Computer Science Department
Duke University
Durham, NC 27708–0129
jeffe@cs.duke.edu
http://www.cs.duke.edu/~jeffe

## Abstract

We present the first nontrivial space-time tradeoff lower bounds for hyperplane and halfspace emptiness queries. Our lower bounds apply to a general class of geometric range query data structures called *partition graphs*. Informally, a partition graph is a directed acyclic graph that describes a recursive decomposition of space. We show that any partition graph that supports hyperplane emptiness queries implicitly defines a halfspace range query data structure in the Fredman/Yao semigroup arithmetic model, with the same space and time bounds. Thus, results of Brönnimann, Chazelle, and Pach imply that any partition graph of size $s$ that supports hyperplane emptiness queries in time $t$ must satisfy the inequality $st^d = \Omega((n/\log n)^{d-(d-1)/(d+1)})$. Using different techniques, we show that $\Omega(n^d/\operatorname{polylog} n)$ preprocessing time is required to achieve polylogarithmic query time, and that $\Omega(n^{(d-1)/d}/\operatorname{polylog} n)$ query time is required if only $O(n \operatorname{polylog} n)$ preprocessing time is used. These two lower bounds are optimal up to polylogarithmic factors. For two-dimensional queries, we obtain an optimal continuous tradeoff between these two extremes. Finally, using a reduction argument, we show that the same lower bounds hold for halfspace emptiness queries in $\mathbb{R}^{d(d+3)/2}$ on a restricted class of partition graphs.

## 1 Introduction

We present the first nontrivial lower bounds on the complexity of data structures that support hyperplane and halfspace emptiness queries. Emptiness query data structures are used to solve several geometric problems including point location [12], ray shooting [2, 12, 24, 27] nearest and farthest neighbor queries [2], linear programming queries [23, 7], depth ordering [5], collision detection [11], and output-sensitive convex hull construction [23, 8].

Most previous range searching lower bounds are presented in the so-called *semigroup arithmetic* model, originally introduced by Fredman [20] and later refined by Yao [31]. In this model, the points are given weights from an additive semigroup, and the goal of a range query is to determine the total weight of the points in a query region. A data structure in this model can be informally regarded as a set of precomputed partial sums in the underlying semigroup. The size of such a data structure is the number of partial sums, and the query time is the number of semigroup additions performed on these partial sums to obtain the answer. (More formal definitions are given in Section 2.) Lower bounds have been established in this model for several types of query ranges [9, 6], in many cases matching the complexities of the corresponding data structures.

Unfortunately, the semigroup model cannot be used to study the complexity of emptiness queries. If the query range is empty, we perform no additions; conversely, if we perform even a single addition, the query range cannot be empty. Similar arguments apply to Tarjan's *pointer machine* model [30], which has been used to derive output-sensitive lower bounds for range *reporting* problems [14]. In fact, the only lower bounds previously known for emptiness queries are trivial. The size of any range searching data structure must be $\Omega(n)$, since it must store each of the points, and the query time must be at least $\Omega(\log n)$ in any reasonable model of computation (such as algebraic computation trees [4] or real RAMs [29]).

*Portions of this research were done while the author was a student in the Computer Science Division, U. C. Berkeley, with the support of a GAANN Fellowship. This work was also supported by the National Science Foundation under grant DMS-9627683 and by the U. S. Army Research Office under grant DAAH04-96-1-0013.



*Computational Geometry 97* Nice France
Copyright 1997 ACM 0-89791-878-9/97/06 ..$3.50

| Space | Preprocessing | Query Time | Source |
|---|---|---|---|
| $O(n^d/\log^d n)$ | $O(n^d/\log^{d-\varepsilon} n)$ | $O(\log n)$ | [10, 25] |
| $O(n)$ | $O(n^{1+\varepsilon})$ | $O(n^{1-1/d})$ | [25] |
| $O(n)$ | $O(n \log n)$ | $O(n^{1-1/d} \operatorname{polylog} n)$ | [21] |
| $n \le s \le n^d/\log^d n$ | $O(n^{1+\varepsilon} + s \log^\varepsilon n)$ | $O(n/s^{1/d})$ | [10, 25] |

**Table 1.** Best known upper bounds for hyperplane emptiness queries.

Our new lower bounds apply to a general class of geometric range query data structures called *partition graphs*. Informally, a partition graph is a directed acyclic graph that describes a recursive decomposition of space. Our model is powerful enough to describe most, if not all[1], known data structures for these problems. Partition graphs have been previously used to study offline range searching problems such as Hopcroft's point-line incidence problem [19, 18].

We summarize our results below. In each of these results and throughout the paper, s denotes space, p denotes preprocessing time, and t denotes worst-case query time. For comparison, the best known upper bounds are listed in Table 1. (For a review of range searching techniques and results, see the surveys by Matoušek [26] and Agarwal [1].)

- $s = \Omega(n)$, $p = \Omega(n \log n)$, and $t = \Omega(\log n)$.
- Any partition graph that supports hyperplane emptiness queries implicitly defines a halfspace range query data structure in the Fredman/Yao semigroup arithmetic model, with the same time and space bounds. Thus, results of Brönnimann, Chazelle, and Pach [6] immediately imply that $st^d = \Omega((n/\log n)^{d-(d-1)/(d+1)})$. This lower bound applies with high probability to a randomly generated set of points.
- Lower bounds on the complexity of Hopcroft's point-hyperplane incidence problem [19, 18] imply the worst case bounds $pt^{(d+2)(d-1)/2} = \Omega(n^d)$ and $pt^{2/(d-1)} = \Omega(n^{(d+2)/d})$. These lower bounds match known upper bounds up to polylogarithmic factors when $d = 2$, $p = O(n \operatorname{polylog} n)$, or $t = O(\operatorname{polylog} n)$ [21, 25]. These results require restrictions on the partition graphs when $d \ge 4$.

All of the previous lower bounds also apply to hyperplane and halfspace counting queries in $\mathbb{R}^d$, and to halfspace emptiness queries in $\mathbb{R}^{d(d+3)/2}$ under some restrictions on the partition graphs.

Lower bounds in the semigroup arithmetic model imply lower bounds for counting queries in the partition graph model. Thus, any partition graph that supports halfspace counting queries satisfies $st^d = \Omega((n/\log n)^{d-(d-1)/(d+1)})$. We also derive the lower bound $st^{d(d+1)/2} = \Omega(n^d)$ for hyperplane queries in the semigroup model, and thus for hyperplane counting queries in the partition graph model as well. This improves the lower bound inherited from hyperplane emptiness queries whenever $s = \Omega(n^{d-1})$ or $t = O(n^{2/d(d+1)})$.

The rest of the paper is organized as follows. Section 2 reviews the definition of the semigroup arithmetic model and states some useful results. In Section 3, we define partition graphs, describe how they are used to answer hyperplane and halfspace queries, and state a few of their basic properties. We prove our new time-space and preprocessing-query tradeoffs for hyperplane emptiness queries in Section 4. In Section 5, we use a reduction argument to derive lower bounds for halfspace emptiness queries. Finally, in Section 6, we offer our conclusions.

## 2 Semigroup Arithmetic

An *additive semigroup* $(S, +)$ is a set S equipped with an associative addition operator $+ : S \times S \to S$. A semigroup is *commutative* if the equation $x + y = y + x$ is true for all $x, y \in S$. A *linear form* is a sum of variables over the semigroup, where each variable can occur multiple times, or equivalently, a homogeneous linear polynomial with positive integer coefficients. A semigroup is *faithful* if any two identically equal linear forms have the same set of variables, although not necessarily with the same set of coefficients. For example, $(\mathbb{Z}, +)$ and $(\{\text{true}, \text{false}\}, \vee)$ are faithful semigroups, but $(\{0, 1\}, + \bmod 2)$ is not faithful.

Let P be a set of n points in $\mathbb{R}^d$, let $(S, +)$ be a faithful commutative additive semigroup, and let $w : P \to S$ be a function that assigns a weight $w(p)$ to each point $p \in P$. For any subset $P' \subseteq P$, let $w(P') = \sum_{p \in P'} w(P)$, where addition is taken over the semigroup. (Since S need not have an additive identity, we assign a special value *nil* to the empty sum.) The range searching problem considered in the semigroup model is to preprocess P so that $w(P \cap q)$ can be calculated quickly for any query range q.

[1]Difficulties in modeling range searching data structures as partition graphs are discussed in [19, Section 3.5].

Let $x_1, x_2, \ldots, x_n$ be a set of $n$ variables over $S$. A *generator* $g(x_1, \ldots, x_n)$ is a linear form $\sum_{i=1}^{n} \alpha_i x_i$, where the $\alpha_i$'s are non-negative integers, not all zero. Given a class $Q$ of query ranges, a *storage scheme* for $(P, Q, S)$ is a collection of generators $\{g_1, g_2, \ldots, g_s\}$ with the following property: For any query range $q \in Q$, there is an set of indices $I_q \subseteq \{1, 2, \ldots, s\}$ and a set of labeled nonnegative integers $\{\beta_i \mid i \in I_q\}$ such that

$$w(P \cap q) = \sum_{i \in I_q} \beta_i g_i(w(p_1), w(p_2), \ldots, w(p_n))$$

holds for *any* weight function $w : P \to S$. The size of the smallest such set $I_q$ is the *query time* for $q$.

We emphasize that although a storage scheme can take advantage of special properties of the semigroup $S$ or the point set $P$, it must work for *any* assignment of weights to $P$. In particular, this implies that lower bounds in the semigroup model do *not* apply to the problem of counting the number of points in the query range, even though $(\mathbb{Z}, +)$ is a faithful semigroup, since a storage scheme for that problem only needs to work for the particular weight function $w(p) = 1$ for all $p \in P$ [9]. For the same reason, even though the semigroup $(\{\text{true}, \text{false}\}, \vee)$ is faithful, the semigroup model cannot be used to prove lower bounds for emptiness queries. Emptiness queries can also be formulated as queries over the one-element semigroup $(\{*\}, * + * = *)$, but this semigroup is not faithful.

For any linear form $\sum_{1=1}^{n} \alpha_i x_i$, call the set of points $\{p_i | \alpha_i \neq 0\}$ its *cluster*. The faithfulness of the semigroup $S$ implies that the union of the clusters of the generators used to determine $w(P \cap q)$ is precisely $P \cap q$:

$$\bigcup_{i \in I_q} \text{cluster}(g_i) = P \cap q.$$

Thus, we can think of a storage scheme as a collection of clusters, such that any set of the form $P \cap q$ can be expressed as the (not necessarily disjoint[2]) union of several of these clusters. The size of a storage scheme is the number of clusters, and the query time for a range $q$ is the minimum number of clusters whose union is $P \cap q$. This is the formulation actually used in [6, 9] to prove lower bounds.

Some of our lower bounds derive from the following result of Brönnimann, Chazelle, and Pach [6].

[2]Whether or not the clusters used to answer a query must be disjoint depends on the semigroup. For example, if the semigroup is $(\mathbb{Z}, +)$, then the clusters must be disjoint; on the other hand, if the semigroup is $(\mathbb{Z}, \max)$, then the clusters can overlap arbitrarily.

**Theorem 2.1 (Brönnimann, Chazelle, Pach).** *Let $P$ be a uniformly distributed set of points in the $d$-dimensional unit hypercube $[0, 1]^d$. With high probability, any storage scheme of size $s$ that supports halfspace queries in $P$ in time $t$ must satisfy the inequality* $st^d = \Omega((n/\log n)^{d-(d-1)/(d+1)})$.

Although lower bounds are known for offline hyperplane searching in the semigroup model [15, 19], we are unaware of any previous results for online hyperplane queries. In particular, Chazelle's lower bounds for simplex range searching [9] do not apply when the ranges are hyperplanes. We easily observe that for a set of points in general position, the smallest possible storage scheme, consisting of $n$ singleton sets, allows hyperplane queries to be answered in constant "time". Much better lower bounds can be obtained by considering degenerate point sets.

**Theorem 2.2.** *Any storage scheme of size $s$ that supports $d$-dimensional hyperplane queries in time $t$ must satisfy the inequality $st^{d(d+1)/2} = \Omega(n^d)$ in the worst case.*

**Proof (sketch):** In two dimensions, we easily observe that the clusters in an optimal storage scheme must consist of maximal colinear subsets of the set of points. The two-dimensional lower bound follows from a construction of $n$ points and $s + 1$ lines with $\Omega(n^{2/3}s^{2/3})$ incidences discovered by Erdős; see [20] or [28, p. 177]. The higher-dimensional results follow from a natural generalization of the Erdős construction to higher dimensions [19, Lemmas 3.5 and 3.10]. We omit further details from this extended abstract. □

## 3 Partition Graphs

A *partition graph* is a directed acyclic (multi-)graph, with one source, called the *root*, and several sinks, called *leaves*. Associated with each non-leaf node $v$ is a set $\mathcal{R}_v$ of *query regions*, satisfying three conditions.

1. $\mathcal{R}_v$ contains at most $\Delta$ query regions, for some constant $\Delta \geq 2$.
2. Every query region is a connected subset of $\mathbb{R}^d$.
3. The union of the query regions in $\mathcal{R}_v$ is $\mathbb{R}^d$.

We associate an outgoing edge of $v$ with each query region in $\mathcal{R}_v$. Thus, the outdegree of the graph is at most $\Delta$. We put no constraints on the indegree. In addition, every non-leaf note $v$ is either a *primal node* or a *dual node*, depending on whether its query regions $\mathcal{R}_v$ are interpreted as a partition of primal or dual space.

```
Preprocess(p):
at each primal node v:
    for each query region R ∈ 𝓡_v:
        if R contains p:
            traverse the edge corresponding to R
at each dual node v:
    for each query region R ∈ 𝓡_v:
        if R intersects p*:
            traverse the edge corresponding to R
at each leaf ℓ:
    add p to P_ℓ
```

(a)

```
Query(h):
at each dual node v:
    for each query region R ∈ 𝓡_v:
        if R contains h*:
            traverse the edge corresponding to R
at each primal node v:
    for each query region R ∈ 𝓡_v:
        if R intersects h:
            traverse the edge corresponding to R
at each leaf ℓ:
    test/count P_ℓ
```

(b)

**Figure 1.** Preprocessing and query algorithms for hyperplane searching.

The query regions associated with primal (resp. dual) nodes are called primal (resp. dual) query regions.

We do not require the query regions to be disjoint. In the general case, we do not require the query regions to be convex, semialgebraic, simply connected, of constant complexity, or even computable in any sense. However, some of our results require the query regions to be constant-complexity semialgebraic sets or constant-complexity polyhedra.

Given a partition graph, we preprocess a set $P$ of points for hyperplane emptiness or counting queries as follows. We preprocess each point $p \in P$ individually by performing a depth-first search of the partition graph, using the query regions to determine which edges to traverse. Whenever we reach a primal node $v$, we traverse the edges corresponding to the query regions in $\mathcal{R}_v$ that contain $p$. Whenever we reach a dual node $v$, we traverse the edges corresponding to the query regions in $\mathcal{R}_v$ that intersect the dual hyperplane $p^*$. Note that the same point may enter or leave a node along several different edges, but we only test the query regions at a node once for each point. For each leaf $\ell$, we maintain the set $P_\ell$ of the points that reach $\ell$.[3] See Figure 1(a).

We answer a hyperplane query almost exactly the same way we preprocess a point: by performing a depth-first search of the partition graph, using the query regions to determine which edges to traverse. The only real difference is that the behavior at the primal and dual nodes is reversed. See Figure 1(b).

At each leaf $\ell$, the query algorithm does something to the corresponding subset of points $P_\ell$. If we are answering a counting query, then the algorithm adds the size of the $P_\ell$ to a running counter, which is output at the end of the search. If we are answering an emptiness query and $P_\ell$ is nonempty, then the algorithm immediately halts and reports that $h$ contains at least one point; if the query algorithm reaches only leaves with $P_\ell$ empty, the algorithm reports that the hyperplane is empty.

By modifying the preprocessing algorithm slightly, we can also use partition graphs to answer halfspace queries. For any hyperplane $h$, let $h^+$ denote its closed upper halfspace and $h^-$ its closed lower halfspace.[4] Recall that the standard duality transformation $(a_1, a_2, \ldots, a_d) \longleftrightarrow x_d + a_d = a_1x_1 + a_2x_2 + \cdots + a_{d-1}x_{d-1}$ preserves incidences and relative orientation between points and hyperplanes: If a point $p$ is above (on, below) a hyperplane $h$, then the dual point $h^*$ is above (on, below) the dual hyperplane $p^*$.

To support halfspace queries, we associate one or more subsets of $P$ with every query region. With each primal region $R \in \mathcal{R}_v$, we associate a subset $P_R$, which contains all the points that reach $v$ and lie inside $R$. With each dual region $R \in \mathcal{R}_v$, we associate two subsets $P_R^+$ and $P_R^-$, which contain all the points that reach $v$ and whose dual hyperplanes lie below and above $R$, respectively. Our modified preprocessing and halfspace query algorithms are shown in Figure 2. Note that the modified preprocessing algorithm can still be used for hyperplane searching.

For purposes of proving lower bounds, the *size* of a partition graph is the number of edges in the graph, the *query time* for a particular hyperplane is the number of edges the hyperplane traverses, and the *preprocessing time* is the total number of edges traversed in the preprocessing phase. We do not consider the complexity of the query regions, the time required in practice to determine which query regions intersect a hyperplane or contain a point, the sizes of the subsets $P_R$, $P_R^+$, $P_R^-$,

[3]These sets are just a convenience for our analysis. In practice, it suffices to record just the size of each set (for counting queries) or whether or not each set is empty (for emptiness queries).

[4]We assume throughout the paper that query halfspaces are closed and that no query halfspace is bounded by a vertical hyperplane. Handling open halfspaces involves only trivial modifications to our query algorithms, which have almost no effect on our analysis. Vertical halfspace queries can be handled either by standard perturbation techniques or by using a lower-dimensional data structure.

```
Preprocess(p):
at each primal node v:
    for each query region R ∈ ℛ_v:
        if R contains p:
            add p to P_R
            traverse the edge corresponding to R
at each dual node v:
    for each query region R ∈ ℛ_v:
        if R intersects p*:
            traverse the edge corresponding to R
        else if R is above p*:
            add p to P_R^+
        else if R is below p*:
            add p to P_R^-
at each leaf ℓ:
    add p to P_ℓ
```

(a)

```
Query(h±):
at each dual node v:
    for each query region R ∈ ℛ_v:
        if R contains h*:
            test/count P_R^±
            traverse the edge corresponding to R
at each primal node v:
    for each query region R ∈ ℛ_v:
        if h intersects R:
            traverse the edge corresponding to R
        else if h± contains R:
            test/count P_R
at each leaf ℓ:
    test/count P_ℓ
```

(b)

**Figure 2.** Modified preprocessing algorithm and query algorithm for halfspaces.

and $P_\ell$, or the time required to maintain and test these subsets.

**Lemma 3.1.** *Every partition graph has the following properties.*

(a) *If a set $P_R$, $P_R^+$, or $P_R^-$ is tested during an emptiness query or counted during a counting query, then the query halfspace contains every point in that set.*

(b) *If a point lies in a query range, it also lies in one of the sets $P_R$, $P_R^+$, $P_R^-$, or $P_\ell$ counted during a counting query. Thus, the output of a counting query is never smaller than the number of points in the query range.*

(c) *An emptiness query never reports that a non-empty query range is empty.*

(d) *Exactly the same edges are traversed during an emptiness query for an empty hyperplane, a counting query for the same hyperplane, and a counting query for either of its halfspaces.*

(e) *Exactly the same edges are traversed during an emptiness query for an empty halfspace and a counting query for the same halfspace.*

**Proof:** These properties either follow directly from definitions or are easily verified by induction. We omit details from this extended abstract. □

We say that a partition graph *supports* a particular type of range query for a given set of points if, after the points are preprocessed, the appropriate query algorithm is always correct. Even though we have a single preprocessing algorithm, a single partition graph need not support all types of queries. However, Lemma 3.1 implies that a partition graph that support one type of query may automatically support other types of queries, with the same or smaller worst-case query time. For example, if a partition graph supports hyperplane or halfspace counting queries, then it also supports hyperplane emptiness queries.

## 4 Hyperplane Emptiness Queries

### 4.1 "Trivial" Lower Bounds

**Theorem 4.1.** *Any partition graph that supports hyperplane emptiness, hyperplane counting, or halfspace counting queries has size $\Omega(n)$, preprocessing time $\Omega(n \log n)$, and worst-case query time $\Omega(\log n)$.*

**Proof:** It suffices to consider hyperplane emptiness queries, since Lemma 3.1 implies that the same lower bounds for the other two types of queries.

Let P be a set of $n$ points all lying on a vertical line $\ell$, and let H be a set of $n$ hyperplanes normal to $\ell$, with each hyperplane just above one of the points in P. Any partition graph that correctly answers hyperplanes queries for P must at least report that every hyperplane in H is empty.

For each point in P, call the hyperplane in H just above it its *partner*. We say that a point is *active* at a particular node $v$ of the partition graph if both the point and its partner reach $v$. We say that a node $v$ *splits* a point $p$ and a hyperplane $h$ if $p$ and $h$ both reach $v$ but no edge out of $v$ is traversed by both $p$ and $h$.

For any primal node $v$ and any query region $R \in \mathcal{R}_v$, there is at most one active point that lies in R whose partner does not intersect R. For any dual node $v$ and any query region $R \in \mathcal{R}_v$, there is at most one active point whose dual hyperplane misses R and whose partner's dual point lies in R. Thus, any node splits at most $\Delta$ points from their partners. Moreover, since every point in P must be split from its partner, there must be at least n query regions, so the partition graph must have at least n edges.

The *level* of a node is its distance from the root. There are at most $\Delta^k$ nodes at level k. At least $n - \sum_{i=0}^{k-1} \Delta^{k+1} \geq n - \Delta^{k+2}$ points are active at some node at level k. In particular, at least $n(1-1/\Delta)$ points are active at level $\lfloor \log_\Delta n - 3 \rfloor$. It follows that at least $n(1-1/\Delta)$ points in P each traverse at least $\lfloor \log_\Delta n - 2 \rfloor$ edges, so the total preprocessing time is at least

$$n(1 - 1/\Delta)\lfloor \log_\Delta n - 2 \rfloor = \Omega(n \log n).$$

Moreover, at least $n(1 - 1/\Delta)$ hyperplanes in H each traverse at least $\lfloor \log_\Delta n - 2 \rfloor$ edges, so the worst-case query time is $\Omega(\log n)$. □

## 4.2 Space/Time Tradeoffs

The next result relates the complexity of range queries in the Fredman/Yao semigroup arithmetic model and counting queries in the partition graph model. Recall that a storage scheme is a "data structure" in the semigroup model.

**Theorem 4.2.** *Let P be a set of n points in $\mathbb{R}^d$. Given a partition graph of size s that supports hyperplane (resp. halfspace) counting queries in time t, we can construct a storage scheme of size $O(s)$ that supports hyperplane (resp. halfspace) queries in time $O(t)$.*

**Proof:** For each query region R and leaf $\ell$, define the subsets $P_R$, $P_R^-$, $P_R^+$, and $P_\ell$ by the preprocessing algorithm in Figure 2. Each of these subsets defines a generator in the semigroup model. There are at most 3s of these generators: at most two for each of the s query regions, plus one for each of the $\leq s$ leaves.

Suppose the partition graph supports hyperplane counting queries. Lemma 3.1(b) implies that for any hyperplane h, the set of points $P \cap h$ is counted as the disjoint union of several sets $P_\ell$, one for every leaf reached by the query algorithm. Since the query algorithm reaches at most t leaves, he set $P \cap h$ is the union of at most t clusters.

If the partition graph supports halfspace counting queries, then for any halfspace $h^\pm$, the points $P \cap h^\pm$ are counted as the disjoint union of at most t subsets $P_R$, at most $\Delta t$ subsets $P_R^\pm$, and at most t subsets $P_\ell$. Thus, the set $P \cap h^\pm$ is the union of at most $(2 + \Delta)t$ clusters. □

This theorem implies that lower bounds for range queries in the semigroup arithmetic model are also lower bounds for the corresponding counting queries in the partition graph model. Theorems 2.1 and 2.2 now immediately imply the following lower bounds.

**Corollary 4.3.** *Let P be a uniformly generated set of n points in $[0,1]^d$. With high probability, any partition graph of size s that supports halfspace counting queries for P in time t satisfies the inequality $st^d = \Omega((n/\log n)^{d-(d-1)/(d+1)})$.*

**Corollary 4.4.** *Any partition graph of size s that supports d-dimensional hyperplane counting queries in time t must satisfy the inequality $st^{d(d+1)/2} = \Omega(n^d)$ in the worst case.*

One way to determine if a hyperplane is empty is by counting the points in its two halfspaces. Thus, any halfspace counting data structure also supports hyperplane emptiness queries. The following result implies that, in our model of computation, the reverse is almost true as well: Any partition graph that supports hyperplane emptiness queries also supports halfspace counting queries, at least over certain semigroups.

**Theorem 4.5.** *Let P be a set of n points in $\mathbb{R}^d$. Given a partition graph of size s that supports hyperplane emptiness queries in time t, we can construct a storage scheme of size $O(s)$ that supports halfspace queries in time $O(t)$.*

**Proof:** Fix a partition graph that supports hyperplane emptiness queries for a set P of n points. For each query region R, define the subsets $P_R$, $P_R^-$, and $P_R^+$ by the preprocessing algorithm in Figure 2.

Each of the subsets $P_R$, $P_R^+$, and $P_R^-$ defines a cluster. There are at most 2s of these clusters, where s is the size of the partition graph. We claim that these clusters define a storage scheme that supports halfspace queries in time at most $\Delta t$, where t is the worst-case time for a hyperplane emptiness query. We emphasize that the partition graph is *not* required to support halfspace queries.

Let h be an empty hyperplane. If we run the counting query algorithm for the upper halfspace $h^+$, the output is the sum of the sizes of several clusters $P_R$ and $P_R^+$. There are at most $\Delta t$ such subsets, where t is the counting query time for the halfspace $h^+$, which by Lemma 3.1(d) is also the emptiness query time for the hyperplane h.

Even though the output of the counting query may be incorrect, we claim that the union of these $\Delta t$ clusters is exactly $P \cap h^+$. Lemma 3.1(a) implies that every point in these clusters is in $h^+$. Lemma 3.1(b) implies that if some point $p \in P \cap h^+$ is not in one of these clusters, then there must be a leaf that is reached while preprocessing p and also reached while querying $h^+$. In that case, Lemma 3.1(d) implies that the same leaf is reached while querying h, which means an emptiness query for h would incorrectly report that h is non-empty.

Lower halfspaces are handled symmetrically. We do not need to consider halfspaces bounded by non-empty hyperplanes, since there is always a halfspace with empty boundary that contains the same points. □

Note that the clusters that form $P \cap h^+$ are not necessarily disjoint. Thus, unlike the storage scheme constructed in Theorem 4.2, the storage scheme we construct here is not correct for all faithful semigroups. It is correct for the semigroups $(\mathbb{Z}, \max)$ and $(\{\text{true}, \text{false}\}, \vee)$, or any other semigroup in which the equation $x + x = x$ always holds.

The following lower bound now follows immediately from Theorem 2.1.

**Corollary 4.6.** *Let P be a uniformly generated set of n points in $[0,1]^d$. With high probability, any partition graph of size s that supports hyperplane emptiness queries for P in time t satisfies the inequality $st^d = \Omega((n/\log n)^{d-(d-1)/(d+1)})$.*

Again, Lemma 3.1 implies that the same lower bound applies to hyperplane counting queries. This improves the lower bound in Corollary 4.4 whenever $s = O(n^{d-1})$ or $t = \Omega(n^{2/d(d+1)})$. Lemma 3.1 also implies that the lower bound applies to halfspace counting queries, giving us an roundabout proof of Corollary 4.3.

## 4.3 Preprocessing/Query Tradeoffs

In a few cases, we can establish tighter tradeoffs by considering preprocessing time instead of space. These tradeoffs follow from offline lower bounds for Hopcroft's problem and its generalizations in an offline model of computation based on partition graphs [19, 18]. A *partitioning algorithm*, given a set of points and hyperplanes as input, constructs a partition graph, preprocesses the points, and queries all the hyperplanes (or equivalently, preprocesses the hyperplanes and queries the points).

**Theorem 4.7.** *Any partition graph that supports line queries in time t after preprocessing time $p = O(n^2)$ satisfies the inequality $pt^2 = \Omega(n^2)$, in the worst case.*

**Proof:** Suppose $p < n^2$, since otherwise there is nothing to prove. Let $m = cp^{3/2}/n$, where c is a constant to be specified later. Note that $m = O(n^2)$, and since $p = \Omega(n \log n)$, we also have $m = \Omega(n^{1/2} \log^{1/2} n)$. There is a set of n points and m lines such that for any partition graph, the total time required to preprocess the n points and correctly answer the m line queries is at least $\alpha n^{2/3} m^{2/3} = \alpha c^{2/3} p$ for some positive constant $\alpha$ [19]. If we choose $c = (2/\alpha)^{3/2}$, the total query time is at least p. Thus, at least one query requires time at least $p/m = \Omega(n/p^{1/2})$. □

Using similar techniques, we can improve this lower bound slightly in three or more dimensions.

**Theorem 4.8.** *Any partition graph, each of whose query regions is a constant complexity polyhedron, that supports d-dimensional plane queries in time t after preprocessing time $p = O(n^d)$ satisfies the inequalities $pt^{(d+2)(d-1)/2} = \Omega(n^d)$ and $pt^{2/(d-1)} = \Omega(n^{(d+2)/d})$ in the worst case. When $d = 3$, these lower bounds hold for arbitrary partition graphs.*

**Proof:** The tradeoffs follow from the offline lower bound $\Omega(n^{1-2/d(d+1)}m^{2/(d+1)} + n^{2/(d+1)}m^{1-2/d(d+1)})$, established in [19] for $d = 3$ and in [18] for $d > 3$, using exactly the same argument as Theorem 4.7. □

Although in general these bounds are far from optimal, there are two interesting special cases that match known upper bounds up to polylogarithmic factors.

**Corollary 4.9.** *Any partition graph, each of whose query regions is a constant-complexity polyhedron, that supports d-dimensional hyperplane emptiness queries after $O(n \operatorname{polylog} n)$ preprocessing time requires query time $\Omega(n^{(d-1)/d}/\operatorname{polylog} n)$ in the worst case.*

**Corollary 4.10.** *Any partition graph, each of whose query regions is a constant-complexity polyhedron, that supports d-dimensional hyperplane emptiness queries in time $O(\operatorname{polylog} n)$, requires preprocessing time $\Omega(n^d/\operatorname{polylog} n)$ in the worst case.*

# 5 Halfspace Emptiness Queries

The space and time bounds for the best known hyperplane (or simplex) emptiness query data structures are only a polylogarithmic factor smaller than the bounds for hyperplane (or simplex) counting queries. The situation is entirely different for halfspace queries. The best known halfspace counting data structure requires roughly $\Omega(n^d)$ space to achieve logarithmic query time [10, 25]; whereas, the same query time can be achieved

|  | Space | Preprocessing | Query Time | Source |
|---|---|---|---|---|
| $d = 2,3$ | $O(n)$ | $O(n \log n)$ | $O(\log n)$ | [13, 3, 16] |
| $d \geq 4$ | $O(n^{\lfloor d/2 \rfloor}/\log^{\lfloor d/2 \rfloor} n)$ | $O(n^{\lfloor d/2 \rfloor}/\log^{\lfloor d/2 \rfloor - \varepsilon} n)$ | $O(\log n)$ | [27] |
|  | $O(n)$ | $O(n^{1+\varepsilon})$ | $O(n^{1-1/\lfloor d/2 \rfloor} 2^{O(\log^* n)})$ | [22] |
|  | $O(n)$ | $O(n \log n)$ | $O(n^{1-1/\lfloor d/2 \rfloor} \operatorname{polylog} n)$ | [27] |
|  | $n \leq s \leq n^{\lfloor d/2 \rfloor}$ | $O(s \operatorname{polylog} n)$ | $O((n \operatorname{polylog} n)/s^{1/\lfloor d/2 \rfloor})$ | [27] |

**Table 2.** Best known upper bounds for halfspace emptiness queries.

with $o(n^{\lfloor d/2 \rfloor})$ space if we only want to know whether the halfspace is empty [27]. Table 2 lists the resource requirements for the best known halfspace emptiness data structures.

In this section, we derive lower bounds on the complexity of halfspace emptiness queries. To prove our results, we use the following reduction argument to transform hyperplane queries into halfspace queries in a higher-dimensional space. (A similar transformation is described in [17].)

Define the function $\sigma_d : \mathbb{R}^{d+1} \to \mathbb{R}^{\binom{d+2}{2}}$ as follows:

$$\sigma_d(x_0, x_1, \ldots, x_d) = (x_0^2, x_1^2, \ldots, x_d^2, \sqrt{2}x_0x_1, \sqrt{2}x_0x_2, \ldots, \sqrt{2}x_{d-1}x_d)$$

This map has the property that $\langle \sigma_d(p), \sigma_d(h) \rangle = \langle p, h \rangle^2$ for any $p, h \in \mathbb{R}^{d+1}$, where $\langle \cdot, \cdot \rangle$ denotes the usual inner product. In a more geometric setting, $\sigma_d$ maps points and hyperplanes in $\mathbb{R}^d$, represented in homogeneous coordinates, to points and hyperplane in $\mathbb{R}^{d(d+3)/2}$, also represented in homogeneous coordinates. For any point $p$ and hyperplane $h$ in $\mathbb{R}^d$, the point $\sigma_d(p)$ is contained in the hyperplane $\sigma_d(h)$ if and only if $p$ is contained in $h$; otherwise, $\sigma_d(p)$ is strictly above $\sigma_d(h)$. Thus, a hyperplane $h$ intersects a point set $P$ if and only if the closed lower halfspace $\sigma_d(h)^-$ intersects the lifted point set $\sigma_d(P)$. In other words, any (lower) halfspace emptiness data structure for $\sigma_d(P)$ is also a hyperplane emptiness data structure for $P$.

Unfortunately, this is not quite enough to give us our lower bounds, since the reduction does not preserve the model of computation. Specifically, the query regions in a partition graph used to answer d-dimensional queries *must* be subsets of $\mathbb{R}^d$. To complete the reduction, we need to show that the $d(d+3)/2$-dimensional partition graph can be "pulled back" to a d-dimensional partition graph.

In order for such a transformation to be possible, we need to restrict the query regions allowed in our partition graphs. A *Tarski cell* is a semialgebraic set defined by a constant number of constant degree polynomial equalities and inequalities. Every Tarski cell has a constant number of connected components, and the intersection of any two Tarski cells is another Tarski cell (with larger constants). A *semialgebraic* partition graph is a partition graph whose query regions are all Tarski cells.

**Theorem 5.1.** *Let $P$ be a set of points in $\mathbb{R}^d$, and let $\hat{P} = \sigma_d(P)$ be the lifted set of points in $\mathbb{R}^{d(d+3)/2}$. Given a semialgebraic partition graph $\hat{G}$ that supports halfspace emptiness queries over $\hat{P}$, we can construct a semialgebraic partition graph $G$ that supports hyperplane emptiness queries over $P$, with (asymptotically) the same space, preprocessing time, and query time bounds.*

**Proof:** We actually prove a stronger theorem, by assuming only that $\hat{G}$ supports emptiness queries for hyperplanes of the form $\sigma_d(h)$. Since no point in $\hat{P}$ is ever below such a hyperplane, any partition graph that supports lower halfspace emptiness queries also supports our restricted class of hyperplane emptiness queries. As noted above, the queries we consider are equivalent to hyperplane emptiness queries over the original point set $P$.

Given $\hat{G}$, we construct $G$ as follows. $G$ has the same set of nodes as $\hat{G}$, but with different query regions. Since each query region $\hat{R}$ in the original partition graph $\hat{G}$ is a Tarski cell, it intersects the algebraic surface $\sigma_d(\mathbb{R}^d)$ in a constant number of connected components $\hat{R}_1, \hat{R}_2, \ldots, \hat{R}_\delta$, where the constant $\delta$ depends on the number and degree of the inequalities that define $\hat{R}$. The query regions in $G$ are the preimages $R_i = \sigma_d^{-1}(\hat{R}_i)$ of these components. See Figure 3.

The edge associated with each d-dimensional query region $R_i$ has the same endpoints as the edge associated with the original query region $\hat{R}$. Thus, there may be several edges in $G$ with the same source and target, but this is perfectly legal. If a query region $\hat{R}$ doesn't intersect $\sigma_d(\mathbb{R}^d)$, then the corresponding edge in $\hat{G}$ is not represented in $G$ at all, so $G$ may not be a connected graph. Nodes in $G$ that are not connected to the root can be safely discarded. The size, preprocessing time, and query time for $G$ are clearly at most a constant factor more than the corresponding resources for $\hat{G}$.

The leaf subsets $P_\ell$ in $G$ cannot be larger than the corresponding subsets $\hat{P}_\ell$ in $\hat{G}$. (They might be smaller,

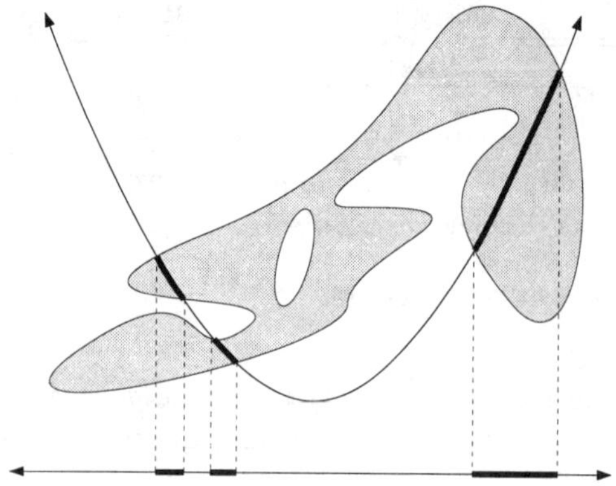

**Figure 3.** Each Tarski cell induces a constant number of lower-dimensional query regions.

but that only helps us.) Similarly, a hyperplane query cannot reach more leaves in G than the corresponding query reaches in $\hat{G}$. It follows that G supports hyperplane emptiness queries — for any hyperplane h, if $\hat{G}$ reports that $\sigma_d(h)$ is empty, G (correctly) reports that h is empty. □

We emphasize that some restriction on the query regions is necessary to prove any nontrivial lower bounds. There is a partition graph of constant size, requiring only linear preprocessing, that supports halfspace emptiness queries in constant time. The graph consists of a single primal node with two query regions — the convex hull of the points and its complement — and two leaves. On the other hand, our restriction to Tarski cells is stronger than necessary. It suffices that every query region intersects (some projective transformation of) the surface $\sigma_d(\mathbb{R}^d)$ in a constant number of connected components.

The following corollaries are now immediate.

**Corollary 5.2.** *Any semialgebraic partition graph that supports d-dimensional halfspace emptiness queries, for any $d \geq 2$, has size $\Omega(n)$, preprocessing time $\Omega(n \log n)$, and worst-case query time $\Omega(\log n)$.*

**Corollary 5.3.** *Any semialgebraic partition graph of size s that supports $d(d+3)/2$-dimensional halfspace emptiness queries in time t satisfies the inequality $st^d = \Omega((n/\log n)^{d-(d-1)/(d+1)})$ in the worst case.*

**Corollary 5.4.** *Any semialgebraic partition graph that supports 5-dimensional halfspace emptiness queries in time t after preprocessing time $p = O(n^2)$ satisfies the inequality $pt^2 = \Omega(n^2)$ in the worst case.*

**Corollary 5.5.** *Any semialgebraic partition graph that supports 9-dimensional halfspace emptiness queries in time t after preprocessing time $p = O(n^3)$ satisfies the inequalities $pt^5 = \Omega(n^3)$ and $pt = \Omega(n^{5/3})$ in the worst case.*

Theorem 5.1 does not imply better preprocessing/query tradeoffs for halfspace emptiness queries in higher dimensions, since the corresponding hyperplane results require polyhedral query regions. Marginally better lower bounds can be obtained in dimensions 14(!) and higher, using the same arguments as in Theorem 4.8, using the lower bounds for offline halfspace emptiness established in [18]. However, since these lower bounds are far from optimal, we omit further details.

The lower bounds in Corollary 5.2 are optimal when $d \leq 3$, and the bounds in Corollary 5.4 are optimal up to polylogarithmic factors.

## 6 Conclusions

We have presented lower bounds on the complexity of hyperplane and halfspace emptiness queries. Our lower bounds apply to a broad class of range query data structures based on recursive decomposition of primal and/or dual space.

The lower bounds we developed for counting queries actually apply to any type of query where the points in the query range are required as the union of several subsets. For example, simplex range searching data structures are typically constructed by composing several levels of halfspace "counting" data structures [25]. To answer a query for the intersection of k halfspaces, the points in the first halfspace are extracted as the (typically disjoint) union of several subsets, and then a $(k-1)$-halfspace query is performed on each subset.

With a few (important!) exceptions, our lower bounds are far from the best known upper bounds, and a natural open problem is to close the gap. In particular, we have only "trivial" lower bounds for four-dimensional halfspace emptiness queries. We conjecture that the correct space-time tradeoffs are $st^d = \Theta(n^d)$ for hyperplanes and $st^{\lfloor d/2 \rfloor} = \Theta(n^{\lfloor d/2 \rfloor})$ for halfspaces. Since these bounds are achieved by current algorithms—exactly for hyperplanes [25], within polylogarithmic factors for halfspaces [27]—the only way to prove our conjecture is to improve the lower bounds.

Our space-time tradeoffs derive from lower bounds for halfspace queries in the semigroup arithmetic model [6], and our preprocessing-query tradeoffs follow from lower bounds for offline range searching problems such as Hopcroft's problem [19, 18]. Any improvements to those lower bounds would improve our results as well.

The best known data structures for d-dimensional hyperplane emptiness queries and 2d- or $(2d+1)$-

dimensional halfspace emptiness queries have the same resource bounds. We conjecture that this is also true of *optimal* data structures for these problems. Is there a reduction from hyperplane queries to halfspace queries that only increases the dimension by a constant factor (preferably two)?

Finally, can our techniques be applied to other closely related problems, such as linear programming queries [23, 7] and ray shooting queries [2, 12, 24, 27]?

## Acknowledgments

I thank Pankaj Agarwal for suggesting this problem for study.

## References

[1] P. K. Agarwal. Range searching. Report CS-1996-05, Dept. Comput. Sci., Duke Univ., 1996. http://www.cs.duke.edu/~pankaj/papers/range-survey.ps.gz. To appear in *The CRC Handbook of Discrete and Computational Geometry*, J. E. Goodman and J. O'Rourke, editors, CRC Press.

[2] P. K. Agarwal and J. Matoušek. Ray shooting and parametric search. *SIAM J. Comput.* 22(4):794–806, 1993.

[3] A. Aggarwal, M. Hansen, and T. Leighton. Solving query-retrieval problems by compacting Voronoi diagrams. *Proc. 22nd Annu. ACM Sympos. Theory Comput.*, pp. 331–340, 1990.

[4] M. Ben-Or. Lower bounds for algebraic computation trees. *Proc. 15th Annu. ACM Sympos. Theory Comput.*, pp. 80–86, 1983.

[5] M. de Berg, M. Overmars, and O. Schwarzkopf. Computing and verifying depth orders. *SIAM J. Comput.* 23:437–446, 1994.

[6] H. Brönnimann, B. Chazelle, and J. Pach. How hard is halfspace range searching? *Discrete Comput. Geom.* 10:143–155, 1993.

[7] T. M. Chan. Fixed-dimensional linear programming queries made easy. *Proc. 12th Annu. ACM Sympos. Comput. Geom.*, pp. 284–290, 1996.

[8] T. M. Chan. Output-sensitive results on convex hulls, extreme points, and related problems. *Discrete Comput. Geom.* 16:369–387, 1996.

[9] B. Chazelle. Lower bounds on the complexity of polytope range searching. *J. Amer. Math. Soc.* 2:637–666, 1989.

[10] B. Chazelle. Cutting hyperplanes for divide-and-conquer. *Discrete Comput. Geom.* 9(2):145–158, 1993.

[11] B. Chazelle, H. Edelsbrunner, L. Guibas, and M. Sharir. Algorithms for bichromatic line segment problems and polyhedral terrains. *Algorithmica* 11:116–132, 1994.

[12] B. Chazelle and J. Friedman. Point location among hyperplanes and unidirectional ray-shooting. *Comput. Geom. Theory Appl.* 4:53–62, 1994.

[13] B. Chazelle, L. J. Guibas, and D. T. Lee. The power of geometric duality. *BIT* 25:76–90, 1985.

[14] B. Chazelle and B. Rosenberg. Simplex range reporting on a pointer machine. *Comput. Geom. Theory Appl.* 5:237–247, 1996.

[15] B. Chazelle. Lower bounds for off-line range searching. *Discrete Comput. Geom.* 17(1):53–66, 1997.

[16] D. P. Dobkin and D. G. Kirkpatrick. Determining the separation of preprocessed polyhedra – a unified approach. *Proc. 17th Internat. Colloq. Automata Lang. Program.*, pp. 400–413. Springer-Verlag, Lecture Notes in Computer Science 443, 1990.

[17] R. Dwyer and W. Eddy. Maximal empty ellipsoids. *Internat. J. Comput. Geom. Appl.* 6:169–186, 1996.

[18] J. Erickson. New lower bounds for halfspace emptiness. *Proc. 37th Annu. IEEE Sympos. Found. Comput. Sci.*, pp. 472–481, 1996.

[19] J. Erickson. New lower bounds for Hopcroft's problem. *Discrete Comput. Geom.* 16:389–418, 1996.

[20] M. L. Fredman. Lower bounds on the complexity of some optimal data structures. *SIAM J. Comput.* 10:1–10, 1981.

[21] J. Matoušek. Efficient partition trees. *Discrete Comput. Geom.* 8:315–334, 1992.

[22] J. Matoušek. Reporting points in halfspaces. *Comput. Geom. Theory Appl.* 2(3):169–186, 1992.

[23] J. Matoušek. Linear optimization queries. *J. Algorithms* 14:432–448, 1993.

[24] J. Matoušek. On vertical ray shooting in arrangements. *Comput. Geom. Theory Appl.* 2(5):279–285, Mar. 1993.

[25] J. Matoušek. Range searching with efficient hierarchical cuttings. *Discrete Comput. Geom.* 10(2):157–182, 1993.

[26] J. Matoušek. Geometric range searching. *ACM Comput. Surv.* 26:421–461, 1994.

[27] J. Matoušek and O. Schwarzkopf. On ray shooting in convex polytopes. *Discrete Comput. Geom.* 10(2):215–232, 1993.

[28] J. Pach and P. K. Agarwal. *Combinatorial Geometry*. John Wiley & Sons, New York, NY, 1995.

[29] F. P. Preparata and M. I. Shamos. *Computational Geometry: An Introduction*. Springer-Verlag, New York, NY, 1985.

[30] R. E. Tarjan. A class of algorithms which require nonlinear time to maintain disjoint sets. *J. Comput. Syst. Sci.* 18:110–127, 1979.

[31] A. C. Yao. On the complexity of maintaining partial sums. *SIAM J. Comput.* 14:277–288, 1985.

# Distribution of Distances and Triangles in a Point Set and Algorithms for Computing the Largest Common Point Sets

Tatsuya Akutsu*
Human Genome Center, Institute of Medical Science, University of Tokyo, Tokyo 108, JAPAN
email:takutsu@ims.u-tokyo.ac.jp

Hisao Tamaki Takeshi Tokuyama
IBM Tokyo Research Laboratory, Yamato, Kanagawa 242, JAPAN
email:{htamaki, ttoku}@trl.ibm.co.jp

**Abstract.** This paper considers the following problem which we call the largest common point set problem (LCP): given two point sets $P$ and $Q$ in the Euclidean plane, find a subset of $P$ with the maximum cardinality which is congruent to some subset of $Q$. We introduce a combinatorial-geometric quantity $\lambda(P,Q)$, which we call the inner product of the distance-multiplicity vectors of $P$ and $Q$, show its relevance to the complexity of of various algorithms for LCP, and give a non-trivial upper bound on $\lambda(P,Q)$. We generalize this notion to higher dimensions, give some upper bounds on the quantity, and apply them to algorithms for LCP in higher dimensions. Along the way, we prove a new upper bound on the number of congruent triangles in a point set in the four-dimensional space.

## 1 Introduction

Matching of point sets in the plane or space is an important problem that arises, in various forms, in computer vision [15] and computational biology [2, 28, 21]. In the simplest form of the problem, we are asked if two finite sets of points $P$ and $Q$ are congruent, i.e., there is a transformation $T$ consisting of translation, rotation, and possibly reflection such that $T(P) = Q$ [5, 25, 4]. A generalization of this question is to ask if $Q$ is congruent to some subset of $P$ (*congruent copy detection* or CCD for short)[7, 13, 24]. We consider in this paper a yet further generalization: we ask for a set $R$ of the maximum cardinality that is congruent to some subset of $P$ and to some subset of $Q$ at the same time. We call such $R$ the *largest common point set* between $P$ and $Q$ and this problem the *largest common point set problem* or LCP. A version of LCP, in which congruence is replaced by similarity (i.e., we allow scaling as well) is previously considered by [16]. We set $n = |P|$, $m = |Q|$ and assume without loss of generality that $n \geq m$.

*Partially supported by the Grant-in-Aid for Scientific Research on Priority Areas, "Genome Informatics", of the Ministry of Education, Science and Culture of Japan.



To solve LCP, we consider two standard schemes for point matching problems, namely *voting* and *alignment*. We consider both deterministic and randomized algorithms based on these schemes. Since the randomization techniques we use are also rather standard, our focus is on combinatorial-geometric analyses of the performance of those algorithms. In particular, we study a quantity which we call the *inner product of distance-multiplicity vectors* of two point sets. This quantity is naturally related to the running time of the algorithms for LCP we consider. We derive a new upper bound on this quantity and hence new upper bounds for LCP.

Given a set $P$ of $n$ points in the plane and a positive real $l$, we define the *multiplicity* of distance $l$ in $P$, denoted by $H_P[l]$, to be the number of ordered pairs $(p, q)$ of $P$ such that the distance between $p$ and $q$ is $l$. The *distance-multiplicity vector* of $P$, denoted by $M(P)$, is the vector $(H_P[l_1], \ldots, H_P[l_\nu])$ where $l_1, \ldots, l_\nu$ is the list of distinct distances with positive multiplicities, sorted in a non-increasing order of their multiplicities in $P$. Erdös [10] asked for theoretical bounds on $\nu = \nu(P)$, the number of distinct distances in $P$, and $H_P[l_1]$, the largest multiplicity of a distance in $P$; currently, the best known upper bound on $H_P[l_1]$ is $O(n^{4/3})$ and the best known lower bound on $\nu(P)$ is $\Omega(n^{4/5})$ [26, 27].

Given two point sets $P$ and $Q$, we define $\lambda(P,Q) = \sum_l H_P[l]H_Q[l]$, where the summation is taken over all distances occurring in $P$ or in $Q$. We call $\lambda(P,Q)$ the "inner product" of the distance-multiplicity vectors $M(P)$ and $M(Q)$, a natural terminology if we regard $M(P)$ as a finite representation of a vector in $(\mathbf{Z}^+)^{\mathbf{R}}$. Since $\lambda(P,Q)$ represents the number of matches between the distances in $P$ and those in $Q$, it is not difficult to imagine that this quantity may play an essential role in the complexity analysis of algorithms based on distance comparisons — the more distance matches we have, the more potential congruences we have to consider. We will see that this is indeed the case for both voting- and alignment-based algorithms.

Let $\lambda(n,m) = \max_{|P|=n,|Q|=m} \lambda(P,Q)$. A trivial upper bound on $\lambda(n,m)$ is $O(n^2m^2)$, while a result of Erdös[10] implies that $\lambda(n,m) \geq \Omega(nm^2\sqrt{\log m})$. The above known bound of $O(n^{4/3})$ on the maximum multiplicity of a distance immediately implies that $\lambda(n,m) = O(n^{4/3}m^2)$. Our result is an upper bound of $O(n^{1.43}m^{1.77})$, which is stronger for all values of $n \leq m^{2.3}$. This bound is proved by extending the technique of Székely, who has recently developed and simplified proofs of the $O(n^{4/3})$ bound on the maximum multiplicity and the $\Omega(n^{4/5})$ bound on the number of distinct distances. We remark that further improving our upper bound, especially the special case of $O(n^{3.2})$ on $\lambda(n,n)$, is arguably difficult, because it would mean improving the best known lower bound $\Omega(n^{4/5})$ on $\nu(P)$, a long standing open problem. Indeed, if there is a point set $P$ with only $O(n^{4/5})$ distinct distances, then $\lambda(P,P)$ is $\Omega(n^{6/5}(n^{4/5})^2) = \Omega(n^{3.2})$.

Similarly to the importance of the distance multiplicities in the plane, the multiplicity of distances and triangles plays an important role in the analysis of LCP algorithms in the three-dimensional space. Here, the multiplicity of a triangle $ABC$ (not necessarily degenerate) in a point set $P$ is defined to be the number of ordered triples $(p,q,r)$ of $P$ such that $pqr$ is congruent to $ABC$. Triangle-multiplicity vectors and their inner-products are defined analogously to distance-multiplicity vectors. Given point sets $P$ and $Q$ in the three-dimensional space, let $\lambda^{(3,1)}(P,Q)$ ($\lambda^{(3,2)}(P,Q)$, resp.) denote the inner-product of the distance-multiplicity (triangle multiplicity, resp.) vectors of $P$ and $Q$. Let $\lambda^{(3,1)}(n,m)$ and $\lambda^{(3,2)}(n,m)$ be defined as before, taking the maximum over $P$ and $Q$ with $|P| = n$ and $|Q| = m$. Our upper bounds are $\lambda^{(3,1)}(n,m) = n^{3/2}m^2(\log^* n)^{O(1)}$ and $\lambda^{(3,2)}(n,m) = \min\{n^{1.8}m^3, n^{4.6875}\}(\log^* n)^{O(1)}$.

The first bound is trivially implied by the known bound on the multiplicity of a distance in three dimensions[8]; the second bound comes from the incidence bound of points and circles (with multiplicities) in the space.

These notions naturally carry over to higher dimensions: for a points set in the $d$-dimensional space, we consider the multiplicity of a $i$-dimensional simplex in the point set, where $1 \leq i \leq d-1$, and define $\lambda^{(d,i)}(P,Q)$ to be the inner-product of the multiplicity vectors of $P$ and $Q$ as before.

We only have a naive (though not entirely trivial) bound of $O(n^{d-1}m^d)$ on $\lambda^{(d,d-1)}$ for $d \geq 5$. For $d = 4$, we have a slightly stronger bound of $O(n^{2.87}m^4)$ on $\lambda^{(4,3)}$. This is proved by showing that, for any set $P$ of $n$ points in the four-dimensional space, the number of triangles in $P$ congruent to a given triangle is at most $O(n^{2.83})$.

From a practical point of view, one might question the relevance of the running time upper bounds of the LCP algorithms obtained from our combinatorial-geometric analyses, because they depend heavily on the exactness of the input and computation. Because of the inevitable errors in the input and computation, several authors formulate point matching problems as that of approximate matching[4, 7, 13]. For example, CCD is formulated as asking for a transformation $T$ that minimizes the directed Hausdorff distance from $T(Q)$ to $P$, i.e., $\max_{q\in Q}\min_{p\in P} dist(T(q),p)$ [7]. Although this is a mathematically clean formulation and adopted by many authors, the time complexity of solving CCD in this model is rather high (the best known upper bound for the two-dimensional case is $O(n^2m^3\log^2 n)$ [7]). Goodrich *et al.* [13] propose to give up exact minimization and to look for a transformation that achieves a Hausdorff distance that is within some constant times the minimum possible, obtaining an algorithm with a reduced running time of $O(n^2m\log^2 n)$ . It is not clear, however, if this approach can be extended to general LCP to produce algorithms with a practical running time. In fact, it is not even clear how we formulate LCP in this model, since we have two quantities to optimize (the subset size and the Hausdorff distance between the subsets).

A more heuristic approach for approximate matching is to take an algorithm for the exact matching model and use its approximate version, where approximate equality is used instead of exact equality to test matches. This is the approach taken by [16] and probably the one preferred in practice. The generalized Hough transformation method[6, 17] commonly used in computer vision may be interpreted as belonging to such an approach. Our position is that it is meaningful to analyze idealized algorithms as long as their approximate versions are used in practice. The hope that such an analysis will tell us something about the performance of the approximate versions is only heuristic, as much as the hope that such approximate versions produce answers of any well-defined correctness.

Table 1 summarizes the algorithmic results of this paper, where we omit logarithmic factors and $K$ denotes the size of the largest common point set. We have included some results on the largest similar point set problem (LSP) and the similar copy detection problem (SCD), where similarity rather than congruence is the condition for matches, which can be obtained in a similar approach. Note that our SCD results improve the $O(n^dm)$ time bound of Rezende and Lee [24] in three or more dimensions (and when specialized to CCD, in two dimensions as well).

The rest of this paper is organized as follows. In Section 2, we review the algorithms based on voting and alignment and see how the inner-product of multiplicity vectors appears in their analysis. The combinatorial-geometric analyses of the inner-products is given in the subsequent sections.

Table 1: Time complexities for LCP problems

| Dimension | 2 | 3 | 4 | $\geq 5$ |
|---|---|---|---|---|
| LCP | $n^{1.43}m^{1.77} + n^2$ | $\min\{n^{1.8}m^3 + n^3, n^{4.7}\}$ | $n^{2.87}m^4 + n^4$ | $n^{d-1}m^d + n^d$ |
| LSP | $n^2m^2$(trivial) | $n^{7/3}m^{8/3} + n^3$ | $n^3m^4 + n^4$ | $n^{d-1}m^d + n^d$ |
| LCP(Rand'zd) | $LCP/K$ | $LCP/K^2$ | NA | NA |
| $\epsilon$-approx | $LCP/K^2$ | $LCP/K^3$ | NA | NA |
| CCD | $\min\{n^{1.43}m^{0.77}, n^{4/3}m\}$ | $n^{1.8}m + \min\{n^{2.5}, n^3m^{-2}\}$ | $n^{2.87}m + n^{3.83}$ | $n^{d-1}m + n^d$ |
| SCD | $n^2m$ [24] | $n^{7/3}m^{2/3} + n^3m^{-2}$ | $n^3m + n^4$ | $n^{d-1}m + n^d$ |

## 2 Algorithms

The goal of this section is to motivate our combinatorial-geometric analysis of the inner product of the multiplicity vectors by showing how it is related to the analysis of several algorithms for LCP. We will consider algorithms based on a voting scheme and algorithms based on an alignment scheme. In the description of the following algorithms, we will consider congruence transformations that consist only of translations and rotations, since reflections can be taken care of by running the algorithm once for $P$ and $Q$ and once for $P$ and the mirror image of $Q$. We will refer to such congruence transformations simply as transformations. We will use $K$ to always denote the size of the largest common point set between $P$ and $Q$.

### 2.1 Voting scheme

In this and the next subsection, we assume that the point sets $P$ and $Q$ are in the plane. Let $dist(x, y)$ be the distance between points $x$ and $y$. Consider pairs $p, p'$ and $q, q'$ of points of $P$ and $Q$, respectively. If $dist(p,p') = dist(q,q')$, these four points determine a unique transformation $T$ such that $T(p) = q$ and $T(p') = q'$. This transformation is denoted by $T[p,p';q,q']$.

For each pair $(p,q)$ of points $p \in P$ and $q \in Q$, and each transformation $T$, we define $mult_{p,q}(T)$ to be the number of pairs $x \in P$ and $y \in Q$ satisfying that $T[p,x;q,y] = T$. If $|T(P) \cap Q| = k$, it is easy to see that $mult_{p,q}(T) = k-1$ for any $q \in T(P) \cap Q$ and $p = T^{-1}q$.

Hence, $\max_{p,q}(\max_T mult_{p,q}(T)) = K-1$ and the transformation achieving this maximum is the transformation that gives the largest common point set.

For each pair $p \in P$ and $q \in Q$, $\max_T mult_{p,q}(T)$ can be computed by considering all matching pairs of edges $(px, qy)$, $x \in P$, $y \in Q$, and letting each pair cast a vote to the transformation $T[p,x;q,y]$. We call this process the *local voting* for pair $(p,q)$.

A similar voting idea is used in the generalized Hough transformation [6], where a vote is cast to a bucket in the transformation space rather than to an individual transformation as in our case. In other words, our scheme may be viewed as a limit of a version of the generalized Hough transformation as the bucket size goes to zero.

In the deterministic version of the algorithm, we need to execute the local voting process for each pair $(p,q)$, with $p \in P$ and $q \in Q$, and take the best result. The total number of votes cast in the algorithm execution is $\lambda(P,Q)$, because each matching pair $(pp', qq')$ of edges contributes exactly one vote, namely in the local voting for the pair $(p,q)$. Thus, $\lambda(n,m)$ bounds the essential term in the running time, with a multiplicative factor of $O(\log n)$ that accounts for the table maintenance for counting a vote.

**Theorem 2.1** *In an efficient implementation, the deterministic voting algorithm computes the two-dimensional LCP in $O((\lambda(n,m) + n^2)\log n)$ time and $O(n)$ space.*

**Proof:** For each $p \in P$, we first sort the set $\{(p,x)|x \in P\}$ of pairs in terms of $dist(p,x)$ and store it in a binary search data structure $D_p(P)$. This needs $O(n \log n)$ time for each $p \in P$. In the local voting for a fixed pair $(p,q)$, we search $dist(q,y)$, for each $y \in Q$, in the data structure $D_p(P)$ in $O(\log n)$ time to obtain the matches in $P$ of the edge $qy$. This needs $O(m \log n)$ time for each of $nm$ local voting processes. The rest of the running time is at most $O(\log n)$ per vote. Thus, the overall running time is $O((\lambda(n,m) + nm^2 + n^2)\log n)$ but we have $\lambda(n,m) = \Omega(nm^2)$ as noted earlier. See Appendix for how to reduce the space complexity to $O(n)$. □

Observe that, in the above deterministic voting algorithm, the optimal transformation $T$ receives the maximum vote of $K-1$ in the voting process for each $(p_i, T(p_i))$, $i = 1, \ldots, K$, where $\{p_i\}$ is the largest common point set. This is a redundancy which is difficult to avoid deterministically. We can reduce the redundancy, however, at the cost of a small failure probability using a standard random sampling procedure: sample a random subset $R$ of $P$ with size $cn/K$, for some sufficiently large $c$, and do local voting only for pairs in $R \times Q$. Here, once we fix a pair $p \in R$ and $q \in R$, the local voting for $(p,q)$ is done exactly as before, scanning the entire sets $P$ and $Q$ (not restricting to $R$). This speeds up the algorithm by a factor of $|P|/|R| = c/K$: the terms $n^2 \log n$ and $nm^2 \log n$ of the deterministic time becomes $|R|n \log n$ and $|R|m^2 \log n$, respectively, and the expected total number of votes cast is $(c/K)\lambda(P,Q)$. The algorithm makes an error only when $R$ is disjoint from the largest common point set. The probability of this is at most $e^{-c}$.

So far, we have assumed that we know the value of $K$, which is not the case in general. However, we can efficiently and with high probability estimate $K$ within a factor of, say, 2, using the randomized approximation scheme described below. (Alternatively, we may appeal to the standard binary search resulting in a slightly larger running time.) We omit the details and only state the result.

**Theorem 2.2** *LCP can be solved in* $O(c(\lambda(n,m)+n^2)K^{-1}\log n)$ *time with probability* $1-2^{-c}$, *if* $K > \log^3 n$.

If we can be satisfied with an approximately optimal solution, that has size at least $(1-\epsilon)K$, we can further reduce the running time by a different sampling strategy. We sample $\tilde{P} \subset P$ with $\alpha n$ points and $\tilde{Q} \subset Q$ with $\beta m$ points, where $\alpha\beta = c \log n / K$ for some sufficiently large constant $c$, and compute the largest common set between $\tilde{P}$ and $\tilde{Q}$.

To see why this sampling approach works, let $T$ be any transformation. For each $p \in P$ such that $T(p) \in Q$, the probability that both $p$ and $T(p)$ are sampled is $\alpha\beta = c\log n/k$. Therefore, if $T$ gives a common point set of size $K$ between $P$ and $Q$, it gives a common point set of expected size $c \log n$ between $\tilde{P}$ and $\tilde{Q}$ and, with high probability, of size at least $(1-\epsilon/2)c\log n$. On the other hand, the probability is also small that any transformation that gives a common point set of size smaller than $(1-\epsilon)K$ between $P$ and $Q$ gives a common point set of size at least $(1-\epsilon/2)K$ between $\tilde{P}$ and $\tilde{Q}$. Thus, the solution of LCP for $\tilde{P}$ and $\tilde{Q}$ gives, with high probability, a near-optimal transformation for $P$ and $Q$ as desired.

For the running time analysis, observe that the expected value of $\lambda(\tilde{P}, \tilde{Q})$ is $\alpha^2\beta^2\lambda(P,Q)$, because the multiplicity of a fixed distance in $P$ ($Q$, resp.) is reduced by a factor of $\alpha^2$ ($\beta^2$, resp.) expectedly and because $\tilde{P}$ and $\tilde{Q}$ are sampled independently. (Note this is a better bound than $\lambda(\alpha n, \beta m)$.) Trading off between $\alpha$ and $\beta$ appropriately to cope with the $n^2 \log n$ pre-processing time of the deterministic voting algorithm, depending on the relative magnitude of $n$ and $m$, we have the following.

**Theorem 2.3** *For any fixed* $\epsilon > 0$, *we can find, with high probability, a common point set between* $P$ *and* $Q$ *of size at least* $(1-\epsilon)K$ *in expected* $O((\lambda(n,m)+n^2)K^{-2}\log^3 n)$ *time.*

### 2.2 Alignment scheme

In the *alignment* scheme, we consider each pair of edges $pp'$ from $P$ and $qq'$ of $Q$ with equal length, align $pp'$ with $qq'$, and count the number of points of $P$ that now coincide with a point of $Q$. In other words, we count the size of the intersection $T(P) \cap Q$, where $T = T[p,p';q,q']$. In the deterministic version, we need to consider $\lambda(P,Q)$ alignments and the counting for each alignment takes $O(m \log n)$ time. Alignments can be enumerated in a manner similar to the voting-based algorithm. Thus, we have a running time bound of $O((\lambda(n,m)m + n^2)\log n)$. This is worse than the voting-based algorithm. However, when we apply random sampling (for exact optimization), the rate of reduction in running time is larger in this case. Sample a subset $R$ of $P$ with size $cn/K$, with sufficiently large $c$ so that $R$ contains at least 2 points of the largest common point set with high probability. Then, we need only to consider each edge $pp'$ in $R$ for alignment. This is in contrast with the voting case, where we need to consider each edge $pp'$ with $p \in R$ and $p' \in P$ for voting. Thus, the first term of the running time is reduced by a factor of $c^2/K^2$ and, when $K$ is close to $m$, is comparable with the voting scheme. Irani and Raghavan [16] apply random sampling to an alignment scheme for LSP; this is reasonable because $K$ is linear in $m$ in the applications they intend.

For CCD, where our goal is to determine if $K = m$, the alignment scheme has an advantage of giving a fast Las Vegas algorithm (i.e. an algorithm that does not risk an error), in contrast to the voting scheme that gives a Monte Carlo randomized algorithm (i.e. an algorithm that has some probability of making an error). The idea is to sample from $Q$ rather than from $P$. The reason we sample from $P$ for the general LCP is because we want to reduce the $n^2 \log n$ term in the deterministic time, which accounts for the preprocessing of distances in $P$ . For CCD, this preprocessing can be avoided and sampling from $Q$ turns out advantageous.

**Theorem 2.4** *CCD can be solved in* $O(n^{4/3} m \log n + n^{4/3}\log^{8/3} n)$ *deterministic time and in* $O(\min\{\lambda(n,m)m^{-1}\log n, n^{4/3}m\log n\} + n^{4/3}\log^{8/3} n)$ *Las Vegas expected time.*

**Proof:** We randomly sample a pair $q, q'$ of points from $Q$. Using the the methods of Agarwal et al. [1], we find all occurrences of the distance $dist(q,q')$ in $P$ in $O(n^{4/3}\log^{8/3} n + k\log n)$ time, where $k$ is the number of occurrences of the distance. In the worst case, we have $k = O(n^{4/3})$, from the bound on the repeated distances, and the expected value of $k$ is $\lambda(P,Q)/m^2$. For each such occurrence $pp'$, we align $qq'$ with $pp'$ and test if the entire $Q$ is matched into $P$, using $O(m \log n)$ time. The total running time is $O(km \log n + n^{4/3}\log^{8/3} n)$, from which the claim follows by plugging in the bounds on $k$. □

### 2.3 Higher dimensions

Both the voting and alignment schemes easily generalize to higher dimensions. We only discuss the voting scheme here. The generalization of the alignment scheme (and especially its specialization to CCD) can be done similarly. We start with the voting scheme in three dimensions.

Let the point sets $P$ and $Q$ be in the three-dimensional space. The local voting is now done fixing a pair of edges $pp'$ from $P$ and $qq'$ from $Q$ of equal length. Each pair of points $x \in P$ and $y \in Q$ such that the triangles $pp'x$ and $qq'y$ are congruent casts a vote to the transformation that matches the triangles. When these triangles are degenerate, i.e., $x$ lies on line $pp'$ and $y$ on line $qq'$, and hence does not determine a transformation, the vote becomes "public", i.e., it conceptually adds a count to every candidate in this particular local voting process. Public votes are counted separately and added to the maximum vote count in the end of the local voting. We do the local voting for every matching pair $(pp', qq')$ of edges and take the transformation with the maximum vote count. It is clear that the maximum vote count is $K - 2$.

The number of local voting processes to be performed is bounded by $\lambda^{(3,1)}(n,m)$ and the number of total votes cast by $\lambda^{(3,2)}(n,m)$. To enumerate matching pairs of edges for which to perform local voting, we use a sorted list of distances in $P$ as we did in the planar case: $O(n^2 \log n)$ preprocessing time and $O(nm^2 \log n)$ query time. To enumerate matching triangles that cast votes, we similarly use a sorted list of triangles: $O(n^3 \log n)$ preprocessing time and $O(\lambda^{(3,1)}(n,m)m \log n)$ query time. Thus, we have:

**Theorem 2.5** *LCP in the three-dimensional space can be solved in time $O((\lambda^{(3,2)}(n,m) + \lambda^{(3,1)}(n,m)m + n^3) \log n)$.*

When we use our current bounds on $\lambda^{(3,2)}(n,m)$ and $\lambda^{(3,1)}(n,m)$ that are roughly $O(n^{1.8}m^3)$ and $O(n^{1.5}m^2)$ respectively, it turns out that the second term is subsumed by the first. The random sampling approaches also work for the three-dimensional space and give a speed up of roughly $K^2$ for exact optimization and $K^3$ for approximate optimization. We omit the details.

The above theorem generalizes to higher dimensions.

**Theorem 2.6** *For any fixed $d \geq 3$, LCP in the $d$-dimensional space can be solved in $O((\lambda^{(d,d-1)}(n,m) + \lambda^{(d,d-2)}(n,m)m + n^d) \log n)$ time.*

Analysis of random sampling schemes for $d \geq 4$ appears difficult if our goal is to achieve the speed-up of $K^{d-1}$, due to the possible degeneracy of the largest common point set.

## 3 Bounding $\lambda(n,m)$

We now turn to the combinatorial-geometric analysis of the inner-product of multiplicity vectors. We start with the planar case.

**Theorem 3.1** $\lambda(n,m) = O(n^{1.43}m^{1.77})$.

We need some lemmas due to Székely [27, 18] (Lemmas 3.1 and 3.2 below), which he used in proving the $\Omega(n^{4/5})$ bound on the number of distinct distances in an arbitrary set of $n$ points.

The *crossing number* of a drawing of a graph $G$ on a plane is the number of intersections of drawn edges (see [22, 23]). Let $cr(G)$ denote the minimum of the crossing numbers over all drawings of $G$.

**Lemma 3.1** *In the graph $G$ with $n$ vertices and $t$ edges, if each edge has at most $y$ edges parallel to it, then $cr(G) = \Omega(t^3/yn^2) - O(y^2n)$*

Recall the notation in the introduction: $\nu(P)$ is the number of distinct distances in $P$, $\mathcal{D}(P) = \{l_1, ..., l_{\nu(P)}\}$ is the set of such distances listed in a non-increasing order of multiplicity, and $H_P[l]$ is the multiplicity of the distance $l$ in $P$. Let $1 \leq k \leq \nu(P)$. We consider the subset $\{l_1, ..., l_k\}$ of $\mathcal{D}(P)$, consisting of the first $k$ entries, and let $f(k) = \sum_{i=1}^k H_P[l_i]$. Seeking an upper bound on $f(k)$, we create $k$ concentric circles of radii $l_1$,..,$l_k$ around each point of $P$. Hence, we have $nk$ circles in total: let $\mathcal{C}(k)$ denote the set of these circles.

The circles in $\mathcal{C}(k)$ have $2f(k)$ incidences with the point set $P$. We delete circles containing at most two points of $P$ and let $\mathcal{C}'(k)$ denote the set of remaining circles. Each circle of $\mathcal{C}'(k)$ contains at least three points of $P$ and these points cut the circle into at least three arcs. If we let $A$ denote the set of these arcs coming from all the circles of $\mathcal{C}'(k)$, then $|A|$ is equal to the number of incidences between the circles of $\mathcal{C}'(k)$ and the points of $P$, and hence, is at least $2f(k) - 2kn$. Define a multigraph $G(k)$ whose vertices are points of $P$, and whose edges are arcs of $A$. Since each pair of at most $kn$ circles of $\mathcal{C}'(k)$ intersect at most twice, the crossing number $cr(G(k))$ of $G(k)$ is at most $k^2n^2$.

The following lemma is due (in a slightly different form) to Székely. The proof is a straightforward adaptation of his proof described in [18].

**Lemma 3.2** *For each $y$, the number of edges in $G(k)$ having at least $y$ edges parallel to them is at most $O(n^2k/y^2 + kn \log n)$.*

**Proof:** Let $M_i$ be the set of pairs $\{u, v\}$ of vertices of $G(k)$ such that there are at least $2^i$ and fewer than $2^{i+1}$ parallel edges connecting $u$ and $v$ in $G(k)$, and let $E_i$ be the set of all edges connecting the pairs from $M_i$. Our goal is to upperbound $\sum_{y/2 \leq 2^i \leq n} |E_i|$.

Let $L_{u,v}$ denote the perpendicular bisector of the segment $uv$ and let $\mathcal{L}_i = \{L_{u,v} \mid \{u,v\} \in M_i\}$. Each edge connecting $u$ and $v$ determines one incidence of the line $L_{u,v}$ with $P$, since the center of a circle through $u$ and $v$ must lie on $L_{u,v}$. On the other hand, we note that one such point-line incidence $(p, L)$, where $p \in S$ and $L \in \mathcal{L}_i$, corresponds to at most $2k$ edges of $E_i$ because there are only $k$ different radii. Therefore, the size of $E_i$ is at most $2kN_i$, where $N_i$ is the number of incidences of $\mathcal{L}_i$ with $P$.

Since each lines of $\mathcal{L}_i$ has at least $2^i$ incidences with $P$ from the definition of $M_i$, we have $N_i \geq 2^i|\mathcal{L}_i|$. On the other hand, Spencer-Szemeredi-Trotter's Theorem[26] (see also [18]) gives the upper bound of $N_i \leq O(|\mathcal{L}_i| + n + n^{2/3}|\mathcal{L}_i|^{2/3})$. Therefore, we have $2^i|\mathcal{L}_i| \leq O(|\mathcal{L}_i| + n + n^{2/3}|\mathcal{L}_i|^{2/3})$ and hence $|\mathcal{L}_i| = O(n^2/2^{3i} + n/2^i)$. Plugging in this bound of $\mathcal{L}_i$ to the Spencer-Szemerédi-Trotter bound again, we conclude that $N_i = O(n^2/2^{2i} + n)$. (Note that we cannot use the inequality $N_i \leq 2^{i+1}|M_i|$ here, because each line of $\mathcal{L}_i$ may correspond to multiple pairs of $M_i$.) Hence, $|E_i| = O(n^2k/2^{2i} + nk)$, from which the lemma follows by summing over $i$. □

**Proposition 3.1** $f(k) = O(\min\{n^{10/7}k^{5/7}, n^2\})$.

**Proof:** $O(n^2)$ is a trivial bound. The $O(n^{10/7}k^{5/7})$ bound needs to be proved for $k < n^{4/5}$, where we can assume that $f(k) \gg kn \log n$.

We construct $G(k)$, set $y = \sqrt{Cn^2k/f(k)}$ for some constant $C$, and delete all edges of multiplicity larger than $y$. From Lemma 3.2, at most $f(k)$ edges are removed, if we set $C$ sufficiently large. Then, we have a graph $G$ with $\Omega(f(k))$ edges, whose edge multiplicity is at most $y$. From Lemma 3.1, $k^2n^2 > cr(G) = \Omega(f(k)^3/yn^2) - O(y^2n)$. Hence, $k^2n^2 = \Omega(f(k)^{3.5}/n^3k^{0.5})$ (the $O(y^2n)$ term is negligible). It follows that $f(k)^{3.5} = O(k^{2.5}n^5)$ and hence $f(k) = O(n^{10/7}k^{5/7})$. □

We are now ready to prove Theorem 3.1. Let $\mathcal{D}(P) = \{l_1, \ldots, l_{\nu(P)}\}$ ($\mathcal{D}(Q) = \{l'_1, \ldots, l'_{\nu(Q)}\}$, *resp.*) denote the set of distinct distances of $P$ ($Q$, resp.) sorted in a non-increasing order of their multiplicities. Note first that

$$\lambda(P,Q) = \sum_{l \in \mathcal{D}(P) \cap \mathcal{D}(Q)} H_P[l]H_Q[l] \leq \sum_{k=1}^{\min(\nu(P),\nu(Q))} H_P[l_k]H_Q[l'_k],$$

due to the monotonicity of the multiplicity sequences. The monotonicity also implies that $H_P[l_k] \leq f(k)/k$ and hence, from the above bound on $f(k)$, that $H_P[l_k] = O(n^{10/7}k^{-2/7})$ for $k \leq O(n^{4/5})$. Similarly, we have $H_Q[l'_k] = O(m^{10/7}k^{-2/7})$ for $k \leq O(m^{4/5})$ and hence:

$$\lambda(P,Q) \leq \sum_{k=1}^{O(m^{4/5})} O(k^{-4/7}n^{10/7}m^{10/7}) = O(n^{10/7}m^{62/35}).$$

Because $n > m$, we can write the bound as $O(n^{1.43}m^{1.77})$.

## 4 Three dimensional case

We now turn to the three-dimensional space. In the following, $O'(f(n))$ will abbreviate $(\log^* n)^{O(1)}f(n)$, i.e., it hides polynomial factors of $\log^* n$. It is known that the number of occurrence of a given distance in a set of $n$ points in the three dimensional space is $O'(n^{3/2})$ [8]. This trivially implies that $\lambda^{(3,1)}(n,m) = O'(n^{3/2}m^2)$. In the following, we derive an upper bound on $\lambda^{(3,2)}(n,m)$.

For each triangle $\Delta$, let $H_P[\Delta]$ denote the multiplicity of $\Delta$ in $P$ (i.e., the number of ordered triples of $P$ forming an occurrence of $\Delta$) and recall that $\lambda^{(3,2)}(P,Q) = \sum_\Delta H_P[\Delta]H_Q[\Delta]$. An upper bound on $H_P[\Delta]$ can be obtained by considering the following circle-point incidence problem. Let $\Delta = ABZ$ be an arbitrary triangle and $h$ be the distance from $Z$ to the line $AB$. For each pair $p, q$ of points of $P$ with $dist(p,q) = dist(A,B)$, the trajectory of the vertex $Z$ of a triangle congruent to $\Delta$ locating $A$ and $B$ at $p$ and $q$, respectively, forms a circle of radius $h$. Thus, if $N = O'(n^{3/2})$ is the multiplicity of distance $dist(A,B)$ in $P$, the multiplicity of $\Delta$ is at most the number of incidences between $N$ circles and $n$ points.

**Lemma 4.1** *Given a set of $N$ circles and $n$ points in the $d$-dimensional space for $d \geq 3$, there are $O(N+n+\min\{N^{2/3}n, Nn^{1/2}, N^{4/5}n^{3/5}\log^* N\})$ circle-point incidences.*

**Proof:** Consider the bipartite graph $H = (V, W, E)$, where $V$ corresponds to the set of circles, $W$ corresponds to the set of points, and $E$ is the edge set, where a vertex of $V$ and a vertex of $W$ are connected with an edge if the corresponding circle and point are incident. Since two circles intersect at most two points (in any dimensional space), the graph does not contain $K_{2,3}$. Thus, we have $O(N + n+\min\{N^{2/3}n, Nn^{1/2}\})$ bound (a Canham-like bound). By projecting the figure to a two-dimensional plane, so that no pair of points overlap, we have an arrangement of ellipses. Now, similarly to [8], we can use cutting theory of arrangements, and obtain the $O(N^{4/5}n^{3/5}\log^* N)$ bound. □

Note: The above $\log^* N$ factor can be slightly improved by using the bounds of Davenport-Shinzel sequences.

Erdös and Purdy [12] gave an $O(n^{3-1/3})$ bound for the number of triangles with a given volume in the space, and asked a question about $H_P[\Delta]$, the number of pairwise congruent triangles. From the above lemma, it immediately follows that $H_P[\Delta] = O'((n^{3/2})^{4/5}n^{3/5}) = O'(n^{1.8})$. We can similarly obtain an $O'(n^{2.2})$ bound on the number of pairwise similar triangles in the space.

Since $H_P[\Delta] = O'(n^{1.8})$, it is immediate to obtain that $\lambda(n,m) = O'(n^{1.8}m^3)$. If $n = m$, the bound becomes $O'(n^{4.8})$. We can obtain another bound which is slightly better if $m$ is large.

**Theorem 4.1** $\lambda^{(3,2)}(n,m) = O'(n^{4.6875})$

**Proof:** We can assume that $n = m$, and further $P = Q$. For each distance $l$, let $\mathcal{T}_l(P) = \{\Delta_1^l, \ldots, \Delta_{k_l}^l\}$ denote the set of distinct triangles of $\mathcal{T}_l$ occurring in $P$, listed in a non-increasing order of their multiplicities. Then,

$$\lambda^{(3,2)}(P,P) = \sum_l \sum_{\Delta \in \mathcal{T}_l} (H_P[\Delta])^2 \leq \sum_l \sum_{k=1}^{k_l} (H_P[\Delta_k^l])^2$$

The number of occurences of $l$ in $P$ is at most $O'(n^{3/2})$, and the worst scenario for our analysis is the case where $O(n^{1/2})$ lengths $l$ such that each occurs $O'(n^{3/2})$ times. So, we assume this happens.

Let us bound $F(k) = \sum_{i=1}^{k} H_P[\Delta_i^l]$ for each $l$. this sum is at most the number of incidences between a family $\mathcal{C}(k)$ of $O'(kn^{3/2})$ circles and $n$ points. However, this family should be considered as a multi-set: If the distance $l$ occurs $s$ times on a line, it may happen that (at most) $\min\{s,k\}$ triangles counted in $F(k)$ define the same circle. Hence, we cannot use the incidence bound in a straightforward manner.

A line containing $s$ points can create at most $s-1$ occurrence of the distance $l$ on the line. From the Spencer-Szemérdi-Trotter theorem, incidence of $m$ lines (in space) and $n$ points is $O(n + m + (nm)^{2/3})$ (although it is usually stated for a configuration in the plane, it is easy to modify it for the space). Hence, if $y \leq n^{0.5}$, the incidence created by the lines containing $\Omega(y)$ points is $O(n^2y^{-2})$. If $y > n^{0.5}$, it is $O(n)$.

Let $\mathcal{C}(k,y)$ denotes the subfamily of $\mathcal{C}(k)$ each of its circle has a multiplicity between $y$ and $2y$. We can assume that $y \leq k$. Let us analyze the incidence between $P$ and $\mathcal{C}(k,y)$. applying Lemma 4.1.

If $y \leq n^{1/4}$ , it is $O'(y(ky^{-1}n^{1.5})^{4/5}n^{3/5})$ bound, which attains the maximum bound $O'(kn^{1.8})$ when $y = k$. If $n^{1/4} \leq y \leq n^{1/2}$, it is $O'(y(ky^{-1}n^2y^{-2})^{4/5}n^{3/5})$, that is $O'(k^{4/5}y^{-7/5}n^{11/5})$, which attains the maximum $O'(k^{4/5}n^{37/20})$ when $y = n^{1/4}$. If $y \geq n^{1/2}$, it is $O'(k^{4/5}y^{1/5}n^{7/5})$, whose maximum is $O'(kn^{7/5})$ when $y = k$.
Hence, $F(k) = O'(kn^{1.8}\log n)$ if $k \leq n^{1/4}$, $O'(k^{4/5}n^{37/20}\log n)$ if $n^{1/4} \leq k \leq n^{1/2}$, and $O'(kn^{7/5}\log n + k^{4/5}n^{37/20}\log n)$ if $k \geq n^{1/2}$. The $\log n$ factor can be removed with a more careful analysis. Since $F(k) = O'(n^{2.5})$ for any $k$, we only need to consider $k$ up to $n^{13/16}$, and the $kn^{7/5}\log n$ term never dominates. From this, we obtain that
$\sum_i (H_P[\Delta_i^l])^2 = O'(n^{(3/5)(13/16)}n^{37/10})$.
Hence, $\lambda^{(3,2)}(P,P) = n^{1/2}\sum_i (H_P[\Delta_i^l])^2 = O'(n^{4.7-1/80}) = O'(n^{4.6875})$. □

Note: The above bound can be slightly refined so that it is better than $O'(n^{1.8}m^3)$ if $m > n^{5/8}$, which we omit in this version.

## 5 Four-dimensional case

### 5.1 Number of congruent triangles

In this subsection, we prove the following theorem:

**Theorem 5.1** *In the given set $P$ of $n$ points in the four-dimensional space, the multiplicity of any given triangle is $O'(n^{65/23}) = O(n^{2.83})$.*

In the following, all geometric objects are in the four-dimensional space unless otherwise stated. We consider triangles congruent to a given triangle $\Delta = ABZ$, whose longest edge $AB$ has length 1, and height $AX$ of the triangle is $h$. We call $AB$ the *base edge* of $\Delta$. We need the following basic fact.

**Proposition 5.1** *The intersection of four mutually distinct three-dimensional unit spheres is either a circle or a set of no more than two points.*

We call two circles $C_1$ and $C_2$ a *distance-circle-pair* if the distance $dist(u,v) = 1$ for any $u \in C_1$ and $v \in C_2$. Note that for any circle $C_1$ of radius $r < 1$, there is a unique circle $C_2$ such that $C_1$ and $C_2$ form a distance-circle pair. To see the existence[12], choose a coordinate system so that $C_1 = \{(x,y,0,0) \mid x^2 + y^2 = r^2\}$ and set $C_2 = \{(0,0,z,w) \mid z^2 + w^2 = 1 - r^2\}$. For uniqueness, take arbitrary four points from $C_1$ and observe that $C_2$ must lie in the intersection of the four unit three-dimensional spheres centered at these four points.

The existence of a distance-circle-pair kills the hope of getting a non-trivial bound on the multiplicity of a distance in four dimensions, because one may place $n/2$ points on each circle of a distance-circle-pair to get $n^2/4$ occurences of the unit distance. Our approach, very roughly speaking, is to bound the number of triangles congruent to $\Delta$ in two parts: the number of those with their base edges occuring from such "heavy" distance-circle-pairs are bounded using the special property of the set of such occurences of unit distances; the number of others are small because the number of unit distances not occurring from "heavy" distance-circle-pairs is small.

Given a point set $P$ of $n$ points, we say that a circle $C$ is *heavy* if it contains more than $n^\alpha$ points of $P$, where $\alpha = 7/23$. Let $N$ denote the number of heavy circles.

**Lemma 5.1** $N = O'(n^{3-5\alpha})$.

**Proof:** By Lemma 4.1 on the number of circle point incidences, we have $Nn^\alpha = O'(N + n + N^{4/5}n^{3/5})$. Hence, $N = O'(n^{3-5\alpha})$ since $\alpha < 0.5$. □

We construct a bipartite graph $G = (U,V,E)$, so that each of $U$ and $V$ corresponds to the point set $P$, and $(u,v) \in E$ if and only if $dist(u,v) = 1$. We consider three subsets $E_0$, $E_1$ and $E_2$ of $E$. An edge $(u,v)$ is in $E_0$ if $u$ is on a heavy circle and $v$ is also on a heavy circle; it is in $E_1$ if $u$ is not on a heavy circle; it is in $E_2$ if $v$ is not on a heavy circle. Clearly, $E = E_0 \cup E_1 \cup E_2$.

**Lemma 5.2** *$|E_1|$ and $|E_2|$ are both $O(n^{(7+\alpha)/4})$.*

**Proof:** First we show that the bipartite graph $(U, V, E_1)$ has no $K_{n^\alpha,4}$ as a subgraph (with the $n^\alpha$ points on the $U$'s side and the four points on the $V$'s side). Suppose otherwise, i.e, that $(U, V, E_1)$ has a subgraph isomorphic to $K_{n^\alpha,4}$. Then the $n^\alpha$ points of $U$ in this subgraph must lie in the intersection of four unit three-dimensional spheres centered at the four points of $V$ in this subgraph. By Proposition 5.1, these $n^\alpha$ points must lie on a circle, making the circle heavy and contradicting the definition of $E_1$. Hence, we have $|U|\binom{|E_1|/|U|}{4} < n^\alpha\binom{|V|}{4}$. Since $|U| = |V| = n$, we have $|E_1| = O(n^{(7+\alpha)/4})$. The bound on $|E_2|$ is similar. □

**Corollary 5.1** *The number of triangles congruent to $\Delta$ whose base edge is in $|E_1 \cup E_2|$ is $O(n^{(11+\alpha)/4})$.*

We next consider triangles associated with edges in $E_0$. We fix a distance-circle-pair $(C_1, C_2)$, both $C_1$ and $C_2$ heavy, and consider the embedding of $\Delta$ so that its base edge is mapped to an edge between $C_1$ and $C_2$. We consider each edge is directed from $C_1$ to $C_2$, so that the copy of $A$ is in $C_1$, and the copy of $B$ is in $C_2$. Recall that $X$ is the point on line $AB$ such that the line segment $ZX$ defines the height $h$ of $\Delta$ and let $X(e)$ denote the image of $X$ when $AB$ is mapped to a directed edge $e$ of unit length. Then, the locus of the opposite vertex $Z$ of $\Delta$ when we transform $AB$ to $e$ forms a two-dimensional sphere $S^{(2)}(e)$ centered at $X(e)$ with radius $h$. Moreover the sphere has $e$ as its normal vector, that is, the normal vector of the three dimensional space containing the two-dimensional sphere. Let us consider the family of spheres $\mathcal{F}(C_1, C_2) = \{S^{(2)}(xy) | x \in C_1, y \in C_2\}$.

**Lemma 5.3** *Consider the family of two-dimensional spheres with radius $h$ containing a fixed circle $C$. Then, the centers of these spheres are located on a circle $C'$. Moreover, the normal of each sphere is tangent to $C'$.*

**Proof:** Let $p$ be the center and $r$ the radius of $C$. Let $L$ be the two-dimensional plane determined by $C$ and let $L'$ be the two-dimensional plane containing $p$ and orthogonal to $L$. Let $S$ be any sphere belonging to the family in the lemma and let $q$ be the center of $S$. Then, $q$ is on $L'$ and at distance $\sqrt{h^2 - r^2}$ from $p$, i.e., it is on a circle $C'$ of radius $\sqrt{h^2 - r^2}$ centered at $p$ and contained in $L'$. The three-dimensional space containing $S$ is determined by the line $pq$ and $L$ and hence is perpendicular to the tangent of $C'$ at $q$. □

**Lemma 5.4** *For any circle $C$, there are at most four spheres in $\mathcal{F}(C_1, C_2)$ that contain $C$.*

**Proof:** Let $F_C$ denote the set of spheres of $\mathcal{F}(C_1, C_2)$ that contain $C$ and assume that $F_C$ is non-empty. By the above lemma, there is some circle $C'$ such that each sphere of $F_C$ has its center on $C'$ and has its normal tangent to $C'$. Let $e = xy$, $x \in C_1$, $y \in C_2$ be an edge such that $S^{(2)}(e) \in F_C$. Then, from the observation preceding the above lemma, the normal of $S^{(2)}(e)$ is $e$ and the center of $S^{(2)}(e)$ is $X(e)$. Therefore, $e$ must be tangent to $C'$ and hence contained in the the two-dimensional plane $L$ containing $C'$. $L$ cannot be coplanar to $C_1$, because if it is then it does not intersect $C_2$ and hence does not contain any edge between $C_1$ and $C_2$. Similarly, $L$ is not coplanar to $C_2$. Therefore, $L$ intersects each of $C_1$ and $C_2$ at no more than two points and hence there are at most four edges between $C_1$ and $C_2$ contained in $L$. Therefore, $S^{(2)}(e) \in F_C$ for at most four edges $e$. □

We are now ready to bound the number of triangles in $P$ congruent to $\Delta$ that have their base edge in $E_0$. We consider the bipartite graph $(E_0, P, T)$, which has vertex set $E_0$ and $P$, and connect $e \in E_0$ and $p \in P$ by an edge of $T$ if and only if $e$ and $p$ forms a triangle congruent to $\Delta$. Consider any $4N + 1$ spheres in $\cup\mathcal{F}(C_1, C_2)$, where union is taken over all distance-circle-pair $(C_1, C_2)$ with both $C_1$ and $C_2$ heavy. Then, there is a distance-circle-pair $(C_1, C_2)$ such that at least five of those spheres are in $\mathcal{F}(C_1, C_2)$. By Lemma 5.4, the intersection of these five circles cannot contain a circle and hence is a set of at most two points. Thus, our bipartite graph does not contain $K_{4N+1,3}$ as a subgraph. Since $|E_0| < n^2$ and $|P| = n$, we have $|T| = O(N^{1/3}n^{7/3})$. Since $N = O'(n^{3-5\alpha})$, $|T| = O'(n^{10/3-5\alpha/3})$.

Putting this together with Corollary 5.1, the number of triangles congruent to $\Delta$ in $P$ is $O'(n^{(11+\alpha)/4} + n^{10/3-5\alpha/3})$. We set $\alpha = 7/23$ to obtain the bound of $O'(n^{65/23})$.

### 5.2 LCP in four dimensional space

Let us consider 3-d simplices in four dimensional space.

**Lemma 5.5** *The maximum number of 3-d simplices in $n$ points in four-dimensional space congruent to a given 3-d simplex $\Gamma$ is $O(n^{2.87})$.*

**Proof:** If we a fix a location of a triangle face $\Delta$ of $\Gamma$, the locus of the opposite vertex $v$ of $\Gamma$ forms a circle. The number of triangles congruent to $\Delta$ in $P$ is $O(n^\gamma)$ for $\gamma = 65/23$. Thus, we have a set $C(\Gamma)$ of $n^\gamma$ circles corresponding to the possible locations of the 3-d simplices.

We would like to use the circle-point incidence to bound the number of 3-d simplices. Unfortunately, different locations of $\Delta$ may create the same circle $C$. So, we cannot use Lemma 4.1 directly.

Suppose that $s$ locations of $\Delta$ can create $C$. These triangles must lie on the plane which contains the center $o$ of $C$, and perpendicular to $C$. Moreover, the set of vertices of the triangles must lie on (at most) three circles $Circ_1(C), Circ_2(C), Circ_3(C)$ centered by $o$. $Circ_1(C)$ contains $s$ vertices, and $Circ_1(C)$ cannot coincide $Circ_1(C')$ if $C \neq C'$: Consider the case where $Circ_1(C)$ has the center at the origin, and contained in the $x$-$y$ plane. If the height of $\Gamma$ with respect to the face $\Delta$ is $h$, $C$ must be the unique circle defined by $x = y = 0, z^2 + t^2 = h^2$.

Let us consider the subset $\mathcal{C}(s)$ of $\mathcal{C}(\Gamma)$ each has multiplicities is equal to or more than $s$ but less than $2s$. Then, for each $C \in \mathcal{C}(s)$, $Circ_1(C)$ contains at least $s$ points of $P$. From Lemma 4.1, at most $O'(ns^{-1} + n^3s^{-5})$ circles contains more than $s$ points. Hence, $\mathcal{C}(s)$ contains $O'(\min\{ns^{-1} + n^3s^{-5}, n^\gamma s^{-1}\})$ circles.

If $s > n^{0.5}$, incidence between $P$ and $\mathcal{C}(s)$ is $O(s(ns^{-1})^{0.8}n^{0.6}) = O(s^{0.2}n^{1.4})$, which is small even if $s = n$. If $s > n^{1/23}$, $n^3s^{-5} < n^\gamma s^{-1}$, and incidence between $P$ and $\mathcal{C}(s)$ is $O(s(n^3s^{-5})^{0.8}n^{0.6})$, which is $O'(n^3s^{-3})$. Hence, there are $O'(n^{66/23})$ incidences created by circles whose multiplicities are at least $n^{1/23}$.

If $s < n^{1/23}$, incidence between $P$ and $\mathcal{C}(s)$ is incidence between $n$ points and $O(n^\gamma s^{-1})$ circles each of which has multiplicity $n^s$; this is $O'(s^{0.2}n^{0.8\gamma+0.6})$. Hence, there are $O'(n^{66/23})$ incidences created by circles whose multiplicities are less than $n^{1/23}$. Hence, the number of $3-d$ simplices is $O'(n^{66/23}) = O(n^{2.87})$. □

The following lemma can be proven similarly to the above one.

**Lemma 5.6** *The maximum number of 3-d simplices in n points in four-dimensional space similar to a given 3-d simplex* $\Gamma$ *is* $O'(n^3)$.

**Theorem 5.2** *LCP in four dimensional space can be solved in* $O((n^{2.87}m^4 + n^4)\log n)$ *time. LSP in four dimensional space can be solved in* $O'((n^3m^4 + n^4)\log n)$ *time. CCD can be solved in* $O(n^{3.83} + n^{2.87}m\log n)$ *time.*

**Proof:** From the bounds of triangles and 3-d simplices, $\lambda^{(4,3)}(n,m) = n^{2.87}m^3$ and $\lambda^{(4,2)}(n,m) = n^{2.83}m^2$. Hence, from Theorem 2.6, we have the bound for LCP. Analysis for LSP is analogous. For CCD, we choose a 3-d simplex $\Gamma$ and its face $\Delta$ from $Q$, and find all possible $O(n^{2.83})$ locations of $\Delta$ in $P$ using $O(n^3)$ time. Next, we find all possible $O(n^{2.87})$ locations of $\Gamma$ in $O(n^{3.83})$ time from all 3-d simplices containing copies of $\Delta$ in $P$. Then, we check whether each of copies of $\Gamma$ is associated with a congruent embedding of $Q$ in $P$ using $O(m\log n)$ time. □

## 6 Higher dimensional case

**Lemma 6.1** *In the d-dimensional space,* $d \geq 5$*, the number of* $(d-1)$*-dimensional simplices in P similar to a given* $(d-1)$*-dimensional simplex is* $O'(n^{d-1})$.

**Proof:**

For $d = 5$, we need special care, which we omit in this version because of space limitation. For $d \geq 6$, given a $(d-1)$-dimensional simplex $\Gamma$, suppose that we fix $(d-1)$ points of $P$ and the $(d-2)$-dimensional simplex $\Delta$ spanned by these points, which is a facet of a similar copy $\Gamma$. Then, the locus of the remained vertex of similar copies of $\Gamma$ forms a circle.

More than one locations of $\Delta$ can have the same circle $C$ as the locus of the remained vertex. Suppose there are $K$ locations of $\Delta$ creating $C$. The set of vertices of copies of $\Delta$ creating $C$ must lie on at most $(d-1)$ $(d-3)$-dimensional spheres which are centered at the center $o$ of $C$ and contained in the $(d-2)$-dimensional plane perpendicular to $C$. If we project these points to one of these spheres, each copy of a face of $\Delta$ create the same $(d-3)$-dimensional simplex on the sphere. Thus, $K$ must be smaller than the number of congruent $(d-3)$-dimensional simplices in the $(d-3)$-dimensional sphere. This is bounded by the number of congruent $(d-4)$-dimensional simplex, since there are at most two $(d-3)$-dimensional congruent simplices sharing a $(d-4)$-dimensional simplex in the sphere. The same $O'(n^{d-4})$ bound as the case of $(d-3)$-dimensional space can be applied, which holds if $d-3 = 3, 4$ as we have seen, and inductively for $d-3 \geq 4$.

So, we have a set of circles with multiplicities, so that the maximum multiplicity is $O'(n^{d-4})$ and the total sum of the multiplicities is $O(n^{d-1})$. The worst scenario for us is that we have a set of $O(n^3)$ circles each of which has multiplicity $O'(n^{d-4})$. From Lemma 4.1, incidence of a set of $O(n^3)$ circles and $n$ points is $O'(n^3)$. Hence, the total incidence considering the multiplicities is $O'(n^{d-1})$. □

**Theorem 6.1** *LSP and LCP in d-dimensional space can be solved in* $O'((n^{d-1}m^d + n^d)\log n)$ *time. for* $d \geq 5$*. SCD and CCD can be solved in* $O'(n^d + n^{d-1}m\log n)$ *time for* $d \geq 5$*, respectively.*

**Proof:** It trivially follows from this lemma that $\lambda^{(d,d-1)}(n,m) = O'(n^{d-1}m^d)$. Note that this bound and the trivial bound of $n^{d-1}m^{d-1}$ on $\lambda^{(d,d-1)}(n,m)$ make the first two terms comparable in the running time upper bound of Theorem 2.6. For SCD (and CCD), we pick a $(d-1)$-dimensional simplex from $Q$, find all $(O'(n^{d-1}))$ similar copies to it in $P$ in $O(n^d)$ time, and check each associated transformation whether it gives an embedding of $Q$ using $O(m\log n)$ time. □

Bounding the number of pairwise congruent (similar) $k$-dimensional simplices in a set of $n$ points in the $d$-dimensional space is a classical problem posed by Moser [19], Erdös and Purdy [12], and Elekes and Erdös [9]. A better answer to this problem for $k = d-1$ and $k = d-2$ will improve the bounds on $\lambda^{(d,d-1)}(n,m)$ and $\lambda^{(d,d-2)}(n,m)$, and hence the bound on the analysis of the algorithms for LCP (or LSP).

Acknowledgement

We wish to thank Jiří Matoušek for communicating Székely's results [27], and also sending us a copy of his unpublished lecture note [18]. We would also like to thank Janos Pach and Pankaj Agarwal for introducing us several useful references.

## References

[1] P. Agarwal, B. Aronov, M. Sharir, and S. Suri, "Selecting Distances in the plane" *Algorithmica* vol 9, pp. 495-514, 1993.

[2] T. Akutsu, "Substructure search and alignment algorithms for three-dimensional protein structures," Proc. SIGAL-41, pp. 1–8, IPSJ, 1994.

[3] T. Akutsu, "On determining the congruency of point sets in higher dimensions," *Proc. ISAAC'94 (LNCS 834)*, pp. 38-55, 1994.

[4] H. Alt, K. Melhorn, H. Wagener and E. Welzl. "Congruence, similarity, and symmetries of geometric objects," *Discrete & Comput. Geom.*, vol. 3, pp. 237-256, 1988.

[5] M. Atkinson, "An Optimal Algorithm for Geometric Congruence" *J. Algorithms* vol. 8, pp. 159-172, 1987.

[6] D. H. Ballard, "Generalizing the Hough Transformation to Detect Arbitrary Shapes", Pattern Recognition, 13-2(1981), pp. 111-122.

[7] P. Chew, M. Goodrich, D. Huttenlocher, K. Kedem, J. Kleinberg, and D. Kravets, "Geometric Pattern Matching under Euclidean Motion" Proc. 5th CCCG pp. 151-156, 1993.

[8] K. Clarkson, H. Edelsbrunner, L. Guibas, M. Sharir and E. Welzl, "Combinatorial complexity bounds for arrangements of curves and spheres," *Discrete & Comput. Geom.*, vol. 5, pp. 99-160, 1990.

[9] G. Elekes and P. Erdös, "Similar configurations and pseudo grids" *Intuitive geometry, Colloq. Math. Soc. Janos Bolyai, 63*, pp. 85 - 104, North-Holland, Amsterdam, 1994.

[10] P. Erdös, "On the set of distances of n points", *Amer. Math. Monthly* vol 53, pp.248-250, 1946.

[11] P. Erdös, "On the set of distances of n points in Euclidean space", *Magyar Tud. Akad. Mat. Kut. Int. Közl.* 5, pp. 165-169, 1960. Also in *Paul Erdös: The Art of Counting (Selected Writing)* (J. Spencer, ed.) pp. 676-679, MIT Press, 1973.

[12] P. Erdös and G. Purdy, "Some extremal problems in geometry", *J. Comb. Theory* vol 10-3, pp. 246-252, 1971.

[13] M. T. Goodrich, J. S. B. Mitchell and M. W. Orletsky, "Practical methods for approximate geometric pattern matching under rigid motions," *Proc. 10th ACM SCG*, pp. 103–112, 1994.

[14] P. J. Heffernan. "Generalized approximate algorithms for point sets congruence," *Proc. Workshop on Algorithms and Data Structures*, pp. 373–384, 1993.

[15] J. E. Hopcroft and D. P. Huttenlocher, Geometric invariance on computer vision, chapter 18, pp.354-374. MIT Press, 1992.

[16] S. Irani and P. Raghavan, "Combinatorial and experimental results for randomized point matching algorithms" *Proc. 12th ACM SCG*, pp. 68-77, 1996.

[17] N. Kiryai, Y. Eldar, and A. M. Bruckstein, "A Probablistic Hough Transform" *Pattern Recognition* Vol 24-4, pp. 303-316, 1991.

[18] J. Matoušek, "Combinatorial and algorithmic geometry", Unpublished lecture notes.

[19] L. Moser, "On the Different Distances Determined by $n$ points" *Amer. Math. Monthly* vol 59, pp. 85-91, 1952.

[20] L. Moser and J. Pach, "Recent Developments in Combinatorial Geometry" in *New Trends in Discrete and Computational Geometry, ed. J. Pach* (1993), Algorithms and Combinatorics 10, Springe Verlag

[21] R. Motwani et al., "Hashing Molecules for Common Substructures" Working paper, 1996.

[22] J. Pach and P. Aggarwal, *Combinatorial Geometry*, Wiley-Interscience Series in Discrete Mathematics and Optimization. John Wiley & Sons, Inc., New York, 1995.

[23] J. Pach and G. Toth, "Graphs Drawn with Few Crossings per Edge" Preprint, 1996.

[24] P. J. de Rezende and D. T. Lee, "Point set pattern matching in $d$-dimensions," *Algorithmica*, vol. 13, pp. 387–404, 1995.

[25] K. Sugihara, "An $n \log n$ algorithm for determining the congruity of polyhedra" *J. Comput. and Sys. Sci.*, vol. 29, pp. 36–47, 1984.

[26] E. Szemerédi and W.T. Trotter, "Extremal problems in discrete geometry," *Combinatorica*, vol. 3, pp. 381–392, 1983.

[27] L. Székely, "Crossing numbers and hard Erdös problems in discrete geometry" Manuscript, 1996.

[28] G. Vriend and C. Sander, "Detection of common three-dimensional substructures in proteins," *PROTEINS: Structure, Function, and Genetics*, vol. 11, pp. 52-58, 1991.



### Appendix: Linear space implementation

The following argument can be extended to the voting algorithms in any dimensions. However, for simplicity, we give the description for the planar case.

We first notice that we process the local voting processes at $p_i, q_j$ in the lexicographic order with respect $i$ and $j$. Hence, we keep the sorted list of $\{dist(p,u) \ : u \in P\}$ only during local voting processes at $p, q_j$ $(j = 1, 2, .., m)$ run.

The space complexity of the voting algorithm is naively $O(nm)$, which is caused because the size of the voting table may become $O(nm)$. We can improve it to $O(n)$ by avoiding to construct the voting table explicitly as follows: At the local voting process, without loss of generality, we assume that $p = q = 0$. The transformation $T$ corresponds to the rotation angle $\tau(T)$, which equals to $\theta(u) - \theta(v)$ if $T$ maps $u$ to $v$, where $\theta(u)$ is the argument of the vector $u$.

Let $A'_l(P)$ be the family of lines consisting of $y = x + \theta(u)$ for all $u \in P$ satisfying $|u| = l$. Let $A''_l(Q)$ be the family of lines consisting of $y = \theta(v)$ for all $v \in Q$ satisfying $|v| = l$. The total number (over all $l$) of lines is at most $n+m$, hence $O(n)$ For a real value $x$, let $g_l(x)$ be the number of segment intersections of $A_l = A'_l(P) \cup A'_l(Q)$ which has $x_0$ as its $x$-coordinate value $x$. Then, the number of the votes collected by a transformation $T$ is $\sum_l g_l(\tau(T))$.

We process the segment intersection computation using Bentley-Ottman's sweep algorithm, with a slight modification that the algorithm share a priority queue controlling the sweeping processes for all $A(l)$, so that it can find the next $x$-coordinate value of the intersection (in precise, lexicographically smallest intersection) among all arrangements.

Using $O(n)$ space, the algorithm maintains the $x$-coordinate value (on the left of the sweep line) which the largest number of intersection points (summed over all $l$) have.

We remark that, in the real implementation, we need not consider segments explicitly.

# RAPID: Randomized Pharmacophore Identification for Drug Design*

P. W. FINN† L. E. KAVRAKI‡ J.-C. LATOMBE§ R. MOTWANI¶ C. SHELTON§
S. VENKATASUBRAMANIAN§ A. YAO||

**Abstract**

This paper describes a randomized approach for finding invariants in a set of flexible ligands (drug molecules) that underlies an integrated software system called RAPID currently under development. An invariant is a collection of features embedded in $\Re^3$ which is present in one or more of the possible low-energy conformations of each ligand. Such invariants of chemically distinct molecules are useful for computational chemists since they may represent candidate pharmacophores. A pharmacophore contains the parts of the ligand that are primarily responsible for its interaction and binding with a specific receptor. It is regarded as an inverse image of a receptor and is used as a template for building more effective pharmaceutical drugs. The identification of pharmacophores is crucial in drug design since the structure of the targeted receptor is frequently unknown, but a number of molecules that interact with the receptor have been discovered by experiments. It is expected that our techniques and the results produced by our system will prove useful in other applications such as molecular database screening and comparative molecular field analysis.

*This research is supported by a grant from Pfizer Central Research.

†Pfizer Central Research, Sandwich, U.K.

‡Department of Computer Science, Rice University, Houston, TX 77005. Partially supported by startup funds from Rice University.

§Department of Computer Science, Stanford University.

¶Department of Computer Science, Stanford University. Partially supported by an Alfred P. Sloan Research Fellowship, an IBM Faculty Partnership Award, an ARO MURI Grant DAAH04-96-1-0007, and NSF Young Investigator Award CCR-9357849, with matching funds from IBM, Mitsubishi, Schlumberger Foundation, Shell Foundation, and Xerox Corporation.

||Department of Computer Science, Princeton University.



*Computational Geometry 97* Nice France



## 1 Introduction

Computational chemists working in the area of structure-based drug design consider both chemical and geometric properties of the interacting molecules when developing new pharmaceutical drugs [8, 28, 33]. The underlying assumption is that drug activity, or pharmacophoric activity, is obtained through the molecular recognition and binding of one molecule (ligand) to a pocket of another, usually larger, molecule (receptor). This assumption is supported by a number of experimental results showing molecules with geometric and chemical complementarity in their active, or binding, conformations [11].

When the three-dimensional structure of the receptor is known, docking methods [10] exploit both the geometric and the chemical information available. However, the geometric structures of relatively few molecules have been obtained via X-ray crystallography or NMR techniques. As a result, computational chemists often try to develop pharmaceutical drugs for receptors whose structure is unknown [9, 37]. The starting point in this case is a collection of ligands that have been experimentally discovered to interact with the considered receptor. By examining the chemical properties and the possible shapes of these ligands, chemists seek to identify a set of features embedded in $\Re^3$ that is contained in some active conformation of each (or most) of the ligands. This is called the *pharmacophore* and it is considered responsible for the observed drug activity. The features of the pharmacophore interact with features of the receptor, while the rest of the ligand acts as a scaffold. Once a pharmacophore has been isolated, it can be used to further improve the activity of pharmaceutical drugs [23].

This paper considers the following problem: Given a set of ligands that interact with the same receptor, find geometric invariants of these ligands, i.e., a set of features embedded in $\Re^3$ that is present in one or more valid conformations of each of the ligands. We refer to this problem as the pharmacophore identification problem since the pharmacophore is such an invariant. Solving this problem requires dealing efficiently with large amounts of spatial data and shape information. Ligand

Figure 1. This ligand is called 1TMN and it is one of the several inhibitors of the protease thermolysin. Figure 1(a) shows the space filling model of 1TMN. In this model a Van der Walls sphere [11] is considered around each atom center. Figure 1(b) shows the corresponding stick model in which only chemical bonds are drawn. The degrees of freedom of ligands include bond lengths, bond angles (angles between two consecutive bonds), and dihedral or torsional angles (angles formed by the first and third of three consecutive bonds, viewed along the axis of the second bond). In practice, only the torsional degrees of freedom are considered since these are the ones that exhibit large variations in their values [11]. Figure 1(c) shows the conformation of 1TMN of Figure 1(a) as a set of points in $\Re^3$. These points may represent atom centers or groups of atoms aggregated to one point endowed with a feature common to all these atoms (e.g., a rigid benzene ring) [37]. We assume that once a conformation is given, one can automatically transform it to a unique collection of points. For the purposes of this paper, a conformation $C$ may represent a valid geometric embedding of a molecule where bonds are retained, or a set of points without bond information.

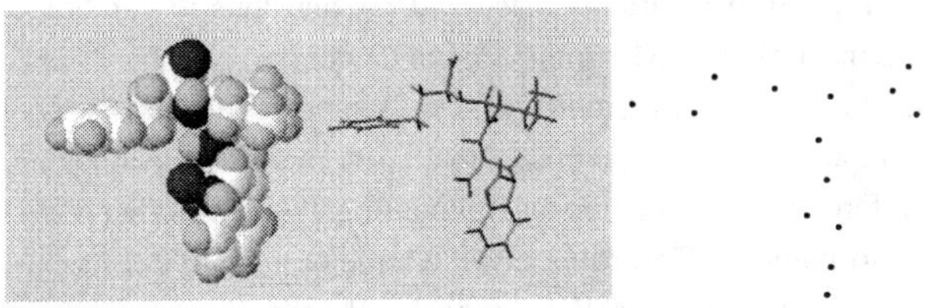

(a) (b) (c)

Figure 1: 1TMN: (a) space-filling, (b) stick, and (c) set-of-points models

molecules are very flexible and can assume many distinct potentially valid conformations. A valid conformation is a rigid spatial realization of the atoms of a molecule whose energy is below a predefined threshold [11]. It is not known, a priori, which of the many possible valid conformations of a ligand is the one that contains the desired pharmacophore.

Besides providing starting points for designing effective pharmaceutical drugs, there are several important applications where geometric invariant identification is useful. For example, the geometric invariants identified by our system can help in the formulation of database queries that may retrieve functionally equivalent, but structurally novel, molecules from molecular databases [44]. Geometric invariants can also suggest alignments of molecules for input to CoMFA (Comparative Molecular Field Analysis) and other 3D QSAR (Quantitative Structure-Activity Relationship) methods [9, 28]. These are methods that predict the activity of hypothetical compounds based on the assayed activity of previously synthesized ones.

This paper describes our efforts to prototype an integrated software system, called RAPID (RAndomized Pharmacophore Identification for Drug design), for addressing the pharmacophore identification problem. We present briefly the overall structure of RAPID in Section 2, and outline related work in Section 3. The two main modules of RAPID, conformational search and identification of invariants, are described in Sections 4 and 5. In this paper, we focus on the latter module and describe in some detail a randomized approach for solving an abstract geometric problem lying at the core of pharmacophore identification – the search for common subsets in a collection of point sets embedded in $\Re^3$. Our system attempts to deal with the numerous complications that arise in real examples. We report in Section 6 some preliminary experimental results. Finally, in Section 7, we conclude with a discussion of some open issues that merit consideration in future work.

## 2 Overview of RAPID

RAPID tries to identify geometric invariants among a collection of small ligand molecules like the molecule shown in

In RAPID, the identification of geometric invariants in a collection of flexible ligands denoted by $M = \{M_1, M_2, \ldots, M_N\}$ is treated as a two-stage process[1] addressing the two following problems:

**Problem 1 (Conformational Search)** *Given a collection of ligands* $M = \{M_1, M_2, \ldots, M_N\}$*, the degrees of freedom of each of them, and an energy function* $E$*, find, for each* $M_i$*, a set of conformations* $\{C_{i1}, C_{i2}, \ldots, C_{ik_i}\}$*, such that* $E(C_{ij}) \leq$ *THRESHOLD and* $d(C_{ij}, C_{il}) >$ *TOLERANCE for* $l \neq j$ *and* $j, l = 1, \ldots, k_i$*. THRESHOLD and TOLERANCE are pre-specified values, while* $d(\cdot, \cdot)$ *is a distance function.*

**Problem 2 (Invariant Identification)** *Given a collection of ligands* $M = \{M_1, M_2, \ldots, M_N\}$*, where each* $M_i$ *has a set of conformations* $\mathcal{C}(M_i) = \{C_{i1}, C_{i2}, \ldots, C_{ik_i}\}$*, determine a set of labeled points* $S$ *in* $\Re^3$ *with the property that for all* $i \in \{1, \ldots, n\}$*, there exists some* $C_{ij} \in \mathcal{C}(M_i)$ *such that* $S$ *is congruent to some subset of* $C_{ij}$*. A solution* $S$*, if it exists, is called an* invariant *of* $M$*.*

Although at this stage the two modules of RAPID work independently, we plan to support their interaction as the system develops. For example, the invariant identification module will be able to request from the conformational search module conformations that contain certain features of the molecule in specified spatial positions. RAPID also needs to deal with a host of complicating factors. In practice, the input may contain ligands that do not bind firmly to a receptor and do not contain the pharmacophore. This requires us to consider a

[1] A third module of RAPID, currently under development, involves the computation of molecular surfaces (see [20, 27]).

relaxation of Problem 2 above, where a geometric invariant need only be present in conformations of some $K$ of the $N$ molecules. Furthermore, an invariant may be perturbed in its spatial layout in distinct ligands' conformations, or there may be a large number of spurious invariants of differing sizes that do not correspond to a pharmacophore. Each of these contributes to a combinatorial explosion in the search process for the invariants.

## 3 Related Work

As far as conformational search is concerned, both systematic and randomized techniques are being investigated [10, 32, 33]. Systematic search methods sample each torsional degree of freedom of the ligand at regularly spaced intervals and minimize all conformations produced [34]. These techniques can be prohibitively expensive [32] and several heuristics are employed to quickly prune away conformations that are "close" to previously generated conformations [43]. Randomized methods work as follows: conformations are obtained by applying random increments to torsional degrees of freedom of the molecule starting with a user-specified initial conformation [22], or with a previously discovered low-energy conformation [14]. Recent articles, which attempt to compare different methods, emphasize the superior quality of the results obtained with randomized methods [22]. Other considered methods produce low-energy conformations, which obey distance constraints, using inverse kinematics [36], algebraic methods [24], or distance geometry techniques [17]. Note that the *protein-folding* problem is also a conformational search problem but its large size prohibits the use of the above techniques (see [18]).

Invariant identification is related to the well-studied problem in geometric optimization [1, 3, 2, 30] of finding common point sets. Determining the congruence of two point-sets in $\Re^3$ is tractable [1, 3] in the absence of complications such as noise. However, invariant identification is more closely related to the problem of identifying the *largest common point set* (LCP). Unfortunately, the LCP problem turns out to exceedingly difficult; in fact, even for $m$ collections of $n$ points on the real line, the LCP cannot be approximated to within an $n^\epsilon$ factor unless $P = NP$, and only weak positive results are known [2, 30]. Of course the problem is polynomially solvable when $m = 2$ [3] but that is not good enough for our purposes. Also, there are constraints in the molecular structures, but there are also various complications as discussed earlier.

In the computational chemistry literature, the most popular algorithms for invariant detection are based on clique-detection. For instance, DISCO [37] initially considers a pair of conformations $c_1$ and $c_2$ belonging to different molecules and constructs a "correspondence graph" $G$. The nodes of $G$ are all node pairs, and an edge is created if the pairs in each of the connected nodes can be matched simultaneously. A clique detection algorithm [13] is then used to find cliques in $G$. These correspond to invariants in $c_1$ and $c_2$, and thus to candidate pharmacophores. Maximum clique detection is NP-hard [21], but this algorithm seems to work well in practice [37, 44]. The approach has been generalized to $n$ conformations by choosing a reference conformation and comparing it with the other $n-1$ conformations. Unfortunately, if a large number of conformations per molecule are to be considered, there is a tremendous blow-up in the number of primitive operations performed by such algorithms [9], rendering them impractical. Given this situation, several new approaches are under development. For instance, one idea is to start with small invariants (2-3 features) and gradually expand them [9]. Our approach is fairly different and involves heavy use of randomized search techniques [39].

## 4 Conformational Search

The goal of conformational search as defined in Problem 1 (Section 2) is to produce a number of distinct low-energy conformations of a given ligand. We proceed as follows. Initially a large number of conformations are generated at random. In contrast with previous randomized search methods, we obtain a random conformation by selecting each degree of freedom from its allowed range according to a user-specified distribution. This distribution is frequently the uniform distribution. However, if some a priori information is available about the preferred values of a particular degree of freedom [31], then the corresponding values are selected according to a distribution that reflects the a priori information (i.e. a Gaussian distribution). An efficient minimizer [12, 40] is then used to obtain conformations at local energy minima. Minimization is the most time-consuming step during conformational search, so we have carefully optimized this procedure.

To obtain a representative set of conformations from our sample, we partition it into sets that reflect geometric similarity as captured by the distance measure DRMS. We define DRMS($C_i$,$C_j$) as the square root of the mean of the squared distances of the corresponding atoms of $C_i$ and $C_j$, after $C_i$ is transformed to $C_j$. This transformation is computed using a basis of three predefined atoms $a_1$, $a_2$, and $a_3$ (see Section 5). We perform a greedy clustering of the conformations by placing a given conformation in an existing cluster if its distance from the "center" of that cluster is less than the predefined value TOLERANCE. If no such cluster is found, a new cluster is created. The center of a cluster is the conformation with the lowest energy in the cluster. We omit further details of the clustering, but note that the implementation employs heuristics for improving the quality of output, e.g., it is ensured the cluster center is close to the average cluster conformation and a post-processing step checks that all cluster centers are sufficiently far apart. While we obtain reasonable experimental results, we are investigating optimizations via algorithms that minimize the maximum intercluster distance [25] and incremental clustering techniques [15].

Our experience with randomized techniques for searching high-dimensional spaces has shown that randomized exploration is superior to systematic exploration when the shape of the underlying space is irregular [29]. The same observation holds for conformational search: a systematic procedure has a higher chance of missing the irregularly shaped basins of attraction of the energy landscape of the molecule (see also [22]). This has been our main motivation for the development of the randomized conformational search procedure described above.

## 5 Identification of Invariants

The set of cluster centers, denoted by $\mathcal{C}(M) = \mathcal{C}(M_1) \cup \ldots \cup \mathcal{C}(M_N)$ is the input for the invariant identification module. Each conformation in $\mathcal{C}(M)$ is now represented as a set of labeled points in $\Re^3$ (see Section 2). We wish to determine a structure $S$ that is congruent to a substructure of some conformation in every molecule. The congruence relation is with respect to 3-D rotations and translations that ensure equality of labels.

Our formulation of the invariant identification problem in Problem 2 (Section 2) assumes noise-free data, specifically that all point positions are known exactly . In practice, atom positions are fuzzy, and it may not be possible to align them exactly. Therefore, we adopt the convention that two points $p_1$ and $p_2$ are said to *match* when $| p_1 - p_2 | \leq \epsilon$, where $\epsilon$ is the *point location error*. Similarly, two triangles are said to be congruent if each point in the first triangle is within $\epsilon$ of its corresponding point in the second.

The invariant identification problem is a variant of the *largest common point set problem (LCP)* in three dimensions, which is the following. Given $s$ point sets $P_1, P_2, \ldots, P_s$ in $\Re^3$, determine the point set of maximum cardinality congruent to some subset of each point set. For convenience, we assume that each point set $P_i$ has cardinality exactly $n$. For arbitrary $s$ and dimension $d$, LCP is hard to approximate within a factor of $n^\epsilon$, for some $\epsilon > 0$ [2]. In the sequel, we consider the following version of LCP, called LCP-$\alpha$: determine a point set $S$ of size $|S| \geq \alpha n$ congruent to some subset of each $P_i, 1 \leq i \leq s$. The motivation for focusing on this subproblem is that it more accurately captures the primary application, where pharmacophores are desired to have a certain minimum size.

We begin by focusing on the case $s = 2$, and later, in Section 5.2 indicate how to generalize for any value of s.

### 5.1 Phase 1: Pairwise Matching

In this section, we focus on the invariant identification problem for two point sets, denoted by MATCH. This problem has been studied extensively in the literature [41, 42, 4]. For the case when $\alpha = 1$, i.e., determining the congruence of two point sets, an algorithm that runs in time $O(n \log n)$ in two and three dimensions was presented in [7]. However, for general $\alpha$, the best known algorithms were obtained by [3]. These algorithms have a worst-case running time of $O(n^{4.6})$ for unknown $\alpha$, and $O(n^{2.6}/\alpha^2)$ (randomized) when $\alpha$ is known, for three dimensions. For two dimensions, the corresponding bounds obtained are $O(n^{3.2})$ and $O(n^{2.2}/\alpha)$. However, these bounds apply only for the noise-free model of point sets. The noisy version of the problem was considered in [5] yielding an $O(n^8)$ algorithm for the two dimensional version.

In this section, we describe two random-sampling schemes for solving LCP-$\alpha$ on noisy data. The first is quite natural, and the second uses more careful random sampling. Note that in these algorithms, we use the notions of congruence and matching that incorporate noise. However, the analysis assumes that the data is exact.

In the sequel, we use the notation $g(n) = \tilde{O}(f(n))$, where $f$ and $g$ are functions, to indicate that $g(n) = O(f(n) \log n)$. Also note that in three dimensions, a unique transformation $T$ (upto reflection) between two point sets $P_1$ and $P_2$ is determined by matching three points $p$, $q$, and $r$ in $P_1$ with three points $s$, $t$, and $u$ in $P_2$. Furthermore, it is known that [19]:

**Proposition 1** *Given a transformation $T$ from $P_1$ to $P_2$, in time $\tilde{O}(n)$ we can determine for each point $p \in P_1$ a corresponding point $p' \in P_2$ such that $|T(p) - p'|$ is minimized.*

The basic random sampling method is as follows.

**BASIC-SAMPLE:** For some constant $c$, perform $(c \log n)/\alpha^3$ iterations of the following sampling process: sample a triplet of points $< p_1, p_2, p_3 >$ randomly from $P_1$; determine three points in $P_2$ congruent to this set; compute the resulting induced transformation and determine the number of points in $P_1$ matching corresponding points in $P_2$; and, if the number exceeds $\alpha n$, declare SUCCESS.

**Theorem 2** *Given a common subset $S$ of size $|S| \geq \alpha n$, the probability that BASIC-SAMPLE fails to declare SUCCESS is at most $1/n$.*

The proof of this theorem is straightforward. Observe that the probability of picking three points that belong to the common substructure is at least $\alpha^3$. Applying a Chernoff bound [39] yields the desired probability.

**Theorem 3** *BASIC-SAMPLE runs in time $\tilde{O}(n^{2.8}/\alpha^3)$ using space $O(n^2)$.*

The running time can be written as $T = n\sum_{i=1}^{1/\alpha^3} \Delta_i$ where $\Delta_i$ is the number of candidate transformations generated during iteration $i$. Let $H_P(t)$ denote the number of triangles congruent to $t$ in point set $P$ (we follow the notation of [3]). Clearly, $\Delta_i = H_{P_2}(t_i)$, where $t_i$ is the triangle induced by the triplet chosen at iteration $i$. In [3], it is shown that $H_P(t) = O(n^{1.8})$. This yields a bound of $\tilde{O}(n^{2.8}/\alpha^3)$ on the running time for BASIC-SAMPLE.

Run-time profiling revealed that BASIC-SAMPLE examines many spurious triples (i.e., those that do not yield a large invariant). We propose the following modification of the random sampling procedure to fix the problem.

**PARTITION-SAMPLE:** For some constant $c$, perform $c \log n$ iterations of the following sampling process: randomly select two subsets $A$ and $B$ of size $1/\alpha$ from $P_1$; also select a subset $C$ of size $1/\alpha$ from $P_2$; store all distances $d(p,q)$, for all $p \in C$ and $q \in P_2 - C$, in a hash table; for every triangle $(a,b,p)$ with $a \in A, b \in B$, and $p \in P - (A \cup B)$, probe for $d(p,a)$ and $d(p,b)$ in the hash table to determine all matching triplets $(c, p_1, p_2)$ with $c \in C$ and $p_1, p_2 \in P_2 - C$; finally, as before, if the resulting transformation induces a match of more than $\alpha n$ points, declare SUCCESS.

**Theorem 4** *Given a common subset $S$ of size $|S| \geq \alpha n$, the probability that PARTITION-SAMPLE fails to declare SUCCESS is at most $1/n$.*

**Proof:** Let the transformation yielding $S$ be $T$. The following statement is clearly true: If $S \cap A \neq \emptyset$ and $S \cap B \neq \emptyset$,then this trial yields a transformation that finds $S$ with probability approximately $\alpha$.

We try all triplets $(a,b,p), a \in A, b \in B, p \in P_1 - A - B$. Since $S$ is large, at least one of these is a triangle wholly contained in $S$. We do not find the corresponding triangle in $P_2$ only if $C$ contains the image of $a$ or $b$ under $T$ (which happens with probability $\leq 1/n$), or if $C$ does not contain any point from $S$, (which happens with probability $(1-\alpha)^{1/\alpha}$).

The probability $f$ that $A \cap S = \emptyset$ is at most $(1-\alpha)^{1/\alpha} \leq e^{-1}$. The probability that a single random trial fails is the probability that $(A \cap S = \emptyset) \vee (B \cap S = \emptyset) \vee ((P_1 - A - B) \cap S = \emptyset)$. For $S \geq 4$, this probability is bounded by $2f + f^2$ which is constant. Therefore, by repeating $c \log n$ times, we obtain a probability of error $F \leq 1/n$.

**Theorem 5** *PARTITION-SAMPLE runs in time $\tilde{O}(n^{3.4}/\alpha^3)$ using space $O(n/\alpha^2)$.*

**Proof:**The number of candidate triangles $t_1, t_2, \ldots, t_k$ generated by iteration from $P_1$ is $n \cdot 1/\alpha \cdot 1/\alpha = n/\alpha^2$. Given a triangle $t_i$, and a point $c \in C$, let $H^c_{P_2}(t_i)$ be the number of triangles in $P_2$ incident on $c$ and congruent to $t_i$. The running time of the algorithm can now be given by the following expression:

$$T(n) = O(n/\alpha^2 + \sum_{c \in C} \sum_{i=1}^{k} nH^c_{P_2}(t_i) + n/\alpha^2 \sum_{c \in C} n^2).$$

The first term is the cost of generating the triangles $t_1, \ldots, t_k$. To estimate $H^c_{P_2}(t_i)$, we need to count the number of triangles in $P_2$ which share the same point $c$ and are congruent to $t_i$. For any such triangle $\{c, p_1, p_2\}$, let $d_1 = ||c - p_1||, d_2 = ||c - p_2||, d_3 = ||p_1 - p_2||$. Let $S_i$ be a sphere centered at $c$ of radius $d_i, i = 1, 2$. Consider any $\{c, p_1, p_2\}$ congruent to $t$. Without loss of generality, assume that $p_i \in S_i, i = 1, 2$. Now, once we fix $p_1$, all such points $p_2$ lie on the intersection of $S_2$ and a circle of radius $d_3$ centered at $p_1$. Therefore, the number of such points (and also the number of congruent triangles fixed at $c$ and $p_1$ is the number of incidences between points of $P_2$ and this fixed–radius circle. Hence, the desired total number of such triangles fixed at $c$ is merely the maximum number of incidences between $n$ circles of fixed radius and $n$ points in $\Re^3$. If we choose an arbitrary direction and project this down to two dimensions, the problem reduces to counting the number of incidences between $n$ ellipses and $n$ points in the plane, which is $O(n^{1.4})$ (from [16]).

Inserting this bound yields:

$$\begin{aligned} T(n) &= O(n^3/\alpha^3 + n\sum_{c \in C} \sum_{i=1}^{k} H^c_{P_2}(t_i)) \\ &= O(n^3/\alpha^3 + n^2/\alpha^3 \cdot n^{1.4}) \\ &= O(n^{3.4}/\alpha^3). \end{aligned}$$

Although the asymptotic running time of PARTITION-SAMPLE is worse than that of BASIC-SAMPLE, experiments (discussed in Section 6) reveal that PARTITION-SAMPLE consistently outperforms BASIC-SAMPLE, generating far fewer spurious triples with an improved degree of success. Intuitively, the modified partitioning favors "large" solutions over small ones. Given a solution $S$ of size $K$ in $P_1$, the probability that each of $A$, $B$, and $P_1 - A - B$ contain at least one point of $S$ increases with $K$.

Additionally, experimental results suggest that the predicted running times are fairly high. In Section 6 we also discuss possible reasons for this, in terms of actual distance distributions among molecules.

**Eliminating Redundant Solutions.** An invariant search algorithm could return many solutions that satisfy a containment relationship with respect to each other. To prevent such redundant solutions from propagating through the complete search procedure, we need to *prune* them. Given invariants $S_1$

and $S_2$, we can check if $S_1 \subseteq S_2$ by invoking the algorithm on these two sets with $\alpha$ set to 1. Clearly, since $\alpha$ is 1, a solution will be produced only if $S_1$ is contained in $S_2$. Essentially, this stage induces many instances of geometric pattern matching, where we wish to determine whether a set of points $P$ is congruent to some subset of a point set $Q$.

**Inaccuracies in Computing the Transformation.** The module at the core of the substructure search algorithm is the following: Given points $p_1, p_2, p_3 \in P_1$, and $p'_1, p'_2, p'_3 \in P_2$, compute a transformation $T$ from $P_1$ to $P_2$ that maps $p_i \rightarrow p'_i$, for $i = 1, 2, 3$. Such a transformation can be computed in many different ways. The simplest such way is: **(i)** align $p_1$ and $q_1$; **(ii)** align the vectors $\overrightarrow{p_1 p_2}$ and $\overrightarrow{q_1 q_2}$; and, **(iii)** align $p_3$ and $q_3$ by a rotation about $\overrightarrow{p_1 p_2}$. This approach over-constrains the fit between the two triangles.

For our purposes we need to compute a transformation that preserves this initial set of correspondences (within distance $\epsilon$), and matches the largest number of points from $P$ to $Q$. Since atom locations are noisy, such a transformation may not be unique, i.e. there may be different transformations that yield maximal but incomparable common subsets. In general, if we consider the six-dimensional space of all rigid transformations in $\Re^3$, the set of transformations satisfying the given correspondences is actually an enclosed volume over which we must maximize the transformation *score*.

If we knew the correspondences between points in $P_1$ and $P_2$, then a "balanced" transformation could be obtained by computing the transformation minimizing DRMS($P'_1$, $P'_2$), $P'_1$ and $P'_2$ being the subsets of $P_1$ and $P_2$ that are matched to each other [6] (see Section 4 for the definition of DRMS). This suggests the following refinement: Use the three-point procedure above to generate a plausible correspondence list, and then use a DRMS minimizing procedure to compute a refined transformation.

However, this approach is more time-consuming, and it is possible that DRMS minimization merely replaces one form of over-constraining by another. Our implementation employs various heuristics to minimize the effect of this problem. One such heuristic is the following: Determine a *seed* transformation $T$ by any of the above methods, and then sample three random pairs from the set of correspondences that $T$ induces, using these pairs to construct a new transformation. Clearly, in a perfect world, we will obtain $T$ again. However, given the inaccuracies in point location, it turns out that some choices of triplets may yield a larger number of correspondences than before.

### 5.2 Phase 2: Multiple Matching

The previous phase yields methods for solving LCP-$\alpha$ on two sets. In Phase 2, candidate solutions obtained from the previous phase are tested against the remaining molecules to determine the invariant. Each MATCH call operates on two conformations. Since each molecule is represented by many conformations, we extend MATCH to two molecules by doing all pair-wise matches between the sets of conformations. Note that comparing a candidate solution against a new conformation may result in $0, 1$, or many solutions, since the solution may decompose into smaller pieces on comparison.

There are various strategies one could use to process multiple molecules. A simple strategy does the obvious linear merging. We take each solution and compare it with the next molecule. We do this for all current solutions, concatenate and prune the results, and repeat with a new molecule. Another approach, which is in some ways less order-sensitive, is a tree-based merge. Here, we run the two-molecule algorithm on distinct pairs of molecules, and recursively combine the results using a binary tree. Both approaches are simple and correct, if point location error is not considered.

In addition to this, we may wish to find an invariant that does not exist in *all* the molecules, but in some fixed number of them. Our current strategy would fail to do this because all candidate solutions are compared against every molecule. We use a modified merging strategy here to keep track of the number of times an invariant fails to match against a molecule, and only reject those which exceed the maximum allowed number of failures.

If an invariant $S$ is contained in a point set $P$, then a call to MATCH($S$, $P$) should return a set of solutions containing $S$. This can easily be checked, which means that we can verify whether an invariant matches against a conformation. Our marking algorithm associates *marks* with each candidate solution. All original conformations are initialized with zero marks. All invariants that result from a MATCH acquire the sum of the marks of the input structures. Every time a solution fails to match against a new molecule (which means that it fails to match with *any* conformation of that molecule), we add a mark to it and continue to propagate it. We reject any solutions for which the number of marks exceeds the prescribed maximum.

## 6 Experimental Results

This section reports experimental results for the algorithms described above. All reported timings are on an SGI Indigo2 with a 175 MHz MIPS R8000 processor and 384MB RAM. Code was written in C/C++, and compiled using SGI CC.

**Input.** In Figure 2, we show four different inhibitors of the protease thermolysin. These molecules fit into the same cavity of thermolysin and by their presence inhibit the activity associated with that cavity. This example was chosen because all the inhibitors have been crystallized with thermolysin and their active conformations are known and recorded in the PDB database (a database of three-dimensional protein structures

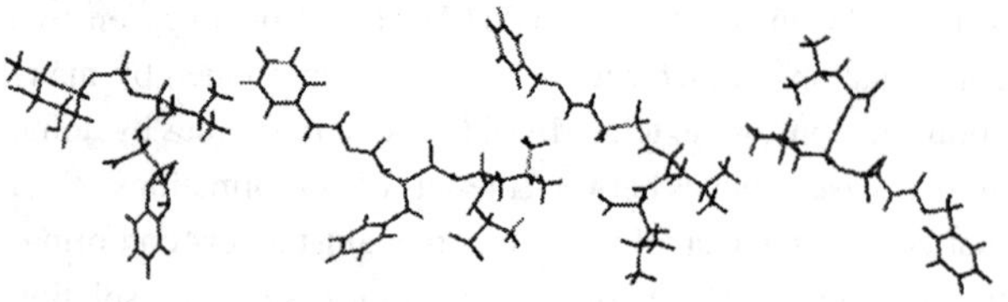

Figure 2: 1TLP, 4TMN, 5TMN, and 6TMN are inhibitors of thermolysin.

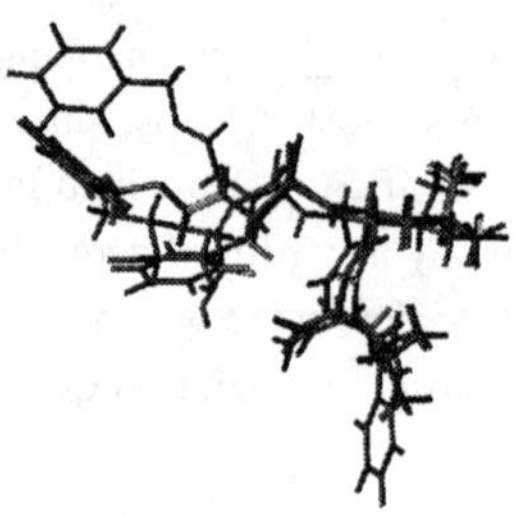

Figure 4: 1TLP, 4TMN, 5TMN, and 6TMN overlapped in their active conformations.

Figure 3: Different clusters of 1TLP.

stored at the Brookhaven National Laboratory). Note that 1TLP has 69 atoms and 10 torsional degrees of freedom, 4TMN has 68 atoms and 15 degrees of freedom, 5TMN has 64 atoms and 13 degrees of freedom, and 6TMN has 63 atoms and 12 degrees of freedom.

**Conformational Search.** Each of the molecules in Figure 2 was run through our conformational search software. A THRESHOLD value of 20 Kcals/mol was used for the energy of the valid conformations. It took 4.3 hours to to produce 4000 valid conformations for each molecule and reduce these to a set of representatives. The timings reported are very reasonable if one takes into account that 10, 15, 13, and 12 dimensional spaces are searched. The number of representatives produced was 850, 1192, 1024, and 955 for 1TLP, 4TMN, 5TMN, and 6TMN, respectively. The TOLERANCE value was set to 1.23 Å for all runs. A few clusters of 1TLP are shown in Figure 3. Since we know the active conformations of all the molecules, we performed a consistency check at the end of conformational search and we confirmed that the active conformations were close to one of the produced representatives.

An important issue in conformational search is to decide the number of valid conformations to produce. At this stage, this number is determined experimentally: we stop producing new conformations when these fall into existing clusters and do not increase significantly the overall number of clusters.

**Identification of Invariants.** To find the invariants in these four molecules, the conformations produced by the first stage were given as input to the search algorithm described in Section 5. For these tests, we considered each non-hydrogen atom to be a separate feature or point. Thus, each conformation had approximately 30 features drawn from 6 feature classes (the feature classes here being the atom types). We define the "solution" to be the overlapping portions of the molecules when aligned as shown in Figure 4. This is the upper right hand T-shaped portion of this diagram. The entire invariant consists of roughly 7 atoms and an additional 7 atoms of "scaffolding," or connecting atoms with no pharmacophore functionality.

Table 1 details the parameters used for invariant identification. The parameters $\epsilon$, $\delta$, and $\alpha$ are as described in Section 5. We set at 5 the number of times the MATCH algorithm is run per comparison, or the maximum number of invariants found per MATCH. These parameters were chosen by experimentation. For example, $\delta$ was selected to be much lower than $\epsilon$. We found that higher values of $\delta$ increased the running time without improving the quality of the solution.

For the sake of brevity, we omit detailed experimental analysis of variants of Multiple Search and use a linear search to compute invariants, with no failures in matching invariants permitted.

In Figure 6, we compare the two schemes experimentally. The running time of the two algorithms (measured in terms of the number of transformations produced) is plotted against $\alpha$ for varying $\alpha$. We see that as $\alpha$ gets smaller, PARTITION-SAMPLE clearly performs far better than BASIC-SAMPLE. At higher values, their behavior is more alike.

We also present in Table 2 a comparison between the two schemes on large data sets, where we vary the number of conformations per molecule. We include the active conformation in our sets. In all cases, the quality of solutions (in terms of the largest solution found) is comparable, and PARTITION-SAMPLE consistently runs faster than BASIC-SAMPLE. When the number of conformations increases, more invariants are produced because some of the added conformations have additional "scaffolding" which also could be matched. Yet, in every case, the invariant of the largest size is the "correct solution." Many non-trivial solutions of size 8 – 14 are also produced.

| Parameter | Search Value | Prune Value |
|---|---|---|
| $\epsilon$ | 1.3 | 3.5 |
| $\delta$ | 0.5 | 1.0 |
| $\alpha$ | 0.3 | 1.0 |

Table 1: Parameters for invariant search trials.

Although we prove worst-case running times for invariant search that are quite high, the algorithms run very quickly in practice. One source of discrepancy is in the analysis of the algorithms, where we use the best-known combinatorial upper bounds on the multiplicity of a fixed triangles induced by a set of points in $\Re^3$. However, these bounds are not tight, and actual molecules tend not to exhibit worst case behavior. In Figure 5, we plot the distribution of inter-atom distances in four of the molecules with which we experimented. For each molecule, the upper-most graph indicated the complete inter-point distance distribution. The lower graphs plot the distributions induced by the points chosen during random sampling (for various values of $\alpha$. Not surprisingly, the distribution becomes flatter as $\alpha$ increases. What is also interesting is that the figures indicate that all these molecules have a bounded diameter, and that the distribution is quite similar to a Poisson distribution. These facts indicate that a more careful analysis incorporating this information might lead to a better running time prediction for the algorithms, and possibly better algorithms.

## 7 Discussion

Our goal is to optimize the modules of the system to perform experiments that involve 5-20 ligands and a large number of conformations per ligand. As far as conformational search is concerned, we are investigating dynamic clustering techniques. By looking at the relative changes in the number of clusters found, we hope that we will be able to automatically end the conformational search stage. For the invariant identification stage, we will try to reduce the dependence of the implementation on the ordering of the input, a phenomenon observed in the current system. Our short-term plans also include the implementation of an algorithm that will extract features from conformations in the way that the computational chemists understand them. Currently, each atom gives rise to a feature that bears as a label the corresponding atom type. Computational chemists, however, have rules to group atoms into features that describe the chemical behavior of parts of the molecule, e.g., hydrophobic parts and positively charged parts. Features will reduce the number of points we consider per conformation and will hopefully help us deal with larger examples.

In Section 5 we mentioned certain combinatorial properties

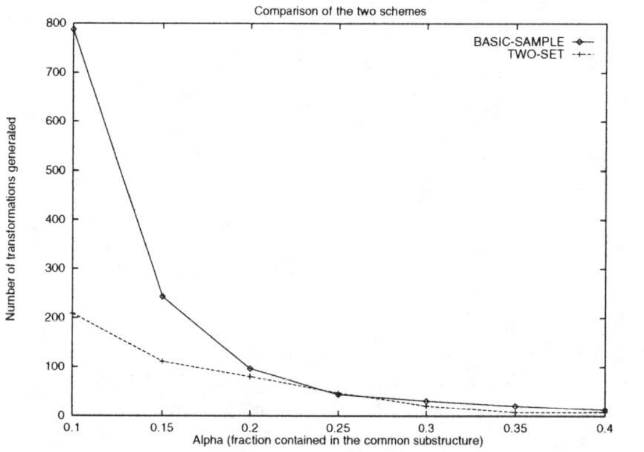


Figure 6: Relative performance of BASIC-SAMPLE and PARTITION-SAMPLE.

of point sets in two and three dimensions. Most of these properties are well understood for exact models i.e models in which all point locations are known perfectly. It would be interesting to determine analogous properties for noisy data sets. These would contribute directly to improving the worst case running time of the algorithms we have presented. In addition, it would be instructive to examine the average case behavior of these algorithms in the light of the remarks made previously about distance distributions in molecules.

## 8 Acknowledgements

The authors would like to thank Piotr Indyk and Dan Halperin for many helpful discussions.

## References

[1] T. Akutsu. On determining the congruence of point sets in higher dimensions. *Lecture Notes in Computer Science 834*, page 38.

[2] T. Akutsu and M. Halldórsson. On the approximation of largest common subtrees and largest common point sets. *Lecture Notes in Computer Science 834*, pages 405–413.

[3] T. Akutsu, H. Tamaki, and T. Tokuyama. Distribution of distances and triangles in a point set and algorithms for computing the largest common point set. *These proceedings*, 1997.

[4] H. Alt and L. Guibas. Discrete Geometric Shapes; Matching, Interpolation and Approximation. A survey. Manuscript 1996.

[5] H. Alt, K. Mehlhorn, H. Wagener, E. Welzl. Congruence, similarity and symmetries of geometric objects. *Discrete Comput. Geom.*, pp. 3:237–256, 1988

[6] K. S. Arun, T. S. Huang, and S. D. Blostein. Least-squares Fitting of Two 3-D Point Sets. *IEEE Trans. Pattern. Anal. Machine Intell.*, volume PAMI-9, No. 5, September 1987.

[7] M. D. Atkinson. An optimal algorithm for geometric congruence. *J. Algorithms* 8 (1987). pages 134–183.

[8] L. Balbes, S. Mascarella, and D. Boyd. A perspective of modern methods in computer-aided drug design. In K. Lipkowitz and D. B. Boyd, editors, *Reviews in Computational Chemistry*, volume 5, pages 337–370. VCH Publishers, 1994.

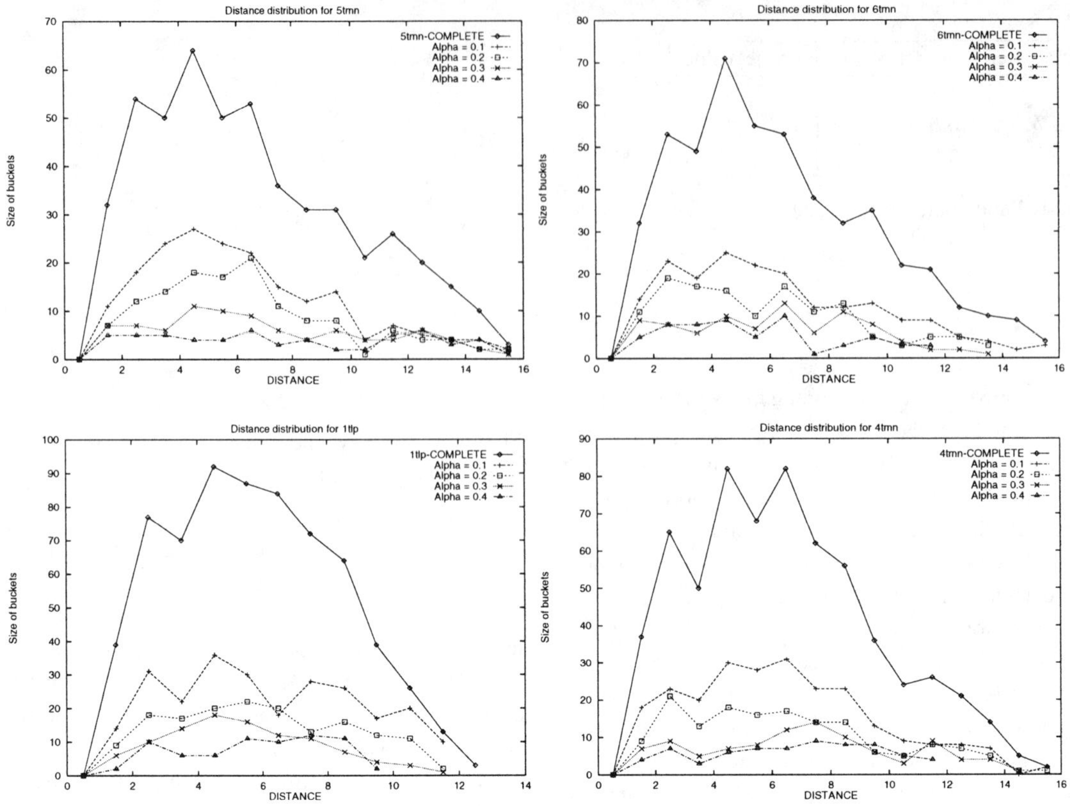


Figure 5: Distance distribution plots for 1TLP, 4TMN, 5TMN, and 6TMN

| Number of | Time to | Histogram of Solution Sizes | | | | | | | | | | | | |
|---|---|---|---|---|---|---|---|---|---|---|---|---|---|---|
| Conformations | Completion (in s) | 4 | 5 | 6 | 7 | 8 | 9 | 10 | 11 | 12 | 13 | 14 | 15 | total |
| BASIC-SAMPLE | | | | | | | | | | | | | | |
| 1 | 4.62 | | 2 | 1 | | | | | | | | 1 | | |
| 11 | 2127.66 | 90 | 34 | 20 | 5 | 6 | 4 | 1 | 1 | 1 | | | | |
| 21 | 24175.42 | 289 | 110 | 46 | 35 | 42 | 18 | 3 | 6 | 2 | 2 | 1 | | |
| PARTITION-SAMPLE | | | | | | | | | | | | | | |
| 1 | 2.85 | | | 1 | | 1 | | | | | 1 | | | |
| 11 | 656.47 | 20 | 9 | 8 | 11 | 11 | 3 | | | | | | | |
| 21 | 8060.00 | 73 | 55 | 53 | 30 | 21 | 25 | 7 | 8 | 1 | 1 | 4 | 1 | |

Table 2: Comparison of BASIC-SAMPLE and PARTITION-SAMPLE

[9] D. Barnum, J. Greene, A. Smellie, and P. Sprague. Identification of common functional components among molecules. To appear in *J. Chem. Inf. Comput. Sci.*, 1996.

[10] J. Blaney and S. Dixon. A good ligand is hard to find: Automated docking methods. *Perspectives in Drug Discovery and Design*, 1:301–319, 1993.

[11] D. B. Boyd. Aspects of molecular modeling. In K. Lipkowitz and D. B. Boyd, editors, *Reviews in Computational Chemistry*, volume 1, pages 321–351. VCH Publishers, 1990.

[12] R. Brent. *Algorithms for finding zeros and extrema of functions without calculating derivatives*. PhD thesis, Stanford University, 1971.

[13] C. Bron and J. Kerbosch. Finding all cliques of an undirected subgraph. *Commn. ACM*, 16:575–577, 1973.

[14] G. Chang, W. Guida, and W. Still. An internal coordinate monte-carlo method for searching conformational space. *J. Am. Chem. Soc.*, 111:4379–4386, 1989.

[15] M. Charikar, C. Chekuri, T. Feder, and R. Motwani. Incremental Clustering. To appear in *Proc. 29th Annual ACM Symposium on Theory of Computing*, May 1997.

[16] K. Clarkson, H. Edelsbrunner, L. Guibas, M. Sharir, and E. Welzl. Combinatorial Complexity Bounds for Arrangements of Curves and Spheres. *Discrete Comput. Geom.* 5:99–160 (1990)

[17] G. Crippen and T. Havel. *Distance Geometry and Molecular Conformation*. Research Studies Press, Letchworth, U.K., 1988.

[18] K. Dill. Folding proteins: Finding a needle in a haystack. *Current Opinion in Structural Biology*, 3:99–103, 1993.

[19] D. Dobkin and R. Lipton. Multidimensional search problems. *SIAM J. Computing* 5:181–186 (1976).

[20] P. Finn, D. Halperin, L. Kavraki, J.-C. Latombe, R. Motwani, C. Shelton, and S. Venkatasubramanian. Geometric manipulation of flexible ligands. In M. Lin and D. Manocha, editors, *LNCS Series – 1996 ACM Workshop on Applied Computational Geometry*. Springer-Verlag, 1996.

[21] M. Garey and D. Johnson. *Computers and Intractability: A Guide to the Theory of NP-Completeness*. Freeman, San Francisco, 1980.

[22] A. Ghose, J. Kowalczyk, M. Peterson, and A. Treasurywala. Conformational searching methods for small molecules: I. study of the sybyl search method. *J. of Computational Chemistry*, 14(9):1050–1065, 1993.

[23] R. Glen, G. Martin, A. Hill, R. Hyde, P. Wollard, J. Salmon, J. Buckingham, and A. Robertson. Computer-aided design and synthesis of 5-substituted tryptamines and their pharmacology at the $5 - HT_{10}$ receptor: Discovery of compounds with potential anti-migraine properties. *J. of Medical Chemistry*, 38:3566–3580, 1995.

[24] N. Go and H. Scherga. Ring closure and local conformational deformations of chain molecules. *Macromolecules*, 3(2):178–187, 1970.

[25] T. F. Gonzalez. Clustering to minimize the maximum intercluster distance. *Theoretical Computer Science*, 38:293–306, 1985.

[26] W. Guida, R. Bohacek, and M. Erion. Probing the conformational space available to inhibitors in the thermolysin active site using monter carlo/energy minimization techniques. *J. of Comp. Chem.*, 13(2):214–228, 1992.

[27] D. Halperin, C. Shelton. A perturbation scheme for spherical arrangements with application to molecular modeling. *This proceedings, 1997*

[28] A. Itai, N. Tomioka, M. Yamada, A. Inoue, and Y. Kato. Molecular superposition for rational drug design. In Kubingi, editor, *3D QSAR in Drug Design: Theory, Methods, and Applications*. ESCOM, 1995.

[29] L. Kavraki. *Random Networks in Configuration Space for Fast Path Planning*. PhD thesis, Stanford University, 1995.

[30] S. Khanna, R. Motwani, and F. F. Yao. Approximation algorithms for the largest common subtree problem. Technical Report STAN-CS-95-1545, Department of Computer Science, Stanford University, 1995.

[31] G. Klebe and T. Mietzener. A fast and efficient method to generate biologically relevant conformations. *J. of Computer-Aided Molecular Design*, 8:583–606, 1994.

[32] A. Leach. A survey of methods for searching the conformational space of small and medium sized molecules. In K. Lipkowitz and D. Boyd, editors, *Reviews in Computational Chemistry*, volume 2, pages 1–47. VCH Publishers, 1991.

[33] T. Lengauer. Algorithmic research problems in molecular bioinformatics. In *IEEE Proc. of the 2nd Israeli Symposium on the Theory of Computing and Systems*, pages 177–192, 1993.

[34] M. Lipton and W. Still. The multiple minimum problem in molecular modeling: Tree searching internal coordinate conformational space. *J. of Computational Chemistry*, 9(4):343–355, 1988.

[35] T. Lybrand. Computer simulation of biomolecular systems using molecular dynamics and free energy perturbation methods. In K. Lipkowitz and D. B. Boyd, editors, *Reviews in Computational Chemistry*, volume 1, pages 295–320. VCH Publishers, 1990.

[36] D. Manocha, Y. Zhu, and W. Wright. Conformational analysis of molecular chains using nano-kinematics. *Computer Application of Biological Sciences (CABIOS)*, 11(1):71–86, 1995.

[37] Y. Martin, M. Bures, E. Danaher, J. DeLazzer, and I. Lico. A fast new approach to pharmacophore mapping and its application to dopaminergic and benzodiazepine agonists. In *J. of Computer-Aided Molecular Design*, volume 7, pages 83–102, 1993.

[38] Y. C. Martin, M. G. Bures, and P. Willet. Searching databases of three-dimensional structures. In K. Lipkowitz and D. B. Boyd, editors, *Reviews in Computational Chemistry*, volume 1, pages 213–256. VCH Publishers, 1990.

[39] R. Motwani and P. Raghavan. *Randomized Algorithms*. Cambridge University Press, 1995.

[40] D. Pierre. *Optimization theory with applications*. Dover, NY, 1986.

[41] P. Raghavan and S. Irani. Combinatorial and experimental results for randomized point matching algorithms. *Proc. 12th. ACM. SCG*, pp. 68-77, 1996.

[42] P. J. Rezende and D. T. Lee. Point set pattern matching in $d$-dimensions. *Algorithmica*, 13, 387-404, 1995.

[43] A. Smellie, S. Kahn, and S. Teig. Analysis of conformational coverage: 1. validation and estimation of coverage. *J. Chem. Inf. Comput. Sci.*, 35:285–294, 1995.

[44] P. Willet. Searching for pharmacophoric patterns in databases of three-dimensional chemical structures. *J. of Molecular Recognition*, 8:290–303, 1995.

# Rotational Polygon Overlap Minimization

Victor J. Milenkovic
University of Miami
Department of Mathematics and Computer Science*

**Abstract**

An effective and fast algorithm is given for *rotational overlap minimization*: given an overlapping layout of polygons $P_1, P_2, P_3, \ldots, P_k$ in a container polygon $C$, translate and rotate the polygons to a layout that minimizes an overlap measure. A (local) overlap minimum has the property that any perturbation of the polygons increases the chosen measure of overlap. Experiments show that the algorithm works well in practice. It is shown how to apply overlap minimization to create algorithms for other layout tasks: *compaction*, *containment*, and *minimal enclosure*. **Compaction**: starting with a non-overlapping layout in a rectangular container, plan a non-overlapping motion that minimizes the length or area of the container. **Containment**: place the polygons into a (possibly non-convex container) without overlapping. **Minimal enclosure**: find a non-overlapping layout inside a minimum-length, fixed-width rectangle or inside a minimum area rectangle. All of these algorithms have important industrial applications.

## 1 Introduction

A number of industries generate new parts by cutting them from stock material: cloth, leather (hides), sheet metal, glass, *etc.* These industries need to generate dense non-overlapping layouts of polygonal shapes. Because fabric has a grain, apparel layouts usually permit only a finite set of orientations. Stripes, plaids, or other patterns on the fabric can further limit the allowed orientations and translations. Nevertheless, apparel manufacturers often allow small rotations (called "tilting"), typically not more than 3 degrees, but still large enough to make a significant difference in cloth utilization. Glass and sheet metal (and sometimes leather) have no grain (or stripes or plaids), and therefore layout on these materials allows arbitrary orientations for the parts.

From the point of view of computational geometry, there are at least four problems related to layout.

**Overlap Minimization** Given an overlapping layout of polygons $P_0$, $P_1$, $P_2$, ..., $P_k$, move the polygons to a layout that minimizes some measure of overlap. Usually, $P_0$ is the unbounded complement $\overline{C}$ of a *container* polygon.

**Compaction** Given a non-overlapping layout of $P_1$, $P_2$, ..., $P_k$, plan a motion that minimizes some measure. For most textile applications, the polygons are inside a rectangular container $R$ with fixed width and the goal is to minimize the length. Both overlap minimization and compaction generate *local* minima.

**Containment** Given polygons $P_1$, $P_2$, ..., $P_k$ and a fixed container $C$, place the polygons into $C$ without overlapping. As in the case of overlap minimization, $P_0$ denotes $\overline{C}$, in which case the goal becomes to find a non-overlapping layout of $P_0$, $P_1$, ..., $P_k$.

**Minimum Enclosure** Find a non-overlapping layout of $P_1$, $P_2$, ..., $P_k$ that minimizes some measure of the container. The most common version in the textile industry is *strip packing*: minimize the length of a rectangular container $R$ of fixed width. It is also useful to minimize the area of $R$.

These four problems are closely related mathematically and all have important applications in industry. Unfortunately, all are NP-hard or PSPACE-hard if $k$ is not a constant. Section 1.3 discusses our previous work on theory, algorithms, and commercially successful software for *translational* versions of the problems. One of the keys to all of these algorithms is a fast technique for *overlap minimization* using a combination of techniques from computational geometry and mathematical programming. In particular, compaction is basically a special case of overlap minimization, and overlap minimization is essentially "half" of the solution to containment and minimum enclosure.

### 1.1 New Results.

This paper gives a new algorithm for *rotational* overlap minimization. Like our translational compaction and overlap minimization algorithms, it finds the minimum through the solution of a sequence of linear programs. Our experiments demonstrate that, like our translational algorithms, the new

*P.O. Box 249085, Coral Gables, FL 33124, http://www.cs.miami.edu/~vjm, vjm@cs.miami.edu. This research was funded by the Textile/Clothing Technology Corporation from funds awarded by the Alfred P. Sloan Foundation, by NSF grant CCR-91-157993, and by a subcontract of a National Textile Center grant to Auburn University, Department of Consumer Affairs.


*Computational Geometry 97* Nice France
Copyright 1997 ACM 0-89791-878-9/97/06 ..$3.50

rotational algorithm has "supernatural" convergence. We say "supernatural" because, unlike standard physically based energy minimization techniques, our algorithm anticipates collisions and overlaps before they happen and therefore converges in fewer steps than the "natural" methods.

For convex $P_i$ and $P_j$, this paper defines OVERLAP($P_i, P_j$) to be the length of the shortest translation of $P_j$ that moves it off $P_i$. For non-convex $P_i$ and $P_j$, we define a more complicated measure for OVERLAP($P_i, P_j$) but one which matches the intuitive notion of overlap. In principle, our algorithm can minimize any monotone (see Section 2.2) differentiable function of the pairwise overlaps. Also, if the user desires, the algorithm can (independently) limit the range of motion or range of angle for each polygon $P_i$. In our experiments, we minimize the overlap measure,

$$\sum_{0 \leq i < j \leq k} \mathrm{OVERLAP}(P_i, P_j).$$

The orientation/position of $P_0 = \overline{C}$, the complement of the container, is fixed, and the other polygons are unconstrained.

This paper also gives new algorithms for rotational compaction and containment which use the new rotational overlap minimization algorithm. The containment algorithm also uses our previous algorithms for translational containment. Previous work [6, 18] has shown that fast algorithms for translational containment and compaction can be used to create algorithms for translational minimum enclosure. The same reasoning works for the rotational case. Implementing rotational versions of these algorithms is a subject for future work. In particular, rotational containment requires the development of robust algorithms for *range intersection*: the intersection of all copies of a polygon as it rotates through a range of angles. The final version of a practical rotational containment algorithm may be different than the one given here, but it will certainly make heavy use of overlap minimization.

The following section discusses the significance of the new work in rotational layout: in theory and in practice. Section 1.3 compares the new algorithms to related work. Finally, Section 1.4 gives a road-map to the rest of the paper.

## 1.2 Significance.

Overlap minimization and compaction take an existing (overlapping or non-overlapping) layout and move the polygons to a *local* minimum of some measure (overlap or container length or area). Analogously, containment and minimum enclosure seek *global* minima.[1] One might think that finding a global minimum is theoretically more difficult or practically more significant than finding a local minimum, but such is not necessarily the case. Translational containment and minimum enclosure are "merely" NP-complete[2] [6, 19], although it is unknown whether rotational containment is in NP. In contrast, compaction and overlap minimization are likely to be PSPACE-hard. Also, among all the algorithms, compaction is by far the most commercially successful algorithm to date. All four problems are theoretically interesting and practically significant.

### 1.2.1 Theoretical Significance.

In his Ph.D. thesis [14], Li proved the PSPACE-completeness of *global compaction*: plan a non-overlapping motion to the global minimum length layout reachable from the current layout. Li used a modification of the hardness proof for the Warehouseman problem [13]. It is also likely that many forms of *local* compaction (the type dealt with in this paper) are also PSPACE-hard, depending on the answers to the following questions. 1) Is it possible to implement a totally mechanical 3D Turing machine? 2) What if parts are only allowed to translate? 3) What about in 2D? A positive answer to any of these questions is equivalent to a proof of PSPACE-hardness for compaction in the corresponding domain. The answer to (1) is certainly yes. Question (2) is likely, and (3) is definitely still open. However, we have shown how to build a translational 2D "counter" [15] which is "powered" by (local) compaction. This means that any "reasonable" compaction algorithm has lower bound running time exponential in $k$. "Reasonable" means the algorithm generates a description of the compaction motion as part of its internal operation. It can easily be shown that all hardness results for compaction also apply to overlap minimization.

### 1.2.2 Practical Significance.

Companies cannot simply generate a few layouts which they heavily reuse to satisfy their needs. At least three forces act to drive companies to generate as many layouts as possible. 1) Producing a greater variety of styles, sizes, and items increases a company's market share. 2) To keep inventories low, companies need to flexibly generate different mixes of sizes, styles, and items to closely track varying demand. 3) Stock dimensions are not standardized: rolls of cloth vary in width from 55 to 60 inches, and the width available might not be known until the day of delivery from the supplier. We know of a single large clothing manufacturer that generates 1500 layouts of 100 or more parts per week–all by hand![3] Nevertheless, the compensation for the layout workers is less that 1% of the money spent on cloth. Thus, if computers are to replace humans, they must generate layouts with nearly the same density. Unfortunately for computers, trained talented humans generate very dense layouts: 89-91% cloth utilization for men's pants, for example.

Compaction is currently the most practically significant algorithm because it can take human-generated layouts as input and improve them. Our translational compaction algorithm [18, 17] can compact an industrial layout of over one hundred clothing parts in 4 or 5 iterations and 1 or 2 minutes on a PC. For some domains, such as intimate apparel, compaction can reduce cloth waste by over 1%, and the average reduction is about 0.5%. For men's pants, the average reduction is about 0.25%. For one manufacturer with \$100 million in cloth cost, compaction saves them about \$250,000 per year. Anticipated yearly savings for customers already under contract[4] is \$2,750,000, and the customer base is growing

[1] Actually, containment only asks if the global minimum overlap is zero or non-zero. Our containment algorithms could be modified to find the actual global minimum overlap layout if the minimum is non-zero, but that is not such an interesting question in practice.

[2] Containment is the feasibility question and minimum enclosure is the optimality question. From the point of view of complexity theory, they are the same question.

[3] Actually, the hand is moving a mouse and using a CAD system which provides some assistance in single-polygon placement: automatically nestling it up against neighboring polygons.

[4] Customers are under contract to Gerber Garment Technologies, which currently holds exclusive license to our compaction

rapidly. Textile manufacturers do not allow arbitrary rotation of the parts because of the grain of the fabric. However, they often allow parts to "tilt" (industry term for a small rotation) a few degrees. Li attempted to add small rotations to translational compaction (Section 1.3). Even though the tilts are small, our experiments on 162 industrial layouts this early version of rotational compaction have shown that rotational compaction saves at least 49% more cloth than translational compaction alone. For industries cutting sheet metal, glass, and (sometimes) leather, the parts can have arbitrary orientations. For these industries, it is essential that compaction and the other layout algorithms can handle rotation and translation.

Compaction is essentially a special case of overlap minimization, but overlap minimization is useful directly as part of *database driven layout* [14, 16]. Companies with large databases of existing layouts can use them as starting point, substitute parts of different style or different but closely matched size, and generate new layouts. This idea has been in the apparel industry for a number of years, and the main obstacle has been the problem of re-laying the parts without overlaps. Clearly, overlap minimization is an excellent starting point, although other techniques might also be needed. We see this as next commercially significant application after compaction. In apparel applications, *rotational* overlap minimization could take advantage of tilting, and it would be essential for glass and sheet metal applications.

In the apparel industry, no layout software has ever replaced a human. Optimal strip packing (a version of minimum enclosure) for $k > 100$ parts would replace humans, but NP-completeness will probably prevent fast enough solutions for $k$ significantly larger than 10. Nevertheless, as Daniels demonstrates in her Ph.D. thesis [6], and we subsequently confirm [7], fast containment algorithms for modest $k$ are an extremely useful tool for building heuristics or approximate algorithms for large values of $k$ (some of our heuristics match human performance a significant fraction of the time). It is clear that *rotational* containment algorithms will similarly permit more powerful layout heuristics in domains with arbitrarily oriented parts.

### 1.3 Related Work.

In the literature, there are essentially four approaches to layout of non-convex polygons: 1) heuristics and meta-heuristics (neural nets, simulated annealing, genetic algorithms), 2) physical simulation, 3) "classical" computational/combinatorial geometry, 4) a combination of computational geometry and mathematical programming (which we use here).

Heuristics and meta-heuristics are limited in theory because they cannot say "no" if there is no solution. Currently, they are limited in practice to falling about 5% behind humans in cloth utilization. See [6, 9, 10, 21] for a survey of published results. As stated in the previous section, heuristics will ultimately be *necessary* to solve large layout problems, but it is clear that they are not *sufficient* by themselves.

Another approach one might take to compaction or layout is to treat the polygons as solid physical objects, add some potential energy functions, and then use numerical methods for simulating physics. However, this approach is very slow in practice for many crowded parts. Experiments have shown that physically based techniques are much slower than our translation compaction algorithm [14, 16] for compacting typical layouts of hundreds of polygons. In recent work [20], we show that a generalization of our translational compaction algorithm can realistically animate 1000 spheres in an hourglass at least 1000 times faster than the latest techniques for true physical simulation of translating 3D solids. We expect the same to be true when rotation is added.

software.

Chazelle [5] introduced the single-polygon containment problem: place $m$-gon $P$ into $n$-gon container $Q$. For convex $Q$, the running time bound is $O(mn^2)$, and for non-convex $P$ and $Q$, $O(m^3n^3(m+n)\log(m+n))$. Avnaim *et al.* [4, 3] improve this to $O(m^3n^3\log(m+n))$. Most recent work deals with finding the largest copy of a convex $P$ that can be placed, which is equivalent to finding the minimum (scaled) enclosure. For convex $Q$ the best running time is $O(mn^2\log n)$ [1], and for non-convex $Q$, $O(m^2n^2)$ [2]. Grinde and Cavalier use an extensive case analysis and linear programming to place a single convex $P$ [11] or convex $P_1$ and $P_2$ [12]. The running time of the first algorithm appears to be $O(m^2n^3)$, and it is not clear what the running time of the second is, but for one of its cases they are able to use parametric programming and find a solution by solving $O(m^4n^4)$ linear programs. This is the only previous result on multi-polygon rotational containment we are aware of.

For multi-polygon *translational* layout, we have had considerable theoretical and practical success using a combination of computational geometry and mathematical programming (CG/MP) [8, 19, 18]. As previously stated, these algorithms can practically solve translational compaction for hundreds of polygons and containment for up to ten polygons. These algorithms heavily rely on a model for compaction/overlap minimization described below. Using the same approach, we have proved a number of theoretical running time bounds including $O((m^2+mn)^{2k}\log n)$ for placing $k$ non-convex $m$-gons into a non-convex $n$-gon and $O(m^{4k-4}\log m))$ for placing them into a minimum area, fixed orientation rectangle [17].[5]

All of our previous translational layout algorithms compute an explicit configuration space, $U_{ij} = P_i \oplus -P_j$, for each pair of polygons, where $\oplus$ is the Minkowski sum. Translational compaction/overlap minimization selects a convex subset $C_{ij} \subseteq U_{ij}$ containing or near to $t_j - t_i$, where $t_i$ is the current translation of $P_i$. (The work defines the overlap of $P_i$ and $P_j$ to be the distance $t_j - t_i$ lies outside $U_{ij}$.) The overlap minimization algorithm sets up a linear program to diminish the distance of each $t_j - t_i$ from $C_{ij}$. Compaction sets up a linear program to keep $t_j - t_i$ inside $C_{ij}$ and minimize the length of the container.

Like the translational algorithms, the rotational algorithms presented in this paper use a CG/MP approach, but it differs considerably in detail. With the addition of a rotation parameter, the configuration space $U_{ij}$ becomes a six-dimensional set with curved surface boundaries. At present, this space is much too complex to compute robustly. Li had previously attempted to use the translational configuration spaces as an approximation to the rotational spaces for rotational compaction [14]. On many inputs, this version of compaction only goes part of the way to a local minimum. The algorithms presented here use a different approach that does not require the configuration spaces to be explicitly constructed. (Nor do they use an extensive case analysis as does Grinde and Cavalier.) Some of the methods used here are standard in the (vast) quadratic programming literature, but we are not aware of any previous work on overlap minimization of non-convex polygons.

[5]This paper also surveys other results in multi-polygon translational layout.

### 1.4 Outline.

Section 2 gives an algorithm for rotational overlap minimization of *convex* polygons. Section 3 generalizes it to non-convex polygons. Section 4 describes how to apply it to accomplish compaction, containment, and minimum enclosure. Section 5 gives experimental results.

## 2 Overlap Minimization

This section gives an algorithm for rotational overlap minimization of *convex* polygons. It first describes a non-linear optimization problem for overlap minimization of convex polygons. This problem is linearized, and interpolation is applied to find an overlap-diminishing step. This step is repeated until the system converges to a local minimum.

### 2.1 Notation

In this paper, lower case letters generally denote scalars, points, and vectors. Upper case letters denote sets of points, usually polygonal regions. Scalar multiplication and vector/point addition are defined as usual. This section employs three vector multiplications:

$$a \cdot b = a_x b_x + a_y b_y, \quad (1)$$

$$a \times b = a_x b_y - a_y b_x, \quad (2)$$

$$ab = (a_x b_x - a_y b_y, a_x b_y + a_y b_x). \quad (3)$$

Equation 1 is the standard dot product; 2 is a 2D scalar cross product; and 3 is the product of complex numbers. The last is very useful for representing rotations: if $u = (\cos\theta, \sin\theta)$, then $up$ is $p$ rotated by angle $\theta$ about the origin. Any operation on points/vectors can be generalized to point-sets. For instance, $P+a = \{p+a \,|\, p \in P\}$ and $uP = \{up \,|\, p \in P\}$.

### 2.2 Non-linear Formulation.

This section describes a non-linear optimization problem which is equivalent to minimizing the overlap of a collection of convex polygons.

Let $P_0, P_1, P_2, \ldots, P_k$ be convex polygons. Let $t_i$ and $u_i = (\cos\theta_i, \sin\theta_i)$ be vector variables representing the translation and rotation of $P_i$, $0 \le i \le k$. Under the notation of the previous section, $u_i P_i + t_i$ represents polygon $P_i$ rotated by $\theta_i$ and translated by $t_i$. Let $L_{ij}$ be the line which *best separates* $u_j P_j + t_j$ from $u_i P_i + t_i$, $0 \le i < j \le k$: 1) $L_{ij}$ is tangent to $u_j P_j + t_j$ and 2) $u_i P_i + t_i$ protrudes the minimum distance $o_{ij}$ into the half-plane containing $u_j P_j + t_j$ (Figure 1). Let $u_{ij}$ be the unit normal vector to $L_{ij}$, pointing into the half-plane containing $u_j P_j + t_j$, and let $d_{ij}$ be the distance from the origin to $L_{ij}$ along the direction $u_{ij}$.

It is easy to prove that $L_{ij}$ is parallel to some edge of $u_i P_i$ or $u_j P_j$ and that $o_{ij} = \text{OVERLAP}(u_i P_i + t_i, u_j P_j + t_j)$, which this paper defines as the length $|t_{ij}|$ of the shortest translation vector $t_{ij}$ such that $u_j P_j + t_j + t_{ij}$ does not overlap $u_i P_i + t_i$.

For $0 \le i < j \le k$, the separation condition can be expressed as,

$$(u_i p_i + t_i) \cdot u_{ij} \le d_{ij} + o_{ij}, \quad (4)$$

for all $p_i \in P_i$, and

$$(u_j p_j + t_j) \cdot u_{ij} \ge d_{ij}, \quad (5)$$

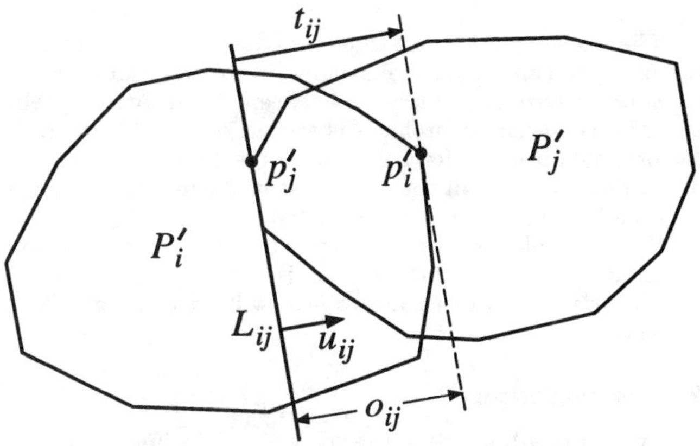


Figure 1: Line $L_{ij}$ best separates overlapping convex polygons $P_i' = u_i P_i + t_i$ and $P_j' = u_j P_j + t_j$. Vertices $p_i' = u_i p_i + t_i$ and $p_j' = u_j p_j + t_j$ are extremal with respect to the direction $u_{ij}$ of minimum overlap $o_{ij}$. (Extremal vertices are not necessarily unique.) Vector $t_{ij}$ is the shortest translation of $P_j'$ that will move it off of $P_i'$.

for all $p_j \in P_j$. Since $P_i$ and $P_j$ are convex polygonal regions, it is sufficient to allow $p_i$ and $p_j$ to range over the *vertices* of $P_i$ and $P_j$, respectively. Since overlaps cannot be negative, we add the constraint $o_{ij} \ge 0$. We refer to these as the *separation constraints.*

Overlap minimization can be expressed as minimizing some *monotone* objective function, $F(o_{ij}, 0 \le i < j \le k)$ on the pairwise overlaps while preserving the separation constraints. A function $F$ is *monotone* in each $o_{ij}$ if increasing any single $o_{ij}$ increases $F$ or leaves it the same. Examples of monotone $F$ are 1) the maximum $o_{ij}$, 2) the sum of the $o_{ij}$, and 3) the sum of the squares of the $o_{ij}$. In our experiments, we use a weighted sum $\sum_{0 \le i < j \le k} w_{ij} o_{ij}$ with non-negative weights.

### 2.3 Linearizing.

This section describes a linearization of the minimization problem of the previous section, resulting in an *overlap minimizing linear program* (OMLP). In this case, $t_i, u_i, u_{ij}, d_{ij}, o_{ij}$ represent *current* (fixed) values and $\Delta t_i, \Delta\theta_i, \Delta\theta_{ij}, \Delta d_{ij}, \Delta o_{ij}$ represent the perturbations of these values. (It is more convenient to use $\Delta\theta_i$ and $\Delta\theta_{ij}$ for the perturbations and $u_i$ and $u_{ij}$ for the unperturbed values.)

Substituting, $t_i + \Delta t_i$ and so forth in Equations 4 and 5 and discarding quadratic terms yields,

$$\begin{aligned}(u_i p_i) \times u_{ij} \Delta\theta_i + u_{ij} \cdot \Delta t_i - (u_i p_i + t_i) \times u_{ij} \Delta\theta_{ij} \\ -\Delta d_{ij} - \Delta o_{ij} \le d_{ij} + o_{ij} - (u_i p_i + t_i) \cdot u_{ij},\end{aligned} \quad (6)$$

$$\begin{aligned}(u_j p_j) \times u_{ij} \Delta\theta_j + u_{ij} \cdot \Delta t_i - (u_j p_j + t_j) \times u_{ij} \Delta\theta_{ij} \\ -\Delta d_{ij} \ge d_{ij} - (u_j p_j + t_j) \cdot u_{ij}.\end{aligned} \quad (7)$$

To keep the overlaps non-negative, $\Delta o_{ij} \ge -o_{ij}$.

The linearized objective is

$$\sum_{0 \le i < j \le k} F_{ij} \Delta o_{ij}, \quad (8)$$

where $F_{ij}$ is the partial derivative of $F$ with respect to $o_{ij}$ at the fixed values of the variables. For our experiments, $F_{ij} = w_{ij}$.

The linearized optimization problem is always feasible because one can set all the $\Delta$-variables to zero. In a neighborhood of zero, the difference between the linearized problem and the original problem is second-order. Therefore if the original optimization has a "negative derivative" in some direction, then so will the linearized problem. To be mathematically precise, we have to include all critical points including unstable equilibria in our definition of "local overlap minimum". As long as the system is not at a critical point, the linearized optimization problem will have a non-trivial solution.

### 2.4 Interpolation.

Because of non-linearities, the solution to the linearized system usually does not necessarily have smaller overlap than the original configuration. However, we can interpolate between the current configuration and the new configuration: for $0 < s < 1$, replace $t_i$ by $t_i + s\Delta t_i$, $\theta_i$ by $\theta_i + s\Delta\theta_i$, and so forth. The linearization ensures that for sufficiently small $s$, the interpolated configuration *will* have smaller overlap.

One can solve exactly for the optimum value of $s$, but the set of equations is messy and at least fourth degree. Our current implementation uses sampling and interval halving. Abusing notation, let $F(s)$ be the overlap for interpolation value $s$. We know that $F'(0) < 0$, and we are seeking $s_{\min}$, the first minimum of $F(s)$ for $0 < s \leq 1$. If we have two sample values $s_a < s_b$ such that $F(s_a) < F(s_b)$, then we know that $s_{\min} < s_b$, and we can "throw away" sample intervals to the right of $s_b$. In our implementation, we alternate between sampling the midpoint of 1) the longest interval and 2) the last (rightmost) interval between samples.

A good value of $s$ yields a new configuration with diminished overlap. The linearization and interpolation are repeated until the system converges to a local overlap minimum. It is highly unlikely that the system would converge to an unstable equilibrium in practice.

## 3 Non-Convex Polygons

One might think that generalizing overlap minimization to non-convex polygons would involve the convex hull, but in fact the trick is to use *inner* convex approximations instead. This section describes how to define OVERLAP$(P, Q)$ for non-convex polygons based on inner convex approximations and to use this definition to generalize the overlap minimization algorithm. Implementation issues are also addressed.

### 3.1 Defining an OVERLAP Measure.

Let $\mathcal{C}_\infty(P)$ be the set of all convex subsets of $P$. Define

$$\text{OVERLAP}_\infty(P,Q) = \max_{C\in\mathcal{C}_\infty(P), D\in\mathcal{C}_\infty(P)} \text{OVERLAP}_\infty(C,D).$$

This measure has two important "intuitive" properties: 1) for convex $P$, $Q$, $\text{OVERLAP}_\infty(P,Q) = \text{OVERLAP}(P,Q)$, and 2) for non-convex $P$, $Q$, $\text{OVERLAP}_\infty(P,Q) = 0$ if and only if $P$ and $Q$ are not overlapping.

We believe that there is a polynomial-time algorithm for calculating $\text{OVERLAP}_\infty$. However, we do not know how to deal with the infinity with respect to the reduction to linear programming. We find it necessary to select a subset $\mathcal{C}(P_i) \subset \mathcal{C}_\infty(P_i)$ for each input polygon $P_i$, and use an approximation to $\text{OVERLAP}_\infty$ which we call "OVERLAP$(P_i, P_j)$",

$$\text{OVERLAP}(P_i,P_j) = \max_{C_i\in\mathcal{C}(P_i), C_j\in\mathcal{C}(P_j)} \text{OVERLAP}(C_i,C_j).$$

This measure still has the two "intuitive" properties if 1) $P_i \in \mathcal{C}(P_i)$ if $P_i$ is convex, and 2) $\cup_{C_i\in\mathcal{C}(P)} C_i = P_i$.

Our initial experiment showed that is good to choose $\mathcal{C}(P_i)$ whose elements overlap. If instead $\mathcal{C}(P_i)$ is chosen to be a *partition* of $P$, such as the triangulation, then OVERLAP$(P_i, P_j)$ can have "unintuitive" local minima. Suppose $L_i$ separates $\mathcal{C}(P_i)$ into two subsets, and suppose the same for $L_j$ and $\mathcal{C}(P_j)$. In this case OVERLAP$(u_iP_i + t_i, u_jP_j + t_j)$ has a valley for all $u_i, t_i, u_j, t_j$ which align $u_iL_i + t_i$ with $u_jL_j + t_j$. Section 5.1 discusses our current choice of $\mathcal{C}(P_i)$.

### 3.2 The Overlap Minimization Algorithm.

Given a choice of $\mathcal{C}(P_i)$ for $P_i$, $0 \leq i \leq k$, it is easy to generalize the algorithm of Section 2. Simply apply the previous algorithm to the set of convex polygons $\cup_{0\leq i\leq k}\mathcal{C}(P_i)$ with two modifications: 1) all the polygons $C_i \in \mathcal{C}(P_i)$ have the *same* translation/rotation variables $t_i$, $u_i$, and 2) each pair of polygons $C_i \in \mathcal{C}(P_i)$ and $C_j \in \mathcal{C}(P_j)$ use the *same* overlap variable $o_{ij}$. We can also drop any constraints between $C_i, C_i' \in \mathcal{C}(P_i)$.

### 3.3 Reducing Running Time.

The rotational overlap minimization algorithm as described in Sections 2 and 3.2 is very inefficient. It creates the constraints of Equations 6 and 7 for each choice $\langle C_i, p_i, C_j, p_j\rangle$, for $0 \leq i < j \leq k$, for $C_i \in \mathcal{C}(P_i)$, for vertex $p_i \in C_i$, for $C_j \in \mathcal{C}(P_j)$, and for vertex $p_j \in C_j$. To reduce avoid considering this huge set of constraint choices ($\Omega(k^2n^4)$ if each $P_i$ has $n$ vertices), we use a *dynamic* approach. The minimization step "guesses" a much smaller set $S$ of constraint choices. If after taking a step it determines that critical constraint choices were missing, it adds them to the set $S$ and starts over.

```
S ← ∅
CreateConstraints()
TakeAStep()
while (UpdateConstraints() > 0)
    Restart()
    TakeAStep()
```

For each $0 \leq i < j \leq k$, the function CreateConstraints() computes a maximally overlapping pair $C_i \in \mathcal{C}(P_i)$ and $C_j \in \mathcal{C}(P_j)$ and adds constraint choices to $S$ of the form $\langle C_i, p_i, C_j, p_j\rangle$, where $p_i' = u_ip_i + t_i \in C_i' = u_iC_i + t_i$ lies on $L_{ij}$ (Section 2.2) and $p_j' = u_jp_j + t_j \in C_j' = u_jC_j + t_j$ lies the maximal distance from $L_{ij}$ on the same side as $C_i$. The function TakeAStep() sets up and solves the OMLP (Section 2.3) and performs interpolation (Section 2.4), using *only* the elements of $S$. The function UpdateConstraints() does the same as CreateConstraints() with the new positions of the polygons and returns the number of *new* constraints it added to $S$. Finally, Restart() moves the polygons back to their starting positions. **NOTE**: this section describes a single overlap minimization step. When UpdateConstraints() returns 0, the set $S$ is verified, and the algorithm has taken a successful step.

## 4 Compaction, Containment, and Minimum Enclosure using Overlap Minimization

This section gives algorithms that use overlap minimization to do compaction and to solve containment and minimum enclosure problems.

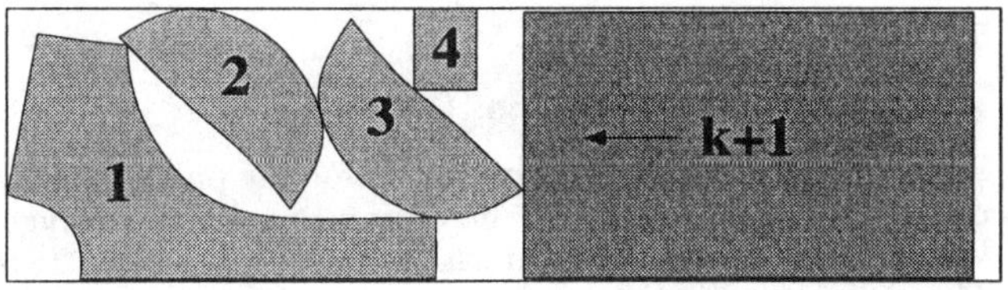


Figure 2: LMLP: Additional term in objective function acts as leftward "force".

### 4.1 Compaction.

To perform compaction using overlap minimization, let $P_0$ be the complement $\overline{R}$ of the rectangular container, and introduce an additional infinite "piston" rectangle $P_{k+1}$ to the right of all the polygons as shown in Figure 2. Start with the overlap minimization linear program (OMLP) for this configuration. Add a term $c \cdot x_{k+1}$ to the objective function, where $c$ is a constant and $x_{k+1}$ is the $x$-coordinate of $t_{k+1}$, the translation of $P_{k+1}$. Finally add constraints so that $P_0$ is fixed and $P_{k+1}$ cannot rotate and can translate only in the $x$-direction. We call this the *length minimizing* linear program (LMLP). Finally, adding non-overlap constraints, $o_{ij} = 0$, $0 \leq i < j \leq k+1$, yields the non-overlap length minimizing linear program (NOLMLP). To compact, iterate the LMLP with a small value of $c$: $c = 0.001$ and $w_{ij} = 1$ $0 \leq i < j \leq k$ (Section 2.3) until convergence. Iterate the NOLMLP to get rid of overlaps.

To minimize the area of the container rather than its length, we use the same strategy with an *area minimizing* linear program (AMLP). In this case, $P_0$ is the complement of the first quadrant, $P_0 = \{(x, y) \,|\, x \leq 0 \text{ or } y \leq 0\}$, and $P_{k+1}$ is the complement of the third quadrant, $P_{k+1} = \{(x, y) \,|\, x \geq 0 \text{ or } y \geq 0\}$. We remove any penalty for overlap between $P_0$ and $P_{k+1}$: in the objective, set the weight $w_{0,k+1}$ of $o_{0,k+1}$ to zero. Translation $t_0$ is fixed at $(0, 0)$ and angles $\theta_0$ and $\theta_{k+1}$ are fixed at zero. The goal is to minimize the area of the rectangle with diagonal $t_0 t_{k+1}$. This rectangle has area $x_{k+1} y_{k+1}$, where $(x_{k+1}, y_{k+1}) = t_{k+1}$. It can be shown [17], that this area can be diminished if and only if its gradient $y_{k+1} \Delta x_{k+1} + x_{k+1} \Delta y_{k+1}$ can be diminished. To finish the creation of the AMLP, we add the terms $c\, y_{k+1} \Delta x_{k+1} + c\, x_{k+1} \Delta y_{k+1}$ to the objective.

### 4.2 Containment.

This section gives an algorithm for rotational containment which uses overlap minimization plus our successful algorithm for translational containment [18].

#### 4.2.1 Range Intersection.

Let $P(\theta)$ denote polygon $P$ rotated by angle $\theta$ about its local origin. Under our previous notation, $P(\theta) = uP$ for $u = (\cos\theta, \sin\theta)$. Let $P[\alpha, \beta]$ denote the *range intersection* of $P$,

$$P[\alpha, \beta] = \bigcap_{\alpha \leq \theta \leq \beta} P(\theta).$$

The range intersection $P[\alpha, \beta]$ has the property that if it cannot be placed in a container, then no rotated copy $P(\theta)$ of $P$ can be placed in the container for any $\theta$ in $[\alpha, \beta]$.

The output of range intersection is a *polycircle*: a planar region bounded by circular arcs and straight line segments. Furthermore, each arc is concentric with the center of rotation. For purposes of applying our current translational containment algorithm, we would have to take an inner polygonal approximation. (Actually, since polycircles are closed under set operations, Minkowski sum, and convex hull, it should be in principle possible to generalize our translational containment algorithm to polycircles because these are the only operations that this algorithm uses.) In the rest of the section, $P[\alpha, \beta]$ denotes an polygonal approximation to $P$. The full paper discusses the issue of selecting an appropriate accuracy for the approximation. The resulting rotational containment algorithm will *not* be approximate because it uses overlap minimization to find the actual solution.

#### 4.2.2 Containment Algorithm.

As stated in Section 1, the goal of containment is to place polygons $P_1, P_2, \ldots, P_k$ into a container $C$. The rotational containment algorithm first establishes an initial *range hypothesis*, $\mathcal{R}^{\text{init}} = \langle [\alpha_i^{\text{init}}, \beta_i^{\text{init}}],\, 1 \leq i \leq k \rangle$, representing the possible angles which a containment solution might have. If $P_i$ has limits on its rotation, $\alpha_i^{\text{init}}$ and $\beta_i^{\text{init}}$ are set to the bounds of this limit. In particular, if $P_i$ is not allowed to rotate, $\alpha_i^{\text{init}} = \beta_i^{\text{init}} = 0$. If $P_i^{\text{init}}$ can rotate arbitrarily, $[\alpha_i^{\text{init}}, \beta_i^{\text{init}}] = [0, 2\pi]$.

The rotational containment algorithm applies two types of operations to a range hypothesis $\mathcal{R} = \langle [\alpha_i, \beta_i],\, 1 \leq i \leq k \rangle$.

- **Evaluate**: attempt to find a solution to containment whose angles fall within these ranges.
- **Subdivide**: for some $i$, split $[\alpha_i, \beta_i]$ into $[\alpha_i, \mu_i]$ and $[\mu_i, \beta_i]$. Create two sub-hypotheses $\mathcal{R}'$ and $\mathcal{R}''$ by replacing $[\alpha_i, \beta_i]$ by either $[\alpha_i, \mu_i]$ or $[\mu_i, \beta_i]$ in $\mathcal{R}$.

#### 4.2.3 Evaluation.

To evaluate a hypothesis $\mathcal{R} = \langle [\alpha_i, \beta_i],\, 1 \leq i \leq k \rangle$, the algorithm runs translational containment on the range intersections $P_1[\alpha_1, \beta_1], P_2[\alpha_2, \beta_2], \ldots, P_k[\alpha_k, \beta_k]$. If there is no solution, then by the property of range intersection, the algorithm can discard the entire hypothesis. If translational containment returns a solution $t_1, t_2, \ldots, t_k$, then the rotational algorithm uses this solution as a starting point for overlap minimization. For $1 \leq i \leq k$, it gives $P_i$ translation $t_i$ and angle $\theta_i = (\alpha_i + \beta_i)/2$. It then runs overlap minimization on $P_1, P_2, \ldots, P_k$. If the minimum overlap is zero, it has found a solution! Otherwise, it subdivides the hypothesis.

#### 4.2.4 Subdivision.

To subdivide the hypothesis, the algorithm selects the polygon $P_i$ which has the largest overlap with its neighbors. Let $\theta_i$ now denote the angle of $P_i$ returned by overlap minimization. If $\theta_i$ is in the interior of the interval $[\alpha_i, \beta_i]$, it subdivides this interval at $\mu_i = \theta_i$. If $\theta_i$ equals $\alpha_i$ or $\beta_i$, it splits the interval at its midpoint. The algorithm then recurses on the two sub-hypotheses.

### 4.3 Minimum Enclosure.

In her Ph.D. thesis [6], Daniels describes how to find a fixed width, minimum length rectangular enclosure for polygons $P_1, P_2, \ldots, P_k$. This method uses containment and compaction. She observes that containment plus binary search can be used to determine the minimum feasible length. Compaction can speed this search by shrinking any feasible length to a local minimum before continuing the binary search.

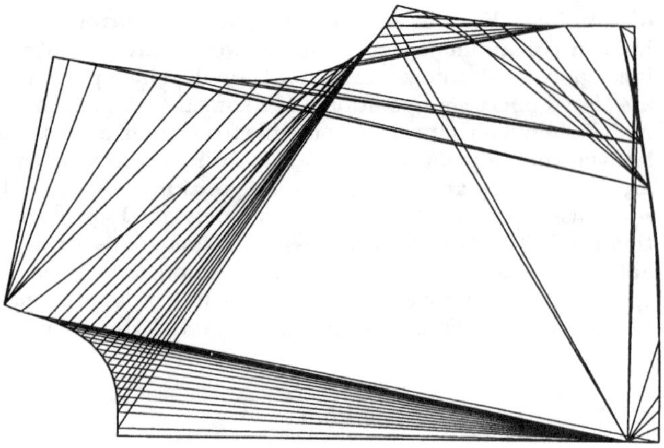

Figure 3: Convex cover of non-convex apparel polygon: 71 vertices, 39 polygons in cover.

Daniels also suggested that the minimum area rectangle can at least be approximated by sampling different widths. In more recent work [18], we show how to use containment directly to solving the following problem without sampling: given area $A$, is there a layout whose enclosing rectangle $R$ has area $\text{AREA}(R) \leq A$? This algorithm can be combined with area compaction to rapidly find a minimum area enclosing rectangle. *All* of these methods work equally well with *rotational* compaction and containment algorithms.

## 5 Experimental Results

We ran our experiments on polygonal parts from apparel manufacturing. Of course, the domains most in need of rotational algorithms are glass and sheet metal. In the full paper, we will also obtain data from manufacturers in these domains and run experiments on these parts. However, the apparel parts are representative of non-convex shapes which arise in manufacturing. All experiments were run on a DEC Alpha 3000/700 (name trademark of Digital Equipment Corporation) using CPLEX 3.0 (product of CPLEX Optimization, Inc.) to solve the linear programs. The standard "optimize" option (simplex method) was used.

### 5.1 Generating Convex Covers

As stated in Section 3.1, it is necessary to generate a subset $\mathcal{C}(P_i) \subset \mathcal{C}_\infty(P_i)$ for each input polygon $P_i$. Our initial experiments showed us that it is good for elements of $\mathcal{C}(P_i)$ to overlap as much as possible. Currently, our system generates a convex partition of $P_i$ in a greedy fashion, and then greedily "fattens" each element of the partition. The partition algorithm repeatedly cuts $P_i$ with a chord. A cut which eliminates two concave vertices is always better than a cut which only eliminates one. A cut which yields two boundary chains with roughly equal number of vertices is better than one which yields chains with very unequal numbers of vertices. The algorithm selects the best cut and recurses on the two halves. To fatten a convex partition element $C_i$, the algorithm greedily adds vertices from $P_i$ to the boundary of $C_i$ as long as the result is a convex subset of $P_i$. Figure 3 shows the result for a typical non-convex apparel part.

### 5.2 Experiments.

#### 5.2.1 Overlap Minimization: One Polygon.

Figure 4 shows overlap minimization for the polygon in Figure 3. The container is the bounding box for the polygon. We applied a fairly large initial rotation and translation. The polygon minimized its overlap with the exterior of the container in 3 iterations. Running times: decomposition, 2.1 seconds; overlap calculations (OC), 1.5 seconds; linear programming (LP), 0.41 seconds. Iterations 1 and 2 required one auxiliary LP each (included in the 0.41 seconds) to dynamically determine constraints (Section 3.3).

#### 5.2.2 Overlap Minimization: Two Polygons.

Figure 5 shows overlap minimization for two copies of the same polygon. The container is chosen to give a tight fit in the $y$-direction. Again, only 3 iterations are required. (If the $y$-dimension is even 1% larger, only 2 iterations are required.) Running times: decomposition, 4.2 seconds; OC, 23.7 seconds; linear programming (LP), 0.93 seconds. Iteration 1 required 2 auxiliary LP calls, and iteration 2 required 4 (included in 0.93 seconds) for dynamic constraint calculation.

Time for decomposition and LP grow linearly, but OC jumps by almost a factor of 10. This happens because we are (naively) calculating the overlap of every $C_1 \in \mathcal{C}(P_1)$ with every $C_2 \in \mathcal{C}(P_2)$ in order to calculate $\text{OVERLAP}(u_1P_1 + t_1, u_2P_2 + t_2)$.

#### 5.2.3 Compaction: 23 Polygons.

Figures 6 and 7 illustrate the even iterations for the LMLP (Section 4.1) applied to compact an actual industrial layout of 23 apparel part polygons. (These figures are rotated 90 degrees to put the piston on the "top".) Iteration 6 is within one part in $10^6$ of optimal. Since it has no overlaps, it is not necessary to apply the NOLMLP. The length (and area) of the layout is diminished by 2.3%. Since the "geometric" overlap calculation (OC) dominates the cost of linear programming, we modified the program to ignore the results of interpolation. This increases the number of constraints added to each linear program, but greatly decreases the number of auxiliary linear programs. To make the cost estimate fair, we ran interpolation anyway, but discarded its output, setting $s = 1$ each time. Running times: 6 essential LP calls (one per iteration), 1.7 minutes; 52 auxiliary LP calls (8.7 per iteration on average), 15 minutes; geometric calculations (decomposition and OC), 333 minutes.

## 6 Conclusions

The full paper will include more experiments, but these three clearly demonstrate that the number of iterations is surprisingly few and that the cost of the *essential* linear programming is very small. The actual translations and orientations of the polygons in each iteration of the 23-polygon example are the output of just *six* linear programs which require less than a total of two minutes to solve. We will also do experiments to compare our method with methods based on physical modeling instead of linear programming. However, it is clear that these physical methods will require many more than six iterations.

Without the use of dynamic constraint constraint calculations, the two-polygon example takes almost 200 seconds, and linear programming dominates the running time. The

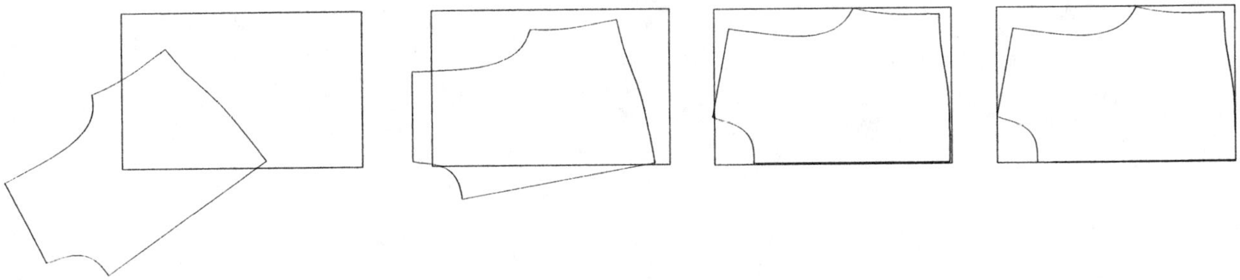

Figure 4: Overlap minimization for one polygon.

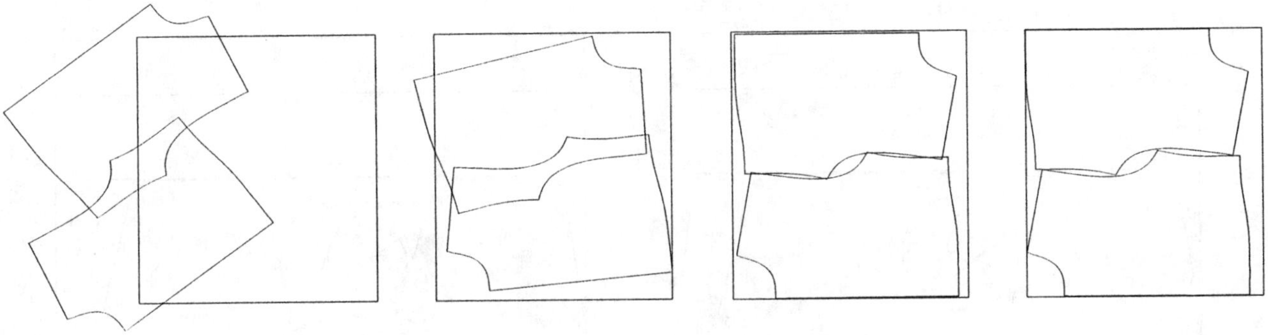

Figure 5: Overlap minimization for two polygons.

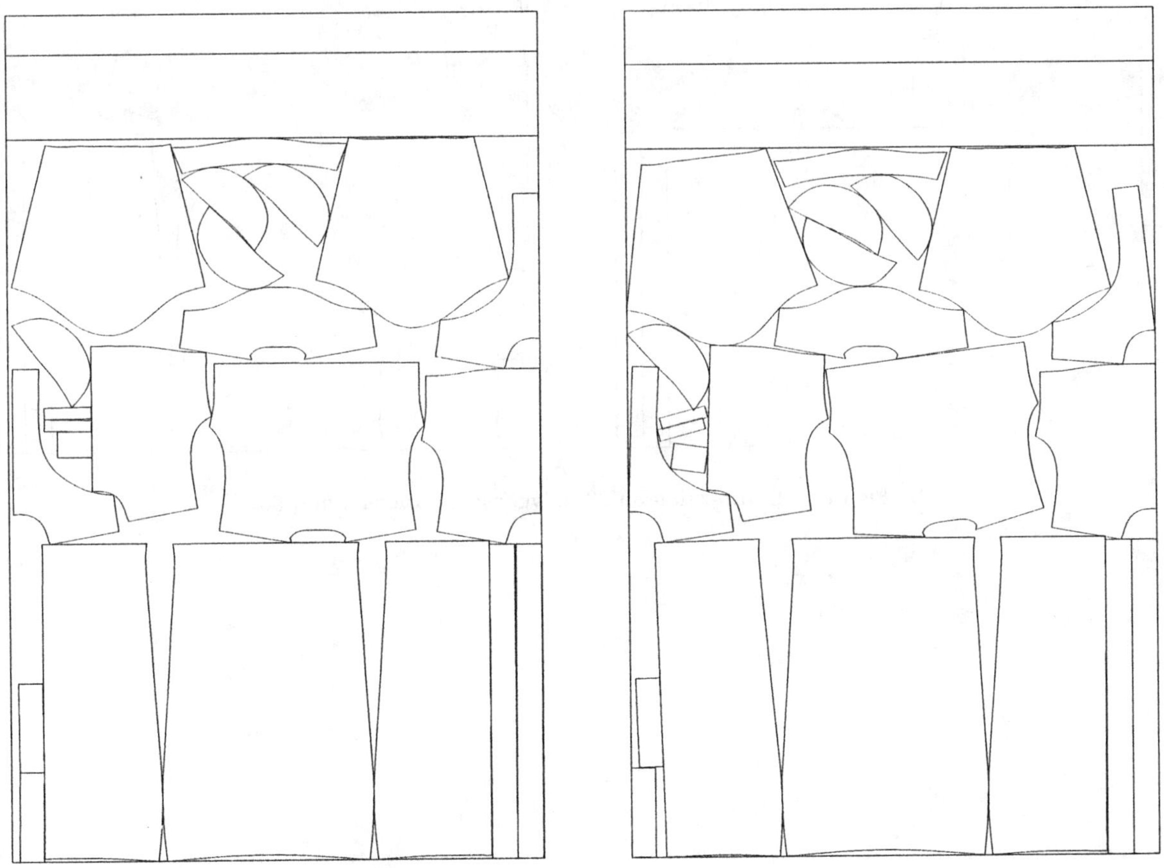

Figure 6: Compaction on 23 polygons: initial layout and iteration 2.

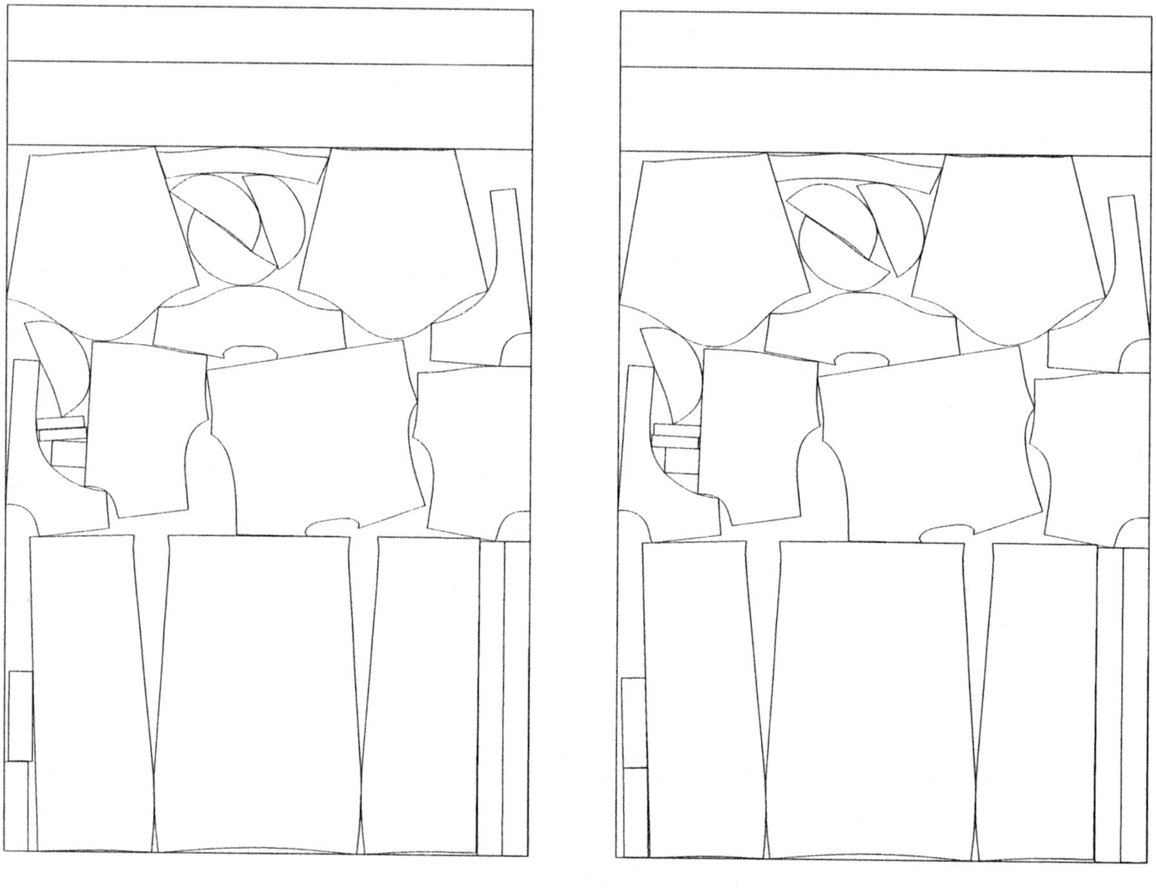

Figure 7: Compaction on 23 polygons: iterations 4 and 6.

23-polygon example would be practically impossible. With dynamic constraint calculations, the cost of linear programming, even including the auxiliary calls, is quite reasonable. Geometric calculations dominate the running time. Currently these are done quite naively, and it is likely they can be sped up by one or two orders of magnitude.

**OPEN QUESTIONS**: Is there a polynomial-time algorithm for $\text{OVERLAP}_\infty$ (Section 3.1)? Is there a way to use it instead of the approximation OVERLAP? Alternatively, what choice of $\mathcal{C}(P_i) \subset \mathcal{C}_\infty(P_i)$ plus what algorithm for overlap calculation gives the fastest algorithm for OVERLAP?

## References


[1] P. Agarwal, N. Amenta, and M. Sharir. Largest placement of one convex polygon inside another. In *Proceedings of the Second Workshop on Algorithmic Foundations of Robotics*, July 1996.

[2] P.K. Agarwal, B. Aronov, and M. Sharir. Motion planning for a convex polygon in a polygonal environment. In preparation., 1992.

[3] F. Avnaim. *Placement et déplacement de formes rigides ou articulées.* PhD thesis, Université de Franche-Comté, France, 1989.

[4] F. Avnaim and J. Boissonnat. Polygon placement under translation and rotation. In *Proceedings of the 5th Annual Symposium on Theoretical Aspects of Computer Science; Lecture Notes in Computer Science (294)*, pages 322–333. Springer-Verlag, 1988.

[5] B. Chazelle. The Polygon Containment Problem. In Preparata, editor, *Advances in Computing Research, Volume 1: Computational Geometry*, pages 1–33. JAI Press, Inc., Greenwich, Connecticut, 1983.

[6] K. Daniels. *Containment algorithms for nonconvex polygons with applications to layout.* Ph.D. thesis, Harvard University, Cambridge, MA, 1995.

[7] K. Daniels and V. Milenkovic. Column-based strip packing using ordered and compliant containment. In *Proc. 1st ACM Workshop on Appl. Comput. Geom.*, pages 33–38, 1996.

[8] K. Daniels and V. J. Milenkovic. Multiple Translational Containment, Part I: An Approximate Algorithm. *Algorithmica, special issue on Computational Geometry in Manufacturing*, In press.

[9] K. A. Dowsland and W. B. Dowsland. Packing Problems. *European Journal of Operational Research*, 56:2 – 14, 1992.

[10] H. Dyckhoff. A typology of cutting and packing problems. *European Journal of Operations Research*, 44:145–159, 1990.

[11] R. Grinde and T. Cavalier. Containment of a Single Polygon Using Mathematical Programming. Technical Report IMSE Working Paper 92-164, The Pennsylvania State University, Department of Industrial and Management Systems Engineering, 1993.

[12] R. B. Grinde and T. M. Cavalier. A new algorithm for the two-polygon containment problem. *Computers and Operations Research*, page to appear, 1996.

[13] J. E. Hopcroft, J. T. Schwartz, and M. Sharir. On the complexity of motion planning for multiple independent objects: P-space hardness of the "Warehouseman's Problem". *Internat. J. Robot. Res.*, 3(4):76–88, 1984.

[14] Z. Li. *Compaction algorithms for nonconvex polygons and their applications.* Ph.D. thesis, Harvard University, Cambridge, MA, 1994.

[15] Z. Li and V. Milenkovic. The complexity of compaction problem. In *Proc. 5th Canad. Conf. Comput. Geom.*, pages 7–11, Waterloo, Canada, 1993.

[16] Z. Li and V. Milenkovic. Compaction and separation algorithms for nonconvex polygons and their applications. *European Journal of Operations Research*, 84:539–561, 1995.

[17] V. Milenkovic. Translational polygon containment and minimal enclosure using linear programming based restriction. In *Proc. 28th Annu. ACM Sympos. Theory Comput. (STOC 96)*, pages 109–118, 1996.

[18] V. Milenkovic and K. Daniels. Translational polygon containment and minimal enclosure using geometric algorithms and mathematical programming. Technical Report 25-95, Center for Research in Computing Technology, Division of Applied Sciences, Harvard University, Cambridge, MA, 1995.

[19] V. J. Milenkovic. Multiple Translational Containment, Part II: Exact Algorithms. *Algorithmica, special issue on Computational Geometry in Manufacturing*, In press.

[20] Victor Milenkovic. Position-based physics: simulating the motion of many highly interacting spheres and polyhedra. In *Computer Graphics*, pages 129–136, 1996. Proc. SIGGRAPH '96.

[21] P. E. Sweeney and E. R. Paternoster. Cutting and Packing Problems: A Categorized, Application-Oriented Research Bibliography. *Journal of the Operational Research Society*, 43(7):691–706, 1992.

# Proximity Problems on Moving Points

Julien Basch Leonidas J. Guibas* Li Zhang

Computer Science Department
Stanford University
Stanford, CA94305
{jbasch,guibas,lizhang}@cs.stanford.edu

**Abstract**

A kinetic data structure for the maintenance of a multidimensional range search tree is introduced. This structure is used as a building block to obtain kinetic data structures for two classical geometric proximity problems in arbitrary dimensions: the first structure maintains the closest pair of a set of continuously moving points, and is provably efficient. The second structure maintains a spanning tree of the moving points whose cost remains within some prescribed factor of the minimum spanning tree.

## 1 Introduction

The goal of this paper is to address a number of proximity problems for points and balls moving continuously in $\mathcal{R}^d$. Examples of such problems are the maintenance of the closest pair and the minimum spanning tree (MST) of the moving objects. In the dynamic simulation of virtual environments, collision detection between the moving objects is frequently important. Complex bodies can be approximated by bounding spheres and knowledge of the closest pair of these spheres at all times can be used to limit where a more elaborate collision-checking algorithm is invoked. In other applications, where the moving objects need to stay in constant communication with each other, the MST might be a desirable way of maintaining a set of point-to-point links among the objects. These are just some examples of the many proximity problems that can be posed for a system of moving bodies — problems which are interesting from both the theoretical and the practical points of view.

Both the closest pair and the MST problems have been extensively studied for sets of points in arbitrary dimensions. It would take us too far afield to survey the known results. What is surprising is that most of these are not directly applicable to the setting of objects moving continuously. Extant algorithms are either *static*, i.e. work for a fixed set of points, or *dynamic*, i.e. allow insertions and deletions to/from the set of points. It is not immediately obvious how to use these classical algorithms to maintain the closest pair and the MST as the points move continuously. In this paper we present algorithms which explicitly take advantage of the continuity or coherence in the motion of the points to gain efficiency. These new algorithms are *kinetic* — they are instances of *kinetic data structures* (KDSs for short), as introduced by Basch, Guibas, and Hershberger [BGH97].

In the kinetic setting, a set of objects is assumed to be continuously changing, or moving. Each object follows a posted *flight plan*, but a plan which can change at any moment through a *flight plan update*. A kinetic data structure maintains a *configuration function* of interest (in our case this will be the closest pair or the MST) by watching for critical events as the objects move. These events correspond to the times at which certain elementary conditions, called *certificates*, switch from being true to being false (an example might be 'points $A$ and $B$ are closer than points $C$ and $D$'). At any one time, the conjunction of all the certificates being watched over by the kinetic data structure proves the combinatorial correctness of its output (for example, $A$ and $B$ are the closest pair). When a certificate fails, the proof structure needs to be modified and the combinatorial description of the configuration function may need to be updated (because now $C$ and $D$ may be the closest pair). A good kinetic data structure will take advantage of the continuity of the object motion to select certificate structures which are easy to update at these critical events; a structure satisfying this condition is called *responsive*. Other desiderata for a kinetic data structure are (1) that it be *efficient*, meaning that the number of events it processes is not much greater than the number of combinatorial changes in the configuration function itself, (2) that it be *compact*, so that the number of active certificates at any one time is roughly linear in the complexity of the moving objects, (3) that it be *local*, so that a flight plan update for any one object affects only a small number of certificates. A kinetic structure which is responsive, efficient, compact, and local is worst-case nearly optimal for the problem at hand. For precise definitions of these concepts the reader is referred to [BGH97].

The results of this paper are based on kinetizing (i.e., maintaining under continuous point motion) a certain type of *multidimensional range search tree* (MDRS tree) [Meh84] used to query the value of some specified function on subsets of the points [BS80]. We apply this MDRS tree to obtain kinetic data structures for the closest pair among $n$ moving points in $\mathcal{R}^d$, the closest pair among a collection of $n$ moving balls, and approximate and exact MSTs for $n$ moving points. Though the structures we propose are fully on-line and work

*Supported in part by National Science Foundation grant CCR-9623851 and by US Army MURI grant DAAH04-96-1-0007.



*Computational Geometry 97* Nice France
Copyright 1997 ACM 0-89791-878-9/97/06 ..$3.50

for arbitrary motions, the bounds we give assume that the motions of the points are algebraic of fixed degree, or that they satisfy certain Davenport-Schinzel type conditions.

In [BGH97] an efficient, compact, responsive and local kinetic data structure is given for the closest pair in $\mathcal{R}^2$. Another event-based algorithm for detecting the collision of moving balls was proposed in [KGS]. For arbitrary dimensions the performance of the algorithms given in this paper is nearly optimal as a function of $n$ (the number of points or balls), but not of $d$ (the dimension — the 'hidden' constant in our bounds is $2^{2d}$). A major open problem is to find kinetic algorithms for proximity problems which do not have this exponential dependence on the dimension.

Section 2 presents our kinetic MDRS tree and shows that it is efficient, compact, responsive, and local, a result which is of independent interest. Section 3 gives a kinetic algorithm for closest pair in $\mathcal{R}^d$ under the Euclidean metric — via a structure which is also efficient, compact, local, and responsive. Our method is based on a certain packing lemma, and extends to moving balls under the assumption that the ratio of the radii of any two balls involved is bounded by a constant. Our results for the MST are not as good. In Section 4 we give a kinetic data structure which maintains the exact MST of the moving points under certain polyhedral metrics (for example, $L_1$ or $L_\infty$). For the $L_2$ metric we can only maintain a tree which is a $(1+\epsilon)$ approximation to the real MST. All of our KDSs for the the MST are compact, local, and responsive, but they are not efficient (in the strict KDS sense).

## 2 Kinetized Multidimensional Range Search Tree

In this section, we study in the kinetic setting the classic range searching problem when the ranges are determined by a fixed set of directions. This version of the range searching problem has received extensive treatment in the literature both in the static and in the dynamic case [Meh84, Mul94]. We focus here on a specific version of this problem that will be well suited for the applications to the closest pair and the MST.

In the generic range searching problem, a set $S$ of $n$ points of $\mathcal{R}^d$ is given, together with a (generally infinite) set of *ranges* which are certain restricted regions of $\mathcal{R}^d$; these ranges define subsets of $S$ of interest. Each point of $S$ is assigned a *weight* from a commutative semigroup, and the semigroup operation allows us to compute the weight of any subset of $S$ induced by a range. The problem then is to compute a data structure $\mathcal{T}$ on $S$ that allows us to find efficiently the weight of any query range.

If each point in the set is assigned a weight from a linearly ordered set, the semigroup setting is general enough to define, for instance, a "min-weight" function such that the min-weight of any range is the list of the $\ell$ elements of smallest weight in that range. In this case, the semigroup operation, which simply merges two sets of size at most $\ell$ each and keeps only the $\ell$ smallest elements, takes time $\Theta(\ell)$, and this factor multiplies all queries and update operations.

For the purpose of this paper, we now focus on a specific version of the range searching problem. It is easily seen that the kinetization described below can be applied to many variants on this kind of range search tree.

### 2.1 A Special Range Searching Problem

Our restricted range searching problem is as follows. We are given a model infinite polyhedral cone $\mathcal{C}_0$ delimited by a constant number $k$ of hyperplanes passing through the origin. The set of ranges are the subsets of $S$ induced by all possible translations of this cone. We denote by $\mathcal{C}_p$ the model cone with apex translated at $p$. For each hyperplane delimiting the model cone, we define an associated *range ordering* on $S$ as the ordering of the projections of the points on the normal to this hyperplane.

For a static set $S$ and a weight function $f$, repeated median partitioning along the range orderings can be used to recursively construct a multidimensional range search (MDRS) tree $\mathcal{T}$ storing the elements of $S$ with the following properties [Meh84]:

- The tree $\mathcal{T}$ can be built in time $O(n \log^k n)$; it has size $O(n \log^{k-1} n)$ and depth $O(\log n)$.
- Each node $\nu$ in $\mathcal{T}$, stores $f(T_\nu)$, where $T_\nu$, called the *span of* $\nu$ is the set of all the elements in the subtree rooted at node $\nu$.
- Each element appears at most $O(\log^{k-1} n)$ times in the tree.
- For each conical range $\mathcal{C}$, there exists a set of $O(\log^k n)$ nodes $N_\mathcal{C}$ such that $S \cap \mathcal{C} = \bigcup_{\nu \in N_\mathcal{C}} T_\nu$, where the union is disjoint. Furthermore, this set of nodes can be found in time $O(\log^k n)$.

Therefore, for any query range $\mathcal{C}$, $f(\mathcal{C})$ can be computed in two steps: first find the set $N_\mathcal{C}$ of $O(\log^k n)$ nodes whose spans are disjoint and whose union is $S \cap \mathcal{C}$. Then compute

$$f(\mathcal{C}) = \sum_{\nu \in N_\mathcal{C}} f(T_\nu),$$

where the $\Sigma$ is to be understood as the generalized semigroup operation to an arbitrary number of values. The query time is thus $O(t_f \log^k n)$, where $t_f$ is the time complexity of the semigroup operation associated with $f$.

In the dynamic setting, we allow the following operations:

- deletion of a point,
- insertion of a point, and
- change of a point weight (this can be viewed as deleting a point and then reinserting it with a new weight)

By relaxing the equal-sized partitioning in the recursive definition of $\mathcal{T}$ to a balanced partitioning and using a careful rotation scheme, the above structure can be maintained and queried within $O(t_f \log^k n)$ worst case time per operation [WL85]. Another way to dynamize $\mathcal{T}$ is to do some local and global rebuilding after every few operations, which gives an amortized bound of $O(t_f \log^k n)$ per operation [Meh84, Ove83], or to use randomized search trees [AS89] and obtain the same bounds in expectation.

### 2.2 Kinetizing the tree $\mathcal{T}$

We now assume that we are given a set $S$ of points whose positions and weights are continuous functions of time; we wish to maintain their MDRS tree $\mathcal{T}$ as time goes on. We also assume that the semigroup operation depends only on the ordering of the weights. This will be the case in our applications, where the weight of a point will be taken to be its inner product with a specific vector, and the semi-group operation will be the min, or some generalized min, as mentioned above.

A kinetic version of the MDRS tree $\mathcal{T}$ is easy to obtain, because the structure of $\mathcal{T}$ depends only on the relative order of the elements of $S$ along the given directions. The kinetized structure consists of:

- A *dynamic* multidimensional search tree $\mathcal{T}$.
- A set of $k+1$ kinetic sorted lists, one for each of the $k$ range orderings and one for the weight ordering. For each sorted list, the kinetic certificates are comparisons between every two adjacent elements in it.
- A global event queue $\mathcal{Q}$, to be used for storing the kinetic structure events; in our case these correspond simply to the times when two adjacent elements in a sorted list exchange their ordering.

There are two types of events which may cause the multidimensional tree to change: one is an exchange in some range ordering among two elements of $S$ and the other is an exchange in their weight ordering. Both types of events can be easily handled by performing a deletion and a re-insertion of the points involved in the event, using the dynamic capabilities of $\mathcal{T}$. In addition, we update the kinetic sorted list involved by exchanging the two elements and rescheduling the events for the newly adjacent pairs in the list — so there are at most 3 events to reschedule. The correctness of this data structure is ensured by the continuity of motion because, for continuously moving points, the sorted lists and all our range and weight orderings can only change when two adjacent points swap order.

**Theorem 1** *The kinetized multidimensional range search tree on n moving points with continuously changing weights is local, compact, and responsive. It can be maintained at a cost of $O(\log^k n)$ per event.*

If the points follow algebraic trajectories of fixed degree, and the weight function is also algebraic of fixed degree, the total number of events will be $O(n^2)$. It is not hard to see that the number of changes the MDRS tree will undergo is $\Theta(n^2)$ in the worst case. Hence, this kinetization is efficient. Note that all range events are external (they change the MDRS structure), but some weight events may not be.

## 3 Maintaining the Closest Pair of Moving Points

We now show how a combination of instances of the kinetic MDRS tree can be used to maintain the closest pair (in the $L_2$ sense) of a set $S$ of moving points in $\mathcal{R}^d$. We take each of the $2^d$ orthants in $\mathcal{R}^d$as a model cone for which we maintain an MDRS tree on the set of points, with a weight for each point that is the projection on the 'diagonal' of this orthant (the line through the origin in the orthant making equal angles with the axes). Such 'quadrant-tilings' have also been previously used for dynamic closest-pair algorithms [Smi92]. What is surprising here is that our rank functions, which are all based on ordering the points along certain fixed directions, can be used to find an exact (not approximate) $L_2$ closest pair. As usual, what helps us is a certain packing lemma.

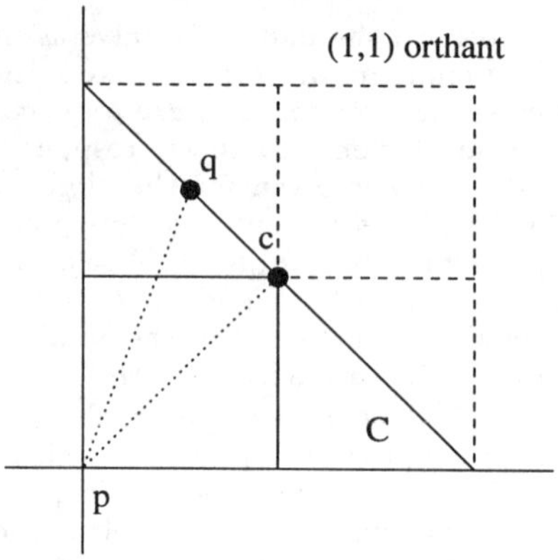


Figure 1: Partitioning of corner

### 3.1 A Packing Lemma

The key observation on the closest pair is that, if we project certain subsets of the points onto certain well-chosen lines and, for each point, we keep a certain number of its nearest neighbors along each of these lines, then the closest pair must be among the pairs thus considered.

In $\mathcal{R}^d$, each vector in an orthogonal basis has an associated normal hyperplane. These $d$ hyperplanes divide the space into $2^d$ cones, called orthants. Formally, if $\sigma = \frac{1}{\sqrt{d}}(\varepsilon_1, \ldots, \varepsilon_d)$ with $\varepsilon_i \in \{-1, 1\}$ (a *diagonal vector*), we define the $\sigma$*-orthant* to be:

$$orth_\sigma = \{(q_1, \ldots, q_d) \in \mathcal{R}^d \mid \forall i \; \varepsilon_i q_i \geq 0\}.$$

We also define the $\sigma$-ordinate of a point $p$ to be the inner product of $p$ with $\sigma$.

For a point $p \in S$ and a diagonal vector $\sigma$, we say that the $2^d - 1$ points of $S \cap orth_\sigma(p)$ that have smallest $\sigma$-ordinates are the $\sigma$*-candidate points* of $p$, and we denote this set by $cand_\sigma(p)$. We say that a pair of points $(p, q)$ is a $\sigma$*-candidate pair* if both $q \in cand_\sigma(p)$ and $p \in cand_{-\sigma}(q)$, and we say that that it is a *candidate pair* if it is a $\sigma$-candidate pair for some $\sigma$.

**Lemma 2** *For any finite set $S \subset \mathcal{R}^d$, the closest pair in $S$ must be a candidate pair. Furthermore, each point $p \in S$ can be part of at most $2^{2d}$ candidate pairs.*

*Proof.* Let $(p, q)$ be the closest pair of $S$, and say that $q$ is in the $\sigma$-orthant of $p$ for some diagonal vector $\sigma$ (Figure 1).

Let $c$ be the projection of $q$ on the line parallel to $\sigma$ passing through $p$. The region $C$ defined by the hyperplanes of the cone and the hyperplane orthogonal to $\sigma$ passing through $c$ can be covered by $2^d - 1$ interior-disjoint hypercubes. If $q$ is not a candidate point of $p$ in the $\sigma$-orthant, there must be at least $2^d$ points (including $p$ itself) lying in $C$. Therefore, there must be two points lying within the same hypercube, which implies that their distance is less than $d(p, c)$ and therefore less than $d(p, q)$, contradicting the hypothesis that $(p, q)$ is the closest pair. By the same argument, $p$ must also be a candidate point of $q$. Thus $(p, q)$ is a candidate pair.

Furthermore, for each point $p$, there are at most $2^d$ candidates in each orthant. Hence, there are at most $2^{2d}$ pairs in total involving $p$. □

### 3.2 Kinetic data structure for the closest pair

We now show how to maintain the candidate pairs with several kinetized MDRS trees. With the addition of a kinetic tournament [BGH97] on the distances defined by these pairs, we will have a KDS for the closest pair.

For each of the $2^d$ diagonal vectors $\sigma$, we build a kinetic MDRS tree $\mathcal{T}$ for $S$ with the $\sigma$-orthant as the model cone and with a weight function $f_\sigma$ that returns the $2^d - 1$ points of smallest $\sigma$-ordinate in the range (as in Section 2.1). With the help of this MDRS tree, the (at most) $2^d$ $\sigma$-candidate points of any point $p$ can be found in time $O(2^d \log^d n)$, as they are simply $f_\sigma(S \cap orth_\sigma(p))$. Similarly, we can decide in the same amount of time whether a point $q$ is a $\sigma$-candidate point of $p$ or not.

Using these facts, we immediately obtain a static algorithm to compute the closest pair in time $O(n \log^d n)$ by first constructing the (static) multidimensional trees, then searching for all candidate points of each point in each orthant, and finally by finding the minimum distance amongst all candidate pairs. The two points realizing this distance are the closest pair by Lemma 2.

Furthermore, by the result of Section 2.1, we can maintain all search trees when the points are moving, as well as, for each point of $S$, the list of its candidate points along each diagonal vector. We will now have $d$ range orderings (each one shared by several MDRS trees), and $2^d$ weight orderings (one for each diagonal vector).

If we were to throw into a tournament tree all pairs $(p, q)$ such that $q$ is a candidate point of $p$, the structure wouldn't be local, as a single point can be the candidate point of arbitrarily many points. This is the reason why the notion of candidate pair was introduced, as each point appears in at most $2^{2d}$ candidate pairs. Hence, in addition to the normal behavior of the MDRS trees, we need to be able to maintain the set of candidate pairs.

**Lemma 3** *A pair $(p, q)$ can become or cease to be a candidate pair only at times when the ordering of $p$ or $q$ changes in one of the $d + 2^d$ (range or weight) orderings.*

*Proof.* Omitted. □

Therefore, in addition to maintaining the MDRS trees, we can proceed as follows to maintain the set of candidate pairs. Let $p$ and $q$ be the points involved in an event; there are two cases.

- *The ordering of $p$, $q$ along one of the axes changes.* In this case, let us assume that $q$ moves from the $\sigma_1$-orthant of $p$ to the $\sigma_2$-orthant ($\sigma_1$, $\sigma_2$ differ by only one coordinate), and oppositely for $q$. The difference between the set of $\sigma_1$-candidate points of $p$ before and after the exchange can be computed by two queries on the MDRS tree for $\sigma_1$ immediately before and after updating it. If a point $r$ disappears from the list of $\sigma_1$ candidate points, the pair $(p, r)$ is removed from the list of candidate pairs (if it was in before). If $r$ appears in this list, we can check whether $p$ is a $-\sigma_1$ candidate point for $r$, and add $(p, r)$ to the list of candidate pairs, if needed. The other three orthants where something happens can be handled similarly.

- *The ordering of $p, q$ changes along a diagonal $\sigma$.* Let us assume that $p$ passes above $q$ along $\sigma$. In this case, we first update the MDRS tree. Then, for each point $r$ such that $(q, r)$ is a $\sigma$-candidate pair, we check whether $q$ is still a $-\sigma$-candidate point of $r$. Finally, we check for all candidate points $r$ of $p$ whether $p$ became a candidate point of $r$. In total, this can be done within $O(2^{2d} \log^d n)$ time. Note that a $-\sigma$ event happens at the same time that takes care of the $-\sigma$-orthant in the same way.

By the above algorithm, we have shown how to maintain the candidate pairs. We just need to build a kinetic tournament on the distances between each candidate pair of points [BGH97]. Note that the number of all the candidate pairs which can ever appear is at most $O(n^2)$. At any time, there is at most a linear number of candidate pairs. The following lemma follows from the results in [BGH97].

**Lemma 4** *If each point follows a degree $s$ algebraic trajectory in $\mathcal{R}^d$, the total number of events in the tournament tree is $O(n\lambda_{2s+2}(n) \log n)$, where $\lambda_{2s+2}(n)$ is the maximum length of a Davenport-Schinzel sequence with parameter $2s + 2$*[1]*.*

This roughly quadratic bound shows that our structure is efficient, since the closest pair can itself change a quadratic number of times. Note also that a point $q$ can be a $\sigma$-candidate for many points but it can be in a candidate pair with only $2^{2d}$ other points. This is what makes the above distance tournament a local KDS in terms of the original data (the MDRS trees are local as seen in the previous section).

**Theorem 5** *The kinetic data structure for the closest pair problem is responsive, efficient, compact, and local.*

### 3.3 Dynamic simulation of moving balls

An interesting dynamic simulation problem is that of simulating the motion of rigid bodies under a physical model. Here, we will describe an application of the above data structure to simulating the motion of balls. In our model, there are moving balls in a bounding volume. Each ball moves in a fixed algebraic trajectory until it hits the boundary, or collides with another ball. Collisions are assumed to be perfectly elastic, or to obey some other standard physical law. Our task is to simulate the motion of balls for a given initial set of motions.

We can, of course, compute the time when a pair of balls may collide, or when a ball hits the boundary and put those events into a global queue. Once a collision happens, we can recompute all the events for the balls involved. However, this asks for $\Omega(n^2)$ event space and $\Omega(n)$ update time per event. But observe that just before two balls collide, they must be the closest pair — so it is enough to worry about collision among the closest pair of balls. For the balls with the same radius, the previous kinetic data structure for the closest pair solves the problem, when applied to the centers of the balls — the closest pair of the centers corresponds to the closest pair of balls. If we add the following two types of events

- the two closest balls collide, and
- a ball hits the boundary,

then we can use the previous structure to continue the simulation.

When a collision happens, we update all the certificates related to the two colliding balls (or the ball colliding with

[1] $The function \lambda_{2s+2}(n)$ is nearly linear in $n$ for any fixed $s$.

the wall). Because of the locality of the data structure, this can be done efficiently. In addition, the data structure takes roughly linear space.

Note that this approach does not work for balls with different radii because we can no longer just maintain the closest pair of centers. However, if the balls are of roughly similar size, say so that $r_{max}/r_{min} \le \beta$, where $r_{max}, r_{min}$ are the maximum and minimum radius of any ball, respectively, then the MDRS tree technique can be adapted, by keeping $(2\beta)^d$ candidates points for each center in each orthant. A similar packing lemma holds and we can show:

**Lemma 6** *In a set of n balls where the maximum radius is $\beta$ times the minimum radius, the closest pair is one of the $2^{2d}\beta^d n$ candidate pairs.*

*Proof.* Omitted. □

By using the same data structure, the moving balls with different but roughly similar radii can be simulated kinetically. More details will be given elsewhere.

## 4 Maintaining the MST of Moving Points

The *Euclidean minimum spanning tree* (EMST or $L_2$-MST) of a set $S$ of $n$ points in $\mathcal{R}^d$ is a spanning tree with minimum total edge length, where an edge length is defined by the Euclidean distance between its endpoints. We will also be interested in MSTs defined by the $L_1$ and $L_\infty$ metrics, and more general polyhedral metrics as well. Given a distance function $\delta$, we define the $\delta$-MST to be the minimum spanning tree according to that distance.

For a constant $\epsilon > 0$, we say that a spanning tree is a $(1+\epsilon)$-EMST if its total edge length is within a factor of $(1+\epsilon)$ of the length of the true EMST. In this section, we will give kinetic data structures maintaining exact MSTs for certain polyhedral metrics and a $(1+\epsilon)$-EMST. We first focus on the structural results needed to approximate the EMST using several MDRS tree auxiliary structures.

### 4.1 Approximating the EMST by a polyhedral MST

To approximate the $L_2$ distance between two points we can project the points onto a line and measure the distance between their projections. If we choose sufficiently many and 'well-spaced' lines, we can make such an approximation arbitrarily good, as stated in the following lemma. This idea, which was also used by Vaidya [Vai84], is the basis of our method for approximating the Euclidean distance by a certain polyhedral metric and thus the EMST by a polyhedral MST. Similar constructions were employed in [AMS92] for the maximum spanning tree problem. However, the technical development given here is different.

In the following, for a point $p \in \mathcal{R}^d$, we denote $\vec{p}$ to be the vector from the origin to $p$. For any given $\alpha$, $0 < \alpha < \pi$, we say that a set of unit vectors $\mathcal{V}$ is *$\alpha$-well-spaced* if it satisfies the following properties:

**Property 1** For any vector $\vec{p} \in \mathcal{R}^d$, there exists $\vec{v} \in \mathcal{V}$ such that the angle between $\vec{p}$ and $\vec{v}$ is no more than $\alpha$, i.e. $\vec{v}\vec{p} \ge \|\vec{p}\| \cos\alpha$, where $\|\vec{p}\|$ is the $L_2$ norm of $\vec{p}$.

**Property 2** The *Voronoi cone* $F_{\vec{v}}$ of $\vec{v} \in \mathcal{V}$, defined by $\{\vec{p} \mid \forall \vec{r} \in \mathcal{V}, \vec{p}\vec{v} \ge \vec{p}\vec{r}\}$, is the intersection of a constant number (dependent on $d$) of halfspaces.

**Property 3** For any $\vec{p}, \vec{q} \in F_{\vec{v}}$, let $\vec{r} = \vec{p} - \vec{q}$, then $\max_{\vec{v} \in \mathcal{V}}(\vec{r}\vec{v}) < \max(\vec{p}\vec{v}, \vec{q}\vec{v})$.

In section 4.2, we will show that in $d$ dimensions, for any $\alpha$, there exists a centrally symmetric set of $\alpha$-well-spaced vectors that has cardinality $O(1/\alpha^{d-1})$. Such a set defines a polyhedral metric $\delta_\mathcal{V}$, with $\delta_\mathcal{V}(p,q) = \max_{\vec{v} \in \mathcal{V}}(\vec{p} - \vec{q})\vec{v}$. Intuitively, the well-spaced properties ensure that (1) the $\delta_\mathcal{V}$ metric is a good approximation of the $L_2$ norm, (2) the function $\delta_\mathcal{V}$ can be computed efficiently with a multidimensional range searching technique, and (3) the $\delta_\mathcal{V}$-MST can be built efficiently with a range searching technique as well.

**Lemma 7** *For any $\epsilon > 0$, there exists a set $\mathcal{V}$ of d dimensional unit vectors, with $|\mathcal{V}| = O(\epsilon^{-(d-1)/2})$ such that, for all $p, q \in \mathcal{R}^d$,*

$$\delta_\mathcal{V}(p,q) \le \|\vec{p} - \vec{q}\| \le (1+\epsilon)\delta_\mathcal{V}(p,q),$$

*where $d_\mathcal{V}$ is the polyhedral metric induced by $\mathcal{V}$.*

*Proof.* Let $\alpha = \epsilon^{1/2}$ and let $\mathcal{V}$ be a corresponding set of $\alpha$-well-spaced vectors, which has the required cardinality by Lemma 11. By Property 1 there exists $\vec{v} \in \mathcal{V}$ such that

$$\vec{v}(\vec{p} - \vec{q}) \ge \|\vec{p} - \vec{q}\| \cos\alpha$$

Since $\cos\alpha \approx 1 - \alpha^2/2$ for small $\alpha$, we have

$$\begin{aligned}(\vec{p} - \vec{q})\vec{v} &\ge \|\vec{p} - \vec{q}\| \cos\alpha \\ &\ge (1 - \epsilon/2)\|\vec{p} - \vec{q}\|.\end{aligned}$$

By the definition of $\delta_\mathcal{V}$, we have $\|\vec{p} - \vec{q}\| \le (1+\epsilon)\delta_\mathcal{V}(p,q)$. In addition, as each vector in $\mathcal{V}$ is a unit vector, we have $\delta_\mathcal{V}(p,q) \le \|\vec{p} - \vec{q}\|$. □

In the rest of this paper, we denote by $\mathcal{V}_\epsilon^d$ the set of well-spaced vectors of Lemma 7, and by $\delta_\epsilon$ the distance function induced by this set.

**Lemma 8** *For any set $S$ in $\mathcal{R}^d$, the $\delta_\epsilon$-MST of $S$ is a $(1+\epsilon)$-EMST of $S$.*

*Proof.* Let $T_1$ be the EMST of $S$ and $T_2$ be its $\delta_\epsilon$-MST. Let $c(T)$, $c'(T)$ to be the total edge length of a tree $T$ under the $L_2$ and $\delta_\epsilon$ metric, respectively. Then, as $T_2$ is a $\delta_\epsilon$-MST, $c'(T_2) \le c'(T_1)$. By Lemma 7, we have

$$c(T_2) \le (1+\epsilon)c'(T_2) \le (1+\epsilon)c'(T_1) \le (1+\epsilon)c'(T_1).$$

i.e., $T_2$ is within $(1+\epsilon)$ of the EMST in the $L_2$ norm. □

Hence, the $(1+\epsilon)$-EMST problem in $\mathcal{R}^d$ can be converted into the MST problem in the $\delta_\epsilon$ metric. To compute the $\delta_\epsilon$-MST, as most efficient geometric MST algorithms do, we first construct a sparse graph on $S$ which is guaranteed to contain all $\delta_\epsilon$-MST edges and then extract the MST from it by standard graph MST algorithms. Here, the sparse graph we are going to use is a variation on the *nearest geographic neighbor graph* [Yao82], also called the *local nearest neighbor graph (LNNG)* [KTI95].

As in the case of the closest pair, we say that, for a given $v \in \mathcal{V}_\epsilon^d$, point $q$ is a *v-candidate point* of $p$ if $q$ is the point of smallest $v$-ordinate in the Voronoi cone of $v$ translated to have its apex at $p$. Two points $p, q$ form a candidate pair if they are candidate points of each other for some vectors in $\mathcal{V}_\epsilon^d$. We define our *LNNG* to be the set of candidate pairs. With an argument similar to that in [Yao82], we have:

**Lemma 9** *The $\delta_\epsilon$-MST of $S$ is a subgraph of the LNNG. The latter has $O(\epsilon^{-(d-1)/2}n)$ edges.*

*Proof.* (By contradiction) Suppose $T$ is the $\delta_\epsilon$-MST of $S$ and edge $pq$ is in $T$ but not in *LNNG*. Assume also that $q$ is not a candidate point of $p$. As $\{F_{\vec{v}} \mid \vec{v} \in \mathcal{V}_\epsilon^d\}$ is a partition of the whole space, there exists $\vec{v} \in \mathcal{V}_\epsilon^d$ such that $q \in \vec{p} + F_{\vec{v}}$. By the assumption that $q$ is not a candidate point of $p$, there is another point $r \in \vec{p} + F_{\vec{v}}$ such that $(\vec{r} - \vec{p})\vec{v} \leq (\vec{q} - \vec{p})\vec{v}$. By Property 3,

$$\max_{\vec{v} \in \mathcal{V}_\epsilon^d}((\vec{q} - \vec{r})\vec{v}) \leq \max((\vec{q} - \vec{p})\vec{v}, (\vec{r} - \vec{p})\vec{v}) .$$

Because both $q$ and $r$ are in $\vec{p} + F_{\vec{v}}$, by the definition of $\delta_\epsilon$, we have

$$\delta_\epsilon(q, r) < \max(\delta_\epsilon(p, r), \delta_\epsilon(p, q)) = \delta_\epsilon(p, q) .$$

Now, delete the edge $pq$ from $T$, disconnecting $T$ into two parts $T_1, T_2$. Suppose $p$ is in $T_1$ and $q$ in $T_2$. If $r$ lies in $T_1$, then we connect $T_1$,$T_2$ by adding $q, r$. Otherwise, we connect by adding $p, r$. Either way, we will get another spanning tree of $S$, but with smaller total edge lengths, contradicting with that $T$ is the $\delta_\epsilon$-MST of $S$. Thus, any edge of $T$ must be in *LNNG*. The bound on the number of edges in *LNNG* follows immediately from the bound on the total number of cones in our construction. □

Hence, to construct the $\delta_\epsilon$-MST, we first build the *LNNG*, then find the MST of the *LNNG* by either a slightly superlinear deterministic algorithm ([GGST86]) or a linear randomized algorithm([KKT95]). Once again, the *LNNG* can be found by the multidimensional range search technique. In order to perform a $d$ dimensional range search, we triangulate each Voronoi cone in constant time into a constant number of sub-cones (Property 2), bounded by $d$ hyperplanes. All the *LNNG* edges can then be detected for each sub-cone in $F_{\vec{v}}$, by building one MDRS tree with $\vec{v}$ as the weight direction and querying each point in each of $O(\epsilon^{-(d-1)/2})$ MDRS trees.

**Theorem 10** *The $\delta_\epsilon$-MST (which is also a $(1+\epsilon)$ EMST) can be found in time $O(\epsilon^{-(d-1)/2} n \log^{d-1} n)$.*

*Proof.* This follows from the above analysis and the complexity bounds for a static multidimensional range search tree. □

## 4.2 Construction of well spaced vectors

In this section, we construct a set of well-spaced vectors as defined in Section 4.1. We simply outline the construction in three dimensions and give indications for its extension to higher dimensions.

Consider a cube with side length $2\sqrt{3}/3$ (small enough so that it fits in the unit sphere) centered at the origin. Partition each of its faces, which are two-dimensional cubes, into a regular $n \times n$ mesh of patches. Define the *center vector* of a patch to be the vector directed from the origin to the center of the patch. We form a vector set $\mathcal{V}$ by normalizing to unit length the center vector of each patch.

It is easy to verify the following facts:

**Lemma 11** *With the above construction,*

1. $|\mathcal{V}| = O(n^2)$.
2. *For any vector $\vec{v}$ whose projection on the cube lies in patch $s$, the angle between $\vec{v}$ and center vector of $s$ is less than* $\arcsin(c/n)$ *for some constant $c$.*
3. *For any vector $\vec{v} \in \mathcal{V}$, the number of vectors in $\mathcal{V}$ whose angle with $\vec{v}$ is less than* $2\arcsin(c/n)$*, is bounded by a constant.*

For any given $\alpha$, we claim that $\mathcal{V}$ is a well spaced vector set for some $n = O(1/\alpha)$. Indeed, Property 1 is ensured by facts 1 and 2 of Lemma 11, plus the fact that $\sin\alpha \approx \alpha$ for small $\alpha$. Again by Lemma 11, we know that, for $\vec{v} \in \mathcal{V}$, any vector in its Voronoi cone has an angle with $\vec{v}$ that is less than $\arcsin(c/n)$. Thus, if $\vec{u} \in \mathcal{V}$ has an angle with $\vec{v}$ that is larger than $2\arcsin(c/n)$, the Voronoi cones of $\vec{u}$ and $\vec{v}$ can't share any plane. By fact 3 of Lemma 11, the complexity of each Voronoi cone is bounded by a constant. Therefore, Property 2 follows also. Property 3 is ensured by the fact that each cone is narrow enough, as argued in [Yao82].

Thus, for any given $\alpha$, we can construct a well spaced vector set in three dimensions. The well spaced vector set in higher dimensions can be constructed in exactly the same way with a partitioning of the faces of the $d$ dimensional hypercube (each one is a $d-1$ dimensional hypercube). The bounds still follow after some simple calculations, which are omitted here. Alas, the hidden constants are exponential in the dimension.

## 4.3 Maintaining a $(1+\epsilon)$-EMST of moving points

In this section, we kinetize the data structure of Theorem 10. By Lemma 9, in order to maintain the $(1+\epsilon)$-EMST, two structures are needed. One is used to maintain the local nearest neighbor graph *LNNG*. The other is used to maintain the MST of a graph under dynamic operations such as insertion and deletion of edges, as well as under continuous changes of the edge lengths involved.

To maintain the *LNNG*, we use our kinetized multidimensional range search tree. We construct an MDRS tree for each weight function and each sub-cone as in Section 2.1. These multidimensional search trees can be maintained kinetically as shown in Section 2.2. Moreover, with this data structure, the *LNNG* (i.e. the set of candidate pairs) can be maintained as in Section 3. Once again, it is the use of candidate pairs instead of candidate points which makes the structure local.

The MST of a graph is completely determined by the ordering of the edges by lengths. Thus, our kinetic data structure will maintain a list of edges ordered by length, and an event queue based on the associated kinetic certificates. There are three types of events that need to be processed in order to maintain the MST of the *LNNG*,

1. the insertion of an edge,
2. the deletion of an edge, and
3. the change of order between two *LNNG* edges in the length-ordered edge list.

The former two events are caused by the dynamic changes in the *LNNG*. We could handle them with the help of any fully dynamic MST algorithm, but such heavy machinery is unnecessary. As a range ordering event doesn't change the ordering of all pairwise distance, it cannot change the structure of the MST. Therefore, the deletion of an MST edge $(p, q)$ can only happen in the following model situation: the point $q$ was the $v$-candidate point of $p$ for some $v$. Suddenly, the $v$-ordinate of $q$ becomes greater than the $v$ ordinate of a point $r$ that also belongs to the Voronoi cone $\vec{p} + F_{\vec{v}}$. In this case, an argument similar to the proof of Lemma 9 shows that $(p, r)$ is the new MST edge that replaces $(p, q)$, and this is also the only case where an edge becomes part of the MST as soon as it is inserted.

If we encounter an event of the third type, this is a signal that the MST may need to be changed. More precisely, if edge $e_1$ was shorter than edge $e_2$ just before they interchange order, this event changes the structure of the MST if and only if, just before this event, (1) $e_1$ is an MST edge, (2) $e_2$ is not an MST edge, and (3) $e_1$ is on the MST path connecting the two endpoints of $e_2$. This last condition can be detected and maintained within $O(\log n)$ time per operation using the link-cut tree data structure of [ST83].

Thus, we have a kinetic data structure to maintain the $(1+\epsilon)$-EMST. The performance of this data structure is analyzed and summarized as follows.

**Number of internal events** There are two types of events. One is caused by a change in (one of) the MDRS trees and the other is caused by the exchange in the length order of two edges in *LNNG*. The number of events of the first and second type is $O(\epsilon^{-(d-1)/2}n^2)$, because we have $O(\epsilon^{-(d-1)/2})$ linear orderings. To bound the number of events of the second type, as in [KTI95], we divide the MST history into $O(n)$ stages with each stage having $O(\epsilon^{-(d-1)/2}n)$ insertions and deletions of edges. Thus, within each stage, there are at most $O(\epsilon^{-(d-1)/2}n)$ edges which ever appear. In addition, for each pair of edges, they can exchange their orders by at most $O(1)$ times for constant degree algebraic point motions. Thus there are at most $O((\epsilon^{-(d-1)/2}n)^2) = O(\epsilon^{-(d-1)}n^2)$ events of third type in each stage. The total is $O(\epsilon^{-(d-1)}n^3)$.

**Compactness** Obviously, the space needed to store all the data structures is $O(\epsilon^{-(d-1)/2}n\log^{d-1}n)$. Therefore, our structure is compact.

**Locality** The structure is also clearly local.

**Responsiveness** Those events caused by maintaining the *LNNG* can be handled in $O(\log^d n)$ time. Those caused by maintaining the MST can be handled in $O(\log n)$ time.

According to above analysis, we have:

**Theorem 12** *The above data structure maintains a $(1+\epsilon)$-EMST of moving points. If the points follow algebraic trajectories of fixed degree, there are $O(\epsilon^{-(d-1)}n^3)$ events. For each event, the structure can be updated within time $O(\log^d n)$. This KDS is compact, local and responsive.*

As mentioned in the introduction, we do not know if the above KDS for the MST is efficient. In any fixed dimension, it is known that the actual number of combinatorial changes to the real EMST for linearly moving points is $O(n^3 2^{\alpha(n)})$ [KTI95], but it is not known whether this bound is tight. Since the $(1+\epsilon)$-EMST is not uniquely defined, the number of combinatorial changes it undergoes has not been studied.

The kinetic data structure shown here can also be used to maintain exact MSTs under polyhedral metrics(for example, $L_1$ or $L_\infty$). It is known that for linear motions the $L_1$- and $L_\infty$-MSTs undergo $O(n^{5/2}\alpha(n))$ changes [KTI95]. Thus our structure is not as efficient as possible even in this case.

## 5 Conclusion and open problems

In this paper, we solved a number of classical proximity problems in the new kinetic setting of [BGH97]. We showed how to maintain a multidimensional range search tree, and how to use it to maintain the closest pair and an approximation of the minimum spanning tree of moving points.

In [BDIZ], an alternate *probabilistic* setting for judging the efficiency of kinetic data structures was proposed. In this setting, one computes the expected number of events processed when a kinetic data structure is run on points drawn independently at random from some prescribed distribution. It is easy to see that our KDS for the closest pair behaves very poorly according to this measure (it processes $\Theta(n^2)$ events in expectation). Can it be modified to process less events in expectation?

The results for the minimum spanning tree are very preliminary, especially as far as efficiency is concerned. A difficulty here is that tight bounds are still unknown for the number of changes of the MST itself for any type of metric. Even for a *fixed* (and even planar) graph with variable edge lengths the problem of kinetically maintaining an MST efficiently seems quite interesting. Finally, a kinetic data structure that can maintain the exact EMST in a responsive, compact, and local manner would also be highly desirable.

## Acknowledgments

We wish to thank Aris Gionis and Tong Zhang for useful discussions, as well as an anonymous referee for helpful suggestions to simplify section 4.1.

## References

[AMS92] Pankaj K. Agarwal, J. Matoušek, and S. Suri. Farthest neighbors, maximum spanning trees and related problems in higher dimensions. *Comput. Geom. Theory Appl.*, 1(4):189–201, 1992.

[AS89] C. Aragon and R. Seidel. Randomized search trees. In *Proc. 30th Annu. IEEE Sympos. Found. Comput. Sci.*, pages 540–545, 1989.

[BDIZ] J. Basch, H. Devarajan, P. Indyk, and L. Zhang. Probabilistic analysis for combinatorial functions of moving points. This volume.

[BGH97] J. Basch, L.J. Guibas, and J. Hershberger. Data structures for mobile data. In *Proc. 8th ACM-SIAM Sympos. Discrete Algorithms*, page to appear, 1997.

[BS80] J. L. Bentley and J. B. Saxe. Decomposable searching problems I: Static-to-dynamic transformation. *J. Algorithms*, 1:301–358, 1980.

[GGST86] H. N. Gabow, Z. Galil, T. Spencer, and R. E. Tarjan. Efficient algorithms for finding minimum spanning trees in undirected and directed graphs. *Combinatorica*, 6:109–122, 1986.

[KGS] Dong-Jin Kim, Leonidas J. Guibas, and Sung-Yong Shin. Fast collision detection among multiple moving spheres. This volume.

[KKT95] D. Karger, P. N. Klein, and R. E. Tarjan. A randomized linear-time algorithm to find minimum spanning trees. *J. ACM*, 42:321–328, 1995.

[KTI95] N. Katoh, T. Tokuyama, and K. Iwano. On minimum and maximum spanning trees of linearly moving points. *Discrete Comput. Geom.*, 13:161–176, 1995.

[Meh84] K. Mehlhorn. *Data Structures and Algorithms 3: Multi-dimensional Searching and Computational Geometry*, volume 3 of *EATCS Monographs on Theoretical Computer Science*. Springer-Verlag, Heidelberg, West Germany, 1984.

[Mul94] K. Mulmuley. *Computational Geometry: An Introduction Through Randomized Algorithms.* Prentice Hall, Englewood Cliffs, NJ, 1994.

[Ove83] M. H. Overmars. *The Design of Dynamic Data Structures*, volume 156 of *Lecture Notes in Computer Science.* Springer-Verlag, Heidelberg, West Germany, 1983.

[Smi92] M. Smid. Maintaining the minimal distance of a point set in polylogarithmic time. *Discrete Comput. Geom.*, 7:415–431, 1992.

[ST83] D. D. Sleator and R. E. Tarjan. A data structure for dynamic trees. *J. Comput. Syst. Sci.*, 26(3):362–381, 1983.

[Vai84] P. M. Vaidya. A fast approximation for minimum spanning trees in $k$-dimensional space. In *Proc. 25th Annu. IEEE Sympos. Found. Comput. Sci.*, pages 403–407, 1984.

[WL85] D. E. Willard and G. S. Lueker. Adding range restriction capability to dynamic data structures. *J. ACM*, 32:597–617, 1985.

[Yao82] A. C. Yao. On constructing minimum spanning trees in $k$-dimensional spaces and related problems. *SIAM J. Comput.*, 11:721–736, 1982.

# Approximate Nearest Neighbor Queries Revisited

Timothy M. Chan*

## Abstract

This paper proposes new methods to answer approximate nearest neighbor queries on a set of $n$ points in $d$-dimensional Euclidean space. For any fixed constant $d$, a data structure with $O(\varepsilon^{(1-d)/2} n \log n)$ preprocessing time and $O(\varepsilon^{(1-d)/2} \log n)$ query time achieves approximation factor $1+\varepsilon$ for any given $0 < \varepsilon < 1$; a variant reduces the $\varepsilon$-dependence by a factor of $\varepsilon^{-1/2}$. For any arbitrary $d$, a data structure with $O(d^2 n \log n)$ preprocessing time and $O(d^2 \log n)$ query time achieves approximation factor $O(d^{3/2})$.

## 1 Introduction

Let $P$ be a set of $n$ point sites in $d$-dimensional space $\mathbb{R}^d$. In the well-known *post office problem*, we want to preprocess $P$ into a data structure so that a site closest to a given query point $q$ (called the *nearest neighbor* of $q$) can be found efficiently. Distances are measured under the Euclidean metric. The post office problem has many applications within computational geometry and from other areas such as data compression, pattern recognition, databases, and statistics.

For $d = 2$, Voronoi diagrams provide an optimal solution to the problem with $O(n \log n)$ preprocessing time, $O(n)$ space, and $O(\log n)$ query time [13]. Unfortunately, even for $d = 3$, no near-linear preprocessing method is known that achieves near-logarithmic query time; the best methods, based on ray shooting [2, 12], require $O((n/m^{1/\lceil d/2 \rceil}) \operatorname{polylog} n)$ query time for an $O(m)$-space structure ($n < m < n^{\lceil d/2 \rceil}$). To obtain better performance, a number of researchers thus turned to an approximate version of the post office problem: instead of a site with the minimum distance to the query point $q$, find a site $s$ whose distance to $q$ is within $c$ times the minimum. We call such an $s$ a *c-nearest neighbor of q* and the number $c > 1$ the *approximation factor*.

For any constant $d$, the approximate post office problem was solved optimally by Arya, *et al.* [5]: an $O(n)$-space structure called the *balanced box-decomposition (BBD) tree* can find $(1+\varepsilon)$-nearest neighbors in $O(\log n)$ time for any fixed $\varepsilon > 0$; this structure can be constructed in $O(n \log n)$ time. Despite its optimality, the main drawback of Arya, *et al.*'s method is the "constant" factors hidden in the big-Oh notation. These factors depend on the parameters $d$ and $\varepsilon$. If $d$ is held fixed and $\varepsilon$ is allowed to vary, then the actual query time is $O(\varepsilon^{-d} \log n)$ in the worst case. The preprocessing does not depend on $\varepsilon$. Even for small values of $d$ such as 3 and 4, obtaining a 1.001-nearest neighbor may be time-consuming with this method.

To obtain better $\varepsilon$-dependence in the query time, one can use a different approximation method due to Clarkson [10], which is based on an earlier randomized method of Arya and Mount [3]. With high probability Clarkson's query algorithm requires only $O(\varepsilon^{(1-d)/2} \log n)$ time, but preprocessing takes $O((\varepsilon^{1-d} \log(\rho/\varepsilon)) n^2)$ time. The parameter $\rho$ is related to the ratio of the distance between the farthest pair of sites to the distance between the closest pair of sites. In Section 2, we obtain a strict improvement of Clarkson's result: the same $O(\varepsilon^{(1-d)/2} \log n)$ query time is obtained but with only $O(\varepsilon^{(1-d)/2} n \log n)$ preprocessing time. This method uses the BBD trees of Arya, *et al.* in a novel way. Further reduction of the $\varepsilon$-dependence by an $\varepsilon^{-1/2}$ factor is possible at the expense of an extra $\log n$ factor in the query time.

The above method, though efficient in low dimensions, is impractical in high dimensions, because constant factors grow exponentially when $d$ varies. This exponential dependence in $d$ is also inherent in Arya, *et al.*'s method and in traditional methods based on grids (bucketing), quadtrees, and $k$-$d$ trees; see Arya, *et al.* [6] for an analysis of a grid method. In some applications such as vector quantization, the dimension $d$ may actually be a function of $n$.

To circumvent the exponential growth problem, one

*Dept. of Math. and Computer Science, University of Miami, Coral Gables, FL 33124, tchan@cs.miami.edu

can demand less and settle for a rough approximation to the post office problem. (In applications where any reasonable metric will do, a rough approximation is just as good.) Bern [7] described an interesting quadtree method to find $O(\sqrt{d})$-nearest neighbors in $O(d2^d \log n)$ time with $O(d8^d n \log n)$ preprocessing time. Alternatively, a randomized method finds $O(d^{3/2})$-nearest neighbors in $O(d \log^2 n)$ query time with high probability, using $O(d4^d n \log^2 n)$ preprocessing time. In Section 3, we improve Bern's technique by removing the exponential factors completely using a deterministic method: we can find $O(d^{3/2})$-nearest neighbors in $O(d^2 \log n)$ query time with $O(d^2 n \log n)$ preprocessing time.

# 2 Fine Approximation in Low Dimensions

In what follows, $\|\cdot\|$ denotes the Euclidean norm. Given $q \in \mathbb{R}^d$ and $r > 0$, $B(q,r)$ denotes the closed ball $\{p \in \mathbb{R}^d : \|p - q\| \leq r\}$. For any point $p \in \mathbb{R}^d$, we use $p_i$ to denote the $i$-th coordinate of $p$, and we use $p'$ and $p''$ to denote the projected points $(p_1, \ldots, p_{d-1})$ and $(p_1, \ldots, p_{d-2})$ respectively. In this section, we assume that the dimension $d$ is a fixed constant and constants in big-Oh notation may depend on $d$.

## 2.1 Preliminaries on BBD trees

Let $P$ be a set of $n$ point sites in $\mathbb{R}^d$. For our purposes, a *BBD tree* for $P$ is a binary tree with $O(n)$ nodes and $O(\log n)$ depth, satisfying the following properties. Each node $v$ of the tree is associated with a cell, $cell(v) \subset \mathbb{R}^d$. Let $P(v) = P \cap cell(v)$. If *root* denotes the root of the tree, then $P(root) = P$. If $left(v)$ and $right(v)$ denote the left and right children of an internal node $v$, then $cell(left(v))$ and $cell(right(v))$ have disjoint interiors, covering $cell(v)$. For simplicity, we assume that no site lies on the boundary of a cell. The cells of the tree obey the following conditions:

1. each cell has constant complexity, and
2. the number of cells with disjoint interior and diameter at least $s$, intersecting a set of diameter $r$, is bounded by $\lceil 1 + Cr/s \rceil^d$ for some fixed constant $C$.

The first condition follows directly from Arya, *et al.*'s construction of the BBD tree [5]; they used cells that are differences of two axis-aligned boxes. The second condition is a consequence of their *packing lemma* [4, 5]. The construction time is $O(n \log n)$.

**Lemma 2.1** *Given $q \in \mathbb{R}^d$ and $r > 0$, one can find $k = O(\log n)$ nodes of the BBD tree, $v_1, \ldots, v_k$, in $O(\log n)$ time, such that* (i) *each site in $B(q,r)$ lies in some $P(v_i)$, and* (ii) *each $P(v_i)$ is contained in $B(q, 2r)$.*

**Proof:** Consider the following algorithm, based on a query algorithm by Arya and Mount for approximate range searching [4]:

**Algorithm** BDD-Query($v$)
1. if $v = nil$ or $cell(v) \cap B(q,r) = \emptyset$ then return $\emptyset$
2. if $cell(v)$ has diameter $< r$ then return $\{v\}$
3. return BDD-Query($left(v)$) $\cup$ BDD-Query($right(v)$)

Let $\{v_1, \ldots, v_k\}$ to be the set of nodes returned by a call to BDD-Query($root$). It is easy to see that (i) holds. To show (ii), observe that for each $v_i$, $cell(v_i)$ intersects $B(q,r)$ and has diameter less than $r$; it follows that each point in $cell(v_i)$ lies in $B(q, 2r)$.

It remains to bound the running time of BDD-Query(). We say that a node $v$ is *expanded* if $cell(v)$ intersects $B(q,r)$ and the diameter of $cell(v)$ is at least $r$. The packing lemma (second condition) ensures that the number of expanded nodes with disjoint interior is bounded by a constant. Since cells of nodes on the same level of the BBD tree have disjoint interiors, the total number of expanded nodes is at most a constant times the depth of the tree, i.e., $O(\log n)$. This bounds the running time as well as the number $k$ of nodes returned by BDD-Query(). □

## 2.2 Two BBD-based data structures

Let $P$ be a set of $n$ point sites in $\mathbb{R}^d$ and let $\varepsilon > 0$ be fixed. We now give a data structure for $P$ such that given a query point $q \in \mathbb{R}^d$, we can quickly report a $(1+\varepsilon)$-nearest neighbor of $q$, i.e., a site $s$ such that $\|s - q\| \leq (1+\varepsilon)\|p - q\|$ for any $p \in P$. The idea is to consider a number of coordinate systems and build BBD trees under each one. (Kapoor and Smid [11] adopted a similar strategy, using range trees, to obtain dynamic data structures for approximate nearest neighbor queries.) We note that each site lies inside a certain narrow cone under some coordinate system. We show that finding a nearest neighbor restricted to such a cone can be solved using approximate range searching via Lemma 2.1.

**Lemma 2.2** *Let $\Delta_1 = \{p \in \mathbb{R}^d : \|p'\| \leq \delta p_d\}$, where $\delta = \sqrt{\varepsilon/8}$. With $O(n \log n)$ preprocessing time and space, one can answer the following query in $O(\log n)$*

*time: given $q \in \mathbb{R}^d$ and $r > 0$, return a site $s$ satisfying the inequality*

$$\min\{\|s-q\|, r\} \leq (1+\varepsilon)\max\{\|p-q\|, r/2\} \quad (1)$$

*for any $p \in P$ with $p - q \in \Delta_1$.*

**Proof:** Construct a BBD tree for the $(d-1)$-dimensional set $P' = \{p' : p \in P\}$ (for simplicity, assume that no two sites have the same projection). For each node $v$ of the tree with cell *cell*$(v)$, sort the point set $P(v) = \{p \in P : p' \in \mathit{cell}(v)\}$ according to the last coordinate and augment the node $v$ with an array storing this sorted list. The space complexity of this augmented BBD tree is $O(n \log n)$, since the tree is of logarithmic depth. The preprocessing time remains $O(n \log n)$, including the sorting step.

Given point $q \in \mathbb{R}^d$ and $r > 0$, we can find a site $s$ with the desired property as follows. Using algorithm BDD-Query() in Lemma 2.1, find $k = O(\log n)$ nodes of the BBD tree, $v_1, \ldots, v_k$, such that (i) each projected site in $B(q', \delta r)$ lies in some $P'(v_i)$, and (ii) each $P'(v_i)$ is contained in $B(q', 2\delta r)$. Now, define $s$ to be a site in $P(v_1) \cup \ldots \cup P(v_k)$ that minimizes $|s_d - q_d|$.

Clearly, $s$ can be found by performing $k = O(\log n)$ binary searches on the sorted lists at nodes $v_1, \ldots, v_k$ in $O(\log^2 n)$ time. To reduce the running time to $O(\log n)$, we employ a standard technique, attributed to Lueker and Willard [13]: for each internal node $v$ of the BBD tree and each point $p \in P(v)$, keep pointers to the successor and predecessor of $p$ in the sorted lists of *left*$(v)$ and *right*$(v)$; this increases preprocessing time and space by at most a constant factor. To answer a query using algorithm BDD-Query(), we only need to perform one binary search at the root; given the position of the query point $q$ in the sorted list at node $v$, we can deduce the position of $q$ in the sorted list at its children in constant time.

Let $p$ be a site with $p - q \in \Delta_1$. It remains to show that inequality (1) holds. If $\|p-q\| > r$, then we are done. Otherwise, since $p - q \in \Delta_1$, we have $\|p'-q'\| \leq \delta|p_d - q_d| \leq \delta\|p-q\| \leq \delta r$. By (i), $p$ must belong to some $P(v_i)$, so that $|s_d - q_d| \leq |p_d - q_d| \leq \|p-q\|$. Furthermore, by (ii), we have $\|s'-q'\| \leq 2\delta r = (\sqrt{\varepsilon/2})r$. Then

$$\begin{aligned}\|s-q\|^2 &= |s_d-q_d|^2 + \|s'-q'\|^2 \\ &\leq \|p-q\|^2 + \varepsilon r^2/2 \\ &\leq (1+2\varepsilon)\max\{\|p-q\|^2, r^2/4\}.\end{aligned}$$

Taking square roots, we get $\|s-q\| < (1+\varepsilon)\max\{\|p-q\|, r/2\}$, implying (1). □

The set $\Delta_1$ is a (spherical) cone of angular diameter $\phi = 2\arctan\delta = \Theta(\delta)$. It is well known that the space $\mathbb{R}^d$ can be covered by $O(\phi^{1-d})$ cones of angular diameter $\phi$; for example, in the plane, such a system of cones can be obtained by rotation over angles of $\phi j$ for $j = 1, \ldots, \lceil 2\pi/\phi \rceil$ (see Yao [16]). Let $\{\Delta_1^{(j)}\}$ be a collection of $O(\phi^{1-d}) = O(\varepsilon^{(1-d)/2})$ rotated copies of $\Delta_1$ covering $\mathbb{R}^d$. We have the following:

**Theorem 2.3** *With $O(\varepsilon^{(1-d)/2} n \log n)$ preprocessing time and space, we can find a $(1+\varepsilon)$-nearest neighbor of a query point $q \in \mathbb{R}^d$ in $O(\varepsilon^{(1-d)/2}\log n)$ time.*

**Proof:** By changing coordinate systems, we can replace $\Delta_1$ with the cone $\Delta_1^{(j)}$ in Lemma 2.2 for each $j$. In the preprocessing, we construct the data structure for each of the $O(\varepsilon^{(1-d)/2})$ cones. The total preprocessing time is $O(\varepsilon^{(1-d)/2} n \log n)$. To answer the query, we first compute a 2-nearest neighbor $t$ of $q$ in $O(\log n)$ time by Arya, *et al.*'s method [5]. Set $r = \|t-q\|$; observe that $\max\{\|p-q\|, r/2\} = \|p-q\|$ for any $p \in P$. Let $s^{(j)}$ be the site returned by the query algorithm of the lemma for the cone $\Delta_1^{(j)}$. Then the site in $\{s^{(j)}\} \cup \{t\}$ closest to $q$ is a $(1+\varepsilon)$-nearest neighbor of $q$. The query time is therefore $O(\varepsilon^{(1-d)/2}\log n)$. □

We can slightly improve the $\varepsilon$-dependence in the time bounds by using a different set of cones and known techniques on planar Voronoi diagrams [13].

**Lemma 2.4** *Let $\Delta_2 = \{p \in \mathbb{R}^d : \|p''\| \leq \delta p_d\}$, where $\delta = \sqrt{\varepsilon/8}$. With $O(n \log n)$ preprocessing time and space, one can answer the following query in $O(\log^2 n)$ time: given $q \in \mathbb{R}^d$ and $r > 0$, return a site $s$ satisfying inequality (1) for any $p \in P$ with $p - q \in \Delta_2$.*

**Proof:** Construct a BBD tree for the $(d-2)$-dimensional set $P'' = \{p'' : p \in P\}$. For each node $v$ of the tree with cell *cell*$(v)$, store the point set $P(v) = \{p \in P : p'' \in \mathit{cell}(v)\}$ and the Voronoi diagram of the two-dimensional point set $\{(p_{d-1}, p_d) : p \in P(v)\}$. The space complexity of this augmented BBD tree is $O(n \log n)$, since the tree is of logarithmic depth. The preprocessing time is $O(n \log^2 n)$ if we use an $O(n \log n)$-time algorithm to construct the Voronoi diagrams. We can reduce the preprocessing time to $O(n \log n)$ if we construct these Voronoi diagrams in a bottom-up fashion, since planar Voronoi diagrams can be transformed into 3-d halfspace intersections, and the intersection of two 3-d convex polyhedra can be be computed in linear time [9].

Given point $q \in \mathbb{R}^d$ and $r > 0$, we follow the proof of Lemma 2.2 and use algorithm BDD-Query() in Lemma 2.1 to find $k = O(\log n)$ nodes of the BBD tree, $v_1, \ldots, v_k$, such that (i) each projected site in $B(q'', \delta r)$

lies in some $P''(v_i)$, and (ii) each $P''(v_i)$ is contained in $B(q'', 2\delta r)$. We then define $s$ to be a site in $P(v_1) \cup \ldots \cup P(v_k)$ that minimizes $|s_{d-1} - q_{d-1}|^2 + |s_d - q_d|^2$. This site $s$ can be found by performing $k = O(\log n)$ point location queries on the Voronoi diagrams at nodes $v_1, \ldots, v_k$. Using an optimal method for planar point location, we can compute $s$ in $O(\log^2 n)$ time. That $s$ satisfies the desired property now follows as in the proof of Lemma 2.2. □

If we ignore the $(d-1)$-st coordinate, then $\Delta_2$ is a cone of angular diameter $\phi$ in $\mathbb{R}^{d-1}$. We can thus cover $\mathbb{R}^d$ with only $O(\phi^{2-d}) = O(\varepsilon^{1-d/2})$ rotated copies $\{\Delta_2^{(j)}\}$ of $\Delta_2$. Lemma 2.4 then implies the following analogue of Theorem 2.3:

**Theorem 2.5** *With $O(\varepsilon^{1-d/2} n \log n)$ preprocessing time and space, we can find a $(1+\varepsilon)$-nearest neighbor of a query point $q \in \mathbb{R}^d$ in $O(\varepsilon^{1-d/2} \log^2 n)$ time.*

### 2.3 Applications

In certain batched applications of the post office problem, we can eliminate the extra $\log n$ factor in Theorem 2.5, by using a simple grid scheme in place of BBD trees. For example, given $n$ red points and $n$ blue points in $\mathbb{R}^3$, we have an $O(\varepsilon^{-1/2} n \log n)$-time algorithm for computing a pair of red and blue points whose distance is at most $1+\varepsilon$ times the closest red-blue pair. This is faster than the known exact methods [1] if $1/\varepsilon$ is within the order of $n^{2/3}$. Similar results can be obtained for an approximation of the Hausdorff distance of two (static) point sets.

## 3 Rough Approximation in High Dimensions

In this section, we assume a real-RAM model of computation that supports integer division and base-2 integer logarithm in unit time. Constants in big-Oh notation do not depend on the dimension $d$.

### 3.1 Preliminaries on balanced quadtrees

We define a *quadtree box* to be a set $B \subseteq [0,2)^d$ of the form

$$B = [\tfrac{a_1}{2^\ell}, \tfrac{a_1+1}{2^\ell}) \times \ldots \times [\tfrac{a_i}{2^\ell}, \tfrac{a_i+1}{2^\ell}) \times [\tfrac{a_{i+1}}{2^\ell}, \tfrac{a_{i+1}+2}{2^\ell}) \times \ldots \times [\tfrac{a_d}{2^\ell}, \tfrac{a_d+2}{2^\ell})$$

for some $\ell \in \mathbb{N}$, $i \in \{0, 1, \ldots, d-1\}$, and integers $a_1, \ldots, a_d$. The number $\ell$ is called the *level* of $B$. We say that $B$ is of *stage* $d\ell + i$. Note that the diameter of $B$ is less than $2^{1-\ell}\sqrt{d}$.

The following lemma, due to Arya, *et al.* [5], when applied recursively, yields a tree of logarithmic depth called the *balanced quadtree*.

**Lemma 3.1** *Given an $n$-point set $P \subset [0,2)^d$, there exists a quadtree box $B$ such that both point sets $P \cap B$ and $P \setminus B$ have cardinality at most $2n/3$. Furthermore, if the points in $P$ have been sorted along each of the $d$ coordinates, then $B$ can be constructed in $O(dn)$ time.*

**Proof:** Define a sequence of quadtree boxes $B_0, B_1, \ldots$ iteratively as follows. Let $B_0 = [0,2)^d$, which is a quadtree box of stage 0. Given a quadtree box $B_{k-1}$ of stage $k-1$ $(k \geq 1)$, write $B_{k-1}$ as the disjoint union of two quadtree boxes of stage $k$. One of the two boxes contains at least $|P \cap B_{k-1}|/2$ of the points of $P$. Let $B_k$ be this box.

Now, consider the smallest index $k$ with $|P \cap B_k| \leq 2n/3$; we know such a $k$ exists since the diameters of $B_0, B_1, \ldots$ converge to 0. Then $|P \cap B_k| \geq |P \cap B_{k-1}|/2 > n/3$ and so $|P \setminus B_k| \leq 2n/3$. This proves the existence of the quadtree box $B$. The proof can be made constructive using integer logarithms to yield the specified time bound, as shown by Arya et al. [5] (they called their algorithm "centroid shrink"). □

For our purposes, a balanced quadtree for the point set $P$ is a binary tree with $O(n)$ nodes and $O(\log n)$ depth, satisfying the following properties. Each node $v$ of the tree stores a quadtree box $B(v)$ and each leaf stores a site. Let $P(v)$ be the set of sites stored in the leaves of the subtree rooted at $v$. If *root* denotes the root of the tree, then $P(root) = P$. If $left(v)$ and $right(v)$ denote the left and right children of an internal node $v$, then $P(left(v)) = P(v) \cap B(v)$ and $P(right(v)) = P(v) \setminus B(v)$. By Lemma 3.1, such a tree exists and can be constructed in $O(dn \log n)$ time after an initial sorting phase of $O(dn \log n)$ time.

### 3.2 A quadtree-based data structure

Let $P$ be a set of $n$ point sites in $\mathbb{R}^d$. Following an approach of Bern [7], we now show that balanced quadtrees can be used to answer approximate nearest neighbor queries on $P$. The idea is to consider a certain set of vectors $\{v^{(j)}\}$ and build balanced quadtrees for the translate $P + v^{(j)}$ for each $j$. Bern showed that using a set of $2^d$ vectors, one can find an $O(\sqrt{d})$-nearest neighbor of any query point. Alternatively, using a set of $O(t \log n)$ random vectors, one can find $O(d^{3/2})$-nearest neighbors with probability $1 - O(1/n^t)$. We note that a set of

$O(d)$ carefully chosen vectors actually suffices to yield approximation factor $O(d^{3/2})$.

We first need some definitions. Given $x \in \mathbb{R}$ and $r > 0$, let $x \operatorname{div} r = \lfloor x/r \rfloor$ and $x \bmod r = x - \lfloor x/r \rfloor r$. We say that two points $p, q \in \mathbb{R}^d$ *belong to the same r-grid cell* if and only if for each $i = 1, \ldots, d$, we have $p_i \operatorname{div} r = q_i \operatorname{div} r$. We say that a point $p$ is *$\alpha$-central in its r-grid cell* if and only if for each $i = 1, \ldots, d$, we have $\alpha r \le p_i \bmod r < (1-\alpha)r$, or equivalently, $(p_i + \alpha r) \bmod r \ge 2\alpha r$.

**Observation 3.2** *Let $p, q \in \mathbb{R}^d$. If $q$ is $\alpha$-central in its r-grid cell and $\| p - q \| \le \alpha r$, then $p$ and $q$ belong to the same r-grid cell.* □

Our key lemma is the following:

**Lemma 3.3** *Suppose $d$ is even. Let*

$$v^{(j)} = (\tfrac{j}{d+1}, \ldots, \tfrac{j}{d+1}) \in \mathbb{R}^d.$$

*For any point $p \in \mathbb{R}^d$ and $r = 2^{-\ell}$ ($\ell \in \mathbb{N}$), there exists $j \in \{0, 1, \ldots, d\}$ such that $p + v^{(j)}$ is $(\frac{1}{2d+2})$-central in its r-grid cell.*

**Proof:** Suppose, on the contrary, that $p + v^{(j)}$ is not $(\frac{1}{2d+2})$-central for any $j = 0, 1, \ldots, d$. Then, for each $j$, there is an index $i(j) \in \{1, \ldots, d\}$ with

$$(p_{i(j)} + \tfrac{j}{d+1} + \tfrac{r}{2d+2}) \bmod r \; < \; \tfrac{r}{d+1},$$

or equivalently,

$$((d+1)2^\ell p_{i(j)} + 2^\ell j + \tfrac{1}{2}) \bmod (d+1) \; < \; 1.$$

By the pigeonhole principle, there exists two distinct indices $j, j' \in \{0, 1, \ldots, d\}$ with $i(j) = i(j')$. Letting $z = (d+1)2^\ell p_{i(j)} + \frac{1}{2}$, we have $(z + 2^\ell j) \bmod (d+1) < 1$ as well as $(z + 2^\ell j') \bmod (d+1) < 1$. This is possible only if $2^\ell j \equiv 2^\ell j' \pmod{d+1}$. Since $2^\ell$ and $d+1$ are relatively prime, we must have $j = j'$: a contradiction! □

We now give a data structure for answering nearest neighbor queries with an $O(d^{3/2})$ approximation factor for even $d$. (For odd $d$, replace $d$ by $d+1$.) If the $L_\infty$ metric is used instead of the Euclidean metric, the approximation factor reduces to $O(d)$.

**Theorem 3.4** *Suppose $d$ is even. Let $c = 4d^{3/2} + 4d^{1/2} + 1$. With $O(d^2 n \log n)$ preprocessing time and $O(d^2 n)$ space, we can find a c-nearest neighbor of a query point $q \in \mathbb{R}^d$ in $O(d^2 \log n)$ time.*

**Proof:** We may assume that the given $n$-point set $P$ is contained in $[0,1)^d$. Let $p^* \in P$ be an exact nearest neighbor of the query point $q$ and let $r = 2^{-\ell}$ ($\ell \in \mathbb{N}$) be such that

$$\tfrac{r}{4d+4} < \| p^* - q \| \le \tfrac{r}{2d+2}.$$

(If $\| p^* - q \| > \frac{1}{2d+2}$, then any site is a $c$-nearest neighbor.)

Suppose that $q$ is $(\frac{1}{2d+2})$-central in its $r$-grid cell, and suppose that a balanced quadtree for $P$ is available. Consider the following query algorithm, which runs in $O(d \log n)$ time and returns a set of $O(\log n)$ sites:

**Algorithm** Quad-Query$(v)$

1. if $v$ is a leaf then return {the site in $v$}
2. if $q \in B(v)$ then
3.     return Quad-Query(*left*$(v)$)
4. else return Quad-Query(*right*$(v)$) $\cup \{s(v)\}$
    where $s(v)$ is any site from $P($*left*$(v))$

We prove that if $p^* \in P(v)$, then Quad-Query$(v)$ contains a $c$-nearest neighbor of $q$. By line 1, the statement is true if $v$ is a leaf. If $v$ is an internal node, we consider four cases:

*Case 1:* $q \in B(v)$, $p^* \in B(v)$. Then $p^* \in P(\textit{left}(v)) = P(v) \cap B(v)$, and inductively, we may assume that the set Quad-Query(*left*$(v)$) from line 3 contains a $c$-nearest neighbor of $q$.

*Case 2:* $q \notin B(v)$, $p^* \notin B(v)$. Then $p^* \in P(\textit{right}(v)) = P(v) \backslash B(v)$, and inductively, we may assume that the set Quad-Query(*right*$(v)$) from line 4 contains a $c$-nearest neighbor of $q$.

*Case 3:* $q \in B(v)$, $p^* \notin B(v)$. Suppose that $B(v)$ is of level $\ell'$. Then the two points $p^*$ and $q$ do not belong to the same $(2^{-\ell'})$-grid cell. However, by Observation 3.2, they belong to the same $r$-grid cell with $r = 2^{-\ell}$. We must have $\ell' > \ell$, implying that the diameter of $B(v)$ is less than $2^{1-\ell'}\sqrt{d} \le r\sqrt{d}$. Then any site $s \in P(\textit{left}(v))$ from the set returned in line 3 is a $c$-nearest neighbor of $q$:

$$\| s - q \| \; < \; r\sqrt{d} \; < \; (4d^{3/2} + 4d^{1/2}) \, \| p^* - q \|.$$

*Case 4:* $q \notin B(v)$, $p^* \in B(v)$. As in Case 3, we can argue that the diameter of $B(v)$ is at most $r\sqrt{d}$. Then the site $s(v) \in P(\textit{left}(v))$ found in line 4 is a $c$-nearest neighbor of $q$:

$$\begin{aligned} \| s(v) - q \| &\le \| s(v) - p^* \| + \| p^* - q \| \\ &< r\sqrt{d} + \| p^* - q \| \\ &< (4d^{3/2} + 4d^{1/2} + 1) \, \| p^* - q \|. \end{aligned}$$

We conclude that Quad-Query(*root*) returns a set of $O(\log n)$ sites containing a $c$-nearest neighbor of $q$, under the assumption that $q$ is $(\frac{1}{2d+2})$-central in its $r$-grid cell. This assumption can be removed as follows. By Lemma 3.3, we know that $q + v^{(j)}$ is $(\frac{1}{2d+2})$-central in its $r$-grid cell for some $j \in \{0, 1, \ldots, d\}$. During preprocessing, build a balanced quadtree for the point set $P + v^{(j)}$ for each $j$. The total preprocessing time for the $d + 1$ trees is $O(d^2 n \log n)$. By answering a query for $q + v^{(j)}$ on the point set $P + v^{(j)}$ for each $j$, we obtain a set of $O(d \log n)$ sites containing a $c$-nearest neighbor of $q$, and we can return the closest point to $q$ in this set. The query time is therefore $O(d^2 \log n)$. □

## 3.3 Applications

One application of our technique is the design of robust dictionaries for high-dimensional point sets. Let $P$ be a set of $n$ points in $\mathbb{R}^d$ such that every two points are of distance at least $\delta$ apart. In a query, we want to find a point in $P$ that is within distance $\varepsilon$ from a given query point. Assuming that $\varepsilon$ is much smaller than $\delta$, we can obtain $O(d^2 \log n)$ query and update time using a simple grid scheme based on Lemma 3.3. Note that the naive grid scheme requires either $\Omega(2^d n)$ space or $\Omega(2^d \log n)$ worst-case query time.

Another application is the construction of sparse spanner graphs. A *t-spanner* of $P$ is a subgraph of the complete Euclidean graph of $P$, with the property that the shortest path length between any pair of points $p, q \in P$ is bounded by $t \| p - q \|$. The number $t > 1$ is called the *stretch factor*. We show how to construct a spanner of size $O(dn \log n)$ with an $O(d^{3/2})$ stretch factor by modifying the proof of Theorem 3.4. Previous algorithms (e.g., [8, 14, 15]) have exponential dependence of $d$ in the running time.

**Theorem 3.5** *Suppose $d$ is even. Let $t = 8d^{3/2} + 8d^{1/2} + 1$. A $t$-spanner with $O(dn \log n)$ edges can be constructed in $O(d^2 n \log n)$ time.*

**Proof:** Given a balanced quadtree for $P \subset [0, 1)^d$, consider the following procedure, which runs in $O(n \log n)$ time and returns a graph with $O(n \log n)$ edges:

**Algorithm** Spanner($v$)
1. if $v$ is a leaf then return $\emptyset$
2. return Spanner(*left*($v$)) $\cup$ Spanner(*right*($v$))
   $\cup$ $\{\{s(v), p\} : p \in P(v)\}$
   where $s(v)$ is any site from $P(\textit{left}(v))$

Given a pair $p, q \in P$, let $r = 2^{-\ell}$ ($\ell \in \mathbb{N}$) be such that

$$\tfrac{r}{4d+4} < \| p - q \| \le \tfrac{r}{2d+2}.$$

Suppose that $q$ is $(\frac{1}{2d+2})$-central in its $r$-grid cell. We prove that if $p, q \in P(v)$, then the shortest path length between $p$ and $q$ in Spanner($v$) is at most $t \| p - q \|$:

*Case 1:* $p, q \in B(v)$. Then $p, q \in P(\textit{left}(v))$, and the claim follows by induction.

*Case 2:* $p, q \notin B(v)$. Then $p, q \in P(\textit{right}(v))$, and the claim follows by induction.

*Case 3:* $p \notin B(v)$, $q \in B(v)$. As in the Case 3 of the proof of Theorem 3.4, we can show that the diameter of $B(v)$ is at most $r\sqrt{d}$. Then the path $p, s(v), q$ in Spanner($v$) has length

$$\begin{aligned} & \| p - s(v) \| + \| s(v) - q \| \\ \le\ & \| p - q \| + 2 \| s(v) - q \| \\ \le\ & \| p - q \| + 2r\sqrt{d} \\ \le\ & (8d^{3/2} + 8d^{1/2} + 1) \| p - q \|. \end{aligned}$$

*Case 4:* $p \in B(v)$, $q \notin B(v)$. Similar to Case 3.

We conclude that Spanner(*root*) contains a path between $p$ and $q$ of length at most $t \| p - q \|$, under the assumption that $q$ is $(\frac{1}{2d+2})$-central in its $r$-grid cell. We can remove this assumption by constructing balanced quadtrees for $P + v^{(j)}$ for $j \in \{0, \ldots, d\}$, using Lemma 3.3, and taking the union of the resulting $d + 1$ graphs. □

It is well known that a minimum spanning tree of a $t$-spanner has weight at most $t$ times the weight of the Euclidean minimum spanning tree. Using a Fibonacci-heap implementation of Prim's algorithm on the spanner graph, we get an $O(d^2 n \log n)$-time algorithm that approximates Euclidean minimum spanning trees to within a factor of $O(d^{3/2})$. By well-known reductions, similar results follow for Euclidean traveling salesman tours and minimum Steiner trees.

For the application to closest pairs, we can reduce the $O(d^2)$ factor in the running time to $O(d)$ using a different algorithm. Details can be found in the full paper.

# 4 Conclusions

We have examined ways of reducing "constant factors" in previous approaches to approximate nearest neighbor queries on a set of $n$ points. For any fixed dimension $d$ and approximation factor $1 + \varepsilon$, we obtain constants that are $O(\varepsilon^{(1-d)/2})$ or $O(\varepsilon^{1-d/2})$. For any dimension and a sufficiently large approximation factor, we obtain constants that are polynomial in the dimension. The time and space bounds of our methods are optimal or near-optimal in terms of $n$.

One open problem for low dimensions is to reduce the $\varepsilon$-dependence even further, if possible, while keeping $O(n \operatorname{polylog} n)$ space and $O(\operatorname{polylog} n)$ query time. Reducing the space complexity to linear in Theorems 2.3 and 2.5 would be interesting. Another problem is to derive our high-dimensional results in Section 3 using only algebraic operations.

## References


[1] P. K. Agarwal, H. Edelsbrunner, O. Schwarzkopf, and E. Welzl. Euclidean minimum spanning trees and bichromatic closest pairs. *Discrete Comput. Geom.*, 6:407–422, 1991.

[2] P. K. Agarwal and J. Matoušek. Ray shooting and parametric search. *SIAM J. Comput.*, 22:764–806, 1993.

[3] S. Arya and D. M. Mount. Approximate nearest neighbor queries in fixed dimensions. In *Proc. 4th ACM-SIAM Sympos. Discrete Algorithms*, pages 271–280, 1993.

[4] S. Arya and D. M. Mount. Approximate range searching. In *Proc. 11th ACM Sympos. Comput. Geom.*, pages 172–181, 1995.

[5] S. Arya, D. M. Mount, N. S. Netanyahu, R. Silverman, and A. Wu. An optimal algorithm for approximate nearest neighbor searching. In *Proc. 5th ACM-SIAM Sympos. Discrete Algorithms*, pages 573–582, 1994.

[6] S. Arya, D. M. Mount, and O. Narayan. Accounting for boundary effects in nearest neighbor searching. *Discrete Comput. Geom.*, 16:155–176, 1996.

[7] M. Bern. Approximate closest-point queries in high dimensions. *Inform. Process. Lett.*, 45:95–99, 1993.

[8] P. B. Callahan and S. R. Kosaraju. Faster algorithms for some geometric graph problems in higher dimensions. In *Proc. 4th ACM-SIAM Sympos. Discrete Algorithms*, pages 291–300, 1993.

[9] B. Chazelle. An optimal algorithm for intersecting three-dimensional convex polyhedra. *SIAM J. Comput.*, 21:671–696, 1992.

[10] K. L. Clarkson. An algorithm for approximate closest-point queries. In *Proc. 10th ACM Sympos. Comput. Geom.*, pages 160–164, 1994.

[11] S. Kapoor and M. Smid. New techniques for exact and approximate dynamic closest-point problems. *SIAM J. Comput.*, 25:775–796, 1996.

[12] J. Matoušek and O. Schwarzkopf. On ray shooting in convex polytopes. *Discrete Comput. Geom.*, 10:215–232, 1993.

[13] F. P. Preparata and M. I. Shamos. *Computational Geometry: An Introduction.* Springer-Verlag, New York, 1985.

[14] J. S. Salowe. Construction of multidimensional spanner graphs, with applications to minimum spanning trees. In *Proc. 7th ACM Sympos. Comput. Geom.*, pages 256–261, 1991.

[15] P. M. Vaidya. A sparse graph almost as good as the complete graph on points in $k$ dimensions. *Discrete Comput. Geom.*, 6:369–381, 1991.

[16] A. C. Yao. On constructing minimum spanning trees in $k$-dimensional spaces and related problems. *SIAM J. Comput.*, 11:721–736, 1982.

# Approximate Shortest Paths and Geodesic Diameters on Convex Polytopes in Three Dimensions*

Sariel Har-Peled†

**Abstract**

Given a convex polytope $P$ with $n$ edges in $\mathbb{R}^3$, we present a relatively simple algorithm that preprocesses $P$ in $O(n)$ time, such that, given any two points $s, t \in \partial P$, and a parameter $0 < \varepsilon \le 1$, it computes, in $O((\log n)/\varepsilon^{1.5} + 1/\varepsilon^3)$ time, a distance $\Delta_P(s,t)$, such that $d_P(s,t) \le \Delta_P(s,t) \le (1+\varepsilon)d_P(s,t)$, where $d_P(s,t)$ is the length of the shortest path between $s$ and $t$ on $\partial P$. The algorithm also produces a polygonal path with $O(1/\varepsilon^{1.5})$ segments that avoids the interior of $P$ and has length $\Delta_P(s,t)$.

Our second related result is: Given a convex polytope $P$ with $n$ edges in $\mathbb{R}^3$, and a parameter $0 < \varepsilon \le 1$, we present an $O(n + 1/\varepsilon^6)$-time algorithm that computes two points $s, t \in \partial P$ such that $d_P(s,t) \ge (1-\varepsilon)\mathcal{D}_P$, where $\mathcal{D}_P = \max_{s,t \in \partial P} d_P(s,t)$ is the geodesic diameter of $P$.

## 1 Introduction

The *three-dimensional Euclidean shortest-path problem* is defined as follows: Given a set of pairwise-disjoint polyhedral objects in $\mathbb{R}^3$ and two points $s$ and $t$, compute the shortest path between $s$ and $t$ which avoids the interiors of the given polyhedral 'obstacles'. This problem has received considerable attention in computational geometry. It was shown to be NP-hard by Canny and Reif [CR87], and the fastest available algorithms for this problem run in time that is exponential in the total number of obstacle vertices (which we denote by $n$) [RS94, Sha87]. The intractability of the problem has motivated researchers to develop polynomial-time algorithms for computing approximate shortest paths and for computing shortest paths in special cases.

In the *approximate three-dimensional Euclidean shortest-path* problem, we are given an additional parameter $\varepsilon > 0$, and the goal is to compute a path between $s$ and $t$ that avoids the interiors of the obstacles and whose length is at most $(1+\varepsilon)$ times the length of the shortest obstacle-avoiding path (we call such a path an $\varepsilon$-approximate path). Approximation algorithms for the three-dimensional shortest path problem were first studied by Papadimitriou [Pap85], who gave an $O(n^4(L + \log(n/\varepsilon))^2/\varepsilon^2)$-time algorithm for computing an $\varepsilon$-approximate shortest path, where $L$ is the number of bits of precision in the model of computation. A rigorous analysis of Papadimitriou's algorithm was recently given by Choi et al. [CSY94]. A different approach was taken by Clarkson [Cla87], whose algorithm computes an $\varepsilon$-approximate shortest path in close to $O(n^2/\varepsilon^4 \log^{O(1)} n)$ time (the complexity of Clarkson's algorithm depends also on an additional parameter).

In a companion paper [HP97], we present an improved solution to an extended variant of the problem: Given a polyhedral environment $V$, and a point $s \in V$, and a parameter $\varepsilon > 0$, preprocess them into a data-structure that supports fast $\varepsilon$-approximate shortest path queries from $s$ to any $t \in V$. The preprocessing takes roughly $O((n^4/\varepsilon^6) \log^2(n)\log^2(1/\varepsilon) + (n^2/\varepsilon^8)\log^5(1/\varepsilon))$ time and the size of the data structure is $O(n^2/\varepsilon^{4+\delta})$, for any $\delta > 0$ (the time complexity of the algorithm depends also on an additional parameter), and a query can be answered in $O(\log(n/\varepsilon))$ time. In Section 5, we present a short review of this and other related results of [HP97].

The problem of computing a shortest path between two points along the surface of a single convex polytope is an interesting special case of the three-dimensional Euclidean shortest-path problem. Sharir and Schorr [SS86] gave an $O(n^3 \log n)$ algorithm for this problem, exploiting the property that a shortest path on a polyhedron *unfolds* into a straight line segment. Mount et al. [MMP87] improved the running time to $O(n^2 \log n)$; their algorithm works for non-convex polyhedra (or polyhedral surfaces) as well. Chen and Han [CH96] gave another algorithm with an improved running time of $O(n^2)$. It is a rather long-standing and intriguing open problem whether the shortest path on a convex polytope can be computed in subquadratic time. This has motivated the problem of finding near-linear algorithms that produce only an approximation of the shortest path. That is, we are given a convex polytope $P$ with $n$ vertices, two points $s$ and $t$ on its surface, and a positive real number $\varepsilon$. Let $\pi_P(s,t)$ denote any shortest path between $s$ and $t$ along the surface of $P$, and let $d_P(s,t)$ denote its length ($\pi_P(s,t)$ is usually, but not always, unique). We want to compute a path on the surface of $P$ between $s$ and $t$ whose length is at most $(1+\varepsilon)d_P(s,t)$.

---

*This work has been supported by a grant from the U.S.–Israeli Binational Science Foundation. This work is part of the author's Ph.D. thesis, prepared at Tel-Aviv University under the supervision of Prof. Micha Sharir.

†School of Mathematical Sciences, Tel Aviv University, Tel Aviv 69978, Israel; sariel@math.tau.ac.il



*Computational Geometry 97* Nice France

The first result in this direction is by Hershberger and Suri [HS95]. They present a simple algorithm that runs in $O(n)$ time, and computes a path whose length is at most $2d_P(s,t)$. Using the algorithm of [HS95], Agarwal et al. [AHPSV96] present a relatively simple algorithm that computes an $\varepsilon$-approximate shortest path (as above, this is a path on $\partial P$ between the points $s,t \in \partial P$ whose length is at most $(1+\varepsilon)d_P(s,t)$), for any prescribed $0 < \varepsilon \leq 1$, where the running time of the algorithm is $O(n\log(1/\varepsilon) + 1/\varepsilon^3)$.

In this paper we extend the analysis of [AHPSV96] to solve two problems involving approximate shortest paths on convex polytopes in $\mathbb{R}^3$. In both problems, our solutions are much faster than the best known algorithms for the corresponding exact problems.

**Approximate shortest paths queries.** In Section 3, we present an alternative to the algorithm of [HS95] that receives as input a convex polytope $P$ with $n$ edges and preprocesses $P$ in $O(n)$ time, such that for any $s,t \in \partial P$, one can compute, in $O(\log n)$ time, a number $\Delta_P(s,t)$, such that $d_P(s,t) \leq \Delta_P(s,t) \leq 8d_P(s,t)$. We then modify the algorithm of [AHPSV96], so that it uses the new algorithm instead of the algorithm of [HS95]. This results in an algorithm with $O(n)$ preprocessing time, such that for any $s,t \in \partial P$ and $0 < \varepsilon \leq 1$, it computes a length $\Delta_P(s,t)$ of an $\varepsilon$-approximate shortest path between $s$ and $t$ on $\partial P$, in $O((\log n)/\varepsilon^{1.5} + 1/\varepsilon^3)$ time. In contrast, the fastest known algorithm for computing a data-structure that supports queries of computing the length of the exact shortest path between any pair of points on $\partial P$ is due to [AAOS96]; it requires $O(n^6 m^{1+\delta})$ space and preprocessing time, for any $\delta > 0$, and answers a query in $O((\sqrt{n}/m^{1/4})\log m)$ time, where $1 \leq m \leq n^2$ (the constants of proportionality depends on $\delta$).

**Approximate geodesic diameter.** We present in Section 4 an algorithm that computes, for a given convex polytope $P$ with $n$ edges in $\mathbb{R}^3$, and a parameter $0 < \varepsilon \leq 1$, two points $s,t \in \partial P$ such that $d_P(s,t) \geq (1-\varepsilon)\mathcal{D}_P$, where $\mathcal{D}_P = \max_{s,t\in\partial P} d_P(s,t)$ is the *geodesic diameter* of $P$. The running time of the algorithm is $O(n + 1/\varepsilon^6)$. In contrast, the fastest known algorithm for the exact geodesic diameter takes $O(n^8 \log n)$ time (see [AAOS96]).

The paper is organized as follows. Section 2 introduces the required terminology and establishes some initial properties, mostly taken or adapted from [AHPSV96]. In Section 3 we describe the algorithm for answering arbitrary two-point approximate shortest-path queries. In Section 4 we derive an algorithm for approximating the geodesic diameter of a convex polytope in $\mathbb{R}^3$. In Section 5, we review a few related results taken from a companion paper [HP97]. We conclude in Section 6 by mentioning a few open problems.

## 2 Preliminaries

We begin with some terminology and some initial observations, most of them taken or adapted from [AHPSV96]. Let $P$ be a convex polytope with $n$ edges in $\mathbb{R}^3$. For a face $f$ of $P$, we denote by $H_f$ the plane passing through $f$. Given a set $A \subseteq \mathbb{R}^3$, and a real number $r \geq 0$, let $A_r$ denote the *Minkowski sum* $A \oplus B_r$, where $B_r$ is a ball of radius $r$ about the origin; that is,

$$A_r = \left\{ x \;\middle|\; \inf_{y\in A} |x-y| \leq r \right\}.$$

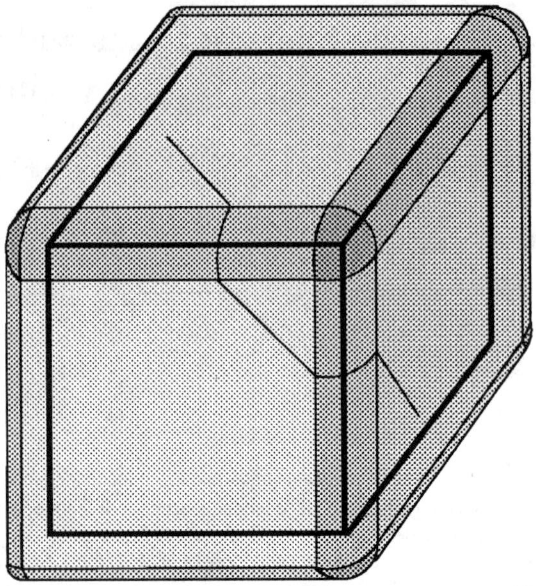

Figure 1: A cube $P$ and the inflated set $P_r$

For a convex polytope $P$ and for $x \in \partial P$, let $F_{P,r}(x) = \left\{ y \;\middle|\; y \in \partial P_r, |x-y| = r \right\}$ be the set of points of $\partial P_r$ 'corresponding' to $x$. See Figure 1 for an example of such an inflated polytope, and see [AHPSV96] for (straightforward) details concerning the structure of $P_r$ and of $F_{P,r}$.

For any plane $H$ that avoids the interior of $P$, the *positive half-space* $H^+$ (resp. *negative half-space* $H^-$) bounded by $H$ is the one containing $P$ (resp. avoiding the interior of $P$); these half-spaces are assumed to be closed. Such a plane $H$ is a *supporting plane* of $P$ if $\partial P \cap H \neq \emptyset$.

Given two points $p,q \in \mathbb{R}^3$ such that $q \neq 0$, we denote by $ray(p,q)$ the ray emanating from $p$ in the direction of $q$; that is, $ray(p,q) = \{p + tq | t \geq 0\}$.

Given two curves[1] $\gamma_1, \gamma_2$ in $\mathbb{R}^3$ that share an endpoint, we denote by $\gamma_1 \,||\, \gamma_2$ the curve resulting from concatenating $\gamma_1$ to $\gamma_2$. For a curve $\gamma \subset \mathbb{R}^3$ and for $a,b \in \gamma$, we denote by $\gamma(a,b)$ the portion of $\gamma$ between $a$ and $b$. An *outer path* of a convex body $K$ is a curve $\gamma$ connecting two points on $\partial K$ and disjoint from the interior of $K$. The length of a curve $\gamma$ is denoted by $|\gamma|$.

The following well-known theorem implies that any outer path of $P$ connecting two points $s,t \in \partial P$ must be of length at least $d_P(s,t)$, where, as above, $d_P(s,t)$ denotes the length of the shortest path between $s$ and $t$ on $\partial P$.

**Theorem 2.1 (see [Pog73])** *Let $F$ be the boundary of a convex body $K$, and $\gamma$ a curve that does not meet the interior of $K$ and connects points $s$ and $t$ on $F$. Then the length of $\gamma$ is at least the shortest distance $d_K(s,t)$ between its endpoints along the surface $F$; it is strictly greater than this distance if the curve does not lie entirely on $F$.*

**Lemma 2.2 (see [AHPSV96])** *Let $P$ be a convex polytope in $\mathbb{R}^3$, let $s,t$ be any two points on $\partial P$, let $0 < \varepsilon < 1$ be an arbitrary parameter, and let $r > 0$. For any $s_r \in F_{P,r}(s)$, $t_r \in F_{P,r}(t)$, there exists an outer path of $P_r$ connecting $s_r$ with $t_r$, whose length is at most $(1+\varepsilon/2)d_P(s,t) + 2\pi r + 100r/\sqrt{\varepsilon}$.*

[1] By a curve we mean the image of a 1-1 continuous mapping of the unit interval $[0,1]$ into $\mathbb{R}^3$.

Lemma 2.2 implies the following:

**Lemma 2.3** *Let $P, Q$ be convex polytopes in $\mathbb{R}^3$, let $0 < \varepsilon < 1$ be an arbitrary parameter, and let $r > 0$. If $P \subset Q \subset P_r$ then for any two points $s, t \in \partial P \cap \partial Q$ we have $d_P(s,t) \leq d_Q(s,t) \leq (1+\varepsilon/2)d_P(s,t) + (2\pi+2)r + 100r/\sqrt{\varepsilon}$.*

*Proof:* Pick two arbitrary points $s_r \in F_{Q,r}(s)$ and $t_r \in F_{Q,r}(t)$. Clearly, $|ss_r| = |tt_r| = r$. Since $s_r - s$ is a normal to $Q$ at $s$, it is also a normal to $P$ at $s$ because $s \in \partial P \cap \partial Q$. This implies that $s_r \in F_{P,r}(s)$, and similarly $t_r \in F_{P,r}(t)$.

Let $\sigma'$ be a shortest path connecting $s_r$ to $t_r$ on $\partial P_r$. By Lemma 2.2 and Theorem 2.1, we have $|\sigma'| \leq (1+\varepsilon/2)d_P(s,t) + 2\pi r + 100r/\sqrt{\varepsilon}$. Let $\sigma = ss_r \,||\, \sigma' \,||\, t_r t$. Clearly, $\sigma \cap int\, Q = \emptyset$, and $|\sigma| \leq (1+\varepsilon/2)d_P(s,t) + (2\pi+2)r + 100r/\sqrt{\varepsilon}$.

By Theorem 2.1,

$$\begin{aligned} d_P(s,t) &\leq d_Q(s,t) \leq |\sigma| \\ &\leq (1+\varepsilon/2)d_P(s,t) + (2\pi+2)r + 100r/\sqrt{\varepsilon}. \end{aligned}$$

■

Let $P \subseteq \mathbb{R}^3$ be a convex polytope. A *point-normal pair* on $P$ is an ordered pair $(p, \eta_p)$ such that (1) $p \in \partial P$, (2) the plane $H(p, \eta_p)$ that passes through $p$ and whose normal is $\eta_p$ is a supporting plane of $P$, and (3) $\eta_p$ is an outward normal to $P$ at $p$. As above, we denote by $H^+(p, \eta_p)$ the closed halfspace containing $P$ that is bounded by $H(p, \eta_p)$. For any $\delta > 0$, we call a collection $S$ of point-normal pairs on $P$ a *$\delta$-dense* set if for any point-normal pair $(q, \eta_q)$ on $P$, there exists a $(p, \eta_p) \in S$ such that $|pq| \leq \delta$, and the angle between $\eta_p$ and $\eta_q$ is at most $\delta$. Let $P(S)$ denote the intersection of all the halfspaces corresponding to elements of $S$, i.e., $P(S) = \bigcap_{(p,\eta_p)\in S} H^+(p, \eta_p)$.

**Lemma 2.4 (See [AHPSV96])** *Given a convex polytope $P$ in $\mathbb{R}^3$ having $n$ edges and contained in a unit ball, then one can construct, in $O(n + (1/\varepsilon^2)\log(1/\varepsilon))$ time, an $\varepsilon$-dense set $S$ of $P$, such that $|S| = O(1/\varepsilon^2)$, and $P \subseteq P(S) \subseteq P_{2\varepsilon^2}$.*

## 3 Approximate Shortest-Path Queries on a Convex Polytope

Let $P$ be a given convex polytope in $\mathbb{R}^3$ with $n$ edges. In this section we present an algorithm that preprocesses $P$ in linear time, into a linear-size data structure, such that given any two query points $s, t \in \partial P$ and a parameter $0 < \varepsilon \leq 1$, one can compute, in $O((\log n)/\varepsilon^{1.5} + 1/\varepsilon^3)$ time, an outer path connecting $s$ to $t$ whose length $\Delta_P(s,t)$ satisfies $d_P(s,t) \leq \Delta_P(s,t) \leq (1+\varepsilon)d_P(s,t)$.

Given two points $s, t \in \partial P$, we assume, without loss of generality, that $s = (0,0,-l)$ and $t = (0,0,l)$. For $r \geq 0$, let $B(r)$ denote the axis parallel cube of side $2r$ centered at the origin. Let $C(r) = \partial B(r) \setminus int\, P$. Let $s(r)$ and $t(r)$ denote the lower and upper intersection points of $C(r)$ with the $z$-axis, respectively. Note that these points are well-defined for $r \geq l$.

**Lemma 3.1** *For any $r \geq l$, there exists a path $\sigma$ between $s$ and $t$ on $\partial P$ such that $\sigma \subset B(r)$, if and only if $s(r)$ and $t(r)$ are connected in $C(r)$.*

*Proof:* If the origin $o$ lies on the boundary of $P$ then $st \subseteq \partial P$ and the lemma follows easily. Otherwise, for any $x \in \partial B(r)$, let $f_r(x)$ denote the point $ray(o,x) \cap \partial P$. It easy to verify, that the function $f_r$ is a continuous bijective deformation of $\partial B(r)$ onto $\partial P$. In particular, given a path

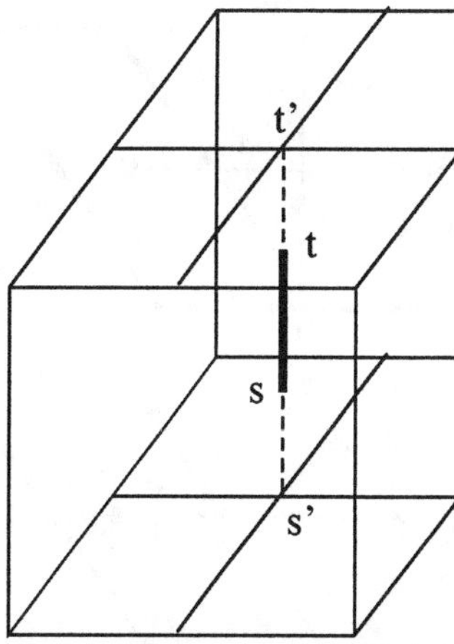


Figure 2: The graph $G$ defined on the boundary of the cube $B(1)$

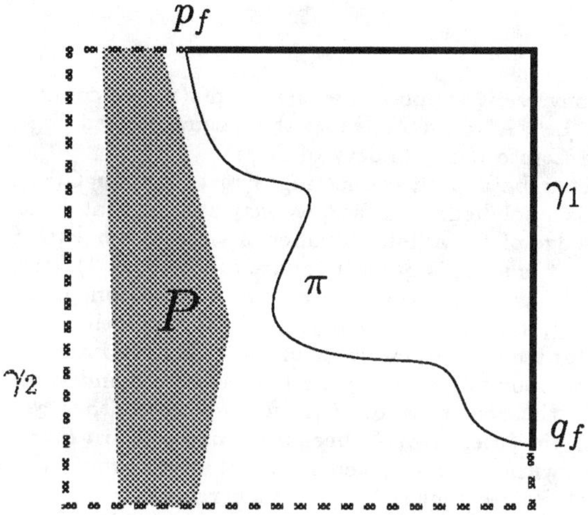


Figure 3: Either $\gamma_1$ or $\gamma_2$ must avoid the interior of $P$.

$\sigma \subset B(r)$ connecting $s$ to $t$ on $\partial P$, then $\sigma^{-1} = f_r^{-1}(\sigma)$ is connected, $\sigma^{-1} \subset C(r)$, and $s' = f_r^{-1}(s), t' = f_r^{-1}(t) \in \sigma^{-1}$, implying that $\sigma^{-1}$ connects $s'$ to $t'$ in $C(r)$.

Conversely, if $s(r)$ and $t(r)$ are connected in $C(r)$, then there exists a path $\gamma$ on $C(r)$ connecting $s(r)$ to $t(r)$. Clearly, $f_r(\gamma)$ is connected, connects $s$ to $t$ on $\partial P$, and $f_r(\gamma) \subset B(r)$.
■

Consider the set $V = \{-1,0,1\} \times \{-1,0,1\} \times \{-1,1\}$ of 18 points on $B(1)$. Let $s' = (0,0,-1)$, $t' = (0,0,1)$. Define a set $E$ of 28 straight segments, such that $vw \in E$ if and only if $v \neq w \in V$, and either $||v-w|| = 1$ or $x(v) = x(w) = \pm 1$, and $y(v) = y(w) = \pm 1$. The resulting graph $G = (V, E)$ is depicted in Figure 2. For $r \geq 0$, let $G(r) = (V, E(r))$, where $vw \in E(r)$ if $vw \in E$ and the segment $T_r(vw)$ does not intersect $int\, P$, where $T_r(a) = ra$ is the linear scaling of $\mathbb{R}^3$ by the factor $r$.

Clearly, $G(r)$ is monotone increasing, that is, $G(r') \subseteq G(r)$, for any $0 \leq r' \leq r$. Moreover, $G(0) = (V, \emptyset)$ and $G(r) = G$, for $r$ sufficiently large.

**Lemma 3.2** *For $r \geq l$, the vertices $s', t'$ in $G(r)$ are connected if and only if $s(r)$ and $t(r)$ are connected in $C(r)$.*

*Proof:* If $s'$ and $t'$ are connected by a path $\sigma$ in $G(r)$, then the path $T_r(\sigma)$ connects $s(r)$ to $t(r)$, and satisfies $T_r(\sigma) \subset$

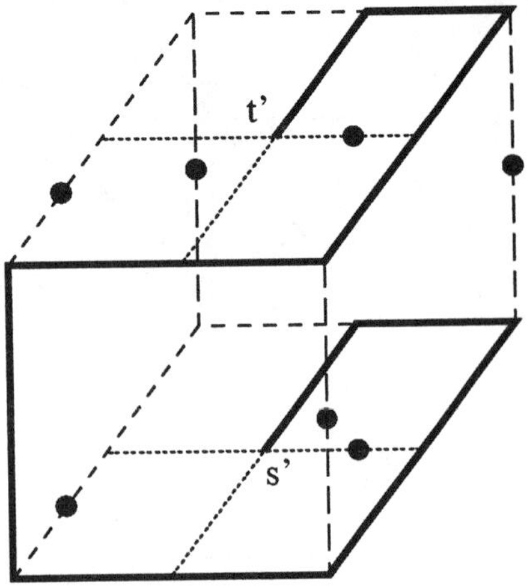


Figure 4: A path realizing the longest possible shortest path (of length 14) in a subgraph $G(r)$.

$C(r)$.

Conversely, suppose that $s(r)$ and $t(r)$ are connected in $C(r)$. Let $G_r = T_r(G)$ denote the natural embedding of the graph $G$ into the boundary of $B(r)$.

Let $\sigma$ be a path connecting $s(r)$ to $t(r)$ in $C(r)$ (since $C(r)$ is a polyhedral surface, we may assume that $\sigma$ intersect each edge of $G_r$ a finite number of times). For each face $f$ of $G_r$ ($f$ is an axis-parallel square of side 1 or 2), such that $\sigma \cap int\, f \neq \emptyset$, let $p_f$ and $q_f$ be two intersection points of $\sigma$ with $\partial f$, such that $int\,\sigma(p_f, q_f) \subseteq int\, f$. Partition $\partial f$ into two Jordan arcs $\gamma_1, \gamma_2$ with disjoint relative interiors such that the points $p_f$ and $q_f$ are their common endpoints.

By the convexity of $P$, it follows that either $\gamma_1$ or $\gamma_2$ avoids the interior of $P$, because $\sigma(p_f, q_f) \cap int\, P = \emptyset$. Assume, without loss of generality, that $\gamma_1$ avoids the interior of $P$ (see Figure 3). We then replace $\sigma$ by the curve $\sigma(s(r), p_f) \mid\mid \gamma_1 \mid\mid \sigma(q_f, t(r))$. We repeat this replacement process for each connected portion of $\sigma \cap int\, f$, until $\sigma \cap int\, f = \emptyset$, and apply this procedure for all faces $f$ of $G_r$. Let $\sigma'$ be the resulting path.

Clearly, $\sigma'$ lies completely on the edges of $G_r$, and it connects $s(r)$ to $t(r)$. Moreover, $\sigma'$ avoids the interior of $P$. This implies that $s(r)$ and $t(r)$ are connected in $G_r$, by a path that avoids the interior of $P$. Thus $s'$ and $t'$ are connected in $G(r)$. ∎

It is easy to verify the following:

**Lemma 3.3** *The length of the shortest path between $s'$ and $t'$ in any subgraph of $G$ in which $s'$ and $t'$ are connected is at most* 18, *where all the horizontal edges of $G$ have weight* 1 *and all the vertical edges have weight* 2.

Actually, the subgraphs $G(\cdot)$ of $G$ have additional conditions imposed on them because of the convexity of $P$. By checking all possible subgraphs of $G$ that comply with these "convexity" conditions, we get the following improvement, which only effects the constants of proportionality in the resulting algorithm.

**Lemma 3.4** *The length of the shortest path between $s'$ and $t'$ in any subgraph $G(r)$ of $G$ in which $s'$ and $t'$ are connected is at most* 14, *where the edges of $G$ have the same weights as above.*

*Moreover, there exist a convex polytope $P$, a pair of points $s, t \in \partial P$ and $r > 0$ such that the length of the shortest path between $s'$ and $t'$ in $G(r)$ is* 14.

The upper bound follows from an exhaustive search through all possible subgraphs $G(r)$ that we performed by a computer program. This can be also verified manually, but it is rather tedious to do so. The lower bound is depicted in Figure 4. The polytope $P$ is the convex hull of all the black dots in that figure (in fact, we have to inflate $P$ a little so that the black dots lie in its interior).

**Lemma 3.5** *Let $r^*$ be the minimal value of $r$ for which $s'$ and $t'$ belong to the same connected component of $G(r)$. Then $2r^* \leq d_P(s,t) \leq 16r^*$.*

*Proof:* Let $\gamma'$ be the path on $C(r^*)$ corresponding to the shortest path between $s'$ and $t'$ in $G(r^*)$. By Lemma 3.4, the length of $\gamma'$ is at most $14r^*$. Let $\gamma = ss(r^*) \mid\mid \gamma' \mid\mid t(r^*)\, t$. Clearly, $|\gamma| \leq 16r^*$. Since $\gamma$ is an outer path of $P$ connecting $s$ to $t$, Theorem 2.1 implies that $d_P(s,t) \leq 16r^*$.

Lemma 3.2 and Lemma 3.1 imply that there exists a path $\sigma' \subset B(r)$ connecting $s$ to $t$ on $\partial P$, if and only if $s'$ and $t'$ are connected in $G(r)$.

Let $\sigma$ be a shortest path between $s$ and $t$ on $\partial P$. We claim that $\sigma$ must intersect $\partial B(r^*)$, for otherwise $s'$ and $t'$ are connected in $G(r^*)$, and this also holds for $r' < r^*$ sufficiently close to it. Hence the vertices $s'$ and $t'$ are connected in $G(r')$, a contradiction to the choice of $r^*$.

If $\sigma$ intersects one of the vertical faces of $B(r^*)$ then its length is at least $2r^*$. If $\sigma$ intersects the top or bottom face of $B(r^*)$, then its length is also at least $(r^* - l) + (r^* + l) = 2r^*$. ∎

**Lemma 3.6** *For an edge $e \in E$, let $r(e)$ denote the minimal value of $r$ for which $e \in E(r)$. Let $e_1, \ldots, e_{28}$ be the edges of $E$ ordered in increasing value of $r(\cdot)$. Then $r^* = r(e_i)$, where $i$ is the smallest index for which $s'$ and $t'$ are connected in the graph $G(r(e_i)) = (V, \{e_1, \ldots, e_i\})$.*

*Proof:* Obvious. ∎

**Lemma 3.7** *A convex polytope $P$ with $n$ edges in $\mathbb{R}^3$ can be preprocessed in linear time, into a linear-size data structure, such that given any two points $s, t \in \partial P$, one can calculate, in $O(\log n)$ time, a value $\Delta_P(s,t)$ satisfying $d_P(s,t) \leq \Delta_P(s,t) \leq 8d_P(s,t)$.*

*Proof:* We construct, in $O(n)$ time, the Dobkin- Kirkpatrick hierarchical decomposition of $P$ [DK85, DK90].

By Lemma 3.6, all we have to do is to compute, for each edge $e = pq \in E$, the value of $r(e)$, which is the minimal value of $r$ for which $e \in G(r)$. Once these values are available, $r^*$ can be obtained in $O(1)$ time. (Recall that the graphs $G(r)$ depend on $s$ and $t$.)

The value of $r(e)$ is the largest $r$ such that $T_r(e) \cap P \neq \emptyset$. Thus, using the hierarchical decomposition of $P$, it is now possible to compute, in $O(\log n)$ time, the two intersection points $x_1, x_2$ between $\partial P$ and the rays induced by $T_r(p)$ and $T_r(q)$ as $r$ varies. Moreover, by locally inspecting $\partial P \cap H$ near $x_1, x_2$, it can be decided whether any of these points is the last intersection point of $T_r(e)$ with $\partial P$, where $H$ is the plane spanned by the origin and $e$. If not, the required intersection point lies between $x_1$ and $x_2$ on $\partial(P \cap H)$. We can find this intersection point by performing an extreme-point query on $P \cap H$. Namely, we are looking for the maximal $r$ for which $l(r) \cap P \neq \emptyset$, where $l(r)$ is the line passing through $T_r(e)$. See Figure 5.

This can be easily done in $O(\log n)$ time if we are given the hierarchical decomposition of the planar polygon $P \cap H$ This two-dimensional hierarchical decomposition (of $P \cap$

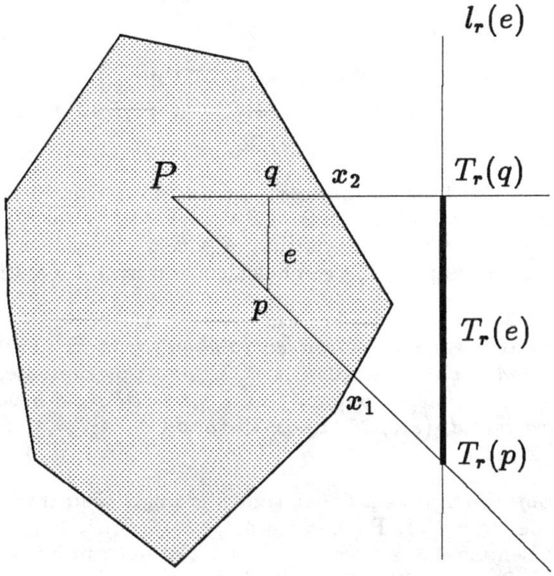


Figure 5: Computing the minimal value for which edge $e$ appears in $G(r)$.

$H$) is stored implicitly in the three-dimensional hierarchical decomposition of $P$, and the query can be performed in $O(\log n)$ time. See [DK90] for details.

Thus, we can compute $r(e)$ in $O(\log n)$ time, for each edge $e$ of $G$. Hence, as already noted, we can calculate $r^*$ in $O(\log n)$ time.

We return $16r^*$ as our approximation value $\Delta_P(s,t)$. By Lemma 3.5, $2r^* \le d_P(s,t) \le 16r^*$, from which lemma follows. ∎

Given a convex polytope $P$ with $n$ edges, two points $s,t \in \partial P$, and a parameter $0 < \varepsilon \le 1$, the algorithm of [AHPSV96] computes an outer path of $P$ connecting $s$ and $t$ and approximating the shortest path, as follows (see [AHPSV96] for more details):

- Compute a crude approximation $d$ satisfying $d_P(s,t) \le d \le 2d_P(s,t)$, using the algorithm of [HS95]. This stage takes $O(n)$ time.
- Compute, in $O(n)$ time, the Dobkin-Kirkpatrick hierarchical decomposition of $P$ [DK85, DK90].
- Using $d$ and the hierarchical decomposition of $P$, compute, in $O((\log n + \log(1/\varepsilon))/\varepsilon^{1.5})$ time, an approximation polytope $Q \supseteq P$ having $O(1/\varepsilon^{1.5})$ faces, such that $s,t \in \partial Q$ and $d_P(s,t) \le d_Q(s,t) \le (1+\varepsilon)d_P(s,t)$.
- Using the algorithm of [CH96], compute, in $O(1/\varepsilon^3)$ time, the shortest path on $\partial Q$ between $s$ and $t$. This is the desired approximating outer path.

A more detailed inspection of the algorithm of [AHPSV96] reveals that any approximation $d$, satisfying $d_P(s,t) \le d \le c \cdot d_P(s,t)$ for some prescribed constant $c > 1$, can be used in the first stage of the algorithm. Thus, by moving the computation of the Dobkin-Kirkpatrick decomposition of $P$ into the preprocessing stage, and by using the algorithm of Lemma 3.7 instead of the algorithm of [HS95], we get:

**Theorem 3.8** *A given convex polytope $P$ with $n$ edges can be preprocessed in $O(n)$ time, such that given any two points $s$ and $t$ on $\partial P$, and a parameter $\varepsilon > 0$, one can construct a polygonal outer path of $P$ between $s$ and $t$, whose length is at most $(1+\varepsilon)d_P(s,t)$, and which consists of $O(1/\varepsilon^{1.5})$ segments, in time $O((\log n)/\varepsilon^{1.5} + 1/\varepsilon^3)$.*

**Remark 3.9** The outer path $\sigma$ of $P$ generated by the algorithm of Theorem 3.8 can be projected onto $\partial P$ in $O(n \log(1/\varepsilon) + 1/\varepsilon^3)$ time, resulting in a path $\sigma'$ on $\partial P$, such that $|\sigma'| \le |\sigma|$. See [AHPSV96]. Since there are (easy) cases where any path on $\partial P$ between $s$ and $t$ has $\Omega(n)$ edges, there is no hope of constructing approximate shortest path on $\partial P$ itself in worst-case sublinear query time.

**Remark 3.10** Given a convex polytope $P$ with $n$ edges, the fastest known algorithm for computing a data-structure that supports queries of computing the length of the exact shortest path between any pair of points on $\partial P$ is due to [AAOS96]; it requires $O(n^6 m^{1+\delta})$ space and preprocessing time, and answers a query in $O((\sqrt{n}/m^{1/4})\log m)$ time, where $1 \le m \le n^2$, $\delta > 0$ (and the constants of proportionality depends on $\delta$).

## 4 Approximating the Geodesic Diameter of a Convex Polytope

In this section we present an efficient algorithm that approximates the geodesic diameter of a convex polytope in $\mathbb{R}^3$.

**Definition 4.1** *(i) Given a convex polytope $P$ in $\mathbb{R}^3$, we define the* geodesic diameter *of $P$ to be*

$$\mathcal{D}_P = \max_{s,t \in \partial P} d_P(s,t).$$

*(ii) Given a subset $S \subseteq \partial P$, we denote by $\Delta_P(S)$ the geodesic diameter of $S$ on the boundary of $P$, namely*

$$\Delta_P(S) = \max_{s,t \in S} d_P(s,t).$$

*(iii) Let $C(P)$ denote the smallest axis-parallel cube bounding $P$, and define the* outer diameter *of $P$ to be $\delta(P) = 6d$, where $d$ is the side length of $C(P)$.*

**Lemma 4.2** *Given a convex polytope $P$ with $n$ edges in $\mathbb{R}^3$, the outer diameter $\delta(P)$ of $P$ can be calculated in linear time. Moreover, $\delta(P)/6 \le \mathcal{D}_P \le \delta(P)$.*

*Proof:* Clearly, $C(P)$, and thus also $\delta(P)$, can be computed in linear time. Let $d = \delta(P)/6$ be the edge length of $C(P)$. Since $\partial P$ must contain two points $s,t$ on opposite facets of $C(P)$, the Euclidean distance between $s$ and $t$ is $\ge d$, implying that $\mathcal{D}_P \ge d_P(s,t) \ge d$. It is also easy to verify that any pair of points on the boundary of $P$ can be connected by an outer path of $P$ of length at most $\delta(P) = 6d$. It follows that $\delta(P)/6 = d \le \mathcal{D}_P \le \delta(P)$. ∎

**Lemma 4.3** *Given a convex polytope $P$ with $n$ edges in $\mathbb{R}^3$ such that $1/6 \le \mathcal{D}_P \le 1$, and a parameter $0 < \varepsilon \le 1$, then a pair of points $s,t \in \partial P$ can be calculated, in $O(n + 1/\varepsilon^6)$ time, such that $d_P(s,t) \ge (1-\varepsilon)\mathcal{D}_P$.*

*Proof:* Let $S$ be an $(\varepsilon/100)$-dense set of $P$. By Lemma 2.4, such an $S$ can be constructed in $O(n + (1/\varepsilon^2)\log(1/\varepsilon))$ time. Let $s,t$ be two points on $\partial P$ realizing the geodesic diameter of $P$. Thus, there are two points $(s', \eta_{s'}), (t', \eta_{t'}) \in S$ such that $|ss'|, |tt'| \le \varepsilon/100$ and $e(\eta_s, \eta_{s'}), e(\eta_t, \eta_{t'}) \le \varepsilon/100$,

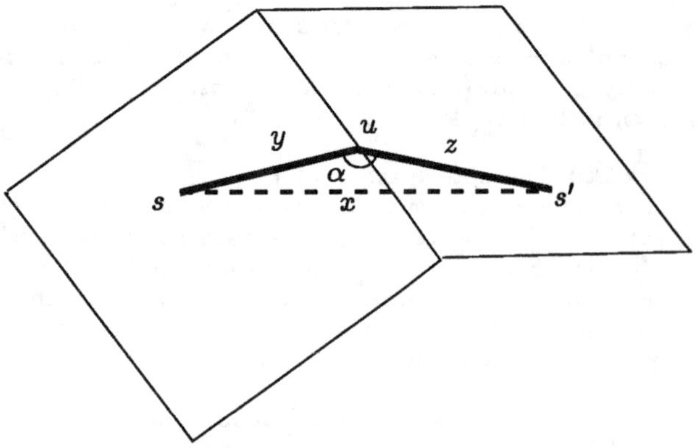


Figure 6: $y+z \le 2x$.

where $\eta_s$, $\eta_t$ are vectors normal to $P$ at $s$, $t$, respectively, and $e(a,b)$ denotes the angle between the vectors $a$ and $b$.

We claim that $d_P(s,s'), d_P(t,t') \le \varepsilon/50$. Indeed, put $x = |ss'|$ and $\psi = e(\eta_s, \eta_{s'})$. Draw the planes $H, H'$ that support $P$ at $s, s'$, and have normals $\eta_s$, $\eta_{s'}$, respectively. Let $W$ be the wedge bounded by $H$ and $H'$ and containing $P$. Let $\sigma$ denote the shortest path on $\partial W$ between $s$ and $s'$. This path consists of two segments $su, us'$, with a common endpoint $u$ lying on $H \cap H'$. Moreover, the angle $\alpha = \measuredangle sus'$ is at least the obtuse dihedral angle between $H$ and $H'$. That is $\alpha \ge \pi - \psi$. See Figure 6.

Put $y = |su|$, $z = |us'|$. By the cosine law,

$$\begin{aligned} x^2 &= y^2 + z^2 - 2yz\cos\alpha = y^2 + z^2 + 2yz\cos(\pi - \alpha) \\ &\ge y^2 + z^2 + 2yz\cos\psi \ge (y+z)^2\cos\psi. \end{aligned}$$

Since $\psi \le \varepsilon/100 \le 1/100$, we have $\cos\psi \ge 1/4$, so $y+z \le 2x$. Since $\sigma = sus'$ is an outer path of $P$ between $s$ and $s'$, we have $d_P(s,s') \le 2|ss'| \le \varepsilon/50$.

Thus,

$$\begin{aligned} \mathcal{D}_P &= d_P(s,t) \le d_P(s,s') + d_P(s',t') + d_P(t',t) \\ &\le \varepsilon/25 + \Delta_P(S). \end{aligned}$$

It follows that $\mathcal{D}_P \le \varepsilon/25 + \Delta_P(S) \le (\varepsilon/2)\mathcal{D}_P + \Delta_P(S)$, or

$$\Delta_P(S) \ge (1 - \varepsilon/2)\mathcal{D}_P. \tag{1}$$

Let $Q = P(S) = \bigcap_{(p,\eta_p)\in S} H^+(p,\eta_p)$. The complexity of $Q$ is $O(1/\varepsilon^2)$, and, by Lemma 2.4, $Q \subset P_r$, for $r = \varepsilon^2/5000$.

Applying Lemma 2.3, with this value of $r$ and with $\varepsilon/2$, we obtain

$$\begin{aligned} d_P(s,t) &\le d_Q(s,t) \\ &\le (1+\varepsilon/4)d_P(s,t) + (1/500)\varepsilon^2 + (\sqrt{2}/50)\varepsilon^{1.5} \\ &\le d_P(s,t) + \left(\frac{1}{4}\varepsilon + \frac{6}{500}\varepsilon^2 + \frac{6\sqrt{2}}{50}\varepsilon^{1.5}\right)\mathcal{D}_P \\ &\le d_P(s,t) + (\varepsilon/2)\mathcal{D}_P, \end{aligned}$$

for any $s,t \in \partial Q \cap \partial P$.

Note that $S \subset \partial Q$. Let $s,t$ be two points in $S$ realizing $\Delta_Q(S)$. The above inequality implies

$$\Delta_P(S) \le \Delta_Q(S) = d_Q(s,t) \le d_P(s,t) + (\varepsilon/2)\mathcal{D}_P.$$

Hence, by (1),

$$\begin{aligned} d_P(s,t) &\ge \Delta_P(S) - (\varepsilon/2)\mathcal{D}_P \\ &\ge (1-\varepsilon/2)\mathcal{D}_P - (\varepsilon/2)\mathcal{D}_P \\ &= (1-\varepsilon)\mathcal{D}_P. \end{aligned}$$

As for the time complexity, the distances from a fixed point $s \in S$ to all other points of $S$ along $Q$ can be calculated in $O(1/\varepsilon^4)$ time by the algorithm of [CH96]. Thus, the pair of points realizing $\Delta_Q(S)$ can be calculated in $O(1/\varepsilon^6)$ time. ■

**Theorem 4.4** *Given a convex polytope $P$ in $\mathbb{R}^3$ with $n$ edges, and a parameter $0 < \varepsilon \le 1$, one can compute, in $O(n + 1/\varepsilon^6)$ time, a pair of points $s,t \in \partial P$ and a value $\Delta$, such that $d_P(s,t) \ge (1-\varepsilon)\mathcal{D}_P$ and $(1-\varepsilon)\mathcal{D}_P \le \Delta \le (1+\varepsilon)\mathcal{D}_P$.*

*Proof:* Compute in linear time the outer diameter $\delta = \delta(P)$. Let $Q = P/\delta$. By Lemma 4.2, $1/6 \le \mathcal{D}_Q \le 1$.

By Lemma 4.3, a pair of points $s,t$ of $\partial Q$ can be calculated in $O(n + 1/\varepsilon^6)$ time, such that

$$d_Q(s,t) \ge (1-\varepsilon)\mathcal{D}_Q. \tag{2}$$

Let $s', t'$ be the points on $\partial P$ corresponding to $s, t$, respectively. Then, by multiplying (2) by $\delta$, it follows that $d_P(s',t') \ge (1-\varepsilon)\mathcal{D}_P$.

Using the algorithm of [AHPSV96], we can compute in $O(n + 1/\varepsilon^3)$ time an $\varepsilon$-approximation $\Delta$ to $d_P(s',t')$, such that $(1-\varepsilon)\mathcal{D}_P \le \Delta \le (1+\varepsilon)\mathcal{D}_P$. ■

**Remark 4.5** The fastest exact algorithm for computing the geodesic diameter of a convex polytope in $\mathbb{R}^3$ is due to Agarwal et al. [AAOS96], and runs in $O(n^8 \log n)$ time.

## 5 Related Results

In this section, we review a few related results from a companion paper [HP97]: we present a new technique how to to use an algorithm that approximates the distances between two points, into an algorithm that construct a data-structure that can be used for $\varepsilon$-approximate shortest path queries from a fixed point, with a query time $O(\log(n/\varepsilon))$.

**Approximate shortest path maps.** The exact algorithms of [SS86, MMP87] receive as input a convex polytope or a polyhedral surface $\mathcal{P}$, and a fixed source point $s$ on $\mathcal{P}$, and compute a map (i.e. a subdivision of $\mathcal{P}$) of complexity $\Theta(n^2)$, that can be used to answer (exact) shortest path queries from $s$ to any point on $\mathcal{P}$ in $O(\log n)$ time (such a query reports the *length* of the shortest path; reporting the path itself might require more time). This raises the problem of computing a map of near-linear size for approximate shortest-path queries from $s$. We show in [HP97] that this is indeed possible: Given a polyhedral surface $\mathcal{P}$ with $n$ edges in $\mathbb{R}^3$, a source point $s \in \mathcal{P}$, and a prescribed $0 < \varepsilon \le 1$, there exists a map (a subdivision of $\mathcal{P}$) of complexity $O((n/\varepsilon)\log(1/\varepsilon))$, such that for any $t \in \mathcal{P}$, one can compute the length of an $\varepsilon$-approximate shortest path between $s$ and $t$ on $\mathcal{P}$ in $O(\log(n/\varepsilon))$ time, by locating $t$ in the map.

We present an algorithm that constructs such an approximation map in $O(n^2 \log n + (n/\varepsilon)\log(1/\varepsilon)\log(n/\varepsilon))$ time, for the case of a polyhedral surface, and in $O((n/\varepsilon^3)\log(1/\varepsilon) + (n/\varepsilon^{1.5})\log(1/\varepsilon)\log n)$ time, for the case of a convex polytope.

**Approximate spatial shortest path maps.** We present in [HP97] a similar result for $\varepsilon$-approximate shortest paths among polyhedral obstacles in $\mathbb{R}^3$. Let $\mathcal{O}$ be a set of polyhedral obstacles with a total of $n$ edges in $\mathbb{R}^3$, $s$ a source point in $\mathbb{R}^3$, and $0 < \varepsilon \leq 1$ a parameter. We show that there exists a spatial subdivision $\mathcal{M}$ of $\mathbb{R}^3$, such that for any $t \in \mathbb{R}^3$, one can compute, in $O(\log(n/\varepsilon))$ time, the length of an $\varepsilon$-approximate shortest path between $s$ and $t$, that avoids the interiors of the obstacles, by performing a spatial point-location query with $t$ in $\mathcal{M}$. The space needed to compute and preprocess $\mathcal{M}$ for spatial point-location is $O(n^2/\varepsilon^{4+\delta})$, for any $\delta > 0$, and the preprocessing time is

$$O\left(\frac{n^4}{\varepsilon^2}\left(\frac{\beta(n)}{\varepsilon^4}\log\frac{n}{\varepsilon} + \log(n\rho)\log(n\log\rho)\right)\log\frac{1}{\varepsilon}\right),$$

where $\rho$ is the ratio of the length of the longest edge in $\mathcal{O}$ to the smallest Euclidean distance between $s$ and any point of the obstacles, $\beta(n) = \alpha(n)^{O(\alpha(n))^{O(1)}}$, and $\alpha(n)$ is the extremely slowly growing inverse of the Ackermann function.

## 6 Conclusions

In this paper we have presented two simple and efficient algorithms for computing approximate solutions to problems involving shortest paths on the surface of a convex polytope in 3-space. We conclude by mentioning a few open problems.

- Can an $\varepsilon$-approximate shortest path between two points on a polyhedral terrain, or on the surface of a nonconvex polyhedron, be computed in time that is near-linear in the number of edges? A recent subquadratic solution has been announced by d Varadarajan and Agarwal [VA96].
- Can the exact shortest path between two points on a convex polyhedron be computed in near-linear time? in subquadratic time?
- A geodesic path on a convex polytope $P$ is a path which is locally optimal. (i.e., it can not be made shorter by a "local" shortcut; in other words, it is a polygonal path with vertices only on edges of $P$, so that the angles between two consecutive segments and the edge of $P$ to which they are adjacent are equal.) Clearly, a shortest path is geodesic. Can one compute a geodesic path between two points on a convex polyhedron in subquadratic time?


## Acknowledgments

The author wishes to thank Pankaj Agarwal and Micha Sharir for helpful discussions concerning the problems studied in this paper and related problems.


## References


[AAOS96] P.K. Agarwal, B. Aronov, J. O'Rourke, and C. Schevon. Star unfolding of a polytope with applications. *SIAM J. Comput.*, 1996, to appear.

[AHPSV96] P.K. Agarwal, S. Har-Peled, M. Sharir, and K. R. Varadarajan. Approximate shortest paths on a convex polytope in three dimensions. Report CS-1996-12, Dept. Comput. Sci., Duke Univ., Durham, NC, 1996.

[CH96] J. Chen and Y. Han. Shortest paths on a polyhedron; Part I: computing shortest paths. *Int. J. Comput. Geom. & Appl.*, 6(2):127–144, 1996.

[Cla87] K. L. Clarkson. Approximation algorithms for shortest path motion planning. In *Proc. 19th Annu. ACM Sympos. Theory Comput.*, pages 56–65, 1987.

[CR87] J. Canny and J. H. Reif. New lower bound techniques for robot motion planning problems. In *Proc. 28th Annu. IEEE Sympos. Found. Comput. Sci.*, pages 49–60, 1987.

[CSY94] J. Choi, J. Sellen, and C. K. Yap. Approximate Euclidean shortest path in 3-space. In *Proc. 10th Annu. ACM Sympos. Comput. Geom.*, pages 41–48, 1994.

[DK85] D. P. Dobkin and D. G. Kirkpatrick. A linear algorithm for determining the separation of convex polyhedra. *J. Algorithms*, 6:381–392, 1985.

[DK90] D. P. Dobkin and D. G. Kirkpatrick. Determining the separation of preprocessed polyhedra – a unified approach. In *Proc. 17th Internat. Colloq. Automata Lang. Program.*, volume 443 of *Lecture Notes in Computer Science*, pages 400–413. Springer-Verlag, 1990.

[HP97] S. Har-Peled. Constructing approximate shortest path maps in three dimensions. manuscript, 1997.

[HS95] J. Hershberger and S. Suri. Practical methods for approximating shortest paths on a convex polytope in $\mathbb{R}^3$. In *Proc. 6th ACM-SIAM Sympos. Discrete Algorithms*, pages 447–456, 1995.

[MMP87] J. S.B. Mitchell, D. M. Mount, and C. H. Papadimitriou. The discrete geodesic problem. *SIAM J. Comput.*, 16:647–668, 1987.

[Pap85] C. H. Papadimitriou. An algorithm for shortest-path motion in three dimensions. *Inform. Process. Lett.*, 20:259–263, 1985.

[Pog73] A. V. Pogorelov. *Extrinsic Geometry of Convex Surfaces*, Volume 35 of *Translations of Mathematical Monographs*. American Mathematical Society, Providence, Rhode Island, 1973.

[RS94] J. H. Reif and J. A. Storer. A single-exponential upper bound for finding shortest paths in three dimensions. *J. ACM*, 41(5):1013–1019, 1994.

[Sha87] M. Sharir. On shortest paths amidst convex polyhedra. *SIAM J. Comput.*, 16:561–572, 1987.

[SS86] M. Sharir and A. Schorr. On shortest paths in polyhedral spaces. *Siam J. Comput.*, 15:193–215, 1986.

[VA96] K.R. Varadarajan and P.K. Agarwal. Approximating shortest paths on a polyhedron. manuscript, 1996.

# $\mathcal{A}$-Shapes of a finite point set

Mahmoud Melkemi*

**Abstract.** In this paper, we develop a family of straight-line graphs which we call $\mathcal{A}$-shape of a finite point set ($\mathcal{A}$ is an arbitrary point set). Derived from the Delaunay triangulation, this family captures the notion of shape hull of a finite point set.

## 1 Introduction

This paper deals with the problem of computing the shape hull of a point set. Apparently, the first work capturing the shape of a point set into a mathematical formalism was that of Edelsbrunner, Kirkpatrick and Seidel [1]. For they propose an elegant concept they call $\alpha$-shape, $\alpha \geq 0$, and which can be viewed as a family of straight-line graphs. The justification of their $\alpha$-shape definition lies essentially in the fact that it is a direct generalization of the convex hull definition. In this context, the convex hull is seen as a "naive" shape and it is reasonable to say that the "desired" shape of a point set may be achieved by an $\alpha$-shape with an appropriate $\alpha$. However, there are situations where this family does not include significant shapes as for example those well modeling sparse and dense patterns in a point set of variable density. Figure 1(a) shows an example of a point set where there is no $\alpha$-shape which can detect the hole $H$ and well model the sparse region. Here, the sparse pattern is constructed so that each "empty" disk passing through two points is of diameter greater than $r$, $r$ being the diameter of the largest circle included in the hole $H$.

In this perspective, we define the shape hull of a finite point set $S$ by a family of straight-line graphs and we refer to it as an $\mathcal{A}$-shape of $S$, where $\mathcal{A}$ is a point set which controls the level of detail reflected by the shape (this concept is a generalization of the work presented in [2]). Conceptually, the $\mathcal{A}$-shape corresponds to a transcription of the **A**-shape of a continuous bounded domain say $D$. By **A**-shape, we simply mean the boundary of some dilation of $D$ with a level controlled by a set of curves **A**. Accordingly, the sets $S$ and $\mathcal{A}$ may be interpreted as samples of the continuous ones $D$ and **A** respectively. As a consequence, the $\mathcal{A}$-shape and **A**-shape are defined via closely related rules.
From this analysis, one can claim the existence of a suitable $\mathcal{A}$ for which the $\mathcal{A}$-shape achieves the desired level of detail. In Figure 1 for example, the illustration (b) shows an $\mathcal{A}$-shape well modeling that of (a). In practice, we suggest a construction of $\mathcal{A}(\alpha,\ t)$ to well model patterns in a point sets with variable densities. The parameter $t$, $0 \leq t \leq 1$, is related to the density measure and $\alpha \geq 0$ controls the level of detail expressed by the shape. This paper is organized as follows. Definitions and properties of $\mathcal{A}$-shape are presented in section 2. The sub-family $\mathcal{A}(\alpha,\ t)$-shape is summarized in section 3. Finally, section 4 presents concluding remarks.

*L.I.G.I.M. Université Claude Bernard Lyon1. 43 boulevard du 11 Novembre 1918, Bat. 710, 69622 Villeurbanne. E-mail : melkemi@ligim.univ-lyon1.fr


*Computational Geometry 97* Nice France


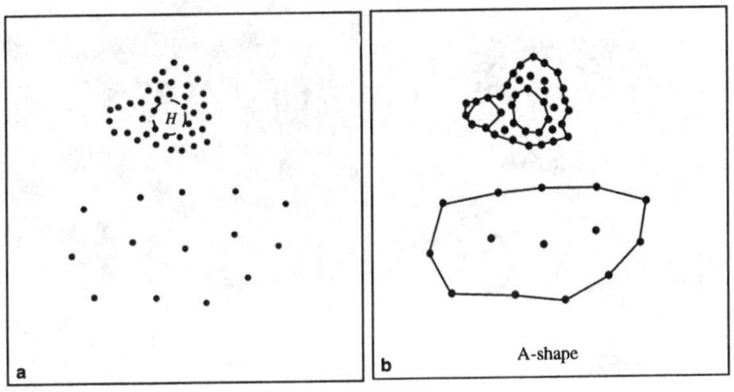


Figure 1: (a) The broken circle is the largest one included in the hole $H$, (b) $\mathcal{A}$-shape of the point set shown in (a).

## 2 $\mathcal{A}$-hull and $\mathcal{A}$-shape of a finite point set

*Notations and Assumption.* Let $S = \{p_1, p_2, ..., p_n\}$ be a set of $n$ distinct points in the plane and let $d(p,\ q)$ be the Euclidean distance between two points $p$ and $q$. $\mathcal{R}(p_i)$, $\mathcal{V}(S)$ and $\mathcal{DT}(S)$ denote, respectively, the Voronoi region of $p_i$, the Voronoi diagram of $S$ and the Delaunay triangulation of $S$. For a finite point set $\mathcal{A}$, we denote $\mathcal{R}(\mathcal{A},\ x)$ the Voronoi region of $x \in S \cup \mathcal{A}$. $D(x,\ r)$ denotes the closed disk of radius $r$ and centered at $x$. If the vertices of $\mathcal{V}(S)$ are labeled from 1 to $m$ then $D_i$ denotes the open Delaunay disk centered at the vertex of label $i$. Throughout this article, we assume that no four points in $S$ are on a common circle and no three points are collinear.

### 2.1 Definitions

Let $\mathcal{A}$ be a finite point set and consider $\mathcal{V}(\mathcal{A} \cup S)$. Let $\mathcal{A}'$ denotes the extremities of Voronoi edges which separate points in sets $S$ and $\mathcal{A}$. $Cdisk(q)$, $q \in \mathcal{A}'$, is the closed complement of the Delaunay disk centered at $q$.

- The $\mathcal{A}$-hull of $S$, $\mathcal{H}(\mathcal{A},\ S)$, is defined as follows.

$$\mathcal{H}(\mathcal{A}, S) = [\bigcap_{p \in \mathcal{A}'} Cdisk(p)] \cap \mathcal{B}(\mathcal{A},\ S)$$

where

$$\mathcal{B}(\mathcal{A},\ S) = \bigcup_{p \in S} \mathcal{R}(\mathcal{A}, p)$$

- A Delaunay edge $[p_i p_j]$ is an $\mathcal{A}$-edge in $\mathcal{DT}(S \cup \mathcal{A})$ if exists a Delaunay circle passing through $p_i$, $p_j$ and a point in $\mathcal{A}$.
- The $\mathcal{A}$-shape of $S$ is $\mathcal{A}$-edges set obtained from $\mathcal{DT}(S \cup \mathcal{A})$.

An example of these notions are shown in Figure 2, the hollow disks are the points of $\mathcal{A}$ and the solid ones represent the points of $S$. The extremities of bold Voronoi edges (see Figure 2(a)) are the points of $\mathcal{A}'$. $\mathcal{B}(\mathcal{A},\ S)$, $\mathcal{A}$-hull and $\mathcal{A}$-shape are shown, respectively, in Figure 2(b), Figure 2(c) and Figure 2(d).

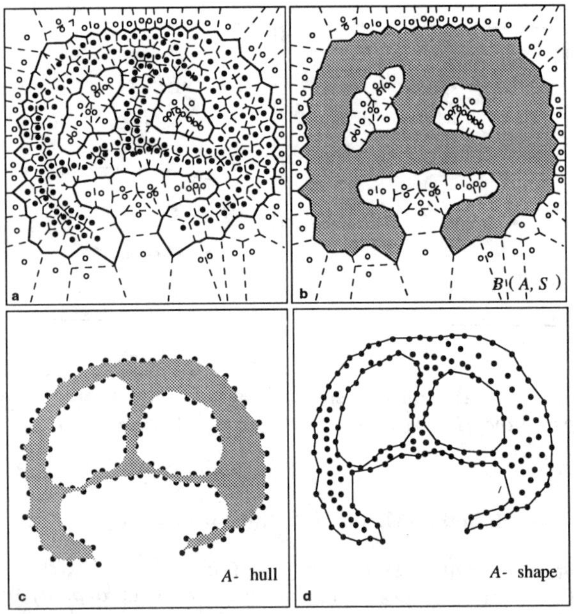


Figure 2: An example of $\mathcal{A}$-hull and $\mathcal{A}$-shape.

### 2.2 A-shape of continuous bounded domains

To understand the $\mathcal{A}$-shape notion, let now consider it in the continuous context. The set $S$ is replaced by bounded domains $F_1, \cdots, F_n$. The **A**-hull of $F$ (where $F$ is the union of $F_1, \cdots, F_n$) is defined as follows.

- Let **A** be a bounded set of curves which verifies : **(i)** the intersection of **A** with interior of $F$ is empty. **(ii)** there exists a set of $\ell$ closed curves $C_1, \cdots, C_\ell$ included in **A** and are boundaries of domains $DOM_1, \cdots, DOM_\ell$ such that $F \subset \bigcup_{i=1}^{i=\ell} DOM_i$.
- We define $\mathbf{A}'$ as the set :
$\mathbf{A}' = \{p \in I\!R^2 - F | d(p, \mathbf{A}) = d(p, F)\}$
- $B(\mathbf{A}, F)$ denotes the domain that contains $F$ and bounded by $\mathbf{A}'$.
- For each $p$ in $\mathbf{A}'$, $Cdisk(p)$ is the closed complement disk of radius $d(p, F)$ centered at $p$
- The **A**-hull of $F$ is the domain :

$$[\bigcap_{p \in \mathbf{A}'} Cdisk(p)] \cap B(A, F)$$

- The **A**-shape of $F$ is the boundary of the **A**-hull of $F$.

An example of $F$, **A**, $\mathbf{A}'$, $B(\mathbf{A},\ F)$ and **A**-hull is depicted in Figure 3. We point out that the sets $S$, $\mathcal{A}$ and $\mathcal{A}'$ can be viewed as samples of $F$, **A** and $\mathbf{A}'$ respectively. $\mathcal{B}(\mathcal{A},\ S)$, $\mathcal{A}$-hull and $\mathcal{A}$-shape are "analogous" to the sets $B(\mathbf{A},\ F)$, **A**-hull and **A**-shape.

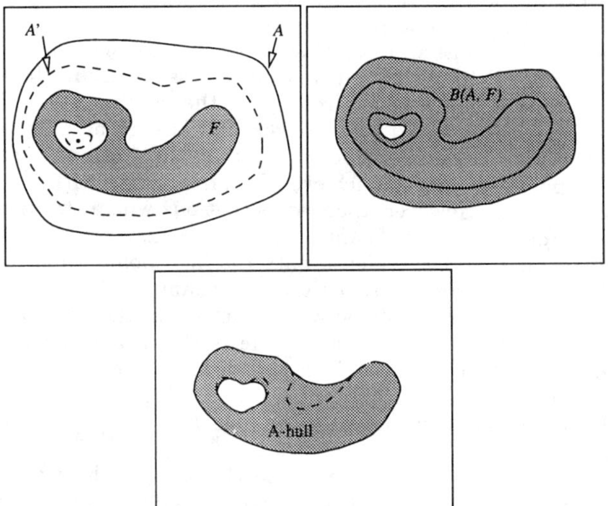


Figure 3: **A**-hull of a domain $F$.

### 2.3 Observations

1. $\mathcal{A}$-shape of $S$ is subset of $\mathcal{DT}(S)$.
2. The set $S$ and its convex hull are elements of the $\mathcal{A}$-shape family.
3. If the vertices of the convex hull of $S \cup \mathcal{A}$ are in $\mathcal{A}$ then $\mathcal{A}$-hull of $S$ is bounded.
4. Consider $[pq]$ a Delaunay edge of $\mathcal{DT}(S)$ where $p$ and $q$ are neighbors across a Voronoi edge which its extremities are labeled $i$ and $j$.
$[pq]$ is an edge of the $\mathcal{A}$-shape, if and only if, $\mathcal{A}$ satisfies the following conditions. **(a)** $\mathcal{A} \cap D_i \cap D_j = \emptyset$. **(b)**

$D_i \cap \mathcal{A} \neq \emptyset$ or $D_j \cap \mathcal{A} \neq \emptyset$. (c) $\forall\, x \in \mathcal{A} \cap D_i$ (respectively $x \in \mathcal{A} \cap D_j$), $\mathcal{A} \cap D(M_x, d(x, M_x)) \cap D_j = \emptyset$ (respectively $\mathcal{A} \cap D(M_x, d(x, M_x)) \cap D_i = \emptyset$). Where $M_x$ is the intersection point between the bisector of $[xp]$ and the Voronoi edge of extremities $i$ and $j$.

Observation 3 gives a sufficient condition which makes $\mathcal{A}$-hull bounded. Under this condition, the $\mathcal{A}$-shape can be seen as the boundary of the $\mathcal{A}$-hull with curved edges replaced by straight edges. Observation 4 states necessary and sufficient conditions which put a Delaunay edge in $\mathcal{A}$-shape. The conditions of this observation are illustrated by the example of Figure 4. Observations and their proofs are detailed in [3] and [4].

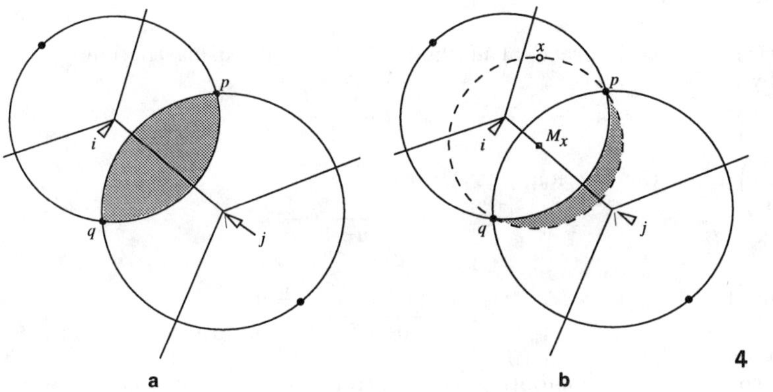


Figure 4: $[pq]$ is in $\mathcal{A}$-shape if the filled domains do not contain any point of $\mathcal{A}$. (a) The filled domain is $D_i \cap D_j$. (b) $\mathcal{A} \cap D_i = \{x\}$, the filled domain is $D(M_x, d(x, M_x)) \cap D_j - D_i \cap D_j$

## 3 $\mathcal{A}(\alpha, t)$-shape family

In this section, we present a construction of a set $\mathcal{A}$ to well model sparse and denser regions in a point set of variable density. This set is named $\mathcal{A}(\alpha,\ t)$, where the parameter $t$ is related to the size of the influence zone associated to each point of $S$. The zone of each point in $S$ is a disk which is large if the point is in a sparse region and small if it is in a denser one. The construction of this set is as follows.
Now $S$ is a point set of variable density, let $p_i^* \in S$ be a point such that $d(p_i,\ p_i^*) =$ minimum of $\{d(p_i,\ p_j),\ j \neq i$ and $j = 1, ..., n\}$. We associate to each $p_i$ the circle centered at $p_i$ and of radius $w_i(t) = d(p_i,\ (1-t)\, p_i + t\, p_i^*)$ where $0 \leq t \leq 1$. Let $d_{pow}(x, P_i) = d^2(x,\ p_i) - w_i(t)^2$ denotes the power distance of $x \in \mathbb{R}^2$ to the circle $P_i = D(p_i,\ w_i(t))$. For a fixed parameter $t$ and positive number $\alpha$, $\mathcal{A}(\alpha, t)$ is the set of points $x$ such that :

- $d_{pow}(x,\ P_i) = d_{pow}(x,\ P_j)$, for $p_i$ and $p_j$ are neighbors in $\mathcal{V}(S)$.
- $d_{pow}(x,\ P_i) = \alpha^2$.
- $d_{pow}(x,\ P_k) \geq \alpha^2$, for $k \neq i$ and $k \neq j$.

Notice that, the points of $\mathcal{A}(\alpha,\ t)$ are on the edges of the power Voronoi diagram and the details captured by $\mathcal{A}(\alpha, t)$-shape decrease for large values of $\alpha$. An example of an $\mathcal{A}(\alpha, t)$-shape is shown in Figure 5.

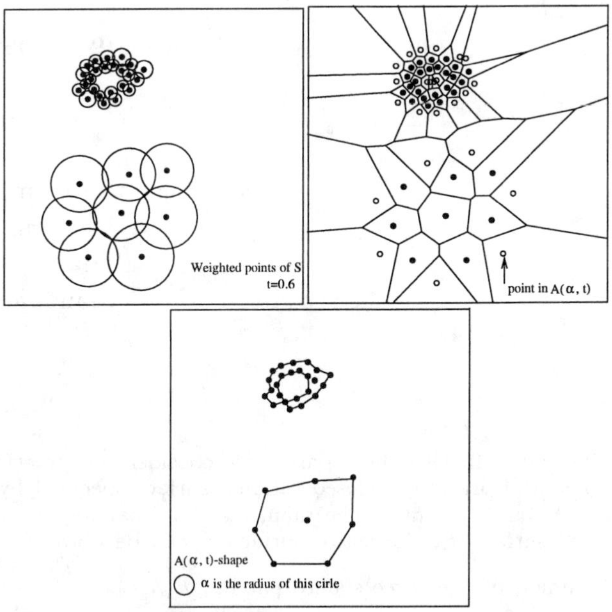


Figure 5: An example of $\mathcal{A}(\alpha,\ t)$-shape

## 4 Conclusion

In this paper, the concept of $\mathcal{A}$-shape of a point set $S$ is established. Because of its close connection with the continuous **A**-shape definition, it has revealed a natural behavior in computing the shape hull of a point set. As an application, when considering the case of sparse-dense regions, a possible construction of appropriate $\mathcal{A}$-shape is presented. An important feature in this notion lies in the fact that there is an $O(n\log n)$ algorithm for constructing the $\mathcal{A}$-shape. For reasons related to space, it is not described here : We merely say that it exploits the fast methods for designing Voronoi diagrams. Of course, as in $\alpha$-shape, the notion of $\mathcal{A}$-shape has a weighted version, and can be extended to three and higher dimensions. These perspectives will be developped in forthcoming works.

## References

[1] H. Edelsbrunner, D.G. Kirkpatrick and R. Seidel. On the shape of a set of points in the plane. *IEEE trans. Inform. Theory* **29**, 551-559 (1983).

[2] M. Melkemi and L. Melkemi. An algorithm for detecting dot patterns. *In : R.A. Earnshaw and J.A. Vince Ed., Computer Graphics*,161-169, Leeds,U.K, Academic Press, (1995).

[3] M. Melkemi. Computing the shape of a planar point set. *LIGIM-TR.01.96 "Submitted to Pattern Recognition Journal"*, (1996).

[4] M. Melkemi. $\mathcal{A}$-shapes of a finite point set. *LIGIM-TR.02.96*,(1996).

# Drawing Closed Rational Surfaces

Jean Gallier

Department of Computer and Information Science
University of Pennsylvania
Philadelphia, PA 19104, USA
jean@saul.cis.upenn.edu

**Abstract.** In this short paper, we consider the practical problem of drawing a closed rational surface specified by a net of control points. The main result is that for any rational surface $F$, the entire surface $F$ can be obtained as the union of the three square patches $F(u,v)$, $F\left(\frac{v}{u}, \frac{1}{u}\right)$, and $F\left(\frac{1}{v}, \frac{u}{v}\right)$, over $[-1,1]\times[-1,1]$. Furthermore, if $F$ is defined by some triangular net $\mathcal{N}$ (consisting of weighted control points), the control nets of six triangular patches corresponding to the above three square patches can be computed very easily from the original control net. In the full paper [9], we illustrate our method on some classical examples (including the "Steiner roman surface"), and we also discuss a method for resolving base points, using some simple "blowing-up" technique involving the computation of new control nets.

## 1 Introduction

In this short paper, we consider the practical problem of drawing a closed rational surface. For example, consider the sphere $F$ specified by the fractions

$$x(u,v) = \frac{2u}{u^2+v^2+1}, \quad y(u,v) = \frac{2v}{u^2+v^2+1},$$

and

$$z(u,v) = \frac{u^2+v^2-1}{u^2+v^2+1}.$$

The first problem is that no matter how large the interval $[r,s]$ is, the trace $F([r,s]\times[r,s])$ of $F$ over $[r,s]\times[r,s]$ is not the trace of the entire surface. In this particular example, we could take advantage of symmetries, but in general, this may not be possible. We could use any of the bijections from $]-1,1[$ to $\mathbf{R}$ to reduce the parameter domain to the square $[-1,1]\times[-1,1]$, but since these maps are at least quadratic, this could triple the total degree of the surface, leading to an impractical method. For example, using the map

$$t \mapsto \frac{t}{1-t^2},$$

the fraction

$$\frac{1}{u^2+uv}$$

becomes

$$\frac{(1-u^2)^2(1-v^2)}{(1-v^2)u^2+(1-u^2)uv}.$$

Recomputing the control net after substitution would also be quite expensive. The reason why the problem is not trivial is that in most CAGD applications, the surface is given in terms of control points rather than parametrically. The key idea is to partition the plane $\mathcal{P} = \mathbf{R}^2$ into regions, in such a way that there exist simple bijections between some fixed closed region of the plane $\mathcal{P}$, say the square $[-1,1]\times[-1,1]$, and some regions in the complement of the open square $]-1,1[\times]-1,1[$ in the projective plane $\widetilde{\mathcal{P}}$. This can be done by dividing the projective plane into 3 regions, the square $[-1,1]\times[-1,1]$, and the two shaded regions as in the diagram below.

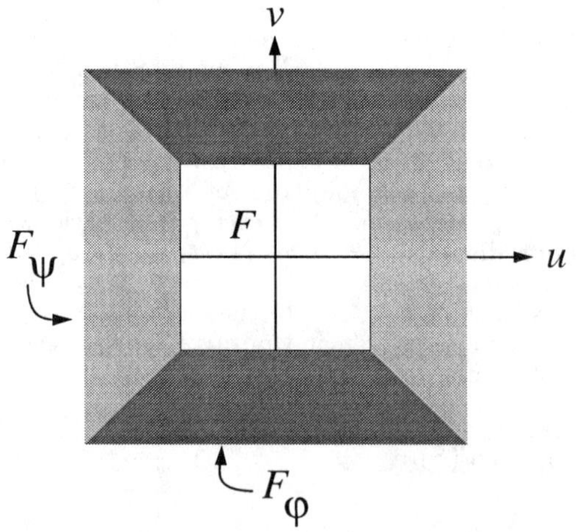


Figure 1: Dividing the projective plane into 3 regions

Then, we can define the bijection

$$\varphi\colon (u,v) \mapsto (v/u, 1/u),$$


*Computational Geometry 97* Nice France

mapping $[-1,1] \times [-1,1]$ onto the subset of the projective plane consisting of the lightly shaded region together with its points at infinity. Similarly, we can define the bijection

$$\psi: (u, v) \mapsto (1/v, u/v),$$

mapping $[-1,1] \times [-1,1]$ onto the subset of the projective plane consisting of the darkly shaded region together with its points at infinity.

The main contribution of this paper is that the trace of the rational surface $F$ can be obtained as the union of the traces of three rational surfaces $F(u, v)$, $F_\varphi(u, v) = F\left(\frac{v}{u}, \frac{1}{u}\right)$, and $F_\psi(u, v) = F\left(\frac{1}{v}, \frac{u}{v}\right)$, over the square $[-1,1] \times [-1,1]$. Actually, since we are assuming that $F$ is defined by a triangular control net $\mathcal{N}$ (as a triangular surface patch), we need to split the square $[-1,1] \times [-1,1]$ into two triangles, and we obtain six triangular patches altogether. There is also a very simple method for computing control nets for these six patches from the control net $\mathcal{N}$ of the original patch. It is important to note that the degree of the two new patches is **not** raised.

The upshot is that in order to render the entire rational surface $F$ it is enough to render $F$, $F_\varphi$, and $F_\psi$, over $[-1,-1] \times [-1,1]$. We will now give an idea of how this result is derived. It turns out that to give a rigorous proof, it is necessary to view rational surfaces as surfaces defined in a suitable projective space in terms of multiprojective maps. An implementation of our method has been written in *Mathematica*, using a subdivision version of the de Casteljau algorithm (see Gallier [9]).

Bajaj and Royappa have also investigated a method for drawing closed rational surfaces (and more generally, rational varieties) [2, 3]. Their method is based on the observation that the four maps

$$(u, v) \mapsto \left(\frac{\sigma_1 u}{1-u-v}, \frac{\sigma_2 v}{1-u-v}\right),$$

with $\sigma_i \in \{-1, 1\}$ for $i = 1, 2$, map the triangle $((1,0),(0,1),(0,0))$ bijectively onto the four quadrants of the plane respectively. However, they do not consider the problem of computing the control nets of the surfaces

$$F\left(\frac{\sigma_1 u}{1-u-v}, \frac{\sigma_2 v}{1-u-v}\right).$$

There is a method for computing control points for these patches, but this method is more complicated than ours. (see Gallier [9]).

Due to the lack of space, we can only sketch how polynomial surfaces and rational surfaces are defined in terms of control points. The deep reason why polynomial triangular surface patches can be effectively handled in terms of control points is that multivariate polynomials arise from multiaffine symmetric maps (see Ramshaw [12], Farin [5, 6], Hoschek and Lasser [11], or Gallier [10]). Recall that rational surfaces are defined by control nets consisting of pairs $\langle a, \lambda \rangle$ where $a \in E$ is a point and $\lambda \neq 0$, called *weighted points*, or vectors $\vec{u} \in \vec{E}$. In order to handle rational surfaces rigorously in terms of control points, it turns out that it is necessary to view rational surfaces as surfaces in some projective space. Fortunately, there is a construction to embed an affine space $E$ into a vector space $\widehat{E}$, in such a way that multiaffine maps extend uniquely to multilinear maps. This construction is described in Berger [4] and is at the heart of the presentation of rational surfaces in Fiorot and Jeannin [7, 8]. However, Fiorot and Jeannin do not use polar forms. We have adapted their approach in the framework of polar forms in Gallier [10]. Technically, a *projective rational surface $F$ of degree $m$* is defined by some symmetric multilinear map $f: (\widehat{\mathcal{P}})^m \to \widehat{\mathcal{E}}$ such that

$$F([u, v, z]) = \mathbf{P}(f)(\underbrace{[u, v, z], \ldots, [u, v, z]}_{m}),$$

for all $[u, v, z] \in \widetilde{\mathcal{P}}$ (all homogeneous coordinates $(u, v, z) \in \mathbf{R}^3$).

## 2 The Main Theorem

As we explained in section 1, the key to drawing completely a closed rational surface is to partition the plane $\mathcal{P} = \mathbf{R}^2$ into regions, in such a way that there exist simple bijections between some fixed closed region of the plane $\mathcal{P}$, say the square $[-1,1] \times [-1,1]$, and some regions in the complement of the open square $]-1,1[\times]-1,1[$ in the projective plane $\widetilde{\mathcal{P}}$. This can be done by dividing the plane into 24 regions. We will use the projective isomorphism $\varphi$ of the projective plane $\widetilde{\mathcal{P}}$ induced by the linear isomorphism of $\mathbf{R}^3$ given by

$$(u, v, w) \mapsto (v, w, u).$$

Choosing the line at infinity $w = 0$, the restriction of this map to the affine plane $\mathcal{P}$ (corresponding to $w = 1$) is the map

$$(u, v, 1) \mapsto (v, 1, u).$$

This is the map from $\mathcal{P}$ to $\widetilde{\mathcal{P}}$ such that

$$(u, v) \mapsto (v/u, 1/u)$$

if $u \neq 0$, and $(0, v) \mapsto (v, 1, 0)$ when $u = 0$.

We also use the projective isomorphism $\psi$ of the projective plane $\widetilde{\mathcal{P}}$ induced by the linear isomorphism of $\mathbf{R}^3$ given by

$$(u, v, w) \mapsto (w, u, v).$$

Choosing the line at infinity $w = 0$, the restriction of this map to the affine plane $\mathcal{P}$ (corresponding to $w = 1$) is the map

$$(u, v, 1) \mapsto (1, u, v).$$

This is the map from $\mathcal{P}$ to $\widetilde{\mathcal{P}}$ such that

$$(u, v) \mapsto (1/v, u/v)$$

if $v \neq 0$, and $(u, 0) \mapsto (1, u, 0)$ when $v = 0$.

Then, $\varphi$ defines a bijection from $[-1,1] \times [-1,1]$ onto the subset of the projective plane consisting of the lightly shaded area (see Figure 1).

Similarly, $\psi$ is a bijection from $[-1,1] \times [-1,1]$ onto the subset of the projective plane consisting of the darkly shaded area (see Figure 1).

We will now use the maps $\varphi$ and $\psi$ to show how the trace of a rational surface $F$ can be obtained as the union of the traces of three rational surfaces over the square $[-1,1] \times$

$[-1, 1]$.[1] The first of these surfaces is $F$ itself, and the two other rational surfaces $F_\varphi$ and $F_\psi$ are easily obtained from $F$. However, depending on the multilinear map $f$ defining $F$, the surface $F_\varphi$ (or $F_\psi$) may have *base points*, that is, points where the multilinears maps $f_\varphi$ (or $f_\psi$) is the zero vector. These are delt with in the full paper [9].

Since we are dealing with triangular surfaces, it is necessary to split the square $[-1,1] \times [-1,1]$ into two triangles, and thus, we will obtain the trace of a rational surface as the union of 6 patches over various triangles in the square $[-1,1]\times[-1,1]$. Letting $a, b, c, d$ be the vertices of the square $[-1,1] \times [-1,1]$ we will consider the following affine frames

$$\Delta bca = ((-1,1),\ (-1,-1),\ (1,1)),$$
$$\Delta dac = ((1,-1),\ (1,1),\ (-1,-1)),$$
$$\Delta bad = ((-1,1),\ (1,1),\ (1,-1)).$$

In particular, a rectangular surface patch defined over the square $[-1,1]\times[-1,1]$ will be treated as the union of two triangular surface patches defined over the triangles $\Delta bca$ and $\Delta dac$.

**Theorem 2.1** *Given an affine space $\mathcal{E}$ of dimension $\geq 3$, for every rational surface $F$ of degree $m$, some surfaces $F_\varphi$ and $G_\psi$ can be defined such that, except for the base points of $F_\varphi$ and $G_\psi$ (if any),*

$$trace(F_\varphi(\Delta bca)) = trace(F(\varphi(\Delta bca))),$$
$$trace(F_\varphi(\Delta dac)) = trace(F(\varphi(\Delta dac)))$$
$$trace(F_\psi(\Delta bca)) = trace(F(\psi(\Delta bca))),$$
$$trace(F_\psi(\Delta dac)) = trace(F(\psi(\Delta dac))).$$

*Furthermore, if the control nets (in $\widehat{\mathcal{E}}$) of the surface $F$ w.r.t. the affine frames $\Delta bca$, $\Delta dac$, and $\Delta bad$, are respectively*

$$\alpha = (\alpha_{i,j,k}),\quad \beta = (\beta_{i,j,k}),\quad \gamma = (\gamma_{i,j,k}),$$

*where $i+j+k=m$, the control nets $\theta^1$ and $\theta^2$ (in $\widehat{\mathcal{E}}$) of the surface $F_\varphi$ w.r.t. the affine frames $\Delta bca$ and $\Delta dac$, and the control nets $\rho^1$ and $\rho^2$ (in $\widehat{\mathcal{E}}$) of the surface $F_\psi$ w.r.t. the affine frame $\Delta bca$ and $\Delta dac$, are given by the equations*

$$\theta^1_{i,j,k} = (-1)^{i+j}\ \beta_{j,k,i},$$
$$\theta^2_{i,j,k} = (-1)^{k}\ \gamma_{i,j,k},$$
$$\rho^1_{i,j,k} = (-1)^{j}\ \gamma_{j,k,i},$$
$$\rho^2_{i,j,k} = (-1)^{i+k}\ \alpha_{k,i,j}.$$

Theorem 2.1 shows that in order to render a rational surface, we just need to compute the control nets $\alpha, \beta, \gamma$ for the surface $F$ w.r.t. the affine frames $\Delta bca$, $\Delta dac$, and $\Delta bad$, since then, the control nets $\theta^1$ and $\theta^2$ (in $\widehat{\mathcal{E}}$) of the surface $F_\varphi$ w.r.t. the affine frames $\Delta bca$ and $\Delta dac$, and the control nets $\rho^1$ and $\rho^2$ (in $\widehat{\mathcal{E}}$) of the surface $F_\psi$ w.r.t. the affine frame $\Delta bca$ and $\Delta dac$, are obtained at trivial cost.

*Remark*: It should be noted that the surface patches associated with the control nets $\alpha$, $\beta$, $\theta^1$, $\theta^2$, $\rho^1$, and $\rho^2$, may overlap in more than boundaries. In fact, there are examples where $\alpha$ and $\beta$ determine the entire surface, and other examples in which $\theta^1$, $\theta^2$, $\rho^1$, and $\rho^2$, determine the entire surface.

[1] While reading Appell's Treatise of Rational Mechanics, we stumbled on the fact that the change of variable $(u, v) \mapsto (1/v, u/v)$ was used by Appell in his solution to a problem of Bertrand (see [1], Tome I, Part III, Chapter XI, page 422-423). Appell explains that he found this "homographic transformation" in 1889.

## 3 Conclusion

We presented an efficient method for drawing closed rational surfaces, defined parametrically or in terms of weighted control points. The method relies on a subdivision version of the de Casteljau algorithm. One of the advantages of our method is that it is incremental. Indeed, the algorithm produces an approximation of the surface as a sequence of control nets. Thus, if we wish to get better accuracy, we can subdivide each control net in the list. We can also achieve a zooming effect by selectively subdividing some subsequences of control nets. Bajaj and Royappa also have investigated methods for drawing closed rational surfaces (and more generally, rational varieties) [2, 3]. However, they do not address the problem of computing control nets. Further comparisons between their method and ours would be interesting. A weakness of our method is that it only applies to rational surfaces. On the other hand, although restricted to rational surfaces, our method is quite efficient. In the full paper [9], we also proposed a method for resolving base points, by computing some refined control nets. Unfortunately, the present version of the method is quite inefficient as soon as the degree becomes greater than 3. Part of the problem is that our method first computes a rectangular control net which is then converted to a triangular net, and this conversion process is very expensive. It would be interesting to compute directly a triangular net, but such a method still eludes us. Alternatively, there may be smarter ways of computing the new control nets.

## References


[1] Paul Appell. *Traité de Mécanique Rationelle, Tome I: Statique, Dynamique du Point.* Gauthier-Villars, sixth edition, 1941.

[2] C. Bajaj and A. Royappa. Triangulation and display of arbitrary rational parametric surfaces. In R. Bergeron and A. Kaufman, editors, *IEEE Visualization '94 Conference.* IEEE, 1994.

[3] C. Bajaj and A. Royappa. Finite representation of real parametric curves and surfaces. *Intl. J. of Computational Geometry and Applications,* pages 313–326, 1995.

[4] Marcel Berger. *Géométrie 1.* Nathan, 1990. English edition: Geometry 1, Universitext, Springer Verlag.

[5] Gerald Farin. *Curves and Surfaces for CAGD.* Academic Press, third edition, 1993.

[6] Gerald Farin. *NURB Curves and Surfaces, from Projective Geometry to practical use.* AK Peters, first edition, 1995.

[7] J.-C. Fiorot and P. Jeannin. *Courbes et Surfaces Rationelles.* RMA 12. Masson, first edition, 1989.

[8] J.-C. Fiorot and P. Jeannin. *Courbes Splines Rationelles.* RMA 24. Masson, first edition, 1992.

[9] Jean H. Gallier. Algorithms for drawing closed rational surfaces. Technical report, University of Pennsylvania, 200 South 33rd Street, Philadelphia, PA 19104, 1996. Submitted for publication.

[10] Jean H. Gallier. Curves and surfaces: Geometry, algorithms, and implementation in Mathematica. Technical report, University of Pennsylvania, 200 South 33rd Street, Philadelphia, PA 19104, 1997. Book in progress.

[11] J. Hoschek and D. Lasser. *Computer Aided Geometric Design.* AK Peters, first edition, 1993.

[12] Lyle Ramshaw. Blossoming: A connect-the-dots approach to splines. Technical report, Digital SRC, Palo Alto, CA 94301, 1987. Report No. 19.

# Fast Collision Detection among Multiple Moving Spheres *

Dong Jin Kim† Leonidas J. Guibas‡ Sung Yong Shin†

**Abstarct**

This paper presents an event-driven approach that efficiently detects collisions among multiple moving spheres of uniform radius. We divide the space containing the spheres into uniform subspaces of cell structure. Each sphere intersecting a subspace is tested against the others intersecting the same subspace for possible collisions. We identify three types of events to detect the sequence of all collisions during our simulation: collision, entering, and leaving. The first type of events is due to actual collisions, and the other two types occur when spheres move from subspace to subspace. By tracing all such events in the order of their occuring times, we are able to simulate $n$ moving spheres with proper collision response[1, 5, 10] in $O(n_c \log n + n_e \log n)$ time with $O(n)$ space after $O(n \log n)$ time preprocessing, where $n_c$ and $n_e$ are the number of actual collisions and that of entering and leaving events during the simulation, respectively. Since $n_e$ depends on the size of subspaces, we adopt the collision model from kinetic theory for molecular gas[3] to determine the subspace size that minimizes simulation time. Experimental results show that collision detection can be done in linear time in $n$ over a large range.

## 1 Introduction

Spheres are widely used in computer graphics and robotics since they are simple and effective to model a variety of objects. They can not only represent objects such as atoms, balls, and sand[9, 12, 13, 14, 15] but also be used to approximate other 3D objects such as desk lamps, piles, and clumps[7, 11]. Particle systems use spheres to represent point masses for modeling complex natural phenomena such as fire, clouds, and water[12, 13, 14, 15]. Molecular graphics use spheres to describe the behavior of molecules composed of atoms[6, 8, 16]. Polyhedral objects are approximately represented by hierarchies of spheres[7, 11]. In order to consider interactions among spheres in such cases, it is a fundemental problem to efficiently detect collisions among spheres. To solve this problem, the number of collision checks must be minimized[1, 5, 10, 17, 18].

*This work was supported in part by Samsung Advanced Institute of Technology(project Research on Multimedia Creation Technology)

†Department of Computer Science, Korea Advanced Institute of Science and Technology, Taejon, Korea 305-701. e-mail:{djkim, syshin}@jupiter.kaist.ac.kr

‡Computer Science Department, Stanford University, Stanford, USA CA 94305. e-mail:guibas@cs.stanford.edu



In physical simulation of moving 3D objects, collision checks are done at a sequence of sampled times called time steps[7, 10, 17, 18]. In order not to miss any collision, the number of steps are increased, or objects are assumed to move sufficiently slowly to hit other objects no more than once between two consecutive time steps[4]. The latter sacrifices generality to restrict the domain of simulation. Meanwhile, the former requires heavy computation time. Assuming that objects are represented as a set of spheres, we propose an efficient event-driven approach for detecting the collisions among multiple moving spheres to avoid excessive collision checks.

In general, the motion of a solid is a mixture of translation and rotation, which makes it hard to predict an exact collision. A sphere is extremely simpler in shape than a general solid, and its motion can be described by translation only. Due to those nice properties, the collision time between two spheres is much easier to compute. In some sense, a simulation of moving spheres may be considered to be too domain-specific. However, its efficiency makes up for the loss of generality. The popularity of bounding spheres and a hierarchy of spheres to approximate a 3D object justify efficient methods for fast collision checks among spheres. Moreover, there are a wide variety of applications of particle modeling for which spheres themselves are massively floating around. The interaction among spheres is important to support those applications.

## 2 Algorithms and Analysis

We take an event-driven approach instead of time-consuming collision check at every time step. In order to localize collision checks, we decompose the space containing moving spheres into a set of uniform subspaces of cell structure. We keep track of the path of every moving sphere and the list of spheres intersecting each subspace. Those paths and lists are changed when a pair of spheres collide and when a sphere in a subspace moves to another subspace, respectively. Their occuring times are kept in a global queue. Thus our approach most closely resembles sweep-line and -plane methods in computational geometry, except that in our approach the dimension being swept over is time. To efficiently detect such events in the order of their occuring times, we first need nice data structure for non-empty subspaces intersecting one or more spheres and candidate pairs of spheres for collision. This idea is further explored in [2] from the theoretical point of view to maintain various attributes of mobile data.

Let $U$ be a sufficiently big cube of volume $D^3$ that contains $n$ moving spheres, where $D$ is the length of edges of the cube. The cube $U$ represents the space in which the moving spheres are simulated. We decompose this space into a set of uniform subspaces of cell structure. Each of the subspaces is a cube of volume $L^3$. For simplicity, $D$ is taken to be a multiple of $L$, that is, $D = mL$. The size $L^3$ of the volume of a subspace is determined to be a constant multiple $c$ of that of a sphere so that each subspace is able to accomodate two or more spheres at a given instance.

Without loss of generality, $U$ is assumed to be contained in the octant in which the three coordinate values of every point are all non-negative. Moreover, a vertex of $U$ is coincident with the origin, and each edge of $U$ is parallel with a principle axis. Let each subspace be denoted by $u(u_x, u_y, u_z)$ for $0 \leq u_x, u_y, u_z < m$ such that

$$u(u_x, u_y, u_z) = [u_x L, (u_x + 1)L] \times [u_y L, (u_y + 1)L] \times [u_z L, (u_z + 1)L]. \quad (1)$$

There are $m^3$ subspaces. However, many of them do not intersect any sphere. Since the size of a subspace is at least twice as large as that of a sphere, each sphere intersects at most eight subspaces. Therefore, there are $O(n)$ non-empty subspaces. Employing a bounded-balanced tree[19], we build a search tree, called a *subspace tree*, to keep those subspaces. A leaf node of the tree represents a non-empty subspace and has a list of spheres intersecting the subspace represented by the node. Search, insert, and delete operations on the subspace tree can be done in $O(\log n)$ time each. As auxiliary data structures, every sphere maintains a list of subspaces intersecting itself. This information is linked to the corresponding leaf nodes of the subspace tree.

In order to efficiently access a subspace in the tree, we assign the unique key to every subspace $u(u_x, u_y, u_z)$ as follows:

$$key(u(u_x, u_y, u_z)) = u_x m^2 + u_y m + u_z, \quad (2)$$

where $0 \leq u_x, u_y, u_z < m$. The function $key(\cdot)$ associates $u(u_x, u_y, u_z)$ with a unique integer in the range from 0 to $m^3 - 1$. It is clear that $key(\cdot)$ is one to one and onto.

Given a sphere with its position and radius, it takes a constant time to find the intersecting subspaces with it. It also takes a constant time to compute the key of each subspace. There are at most eight subspaces intersecting a sphere. Thus, we can compute all keys for a sphere in a constant time. It takes $O(\log n)$ time to insert and update each subspace. Since there are a maximum of eight subspaces to handle for a sphere, each sphere needs $O(\log n)$ time. Hence, it takes $O(n \log n)$ time to initially construct a subspace tree. Moreover, the number of leaf nodes is $O(n)$ which immediately gives $O(n)$ space to keep the tree[19].

We identify three types of events to detect the sequence of collisions during a simulation: collision, entering, and leaving. The first type of events is due to actual collisions, and the other two types occur when spheres enter new subspaces and leave the current subspace(s), respectively. Allowing penetration with other spheres, we can compute the times for candidate collision events of each sphere with the others in the same subspace as well as those candidates for leaving and entering subspaces. Under this condition, the candidate events seem to have nothing to do with actual ones. However, given all such events for $n$ spheres, no penetration occurs until the earliest among them. Therefore, an actual event must occur at the time of the earliest candidate, and no others do by this time. Assuming that the subspace tree is correctly updated to reflect the actual event, we can again catch the next earliest event in the same manner. Repeating this process, all the events can be generated in their time sequence.

Every sphere has a constant number of candidate events: $O(c)$, $O(1)$, and $O(1)$ candidates for collision, entering, and leaving, respectively. Thus, the total number of candidates are $O(n)$. We again adopt a bounded-balanced tree to maintain all candidate events. A leaf node of the event tree represents a candidate event. Its occuring time is used for the search key. Occuring times may not to be unique. However, we can always use sphere labels and event types in addition to those times to break any ties. The candidate events are arranged in the ascending order of their keys. By a similar analysis for the subspace tree, we can initially construct the event tree in $O(n \log n)$ time with $O(n)$ space.

According to the type of the current event, the change of distribution of the spheres over the subspaces together with its resulting change of candidate events are correctly updated in the data structures such as subspace and event trees. This change includes:

1. the candidate events of each sphere that is involved in the current event.

2. the subspaces of each sphere of the current event.

Suppose that the current event is caused by an actual collision between two spheres. Then, the portion of the moving path of each of those spheres after the collision is changing due to the collision response of one sphere to the other. Collision response is beyond the scope of this paper although it itself is very interesting. There are several models for collision response available in computer graphics literature[1, 5, 10]. By renewing the paths of two colliding spheres, we can compute the new occurring times of the candidate events of each sphere. They include the times for candidate collisions with the other spheres in the same subspace(s) as well as for the candidate entering and leaving events of the sphere. There are a pair of spheres that are involved in a collision. Moreover, every sphere intersects at most eight subspaces, each of which may have a constant number $c$ of intersecting spheres. Therefore, every collision causes a constant number of candidate events to be renewed. Since each of those events needs $O(\log n)$ time to be updated in the event tree, it takes $O(\log n)$ time to handle a collision event.

Now, consider the case that the current event occurs due to a sphere entering a subspace. The subspace has the entering sphere as a new member in its list of intersecting spheres. That sphere needs adding to the list for the subspace in the subspace tree, which takes $O(\log n)$ time. Accordingly, the new candidate events for collisions between the sphere and the others in the subspace are added to the event tree. This can also be done in $O(\log n)$ time since the subspace intersects a constant number of spheres. We finally renew the candidate event time for entering another subspace in $O(\log n)$ time. Therefore, it takes $O(\log n)$ time to handle an entering event.

Finally, suppose that the current event is caused by a sphere leaving a subspace. Then, that sphere needs removing from the subspace, and the candidate events for collisions between the sphere and the others in the subspace are also required to be removed. Clearly, it takes $O(\log n)$ time to update the subspace and event trees to reflect those removals, respectively. It also takes $O(\log n)$ time to renew the candidate event time of the sphere for leaving. Therefore, we need $O(\log n)$ time to handle a leaving event.

A brute-force approach needs $O(n_t n^2)$ operations for collision detection, where $n_t$ denotes the number of time steps during a simulation. Let $n_c$ and $n_e$ be the number of actual collisions and that of actual entering and leaving events for the simulation, respectively. Since it takes $O(\log n)$ time to handle an event regardless of its type, we need $O(n_c \log n +$

$n_e \log n)$ operations for the whole simulation of $n$ moving spheres after $O(n \log n)$ initial investment for constructing subspace and event trees. It is clear that $n_c$ cannot be further reduced. However, $n_e$ is a function of the size of subspaces, and thus needs more attention to improve.

## 3 Determining the Size of Subspaces

For a given $n$, $D$, a radius of spheres, and their mean velocity, we can choose $L$ by solving the following equation:

$$6vTL^{-2}\left[2n\frac{(L+2r)^2}{D^3}\left\{c_1\pi r^2 n_0 L^2 + c_3(L-r)\right\} - c_4\right] = 0$$

This equation is derived using the collision model of kinetic theory for molecular gas[3]. The obtained subspace size is not optimal. However, experiments support that it is a good approximation of the optimal subspace size.

## 4 Experimental Results

In order to show the efficiency of the proposed event-driven approach, experiments are performed on an IRIS Indigo2 machine(128MHz, R10,000CPU, 128MB) for multiple spheres of uniform radii freely moving in a given box. Assuming that a gravity force is zero, the motion of each sphere is given by its initial position and velocity. A path of a sphere is changed whenever it undergoes a collision with other spheres or collides with a face of the box.

Varying the number of spheres for a fixed radius and mean speed while optimizing the subspace size, we measure actual running times(Figure 1). The vertical axis gives an actual running time for a unit time of simulation of given data. As plotted in the figure, it is nearly linear in $n$ over a large range.

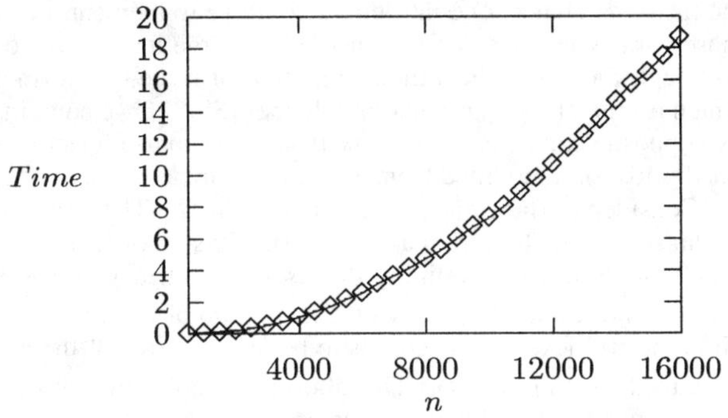


Figure 1: The radius of spheres is 1.5, their mean speed is $10\sqrt{3}$, and the size of the space is $200\times200\times200$,

## 5 Conclusions

This paper proposes an event-driven approach for efficient collision detection among multiple moving spheres of uniform size. The proposed approach enables us to simulate in real time several thousands of spheres freely moving in a 3D box. Although we assume that all spheres are of uniform size, our approach can also be applicable to spheres of non-uniform radii if the size of the largest sphere is a small constant multiple of that of the smallest one.


### Acknowledgements

We would like to thank Mr. Chan-Su Shin for his careful reading.


## References


[1] David Baraff. Curved surfaces and coherence for non-penetrating rigid body simulation. *Computer Graphics*, 24(4):19–28, August 1990.

[2] Julien Basch, Leonidas J. Guibas, and John Hershberger. Data structures for mobile data. In *ACM-SIAM Symposium on Discrete Algorithms*, 1997(to appear).

[3] Feynmann, Leighton, and Sands. *The Feynmann Lectures on Physics*, volume 1, pages 43.1–43.4. Addison-Wesley, 1963.

[4] Stefan Gottschalk, Ming Lin, and Dinesh Manocha. Obbtree: A hierarchical structure for rapid interferece detection. *Computer Graphics Proceedings, Annual Confernece Series*, pages 171–180, 1996.

[5] James K. Hahn. Realistic animation of rigid bodies. *Computer Graphics*, 22(4):299–308, August 1988.

[6] Dan Halperin and Mark H. Overmars. Spheres, molecules, and hidden surface removal. In *Proc. 10th Annual ACM Symposium on Computational Geometry*, pages 113–122, 1994.

[7] Philip M. Hubbard. Approximating polyhedra with spheres for time-critical collision detection. *IEEE Transactions of Graphics*, 15(3):179–210, July 1979.

[8] Cyrus Levinthal. Molecular model-building by computer. *Scientific American*, pages 42–52, June 1966. 213(6).

[9] Victor J. Milenkovic. Position-based physics: Simulating the motion of many highly interacting spheres and polyhedra. *Computer Graphics Proceedings, Annual Conference Series*, pages 129–136, 1996.

[10] Matthew Moore and Jane Wilhelms. Collision detection and response for computer animation. *Computer Graphics*, 22(4):289–298, August 1988.

[11] Joseph O'Rourke and Norman Badler. Decomposition of three-dimensional objects into spheres. *IEEE Transactions on Pattern Analysis and Machine Intelligence*, PAMI-1(3):295–305, July 1979.

[12] W. T. Reeves. Particle systems - a technique for modeling a class of fuzzy objects. *ACM Transaction on Graphics*, 2(2):91–108, April 1983.

[13] Craig W. Reynolds. Flocks, herbs, and schools: A distributed behavioral model. *Computer Graphics*, 21(4):25–34, 1987.

[14] Karl Sims. Particle animation and rendering using data parallel computation. *Computer Graphics*, 24(4):405–413, 1990.

[15] Richard Szeliski and David Tonnesen. Sufrace modeling with oriented particle systems. *Computer Graphics*, 26(2):185–194, July 1992.

[16] G. Turk. Interactive collision detection for molucular graphics. Technical report, 90-014, Dept. of Computer Science, Univ. of North Calolina at Chapel Hill, Jan. 1990.

[17] Vincent Hayward, Stéphane Aubry, André Foisy, and Yasmine Ghallab. Efficient collision prediction among many moving objects. *The International Journal of Robotics Research*, 14(2):129–143, 1995.

[18] Robert Webb and Mike Gigante. Using dynamic bounding volume hierarchies to improve efficiency of rigid body simulation. *Visual Computing*, pages 825–842, 1992. (CGI Proc).

[19] Dan E. Willard and George S. Lueker. Adding range restriction capability to dynamic data structure. *Journal of The Association for Computing Machinery*, 32(3):pp. 597–617, July 1985.

# Robust and Efficient Evaluation of Functionals on Parametric Surfaces

Leif Kobbelt

*Computer Sciences Department (IMMD IX), University of Erlangen - Nürnberg*
*Am Weichselgarten 9, 91058 Erlangen, Germany*
kobbelt@informatik.uni-erlangen.de

Quality control is an important issue in computational geometry. Since analytical properties like the differentiability of a boundary representation are often not sufficient to fully judge the quality of an object, additional criteria measured in terms of scalar valued functionals have to be verified. Energy functionals in the context of *fairing* are just one example where the global smoothness of a surface is rated by one single value, e.g., the total amount of (squared) curvature. More "industrial" examples are the computation of *physical properties* like the surface area or the volume of an object, its center of gravity, or the moments of inertia.

Let an object $X$ of arbitrary topology be given by a closed boundary representation $\Phi : \{\Omega_i\} \to \mathbf{R}^3$. The functionals we consider, have the form

$$P(X) := \sum_i \int_{\Omega_i} F(\Phi, \Phi_u, \Phi_v) \, du \, dv \tag{1}$$

where $F : \mathbf{R}^3 \to \mathbf{R}$ does not depend on the geometry of $X$. Hence, once the functional $P$ characterizing the property under investigation is found, the problem reduces to the numerical approximation of a bivariate integral. A closed form solution for $P$ is usually not known – even if $X$ is a polyhedron with $\Phi$ being piecewise linear.

In this paper we suggest solutions for the key steps of an algorithm which performs the task of approximately evaluating (1) on a parametric surface. The efficiency of the algorithm results from the adaption of powerful quadrature techniques to the particular geometric setting. The robustness of the algorithm is due to exact arithmetic operations.

## Non-uniform Romberg-quadrature

Iterative algorithms to approximately evaluate (multivariate) integrals use an *adaptive* refinement approach. The integrand function is sampled on an initial uniform grid and a simple cubature formula is applied. A posteriori error estimates decide whether the grid has to be refined in order to meet prescribed error tolerances[Str71]. The different approximations to the exact solution which one obtains by this iterative process can be subject to an extrapolation scheme that allows to further reduce the approximation error [Sto83].

Romberg-quadrature [Bul67] efficiently computes the Richardson-extrapolation by a Neville-type scheme. If a polynomial expansion of the approximation error as a function of the grid's step width $h$ exists, the value of this polynomial at $h = 0$ can be computed by extrapolating approximate results $T_i$ with step widths $h_i$. Romberg-quadrature specifically uses the trapezoid quadrature rule and the step widths $h_i = 2^{-i}$. From $T_{i,i} := T_i$ we recursively compute a tableau of values $T_{i,j}$ by

$$T_{i,j} := \frac{4^{j-i}\, T_{i+1,j} - T_{i,j-1}}{4^{j-i} - 1}. \tag{2}$$

[Bul66] shows that the error of the extrapolated value $T_{0,k}$ with respect to the exact solution $T_\infty$ can be bounded by

$$|T_{0,k} - T_\infty| \le \sigma\, |T_{1,k} - T_{0,k-1}| \tag{3}$$

for an appropriate $0 < \sigma \le 1$.

The major draw-back of this approach is the exponentially increasing number of function evaluations that are necessary to obtain an approximation on the next refinement level. Slightly reducing the prescribed error tolerance may increase the running time by the factor 4. Hence, one introduces *selective* refinement, i.e., only those parameter regions are refined where large local errors occur. Although this contradicts the assumption of nested uniform grids which is crucial for extrapolation schemes [Slu82], we combine the two opposite concepts in a scheme that allows to use selectively refined grids for a modified Romberg-quadrature.

Consider a triangular parameter domain $\Omega$. The exact trapezoidal sums $T_{i,i}$ stem from a uniform subdivision of $\Omega$ into $4^i$ sub-triangles. Selectively refining only *some* sub-triangles yields an approximate value $\widetilde{T}_{i+1,i+1}$ which is the sum of local results from different grid levels. Our goal is to bound the error of the extrapolated values $\widetilde{T}_{0,k} \approx T_{0,k}$. This allows the application of (2) with (3) replaced by a modified error estimate.

Due to the exponential increase of parameter triangles, most of the effort is spent on the finest level. Reducing the complexity on this level has the heaviest impact on the total running time of the algorithm. The geometric background of the functionals ensures a regular behavior of the integrand function. Since the local convergence of the trapezoid rule is $O(h^4)$, the probability that a selective refinement algorithm picks a triangle from level $i+1$ before all level $i$ triangles are subdivided, is low (the local error would have to differ by more than the factor 16).

Assume that the values $T_{i,i}$ with $i = 0, \ldots, k-1$ are exact and $\widetilde{T}_{k,k} = T_{k,k} + \varepsilon$ since not all triangles from generation $k-1$ have been subdivided yet. Applying (2) yields

$$\widetilde{T}_{0,k} = T_{0,k} + \left( \prod_{i=1}^{k} \frac{4^i}{4^i - 1} \right) \varepsilon.$$



*Computational Geometry 97* Nice France

Hence, no matter what $k$, $\widetilde{T}_{0,k}$ deviates from $T_{0,k}$ by less than $\frac{3}{2}\,\varepsilon$ and the analogue holds for $\widetilde{T}_{1,k}$. Taking into account that $\widetilde{T}_{0,k-1} = T_{0,k-1}$, we have

$$|\widetilde{T}_{0,k} - T_\infty| \;\leq\; \sigma\,|\widetilde{T}_{1,k} - \widetilde{T}_{0,k-1}| + 3\,\varepsilon \qquad (4)$$

This estimate allows to use the result $\widetilde{T}_{0,k}$ extrapolated from a selectively refined grid to approximate the exact solution $T_\infty$. The bound (4) gives an estimate of the current approximation error while $\varepsilon$ is bounded by summing the local approximation errors over those parameter triangles which are still from generation $k-1$. Inverse Richardson-extrapolation allows to compute the difference $|\widetilde{T}_{1,k} - \widetilde{T}_{0,k-1}|$ without actually generating the table entries $\widetilde{T}_{j,k}$ [Kob92].

## Exact Arithmetics

The reliable evaluation of algebraic expressions is an important issue in current research on computational geometry [For96], [Gui96]. Many computational geometers have proposed algorithms to perform exact arithmetic operations. [For93] and [She96] observe that most geometric *predicates* can be formulated as determinants. [Dek71] and [FW93] consider more elementary operations which allow to build code for the exact evaluation of fairly general formulae (assuming exact floating-point input data). However, the complexity of these solutions increases faster than the complexity of the formulae themselves.

When evaluating the functionals (1) either by a uniform or a selective refinement approach, the most unstable operation is the summation (*destillation*) of the local results. Both approaches, yield an increasing number of *local* estimates which have to be added up in order to obtain the *global* result. Eventually, one has to accumulate a huge number of local approximations, each within some prescribed tolerance to the exact value. Obviously, the rounding errors during this step can lead to invalid answers because they are usually not covered by error bounds emerging from numerical integration theory.

In practice one encounters situations where further refinement of the grid makes the approximation of the integral worse because of this effect. Consider a selective refinement cubature algorithm. The current approximation $T$ is stored as the accumulated sum of the local values $T_i$. Local refinement means replacing one local value $T_i$ by $T'_{i,1}, \ldots T'_{i,r}$ and updating

$$T \;\leftarrow\; T - T_j + T'_{j,1} + \ldots + T'_{j,r}.$$

Hence, with every refinement step, the uncertainty of $T$ increases. Notice that during the computation of the integral, this update operation might be performed several million times! Otherwise, if the sum $T$ is not updated in every step but only computed once in the end, the current approximation is not available for extrapolation and error estimation.

There are several approaches to *destillation* which turns out to be an important step in almost any numerical algorithm. A whole family of algorithms perform a more or less strict pre-sorting step to order the elements by their magnitude and compute the sum by exploiting this special configuration [Kul81, Pri92]. These algorithms always have a $O(n \log n)$ time complexity. Another draw-back is that in order to apply such algorithms, all $n$ elements have to be stored and cannot be processed on the fly.

A second class of algorithms uses special high precision fixed point arithmetics to compute the sum and project the result back into floating point representation [Boh90]. Those approaches are usually machine dependent – in the extreme they are designed for hardware implementation. More portable algorithms should be formulated relative to a specific floating point arithmetics standard (e.g. IEEE 754) and can work on any computer providing this standard [Kob94].

We present a simple algorithm which allows to compute arbitrarily large sums without any rounding errors by just using standard IEEE arithmetic. The scheme has linear time complexity $O(n)$ with a small constant $1+\varepsilon$ and does not depend on custom designed long integer arithmetics. It is an improved version of [Kob94]. Its software implementation only needs a small amount of intermediate storage and its simplicity makes it also a candidate for hardware realization.

The basic idea of the proposed algorithm can be described as follows. A floating point number is given by a sign, a $p$-bit mantissa, and an exponent (all being integer). It can be observed that the addition (or subtraction) of two floating point numbers does not produce any roundoff error if both have the same exponent and both mantissae have the same parity. For subtraction this is trivial since subtraction on the same scale is simply integer arithmetic plus normalization. Adding two $p$-bit integers yields a $(p+1)$-bit integer which is subject to rounding (back to $p$-bits). However, since both addends have the same parity, their sum will be an even number and the rounding step cuts off a zero.

We can use the exponent of a floating point number combined with its mantissa's parity bit as a hashing key. Storing all incoming addends into a hash table, we know that if a collision occurs, the two numbers can be processed without rounding error. The table has to be large enough to hold two times the number of valid exponents of floating point numbers. In IEEE standard this means, we need 2 KByte (single precision) or 32 KByte (double precision) to store the whole accumulator table.

```
insert(value x,table t)
  { i = parity(x) + 2 * exponent(x)
    while (t[i] != 0.0)
      { x    = x + t[i]
        t[i] = 0.0
        i    = parity(x) + 2 * exponent(x) }
    t[i] = x }
```

The hashing takes constant time per access: Parity and exponent of a floating point number $x$ can be isolated by masking out the corresponding bit-fields. Once all addends are inserted, we have a table, with a few remaining entries orderd according to their exponent's magnitude. In [Dek71] it is shown that for two floating point numbers $a$ and $b$ with $|a| \geq |b|$ the rounding error of the floating point operation $a+b$ can be computed exactly (in floating point arithmetic) by $b - ((a+b) - a)$. This remains true as long as the exponent of $a$ is not smaller than the exponent of $b$. Hence we can add the hashtable's two top entries and store the result as well as the exact rounding error back into the table. By repeating this operation, we obtain the best approximation to the true sum which is representable as a floating point number together with an expansion of the approximation error (the remaining entries).

```
sum(table t)
  { i = search_downwards_from(TOP)      % first
    j = search_downwards_from(i-1)      % second
    while (j>=BOTTOM)
      { if (t[i] + t[j] != t[i])
          { a = t[i] + t[j]
            b = t[j] - (a - t[i])
            delete t[i],t[j]
            insert(a,t)
            insert(b,t)
            i = search_downwards_from(i+3) }
        else
          { i = j }
        j = search_downwards_from(i-1) }
    i = search_downwards_from(TOP)
    j = search_downwards_from(i-1)
    return t[i]+t[j] }
```

The destillation algorithm has linear complexity since no auxilliary additions are necessary. To sum up $n$ numbers, a minimum of $n-1$ additions is always necessary. While hashing the addends we occasionally perform additions and end up with $m$ remaining numbers in the table. Hence, $n-m$ additions have taken place so far. The only computational overhead stems from the evaluation of the hashing function. The number $m$ does not depend on $n$ but rather on the orders of magnitude which are covered by the addends. Summing up the remaining numbers is an $O(m)$ process with $m \ll n$.

### Reformulation of Physical Properties as Functionals

The Divergence Theorem provides a tool to transform the volume based physical properties into proper integral expressions of the form (1):

$$\int_V \nabla G \, dx\, dy\, dz = \int_{\partial V} G^T N \, d\sigma \qquad (5)$$

where $N : \partial V \to \mathbf{R}^3$ is the unit-length outer normal vector on $\partial V$ and $G : \mathbf{R}^3 \to \mathbf{R}^3$ is an arbitrary vector field (cf. [TS80]). Replacing $N$ by the cross product of the partial derivatives of $\Phi$ and transforming (5) into an integral over the parameter domain $\Omega = \bigcup_i \Omega_i$ yields

$$\int_V \nabla G \, dx\, dy\, dz = \int_\Omega G(\Phi)^T \left[\Phi_u \times \Phi_v\right] du\, dv.$$

which reduces the reformulation of physical properties to the construction of a function $G$ such that $\nabla G$ fits to the volume integral which defines the property.

The volume of an object is given by the integral over $\nabla G = 1$ and a suitable (symmetric) choice is $G := \frac{1}{3}(x, y, z)$. Analogously, we may chose $G := \frac{1}{2}(x^2, 0, 0)$ for the center of gravity's $x$-coordinate ($\nabla G = x$) and for the other coordinates respectively. To compute the moments of inertia with respect to the $z$-axis, we use $G := \frac{1}{3}(x^3, y^3, 0)$, i.e., $\nabla G = x^2 + y^2$, and for the $x$- and $y$-axis respectively.

### Conclusion

The ingredients presented in this paper yield a framework to implement a robust and efficient algorithm for the numerical evaluation of functionals on surfaces. Efficiency is obtained through a modification of the Romberg-quadrature which does not require to subdivide the domain uniformly but allows to use a selective refinement strategy. Volume based physical properties are reformulated as surface integrals which strongly reduces the complexity. The implementation becomes reliable and robust by including an algorithm to sum up the intermediate results without rounding errors. This exact arithmetic does not affect the performance of the algorithm significantly but it prevents the quadrature algorithm from accumulating rounding errors. This is important for engineering applications since rounding errors are not covered by error bounds emerging from numerical integration theory.

### References


[Boh90] G. Bohlender, *What do we need beyond IEEE-arithmetics?*, Comp. Arith. and Self-validating Numerical Methods, pp. 1–32, Academic Press, New York 1990

[Bul66] R. Bulirsch / J. Stoer, *Asymptotic Upper and Lower Bounds for Results of Extrapolation Methods*, Numer. Math. 8(1966) pp. 93-104

[Bul67] R. Bulirsch / J. Stoer, *Numerical Quadrature by Extrapolation*, Numer. Math. 9(1967) pp. 271-278

[Dek71] T. Dekker, *A floating-point technique for extending the available precision*, Numerische Mathematik 18 (1971), pp. 224–242

[Gui96] L. Guibas, *Implementing Geometric Algorithms Robustly*, Proc. Workshop on Applied Computational Geometry, pp. 15–22, Springer Verlag, 1996

[For93] S. Fortune, *Progress in Computational Geometry*, Directions in Geometric Computing, R.R. Martin ed. pp. 81-128, Information Geometers Ldt. 1993

[For96] S. Fortune, *Robustness issues in geometric algorithms*, Proc. Workshop on Applied Computational Geometry, pp. 9–14, Springer Verlag, 1996

[FW93] S. Fortune / J. Van Wyk, *Efficient Exact Arithmetics for Computational Geometry*, Proc. of the 9th annual symposium on Computational Geometry, pp. 163–172, ACM 1993

[Kob92] L. Kobbelt, *Iterative Berechnung metrischer Eigenschaften*, Thesis, University of Karlsruhe, Germany

[Kob94] L. Kobbelt, *A fast dot-product algorithm with minimal rounding errors*, Computing 52 (1994), pp. 355–369, Springer Verlag 1994

[Kul81] U. Kulisch / W. Miranker, *Computer Arithmetic in Theory and Practice*, Academic Press, New York, 1981

[Pri92] D. Priest, *On properties of floating point arithmetics*, Thesis, University of California at Berkeley, 1992

[She96] J. Shewchuk, *Adaptive precision floating point arithmetic and fast robust geometric predicates*, Report CMU-CS-96-140, Carnegie Mellon University, Pittsburgh, PA

[Slu82] A. van der Sluis, *Asymptotic expansions for quadrature errors over a simplex*, Numerical Integration, proceedings of the conference held at the Mathematisches Forschungsinstitut Oberwolfach Birkhäuser Verlag 1982 pp. 222-240

[Sto83] Josef Stoer, *Einführung in die Numerische Mathematik I*, Springer-Verlag 1983

[Str71] A. H. Stroud, *Approximate Calculation of Multiple Integrals*, Prentice-Hall 1971

[TS80] Timmer / Stern, *Computation of global geometric properties of solid objects*, Computer Aided Design 12(1980) pp. 301-304

# GASP-II: A Geometric Algorithm Animation System for an Electronic Classroom

Maria Shneerson
Department of Applied Mathematics
The Weizmann Institute of Science
Rehovot, Israel, 76100

Ayellet Tal
Department of Electrical Engineering
Technion – Israel Institute of Technology
Haifa, Israel, 32000

**Abstract**

This paper describes GASP-II, an algorithm animation system for geometric computing, which is particularly suitable for a distributed electronic classroom. The system allows the creation, presentation and interactive exploration of 3D geometric algorithms. GASP-II can also be used as a visual debugger for geometric computing.

## 1 Introduction

Geometric algorithms can be highly complex and difficult to grasp. Algorithm animations can provide the best way to help explain algorithms and to visually debug code. We present in this paper a system, GASP-II, that focuses on algorithms with a geometric flavor, which is accessible to even naive users.

Most algorithm animation systems (e.g., Balsa [5], Balsa-II [2], Tango [9], Zeus [3], CAT [4]) are designed for fundamental algorithms. Their support for 3D geometric algorithms is limited. There exist a few special purpose systems for geometric computing that come to ease the unusual difficulty of the implementation cycle (e.g., [7], [8], [6]). The Computational Geometry Workbench [6] and the XYZ GeoBench [8] support algorithm animations, but this is not their chief goal, and thus creating an animation is still not trivial. Mocha's [1] major objective is to provide algorithm animations over the World Wide Web. It focuses on 2D algorithms. GASP [10] goes beyond these systems by providing a rich set of 3D visualization and animation tools.

While GASP is mainly oriented towards debugging, GASP-II focuses on education. It allows the presentation and exploration of 3D geometric algorithms in a distributed electronic classroom. With GASP-II, views of the same running algorithm may reside on different machines, that can be distributed in various geographic locations. A fundamental difference between GASP-II and most other electronic classroom systems is that GASP-II is interactive. While in most other systems students are completely passive and the system simply displays the same X window on multiple machines, GASP-II allows the students to actively participate in the process of learning by interacting with the algorithm and collaborating with the instructor.

Each student can choose a different *style* for the animation which is suitable to the student's taste. This is done through a comfortable user interface without any need to write or modify any code. During a lecture, each student views a different animation of the same running algorithm. GASP-II also provides every student with a control panel for starting, pausing, stopping, rewinding and controlling the speed of the animation. It is thus possible to rewind and rerun the difficult parts of the algorithm until they are fully understood.

To support debugging, GASP-II provides a rich library of high-level operations to create an animation. During debugging, it is possible to run three communicating processes, GASP-II, a standard debugger, and the user's program simultaneously. The single stepping capability of the debugger combined with the ability to move both forwards and backwards in GASP-II makes it easy to spot bugs visually that would be nearly impossible to find with a standard ASCII debugger.

GASP-II is designed so that each machine is in charge of the graphical work displayed on it. It allows each student to have a full control over the animation. In addition, potential network traffic bottlenecks are avoided since only a relatively small amount of short messages are sent over the network. Moreover, in contrast to the X approach, the least powerful machine does not determine the rate of the whole class.

The next section describes how GASP-II can be used in an electronic classroom. In section 3 we discuss the programmer interface. Section 4 discusses implementation issues. We conclude in section 5.

## 2 End-User Perspective

GASP-II supports two modes of studying – *self-study* and *guided-study*. In the self-study mode, the lecture becomes similar to studying a chapter from an interactive electronic textbook. In the guided-study mode, the students must follow the instructor throughout the lecture. It is up to the instructor to decide when, during the lecture, a self-study mode is appropriate and when the students should be guided by the instructor. In the latter case, GASP-II ensure that all the views are synchronized with the instructor's view.

An unlimited number of students can be viewing the same algorithm, and they need not be at the same geographical location. Fig. 1 shows three snapshots of screens taken at

[0]This work has been supported in part by the Israeli Ministry of Science, Eshkol Grant 0554-2-96.


*Computational Geometry 97* Nice France

the same time. Each screen consists of three windows: the *Control Panel* through which the animation is controlled, the *Algorithm Window* in which the animation runs, and a *Text Window* in which the algorithm is explained.

**The Control Panel** (on the top left of Fig. 1(a)) contains *general controls* and *study-mode controls.* Exploring the animation is performed through the VCR-like general controls. The viewer can run the algorithm at varying speeds and "rewind" the algorithm in order to observe the confusing parts of the algorithm multiple times - an invaluable tool both for education and for debug.

The instructor's control panel differs from the student's control panel in the study-mode controls. The instructor can choose between a self-study mode and a guided-study mode. In the latter case, the general controls at the students' screens are disabled, and the students must follow the instructor in the course of the lecture. GASP-II supports an electronic discussion between the instructor and the students. An electronic discussion is more than a text-exchange, since it is accompanied by a 3D animation of the algorithm. The instructor controls the electronic discussion through this panel.

The student's control panel supports two levels of synchronization. With *instructor unit* synchronization, the animation viewed by the student shows the same phase of the algorithm as the instructor's. With *instructor home position* synchronization, the current camera positions on both screens are the same. The student can then observe the scene in the exact same way as the instructor.

**The Algorithm Window** (on the right hand side of Fig. 1(a)) is where the algorithm is observed. Different views of the same algorithm animation are shown in Fig. 1(a) – 1(b). In particular, different colors and transparency values were chosen by the first student. In Fig. 1(c) a different phase of the algorithm is being animated. The student rewinded the algorithm to a previous phase.

End-users can modify the default animation created by the system in order to accommodate their personal taste and to enhance their understanding. The preferred animation is selected via a style panel, without any need to write code. An important advantage of GASP-II over other animation systems is that different graphical views should not be programmed beforehand. GASP-II can automatically generate many animations for the same algorithm.

**The Text Window** (on the bottom left of Fig. 1(a)) is used to elucidate the events and to direct the viewer's attention to specific details. Every logical unit of the algorithm is associated with a piece of text which explains the events occurring during this phase.

## 3 Programmer Perspective

The main principle behind the model of GASP-II is that programmers should not be concerned with the presentation aspects of the animation and can choose to be isolated from any decisions of a graphical nature. Instead, programmers should concentrate solely on the contents of the animation. How the algorithm is visualized is left for the system to automatically determine. GASP-II is capable of doing so because it has knowledge about the geometric domain and the proper way to visualize it.

Our framework follows the GASP approach in which the the important logical phases of the algorithm, its *atomic units*, should be identified. Each atomic unit groups together the geometric operations that belong to the same logical phase of the algorithm, and GASP-II animates these

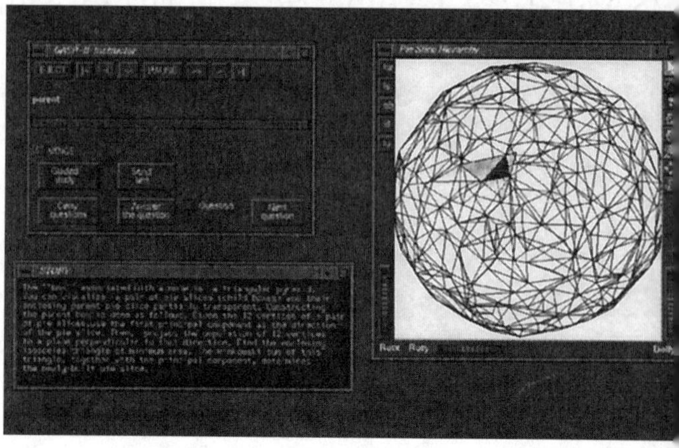

(a) The Instructor's Screen

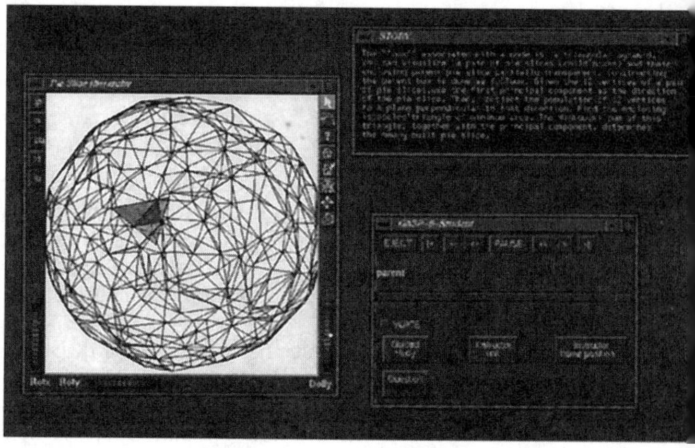

(b) A Student's Screen

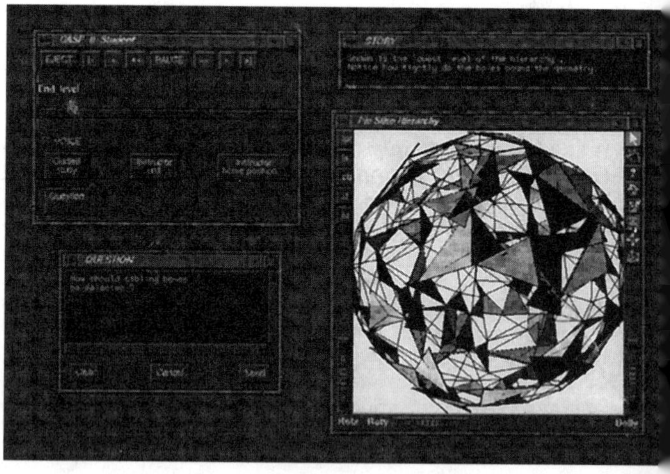

(c) Another Student's Screen

Figure 1: Snapshots of Screens

operations as a single unit. Each atomic unit consists of one or more calls to functions in the GASP-II library.

This library provides a rich set of functions which enable the programmer to write animations with minimal effort. The library contains functions which animate the creation, modification, and removal of geometric objects in 2D as well as in 3D. Each such function generates a smooth animation which is appropriate for the required operation.

The application transfers to GASP-II only parameters which are relevant to the geometry (e.g., the vertices and the faces of a polyhedron). No parameters of a graphical nature (e.g., colors, number of frames, the rate of the animation, the way each operation is animated) are required. GASP-II determines all these parameters. This is the major strength of the system, that contains sufficient knowledge regarding the appropriate visualization being generated.

Should the users wish to modify and extend various visualization aspects of the animation, they can change a few parameters in the style panel. Even if the style is modified, the animation is still generated automatically by the system, only a different animation will be generated. The style affects the animation, not the implementation!

## 4 Implementation

GASP-II's architecture is illustrated in Fig. 2. Two processes, which communicate with each other through messages, are running on the instructor's machine. Process 1 consists of the application code and the GASP-II library. Process 2 is responsible for executing the animation on the instructor's machine and for sending instructions regarding the animation to the students' machines. The process running on each of the students machines is in charge of executing the animation and handling the student's input.

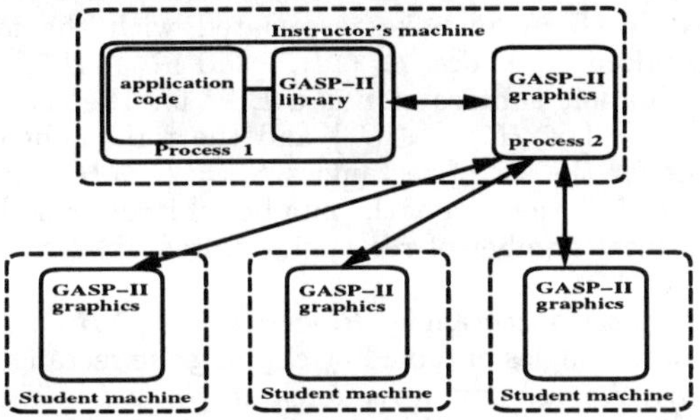


Figure 2: GASP-II's Architecture

The application initiates a call to a GASP-II function. Process 1 prepares messages containing the relevant information regarding this operation, and sends them to Process 2. Utilizing a hand-shaking approach, Process 1 then waits for a reply. Upon receiving the messages, Process 2 consults its style information and executes the animation. It then broadcasts the messages to the students' machines, and immediately sends an acknowledgment to Process 1 without waiting for a response. Process 1 can then return to the application's code. The application now proceeds executing the algorithm. Meanwhile, on each of the students' machines, an animation is executed.

The hand-shaking approach used between the processes on the instructor's machine, comes to facilitate debugging, when the system is used by a single user. It allows the user to visualize the scene at the time when the calls occur. Since rendering is done within an event mainloop, it is otherwise difficult to return to the application after each call.

In a distributed environment, the above architecture has a few important advantages. First, the graphical work is distributed among all participants in the classroom. Second, this architecture allows the students to choose their own styles for a specific animation and view completely different animation for the same algorithm. This is so because the messages received by the students' machines do not specify how the animation should be carried out, and thus each machine is free to determine the appropriate style. Third, since the information sent is very high level, the network traffic is very low. Fourth, if synchronization is required, it can be easily handled. Fifth, if a failure occurs at a student machine, it does not affect the other machines. Sixth, the system fully protects the algorithm code which is never sent over the network. Finally, since each machine can specify in the style panel parameters that fit this machine, machines on the lower-end do not slow down the whole class.

## 5 Conclusion

GASP-II is an algorithm animation system for geometric computing. It is particularly designed for use in an electronic classroom, yet is also suitable as a debugging tool. With GASP-II, programmers can produce "quick-and-dirty" animations by annotating their code with very short excerpts of C code. Users can experiment with various animations by changing the system's default style through a style panel, without ever modifying or compiling the code. Students can experiment with a finished animation and collaborate with other users.

We believe that animations can be used as a central part of teaching computational geometry, both for demonstrating algorithms, and for accompanying programming assignments. Finally, many possibilities exist in making an electronic book out of GASP-II.

## References

[1] J.E. Baker, I.F. Cruz, G. Liotta and R.Tamassia. Algorithm Animation over the World Wide Web. *Proc. Int. Workshop on Advanced Visual Interfaces (AVI '96)*, pages 203-212, 1996.

[2] M.H. Brown. *Algorithm Animation*. MIT Press, 1988.

[3] M.H. Brown. Zeus: A system for algorithm animation and multi-view editing. *Computer Graphics*, 18(3):177–186, May 1992.

[4] M.H. Brown and M.A. Najork. Collaborative Active Textbooks: A Web-Based Algorithm Animation System for an Electronic Classroom. *SRC Research Report 142*, May 1996.

[5] M.H. Brown and R. Sedgewick. Techniques for algorithm animation. *IEEE Software*, 2(1):28–39, Jan 1985.

[6] P. Epstein, J. Kavanagh, A. Knight, J. May, T. Nguyen, and J.-R. Sack. A workbench for computational geometry. *Algorithmica*, 11(4):404–428, April 1994.

[7] K. Mehlhorn and S. Naher. Leda: a library of efficient data types and algorithms. *Technical Report A 04/89, FB10*, Universitat des Saarlandes, Saarbruken, 1989.

[8] P. Schorn. *Robust Algorithms in a Program Library for Geometric Computation*. PhD thesis, Informatik-dissertationen eth zurich, 1992.

[9] J. Stasko. Tango: A framework and system for algorithm animation. *IEEE Computer*, September 1990.

[10] A. Tal and D.P. Dobkin. Visualization of Geometric Algorithms. *IEEE Transactions on Visualization and Computer Graphics*, 1:194–204, 1995.

# Practical Techniques for Constructing Binary Space Partitions for Orthogonal Rectangles

Pankaj K. Agarwal* T. M. Murali† Jeffrey Scott Vitter‡
Center for Geometric Computing
Department of Computer Science, Duke University
Box 90129, Durham, NC 27708-0129
Email: {pankaj,tmax,jsv}@cs.duke.edu

## Abstract

We present the first systematic comparison of the performance of algorithms that construct Binary Space Partitions for orthogonal rectangles in $\mathbb{R}^3$. We compare known algorithms with our implementation of a variant of a recent algorithm of Agarwal et al. [1]. We show via an empirical study that their algorithm constructs BSPs of near-linear size in practice and performs better than most of the other algorithms in the literature.

## 1 Introduction

The Binary Space Partition (BSP) [3, 5] is a versatile and popular data structure, with applications in many problems—hidden-surface removal, global illumination, shadow generation, solid geometry, geometric data repair, ray tracing, network design, and surface simplification; see [1] for a detailed list of references.

The efficiency of most BSP-based algorithms depends on the size and/or the depth of the BSP (we formally define the size of a BSP later). Therefore, several algorithms have been developed to construct BSPs of small size and depth; see [1] for a list of references.

Recently, Agarwal et al. [1] developed an algorithm that constructs a BSP for orthogonal rectangles in $\mathbb{R}^3$ when most rectangles have aspect ratio bounded by a constant. We have implemented their algorithm to study its performance on "real" data sets. We have also systematically compared its performance to that of various existing algorithms. We show that the algorithm of Agarwal et al. is indeed practical: it constructs a BSP of near-linear size on real data sets (the size varies between 1.5 and 1.8 times the number of input rectangles). This algorithm performs better than not only Paterson and Yao's algorithm [4] but also most heuristics described in the literature [2, 3, 7].

To compare the different algorithms, we measure the size of the BSP each algorithm constructs and the total number of pieces into which the rectangles are partitioned by the BSP. The size measures the storage needed for the BSP.

*Support was provided by National Science Foundation research grant CCR-93-01259, by Army Research Office MURI grant DAAH04-96-1-0013, by a Sloan fellowship, by a National Science Foundation NYI award and matching funds from Xerox Corp, and by a grant from the U.S.-Israeli Binational Science Foundation.

†This author is affiliated with Brown University. Support was provided in part by National Science Foundation research grant CCR-9522047 and by Army Research Office MURI grant DAAH04-96-1-0013.

‡Support was provided in part by National Science Foundation research grant CCR-9522047, by Army Research Office grant DAAH04-93-G-0076, and by Army Research Office MURI grant DAAH04-96-1-0013.



## 2 Geometric Preliminaries

A *binary space partition* $\mathcal{B}$ for a set $S$ of pairwise-disjoint rectangles in $\mathbb{R}^3$ is a tree defined as follows: Each node $v$ in $\mathcal{B}$ represents a box (rectangular parallelepiped) $\mathcal{R}_v$ and a set of rectangles $S_v = \{s \cap \mathcal{R}_v \mid s \in S\}$ that intersect $\mathcal{R}_v$. The box associated with the root is $\mathbb{R}^3$ itself. If $S_v$ is empty, then node $v$ is a leaf of $\mathcal{B}$. Otherwise, we partition $\mathcal{R}_v$ into two boxes by an orthogonal *cutting plane* $H_v$. At $v$, we store the equation of $H_v$ and $\{s \cap H_v \mid s \in S_v\}$, the subset of rectangles in $S_v$ that lie in $H_v$. If we let $H_v^+$ be the positive halfspace and $H_v^-$ the negative halfspace bounded by $H_v$, the boxes associated with the left and right children of $v$ are $\mathcal{R}_v \cap H_v^-$ and $\mathcal{R}_v \cap H_v^+$, respectively. The left subtree of $v$ is a BSP for the set of rectangles $S_v^- = \{s \cap H_v^- \mid s \in S_v\}$ and the right subtree of $v$ is a BSP for the set of rectangles $S_v^+ = \{s \cap H_v^+ \mid s \in S_v\}$. The size of $\mathcal{B}$ is the sum of the number of interior nodes in $\mathcal{B}$ and the total number of rectangles stored at all the nodes in $\mathcal{B}$.[1]

Given a set of rectangles $R$, let $R_B = \{s \cap B \mid s \in R\}$ be the set of rectangles obtained by clipping the rectangles in $R$ within $B$. We say that a rectangle in $S_B$ is *free* if none of its edges lies in the interior of $B$; otherwise it is *non-free*. A *free cut* is a cutting plane that does not cross any rectangle in $S$ and that either divides $S$ into two non-empty sets or contains a rectangle in $S$.

## 3 The Agarwal et al. Algorithm

In this section, we describe a variant of the algorithm of Agarwal et al. [1] that we have implemented. In our implementation, we have modified their algorithm slightly in order to improve its performance. We call this algorithm Rounds.

A box $B$ in $\mathbb{R}^3$ has six faces—*top, bottom, front, back, right,* and *left.* We assume, without loss of generality, that the back, bottom, left corner of $B$ is the origin (i.e., the back

[1] We do not store the leaves of $\mathcal{B}$ explicitly since all the information about a leaf is captured by its parent and the cutting plane at the parent.

face of $B$ lies on the $yz$-plane). We say that a rectangle $r$ in $S_B$ is *long* with respect to a box $B$ if none of the vertices of $r$ lie in the interior of $B$. Otherwise, $r$ is said to be *short*. We can partition long rectangles into three classes: a rectangle $s$ that is long with respect to $B$ belongs to the *top class* if two parallel edges of $s$ are contained in the top and bottom faces of $B$. We similarly define the *front* and *right* classes. For a set of points $P$, let $P_B$ be the subset of $P$ lying in the interior of $B$.

The algorithm proceeds in rounds. At the beginning of the $i$th round, where $i > 0$, the algorithm has a top subtree $\mathcal{B}_i$ of the BSP for $S$. Let $Q_i$ be the set of boxes associated with the leaves of $\mathcal{B}_i$ containing at least one rectangle. The initial tree $\mathcal{B}_1$ consists of one node and $Q_1$ consists of one box that contains all the input rectangles. If $Q_i$ is empty, we are done. Otherwise, in the $i$th round, for each box $B \in Q_i$, we construct a top subtree $\mathcal{T}_B$ of the BSP for the set $S_B$ and attach it to the corresponding leaf of $\mathcal{B}_i$. This gives us the new top subtree $\mathcal{B}_{i+1}$. Thus, it suffices to describe how to build the tree $\mathcal{T}_B$ on a box $B$ during a round.

Let $F \subseteq S_B$ be the set of rectangles that are long with respect to $B$. Set $f = |F|$, and let $k$ be the number of vertices of rectangles in $S_B$ that lie in the interior of $B$ (note that each such vertex is a vertex of an original rectangle in $S$). As in the Agarwal et al. algorithm, we choose a parameter $a = 2^{\sqrt{\log(f+k)}}$, which remains fixed throughout the round. In a round, we partition $B$ using a sequence of cuts in two stages, the separating stage and the dividing stage. The separating stage divides $B$ into a set of boxes $\mathcal{C}$ such that for each box $C \in \mathcal{C}$, $F_C$ contains only two classes of rectangles. The dividing stage further refines each such box $C$ until a new round is to be started in the resulting boxes. We now describe each stage in detail.

***Separating Stage:*** Assume without loss of generality that the longest edge of $B$ is parallel to the $x$-axis. The rectangles in $F$ that belong to the front class can be partitioned into two subsets: the set $R$ of rectangles that are vertical (and parallel to the right face of $B$) and the set $T$ of rectangles that are horizontal (and parallel to the top face of $B$). Let $e$ be the edge of $B$ that lies on the $z$-axis and $e'$ be the edge of $B$ that lies on the $y$-axis. The intersection of each rectangle in $R$ with the back face of $B$ is a segment parallel to the $z$-axis. Let $\bar{r}$ denote the projection of this segment onto the $z$-axis, and let $\bar{R} = \{\bar{r} \mid r \in R\}$. Let $z_1 < z_2 < \cdots < z_{k-1}$ be the endpoints of intervals in $\bar{R}$ that lie in the interior of $e$ but not in the interior of any interval of $\bar{R}$. Similarly, for each rectangle $t$ in the set $T$, we define $\bar{t}$ to be the projection of $t$ onto the $y$-axis, and $\bar{T} = \{\bar{t} \mid t \in T\}$. Let $y_1 < y_2 < \cdots < y_{l-1}$ be the endpoints of intervals in $\bar{T}$ that lie in the interior of $e'$ but not in the interior of any interval of $\bar{T}$.

We divide $B$ into $kl$ boxes by drawing the planes $z = z_i$ for $1 \le i < k$ and the planes $y = y_j$ for $1 \le j < l$. We refer to these cuts as *$\alpha$-cuts*. Let $\mathcal{C}$ be the set of boxes into which $B$ is partitioned in this manner.

***Dividing Stage:*** We refine each box $C$ in $\mathcal{C}$ by applying cuts as described below. Let $V_C$ be the set of vertices of rectangles in $S_C$ that lie in the interior of $C$. We recursively invoke the following steps until $|F_C| + 2a|V_C| < (f + ak)/a$ and $S_C$ does not contain any free rectangles.

1. If $C$ has any free rectangle, we use the free cut containing that rectangle to split $C$ into two boxes.
2. If the rectangles in $F_C$ belong to two classes, we make
   (i) either one cut that partitions $C$ into two boxes $C_1$ and $C_2$ so that for $i = 1, 2$ $|R_{C_i}| + w|V_{C_i}| \le 2(|R_C| + w|V_C|)/3$, or
   (ii) at most two parallel cuts that divide $C$ into three boxes $C_1, C_2$, and $C_3$ such that $|R_{C_2}| + w|V_{C_2}| \ge (|R_C| + w|V_C|)/3$ and such that all rectangles in $R_{C_2}$ belong to the same class (these two cuts are unique).
3. If $F_C$ has only one class of rectangles, let $g$ be the face of $C$ that contains exactly one of the edges of each rectangle in $R_C$. Let $P$ be the set of those vertices of the rectangles in $R_C$ that lie in the interior of $g$. We use a plane that is orthogonal to $g$ to partition $C$ into two boxes $C_1$ and $C_2$ so that $|P \cap C_i| + w|V_{C_i}| \le 2(|P| + w|V_C|)/3$, for $i = 1, 2$.

In Steps 2i and 3, if there are many planes that satisfy the conditions on the cuts, we use the plane that intersects the smallest number of rectangles in $S_C$.

## 4 Other Algorithms

In this section we discuss our implementation of some algorithms available in the literature for constructing BSPs. All the algorithms work on the same basic principle: examine all the planes containing the rectangles in $S_B$ and determine how "good" each plane is. Split $B$ using the "best" plane and recurse. Our implementation refines the original descriptions of these algorithms in two respects: (i) At a node $B$, we first check whether $S_B$ contains a free rectangle; if it does, we apply the free cut containing that rectangle.[2] (ii) If there is more than one "best" plane, we choose one of these planes based on some simple criteria. To complete the description of each technique, it suffices to describe how it measures how "good" a candidate plane is.

For a plane $\pi$, let $f_\pi$ denote the number of rectangles in $S_B$ intersected by $\pi$, $f_\pi^+$ the number of rectangles in $S_B$ completely lying in the positive halfspace defined by $\pi$, and $f_\pi^-$ the number of rectangles in $S_B$ lying completely in the negative halfspace defined by $\pi$. We also define the *occlusion factor* $\alpha_\pi$ to be the ratio of the total area of the rectangles in $S_B$ lying in $\pi$ to the area of $\pi$ (when $\pi$ is clipped within $B$), the *balance* $\beta_\pi$ to be the ratio $\min\{f_\pi^+, f_\pi^-\}/\max\{f_\pi^+, f_\pi^-\}$ between the number of polygons that lie completely in each halfspace defined by $\pi$, and $\sigma_\pi$ to be the *split factor* of $\pi$, which is the fraction of rectangles that $\pi$ intersects, i.e., $\sigma_\pi = f_\pi/|S_B|$. We now discuss how each algorithm measures how good a plane is.

ThibaultNaylor: We discuss two of the three heuristics that Thibault and Naylor [7] present (the third performed poorly in our experiments). Below, $w$ is a positive weight that can be changed to tune the performance of the heuristics.

1. Pick a plane the minimizes the function $|f_\pi^+ - f_\pi^-| + w f_\pi$.
2. Maximize the measure $f_\pi^+ \cdot f_\pi^- - w f_\pi$.

[2] Only Paterson and Yao's algorithm [4] originally incorporated the notion of free cuts.

In our experiments, we use $w = 8$, as suggested by Thibault and Naylor [7].

Airey: Airey [2] proposes a measure function that is a linear combination of a plane's occlusion factor, its balance, and its split factor: $0.5\alpha_\pi + 0.3\beta_\pi + 0.2\sigma_\pi$.

Teller: Let $0 \leq \tau \leq 1$ be a real number. Teller [6] chooses the plane with the maximum occlusion factor $\alpha_\pi$, provided $\alpha_\pi \geq \tau$. If there is no such plane, he chooses the plane with the minimum value of $f_\pi$. We use the value $\tau = 0.5$ in our implementation, as suggested by Teller.

PatersonYao: For a box $B$, let $s_x$ (resp., $s_y, s_z$) denote the number of edges of the rectangles in $S_B$ that lie in the interior of $B$ and are parallel to the $x$-axis (resp., $y$-axis, $z$-axis). We define the measure of $B$ to be $\mu(B) = s_x s_y s_z$. We make a cut that is perpendicular to the smallest family of edges and divides $B$ into two boxes, each with measure at most $\mu(B)/4$.

## 5 Experimental Results

We have implemented the above algorithms and run them on the following data sets containing orthogonal rectangles:[3]

1. the `Fifth` floor of Soda Hall containing 1677 rectangles,
2. the `Entire` Soda Hall model with 8690 rectangles,
3. the Orange United Methodist `Church` Fellowship Hall with 29988 rectangles,
4. the Sitterson Hall `Lobby` with 12207 rectangles, and
5. `Sitterson` Hall containing 6002 rectangles.

We performed our experiments on a Sun SPARCstation 5 running SunOS 5.5.1 The table below displays the size of the BSP and the total number of times the rectangles are fragmented by the cuts made by the BSP.

| | Fifth | Entire | Church | Lobby | Sitt. |
|---|---|---|---|---|---|
| #rectangles | 1677 | 8690 | 29988 | 12207 | 6002 |
| Size of the BSP | | | | | |
| Rounds | 2744 | 14707 | 45427 | 22225 | 9060 |
| Teller | 2931 | 14950 | 33518 | 13911 | 7340 |
| PatersonYao | 3310 | 22468 | 56868 | 30712 | 20600 |
| Airey | 3585 | 24683 | 41270 | 21753 | 19841 |
| ThibaultNaylor1 | 6092 | 32929 | 65313 | 25051 | 10836 |
| ThibaultNaylor2 | 3235 | 20089 | 58175 | 23159 | 12192 |
| Number of Fragments | | | | | |
| Rounds | 113 | 741 | 838 | 475 | 312 |
| Teller | 301 | 1458 | 873 | 514 | 153 |
| PatersonYao | 449 | 5545 | 12517 | 9642 | 6428 |
| Airey | 675 | 7001 | 5494 | 5350 | 8307 |
| ThibaultNaylor1 | 1868 | 10580 | 13797 | 3441 | 1324 |
| ThibaultNaylor2 | 262 | 2859 | 6905 | 1760 | 1601 |

Examining this table, we note that, in general, the number of fragments and size of the BSP scale well with the size of the data set. For the Soda Hall data sets (`Fifth` and `Entire`), algorithm Rounds creates the smallest number of fragments and constructs the smallest BSP. For the other three sets, algorithm Teller performs best in terms of BSP size. However, there are some peculiarities in the table. For example, for the `Church` data set, Rounds creates a smaller number of fragments than Teller but constructs a larger BSP. We believe that this difference is explained by the fact that the 29998 rectangles in the `Church` model lie in a total of only 859 distinct planes. Since Teller makes cuts based on how much of a plane's area is covered by rectangles, it is reasonable to expect that the algorithm will "place" a lot of rectangles in cuts made close to the root of the BSP, thus leading to a BSP with a small number of nodes.

The time taken to construct the BSPs also scaled well with the size of the data sets. Rounds took 11 seconds to construct a BSP for the `Fifth` floor of Soda Hall and about 4.5 minutes for the `Church` data set. Typically, PatersonYao took about 15% less time than Rounds while the heuristics (Airey, ThibaultNaylor, and Teller) took 2-4 times as much time as Rounds to construct a BSP.

[3]We discarded all non-orthogonal polygons from these data sets. The number of such polygons was very small.

## 6 Conclusions

Our comparison indicates that Rounds and Teller are the best algorithms for constructing BSPs for orthogonal rectangles in $\mathbb{R}^3$. We plan to extend our algorithm to construct BSPs for triangles in $\mathbb{R}^3$. One algorithm we intend to implement is to enclose each triangle in an orthogonal bounding box and construct a BSP for the bounding boxes. We also plan to compare the different BSPs in terms of the time they take to answer standard visibility queries like ray shooting and line stabbing.

**Acknowledgments** We would like to thank Seth Teller for providing us with the Soda Hall data set created at the Department of Computer Science, University of California at Berkeley. We would also like to thank the Walkthrough Project, Department of Computer Science, University of North Carolina at Chapel Hill for providing us with the data sets for Sitterson Hall, the Orange United Methodist Church Fellowship Hall, and the Sitterson Hall Lobby.

## References

[1] P. K. Agarwal, E. F. Grove, T. M. Murali, and J. S. Vitter, Binary space partitions for fat rectangles, *Proceedings of the 37th IEEE Annual Symposium on foundations of Computer Science (FOCS '96)*, October 1996.

[2] J. M. Airey, *Increasing Update Rates in the Building Walkthrough System with Automatic Model-space Subdivision and Potentially Visible Set Calculations*, Ph.D. Thesis, Dept. of Computer Science, University of North Carolina, Chapel Hill, 1990.

[3] H. Fuchs, Z. M. Kedem, and B. Naylor, On visible surface generation by a priori tree structures, *Comput. Graph.*, 14 (1980), 124–133. Proc. SIGGRAPH '80.

[4] M. S. Paterson and F. F. Yao, Optimal binary space partitions for orthogonal objects, *J. Algorithms*, 13 (1992), 99–113.

[5] R. A. Shumacker, R. Brand, M. Gilliland, and W. Sharp, Study for applying computer-generated images to visual simulation, Report AFHRL-TR-69-14, U.S. Air Force Human Resources Lab., 1969.

[6] S. J. Teller, *Visibility Computations in Densely Occluded Polyhedral Environments*, Ph.D. Thesis, Dept. of Computer Science, University of California, Berkeley, 1992.

[7] W. C. Thibault and B. F. Naylor, Set operations on polyhedra using binary space partitioning trees, *Comput. Graph.*, 21 (1987), 153–162. Proc. SIGGRAPH '87.

# Assembling Objects Through Parts Correlation

Pamela J. Vermeer
Washington and Lee University
Lexington, Virginia 24450 USA
pvermeer@wlu.edu http://www.wlu.edu/~pvermeer

## 1 Introduction

When designing a complex mechanism or structure, it is often easiest to construct the object component by component. The simpler individual components can then be pieced together by correlating geometric elements of pairs of objects. In this paper, we present a solution to the problem of assembling four parts by matching points and planes in pairs of parts. This problem would be simple if the parts could be assembled by fixing one part and adding the others sequentially. In the situation we consider, however, after fixing one part, the remaining three parts must be positioned simultaneously. To do this requires solving an interlocked system of nonlinear equations with up to sixteen different solutions. Since we are interested in obtaining all possible positionings of the four objects which satisfy the given constraints, an analytic approach is taken to solve the resulting system of equations.

We first discuss how our problem arises in the context of geometric constraint solving and we establish the framework of the required matching, that is, how the four base objects are related to each other. This basic framework leads to a suite of problems which must be solved, each corresponding to a different type of matching between the four parts. We next develop a general solution method for finding all the possible positionings of the four components which satisfy a particular matching. We finish with a discussion of an implementation in progress of a program to allow interactive solving of these constraint problems.

## 2 Problem Framework

The problem considered in this paper arises from the graph-based approach to spatial constraint solving. In the analysis phase of that method, geometric entities (or *geoms*) which can be placed relative to one another are formed into groups called *clusters*. Each geom begins with three degrees of freedom, so placing a new geom requires that it be constrained by three other previously fixed geoms. Thus to begin a cluster, there must be at least one set of three pairwise constrained geoms. These three geoms can be positioned in a



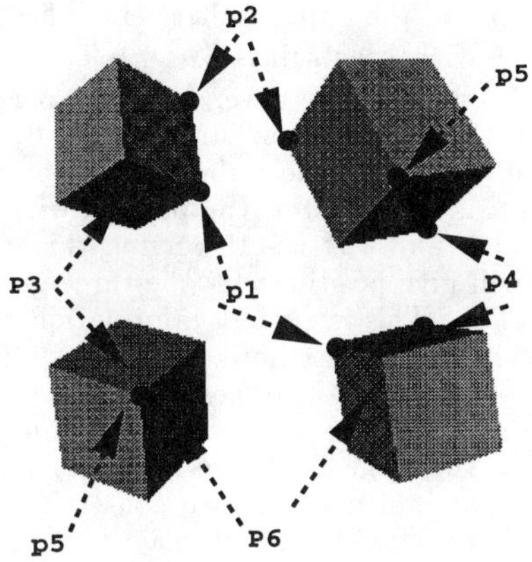


Figure 1: Four boxes to be assembled into a single unit

standard way depending on the types of geoms involved and the constraints between them. A new geom or another cluster can be added to a cluster if there are three constraints between the existing cluster and the new geom or cluster to be added. This process continues until no new clusters can be formed and no additional geoms can be added to any of the existing clusters.

Once the initial clusters have been formed, *cluster merging* takes place. Having multiple constraints may have allowed a geom to be part of more than one cluster. Cluster merging combines clusters which share a geometric entity in such a way that the result is rigid in $\Re^3$, up to translation and rotation as an entire body. Further details about cluster formation and merging in spatial constraint solving can be found in[DH95].

In the problem we are working on, we begin with four objects. In order to assemble these objects into a single object, each object must be matched with every other object. That is, three different geoms of the first object must be chosen such that one of them is the same as some geom on the second object, another is the same as some geom on the third object, and the third is the same as some geom on the fourth object. Then two more geoms on the second object must be chosen so that the first matches a geom on the third object, and the third selected geom matches a geom on the fourth object. Finally, a geom on the third object must be selected to match a geom on the fourth object. With this framework, the problem becomes one of merging four clusters, where each cluster shares exactly one geom with each other cluster. Each cluster corresponds to the geoms

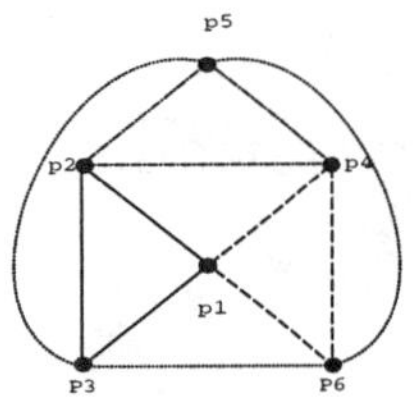


Figure 2: Cluster abstraction of cube assembly problem

selected on one of the objects.

For example, consider Figure 1. Here each unit cube is one of the objects to be assembled into a larger unit. The cube in the upper left can be reduced to the cluster $\{p_1, p_2, P_3\}$ with constraints $d_{12} = 1$, $d_{13} = 0$, and $d_{23} = -1$ (assuming outward normals on the object surfaces). The cube in the upper right is simplified to the cluster $\{p_2, p_4, p_5\}$ with constraints $d_{24} = \sqrt{3}$, $d_{25} = \sqrt{2}$, and $d_{45} = \sqrt{2}$. The cube in the lower left is represented by the cluster $\{P_3, p_5, P_6\}$ with constraints $d_{35} = 0$, $\alpha_{36} = \pi/2$, and $d_{56} = 0$. Finally, the cube in the lower right can be thought of as the cluster $\{p_1, p_4, P_6\}$ with constraints $d_{14} = 1$, $d_{16} = 0$, and $d_{46} = -1$. Solving this cluster merging problem will give positions for the three geometric entities on each cube. This is enough information to be able to rigidly place the cubes together.

The abstraction of the problem to cluster merging can be visualized as in Figure 2. Each of the four triangles in the figure with the same line style for edges represents one cluster. The lines indicate a known constraint between the geoms. The clusters are joined at exactly one geometric element; this indicates a matching between geoms in different clusters. Because of the octahedral nature of this abstract representation, we sometimes refer to the problem of matching objects in this way as *the octahedron problem.* In fact, if the geometries are all points and the constraints all positive distances between the points, the final configuration of the cluster geoms is an octahedron.

If one cluster is fixed in space, the other three clusters cannot be added sequentially because there are only two known constraints between the fixed cluster and any unfixed cluster. This means that to obtain a rigid configuration of the four clusters as a single object, the remaining three clusters must be placed simultaneously. In the rest of the paper, we present a solution to the problem of fixing the first cluster and then positioning the other three clusters with respect to that cluster.

### Notation

Throughout the paper we use the following notation: $g_i$ refers generically to either a point or a plane, while $p_i$ is a point and $P_i$ is a plane. Any constraint between two points or between a point and a plane is a signed distance $d_{ij}$, where the sign of the distance between a point and a plane is positive if the point is on the side of the plane towards which the normal points and is negative otherwise. Any constraint between two planes is an angle $\alpha_{ij}$

## 3 A General Solution

An initial analysis of the problem in its abstract formulation as six geometric elements, each of which is either a point or a plane, comprising four clusters yields ten distinct cases which must be solved. These range from all clusters containing only points to all clusters containing only planes.

The following steps outline the solution process for any of the subcases.

1. Fix the coordinate system using the first cluster. This is done by explicitly positioning the three elements in the cluster. We now consider only the three remaining clusters.
2. For each of the three remaining unfixed elements, establish a parametric representation of the point, or of the normal direction in the case of a plane. This representation is formulated from the interactions between the unfixed element and the fixed element in each cluster in which the unfixed element appears.
3. Use the remaining constraint in each cluster between two unfixed elements to establish three algebraic equations in the three unknown parameters of the representation in step 2.
4. Solve the system of equations.
5. Substitute the parameter values into the representation in step 2 to determine the position of the three remaining elements.

In the following sections, these steps are explained in greater detail. We note that the case of six points is solved in its entirety in Nanua *et. al.* [NWM90]

### 3.1 Fixing the First Cluster

The first cluster is fixed by establishing a coordinate system based on its geometric elements and their relative positions. In every case in which there is at least one plane and one point, we make a cluster with a plane and a point the first cluster. We assume that this point, $p_1$, lies on the plane, $P_3$; this requires that we adjust any distance constraints between the plane $P_3$ and the other points in the system by $-d_{13}$. We then fix the first cluster by making $P_3$ the $xy$-plane, and $p_1$ the origin. The remaining point or plane completes the coordinate system.

In the last case, all six geometric elements are planes, and the only constraints are angles between the planes. This situation is underconstrained, as is the case of five planes and one point, but we can find necessary normal directions for the planes. In this case, we make the origin the intersection point of the three planes in the first cluster, and then proceed as before.

### 3.2 Determining the Parametric Representation

The first step of the placement process completely determines the positions of the three geometric elements $g_1$, $g_2$, and $g_3$ of the first cluster. The three remaining clusters have one fixed element (the element which is also in the first cluster), and two unfixed elements. Each unfixed element is in two of the remaining clusters, so has a relationship with two fixed elements. We can use the constraints between the unfixed element and these two fixed elements to establish a parametric representation of the unfixed element. For a point, the parameterization is the set of possible positions for the point as determined from the two fixed geometries. For a plane, the parameterization is the set of the possible directions of the normal vector. As an example, consider how the paramateric representation of an unfixed point relative to a fixed point and plane is determined.

Let $p_1$ and $P_3$ be the fixed entities, and $p_4$ be the unfixed point. The distance constraint between $p_4$ and $P_3$ means that $p_4$ must lie on a plane parallel to $P_3$, and the distance constraint between $p_4$ and $p_1$ constrains $p_4$ to lie on a sphere. The intersection of these two surfaces is a circle of form

$$(A + B\cos u,\ C\sin u,\ D)$$

When all the possible combinations of points and planes in the fixed and unfixed roles are explored, we find that the interactions lead to a parametric representation of the unfixed geometry which is one of three forms:

$$(A + B\cos u,\ C\sin u,\ D) \tag{1}$$
$$(A + B\sin u,\ C\cos u,\ D + E\sin u) \tag{2}$$
$$(A,\ u,\ B) \tag{3}$$

### 3.3 Forming the Equations

In the previous step, a parametric representation was determined for each unfixed point or plane normal. We now use these representations and the constraint between the two unfixed geometries in each of the three unfixed clusters to set up three equations in the three parameters. There are three different equations, depending on whether the two unfixed geometries are both points, both planes, or one point and one plane.

The first equation comes from the distance between two points. If the two points are $p_i : (x_i, y_i, z_i)$ and $p_j : (x_j, y_j, z_j)$, then the equation is

$$(p_i - p_j) \cdot (p_i - p_j) = {d_{ij}}^2 \tag{4}$$

The second equation comes from the angle between two planes. If the plane unit normals are $n_i : (x_i, y_i, z_i)$ and $n_j : (x_j, y_j, z_j)$, then the equation is

$$n_i \cdot n_j = \alpha_{ij} \tag{5}$$

The final equation comes from the distance between a point and a plane. If the point is $p_i : (x_i, y_i, z_i)$ and the plane has normal $n_j : (x_j, y_j, z_j)$ and distance to the origin $l_j$, then the equation is

$$n_j \cdot p_i = d_{ij} + l_j \tag{6}$$

To set up the actual equations to solve, we choose $u$ as the parameter for the unfixed geometry $g_4$, $v$ for $g_5$, and $w$ for $g_6$. We then compute the coefficients of each parametric representation, and substitute the form into the appropriate equation above. Next, we make the standard rational parametric substitutions for the trigonometric expressions, and simplify. The result in all ten cases is that each equation has total degree four, and the degree in each of the two variables is two. Each equation has mixed terms, and so it is not possible to solve the three equations by simple substitution.

### 3.4 Solving the Equations

Because we desire all possible solutions to the assembly problem, the system of equations is solved symbolically as far as possible. When we perform elimination on the equations derived in the previous step, we obtain a polynomial of degree 16 in the remaining variable. We must rely on numerical methods to find solutions to this polynomial For each solution of the polynomial, there is at most one solution for the remaining two variables which satisfy all three equations. Thus there are at most 16 different positions in real space which satisfy the given constraints.

In the final step, having obtained values for each of the parameters $u$, $v$, and $w$, we substitute into the original parametric forms to obtain the positions of the points and the directions of the normals. In addition, except in the final two cases, we are able to determine the distance of each plane from the origin, since we either have the distance directly as a constraint or we know the distance from some fixed point to the plane and can compute the distance from the origin using that information.

## 4 Implementation

We have implemented a library which solves all ten cases of this problem, making distance assumptions in the last two cases to make them well-constrained. This library uses Maple to carry out the necessary numerical solving processes. We have also implemented a graphical user interface front-end to the assembler library using Open Inventor. This interface displays four objects and allows the user to interactively select a matching between points and planes of the four objects which correspond to well-constrained matching problem described in the paper. When a valid matching is selected, the program extracts the positions of the selected points and planes, and computes the distances and angles to input to the library.

The back-end of the interface uses the new positions of the geometries generated by the solver to compute transformations which take the geometries from their original positions to their new positions. The corresponding transformation is applied to each object and the assembled object will be displayed. A major advantage to our method over numerical solving of the equation system is seen by browsing through all the returned solutions: A numerical solution returns only one solution, which may be a solution in which the four objects intersect each other. Since our method returns all solutions, if there is a solution in which the objects do not intersect, it will be found and displayed.

## 5 Acknowledgements

Thanks to John Hills and Stephen Williard who developed and programmed the graphical user interface to the assembler library, and to Jean-Guy Speton who assisted in programming the assembler library. This work was funded in part by NSF grant number CCR-9505745, by the Washington and Lee University Glenn Grant program, and by the Washington and Lee University R. E. Lee Research Scholars program.

## References

[DH95] D.-Z. Du and F. Hwang, editors. *Computing in Euclidean Geometry*, pages 266–298. World Scientific Publishing Co Pte Ltd, 2 edition, 1995.

[HV95] Christoph M. Hoffmann and Pamela J. Vermeer. A spatial constraint problem. In *Computational Kinematics Workshop*, Nice, France, Sept. 1995.

[NWM90] P. Nanua, K. J. Waldron, and V. Murthy. Direct kinematic solution of a stewart platform. *IEEE Transactions on Robotics and Automation*, 6(4):438–444, Aug. 1990.

Full paper: www.wlu.edu/pvermeer/papers/SoCG.html

# A Practical Evaluation of Kinetic Data Structures

Julien Basch    Leonidas J. Guibas*    Craig D. Silverstein†    Li Zhang

Computer Science Department
Stanford University
Stanford, CA94305
{jbasch,guibas,csilvers,lizhang}@cs.stanford.edu

## 1 Introduction

In many applications of computational geometry to modeling objects and processes in the physical world, the participating objects are in a state of continuous change. Motion is the most ubiquitous kind of continuous transformation but others, such as shape deformation, are also possible. In a recent paper, Basch, Guibas, and Hershberger [BGH97] proposed the framework of *kinetic data structures* (KDSs) as a way to maintain, in a completely on-line fashion, desirable information about the state of a geometric system in continuous motion or change. They gave examples of kinetic data structures for the maximum of a set of (changing) numbers, and for the convex hull and closest pair of a set of (moving) points in the plane. The KDS framework allows each object to change its motion at will according to interactions with other moving objects, the environment, etc.

We implemented the KDSs described in [BGH97], as well as some alternative methods serving the same purpose, as a way to validate the kinetic data structures framework in practice. In this note, we report some preliminary results on the maintenance of the convex hull, describe the experimental setup, compare three alternative methods, discuss the value of the measures of quality for KDSs proposed by [BGH97], and highlight some important numerical issues.

## 2 Kinetic Data Structures

Given a set of continuously moving points in the plane, we wish to maintain some *configuration function* of interest, dependent on the positions of the points. This function could be, for example, the minimum distance between all pairs of the points. Even though this distance changes continuously, it can be derived from a combinatorial function $F$ — describing the identity of the pair of points that realize this distance — that changes at discrete times only. A kinetic data structure maintains the combinatorial function $F$ through a list of *certificates* whose validity at a particular instant in time implies the correctness of $F$ at that instant. These certificates are typically low degree algebraic inequalities involving a few of the moving points, such as "point $p$ is to the left of point $q$," or "the distance $d(p,q)$ is greater than the distance $d(p,r)$."

In the KDS framework each point is assumed to follow a particular *flight plan*. Typically we will assume that these motions are piecewise algebraic of low degree. Under this constraint, for a given certificate $c$, we can calculate the next future time $t(c)$ at which $c$ becomes false by solving for "the next larger" root of an algebraic equation,. All these "certificate failure times," or *events* for short, are kept in an event queue associated with the combinatorial function $F$. Since $F$ cannot change between events, we can perform a dynamic simulation of the system by jumping from event to event. Processing an event involves recomputing a new certificate list — since at least one certificate is now false — that re-validates (the same or a modified) $F$ at the current time, and appropriately updating the event queue. If $F$ changes at the time of an event, the event is called *external*. Otherwise the event is called *internal*. For a KDS to be of good quality, the following criteria should be met:

- the certificate list does not change too much when an event occurs (*responsiveness*);
- the overhead of internal events with respect to external events is reasonable (*efficiency*);
- the KDS itself is of small size, typically linear or slightly superlinear (*compactness*); and
- each point is involved in only a small number of certificates (*locality*).

Precise ways to make these criteria quantitative are discussed in [BGH97]. To elaborate on the second condition, we say a KDS is *efficient* if, over a given class of allowed motions, the ratio of the worst case number of total (internal plus external) events to the worst case number of external events is small (as a function of the number of points involved). Note that in this definition of efficiency, as originally proposed in [BGH97], we are comparing two worst-case scenarios, even though these may arise for different configurations of moving points.

The *locality* condition is important when points change their motion law, something we call a *flight plan update*.

*Supported in part by National Science Foundation grant CCR–9623851 and by US Army MURI grant DAAH04–96–1–0007.

†Supported by the Department of Defense, with partial support from NSF Award CCR–9357849, with matching funds from IBM, Mitsubishi, Schlumberger Foundation, Shell Foundation, and Xerox Corporation.



*Computational Geometry 97* Nice France
Copyright 1997 ACM 0-89791-878-9/97/06 ..$3.50

This is common in typical simulations, where for instance points may bounce off walls or off each other. When a point changes motion, the failure times of all certificates dependent on that point must be updated. The locality condition ensures that this operation can be performed within reasonable time bounds.

## 3 Three ways to maintain a convex hull

Along with the method proposed in [BGH97], we implemented two other methods for comparison: the Delaunay triangulation and a less sophisticated method, dubbed "brute force." All these methods share the traits that they are event driven and exact.

BGH97-CH : The principle of the kinetic data structure proposed in [BGH97] for the maintenance of the convex hull is the following: the set of points is divided arbitrarily into a red half and a blue half, and the convex hull of each half is computed and maintained recursively. The convex hull of the whole is then certified by the slope ordering between all red and blue convex hull edges and by the orientation of a set of bichromatic triangles. If the motion is algebraic of degree $k$, it is possible to compute the time at which two edges become parallel, and the time at which a triangle becomes degenerate, by solving in each case an equation of degree $2k$. In [BGH97], it is shown that, with some small adaptation, this kinetic data structure is both local and efficient.

DELAUNAY : An alternate way to maintain the convex hull of moving points is to maintain their Delaunay triangulation, whose edges form a superset of the convex hull. This configuration is especially straightforward to kinetize, as its set of edges, together with "InCircle" tests that certify "local Delaunayhood," is a correct certificate structure [GMR91]. As a way to maintain the convex hull, this KDS is not local (a given point can have an arbitrarily high degree in the underlying graph) and is not known to be efficient (the best known upper bounds for the number of changes to the DT when points move along algebraic trajectories is roughly cubic). If the motion of each point can be described by algebraic pieces of degree $k$, the event times for the Delaunay certificates will be roots of polynomials of degree $4k$.

BRUTE-CH : There is a simple, brute force data structure for maintaining a convex hull. We simply calculate the time at which each point will hit (or leave) the convex hull, assuming its current motion remains unchanged. When a point on the hull has a flight plan update, we must recalculate the event times of all the other points. (When a non-hull point updates its flight plan, we need only recalculate its own event time.) Furthermore, whenever a point enters or leaves the convex hull, all events must be rescheduled. Hence, as one flight plan update causes the rescheduling of up to $\Omega(n)$ events, this structure is not local.

## 4 Numerical issues

The major problem in implementing robust kinetic data structures is the approximate numerical computation of roots of polynomials. As the same issues arise for any implementation of a line sweep algorithm on curves (for which we are not aware of any general implementation), the solution we propose is of independent interest.

To focus on the specific problems posed by a KDS, we make the assumption that the value of a polynomial at a given point (time) can be computed exactly, but the roots of a polynomial are computed with a small error. In this setting there are two problems we must address: (1) events can get out of order and (2) an event that really happens in the future can appear to have already happened, or vice versa. We discuss the second problem, which occurs frequently; we never observed an occurrence of the first problem.

To be concrete, assume that at time $t = t_0$ we are processing the event "$(abcd)$ are cocircular". Once we have processed this event, we need to schedule another event for the same four points, because they may become cocircular again. The associated polynomial $P(t)$ has a root at $t = t_0$. Because of numerical error, however, this root might appear to be at some $t_\epsilon > t_0$, and the event that was just processed will be improperly rescheduled. To avoid this problem, we discard the root closest to $t_0$ before finding the next event time.

Using this method, we were able to run reliable simulations involving 50,000 points, or about 400,000 events for the BGH97-CH KDS.

## 5 Experimental setup

In order to remain independent of specific implementation details, the cost of a KDS was taken to be a weighted sum of the number of polynomial equations solved. This statistic is motivated by the fact that most of the time spent by the data structures we consider (over 80 percent, for DELAUNAY ) is spent solving equations used in scheduling events. We weigh each equation by the time it takes to solve: degree 1 equations (which are dominated by division) take 1 time unit, degree 2 equations (which are dominated by square root) take 4 time units, and degree 4 equations (which require either many square and cube roots, or an iterative algorithm) take 80 time units. These ratios were obtained empirically by repeatedly solving several random equations of each type (on a 90 MHz. Pentium; a 167 Mhz. UltraSparc-I; and an 8 processor, 250 MHz. MIPS R4400).

We ran the three methods described in section 3 with $n$ ranging from 10 to 10000 points with initial positions and speeds chosen independently at random in the unit square.[1] The first set of experiments was intended to measure the efficiency on average (as opposed to worst case) of the different kinetic data structures. For this purpose, we let the simulation run until the convex hull stabilized and there was no more events in the event queue (Figure 1). The second set of tests was designed to judge of the importance of the locality requirement for a KDS: we introduced walls against which the points bounced, and let the simulation run until $n$ bounces had occurred (Figure 2). Since each bounce requires rescheduling every event associated with the bouncing point, non-local structures are expected to perform poorly. For each test, the result was averaged over five runs.

All tests were conducted on an SGI using an 8 processor MIPS R4400 at 250MHz., running IRIX 6.2. The code was written in C++ and compiled with the SGI `CC` compiler with the `-O2 -mips2` optimization options.

## 6 Results

This study allowed us a close examination of the practical value of the quality measures for KDSs proposed by [BGH97] and highlighted some important issues that did not arise in

[1] Alternative distributions, such as the uniform unit disk and the Gaussian, gave qualitatively similar results.

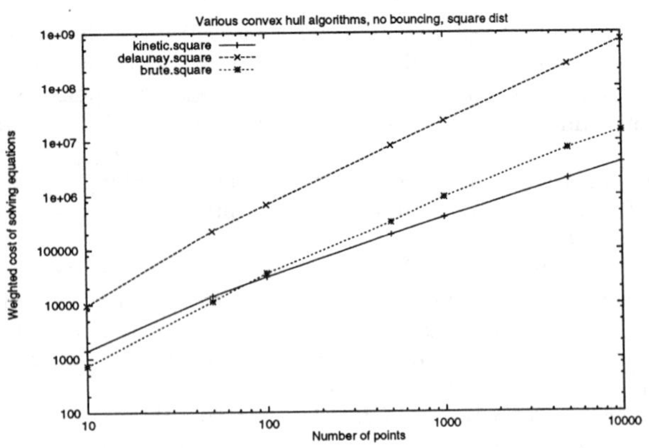


Table 1: Cost (weighted sum of the number of equations solved) of the three different methods that maintain the CONVEX HULL, when the points move along an unchanging, linear flight plan.

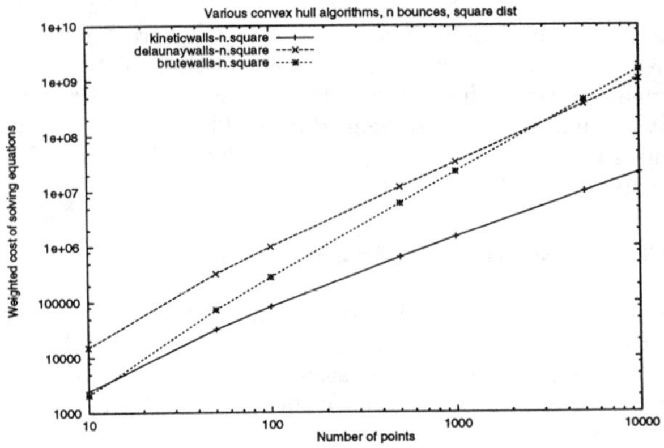


Table 2: Cost (weighted sum of the number of equations solved) of the three methods for maintaining the convex hull, when the points bounce inside a square. The simulation was run until there were $n$ bounces ($n$ is the number of points).

the theoretical study. The first important conclusion, however, is that the implementation of kinetic data structures did not pose any major problem and the framework is readily practical. It provides a solid paradigm for dynamic simulations that avoids approximation due to arbitrary time discretization.

On the side of efficiency, the number of equations solved can be seen to be roughly linear for BGH97-CH , roughly $\Theta(n^{3/2})$ for both DELAUNAY and BRUTE-CH . The worst case is therefore far from being attained, but, as the number of external events itself is extremely low (about $\log^2 n$), the efficiency ratio is in all cases rather high. It therefore makes more sense to compare the relative performance of the different methods. In this sense, BGH97-CH is vastly superior to Delaunay but only marginally better than BRUTE-CH for the numbers considered (Figure 1).

Simulations with bouncing were performed to test the value of the locality requirement. We mentioned earlier that DELAUNAY was not a local KDS in the worst case. However, in this context, as the expected degree of a vertex is constant, the performance of DELAUNAY doesn't degrade with bouncing. On the other hand, BRUTE-CH , which is non-local at all points, has a quadratic growth due to the intense rescheduling necessary upon each bouncing. Finally, BGH97-CH handles bouncing gracefully as expected (Figure 2).

There is one particular aspect of a KDS that is not taken into account in the framework of [BGH97]: it is the cost of solving a single equation. The huge cost of solving degree 4 equations for DELAUNAY is the only reason why this method performs so poorly for $n$ up to a hundred. In this range, the number of equations solved is comparable with that of BGH97-CH . This suggests that it would be worthwhile to maintain some arbitrary triangulation instead of the Delaunay. An arbitrary triangulation can be maintained using only collinearity tests and might well perform better than BGH97-CH for small $n$.

## 7 Conclusion

In this paper, we reported the implementation of and experimentation with several data structures for maintaining the convex hull of points continuously moving in the plane. Our primary finding is that kinetic data structures do not pose major implementation problems and perform well on several natural distributions. For a large number of points, alternative data structures suffer from non-locality or the expensive root-finding operations they perform. However, all methods stand to gain from more careful implementation. Particular advantage can be gained by using simpler data structures when the number of points is small.

Several other kinetic data structures can be imagined for maintaining the convex hull. As mentioned above, it is enough to maintain any triangulation, not just the Delaunay one. These alternative data structures can be evaluated both in the theoretical framework of [BGH97], in a probabilistic setting [BDIZ], and in the experimental framework of this paper. We will report in a forthcoming paper on experiments with kinetic data structures for maintaining other configuration functions, such as the closest pair and the maximum.

A program demonstrating kinetic data structures is available at `http://graphics.stanford.edu/~jbasch/demokin`.

## Acknowledgments

We wish to thank Aris Gionis and Piotr Indyk for useful discussions.

## References

[BDIZ] J. Basch, H. Devarajan, P. Indyk, and L. Zhang. Probabilistic analysis for combinatorial functions of moving points. This volume.

[BGH97] J. Basch, L. J. Guibas, and J. Hershberger. Data structures for mobile data. In *8th Symposium on Discrete Algorithms*, pages 747–756, 1997.

[GMR91] L. Guibas, J. S. B. Mitchell, and T. Roos. Voronoi diagrams of moving points in the plane. In *Proc. 17th Internat. Workshop Graph-Theoret. Concepts Comput. Sci.*, volume 570 of *Lecture Notes in Computer Science*, pages 113–125. Springer-Verlag, 1991.

# Guaranteed-Quality Delaunay Meshing in 3D (Short Version)

L. Paul Chew *
Department of Computer Science
Cornell University
Ithaca, NY 14853

## 1 Introduction

The main contribution of this paper is a new mesh generation technique for producing 3D tetrahedral meshes. Like many existing techniques, this one is based on the Delaunay triangulation (DT). Unlike existing techniques, this is the first Delaunay-based method that is mathematically guaranteed to avoid *slivers*. A *sliver* is a tetrahedral mesh-element that is almost completely flat. For example, imagine the tetrahedron created as the (3D) convex hull of the four corners of a square; this tetrahedron has nicely shaped faces — all faces are 45 degree right-triangles — but the tetrahedron has zero volume. Slivers in the mesh generally lead to poor numerical accuracy in a finite element analysis.

The Delaunay triangulation (DT) has been widely used for mesh generation. In 2D, the DT maximizes the minimum angle for a given point set; thus, small angles are avoided. There is no analogous property involving angles in 3D. We make use of the *Empty Circle Property* for the DT of a set of point sites: the circumcircle of each triangle is empty of all other sites. In 3D, the analogous property holds: the circumsphere of each tetrahedron is empty of all other sites. The Empty Circle Property can be used as the definition of the DT.

There is a vast literature on mesh generation with most of the material emanating from the various applications communities. We refer the reader to the excellent survey by Bern and Eppstein [BE92]. We consider here only work related to the topic of mesh generation with mathematical quality guarantees. Chew [Che89] showed how to use the DT to triangulate any 2D region with smooth boundaries and no sharp corners to attain a mesh of uniform density in which all angles are greater than 30 degrees. An optimality theorem for meshes of nonuniform density was developed by Bern, Eppstein and Gilbert [BEG94] using a quadtree-based approach. Ruppert [Ru93] later showed that a modification of Chew's algorithm could also attain the same strong results as Bern, Eppstein and Gilbert. More recently, Chew [Che93] extended his own work to non-uniform density meshes for 2D curved surfaces embedded in 3D. Mitchell and Vavasis [MV92] used octrees to extend the Bern-Eppstein-Gilbert ideas to 3D.

*This work was supported by DARPA under contract N00014-96-1-0699 and by the Cornell Theory Center which receives funding from its its Corporate Research Institute, NSF, New York State, DARPA, NIH, and IBM Corporation.


*Computational Geometry 97* Nice France


There are several reasons why the Delaunay-based technique presented here might be preferred over the octree technique. For instance, the Delaunay-based technique produces a known bound on the aspect ratio of the tetrahedra while the mathematical theory of Mitchell and Vavasis proves the existence of such a bound for their octree-based technique without actually determining its numerical value. In addition, octree-based algorithms tend to produce meshes with some features that are undesirable in practice: the worst elements are along the boundary of the domain — typically, this is where the user would like to have the highest quality elements — and the mesh has a clear orientation in the sense that one can usually determine the orientation of the octree axes by examining the resulting mesh.

Due to space limitations, most of the proofs in this paper have been either condensed or eliminated. A more complete version of this paper is available as a tech report.

## 2 A Simple 2D Technique

We start with a simple 2D meshing technique to illustrate some of the ideas and to motivate the development of the 3D meshing algorithm. This technique is similar to that used in [Che89, Che93], although for ease of presentation, the version used here makes several simplifying assumptions about the region to be meshed. These simplifying assumptions are not strictly necessary and will be dropped in later sections. We assume that

1. the region to be meshed is convex,
2. the goal is to create a mesh with constant density, and
3. the boundary of the region has already been subdivided in such a way that all boundary edges have lengths between 1 and $\sqrt{3}$ units and no two vertices are closer than 1 unit.

The size of the *unit* here controls the size of the elements in the resulting mesh. The first assumption can easily be dropped, but by restricting ourselves to convex regions we avoid having to discuss Delaunay edges that occur outside the region.

The assumptions about the boundary subdivision imply something about the shape of the region to be meshed.

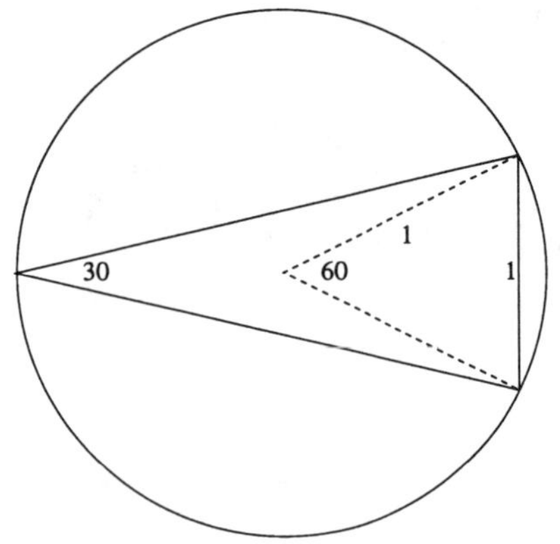


Figure 1: The minimum angle is 30 degrees.

In particular, such a subdivision cannot be obtained if the region-to-be-meshed has a boundary angle of less than 30 degrees. In practice, boundary angles less than 30 degrees can be removed from the problem region and triangulated in a postprocessing step; this type of "lopping off" technique is used in [BEG94, Ru93].

```
Algorithm A:
Form the DT of the boundary points
while there is a triangle with circumradius > 1 do
    Add the triangle's circumcenter to
     the set of sites and update the DT
```

**Theorem 1** *Algorithm A halts, producing a mesh in which all triangles have angles between 30 and 120 degrees.*

*Proof*: First, note that no two vertices are ever closer than 1 unit; this holds initially and continues to hold throughout the algorithm. Second, note that Algorithm A must halt: the region has to fill up after some finite time since the vertices are all 1 unit apart. When the algorithm halts, each triangle has a circumcircle with radius less than 1 and each edge has length $\geq 1$. It's easy to see that the smallest angle occurs when the shortest possible edge is used in the largest possible circumcircle (see Figure 1). In this case, the central angle is easily seen to be 60 degrees and the corresponding angle of the triangle is 30 degrees. The 30 degree bound for the minimum angle immediately implies the 120 degree bound for the maximum angle. □

The 2D algorithm, Algorithm A, can be generalized in a fairly straightforward way to produce a 3D algorithm: Algorithm B.

**Theorem 2** *Algorithm B produces a mesh in which all triangular faces of tetrahedra have angles between 30 and 120 degrees.*

*Proof*: The proof requires some some fairly strong assumptions about the boundary and is fundamentally similar to the proof for Algorithm A. □

Unfortunately, knowing that a tetrahedron has nicely-shaped faces is not enough to know that it's not a sliver. The tetrahedron created from the (3D) convex hull of a square has faces that are very nicely-shaped, but the tetrahedron is the worst possible kind of sliver, with no volume at all.

## 3 Randomness and Approximate Circumcenters

Let us reconsider the 2D technique. Suppose we don't compute exact circumcenters. Instead, we'll choose a point randomly within a ball of radius $f < 1$ about the true circumcenter. If we do this, it's easy to see that the minimum edge length is $1 - f$. Some simple geometry shows that a chord of length $1 - f$ in a circle of radius 1 has an opposite angle of $\arcsin(\frac{1-f}{2})$. This is the minimum angle that occurs in a 2D mesh when we approximate circumcenters with points within $f$ of the true circumcenter. Thus, $f = 0$ gives us a minimum angle of 30 degrees, as expected; $f = \frac{1}{2}$ gives us a minimum angle of about 14.5 degrees.

This technique gives us some hope for 3D. In Algorithm B, we had no choice in where to place each new vertex; if some new vertex created a sliver, we were stuck with it. But by choosing a new vertex from somewhere within an interior ball, we might be able to position the vertex in such a way that all slivers are avoided. Ideally, one might hope to calculate exactly where the new vertex should be placed within this interior ball, but this approach has not been successful, since the correct placement depends on the positions of potentially many surrounding vertices. Fortunately, randomness comes to our rescue: we don't need to determine exact placement as long as we can estimate the probability of finding a good placement when choosing random positions within the interior ball.

As a concrete basis for calculations, we take $f = \frac{1}{2}$; this is not necessarily the best value for $f$. Note that, with $f = \frac{1}{2}$, we expect that no two vertices will be closer that 0.5.

As before, we make some fairly strong assumptions about the region to be meshed. We assume that

1. the region to be meshed is a convex polyhedron,
2. the goal is to create a mesh with constant density, and
3. the boundary of the region has already been subdivided into triangles in such a way that all circumcircles of boundary triangles have diameter $\leq \sqrt{3}/2$ units and no two vertices are closer than 0.5 units.

```
Algorithm C:
Form the 3D DT of the boundary points
while there is a tetrahedron with circumradius > 1 d
   Randomly pick a point within 0.5 units of the
    circumcenter, repeating until we find one that
    does not form a sliver
   Add this point to the set of sites and update
    the DT
```

The exact definition of *sliver* as used in Algorithm C is determined by the requirements of the proof of the following theorem.

**Theorem 3** *Algorithm C halts, producing a mesh in which all triangular faces of tetrahedra have angles between* $\arcsin(\frac{1}{4})$ *(about 14.5 degrees) and about 151 degrees. There are no slivers in the resulting mesh except possibly for tetrahedra that have all their vertices on the boundary.*

*Proof*: Assuming that the algorithm halts, it's easy to see that the bound on the face angles holds and that the only possible slivers are those that appear due to the initial boundary subdivision. The bound on face angles follows from the maximum circumsphere size of 1 and the minimum edge size of 1/2. The only remaining part of the proof is to show that the algorithm halts. In other words, we need to show that,

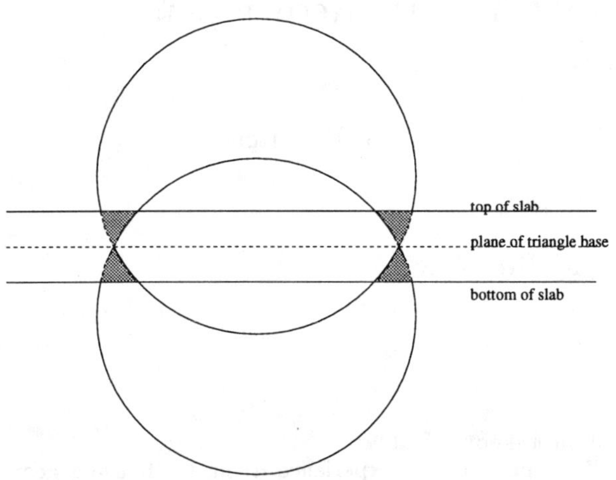


Figure 2: The fourth point of the sliver must lie in the grey region near the circumcircle of the base triangle.

when we randomly pick a point within 0.5 units of the circumcenter, we will eventually find one that forms no slivers.

First, observe that there are some slivers that we don't have to worry about. Any sliver that has a large circumsphere (with radius $> 1$) is caught later in the algorithm and the corresponding tetrahedron is destroyed. With this observation, we define a *sliver* as used in Algorithm C to be a tetrahedron that has a unit-radius circumsphere and that has height less than $\varepsilon$ where $\varepsilon$ is a value to be determined later in the proof. Note that new tetrahedra all use our new near-circumcenter point as their apex. We refer to the remaining three vertices of a new tetrahedron as the *base triangle* for for that tetrahedron. Height is measured from the plane of this base triangle.

The remainder of the proof is based on the observation that slivers have a very special shape. Given a base triangle, the apex vertex of a sliver can't be just anywhere. To be a sliver, the apex vertex must lie almost within the plane of the base triangle. Thus, a base triangle determines a slab (i.e., a thickened plane) of thickness $2\varepsilon$ with the property that an apex vertex chosen outside the slab cannot possibly form a sliver with the base triangle. Furthermore, the apex vertex must lie near the circumcircle of the base triangle. The exact shape is a ring with an hourglass cross-section; see Figure 2. Thus each potential base-triangle has just a small volume of space (called the *disallowed region*) where a forth vertex could make a sliver with that base-triangle.

Note that there are just a finite number of potential base triangles and consider the volume covered by the disallowed regions of all the potential base triangles. By choosing $\varepsilon$ sufficiently small, we can ensure that all of these disallowed regions fail to completely cover the *picking-sphere* (the inner sphere of radius 0.5 about the true circumcenter). We can force some fraction of the volume within the picking-sphere to be uncovered; thus, the picking process must halt in constant expected time. □

Currently, the value of $\varepsilon$ needed for the above proof is about 0.00058. This value is a fairly crude estimate and follows from a number of rather crude bounds on geometric quantities. Several of these bounds are multiplied together and/or cubed; thus any over-estimation can become quite significant. Rough experimental evidence indicates that a better estimate is $\varepsilon = 0.1$. Better estimates should be available when the algorithm is fully implemented.

## 4 Creating a Practical Algorithm

There are three problems that need to be resolved: (1) as presented in Algorithm C, the boundary is completely meshed into *very* nice triangles during an unspecified (and possibly impractical) preprocessing step, (2) if initially there are some slivers whose vertices are all on the boundary then these slivers can remain in the final mesh, and (3) Algorithm C produces a constant density mesh while practitioners generally want control of mesh density, with small tetrahedra in "interesting" areas for accuracy and large tetrahedra elsewhere for computational efficiency. For the third problem, a relatively small modification of the algorithm allows varying density, but this leads to large changes in the bounds used in the proofs. We briefly discuss resolutions for the first two problems. Full explanations, with proofs, would significantly lengthen this paper.

Boundary preprocessing is unnecessary; the boundary meshing can be done during the tetrahedral meshing. The idea is to split a boundary triangle by adding its circumcenter whenever it becomes clear that a boundary triangle is too large.

This leaves the problem of how to get rid of slivers whose vertices are all boundary vertices. The idea here is to allow approximate circumcenters of boundary triangles, picking a random point near the true circumcenter. By adjusting $\varepsilon$ appropriately one can show that disallowed *areas* along the boundary can be avoided using a constant expected number of picks.

## References


[BE92] M. Bern and D. Eppstein, Mesh Generation and Optimal Triangulation, *Computing in Euclidean Geometry*, edited by F. K. Hwang and D.-Z. Du, World Scientific, 1992. Also appears as Tech Report CSL-92-1, Xerox PARC, March 1992.

[BEG94] M. Bern, D. Eppstein, and J. R. Gilbert, Provably Good Mesh Generation, *Journal of Computer and System Sciences* 48, 384–409, 1994. An earlier version appears in *Proceedings of the 31st IEEE Symposium on the Foundations of Computer Science*, 231-241, 1990.

[Che89] L. P. Chew, *Guaranteed-Quality Triangular Meshes*, Department of Computer Science Tech Report TR 89-983, Cornell University, 1989.

[Che93] L. P. Chew, Guaranteed-Quality Mesh Generation for Curved Surfaces, *Proceedings of the Ninth Symposium on Computational Geometry* (1993), ACM Press, 274–280.

[MV92] S. A. Mitchell and S. A. Vavasis, Quality Mesh Generation in Three Dimensions, *Proceedings of the Eighth Annual Symposium on Computational Geometry*, 212-221, ACM Press, 1992. Full version appears as Department of Computer Science Tech Report TR 92-1267, Cornell University, 1992.

[Ru93] J. Ruppert, A new and simple algorithm for quality 2-dimensional mesh generation, *Proceedings of the 4th ACM-SIAM Symposium on Discrete Algorithms*, 83–92, 1993.

# Geometric Decision Trees for Optical Character Recognition

(Extended Abstract*)

George N. Sazaklis[†] Esther M. Arkin[‡] Joseph S. B. Mitchell[§]

Steven S. Skiena[¶]

State University of New York, Stony Brook, NY 11794

**Abstract**

A fundamental problem in computer vision is identifying which of a given set of geometric models is present in an image. We consider an approach to model recognition based on computing efficient strategies (*decision trees*) for "probing" a scanned image of a typeset document, in order to perform fast and effective optical character recognition (OCR). We consider a "probe" to be a simply computed local operator that can be applied to discriminate between two sets of possible models. By carefully constructing effective probes, and assembling them into a geometric decision tree, we have devised, implemented, and compared a variety of methods to perform OCR. In this paper, we present algorithms for probing strategies and decision tree construction, and we report experimental results on the effectiveness of these algorithms in identifying English characters and numerals in scanned images of printed pages of text. These algorithms are implemented as part of a system used by a document processing company (Syngen Corp.).

## 1 Introduction

The field of optical character recognition (OCR) strives to build systems enabling a machine to "read" pages of printed text that have been scanned into a digital file format. Extensive research has been invested in OCR since the 1960's, but the problem is still largely unsolved. Many commercial OCR packages exist, but most of them encounter problems in multi-font or low-quality documents. Moreover, most approaches are based on statistical decision theory that requires a set of training data to "tune" the classifier. However, such an extensive training set is not always available, as when one is scanning a book, or working on a document with an unknown font.

We report on our experience doing OCR using geometric probes, arranged in a *decision tree classifier*. Each probe consists of a simple local operator (e.g., summing the intensity values of the set of pixels that lie within a selected small box), and a comparison of the resulting value with threshold values that determine which branch of the decision tree is to be taken next.

This work can be considered a follow-up to Arkin et al. [1, 2], who studied the geometric decision tree problem. Here, we investigate the practical aspects of applying geometric decision trees to a real problem (OCR) and provide the experimental analysis that permits them to be further developed for other applications.

The "probing paradigm" has been applied to model-based object recognition; see [6], and references cited therein. In effect, the probing schemes serve to "factor out" the effect of translation and rotation, reducing the final decision problem to that of this paper. For previous work on the decision tree classifier paradigm, see [3, 4, 11]. For general background on the extensive field of OCR, we refer the reader to the survey by Mori, Suen and Yamamoto [7]. The method we are employing most closely resembles the "peephole" method that was among the first approaches to the OCR problem [5].

### Main Results

(1). We formalize the decision tree methodology, as applied to OCR, and describe and analyze heuristic algorithms designed to produce decision trees having low height. In particular, we conduct an experimental investigation into the methods based on greedy set cover heuristics, which were introduced in previous theoretical work of Arkin et al. [1, 2].

(2). Our decision-tree construction algorithm is based on an error model, which provides us with the probabilistic information required to assess the reliability of local probes. A detailed description of our error model is provided in the full paper.

(3). We develop a method of *verifying* a hypothesis given by a recognizer, in order to make more stringent demands on the level of certainty. Our verifier is based on set cover heuristics, and uses a select set of probes in order to raise the recognition confidence.

(4). We report on experimental results and on the practical effectiveness of geometric decision trees for use in scanned document processing. Our implementation is part of a system being tested now for use within a full-scale production environment to scan and read consumer surveys, financial account documents, and other forms.

---

*Full paper at http://ams.sunysb.edu/~jsbm/jsbm.html

[†]sazaklis@cs.sunysb.edu; Dept. Computer Science. Supported in part by a grant from Syngen Corp. and by SPIR, College of Engineering and Applied Science, SUNY Stony Brook.

[‡]estie@ams.sunysb.edu; Dept. Applied Math & Statistics.

[§]jsbm@ams.sunysb.edu; Dept. Applied Math & Statistics.

[¶]skiena@cs.sunysb.edu; Dept. Computer Science.



*Computational Geometry 97* Nice France

## 2 Methodology

We assume that each model in the library $S$ is given in a fixed position, orientation, and scale, as a greyscale array of pixels. From the set of models, we construct a decision tree (a *probe tree*), which can be used to perform recognition of scanned images. We assume that the scanned image has been segmented into subimages that each contain exactly one instance of one model. While segmentation can itself be a challenging task, for this work we chose to concentrate on classification and keep the segmentation issue separate.

As in [1, 2], we consider a *probe* to be a function that maps an image to a real number, its *outcome*, typically by computing a simple local function of the image values in one or more selected locations or neighborhoods within the image. The probe outcome must be simple to compute. By normalization, we can assume that all outcome values are in [0, 1]. The outcome of a probe when applied on a model is a random variable. The support interval of its distribution is determined by the error model and is called the *outcome interval.*

At each node $v$ of the probe tree there is an associated subset $S_v \subseteq S$ of the models. In the tree construction algorithm, our goal is to partition $S_v$ into roughly equal-size subsets by means of selecting an effective probe, $\pi_v$, to apply at node $v$. We determine one or more *classification thresholds* for each probe that we use. The classification thresholds partition [0, 1] into intervals, each of which corresponds to a child of the current node. While the thresholds partition the real numbers, we may not end up with a partition of the models $S_v$: Some models may be assigned to two or more children of $v$. In particular, since the outcome interval $I_m$ for a model $m$ contains the set of values that $\pi_v$ might produce, $m$ is included in any child whose interval overlaps with $I_m$. The optimal classification intervals for each candidate probe, together with the model subsets for each child are computed using dynamic programming. The "best" split is a set of $k$ points (with $k < 5$, typically) in the interval arrangement that generate $k+1$ children, while minimizing the model overlap between different children. We have found that minimizing the objective function $\sum_{k=1}^{n} |S_i|^3$ is an effective means of minimizing model overlap, where $S_i$ denote the model sets of the children. (While here we use a sum of cubes, any sum of convex functions will suffice). A dynamic programming algorithm ($O(n^2)$) is used to select the optimal classification intervals.

We then apply a *greedy* strategy, selecting the probe whose corresponding split of the models is the most balanced. The greedy heuristic for constructing decision trees was discussed and analyzed in [2], where it was shown that the greedy strategy to split a set of models produces a tree within a $\lceil \lg k \rceil$ factor of optimal, when we have $|M| = k$ models. While greedy probes produce short trees, there is a tradeoff between partitioning and accuracy. Currently, we choose the most reliable probe among those that are greedy enough (have balanced children).

The resulting decision tree can then be used for character classification. As we see our tree growing strategy is top-down, because we want to use only local information to make a decision on a tree node.

During classification, we descend to the child (or children) of $v$ whose interval contains the outcome of probe $\pi_v$.

Currently we use four distinct types of probes: *Bounding Box* probes measure the width or the height of the bounding box of the character. *Single-pixel* probes consist of a lone pixel, while *Rectangle Probes* are a generalization of single-pixel probes, with the probe calculating the average intensity over a rectangular area. Finally, *Pixel Set* probes are even more general. They may consist of any collection of pixels, equipped with signs (to specify whether the pixel intensity has to be inverted before taking the average). The *Pixel Set* type can be used to "home in" on where the essential geometrical differences are between two shapes.

## 3 Verifier

Because of distortions in the input image, a local operator may make mistakes during recognition. We can overcome this danger, however, if we use the redundancy present in the character image, to *verify* the classifier's suggestion.

The concept of *verification* is to use additional probes in order to obtain further evidence supporting the suggestion made by the decision tree.

To verify a suggestion $m$, made by the probe tree, we apply a sequence of probes to the sample, that can distinguish between model $m$ and any other model in our alphabet. We say a probe $p$ *covers* model $c$ against $c'$, if and only if it can distinguish between $c$ and $c'$. For probe $p$ to be used as a verification probe between $c$ and $c'$, it has to cover $c$ against $c'$ and be independent (disjoint) of any probes used so far in the decision tree for $c$, so that it does not repeat any errors made by the tree probes.

To build our verifier, we construct the so called *ambiguity graph.* The nodes are the models, while there is an edge between two models, if they are not discriminated by the current set of verifying probes. The edge is labeled by those probes that can do the discrimination, and break the ambiguity. This way, the verifier construction problem is transformed into a set covering problem. To construct the verifier with the minimum number of probes, we apply the greedy heuristic, selecting the probe that breaks the highest number of unresolved ambiguities.

As a different verifier variety, we can have a separate set of probes for each model, producing many set covers, independent from one another. The second flavor is called the "multiple cover" verifier.

Finally, we can ask for multiple coverage ("$C$-coverage") to increase the confidence of the final decision, where $C$ is a parameter that determines the minimum number of probes needed to break the ambiguity between two models.

## 4 Experimental Results

Our experiments were done on two fonts, OCR-font and Times-Roman 10pt. The OCR-font data was *real* data provided to us by Syngen Corp. As the OCR-font was designed to be machine-readable, it draws the numbers using only thick, constant width, horizontal and vertical lines, except "7", which has a tilted stroke. The experiment was conducted on 5183 number sequences, each segmented into 7 numerals, for a total of 36281 characters. The results for each tree individually without the verifier and for the system as a whole (with verification) appear in Table 1.

| | Rec. Rate | Error Rate | Rejections |
|---|---|---|---|
| Tree1 Only | 99.94 % | 0.05 % | 0.01 % |
| Tree2 Only | 99.96 % | 0.03 % | 0.00 % |
| Whole system | 99.87 % | 0.00 % | 0.13 % |

Table 1: Rates for OCR-font

| Decision Tree | Verifier | Originals | | Photocopy | | 2nd Generation | | Fax | |
|---|---|---|---|---|---|---|---|---|---|
| | | Rec. | Error | Rec. | Error | Rec. | Error | Rec. | Error |
| Single Pixel | Y | 99.87 % | 0.00 % | 96.20 % | 0.00 % | 72.57 % | 0.00 % | 42.95 % | 0.07 % |
| | N | 99.90 % | 0.10 % | 98.87 % | 1.12 % | 92.63 % | 6.83 % | 82.73 % | 16.48 % |
| Pixel Set | Y | 99.15 % | 0.00 % | 94.03 % | 0.00 % | 80.99 % | 0.00 % | 55.32 % | 0.20 % |
| | N | 99.84 % | 0.13 % | 98.13 % | 1.85 % | 94.04 % | 5.74 % | 76.17 % | 19.76 % |

Table 2: Experimental results for the Times-Roman font, both with ("Y") and without ("N") the verifier.

| Trees | 1-cover Ambiguity Graph | | 2-cover Ambiguity Graph | | 1-cover Mult. Verifiers | | 2-cover Mult. Verifiers | |
|---|---|---|---|---|---|---|---|---|
| | CR | WA | CR | WA | CR | WA | CR | WA |
| **Original Print** | | | | | | | | |
| Single Pixel | 0.03 % | 0.00 % | 0.12 % | 0.00 % | 0.02 % | 0.00 % | 0.30 % | 0.00 % |
| Pixel Set | 0.43 % | 0.00 % | 0.68 % | 0.00 % | 1.89 % | 0.00 % | 0.77 % | 0.00 % |
| **Photocopy (1st)** | | | | | | | | |
| Single Pixel | 2.48 % | 0.00 % | 5.42 % | 0.00 % | 4.92 % | 0.00 % | 5.82 % | 0.00 % |
| Pixel Set | 4.16 % | 0.00 % | 6.20 % | 0.00 % | 7.39 % | 0.00 % | 8.55 % | 0.00 % |
| **Photocopy (2nd)** | | | | | | | | |
| Single Pixel | 9.09 % | 0.96 % | 34.79 % | 0.03 % | 21.56 % | 0.00 % | 24.00 % | 0.00 % |
| Pixel Set | 10.67 % | 0.07 % | 13.81 % | 0.00 % | 40.37 % | 0.00 % | 24.92 % | 0.00 % |

Table 3: Verification rates for different strategies on originals, copies, and second-generation copies.

Figure 1: Left to right: Original, first-generation copy, second-generation copy, faxed.

Times-Roman, on the other hand, presents a greater challenge. As a font, it is considered non-trivial to be processed by OCR systems, as it has stroke width variability and serifs. We chose a size of 10pt, as it is more challenging than the more commonly used size of 12pt.

Printed pages with uppercase, lowercase letters as well as numbers were scanned on a flatbed Ricoh IS-60 scanner at 300 dpi, as 8-bit greyscale images. We also scanned first and second-generation photocopies, as well as faxed images to test our system on badly distorted documents. In Figure 1 we present typical examples of characters for our 4 document quality categories. We see that the probe tree has to deal with quite disparate distortions on the 4 documents: On photocopied images, individual pixels are flipped, smudges make recognition difficult, while defects accumulate in higher generations. Faxed data is the most challenging, since many characters are elongated vertically, due to the mechanical inaccuracies of the sending device, which scans the page; the image also suffers from ink dispersion that occurs at the receiving end.

Character models and probe trees were constructed as was outlined previously. Table 2 displays the recognition and error rates for different choices of probe trees and sources of scanned images. The remaining percentage in each case is due to rejections either from the tree or the verifier. The sample size for these experiments was 15692 characters, distributed equally among the 26 uppercase, the 26 lowercase and the 10 numerals.

Table 3 reports our verification rates for different verifier choices, different probes and sources of scanned data. In each table cell, two error measures are shown: The percentage of Correct but Rejected (CR) samples as well as the percentage of Wrong but Accepted (WA) samples. Our verifiers tend to be aggressive with rejections, since the cost of accepting a wrong classification is much higher than rejecting a correct one.

## References


[1] E. Arkin, M. Goodrich, J. Mitchell, D. Mount, C. Piatko, and S. Skiena. Point probe decision trees for geometric concept classes. In *Proc. 3rd WADS*, pp. 95–106, 1993.

[2] E. Arkin, H. Meijer, J. Mitchell, D. Rappaport, and S. Skiena. Decision trees for geometric models. In *Proc. 9th Annu. ACM Sympos. Comput. Geom.*, 369–378, 1993.

[3] L. Breiman, J. Friedman, R. Olshen, and C. Stone. *Classification and Regression Trees.* Wadsworth, 1984.

[4] P. Chou. Optimal partitioning for classification and regression trees. *IEEE Transactions on Pattern Analysis and Machine Intelligence*, 13(4):340–354, 1991.

[5] ERA. An electronic reading automaton. *Electronic Engineering*, pages 189–190, April 1957.

[6] R. Freimer, S. Khuller, J. Mitchell, C. Piatko, K. Romanik, and D. Souvaine. Localizing an object with finger probes. In R. Melter and A. Wu, eds., *Proc. of Vision Geometry III*, pp. 272–283, SPIE, Nov., 1994.

[7] S. Mori, C. Suen, and K. Yamamoto. Historical review of OCR research and development. *IEEE Proceedings*, 80:1029–1058, July 1992.

[8] S. Mori, K. Yamamoto, and M. Yasuda. Research on machine recognition of handprinted characters. *IEEE Trans. PAMI*, 6:386–405, 1984.

[9] G. Nagy. State of the art in pattern recognition. *Proc. IEEE*, 56:836–860, 1968.

[10] K. Romanik and C. Smith. Testing geometric objects. *Comput. Geom. Theory Appl.*, 4:157–176, 1994.

[11] S. Safavian and D. Landgrebe. A survey of decision tree classifier methodology. *IEEE Transactions on Systems, Man, and Cybernetics*, 21(3):660–674, 1991.

# A Wavelet Approach to Foveating Images

Ee-Chien Chang and Chee K. Yap*
Courant Institute of Mathematical Sciences
New York University
New York, NY 10012
U.S.A.

**Abstract**

Motivated by applications of foveated images in visualization, we introduce the **foveation transform** of an image. We study the basic properties of these transforms using the multiresolution framework of Mallat. We also consider practical methods of realizing such transforms. In particular, we introduce a new method for foveating images based on wavelets. Preliminary experimental results are shown.

## 1 Introduction

Conventional images have uniform resolution. **Foveated images** which have non-uniform resolution arise in biological vision. In figure 1 we show a foveated version of an uniform image. The process of going from an uniform image to figure 1 is called "foveation".

One of the most interesting forms of foveated images is based on the complex logarithm function. Such **logmap images** were studied by Rojer and Schwartz [10] and others. The complex logmap is a model consistent with empirical data on the mapping from primate retina to the visual cortex [11, 12]. This neuro-physiological model goes back to the pioneering (and the Nobel prize) work of Hubel and Wiesel [5, 6]. Rojer [9] demonstrated many favorable properties of such images. Perhaps the most striking fact is that the data density in such images grows logarithmically with the diameter of the visual field (as opposed to quadratically in the case of uniform images). Such low density images have been exploited in applications such as video phones [14]. In this paper, we use the term "foveated" to refer to any variable resolution image, not just those with the geometry of the logmap or having a definite "foveal region".

**Visualization Applications.** Foveated images have been exploited in computer vision, (e.g., [4, 13]), especially in the context of **active vision**[1, 2]. But they are also useful in **visualization**, although this aspect is less explored. For visualization, we must compensate for the loss of peripheral resolution in foveated images by rapidly presenting images that are foveated at points of user interest. We could use eye-tracking technology [7] to follow the eye and to present the corresponding foveated images. In principal, if the foveation uses the correct parameters and eye-tracking is good, a user should not be able to detect any difference from viewing a uniform image. A system with such properties has been reported by [8]. But in our visualization applications, we can dispense with eye-tracking — the visualizer may be counted upon to cooperate with the image server. For instance, the visualizer can indicate moving areas of interest in an image via a standard mouse input. Because of the rapid eye activity in normal visualization, we call this approach **active visualization.**

**Thinwire Visualization.** We describe one application, viz., in "thin-wire" models of computation. Here, we have a **server** who acts as an image-server and a **client** who wishes visualize images from the server. The images may be very large (e.g., a terrain or map) and have high resolution. The server has large computational resources but there is only a thin wire (such as the Internet) connecting the server and client. A basic problem in this scenario is the construction foveated images from uniform images. Our paper focuses on this problem.

## 2 Foveation Theory

There are various practical methods to foveate an image. We would like to provide the mathematical formulation of this process. This would allow us to evaluate the various methods.

As usual, one can begin with one-dimensional images or functions. Given an image $I(x)$, we would like to define a foveated image $I_0(x)$ A **foveation** is determined by a **weight function** and a **scaling function**,

$$w : \mathbb{R} \to \mathbb{R}_{\geq 0}, \quad s : \mathbb{R} \to \mathbb{R}.$$

We require the weight function to be positive except at finitely many values (so $w(x)^{-1} < \infty$ at all but finitely many $x$). The scaling function is analogous to the "father wavelet" usually denoted $\phi(x)$ in wavelet theory. We define

$$I_0(x) := \int I(t) C^{-1}(x) s\left(\frac{t-x}{w(x)}\right) dt,$$

where $C(x) := \| s\left(\frac{\cdot - x}{w(x)}\right) \|$. The **weighted translation** (or,

---

*yap@cs.nyu.edu, supported by an National Science Foundation Grant #CCR-9402464.

Figure 1: Blended image with 2 gaze points (on the extreme left and extreme right faces with rates $\frac{1}{20}$ and $\frac{1}{30}$ respectively). The image is obtained using the definition of foveation and blending of two standard weight functions.

Figure 2: Same as figure 1 but is is obtained using wavelet foveation.

$w$-translation) of $s$ by $x$ is defined to be

$$s_x(t) := C^{-1}(x) s\left(\frac{t-x}{w(x)}\right). \quad (1)$$

Now the foveated image could be rewritten as

$$I_0(x) = \langle I, s_x \rangle.$$

A **standard weight function** is one of the form

$$w_{\text{std}}(t) = w_{\alpha,\beta,\gamma}(t) := \alpha|t-\gamma| + \beta,$$

where $\alpha, \beta, \gamma$ are fixed constants and $\alpha, \beta \geq 0$. We call $\alpha$ the **rate**, $\beta$ the **foveal resolution** and $\gamma$ the **gaze point**. Intuitively, this weight function defines the point $t = \gamma$ as the fovea with maximum resolution $\beta$. The resolution decreases linearly with the distance from the gaze point, at the rate $\alpha$.

We can obtain new weight functions by combining several given weight functions. The following definition corresponds to an effective method for blending images: given weight functions $w_1$ and $w_2$, their **blended weight function** $w$ is the pointwise minimum of the two functions: $w(x) = \min\{w_1(x), w_2(x)\}$.

## 3 Weighted Distance

For a function $I$, we want to measure how far $I$ is from the foveation $I_0$. It is not useful to use the usual $\|I - I_0\|$ norm as we want to place less emphasis on $I(x) - I_0(x)$ when $x$ is far away from the "foveal region". Define for a function $f$,

$$\|f\|_w = \left(\int (w(t))^{-1} |f(t)|^2 \, dt\right)^{1/2}. \quad (2)$$

Then the **$w$-distance** between $f$ and $g$ is defined to be $d_w(f,g) := \|f-g\|_w$. Let $L(t)$ be a function such that $dL(t) = \frac{dt}{w(t)}$. If $L^{-1}(t)$ is well-defined then

$$\int |f(t)|^2 (w(t))^{-1} dt = \int |f(L^{-1}(y))|^2 dy. \quad (3)$$

$L(t)$ and $L^{-1}(t)$ for the standard weight functions with $\beta = 1$ is are given by

$$L(t) = \text{sg}(t) \cdot \ln(\alpha|t| + 1), \qquad L^{-1}(t) = \text{sg}(t)(e^{\alpha|t|} - 1).$$

## 4 Multiresolution Foveation

For any weight function $w$ and $n \in \mathbb{Z}$, let $w_n(t) := 2^{-n} w(t)$. (An alternative definition is $w_n(t) := w(2^{-n}t)$). Define $I_n$ to be the foveation using the weight function $w_n$. Then $\{I_n\}$ is now a sequence of increasingly better foveation as $n$ increases.

## 5 Practical Foveation Methods

Computing $I_0$ from its definition is computationally inefficient. We consider practical methods to approximate $I_0$. There are two general approaches. The first approach is based on **sampling**. Previous papers on foveation can be put under this approach. This is computationally efficient (usually implemented with lookup tables), but lack flexibility when different weight functions are required. In this paper we introduce a second approach. It exploits any multiresolution representation of an image of $I$: to create a foveated image from $I$, we 'cut' the necessary pieces from each level of the multiresolution representation and 'glue' them together. There is considerable freedom in choosing the multiresolution representation (e.g., different kinds of wavelet transforms) and in how we cut and paste.

### 5.1 Foveation via Wavelets

We give an approximation of foveation using wavelets. Let $\psi$ be the mother wavelet and $\phi$ be the corresponding father wavelet. Let $\psi_{m,n}$ be the translated and dilated version of $\psi$. Specifically, $\psi_{m,n}(t) = \psi(2^{-m}t - n)$. Assume that $\{\psi_{m,n}\}$ form an orthonormal basis for $L_2(\mathbb{R})$. The approximation is as follows.

$$I^{\text{wav}} = \sum_{(m,n) \in B(w)} \langle \psi_{m,n}, I \rangle \, \psi_{m,n}(x),$$

where $B(w)$ is a discrete set depending on the weight function $w$,

$$B(w) = \{(m,n) | 2^m \geq w(n2^m)\}.$$

For the standard weight $w = \alpha|t| + \beta$, the set $B(w)$ is $\{(m,n)|2^m \geq \alpha|n|2^m + \beta\}$, which is $\{(m,n)|2^{-m}c_1 + |n| \leq r\}$

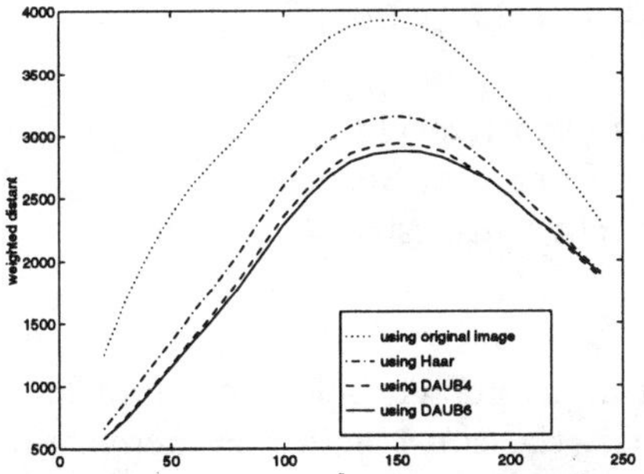


Figure 3: The weighted distance $d_w(I, I_0)$ and $d_w(I^{\text{wav}}, I_0)$ where $I^{\text{wav}}$ is computed using different wavelets. The weight function used is $w(t) = \frac{1}{r}\|t\|_\infty$.

where $c_1 = \alpha^{-1}\beta$ and $r = \alpha^{-1}$. This means that we just have to take at most $r$ coefficients from each level. If the foveal resolution $\beta$ is greater than $2^m$, then no coefficients from level $m$ is in $B(w)$ ($m$ could be a negative integer). This simple structure is useful for fast reconstruction.

We use the weighted distance to measure the quality of approximations using different wavelets. Figure 3(a) shows experimental results of this comparison. The two dimensional weighted distance is

$$d_w(f,g) := \int\int w(x,y)^{-2}\|f(x,y) - g(x,y)\|^2 dx\, dy.$$

### 5.2 Bandwidth reduction

Our techniques show promise for thin-wire visualization. Figure 4 illustrates experimental results on the reduction in number of coefficients and number of bytes (note that a coefficient may take more than one byte) needed to represent foveated images at various rates. The number of bytes is computed after a lossless compression. The original image used is the uniform version of figure 1.

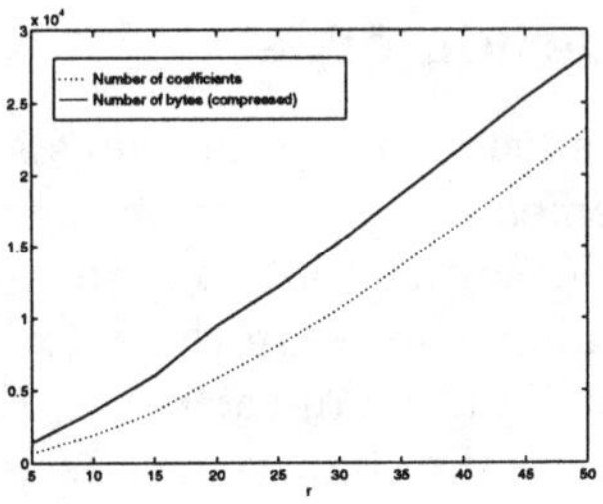


Figure 4: Size of original image is $375 \times 469$ pixels with 256 gray levels. When $r = 20$, on average, the image uses 0.43 bit per pixel. When $r = 50$, the image uses 1.28 bit per pixel.

## 6 Further Work

We are currently looking at the mathematical analysis of multiresolution foveation and comparison of various foveation approaches using our formulation. On the practical side, we are implementing active visualization of a large (for example 4000 by 4000 pixels) image across a network. We are also looking into generalization to 3-d, that is, volume data.

## References


[1] J. Aloimonos, I. Weiss, and A. Bandyopadhyay. Active vision. *Intl. J. Computer Vision*, 2:333–356, 1988.

[2] R. Bajcsy. Active perception. *IEEE Proceedings*, 76(8):996–1005, 1988.

[3] Benjamin B. Bederson. *A miniature space-variant active vision system: Cortex-I.* PhD thesis, Computer Science, Courant Institute, New York University, 1992.

[4] C. Colombo, M. Rucci, and P. Dario. Integrating selective attention and space-variant sensing in machine vision. In Jorge L.C. Sanz, editor, *Image Technology: Advances in Image Processing, Multimedia and Machine Vision*, pages 109–128. Springer, 1996.

[5] D.H.Hubel and T.N.Wiesel. Receptive fields and functional architecture of monkey striate cortex. *J. Physiol. (Lond.)*, 195:215–243, 1968.

[6] D.H.Hubel and T.N.Wiesel. Sequence regularity and geometry of orientation columns in the monkey striate cortex. *J. Comp.Neurol.*, 158:267–293, 1974.

[7] R. Jacob. Emerging user-interface media: potentials and challenges – eye tracking. In *Tutorial, SIGGRAPH '89*, 1989.

[8] Philip Kortum and Wilson S. Geisler. Implementation of a foveated image coding system for image bandwidth reduction. In *Human Vision and Electronic Imaging, SPIE Proceedings Vol. 2657*, pages 350–360, 1996.

[9] Alan S. Rojer. *Space-variant computer vision with a complex-logarithmic sensor geometry.* PhD thesis, Computer Science, Courant Institute, New York University, 1989.

[10] Alan S. Rojer and E.L. Schwartz. Design considerations for a space-variant visual sensor with a complex-logarithmic sensor geometry. In *10th ICPR*, volume 2, pages 278–285, 1990.

[11] E.L. Schwartz. The development of specific visual projections in the monkey and the goldfish: outline of a geometric theory of receptotopic structure. *J.Theoret.Biol.*, 69:655–685, 1977.

[12] E.L. Schwartz. Spatial mapping in primate sensory projection: analytic structure and relevance to perception. *Biological Cybernetics*, 25:181–194, 1977.

[13] E.L. Schwartz, D.N. Greve, and G. Bonmassar. Space-variant active vision: Definition, overview and examples. *NeurNet*, 8(7–8):1297–1308, 1995.

[14] R.S. Wallace, B.B. Bederson, and E.L. Schwartz. Voice bandwidth visual communication through logmaps: The Telecortex. In *Workshop on Applications of Computer Vision*, pages 4–10, 1992.

# Good orders for incremental (re)construction*

Jack Snoeyink
Dept. of Computer Science
University of British Columbia

Marc van Kreveld
Dept. of Computer Science
Utrecht University

## Abstract

Voronoi diagrams, Delaunay triangulations, and vertical decompositions in the plane are structures that are canonically defined for a set of $n$ points or line segments. Construction requires $\Theta(n \log n)$ time in the worst case. We show that an order exists in each case such that construction takes only linear time. The order can be determined in linear time from the Voronoi diagram, Delaunay triangulation, or vertical decomposition. This result has applications in compression and decompression of data, and as a consequence transmission overhead can be reduced. The method is simple and our implementation can be seen at `http://www.cs.ubc.ca/spider/snoeyink/terrain/Demo.html`

## 1 Introduction

The best-known geometric structures in computational geometry—convex hulls, Voronoi diagrams, Delaunay triangulations, and vertical decompositions—can all be constructed in the plane in $O(n \log n)$ time, which is optimal. The lower bounds can be obtained by reduction from sorting. Presorting by $x$-coordinate doesn't help in the case of Delaunay triangulations, Voronoi diagrams, and 3D convex hulls; the lower bound remains $\Omega(n \log n)$ time [5, 13].

This paper shows that an ordering of a set of $n$ points exists such that the Delaunay triangulation can be constructed from this order in only linear time. The same result holds for Voronoi diagrams, trapezoidal decompositions of a set of $n$ line segments in the plane, and 3D convex hulls. The result is based on Kirkpatrick's hierarchical point location scheme, often called *triangulation refinement* [8].

In addition to being simple, our method has two characteristics that allow progressive expansion: the rough structure can be known before all data has been processed. First, the construction is incremental; when $k$ vertices have been processed, the Delaunay triangulation, Voronoi diagram, or convex hull of those $k$ points is constructed in $O(k)$ time. Second, the scheme is flexible enough to incorporate importance measures so that "important vertices" are included early in the order.

This gives an interesting application to storing geometric models and transmitting them over a network. The Delaunay triangulation is widely used for mesh generation [2], and as a model for terrains in geographic information systems, where vertices have a third coordinate to represent height [10]. Although triangulated terrain models are attractive as an adaptive data structure that represents a surface, the overhead of storing triangles and their neighbor relations is high, especially when compared with gridded elevation data, which only stores $z$-coordinates at regularly sampled locations. A Delaunay triangulation can be compressed by omitting all structural information (edges and triangles), leaving only the vertices. By our result, the reconstruction takes only linear time if the vertices are stored in the determined order. Since the reconstruction is incremental, the rough structure of the terrain can be visualized before all the data has been read or transmitted. This type of progressive expansion is also of interest to computer graphics [4, 7].

## 2 Computing an incremental construction order

The structures listed in the previous section have randomized incremental construction algorithms that run in expected $O(n \log n)$ time, where the expectation is taken over random choices in the algorithm for a worst-case input distribution. In each case, the expected cost of all changes to the data structure is linear, but the expected time to locate where the changes must occur is $\Theta(n \log n)$ [1, 6, 9, 12]. Our data order must ensure that the cost of change actually is linear (not just in expectation) and that the location time is also linear in $n$. Fortunately, this is easily done.

In this section, we use the computation of a Delaunay triangulation of $n$ points $P$ in the plane to give an example of the steps of our approach. Our first algorithm

*This research is supported by the NSERC and by the ESPRIT IV LTR Project No. 21957 (CGAL).


*Computational Geometry 97* Nice France

produces a permutation of the data (in reverse order); our second takes this permutation and reconstructs the graph.

**Order Determination Algorithm**

O1. **Bound the problem.** Extend the triangulation to the entire plane by adding three vertices "at infinity" on the positive $x$- and $y$-axes, and the half-line $y = x$ with negative values. These vertices are joined to the convex hull vertices.

O2. **Independent set.** Find a maximal (not maximum) independent set of size at least 1/5 of the vertices—no two vertices in the set are adjacent [3, 11]. (Low degree is useful, but not mandatory.)

O3. **Delete and locate data.** Delete the vertices of the independent set and repair the remaining triangulation, noting which new triangles now contain the deleted vertices. There will be at most one deleted vertex per triangle.

O4. **Traverse graph & output permutation.** Enumerate triangles by any canonical method: depth-first, breadth-first, etc. Output the deleted vertices in the reverse of the order that their triangles are visited by the enumeration.

O5. **Output end-of-phase signal.** Place a marker in the output sequence to signal that a phase has ended and a new one will begin.

O6. **Repeat from O2.** Continue until no finite vertex remains.

As in Kirkpatrick's hierarchical scheme, a constant fraction of the vertices is removed in each phase. Therefore, only a logarithmic number phases is needed in the algorithm. There are several ways to signal phase boundaries that do not require explicit markers. One could use independent sets of size exactly $\lfloor n/5 \rfloor$ instead of a maximal independent set, or one can make sure the first point of one phase does not fall in a triangle visited after the last point of another.

**Lemma 1** *Given the Delaunay triangulation of $n$ points, the Order Determination Algorithm can be implemented to run in $O(n)$ time.*

## 3 Reconstruction from the order

If we now reverse the ordered list of points and reconstruct the Delaunay triangulation, our traversal can accomplish a linear amount of point location in linear time.

**Reconstruction Algorithm**

R1. **Initialize.** Start with the "triangulation" of the three vertices at infinity.

R2. **For each phase do R3–R4.** A phase is identified by an end-of-phase signal.

R3. **Traverse and locate.** Visit the triangles by the same traversal as in O4 and link vertices to the triangles that contain them as they are found.

R4. **Insert entire phase.** Update the Delaunay triangulation by incremental insertion and edge flips for each new vertex.

**Lemma 2** *The Reconstruction Algorithm for the Delaunay triangulation takes $O(k)$ time after reading $k$ points.*

## 4 Vertical decompositions of line segments

The method outlined in the previous sections applies directly to 3D convex hulls and Voronoi diagrams. The case of vertical decompositions of set of line segments in the plane is slightly different. Here we wish to determine an order on the line segments, not their endpoints.

First add long horizontal segments above and below the trapezoidal decomposition. Then we extend the vertical decomposition of which the construction order is determined by adding a connection (shown dashed) between the endpoints of two segments that bound the same trapezoid, but aren't connected by a vertical extension. Two line segments are independent if they are

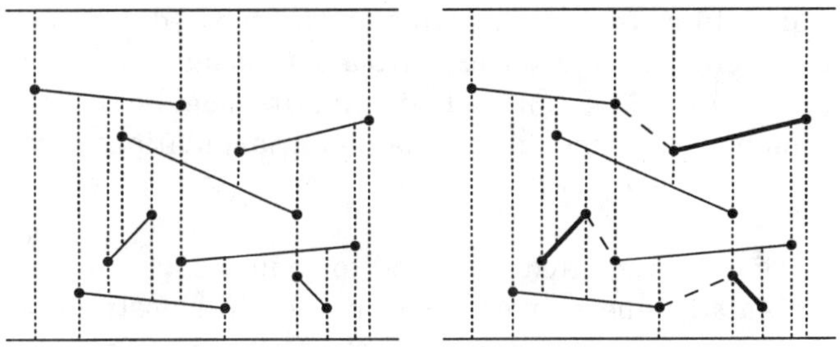

not connected by a vertical extension or an extra connection. By considering the graph where the segments are nodes and the vertical extensions and connections are the arcs, we have a planar graph of which an independent set (shown fat) can be determined. During the reconstruction, we determine of the segments in one phase where they should be inserted by locating the trapezoid that contains their left endpoint. More details are in the full paper.

## 5 Experiments

An important issue is how to determine the maximal independent set. The reconstruction time drops if the independent sets are larger. By the four-color theorem, an independent set of size 1/4 exists, but it takes quadratic time and a case analysis with 633 cases to find the four-coloring [11]. Much simpler linear time algorithms exist that guarantee an independent set of size 1/5 of the total number of vertices [3, 11]. In the full paper we show that

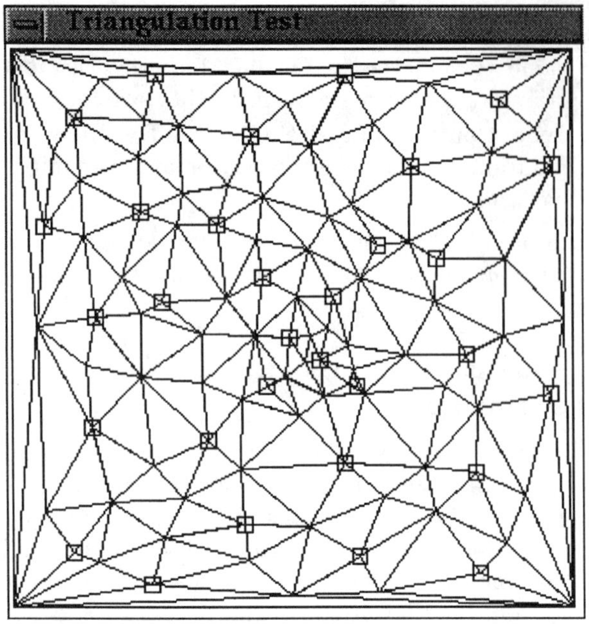


Figure 1: An independent set with 31 vertices out of a triangulation with 100 vertices.

a simple greedy algorithm guarantees 1/6 of the total using degree at most 9: Delete vertices of degree greater than 9 from the triangulation, then repeatedly choose a minimum degree vertex $v$ from the remaining graph, add $v$ to the MIS, and delete $v$ and its neighbors. Preliminarly experiments on Delaunay triangulations show an MIS of 30% is often found.

We have presented a method to compute an order on $n$ points or line segments, so that the (re)construction of canonical planar diagrams—the Delaunay triangulation, Voronoi diagram, 3D convex hull, and trapezoidal decomposition—takes only linear time. The method also extends to constrained Delaunay triangulations.

Some interesting difficulties arise when the data is not in general position because the diagram may no longer be canonical. Different diagrams may lead to different traversals on reconstruction. For trapezoidations the diagram can be made canonical by geometric perturbation methods, but for the Delaunay triangulation this appears more difficult—most perturbation methods rely on the ordering of vertices. In the full paper, we show that degenerate Delanay triangulations can be handled by ensuring that the traversal order is preserved.

## References


[1] M. de Berg, M. van Kreveld, M. Overmars, and O. Schwarzkopf. *Computational Geometry – Algorithms and Applications.* Springer-Verlag, Berlin, 1997.

[2] M. Bern and D. Eppstein. Mesh generation and optimal triangulation. In F. K. Hwang and D.-Z. Du, editors, *Computing in Euclidean Geometry.* World Scientific, March 1992.

[3] N. Chiba, T. Nishizeki, and N. Saito. A linear 5-coloring algorithm of planar graphs. *Journal of Algorithms*, 2:317–327, 1981.

[4] M. Deering. Geometry compression. *Computer Graphics*, pages 13–20, 1995. Proc. SIGGRAPH '95.

[5] H. Djidjev and A. Lingas. On computing the Voronoi diagram for restricted planar figures. In *WADS '91*, LNCS 519, pages 54–64. Springer-Verlag, 1991.

[6] L. J. Guibas, D. E. Knuth, and M. Sharir. Randomized incremental construction of Delaunay and Voronoi diagrams. *Algorithmica*, 7:381–413, 1992.

[7] H. Hoppe. Progressive meshes. *Computer Graphics*, pages 99–108, 1996. Proc. SIGGRAPH '96.

[8] D. Kirkpatrick. Optimal search in planar subdivisions. *SIAM Journal on Computing*, 12:28–35, 1983.

[9] K. Mulmuley. *Computational Geometry: An Introduction Through Randomized Algorithms.* Prentice-Hall, Englewood Cliffs, N.J., 1993.

[10] T. K. Peucker, R. J. Fowler, J. J. Little, and D. M. Mark. The triangulated irregular network. In *Amer. Soc. Photogrammetry Proc. Digital Terrain Models Symposium*, pages 516–532, 1978.

[11] N. Robertson, D. P. Sanders, P. Seymour, and R. Thomas. Efficiently four-coloring planar graphs. In *Proc. 28th ACM STOC*, pages 571–575, 1996.

[12] R. Seidel. Backwards analysis of randomized geometric algorithms. In J. Pach, editor, *New Trends in Discrete and Computational Geometry*, volume 10 of *Algorithms and Combinatorics*, pages 37–68. Springer-Verlag, 1993.

[13] R. Seidel. A method for proving lower bounds for certain geometric problems. In G. T. Toussaint, editor, *Computational Geometry*, pages 319–334. North Holland, Amsterdam, 1985.

# Algorithmic Tolerances and Semantics in Data Exchange

T. J. Peters*
Computer Science & Eng. U-155
University of Connecticut
Storrs, CT 06269-3155 USA
tpeters@eng2.uconn.edu

N. F. Stewart†
Dep. IRO
Univ. de Montréal
Montréal, Qc, Canada H3C 3J7
stewart@iro.umontreal.ca

D. R. Ferguson
Boeing
Seattle, WA 98124-2207 USA
drf@espresso.rt.cs.boeing.com

P. S. Fussell
Alcoa Technical Center
Alcoa Center, PA 15069 USA
fussell_ps@atc.alcoa.com

**Abstract**

Within industrial contexts, a common view of the level of confidence needed in CAD data transfer processes may be expressed as *"I don't mind if the [CAD data-transfer] algorithm sometimes has difficulties, provided I get a warning message that tells me something has gone wrong."* A particularly annoying, even mystifying, CAD data exchange problem occurs in the so-called 'round-trip' problem. There, an acceptable model is transferred from an originating system to a neutral format and then back to the original system, only to find that serious new flaws now appear in the model within the originating system.

## 1 Introduction

The question of ensuring the integrity of data exchanges between companies is of great importance in Computer-Aided Design (CAD). Indeed, the problem is serious enough, and important enough, to have spawned a fairly large industry dedicated to its solution [5].

One integrity issue is the desire for consistency of topology and geometry data. Design models contain symbolic and numeric data types expressing, respectively, topology and geometry, with some level of interaction between the two. The symbolic data is exact, whereas the numeric data suffers the errors of floating point arithmetic. The unavoidable nature of this disparity is well recognized; attempts to resolve it are based upon the use of tolerance values. When we say that the geometry and topology are inconsistent, we mean, for example, that the symbolic data indicates that two geometric elements are adjacent, yet the corresponding floating point data implies that they are disjoint.

Particularly vexing (sometimes mystifying, even to seasoned CAD practitioners) is the 'round-trip' data exchange problem. This describes the following exchange pathology. The model has topological and geometric consistency in its native system. It is moved to an exchange representation (like STEP) [6], where the consistency is preserved. However, upon being transferred back to the native system, the model may no longer be consistent. We illustrate how and why this can happen, and indeed, why it is even likely for some class of problems.

## 2 Related Work

Some evidence [5] indicates that model transfer can be improved when the STEP default tolerances are replaced by ones more closely adapted to those of the input system. In a similar vein [5], several illustrative examples of topology and geometry inconsistency were created, and labeled as 'gap problems', where two entities are adjoint by their topological data, yet their geometric data indicates they are disjoint.

New intersection algorithms [2, 3] incorporate tolerance intervals that are adaptive to the numerical errors compounded by extensive geometric computations.

Backward error analysis [1] has been performed relative to the errors for certain geometric algorithms.

## 3 Example

Consider the simple illustrative geometry of Figure 1. Vertex V0 serves as a trimming endpoint for edge E0, with V0 being within tolerance of E0.

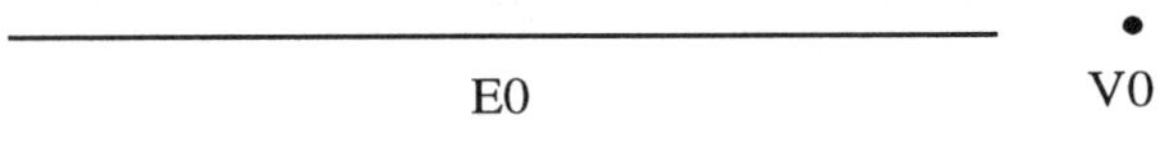


Figure 1: Simple Illustrative Geometry

Now, during model transfer, E0 could be perturbed sufficiently so that VO was no longer within tolerance of E0. For example, suppose the line segment indicated was represented in the native system by an abstract data type having two

*This author acknowledges, with appreciation, partial funding for this work received under National Science Foundation Grant Number MII-9308346 and the hospitality of the University of Montréal during sabbatical leave. The views expressed herein are of this author, not of the Foundation.

†The research of this author was supported in part by a grant from the Natural Sciences and Engineering Research Council of Canada



*Computational Geometry 97* Nice France
Copyright 1997 ACM 0-89791-878-9/97/06 ..$3.50

geometric points, with the metric used for tolerance comparisons in the originating system being the Manhattan (also known as the taxi-cab) metric. For points $p_0 = (x_0, y_0)$ and $p_1 = (x_1, y_1)$, the Manhattan metric is given by the equation

$$d_M(p_0, p_1) = |x_0 - x_1| + |y_0 - y_1| \quad .$$

Suppose also that the neutral transfer representation called for lines to be represented in the format of an originating point and a directional unit vector with a separate scalar for its magnitude. In the conversion of the original line representation into the neutral format, some numerical error could occur, thereby perturbing E0.

Suppose V0 was transferred exactly (via faithful bit transfers) and the tolerance values for the neutral format were identical to that of the originating system, but that the metric used to test for a point being on an edge within the neutral format was the Euclidean metric, whose convex unit sphere contains that of the Manhattan metric. For points $p_0$ and $p_1$, the Euclidean metric is given by the equation

$$d_E(p_0, p_1) = \sqrt{(x_0 - x_1)^2 + (y_0 - y_1)^2} \quad .$$

The Euclidean metric, by virtue of its larger unit sphere, could still have the perturbed edge within tolerance of V0, as is indicated in more detail in Section 4 and Figure 2. Hence, the model could be topologically consistent in both its original form and in its neutral transferred format. However, since E0 has been perturbed, when the model is transferred back to the originating system, the more sensitive Manhattan metric could determine that the tolerances were violated, as shown in Figure 2.

This illustrative example appears to rely crucially upon the differences between the Manhattan and Euclidean metrics. However, a moment's reflection shows that even if both metrics were Euclidean, then merely slight differences in their coded implementations (for instance, use a squaring operation versus explicitly multiplying an argument by itself) could lead to similar inconsistencies. Furthermore, the first author has reviewed many instances of commercial CAD/CAM implementations where the two cited metrics were used interchangeably, with no provision for their differing characteristics. These differences are symptomatic of semantic inconsistency across module interfaces. In this example, the expression of each individual metric is mathematically valid, yet the ensuing communication does not preserve the intended meaning.

A simplistic approach to resolution might be just to perturb the co-ordinates of V0 slightly. Alternatively, for the simple geometry depicted, a slight linear extrapolation of E0 might be appropriate. While these approaches might resolve this particularly simple geometry, it is clear that neither approach is extensible to general situations. Note, that for more complex models, the point perturbation 'solution' could be fraught with all the difficulties of changing one item in a highly interrelated constraint network, and the linear extrapolation 'solution' would be totally inapplicable for curvilinear geometry.

## 4 More Complex Geometry

Figure 2 depicts the original curve as a dotted line, shown to be within the tolerance neighborhood about a point relative to the Manhattan metric, shown as a dashed diamond. The perturbed curve and the Euclidean tolerance neighborhood are shown with solid display.

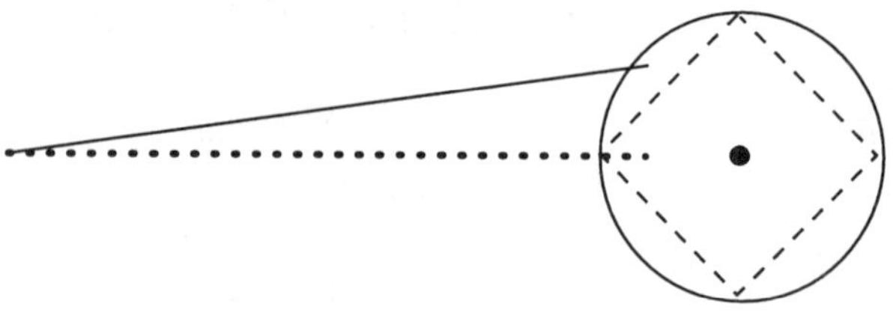

Figure 2: Two Tolerance Neighborhoods

The conversion of the line representation could be implemented in various ways, but the floating point operations invoked contribute to the error incurred. The representation conversion for the line segment is particularly simple. Yet the ideas expressed are indicative of more complex geometry. Consider a trimmed Non-Uniform Rational B-Spline (NURBS) curve represented in one system, with the intent to transfer it to another system. The originating data would consist of the trimming point and an abstract data type for the NURBS curve. The condition that the trimming point was 'on' the NURBS curve would typically be expressed by the distance between the point and the curve being less than some tolerance. Even if one were to suppose that the two communicating systems had identical tolerances and metrics, problems could occur. For instance, suppose the receiving system could only represent Bézier curves. Then, the exchange would require a conversion of the NURBS representation into its best Bézier approximant. The errors induced with that conversion, both due to the curve approximation and to floating point arithmetic, could be such that the distance from the trimming point to the curve would exceed the tolerance, implying that the trimming point was no longer 'on' the curve. The situation is further compounded when the communicating systems invoke differing tolerance values. Furthermore, this illustrative example considered only a dedicated transfer from one system to another, whereas, in practice usually additional conversions are necessary to and from a neutral intervening format.

## 5 Software Design Issues

The tolerance neighborhoods (Manhattan and Euclidean) depicted in Figure 2 model the situation where geometric co-ordinates are known with sufficient uncertainty that they could lie anywhere within the depicted tolerance neighborhoods. For ease of discussion, consider the location of the individual point to be fixed, as is consistent with our discussion of the transfer example. Then the neighborhood about the point can be considered the region in which there exists non-zero probability for non-void intersection with the line segment. To fully model the underlying semantics would require assigning a probability distribution over this neighborhood. If one were to assume simplistically that the probability distribution were uniform, the Euclidean neighborhood's greater area reflects a more pessimistic view of the uncertainty of the data than does the smaller Manhattan neighborhood.

Our critical issue is to emphasize that these 'hidden assumptions' must be explicated and their choices justified, as these seemingly innocent choices may have unintended implications. Resolving all such concerns is essential for establishing fully consistent, logically correct, formally verifiable semantics for geometric modeling. The cost to do so will be high. But it is our judgement that the failure to do so will be much higher, where the fourth author provides perspective on recent supporting evidence from Alcoa:

> About 18% of our best CAD engineers' and designers' time is spent on re-asserting the consistency of our models (these problems require too much insight for our medium skilled designers to quickly solve). At Alcoa, we expect that proper semantics could help us reclaim a substantial portion of time of our best people, but, more importantly, we will more effectively communicate with our suppliers and customers.

The fundamental difference for the level of consistency seen for the original, neutral and final models arises from two causes

- the differing use of 'metric', and
- the separation distance being near the tolerance limit.

Within *a single* system, adherence to contemporary software engineering principles can be effective as a disciplined means to enforce consistent semantics. For instance, the use of a metric should be standardized within one system. It should be extracted as a separate mathematical utility. All software engineers should then access this code, when needing such a utility. Even in the presence of such effective technical steps, it remains imperative to have ongoing internal education amongst software engineers to inform them of the importance of adhering to these standards while implementing deviating code only under carefully controlled, approved and documented circumstances. This does nothing to help legacy code which may fail to conform to these standards and the re-engineering of such code can be an expensive undertaking.

The demands for ensuring such semantic consistency become significantly more complex when one considers the issue outside the controlled bounds of writing software for only one system. Those more complex issues will be discussed in the next section.

## 6 Next Steps

The round-trip problem, as illustrated, was due to semantic differences across the exchange interface and to the separation distance being near the tolerance limit. The round-trip problem was chosen as one of the simplest transfer problems to illustrate the need for consistent semantics. Clearly, the problems become even more complex for exchange amongst multiple CAD systems.

For ease of exposition, our example deliberately focused upon the use of the same tolerance values. The problem becomes more complex when many different tolerance values are used, as is reasonable in a large system, where the tolerance values are locally adapted to specific algorithmic needs. It then seems appropriate to rationalize the interplay amongst all such values. An assessment of the success of past practice in isolating all such subtle dependencies has been pessimistically expressed [4]:

> System tuning by adjusting tolerances creates a subtle coupling of otherwise disjointed subprocesses within the system. Tolerance matching is then required. This is difficult or impossible to document. To make matters worse, tolerances are the most tempting control handle available for software maintenance. A programmer may fix a problem by changing a tolerance slightly without fully understanding the implications of the change.
>
> Among users, there is likely to be little guidance beyond a few rules of thumb.

However, we do not view this pessimism as reason to despair. Rather, we expect that the failures to date can be somewhat overcome by a more abstract mathematical approach to these problems. The combinatorial complexity certainly argues that simplistic methods will not suffice. The history of developing successful compilers for programming languages should prove instructive. Compiler theory must address issues of significant combinatorial complexity and it was necessary to develop correspondingly rich mathematical semantics for formally verifiable compilers. Such a development of new semantical theory for CAD tolerances is likely to be a long term project, but there are some helpful steps that can be taken in the interim.

Different conceptual approaches to algorithms (such as Manhattan versus Euclidean) could be unified. This might be particularly helpful, if even differences of implementation (squaring function versus repeated multiplication) could be avoided. This might suggest, (at least in the CAD arena) creation of some standardized geometrical libraries. This would be offering a solution by means of 'shared code', only a slightly more sophisticated approach than requiring homogeneous design environments for data exchange. Even shared code would not address the question as to slightly different results on different architectures. The use of such libraries would require extensive re-writing of existing commercial CAD systems and such effort may be economically prohibitive. While such retroactive fixes may not be feasible, that software architecture ideal should serve as a guide to future CAD software design and development.

A more immediate approach may provide a partial response to the issue raised in the abstract, above, relative to user notification. When computed values come 'close' to the tolerance limits of the communicating CAD system and/or of the transfer software, this situation should be flagged *before any transfer* is executed. It would require an interface to internal tolerance values used in both the communicating CAD systems and the transfer software. The result could be a warning, 'near tolerance–may require user intervention'. So warned, the user could then decide to take responsive action.

## References


[1] Desaulniers, H., and Stewart, N. F., *Robustness of numerical methods in geometric computation when problem data is uncertain*, CAD, 25 (9), pp. 539 - 545.

[2] Hu, C.-Y., Patrikalakis, N. M., and Ye, X., *Robust interval solid modeling, Part I: representations*, CAD, 28, 1996, pp. 807–818.

[3] Hu, C.-Y., Patrikalakis, N. M., and Ye, X., *Robust interval solid modeling, Part II: boundary evaluation*, CAD, 28, 1996, pp. 819–830.

[4] NURBS White paper, Automotive Industry Action Group, Southfield, MI, 1991.

[5] PDES, Inc. Geometric Accuracy Team, Interim Report, 24 July, 1996.

[6] 'STEP on a page', http://www.nist.gov/sc5/soap/

# A General Method for Partial Point Set Matching

Michiel Hagedoorn and Remco Veltkamp
Utrecht University, Department of Computing Science
Padualaan 14, 3584 CH Utrecht, The Netherlands
mh@cs.ruu.nl, Remco.Veltkamp@cs.ruu.nl

## 1 Introduction

In applications such as stereo matching, content-based image retrieval, object recognition, and radiotherapy alignment, a major problem is finding a transformation which matches part of a point set $A$ to some part of another point set $B$. For example, $A$ could consist of points extracted from the left image, query image, object model, or patient body data at the planning stage, while the other point set $B$ is formed by points from the right image, database image, range data, or patient body points at the radiation stage.

In this short communication, we will present a method for finding transformations which map some subset of $A$ within a specified Hausdorff distance of $B$, while deciding whether such transformations exists. This method is general in the sense that it works for points of arbitrary dimension, and various classes of transformations, including scaling, general linear transformation, rotation, translation, scaling and translation, affine transformation, and rigid motion.

## 2 Partial Matching

Let $V$ be a finite-dimensional real vector space with a corresponding metric $d : V \times V \to \mathbb{R}$. A frequently used measure of similarity between finite point sets $A, B \subseteq V$ is the Hausdorff distance. Consider for each point $a \in A$ the distance to its closest point in $B$. Now define the *directed* Hausdorff distance $\vec{h}(A, B)$ to be the maximum over all these distances. The (undirected) Hausdorff distance $h$ is given by $h(A, B) = \max\{\vec{h}(A, B), \vec{h}(B, A)\}$. Having a small Hausdorff distance $h(A, B)$, means that each point of $A$ must be close to some point of $B$ and vice versa.

The *partial* Hausdorff distance measures distance between subsets of $A$ and $B$. The directed partial Hausdorff distance with rank $k$ of $A$ relative to $B$, denoted by $\vec{h}_k(A, B)$, can be obtained by sorting the distances of points in $A$ to their closest points in $B$ in increasing order, and taking the $k$-th distance. The (undirected) partial Hausdorff distance is defined as $h_k(A, B) = \max\{\vec{h}_k(A, B), \vec{h}_k(B, A)\}$. Unlike the Hausdorff distance, the partial Hausdorff distance is not always a metric on $\wp(V)$.

For some class of transformations $\mathcal{F}$, finite sets $A, B \subseteq V$, a natural number $k$ and a real $\epsilon > 0$, define the *partial Hausdorff decision problem* as follows:

**Problem 1** *Is there an $f \in \mathcal{F}$ such that $h_k(f(A), B) \leq \epsilon$?*

The *directed* partial Hausdorff decision problem can be defined using $\vec{h}_k$ instead of $h_k$.

Solutions to problems concerning the Hausdorff distance have been presented for various types of transformations. In [1], Huttenlocher *et al.* describe an algorithm for finding the minimal Hausdorff distance under the group of translations in the plane in $O(n^3 \log n)$ time. In [3], Chew *et al.* present a sweep-line algorithm which can be modified to solve Problem 1 for the group of rigid motions in the plane, having an $O(n^5 \log n)$ time complexity.

The mentioned algorithms have high worst-case complexities, and can't be expected to be efficient for the average case either. In [2], Rucklidge and Huttenlocher take an alternative approach to Problem 1 by traversing a multi resolution pyramid built on transformation space. In principle, their method is applicable to many classes of transformations. It performs well for translation or translation combined with scaling, but is less suitable for affine transformations and rigid motion in two or more dimensions. This fact inspired us to devise a more general method.

The next section deals with this method. The subsequent section illustrates our method for specific classes of transformations, including rotation, rigid motion, and affine transformation.

## 3 General Method

First, we describe a general method for solving the directed partial Hausdorff distance decision problem. After that, we show how to extend it to the undirected case, solving Problem 1.

Let $\mathcal{F}$ be a set of functions (transformations) on $V$. We assume there exists a mapping from some finite-dimensional real vector space $V_{\mathcal{F}}$ (configuration space) onto $\mathcal{F}$. For $x \in V$, we define the function $\tau_x : \mathcal{F} \to \mathcal{V}$ by $\tau_x(f) = f(x)$. We assume all functions $\tau_x$ are continuous.

Defining $X^\epsilon = \bigcup_{x \in X}\{y \in V \mid d_V(y, x) \leq \epsilon\}$ allows us to make the following observation:

**Observation 1** $\vec{h}_k(A, B) \leq \epsilon$ *iff there exist at least $k$ points $a \in A$ such that $a \in B^\epsilon$.*

```
PARTIAL-MATCH(F)
1  if γ(F) ≤ δ
2     then return D(F)
3  if D(F)
4     then Cover F with sets {F_1, ..., F_c}
5            such that diam(F_i) < diam(F).
6          for i ← 1 to c
7          do if PARTIAL-MATCH(F_i)
8                then return TRUE
9  return FALSE
```

Algorithm 1: Decision Procedure

Let $F \subseteq \mathcal{F}$. Observe that if $f(a) \in B^\epsilon$ for some $f \in F$, the intersection of $\tau_a(F)$ and $B^\epsilon$ is not empty. If we apply this idea to a point set $A$ instead of a single point $a$, we can use Observation 1 to come at the following lemma:

**Lemma 1** *Let $F \subseteq \mathcal{F}$. If $F$ contains a transformation $f$ such that $\vec{h}_k(f(A), B) \leq \epsilon$, then there are at least $k$ points $a \in A$ such that $\tau_a(F) \cap B^\epsilon \neq \varnothing$.*

Let $D(F)$ be the predicate that is true exactly when there are at least $k$ points $a \in A$, such that $\tau_a(F) \cap B^\epsilon \neq \varnothing$. Lemma 1 tells us that whenever there is an $f \in F$ such that $\vec{h}_k(f(A), B) \leq \epsilon$, the predicate $D(F)$ must be true. Define the function $\gamma$ on bounded sets $F \subseteq \mathcal{F}$ by $\gamma(F) = \max_{a \in A} \operatorname{diam}(\tau_a(F))$, where $\operatorname{diam}(X) = \sup_{x,y \in X} d_V(x, y)$. The predicate $D$ and the function $\gamma$ are used in the following lemma:

**Lemma 2** *If there is an $f \in F$ such that $\vec{h}_k(f(A), B) > \epsilon + \gamma(F)$, then $D(F)$ is false.*

Lemma 1 and 2 lead to Algorithm 1, which solves the directed partial Hausdorff decision problem for bounded $F \subseteq \mathcal{F}$. Algorithm 1 can be used with any desired accuracy $\delta > 0$.

An important observation is that at the moment that our algorithm returns TRUE, all mappings in $f$ in the current $F$ are such that $\vec{h}_k(f(A), B) \leq \epsilon + \delta$. This means that if some $f \in F$ exists for which $\vec{h}_k(f(A), B) \leq \epsilon$, the algorithm is guaranteed to find some $f \in F$ for which $\vec{h}_k(f(A), B) \leq \epsilon + \delta$.

**Theorem 1** *If there is an $f \in F$ such that $\vec{h}_k(f(A), B) \leq \epsilon$, then Algorithm 1 will return* TRUE. *If for all $f \in F$, $\vec{h}_k(f(A), B) > \epsilon + \delta$, then Algorithm 1 will return* FALSE.

Algorithm 1 can be modified to work for the undirected problem by replacing $D(F)$ with a stronger predicate $U(F)$, which can be obtained from $D(F)$ by demanding that, in addition, at least $k$ points $b \in B$ belong to $\cup_{a \in A} \tau_a(F)^\epsilon$.

Suppose $\mathcal{F}$ is a group of transformations on $V$ that leaves distance $d$ invariant, with a continuous inversion mapping $i : f \mapsto f^{-1}$. Then, $\vec{h}_k(B, f(A)) = \vec{h}_k(f^{-1}(B), A)$, meaning we can write the predicate $U(F)$ as the conjunction of $D(F)$ and $D(F^{-1})$ (the latter with $A$ and $B$ reversed). Consider, as an example, the group of rigid motions combined with the Euclidean distance.

## 4 Transformation Classes

In this section we explain how to apply the general method outlined above to several types of transformations. For clarity, we will only focus on its directed version.

Let $V = \mathbb{R}^d$. First, we will deal with the set of translations ($\mathcal{F}$) on $\mathbb{R}^d$ and the $L_\infty$ distance ($d$). Every vector $v \in \mathbb{R}^d$ defines a unique translation $t : x \mapsto v + x$. We can apply Algorithm 1 for rectangular regions $F \subseteq \mathcal{F}$. The computation of a finite cover (line 4) can be done simply by splitting $F$ into several smaller rectangular regions $F_i$. Testing the predicate $D(F)$ reduces to performing evaluations $\tau_a(F) \cap B^\epsilon \neq \varnothing$, or equivalently $\tau_a(F)^\epsilon \cap B \neq \varnothing$, for a number of points $a \in A$. For rectangular regions $F$, we get rectangular $\tau_a(F)^\epsilon$. Whether such a rectangle contains some point in $B$ or not can be determined efficiently using a kd-tree or a range tree. Alternatively, for square $F \subseteq \mathcal{F}$, every $\tau_a(F)^\epsilon$ is square. This means we can evaluate $D(F)$ efficiently by probing an $L_\infty$-distance transform built on $B$.

Also for scaling, linear transformations, scaling and translation, and affine transformation we can make a mapping from configuration space $V_\mathcal{F}$ to transformation space $\mathcal{F}$ such that each rectangular $F \subseteq \mathcal{F}$ results in a rectangular $\tau_a(F)$. This means that for these transformations we can use kd-trees or range trees just like for translation.

More interesting applications are rotation and rigid motion, under the Euclidean distance. For these instances, choosing a convenient mapping from configuration space $V_\mathcal{F}$ to $\mathcal{F}$ is less trivial. This is due to the properties of rotation. It would be nice to have a mapping $V_\mathcal{F} \to \mathcal{F}$ such that each region $\tau_a(F)$ has a simple shape for a simple-shaped $F$. This would allow some data structure for the efficient evaluation of $\tau_a(F)^\epsilon \cap B \neq \varnothing$.

A rotation in the plane can be represented by a single angle. Choosing $F$ to be a closed interval in $\mathbb{R}$, gives a $\tau_a(F)$ which is a circular arc. Rotations in $d \geq 3$ dimensions however, are determined by $\frac{1}{2}d(d-1)$ angles. Using this mapping $V_\mathcal{F} \to \mathcal{F}$, a rectangular region $F \in \mathcal{F}$ results in a $\tau_a(F)$ which is not easy to use. However, we can bound these shapes with blocks whose sizes go to zero as $F$ converges to any point. The intersection of such a block and the surface of a sphere, resulting in a ball-segment still fully contains the corresponding $\tau_a(F)$. If we replace each $\tau_a(F)$ by such a ball-segment we get correct results. Now we can use an efficient procedure on a kd-tree to test whether or not a ball-segment intersects $B^\epsilon$.

Rigid motions, being products of rotations with translations, can be dealt with by using the Minkowski sum of ball-segments resulting from rotation and rectangular regions resulting from translation, and using these sums instead of $\tau_a(F)$.

## 5 Conclusion

We presented a general technique for solving both directed and undirected versions of the partial Hausdorff decision problem under several types of transformations in various dimensions. The basic algorithm also finds transformations which map some subset of $A$ within some specified distance of some subset of $B$, if these exist.

We implemented the algorithm for vector spaces of arbitrary dimension and the following choices of $\mathcal{F}$: scaling, linear transformation, rotation, translation, scaling and translation, affine transformation, and rigid motion. The implementation shows good results for all of these groups in the plane. In three dimensions, good results have been obtained for translation, translation combined with rotation around a fixed axis, and pure rotation. In general, our algorithm processes far less regions $F$ than the method given in [2].

Figure 1 shows an image of the Mir space station. Edge detection and thresholding yields a model point set $A$, the

Figure 1: Model image.

Figure 2: Target image.

black dots in figure 3. Similarly, figure 2 yields a target point set $B$, the grey dots in figure 3. Note that these images are taken from different viewpoints, and are not just rotated copies of one another. Figure 3 shows the resulting match under rigid motion and the directed partial Hausdorff distance, so that most of the black dots lie close to a grey dot.

Without much trouble, our approximate decision algorithm can be converted into an approximate optimisation algorithm, which finds transformations that lie within a specified distance from an actual minimum. Advantages of our method over other existing methods lie in its relative efficiency, generality and simplicity.

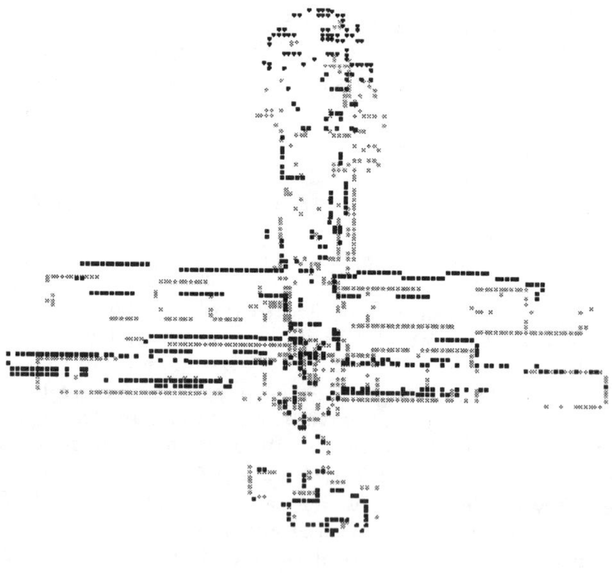

Figure 3: Match.

## Acknowledgement

This research was partially supported by the ESPRIT IV LTR Project No. 21957 (CGAL).

## References

[1] Daniel P. Huttenlocher, Klara Kedem, and Micha Sharir. The upper envelope of Voronoi surfaces and its applications. *Discrete Computational Geometry*, 9:267–291, 1993.

[2] Daniel P. Huttenlocher and William J. Rucklidge. A multi-resolution technique for comparing images using the Hausdorff distance. Technical Report 92-1321, Cornell University, 1992.

[3] L. Paul Chew, Michael T. Goodrich, Daniel P. Huttenlocher, Klara Kedem, Jon M. Kleinberg, and Dina Kravets. Geometric pattern matching under Euclidean motion. In *Fifth Canadian Conference on Computational Geometry*, pages 151–156, 1993.

# An Interactive Environment for Visualizing and Animating Algorithms


Jürgen Döllner, Klaus Hinrichs, Hermann Spiegel

FB 15, Institut für Informatik, Westfälische Wilhelms-Universität
Einsteinstr. 62, D - 48149 Münster, Germany
E-Mail: {dollner | khh | spiegel}@uni-muenster.de



## Abstract

This communication describes an interactive environment for visualizing and animating geometric algorithms and geometric data. In contrast to traditional systems for algorithm animation, our approach is based on a new paradigm for modeling geometry and behavior, the *geometry graphs* and *behavior graphs*. They allow developers to specify complex geometric representations of algorithm-specific geometric data, to relate time- and event-dependent processes to geometric data, and to add interactive capabilities to both animation and geometry components. This paradigm reduces the complexity to specify the visualization and animation of an algorithm significantly. Visualization, animation and algorithms are glued together by an interactive environment based on [incr Tcl], an object-oriented extension to Tcl/Tk. The system has been used successfully for interactive animations of computational geometry algorithms, to produce high-quality videos, and as a graphics kernel for 3D applications.



**Keywords:** Algorithm Animation, Geometric Modeling, Visualization, Computer Graphics, Simulation.


## Introduction

Computational geometry benefits from algorithm visualization and animation because these techniques allow us to explore how algorithms work, and to see how and when these algorithms transform geometric data. Traditional algorithm animation systems are faced by several limitations. Most systems require that an algorithm represents its geometric data in terms of built-in geometric entities which leads to data duplication and a loss of application semantics in the visualization subsystem. Another limitation results from a tight coupling to a specific rendering toolkit. In this case, we cannot exchange, for example, a fast real-time graphics engine (e.g., used for interactive simulations) with a high-quality photorealistic renderer (e.g., used for video production).

The description of time- and event-dependent behavior of algorithms and geometric data is of considerable complexity. Most (procedural) approaches to algorithm animation cannot reflect the dynamic relationships between algorithms and geometric data in an object-oriented fashion.



In addition, algorithm animation is still limited with respect to 3D interaction capabilities since 3D interaction requires input techniques which are typically less developed in the underlying low-level 3D graphics systems. However, 3D interaction is the key to interactive algorithm animations which offer new strategies for exploring the dynamic behavior of algorithms (e.g., modification of input data in real-time).

Another more technical issue is the integration of algorithm libraries into the animation system. If an animation system forces the re-coding of an algorithm, or if it does not link to state-of-the-art user interface technology, developers will not use it. There are also competing architecture requirements: graphics and algorithms should be compiled in order to be fast, whereas interpretative languages make animation design, algorithm test, and user interface development easier. Compared to existing systems for algorithm animation (e.g., GASP [12] or Obliq-3D [8]), our system focusses on an explicit time management, 3D interaction techniques, and an extendible 3D graphics kernel.

## System Architecture

The architecture of our system consists of two layers, the MAM modeling layer, and the VRS rendering layer (Fig. 1). The *Modeling and Animation Machine* [3] is responsible for the geometric modeling and the modeling of animation and interaction. The *Virtual Rendering System* [2] provides a uniform object-oriented interface to different low-level 3D rendering toolkits, e.g. OpenGL, Radiance, POVRay and RenderMan. MAM/VRS operates on shared geometric and graphics objects which represent algorithm (i.e., application) data. Objects from all subsystems can be accessed by an [incr Tcl] [11] shell.

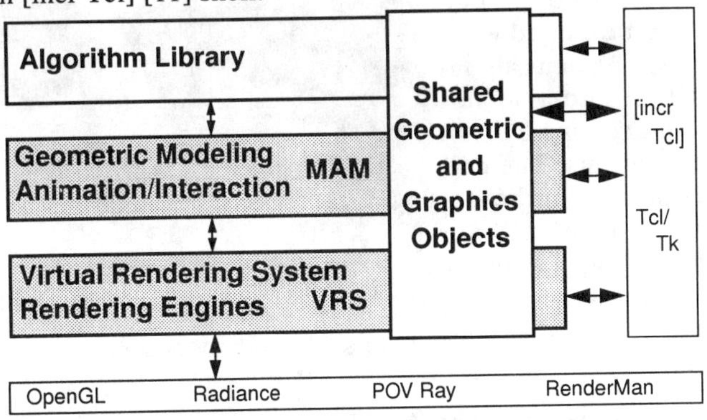


Figure 1: MAM/VRS System Architecture.

## Integration of Geometric Data

One of the key differences to existing systems is that we explicitly encourage the usage of algorithm-specific data types in the modeling and rendering subsystems. There is no notion of "built-in" data types (although a large variety of classes is supplied) since MAM/VRS allows developers to integrate their own data types by inheritance. As a consequence, data duplication is avoided since geometric data need not be converted into built-in graphics primitives, and semantics of data types are available both to animation and rendering. If new algorithm-specific types are integrated into MAM/VRS, they can use the animation features and the scripting language. Geometric and graphics objects are associated with *geometry nodes* which are organized in directed acyclic *geometry graphs*.

In our experience, an integrated solid modeler has been a valuable tool for algorithm animation. For example, it allows to explore polyhedral objects using solid operators.

Example: Regular Polyhedron with Hidden Symmetries

The polyhedral realization of Coxeter's {4,6|3} exposes hidden symmetries [1]. To inspect the geometric structure of the polyhedron, we project the edges of the polyhedron onto a canvas (Fig. 2), and we allow the user to split the polyhedron by an arbitrary plane. During the rotation of the polyhedron, its projected edges are updated.

Figure 2: Polyhedron and Edge Projection. Splitted Polyhedron.

Solid modeling operations are essential to achieve constructive geometric results.

Example: 3D Voronoi Cells

We use an interactive application to construct and explore 3D Voronoi cells. The cells are created when new 3D points are inserted into a default solid volume by split operations. The cell complex can be decomposed interactively, and a clipping plane can be used to extract cross sections (Fig. 3).

Figure 3: 3D Voronoi Cell Inspector.

## Animation of Algorithms

Algorithm animations are specified by directed acyclic graphs called *behavior graphs*. They control the flow of time and events, constrain geometric and graphics objects, and can be used to code story books. Geometry and behavior graphs are tightly coupled since both are associated with shared geometric and graphics objects (Fig. 4). Behavior nodes calculate, distribute and transform model time and time intervals. Each geometric object has its time requirement and gets synchronized when a new point in time (of its local model time) is reached.

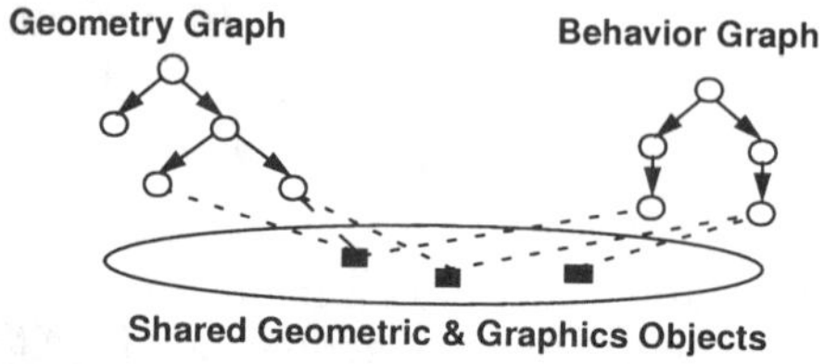


Figure 4: Object sharing between the graphs.

For example, to translate a 3D point onto a paraboloid, we associate the point with a behavior node which constrains its position, and assign a time requirement (i.e, the desired duration) to the behavior node. If we activate the behavior node, a time process is started and sends synchronization events to the behavior node which in turn calculates and updates the position of the point.

Example: Constructing a Voronoi Diagram

For a set of *n* points in a d-dimensional Euclidean space, the arrangement induced by these sites is defined by the *n* hyperplanes that are tangential to the *(d+1)*-dimensional unit parabola at the points lifted to the parabola [5]. For *d=2* we can visually construct the 2-dimensional Voronoi diagram as follows: (1) the points are lifted from the *xy*-plane onto the unit paraboloid, (2) planes that are tangential to the lifted points on the paraboloid are faded in, (3) the plane and the paraboloid are faded out, and finally the Voronoi diagram is exposed (Fig. 5) by looking from above into the open paraboloid.

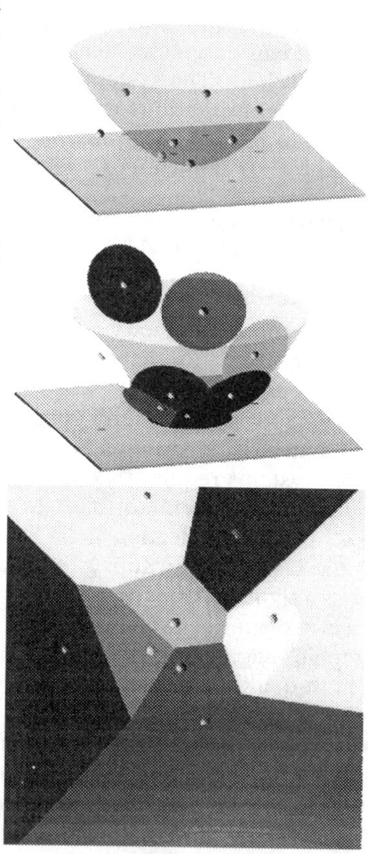

Figure 5: Tangential Planes Constructing the Voronoi Diagram.

For each point we create a sphere object and a tangential plane object. In (1), we constrain the sphere position of each site by a translation constraint behavior node. The speed is specified relative to the distance to the paraboloid, and expressed by the time requirements of the behavior nodes. The growing of the tangential planes is represented by behavior nodes which scale the facet. Fade-in and fade-out effects are achieved by constraining the transmission attributes of the *xy*-plane and the paraboloid.

We observed that the time management facilitates the design of large and complex animations since we can use a variety of basic behavior types which can be combined and specialized, instead of purely functional descriptions [3]. For example, time manipulator nodes can be mixed arbitrarily with constraint behavior nodes. For example, to slow down the lifting of a sphere close to its final position, we prefix its behavior node with a "creep" behavior node.

## Interactive Animations

Interaction nodes control the flow of events in behavior graphs and manipulate associated shared geometric and graphics objects. Interaction nodes are specialized behavior nodes and contained in behavior graphs. They can be combined with time-related behavior nodes and form part of animations. For example, while a sweep plane is moving, the 3D scene can be rotated by a trackball.

Interaction can be used in algorithm animations to configure geometric data, to control and design the animation steps, and to observe the execution of the algorithm. One limitation of many (algorithm) animation systems is that they do not offer *three-dimensional* and *precise* interaction mechanisms. For example, during the motion of a point aligned to a plane, the interaction mechanism must ensure that the mouse position and the associated 3D position coincide, and that the point remains constrained to the plane.

MAM/VRS offers a collection of interaction classes which can be customized to algorithm-specific geometric data, (e.g. 3D translators constrained to lines or planes) and selection mechanisms for arbitrarily contoured selection regions. The orthogonality of geometry nodes and behavior nodes, and their association to a pool of shared application objects provides a clear organization scheme for animations.

### Example: Interactive Animation of Fortune's Algorithm

Fortune [6] developed an optimal algorithm for constructing the Voronoi diagram by sweeping the plane with a parabolic front [9]. In our animation we create for each point a sphere and a cone; the cone is erected upon the point. The user can drag interactively the points, and see how the Voronoi diagram changes. The interaction handler must guarantee that each point is not dragged outside the *xy*-plane. The camera can be animated to obtain the "best view" from $z = -\infty$. Additionally, the parabolic front is obtained by intersecting a 45-degree plane with the cones. The user can sweep the plane along the *x*-axis. Fig. 6 (left) shows the 3D configuration; Fig. 6 (right top) shows the parabolic front, and Fig. 6 (right bottom) exposes the Voronoi diagram (the semi-transparent circular disk which embeds the given points is provided as help for orientation).

## Implementation

The system is implemented in C++. We provide a C++ and an [incr Tcl] application programming interface. The mapping between C++ and [incr Tcl] is based on [7]. Bindings for different user interface toolkits are available (e.g., Motif, ViewKit, Tk [10]). Algorithm animations can be previewed

Figure 6: Interactive Animation of Fortune's Algorithm.

and designed with the OpenGL rendering engine. To produce a high-quality video, MAM/VRS supports among different renderers Pixar's RenderMan. A native Java implementation is under development.

Some animations and more details about MAM/VRS can be found at `http://wwwmath.uni-muenster.de/~mam`.

## References


1. J. Bokowski, J. M. Wills: *Regular Polyhedra with Hidden Symmetries*, The Mathematical Intelligencer, **10**(1), 27-32 (1988).
2. J. Döllner, K. Hinrichs: *The Virtual Rendering System - A Toolkit for Objet-Oriented 3D Rendering*, EduGraphics/CompuGraphics, Combined Proceedings, 309-318, 1995.
3. J. Döllner, K. Hinrichs: *Object-Oriented 3D Modeling, Animation and Interaction*, The Journal of Visualization and Computer Animation, **8**(1), 33 - 64 (1997).
4. J. Döllner: *Object-Oriented 3D Modeling, Animation and Interaction*, PhD Thesis, University of Münster, 1996.
5. H. Edelsbrunner, R. Seidel: *Voronoi diagrams and arrangements*, Discrete & Computational Geometry, 1, 25-44 (1986).
6. S. Fortune: *A Sweepline Algorithm for Voronoi Diagrams*, Algorithmica **2**, 153-174 (1987).
7. W. Heidrich, P. Slusallek, H.-P. Seidel: *Using C++ Class Libraries from an Interpreted Language*, Technology of Object-Oriented Languages & Systems, TOOLS 14 (Santa Barbara), 397-408, 1994.
8. M. A. Najork, M. H. Brown, *Obliq-3D: A High-Level, Fast-Turnaround 3D Animation System*. IEEE Transactions on Visualization and Computer Graphics, **1**(2), 175-192 (1995).
9. J. O'Rourke. Computational Geometry in C, Cambridge University Press, 1994.
10. J. K. Ousterhout, *Tcl and the Tk Toolkit*, Addison-Wesley, 1994.
11. M. J. McLennan, *[incr Tcl]: Object-Oriented Programming in Tcl,* Proceedings of the Tcl/Tk Workshop 1993, University of California, http://www.wn.com/biz/itcl.
12. A. Tal and D. Dobkin *Visualization of Geometric Algorithms*, IEEE Transactions on Visualization and Computer Graphics, **1**(2), 194-204 (1995).

# Classical Computational Geometry in GeomNet*

Gill Barequet† Stina S. Bridgeman‡ Christian A. Duncan† Michael T. Goodrich†
Roberto Tamassia‡

## Abstract

In this paper we present *GeomNet*, a system for performing distributed geometric computing over the Internet. We also provide several examples of actual geometric algorithms that our system already supports. Application domains for GeomNet include collaborative research and distance education.

**Keywords:** geometric computing, Internet computing, cooperative computing, visualization.

## 1 Introduction

Researchers in computational geometry are developing a host of efficient algorithms for solving geometric problems, and many are taking the design of these algorithms all the way to working implementations. This implementation component is no small task, for computational geometry programs must deal with many issues that are not faced in other software applications. This is largely due to the fact that geometric data typically involves a non-trivial combination of numerical and combinatorial relationships, and much care must be taken to develop formats for representing geometric data and for performing robust and reliable computations using such representations. Fortunately, there is a rich, growing literature directed at developing techniques for dealing with such robustness issues [1, 7, 9, 10, 11, 12, 13, 15, 18, 21]. Moreover, fundamental geometric algorithms are being redesigned taking into account robustness and arithmetic precision issues [9, 16]. Another important aspect of software development in computational geometry is the visualization of geometric objects and the animation of geometric algorithms, with the dual purpose of demonstration and debugging [14].

### 1.1 Collections of Geometric Software

Collections of geometric software can be found on the Internet at Amenta's "Directory of Computational Geometry Software" site,[1] and a host of applications of such software can be found at Eppstein's "Geometry in Action" site.[2] Indeed, Erickson has an even more extensive collection of geometric software at his "Computational Geometry Software" site,which is actually just a small part of his extensive directory of "Computational Geometry Pages."[3] In addition, there is a large collaborative effort underway to develop an extensive Computational Geometry Algorithms Library (CGAL)[4] [8, 20], which itself is built upon the successful Library of Efficient Data Types and Algorithms (LEDA)[5] [6, 17]. This effort is directed at building a large collection of C++ routines for solving computational geometry problems. Therefore, in spite of the difficulties presented by geometric algorithm implementation, there is now a rich collection of robust implementations of geometric algorithms, and this collection should grow significantly in the years ahead.

Unfortunately, each package of geometric software typically creates its own unique, incompatible data format. Unlike many other domains such as numerical analysis, which have simple, standard notions for describing problems (e.g., using matrices), there are no standard mechanisms for fully representing geometric data. The computational geometry research community could, of course, spend a considerable amount of effort on designing a standard language for describing geometric data, but this does not seem likely. Such an effort would necessitate the encoding of all possible relationships between the numerical and combinatorial components of geometric data, which would be a Herculean task.

We feel that practitioners and researchers can take better advantage of the latest developments in computational geometry if implementations of geometric algorithms are made widely available on the Internet and are fully usable without the learning of new file for-

*The work of all five authors is supported in part by the U.S. Army Research Office under grant DAAH04–96–1–0013. In addition, work of the second author is supported in part by a National Science Foundation Graduate Fellowship, work of the fourth author is supported in part by the National Science Foundation under grant CCR–9625289, and work of the fifth author is supported in part by the National Science Foundation under grant CCR–9423847.

†Center for Geometric Computing, Dept. of Computer Science, Johns Hopkins University, Baltimore, MD 21218. E-mail: `[barequet|duncan|goodrich]@cs.jhu.edu`

‡Center for Geometric Computing, Dept. of Computer Science, Brown University, Providence, RI 02912. E-mail: `[ssb|rt]@cs.brown.edu`



*Computational Geometry 97* Nice France


[1]`http://www.geom.umn.edu/software/cglist/`
[2]`http://www.ics.uci.edu/~eppstein/geom.html`
[3]`http://www.cs.duke.edu/~jeffe/compgeom/`
[4]`http://www.cs.ruu.nl/CGAL/`
[5]`http://www.mpi-sb.mpg.de/LEDA/leda.html`

mats, calling sequences, or class hierarchies in software libraries. In this paper we present a system designed to address these issues, give examples of applications supported by the current prototype implementation, and discuss further extensions.

## 2 The GeomNet Vision

We envision an "Internet computing" framework, which we call *GeomNet*, where a family of *geometric computing servers* execute a variety of geometric algorithms on behalf of remote clients, which can be either users interacting through a Web browser interface, or application programs connecting directly through sockets. Such an environment could be used in a variety of scenarios, including execution of algorithms on behalf of remote clients, checking geometric structures for consistency, experimental study and comparison of algorithms, design of new algorithms through plug-and-play integration of existing algorithms, demonstrations in an educational setting, interactive electronic books and manuals, and electronic commerce for CAD and GIS companies.

### 2.1 Cooperative Computing

The goal for GeomNet is a networked geometric computing environment whose functionality is independent of whether the user is interfacing with it through the Internet on a notebook computer or locally on a high-end graphics workstation. Thus, our system is based on a layered client-server architecture. The system (with minimal user guidance) forwards difficult computations to a compute server, converts between various file formats, performs a requested computation, and sends the results back to the user. We refer to such negotiations between the client and server as *cooperative computing*, with the following scenarios serving as typical levels of computational commitment on the part of the client:

**Level 0**. The client has minimal computational and graphical resources. In this case, all computation, including rendering, must be performed on the server and transferred to the client as images or display primitives.

**Level 1**. The client has some computational and graphical resources but limited network bandwidth. In this case, non-CPU computations, such as those pertinent to the graphical user interface, should be performed on the client, while CPU-intensive computations should be performed by the server or another powerful computer on the network.

**Level 2**. The client has workstation-quality computational and graphical resources and high network bandwidth. In this case all but the most CPU-intensive computations should be performed on the client, with very CPU-intensive computations still performed by the server or another powerful computer on the network.

**Level 3**. The client has significant computational and graphical resources (but not necessarily very high network bandwidth). In this case the client is better off performing the entire computation locally (with the server performing only code-shipping services).

Levels 0 and 3 are well-developed paradigms in distributed computing, with Level 0 cooperative computing fitting the X network-transparent window system, and Level 3 cooperative computing fitting the standard Java computing paradigm. Our intention, however, is to define a system that can negotiate with a client to determine the most efficient level of cooperation. The intermediate levels also allow the integration of existing implementations not written in Java as well as enabling users to access code that the authors may not wish to make available, even in compiled form. Thus, our design of GeomNet is concerned mostly with Level 1 and Level 2 cooperative computing.

### 2.2 System Goals

Specific goals of the GeomNet system embodying our vision of a cooperative computing environment include:

1 *Security*: both client and server should be safe from intruder attacks;

2 *Code protection*: code should be protected from software piracy as much as possible;

3 *Authoring*: in order to promote the development of a user community, both new and legacy implementations should be easily integrated into the system;

4 *Efficiency*: computations should be performed quickly and require minimum communication costs;

5 *Correctness*: the system should provide "certificates" of the correctness of its computations; and

6 *Cooperation*: users with all levels of computing power should be able to easily perform interesting geometric computations and visualizations.

Our GeomNet system currently satisfies each of the first four goals to some degree. It provides security inherited from the Java virtual environment, code protection between client and server and between separate components of the server, authoring through a modular design allowing even non-Java packages to be integrated with a simple application "wrapper" mechanism, and efficiency through the integration of well-designed implementations of fast algorithms. With further integration of applications we hope that it will soon also satisfy the latter two design goals as well. Still, even in its current form, the GeomNet environment provides a powerful tool for those applications that interact with geometric data or can benefit from geometric visualizations of non-geometric data.

In addition, we feel that one of the prime strengths of GeomNet is that it complements rather than replaces the traditional model of supporting the development of geometric computing applications by providing software libraries. Indeed, for GeomNet to reach its full potential we need to be able to integrate as many tools for performing geometric computations as is possible (and allowed). A prototype implementation of GeomNet is available on the Internet through the support of the Center for Geometric Computing,[6] with prototype (pre-alpha release) servers available at Brown's GeomNet site[7] and the GeomNet site[8] at Johns Hopkins. New releases are planned throughout 1997.

[6] `http://www.cs.brown.edu/cgc/`

[7] `http://loki.cs.brown.edu:8081/geomNet/`

[8] `http://www.cgc.cs.jhu.edu/geomNet/`

## 3 Applications

Having covered some of the features of GeomNet, let us now discuss some example applications.

**Fundamental Algorithms**. One of the prime applications that we envision is for GeomNet to be used to invoke implementations of fundamental geometric algorithms. We have therefore included in our prototype implementations of several such fundamental algorithms, including 2-D and *d*-D convex hulls, Delaunay triangulations and Voronoi diagrams, and width of a *d*-D point set. We have developed several interface mechanisms for invoking these routines. The simplest interface is the interactive applet interface designed to gain intuition about geometric algorithms as well as for testing GeomNet servers. For handling large geometric data files we provide an HTML forms interface, and for production-level usage we provide a socket interface which may be invoked by an application program running on the client. The output from the server can be immediately sent to an interactive display system running on the client or piped into a client application.

**Graph Drawing Server**. The Graph Drawing tool can be used for drawing graphs for documents or user applications, studying and comparing graph drawing algorithms, translating between formats for describing graphs and their drawings, creating a database of graphs occurring in user applications, and providing demonstrations in an educational setting. The current prototype supports three algorithms for orthogonal drawings of general graphs, and two specialized algorithms for acyclic digraphs and series-parallel digraphs. The user is free to specify any input/output format independently of the algorithm requested.

**Geometric Algorithm Animation**. We are incorporating geometric algorithm animation capabilities into GeomNet (at cooperative computing levels 1 and 2) by extending our previous work on the Mocha model for Web-based algorithm animation [2, 3, 4, 5]. Within the GeomNet client-server architecture, Mocha optimally partitions the software components of a typical algorithm animation system. A variety of animations of fundamental 2-D geometric algorithms implemented within LEDA are available through Mocha.

## 4 Conclusion

We have presented the GeomNet architecture for performing geometric computing over the Internet. We anticipate the availability of a significant variety of applications based on the GeomNet architecture, mainly due to the small amount of effort required for integrating new applications into the modular system. It is our feeling that such a system provides the accessibility of geometric algorithms while overcoming the usual problems of lack of knowledge of where to look for such algorithms, differences in interfaces and in file formats, and unavailability of sufficient computer resources. All these problems are handled by GeomNet, which we consider as a significant step towards sharing geometric data and computing resources.

## References


[1] F. Avnaim, J.-D. Boissonnat, O. Devillers, F. Preparata, and M. Yvinec. Evaluation of a new method to compute signs of determinants. In *Proc. 11th Ann. ACM Symp. Comp. Geom.*, C16–C17, 1995.

[2] J. E. Baker, I. F. Cruz, G. Liotta, and R. Tamassia. A new model for algorithm animation over the WWW. *ACM Comp. Surv.*, 27(4):568–572, 1995.

[3] J. E. Baker, I. F. Cruz, G. Liotta, and R. Tamassia. Algorithm animation over the World Wide Web. In *Proc. Int. Workshop on Advanced Visual Interfaces (AVI '96)*, 203–212, 1996.

[4] J. E. Baker, I. F. Cruz, G. Liotta, and R. Tamassia. Animating geometric algorithms over the Web. In *Proc. 12th Ann. ACM Symp. Comp. Geom.*, C3–C4, 1996.

[5] J. E. Baker, I. F. Cruz, G. Liotta, and R. Tamassia. The Mocha algorithm animation system. In *Proc. Int. Workshop on Advanced Visual Interfaces (AVI '96)*, 248–250, 1996.

[6] C. Burnikel, J. Könnemann, K. Mehlhorn, S. Näher, S. Schirra, and C. Uhrig. Exact geometric computation in LEDA. In *Proc. 11th Ann. ACM Symp. Comp. Geom.*, C18–C19, 1995.

[7] H. Edelsbrunner and E. P. Mücke. Simulation of simplicity: a technique to cope with degenerate cases in geometric algorithms. *ACM Trans. Graph.*, 9:66–104, 1990.

[8] A. Fabri, G.-J. Giezeman, L. Kettner, S. Schirra, and S. Schoenherr. The CGAL kernel: A basis for geometric computation. In *Applied Computational Geometry (Proc. WACG '96)*, Lecture Notes in Computer Science. Springer-Verlag, 1997.

[9] S. Fortune. Numerical stability of algorithms for 2-d Delaunay triangulations. *Int. J. Comp. Geom. Appl.*, 5(1):193–213, 1995.

[10] S. Fortune and V. Milenkovic. Numerical stability of algorithms for line arrangements. In *Proc. 7th Ann. ACM Symp. Comp. Geom.*, 334–341, 1991.

[11] D. H. Greene and F. F. Yao. Finite-resolution computational geometry. In *Proc. 27th Ann. IEEE Symp. Found. Comp. Sci.*, 143–152, 1986.

[12] L. Guibas and D. Marimont. Rounding arrangements dynamically. In *Proc. 11th Ann. ACM Symp. Comp. Geom.*, 190–199, 1995.

[13] L. J. Guibas, D. Salesin, and J. Stolfi. Epsilon geometry: building robust algorithms from imprecise computations. In *Proc. 5th Ann. ACM Symp. Comp. Geom.*, 208–217, 1989.

[14] A. Hausner and D. Dobkin. Making geometry visible: An introduction to the animation of geometric algorithms. In J.-R. Sack and J. Urrutia, editors, *Handbook on Computational Geometry*, North Holland, 1997. To appear.

[15] C. M. Hoffmann, J. E. Hopcroft, and M. S. Karasick. Towards implementing robust geometric computations. In *Proc. 4th Ann. ACM Symp. Comp. Geom.*, 106–117, 1988.

[16] G. Liotta, F. P. Preparata, and R. Tamassia. Robust proximity queries in implicit Voronoi diagrams. TR CS-96-16, Center for Geometric Computing, CS Dept., Brown Univ., Providence, RI, 1996.

[17] K. Mehlhorn and S. Näher. LEDA: a platform for combinatorial and geometric computing. *Commun. ACM*, 38:96–102, 1995.

[18] V. Milenkovic. Double precision geometry: a general technique for calculating line and segment intersections using rounded arithmetic. In *Proc. 30th Ann. IEEE Symp. Found. Comp. Sci.*, 500–505, 1989.

[19] T. Munzner, S. Levy, and M. Philips. Geomview: A system for geometric visualization. In *Proc. 11th Ann. ACM Symp. Comp. Geom.*, C12–C13, 1995.

[20] M. H. Overmars. Designing the computational geometry algorithms library CGAL. In *Applied Computational Geometry (Proc. WACG '96)*, Lecture Notes in Computer Science. Springer-Verlag, 1997.

[21] C. K. Yap. A geometric consistency theorem for a symbolic perturbation scheme. *J. Comp. Syst. Sci.*, 40:2–18, 1990.

# Empirical Testing of Algorithms for Variable-Sized Label Placement


*Jon Christensen*
Painted Word
124 Mt. Auburn St.
Cambridge, MA 02138
jon_christensen@paintedword.com

*Stacy Friedman*
Oracle Corp.
500 Oracle Parkway
Redwood Shores, CA 94065
sfriedma@us.oracle.com

*Joe Marks*
MERL — A Mitsubishi Electric Research Laboratory
201 Broadway
Cambridge, MA 02139
marks@merl.com

*Stuart M. Shieber*
Harvard University
33 Oxford St.
Cambridge, MA 02138
shieber@deas.harvard.edu


The development of methods for placing labels on map features is a central problem in automated cartography. A well-studied problem of this sort is the *point-feature label placement* (PFLP) problem: Given a set of points and a set of candidate label positions adjacent to each point find the choice of label position for each point that minimizes the total number of label-label and label-point overlaps. Sample labelings of this sort are depicted in Figure 1. This NP-hard problem has been attacked by a number of researchers. Thorough empirical testing has shown that a heuristic algorithm based on simulated annealing (SA) outperforms all previously published practical algorithms for this problem [1].

Recently, Wagner and Wolff [3] have explored a variant of the PFLP problem. In their problem, which we call *variable-sized point-feature label placement* (VPFLP), the goal is to find the largest label scale at which the set of points can be labeled with zero overlaps; that is, only perfect labelings are allowed, but the size of the labels is allowed to vary uniformly. (See Figure 1.) Wagner and Wolff present an algorithm for the specific case of VPFLP in which each point has four candidate label positions (4-VPFLP). Wagner and Wolff do not compare their own algorithm for this problem with those of other researchers. In this note, we report on empirical testing of Wagner and Wolff's algorithm (which we will refer to as WW) for 4-VPFLP with the SA algorithm. We find that SA performs essentially identically to WW, though it is slower. However, the generality of the SA algorithm (explored, for instance, by Edmondson et al. [2]) means that, unlike WW, SA can be applied to a far broader range of VPFLP variants. Finally, we address the question of whether the clever preprocessing method used in WW might be advantageously applied to PFLP by using it as a preprocessing step for SA, and show that it provides no advantage.

The WW algorithm for VPFLP works by performing a simple binary search over scales; at each candidate scale, a test is performed to determine whether a perfect labeling exists. To perform this *admissibility test*, any algorithm for PFLP may be used, but Wagner and Wolff take advantage of the fact that finding a good but imperfect solution to this embedded PFLP problem is profligate; they propose instead a very fast heuristic that determines quickly whether a perfect labeling exists. Their admissibility test is particularized to the case where each point has only four candidate label positions. It works by (i) a *preprocessing phase* to eliminate candidate positions that cannot occur in a perfect labeling; (ii) a *greedy heuristic phase* to eliminate all but two candidate positions for each point, thereby reducing the problem to 2-PFLP; (iii) a *solution phase* to solve the 2-PFLP problem by reduction to 2-SAT and solution of the 2-SAT problem. Wagner and Wolff claim that their admissibility test performs well in solving 4-VPFLP problems.

We tested the WW algorithm for 4-VPFLP in comparison with binary search using other admissibility tests, in particular: the SA variant mentioned above, a random-descent method (DE) also proposed by Christensen et al. [1], and a baseline random-placement method (RND). Figure 2 shows the result of these tests on a set of problems at a wide range of densities generated randomly according to the methods of Christensen et al. [1]. Label density, varying along the $x$ axis, is given by the number of labels of size $30 \times 7$ on a map of dimensions $792 \times 612$, ranging from 100 to 1000 in multiples of 100. At each label density, for each method tested, runs were made on 50 maps. Scale is presented on the $y$ axis by the computed maximal scale relative to a benchmark "space-filling scale", the scale at which the area of all the labels just equals the area of the map. This serves as a very weak upper bound on the optimal scale. The box plot shows the mean value over the 50 runs with a crosshair. The top middle and bottom lines in the box mark the 75th, 50th (median), and 25th percentiles, and the top and bottom extensions mark the 90th and 10th percentiles.

The graph shows the clear and unsurprising superiority of WW and SA over the two alternatives; the repeated use of an admissibility test in each VPFLP problem means that small improvements in a PFLP algorithm are leveraged into large improvements in a VPFLP algorithm. Figure 3 graphs the average ratio of SA label size to WW label size at different densities. The ratio stays extremely close to 1, with neither algorithm clearly superior. The differences are not



*Computational Geometry 97* Nice France

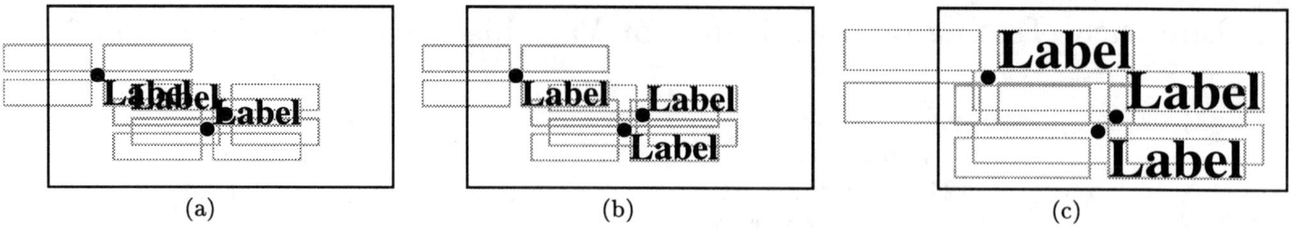


Figure 1: Examples of fixed- and variable-size label placement. A map of three points (a) shows the four potential label positions for each point and depicting an especially poor choice of label positions. Viewed as a fixed-size label placement problem, the points can be labeled well as in (b). Viewed as a variable-size label placement problem, the size of the labels may be increased by some 40 percent and still labeled without overlap (c) by moving the label of the upper left point.

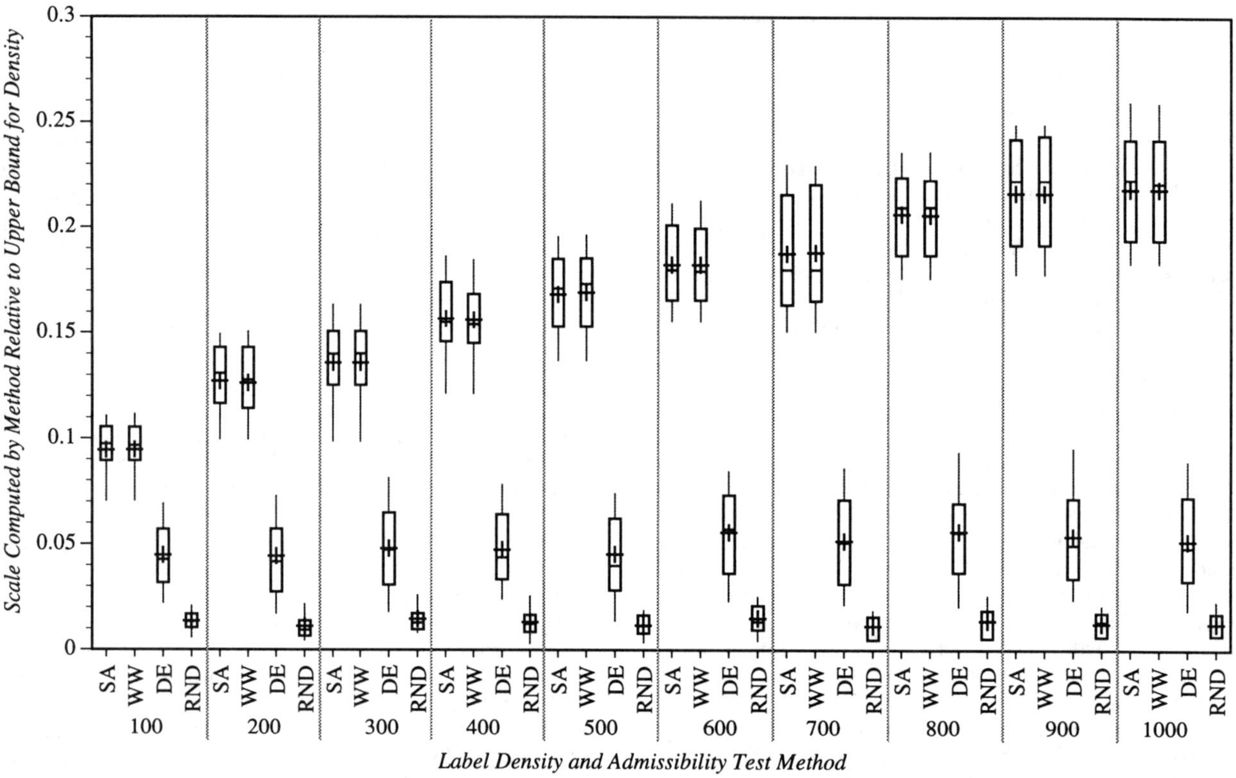


Figure 2: Box plot of four admissibility test methods used to solve 4-VPFLP at various densities.

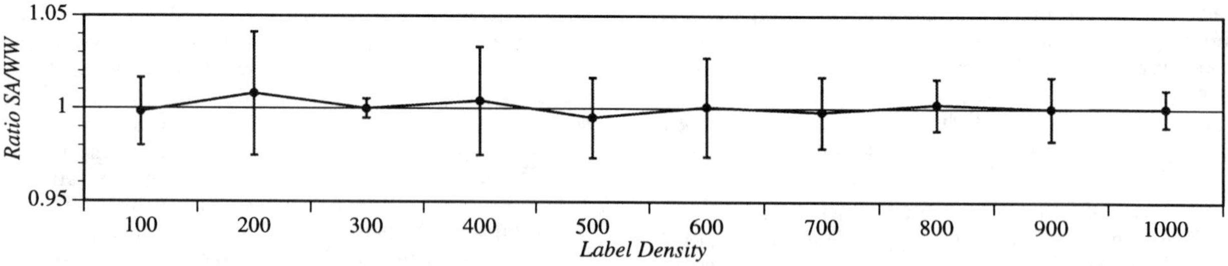


Figure 3: Mean and standard error of ratios of SA-computed scale to WW-computed scale for 4-VPFLP.

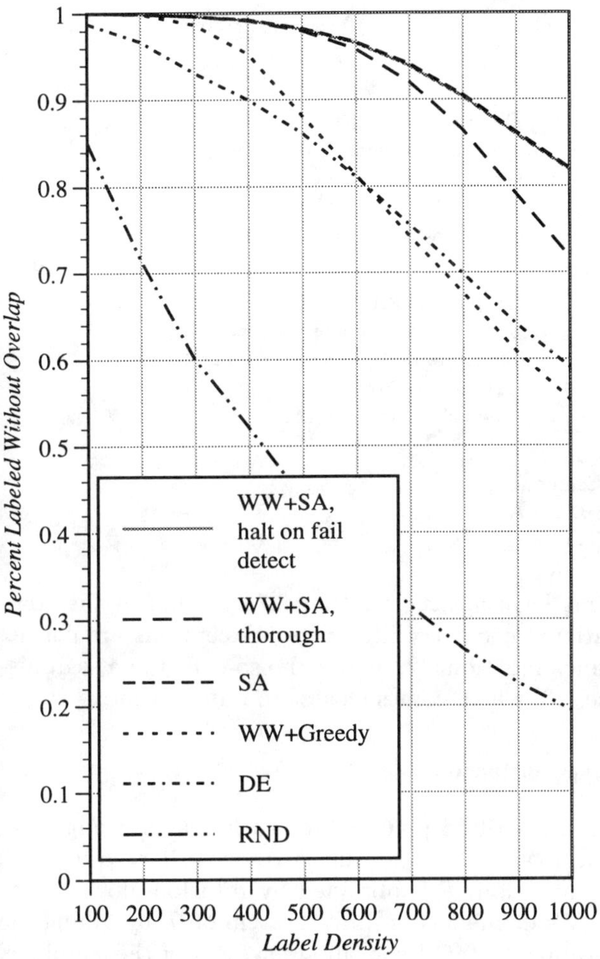


Figure 4: Performance of various algorithms on PFLP problems at various densities, including SA with and without WW preprocessing.

significant according to a one-tail paired two-sample t-test over all 500 maps tested (t = .42, df=499, P =.337). Similar significance levels obtain for tests at each label density.

On average, the SA-based VPFLP algorithm required about three times more time than the WW-based one. This is not surprising, as the WW admissibility test is designed to fail quickly for PFLP problems with no perfect labeling, whereas SA must run through an entire annealing process before it can test whether the final labeling generated is perfect. Thus, for the 4-VPFLP problem, WW is preferable to SA.

On the other hand, the fact that the SA performance is essentially identical to that of WW means that SA is as good a method for determining admissibility as WW for the four-candidate variant. We can thus expect that for other versions of VPFLP for which WW cannot be applied, such as versions with more than four candidate positions per point, a priori preferences among positions, or with linear and area features, SA will perform extremely well. As extensions beyond the capability of WW are usual in cartographic applications, the efficacy of SA for variable-sized label placement problems is welcome news.

The preprocessing phase of WW is conservative in the sense that if a perfect labeling of the original set of candidate positions exists, then one will still exist after the preprocessing phase has eliminated its candidates. A natural question is whether the preprocessing phase might therefore be used to eliminate some potential candidate positions as a preprocess to SA so that SA might find better solutions to the PFLP problem. Of course, if the preprocess reports that no perfect labeling exists (as it well might), we need to decide what to do. Figure 4 show the result of a test of algorithms using the preprocess to augment SA. (The methodology is identical to that used by Christensen et al. [1] so that the results are directly comparable.) The line marked "WW+SA halt on fail detect" stops eliminating candidate positions as soon as the preprocess concludes that no perfect labeling exists. The line marked "WW+SA, thorough" continues to eliminate candidate positions thereafter. This more aggressive use of the preprocess when used to prune candidate positions for the SA algorithm is clearly inferior. The former method, though it does not perform better than SA, performs only very slightly worse, and provides some slight speedup. Thus, the use of the preprocessing does not seem to provide a significant advantage to SA.

In summary, the empirical study reported on here demonstrates that SA performs as well as the WW 2-SAT heuristic on 4-VPFLP though it is considerably slower. Use of the WW preprocessing heuristic does not seem to benefit SA. The generality of SA makes it the method of choice for variable-sized label placement problems beyond 4-VPFLP.

## Acknowledgements

This paper is based upon work supported in part by the National Science Foundation under Grant Number IRI-9350192 and by a gift from MERL — A Mitsubishi Electric Research Laboratory.

## References

[1] Jon Christensen, Joe Marks, and Stuart M. Shieber. An empirical study of algorithms for point feature label placement. *Transactions on Graphics*, 14(3), July 1995.

[2] Shawn Edmondson, Jon Christensen, Joe Marks, and Stuart M. Shieber. A general cartographic labeling algorithm. *Cartographica*, 1997. To appear.

[3] Frank Wagner and Alexander Wolff. A practical map labeling algorithm. To appear in the *Journal of Computational Geometry*.

# Voronoi Diagrams for Direction-Sensitive Distances*

Oswin Aichholzer[†]
Franz Aurenhammer
Institute for Theoretical Computer Science
Graz University of Technology
Graz, Austria
e-mail: {oaich,auren}@igi.tu-graz.ac.at

Danny Z. Chen[‡]
Department of Computer Science & Engineering
University of Notre Dame
Notre Dame, IN 46556, U.S.A.
e-mail: chen@cse.nd.edu

D. T. Lee[§]
Department of Electrical and Computer Engineering
Northwestern University
Evanston, IL 60208, U.S.A.
e-mail: dtlee@ece.nwu.edu

Asish Mukhopadhyay
Department of Computer Science
Indian Institute of Technology, Kanpur
UP-208016, India
e-mail: am@iitk.ernet.in

Evanthia Papadopoulou
IBM Watson Research Center
Yorktown Heights, NY 10598, U.S.A.
e-mail: evanthia@watson.ibm.com

## 1 Introduction

Most computational geometry research on planar problems assumes that the underlying plane is perfectly 'flat', in the sense that movement between any two points on the plane always takes the same cost as long as the Euclidean distance between the two points is the same. In real environments, distances may depend on the direction one moves along [10], or even may be influenced by local properties of the plane [8]. These situations sometimes can be modeled by considering a piecewise-linear surface as the underlying 'plane', and measuring distances therein; see e.g. [7]. In fact, many distance problems on non-flat planes are hard to deal with from the computational geometry point of view.

We study distance problems for the basic case of a 'tilted' plane in three-space. In this model, the cost of moving depends not only on the Euclidean distance but also on how much upwards or downwards the movement has to travel, simulating the situation when driving a vehicle on the tilted plane. Direction-sensitive distances and, in particular, their induced Voronoi diagrams have not received much attention in the computational geometry literature. Exceptions are convex distance functions [2, 5] and the so-called boat-sail distance [10] which comes closest to our concept.

*Part of this work was carried out when the authors were visiting the Center for Applied Science and Engineering and the Institute of Information Science, Academia Sinica, Nankang, Taiwan, June – July 1996.

[†]The research of this author was supported by the Austrian Spezialforschungsbereich F003, Optimierung und Kontrolle.

[‡]The research of this author was supported in part by the National Science Foundation under Grant CCR-9623585.

[§]The research of this author was supported in part by the National Science Foundation under the Grant CCR-9309743, and by the Office of Naval Research under the Grants No. N00014-93-1-0272 and No. N00014-95-1-1007.



## 2 Skew distances

Let $T$ be a tilted plane in $\mathcal{R}^3$ such that the angle between $T$ and the $xy$-plane is $\alpha$, with $0 \leq \alpha \leq \pi/2$. By convention, $T$ is obtained by rotation along the $x$-axis. We define a coordinate system on $T$ by taking the accordingly rotated coordinate system of the $xy$-plane. Based on this system, a point $p$ on $T$ is described by its coordinates $x(p)$ and $y(p)$. The Euclidean distance in $\mathcal{R}^3$ between $p, q \in T$ then can be expressed as $d_2(p, q)$ $= ((x(p) - x(q))^2 + (y(p) - y(q))^2)^{1/2}$.

A simple distance function on $T$ is obtained by taking, for points $p, q \in T$, their Euclidean distance plus their signed difference in height, that is, $d_2(p,q)+(h(q)-h(p))$. The latter term is equal to $(y(q) - y(p)) \cdot \sin \alpha$, which means that the slope of $T$ affects the $y$-difference of $p$ and $q$ by a factor in the range $[0, 1]$. To obtain a richer, but still realistic class of distance functions on $T$, we define the *skew distance* from $p$ to $q$ as

$$d(p,q) = d_2(p,q) + k \cdot (y(q) - y(p)),$$

where $k \geq 0$ is a constant. The parameter $k$ has a nice physical interpretation. Imagine a ball moving on $T$, and let $\frac{1}{k}$ be the frictional coefficient on $T$. For $k < 1$, friction dominates gravity and the ball sticks on the skew plane. Friction and gravity balance out if $k = 1$. For $k > 1$, gravity dominates friction and the ball rolls downhill. The last case also models the assumption that energy is gained when a (electric-driven) vehicle moves downhill. Note that $d$ is non-symmetric as $d(p,q) \neq d(q,p)$ unless $y(p) = y(q)$ or $k = 0$, but $d$ still obeys the triangle inequality.

For $k > 1$, $d(p,q)$ may be negative. Moreover, $d(p,q)$ may *decrease* when a point $q$ below $p$ is moved downwards $T$. Let

$$L_0(p) = \{a \mid d(p,a) = 0\}.$$

Clearly, for $k < 1$, $L_0(p) = \{p\}$. For $k = 1$, $L_0(p)$ is the vertical ray emanating from $p$ and extending below $p$. For $k > 1$, $L_0(p)$ is composed of two rays of slopes $\pm 1/(\sqrt{k^2 - 1})$, emanating from and extending below $p$. We refer to these two rays as the *0-rays* of $p$.

As $d$ is non-symmetric, two different 'unit circles' with $p$ as the fixed center can be defined, $\sigma(p) = \{a \mid d(p,a) = 1\}$ and $\sigma'(p) = \{a \mid d(a,p) = 1\}$. However, $\sigma'(p)$ is just the reflection of $\sigma(p)$ through the horizontal line passing through $p$. We adopt the convention of considering only the 'outgoing' skew distance in the definitions of geometric structures like $\sigma(p)$, $L_0(p)$, etc., keeping in mind that their 'incoming' versions can be obtained by reflection.

**Lemma 1** *For $k > 0$, $\sigma(p)$ is a conic with focus $p$, directrix the horizontal line at $y$-distance $1/k$ above $p$, and eccentricity $k$. Thus $\sigma(p)$ is an ellipse for $0 < k < 1$, a parabola for $k = 1$, and a hyperbola for $k > 1$.*

## 3 Skew Voronoi diagrams and related structures

Let $S$ be a set of $n$ point sites on the tilted plane $T$. The *skew Voronoi diagram* of $S$, for short $SV(S)$, associates each site $p \in S$ with the region

$$reg(p) = \{a \mid d(p,a) < d(q,a), \forall q \in S\}.$$

Figure 1 depicts a skew Voronoi diagram for six sites. The dashed line segments indicate the upper envelope of their 0-rays. Note the emptiness of $reg(q)$ and $reg(t)$.

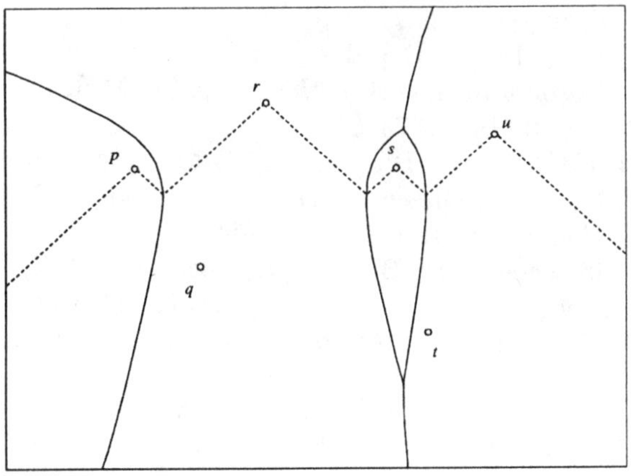


Figure 1: Skew Voronoi diagram for $k = 1.5$

Of course, $SV(S)$ depends on the 'slope parameter' $k$. The case $k < 1$ leads to known structures: According to Lemma 1, the skew distance $d$ then is a *convex distance function* as defined in [2], and Voronoi diagrams for convex distance functions are well-studied objects; see e.g. [5].

Skew Voronoi diagrams have another nice and helpful interpretation for general $k$. Consider the *bisector* of two sites $p, q \in S$, $b(p,q) = \{a \mid d(p,a) = d(q,a)\}$. It can be described by the equation

$$d_2(p,a) - k \cdot y(p) = d_2(q,a) - k \cdot y(q).$$

Without loss of generality, let $S$ be contained in the $y$-positive halfplane of $T$. For each site $p \in S$, define $C(p)$ as the (Euclidean) circle with center $p$ and radius $k \cdot y(p)$. Then $b(p,q)$ is just the set of points equidistant under the Euclidean distance from the two circles $C(p)$ and $C(q)$. We hence conclude:

**Theorem 3.1** *Let $S$ be a set of point sites in the $y$-positive halfplane of $T$, and let $C(S) = \{C(p) \mid p \in S\}$. Then $SV(S)$ is the Euclidean closest-point Voronoi diagram of $C(S)$.*

Various properties are known for the Voronoi diagram of circles in the plane; see [3, 6, 9]. Its edges are hyperbolic arcs, its regions are star-shaped, and its size is linear. Among the available techniques for its construction, the $O(n \log n)$ time plane-sweep algorithm in [3] is reasonably simple to implement. Clearly, $C(S)$ can be computed from $S$ in $O(n)$ time. So, for any fixed $k \geq 0$, the skew Voronoi diagram can be computed in time $O(n \log n)$ and space $O(n)$.

Another kind of direction-sensitive diagram, the so-called *Voronoi diagram in a river*, has been investigated in [10]. The underlying distance function is the time required to reach a point on the river by a boat starting from a site (and fighting against a constant river flow). Despite apparent similarity, the 'boat-sail' distance is different from the skew distance. For example, if the river is slower than the boat then the river Voronoi diagram and the corresponding Euclidean Voronoi diagram are combinatorially the same. If the river is faster than the boat then all regions start at their respective sites.

Let us now give a brief catalogue of properties of $SV(S)$ in dependence of $k$. In fact, for $k < 1$, $SV(S)$ has a combinatorial structure identical to the Euclidean Voronoi diagram of some set $S'$ of point sites. This can be seen by applying an affine transformation that takes $S$ into $S'$, and the elliptic circles defined by the skew distance into Euclidean circles[1]. Observe that the convex hulls of $S$ and $S'$ are combinatorially the same, so exactly the sites lying on the convex hull $S$ have unbounded regions in $SV(S)$. Still, $SV(S)$ and the Euclidean Voronoi diagram of $S$ are combinatorially different, in general. For example, two sites which are

[1]Thanks go to Olivier Devillers for pointing out this property.

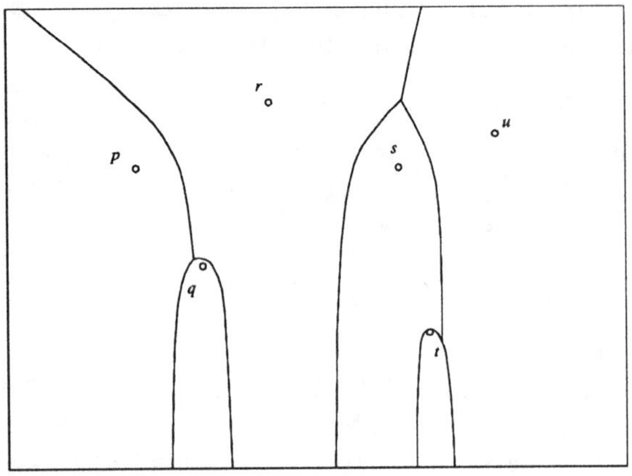


Figure 2: Skew Voronoi diagram for $k = 1$.

neighbored in the $x$-sorting of $S$ but have non-adjacent Euclidean regions will have adjacent skew regions, provided $k$ is sufficiently close to 1; see the case $k = 1$.

The case $k > 1$ is the most interesting one from the geometric point of view. Let the *0-envelope*, $E_0(S)$, be defined as the upper envelope of the 0-rays for all $p \in S$. From the triangle inequality we get $reg(p) \neq \emptyset$ if and only if $p \in E_0(S)$. As $p \in reg(p)$ for each non-empty region, $E_0(S)$ cuts each such region into an upper and a lower part. This implies that the edges of $SV(S)$ lying above and below $E_0(S)$, respectively, have the structure of a forest[2]. Using the interpretation of $SV(S)$ as a Voronoi diagram for circles, sites with unbounded regions can be characterized: A region $reg(p)$ is unbounded if and only if $p$ lies on both $E_0(S)$ and the upper convex hull of $S$.

For $k = 1$, all regions of $SV(S)$ are unbounded and extend to minus infinity in $y$-direction. Their left-to-right order reflects the $x$-sorted order of $S$.

See Figure 1 for an example of a 0-envelope and the corresponding skew Voronoi diagram. For the same set of sites, the diagram that results from setting $k = 1$ is shown in Figure 2. Some pictures of skew Voronoi diagrams for different values of $k$ and some animated sequences for steadily increasing $k$ can be obtained from `http://www.cis.tu-graz.ac.at/igi/oaich/skewvd`.

To get an output-sensitive algorithm for $k > 1$, sites with empty regions can be pruned away by first constructing $E_0(S)$. By a coordinate transformation, the latter task is essentially equivalent to finding all the $m$ maximal dominating elements among $S$, which can be done in $O(n \log m)$ time and $O(n)$ space [4]. This gives an algorithm with the same complexity for constructing $SV(S)$ if $m$ regions are non-empty, which is optimal by reduction to the planar convex hull problem.

A generalization of the skew Voronoi diagram problem, still solvable in time $O(n \log n)$, is the following: Given a set $S$ of $n$ sites, let $S' = S$, compute $SV(S')$, and call it the *layer*-1 *diagram*, $SV_1(S)$. Next, remove from $S'$ all sites with non-empty regions in $SV_1(S)$ (that is, all $p \in E_0(S')$), and repeat the computation on $S'$, which next gives $SV_2(S)$, until $S' = \emptyset$. Note that the collection of 0-envelopes for the shrinking set $S'$ can be computed in overall time $O(n \log n)$ using the max-dominance algorithm by [1].

$SV_1(S)$ and $SV_2(S)$ can be used to solve, in time $O(n \log n)$, the all nearest neighbors problem under the skew distance: For each $p \in S$, find a site $q \neq p$ such that $d(q,p)$ is minimized. Note that $q$ may *not* be any of the Voronoi neighbors of $p$ in $SV(S)$. For example, when $E_0(S)$ contains only one site $p$ of $S$, then $p$ has no Voronoi neighbor in $SV(S)$ but still has a closest $q \neq p$. However, it can be shown that $q$ has a non-empty region in $SV_2(S)$ then.

[2]Thanks go to Günter Rote for pointing out the forest property.

## References


[1] Atallah, M.J., Kosaraju, S.R., *An efficient algorithm for maxdominance, with applications*, Algorithmica 4 (1989), 221–236.

[2] Chew, L.P., Drysdale, R.L., *Voronoi diagrams based on convex distance functions*, Proc. 1st Ann. ACM Symp. Computational Geometry, 1985, 235–244.

[3] Fortune, S., *A sweepline algorithm for Voronoi diagrams*, Algorithmica 2 (1987), 153–174.

[4] Kirkpatrick, D.G., Seidel, R., *Output-size sensitive algorithms for finding maximal vectors*, Proc. 1st Annu. ACM Sympos. Computational Geometry, 1985, 89–96.

[5] Klein, R., *Concrete and Abstract Voronoi Diagrams*, Lecture Notes in Computer Science 400, Springer, 1989.

[6] Lee, D.T., Drysdale, R.L., *Generalization of Voronoi diagrams in the plane*, SIAM J. Computing 10 (1981), 73–87.

[7] Mitchell, J.S.B., Mount, D.M., Papadimitriou, C.H., *The discrete geodesic problem*, SIAM J. Computing 16 (1987), 647–668.

[8] Mitchell, J.S.B., Papadimitriou, C.H., *The weighted region problem: Finding shortest paths through a weighted planar subdivision*, Journal of the ACM 38 (1991), 18–73.

[9] Sharir, M., *Intersection and closest-pair problems for a set of planar discs*, SIAM J. Computing 14 (1985), 448–468.

[10] Sugihara, K., *Voronoi diagrams in a river*, Int'l. J. Computational Geometry & Applications 2 (1992), 29–48.

# Dynamic visibility in polygonal scenes with the visibility complex

RIVIÈRE Stéphane *

## 1 Introduction

This paper proposes methods for efficient maintaining of the view around a point (i.e., the visibility polygon) in static and dynamic polygonal scenes in the plane. The algorithms presented use the *visibility complex*, that we have adapted for polygonal scenes (composed of disjoint simple polygons) to take advantage of the spatial and temporal coherence.

The first algorithm shows how to maintain the view around a moving point in $O(\log^2 v)$ time at each visibility change ($v$: current size of the view), improving the $O(\log^2 n)$ time bound ($n$: number of polygon vertices) of [GS96], and is well-suited for small consecutive moves. The second algorithm maintains the view around a point moving along a known trajectory: After a preprocessing in $O(v \log v)$ time, the view is updated in $O(\log v)$ time at each change of visibility. Finally, we show how to maintain the visibility complex for a dynamic scene (i.e., a scene in which the polygon vertices are moving) and how to maintain the view around a moving point in a dynamic scene.

## 2 The visibility complex

Visibility computations involve determining the object seen along directions of vision, that is along maximal free line segments (line segments of maximal length in free space), that we also call *rays*. Recomputing each time the object seen along a ray can be too time consuming for solving some visibility problems. One solution to this issue is to classify rays according to their visibility: to find the object seen along a ray, we then just have to identify the set of rays that contains the given ray and to read the visibility properties of this set.

Pocchiola and Vegter [PV96] devised a new data structure, the visibility complex, which represents sets of rays having the same visibility properties. A ray going from an object $O_l$ to another object $O_r$ being characterized by its label $(O_l, O_r)$, the visibility complex is the quotient space of the space of rays under the following relation $\sim$: $r_1 \sim r_2$ iff $r_1$ can be moved continuously to $r_2$ while keeping the same label.

We have adapted here this structure, created initially for scenes of convex curved objects, for polygonal scenes: Each side of a polygon is a distinct object. Then the complex is composed of three types of elements: faces (2D components), edges (1D components representing rays passing through a polygon vertex), and vertices (0D components representing rays passing through two polygon vertices).

These elements are best handled by means of a duality relation (figure 1). We consider the duality relation that maps the line $l : y = ax + b$ of the scene into the point $l^* : (a, b)$ in the dual space. This duality maps the set of lines passing through a point $p : (a, b)$ of the scene into the dual curve $p^* : y = -ax + b$. An edge $e$ of the complex is contained in a dual curve $p^*$: We say that $p$ is the point associated with $e$. A vertex $v$ of the complex is the intersection of two dual curves $p^*$ and $q^*$; in the scene, $(p, q)$ is an edge of the visibility graph, and we say that the ray $v^*$ containing $(p, q)$ is the ray associated with $v$.

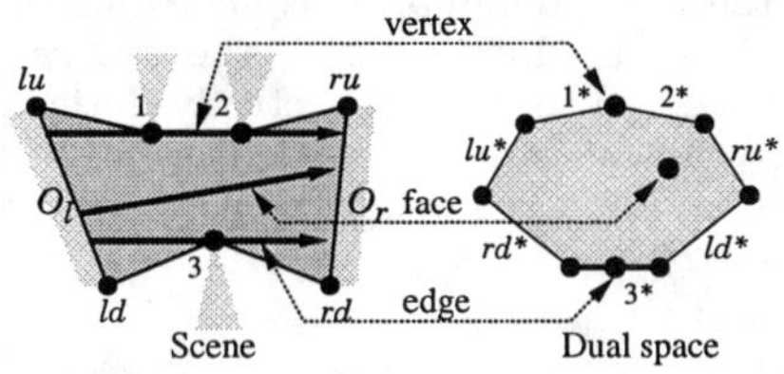


Figure 1: Visibility complex in dual space.

The visibility complex of polygonal scenes has a size $O(k)$ ($k$ size of the visibility graph) and can be computed in optimal $O(n \log n + k)$ time and $O(n)$ working space with the algorithm of Rivière [Riv97].

Pocchiola and Vegter [PV96] have shown how to use the visibility complex to compute a view around a point. We recall here the algorithm for we reuse its results later.

## 3 Computing a view around a point

Computing the view around a point of view $p_v$ amounts to computing changes of visibility occurring along a ray pivoting around $p_v$. A change of visibility occurs when the ray is leaving the face $f$ of the visibility complex to which it belongs, and is entering into a new face $f'$.

*iMAGIS-GRAVIR/IMAG - BP 53
38041 GRENOBLE CEDEX 09 - FRANCE
e-mail: Stephane.Riviere@imag.fr
www: http://www-imagis.imag.fr/Membres/Stephane.Riviere
iMAGIS is a joint project of CNRS/INRIA/INPG/UJF

In the dual space, the set of rays passing through $p_v$ is a dual curve $p_v^*$ which passes through a set of faces of the visibility complex. The dual curve passes from one face $f$ to another face $f'$ by crossing an edge $e$ separating the two faces; the point $p$ associated with $e$ is then a point of the view. The algorithm for computing the view computes the list of edges of the complex crossed by the dual line of the point of view. Depending on the method used to find the crossed edges, the view can be computed in $O(v \log n)$ time, or in $\Omega(v)$ and $O(vn)$ time ($v$ size of the view).

With this list of crossed edges, we can maintain the view when the point of view is moving in the scene.

## 4 Maintaining a view around a moving point

Let $p_v$ be a point moving in the scene. The view around $p_v$ changes when $p_v$ crosses an extended edge of the visibility graph (i.e., the ray containing the edge): A point of the view becomes invisible, or a new point becomes visible, or $p_v$ crosses the non-extended edge. In the last case, although the view does not change, we still classify this event as a visibility change: It corresponds to a visibility event in the algorithms that we describe.

By extending each edge of the visibility graph in free space, we obtain a partition of free space into regions of constant visibility: The view around a point is the same for all points inside the same *visibility region*. To maintain the view around $p_v$, we do not compute all the $O(n^4)$ visibility regions, but only the limits of the visibility region that contains $p_v$.

When $p_v$ is moving, we check whether $p_v$ has left its visibility region; if so, we search the side of the visibility region crossed by $p_v$ and compute the limits of the region incident to that side. We repeat the process until finding the new visibility region containing $p_v$.

We compute the limits of the visibility region of $p_v$ by using the list of edges of the complex crossed by the dual curve $p_v^*$. Let us consider a crossed edge $e = (v_l, v_r)$ whose extremities are the vertices $v_l$ and $v_r$. If $p_v^*$ passes through one of these vertices, then the list of crossed edges — and therefore the view around $p_v$ — changes. Let $p$ be the point associated with $e$. In the scene, the two rays $v_l^*$ and $v_r^*$ associated with $v_l$ and $v_r$ respectively, form a wedge that contains $p_v$ and no other (extended) edges of the visibility graph incident to $p$: The wedge contains the visibility region of $p_v$ (figure 2).

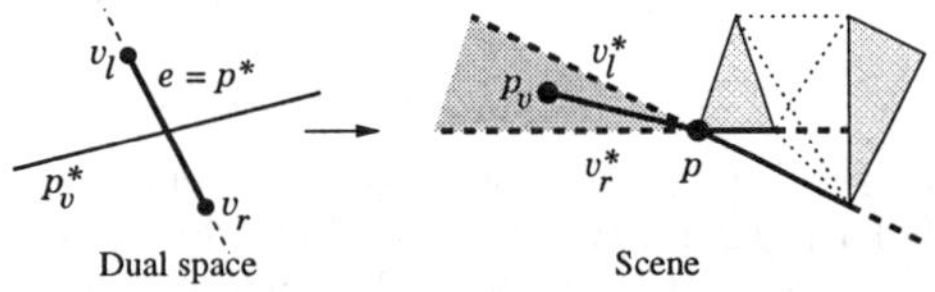


Figure 2: Wedge of visibility given by a point of the view.

More precisely, each wedge is made of an upper line and a lower line. The limits of the visibility region of $p_v$ is then composed of the lower envelope of the set of upper lines and of the upper envelope of the set of lower lines (figure 3).

Once the view around $p_v$ is computed with the visibility complex, the limits of the visibility region of $p_v$ can be computed in $O(v \log v)$ time ($v$ size of the view).

When the point of view is moving from $p_v$ to $p'_v$, we can check in logarithmic time whether the point has left its region and, if it is the case, find which line of the upper (resp. lower) has been crossed by $(p_v, p'_v)$. In the visibility

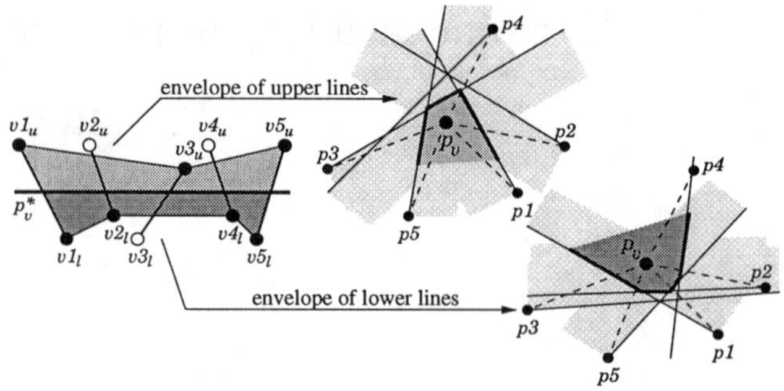


Figure 3: Computing the limits of the visibility region.

complex, the vertex $v$ corresponding to the crossed line is swept by the dual curve $p_v^*$. Changes in the list of crossed edges (and therefore in the view) are local: They concern only the edges around the swept vertex and are done in constant time.

We must then update the limits of the visibility region. The changes are local too: We must remove (resp. insert) at most two lines in the set of lines, and the envelopes can be maintained by adapting the algorithms of Overmars and van Leeuwen [OvL81] or of Dobrindt and Yvinec [DY93] for dynamically maintaining a convex envelope.

**Proposition 1** *After computing the initial view around a point $p_v$ and a preprocessing step in time $O(v_0 \log v_0)$ ($v_0$ size of view at $t = 0$), the view around a moving point can be maintained in $O(\log^2 v)$ time (resp. randomized $O(\log v)$ time).*

We suppose now that the point of view is moving along a known trajectory, given for example by its position $p_v(t)$, $t \in [0, 1]$. We do not need to compute all the limits of each visibility region crossed by the trajectory, but only the limits cut by the trajectory.

We suppose that we can compute in constant time the point of intersection of the trajectory with a line, and that we can compare in constant time two points on the trajectory (i.e., establish which one is before the other). It allows us to use a priority queue in which lines are ordered according to the position of their intersection point with the trajectory, so that we can pass the limits of visibility in order.

To initialize the priority queue, we first compute the view around $p_v(0)$, and for each edge of the complex crossed by $p_v^*(0)$, we check whether the upper line (resp. lower line) intersects the trajectory and — if it is the case — put it in the priority queue. Then, at each step we consider the first line of the priority queue, line which indicates the limit of visibility being crossed. We update (locally) the view as in the previous algorithm, consider the (at most two) new lines that limit the region, and put them in the priority queue if they cut the trajectory.

**Proposition 2** *After the computation of the initial view around a point $p_v$ and a preprocessing step done in $O(v_0 \log v_0$ time, the view along a trajectory can be maintained in $O(\log v)$ time at each change of visibility.*

## 5 Dynamically maintaining the visibility complex

We now show how the visibility complex can be maintained when the polygon vertices are moving in the scene. More precisely, each of the $n$ polygon vertices is defined by its position $p_i(t)$ for $t \in [0,1]$, and we want to maintain the topological structure of the complex from $t=0$ to $t=1$. Modifications in the complex arise at *elementary visibility changes*. They occur when three points $p_1$, $p_2$, and $p_3$ such that $(p_1,p_2)$ and $(p_2,p_3)$ are edges of the visibility graph, change their relative positions: $p_2$ previously below (resp. above) $(p_1,p_3)$ becomes above (resp. below) $(p_1,p_3)$ (figure 4a).

Corresponding changes in the visibility complex are local: They occur around the vertices $v_{12}$ and $v_{23}$ corresponding to $(p_1,p_2)$ and $(p_2,p_3)$ respectively (figure 4b), and can be made in constant time (there are 50 cases to consider).

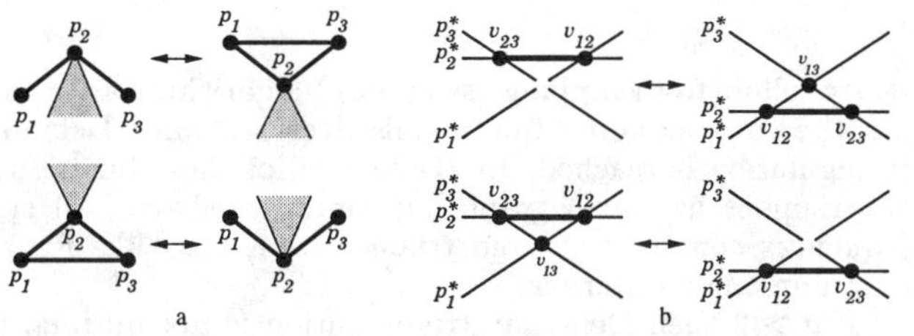

Figure 4: Elementary visibility changes. **a.** In the scene. **b.** In the visibility complex.

To maintain the complex from $t=0$ to $t=1$, we must handle all the visibility changes in order. We compute for each edge $e_j = (v_{ij}, v_{jk})$ $(v_{ij} = p_i^* \cap p_j^*)$ of the complex a "date of death" $t_d$ that is the time when the three corresponding points $p_i(t_d)$, $p_j(t_d)$, and $p_k(t_d)$ in the scene will be collinear. Then we put the edges in a priority queue and handle the visibility changes in order. At each visibility change, the set of dates of death can be updated in constant time and the priority queue can be updated in $O(\log k) = O(\log n)$ time ($k$, size of the visibility graph, verifies $k = O(n^2)$). If dates of death can be computed in constant time, then:

**Proposition 3** *After a preprocessing step in $O(k_0 \log n)$ time ($k_0$ size of the visibility complex at $t=0$), the visibility complex of a moving polygonal scene can be maintained in $O(\log n)$ time at each visibility change.*

## 6 Maintaining a view around a moving point among moving polygons

We finally show how to maintain a view around a point when the point of view and the obstacles are moving. In this context, using the data structures shown before for maintaining a view (dynamic convex envelope or priority queue) is not well-suited: At each change in the visibility complex, these structures must be updated and the modifications may not be local. Thus we integrate the visibility changes around the point of view with the visibility changes of the visibility complex.

For each vertex $v_{ij}$ of the complex extremity of an edge crossed by the dual curve $p_v^*$ we also compute a "date of death", that is the date $t_d$ when $p_v^*(t_d)$ passes through $v_{ij}(t_d)$, that is when in the scene $p_v(t_d)$, $p_i(t_d)$, and $p_j(t_d)$ are collinear, and put all these events in the priority queue.

When an elementary visibility change occurs in the complex, some dates of death related to the view around $p_v$ must also be updated. However, these changes are local, and if for each edge put in the priority queue we add a flag indicating whether the edge is crossed by the dual curve of the point of view, all modified dates of death can be updated in constant time.

**Proposition 4** *After a preprocessing step in $O(k_0 \log n)$ time ($k_0$ size of the visibility complex at $t=0$), the view around a moving point among moving polygons can be maintained in $O(\log n)$ time each time a visibility change occurs in the scene or in the view.*

## 7 Conclusion

We have shown two output-sensitive algorithms for maintaining the view around a moving point in a fixed scene. These algorithms use the visibility complex so that local changes of visibility in the scene yield also "local" recomputations. They are particularly efficient in large occluded environments, where the size of the visibility complex is rather linear than quadratic (in the number of polygon vertices) and the size of the view is much smaller that the size of the scene. They are relatively simple and have been implemented.

Then we have shown an algorithm for maintaining the visibility complex for dynamic scenes which can be used in global visibility computations. The algorithm handles the changes of visibility according to their date of occurrence. We are currently studying whether visibility changes happening in different hidden parts of the scene could be handled independently in any order. We are also studying algorithms for maintaining the complex upon insertion or deletion of polygons in a scene, algorithms which could reuse (for a first study) our algorithms for dynamic visibility.

Finally, we have shown an algorithm for maintaining the view around a point moving in a dynamic scene, algorithm more suited to the case when most of the objects of the scene are seen at some point in time from the point of view.

## References


[DY93] K. Dobrindt and M. Yvinec. Remembering conflicts in history yields dynamic algorithms. In *Proc. 4th Annu. Internat. Sympos. Algorithms Comput. (ISAAC 93)*, volume 762 of *Lecture Notes in Computer Science*, pages 21–30. Springer-Verlag, 1993.

[GS96] S. Ghali and A.J. Stewart. Incremental update of the visibility map as seen by a moving viewpoint in two dimensions. In *Proc. Eurographics Workshop on Comp. Anim. and Simul.*, pages 3–14, 1996.

[OvL81] M. H. Overmars and J. van Leeuwen. Maintenance of configurations in the plane. *J. Comput. Syst. Sci.*, 23:166–204, 1981.

[PV96] M. Pocchiola and G. Vegter. The visibility complex. *Internat. J. Comput. Geom. Appl.*, 6(3):279–308, 1996. Special issue devoted to ACM-SoCG'93.

[Riv97] S. Rivière. *Visibility computations in 2D polygonal scenes*. Ph.D. thesis, Université Joseph Fourier, Grenoble, France, 1997.

# PROPERTIES OF THE DELAUNAY TRIANGULATION

Oleg R. Musin

Dept. of Cartography and Geoinformatics
Moscow State University, Moscow, 119899, Russia
E-MAIL: musin@gislab.geogr.msu.su

## 1 INTRODUCTION

Some of the most well-known names in Computational Geometry are those of two prominent Russian mathematicians: Georgy F. Voronoi (1868 – 1908) and Boris N. Delaunay (1890 – 1980). Their considerable contribution to the Number Theory and Geometry is well known to the specialists in these fields. Surprisingly, their names (their works remained unread and later re-discovered) became the most popular not among "pure" mathematician, but among the researchers who used geometric applications. Such terms as "Voronoi diagram" and "Delaunay triangulation" are very important not only for Computational Geometry, but also for Geometric Modeling, Image Processing, CAD, GIS etc.

Delaunay triangulation is used in numerous applications. It is widely used in plane and 3D case. A natural question may arise: why this triangulation is better than the others. Usually the advantages of Delaunay triangulation are rationalized by the max-min angle criterion and other properties [1,2,5,10,11,12]. The max-min angle criterion requires that the diagonal of every convex quadrilateral occurring in the triangulation "should be well chosen" [12], in the sense that replacement of the chosen diagonal by the alternative one must not increase the minimum of the six angles in the two triangles making up the quadrilateral. Thus the Delaunay triangulation of a planar point set maximizes the minimum angle in any triangle. More specifically, the sequence of triangle angles, sorted from sharpest to least sharp, is lexicographically maximized over all such sequences constructed from triangulation of $S$.

We defined several functionals on the set of all triangulations of the finite system of sites in $\mathbf{R}^d$ attaining global minimum on the Delaunay triangulation (DT). First we consider a so called "parabolic" functional and prove that it attains its minimum on DT in all dimensions. It could be used as an equivalent definition for DT. Secondly we treat "mean radius" functional (the mean of circumradii of triangles) for planar triangulations. Thirdly we treat a so called "harmonic" functional. For a triangle this functional equals the ration of the sum of squares of sides over area. Finally, we consider a discrete analogue of the Dirichlet functional. Actually in all these cases the optimality of DT in 2D directly follow from flipping (swapping) algorithm: after each flip the corresponding functionals decrease until Delaunay triangulation is reached. In 2D case all of these functionals on triangles are lexicographically minimized over all such sequences constructed from triangulation of $S$ like for the max-min angle criterion.



*Computational Geometry 97* Nice France
Copyright 1997 ACM 0-89791-878-9/97/06 ..$3.50

If $d > 2$ then Delaunay triangulation is not optimal for the functionals "mean radius", "harmonic" and "Dirichlet". ¿From this point of view the usage of DT in dimensions $d > 2$ may be nonappropriate. Thus the problem of finding "good" triangulations for this functionals in higher dimensions is opened and more detailed consideration is necessary.

## 2 OPTIMALITY DELAUNAY TRIANGULATIONS FOR PARABOLIC FUNCTIONAL

Let $t$ be a triangulation of the set $S$ in $\mathbf{R}^d$, $\Delta_i$ denotes the $i$-th $d$-simplex of $t$ and $x_{ij} \in \mathbf{R}^d$, $j = 0, 1, \ldots, d$ are its vertices . Let

$$V(t) = \sum_i (x_{i0}^2 + \ldots + x_{id}^2) vol(\Delta_i),$$

where $vol(\Delta_i)$ is volume (area for $d = 2$) of the simplex $\Delta_i$. We call functional $V$ *parabolic.*

The parabolic functional induces an order on triangulations of the set $S$ by the rule: $t_1 > t_2$ iff $V(t_1) > V(t_2)$. The value of $V$ depends on the choice of the origin. If we move the origin to $x_0$ then this order does not change i.e.

$$V(t_1) > V(t_2) \text{ iff } V(t_1, x_0) > V(t_2, x_0).$$

The main result for functional $V$ is the following:

**Theorem 1** *The functional $V(t)$ achieves its minimum if and only if $t$ is the Delaunay triangulation*

A simple proof of these theorems follows from paroboloid construction of the DT found by Voronoi [13] and rediscovered only in 1979 for a sphere (K.Brown), and later also for a paraboloid.

Let us consider an arbitrary triangulation $t$ of the set $S$ and "lift" it onto the paraboloid in $\mathbf{R}^{d+1}$, i.e. let us build a polyhedral surface in $\mathbf{R}^{d+1}$ connecting corresponding vertices on the paraboloid. Note that the functional $V$ up to a constant equals to the volume of the solid body below this surface. Thus, the minimum of $V$ is attained on the Delaunay triangulation.

Let us consider another functional for triangulations:

$$C2(t) = \sum_i ||c(\Delta_i)||^2 vol(\Delta_i),$$

where $c(\Delta_i)$ is the center (barycenter) of the $\Delta_i$, $c(\Delta_i) = \sum_j x_{ij}/(d+1)$ and $||\cdot||$ is Euclidean norm.

By direct calculation (it is sufficient to check the formula on a simplex) it is possible to prove that

$$(d+1)^2 C2(t) + V(t) = (d+1)(d+2) \int\limits_{CH(S)} ||x||^2 dx,$$

where $CH(S)$ is convex hull of set $S$ in $\mathbf{R}^d$.

¿From Theorem 1 and this formula directly follows that:

**Theorem 2** *The functional C2(t) on triangulations of the set S achieves its maximum if and only if t is the Delaunay triangulation.*

The functional $V$ may be useful in analysis of algorithms and construction of DT. For example, the proof that sequence of "local transformation" in [4] does not cycle easily follows from consideration of the functional $V(t)$. Indeed, this functional decreases after each local transformation used in the algorithm. In [4] quite complicated proof is given for it.

## 3 MEAN RADIUS OF TRIANGULATION

Let $t$ be a triangulation of $n$ sites $S$ in the plane. Assume that each triangle $\Delta_i$ of this triangulation is related to the radius of its circumcircle $R_i$. Thus every triangulation $t$ is related to the set $\{R_{\Delta_1}, ..., R_{\Delta_k}\}$ of circumradii of triangles $\Delta_i \in t$. The numbers of triangles for any two triangulations of $S$ are equal, so it is possible to compare sets of radii for different triangulations. In particular, it is possible to compare sums of radii: $\sum R_{\Delta_i}$ or power sums: $\sum R^a_{\Delta_i}, a > 0$. It seems that triangulation having minimal sum of radii is "better", because all its triangles in "average" are nearer to the regular triangles.

**Theorem 3** *The functional $R(t,a) = \sum R^a_{\Delta_i}, a > 0$ attains its minimum iff t is Delaunay triangulation.*

This theorem for convex quadrilateral is some kind of a four-vertex theorem for polygon [9] and follows from lemma.

**Lemma** *Let ABCD is convex quadrilateral and $t_{BD}$ its Delaunay triangulation. Then one of the following five relations takes place:*

$(1) R_{ABD} < R_{ABC}, R_{ABD} < R_{ADC}$,
$R_{BCD} < R_{ABC}, R_{BCD} < R_{ADC}$;
$(2) R_{ABD} < R_{ABC} < R_{BCD} < R_{ADC}$;
$(3) R_{ABD} < R_{ADC} < R_{BCD} < R_{ABC}$;
$(4) R_{BCD} < R_{ABC} < R_{ABD} < R_{ADC}$;
$(5) R_{BCD} < R_{ADC} < R_{ABD} < R_{ABC}$,
*where $R_{XYZ}$ means radius of circumcircle of triangle XYZ.*

*I do not know how to generalize theorem 3 to higher dimensions.*

## 4 HARMONIC INDEX OF TRIANGULATION

This index has its origin in the theory of the so called "harmonic maps".

For polygon $P$ its *harmonic index*

$$hrm(P) = \sum a_i^2/S(P),$$

where $a_1, \ldots, a_n$ are the lengths of sides of $P$ and $S(P)$ is its area. This index is the same for similar polygons.

We will call $n$-gon $P$ *harmonic* if the $hrm(P)$ achieves its minimum for $P$. It is easy to prove that harmonic triangle is equiangular. The same result holds for arbitrary $n$.

*The polygon P is harmonic iff P is regular.*

The proof of this proposition is very close to proof of minimality properties of the circle in book [3] (more accurately, isoperimetric unequality for polygons).

For planar triangulation $t$ of a set $S$ let denote by $hrm(t)$ (harmonic index of triangulation $t$) the sum of $hrm$ of its triangles:

$$hrm(t) = \sum_{\Delta_i \in t} hrm(\Delta_i)$$

. Our main result for harmonic index of planar triangulation is following:

**Theorem 4** *Harmonic index hrm(t) of triangulation t of S achieves its minimum if and only if t is the Delaunay triangulation of S.*

The harmonic functional for triangle attains its minimum if this triangle is equiangular. Usually, a triangulation is regarded as "good" for different purposes if its triangles are nearly equiangular. The harmonic index of triangulation $t$ achieves its global minimum if $t$ is a part of a regular triangular lattice. In some sense, theorem 3 shows that the Delaunay triangulation is as close as possible to equiangular triangulation.

We have proposed the generalization of harmonic index for polygons. For polygon $P$ let

$$hrm(P,k) = \sum a_i^{2k}/S(P)^k$$

, where $k$ is some real number. If $k \geq 1/2$ then $hrm(P,k)$ achieves its minimum for regular polygon. In the case $k = 1/2$ this result gives classic isoperimetric unequality for polygons [3].

It is easy to give an example of a convex quadrilateral $P$ when its triangulation $t$ minimizing the functional $hrm(t,k), k \neq 1$ is not Delaunay triangulation. In this connection the following problem is open: *"For fixed $k \geq 1/2$ find an efficient algorithm for construction of a triangulation t of S minimizing the functional hrm(t,k)."*

Let us consider harmonic index in $d$-dimension. It is clear, that right extension of $hrm$ for polyhedron $P$ is following:

$$hrm(P) = \sum F_i^d/Vol^{d-1},$$

where $F_i$ are the volumes of facets of $P$ and $Vol$ is volume of $P$.

As in 2D case a harmonic tetrahedron (simplex) is regular. It seems to us that all Platonic solids are harmonic. For general case it need to study properties of harmonic polytops more carefully.

It is easy to construct $d+2$-vertices, $d > 2$, polyhedron that its Delaunay triangulation does not gives minimum of $hrm$. Open problem is *to study "harmonic" triangulation and find efficient algorithm for its construction*

## 5 DIRICHLET FUNCTIONAL FOR TRIANGULATIONS.

Let $S = \{x_i\}$ be a set of $n$ sites in $\mathbf{R}^d$, each associated with a real number $y_i$. Denote by $Y$ the set of these numbers, i.e. $Y = (y_1, ..., y_n)$. There are a lot of different problems in GIS, Geology, Topography, CAD/CAM etc., where we need to construct a surface in $\mathbf{R}^{d+1}$ corresponding to this dataset. The main problem is the following: to find a function $y = f(x)$, such that $f(x_i) = y_i$. One of the oldest and the most famous methods is modeling by triangulation. If we have some triangulation of $S$ then for a set of data $Y$ there is only one method to construct a piecewise linear function (polyhedral surface) on this triangulation. Usually Delaunay triangulation is used for this purpose.

One of the criteria that triangulation is suitable for the modeling is that area (volume if $d > 2$) of the polyhedral surface $y = f(x)$) in $\mathbf{R}^{d+1}$ is minimized on this triangulation. Let us introduce the following functional:

$$SV(t, Y) = \sum_i vol(\Delta_i(Y)),$$

where $(\Delta_i(Y)$ denotes the simplex on the surface $y = f(x)$ in $\mathbf{R}^{d+1}$ related to $(\Delta_i$.

It turns out that on the plane for "small" $y_i$ the triangulation minimizing this functional does not depend on $Y$ and it is DT.

**Theorem 5** *For every set of points $S$ in the plane there exists $\epsilon > 0$ such that if for all $i$ : $|y_i| < \epsilon$, then functional $SV(t, Y)$ achieves its minimum iff $t$ is Delaunay triangulation.*

It is easy to give an example of $n > 3$ points on the plane and some set $Y$ when the minimum of the functional $SV$ is not attained on DT.

In this connection we state an open problem: *to find an algorithm for constructing triangulation that gives minimum of the functional SV for an arbitrary set of sites Y.*

One more minimum criterion is a discrete analogue of the Dirichlet functional: $\int grad^2 f(x) dx$. For interval $(d = 1)$, a spline $deg = 2k - 1$ is a function $y = f(x)$ such that $f(x_i) = y_i$ and

$$\int_a^b [f^{(k)}(x)]^2 dx = min.$$

For $k = 1, d > 1$ and piecewise linear function $f$ we get

$$DF(t, Y) = \int_{CH(S)} grad^2 f(x) dx =$$

$$= \sum_i vol^2(\Delta_i(Y))/vol(\Delta_i) - vol(CH(S)).$$

Triangulation minimizing functional $DF$ may be called discrete spline triangulation (DST). For the plane DST does not depend upon $Y$ and it is DT.

**Theorem 6** *(S.Rippa [11]) If $d = 2$ then $DF(t, Y)$ achieves its minimum iff $t$ is DT.*

The proof of the theorems 5 and 6 directly follows from the fact that these functionals on quadrilaterals are minimized if $t$ is DT. The proof also follows from some general result that is given below.

Let $S$ be a set of $d+2$ points $x_1, ..., x_{d+2}$ in $\mathbf{R}^d$. C.Lawson [6] proved that there are at most two distinct triangulations of the convex hull of $S$. Suppose $S$ admits two triangulations $t_1$ and $t_2$, and $Y = (y_1, ..., y_{d+2})$ is a set numbers corresponding to $x_1, ..., x_{d+2}$ as above. Let $B(Y, S) = DS(t_1, Y) - DS(t_2, Y)$. Note $B(Y, S)$ is a quadratic form depending on $Y$.

**Theorem 7** *The optimal (DST) triangulation for $n = d+2$ does not depend on $Y$.*

This theorem follows from the fact that

$$B(Y) = const(S) L^2(Y),$$

where $L(Y)$ is some linear form on $Y$.

For $d > 2$ it is easy to give an example when triangulation is DST, but not DT. We state an open problem: *can DST depend on Y if $n > d + 2$ ?*

REFERENCES


1. D'Azevedo, E.F. and Simpson, R.B. On optimal interpolation triangle incidences, SIAM J. Sci. Statist. Comput., vol. 10, No. 6 , pp. 1063–1075, 1989.
2. D'Azevedo, E.F. Optimal triangular mesh generation by coordinate transformation, SIAM J. Sci. Statist. Comput., vol. 12 , No. 4, pp. 755–786, 1991.
3. Blaschke, W. Kreis und Kugel, Leipzig, 1916.
4. Joe, B. Construction of three – dimensional Delaunay triangulation using local transformation, Computer Aided Geometric Design, No. 8, pp. 123-142, 1992.
5. Lawson, C.L. Software for $C^1$ surface interpolation, in: J.R. Rice, ed., Mathematical Software III, Academic Press, New York, pp. 161-194, 1977.
6. Lawson, C.L. Properties of n-dimensional triangulations, Computer Aided Geometric Design, No. 3, pp. 231-246, 1986.
7. Musin, O.R. Delaunay triangulation and optimality, ARO Workshop Comp. Geom., Raleigh (North Carolina), pp. 37-38, 1993.
8. Musin, O.R. Index of harmony and Delaunay triangulation, Symmetry: Culture and Science, Vol. 6, No. 3, pp. 389-392, 1995.
9. Musin, O.R. Four-vertex theorem for polygon, Kvant, No 2, 1997 (in Russian).
10. Rajan V.T. Optimality of the Delaunay triangulation in $R^d$, Proc. 7th Annu. ACM Sympos. Comput. Geom., pp. 357–363, 1991.
11. Rippa, S. Minimal roughness property of the Delaunay triangulation, Computer Aided Geometric Design, No. 7, pp. 489-497, 1990
12. Sibson, R. Locally eguiangular triangulations, Comput. J., Vol.21, No. 3, pp. 243-245, 1978.
13. Voronoi, G.F. Nouvelles applications des parametres continus a la theorie des formes quadratiques, J. Reine u. Angew. Math., 34 , 198-287, 1908

# New Competitive Strategies for Searching in Unknown Star-Shaped Polygons

Jae-Ha Lee Chan-Su Shin Jae-Hoon Kim Sung Yong Shin
Kyung-Yong Chwa

Dept. of Computer Science
Korea Advanced Institute of Science and Technology
email: {jhlee,cssin,jhoon,syshin,kychwa}@jupiter.kaist.ac.kr

## Abstract

We consider searching problems in robotics that a robot has to find a path to a target by traveling in an unknown star-shaped polygon $\mathcal{P}$. The goal is to minimize the ratio of the distance traveled by the robot to the length of the shortest start-to-target path. Let $s$ be a starting point in $\mathcal{P}$. We first present a competitive strategy to find a path from $s$ to the closest kernel point $k$ of $\mathcal{P}$. The length of the path that the robot generates is less than $1 + 2\sqrt{2}$ ($< 3.829$) times the distance from $s$ to $k$, which improves the best previous bound 5.331 [3]. Second, given a specified target $t$ in $\mathcal{P}$, we present a competitive strategy to find a path from $s$ to $t$ whose length does not exceed 17 times the length of the shortest $s$-$t$ path.

## 1 Introduction

A typical navigation problem in robotics is to search for a target point in an "unknown" polygon. Searching in the unknown polygon means that a robot has no priori information about the geometry of the polygon and the location of the target. Thus, the robot with the vision system must make a decision as it moves around depending only on what it has seen so far. In that sense, these problems are "on-line", and thus it is natural to use the *competitive ratio* [9] to measure the performance of the robot's strategy. In particular, we wish to minimize the ratio between Euclidean ($L_2$) distance traveled by the robot and the length of the (off-line) shortest start-to-target path in the polygon. The competitive ratio is the worst case ratio achieved over all possible problem instances. A strategy is called *competitive* if its competitive ratio is constant.

In recent years, the competitive searching in unknown polygons has been intensively studied by many researchers [1, 6, 2, 8, 3] in computational geometry. Since it is impossible to competitively find a target in general polygons [6, 8], most of the work has focused on restricting the classes of polygons for which constant competitive ratios can be achieved, most notably Streets[6, 8] and $\mathcal{G}$-street [2].

As another class of polygons, we consider the class of *star-shaped polygons*. A polygon $\mathcal{P}$ is *star-shaped* if there is some point inside $\mathcal{P}$ that can see the entire boundary of $\mathcal{P}$. A set of those points is called a *kernel* of $\mathcal{P}$, denoted by $\mathcal{K}(\mathcal{P})$. In this paper, we consider two searching problems in a star-shaped polygon $\mathcal{P}$: From a given starting point $s$ in $\mathcal{P}$, (1) one is to search for the closest point $k \in \mathcal{K}(\mathcal{P})$, called *kernel searching problem*, and (2) the other is to search for a given target point $t$ in $\mathcal{P}$, called *target searching problem*.

For the kernel searching problem, Icking and Klein [3, 7] presented a 5.331 ($< 4\sqrt{2}$)-competitive strategy to find a $s$-$k$ path, and also gave a lower bound of $\sqrt{2}$. Here, note that the shortest path from $s$ to $k$ is a line segment because $k$ is a kernel point of $\mathcal{P}$. Their strategy generates the path that consists of $O(n)$ pieces of hyperbolic or elliptic arcs (including circular arcs and line segments), where $n$ denotes the number of vertices of $\mathcal{P}$. In this paper, we present a new strategy that generates a $s$-$k$ path whose length is bounded by $2\sqrt{2}+1$ ($< 3.829 < 3\sqrt{2}$) times the distance from $s$ to $k$, i.e., 3.829-competitive.

For the target searching problem, we present a 17-competitive strategy to find a path from $s$ to $t$ in the star-shaped polygons. Recently, Lopez-Ortiz and Schuierer [5] obtained 14.5 competitive algorithm and proved that the lower bound is 9.

## 2 Preliminaries

In this section, we give definitions and notation used throughout this paper; most of them has been introduced in [3, 8].

Let $\mathcal{P}$ be a star-shaped polygon with a given starting point $s$ in the interior or on the boundary. A robot is required to travel from $s$ to a target point: It may be either the closest kernel point $k$ or a specified point $t$.

Suppose that the robot is currently located at a point $p$ inside $\mathcal{P}$. We define a *cave* to be a connected chain of the boundary of $\mathcal{P}$ such that the robot sees the two end points of the chain but no other points between them. These two points are on the same line of sight from $p$. We call the one closer to $p$ the "mouth" of the cave. Each cavemouth $v$ is a reflex vertex of $\mathcal{P}$, and the cave associated with $v$ lies either to the left or right of a ray $pv$. Accordingly, we refer to $v$ as being *left* or *right cavemouths*. Moreover, for a left(right) cavemouth $v$, $\mathcal{K}(\mathcal{P})$ should be contained in the right(left) side of the ray $pv$. In other words, a cavemouth and the robot's position define an inner halfplane where $\mathcal{K}(\mathcal{P})$ is contained. Using the fact that $\mathcal{P}$ is star-shaped, one can easily verify that all left cavemouths lie to the left of all right cavemouths. Furthermore, we can define the left-to-right ordering among the cavemouths. We define $c_l$ to be the rightmost one among the left cavemouths, and $c_r$ to be the leftmost one among the right cavemouths.


*Computational Geometry 97* Nice France
Copyright 1997 ACM 0-89791-878-9/97/06 ..$3.50

## 3 New Competitive Strategy Searching for $\mathcal{K}(\mathcal{P})$

At the current position $p$ of the robot, if there are caves with respect to $p$, then it has to proceed in the direction being able to see the inside of the caves. To determine the direction, we define two wedges $G(p)$ and $B(p)$ at the point $p$, which were first defined in [3]. Those wedges represent regions in $\mathcal{P}$ where $\mathcal{K}(\mathcal{P})$ is contained, and the robot walks into $G(p)$ or $B(p)$.

More precisely, let $G(p)$ denote a *gaining wedge* that is an intersection of the inner halfplanes defined by $p$ and each cavemouth. Let $B(p)$ denote a *bounding wedge* that is an intersection of the inner halfplanes induced by all edges of $\mathcal{P}$ that are visible from and collinear with $p$. If there are no halfplanes associated with $G(p)$ and $B(p)$, both of the wedges are defined as $\mathcal{P}$ itself. For example, at the point $p_2$ on the robot's path in Figure 2, $G(p_2)$ is a wedge bounded by two rays, $p_2v_6$ and $p_2v_1$, and $B(p_2)$ is a right halfplane of a ray $p_2v_5$.

Since $\mathcal{K}(\mathcal{P}) \subseteq G(p) \cap B(p)$, if $G(p) \neq \mathcal{P}$, that is, there are any caves with respect to $p$, then the robot walks into $G(p) \cap B(p)$to see the caves, and moreover it never moves out of $G(p) \cap B(p)$. Thus, the visible region grows monotonically as the robot walks into $G(p) \cap B(p)$, and eventually it reaches a kernel point in $\mathcal{K}(\mathcal{P})$.

In order to decide a direction to move, the robot uses a local coordinate system, like sailor's compass. At the starting point $s$, the positive $y$-axis(like magnetic north), denoted by the $y^+$-axis, is determined as an arbitrary ray which is passing through $s$ and entering $G(s)$. Then, all left(right) cavemouths of $s$ has negative(positive) $x$-coordinates. As the robot proceeds, the coordinate system is translated so that the robot is always at the origin.

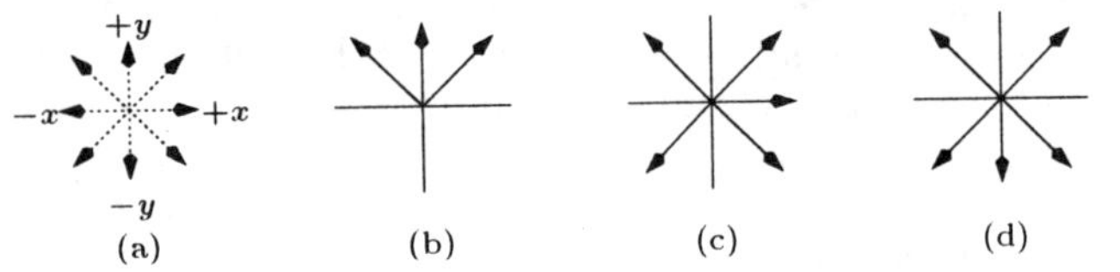


Figure 1: Candidates for the robot's direction. The octants are called 1-octant, $\cdots$, 8-octant from the $x^+$-axis in the counterclockwise order.

Our strategy permits the robot to move only in one of eight directions determined by representative vectors $(\pm1, \pm1)$, $(\pm1, 0)$ and $(0, \pm1)$ (see Figure 1 (a)), which are denoted by $\pm x \pm y$, $\pm x$ and $\pm y$, respectively. The robot proceeds bit by bit, in fact, infinitesimally, while it updates two cavemouths $c_l$ and $c_r$, and chooses one of the eight directions for the next movement. However, exceptionally, if the movement in the determined direction makes the robot leave out of $B(p)$, it moves along the boundary of $B(p)$ that is contained in $G(p)$.

We denote by *Right* the set of vertors that is defined as follows: If $c_r$ with respect to $s$ is in the 1-quadrant of $s$, then $Right = \{\pm x \pm y, +x\}$ (see Figure 1 (c)), else (if $c_r$ is in the 4-quadrant) $Right = \{\pm x \pm y, -y\}$ (see Figure 1 (d)). Notice that all right cavemouths of $s$ have positive $x$-coordinates, and thus the cavemouth $c_r$ lies either in the 1-quadrant or the 4-quadrant of $s$. According to the position of $c_l$ of $s$, we similarly define the set of vectors, denoted by *Left*, as follows: $Left = \{\pm x \pm y, -x\}$ or $Left = \{\pm x \pm y, -y\}$.

We determine the robot's direction at each position $p$ by the following rule. If both $c_r$ and $c_l$ exist, then the robot should move in the one of three directions, $\{+y, \pm x + y\}$, depending on the $y$-coordinates of $c_r$ and $c_l$ (see Figure 1 (b) and **case 1.1** $\sim$ **case 1.3** in the strategy **SFK** ). Otherwise, i.e., either $c_r$ or $c_l$ exists (but not both) (see **case 2** and **case 3**), it should move in the one of five directions in *Right* or *Left*, respectively. The detail description is summarized in the strategy **SFK**. An angle $\angle pqr$ denotes a counterclockwise one from the ray $qp$ to the ray $qr$.

**Strategy Searching-For-Kernel (SFK)**
$p = s$; (* $p$ is the robot's current position *)
determine the $y^+$-direction, and *Left* and *Right*
**while** there are any caves at $p$ **do**
(* $p$ is always the origin of coordinate system *)
compute $c_l$ and $c_r$ at $p$;
(* determine the direction vector $d$ as follows *)
**if** both $c_l$ and $c_r$ exist (* **case 1** *)
**then**
**case 1.1** $(c_l.y)(c_r.y) > 0$
$d = +y$
**case 1.2** $c_l.y < 0$ and $c_r.y > 0$
$d = -x+y$
**case 1.3** $c_l.y > 0$ and $c_r.y < 0$
$d = +x+y$
**else if** only $c_r$ exists (* **case 2** *) **then**
$d = arg\min_{q' \in Right}\{\angle c_r p q' | \angle c_r p q' \geq \frac{\pi}{4}\}$
**else** (* only $c_l$ exists: **case 3** *)
$d = arg\min_{q' \in Left}\{\angle q' p c_l | \angle q' p c_l \geq \frac{\pi}{4}\}$
**end-if**
**if** $d$ makes the robot move out of the $B(p)$
**then** move along the boundary of $B(p)$ in $G(p)$
**else** move in the $d$ direction
**end-if**
**end-while**
move straight to the closest kernel point $k$
**end-strategy**

Figure 2 illustrates an example for the robot to find a path from $s$ to $k$ by **SFK**.

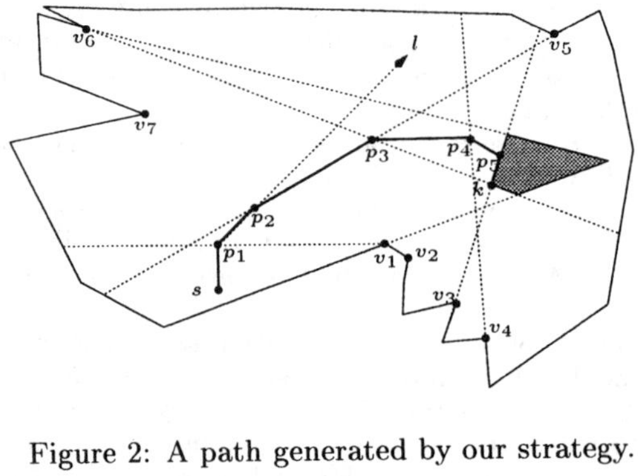


Figure 2: A path generated by our strategy.

## 4 Analysis of the Competitive Strategy SFK

Let $P$ be a path from $s$ to $k$. Let $len(P)$ denote the length of $P$ and $|qr|$ denote the length of a line segment $qr$. A *convex chain* $C = (p_1, p_2, \cdots, p_m)$ of $P$ is a maximal subpath such that $p_i$ and $p_j$ are visible from each other for all $1 \leq i, j \leq m$. Then, the path $P$ can be considered as a sequence of convex chains $C_1, C_2, \cdots, C_M$ as shown in Figure 3 (a). Two adjacent chains $C_i$ and $C_{i+1}$ share exactly one edge of $P$. Let $q$ and $r$ be any two points on a convex chain $C$. Note here that they are not necessarily $p_j$'s of $C$. A *carving* operation for $q$ and $r$ is to replace the line segments on $P$ between $q$ and $r$ with a line segment $qr$ (see Figure 3 (b)). Obviously, the segment $qr$ does not intersect with $P$ except at $q$ and $r$. A *carved path* is defined to be a path obtained by a sequence of carving operations. Then, we can easily obtain the following lemma.

**Lemma 1** *Let $P$ be a path and $P^c$ be a carved path of $P$. Then, $len(P^c) \leq len(P)$.*

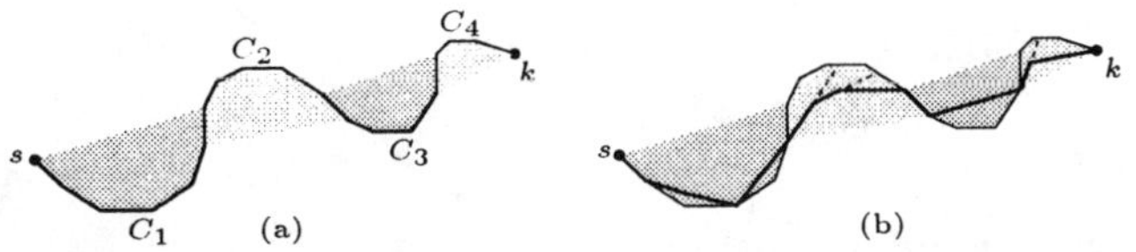


Figure 3: (a) A path $P$ consists of four maximal chains. (b) A carved path of $P$ is represented as thick lines.

**Outline of the Analysis:** Let $P_R$ be a path from $s$ to $k$ generated by our strategy **SFK**. To analyze $len(P_R)$, we first define three basic types of paths from $s$ to $k$, denoted by $P_1$, $P_2$, and $P_3$, and then show in Lemma 2 that $len(P_i) \leq (2\sqrt{2}+1)|sk|$ for each $i$. Next, we can show that $len(P_R)$ is bounded by the length of a carved path of $P_i$ for some $i$. Then, by Lemma 1, $len(P_R) \leq \max_i\{len(P_i)\}$. From Lemma 2, it follows that $len(P_R) \leq (2\sqrt{2}+1)|sk| < 3.829|sk|$. For the complete proof, refer to the full version of this communication.

**Theorem 1** *Strategy* **SFK** *is* $2\sqrt{2}+1$ $(< 3.829)$*-competitive.*

**Basic Paths:** Now, we define three basic paths $P_1$, $P_2$, and $P_3$ from $s$ to $k$. A basic path $P_1$ is defined as shown in Figure 4 (a). The path starts at $s$ in the $-x+y$ direction, turns to the $+x+y$ direction at the $x$-axis, turns to the $+x$ direction when it meets a ray with the $-x+y$ direction from $k$, turns to the $+x-y$ direction at the $y$-axis, follows along the line $sk$ towards $k$ when it meets $sk$, and finally ends at $k$. Other basic paths $P_2$ and $P_3$ are shown in Figure 4 (b) and (c), respectively. The path $P_2$ starts at $s$ and moves up until meets the $x$-axis, turns to the $+x+y$ direction at the $x$-axis, turns to the $+x-y$ direction at the $y$-axis, follows along the line $sk$, and ends at $k$. The path $P_3$ is a monotone path with respect to two vectors whose angle is exactly $\frac{\pi}{4}$. Figure 4 (c) illustrates the case that $P_3$ is monotone with respects to two vectors $+y$ and $+x+y$.

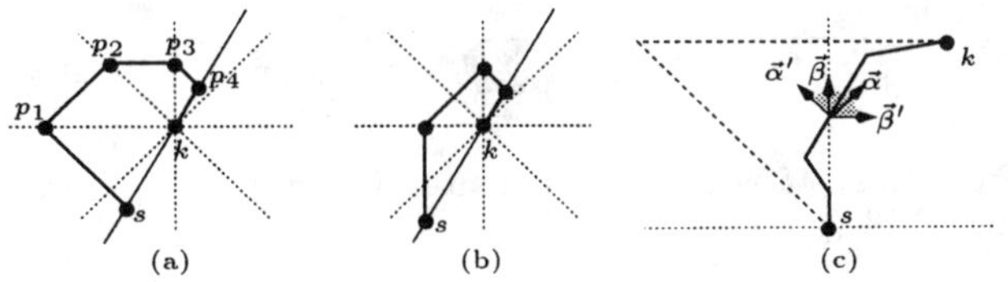


Figure 4: Basic paths. (a) $P_1$. (b) $P_2$. (c) $P_3$.

**Lemma 2** *$len(P_1) \leq 3.746\,|sk|$, $len(P_2) \leq (2\sqrt{2}+1)\,|sk|$, and $len(P_3) \leq 2.613\,|sk|$.*

## 5 A Competitive Strategy Searching for Targets in $\mathcal{P}$

In this section, we present a 17-competitive strategy searching for a specified target $t$ in a star-shaped polygon $\mathcal{P}$. Before describing our strategy, we begin with a fundamental fact about the shortest paths in $\mathcal{P}$.

**Lemma 3** *If $t$ is contained in a cave $C$ of $\mathcal{P}$, then the shortest path from $s$ to $t$ must touch the mouth of $C$.*

By Lemma 3, the shortest path from $s$ to $k$ is a chain of line segments joined at reflex vertices of $\mathcal{P}$. Let $d(\cdot,\cdot)$ denote the length of the shortest path between two points in $\mathcal{P}$. Let $D(p,\delta)$ be the set of points $q$ in $\mathcal{P}$ such that $d(p,q)$ is less than or equal to $\delta$.

**Definition 1** *Given a positive constant $\delta$, a* left path $L_\delta = (c_l^0, c_l^1, \cdots, c_l^{\sigma_l})$ *is a maximal path satisfying the following three conditions:*

(1) $c_l^0 = s$,

(2) $c_l^i \in D(s,\delta)$ *for each* $i$, *and*

(3) $c_l^i$ *is the rightmost one among left cavemouths with respect to* $c_l^{i-1}$ *for* $1 \leq i \leq \sigma_l$.

*A* right path $R_\delta = (c_r^0, c_r^1, \cdots, c_r^{\sigma_r})$ *is defined similarly.*

Combining Lemma 3 with the second condition, one can see that $c_l^i$ is in $D(c_l^{i-1}, \delta - d(s, c_l^{i-1}))$. This fact immediately implies that the length of $L_\delta$ is no more than $\delta$. The same property holds for $R_\delta$. Using the fact that $\mathcal{K}(\mathcal{P})$ lies between $c_l$ and $c_r$, we can prove the following key lemma.

**Lemma 4** *For any positive constant $\delta$, every point in $D(s,\delta)$ is visible from some point either on $L_\delta$ or on $R_\delta$.*

Our strategy is to apply the doubling technique of [1] to $L_\delta$ and $R_\delta$.

**Theorem 2** *Our strategy produces a path from $s$ to $t$ whose length is bounded by $17d(s,t)$.*

For the target searching problem, the same strategy was obtained independently by Lopez-Ortiz and Schuierer [4]. Recently, they improve the competitive ratio to 14.5 and proves the lower bound is 9 [5].

## References


[1] R. A. Baeza-Yates, J. C. Culberson, and G. J. E. Rawlins. Searching in the plane. *Information and Computation*, (106):234–252, 1993.

[2] A. Datta and Ch. Icking. Competitive searching in a generalized street. In *Proc. 10th Annu. ACM SoCG*, pages 175–182, 1994.

[3] Christian Icking and Rolf Klein. Searching for the kernel of a polygon: A competitive strategy. In *Proc. 11th Annu. ACM SoCG*, pages 258–266, 1995.

[4] A. Lopez-Ortiz. On the target searching in bounded and unbounded domains. Ph.D. thesis, University of Waterloo, 1996.

[5] A. Lopez-Ortiz and S. Schuierer. Position-Independent near optimal searching and on-line recognition in star polygons. In *this Proceeding*.

[6] R. Klein. Walking an unknown street with bounded detour. In *Proc. 32nd Annu. IEEE FOCS*, pages 304–313, 1991.

[7] R. Klein. personal communication. 1996.

[8] J. M. Kleinberg. On-line search in a polygon. In *Proc. 5th ACM SODA*, pages 8–15, 1994.

[9] C. H. Papadimitriou and M. Yannakakis. Shortest paths without a map. In *Proc. 16th ICALP*, volume 372 of *Lecture Notes in Computer Science*, pages 610–620. Springer-Verlag, 1989.

# Exact Primitives for Smallest Enclosing Ellipses *

Bernd Gärtner and Sven Schönherr

Institut für Informatik, Freie Universität Berlin
Takustr. 9, D-14195 Berlin, Germany
email: {gaertner,sven}@inf.fu-berlin.de

## Introduction

The problem of finding the unique closed ellipsoid of smallest volume enclosing an $n$-point set $P$ in $d$-space (known as the *Löwner-John ellipsoid* of $P$ [5]) is an instance of convex programming and can be solved by general methods in time $O(n)$ if the dimension is fixed [12, 6, 3, 1]. The problem-specific parts of these methods are encapsulated in *primitive operations* that deal with subproblems of constant size.

We derive explicit formulae for the primitive operations of Welzl's randomized method [12] in dimension $d = 2$. Compared to previous ones [9, 7, 8], these formulae are simpler and faster to evaluate, and they only contain rational expressions, allowing for an exact solution.

## Primitive Operations

For a finite point set $P$ in the plane, not all points on a line, we denote by $\text{ME}(P)$ the smallest enclosing ellipse of $P$. An inclusion-minimal set $S \subseteq P$ with $\text{ME}(S) = \text{ME}(P)$ is a *support set* of $P$. Any support set satisfies $|S| \leq 5$ and $\text{ME}(S) = \overline{\text{ME}}(S)$, where $\overline{\text{ME}}(S)$ denotes the smallest ellipse with all points of $S$ on the boundary. In general, if some ellipse exists with a set $B$ on its boundary, then also $\overline{\text{ME}}(B)$ exists and is unique [12].

Given $P$, Welzl's algorithm computes a support set $S$ of $P$, provided the following primitive operation is available.

> Given $B \subseteq P$, $3 \leq |B| \leq 5$, such that $\overline{\text{ME}}(B)$ exists, and a query point $q \in P \setminus B$, decide whether $q$ lies inside $\overline{\text{ME}}(B)$.

We call this operation the *in-ellipse test.* As we will see, the case $|B| = 4$ presents the actual difficulty. Our method is based on the concept of *conics.*

## Conics

A *conic* $\mathcal{C}$ in *linear form* is the set of points $p = (x, y)^T \in \mathbb{R}^2$ satisfying the quadratic equation

$$\mathcal{C}(p) := rx^2 + sy^2 + 2txy + 2ux + 2vy + w = 0, \quad (1)$$

*supported by the ESPRIT IV LTR Project No.21957 (CGAL)



$r, s, t, u, v, w$ being real parameters. $\mathcal{C}$ is invariant under scaling the vector $(r, s, t, u, v, w)$ by any nonzero factor. After setting

$$M := \begin{pmatrix} r & t \\ t & s \end{pmatrix}, \quad m := \begin{pmatrix} u \\ v \end{pmatrix}, \quad (2)$$

the conic assumes the form $\mathcal{C} = \{p^T M p + 2p^T m + w = 0\}$. If a point $c \in \mathbb{R}^2$ exists such that $Mc = -m$, $\mathcal{C}$ is symmetric about $c$ and can be written in *center form* as

$$\mathcal{C} = \{(p - c)^T M (p - c) - z = 0\}, \quad (3)$$

where $z = c^T M c - w$. If $\det(\mathcal{C}) := \det(M) \neq 0$, a center exists and is unique. Conics with $\det(\mathcal{C}) > 0$ define *ellipses.*

By scaling with $-1$ if necessary, we can w.l.o.g. assume that $\mathcal{C}$ is *normalized*, i.e. $r \geq 0$. If $\mathcal{E}$ is a normalized ellipse, $q$ lies inside $\mathcal{E}$ iff $\mathcal{E}(q) \leq 0$.

If $\mathcal{C}_1, \mathcal{C}_2$ are two conics, the *linear combination*

$$\mathcal{C} := \lambda \mathcal{C}_1 + \mu \mathcal{C}_2, \quad \lambda, \mu \in \mathbb{R}$$

is the conic given by $\mathcal{C}(p) = \lambda \mathcal{C}_1(p) + \mu \mathcal{C}_2(p)$. If $p$ belongs to both $\mathcal{C}_1$ and $\mathcal{C}_2$, $p$ also belongs to $\mathcal{C}$.

Now we are prepared to describe the in-ellipse test, for $|B| = 3, 4, 5$.

### In-ellipse test, $|B| = 3$

It is well-known [11, 7, 8] that $\overline{\text{ME}}(\{p_1, p_2, p_3\})$ is given in center form (3) by

$$c = \frac{1}{3} \sum_{i=1}^{3} p_i, \quad M^{-1} = \frac{1}{3} \sum_{i=1}^{3} (p_i - c)(p_i - c)^T, \quad z = 2.$$

From this, $M$ is easy to compute. Query point $q$ is inside $\overline{\text{ME}}(B)$ iff $(p - c)^T M (p - c) - z \leq 0$.

### In-ellipse test, $|B| = 4$

$\overline{\text{ME}}(B)$ is some conic through $B = \{p_1, p_2, p_3, p_4\}$, and any such conic is a linear combination of two special conics $\mathcal{C}_1, \mathcal{C}_2$ through $B$ [10], see Figure 1.

To see that these are indeed conics, consider three points $q_1 = (x_1, y_1), q_2 = (x_2, y_2), q_3 = (x_3, y_3)$ and define

$$[q_1 q_2 q_3] := \det \begin{pmatrix} x_1 - x_3 & x_2 - x_3 \\ y_1 - y_3 & y_2 - y_3 \end{pmatrix}.$$

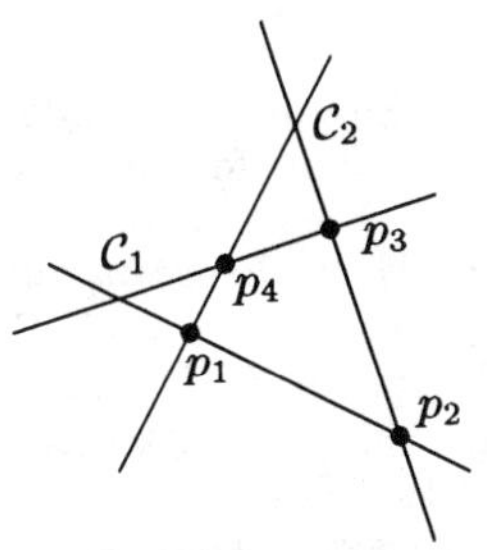


Figure 1: Two special conics through four points

$[q_1q_2q_3]$ records the orientation of the point triple; in particular, if the points are collinear, then $[q_1q_2q_3] = 0$. This implies

$$\mathcal{C}_1(p) = [p_1p_2p][p_3p_4p], \quad \mathcal{C}_2(p) = [p_2p_3p][p_4p_1p],$$

and these turn out to be quadratic expressions as required in the conic equation (1), easily computable from the points in $B$.

Now, given the query point $q$, we determine the (unique [10]) conic $\mathcal{C}_0$ through the five points $B \cup \{q\}$. We get $\mathcal{C}_0 = \lambda_0\mathcal{C}_1 + \mu_0\mathcal{C}_2$, with $\lambda_0 = \mathcal{C}_2(q), \mu_0 = -\mathcal{C}_1(q)$. In the sequel we assume that $\mathcal{C}_0$ is normalized.

**Case 1.** $\det(\mathcal{C}_0) \leq 0$, i.e. $\mathcal{C}_0$ is not an ellipse. Then exactly one of the following statements holds.
(i) $q$ lies inside any ellipse through $B$.
(ii) $q$ lies outside any ellipse through $B$.

To prove this, assume there are two ellipses $\mathcal{E}, \mathcal{E}'$ through $B$, with $\mathcal{E}(q) \leq 0$ and $\mathcal{E}'(q) > 0$. Then we find $\lambda \in [0,1)$ such that $\mathcal{E}'' := (1-\lambda)\mathcal{E} + \lambda\mathcal{E}'$ satisfies $\mathcal{E}''(q) = 0$, i.e. $\mathcal{E}''$ goes through $B \cup \{q\}$. Hence $\mathcal{E}''$ equals $\mathcal{C}_0$ and is not an ellipse. On the other hand, the convex combination of two ellipses is an ellipse again, a contradiction.

Thus, it suffices to test $q$ against *any* ellipse through the four points to obtain the desired result. Let

$$\alpha = r_1s_1 - t_1^2, \; \beta = r_1s_2 + r_2s_1 - 2t_1t_2, \; \gamma = r_2s_2 - t_2^2,$$

$r_i, s_i, t_i$ the parameters of $\mathcal{C}_i$ in the linear form (1). Then $\mathcal{E} := \lambda\mathcal{C}_1 + \mu\mathcal{C}_2$ with $\lambda = 2\gamma - \beta, \mu = 2\alpha - \beta$ defines such an ellipse. This follows from the fact that

$$\det(\mathcal{E}) = (4\alpha\gamma - \beta^2)(\alpha + \gamma - \beta),$$

and both factors can be shown to have negative sign if the $p_i$ are in convex position (which holds because we know that $\overline{\text{ME}}(B)$ exists) and in (counter)clockwise order (which can be achieved in a preprocessing step)[4].

**Case 2.** $\det(\mathcal{C}_0) > 0$, i.e. $\mathcal{C}_0$ is an ellipse $\mathcal{E}$. We need to check the position of $q$ relative to $\mathcal{E}^* = \overline{\text{ME}}(B)$, given by

$$\mathcal{E}^* = \lambda^*\mathcal{C}_1 + \mu^*\mathcal{C}_2,$$

with unknown parameters $\lambda^*, \mu^*$. In the form of (1), $\mathcal{E}$ is determined by $r_0, \ldots, w_0$, where $r_0 = \lambda_0 r_1 + \mu_0 r_2$. By scaling the representation of $\mathcal{E}^*$ accordingly, we can also assume that $r_0 = \lambda^* r_1 + \mu^* r_2$ holds. In other words, $\mathcal{E}^*$ is obtained from $\mathcal{E}$ by varying $\lambda, \mu$ along the line $\{\lambda r_1 + \mu r_2 = r_0\}$. This means,

$$\begin{pmatrix} \lambda^* \\ \mu^* \end{pmatrix} = \begin{pmatrix} \lambda_0 \\ \mu_0 \end{pmatrix} + \tau^* \begin{pmatrix} -r2 \\ r1 \end{pmatrix}. \tag{4}$$

for some $\tau^* \in \mathbb{R}$. Define

$$\mathcal{E}^\tau := (\lambda_0 - \tau r_2)\mathcal{C}_1 + (\mu_0 + \tau r_1)\mathcal{C}_2, \; \tau \in \mathbb{R}.$$

Then $\mathcal{E}^0 = \mathcal{E}, \mathcal{E}^{\tau^*} = \mathcal{E}^*$. The function $g(\tau) = \mathcal{E}^\tau(q)$ is linear, hence we get

$$\mathcal{E}^*(q) = \tau^* \frac{\partial}{\partial\tau}\mathcal{E}^\tau(q)\Big|_{\tau=0} = \rho\,\tau^*,$$

where $\rho = \mathcal{C}_2(q)r_1 - \mathcal{C}_1(q)r_2$. Consequently, $q$ lies inside $\overline{\text{ME}}(B)$ iff $\rho\tau^* \leq 0$.

The following Lemma is proved in [2], see also [8].

**Lemma** *Consider two ellipses $\mathcal{E}_1, \mathcal{E}_2$, and let*

$$\mathcal{E}^\lambda = (1-\lambda)\mathcal{E}_1 + \lambda\mathcal{E}_2$$

*be their convex combination, $\lambda \in (0,1)$. Then $\mathcal{E}^\lambda$ is an ellipse satisfying* $\text{Vol}(\mathcal{E}^\lambda) < \max(\text{Vol}(\mathcal{E}_1), \text{Vol}(\mathcal{E}_2))$.

Since $\mathcal{E}^\tau$ is a convex combination of $\mathcal{E}$ and $\mathcal{E}^*$ for $\tau$ ranging between 0 and $\tau^*$, volume of $\mathcal{E}^\tau$ decreases as $\tau$ goes from zero to $\tau^*$, hands

$$\text{sgn}(\tau^*) = -\text{sgn}\left(\frac{\partial}{\partial\tau}\text{Vol}(\mathcal{E}^\tau)\Big|_{\tau=0}\right).$$

If $\mathcal{E}^\tau$ is given in center form (3), its area is

$$\text{Vol}(\mathcal{E}^\tau) = \frac{\pi}{\sqrt{\det(M/z)}},$$

as can be seen by choosing the coordinate system according to the principal axes of $E$, such that $M$ becomes diagonal. Consequently,

$$\text{sgn}\left(\frac{\partial}{\partial\tau}\text{Vol}(\mathcal{E}^\tau)\Big|_{\tau=0}\right) = -\text{sgn}\left(\frac{\partial}{\partial\tau}\det(M/z)\Big|_{\tau=0}\right).$$

Recall that if $M, m$ collect the parameters of $\mathcal{E}^\tau$ as in (2), $c = M^{-1}m$ being its center, we get $z = c^TMc - w = m^TM^{-1}m - w$, where $M, m, w$ depend on $\tau$ (which we omit in the sequel, for the sake of readability). Noting that

$$M^{-1} = \frac{1}{\det(M)}\begin{pmatrix} s & -t \\ -t & r \end{pmatrix},$$

we get

$$z = \frac{1}{\det(M)}(u^2s - 2uvt + v^2r) - w.$$

Let us introduce the following abbreviations.

$$d := \det(M), \quad Z := u^2s - 2uvt + v^2r.$$

With primes ($d', Z'$ etc.) we denote derivatives w.r.t. $\tau$. Now we can write

$$\frac{\partial}{\partial\tau}\det(M/z) = (d/z^2)' = \frac{d'z - 2dz'}{z^3}. \tag{5}$$

Since $d(0), z(0) > 0$ (recall that $\mathcal{E}$ is a normalized ellipse), this is equal in sign to

$$\delta := d(d'z - 2dz'),$$

at least when evaluated for $\tau = 0$, which is the value we are interested in. Furthermore, we have

$$\begin{aligned} d'z &= d'(\frac{1}{d}Z - w) = \frac{d'}{d}Z - d'w, \\ dz' &= d(\frac{Z'd - Zd'}{d^2} - w') = \frac{Z'd - Zd'}{d} - dw'. \end{aligned}$$

Hence

$$\begin{aligned} \delta &= d'Z - dd'w - 2(Z'd - Zd' - d^2w') \\ &= 3d'Z + d(2dw' - d'w - 2Z'). \end{aligned}$$

Rewriting $Z$ as $u(us - vt) + v(vr - ut) = uZ_1 + vZ_2$, we get

$$\begin{aligned} d &= rs - t^2, & Z_1' &= u's + us' - v't - vt', \\ d' &= r's + rs' - 2tt', & Z_2' &= v'r + vr' - u't - ut', \end{aligned}$$

$$Z' = u'Z_1 + uZ_1' + v'Z_2 + vZ_2'.$$

For $\tau = 0$, all these values can be computed directly from $r(0), \ldots, w(0)$ (the defining values of $\mathcal{E}$) and their corresponding primed values $r'(0), \ldots w'(0)$. For the latter we get $r'(0) = 0, s'(0) = r_1s_2 - r_2s_1, \ldots, w'(0) = r_1w_2 - r_2w_1$. We obtain that $q$ lies inside $\overline{\text{ME}}(B)$ iff sgn $(\rho\,\delta(0)) \leq 0$.

### In-ellipse test, $|B| = 5$

In Welzl's algorithm, $B$ attains cardinality 5 only if before, a test '$p$ inside $\overline{\text{ME}}(B \setminus \{p\})$ ?' has been performed (with a negative result), for some $p \in B$. In the process of doing this test, the unique conic (which we know is an ellipse $\mathcal{E}$) through the points in $B$ has already been computed, see previous section. Now we just 'recyle' $\mathcal{E}$ to conclude that $q$ lies inside $\overline{\text{ME}}(B)$ iff $\mathcal{E}(q) \leq 0$.

### Implementation

We have implemented the in-ellipse tests as subroutines of Welzl's method with move-to-front heuristic [12], without any tuning.[1] On a Sun SPARC-station 20, using rational arithmetic over integers of arbitrary length provided by LEDA[2], the algorithm takes 220 seconds to compute $\text{ME}(P)$, $P$ a set of 10,000 points with random 32-bit integer coordinates. Under floating-point arithmetic, the computing time drops to 2 seconds, but the result might be incorrect. This gap (suggesting successful usage of floating-point filters and other techniques to combine fast arithmetic with exact computation) is explained by the fact that numbers get large under rational arithmetic. If the input coordinates are $k$-bit integers, an exact evaluation of $\delta(0)$ in case of $|B| = 4$ (which is the most expensive operation) requires $30k + O(1)$ bits of precision in the worst case.

The output of the algorithm is a support set $S$. In addition, for $|S| \neq 4$, our method determines $\text{ME}(P) = \text{ME}(S) = \overline{\text{ME}}(S)$ explicitly. For $|S| = 4$, the value $\tau^*$ defining $\overline{\text{ME}}(S)$ via (4) appears among the roots of (5); a careful analysis [7, 8] reduces this to a cubic polynomial in $\tau$, thus an exact symbolic representation or a floating-point approximation of $\tau^*$ and $\overline{\text{ME}}(S)$ can be computed in a postprocessing step.

[1] A tuned version will become part of the CGAL library, see http://www.cs.ruu.nl/CGAL/

[2] See http://www.mpi-sb.mpg.de/LEDA/leda.html

### References


[1] I. Adler and R. Shamir. A randomization scheme for speeding up algorithms for linear and convex quadratic programming problems with a high constraints-to-variables ratio. *Math. Programming*, 61:39–52, 1993.

[2] L. Danzer, D. Laugwitz, and H. Lenz. Über das Löwnersche Ellipsoid und sein Analogon unter den einem Eikörper eingeschriebenen Ellipsoiden. *Arch. Math.*, 8:214–219, 1957.

[3] M. E. Dyer. A class of convex programs with applications to computational geometry. In *Proc. 8th Annu. ACM Symp. on Computational Geometry*, pages 9–15, 1992.

[4] B. Gärtner and S. Schönherr. Smallest enclosing ellipses – fast and exact. Technical report, Free University Berlin, 1997.

[5] F. John. Extremum problems with inequalities as subsidiary conditions. In *Studies and Essays presented to R. Courant on his 60th Birthday*, pages 187–204. Interscience Publishers, NY, 1948.

[6] J. Matoušek, M. Sharir, and E. Welzl. A subexponential bound for linear programming. In *Proc. 8th Annu. ACM Sympos. Comput. Geom.*, pages 1–8, 1992.

[7] M. J. Post. Computing minimum spanning ellipses. Technical Report CS-82-16, Department of Computer Science, Brown University, 1982.

[8] S. Schönherr. Berechnung kleinster Ellipsoide um Punktemengen. Diploma thesis, Free University Berlin, 1994.

[9] B. W. Silverman and D. M. Titterington. Minimum covering ellipses. *SIAM J. Sci. Statist. Comput.*, 1:401–409, 1980.

[10] B. Spain. *Analytical Conics.* Pergamon Press, 1957.

[11] D. M. Titterington. Optimal design: some geometrical aspects of D-optimality. *Biometrika*, 62(2):313–320, 1975.

[12] E. Welzl. Smallest enclosing disks (balls and ellipsoids). In H. Maurer, editor, *New Results and New Trends in Computer Science*, volume 555 of *Lecture Notes in Computer Science*, pages 359–370. Springer-Verlag, 1991.

# Piecewise Linear Approximation of Bézier-Curves*

Helmut Alt† Emo Welzl‡ Barbara Wolfers§

While geometric objects are often designed and represented as smooth curves or surfaces, many efficient algorithms (e.g. rendering tools) require their input in piecewise linear form. Therefore it becomes necessary to efficiently approximate objects of the first type by objects of the second type. We will investigate this problem for Bézier-curves in two or higher dimensions, i.e. parameterized curves of the form

$$C(t) = \sum_{i=0}^{n} B_{i,n}(t)\, p_i, \qquad t \in [0,1],$$

where $p_0, p_1, \ldots p_n \in \mathbb{R}^d$ are the *control points* and

$$B_{i,n}(t) = \binom{n}{i} t^i (1-t)^{n-i}$$

are the *Bernstein polynomials.* We consider polygonal approximations $\hat{C}$ of $C$ with vertices on the curve $C$. That is, the vertices are $C(t_0), C(t_1), \ldots C(t_k)$ where $0 = t_0 < t_1 < \ldots < t_k = 1$ is a suitably chosen subdivision of the unit interval. We analyze the quality of this approximation, if the subdivision is pre-set (independently of the control points). In this way, the necessary coefficients $B_{i,n}(t_j)$ can be precomputed and stored in a table, and the vertices of the approximation can be easily retrieved as linear combinations of the control points.

Observe that $\hat{C}$ has a parameterization

$$\hat{C}(t) = \sum_{i=0}^{n} \hat{B}_{i,n}(t)\, p_i, \qquad t \in [0,1],$$

where the $\hat{B}_{i,n}$'s are piecewise linear approximations of the $B_{i,n}$ with $\hat{B}_{i,n}(t_j) = B_{i,n}(t_j)$ for $i = 0, 1, \ldots n$ and $j = 0, 1, \ldots k$ (see e.g. Figure 1). More specifically,

$$\hat{B}_{i,n}(t) = B_{i,n}(t_j) + (t - t_j)\frac{B_{i,n}(t_{j+1}) - B_{i,n}(t_j)}{t_{j+1} - t_j}$$

for $t \in [t_j, t_{j+1}]$. As a measure of its quality we use the distance function $d(C, \hat{C}) = \max_t \|C(t) - \hat{C}(t)\|$. This distance measure is related to the *Fréchet-distance* between curves (see [1]), where in addition the minimum is taken over all possible monotone parameterizations of the curves.

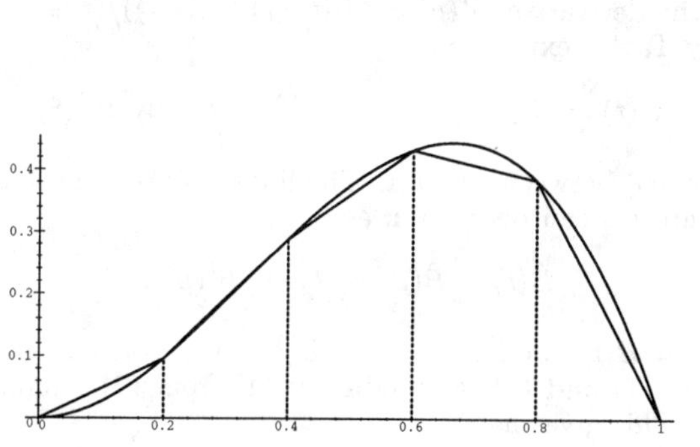


Figure 1: Piecewise linear approximation $\hat{B}_{2,3}$ of the Bernstein polynomial $B_{2,3}$

We have

$$d(C, \hat{C}) = \|\sum_{i=0}^{n}(B_{i,n}(\tau) - \hat{B}_{i,n}(\tau))\, p_i\| = \|\sum_{i=0}^{n} \alpha_i p_i\|$$

for some $\tau \in [0,1]$ and $\alpha_i = B_{i,n}(\tau) - \hat{B}_{i,n}(\tau)$ for $i = 0, \ldots, n$. $\sum_{i=0}^{n} \alpha_i = 0$, since $\sum_{i=0}^{n} B_{i,n}(t) = \sum_{i=0}^{n} \hat{B}_{i,n}(t) = 1$ for all $t$.

Let $\alpha = max_{i,t}|B_{i,n}(t) - \hat{B}_{i,n}(t)|$ be the maximum error in the approximations of the Bernstein polynomials, i.e. $|\alpha_i| \le \alpha$ for $i = 0, \ldots, n$. Let $\beta_i = \alpha_i/\alpha$ for $i = 0, \ldots, n$. Then

*This research was partially supported by the ESPRIT IV LTR Project No. 21957 (CGAL).

†Institut für Informatik, Freie Universität Berlin, Takustr. 9, D-14195 Berlin, Germany, <alt@inf.fu-berlin.de>.

‡Institut für Theoretische Informatik, ETH Zürich, CH-8092 Zürich, Switzerland, <emo@inf.ethz.ch>.

§mental images GmbH & Co. KG, Fasanenstr. 81, D-10623 Berlin, Germany, <wolfers@mental.com>.



*Computational Geometry 97* Nice France

we want an upper bound on

$$\alpha \| \sum_{i=0}^{n} \beta_i p_i \|$$

where $|\beta_i| \leq 1$ for $i = 0, ..., n$ and $\sum_{i=0}^{n} \beta_i = 0$.

For a given point set $P = \{p_0, p_1, \ldots p_n\}$, the set

$$\text{bzh}\, P := \{\sum_{i=0}^{n} \lambda_i p_i \mid \sum_{i=0}^{n} \lambda_i = 0, \text{ and } |\lambda_i| \leq 1, \text{ for all } i\}$$

has interesting relations to the so-called $k$-set problem. In fact, it can be shown that this set is a convex polytope, whose vertices have a one-to-one correspondence to subsets of $P$ of cardinality[*1] $\lfloor (n+1)/2 \rfloor$ that can be separated from their complement in $P$ by a hyperplane. Moreover, it follows from the analysis of bzh $P$ that

$$\max\{\|p\| \mid p \in \text{bzh}\, P\} < \lfloor \frac{(n+1)}{2} \rfloor \text{diam}\, P,$$

where diam $P$ is the diameter of $P$.

For an analysis of the factor $\alpha$ (see e.g. [2]), fix some $B = B_{i,n}$ and some $j$, $0 \leq j \leq k-1$. Let $\mu \in [t_j, t_{j+1}]$ such that $|B(\mu) - \hat{B}(\mu)|$ is maximized in this interval, which implies that the derivative $B'(\mu) = (B(t_{j+1}) - B(t_j))/(t_{j+1} - t_j)$.

By Taylor expansion

$$B(t) = B(\mu) + (t-\mu)B'(\mu) + (t-\mu)^2 B''(\xi) \tag{1}$$

for some $\xi$ between $t$ and $\mu$. The linear approximation $\hat{B}$ in this interval can be written as

$$\hat{B}(t) = \hat{B}(\mu) + (t-\mu)B'(\mu)\ . \tag{2}$$

$\hat{B}(t_j) = B(t_j)$ and $\hat{B}(t_{j+1}) = B(t_{j+1})$. Now substitute $t_j$ for $t$ in (1) and (2), and subtract (1) from (2) (similar for $t_{j+1}$). This gives us

$$\hat{B}(\mu) - B(\mu) = (t_j - \mu)^2 B''(\xi_j) = (t_{j+1} - \mu)^2 B''(\xi_{j+1})$$

for some $\xi_j$ and $\xi_{j+1}$ with $t_j < \xi_j < \mu < \xi_{j+1} < t_{j+1}$. By exploiting the fact that one of $|t_j - \mu|$ and $|t_{j+1} - \mu|$ does not exceed $(t_{j+1} - t_j)/2$, the error in $\mu$ is bounded by

$$\left(\frac{t_{j+1} - t_j}{2}\right)^2 \max_{\xi \in [t_j, t_{j+1}]} |B''(\xi)|\ . \tag{3}$$

The second derivative of the Bernstein polynomials can be written as ([3])

$$B''_{i,n} = n(n-1)(B_{i-2,n-2} - 2B_{i-1,n-2} + B_{i,n-2})$$

which shows that $\max_{t \in [0,1]} |B''(t)| < 2n(n-1)$.

Let us first assume that the $t_i$'s are evenly spaced with distance $\delta = 1/k$.

**Lemma 1** *In the equidistant (in parameter space) approximation with step size $\delta$ of a Bézier-curve of degree $n$, the error is bounded by*

$$\frac{1}{4}\delta^2 n(n-1) \lfloor \frac{n+1}{2} \rfloor \text{diam}\, P,$$

*where $P$ is the set of control points.* □

[*1] Note that the cardinality of $P$ is $n+1$.

A closer look at (3) suggests a more adaptive step size which ideally makes the expression in (3) constant. More precisely, since we have to guarantee the bound for all Bernstein polynomials, we have to ensure that

$$\left(\frac{t_{j+1} - t_j}{2}\right)^2 \max_{\xi \in [t_j, t_{j+1}],\ 0 \leq i \leq n} |B''_{i,n}(\xi)|$$

stays below some given threshold. So for large values of the second derivative a small and for small values a large step size should be used. We analyzed this relationship in more detail and developed a method for choosing an adaptive sequence of breakpoints.

We have run experiments both for equidistant and adaptive step size. Here are the results for three representative curves with the following control point sequences (see Figures (2 - 4)).

curve 1: (0,0),(1,3),(0,1),(0,2),(4,0),(2,3).
curve 2: (1,1),(2.5,3),(5.5,1),(5.5,3.5),(8,2.5),(10,4.5).
curve 3: (1,1),(2,4.2),(6,4.2),(7,1).

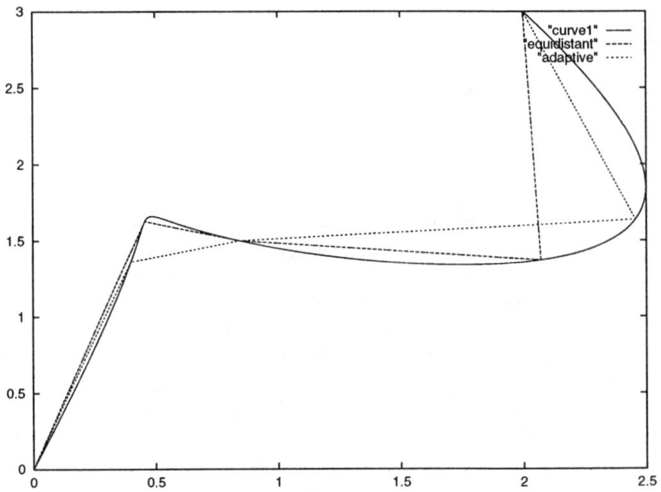


Figure 2: Curve 1.

Table 1 gives for all three curves in the case of 100 equidistant breakpoints a comparison between the bound on the error in the approximation derived in the analysis according to the formula in Lemma 1 and the actual value. We see that our formula is too pessimistic by a factor between approximately 3.5 and 15. One of the reasons is that Lemma 1 takes into account only the diameter of the set of control points, not the actual configuration (an improvement would be possible if we substitute $\max\{\|p\| \mid p \in \text{bzh}\, P\}$ for the factor $\lfloor (n+1)/2 \rfloor$ diam $P$).

| | Curve 1 | Curve 2 | Curve 3 |
|---|---|---|---|
| bound | 0.0067 | 0.0145 | 0.0018 |
| distance | 0.0019 | 0.0010 | 0.00032 |

Table 1: Derived bound and actual distance, $\delta = 0.01$ .

Table 2 gives the same information as Table 1 for the case of adaptive step size.

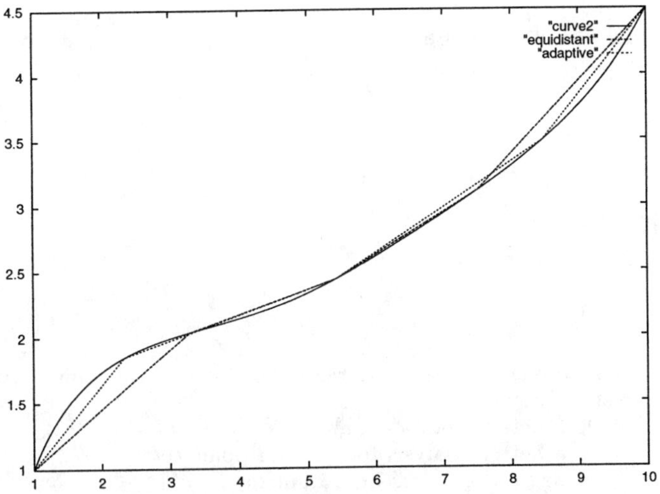


Figure 3: Curve 2.

| | Curve 1 | Curve 2 | Curve 3 |
|---|---|---|---|
| bound | 0.00219 | 0.00473 | 0.00109 |
| distance | 0.00070 | 0.00038 | 0.00053 |

Table 2: Derived bound and actual distance, adaptive subdivision for $k = 100$.

Finally, Table 3 gives the result of comparing the performance of equidistant vs. adaptive step size. For Curves 1 and 2 adaptation really pays yielding results nearly three times as good as the ones for equidistant breakpoints. For Curve 3 adaptation has a negative effect. The reason is that Curve 3 is bent more strongly in its middle part where the adaptive breakpoints are more sparse. However, more importantly, when comparing the first lines of Tables 1 and 2 we can see that the *guaranteed* upper bound is by a factor of 2-3 better in the adaptive case. This means that in our application for a given error tolerance we can choose a polygonal chain of correspondingly less segments.

| k | 4 | 6 | 8 | 10 | 20 | 40 | 100 |
|---|---|---|---|---|---|---|---|
| Curve1 | 0.47 | 0.43 | 0.44 | 0.46 | 0.37 | 0.37 | 0.36 |
| Curve2 | 0.52 | 0.46 | 0.43 | 0.41 | 0.38 | 0.37 | 0.36 |
| Curve3 | 1.22 | 1.40 | 1.44 | 1.59 | 1.72 | 1.29 | 1.60 |

Table 3: Error ratio: adaptive / equidistant.

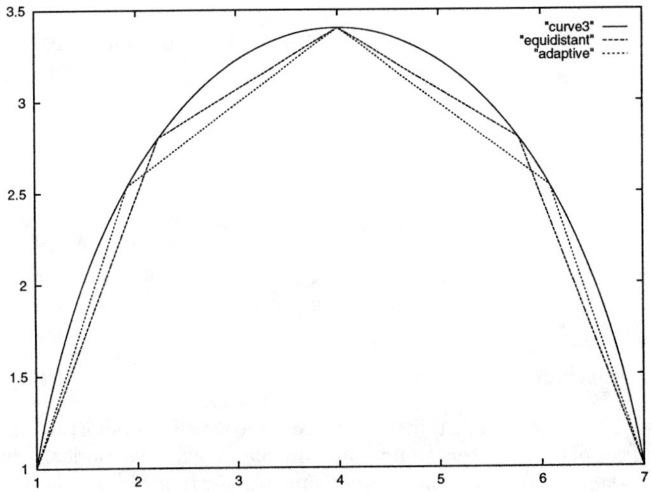


Figure 4: Curve 3.

## References

[1] H. ALT AND M. GODAU, Computing the Fréchet distance between two polygonal curves, *International Journal of Computational Geometry and Applications* **5** (1995) 75-91.

[2] C. DE BOOR, *A Practical Guide to Splines*, Springer-Verlag, (1978).

[3] G. FARIN, *Curves and Surfaces for Computer Aided Geometric Design*, 3rd Ed., Academic Press, (1992).

[4] J. D. FOLEY, A. VAN DAM, S. K. FEINER, AND J. F. HUGHES, *Computer Graphics – Principles and Practice*, Addison-Wesley, (1996).

# On Exact Solutions for the Rectilinear Steiner Tree Problem

Ulrich Fößmeier * Michael Kaufmann *

**Abstract**

The rectilinear Steiner tree problem asks for a shortest tree connecting given points in the plane with rectilinear distance. To find exact solutions for such problems is very challenging and could be done only for problems of very small size. The best theoretically analysed algorithms for this problem have a running time of $O(n^2 \cdot 2.62^n)$ (Ganley/Cohoon) resp. $n^{O(\sqrt{n})}$ (Smith). The best implementations use Branch & Bound and can solve random problems of size 35 (Salowe/Warme) within a day, while the algorithm of Ganley/Cohoon can only solve problems of size 27. In this paper we improve the theoretical worst-case time bound to $O(n \cdot 2.38^n)$, for random problem instances we prove a running time of less than $2^n$; implementations show drastic improvements.

## 1 Introduction

Let $S = \{s_0, .., s_{n-1}\}$ be a set of points in the plane which are called *terminals*. A *Steiner tree* is a tree in the plane which contains $S$. The Steiner tree problem is to find a Steiner tree of minimum length (SMT). It is one of the most famous combinatorial problems next to the traveling salesman problem. There are several versions of the problem, extensively described in the literature [5, 6, 8]. In the *rectilinear* Steiner tree problem, the distance between two points is the sum of the differences of their $x-$ and $y-$coordinates. This classical combinatorial problem received new importance in the development of techniques for VLSI routing [9, 10]. In this area, it is important to achieve optimum solutions. One percent increase of the wire length could mean a considerable loss of performance of the corresponding chip. The Steiner tree problem is *NP*-complete [5], so many approximation algorithms for SMTs appeared in the last two decades [1, 2, 11, 14]. The known exact algorithms perform extremely poorly from the analytical point of view as well as from the implementation side. We show how to solve this problem exactly in a 'reasonable' computing time. Our results are:

1. a better worst-case analysis
2. a better analysis for random point sets
3. drastic practical improvements.

The terminals have to be connected via a tree $T$ consisting of horizontal and vertical segments. $T$ may contain other points than the terminals, the so-called *Steiner points.* $T$ is called a *full Steiner tree* if all terminals are leaves in $T$. At terminals with degree more than one we can split $T$ into edge-disjoint full Steiner trees, which have only the split terminals in common. Those trees are called *full components* in the following.

All previous approaches to compute rectilinear SMT's (see [4] and [8] for a review) use a two-phases scheme due to P. Winter: First a sufficient number of full components is computed and in the second phase an SMT is constructed using dynamic programming or Branch & Bound. For both approaches the number of full components constructed in phase 1 is the main parameter for the total running time; thus all algorithms try to minimize this number.

The experiences in [4] and [8] show that the dynamic programming approach leads to the best theoretic results, while a Branch & Bound algorithm [12] has the best practical behaviour. The theoretically best algorithm of Smith ([13]) with a complexity of $n^{O(\sqrt{n})}$ is highly impractical because of the large constant in the exponent. The most advanced work on dynamic programming by Ganley/Cohoon [4] gives simple algorithms that perform quite poorly in practice. Here SMTs are constructed for all possible subsets of $S$ by putting together a full component $FC$ and a previously computed SMT $ST$ for a smaller subset if $FC$ and $ST$ have exactly one terminal in common. If two trees for the same subset are constructed then only the shorter tree is stored. In the $i$-th iteration SMTs for subsets of size $i$ are constructed and the algorithm terminates after iteration $n$.

The total time complexity is $\sum_{m=3}^{n} \binom{n}{m} \cdot m \cdot f(m)$, where $f(m)$ denotes the number of full components within a terminal set of size $m$. In [4], $f(m)$ is bounded by $1.62^m$ and a bound for the running time of $O(n^2 \cdot 2.62^n)$ could be proved. The known Branch & Bound algorithms [12] can only guarantee a double exponential running time. We improve these results by considerably reducing the number of necessary full components for both approaches and we also show how to reduce the number of all subsets of $S$ to be considered for the dynamic programming.

*Universität Tübingen, Wilhelm-Schickard-Institut, Sand 13, 72076 Tübingen, Germany, email: { foessmei / mk } @informatik.uni-tuebingen.de



*Computational Geometry* 97 Nice France

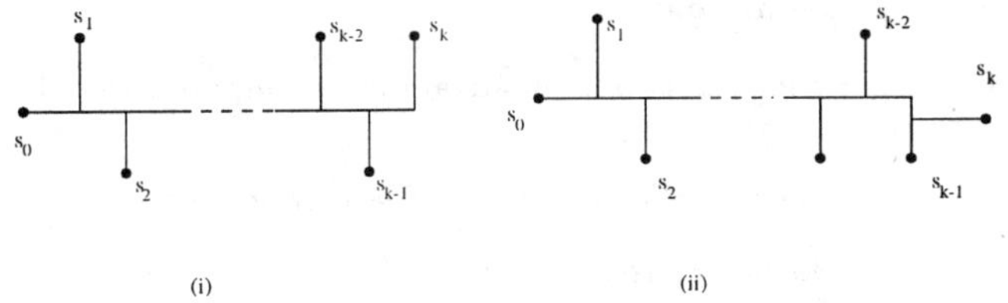


Figure 1: Two shapes of a full Steiner component

## 2 Basics and Full Components

A classical Lemma of Hwang [7] restricts the topology of the full components:

**Lemma 1** *Let $n > 4$. Suppose that the SMT of a given set of terminals is a full Steiner tree. Then there exists an SMT that either consists of a single line with $n-1$ alternating incident segments, or of a corner with $n-3$ alternating segments being incident to one leg and a single segment being incident to the other leg.*

In Figure 1 we show the two shapes described in Lemma 1. The single middle line is called *Steiner chain*, the only terminal on the Steiner chain is the *root* of the full component. Note that the second shape is similar to the first one, except of the last leg which is oriented in the same way as the Steiner chain. From now on we only consider full components of type (i) with a horizontal Steiner chain having the root at the left end. The number of components of type (ii) can be assumed not to exceed the number of components of type (i) ([3]).
At the beginning of phase 1 of the algorithm every subset of the terminals is computed that has one of the shapes of Fig. 1; this set of full components is sufficient to construct a Steiner minimum tree.

We describe the two most important tests to decrease the number of full components:
We say that a full component is defining *empty rectangles* if for every pair of consecutive terminals of the component the smallest rectangle containing them does not contain any other terminal of the given set. From [2] we know

**Lemma 2** *For any set of terminals there is an SMT using only full components defining empty rectangles.*

**Corollary 1** *When we fix the set of terminals above (below) the Steiner chain, then the set of terminals below (above) the chain is determined automatically, namely the terminals with the smallest distance to the chain in the corresponding vertical stripes.*

Using Corollary 1, we can prove

**Lemma 3** *For $n$ terminals, there are only $O(2^{n/2}) \approx O(1.42^n)$ full components defining empty rectangles.*

A full component $FC$ with Steiner points $c_1, \ldots, c_k$ is called a *tree star* if the minimum spanning tree of the point set $S \cup \{c_1, \ldots, c_k\}$ contains every edge of $FC$ (i.e. the connections of terminals to Steiner points and the connections between two Steiner points) and if $FC$ defines empty rectangles. In [3], we show

**Lemma 4** *For any set of terminals $S$ there is an SMT containing only tree stars as full components.*

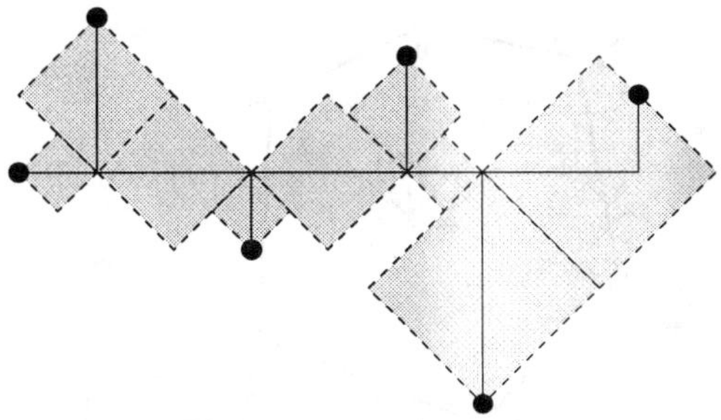

Figure 2: The empty regions of a tree star.

**Corollary 2** *In a tree star the triangles defined by 45°-diagonals on both sides of each segment do not contain any other terminal of $S$ (see Fig. 2).*

We bound the number of tree stars in the worst case and for random point sets:

**Theorem 1** *In a rectilinear Steiner tree problem with $n$ terminals there are $O(1.384^n)$ tree stars.*

**Theorem 2** *The number of tree stars in a random problem instance is $\leq O(n \cdot 2^{\sqrt{n \log n}})$.*

Note that in [12] only a bound of $O(n^2 \cdot 2^k \cdot k^k)$ for full components of size $k$ has been achieved. We have shown that the concept of tree stars can reduce the number of full components to be considered; nevertheless the concept of tree stars does not lead to a polynomial number of necessary full components: In [3] we give an RST problem with $\Omega(1.32^n)$ tree stars. In practical examples however, the number of tree stars is really small: In [12] a very large number of examples is considered and it can be seen that the number of their full components is about $4n$ (linear!) for most of the examples. Our implementation can delete about half of them, therefore we have to deal with about $2n$ full components for randomly generated examples.

## 3 Phase 2: The Merge – Step

In our previous analysis of the running time, we assumed that for *all* subsets $R$ an SMT($R$) will be constructed. But we only need to compute the subsets $R$ that could be extended to an SMT for the total point set $S$ by subsequently adding other full components. There are many subsets $R$ that do not fulfill this condition and that can be excluded: If e.g. two terminals in $R$ lie at opposite corners and the terminals between them are not in $R$, the tree SMT($R$) will obviously not appear as a part of SMT($S$) and we do not need to consider $R$. We now make this observation more precise.
Let $Q(a, b)$ be the axisparallel quadrangle defined by the points $a, b$ such that $a$ and $b$ lie at opposite sides of $Q(a, b)$ and the center of $Q(a, b)$ lies in the middle of the segment $[ab]$. We define $EQ$ to be the graph $(V, E)$ with $V$ to be the set of terminals and $E = \{(a, b); Q(a, b)$ does not contain any terminals$\}$. Fig. 3 shows an example. The black vertices in Fig. 3 show a subset of terminals that the algorithm should not consider (one of the terminals lies inside the fat cycle, the other one outside). We can prove

**Lemma 5** *Let $R$ be any subset of the terminals, $\overline{R} = V - R$ and $EQ(\overline{R})$ is the induced subgraph of $\overline{R}$. If $EQ(\overline{R})$ contains*

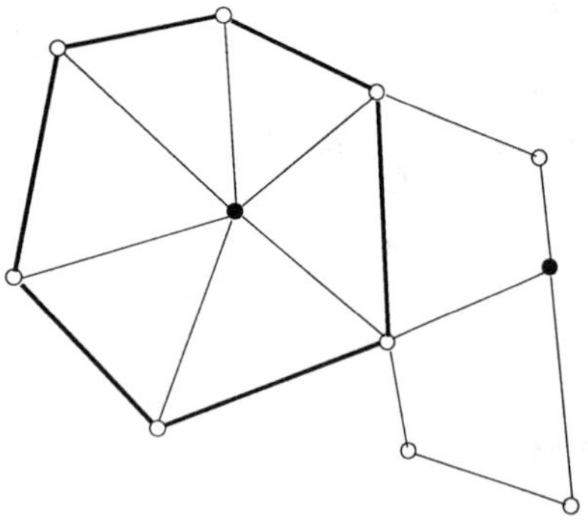

Figure 3: An example for a 'non reasonable' subset

*a simple cycle such that there are vertices of $R$ inside and outside of the cycle then we do not need to consider the subset $R$ of terminals.*

In [3] we show that an algorithm that makes use of this property has to consider less than $\alpha^n$ subsets of terminals in random problem instances for $\alpha < 2$. Together with Theorem 2, we can only prove that for random terminal sets the running time is smaller than $\beta^n$ for a constant $\beta < 2$. There are some worst-case examples where the number of subsets cannot be reduced by the technique of graph $EQ$, but for these cases only a small fraction of them will be considered by other reasons. We suspect that only a polynomial number of subsets might be constructed. But this is not part of the graph concept. The examples and their impacts deserve further investigations.

We summarize our results by the following algorithm:

(1) for $m = 2$ to $|S|$ do<br>
(2) $\quad L(m) := \{T \in F(S); |T| = m\}$;<br>
(3) $\quad F(m) := L(m)$;<br>
(4) od;<br>
(5) for $m = 2$ to $|S|$ do<br>
(6) $\quad$ for $i = 2$ to $|S|$ do<br>
(7) $\qquad$ For all $A \in F(i)$ and $B \in L(m)$ with $|A \cap B| = 1$ and<br>
(8) $\qquad$ component $A$ does not cross any edge $\in EQ(\overline{A \cup B})$<br>
(9) $\qquad\quad$ do $R := A \cup B$;<br>
(10) $\quad$ if $R \notin L(m+i-1)$ then append $R$ in $L(m+i-1)$;<br>
(11) $\qquad l(R) := min\{l(R), l(A) + l(B)\}$;<br>
(13) return $l(entry(L(|S|)))$;

The $F$-sets contain the set of full components computed in phase 1. In the $L(i)$-sets we only store sets of size $i$ that could have been constructed by combining full components. So by far, not all subsets of size $i$ will be considered. Note that the first condition in line (7) means that sets $A$ and $B$ can be combined, and the second in line (8) concerns the conditions given by Lemma 5.

We summarize the theoretical analyses and combine the last results from this section, Theorem 2 and 1:

**Theorem 3** *We can solve the rectilinear Steiner tree problem for n points in the plane in time $O(n \cdot 2.384^n)$ in the worst case, and for n randomly distributed points we only need $O(\beta^n)$ time, with $\beta < 2$.*

## 4 Implementations

Our algorithm has two main advantages compared with former algorithms:

1. We use a smaller number of full components.
2. We do not consider all possible subsets of the set of terminals.

Enhanced with several heuristics, our dynamic programming implementation can solve problem instances of size 35 in one day and has thus a comparable power as the Branch & Bound algorithm in [12]. The main bottleneck for a better performance of the dynamic programming approach is the space-consuming and slow phase 2. To illustrate the importance of a careful phase 1 (leading to a small number of full components) also for a Branch & Bound algorithm we combine our phase 1 with phase 2 of Salowe and Warme and achieve tremendous improvements. Reducing the number of full components by about one half, we can increase the range of solvability for random problem instances from size 35 [12] to about size 55 as indicated by first experiments.

## References


[1] S. Arora, *Polynomial Time Approximation Schemes for Euclidian TSP and other Geometric Problems.* In Proc. of 37th IEEE Symp. on Foundations of Computer Science, 2–11, 1996.

[2] P. Berman and V. Ramaiyer. *Improved approximations for the Steiner tree problem.* Proc. of 3d ACM-SIAM Symp. on Discrete Algorithms, 325–334, 1992.

[3] U. Fößmeier and M. Kaufmann. *On Exact Solutions for the Rectilinear Steiner Problem. Part I: Theoretical results,* Tech. Report WSI-96-09, Tübingen, 1996.

[4] Ganley J. L., J. P. Cohoon, *Optimal rectilinear Steiner minimal trees in $O(n^2 2.62^n)$ time,* Proc. 6th Canad.Conf.Comput.Geom., 308-313, 1994.

[5] M. R. Garey, D. S. Johnson. *The Rectilinear Steiner Problem is NP-Complete.* SIAM J. Appl. Math., 32, 826-834, 1977.

[6] M. Hanan. *On Steiner's Problem with Rectilinear Distance.* SIAM J. Appl. Math.,14, 255-265, 1966.

[7] F. K. Hwang. *On Steiner Minimal Trees with Rectilinear Distance.* SIAM J. Appl. Math.,30, 104-114, 1976.

[8] F. K. Hwang, D. S. Richards and P. Winter. *The Steiner Tree Problem.* Annals of Disc. Math. 53, North-Holland, 1992.

[9] B. Korte, H. J. Prömel, A. Steger. *Steiner Trees in VLSI-Layouts.* In Korte et al.: Paths, Flows and VLSI-Layout, Springer, 1990.

[10] Th. Lengauer. *Combinatorial Algorithms for Integrated Circuit Layout.* John Wiley, 1990.

[11] D. Richards. *Fast Heuristic Algorithms for Rectilinear Steiner Trees.* Algorithmica, 4, 191-207, 1989.

[12] J. S. Salowe and D. M. Warme. *Thirty-Five Point Rectilinear Steiner Minimal Trees in a Day.* Networks Vol. 25, 69-87, 1995.

[13] W. D. Smith. *How to find Steiner minimal trees in Euclidian d-space.* Algorithmica 7, 137-177, 1992.

[14] A. Z. Zelikovsky. *An 11/8-approximation Algorithm for the Steiner Problem on Networks with Rectilinear Distance.* In Sets, Graphs and Numbers. Coll. Math. Soc. J. Bolyai 60: 733-745, 1992.

# Clusters of Stars

Ileana Streinu*

**Abstract**

We solve two open problems posed by Goodman and Pollack[GP84] about sets of signed circular permutations (clusters of stars) arising from generalized configurations of points: recognition and efficient reconstruction (drawing). As a biproduct we get an $\mathcal{O}(n^2)$ space data structure constructible in $\mathcal{O}(n^2)$ time, representing the order type of a (generalized) configuration of points and from which the orientation of each triple can be found in constant time, a problem posed in [EHN].

## 1 Introduction and motivation

Imagine that robots equiped with cameras are sending visibility information from an unknown environment. The goal is to reconstruct a global model of what the environment might be like. In a simplified model, assume that the observers are placed at the vertices of a polygon, with each observer sending information about the combinatorial structure of its *visibility polygon*: an ordered list of vertices and edges. The receiver will verify the consistency of the data and *draw* a model matching the given combinatorial information. As a first approximation one is interested in reconstructing a pseudo-polygon (see [OS96]) without regard as to whether its sides can or cannot be stretched into straight lines, since there is evidence that stretchability is considerably harder to achieve. In this paper we conclude the work started in [OS96] by providing efficient algorithms for recognizing and drawing clusters of stars, from which the pseudo-polygon drawing problem is a simple consequence. This way we also answer two problems posed by Goodman and Pollack in [GP84]: finding conditions for a cluster of stars to fit into a generalized configuration of points, and finding a simple procedure for drawing the generalized configuration of points without going through a double-dualizing process.

*Dept. of Computer Science, Smith College, Northampton, MA 01063, USA. streinu@cs.smith.edu.



*Computational Geometry 97* Nice France



## 2 Definitions, Problems, Results

We refer to [OS96], [GP84], [Bj93] for definitions of pseudo-line (p-line) arrangements, generalized configurations of points, clusters of stars, local (or $i$-) sequences, allowable sequences and wiring diagrams. We assume general position throughout.

Given a finite set of planar points $\{p_1, \cdots, p_n\}$, and a direction for each line $l_{ij}$ through the points $p_i$ and $p_j$, we associate a *directed cluster of stars* by recording $\forall i$ the circular sequence of all the directed lines $l_{ij}$ around vertex $p_i$. This doubly infinite periodic sequence is fully characterized by a half-period. A signed permutations (called a star permutation) of indices of points different from $i$ suffices to represent it: indices occur positively or negatively, depending on whether the corresponding line is encountered in a direction coinciding or not to its orientation. We will also work with undirected clusters of stars as in [OS96], where no predefined orientation of the connecting lines $l_{ij}$ is assumed. The concepts can be introduced for generalized configurations of points as well. For the example in Fig.1 the directed star permutations can be taken as: $1 : 42\bar{3}$, $2 : 1\bar{3}\bar{4}$, $3 : 4\bar{2}\bar{1}$ and $4 : 13\bar{2}$. The undirected stars are similar but signless.

Goodman and Pollack's first problem is: *Under what conditions can one complete a picture such as Fig.1a to a generalized configuration, with each $j$-line radiating from point $i$ connecting to the $i$-line radiating from point $j$ in such a way that the directions match?* The goal is to find conditions verifyable in low polynomial time. If for every pair of points $(i, j)$ we knew that the direction of the underlying pseudo-line was, say, from $i$ to $j$, then the information in the cluster of stars would be equivalent to the information in a chirotope, which can be verified by looking at all 5-tuples and checking in constant time that the orientation of all its triples is compatible to the sign patterns re-

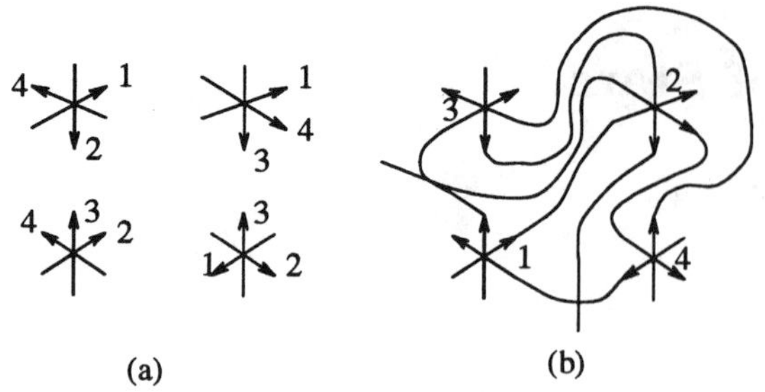


Figure 1: A cluster of stars and a generalized configuration of points from it.

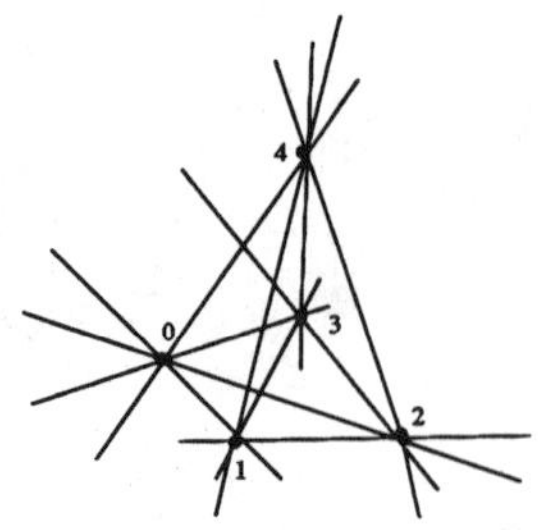

Figure 2: Clusters of stars for Problem 1.

sulting from the Grassman-Plücker relationships (see [Bj93]). This can be done naively in $\mathcal{O}(n^5)$ time. Trying all possible ways of orienting the pairs, and checking the chirotope conditions for each would be exponential. We will give a simple $\mathcal{O}(n^2)$ time algorithm to carry on this verification.

One way of drawing the underlying p-line arrangement of a generalized configuration of points as a wiring diagram is described in [GP84]. It works by using duality twice. First they construct the dual arrangement as a wiring diagram, then apply Levy's Enlargement Lemma to connect the vertices of the arrangement by p-lines, then record the order of all the p-lines (from the wiring diagram and from Levy's lemma) at $x_{-\infty}$ and use that as the allowable sequence for a dual arrangement. The process is quite complicated and because it uses Levy's Lemma, it is not algorithmic. Hence the second open problem from [GP84]: *Given a cluster of stars, realize it as a generalized configuration of points without going through the double dualizing procedure.* Here we provide an $\mathcal{O}(n^2)$ algorithm for drawing the connecting p-lines of a generalized configuration of points. Along the way we answer a problem mentioned in [EHN]: *Find an $\mathcal{O}(n^2)$ space data structure, constructible in $\mathcal{O}(n^2)$ time, for storing the order type of a configuration of points, from which the orientation of each triplet of points can be found in constant time.* We have realized recently that this data structure has also been defined in Knuth[Kn92], page 16, but its usage seems to be novel.

## 3 Realizability of Clusters of Stars as Generalized Configurations of Points

We identify two variants of the problem, corresponding to the undirected and directed cluster of stars problem. We give an efficient algorithm for recognizing (undirected) clusters of stars and show that the original problem of Goodman and Pollack (the directed stars) can be reduced to it at the expense of an additional $\mathcal{O}(n^2)$ factor.

**Problem 1.** *(Undirected) cluster of stars.* Given $n+1$ star permutations, the first one $\pi_0$ being the identity of $\{1, \cdots, n\}$ and each $\pi_i$ starting with 0, find necessary and sufficient conditions for them to be the cluster of star permutations of a generalized planar configuration of points with undirected underlying p-lines, so that the two points 0 and 1 form an edge of the convex hull.

For example, the set of star permutations: $0:1234$, $1:0234$, $2:0143$, $3:0142$, $4:0132$ can be realized as in Fig.2.

**Problem 2.** *Directed cluster of stars.* Given a set of signed star permutations, find an efficient algorithm to decide when they can fit into a generalized configuration of points with matching directions.

We solve Problem 2 using Problem 1 as follows. For all the pairs of points, associate an instance of Problem 1 where the two points are the distinguished edge of the convex hull. Ignore the signs for the time being. If the instance doesn't have a solution, try next pair. If it does, then check the signs (directions) for compatibility. If they do not match for any of the pairs for which the problem 1 has a solution, then problem 2 does not have a solution. As an aside we should mention that solutions for Problems 1 and 2, when they exists, are unique with respect to the order type.

The characterization needed to solve Problem 1 is based on a relationship $\prec$ on the set of pairs $ij$ (where $ij = ji$): if in the permutation $k$, $i$ occurs somewhere before $j$ (and after 0), then define $ik \prec jk$.

**Theorem 3.1** *Given a cluster of stars, it is realizable as a generalized configuration of points iff the following property (*the $\Delta$-property*) holds:*

*For any triple $0 < i < j < k \leq n-1$, exactly one of the following is true:*

1. $jk \prec ik \prec ij$, *or*
2. $ij \prec ik \prec jk$.

**Proof.**(Sketch) The necessity is straightforward. For the sufficiency we will prove by a double induction a dual statement: *Given a cluster of stars satisfying the $\Delta$-property, there exists a p-line arrangement with the p-lines numbered $0, \cdots, n$ at $x_{-\infty}$ and whose local sequences are the given permutations of the stars.* The proof can be turned into a verification and reconstruction algorithm, which will attempt to construct the arrangement just like the classical [EOS] algorithm. A careful implementation turns this into an $O(n^2)$ time algorithm. We omit the details.

## 4 Efficiently drawing a generalized configuration of points

We are given a cluster of stars as in Problem 1 and assume it satisfies the Δ-condition. Associate a directed graph $G$ with vertices labeled by pairs $ij$ with $i, j \neq 0$ and an arc $ij \rightarrow ik$ whenever $ij \prec ik$. The Δ condition implies, among others, that the digraph is acyclic. Perform a topological sort on it. This will be the order of the "slopes" of the connecting p-lines in the final drawing, where the vertex 0 is at infinity and the p-lines through 0 are verticals.

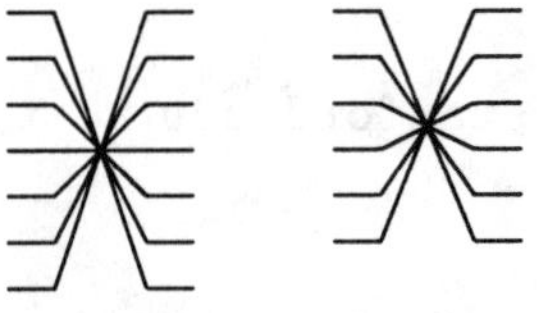


Figure 3: A switch at a vertical reference line.

The drawing is done using $\binom{n}{2}$ horizontal reference lines and $n$ vertical reference lines. The horizontal lines are labeled top-down by the pairs $ij$ in the order provided by the topological sort. The vertical lines are labeled left to right by the point indices $1, 2, \cdots, n$. All pseudo-lines labeled with pairs containing $k$ will switch at the $k$th vertical (on one or between two horizontals, see Fig.3) and will not cross before or after the $k$th vertical reference line. This implies that their relative ordering stays the same before and after the switch, even though they may move up and down on the horizontal reference lines, generating other crossings.

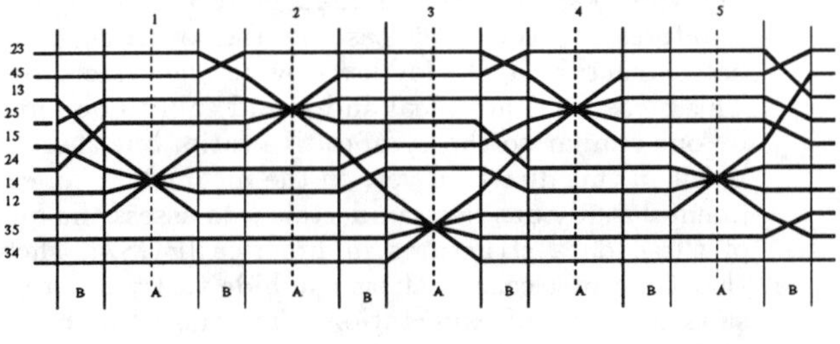


Figure 4: A generalized configuration of points drawn by the algorithm. Type $A$ and type $B$ regions are represented.

The main ingredient of the algorithm is a function to compute how many p-lines are above a switch point. We can do this by counting the number of inversions in its star permutation, which yields $\mathcal{O}(n \log n)$ time by sorting. To improve this to linear time we use an efficient data structure. It is based on having an inverse permutation for each star permutation. The $i$th inverse permutation stores in its $j$th position the *rank* of the p-line $l_{ij}$ around vertex $i$, counted from the p-line $l_{i0}$. It is straightforward that this additional data occupies $\mathcal{O}(n^2)$ space and can be computed in $\mathcal{O}(n^2)$ time. From it we can answer in constant time questions about the orientation of triples of points, as well as compute the number of p-lines above/below points in linear time.

To complete the drawing we divide it into two types of vertical slabs: type $A$ and type $B$. The type $A$ slabs are centered at the vertical lines corresponding to the $n$ points. The $B$ slabs are between type $A$ slabs. In the $A$ slabs we place the "spiders" from Fig.3. In the type $B$ slabs we draw a matching. The whole computation takes $\mathcal{O}(n^2)$ time. As for the drawing per se, although it looks like being on an $\mathcal{O}(n^3)$ grid, it can be done in $\mathcal{O}(n^2)$, too. Figure 4 is a complete example of the output of our algorithm on the allowable sequence $23, 45, 13, 25, 15, 24, 14, 12, 35, 34$ derived from the cluster $1 : 03542$, $2 : 03541$, $3 : 02154$, $4 : 05213$ and $5 : 04213$. The inverse permutations are: $1 : 0 * 4132$, $2 : 04 * 132$, $3 : 021 * 43$, $4 : 0324 * 1$ and $5 : 03241*$.

The drawing algorithm has been implemented using Leda and is accessible via ftp from:
cs.smith.edu/pub/streinu/code/cluster.cc.

**Acknowledgements**. We are grateful to Joe O'Rourke for many stimulating discussions, and to Jorjeta Jetcheva for implementing the drawing algorithm.

### References


[Bj93] A. Björner, M. Las Vergnas, B. Sturmfels, N. White and G. Ziegler. *Oriented Matroids*, Cambridge University Press, 1993

[EOS] H. Edelsbrunner, J. O'Rourke and R. Seidel. Constructing arrangements of lines and hyperplanes with applications SIAM J. Comput. 15, (1986), 341-363.

[EHN] H. Everett, F. Hurtado and M. Noy. *Stabbing Information of a Simple Polygon* 8th Canadian Conference of Computational Geometry, Ottawa, to appear, August 1996.

[GP84] J.E. Goodman and R. Pollack. *Semispaces of configurations, cell complexes of arrangements*, *J. of Combinatorial Theory, Series A*, 37:257–293, 1984.

[GP93] J.E. Goodman and R. Pollack. *Allowable sequences and Order Types in Discrete and Computational Geometry*, in *New Trends in Discrete and Computational Geometry*, J. Pach (ed.), Springer Verlag, 1993

[Kn92] D.E. Knuth. *Axioms and Hulls*, Springer Verlag, New York, 1992.

[OS96] J. O'Rourke and I. Streinu. *Vertex-Edge Pseudo-Visibility Graphs: Characterization and Recognition*, Smith College Tech report, 1996. Also these proceedings.

# Probabilistic Analysis for Combinatorial Functions of Moving Points*

Li Zhang[†] Harish Devarajan Julien Basch[‡] Piotr Indyk[§]

Computer Science Department
Stanford University
Stanford, CA94305
{lizhang,harish,jbasch,indyk}@cs.stanford.edu

**Abstract**

We initiate a probabilistic study of configuration functions of moving points. In our probabilistic model, a *particle* is given an initial position and a velocity drawn independently at random from the same distribution $\mathcal{D}$. We show that if $n$ particles are drawn independently at random from the uniform distribution on the square, their convex hull undergoes $\Theta(\log^2 n)$ combinatorial changes in expectation, their Voronoi diagram undergoes $\Theta(n^{3/2})$ combinatorial changes, and their closest pair undergoes $\Theta(n)$ combinatorial changes.

## 1 Introduction

Given a set of $n$ points, what is the description complexity of their convex hull? In our world, this question is understood with an implicit "in the worst case", and the answer is $n^{\lfloor d/2 \rfloor}$ where $d$ is the dimension of the underlying space. This is not entirely satisfactory, as this description complexity can vary tremendously depending on the positions of the points. Another approach is to look at the expected description complexity when the points are drawn from a given distribution. This type of analysis, initiated by Rényi and Sulanke [RS63] and pursued by others [Efr65, Ray70, Dwy90], gets its value from the fact that this expectation is in general much smaller than in the worst case, and, more importantly, in that it often allows one to design algorithms that have expected running times against which worst case aware algorithms cannot compete [Sei86, BCL90, Dwy91]. For instance, the convex hull of $n$ points drawn independently uniformly at random from a $d$-dimensional hypercube has expected complexity $O(\log^{d-1} n)$, and can be computed in expected linear time.

In parallel, in the past decade, a number of papers have considered a setting where points are allowed to move along low degree algebraic trajectories [OW84, Ata85, GMR91, KTI95, BGH97]. Different questions have been asked in this context. In particular, Atallah [Ata85], studied the number of times the combinatorial description of the convex hull or closest pair can change, in the worst case ("dynamic computational geometry"). More recently, Basch, Guibas, and Hershberger [BGH97] have designed *kinetic* data structures to maintain these attributes in an *online* setting, measuring the quality of a kinetic data structure by the ratio of the worst case number of changes to the configuration of interest, to the worst case number of changes to the data structure itself, for low degree algebraic motions. This measure is not ideal and would gain to be replaced by one similar to the competitive ratio, but there is no result in this direction yet. In the meanwhile, an experimental study has been undertaken to assess the quality of these data structures in practice [BGSZ], showing that the worst case analysis can hide vastly different results in terms of expectation when the point positions and speeds are drawn at random from some distributions. It is this study that motivated the present paper.

Indeed, in the light of previous works on static problems, it is natural to study theoretical bounds for the expected number of changes of combinatorial functions of moving points, when these points are drawn from prescribed distributions, as well as expected time bounds for kinetic data structures that maintain these combinatorial functions.

We report here tight bounds on the expected number of changes to the convex hull, the Delaunay triangulation, and the closest pair of points in the plane for a distribution uniform on a unit square. We also give related (not tight) results, and analyze the expected run-

*The authors' names appear in random order.

[†]Supported in part by National Science Foundation grant CCR-9623851 and by US Army MURI grant DAAH04-96-1-0007.

[‡]Supported in part by the Okawa Foundation.

[§]Supported in part by ARO MURI Grant DAAH04-96-1-0007 and by NSF Award CCR-9357849, with matching funds from IBM, Mitsubishi, Schlumberger Foundation, Shell Foundation, and Xerox Corporation.

ning time of one of the algorithms proposed in [BGH97] for the maintenance of the convex hull.

## 2 Probabilistic Setting

As a point in the plane cannot move without losing its identity, we call the objects of our study *particles*. A particle $X$ is given an initial position $X_0$ at time $t = 0$ and a constant velocity $V_X$, so that it has a position $X(t) = X_0 + tV_X$ at any time $t \geq 0$. Given a distribution $\mathcal{D}$ in $\mathcal{R}^d$, we say that a particle is drawn from $\mathcal{D}$ if its initial position and velocity are each drawn independently at random from $\mathcal{D}$. In [BGSZ], a number of particles were independently drawn at random this way, and the number of events of several kinetic data structures, for several underlying distributions, were studied experimentally.

Using an appropriate space-time transformation, it is easy to see that the above model is combinatorially equivalent to a model where a particle is given an initial position $X_0$ and a final position $X_1$ independently drawn at random from $\mathcal{D}$. In this setting, the position of $X$ at time $t \in [0, 1]$ is given by $X(t) = (1 - t)X_0 + tX_1$ ($X$ moves at a constant speed from $X_0$ to $X_1$). The proofs are easier to carry out in this second model.

## 3 Results

In this communication, we report some results in the plane, when points are drawn from the uniform distribution on the unit square.

**Theorem 1** *If $n$ particles are drawn independently at random from the uniform distribution on the unit square, then the expected number of changes to the following combinatorial functions is:*

| | |
|---|---|
| *Closest pair* | $\Theta(n)$ |
| *Delaunay triangulation* | $\Theta\left(n^{3/2}\right)$ |
| *Convex hull* | $\Theta\left(\log^2 n\right)$ |

Some comparisons will help put these results in perspective. In the case of the convex hull, its expected size in the static setting $\Theta(\log n)$, while the worst case number of changes in the dynamic setting is $\Theta\left(n^2\right)$. On the other hand, the closest pair has a static description complexity of $O(1)$, and a worst case number of changes of $\Theta\left(n^2\right)$, but the expected number of changes seems surprisingly high. Finally, the static Delaunay triangulation always has linear size, while no tight bounds are known for the number of changes in the worst case when points move (there is a trivial lower bound of $\Omega(n^2)$ and an upper bound of $O(n^3)$). It is pleasant to see that the expected value was rather easy to obtain.

The probabilistic analysis can be applied to all configuration functions that are typically looked at in the setting of moving points, and to the kinetic data structures that are used to maintain them.

For instance, as the minimum spanning tree is a subgraph of the Delaunay triangulation, the result on the Delaunay triangulation coupled with a standard batching argument imply that, in our model, the MST in the plane changes $O(n^{5/2})$ times on average. Although this is probably not a tight bound, it is to be compared with the best known worst case upper bound of $O\left(n^3 2^{\alpha(n)}\right)$ [KTI95].

The furthest pair, on the other hand, is always constituted of two points that are on the convex hull. From this and Devroye's moment inequalities [Dev83], we deduce that the number of changes to the furthest pair is polylogarithmic in expectation, for the square distribution.

In [BGH97], a kinetic data structure is described to maintain the convex hull of a set of moving points. This data structure divides the point set arbitrarily into a blue and a red half, recursively computes the value and red convex hulls, and maintains a set of certificates between edge-vertex pairs and edge-edge pairs to certify the convex hull of the whole set. Our convex hull result directly imply:

**Theorem 2** *If $n$ particles are drawn independently at random from the uniform distribution on the unit square, then the kinetic data structure of [BGH97] processes a linear number of events in expectation to maintain the convex hull.*

Hence, whatever the worst case number of changes to the Delaunay triangulation turns out to be, this type of result suggest strongly that it is not a good idea to use it to maintain the convex hull. Our probabilistic framework provides a way that is different from and complementary to [BGH97] for analysing and comparing kinetic data structures.

We now review several kinetic data structures for the maintenance of the closest pair. The Delaunay triangulation can be used to maintain the closest pair, and will do it in roughly $n^{3/2}$ time in our probabilistic model. How do other methods compare? A method is to maintain the $L_1$ Delaunay triangulation [CD85], as one of its edges is the closest pair, but our method shows that it undergoes roughly the same number of changes as the $L_2$ Delaunay. Another method was proposed in [BGH97] and modified in [BGSZ], whose average running time is experimentally also roughly $\Theta\left(n^{3/2}\right)$. At last, one may consider a more straightforward algorithm, which cuts the square (say) into $n$ cells and tracks every particle as it goes from cell to cell [KSG]: although extremely bad in the worst case, it is easy to see that the average case of this algorithm in our probabilistic model is also precisely $\Theta\left(n^{3/2}\right)$. There is probably a good reason for that.

## 4 Some proof ideas

The basic idea behind the proofs of the three results of theorem 1 is as follows. Firstly, we recall the approach of Rényi and Sulanke [RS63] for computing the expected size of the static convex hull. The idea is to consider a given pair of random points $(P, Q)$, and to compute the probability that this pair forms an edge of the convex hull, i.e. that all other points $R_1, \cdots, R_n$, whose positions are chosen independently, are on the same side (say to the right) of the line passing through $(P, Q)$. The positions of the other points with respect to the line $(PQ)$, although not independent (as they all depend on $P, Q$), are independent conditionally on $P, Q$. The recipe, then, for obtaining the required probability, is to compute the probability density of the line defined by two random points, and integrate $(1 - G(\ell))^n$ with respect to this density, where $G(\ell)$ is the probability that a point lies to the left of line $\ell$.

When points start to move, we can use basically the same approach, but we now need three points to characterize the combinatorial event that the convex hull changes. Hence, the question becomes: Given three particles $P, Q, R$, what is the probability that, when they become collinear, they are on the convex hull ? Conditioning on the position of these three points at that time, we are left with the following two problems:

1. what is the probability that, at a given time $t$, a particle is to the left of a line $\ell$ ?
2. what is the probability density of the time and line on which three particles are collinear ?

For the closest pair and the Delaunay triangulation, these questions are replaced by similar questions that involve four points instead of three. The proof of Theorem 1 requires a detailed case analysis for different regions of the unit square. To our knowledge, this case analysis is unavoidable.

## 5 Conclusion

Further work is called for to investigate the probabilistic behavior of the number of changes of geometric structures on moving points. In particular, the results in this paper could be generalized to arbitrary dimensions, to more general distributions, and to algebraic motion within an appropriate probabilistic model. We also hope that some problems that are extremely difficult to analyse in the worst case will be amenable to our analysis to obtain tight answers. The minimum spanning tree is the most interesting of these problems (in part due to its connection with the $k$-level)


## Acknowledgments

We would like to express our gratitude to Alexei Novikov and Suresh Venkatasubramanian for several very useful discussions, and Leo Guibas for his constant encouragements.


## References


[Ata85] M. J. Atallah. Some dynamic computational geometry problems. *Comput. Math. Appl.*, 11:1171–1181, 1985.

[BCL90] J. L. Bentley, K. L. Clarkson, and D. B. Levine. Fast linear expected-time algorithms for computing maxima and convex hulls. In *Proc. 1st ACM-SIAM Sympos. Discrete Algorithms*, pages 179–187, 1990.

[BGH97] J. Basch, L. J. Guibas, and J. Hershberger. Data structures for mobile data. In *8th Symposium on Discrete algorithms*, pages 747–756, 1997.

[BGSZ] J. Basch, L. J. Guibas, C. Silverstein, and L. Zhang. A practical evaluation of kinetic data structures. This volume.

[CD85] L. P. Chew and R. L. Drysdale, III. Voronoi diagrams based on convex distance functions. In *Proc. 1st Annu. ACM Sympos. Comput. Geom.*, pages 235–244, 1985.

[Dev83] L. P. Devroye. Moment inequalities for random variables in computational geometry. *Computing*, 30:111–119, 1983.

[Dwy90] R. A. Dwyer. Kinder, gentler average-case analysis for convex hulls and maximal vectors. *SIGACT News*, 21(2):64–71, 1990.

[Dwy91] R. A. Dwyer. Higher-dimensional Voronoi diagrams in linear expected time. *Discrete Comput. Geom.*, 6:343–367, 1991.

[Efr65] B. Efron. The convex hull of a random set of points. *Biometrika*, 52(3):331–343, 1965.

[GMR91] L. Guibas, J. S. B. Mitchell, and T. Roos. Voronoi diagrams of moving points in the plane. In *Proc. 17th Internat. Workshop Graph-Theoret. Concepts Comput. Sci.*, volume 570 of *Lecture Notes in Computer Science*, pages 113–125. Springer-Verlag, 1991.

[KSG] D.-J. Kim, S.-Y. Shin, and L. J. Guibas. Fast collision detection among multiple moving spheres. This volume.

[KTI95] N. Katoh, T. Tokuyama, and K. Iwano. On minimum and maximum spanning trees of linearly moving points. *Discrete Comput. Geom.*, 13:161–176, 1995.

[OW84] T. Ottmann and D. Wood. Dynamical sets of points. *Comput. Vision Graph. Image Process.*, 27:157–166, 1984.

[Ray70] H. Raynaud. Sur l'enveloppe convexe des nuages de points aléatoires. *J. Appl. Prob.*, 7:35–48, 1970.

[RS63] A. Rényi and R. Sulanke. Über die konvexe Hülle von $n$ zufällig gerwähten Punkten I. *Z. Wahrsch. Verw. Gebiete*, 2:75–84, 1963.

[Sei86] R. Seidel. Constructing higher-dimensional convex hulls at logarithmic cost per face. In *Proc. 18th Annu. ACM Sympos. Theory Comput.*, pages 404–413, 1986.

# Position-Independent Near Optimal Searching and On-line Recognition in Star Polygons*

Alejandro López-Ortiz† Sven Schuierer‡

## Abstract

We study the problem of on-line searching for a target inside a polygon. In particular we propose a strategy for finding a target of unknown location in a star polygon with a competitive ratio of 12.72. This makes star polygons the first non-trivial class of polygons known to admit constant competitive searches independent of the position of the target. We also provide a lower bound of 9 for the competitive ratio of searching in a star polygon —which is close to the upper bound. A similar task consists of the problem of on-line recognition of star polygons for which we present the first strategy with a constant competitive ratio *including* negative instances. We also show, by means of a new lower bound, that it is not posssible to optimally walk into the kernel on the $L_1$ metric.

## 1 Introduction

In the past years on-line searching has been an active area of research in Computer Science (e.g. [1, 3, 6, 7, 11]). In its full generality, an on-line search problem consists of an agent or robot searching for a target on an unknown terrain. In the worst case a search by a robot on a general domain can be arbitrarily inefficient. However, as it is to be expected, strategies can be improved depending on the type of terrain and the searching capabilities of the robot.

*This research is partially supported by the DFG-Project "Diskrete Probleme", No. Ot 64/8-1.

†Research Scientist, Open Text Corp., 180 Columbia St. W., Waterloo, Ontario CANADA N2L 3L3, e-mail: alopez-o@daisy.UWaterloo.ca

‡Institut für Informatik, Universität Freiburg, Am Flughafen 17, D-79110 Freiburg, FRG, e-mail: schuiere@informatik.uni-freiburg.de



The robot is assumed to be equipped, as it is standard in the field, with an on-board vision system that allows it to see its local environment. The performance of an on-line search strategy is measured by comparing the distance traveled by the robot with the length of the shortest path from the starting point $s$ to the target location $t$. The ratio of the distance traveled by the robot to the optimal distance from $s$ to $t$ is called the *competitive ratio* of the search strategy.

There are several known classes of polygons admitting searches for some targets at constant competitive ratios, most notably *streets* [6], *$\mathcal{G}$-streets* [3, 10], *HV-streets* [2] and *$\theta$-streets* [2]. However, the existence of a constant competitive searching strategy for these classes of polygons is strongly dependent on the position of the target.

A natural question is to find a class of polygons which the robot may search at a constant competitive ratio independently of the position of the target.

In this paper we present the first non-trivial class of polygons which admits constant competitive-ratio position-independent target searching, namely star polygons. We devise a 12.72-competitive algorithm for target searching in star polygons and prove a lower bound of 9 for the competitive ratio of any search strategy for star polygons. We also use this strategy to construct the first constant competitive algorithm for recognition of star polygons. That is, given a polygon, the robot follows a path that proves or disproves that the polygon is a star and the path is no more than a constant times longer than a shortest path with the same property. Furthermore, such a path leads into the kernel in a constant competitive ratio as well. We also improve the $\sqrt{2}$ lower bound for walking into the kernel of a star polygon to $\sim 1.48$.

## 2 Definitions

We denote the visibility polygon of a point $p$ inside a polygon $P$ by $V_P(p)$. If the robot does not start in the kernel of $P$, then there are regions in $P$ that cannot be seen by it. The connected components of $P \setminus V_P(p)$ are called **pockets**.

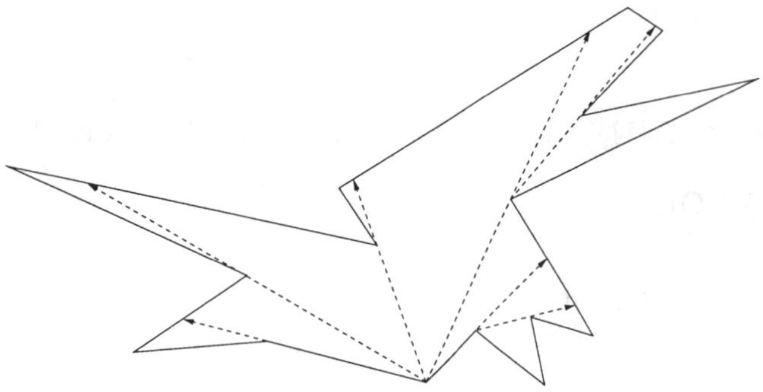

Figure 1: Extended pocket edge search.

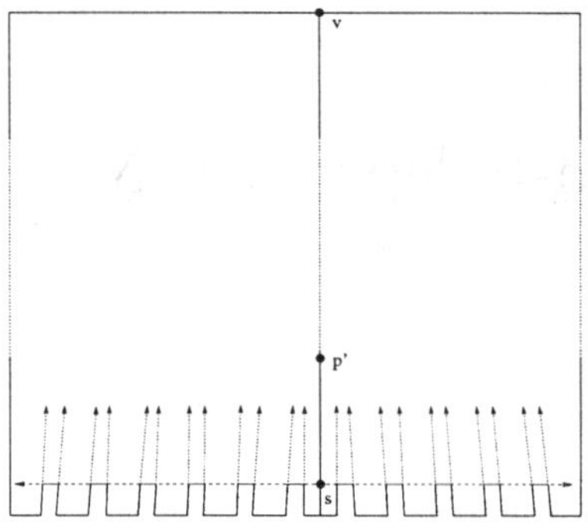


Figure 2: Lower bound for searching for a target.

A **pocket edge of** $p$ is a ray emanating from $p$ which contains a window to a pocket. Each pocket edge passes through at least one reflex vertex of the polygon, which is also an end point of the window associated to the pocket edge. This reflex vertex is called the **entrance point** of the pocket. A pocket edge is said to be a left (right) pocket edge if the pocket defined by it lies to left (right resp.) of the directed pocket edge emanating from $p$.

Since a point in the kernel of $P$ sees all the points in $P$, in particular $p$, a pocket of $V_P(p)$ does not intersect the kernel of $P$ which implies the following observation.

**Observation 1** *The kernel lies to the right of all left pocket edges and to the left of all right pocket edges.*

## 3 Target Searching in Star Polygons

There are many similarities between searching for the target and searching for the kernel in a star polygon. However, note that, when searching for a target, it is easy to construct examples for which it is not an efficient strategy to first go to the kernel and from there move to the target.

Consider the set of pocket edges seen by the robot from the starting position. Given a polygon $P$, an **extended pocket edge** from a point $s$ is a polygonal chain $q_0, q_1, q_2, \ldots, q_k$ such that $q_0 = s$, and each of $q_i$ is a reflex vertex of $P$, save possibly for $q_k$. Furthermore $q_{k-2}$, $q_{k-1}$ and $q_k$ are collinear and form a pocket edge with $\overline{q_{k-1}q_k}$ as associated window into the pocket. If $\overline{q_{k-2}q_k}$ is a left (right) pocket edge, then each of $\angle q_{i-1}q_iq_{i+1}$ is a counterclockwise (clockwise) reflex angle.

The strategy proposed for finding a target traverses extended pocket edges in increasing length order, and alternates between left and right extended pocket edges. As each pocket edge splits the polygon in two parts and since the polygon is a star, it follows that at least one of the two parts is now completely explored. Eventually the entire polygon is explored.

To obtain a constant competitive ratio, the robot sets, at step $i$ the maximum length $d_i$ of the next pocket edge to be traversed to $d_i = kd_{i-1}$ for fixed $k > 1$. The robot then considers all pocket edges of lesser length and among those it selects the leftmost or rightmost depending on the current state of the alternation.

**Theorem 1** *There exists an extended pocket edge strategy for searching for a target inside a star polygon with a competitive ratio of at most 14.5. Moreover, a generalization of the extended pocket edge strategy results in a strategy that has a competitive ratio of* 12.72.

For the lower bound on the same problem we use the following theorem.

**Theorem 2 ([8, Theorem 2.2])** *Any on-line search strategy on the line for a target at a distance of at most $D$ is at least $(9 - f(D))$-competitive, where $f(n) \leq 24/\log_4 n$, for sufficiently large $D$.*

Applying the preceding Theorem to Figure 2, where the width of the construction is $n$ and the height is $n^2$ outlines the argument for Theorem 3.

**Theorem 3 ([8])** *Any strategy for searching for a target inside a star polygon is at least 9-competitive.*

## 4 Recognition of Star Polygons

For the on-line star recognition problem, we assume that given a polygon, the robot aims to determine if it is star shaped. Similarly to target searches, the competitive ratio is given by the quotient between the shortest path that proves or disproves that a given polygon is a star and the distance traversed by the robot.

The next theorem shows that kernel searches are strictly worse than $\sqrt{2}$-competitive. This result stands out against several other lower bounds for searching in simple domains, for which a robot can find an optimal path on-line for the $L_1$ metric [7].

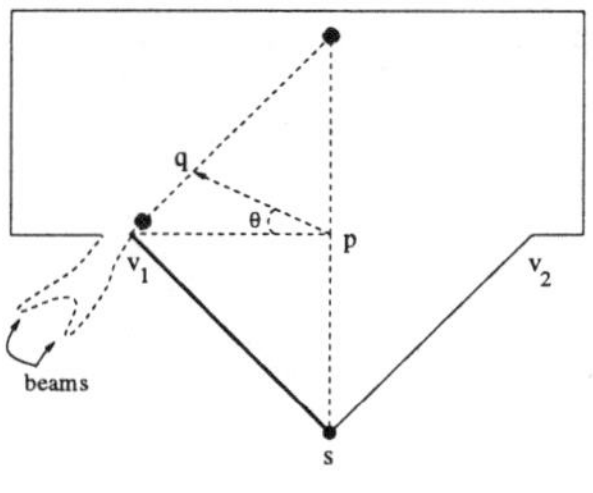


Figure 3: Lower bound configuration.

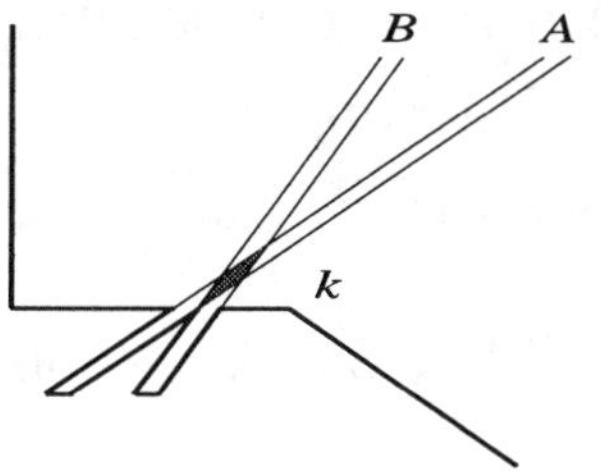


Figure 4: Progressively thinner beams.

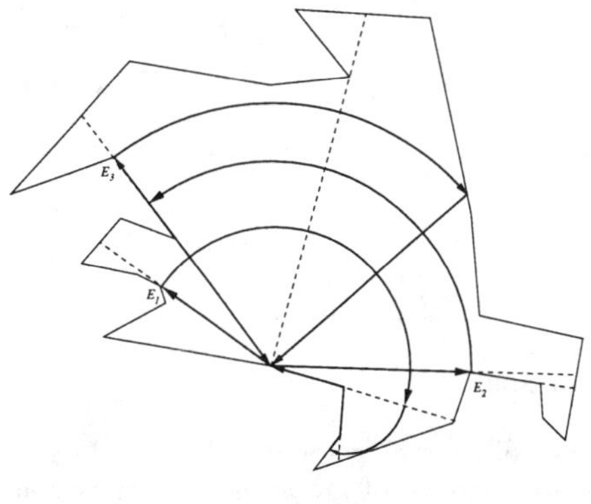

Figure 5: Recognizing a polygon.

Recall that the visibility region of a subset $B$ of a polygon is the set of all points in the polygon which see all points in $B$. Given the current position of the robot $p$ and a pocket $B$ with respect to that point, the visibility region of $B$ forms a **beam**. Clearly the kernel lies in the intersection of all beams. This allows us to construct a polygon like that of Figure 3 with the kernel on either of the two areas indicated by the bold circles.

**Theorem 4** *Searching for the kernel of a polygon is at least* $1/2 + 3/8\sqrt{2} + (2+\sqrt{2})/8\sqrt{10\sqrt{2}-13} \sim 1.48$*-competitive.*

It is possible to walk into the kernel of a given star polygon using a no worse than $\sqrt{4+(2+\pi)^2} \sim 5.5$-competitive strategy [5]. However, it is unclear if the same algorithm applied to a general polygon would terminate at a constant competitive ratio for negative instances. A modification of the target searching strategy of Theorem 1 can be used for this purpose. This modified strategy also traverses extended pocket edges but moves from a left pocket edge to a right pocket edge using a circular arc —not unlike a windshield wiper (see Figure 5). After each circular "swipe" the robot has explored the entire circular sector as well as the pockets associated to the two extended pocket edges traversed.

**Theorem 5** *There exists a* 46.35*-competitive strategy that identifies if a polygon is or is not a star.*

**Theorem 6** *Recognition of a star polygon, including negative instances, is at least* 9*-competitive.*

## 5 Conclusions

We have outlined a strategy for on-line searching in a star polygon. This strategy has constant competitive ratio independent of the starting position of the robot and the position of the target, in contrast to other classes of polygons for which both the position of the target and the starting position are heavily limited. We have also presented a lower bound of 9 for on-line searching in a star polygon. For the case of on-line recognition of a star polygon we present a strategy with a constant competitive ratio as well as a lower bound, both for positive an negative instances. Finally, we present a lower bound of 1.48 for walking into the kernel, which improves on the best previously known lower bound of $\sqrt{2}$.

## References

[1] R. Baeza-Yates, J. Culberson and G. Rawlins. "Searching in the plane", *Information and Computation*, Vol. **106**, (1993), pp. 234-252.

[2] A. Datta, Ch. Hipke, and S. Schuierer. "Competitive searching in polygons—beyond generalized streets", in *Proc. Sixth Annual International Symp. on Algorithms and Computation*, pages 32–41. LNCS 1004, 1995.

[3] A. Datta and Ch. Icking. "Competitive searching in a generalized street", *Proc. 10th ACM Symp. on Computational Geometry*, (1994), pp. 175-182.

[4] S. Gal. *Search Games*, Academic Press, 1980.

[5] Ch. Icking and R. Klein. "Searching for the kernel of a polygon. A competitive strategy", *Proc. 11th ACM Symp. on Computational Geometry*, (1995).

[6] R. Klein. "Walking an unknown street with bounded detour", *Computational Geometry: Theory and Applications*, Vol. **1**, (1992), pp. 325-351.

[7] J. Kleinberg. "On-line search in a simple polygon", *Proc. of 5th ACM-SIAM Symp. on Discrete Algorithms*, (1994), pp. 8-15.

[8] A. Lopez-Ortiz. "On-line target searching in bounded and unbounded domains", Ph.D. Thesis, University of Waterloo, 1996.

[9] A. Lopez-Ortiz and S. Schuierer. "Going home through an unknown street", *Proc. of 4th Workshop on Data Structures and Algorithms*, 1995, LNCS 955, Springer-Verlag, pp. 135-146.

[10] A. López-Ortiz and S. Schuierer. "Generalized streets revisited", In J. Diaz and M. Serna, editors, *Proc. 4th European Symp. on Algorithms*, LNCS 1136, pages 546–558. Springer Verlag, 1996.

[11] A. Lopez-Ortiz and S. Schuierer. "Walking streets faster", *Proc. of 5th Scandinavian Workshop in Algorithmic Theory*, 1996, LNCS 1097, Springer-Verlag, pp. 345-356.

# Searching for Mobile Intruders in a Polygonal Region by a Group of Mobile Searchers*

(Extended Abstract)

Masafumi Yamashita[†] Hideki Umemoto[‡] Ichiro Suzuki[§] Tsunehiko Kameda[¶]

## 1 Introduction

The problem of searching for mobile intruders in a polygonal region by a group of mobile searchers is considered. A searcher can move continuously inside a polygon, holding a flashlight that emits a single ray of light whose direction can be changed continuously. The visibility of a searcher at any time instant is limited to the points on the ray. The intruders can also move continuously with unbounded speed.

Previous work on this type of "pursuit-evasion" problem includes (1) the graph search problem [5, 8] of searching for mobile intruders in a graph (or "clearing" an initially contaminated graph) using as few searchers (i.e., pebbles that can be moved along the edges) as possible, (2) the polygon search problem [1, 4, 10] of searching for mobile intruders in a polygonal region by one or more searchers having various degrees of visibility, (3) the searchlight scheduling problem [9] of computing a search schedule for detecting mobile intruders in a polygonal region using a group of stationary searchers each emitting a single ray of light, and (4) the two-guard problem [3] that asks if it is possible for two mobile guards to simultaneously walk along the two boundary chains between two points on the boundary of a simple polygon $P$ in such a way that they are always mutually visible. Note that the polygon search problem is solvable for a polygon $P$ if the two-guard problem is solvable. Also recently an experimental study has started at Oak Ridge National Laboratory on the problem of searching for intruders in a given region using multiple mobile robots [7]. The goal is to derive effective heuristic algorithms for coordinating the motion of the robots so that all intruders will be found regardless of their move.

In this paper we discuss the problem of finding mobile intruders in a simple polygon $P$ using as few searchers as possible. The main results are the following. Let $ps(P)$ be the number of searchers necessary and sufficient to search a simple polygon $P$. We call $ps(P)$ the *polygon search number* of $P$. We present upper and (worst case) lower bounds on $ps(P)$ using the following four different measures of the shape complexity of $P$: (1) The number $n$ of edges, (2) the number $r$ of reflex vertices, (3) the "bushiness" $b$, which is defined as the minimum number of triangles that share no edges with $P$ over all triangulations of $P$, and (4) the size $g$ of a minimum guard set, where a finite set $\mathcal{G}$ of points in $P$ is called a *guard set* of $P$ if every point $x \in P$ is visible from some point $y \in \mathcal{G}$. (The goal of the art gallery problem [6] is to find a smallest such $\mathcal{G}$.)

Specifically, we prove the following upper bounds on $ps(P)$ in terms of these four measures:

- $ps(P) \le 1 + \lfloor \log_3(n-3) \rfloor$.
- $ps(P) \le 1 + \lfloor \log_3 r \rfloor$.
- $ps(P) \le 1 + \lfloor \log_3(2b+1) \rfloor$.
- $ps(P) \le 2 + \lceil \log_2 g \rceil$.

We then show that these bounds are either tight (for the case of measure $b$) or almost tight (for the other three measures) in the worst case by presenting, for each natural number $s \ge 2$, a polygon $P$ with $ps(P) = s$ such that

$$\begin{aligned} ps(P) &= \log_3(n+1) = \log_3(2r+3) = 1 + \lfloor \log_3(2b+1) \rfloor \\ &= 1 + \log_3(2g-1). \end{aligned}$$

(Precisely, we then extend this construction and prove similar lower bounds for any $n \ge 3$, $r \ge 1$, $b \ge 1$ and $g \ge 1$.) The upper bounds are obtained either by a reduction of the polygon search problem to the graph search problem (see above), or by a decomposition of the given instance into a number of smaller related search problems called *chasing*, *shooing*, and *corridor search problems*, that deal with regions with "entrances." (We omit the details in this abstract due to the space limitation.) The decomposition is often done by repeated applications of a search strategy called *the one-way sweep strategy (OWSS)* [9]. In this abstract, we present only the key ideas; all the details and the missing proofs can be found in the full paper.

## 2 The Upper Bounds

### 2.1 The Number of Edges and the Bushiness

Our upper bounds on the search number $ps(P)$ in terms of the number of edges $n$ and the bushiness $b$ are obtained using the concept of *graph search*.

---

*This work was supported in part by a Scientific Research Grant-in-Aid from the Ministry of Education, Science and Culture of Japan under Grant 07243219 and 07680360, the National Science Foundation under Grant IRI-9307506, the Office of Naval Research under Grant N00014-94-1-0284, the Natural Sciences and Engineering Council of Canada, and an endowed chair supported by Hitachi Ltd. at Faculty of Engineering Science, Osaka University.

[†]Department of Electrical Engineering, Faculty of Engineering, Hiroshima University, Kagamiyama, Higashi-Hiroshima 739, Japan. mak@se.hiroshima-u.ac.jp

[‡]Technology Center, Anritsu Corporation, Onna, Atsugi 243, Japan. umemoto@accpd3.anritsu.co.jp

[§]Department of Electrical Engineering and Computer Science, University of Wisconsin – Milwaukee, P.O. Box 784, Milwaukee, WI 53201, U.S.A. suzuki@blatz.cs.uwm.edu

[¶]School of Computing Science, Simon Fraser University, Burnaby, B.C., V5A 1S6, Canada. tiko@cs.sfu.ca

The "edge search" version of the graph search problem is the following [5]: Given a connected graph in which all edges are initially "contaminated," make all edges simultaneously "clear" by a sequence of the following basic operations: (1) placing a searcher on a vertex, (2) removing a searcher from a vertex, and (3) moving a searcher from one vertex to another along an edge. A contaminated edge $e = (u, v)$ becomes clear if we place a searcher on $u$ as a guard and move a second searcher along $e$ from $u$ to $v$. If all edges incident on $u$ other than $e$ are already clear, however, we can dispense with the second searcher and clear $e$ simply by moving the guard along $e$ from $u$ to $v$. A clear edge remains clear until a searcher-free path to a contaminated edge is formed, in which case the edge is recontaminated. A *search strategy* is a sequence of the basic operations that clears an initially contaminated graph. The *graph search number* $es(G)$ of a graph $G$ is the minimum number of searchers for which a search strategy exists.

**Lemma 1** [5] *Let $T$ be a tree with $m \geq 2$ vertices. Then $es(T) \leq 1 + \lfloor \log_3(m-1) \rfloor$.*

Lemma 2 given next is a key lemma that relates the polygon search number $ps(P)$ and the graph search number $es(T)$. The proof is by induction on $es(T)$, with a careful use of OWSS [9] in the induction step. Details are omitted in this abstract.

**Lemma 2** *Let $T$ be the dual tree of a triangulation of a simple polygon $P$. Then $ps(P) \leq es(T)$.*

Using these lemmas, we can prove the next theorem that gives an upper bound on $ps(P)$ in terms of $n$.

**Theorem 1** *For any simple polygon $P$ with $n \geq 4$ edges, $ps(P) \leq 1 + \lfloor \log_3(n-3) \rfloor$.*

**Proof** By Lemmas 1 and 2, $ps(P) \leq es(T) \leq 1 + \lfloor \log_3(m-1) \rfloor$, where $T$ and $m$ are the dual graph of a triangulation of $P$, and the number of vertices of $T$, respectively. Since $m = n - 2$, we have $ps(P) \leq 1 + \lfloor \log_3(n-3) \rfloor$. □

The *reduction* of a graph $G$ is the graph $G'$ obtained from $G$ by repeatedly replacing any degree-two vertex $u$ and its two incident edges by a single edge connecting the two neighbors of $u$, until no degree-two vertices remain. It is known that:

**Lemma 3** [5] *A graph and its reduction have the same search number.*

One of the main results of this section is the next theorem.

**Theorem 2** *For any simple polygon $P$ with bushiness $b$, $ps(P) \leq 1 + \lfloor \log_3(2b+1) \rfloor$.*

**Proof** Let $T$ be the dual tree of a thin triangulation of $P$, and $T'$ be the reduction of $T$. Since $T'$ has $b$ degree-three vertices and no degree-two vertices, there are $2b+2$ vertices in $T'$. Therefore, by Lemmas 1, 2 and 3, $ps(P) \leq es(T) = es(T') \leq 1 + \lfloor \log_3(2b+1) \rfloor$. □

### 2.2 The Number of Reflex Vertices and the Size of a Minimum Guard Set

Let us consider the number $r$ of reflex vertices of $P$. If $r = 0$, then $P$ is convex and therefore $ps(P) = 1$. For the case $r \geq 1$ the following theorem gives an upper bound on $ps(P)$ in terms of $r$. The proof is by an explicit construction of a search schedule based on recursive applications of OWSS [9], instead of a reduction to the graph search problem.

**Theorem 3** *For any simple polygon $P$ with $r \geq 1$ reflex vertices, $ps(P) \leq 1 + \lfloor \log_3 r \rfloor$.*

**Outline of Proof** It suffices to prove the inequality for all $r$ of the form $r = 3^s - 1$ for some natural number $s \geq 1$. We use induction on $s$. For the case $s = 1$ (or $r = 2$), we show that every simple polygon with two reflex vertices is searchable by one searcher by explicitly constructing a search schedule. For the induction step, we assume that every simple polygon with at most $3^s - 1$ reflex vertices is searchable by $s$ searchers and construct a search schedule of $s+1$ searchers for any given simple polygon $P$ with $3^{s+1} - 1$ reflex vertices. We do so by repeatedly applying OWSS [9] in such a way that any region $R_i$ not visible from the pivot of OWSS that contains $3^s$ or more reflex vertices is further decomposed by the same procedure. We then carefully examine the structure of the decomposition and show how $P$ can be searched using $s+1$ searchers. The details will be given in the full paper. □

The upper bound of $1 + \lfloor \log_3 r \rfloor$ on $ps(P)$ in terms of $r$ is at least as strong as that $(1 + \lfloor \log_3(n-3) \rfloor)$ in terms of $n$ we derived in the last subsection, since $r \leq n-3$ holds in general. On the other hand, the upper bounds in terms of $r$ and $b$ are incomparable.

Lemma 4 given next is a key lemma for computing an upper bound on $ps(P)$ in terms of the size $g$ of a minimum guard set of $P$. The proof is by induction similar to that used in the proof of Theorem 3. However, this time we need to take into account the effect of guards that are located outside of the region under consideration—for a guard set $\mathcal{G}$ of $P$ and some region $R \subset P$, $R \cap \mathcal{G}$ may not be a guard set of $R$. We omit the details in this abstract.

**Lemma 4** *Let $P$ be a simple polygon, and $\mathcal{G} \subseteq P$ a nonempty set of points in $P$ of size $|\mathcal{G}| = g \leq 2^s - 1$ for some natural number $s \geq 1$. If $P$ is weakly visible from $\mathcal{G}$ and an edge $e$, then $P$ is searchable by $s+1$ searchers.*

The next theorem follows from Lemma 4.

**Theorem 4** *Let $P$ be a simple polygon having a guard set of size $g$. Then $ps(P) \leq 2 + \lfloor \log_2 g \rfloor$.*

## 3 The Lower Bounds

In this section we present a series of simple polygons $P(i)$, $i \geq 1$, such that $ps(P(i)) = i+1$, that provide lower bounds on the polygon search number in terms of the four parameters discussed in the previous section.

$P(1)$ is given in Figure 1(a). $P(2)$ is shown in Figure 1(b), and is constructed by "pasting" three copies of $P(1)$ (of which one is drawn backwards) to the three "hooks" of a fourth copy of $P(1)$ (drawn at the center). Analogously, $P(i+1)$ is recursively constructed from three copies of $P(i)$ and a copy of $P(1)$: the three copies of $P(i)$ (of which one is drawn backwards) are pasted to a copy of $P(1)$ (drawn at the center) at its three hooks. Figure 1(c) shows $P(3)$ that consists of three copies of $P(2)$ and a copy of $P(1)$.

The following lemma gives the values of the four measures of shape complexity of $P(i)$.

**Lemma 5** *Let $n_i, r_i, b_i$ and $g_i$ be the number of edges, the number of reflex vertices, the bushiness, and the size of a minimum guard set of $P(i)$, respectively. Then $n_i = 3^{i+1} - 1$, $r_i = 3(3^i - 1)/2$, $b_i = (3^i - 1)/2$, and $g_i = (3^i + 1)/2$.*

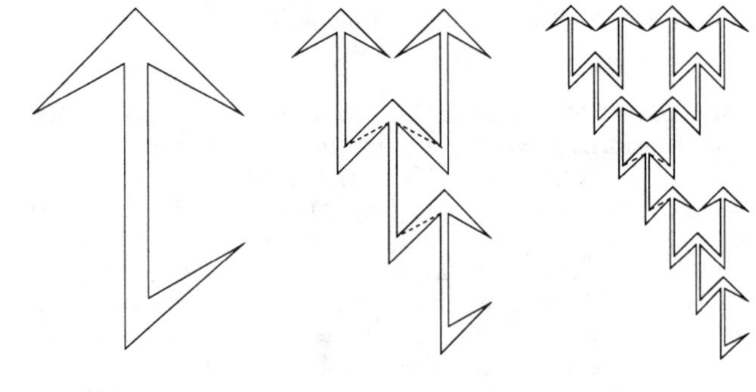


Figure 1: Simple polygons $P(1)$, $P(2)$ and $P(3)$.

The next lemma states that $i+1$ searchers are necessary and sufficient to search $P(i)$. The proof is rather technical, and is by induction on $i$. Details are omitted in this abstract.

**Lemma 6** *For any $i \geq 1$, $ps(P(i)) = i+1$.*

By Lemmas 5 and 6, we obtain:

**Theorem 5** *For any natural number $s \geq 2$, there is a simple polygon $P$ with $ps(P) = s$ such that*

$$\begin{aligned} ps(P) &= \log_3(n+1) = \log_3(2r+3) = 1 + \log_3(2b+1) \\ &= 1 + \log_3(2g-1), \end{aligned}$$

*where $n$, $r$, $b$ and $g$ are the four measures of shape complexity for $P$ introduced earlier.*

Theorem 5 applies to only certain values of $n$, $r$, $b$ and $g$. The next theorem extends this result to all values of $n$, $r$, $b$ and $g$. The proof is by construction of polygons through additions of extra vertices and edges to some $P(i)$.

**Theorem 6**

- *For any $n \geq 3$, there exists a simple polygon $P$ with $n$ edges such that $ps(P) \geq \lfloor \log_3(n+1) \rfloor$.*
- *For any $r \geq 1$, there exists a simple polygon $P$ with $r$ reflex vertices such that $ps(P) \geq \lfloor \log_3(2r+3) \rfloor$.*
- *For any $b \geq 1$, there exists a simple polygon $P$ with bushiness $b$ such that $ps(P) = 1 + \lfloor \log_3(2b+1) \rfloor$.*
- *For any $g \geq 1$, there exists a simple polygon $P$ with a minimum guard set of size $g$ such that $ps(P) \geq 1 + \lfloor \log_3(2g-1) \rfloor$.*

## 4 Concluding Remarks

We presented upper and (worst case) lower bounds on the polygon search number $ps(P)$ of a simple polygon $P$ in terms of four measures of shape complexity. The gap between the upper and lower bounds given in terms of measures $n$ and $r$ are at most one for all values of $n$ and $r$, whereas those given in terms of bushiness $b$ match. So these bounds are either tight or almost tight. For measure $g$, the upper bound is up to a factor of two greater than the lower bound, and unfortunately, we were not able to bound the gap by a constant. We believe that a more elaborate decomposition of a given polygon into subregions could yield a better upper bound on $ps(P)$, and we conjecture $ps(P) \leq 2 + \lceil \log_3 g \rceil$.

For the case where searchers have 360° visibility, a tight worst case *asymptotic* bound of $ps(P) = \Theta(\log n)$ for a simple polygon $P$ with $n$ edges has been proved by de Berg and Klein [2] and LaValle, et al. [4]. For such searchers we recently proved a tight worst case bound of $ps(P) = 1 + \lfloor \log_3(2b+1) \rfloor$ for a simple polygon $P$ with bushiness $b$.

The problem of computing $ps(P)$ for searchers with 360° visibility is NP-hard if $P$ can contain holes [4], and the same result can be proved for searchers whose visibility is restricted to a single ray. (Known bounds on $ps(P)$ of a polygon $P$ with $n$ edges and $h$ holes include $\Omega(\log n + \sqrt{h})$ and $O(\log n + h)$ for searchers with 360° visibility [4], and an upper bound of $1 + h + \lfloor \log_3(n+2h-3) \rfloor$ even for searchers whose visibility is restricted to a single ray [11].) For the case where the regions do *not* contain holes, the complexity of computing $ps(P)$ for an arbitrarily given simple polygon $P$ is still open for either type of searcher mentioned above. Indeed, the polygon search number $ps(P)$ for such polygons cannot be bounded from below by either one of the four measures discussed in this paper (i.e., there are polygons with high shape complexity that require only a few searchers), and discovering a new measure of shape complexity that can be used to compute $ps(P)$ of a simple polygon $P$ remains as an open problem.

## References


[1] D. Crass, I. Suzuki and M. Yamashita, "Searching for a mobile intruder in a corridor – The open edge variant of the polygon search problem," *International J. Computational Geometry and Applications 5*, 4, 1995, 397–412.

[2] M. de Berg and R. Klein, private communication, 1993.

[3] C. Icking and R. Klein, "The two guards problem," in *Proc. 7th ACM Symp. on Computational Geometry*, 1991, 166–175.

[4] S.M. LaValle, D. Lin, L.J. Guibas, J.-C. Latombe and R. Motwani, "Finding an unpredictable target in a workspace with obstacles," in *Proc. IEEE International Conf. Robotics and Automation*, April 1997, to appear.

[5] N. Megiddo, S.L. Hakimi, M.R. Garey, D.S. Johnson and C.H. Papadimitriou, "The complexity of searching a graph," *J. ACM 35*, 1, 1988, 18–44.

[6] J. O'Rourke, *Art Gallery Theorems and Algorithms*, Oxford University Press, New York, Oxford, 1987.

[7] L.E. Parker, "Behavior-based cooperative robotics applied to multi-target observation," in *Proc. Dagstuhl Seminar on Modeling and Planning for Sensor-Based Intelligent Robot Systems*, Dagstuhl, Germany, September 1996.

[8] T.D. Parsons, "Pursuit-evasion in a graph," in *Theory and Applications of Graphs*, Y. Alavi and D. Lick (eds.), Lecture Notes in Mathematics 642, Springer-Verlag, Berlin, 1976, 426–441.

[9] K. Sugihara, I. Suzuki and M. Yamashita, "The searchlight scheduling problem," *SIAM J. Computing 19*, 6, 1990, 1024–1040.

[10] I. Suzuki and M. Yamashita, "Searching for a mobile intruder in a polygonal region," *SIAM J. Computing 21*, 5, 1992, 868–888.

[11] H. Umemoto, *Searching a Polygon by 1-Searchers*, Master's Thesis, Department of Electrical Engineering, Hiroshima University, 1995.

# A polynomial time solution for labeling a rectilinear map

Chung Keung Poon    Binhai Zhu    Francis Chin

## 1 Introduction

Map labeling is an important problem in Geographical Information Systems (GIS). This problem is well studied in the area of cartography, especially using heuristics and logic programming [Yo72, Imh75, Jon89, DF92, CMS93, BC94]. The study of map labeling problem from the algorithmic point of view starts very recently. In general, map labeling problems are hard and two discrete versions are shown to be NP-complete [FW91, KR92]. In spite of these negative results it was shown that by using 2SAT one can obtain good approximate algorithms and heuristics in labeling a set of sites (points) [FW91, Wag94, WW95]. The main reason is that if for each site only two labels are allowed then one can naturally formulate the problem as 2SAT, which can be solved in linear time [EIS76].

In a recent paper [DMMMZ97], several generalized versions of map labeling problems are studied. For example, in [DMMMZ97], constant approximation algorithms are proposed to label a set of sites with uniform squares along any directions and with uniform circles. Also in [DMMMZ97], different from all previous research in map labeling (which only studies the problem of labeling a set of sites or points), the problem of labeling a rectilinear map which consists of a set of rectilinear, i.e., either horizontal or vertical, segments and each segment is allowed to use three natural labels, is studied. If all segments are horizontal then the problem can be solved in optimal $\Theta(n \log n)$ time. For the general rectilinear map labeling problem a factor-2 approximate solution and a *Polynomial Time Approximation Scheme* are also proposed.

The Authors may be contacted at: C.K. Poon, ckpoon@cs.cityu.edu.hk, B. Zhu, bhz@cs.cityu.edu.hk, both with Dept. of Computer Science, City University of Hong Kong, and F. Chin, chin@cs.hku.hk, Dept. of Computer Science, University of Hong Kong. This work is supported by the research grant No.9030526 from City Univ of HK and a RGC grant (HKU 541/96E).



In this paper, we show that the general rectilinear map labeling problem is polynomially solvable by a nontrivial and an indirect use of 2SAT. This solves the open problems posed in [DMMMZ97]. In some sense we extend the use of 2SAT in map labeling even further compared with what have been done in [FW91, WW95]. We believe this technique can be used to solving other geometric packing problems.

## 2 The 3-position Rectilinear Segment Labeling

In this section, the problem of labeling $n$ rectilinear (i.e either horizontal or vertical) line segments using 3 candidate labels [DMMMZ97] is re-studied. Note that each line segment can be labeled in three possible ways, either on the line segment or at one of its two sides. For the simplicity of presentation, we only deal with labeling line segments which are either horizontal or vertical but the idea also works for non-rectilinear line segments as long as they do not intersect and do not share endpoints.

First of all we will study the corresponding *decision* problem. Given a map consisting of $n$ disjoint rectilinear line segments, and a positive number $B$, the *decision* problem for *3-position Rectilinear Segment Labeling (3RSL)* is to determine whether there exists a placement for $n$ non-overlapping rectangular valid labels, each attached to a distinct line segment, of width $B$ and of length same as its line segment. Our algorithm actually finds a labeling when there exists one. Given such an algorithm, it is not difficult to solve the original *maximization* problem (i.e. finding a map labeling which maximizes $B$) since the number of possible values which the optimal solution can take is bounded and can be computed without too much difficulty.

For a horizontal line segment, a *valid* label can be above/on/below the line segment. Assume the labels are rectangular in shape, having the same length as the line segment and a fixed width $B$. The rectilinear area around the line segment is then divided into four regions, two of them, $r_1$ and $r_2$, are above and two of them, $r'_1$ and $r'_2$, are below the line segment, each of which is of width $B/2$ and of the same length as the line segment (Figure 1 (a)). A label above the line segment *(above-line label)* will occupy regions $r_1$ and $r_2$; regions $r_2$ and $r'_1$ are for the label on the line segment *(on-the-line la-*

*bel)*, while $r'_1$ and $r'_2$ are for the label below the line segment *(below-line label)*. These three ways of labeling can be represented by truth assignment of two boolean variables, $x_1$ and $x_2$. The true/false value of $x_1$ specifies whether $r_1$ or $r'_1$ is occupied for labeling. To be specific, $x_1 = 1$ means $r_1$ is occupied for labeling and not $r'_1$, while $\overline{x_1} = 1$ means $r'_1$ is occupied and not $r_1$. Similarly $x_2 = 1$ means $r_2$ is used for labeling but not $r'_2$ and $\overline{x_2} = 1$ means $r'_2$ is used but not $r_2$.

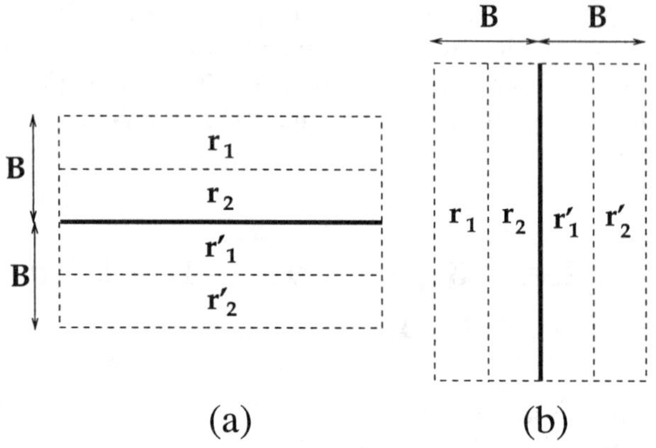


Figure 1: Region Partitions for 3-position Labeling.

With this assumption, the above-line label can be represented by having $x_1 = 1$ and $x_2 = 1$, the on-the-line label by $\overline{x_1} = 1$ and $x_2 = 1$, and the below-line label by $\overline{x_1} = 1$ and $\overline{x_2} = 1$. These three valid ways of labeling is implied by a 2-literal clause $(\overline{x_1} \vee x_2)$ or $(x_1 \rightarrow x_2)$, the satisfaction of which enforces the existence of a valid label on the line segment. On the other hand, if there exists one valid labeling then $(\overline{x_1} \vee x_2)$ is satisfied.

The labeling of a vertical line segment can be dealt with similarly. There are again four regions of width $B/2$, regions $r_1$ and $r_2$ are on the left of the line and regions $r'_1$ and $r'_2$ on its right (Figure 1 (b)). The labeling of the vertical line segment can also be represented by the truth assignment of two boolean variables $x_1$ and $x_2$ that satisfies clause $(\overline{x_1} \vee x_2)$.

Let the line segments be denoted by $s_i$, $1 \leq i \leq n$ and their associated four regions, by $r_{i1}\ r_{i2}, r'_{i1}\ r'_{i2}$, $1 \leq i \leq n$. We say two line segments $s_i$ and $s_j$ *interact* if and only if any of the four regions of $s_i$ overlap with those of $s_j$, i.e., $R_i \cap R_j \neq \phi$ where $R_i = r_{i1} \cup r_{i2} \cup r'_{i1} \cup r'_{i2}$ and $R_j = r_{j1} \cup r_{j2} \cup r'_{j1} \cup r'_{j2}$.

Consider the example in Figure 2, the interactions are ($s_1$, $s_2$), ($s_1$, $s_4$), ($s_1$, $s_5$), ($s_5$, $s_6$), ($s_6$, $s_7$), ($s_6$, $s_8$) and ($s_7$, $s_8$). We want to ensure that the labels of the interacting line segments are valid and non-overlapping.

When two line segments $s_i$ and $s_j$ interact, the placements of the labels for $s_i$ and $s_j$ are not independent and they cannot overlap. Depending on how $R_i$ and $R_j$ intersect each others, different 2-literal clauses on variables, $x_{i1}$, $x_{i2}$, $x_{j1}$,$x_{j2}$, are generated, whose satisfaction

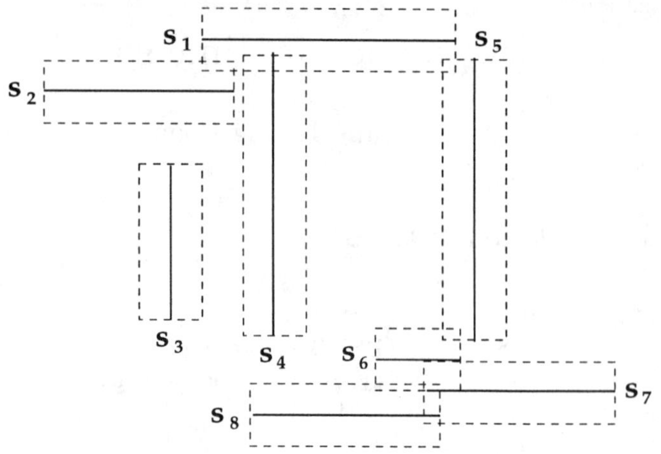


Figure 2: An Example of Interactions Among Rectilinear Segments.

implies the non-overlapping of the labels. For example, overlapping $r_{i1}$ and $r'_{j2}$ implies the generation of clause $(\overline{x_{i1} \wedge \overline{x_{j2}}})$, i.e., $(\overline{x_{i1}} \vee x_{j2})$. As $x_{i1} = 1$ means $r_{i1}$ is occupied and $\overline{x_{j2}} = 1$ means $r'_{j2}$ is occupied, the satisfaction of the clause $(\overline{x_{i1} \wedge \overline{x_{j2}}})$ ensures that both $r_{i1}$ and $r'_{j2}$ will not simultaneously be used for labels.

The following example also illustrates that if two line segments $s_i$ and $s_j$ are too close to each other, the satisfaction of the generated 2-literal clauses through overlapping regions ensures that some placement of labels on $s_i$ and $s_j$ might not be possible.

**Example:** Let $s_i$ and $s_j$ be two horizontal line segments facing each other with $s_i$ above $s_j$ and more than $B/2$ but less than distance $B$ apart. We have $C = (\overline{x_{i2}} \vee \overline{x_{j1}})$ $(x_{i1} \vee \overline{x_{j1}})$ $(x_{i1} \vee \overline{x_{j2}})$ $(x_{i2} \vee \overline{x_{j2}})$ $(x_{i2} \vee x_{j1})$ as $r_{i2}$ and $r_{j1}$, $r'_{i1}$ and $r_{j1}$, $r'_{i1}$ and $r_{j2}$, $r'_{i2}$ and $r_{j2}$, $r'_{i2}$ and $r'_{j1}$ overlap. The satisfaction of $C$ with $(\overline{x_{i1}} \vee x_{i2})$ and $(\overline{x_{j1}} \vee x_{j2})$ ensures that $x_{i2} = 1$ and $x_{j1} = 0$. This conclusion matches with the interpretation that below-line label for $s_i$ and above-line label for $s_j$ are not possible.

## 2.1 The Algorithm

The following algorithm solves the 3RSL problem:

**Step 0:** Compute all the interactions between the labeling rectangles, each with width $B$, of different segments.

**Step 1:** Construct a set of 2-literal clauses, one for each line segment,

$C_s = \{(\overline{x_{i1}} \vee x_{i2}) \mid 1 \leq i \leq n\}$.

**Step 2:** Construct a set of clauses for each interaction of two line segments. Assume $s_i$ and $s_j$ interact,

$C_{ij} = \{(\overline{a} \vee \overline{b}) \mid a = x_{ip}(\overline{x_{ip}}),\ b = x_{jq}(\overline{x_{jq}})$ if $r_{ip}(r'_{ip})$ and $r_{jq}(r'_{jq})$ overlap$\}$

Let $C_I$ be a collection of $C_{ij}$ where $s_i$ and $s_j$ interact.

**Step 3:** Check if there exists a truth assignment of variables $\{x_{i1}, x_{i2} \mid 1 \leq i \leq n\}$ satisfying all clauses in $C_s$ and $C_I$. If there is such an assignment, then there exists a feasible solution for the 3RSL problem. The placement of label for each line segment $s_i$ can be determined from the truth assignments of $x_{i1}$ and $x_{i2}$.

**Theorem 1.** The 3RSL problem for $n$ line segments can be solved in $O(I+n \log n)$ time where $I$ is the number of pairwise line segment interactions and is bounded by $O(n^2)$.

In practice, the width of the label normally should not be larger than the length of the line segment. If that is the case, the 3RSL problem can be solved in $O(n \log n)$ time which matches the $\Omega(n \log n)$ lower bound [DMMMZ97] and is hence optimal.

**Corollary 2.** The 3RSL problem for $n$ line segments can be solved in $O(n \log n)$ time if each line segment is of length at least $B$.

## 3 The $k$-position Rectilinear Segment Labeling

In this section, it is assumed that the label of a line segment can be placed in $k$ possible ways. Each label is still of width $B$ and of the same length as the line segment. We show that the same method for solving the 3-position RSL problem can be used to solve the $k$-position RSL problem.

**Theorem 3.** The $k$-position RSL problem for $n$ line segments can be solved in $O(k(I+n)+n \log n)$ time where $I$ is the number of pairwise line segment interactions.

## 4 The Maximum Width Rectilinear Segment Labeling

Given $n$ rectilinear line segments, the *Maximum Width Rectilinear Segment Labeling (MWRSL)* problem is to find the maximum $B$, such that there exists a placement for $n$ non-overlapping rectangular valid labels, each attached to a distinct line segment, of width $B$ and of length same as its line segment. We have the following results.

**Theorem 4.** The 3-position MWRSL problem for $n$ line segments can be solved in $O(n^2 \log n)$ time.

**Corollary 5.** Assume $k < n$, the $k$-position MWRSL problem for $n$ line segments can be solved in $O(kn^2 \log n)$ time.

## References


[BC94] D. Beus and D. Crockett. Automated production of 1:24,000 scale quadrangle maps. In *Proc. 1994 ASPRS/ACSM Annual Convention and Exposition*, volume 1, pages 94–99, 1994.

[CMS93] J. Christensen, J. Marks, and S. Shieber. Algorithms for cartographic label placement. In *Proc. 1993 ASPRS/ACSM Annual Convention and Exposition*, volume 1, pages 75–89, 1993.

[DF92] J. Doerschler and H. Freeman. A rule-based system for cartographic name placement. *CACM*, 35:68–79, 1992.

[DMMMZ97] S. Doddi, M. Marathe, A. Mirzaian, B. Moret and B. Zhu. Map labeling and its generalizations. In *Proc. 8th ACM-SIAM Symp. on Disc. Alg.*, pages 148-157, 1997.

[EIS76] S. Even, A. Itai and A. Shamir. On the complexity of timetable and multicommodity flow problem. *SIAM J. Comput.*, 5:691–703, 1976.

[FW91] M. Formann and F. Wagner. A packing problem with applications to lettering of maps. In *Proc. 7th Annu. ACM Sympos. Comput. Geom.*, pages 281–288, 1991.

[Imh75] E. Imhof. Positioning names on maps. *The American Cartographer*, 2:128–144, 1975.

[Jon89] C. Jones. Cartographic name placement with Prolog. *Proc. IEEE Computer Graphics and Applications*, 5:36–47, 1989.

[KR92] D. Knuth and A. Raghunathan. The problem of compatible representatives. *SIAM J. Disc. Math.*, 5:422–427, 1992.

[PS85] F.P. Preparata and M.I. Shamos. *Computational Geometry: An Introduction*. Springer-Verlag, 1985.

[Wag94] F. Wagner. Approximate map labeling is in $\Omega(n \log n)$. *Inform. Process. Lett.*, 52:161–165, 1994.

[WW95] F. Wagner and A. Wolff. Map labeling heuristics: Provably good and practically useful. In *Proc. 11th Annu. ACM Sympos. Comput. Geom.*, pages 109–118, 1995.

[Yo72] P. Yoeli. The logic of automated map lettering. *The Cartographic Journal*, 9:99–108, 1972.

# Time-Series Similarity Problems and Well-Separated Geometric Sets

Béla Bollobás*
bollobas@IAS.EDU

Gautam Das†
dasg@mathsci.msci.memphis.edu

Dimitrios Gunopulos‡
gunopulo@almaden.ibm.com

Heikki Mannila§
Heikki.Mannila@cs.helsinki.fi

**Abstract**

Given a pair of nonidentical complex objects, defining (and determining) how similar they are to each other is a nontrivial problem. In data mining applications, one frequently needs to determine the similarity between two time series. We analyze a model of time-series similarity that allows outliers, and different scaling functions. We present deterministic and randomized algorithms for computing this notion of similarity. The algorithms are based on nontrivial tools and methods from computational geometry. In particular, we use properties of families of well-separated geometric sets. The randomized algorithm has provably good performance and also works extremely efficiently in practice.

## 1 Introduction

Being able to measure the *similarity* between objects is a crucial issue in many data retrieval and data mining applications; see [9] for a general discussion on similarity queries. Typically, the task is to define a function $Sim(X, Y)$, where $X$ and $Y$ are two objects of a certain class, and the function value represents how "similar" they are to each other. For complex objects, designing such functions, and algorithms to compute them, is by no means trivial.

*Time series* are an important class of complex data objects: they arise in many applications. Examples of time series databases are stock price indices, volume of product sales, telecommunication data, one-dimensional medical signals, audio data, and environmental measurement sequences. In *data mining* applications, it is often necessary to search within a series database for those series that are similar to a given query series. This primitive is needed, for example, for prediction and clustering purposes. While the statistical literature on time-series is vast, it has not studied similarity notions that would be appropriate for data mining applications.

In this paper we present and analyze a similarity function for time series. We present new deterministic and randomized algorithms that are both practical *and have* provable performance bounds. Interestingly, these algorithms are based on several subtle geometric properties of the problems, in particular properties of certain *well-separated* geometric sets, which are interesting in their own right. In addition to our theoretical analysis, we present several implementation results.

Next we describe the similarity notion used here; related work is considered at the end of the next section.

Suppose we are given two sequences $X = x_1, \ldots, x_n$ and $Y = y_1, \ldots, y_n$. A simple starting point would be to measure the similarity of $X$ and $Y$ by using their $L_p$-distance as points of $R^n$. For time series, this way of measuring similarity or distance is not appropriate, since the sequences can have outliers, different scaling factors, and baselines.

*Outliers* are values that are measurement errors and should be omitted when comparing the sequence against others. A grossly outlying value can cause two otherwise identical series to have large distance. A more reasonably similarity notion is based on the *longest common subsequence* concept, where intuitively $X$ and $Y$ are considered similar if they exhibit similar behavior for a large part of their length. More formally, let $X' = x_{i_1}, \ldots, x_{i_l}$ and $Y' = y_{j_1}, \ldots, y_{j_l}$ be the longest subsequences in $X$ and $Y$ respectively, where (a) for $1 \leq k \leq l-1$, $i_k < i_{k+1}$ and $j_k < j_{k+1}$, and (b) for $1 \leq k \leq l$, $x_{i_k} = y_{j_k}$. We define $Sim(X, Y)$ to be $l/n$. Note that it is not necessary for the two given sequences to have the same length, because the shorter sequence can always be padded with dummy numbers.

There are several shortcomings in the above model. The two sequences may have different *scaling factors* and *baselines*. For example, two stock indices could be es-

*Institute for Advanced Study, Princeton, N.J. 08540 AND University of Memphis, Department of Mathematical Sciences, Memphis, TN 38152, USA

†University of Memphis, Department of Mathematical Sciences, Memphis, TN 38152, USA

‡IBM Almaden RC K55/B1, 650 Harry Rd., San Jose CA 95120, USA

§University of Helsinki, Department of Computer Science, P.O. Box 26, FIN-00014 Helsinki, Finland

sentially similar (i.e. they react similarly to changing market conditions) even though one fluctuates near \$30 while the other fluctuates near \$100. In practice there should also be some allowable *tolerance* when comparing elements from both sequences (even after one sequence has been appropriately scaled or otherwise transformed to resemble the other sequence). The following similarity function overcomes these problems.

**The similarity function $Sim_{\epsilon,\delta}(X,Y)$:**

Let $\delta > 0$ be an integer constant, $0 < \epsilon < 1$ a real constant, and $f$ a linear function (of the form $f : y = ax+b$) belonging to the (infinite) family of linear functions $\mathcal{L}$. Given two sequences $X = x_1, \ldots, x_n$ and $Y = y_1, \ldots, y_n$, let $X' = (x_{i_1}, \ldots, x_{i_l})$ and $Y' = (y_{j_1}, \ldots, y_{j_l})$ be the longest subsequences in $X$ and $Y$ respectively such that

1. for $1 \le k \le l-1$, $i_k < i_{k+1}$ and $j_k < j_{k+1}$,
2. for $1 \le k \le l$, $|i_k - j_k| \le \delta$, and
3. for $1 \le k \le l$, $y_{j_k}/(1+\epsilon) \le f(x_{i_k}) \le y_{j_k}(1+\epsilon)$.

Let $S_{f,\epsilon,\delta}(X,Y)$ be defined as $l/n$. Then $Sim_{\epsilon,\delta}(X,Y)$ is defined as $\max_{f\in\mathcal{L}}\{S_{f,\epsilon,\delta}(X,Y)\}$.

Thus, when $Sim_{\epsilon,\delta}(X,Y)$ is close to 1, the two sequences are considered to be very similar. The constant $\delta$ ensures that the positions of each matched pair of elements are not too far apart. In practice this is a reasonable assumption, and it also helps in designing very efficient algorithms.

The linear function (or transformation) $f$ allows us to detect similarity between two sequences with different base values and scaling factors. Note than an algorithm trying to compute the similarity will have to find out *which* linear function to use (i.e. the values of $a$ and $b$) that maximizes $l$. The tolerance $\epsilon$ allows us to "approximately" match an element in $X$ (after transformation) with an element in $Y$. We observe that our similarity function is not necessarily symmetric; a symmetric definition of similarity is $\max\{Sim_{\epsilon,\delta}(X,Y), Sim_{\epsilon,\delta}(Y,X)\}$.

Given two sequences of length $n$, the longest common subsequence can be found in $O(n^2)$ time by a well known dynamic programming algorithm [3]; This algorithm can easily be modified to compute $S_{f,\epsilon,\delta}(X,Y)$ in $O(n\delta)$ time where $f$ is a given linear function. Note that this is essentially linear time. We refer to this algorithm as the LCSS algorithm.

To design algorithms to compute $Sim_{\epsilon,\delta}(X,Y)$, the main task is to locate a finite set of all fundamentally different linear transformations and run LCSS on each one.

Here we present algorithms that are based on a thorough analysis of the geometric properties of the problem. These algorithms have provable performance bounds, and some of them are very efficient in practice.

## 2 Main results

We summarize our results below.

1. **Cubic-time exact algorithm:**

   $Sim_{\epsilon,\delta}(X,Y)$ can be computed in $O(n^3\delta^3)$ time.

**Proof :** (Sketch)

Let $V = \{(x_i, y_j) | x_i \in X, y_j \in Y, |i-j| \le \delta\}$ be a set of points in the $xy$-plane. Clearly $|V| = O(n\delta)$. For every point $p = (x_i, y_j) \in V$, consider a vertical line segment $L_p = (p_1, p_2)$, where $p_1 = (x_i, y_j(1+\epsilon))$ and $p_2 = (x_i, y_j/(1+\epsilon))$. Let $R = \{L_p | p \in V\}$. Consider the linear transformation $f : y = ax + b$. Let $R_f$ be the segments of $R$ stabbed by the infinite line $f$. Let $V_R$ be the set of end points of all vertical segments in $R$. Thus $|V_R| = 2|V| = O(n\delta)$. Consider any linear function $f$ that does not pass through any point in $V_R$. Clearly if we perturb $f$ slightly, $R_f$ will not change. Recall that $\mathcal{L}$ is the (infinite) family of all linear functions. Let $\mathcal{L}'$ be the family of linear functions such that $f' \in \mathcal{L}'$ iff $f'$ passes through two points of $V_R$. It is easy to see that for *any* linear function $f \in \mathcal{L}$, there is a function $f' \in \mathcal{L}'$ such that $V_f = V_{f'}$.

Thus, our algorithm first computes $\mathcal{L}'$ (whose size is $O(n^2\delta^2)$), then runs LCSS on each function, and finally outputs the normalized length of the longest sequence found. This gives a total running time of $O(n^3\delta^3)$. □

2. **Quadratic-time approximation algorithm:**

   Let $0 < \beta < 1$ be any desired small constant. A linear transformation $f$ and the corresponding $S_{f,\epsilon,\delta}(X,Y)$ can be computed in $O(n^2\delta^2 + n\delta^3/\beta^2)$ time such that $Sim_{\epsilon,\delta}(X,Y) - S_{f,\epsilon,\delta}(X,Y) \le \beta$.

3. **Linear-time randomized approximation algorithm:**

   Let $0 < \beta < 1$ be any desired small constant. A linear transformation $f$ and the corresponding $S_{f,\epsilon,\delta}(X,Y)$ can be computed in $O(n\delta^3/\beta^2)$ running time such that $Sim_{\epsilon,\delta}(X,Y) - S_{f,\epsilon,\delta}(X,Y) \le \beta$ with high probability.

We note that the randomized approximation algorithm is very simple: it randomly selects a constant number of linear functions in $\mathcal{L}'$, and runs LCSS for each one.

While developing these algorithms we discovered several geometric results which are of independent interest. We describe some of these results below.

Let $S_i \Delta S_j$ represent the *symmetric-difference* between two sets. Let $V$ be a finite set and $2^V$ be its power set. Let $k > 0$ be an integer. A family of finite sets $\mathcal{S} \subseteq 2^V$ is *k-separated* if for all $S_i, S_j \in \mathcal{S}$, $|S_i \Delta S_j| > k$. It is known that there exist a $k$-separated family $\mathcal{S}$ where $|\mathcal{S}| = 2^{\alpha n}$ where $\alpha$ depends on $k/n$ [4]. However, we can get much better upper bounds if we only consider certain kinds of geometric sets.

Consider the following set system. Let $R$ be a fixed set of $n$ line segments on the plane. Given any infinite line $L$, let $R_L$ be the set of line segments of $R$ intersected (or *stabbed*) by $L$. (If $R_L = R$, then $L$ is known as a *transversal* of $R$.) Let the family $\mathcal{S}$ consist of the distinct stabbed sets $R_L$ of $R$, for all possible infinite lines $L$. The following result and its corollaries are crucial for our algorithms.

4. **$k$-separated stabbed sets:**

   The maximum size of a $k$-separated family $\mathcal{S}' \subseteq \mathcal{S}$ is $\Theta(n^2/k^2)$.

In particular, if $k = \beta n$ for some constant $0 < \beta < 1$, the size of any $k$-separated family is $O(1)$. It is this property that is used in our approximation algorithms.

We note here that the last result can also be derived from the cutting theorem of [6, 5], but our proof is simpler and direct.

**Related work:** There has been some recent work on the problem of defining similarity between time series; due to lack of space we only give some references. The problem was introduced to the data mining community by papers [1, 8]. The similarity measures considered in [2, 7, 12] use the concept of *longest common subsequence.* In [1], a *fingerprint* method is used, where the discrete Fourier transform is employed to reduce the dimensions of each sequence, and the resulting fingerprints are compared. In [10], *feature extraction* techniques are used; see [11] for a general discussion on fingerprinting techniques.

## 3 Experimental results

We experimented with the exact algorithm and the randomized approximation algorithm using three collections of sequences. One consisted of quarterly indicators of the status of the Finnish economy (67 sequences, 85 points), another of measurements of traffic data, error counts and call counts at 15 minute intervals (17 different phone lines, i.e., 51 sequences, 478 points each), and the third one about stock prices at the NYSE. Especially in the phone line data outliers are truly a problem: there are values in the sequence that differ by a factor of 2–5 from all the other values. In most cases these are outliers, but in some cases not; removing them permanently from the data is not possibly.

Overall, the algorithms behaved as predicted by the theoretical analysis. The exact $O(n^3)$ algorithm was far too slow to use for any but the smallest sequences and displacements. The randomized algorithm proved to be very efficient. Moreover, it produced approximations to the true similarity that are very close to the correct values. For example, in Table 1 we see that for varying $\epsilon$, the randomized algorithm got to within 1 from the true optimum already after 500 randomly chosen linear functions. Table 2 shows the time needed for this analysis. The randomized algorithm works extremely fast, checking 500 random linear functions in about 6.5 seconds, and this time is almost independent of the displacement. The results for the other sequence types were similar, and they are omitted from this version of the paper.

| | | | K | | | | | |
|---|---|---|---|---|---|---|---|---|
| $\epsilon$ | $\delta$ | LCSS | 10 | 50 | 100 | 200 | 500 | 1000 |
| 0.10 | 0 | 85 | 80.6 | 83.1 | 83.4 | 83.2 | 84.0 | 84.0 |
| 0.10 | 1 | 85 | 79.2 | 83.2 | 83.7 | 83.9 | 84.7 | 84.9 |
| 0.05 | 0 | 81 | 77.7 | 80.1 | 80.2 | 80.2 | 80.5 | 80.9 |
| 0.05 | 1 | 81 | 78.1 | 80.3 | 80.3 | 80.6 | 81.0 | 80.9 |
| 0.01 | 0 | 56 | 49.5 | 53.4 | 53.8 | 54.7 | 55.3 | 55.3 |
| 0.01 | 1 | 57 | 50.2 | 52.5 | 53.6 | 54.4 | 54.8 | 55.6 |

Table 1: True similarity between sequences and the length of longest common subsequence found by using $K$ randomly chosen linear functions; averages over 10 trials. Data: two series of 85 points about the Finnish national economy.

| | | | | K | | | | |
|---|---|---|---|---|---|---|---|---|
| $\epsilon$ | $\delta$ | LCSS | exact | 10 | 50 | 100 | 500 | 1000 |
| 0.10 | 0 | 85 | 361 | 0.16 | 0.67 | 1.3 | 6.5 | 13.0 |
| 0.10 | 1 | 85 | 3211 | 0.15 | 0.68 | 1.3 | 6.5 | 13.0 |
| 0.05 | 0 | 81 | 358 | 0.15 | 0.67 | 1.4 | 6.5 | 13.0 |
| 0.05 | 1 | 81 | 3236 | 0.14 | 0.57 | 1.1 | 5.6 | 11.3 |
| 0.01 | 0 | 56 | 364 | 0.15 | 0.67 | 1.3 | 6.5 | 13.0 |
| 0.01 | 1 | 57 | 3269 | 0.11 | 0.58 | 1.1 | 5.6 | 11.2 |

Table 2: Time (measured on a Pentium 32MB machine) used by the exact algortithm and the randomized approximate algorithm (for $K$ randomly chosen linear functions; averages over 10 trials). Data: as above.

## References


[1] R. Agrawal, C. Faloutsos and A. Swami. Efficient Similarity Search in Sequence Databases. In *Proc. of the 4th Intl. Conf. on Foundations of Data Organization and Algorithms (FODO'93)*, 1993.

[2] R. Agrawal, K.-I. Lin, H. S. Sawhney and K. Shim. Fast Similarity Search in the Presence of Noise, Scaling, and Translation in Time-Series Databases. In *Proc. of the 21st Intl. Conf. on Very Large Data Bases (VLDB'95)*, pp 490–501.

[3] A. V. Aho. Algorithms for Finding Patterns in Strings. In *Handbook of Theoretical Computer Science, Volume A: Algorithms and Complexity*, Elsevier, 1990, pp. 255–400.

[4] B. Bollobas. Combinatorics. Cambridge University Press, 1986.

[5] B. Chazelle. Cutting hyperplanes for divide-and-conquer. In *Discrete Comput. Geom.*, 9, 2, 1993, pp. 145-158.

[6] B. Chazelle and E. Welzl. Quasi-optimal range searching in spaces of finite VC-dimension. In *Discrete Comput. Geom.*, 4, 1989, pp. 467-489.

[7] G. Das, D. Gunopulos and H. Mannila. Finding Similar Time Series. Manuscript, 1996.

[8] C. Faloutsos, M. Ranganathan and Y. Manolopoulos. Fast Subsequence Matching in Time-Series Databases. In *SIGMOD'94*, 1994.

[9] H.V. Jagadish, A. O. Mendelzon and T. Milo. Similarity-Based Queries. In *Proc. of 14th Symp. on Principles of Database Systems (PODS'95)*, 1995, pp. 36–45.

[10] H. Shatkay and S. Zdonik. Approximate Queries and Representations for Large Data Sequences. In *ICDE'96*, 1996.

[11] D. A. White and R. Jain. Algorithms and Strategies for Similarity Retrieval. Technical Report VCL-96-101, Visual Computing Laboratory, UC Davis, 1996.

[12] N. Yazdani and Z. M. Ozsoyoglu. Sequence Matching of Images. In *Proc. of the 8th Intl. Conf. on Scientific and Statistical Database Management*, 1996, pp. 53–62.

# The Area Bisectors of a Polygon and Force Equilibria in Programmable Vector Fields*

Karl-Friedrich Böhringer[†] Cornell University
Bruce Randall Donald[‡] Cornell University
Dan Halperin[§] Tel Aviv University

## Abstract

We consider the family of area bisectors of a polygon (possibly with holes) in the plane. We say that two bisectors of a polygon $P$ are *combinatorially distinct* if they induce different partitionings of the vertices of $P$. We show that there are simple polygons with $n$ vertices that have $\Omega(n^2)$ combinatorially distinct area bisectors (matching the obvious upper bound), and we present an output-sensitive algorithm for computing an explicit representation of all the bisectors of a given polygon. Our study is motivated by the development of novel, flexible feeding devices for parts positioning and orienting. The question of determining all the bisectors of polygonal parts arises in connection with the development of efficient part positioning strategies when using these devices.

## 1 Introduction

Let $P$ be a polygon in the plane, possibly with holes, and having $n$ vertices in total. We denote by $V$ the set of vertices of $P$. For a directed line $\lambda$ in the plane, we denote by $h_l(\lambda)$ (resp. $h_r(\lambda)$) the open half-plane bounded by $\lambda$ on the left- (resp. right-) hand-side of $\lambda$. The line $\lambda$ is an area bisector of $P$ if the area of $P \cap h_l(\lambda)$ is equal to the area of $P \cap h_r(\lambda)$.

A line $\lambda$ partitions $V$ into three sets (two of which may be empty): $V \cap h_l(\lambda)$, $V \cap \lambda$, and $V \cap h_r(\lambda)$. We say that two area bisectors of $P$ are *combinatorially distinct* if the partitioning of $V$ as above induced by the two bisectors is different. We say that two area bisectors of $P$ are *combinatorially equivalent* if they induce the same partitioning of $V$. We assume that the polygon $P$ is connected, and non-degenerate in the sense that the complement of $P$ has the same boundary as $P$.

An obvious upper bound on the number of distinct area bisectors of a polygon with $n$ vertices is $O(n^2)$—each combinatorial equivalence class of area bisectors is determined by a pair of vertices of the polygon. In Section 4 we show that a polygon with $n$ vertices can have $\Omega(n^2)$ distinct area bisectors. (Note that the polygon in our construction is *simple*.)

We devise an output-sensitive algorithm for computing an explicit representation of all the area bisectors of a given polygon, by constructing the *bisector curve* $\beta$ defined in a plane dual to the plane containing the polygon: the curve $\beta$ is the union of points dual to area bisectors in the primal plane. Our algorithm proceeds by constructing the *zone* of the curve $\beta$ in an *arrangement of lines* [5] in the dual plane, where each line of the arrangement is the dual of a vertex of the polygon. A sketch of the algorithm is given in Section 5.

Area bisectors were considered by Díaz and O'Rourke [4]. However, their focus is on the continuous version of the *ham-sandwich cut* problem, and of a problem they introduce of *orthogonal four-sections*; see [4] for more details. The problem that we study here can be viewed as a continuous version of the well-known *k-set* problem [5].

## 2 Motivation: Programmable Vector Fields

Programmable vector fields can be used to control a variety of flexible planar part feeders. These devices often exploit exotic actuation technologies such as arrayed, microfabricated motion pixels [2] or transverse vibrating plates [1]. These new automation designs promise great speed, flexibility, and dexterity—we believe they may be employed to orient, singulate, sort, and feed parts. However, since they have only recently been invented, programming and controlling them for manipulation tasks is challenging.

In [3], we devise a technique for analyzing programmable vector fields called *equilibrium analysis*, lower bounds (i.e., impossibility results) on what the devices *cannot* do, and results on a classification of control strategies yielding design criteria by which well-behaved manipulation strategies may

*Work on this paper by K.-F. Böhringer and B. R. Donald has been supported in part by NSF under grants No. IRI-8802390, IRI-9000532, IRI-9201699, and by a Presidential Young Investigator award to B. Donald, in part by NSF/ARPA Small Grant for Exploratory Research No. IRI-9403903, and in part by the Air Force Office of Sponsored Research, the Mathematical Sciences Institute, Intel Corporation, and AT&T Bell labs. Work on this paper by D. Halperin has been supported by an Alon Fellowship and by ESPRIT IV LTR Project No. 21957 (CGAL).

[†]Robotics & Vision Lab, Dept. of Comp. Sc., Cornell Univ. Current address: ALPHA Lab, Dept. of Ind. Eng. and Op. Research, Univ. of CA, Berkeley. Email: karl@ieor.berkeley.edu.

[‡]Robotics & Vision Laboratory, Department of Computer Science, Cornell University. Email: brd@cs.cornell.edu.

[§]Dept. of Comp. Sc., Tel Aviv Univ., Tel Aviv 69978, ISRAEL. halperin@math.tau.ac.il. Part of the work was carried out while D. H. was at the Robotics Lab, Dept. of Comp. Sc., Stanford University.


*Computational Geometry 97* Nice France

be developed. Equilibrium analysis is a fundamentally geometric problem. To illustrate, let us make the following assumptions. Suppose that (1) Each motion pixel is very small relative to the part we wish to manipulate, and hence a dense array of pixels may be modeled as a 2-dimensional vector field. (2) The part to be manipulated is essentially laminar. (3) The part is connected.

A squeeze field is defined by making all the actuators push normally towards a common *squeeze line*. In [3], we show that under certain dynamic and mechanical assumptions, we can compute a sequence of squeeze fields guaranteed to bring any part into unique moment equilibrium. No sensing is required, but a clock is required to switch between fields.

In order to be in equilibrium, the forces and moments must balance. Force equilibrium, under our assumptions, is equivalent to: *the squeeze line must be an area bisector of the polygon.* The number of equilibria critically impacts both the complexity of plan generation, and plan size. Therefore, it is important to bound the number of equilibria for a given part or class of parts. In [3] we show that every convex part has at most $O(n)$ equilibria, and every non-convex part has $O(rn^2)$ equilibria ($r$ is the maximum number of edges of $P$ intersected by any line), under squeeze fields.

In this paper, we show how to analyze the number of area bisectors, and hence bound the number of force equilibria. Our algorithm for computing area bisectors can be used as a preliminary step in designing alignment plans (see Section 6).

## 3 Properties of Area Bisectors

In this section we state several properties of area bisectors of polygons. The proofs can be found in the full version of the paper.[1]

**Lemma 3.1** *Let $P$ be a non-degenerate polygon with $n$ vertices. (1) There exist $O(n^2)$ combinatorially distinct ways in which a line can partition $P$. (2) Let $A$ and $B$ be the intersections of an area bisector $\lambda$ with the boundary of the convex hull of $P$. As the slope of $\lambda$ varies from $-\infty$ to $+\infty$, $A$ and $B$ progress monotonically counterclockwise on the boundary of the convex hull of $P$. (3) For every slope $\bar{x}$ there exists a unique bisector $\lambda$ of $P$ with slope $\bar{x}$.*

**Lemma 3.2** *Let $P$ be a polygon with $n$ vertices. Let $s$ be a point in $\mathbb{R}^2$ and let $\lambda$ be a line that intersects $r$ edges of $P$. The area bisectors of $P$ that are combinatorially equivalent to $\lambda$ and pass through $s$ are determined by the roots of a polynomial equation of degree $r$.*

In the full version of the paper we show that the bisectors of a polygon $P$ can be described by a piece-wise algebraic curve, where each piece is described by a polynomial whose degree depends on the number of edges of $P$ intersected by the corresponding bisectors.

## 4 Lower Bound

As argued above, a polygon with $n$ vertices can have at most $O(n^2)$ combinatorially distinct area bisectors. In this section we give an example of a simple polygon with $n$ vertices where the bound $\Omega(n^2)$ is attained.

Consider Figure 1. All the vertices $v_i, v_i', u_i$ and $u_i'$ lie on a circle whose center is at $c$. The vertices $w_j$ lie very close

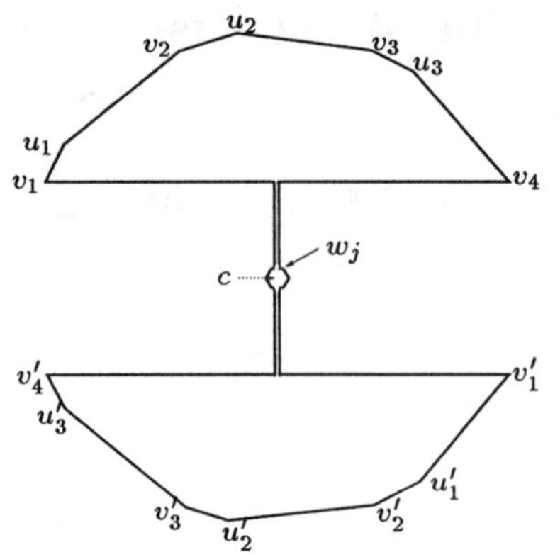


Figure 1: A simple polygon with $n$ vertices that has $\Omega(n^2)$ combinatorially distinct bisectors.

to $c$ on a small circle whose center is $c$ as well, along two convex polygonal chains.

We fix an integer $m$ (that we will determine later; for the polygon in the figure $m = 3$). The distance between the vertices $v_i$ and $v_{i+1}$ is the same for $i = 1, \ldots, m$, and it is the same as the distance between $v_i'$ and $v_{i+1}'$ for $i = 1, \ldots, m$. The area of all triangles $v_i u_i v_{i+1}$ for $i = 1, \ldots, m$ is the same and is equal to the area of all triangles $v_i' u_i' v_{i+1}'$ for $i = 1, \ldots, m$. There are $2m$ vertices $w_j$ near $c$ and they are equally spaced on a small circle centered at $c$. As can be easily verified, for every pair of vertices $v_i$ and $v_i'$, there is a bisector passing through these points that passes also through the center point $c$. We next claim that as we rotate the bisector from $v_i$ to $v_{i+1}$ it will move off the center $c$ and sweep $m$ vertices $w_j$. The reason is that the angle $\angle u_i v_i v_{i+1}$ is larger than the angle $\angle u_i' v_i' v_{i+1}'$. Hence, as the bisector rotates, it will proceed 'faster' on the bottom part of our polygon than on the top part and therefore will sweep half of the vertices $w_j$ on its way. Finally $m$ is chosen such that (roughly) $n = 6m+8$. The number of distinct area bisectors is evidently $\Omega(m^2) = \Omega(n^2)$.

## 5 Output-Sensitive Algorithm

It is convenient to study the algorithmic problem in a dual plane: a line $y = 2\bar{x}x - \bar{y}$ in the primal plane is transformed into the point $(\bar{x}, \bar{y})$ in the dual plane. A point $(x, y)$ in the primal plane is transformed into the line $\bar{y} = 2x\bar{x} - y$ in the dual. The dual of an object $o$ will be denoted by $o^*$. If $O$ is a set of objects in the plane, $O^*$ will denote the set of dual objects.

Let $P$ be a polygon with $n$ vertices as defined in the Introduction, namely connected, non-degenerate and possibly with holes. In the dual plane every vertex $v$ of $P$ is transformed into a line $v^*$ which is the collection of all points dual to lines in the primal plane that pass through $v$.

For any given direction there is a unique area bisector. We denote the oriented bisector of $P$ that makes an angle $\theta$ with the positive $x$-axis by $B(\theta)$, and (because of symmetry) confine ourselves to the range $[-\pi/2, \pi/2)$ for $\theta$. We denote the collection of points dual to area bisectors of $P$ in that range by $\beta$. Note that any $\theta$ (besides $-\pi/2$) corresponds to an $\bar{x}$-coordinate in the dual plane.

The curve $\beta$ is a piece-wise algebraic and $\bar{x}$-monotone curve (this is proved in the full version of the paper). We

[1] Forthcoming as a Technical Report, Computer Science Department, Cornell University.

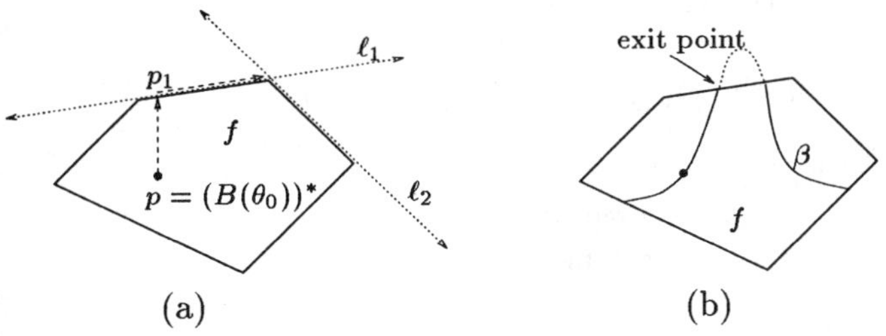


Figure 2: Ray shooting to determine the face $f$ containing $p$ (a), and then finding the maximal pieces of $\beta$ inside $f$ and its exit points from $f$ (b).

call $\beta$ the *bisector curve* of $P$, as it gives a complete specification of all the area bisectors of the polygon $P$. We denote by $\kappa$ the number of maximal connected algebraic pieces in $\beta$, where the function describing each piece is defined by the fixed set of edges that the corresponding set of bisectors cross. In this section we describe an output-sensitive algorithm to compute $\beta$. Since we aim for output-sensitivity, we cannot afford to compute the entire arrangement $\mathcal{A}(V^*)$ whose complexity may be $\Omega(n^2)$. We will discover the maximal pieces of $\beta$ in their order along $\beta$, using two primitive operations: *ray shooting* among the lines $V^*$, and intersection of an algebraic curve with a straight line.

We choose an arbitrary direction $\theta_0 \in [-\pi/2, \pi/2)$ and look for the area bisector of $P$ in that direction. This requires $O(n \log n)$ time since the polygon may have holes. Next, we obtain the set of edges crossed by $B(\theta_0)$. We denote by $E(\theta)$ the set of edges crossed by $B(\theta)$. The set $E(\theta_0)$ determines a function $\rho := \rho(\theta_0)$ describing the bisector curve $\beta$ in a neighborhood of $\theta_0$, as long as the set of edges crossed by the bisector does not change. In the dual plane the function $\rho$ describes the curve $\beta$ as long as we do not leave the face of $\mathcal{A}(V^*)$ which contains the point $p := (B(\theta_0))^*$.

Our next step is to construct the face $f$ that contains the point $p$ in $\mathcal{A}(V^*)$. We describe this procedure assuming $f$ is bounded; the extension to unbounded faces is straightforward. We prepare in advance a data structure $R(V^*)$ that supports efficient ray shooting among the lines $V^*$. We shoot a ray from $p$ in the upward vertical direction and identify the line $\ell_1 \in V^*$ supporting the edge of $f$ above $p$. See Figure 2(a). We next proceed in clockwise direction along the boundary of $f$. From the point $p_1 \in \ell_1$ we shoot a ray in $\mathcal{A}(V^*)$ along $\ell_1$, and identify the line $\ell_2$ supporting the next edge on the boundary of $f$, and so on until we have returned to $\ell_1$ and thus have identified the entire face $f$.

Now we determine the maximal connected pieces of $f \cap \beta$. For each edge on boundary of $f$ we compute the intersection of its supporting line $v^*$ with $\beta$. This computation is equivalent to finding the bisectors that pass through the vertex $v$, and intersecting the edges in $E(\theta_0)$. By Lemma 3.2, this reduces to solving a polynomial equation of degree $r$, where $r$ is the number of edges in $E(\theta_0)$. We denote the time required to find these roots by $\psi(r)$.

We order the resulting intersections along the $\bar{x}$-axis. Since the curve $\beta$ is $\bar{x}$-monotone, this ordered list of intersections provides a description of the curve $\beta$ inside $f$, and indicates what are the neighboring faces that $\beta$ crosses. We mark each of these additional faces by the point where $\beta$ crosses out of $f$. We call each such point an *exit point*. See Figure 2(b).

Since $f$ has already been constructed, we know for each exit point of $\beta$ the line that contains it. Therefore we can construct each new face using ray shooting queries and proceed as above. We keep a data structure that describes all the faces of $\mathcal{A}(V^*)$ that have already been constructed so that we do not construct the same face twice.

The algorithm stops when we have identified all the intersection points of $\beta$ with lines in $V^*$, and so we have also identified the *zone* of $\beta$ in $\mathcal{A}(V^*)$, namely all the face of $\mathcal{A}(V^*)$ crossed by $\beta$.

Further details on the algorithm can be found in the full version of the paper. We summarize the algorithmic result in the following theorem.

**Theorem 5.1** *Let $P$ be a non-degenerate polygon (possibly with holes) with $n$ vertices, and such that any line crosses at most $r$ edges of $P$. For any $\varepsilon > 0$ we can find a complete specification of the area bisectors of $P$ in time $O(\kappa^{2/3} n^{2/3+\varepsilon} + \kappa\alpha(\kappa)\psi(r))$, where $\kappa$ and $\psi(r)$ are as defined above, and $\alpha()$ is the functional inverse of Ackermann's function. If $P$ is rectilinear, then the algorithms runs in time $O(\kappa^{2/3} n^{2/3+\varepsilon})$. The space required by the algorithm is $O(\kappa^{2/3} n^{2/3+\varepsilon})$.*

## 6 Moment Equilibria and Alignment Plans

As described in Section 2, bisectors correspond to force equilibria of $P$ in a squeeze field. For *total* equilibrium, the moment acting on the part has to be taken into account as well. In particular, not all of the force equilibrium configurations will be moment equilibria. For each maximal piece $b$ of the bisector curve $\beta$ there exists only a finite number of moment equilibria (we omit the proof here):

**Lemma 6.1** *Let $P$ be a polygon whose interior is connected. Let $\lambda$ be a bisector of $P$ that intersects $r$ edges of $P$. There exist $O(r)$ lines $\lambda'$ that are combinatorially equivalent to $\lambda$ such that $P$ is in total equilibrium when $\lambda'$ coincides with the center line of a squeeze field.*

It follows that a squeeze field induces a finite number of total equilibria on a polygonal part $P$. In [3] we show how to exploit this finiteness property to automatically generate *alignment plans* that bring $P$ into a unique (up to symmetry) orientation, by cascading squeeze fields that systematically reduce the possible orientations of $P$.

## References


[1] K.-F. Böhringer, V. Bhatt, and K. Y. Goldberg. Sensorless manipulation using transverse vibrations of a plate. In *Proc. IEEE Int. Conf. on Robotics and Automation (ICRA)*, pages 1989 – 1996, Nagoya, Japan, May 1995.

[2] K.-F. Böhringer, B. R. Donald, and N. C. MacDonald. Single-crystal silicon actuator arrays for micro manipulation tasks. In *Proc. IEEE Workshop on Micro Electro Mechanical Systems (MEMS)*, San Diego, CA, Feb. 1996.

[3] K.-F. Böhringer, B. R. Donald, and N. C. MacDonald. Upper and lower bounds for programmable vector fields with applications to MEMS and vibratory plate parts feeders. In J.-P. Laumond and M. Overmars, editors, *Algorithms for Robotic Motion and Manipulation*. A. K. Peters, Wellesley, MA 02181, 1997.

[4] M. Díaz and J. O'Rourke. Ham-sandwich sectioning of polygons. In *Proc. 2nd Canadian Conference on Computational Geometry*, pages 98–101, Ottawa, 1990.

[5] H. Edelsbrunner. *Algorithms in Combinatorial Geometry*, volume 10 of *EATCS Monographs on Theoretical Computer Science*. Springer Verlag, Heidelberg, Germany, 1987.

# Efficient Breakout Routing in Printed Circuit Boards

John Hershberger

Mentor Graphics

1001 Ridder Park Drive

San Jose, CA 94301

Subhash Suri*

Department of Computer Science

Washington University

St. Louis, MO 63130

**Abstract**

Breakout routing is a single-layer wire-routing problem in which each of a set of pins must be connected to one of a set of vias, but no matching is prespecified. We propose a network-flow approach to breakout routing in which the wiring grid is modeled by a more compact graph. Our graph is a factor of $\Theta(\kappa^2)$ smaller than the wiring grid, where $\kappa$ is the ratio of via spacing to pin spacing, which improves both the space and run time efficiency of the flow computation. A flow in the compact graph can be transformed into a wire layout, and vice versa.

## 1 Introduction

Breakout routing is a crucial problem in routing printed circuit boards (PCBs) and multichip modules (MCMs). The problem arises because a typical PCB or MCM has multiple layers available for routing, but electronic components are often connected to only a single layer of the board. Completing the routing requires that these surface-mounted components be connected to other layers by vias. The most efficient way to do this is to place a set of breakout vias near the component, connect each surface-mounted pin to a unique via, and then finish routing the connections from via to via using all the routing layers. Breakout routing is qualitatively different from ordinary PCB routing: the connections from pins to breakout vias lie in a single routing layer, whereas ordinary routing may use multiple layers to cross one wire over another. The pairing of pins with breakout vias is not fixed: any pin may be connected to any via, because the connections specified by the circuit schematic will be routed between pairs of vias.

Among the previous work, the two methods most relevant to our approach are pattern routing in a grid and network flow. In pattern routing, vias lie in a rectangular grid, pins are located on its periphery, and one connects pins to vias with a regular pattern of wires. This method is guaranteed to find at least $1/\sqrt{2}$ of the maximum possible number of connections [7], but it is limited to the case of a fully populated via grid. The network flow methods always find a feasible routing if one exists, but appear to be more popular in theory [2, p. 625] than in practice, perhaps because of their large asymptotic complexity. One big disadvantage of the "vanilla" network flow method is that the size of the grid graph can be quadratic in the number of pins.[1]

Our method can be viewed as combining the best features of pattern routing in a grid and network flow. Like pattern routing in a grid, it exploits the fact that vias lie on a grid, but it allows some sites in the via grid to be inactive. It uses a network flow framework and always finds an optimal number of routes, but it does not explicitly build the entire grid graph. In particular, suppose that $n$ breakout vias lie within a rectangular grid of size $p \times q$, and $m$ component pins lie on the boundary of the rectangle, where $m \leq n \leq pq$. Our algorithm computes network flow in a network of size $O(pq)$, whereas the conventional approach will construct a grid graph of size at least $\Omega(pq\kappa^2)$, where $\kappa + 1$ is the ratio of via spacing to pin spacing. (In Figure 1, $p = q = 5$, $m = 6$, and $\kappa = 3$.) We believe that the improvement by a factor of $\kappa^2$ in running time and, more significantly, in space can make our approach a viable candidate for breakout routing in practice.

## 2 Preliminaries and the Combinatorial Wire Flow

Consider a $(\kappa+1) \times (\kappa+1)$ square, partitioned into $(\kappa+1)^2$ unit squares by a square grid, for some nonnegative integer $\kappa$. Let us call the big square a *via cell*, its four corners *via sites*, the unit grid the *wire grid*, and vertices of the unit grid *grid points*.[2] The parameter $\kappa$ is called the *via separation*. Figure 1 shows an example.

A *wire element* is an edge of a unit square in the wire grid. A *wire path* is a simple (non-self-intersecting) path made up of wire elements. In this paper, all our wire paths will connect a pin location to a via location, and we assume throughout that these paths are *directed* from the pin to the via. Each via can either act as a *sink* node for a wire path, or as a *transit* node, but not both. If a via can act only as a sink, and not a transit node, then our algorithm and method still apply, with only a minor modification in the

*Research supported in part by National Science Foundation Grant CCR-9501494



*Computational Geometry 97* Nice France



[1] Recent work by Yu, Darnauer, and Dai [8] uses network flow on a graph derived from a constrained Delaunay triangulation. The resulting flow may not correspond to a feasible routing, but the algorithm is applicable to a broader class of problems than the grid-based approach.

[2] Thus, a via site is also a grid point.

construction.

A *matching* between a set of pins $P = \{p_1, p_2, \ldots, p_m\}$ and a set of vias $V = \{v_1, v_2, \ldots, v_n\}$ is a pairing $\{(p_1, v_1), (p_2, v_2), \ldots, (p_k, v_k)\}$ where each pin or via appears at most once. A *complete matching* is a matching where each pin appears exactly once; that is, $k = m$. A matching is *embeddable* if each matched pair can be joined by a wire path in the wire grid so that all paths are pairwise non-intersecting. In this paper, we will be interested only in embeddable matchings, and so we assume in the following that a matching is always embeddable, and that each wire path is directed from its source pin to its sink via. We say that a wire element $x$ *enters* a cell $c$ if $x$'s head is on the boundary of $c$ but its tail is outside $c$. Similarly, $x$ *exits* a cell $c$ if $x$'s tail is on the boundary of $c$ but its head is outside $c$.

Let us define the *core* of a cell $c$ to be the $\kappa \times \kappa$ subgrid in the center of $c$. The *boundary* of $c$ is the square defining $c$. In Figure 1 (a), the central $3 \times 3$ grid is the core of the via cell. Let us discuss the combinatorial properties of a wire layout with respect to the core of a cell. The main property of a core that we need in the following discussion is that a core has no vias, and therefore no wire terminates inside a core.

Let $\mathcal{M}$ be a fixed matching, and consider the core $b$ of a via cell $c$. Let $e_i(b)$, for $1 \le i \le 4$, denote the edges of the square $b$ in clockwise order, starting at the left edge. Let $f_i(b)$ denote the number of wire elements in $\mathcal{M}$ that enter the core $b$ through edge $e_i(b)$, and let $g_i(b)$ denote the number of wire elements in $\mathcal{M}$ that leave $b$ through $e_i(b)$, where $1 \le i \le 4$. The following two constraints are easily established:

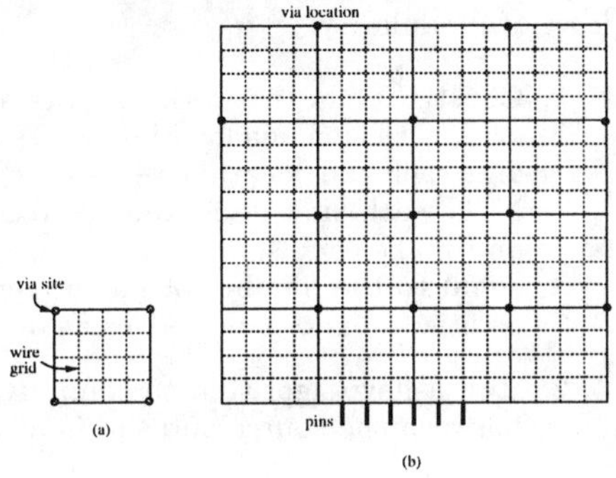


Figure 1: (a) A via cell; $\kappa = 3$. (b) A tiling of via cells. Pins are shown as thick rectangles along the bottom row of the printed circuit board. Dark circles represent location of actual vias.

1. [Capacity Constraint] $f_i(b) + g_i(b) \le \kappa$.
2. [Wire Conservation] $\sum_{i=1}^{4} f_i(b) - \sum_{i=1}^{4} g_i(b) = 0$.

The capacity constraint and wire conservation are clearly necessary for a combinatorial wire flow to be embeddable, though not always sufficient. It turns out that the maximum number of wire elements entering $b$ in a legal embedding can vary from $\kappa$ to $2\kappa$ depending on the topology of wires. Our next lemma gives a full characterization of this combinatorial fact.

Let $x_i = f_i - g_i$ denote the net flow of wires into $b$ through the edge $e_i$. The edge $e_i$ is called a *surplus edge* if $x_i > 0$, a *deficit edge* if $x_i < 0$, and neutral otherwise. We define the *type* of a square $b$ depending on the number of surplus edges:

**Type I:** $b$ has exactly one surplus edge or exactly one deficit edge.

**Type II:** $b$ has two surplus edges sharing a common vertex.

**Type III:** $b$ has two surplus edges that are opposite (non-adjacent) edges of $b$.

**Lemma 2.1** *The maximum wire inflow into a square core $b$ satisfies the following bounds:*

$$\sum_{i=1}^{4} \max(f_i - g_i,\ 0) \le \begin{cases} \kappa & \text{if } b \text{ is Type I or II} \\ 2\kappa & \text{if } b \text{ is Type III and } \kappa \text{ is even} \\ 2\kappa - 1 & \text{if } b \text{ is Type III and } \kappa \text{ is odd.} \end{cases}$$

*These bounds are tight.*

## 3 A Network Flow Formulation

In our construction, the following *diamond network* will play a crucial role; see Figure 2. The four nodes on the outer cycle are labeled $a_1, a_2, a_3, a_4$; they are joined by edges of bidirectional capacity $\kappa_1 = \lfloor \kappa/2 \rfloor$. Whenever an edge is drawn undirected, it should be interpreted as having the same capacity in both directions. In addition, there is a central node $a_0$, joined to the four other nodes by edges of bidirectional capacity $\kappa_2 = \kappa - 2\lfloor \kappa/2 \rfloor$. The central node $a_0$ also has a node capacity constraint $\sigma(a_0) = 1$; other nodes have no capacity constraints.

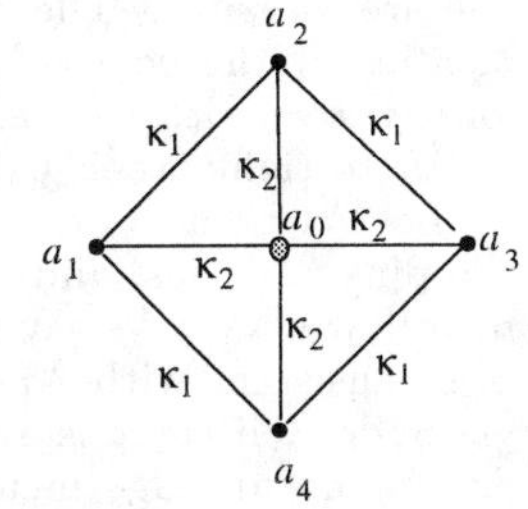


Figure 2: The diamond network for the core of a cell.

**Lemma 3.1** *The diamond network admits a flow for each of the following cases.*

1. *Exactly one node on the cycle, say $a_1$, is a supply node with $\sigma(a_1) \le \kappa$.*
2. *Two adjacent nodes on the cycle, say $a_1$ and $a_2$, are supply nodes with $\sigma(a_1) + \sigma(a_2) \le \kappa$.*
3. *Two opposite nodes on the cycle, say $a_1$ and $a_3$, are supply nodes with $\sigma(a_1) + \sigma(a_3) \le 2\kappa - \kappa_2$.*
4. *Three nodes on the cycle, say $a_1$, $a_2$, and $a_3$, are supply nodes with $\sigma(a_1) + \sigma(a_2) + \sigma(a_3) \le \kappa$.*

A via cell is modeled using the network shown in Figure 3. We call this a *network element*. The nodes $v_1, v_2, v_3, v_4$ represent the four via sites; the central diamond represents the core $b$ of the cell; and the nodes $u_1, u_2, u_3, u_4$ act as connections between the core and the cell boundary. For ease of reference, we will often use the term *connection nodes* for the $u_i$ nodes, *connection edges* for the edges $(u_i, a_i)$, and *boundary edges* for the edges $(v_i, u_j)$.

The via nodes $v_i$ and the middle diamond node $a_0$ have node capacity 1; the connection nodes $u_i$ have capacity $\kappa$ each; all other nodes have no capacity constraint. Let $\mathcal{N}$ denote the network corresponding to the entire via grid. (See Figure 4.) We complete the network by adding two artificial nodes $s$ and $d$, called *supply* and *demand* nodes. We join each via location to $d$ with a unit-capacity edge. We join $s$ to each node on the boundary of the via grid.

The capacity of the edge joining $s$ to $u$ is the number of "pins" on the portion of the boundary corresponding to $u$: if $u$ is a via site, then the capacity is at most 1; otherwise, the capacity is the number of pins lying between the via neighbors of $u$. Finally, we set $\sigma(s) = |P|$ and $\sigma(d) = -|P|$. (For example, in Figure 1, there are 9 nodes on the bottom row, of which five are via sites. The capacities of the edges that join these nodes to $s$ are, in left to right order, $0, 0, 0, 3, 1, 2, 0, 0, 0$.)

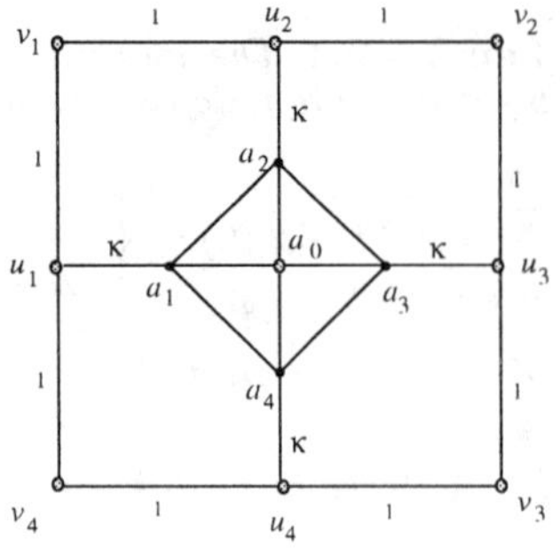


Figure 3: Network for a via cell.

Let $\mathcal{M}$ be a matching of the pins and vias, and let $\phi$ be a flow in $\mathcal{N}$. We say that $\phi$ is consistent with $\mathcal{M}$ at a via node $v_i$ if there is a unit flow along an edge incident to $v_i$ if and only if there is a wire path in $\mathcal{M}$ incident to $v_i$ along that edge. We also say that $\phi$ is consistent with a cell $c$ if $\phi(u_i, a_i) = f_i$, and $\phi(a_i, u_i) = g_i$, where $f_i$ and $g_i$, for $1 \leq i \leq 4$, are the number of wire elements entering and leaving the core of $c$ through the edge $e_i$. Finally, we say that $\phi$ is *consistent* with $\mathcal{M}$ if it is consistent with each cell and each via.

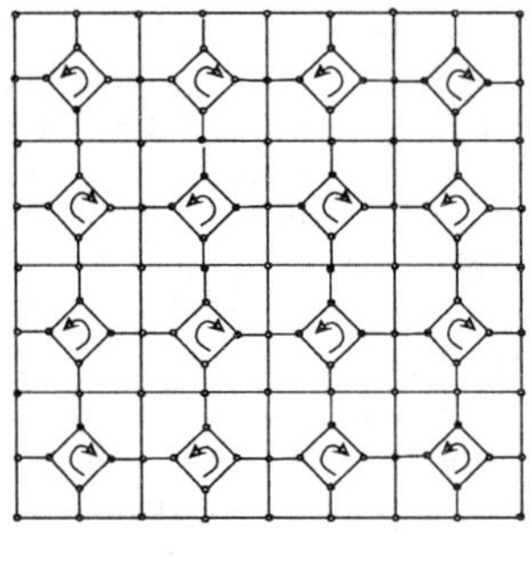

Figure 4: The network $\mathcal{N}$.

**Lemma 3.2** *Given a matching $\mathcal{M}$, there is always a flow in $\mathcal{N}$ consistent with $\mathcal{M}$.*

## 4 Converting a Flow into a Wire Layout

In this section, we show how to construct a matching consistent with a given flow in $\mathcal{N}$. Our algorithm consists of three steps: cell orientation and tail-ordering, wiring inside a core, and wiring in the cell. Due to lack of space, we sketch the ideas of these steps only briefly.

We orient each cell $c$ either clockwise or counterclockwise, which enforces a direction on each boundary edge of $c$. If an edge $(a, b)$ is oriented from $a$ to $b$, then our wiring layout pushes all wires on $(a, b)$ to the $b$ end. We assign the orientations in a checkerboard fashion, alternating between clockwise and counterclockwise. This ensures that the direction enforced on an edge by its two neighboring cells is the same.

Let $u$ be a connection node in a cell $c$, and let $v$ be the via site at the head of $u$'s edge, as determined by $c$'s orientation. Cell $c$ has a *tail violation* at $u$ if there is positive (or negative) flow from $c$'s core to $u$ and from $u$ to $v$. A flow is *tail-ordered* if no cell has a tail violation. The full paper shows that a flow with a tail violation may not be embeddable, but that any flow can be made tail-ordered by local modifications.

For wiring in a core, consider the network element corresponding to a cell $c$. We are interested in $\phi$'s restriction to the subgraph spanned by $c$'s diamond and the connection nodes—this subgraph corresponds to the core $b$ of the cell $c$. We will show how to compute a layout consistent with the flow in this subgraph.

Since $\phi$ is anti-symmetric, each connection edge $(u_i, a_i)$ has flow in only one direction. Thus, we can use the direction of flow in $(u_i, a_i)$ to label the corresponding core boundary edge. We say that the edge $e_i$ is surplus if $\phi(u_i, a_i) > 0$; it is deficit if $\phi(a_i, u_i) < 0$; and neutral otherwise. The core $b$ can now be classified as Type I, II, or III, depending on which of its edges are surplus, exactly as in Section 2. We show that wiring is possible in each of these cases, under the further restriction that wiring must be consistent with the cell orientation of $c$.

Once the core wiring is complete, the only remaining wires are the ones that correspond to flow along boundary edges of a network element. It suffices to consider a side of a cell, and show how the wires from its two neighboring cores can be augmented with the boundary wires. The full paper describes this in detail.

## 5 Algorithm and Complexity Analysis

All the pieces of our breakout routing algorithm are in place now, and the algorithm can be described as follows at a high level of modularity:

1. Construct the network $\mathcal{N}$ corresponding to the via grid, with each via location represented by a sink node, and each pin location represented by a source node. Make each sink node a demand node with value one, and make each source node a supply node with value equal to the number of pins it represents. Assign capacities to nodes and edges as described in Section 3.

2. Run a flow algorithm to compute a maximum flow $\phi$ between the source and sink nodes [2].

3. Modify the flow, if necessary, to make it tail-ordered.

4. Use the tail-ordered flow $\phi$ to compute a wire layout consistent with $\phi$ in each cell.

## References


[1] W.-T. Chan and F. Y. L. Chin. Efficient algorithms for finding disjoint paths in grids. In *Proceedings of ACM-SIAM Symposium on Discrete Algorithms*, pages 454–463, 1997.

[2] T. H. Cormen, C. E. Leiserson, and R. L. Rivest. *Introduction to Algorithms.* The MIT Press, Cambridge, Mass., 1990.

[3] W. W.-M. Dai, R. Kong, and M. Sato. Routability of a rubber-band sketch. In *Proceedings of 28th IEEE/ACM Design Automation Conf.*, pages 45–48, 1991.

[4] F. M. Maley. *Single-Layer Wire Routing and Compaction.* The MIT Press, Cambridge, Mass., 1990.

[5] V. P. Roychowdhury, J. Bruck, and T. Kailath. Efficient algorithms for reconfiguration in VLSI/WSI arrays. *IEEE Transactions on Computers*, pages 480–489, 1990.

[6] R. E. Tarjan. *Data Structures and Network Algorithms*, volume 44 of *CBMS-NSF Regional Conference Series in Applied Mathematics.* Society for Industrial Applied Mathematics, 1983.

[7] M.-F. Yu and W. W.-M. Dai. Single-layer fanout routing and routability analysis for ball grid arrays. In *Proceedings of IEEE/ACM International Conf. on CAD*, pages 581–586, 1995.

[8] M.-F. Yu, J. Darnauer, and W. W.-M. Dai. Interchangeable pin routing with application to package layout. In *Proceedings of IEEE/ACM International Conf. on CAD*, pages 668–673, 1996.

# Voronoi Diagram in Statistical Parametric Space by Kullback-Leibler Divergence

Kensuke Onishi* Hiroshi Imai†


## stract

ronoi diagram has been a main theme in computational geom-y, and the theory of generalized Voronoi diagrams for various plications in robotics, VLSI CAD, etc., has been developed in ms of arrangements, Davenport-Schinzel sequences and lower velopes. In this paper, we propose a new direction of research wards introducing discrete proximity structures in statistical rametric spaces by Voronoi diagrams defined by statistically eaningful distance, partially based on information geometry mari [1]), and the Voronoi diagram in the upper half space is nstructed for canonical normal distributions by revealing its ation with the Euclidean Voronoi diagram.

This paper investigates the statistical parametric space of rmal distributions by adopting the Kullback-Leibler divergence a distance to generate the Voronoi diagram. The Kullback-ibler divergence is the most fundamental divergence in infor-ation theory (e.g., see [3, 5]), and similarity of the diagram tained from this divergence with that in [6, 7] is shown. Due relative simplicity, the Kullback-Leibler divergence allows us compute the Voronoi diagram for general normal distribution. nearization technique as well as lower envelope arguments is lly made use of in deriving bounds of this paper. Voronoi di-rams for other divergences in the statistical parametric space probability distribution of a discrete variable taking $d$ values e also touched upon.


## Introduction

omputational geometry mostly is based on Euclidean space and uclidean geometry. Suppose that the Euclidean Space is re-aced Riemannian Space. We consider the Voronoi diagram hich is a main target of computational geometry in the Rie-annian Space.

In this paper specially suppose that there are $n$ points, each oint correspond to a probability distribution, in statistical para-etric space which is regarded as Riemannian space. As we fine Voronoi diagram in the statistical parametric space, we tain the statistical "distance". The "distance" may be the uclidean distance in Euclidean space,

One choice of the statistical "distance" is a metric induced Fisher information matrix [1] We introduced the Voronoi di-gram by the above metric for statistical metric space of nor-al distributions in [6] and high dimensional normal distribution hose variance-covariance matrix is $\sigma^2 I$($I$ is identity matrix) in ].

Another choice of the "distance" is the Kullback-Leibler di-ergence, which is very famous and useful for probability density nctions in information theory. The divergence does not satisfy e symmetric property of metric. However we can regard the vergence as a kind of convex distance function [2]. Based on is idea, we define the Voronoi diagram in this space by the ullback-Leibler divergence (section 2.1). In section 2.2, we an-yze the number of faces of this Voronoi diagram. We obtain the following interesting results: We prove that the number of faces is bounded by $O(n)$ for $n$ point set in the case of a parametric space of 1-dimensional normal distribution. It is bounded by $O(n^{\lfloor (d+1)/2 \rfloor})$ for $n$ point set by the method of linearization in the case of special high dimensional normal distribution. In the later section we deal with more general case. The bound becomes $O(n^d)$ for $n$ point set by the method of linearization in the case that the variance-covariance matrix is diag$[{\sigma_1}^2, \ldots, {\sigma_d}^2]$. Moreover the number of faces is bounded by $O(n^3)$ in the case of statistical parametric space of 2-dimensional normal distribution. These analysis shows that the Voronoi diagram for the Kullback-Leibler divergence is the lower envelope of nice quadratic surfaces which can be regarded as part of lower envelope of hyperplanes.

In this paper, we assume that real RAM for model of computation, which can especially calculate the logarithmic functions in constant time (e.g., see [8]).


*Graduate School of Science and Technology, Kobe University, Kobe, pan. Email: onishi@math.s.kobe-u.ac.jp

†Department of Information Science, University of Tokyo, Japan. mail: imai@is.s.u-tokyo.ac.jp





*Computational Geometry 97* Nice France
Copyright 1997 ACM 0-89791-878-9/97/06 ..$3.50


## 2 Statistical Parametric Space

In this section we discuss about "distance", instead of the Euclidean distance, on statistical parametric space. In fact, the Euclidean distance does not have invariance necessary in statistics. By the "distance" we define our Voronoi diagrams and analyze them. Firstly, we define statistical parametric space.

**Definition 1** *For probability density function $p(x;\xi)$, let $x$ be a random variable. If the probability distribution is determined by $[\xi] = [\xi_1, \ldots, \xi_d]$, then the space of $[\xi]$ is called* statistical parametric space.

In this paper we consider as a "distance" in the space the Kullback-Leibler divergence in information theory.

### 2.1 Kullback-Leibler Divergence

Firstly, we define the Kullback-Leibler divergence.

**Definition 2** [3, 5] *For probability density function $p(x), q(x)$, the Kullback-Leibler divergence $D_K$ from $p(x)$ to $q(x)$ is defined as follows:*

$$D_K(p(x), q(x)) = \int p(x) \log \frac{p(x)}{q(x)} dx.$$

There are the following properties of this divergence.

- $D_K(p(x), q(x)) \geq 0$ (this fact is known also as a Shannon's inequality).
- $D_K(p(x), q(x)) = 0$ iff $p(x) = q(x)$,
- If $p(x) \neq q(x)$, then it is not always realized that $D_K(p(x), q(x)) = D_K(q(x), p(x))$.

Therefore the divergence does not satisfy the symmetry property of distance. However as we consider Voronoi diagrams, we may recall the following observation: "For two points in the statistical parametric space and we are interested in (this is why the divergence is not called a distance) the set of points from which the divergence are equal, i.e. we take no thought of the failure of symmetry property of divergence."

**[Remark]** In other words, the divergence is considered as a convex distance function in [2]. In [2] a convex distance function from generators to any point is treated, but in this paper we consider divergence from any point to generators (reflective version) in a converse manner.

Consequently we can define a Voronoi diagram in the statistical parametric space.

**Definition 3** *Let $P = \{\tilde{p}^{(1)}, \ldots, \tilde{p}^{(n)}\}$ be a point set in statistical parametric space, i.e. each point $\tilde{p}^{(\alpha)}$ represent a probability density function, which point is called* generator. *The* Voronoi region $\mathrm{Vor}(\tilde{p}^{(\alpha)})$ *of $P$ is defined as follows:*

$$\mathrm{Vor}(\tilde{p}^{(\alpha)}) = \{x \in \mathbb{R}^d | D_K(x, \tilde{p}^{(\alpha)}) \leq D_K(x, \tilde{p}^{(\beta)}), \forall \beta \neq \alpha\}, \quad (1)$$

*where $D_K(x, y)$ is the Kullback-Leibler divergence from $x$ to $y$. The whole of Voronoi region for $P$ partions the statistical parametric space, which is called* Voronoi diagram *for the Kullback-Leibler divergence.*

In the later this paper we denote generator by $\tilde{p}$. Thus when we consider the Kullback-Leibler divergence $D_K(p^{(1)}, p^{(2)})$, the first point $p^{(1)}$ is a variable and the second is a generator which is given in the statistical parametric space.

### 2.2 Normal Distribution Case

In this section we consider normal distributions. Firstly, we define normal distribution. The normal distribution is defined by the following probability density function.

$$p(x; \xi) = \frac{1}{\sqrt{2\pi}\sigma} \exp\left\{-\frac{(x-\mu)^2}{2\sigma^2}\right\}, \quad [\xi] = [\mu, \sigma], \sigma > 0, \quad (2)$$

where $x \in \mathbb{R}$ is a random variable, $\mu_i \in \mathbb{R}$ is a mean, $\sigma$ is a variance.

Secondly, we calculate Kullback-Leibler divergence from (2). Suppose $p^{(1)}(x; \mu^{(1)}, \sigma^{(1)}), p^{(2)}(x; \mu^{(2)}, \sigma^{(2)})$ are points in the statistical parametric space. The divergence is expressed as follows:

$$D_K\left(p^{(1)}, p^{(2)}\right) = \log \frac{\sigma^{(2)}}{\sigma^{(1)}} + \frac{\left(\sigma^{(1)}\right)^2 + \left(\mu^{(1)} - \mu^{(2)}\right)^2}{2(\sigma^{(2)})^2} - \frac{1}{2}.$$

By this divergence and **Definition 3**, we can define the Voronoi diagram in statistical parametric space. The point set $(X, Y)$ which has the same divergence to $\tilde{p}^{(1)}(\mu^{(1)}, \sigma^{(1)})$ and $\tilde{p}^{(2)}(x; \mu^{(2)}, \sigma^{(2)})$ can be calculated as follows:

$$\begin{cases} X = \dfrac{\mu^{(1)} + \mu^{(2)}}{2}, & \sigma^{(1)} = \sigma^{(2)}, \\ (X - q)^2 + Y^2 = r, & \sigma^{(1)} \neq \sigma^{(2)}, \end{cases} \quad (3)$$

where $q, r$ are dependent on $\mu^{(i)}, \sigma^{(i)}$ which is called "perpendicular bisector". This curve becomes Voronoi edge in this space.

Comparing the equation (3) and the equation of perpendicular bisector by Fisher information metric [6], we can find some similarity.

- If $\sigma^{(1)} = \sigma^{(2)}$, these becomes the same equation.
- If $\sigma^{(1)} \neq \sigma^{(2)}$, each equation is of type $(X - p)^2 + Y^2 = R^2$ (both of equations become geodesic [6, 7]).

We state the complexity of edge number of this Voronoi diagram.

**Lemma 4** *There exists at most one point which has the same divergence to three points in the statistical parametric space.*

**Proof.** Consider three distinct points in this space. For each pair of points we can get the "perpendicular bisector", but these curves meet at most a point. Because each circle is considered as a part of circle, the circle crosses at two points. However the center of the circle is on $\sigma = 0$ and this space is defined only $\sigma > 0$. Thus at most one intersection is included in this space. □

By this lemma, we can get the complexity of the number of Voronoi edge in the parametric space of normal distribution.

**Theorem 5** *The number of Voronoi edges for $n$ points set is bounded by $O(n)$ in the statistical parametric space of normal distribution.*

**Proof.** Each Voronoi region is connected, the dual structu becomes planar graph. Moreover the Voronoi diagram is a sul division of the plane and the above property is proved. Th from famous theorem of planer graph, the edge number is most $3n - 6$. Since the edge number of dual structure is same the edge number of Voronoi diagram, the number is bounded $O(n)$.

### 2.3 High Dimensional Normal Distribution Case

In this section we consider the parametric space of high dime sional normal distribution. The high dimensional normal distr bution is defined:

$$p(x|\xi^{(\alpha)}) = \frac{1}{\sqrt{2\pi}^d |\Sigma^{(\alpha)}|} \cdot \exp\left\{-\frac{1}{2}(x - \mu^{(\alpha)}) \left(\Sigma^{(\alpha)}\right)^{-1} {}^t(x - \mu^{(\alpha}\right.$$

(

where $x \in \mathbb{R}^d$ is a $d$-dimensional random variable, $\mu_i \in \mathbb{R}^d$ is mean, $\Sigma^{(\alpha)}$ is a $d$-dimensional variance-covariance matrix who $(i, j)$-element is $(\sigma_{ij}^{(\alpha)})^2$ (note that this matrix is symmetric $\alpha = 1, \ldots, n$, and the statistical parametric space

$$\xi = [\mu_1, \ldots, \mu_d, \sigma_{11}, \sigma_{12}, \ldots, \sigma_{1d}, \sigma_{22}, \ldots, \sigma_{2d}, \sigma_{33}, \ldots, \sigma_{dd}]$$

is of dimension $d + d(d+1)/2 = d(d+3)/2$. We consider thr cases by the type of variance-covariance matrix $\Sigma$:

- $\Sigma$ is an identity matrix times constant number,
- $\Sigma$ is a diagonal matrix,
- $\Sigma$ is a symmetric matrix.

#### 2.3.1 Identity Matrix Type

In this case, the Kullback-Leibler divergence from $p^{(1)}(x; \mu_1^{(1)}, \ldots, \mu_d^{(1)}, \sigma^{(1)})$ to $p^{(2)}(x; \mu_1^{(2)}, \ldots, \mu_d^{(2)}, \sigma^{(2)})$ becom as follows:

$$D_K(p^{(1)}, p^{(2)})$$
$$= \frac{1}{2}\sum_{i=1}^{d}\left(\frac{(\mu_i^{(1)})^2 - (\mu_i^{(2)})^2}{\sigma^{(2)}}\right)^2 + \frac{d}{2}\left(\left(\frac{\sigma^{(1)}}{\sigma^{(2)}}\right)^2 - 2\log\frac{\sigma^{(1)}}{\sigma^{(2)}} -\right.$$

Furthermore the "perpendicular bisector" to two points this space is expressed as follows:

$$(X_1 - a_1)^2 + \cdots + (X_d - a_d)^2 + dY^2 = b \quad (a_i, b \in \mathbb{R}).$$

This equation is a kind of ellipse. We can use linearizatio technique in this case.

Now we explain the linearization technique. The linearizatio is a method that the original structure is embedded in high dimension so that it is a structure in the high dimension. E this method, the complexity is bounded by the Voronoi diagra in such high dimension.

Thus we consider the following transform:

$$\mathcal{X}_i := X_i, \quad \mathcal{Y} := {X_1}^2 + \cdots + {X_d}^2 + dY^2.$$

By this transformation, the above Kullback-Leibler divergen becomes linear (a part of $\log \sigma^{(1)}$ is lifting up a same volun for each generator). Moreover above "perpendicular bisecto becomes

$$2a_1\mathcal{X}_1 + \cdots + 2a_d\mathcal{X}_d + \mathcal{Y} = 0.$$

From the upper bound theorem of polytope, we can get the fo lowing theorem.

**Theorem 6** *Suppose high dimensional normal distribution wh probability function is (4) and the variance-covariance matrix $\sigma^2 I_d$. The number of the faces of Voronoi diagram for $n$ poi set in this statistical parametric space is bounded by $O(n^{\lfloor\frac{d+1}{2}\rfloor}$*

### 3.2 Diagonal Matrix Type

ext we consider the probability density function of high dimen-onal normal distribution is expressed as (4).

Secondly we state only the variance-covariance matrix is ex-essed as follows:

$$\Sigma = \operatorname{diag}\left[(\sigma_1)^2, (\sigma_2)^2, \cdots, (\sigma_d)^2\right], \qquad (5)$$

here the parametric space becomes $(2d)$-dimensional space $[\xi] = {}_1, \ldots, \mu_d, \sigma_1, \ldots, \sigma_d] = [X_1, \ldots, X_d, Y_1, \ldots, Y_d]$.

Now calculate the divergence from $p^{(1)}(x; \mu_1^{(1)}, \ldots, {}^{1)}_{d}, \sigma_1^{(1)}, \ldots, \sigma_d^{(1)})$ to $p^{(2)}(x; \mu_1^{(2)}, \ldots, \mu_d^{(2)}, \sigma_1^{(2)}, \ldots, \sigma_d^{(2)})$.

$$D_K(p^{(1)}, p^{(2)}) = \frac{1}{2}\sum_{i=1}^{d}\left\{\frac{(\sigma_i^{(1)})^2}{(\sigma_i^{(2)})^2} + \left(\frac{\mu_i^{(2)} - \mu_i^{(1)}}{\sigma_i^{(2)}}\right)^2 - 2\log\frac{\sigma_i^{(1)}}{\sigma_i^{(2)}} - 1\right\}.$$

By this Kullback-Leibler divergence and **Definition 3**, define e Voronoi diagram in this parametric space. The "perpendic-ar bisector", which becomes a facet in this space, is expressed follows:

$$a_1\left\{(X_1 - b_1)^2 + Y_2^2\right\} + \cdots + a_d\left\{(X_d - b_d)^2 + Y_d^2\right\} = c$$

r $a_i, b_i c \in \mathbb{R}^d$. This surface is a quadratic surface, but we can-ot decide the form of the surface because the form depends on e point. However by linearization we can obtain the complexity faces of Voronoi diagram in this space.

In this case we use the following linearization.

$$\mathcal{X}_i := X_i = \mu_i, \quad \mathcal{Y}_i := {X_i}^2 + {Y_i}^2 = {\mu_i}^2 + {\sigma_i}^2.$$

y this transformation, above Kullback-Leibler divergence be-mes linear (a part of $\log \sigma_i^{(1)}$ is lifting up for each generator). oreover above "perpendicular bisector" becomes

$${}_1\mathcal{Y}_1 - 2a_1b_1\mathcal{X}_1) + \cdots + (a_d\mathcal{Y}_d - 2a_db_d\mathcal{X}_d) = c \quad (a_i, b_i, c \in \mathbb{R}).$$

hus this structure is considered as the geometrical structure in e $(2d)$-dimensional space.

This surface is regarded as a hyperplane in the $(2d)$-dimensional ace. The complexity of number of faces in this Voronoi dia-am is based on the number of faces in the Voronoi diagram $(2d)$-dimensional space. Therefore we can prove the following eorem.

**heorem 7** *Suppose high dimensional normal distribution whose obability function is (4) and the variance-covariance matrix is ). The number of faces of Voronoi diagram for n point set in is statistical parametric space is bounded by $O(n^d)$.*

### 3.3 Symmetric Matrix Type (2-dimensional)

he later of this section we discuss about general 2 dimensional rmal distribution whose variance-covariance matrix $\Sigma$ is not stricted and is given by $\begin{pmatrix} (\sigma_{11})^2 & (\sigma_{12})^2 \\ (\sigma_{12})^2 & (\sigma_{22})^2 \end{pmatrix}$. In this case, the rametric space has 5 dimensional. The parametric space is ven by

$$[\xi] = [\mu_1, \mu_2, \sigma_{11}, \sigma_{12}, \sigma_{22}].$$

Firstly the calculate Kullback-Leibler divergence from ${}_1^{(1)}, \mu_2^{(1)}, \sigma_{11}^{(1)}, \sigma_{12}^{(1)}, \sigma_{22}^{(1)})$ to $(\mu_1^{(2)}, \mu_2^{(2)}, \sigma_{11}^{(2)}, \sigma_{12}^{(2)}, \sigma_{22}^{(2)})$, where the riance-covariance matrix is expressed as

$$\Sigma = \begin{pmatrix} (\sigma_{11}^{(1)})^2 & (\sigma_{12}^{(1)})^2 \\ (\sigma_{12}^{(1)})^2 & (\sigma_{22}^{(1)})^2 \end{pmatrix}, \quad \Sigma' = \begin{pmatrix} (\sigma_{11}^{(2)})^2 & (\sigma_{12}^{(2)})^2 \\ (\sigma_{12}^{(2)})^2 & (\sigma_{22}^{(2)})^2 \end{pmatrix}.$$

The Kullback-Leibler divergence is expressed as follows:

$$D_K(p^{(1)}, p^{(2)}) = \frac{1}{2}\log\frac{|\Sigma'|}{|\Sigma|} + \frac{(\sigma_{22}^{(2)})^2}{2|\Sigma'|}\left\{(\sigma_{11}^{(1)})^2 + (\mu_1^{(1)} - \mu_1^{(2)})^2\right\}$$
$$+\frac{(\sigma_{11}^{(2)})^2}{2|\Sigma'|}\left\{(\sigma_{22}^{(1)})^2 + (\mu_2^{(1)} - \mu_2^{(2)})^2\right\}$$
$$-\frac{(\sigma_{12}^{(2)})^2}{|\Sigma'|}\left\{(\sigma_{12}^{(1)})^2 + (\mu_1^{(1)} - \mu_1^{(2)})(\mu_2^{(1)} - \mu_2^{(2)})\right\} - 1$$

The point set $(X_1, X_2, Y_{11}, Y_{12}, Y_{22})$ that has the same diver-gence to two points is represented by

$$\sum_{i=1}^{2}\left\{a_i((X_i)^2 + (Y_{ii})^2) + b_iX_i\right\} + c_1(X_1X_2 + (Y_{12})^2) + c_2 = 0,$$

where $a_i, b_i, c_i \in \mathbb{R}$.

The point set is expressed above but by linearization method we can state the complexity of Voronoi diagram. The following is transformation for linearization.

$$\mathcal{X}_i^1 := X_i, \mathcal{X}_i^2 := (X_i)^2 + (Y_{ii})^2, \mathcal{X}_{12} = X_1X_2, \mathcal{Y}_{ij} := Y_{ij}.$$

By this linearization the whole of the structure is embedded in 6-dimensional space and above "perpendicular bisector" can be transformed.

$$\sum_{i=1}^{2}\left\{a_i\mathcal{X}_i^2 + b_i\mathcal{X}_i^1\right\} + c_1(\mathcal{X}_{12} + \mathcal{Y}_{12}) + c_2 = 0 \quad (a_i, b_i, c_i \in \mathbb{R})$$

Thus since the structure is embedded in the 6 dimensional space, we can get next complexity.

**Theorem 8** *The face number of Voronoi diagram for n point set in the statistical parametric space of general 2 dimensional normal distribution is bounded by $O(n^3)$.*

## 3 Conclusion and Remarks

In this paper, we have analyzed the Voronoi diagram in the statistical parametric space of normal distribution.

It is possible that lower envelope of a kind of surface regardes as Voronoi diagram. For example in normal distribution we only consider the following function for each generator.

$$(X - \mu)^2 + Y^2 - (\sigma)^2 \log Y = (\sigma)^2(1 + Z - \log\sigma) \quad (Z > 0)$$

This function becomes 0 at $(\mu, \sigma)$ and if increasing $Z$, then the region is increasing. As the lower envelope of these $n$ functions project in the original space, we can get the Voronoi diagram of parametric space of normal distribution. From above fact we use the theorem of [4], then we can also bound the complexity of the edge number.


## Acknowledgment

The work of the second author was supported in part by the Grant-in-Aid of the Ministry of Education, Japan.


## References


[1] S. Amari *Differential Geometrical Methods in Statistics*, Lecture Notes in Statistics. 28, Springer-Verlag, New York, 1985.

[2] L. P. Chew and R. L. (Scote) Drysdale. Voronoi Diagrams Based on Convex Distance Function. *Proc. of ACM Computational geometry*, Baltimore, Maryland, pp. 235-244, 1985.

[3] T. M. Cover and J. A.Thomas. *Elements of Information Theory.* Wiley-Interscience Publication, New York, 1991.

[4] H. Edelsbrunner. *Algorithms in Combinatorial Geometry.* Springer-Verlag, Berlin, 1987.

[5] T. Han and K. Kobayashi. *Mathematical Information and Code* (in Japanese). Iwanami Publisher, Japan, 1994.

[6] K. Onishi and N. Takayama. Construction of Voronoi Diagram on the Upper Half-plane. *IEICE Trans. on Fundamentals,* Vol.E79-A, No.4, pp. 533-539, 1996.

[7] K. Onishi. Voronoi Diagram in the $d$-dimensional Hyperbolic Space. *Proc. of the International Symposium on Combinatorics and Applications,* Nankai University, Tianjin, pp. 314-324, 1996.

[8] F. P. Preparata and M. I. Shamos. *Computational Geometory.* Springer-Verlag, New York, 1985.

# Computing the Minimum Diameter for Moving Points: An Exact Implementation Using Parametric Search*

Jörg Schwerdt† Michiel Smid† Stefan Schirra‡

## 1 Introduction

Parametric Search, developed by Megiddo [8], is a powerful algorithmic technique that can be used to solve a large variety of geometric optimization problems, see e.g. [1]. Although this technique is ingenious, it is, in general, hard to implement. In this paper, we report on the implementation of (a practical variant of) an algorithm that is based on parametric search, and that is due to Gupta *et al.* [7]. As far as we know, this is the first implementation of a parametric search algorithm.

The algorithm of [7] solves the following problem. We are given a set of $n$ points in the plane that are moving at constant but possibly different velocities. The diameter of the points at time $t$ is the largest Euclidean distance among all pairs of points at time $t$. Our goal is to compute the time $t^*$ at which the diameter is minimum.

This problem can be solved trivially as follows. For each pair of points, the square of their distance defines a quadratic function in $t$, i.e., a parabola. The minimum diameter is obtained by finding the lowest point on the upper envelope of these $\binom{n}{2}$ parabolas. This gives an algorithm with running time $O(n^2 \log n)$, and using $O(n^2)$ space. It was shown in [7], that the problem can be solved, using parametric search, in $O(n \log^3 n)$ time, using $O(n)$ space. The latter algorithm is complicated and not practical; it uses e.g. an optimal parallel sorting algorithm. Therefore, we have implemented a variant of it, which has running time $O(n \log^6 n)$.

We implemented this algorithm in C++ using LEDA [9, 10] and CGAL [5, 6]. In our first implementation we used the number type `double` in all numerical computations. For many problem instances the implementation could not find the minimum diameter. During parametric search an interval containing $t^*$, the time at which the diameter is minimum, is maintained and repeatedly made smaller until it is finally shrunken to $t^*$. For many problem instances the interval does not collapse to a single point due to rounding errors and a fairly small interval was reported in the end. We got rid of these precision problems by using the number type `real` [3, 4] from LEDA [9] in the numerical computations, cf. Section 4.

Our implementation of the algorithm of Gupta et al. [7] shows that parametric search is not at all a purely theoretical and totally impractical technique in algorithm design in general and computational geometry in particular.

---

* This paper is based on the first author's Master's Thesis [12].

† Fakultät für Informatik, Otto-von-Guericke-Universität Magdeburg, Universitätsplatz 2, D-39106 Magdeburg, Germany. E-mail: {schwerdt,michiel}@cs.uni-magdeburg.de.

‡ Max-Planck-Institut für Informatik, Im Stadtwald, D-66123 Saarbrücken, Germany. E-mail: stschirr@mpi-sb.mpg.de. This work was (partially) supported by the ESPRIT IV LTR Project No. 21957 (CGAL).



*Computational Geometry 97* Nice France



## 2 Applying Parametric Search

We assume that the reader is familiar with the parametric search technique, see [8, 1, 7]. Recall that $t^*$ denotes the time at which the diameter of the moving points is minimum. In order to apply parametric search, we need a sequential algorithm $A$ and a parallel algorithm $B$, each of which decides if a given $t \in \mathbb{R}$ is smaller, larger, or equal to $t^*$.

Let $Z_{pq}(t)$ denote the square of the Euclidean distance between the moving points $p$ and $q$. Note that $Z_{pq}(t)$ is a parabola. Let $t^*_{pq}$ be the time at which $p$ and $q$ are closest. Finally, let $D(t)$ denote the diameter at time $t$.

**Lemma 2.1** *We have $t < t^*$ iff $t < t^*_{pq}$ for all points $p$ and $q$ such that $Z_{pq}(t) = D^2(t)$. A similar claim can be used to decide if $t > t^*$.*

Based on this lemma, algorithms $A$ and $B$ do the following. They first compute the positions of the points at time $t$. Then they compute all diametral pairs, i.e., all pairs $p, q$ such that $Z_{pq}(t) = D^2(t)$. Finally, for each such pair, they check if $t < t^*_{pq}$, $t > t^*_{pq}$, or $t = t^*_{pq}$. Algorithm $A$ can be implemented in $T_A(n) = O(n \log n)$ time, whereas algorithm $B$ can be implemented on a PRAM such that it takes $T_B(n) = O(\log n)$ time using $P(n) = n$ processors.

Both $A$ and $B$ perform computation and comparison steps. Each computation step only involves the basic operations $+, -, *$, and $/$, whereas each comparison step involves determining the sign of a polynomial at the value $t$. The coefficients of such a polynomial depend on the moving points only, and, as we will see later, its degree is at most three.

The parametric search algorithm computes $t^*$ by sequentially simulating the parallel algorithm $B$ on the unknown value $t^*$. Comparisons are resolved by running algorithm $A$ on the roots of the corresponding polynomials. As is shown in [8], the total time for finding $t^*$ is bounded by $O(P(n)T_B(n) + T_A(n)T_B(n) \log P(n)) = O(n \log^3 n)$.

## 3 The Implementation

The implementation of the sequential algorithm $A$ does the following. First, it calls a LEDA routine to sort the points—at a given time $t$—by their $x$-coordinates. Then the convex hull of the points is constructed by running Andrew's modification of Graham's Scan (see [11]) on the sorted sequence. Next, the rotating calipers algorithm is used to compute all diametral pairs of the convex hull. Finally, for each diametral pair $p, q$, it is decided if $t < t^*_{pq}$, $t > t^*_{pq}$, or $t = t^*_{pq}$. The entire algorithm has running time $O(n \log n)$.

Since the optimal parallel algorithm $B$ is complicated, we decided to implement a simpler, and slower, variant. This algorithm does the following.

Using parallel merge-sort in a bottom-up fashion, the points—at time $t$—are sorted by their $x$-coordinates. This algorithm consists of $O(\log n)$ stages. At the beginning of stage $i$, we have $n/2^i$ sorted sequences, each of length $2^i$. For each element $x$, there is one processor that uses binary search to find the rank of $x$ in the neighboring sequence. Given these ranks, we can merge pairs of neighboring sequences. The entire sort algorithm takes $O(\log^2 n)$ time, using $n$ processors. The polynomials that occur during comparison steps have degree one.

The convex hull of the sorted sequence is computed by parallel bottom-up merging. Again, this algorithm consists of $O(\log n)$ stages. At the beginning of stage $i$, we have the convex hulls of the leftmost $2^i$ points, the next $2^i$ points, and so on. For each pair of neighboring convex hulls, there is one processor that computes their upper and lower tangents using binary search. For resolving a comparison, nine cases have to be distinguished (see [11, Figure 3.21]), eight of which lead to a polynomial in $t$ of degree two. Unfortunately, one case (concave–concave in [11, Figure 3.21]) leads to a polynomial of degree three. In this case, we do not proceed as in [11], but change the tangent finding algorithm, thereby slowing it down (see [12] for details). As a result, one stage may take $O(\log^3 n)$ parallel time. The entire parallel convex hull algorithm takes $O(\log^4 n)$ time using $n$ processors. All polynomials that occur have degree two.

Given the convex hull, all diametral pairs are obtained as follows. For each hull edge $e = (p, q)$, there is one processor that uses binary search to find the at most two points $a$ and $b$ that are at maximum distance from the line through $e$. (The distances of the hull vertices to this line form a uni-modal sequence.) Then the four pairs $(p, a)$, $(p, b)$, $(q, a)$, and $(q, b)$ are inserted into appropriate positions of an array. All this takes $O(\log n)$ time using $n$ processors. The polynomials that occur have degree two.

At this point, we know that the array contains all diametral pairs. By playing a tournament in parallel, level by level, these pairs are found in another $O(\log n)$ time using $n$ processors. Again, the polynomials that are involved have degree two. Finally, given all diametral pairs, it can be decided in $O(1)$ time using $n$ processors, whether $t < t^*$, $t > t^*$, or $t = t^*$. The comparisons in this final step do not depend on $t^*$ and, hence, no polynomials are involved.

The entire algorithm $B$ takes $O(\log^4 n)$ time and uses $n$ processors. Moreover, the polynomials that arise at comparison steps all have degree at most two.

The sequential simulation of $B$ on the value $t^*$ is straightforward, and the total time for finding $t^*$ is bounded by $O(n \log^6 n)$.

## 4 Computing Exact Results

Due to precision problems our implementation with `doubles` often could not find $t^*$, but reported an interval only. Fig. 1 gives an example. The reported interval might be a sufficiently good approximation to $t^*$ and in our examples it usually was. However, when we got the result of the implementation with `double` arithmetic we had no idea how good an approximation we had got. Even worse, we did not really know whether the fact that we got an interval and not a single time was caused by precision problems or a bug in the implementation.

| | |
|---|---|
| *exact*: | $-\frac{507777}{587485} + \frac{1}{1174970}\sqrt{1239667408036}$ |
| `double`: | ( 0.0832666379676957336. . . , 0.0832783171914589365. . . ) |
| `real`: | 0.0832783171914590203. . . |

Figure 1: Results for $t^*$.

Replacing double arithmetic by exact rational arithmetic is not sufficient, because quadratic irrationals, real numbers of the form $A + B\sqrt{C}$, where $A$, $B$, and $C$ are rational, arise in the computations (as roots of the polynomials). Fortunately the number type `real` [3, 4] in LEDA [9] provides exact computation for arithmetic expressions involving $+, -, \cdot, /$ and $\sqrt{\ }$. To be more precise, it provides exact comparisons of `reals` and hence the use of `reals` ensures correct control flow in the implementation: All decisions made in branching steps are correct, i.e., as if the computation would have been done with infinite precision.

In C++-code `reals` are very easy to use, they can be used exactly like `doubles` [3]. The switch from `double` to `real` in our code was simplified by the use of a preliminary version of the CGAL kernel [6], which provides geometric objects, especially points, parametrized by number types.

In addition, we wrote a number type which maintains quadratic irrationals symbolically in the form $A + B\sqrt{C}$ using LEDA `rationals` to represent $A$, $B$, and $C$. In the implementation all arithmetic operations are over elements from the same real quadratic field. In addition to arithmetic operations in a real quadratic field we had to provide a comparsion operation for quadratic irrationals from different extension fields. This comparison is made by repeated squaring.

## 5 Running Times

In our implementation we used the newest version of the LEDA number type `real` which incorporates recent results on separation bounds [2]. The old version was much too slow. For the old version it would have been necessary to provide reasonable separation bounds calculated by hand. The new version automatically finds such good bounds.

Fig. 2 shows ranges of running times in seconds of our `double` and `real` version for "random" point sets of size $n \in \{500, \ldots, 3000\}$ on a SUN SPARCstation 4. In the examples a moving point $p$ was created by randomly choosing four integers $x, y, v_x, v_y$ in the range $[-500, 500]$. The position of point $p$ at time $t$ is then $p(t) = (x, y) + t\,(v_x, v_y)$. For each value of $n$ we created 10 different example point sets. It turned out that the exact `real` version is about 6 times

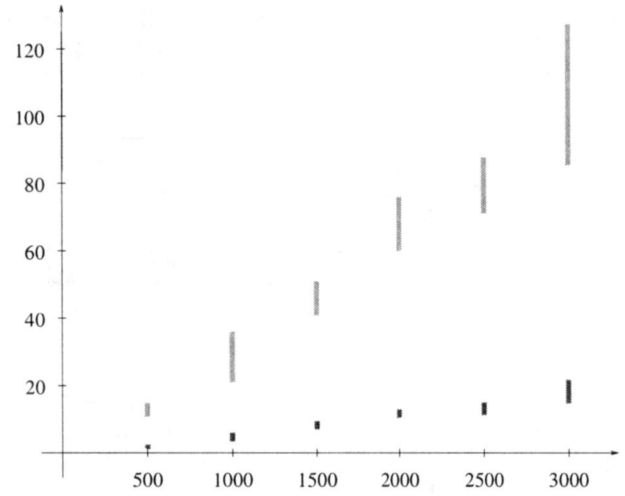


Figure 2: Running times in seconds for **real** version ▮ and **double** version ▮ .

slower than the inherently inexact **double** version, which returned time $t^*$ in less than 20% of our test examples. The implementation with symbolic quadratic irrationals is about 40 times slower than the double version.

## References


[1] P.K. Agarwal, M. Sharir and S. Toledo. *Applications of parametric searching in geometric optimization.* J. Algorithms **17** (1994), pp. 292–318.

[2] C. Burnikel, R. Fleischer, K. Mehlhorn, and S. Schirra. A strong and easily computable separation bound for arithmetic expressions involving square roots. In *Proceedings of the 8th ACM-SIAM Symposium on Discrete Algorithms*, 1997, pp. 702–709.

[3] C. Burnikel, J. Könemann, K. Mehlhorn, S. Näher, S. Schirra, and C. Uhrig. Exact geometric computation in LEDA. In *Proceedings of the 11th ACM Symposium on Computational Geometry*, 1995, pp. C18–C19.

[4] C. Burnikel, K. Mehlhorn, and S. Schirra. *The LEDA class real number.* Technical Report MPI-I-96-1-001, Max-Planck-Institut für Informatik, Saarbrücken, January 1996.

[5] CGAL Project. cf. **http://www.cs.ruu.nl/CGAL/**

[6] A. Fabri, G.-J. Giezeman, L. Kettner, S. Schirra, and S. Schönherr. The CGAL kernel: A basis for geometric computation. In M.C. Lin and D. Manocha, editors, *Applied Computational Geometry: Towards Geometric Engineering,* Proceedings First ACM Workshop on Applied Computational Geometry, 1996, Springer LNCS 1148, pp. 191–202.

[7] P. Gupta, R. Janardan and M. Smid. *Fast algorithms for collision and proximity problems involving moving geometric objects.* Computational Geometry: Theory and Applications **6** (1996), pp. 371-391.

[8] N. Megiddo. *Applying parallel computation algorithms in the design of serial algorithms.* Journal of the ACM **30** (1983), pp. 852–865.

[9] K. Mehlhorn, S. Näher, and C. Uhrig. *The LEDA User Manual*;
cf. **http://www.mpi-sb.mpg.de/LEDA/leda.html**

[10] K. Mehlhorn and S. Näher. *LEDA: A platform for combinatorial and geometric computing.* Communications of the ACM **38** (1995), pp. 96–102.

[11] F.P. Preparata and M.I. Shamos. *Computational Geometry, an Introduction.* Springer–Verlag, 1985.

[12] J. Schwerdt. *Das Diameterproblem einer bewegten Punktemenge: eine Implementierung mit Hilfe von Parametric Search.* Master's Thesis (Diplomarbeit), Universität des Saarlandes, Saarbrücken, 1996.

# Sweeping Lines and Line Segments with a Heap


Julien Basch  Leonidas J. Guibas*  G.D. Ramkumar

Computer Science Department
Stanford University
Stanford, CA 94305
{jbasch,guibas}@cs.stanford.edu

Research & Development Division
Hitachi America Ltd.
3101, Tasman Drive, MS: 120
Santa Clara, CA 95054
ramkumar@hitachi.com



## Abstract

Given $n$ line segments in the plane, the Bentley-Ottmann sweep maintains the exact ordering of the intersections of the segments with a vertical line, as this line sweeps the plane from left to right. To accomplish this, every intersection between two segments must be processed, and the running time of the sweep can be $\Omega(n^2)$. In this paper, it is shown how a heap on the intersections can be maintained during the sweep. This new type of sweep processes $O(n \log^2 n)$ intersections when sweeping over lines and $O(n\sqrt{n \log n})$ intersections when sweeping over line segments. A lower bound of $\Omega(n \log n)$ is also established.


## 1 Introduction

One of the common introductory problems in geometric algorithms is that of finding all pairwise intersections in a family of line segments in the plane. The first non-trivial solution to this problem was given by Bentley and Ottmann [BO79], who introduced in 1979 the now familiar line sweep paradigm. A vertical line is moved ('swept') from the far left to the far right of the plane, and the exact order of the intersections with the segments is maintained as the line sweeps along. All vertices in the arrangement of segments are detected in this process.

From the point of view of the sweep line, the segments can be viewed as points which move up and down the line. The Bentley-Ottmann algorithm requires that we maintain the full sorted list of these points along the line and update it as new points are inserted (segment appearance), old ones are deleted (segment disappearance), and adjacent segments interchange position (intersection). In the same vein, we may be interested in maintaining other order-related information of these points along the sweep line, for example the topmost point, the $k$-th point from the top, etc. Since these variants do not require full sorting of the points during the sweep line, we may hope for an algorithm more efficient than a normal sweep. Indeed, if we just want to know the topmost point at all times, we can merely compute the upper envelope of the line segments, and it is known how to do this in $O(n \log n)$ time for an input of size $n$ [Her89], irrespective of how many intersections the segments may have.

But can we actually compute the upper envelope of the segments using a line sweep method? One possibility is to put the segments intersecting the sweep line in a binary heap, where the priority of such a segment is the vertical coordinate of the intersection with the sweep line. At any given time, the validity of the heap structure can be guaranteed by one comparison per edge in the heap. The set of comparisons is called the *certificate list* of the structure. As the priorities of the heap elements change continuously with time, a comparison will fail exactly when the two segments involved intersect. For each edge of the heap, this time can be computed and put into an event queue, and the sweep can proceed from event to event exactly as in the case of the Bentley-Ottman sweep. The processing of an event requires a swap between the two intersecting segments in the heap. It also requires the descheduling and rescheduling of at most four events corresponding to the edges affected by the swap (Figure 1). A segment insertion in the sweep line is processed by inserting the segment into the heap, and rescheduling all events corresponding to edges disturbed by this operation.

What we have described here is a *kinetic heap*, an example of a *kinetic data structure* (a data structure for continuously changing data) as introduced by Basch, Guibas, and Hershberger [BGH97]. The kinetic heap is a data structure for maintaining the maximum of a collection of continuously changing numbers, where insertion of a new number into the collection, and dele-


*Supported in part by National Science Foundation grant CCR-9623851 and by US Army MURI grant DAAH04-96-1-0007.

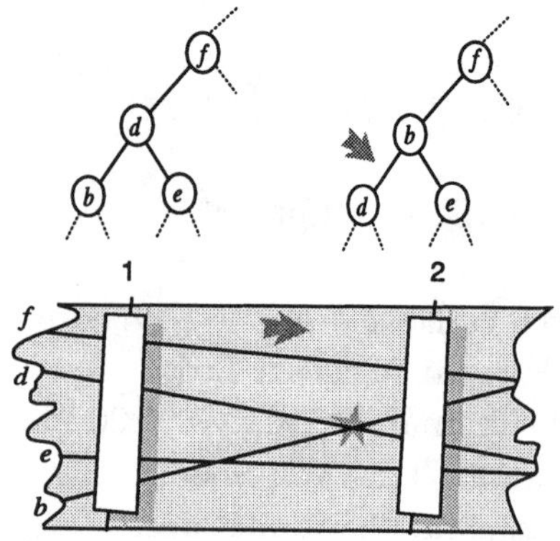


Figure 1: At the first position sweep line this figure, segments $b$ and $e$ are not in the same subtree, hence their intersection is not scheduled as an event. The first intersection to be scheduled is $bd$. At this point, the heap property is maintained by a swap between $b$ and $d$. The events corresponding to the four edges around this pair are descheduled before the swap and rescheduled with the new identities of the lines involved after the swap

tion of an old number from it, are also allowed. Two other data structures have been proposed for the same purpose: the *heater* [BGR96], and the *kinetic tournament* [BGH97].

We measure the performance of a kinetic data structure by the number of events that it needs to process. On a set of $n$ segments in the plane, the heater is a randomized structure that processes $O(n\alpha(n)\log n)$ events in expectation, while the kinetic tournament achieves the same bounds in the worst case.

The kinetic heap is arguably the most natural kinetic structure for maintaining the maximum. Its analysis, however, appears to be much less natural than that of its competitors. The primary reason for this state of affairs is that there is a complicated dependence between the exact structure of the heap at a given time and the prior history of the sweep. Nevertheless, we can show that if the sweep is done over entire lines (or segments with the same $x$-projection), then the number of events is $O(n\log^2 n)$ in the worst case. If the sweep is over line segments, we can only prove the weaker upper bound of $O(n\sqrt{n\log n})$. The best lower bound we can prove is $\Omega(n\log n)$ in both cases.

## 2 Notation

In what follows, we say that a vertex is *red* if it is processed as an event during the sweep, and *green* otherwise. Thus, a vertex is red if the two segments that define it are parent and child in the heap at the time they intersect. We are interested in bounding the number of red vertices. For a time $t$, corresponding to a position of the sweep line, we denote by $t^-$ and $t^+$ the times immediately before and after $t$.

We assume that the lines or segments are in general position, so that no two events (intersections or endpoints) occur at the same time, and no three lines or segments intersect at a common point. We write $a <_t b$ to mean that the line or segment $a$ is below $b$ at time $t$.

## 3 Sweeping over lines

We now consider a binary heap sweeping over a set of $n$ infinite lines. The heap structure remains fixed during the sweep, even though the node contents change over time. The *level* of a line $a$ at time $t$, denoted by $\lambda_t(a)$, is the distance from the node containing $a$ at $t$ to the bottom level. Thus the level of a line is an integer between 1 and $\lceil \lg n \rceil$. We denote by $\hat\lambda_t(a)$ the highest level ever attained in the heap by line $a$, from the initial time until time $t$.

**Lemma 3.1** *If $ab$ is a red vertex with $b <_{t^-} a$, then $\hat\lambda_t(a) \geq \hat\lambda_t(b) + 1$.*

**Proof:** Consider the last time $\tau$ before $t$ when $b$ was at its maximum level $\hat\lambda_t(b)$. At time $\tau$, line $b$ was in some node $\nu$ of of the heap and from then on $b$ only moved within the subtree $U$ rooted at $\nu$. If $b$ is in $\nu$ at $t^-$, then $a$ is in the parent of $\nu$ at that time, and the claim trivially holds.

Otherwise, as $b <_\tau a$, line $a$ could not be in $U$ at $\tau$. Therefore, the path followed by $a$ in the heap from $\tau$ to $t^-$ goes from the complement of $U$ into $U$, and thus passes through the parent of $\nu$. The conclusion follows. □

**Theorem 3.2** *The number of red vertices encountered during the maintenance of a standard binary heap in a sweep over $n$ lines is $O(n\log^2 n)$.*

**Proof:** We use a potential function argument. For a line $a$ in the heap, define its potential $\phi_t(a)$ at time $t$ to be

$$\phi_t(a) = \hat\lambda_t(a)(\lambda_t(a) - \hat\lambda_t(a)).$$

The potential of the entire heap is the sum of the potentials of all the lines. Note that this potential is 0 before the start of the sweep (for every $a$, $\lambda_t(a) = \hat\lambda_t(a)$); it is non-positive for all $t$ (because $\lambda_t(a) \leq \hat\lambda_t(a)$) and at the end of the sweep its absolute value is at most $O(n\log^2 n)$. We now show that the occurrence of a red vertex decreases the total potential by at least 1.

Consider a swap at time $t$ between a parent $a$ and a child $b$ in the heap. The potential of any line other than $a$ or $b$ doesn't change. Also, the quantity $\hat\lambda_{t^-}(a)$ doesn't change, so that the potential change for $a$ is $\phi_{t^+}(a) - \phi_{t^-}(a) = -\hat\lambda_{t^-}(a)$.

There are two cases for $b$: either it breaks its current $\hat\lambda$ record (i.e. $\lambda_{t^-}(b) = \hat\lambda_{t^-}(b)$), or it does not. In the first case, $\hat\lambda_{t^-}(b) = \lambda_{t^-}(b)$ and $\hat\lambda_{t^+}(b) = \lambda_{t^+}(b)$, so $\phi_{t^-}(b) = \phi_{t^+}(b) = 0$, and $b$'s potential doesn't change. The decrease in potential is therefore $\hat\lambda_{t^-}(a)$, which is at least 1 because $a$ is not on the bottom level before the swap.

In the second case, the potential of $b$ increases by $\hat{\lambda}_{t^-}(b)$, and the net potential change is $\hat{\lambda}_{t^-}(b) - \hat{\lambda}_{t^-}(a)$, which is at most $-1$ by Lemma 3.1. □

## 4 Sweeping over line segments

The analysis used in the previous section completely breaks down in the case of line segments, as Lemma 3.1 no longer hold. Instead, we use another potential argument reminiscent of the one used independently by several authors [ELSS73, Gus79, EW85] for proving upper bounds on the $k$-level of an arrangement of lines.

**Lemma 4.1** *The number of red vertices processed during the maintenance of a standard binary heap in a sweep over $n$ line segments is $O(n\sqrt{n \log n})$.*

**Proof:** Let us order the $n$ segments by increasing slope, and denote by $r(s)$ the rank of a segment $s$ in this ordering. Let $H(\nu)$ be the subtree rooted at a node $\nu$, and let $r_t(\nu)$ be the rank of the segment in node $\nu$ at time $t$.

For a node $\nu$ in the heap, define its potential $\phi_t(\nu)$ at time $t$ to be:

$$\phi_t(\nu) = \sum_{\mu \in H(\nu)} r_t(\mu).$$

In other words, the potential of a node is the sum of the ranks of all segments in the subtree rooted at that node. The potential of the entire heap is the sum of the potentials of all the nodes of the heap. It is 0 at both ends of the sweep (when the heap is empty).

If two nodes $\mu, \nu$ are parent and child, a swap of their contents at $t$ changes the potential by $r_{t^-}(\mu) - r_{t^-}(\nu)$, which is at most $n$. In a binary heap, a segment insertion/deletion is implemented by $O(\log n)$ swaps and an insertion/deletion at the bottom of the heap. This sequence of operation change the potential by $O(n \log n)$. Hence, during the entire sweep, the total potential *increase* due to endpoints is $O(n^2 \log n)$.

Let $R$ be the set of red vertices. When $v \in S$ is processed, the potential can only decrease, as it is the higher ranked segment that goes up one level in the heap, and it will always remain non-negative. Hence, if we denote by $\delta(v)$ the absolute difference in rank between the two segments that define a vertex $v$, we have:

$$\sum_{v \in R} \delta(v) \leq O(n^2 \log n).$$

Therefore, for any fixed $B$, the number of vertices $v \in S$ with $\delta(v) \geq B$ is $O(\frac{n^2 \log n}{B})$. Moreover, there are at most $nB$ vertices with $\delta(v) \leq B$. Choosing appropriately $B = \sqrt{n \log n}$ gives the desired bound. □

## 5 Lower bound

We describe an arrangement of $n$ lines where the number of red vertices seen while sweeping with a heap is $\Omega(n \log n)$. Place each line so that it is tangent to an upward facing parabola. Each line appears on the upper envelope exactly once. In particular, a line that starts at the bottom of the heap has to appear at the root at the time it is on the upper envelope. This cannot be done in less than $\log n$ swaps. As there are $n/2$ lines that start at the bottom of the heap. As a swap increases the level of only one line at a time, the total number of swaps generated by this arrangement is $\Omega(n \log n)$.

## 6 Conclusion

In this paper, we established bounds on the number of events that need to be processed to maintain a heap while sweeping over an arrangement of lines and line segments. As the kinetic heap is the most natural structure to maintain the extremum of a set of points moving along a line, we would like to see the bounds tightened and extended to the case of algebraic curves or arcs of low degree.

## References


[BGH97] J. Basch, L. J. Guibas, and J. Hershberger. Data structures for mobile data. In *8th Symposium on Discrete Algorithms*, pages 747–756, 1997.

[BGR96] J. Basch, L.J. Guibas, and G.D. Ramkumar. Reporting red-blue intersections between connected sets of line segments. In J. Diaz and M. Serna, editors, *Algorithms — ESA '96*, LNCS 1136, pages 302–319, sep 1996.

[BO79] J. L. Bentley and T. A. Ottmann. Algorithms for reporting and counting geometric intersections. *IEEE Trans. Comput.*, C-28:643–647, 1979.

[ELSS73] P. Erdős, L. Lovász, A. Simmons, and E. Straus. Dissection graphs of planar point sets. In J. N. Srivastava, editor, *A Survey of Combinatorial Theory*, pages 139–154. North-Holland, Amsterdam, Netherlands, 1973.

[EW85] H. Edelsbrunner and E. Welzl. On the number of line separations of a finite set in the plane. *J. Combin. Theory Ser. A*, pages 15–29, 1985.

[Gus79] D. Gusfield. Bounds for the parametric spanning tree problem. In *Proc. Humbolt Conf. on Graph Theory, Combinatorics and Computing*, pages 173–183, 1979.

[Her89] J. Hershberger. Finding the upper envelope of $n$ line segments in $O(n \log n)$ time. *Inform. Process. Lett.*, 33:169–174, 1989.

# 6th Annual Video Review of Computational Geometry

The following notes give brief descriptions of the video segments that make up the accompanying video proceedings. As in previous years, the videos use images to enhance our understanding of geometric ideas. In some cases this involves the illustration of algorithms in action. In others, we see applications of geometric algorithms in other disciplines. As the field matures, we see the videos serving as valuable adjuncts in each of these functions.

The seven video segments were selected by a video program committee meeting in Vancouver on February 28th, 1997. The segments were judged based on their clarity, quality, and contribution to the field. Some accepted videos have been revised to reflect the comments of the committee.

We thank the members of the Video Program Committee for their help in evaluating the entries. The members of the committee were Alain Fournier (University of British Columbia), Mark Keil (University of Saskatchewan), David Salesin (University of Washington), Thomas C. Shermer (Simon Fraser University), and Jack Snoeyink (University of British Columbia). We thank the Computer Science Department at the University of British Columbia for providing support in the creation of the final video.

Jack Snoeyink
Video Review Chair

# The Bisector Surface of Freeform Rational Space Curves *

*Gershon Elber*
Department of Computer Science
Technion, Israel Institute of Technology
Haifa 32000, Israel
Email: gershon@cs.technion.ac.il

*Myung-Soo Kim*
Department of Computer Science
POSTECH
Pohang 790-784, South Korea
Email: mskim@vision.postech.ac.kr

**Abstract**

Given a point and a rational curve in the plane, their bisector curve is rational [3]. However, the bisector of two rational curves in the plane is not rational, in general [4]. Given a point and a rational *space* curve, the *bisector surface* is a rational ruled surface. Moreover, given two rational space curves, the bisector surface is rational (except for the degenerate case in which the two curves are coplanar). The last two cases are demonstrated in this video.

**Key Words:** Bisector, Voronoi surface, Medial surface, skeleton.

Given two objects in the plane or space, their bisector curve or surface is defined as the set of points which are equidistant from the two objects. While the bisector curve or surface can never self intersect, in this work, we consider rational representations of curves and surfaces that *contain* the exact bisector curve and/or surface. There are some redundant surface regions in the rational form that is presented that self intersect and hence must be eliminated. For brevity reasons, we denote in this work, those rational forms as the bisector curves and/or surfaces. The medial axis and medial surface are also closely related to the bisector curve and surface; that is, given an object in the plane or space, the medial axis or surface is defined as the set of interior points of the object which have minimum distance from at least two different points on the boundary of the object.

Bisector surfaces and medial surfaces have many important applications in engineering [1, 5, 6, 8]. (See Sherbrooke et al. [8] for a detailed survey.) However, their construction is non-trivial except for some special cases. Dutta and Hoffmann [1] considered the bisectors for simple surfaces such as natural quadrics and toroidal surfaces. For these special types of surfaces, the bisector surfaces have simple closed-form representations. For general algebraic surfaces, Hoffmann et al. [5, 6] formulated the bisector surface (called the Voronoi surface) using a simultaneous system of non-linear polynomial equations. The solution scheme is based on the *dimensionality paradigm* [5], which requires a preprocessing step that determines the global topological structure of the solution space.

Approximation of bisectors is computationally quite expensive. Therefore, it is desirable to classify special cases in which the bisectors have exact representations or simple approximations. Farouki and Johnstone [3, 4] investigated the bisector problem for planar rational curves. Given a point and a rational curve in the plane, the bisector curve is rational [3]. However, the bisector of two polynomial or rational curves in the plane is algebraic, but not rational, in general [4]. In practice, numerical tracing techniques are required to approximate the bisector curve. This video considers the bisector problem for space points and curves and shows that the added extra dimension of the 3D space actually alleviates the computational difficulty! That is, given a point and a space curve, their bisector is a rational ruled surface. Moreover, given two space curves, their bisector surface is rational (except for the two dimensional degenerate case in which the two curves are coplanar).

Given two $C^1$-continuous space curves, $C_1(s)$ and $C_2(t)$, consider the necessary conditions for the two points $C_1(s_0)$ and $C_2(t_0)$ to generate a bisector point $P$:

1. Point $P$ must be located in the normal plane $L_1(s_0)$ of $C_1(s)$ at $C_1(s_0)$.
2. Similarly, point $P$ must also be located in the normal plane $L_2(t_0)$ of $C_2(t)$ at $C_2(t_0)$.
3. Moreover, point $P$ is also located in the bisector plane $L_{12}(s_0, t_0)$ which is the set of equidistant points from $C_1(s_0)$ and $C_2(t_0)$.

When the three planes: $L_1(s_0)$, $L_2(t_0)$, and $L_{12}(s_0, t_0)$, are in general position (i.e., no two of them are parallel to each other), there exists a unique intersection point $P$ of the three planes.

When the curves $C_1(s)$ and $C_2(t)$ are rational 3-space curves, the planes $L_1(s)$, $L_2(t)$, and $L_{12}(s, t)$ can be represented as implicit equations (of $x, y, z$) with rational coefficients in $s$ and $t$:

$$\begin{aligned} L_1(s): &\quad a_1(s)x + b_1(s)y + c_1(s)z = d_1(s), \\ L_2(t): &\quad a_2(t)x + b_2(t)y + c_2(t)z = d_2(t), \\ L_{12}(s,t): &\quad a_{12}(s,t)x + b_{12}(s,t)y + c_{12}(s,t)z = d_{12}(s,t), \end{aligned}$$

where all the coefficients are rational functions of $s$ and $t$. Based on Cramer's rule, it is quite straightforward to show that $x, y, z$ are rational functions of these coefficients; therefore, they are rational functions of $s$ and $t$. As a result, we can represent the bisector surface $P(s, t)$ as a rational surface of $s$ and $t$.

*This research was supported in part by the Korean Ministry of Science and Technology under Grants 95-S-05-A-02-A and 95-S-05-A-02-C of STEP 2000, and by the Electronics and Telecommunications Research Institute of Korea.

When one of the two curves (say, $C_1(s)$) degenerates to a single point $Q$, the orthogonal plane $L_1(s)$ is not defined. Let $L_{12}(t_0)$ denote the bisector plane between points $Q$ and $C_2(t_0)$. Then, we have the following plane equations of $L_2(t)$ and $L_{12}(t)$:

$$L_2(t): \qquad a_2(t)x + b_2(t)y + c_2(t)z = d_2(t)$$
$$L_{12}(t): \qquad a_{12}(t)x + b_{12}(t)y + c_{12}(t)z = d_{12}(t),$$

where all the coefficients are rational functions of $t$. The common solution of the above two equations is the intersection line of two planes: $L_2(t)$ and $L_{12}(t)$. That is, each parameter $t$ contributes a line to the bisector surface; thus the bisector surface becomes a ruled surface. The ruling direction $N(t)$ is given by the cross product of the normals of $L_2(t)$ and $L_{12}(t)$:

$$N(t) = (a_2(t), b_2(t), c_2(t)) \times (a_{12}(t), b_{12}(t), c_{12}(t)),$$

which is rational. To represent the ruled bisector surface as a rational surface, we need to construct a rational directrix curve on the surface. Let $L_1(t)$ be the plane which passes through the given point $Q$ and is orthogonal to the ruled direction $N(t)$ at $t$:

$$L_1(t): \qquad a_1(t)x + b_1(t)y + c_1(t)z = d_1(t),$$

where $(a_1(t), b_1(t), c_1(t)) = N(t)$ and $d_1(t) = \langle N(t), Q\rangle$. All the coefficients of $L_1(t)$, $L_2(t)$, and $L_{12}(t)$ are rational; therefore, their common intersection point is also rational in $t$. The trace of these intersection points generates a rational directrix curve on the ruled bisector surface. Since the indicatrix curve $N(t)$ is also rational, the bisector surface is a rational ruled surface.

The computation of the bisector surface is efficient as it requires the symbolic solution of a small (3 by 3) linear system. The symbolic manipulation involves the summation and product of different piecewise polynomial and rational functions. Overall, and for all the bisector surfaces in this video, the computation of a single bisector surface on a high-end workstation takes a fraction of a second only. A wire-frame animation of the same video can, in fact, be computed and animated in real-time on a high-end workstation, at the rate of several frames per second.

Given two rational space curves, the existence of a rational bisector surface has great potential in surface design as well as in conventional engineering applications of medial surfaces. It is easy and quite intuitive to control the geometric shape of a (bisector) surface by changing the shape and orientation of the two base curves or to control the shape of a ruled (bisector) surface with one base curve and a base point. Motivated by the efficiency of the bisector surface computation, we propose the possibility of using the bisector surface construction as another freeform modeling construction scheme in traditional modeling environments, in a similar way to the sweep surface or surface of revolution construction schemes.

Clearly not all bisectors have rational representations. Questioning the varieties that can benefit from the presented symbolic approach is natural. Moreover, the extension of the proposed scheme to higher dimensions is feasible. In $\mathbb{R}^3$, the bisectors of the curve–point, curve–curve, and surface–point cases all have bivariate rational forms.

This video consists of a sequence of nearly 4000 frames, with each frame containing a bisector surface computed from different and continuously changing curves and points. The bisectors were computed using the symbolic tools of the IRIT [7] solid modeling system, developed at the Technion, Israel. The resulting bisector surfaces were then rendered transparently using a rendering tool of the same modeling system. Figure shows one example of a bisector surface computed to two quadratic Bspline space curves.

This text is an abstracted version of Reference [2].

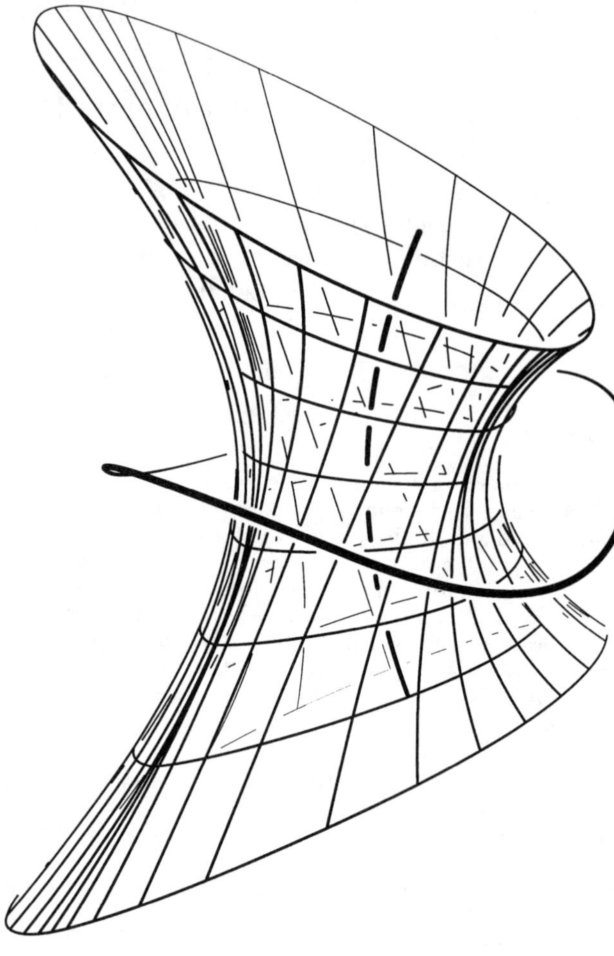

Figure 1: A bisector surface of two quadratic Bspline space curves. The bisector surface is order seven by seven.

## References


[1] Dutta, D., and Hoffmann, C., (1993), "On the Skeleton of Simple CSG Objects," *ASME J. of Mechanical Design*, Vol. 115, pp. 87–94.

[2] Elber, G., and Kim, M. S. (1996), "The Bisector Surface of Freeform Rational Space Curves," Center for Intelligent Systems Tech. Report, CIS 9619, Computer Science Department, Technion.

[3] Farouki, R., and Johnstone, J., (1994), "The Bisector of a Point and a Plane Parametric Curve," *Computer Aided Geometric Design*, Vol. 11, No. 2, pp. 117–151.

[4] Farouki, R., and Johnstone, J., (1994), "Computing Point/Curve and Curve/Curve Bisectors," *Design and Application of Curves and Surfaces: Mathematics of Surfaces V*, R.B. Fisher (Ed.), Oxford University Press, pp. 327–354.

[5] Hoffmann, C., (1990), "A Dimensionality Paradigm for Surface Interrogations," *Computer Aided Geometric Design*, Vol. 7, No. 6, pp. 517–532.

[6] Hoffmann, C., and Vermeer, P., (1991), "Eliminating Extraneous Solutions in Curve and Surface Operations," *Int'l J. of Computational Geometry and Applications*, Vol. 1, No. 1, pp. 47–66.

[7] IRIT 6.0 User's Manual, February 1996, Technion. http://www.cs.technion.ac.il/~irit.

[8] Sherbrooke, E., Patrikalakis, N., and Brisson, E., (1996), "An Algorithm for the Medial Axis Transform of 3D Polyhedral Solids," *IEEE Tans. Visualization and Computer Graphics*, Vol. 2, No. 1, pp. 44–61.

# 3D Visibility made visibly simple: an introduction to the Visibility Skeleton

Frédo Durand, George Drettakis, Claude Puech
iMAGIS/GRAVIR-IMAG*

Many problems in computer graphics require accurate global visibility information, such as the view from a given point, the mutual visibility of two objects (e.g. lighting simulation), the limits of umbra and penumbra, etc. Previous approaches have used coarse approximations or have been complicated to implement, numerically unstable and very expensive both in storage and time. We present a new data-structure, the *Visibility Skeleton*, a multi-purpose tool which can answer many types of queries. It is easy to implement and has been tested with scenes of a few hundred polygons. We present the visual events which encode all visibility changes for polygonal scenes and sketch a construction algorithm as well as queries such as the computation of the part of a polygon visible from a vertex of the scene and the limits of umbra and penumbra between two polygons.

## 1 Introduction and Previous Work

The computation of accurate visibility information has been addressed in computer graphics, computational geometry and computer vision. In computational geometry, researchers have described algorithms using line sets and Plücker coordinates (see e.g. [Pel90]).

In computer graphics, Teller [Tel92] has used these concepts to compute the limits of penumbra cast by an area light source and Drettakis and Fiume [DF94] and Stewart and Ghali [SG94] have computed *discontinuity meshes* which are the limits of umbra and penumbra together with *back-projections* which encode the topological aspect of the light source for a region of the scene where it does not change.

In computer vision researchers have long been interested in computing the aspect graph which is a partition of the viewing space according to the topological aspect of a view (see e.g. [GM90, PD90]). To compute such a partition, Plantinga and Dyer [PD90] define a data-structure called the *asp* which is a subdivision of the 4 dimensional space of lines for orthographic projection, and a subdivision of the 5 dimensional space of rays for the perspective projection.

*BP 53, 38041 Grenoble Cedex 09, France Frederic.Durand@imag.fr, http://w3imagis.imag.fr
iMAGIS is a joint project between CNRS/INRIA/INPG/UJF


*Computational Geometry 97* Nice France
Copyright 1997 ACM 0-89791-878-9/97/06 ..$3.50

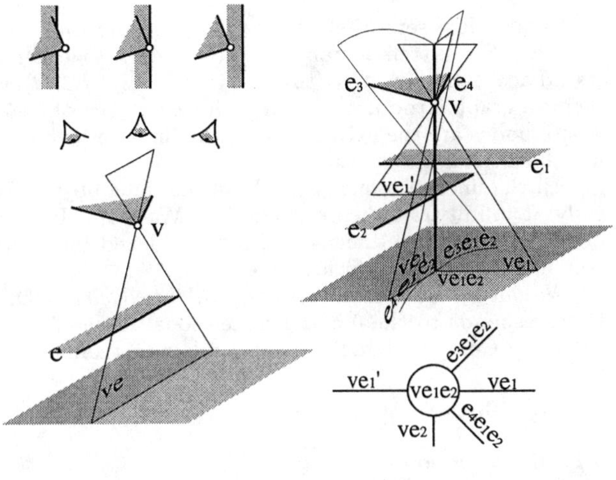


(a) (b)

Figure 1: (a) $EV$ critical line swath. (b) $VEE$ extremal stabbing line and its adjacent critical line swaths; below, the graph induced.

In 2D, Pocchiola and Vegter have defined a data-structure called the *visibility complex* (see e.g. [PV96, DP95]) which encodes all the visibility information of a scene. This data structure has been used for lighting simulations in flatland [DORP96], and a 3d generalisation has been proposed but not implemented [DDP96].

In this video we introduce a simplification of the 3d Visibility complex which is easy to implement since we only deal with the 0 and 1-dimensional faces.

## 2 The Visibility Skeleton

Visibility changes can be characterized by *critical line sets* or *line swaths* and by *extremal stabbing lines*.

When the eyepoint traverses a line swath the topology of the view changes. For example, in fig 1(a), the vertex $v$ is initially projected onto the floor, and subsequently projected onto the polygon adjacent to the edge $e$.

A line swath (or critical line set) is a 1 dimensional set of lines defined by its generators, ($v$ and $e$ in fig 1(a)). In addition there are also $EEE$ swaths which are ruled quadrics.

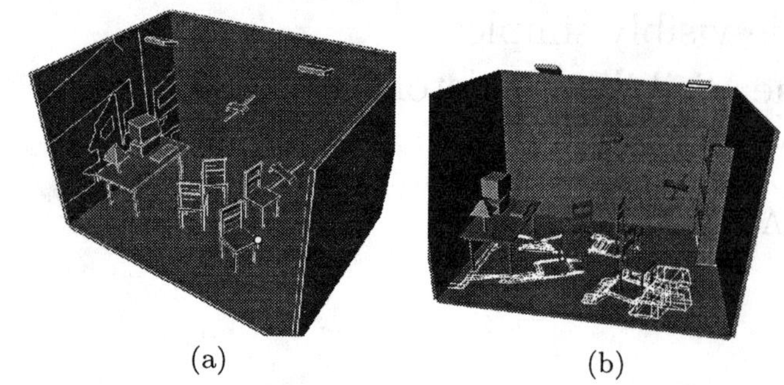

Figure 2: (a) Part of the left wall seen from the vertex on the chair (b) Limits of umbra and penumbra (*discontinuity mesh*) cast by the upper right lamp onto the floor.

Critical line sets meet at *extremal stabbing lines.* In fig 1(b), a $VEE$ extremal stabbing line is shown together with its adjacent line swaths (three $EV$ and two $EEE$). We define a graph structure (shown in the lower part of fig 1(b)) whose nodes are the extremal stabbing lines and whose arcs are the critical line swaths.

All the lines of a line swath have the same objects (possibly at infinity) at their extremities. We store the arcs of our graph in a two-dimensional array indexed by their extremities to perform efficient queries.

We call this data-structure the *Visibility Skeleton* [DDP97]. It corresponds to the 0 and 1-dimensional faces of the 3D-Visibility Complex [DDP96]; this explains its name.

## 3 Algorithms

### 3.1 Construction

The potential nodes of the Visibility Skeleton are enumerated using nested loops on their generators. They are subsequently tested for occlusion using a ray-caster which also finds their extremities. Each node is then inserted in the graph. For each adjacent arc, we check if it has already been created. If not, it is created. Otherwise the arc is split if appropriate. The node can then been attached to the arc.

This algorithm has $O(n^5)$ time complexity, but this is a very pessimistic bound, and simple culling techniques allow the practical running time to grow about quadratically in the few examples we have tested. See [DDP97] for more details.

### 3.2 Queries

To compute the part of a polygon $p$ seen by a vertex $v$, we consider the arcs for which $p$ is an extremity and $v$ is a generator. An example is shown fig 2(a).

The limits of umbra and penumbra cast by a polygon considered as a light source upon another polygon of the scene correspond to the arcs whose extremities are those two polygons, or the arcs which have a generator on the light source and an extremity on the other polygon. An example is shown fig 2(b).

## 4 The video

The video introduces the notions of critical line sets and extremal stabbing lines. The construction algorithm is sketched and some queries are demonstrated. The initial scene has been chosen since, despite being simple, it includes a wide range of types of critical line swaths and extremal stabbing lines, while the second scene (fig 2) is a typical computer-graphics scene of 450 polygons.

The construction of the skeleton and the queries were implemented in C++ on a SGI workstation using the library Open Inventor for visualization.

The video was recorded in real time and transfered on a Macintosh where the editing was performed with Adobe Premiere and Photoshop. Graphs and arrays were added manually in Premiere, as were animations such as the arrows for ray casting.

## Author/Title Index


[DDP96] Frédo Durand, George Drettakis, and Claude Puech. The 3d visibility complex, a new approach to the problems of accura te visibility. In X. Pueyo and P. Shroeder, editors, *Proceedings of 7th Eurographics Workshop on Rendering in Porto, Portugal (Rendering Techniques '96, Springer Verlag)*, pages 245–257. Eurographics, June 1996.

[DDP97] Frédo Durand, George Drettakis, and Claude Puech. The visibility skeleton: A powerful and efficient multi-purpose global visibility tool. In *Computer Graphics Proceedings, Annual Conference Series:* SIGGRAPH '97 (Los Angeles). ACM, August 1997.

[DF94] George Drettakis and Eugene Fiume. A fast shadow algorithm for area light sources using back projection. In *Computer Graphics Proceedings, Annual Conference Series:* SIGGRAPH '94 (Orlando, FL), pages 223–230. ACM SIGGRAPH, New York, July 1994.

[DORP96] Frédo Durand, Rachel Orti, Stéphane Rivière, and Claude Puech. Radiosity in flatland made visibly simple. In *video 12th Annu. ACM Sympos. Comput. Geom.*, 1996.

[DP95] Frédo Durand and Claude Puech. The visibility complex made visibly simple. In *video 11th Annu. ACM Sympos. Comput. Geom.*, 1995.

[GM90] Ziv Gigus and Jitendra Malik. Computing the aspect graph for the line drawings of polyhedral objects. *IEEE Trans. on Pat. Matching & Mach. Intelligence*, 12(2), February 1990.

[PD90] H. Plantinga and C. R. Dyer. Visibility, occlusion, and the aspect graph. *Internat. J. Comput. Vision*, 5(2):137–160, 1990.

[Pel90] M. Pellegrini. Stabbing and ray shooting in 3-dimensional space. In *Proc. 6th Annu. ACM Sympos. Comput. Geom.*, pages 177–186, 1990.

[PV96] M. Pocchiola and G. Vegter. The visibility complex. *Internat. J. Comput. Geom. Appl.*, 96. special issue devoted to ACM-SoCG'93.

[SG94] A. James Stewart and Sherif Ghali. Fast computation of shadow boundaries using spatial coherence and backprojections. In Andrew Glassner, editor, *Proceedings of SIGGRAPH '94 (Orlando, Florida, July 24–29, 1994)*, Computer Graphics Proceedings, Annual Conference Series, pages 231–238. ACM SIGGRAPH, ACM Press, July 1994. ISBN 0-89791-667-0.

[Tel92] Seth J. Teller. Computing the antipenumbra of an area light source. *Computer Graphics*, 26(4):139–148, July 1992. Proceedings of SIGGRAPH '92 in Chicago (USA).

# Application of an Effective Geometric Clustering Method to the Color Quantization Problem

Mary Inaba*, Hiroshi Imai*, Motoki Nakade* and Tatsurou Sekiguchi*

## 1 Introduction

Geometric clustering is the grouping of similar objects in some high dimensional space with an metric defined appropriately for similarity. The color quantization problem in computer graphics is a geometric clustering problem in the three-dimensional space, typically the RGB-space with Euclidean or $L_1$ metric, and there have been proposed many methods [1, 6, 9].

The geometric clustering problem for a set $S$ of $n$ points $\boldsymbol{x}_i$ in the $d$-dimensional space such that each cluster of points has its representative point, not necessarily in $S$, is stated in general as a problem of finding a $k$-clustering of $S$ into $S_j$ $(j = 1, \ldots, k)$ minimizing

$$\text{Intercluster-cost}\{\ \text{Intracluster-cost}(S_j) \mid j = 1, \ldots, k\}$$

where

- the Intracluster-cost of a cluster $S_j$ is typically a variance-based cost $V(S_j) = \sum_{p_i \in S_j} \|\boldsymbol{x}_i - \bar{\boldsymbol{x}}(S_j)\|_2^2$, or an $L_1$-type cost $\sum_{p_i \in S_j} \|\boldsymbol{x}_i - \hat{\boldsymbol{x}}(S_j)\|_1$, where $\|\cdot\|_p$ is the $L_p$ norm, and $\bar{\boldsymbol{x}}(S_j)$ is the centroid of points in $S_j$, i.e., $\frac{1}{|S_j|}\sum_{p_i \in S_j} \boldsymbol{x}_i$, and $\hat{\boldsymbol{x}}(S_j)$ is the medoid (median) of $S_j$, and
- the Intercluster-cost function is mostly either the summation or the maximum.

For the color quantization problem, the dimension $d$ is three, and for a pseudo-color display of depth 8bit with $1000 \times 1000$ pixels, $n$ is a million and $k = 256$. As for the clustering criteria, the summation is used mostly as the Intercluster-cost function. As the Intracluster-cost, Wan, Wong, Prusinkiewicz [9] used the variance-based cost and presented a top-down $k$-clustering algorithm which recursively uses a 2-clustering algorithm finding an optimum 2-clustering dissected by a hyperplane perpendicular to an axis. Heckbert [1] used the $L_1$-type cost and presented an algorithm which has similar structure as above. These are implemented in `rlequant` of Utah Raster Toolkit, and `ppmquant` of X11 or `tiffmedian` of Tiff Soft. Although these implementations run rather fast in practice, roughly speaking in $O(n \log n)$ time, there is no theoretical guarantee about how good their $k$-clustering solution is.

Inaba, Imai and Katoh [6, 7] (see also [2]) have proposed a randomized geometric clustering algorithm with promising experimental results. The algorithm makes use of the random sampling technique, and provides a theoretical guarantee for the approximation ratio to the exact clustering cost for the 2-clustering case (this may be extended for general $k$, although bounds may become looser). One of its features is that 2-clusterings dissected by planes not necessarily perpendicular to some axis are treated, which has many nice implications.

Furthermore, Inaba and Imai [4] have investigated a $k$-means algorithm, which is a well-known local improvement algorithm for this geometric clustering problem, and relate it to some mathematical programming problem called geographic optimization problem by Iri, Murota, Ohya [3]. Inaba and Imai [5] further considers an improvement of the $k$-means algorithm via overrelaxation, motivated from the viewpoint of mathematical programming.

In this paper with companion video, an algorithm animation of the randomized geometric clustering algorithm is given. Applications of this algorithm together with the (enhanced) $k$-means algorithm are further shown by using several types of pictures. This part is concerned with three-dimensional clustering problems, and 3D animations are fully demonstrated. From these, features of typical color quantization problems are visually revealed, and some advantages of our methods are explained. In the sequel, this paper describes backgrounds of the algorithms with video explanations.

## 2 Randomized 2-clustering algorithm

Now the problem is, for a set of $n$ points $\boldsymbol{x}_i$ in the $d(=3)$-dimensional space, to find a $k$-clustering of $S$ into $S_j$ $(j = 1, \ldots, k)$ minimizing $\sum_{j=1}^{k} V(S_j)$ for the variance-based cost $V(S_j)$ as treated in [6]. For this problem, an optimal partition is generated by the Voronoi Diagram for the centroid of each partition, and is called the Voronoi partition.

For this problem, we have proposed the following 2-clustering randomized algorithm in [6].

1. Sample a subset $T$ of $m$ points from $S$ at random;
2. For every linearly separable 2-clustering $(T_1, T_2)$ of $T$, execute the following:
   (a) Compute the centroids $t_1$ and $t_2$ of $T_1$ and $T_2$;
   (b) Find a 2-clustering $(S_{t_1}, S_{t_2})$ of $S$ divided by the perpendicular bisector of $\overline{t_1 t_2}$;
   (c) Compute the centroids $s_1$ and $s_2$ of $S_{t_1}$ and $S_{t_2}$;
   (d) Find a 2-clustering $(S_1, S_2)$ of $S$ divided by the perpendicular bisector of $\overline{s_1 s_2}$;
   (e) Compute the value of $V(S_1) + V(S_2)$;
3. Output the 2-clustering of $S$ with minimum value.

The main idea of this randomized algorithm is to use all pairs of centroids of linearly separable 2-clusterings for the sampled point set $T$ to obtain $S_{t_1}$ and $S_{t_2}$, then perform local improvement in (2.c) and (2.d). Even without these, we have the following:

*Department of Information Science, University of Tokyo, Tokyo 113, Japan. E-mail: {mary,imai,nakade,cocoa}@is.s.u-tokyo.ac.jp



*Computational Geometry 97* Nice France

**Theorem 1** (**[6]**) *When there is an optimum 2-clustering* $(S_1^*, S_2^*)$ *of* $S$ *such that* $|S_1^*|,\ |S_2^*| \geq \gamma|S| = n$ *for a constant* $\gamma$ $(0 < \gamma \leq 1/2)$*, this randomized algorithm, sampling* $m$ *points, finds a 2-clustering whose expected ratio is* $1 + O(1/m)$ *from the optimal value with high probability in* $O(m^d n)$ *time.* □

In [7], results of computational experiments are shown for this algorithm, and it is shown that in practice a very small set of sampled points provides a good clustering. Experimentally, even sixteen samples from a huge ground set are shown to be sufficient. This may be because the above-mentioned algorithm performs a one-step improvement in step (2.d), whose fact is not utilized in obtaining the theoretical bound in the theorem above.

In the companion video, this algorithm is explained with animation. Then, the color space features of typical pictures are shown by examples. It is observed in some examples that, in the RGB color space, points are distributed in only neighboring region to the diagonal connecting two points representing white and black. Even with such a biased distribution of points, the above 2-clustering algorithm can adjust itself since dissecting planes used are independent of the coordinate axes unlike [1, 9].

## 3 Recursive method for $k$-clustering

A $k$-clustering can be obtained by applying the 2-clustering algorithm described above in a top down recursive manner. We used a heap to determine a subset with maximum cost to be divided in the next step.

In the companion video, this stage is shown in detail. In it, as the number of clusters increases, the corresponding colored images are shown, which reveals the importance of color quantization. In this example, $k = 32$ and hence via our clustering algorithm the original picture of more than 40000 colors are displayed only using 32 colors without losing its essence. To represent point distributions in the animation, the RGB space is divided into $16 \times 16 \times 16$ meshes and each nonemptry mesh containing at least one point is denoted by a cube. Also, an animation of how representative points of clusters are derived is shown. We also put the original image on the right top, and the generated image at each stage on the right bottom.

## 4 $k$-means and extensions

The $k$-means algorithm is an iterative improvement algorithm for the $k$-clustering problem (e.g., see [8]). It starts with an initial $k$-clustering and then iterate to move each representative point to the centroid of each cluster in the Voronoi partition generated by the current representative points, until a local minimum solution is found. It can be described as follows:

1. Find an initial $k$-clustering $S_j$ $(j = 1, \ldots, k)$ for $n$ points $\boldsymbol{x}_i$ $(i = 1, \ldots, n)$ by some method;
2. **repeat**

   Compute the centroid $\bar{\boldsymbol{x}}(S_j)$ of each cluster;

   Update the $k$-clustering to the Voronoi partition by the Euclidean Voronoi diagram of $\bar{\boldsymbol{x}}(S_j)$;

   **until** a local minimum is found.

For this kind of algorithm, the initial solution plays an important role, as shown in [7], and the $k$-clustering obtained by the top-down recursive application of the 2-clustering in [6] is very suitable.

Furthermore, relating this $k$-means algorithm to a geographic optimization problem in [3], a step of the $k$-means can be regarded as a Newton step in minimizing the clustering cost function, and therefore naturally performing a line search (one-dimensional minimization along the search direction) is considered. However, in the clustering problem, evaluating the function even at points along the line is costly, and Inaba and Imai [5] proposes a simple overrelaxation scheme. In the companion video, animations of $k$-means algorithm for a planar case and an RGB-space case are shown. Here, the traces of centroids are also shown and the overrelaxation part is also touched upon.

## 5 Environments creating the animation

The clustering algorithm for the color quantization problem was implemented in C language on SPARC station 20. Each frame of the animation was generated by VC++ program, OpenGL on Windows NT, and merged to animation by Media Studio VE of Canopus on Windows 95.


## Acknowledgment

The work was supported in part by the Grant-in-Aid for Scientific Research of the Ministry of Education, Japan.


## References


[1] P. Heckbert: Color image quantization frame buffer display. *ACM Trans. on Computer Graphics*, Vol.16, No.3 (1982), 297–304.

[2] S. Hasegawa, H. Imai, M. Inaba, N. Katoh and J. Nakano: Efficient algorithms for variance-based $k$-clustering. *Proc. 1st Pacific Conf. on Computer Graphics and Applications*, World Scientific, 1993, 75–89.

[3] M. Iri, K. Murota and T. Ohya: A fast Voronoi-diagram algorithm with applications to geographical optimization problems. *Proc. 11th IFIP Conf. on System Modelling and Optimization*, Lecture Notes in Control and Information Science, Vol.59, Springer-Verlag, 1984, 273–288.

[4] H. Imai and M. Inaba: Geometric clustering with applications. *Zeitschrift für Angewandte Mathematik und Mechanik (ZAMM)*, ICIAM95 Proceedings, Issue 3, "Applied Atochastics and Optimization", 1996, 183–186.

[5] M. Inaba and H. Imai, in preparation.

[6] M. Inaba, H. Imai and N. Katoh: Applications of weighted Voronoi diagrams and randomization to variance-based $k$-clustering. *Proc. 10th ACM Symp. on Computational Geometry*, 1994, 332–339.

[7] M. Inaba, H. Imai, and N. Katoh: Experimental results of randomized clustering algorithm. *Proc. 12th ACM Symp. on Computational Geometry*, 1996, C1-C2.

[8] S. Z. Selim, M. A. Ismail: $K$-means-type algorithms: A generalized convergence theorem and characterization of local optimality. *IEEE Trans. Pattern Anal. Mach. Intell.*, 1984, 81–87.

[9] S. J. Wan, S. K. M. Wong and P. Prusinkiewicz: An algorithm for multidimensional data clustering. *ACM Trans. on Mathematical Software*, Vol.14, No.2 (1988), 153–162.

# Animating the Polygon-Offset Distance Function*

Gill Barequet† Amy Briggs‡ Matthew Dickerson‡ Cristian Dima‡ Michael T. Goodrich†

## 1 Introduction

In this video, we use *The Geometer's Sketchpad*$^{TM}$ to describe and animate a new *polygon-offset* distance function, illustrating how offset polygons grow, how they differ from scaled polygons, how distance is measured using offset polygons, why the function does not satisfy the triangle inequality, and also what a Voronoi boundary based on this distance function looks like.

Offset polygons were formally defined and used in [BBDG] in solutions to a number of geometric problems with applications to robot localization and geometric tolerancing. A similar construct was also explored by Aicholzer et al. [AAAG, AA] and given the name *straight skeletons*. In [BDG], a *polygon-offset distance function* was first formally defined based on these offset polygons (or straight skeletons) in the case when the polygon is convex. Some properties of this distance function were explored in that paper, as was the Voronoi diagram based on this distance function. Figure 1 shows a sample polygon $P$ along with two offset polygons, one inside $P$ and one outside $P$. Note that the inner offset has fewer edges than $P$.

The *centroid* of a polygon $P$ is the medial axis vertex where the inner offset polygon collapses to a single point. Note that in the case of a "degenerate" polygon with two long parallel edges, the inner offset may collapse to an edge rather than a point. We thus assume a general position assumption with no degenerate polygons. Distance in this offset polygon distance function is now defined as follows: to

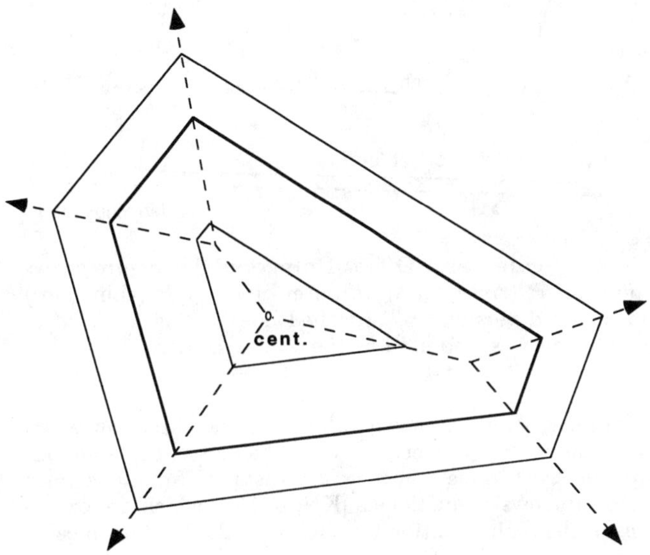


Figure 1: **Polygon with Inner and Outer Offsets.** The figure shows a convex polygon (thick lines) with its medial axis lines extended as rays to infinity(dashed lines) and an inner and outer offset polygon (thin lines).

measure distance from a point $p$ at the centroid of polygon $P$, to any point $q$, we find the smallest offset $P^*$ of $P$ which contains $q$. Let $d_u$ be the Euclidean distance from $p$ to the nearest edge of $P$, and $d^*$ be the Euclidean distance from $p$ to the nearest edge of $P^*$. Then the offset polygon distance from $p$ to $q$ is given by $d^*/d_u$. Note that $d_u$ is the distance the polygon must be offset inward to collapse to the centroid. Note also that the distance normalizes so that $d(p,p) = 0$, and $d(p,q) = 1$ for any point $q$ on $P$.

## 2 Why a Polygon-Offset Distance Function

The Voronoi diagram of a set $S \subset \mathbb{R}^2$ is a well-known, powerful tool for handling a host of geometric problems dealing with distance relationships (see, e.g., [OBS]). Voronoi diagrams have been used extensively, for example, for solving nearest-neighbor, furthest-neighbor, and matching problems in many contexts. The underlying distance function used to define Voronoi diagrams is typically either the usual Eu-

*Work by the first author has been supported in part by the U.S. Army Research Office under Grant DAAH04-96-1-0013. Work by the third author has been supported by the National Science Foundation under Grant CCR-93-1714. Work by the fifth author has been supported by NSF grant CCR-9625289 and by U.S. ARO grant DAAH04-96-1-0013.

†Center for Geometric Computing, Dept. of Computer Science, Johns Hopkins University, Baltimore, MD 21218. E-mail: [barequet|goodrich]@cs.jhu.edu

‡Dept. of Mathematics and Computer Science, Middlebury College, Middlebury, VT 05753. E-mail: [briggs|dickerso|dima]@middlebury.edu

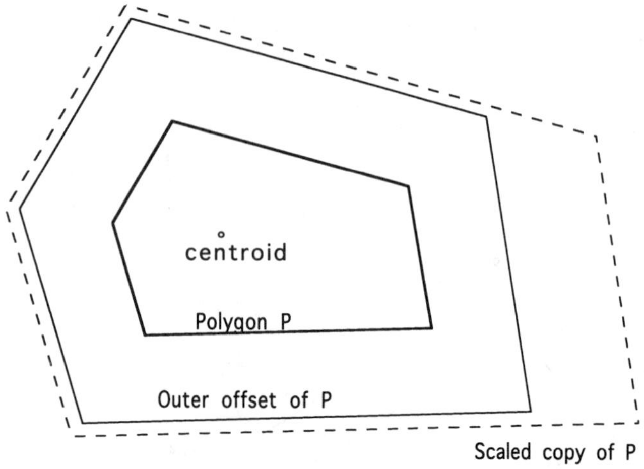


Figure 2: **Scaled and Offset Polygons.** The figure shows a polygon $P$ (thick lines) an outer offset of $P$ (thin lines) and a scaled version of $P$ (dashed lines). The *centroid* of the polygon was also used as the scaling center.

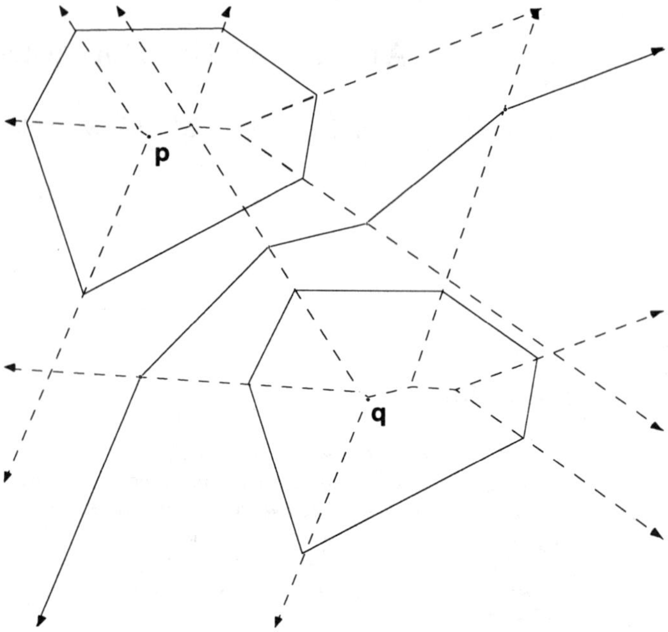


Figure 3: **Voronoi Boundary.** The figure shows two points $p$ and $q$, a copy of a polygon $P$ placed with its *centroid* on each point, and the polygon-offset distance function Voronoi polygonal bisector between the two points.

clidean metric, or more generally a distance function based upon one of the $L_p$ metrics. There has also been some interesting work done using convex distance functions (also called Minkowski functionals [KN, p. 15]), which are extensions of the scaling notion for circles (in the Euclidean case) to convex polygons. Figure 2 shows a polygon $P$ as well as scaled and offset polygons based on $P$. We feel that defining distance in terms of an offset from a polygon is actually more natural than scaling in many applications, including those dealing with manufacturing processes. This is because the relative error of the production tool (a milling head, a laser beam, etc.) is independent of the location of the produced feature relative to some artificial reference point (the origin). Therefore it is more likely to allow (and expect) local errors bounded by some tolerance, rather than scaled errors relative to some (arbitrary) center.

We thus investigate distance functions based on *offsetting* convex polygons, where the distance is measured along the infinitely extended medial-axis of such a polygon. While the scaling operation shifts each edge of the polygon *proportionally* to its distance from the origin, the offset operation shifts all the edges by the *same amount.* Offset polygons are therefore not homothetic copies of the original polygon (unless the original polygon is regular). We are interested in the investigation of basic properties of polygon-offset distance functions, with particular attention paid to how they may be used in the definition of Voronoi diagrams.

## 3 Summary of Video

This video illustrates this distance function, animating its behavior for several different polygons, and showing some of its properties. It is shown how the distance function differs from its relatives, the Minkowski distance functions (or scaled polygon convex distance functions). For example, the offset polygon distance function does not obey the normal triangle inequality. In fact, for colinear points, it obeys a reverse inequality.

The video ends by showing the Voronoi boundaries of two and three sites in the plane. The boundary is first shown growing along expanding offset polygons, and then shown as it would be constructed from the medial axis lines and bisectors of polygon edges.

All the animations were done using *The Geometer's Sketchpad*$^{TM}$, version 3.00 by *Key Curriculum Press*, which provides an excellent environment for geometric investigation.

## References

[AAAG] O. Aicholzer, F. Aurenhammer, D. Alberts, and B. Gärtner, "A novel type of skeleton for polygons" *Journal of Universal Computer Science* **1** (1995) 752–761.

[AA] O. Aicholzer, F. Aurenhammer, "Straight skeletons for general polygonal figures in the plane", *Proceedings of the Second COCOON Conference: Springer Lecture Notes in Computer Science 1090* (1996) 117–126.

[BBDG] G. Barequet, A. Briggs, M. Dickerson, M. Goodrich, "Offset polygon annulus placement problems", manuscript.

[BDG] G. Barequet, M. Dickerson, M. Goodrich, "Voronoi diagrams for Polygon-Offset Distance Functions", manuscript.

[KN] J.L. Kelley and I. Namioka, Linear Topological Spaces, Springer Verlag, 1976.

[OBS] A. Okabe, B. Boots, and K. Sugihara, Spatial Tessellations: Concepts and Applications of Voronoi Diagrams, John Wiley & Sons, Chichester, England, 1992.

# A Triangulation-Based Object Reconstruction Method*

Fausto Bernardini    Chandrajit L. Bajaj

Jindong Chen†    Daniel R. Schikore

Department of Computer Sciences
Purdue University ‡

## 1 Introduction

Reconstructing the shape of a 3D object from a digital scan of its surface has a range of applications, such as reverse engineering, authoring 3D synthetic worlds, shape analysis, 3D faxing and tailor-fit modeling.

Input data might come in different forms, depending on the scanning device used. It is usually comprised of the location $(x_i, y_i, z_i)$ of points on the surface of the object, and at times additional topological and geometric information, as well as measures of other physical properties. The sampling provided by recent scanning devices (such as the *laser range scanner*) is *dense*, in the sense that the resolution is much smaller than the size of shape features of interest. Often multiple scans are required to capture the entire object's surface. We make no assumptions on spatial relations among sample points, and assume that the input is a large, but unorganized, collection of measurements.

Our goal is to reconstruct a boundary representation of the object, based on implicit polynomial surface patches of low degree, that has the desired geometric continuity and approximates the data within a user-specified parameter $\varepsilon$. For a discussion of related prior work the reader is referred to [6].

In [1], we presented a method based on alpha-shapes, to build an initial piecewise-linear reconstruction, followed by an incremental, adaptive piecewise polynomial fitting of the signed distance function defined by the alpha-shape. The method relied on the user to select a good $\alpha$-value. The final reconstructed model was represented as a collection of $C^1$-smooth implicit algebraic patches. A more up-to-date, detailed description of the algorithm can be found in [2].

In this paper, and the accompanying video presentation, we present a novel reconstruction technique (more details can be found in [6]), capable of handling piecewise-smooth objects. While we still use alpha-shapes and implicit algebraic patches in the new method, there are significant differences with our previous method. First, we devised an automatic selection method to find an optimal $\alpha$-value. Second, we developed a heuristic-based approach, to improve the quality of the selected (regularized) alpha-shape. Third, instead of approximating the signed-distance function, we apply a mesh-reduction algorithm followed by an interpolatory A-patch fitting scheme.

*This research was partially supported by AFOSR grant F-49620-94-1-0080, NSF grant CCR 92-22467 and ONR grant N00014-94-1-0370

†Present address: Corporate Research & Technology, Xerox. Email: `jchen@cp10.es.xerox.com`.

‡West Lafayette, Indiana 47907-1398 USA

`{bajaj,fxb,jdc,drs}@cs.purdue.edu`
`http://www.cs.purdue.edu/research/shastra`



*Computational Geometry 97* Nice France



## 2 Algorithm overview

An example of the reconstruction process is shown in Figure 1. Our algorithm is based on the following three phases:

**1.** Build an initial triangle mesh that interpolates all data points, approximating the object shape (Figure 1 (a)-(c)). Our approach is based on (regularized) alpha-shapes [8], and is capable of automatically selecting an optimal $\alpha$-value and improving the resulting mesh in areas of insufficient sampling.

The resulting triangle mesh can be used to estimate normals at smooth vertices (by averaging the normals of incident triangles) and to detect sharp features (by looking at the dihedral angle formed by two adjacent triangles). Notice that for the dense surface sampling that we are interested in, these estimates are usually quite accurate. The use of more complex and accurate sharp-feature detection strategies will be investigated in the future.

**2.** Simplify the mesh to reduce the number of triangles, while guaranteeing good aspect-ratio of triangles, bounded distance of the data points from the reduced mesh, and feature preservation (Figure 1 (d)). The technique used in our paper has been extended from [4]. The edges and vertices of the reduced mesh are "tagged" as either *smooth* (the surface is $C^1$ continuous across it) or *sharp* (only $C^0$ continuity), and vertices are classified according to the type of incident edges and the number and type of estimated vertex normals (Figure 1 (e)).

**3.** The reduced mesh is used as the starting point for a polynomial-patch data fitting. For every triangle, we build an implicit Bernstein-Bézier patch of low degree which interpolates the vertices and vertex normals (if defined) and least-squares approximates data points in its vicinity. The

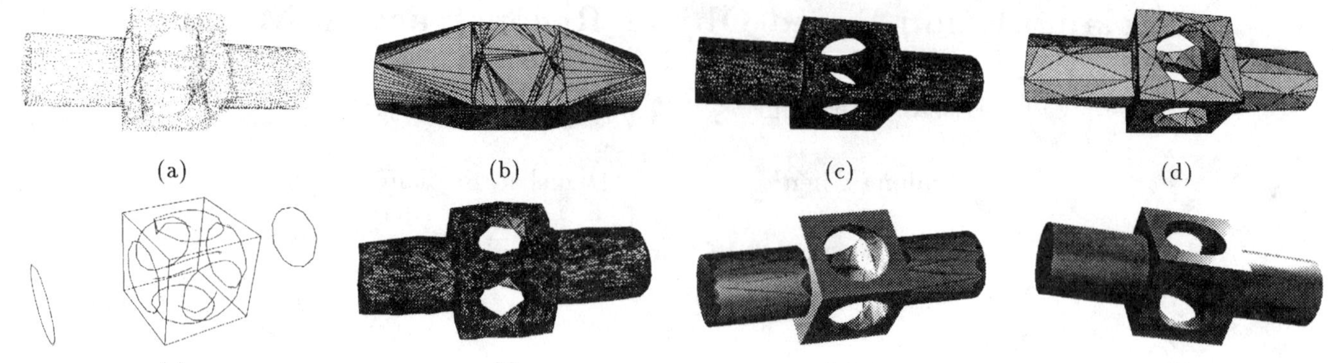

Figure 1: The complete reconstruction process. (a) Point sampling. (b) 3D Delaunay triangulation. (c) Alpha-solid. (d) Simplified mesh. (e) Sharp features. (f) Support mesh. (g) A-patches. (h) Reconstructed model.

algebraic patches used (cubic A-patches [3]) allow a simple formulation of $C^1$ continuity constraints between adjacent patches, and have been extended to allow the modeling of sharp features such as linear sharp edges, piecewise-planar curved creases and sharp corners (Figure 1 (f)-(h)).

## 3 Shape Reconstruction Using Alpha-Shapes

One of the most difficult problems of surface reconstruction from unorganized points is understanding how to connect the points so as to form a surface that has the same topological (e.g. number of handles) and geometric (e.g. depressions and protrusions) characteristics of the original. In particular, we look at the following problem ($D(K, B)$ is a suitably defined "distance" measure):

*Let $B$ be the boundary of a solid $M$, and $S \subset B$ a finite set of points (sampling). Construct a (geometric) simplicial complex $K$, such that $K^{(0)} = S$, $K$ is homeomorphic to $B$, and $D(K, B) < \varepsilon$.*

We call $\rho$-dense a sampling such that any sphere of radius $\rho$, centered on the surface of the object, contains at least one sample point. We prove in [5] that for a manifold whose radius of curvature and "feature-size" are larger than $\rho$, a $\rho$-dense sampling suffices to guarantee a homeomorphic and error-bounded reconstruction using alpha-shapes. While the theorems above give us sufficient conditions for a sampling to allow a faithful reconstruction using alpha-shapes, in practice one has to deal with less than ideal scans.

In general, i.e. when the conditions of the theorems above are not satisfied, an alpha-shape is a non-connected, mixed-dimension polytope. We define the alpha-solid to be the union of all tetrahedra in the 3D Delaunay triangulation $\mathcal{T}$ of $S$ that are contained within a continuous envelope of alpha-shape triangles. The alpha-solid is clearly a homogeneously three-dimensional object. Observe also that it can be computed very efficiently from the underlying triangulation, by simply traversing the adjacency graph. Varying $\alpha$, one obtains a finite collection of different alpha-solids, ranging from the empty set for sufficiently small values of $\alpha$, to the convex hull of the set of points for $\alpha$ large enough. We can perform a binary search for the minimal $\alpha$-value such that all points lie on the boundary or in the interior of the alpha-solid, and such that the alpha-solid is a closed, connected manifold. This search takes $O(\log n)$ time (where $n$ is the size of $S$) because the number of possible $\alpha$-values is bounded by the number of simplices in $\mathcal{T}$, which is polynomial in $n$. Observe also that the convex hull satisfies the three properties above, so the algorithm always terminates successfully. For a sufficiently dense and uniform sampling of the object boundary the alpha-solid selected by this strategy is a good approximation of the object's shape. However, small, concave features may be occluded by unwanted tetrahedra, and some of the sampled points might lie in the interior of the alpha-solid.

Our criterion for improving the initial alpha-solid is based on the search for a subset of tetrahedra whose boundary interpolates all the points, and that maximizes the "smoothness" of the triangle mesh, defined as $\sum_{e_i} |\pi - \gamma_i|$, over all edges of the triangle mesh, where $\gamma_i$ is the dihedral angle formed by the two triangles incident on edge $e_i$.

Searching for the global optimum for this optimization problem would be clearly computationally expensive, but in practice our alpha-solid is already a good approximation of the optimal polyhedron, and we only need to modify it where concave, high curvature features are present. We therefore resort to a simple greedy strategy, similar to the "sculpturing" approach proposed by Boissonnat [7]. However, we apply the iterative removal of tetrahedra only to locally improve the alpha-solid, rather than as a global strategy to extract an interpolating mesh from the 3D Delaunay triangulation. Figure 2 and Table 1 illustrate some examples of alpha-solids computed with the technique described above.

## 4 Mesh Simplification

Surface mesh simplification refers to a general category of techniques designed to generate compact, adaptive approximations of dense tessellated surfaces. A discussion of previous work in the field can be found in [4].

In this work, we adapt and improve the method of [4] to handle explicit edge feature detection and preservation. The resulting algorithm is able to maintain a strict bound on the distance between the original mesh and the surface mesh, in addition to maintaining sharp features in the reconstructed triangulation.

The simplification algorithm follows the basic strategy of other "vertex deletion" schemes, and is based on accumulated error bounds which are propagated from the original surface mesh through the successive simplified meshes produced by point deletion. A compact error representation

| Object | Number of Points | Alpha-solid Time | Number of tetrahedra | Removed tetrahedra | Number of Triangles |
|---|---|---|---|---|---|
| Femur | 9807 | 1.5 | 36182 | 3704 | 19610 |
| Tibia | 9200 | 1.4 | 33232 | 2172 | 18396 |
| Fibula | 8146 | 1.1 | 30876 | 2896 | 16288 |
| Patella | 2050 | 0.3 | 7536 | 683 | 4096 |
| CSG Solid | 13040 | 2.5 | 42507 | 2473 | 26088 |
| Club | 16864 | 4.1 | 58657 | 754 | 33142 |
| Bunny | 33123 | 19.6 | 127607 | 3761 | 66224 |
| Mannequin | 10392 | 2.1 | 35383 | 2077 | 19802 |

Table 1: Results of alpha-solid reconstruction. The table shows for each object, from left to right: (1) The number of points in the sampling; (2) The time, in minutes, required by the alpha-solid computation (including 3D Delaunay triangulation, computation of family of alpha-shapes, automatic selection of $\alpha$-value, improvement by local sculpturing). All computations were carried out on a SGI Indigo2, with a 250MHz MIPS 4400 CPU; (3) The number of tetrahedra in the initial alpha-solid; (4) The number of tetrahedra removed by the heuristic; (5) The number of triangles in the boundary of the final reconstructed model.

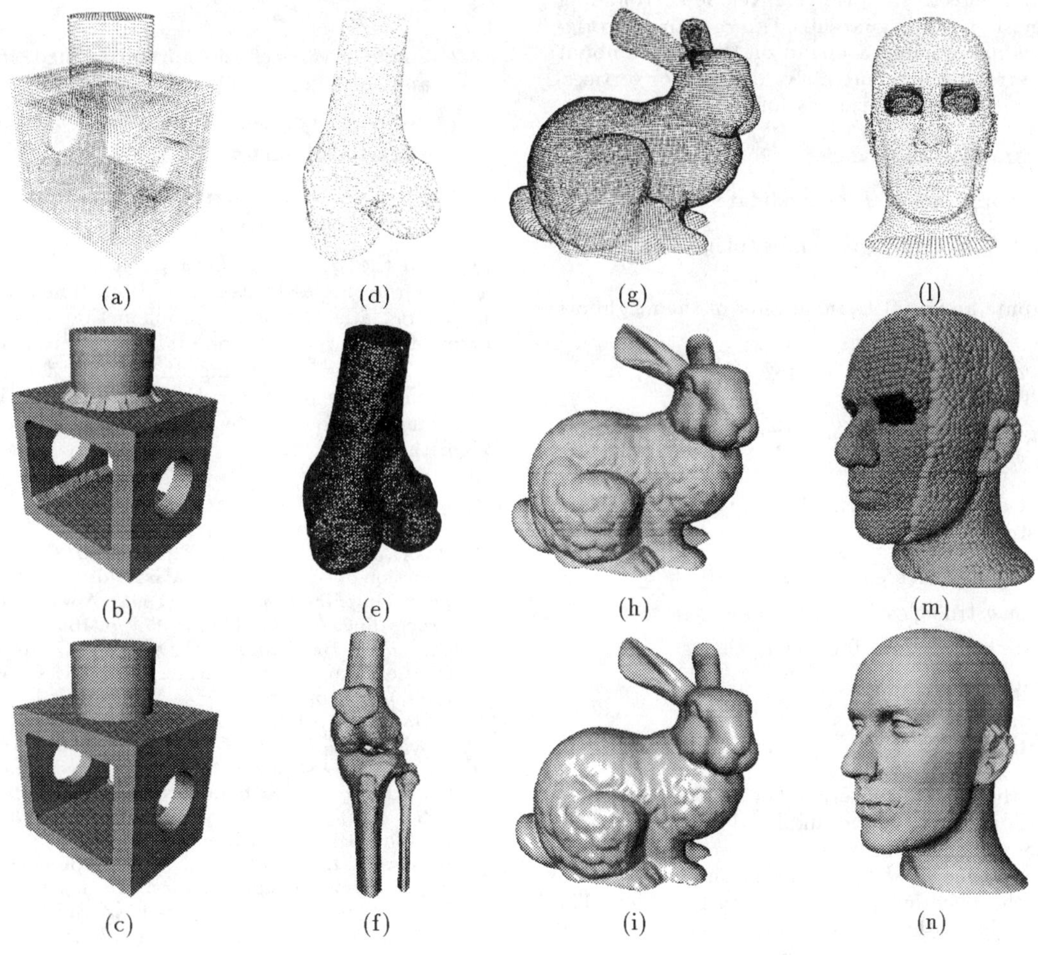

Figure 2: Example of reconstruction using alpha-shapes. (a)-(c) Random sampling of a solid generated with a commercial modeler. (a) Point sampling. (b) Selected alpha-shape. (c) Improved via local smoothness heuristic. (d)-(f) Reconstruction from isocontours of a CT scan (Visible Human Project). (d) Points. (e) Alpha-solid. (f) Reconstructed knee model. (g)-(i) Reconstruction from laser range data. (g) Data points. (h) Improved alpha-solid. (i) Phong rendering of the alpha-solid. (l)-(m) Use of weighted alpha-shapes for reconstruction from multiresolution scans. (l) Sampling. (m) Weighted points. (n) Alpha-solid.

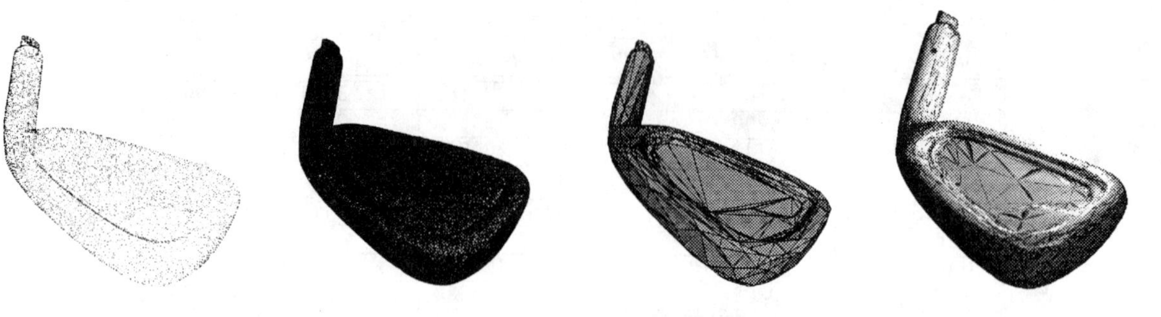

(a) (b) (c) (d)

Figure 3: Reconstruction and piecewise-smooth A-patch fitting. (a) Scanned points. (b) alpha-solid. (c) Decimated mesh. (d) Reconstructed model (different colors identify different surface patches).

consisting of two scalar error bounds per triangle is used. The error values correspond to a bound on the error (geometric displacement) toward the outside (inside) of the object. These bounds effectively form an envelope surrounding the simplified mesh which is guaranteed to contain the original surface, thus maintaining a bound on the total amount of accumulated error through successive deletion of vertices. The algorithm can be summarized as follows:

1. Initialize errors on all triangles to 0
2. Initialize priority queue $P$ of candidate vertices $v_i$
   (a) Classify $v_i$ according to number of incident "sharp" edges
   (b) Compute an initial triangulation of the neighbors of $v_i$
   (c) Perform edge flipping to lower the error in the new triangulation
   (d) Assign priority based on introduced error associated with $v_i$
3. While next candidate vertex $v$ from $P$ does not violate error constraints
   (a) Delete $v$ and incident triangles
   (b) Add new triangles
   (c) Update error values for new triangles
   (d) Update $P$

## 5 A-patch Fitting

Our A-patch fitting scheme interpolates the vertices (and estimated surface normals) of the simplified mesh computed as described above, and approximates the remaining data points. Features tagged as *sharp* during mesh simplification are retained in the resulting piecewise-smooth model. The fitting process begins with the construction of a tetrahedral mesh to act as support for the A-patches. Then, weights for each patch are set to interpolate vertices and sharp features, and least-squares approximate the remaining point. Finally, a fairing and fitting optimization can be applied to improve the quality of the reconstructed model.

The scheme used here follows Bajaj, Chen and Xu [3], and consists basically in building one tetrahedron for each side of every triangle of the mesh (*face-tetrahedra*) and four edge-tetrahedra to close the gaps.. Clough-Tocher splitting is only used in special cases. This scheme can be extended to accommodate sharp features of the following type:

1. Sharp corners (multiple normals defined, for example the corner of a cube).
2. Singular vertices (no normal defined, for example the apex of a cone).
3. Straight edges (two normals, one for each side, defined at each endpoint).
4. Planar or piecewise-planar curved ridges (two normals at each endpoint).

Some of the weights of each patch are only constrained in sign by the single-sheeted condition. Their value can be set by solving a least-squares problem that minimizes the local error-of-fit to the data points. An additional optimization step to improve the fitting and obtain a better fairing can be applied as a postprocessing to the piecewise surface.

Figure 3 shows an example of reconstruction using this algorithm.

## References


[1] Bajaj, C., Bernardini, F., and Xu, G. Automatic reconstruction of surfaces and scalar fields from 3D scans. In *Computer Graphics Proceedings* (1995), Annual Conference Series. Proceedings of SIGGRAPH 95, pp. 109–118.

[2] Bajaj, C., Bernardini, F., and Xu, G. Reconstructing surfaces and surfaces on surfaces from 3D scans. *Algorithmica* (1997). To appear.

[3] Bajaj, C., Chen, J., and Xu, G. Modeling with cubic A-patches. *ACM Trans Graph. 14*, 2 (1995), 103–133.

[4] Bajaj, C., and Schikore, D. Error-bounded reduction of triangle meshes with multivariate data. In *Proceedings of SPIE Symposium on Visual Data Exploration and Analysis III* (Jan. 1996), SPIE, pp. 34–45.

[5] Bernardini, F., and Bajaj, C. Sampling and reconstructing manifolds using alpha-shapes. Tech. Rep. CSD-TR-97-013, Department of Computer Sciences, Purdue University, 1997. Revised version submitted for publication.

[6] Bernardini, F., Bajaj, C., Chen, J., and Schikore, D. Automatic reconstruction of 3D CAD models from digital scans. Tech. Rep. CSD-TR-97-012, Department of Computer Sciences, Purdue University, 1997. Submitted for publication.

[7] Boissonnat, J.-D. Geometric structures for three-dimensional shape representation. *ACM Trans. Graph. 3*, 4 (1984), 266–286.

[8] Edelsbrunner, H., and Mücke, E. P. Three-dimensional alpha shapes. *ACM Trans. Graph. 13*, 1 (Jan. 1994), 43–72.

# Approximating Weighted Shortest Paths on Polyhedral Surfaces*

Mark Lanthier, Anil Maheshwari and Jörg-Rüdiger Sack
School of Computer Science
Carleton University
1125 Colonel By Drive
Ottawa, ON, CANADA K1S 5B6
E-mail: {lanthier,maheshwa,sack}@scs.carleton.ca

## 1 Introduction

Shortest path problems are among the fundamental problems studied in computational geometry. In this video, we consider the problem of computing a shortest cost path between two points $s$ and $t$ on a (possibly non-convex) polyhedral surface $\mathcal{P}$. The surface is composed of triangular regions (faces) in which each region has an associated positive weight. The cost of travel through each region is the distance traveled times its weight.

The computation of Euclidean shortest paths on non-convex polyhedra has been investigated by [5, 2, 1]; currently, the best known algorithm due to Chen and Han [2] runs in $\mathcal{O}(n^2)$ time. Mitchell and Papadimitriou [6] introduced the Weighted Region Problem and an algorithm that computes a shortest weighted cost path between two points in a planar subdivision; it requires $\mathcal{O}(n^8 \log n)$ time in the worst case. They state that their algorithm applies to non-convex polyhedral terrains with modifications.

Most shortest path applications demand simple and efficient algorithms to compute approximate shortest paths as opposed to a complex algorithm that computes an exact path. Polyhedra arising in these applications approximate real surfaces and thus an approximate path will typically suffice. Our interest is also motivated by our research and development on a parallel system for GIS and spatial modeling. For a more detailed explanation of the contents of this video, see [3] or [4].

[1] Research is supported by ALMERCO Inc. and the Natural Sciences and Engineering Research Council of Canada.



## 2 Video Contents

The video describes several schemes that allow the computation of approximate shortest paths on polyhedral surfaces in both the weighted and unweighted scenarios. The schemes are based on adding Steiner points along the polyhedral edges and interconnecting them across each face thereby building weighted graphs. The schemes vary in the way in which Steiner points and edges are added. An approximate shortest path for the polyhedral surface is obtained by applying Dijkstra's shortest path algorithm on the weighted graph. As more and more Steiner points are added, the path accuracy increases (bounds on the accuracy have been established in the accompanying paper). We show that with a constant number (around six) of Steiner points per polyhedral edge, the path cost converges to a near-optimal value.

We have implemented these schemes as well as Chen and Han's [2] shortest path algorithm. Each of the schemes is described in the video and we also give experimental results that show how well they perform on a test suite of terrain data. In the unweighted case, we compare the path accuracy and algorithm running time with our implementation of Chen and Han's algorithm. We present graphs showing that our schemes run (much) faster and obtain near-optimal path accuracy. Since we do not have an implementation of an algorithm to compute exact weighted shortest paths, we are unable to make a comparison of our weighted paths. We do show however, that our weighted paths exhibit similar convergence behaviour as in the unweighted case. We also give results showing the effects on the approximated paths when the terrain becomes more "spiky".

## 3 Experimental Testing

Our testing was performed on a special class of non-convex polyhedra called a *Triangular Irregular Network* (TIN). This is a typical terrain model used in Geographical Information Systems (GIS) to represent a 2.5-d surface in which each vertex is given a height value. One of the main difficulties in presenting experimental results is to choose *benchmark* TINs. It is conceivable that different TIN characteristics could affect the performance of an algorithm. We have attempted to accommodate different characteristics by performing our tests with TINs that have different sizes (i.e., 1012, 5000 and 10,082 faces), height characteristics (i.e., smooth or spiky (modeled by accentuating the heights)), and data sources (i.e., random or sampled from Digital Elevation Models (DEM)).

For each TIN, we computed a set of 100 randomly selected pairs of vertices. We then tested each of the approximation schemes. The tests were performed in iterations based on the number of Steiner points per edge. In each iteration, we computed a graph and then we computed the path cost between each of the 100 vertex pairs and then obtained an "average path cost" and "computation time" for these pairs. Each scheme was tested for both weighted and non-weighted scenarios.

## 4 Video Preparation

The video was created using Microsoft Powerpoint '97. The video contains some terrain images which were obtained from snapshots of our implementation which used XGL and Motif on a SUN Sparc 20. The authors thank Carleton University's Instructional Media Center for technical support.

## References


[1] P.K. Agarwal, B. Aronov, J. O'Rourke, and C. Schevon, "Star Unfolding of a Polytope with Applications", to appear in *SIAM Journal on Computing*, 1997.

[2] J. Chen and Y. Han, "Shortest Paths on a Polyhedron", *Proceedings of the 6th ACM Symposium on Computational Geometry*, 1990, pp. 360-369.

[3] M. Lanthier, A. Maheshwari and J.-R. Sack, "Approximating Weighted Shortest Paths on Polyhedral Surfaces", *Technical report, School of Computer Science, Carleton University*, 1996.

[4] M. Lanthier, A. Maheshwari and J.-R. Sack, "Approximating Weighted Shortest Paths on Polyhedral Surfaces", in this proceedings, *13th Annual ACM Symposium on Computational Geometry*, June 1997.

[5] J.S.B. Mitchell, D.M. Mount and C.H. Papadimitriou, "The Discrete Geodesic Problem", *SIAM Journal of Computing*, **16**, August 1987, pp. 647-668.

[6] J.S.B. Mitchell and C.H. Papadimitriou, "The Weighted Region Problem: Finding Shortest Paths Through a Weighted Planar Subdivision", *Journal of the ACM*, **38**, January 1991, pp. 18-73.

# Back-face Computation of Polygon Clusters*

Subodh Kumar — Johns Hopkins University†

Dinesh Manocha, William Garrett, Ming Lin — University of North Carolina‡

## 1 Introduction

Polygons whose normals point away from a view-point are called *back-facing*. Given a solid polyhedron with consistently oriented normals, back-facing polygons are not visible from the view-point. By eliminating these polygons quickly, applications that need to perform computations only on the visible polygons can reduce the working set by approximately half in general. For example, it is common for a polygon renderer to *cull* away any polygon that is back-facing and avoid drawing them, thus improving the frame rate. Many high end graphics systems have an hardware implementation of back-face culling. Computation of back-facing polygons has been used for shadow generation [3] and collision detection [1]. At the same time, these algorithms can be used to compute the front-facing polygons, and therefore, the silhouettes.

The simplest algorithms for back-face computation involve computing the dot product of the polygon normal with a vector from the view-point to any point on the polygon. If the dot-product is positive, the polygon is back-facing. The complexity of these standard algorithms is linear in the number of polygons. More recently, Kumar et al. have presented a sub-linear algorithm for back-face computation [6]. It is based on using a hierarchical clustering of the polygons. This video explains that algorithm and demonstrates its performance on different models. In practice, the algorithm takes almost constant time to compute if a cluster is backfacing, utilizing spatial and temporal coherence. It has three main components: cluster formation, spatial partition and view-point tracking. The input polygonal model is partitioned into a hierarchy of clusters based on both similarity of orientation and physical proximity of the polygons.

*Supported in part by ARPA ISTO Order No. A410, NSF Grant No. MIP-9306208, Alfred P. Sloan Foundation Fellowship, ARO Contract P-34982-MA, NSF Grant CCR-9319957, ONR Contract N00014-94-1-0738, ARPA Contract DABT63-93-C-0048 and NSF/ARPA Center for Computer Graphics and Scientific Visualization

†Department of Computer Science, Johns Hopkins University, Baltimore, MD 21218-2694. Email: subodh@cs.jhu.edu. Web: http://www.cs.jhu.edu/~subodh.

‡Department of Computer Science, University of North Carolina, Chapel Hill, NC 27599-3175.
Email: {manocha,garrett,lin}@cs.unc.edu.
Web: http://www.cs.unc.edu/~{manocha,garrett,lin}.



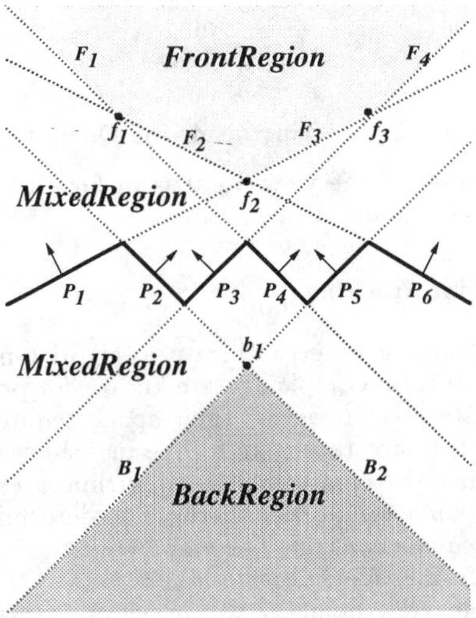


Figure 1: Different Regions of a Cluster

## 2 Spatial Partition

The space is partitioned into *BackRegion*, *FrontRegion* and *MixedRegion* with respect to each cluster. All polygons in a cluster are back-facing (front-facing) if the view-point is in the *BackRegion* (*FrontRegion*) of that cluster.

Note that there is a separate spatial partition for each polygon cluster. We illustrate these *Regions* for a collection of lines in 2-D in Fig. 1. The *BackRegion* of a cluster is an open convex set consisting of boundary planes, and defined as the intersection of (back) half-spaces defined by the polygons in the cluster. If the view-point $\mathbf{V}$ lies in the *BackRegion*, then *all* the polygons in the cluster $\mathcal{C}$ are back-facing. Similarly, if $\mathbf{V}$ lies in the *FrontRegion*, all the polygons in that cluster are front-facing. Finally, if the view-point lies in the *MixedRegion*, some of the polygons are back-facing and the others are front-facing.

Each *Region* is further decomposed into *Query* cells, each defined by three bounding planes. Each *Query* cell is a convex regions bounded by *three* planes. As a result, a point-containment query for such cells can be answered in constant time.

The decomposition of *BackRegion*(the same formulation applies to *FrontRegion*) is done by computing a center-point in *BackRegion*. Planes passing through this center-point and the boundary lines of *BackRegion* partition it into disjoint cells. It is easier to understand this partitioning in 2-D. Fig. 2 shows one such partitioning. The lightly shaded region shows the *BackRegion* of a cluster. **CB** is the center-point. For each edge $\mathbf{B}_j = \mathbf{b}_i\mathbf{b}_j$ on the boundary of *BackRegion*, the query cell $\mathsf{BQ}_j$ is the open cone with **CB** as its apex. These cells have the following property: if the view-point lies in $\mathsf{BQ}_j$, it is sufficient to test against the line

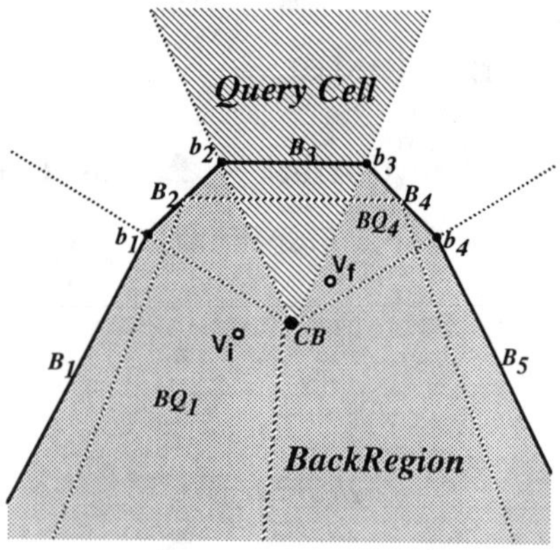


Figure 2: Tracking inside the Query Cells

$B_j$ to determine if the view-point is in fact in *BackRegion*. The 3-D case is a simple extension.

## 3 View-point Tracking

In computational geometry literature, algorithms with logarithmic asymptotic complexity are known for point location in convex sets [4]. However, their space requirements and constant factors are rather high. We instead use a tracking based algorithm. This tracking algorithm uses the *Query* cells and exploits temporal coherence to determine which of the three *Regions* contains the view-point.

At each frame the algorithm starts at the *Query* cell that contained the view-point in the previous frame and incrementally locates the *Query* cell containing it in the current frame. We start at a random cell in the very first frame.

The algorithm uses the pre-computed cluster descriptions to locate the view-point in the corresponding *Region* for each cluster. For each frame, the algorithm first tests whether the view-point continues to be contained in the same cell(s) (as the previous frame). Each test involves computing the position of the view-point with respect to each bounding plane. If the view-point fails a bounding plane test, the algorithm *walks* to the neighboring cell adjoining that plane, as shown in Fig. 2 and the video. (Red cell, as shown in the video, is the tracked cell.) This walk is repeated until the algorithm finds the cell containing the view-point. If the resulting cell lies in the *MixedRegion*, the algorithm is applied recursively to each child of that cluster. Some extra processing is required whenever the algorithm moves from *BackRegion* or *FrontRegion* to *MixedRegion* (or vice-versa).

The running time of the algorithm is a function of the number of cells traversed per frame. Due to coherence, this number is typically a small constant. Thus, if a cluster consists of $k$ sub-clusters, the expected time to traverse all sub-clusters is $O(k)$. On the other hand if we do not assume coherence, the expected number of cells traversed in an cluster with $m$ equi-sized cells, is $O(\sqrt{m})$, if the cells uniformly partition the space. This is because the average length of the shortest path from one cell to the other is $O(\sqrt{m})$. For a model with $n$ polygons, the total traversal time for a $k$-way hierarchy is $O(\sqrt{nk})$ in that case.

## 4 Video Description

The overall performance of this algorithm is a function of number of clusters, the height of the cluster tree, and the motion of the view-point between successive frames. The performance is nearly independent of the number of polygons in a cluster. We have implemented the algorithm and employed it to perform back-face culling for rendering. In practice, it improves the overall frame rate by 40 – 75% as compared to hardware back-face culling.

All the frames in the video were captured in real time from an SGI Indigo 2/Extreme workstation. The first segment of the video includes a visualization of the basic components of the algorithm. It demonstrates a single level of polygon clustering. The spatial partitioning with respect to each cluster is shown by using transparent planes. The planes themselves have been truncated for better visualization. The *Query* cell containing the view-point in the simulation is highlighted to illustrate the coherence. Since the view-point does not move significantly between successive frames, the cell containing the view-point in the new frame is often adjacent to the one in the previous frame, if not the same.

The second segment of the video demonstrates the performance of the algorithm by comparing it with that of the linear algorithm implemented in hardware. Even though we implemented the [6] algorithm in software, it is able to outperform the hardware implementation available on the SGI workstations.

## References


[1] W. Bouma and G. Vanecek. Velocity-based collision detection. In A. Paeth, editor, *Graphics Gems V*, pages 380–385, Academic Press, 1995.

[2] J.H. Clark. Hierarchical geometric models for visible surface algorithms. *Communications of the ACM*, 19(10):547–554, 1976.

[3] F. C. Crow. Shadow algorithms for computer graphics. *ACM Computer Graphics*, 11(3):242–248, 1977. (SIGGRAPH Proceedings).

[4] D. P. Dobkin and D. G. Kirkpatrick. Fast detection of polyhedral intersection. In *Proc. 9th Internat. Colloq. Automata Lang. Program.*, volume 140 of *Lecture Notes in Computer Science*, pages 154–165. Springer-Verlag, 1982.

[5] H. Fuchs, Z. Kedem, and B. Naylor. On visible surface generation by a priority tree structure. In *Proceedings of ACM SIGGRAPH*, pages 124–133, 1980.

[6] S. Kumar, D. Manocha, B. Garrett, and M. Lin. Hierarchical back-face culling. In *7th Eurographics Workshop on Rendering*, pages 231–240, Porto, Portugal, 1996.

[7] M. Newell, R. Newell, and T. Sancha. A new solution to the hidden surface problem. *Proc. ACM Ann. Conf.*, pages 443–448, 1972.

[8] R. Schumacker, B. Brand, M. Gilliland, and W. Sharp. Study for applying computer-generated images to visual generation. Technical report, AFHRL-TR-69-74, US Air Force Human Resources Lab, 1969.

[9] I. Sutherland, R. Sproull, and R. Schumaker. A characterization of ten hidden-surface algorithms. *Computing Surveys*, 6(1):1–55, 1974.

# Index